NONLINEAR FINITE ELEMENTS FOR CONTINUA AND STRUCTURES

NONLINEAR FINITE ELEMENTS FOR CONTINUA AND STRUCTURES

Second Edition

Ted Belytschko
Department of Mechanical Engineering
Northwestern University, USA

Wing Kam Liu
Department of Mechanical Engineering
Northwestern University, USA

Brian Moran
Physical Sciences and Engineering Division
King Abdullah University of Science and Technology
Kingdom of Saudi Arabia

Khalil I. Elkhodary
Department of Mechanical Engineering
The American University in Cairo, Egypt

This edition first published 2014

Registered Office
John Wiley & Sons, Ltd, The Atrium, Southern Gate, Chichester, West Sussex, PO19 8SQ, United Kingdom

For details of our global editorial offices, for customer services and for information about how to apply for permission to reuse the copyright material in this book please see our website at www.wiley.com.

Library of Congress Cataloging-in-Publication Data

Belytschko, Ted, 1943–
Nonlinear finite elements for continua and structures / Ted Belytschko, Wing Kam Liu, Brian Moran, Khalil I. Elkhodary. – Second edition.
pages cm
Includes bibliographical references and index.
ISBN 978-1-118-63270-3 (pbk.)
1. Finite element method. 2. Continuum mechanics. 3. Structural analysis (Engineering) I. Liu, W. K. (Wing Kam) II. Moran, B. (Brian), 1958– III. Elkhodary, Khalil I. IV. Title.
TA347.F5B46 2014
620′.001515355–dc23

2013023511

A catalogue record for this book is available from the British Library.

Set in 10/12pt Times by SPi Publisher Services, Pondicherry, India

1 2014

On the cover: Fracture-process profile of a Ti-modified 4330 steel, shown in the unloaded configuration. An initial microvoid volume fraction of 0.04% is simulated using the multiresolution continuum theory described in chapter 12 with about 270 million 8-node elements and approximately 3 billion degrees of freedom.

To our families

Contents

Foreword

This book provides a comprehensive introduction to the theory of nonlinear finite element analysis and its various implementation strategies. It is intended for beginning graduate students studying the areas of mechanical engineering, civil engineering, applied mathematics, engineering mechanics and materials science.

The authors provide a wide selection of material to suit the needs and preferences of many instructors. This second edition provides a thorough coverage of the key topics for the benefit of students, practitioners and developers of the nonlinear finite element software. Since 2001, I have taught the first edition of this book as a textbook to students at Columbia University, RPI and engineers at the Knolls Nuclear Laboratory. Students absolutely loved it due to the fact that difficult concepts are explained in a way that engineers and graduate students can easily comprehend.

The second edition of the book comes with a solutions manual for the benefit of instructors and students. The solutions are accompanied by MATLAB® and FORTRAN codes for select computer problems, to facilitate and accelerate the implementation of the more advanced concepts and algorithms covered in the book.

The second edition also includes three new chapters, which offer a concise introduction to some of the cutting-edge methods that have evolved in recent years in the field of nonlinear finite elements, namely: the eXtended Finite Element Method (XFEM), Multiresolution Continuum Theory (MCT) for multiscale microstructures, and dislocation-density-based crystalline plasticity. With these timely additions, the book will now reach a broader community at the forefront of research.

In summary, the book has proven over the years to be an excellent reference on nonlinear finite elements for instructors, students, engineers and researchers. It can be easily adopted as a graduate textbook or used by students for self-study. With the addition of a solutions manual, along with the relevant computer codes, the learning curve for some of the most advanced concepts in nonlinear finite analysis could be greatly shortened. Finally, the additional three chapters add the new dimension of cutting edge research.

Jacob Fish
Columbia University, USA

Preface

The objective of this book is to provide a comprehensive introduction to the methods and theory of nonlinear finite element analysis. We have focused on the formulation and solution of the discrete equations for various classes of problems that are of principal interest in applications of the finite element method to solid mechanics, the mechanics of materials, and structural mechanics. The core topics are presented first, which include: the discretization by finite elements of continua in one dimension and in multi-dimensions; the formulation of constitutive equations for nonlinear materials and large deformations; and procedures for the solution of the discrete equations, including considerations of both numerical and physical instabilities. More specialized applications are then presented. These include: the treatment of structural and contact-impact problems; representation of weak and strong discontinuities that evolve in failing solids; and mechanism-based modeling of material nonlinearities, illustrating advanced treatments of their multiscale aspects and microstructural origins. These are the topics which are of relevance to industrial and research applications and which are essential to those in the practice, research, and teaching of nonlinear finite elements.

The book has a mechanics style rather than a mathematical style. Although it includes analyses of the stability of numerical methods and the relevant partial differential equations, the objective is to teach methods of finite element analysis and the properties of the solutions and the methods. Topics such as proofs of convergence and the mathematical properties of solutions are not considered.

In the formulation of the discrete equations, we start with the governing equations based on the mechanics of the system, develop a weak form, and use this to derive the discrete equations. Weak forms and the discrete equations are developed for Lagrangian, arbitrary Lagrangian, and Eulerian meshes, for in the simulation of industrial processes and research, problems with large deformations that cannot be treated by Lagrangian meshes are becoming more common. Both the updated Lagrangian and the total Lagrangian approaches are thoroughly described.

Since a fundamental understanding of the equations requires substantial familiarity with continuum mechanics, Chapter 3 summarizes the continuum mechanics which is pertinent to the topics in this book. The chapter begins with a basic description of motion with an emphasis on rotation. Strain and stress measures are described along with transformations between them, which are later generalized as push-forward and pullback operations. The basic conservation laws are described in both so-called Eulerian and Lagrangian descriptions. Objectivity, often known as frame invariance, is introduced.

Chapter 4 describes the formulation of the discrete equations for Lagrangian meshes. We start with the development of the weak forms of momentum balance and use these to develop the discrete equations. Both the total Lagrangian and the updated Lagrangian formulations are thoroughly described, and methods and approaches for transforming between these formulations are discussed. Examples are given of the development of various elements in two and three dimensions.

Chapter 5 treats constitutive equations, with particular emphasis on the aspects of material models that are relevant to the treatment of material nonlinearities and large deformations.

Solution procedures and analyses of stability are described in Chapter 6. Both explicit and implicit integration procedures are described for transient processes and solutions; continuation procedures for equilibrium problems are considered. Newton methods and the linearization procedures required for the construction of the Newton equations are developed. In the solution of nonlinear problems, the stability of the numerical procedures and of the physical processes is crucial. Therefore, the theory of stability is summarized and applied to the determination of the stability of solutions and numerical procedures. Both geometric and material stability are considered.

Chapter 7 deals with arbitrary Lagrangian Eulerian methods. This chapter also provides the tools for Eulerian analysis. Numerical techniques needed for this class of meshes, such as upwinding and the SUPG formulation, are described.

Chapter 8 deals with element technology, the special techniques which are needed for the successful design of elements in constrained media problems. Emphasis is placed on the problem of incompressible materials but the techniques are described in a general context. One-point quadrature elements and hourglass control are also described.

Chapter 9 is devoted to structural elements, particularly shells and beams; plates are not treated separately because they are special cases of shells. We emphasize continuum-based structural formulations because they are more easily learned and more widely used for nonlinear analysis. The various assumptions are carefully studied and continuum-based formulations for beams and shells are developed. Much of this chapter rests heavily on the preceding chapters, since continuum-based elements can be developed from continuum elements with minor modifications. Therefore, topics such as linearization and material models are treated only briefly.

Contact-impact is described in Chapter 10. Contact-impact is viewed as a variational inequality, so that the appropriate contact inequalities are met in the discrete equations. Both displacement-based and velocity-based formulations are described. Attention is focused on the nonsmooth character of contact-impact and its effect on solution procedures and simulations.

Chapter 11 covers the modeling of strong and weak discontinuities. An overview of methods in classical finite elements is provided as a historical introduction. The chapter focuses on using the extended finite element method (XFEM) to model discontinuities with non-conforming meshes. For strong discontinuities the emphasis is on modeling fracture, with extensions to other problems. For weak discontinuities emphasis is on material interfaces, but the developments presented are easily extendable to other weak discontinuities. The discussion begins with the 1D formulation and then builds to multiple dimensions. Discussions are included for both implementation and integration of XFEM as well as a brief overview of the level set method, which is often coupled with XFEM. The chapter concludes with an example.

The role of material microstructure in defining material nonlinearities is introduced in Chapter 12. Emphasis is made on the *multiresolution continuum theory*, a multiscale mechanics theory for the large deformation of heterogeneous materials. Its aim is to link the mechanics of solids to materials science. The theory is developed from variational principles and discretized for finite element implementation. Representative volume elements (RVEs) and their role in developing mechanism-based multiscale constitutive formulations are then discussed and integrated in the multiresolution framework.

RVE modeling of single crystals by finite elements is discussed in Chapter 13, as an example of mechanism-based modeling of non-linear materials. From materials science, the crystallographic description of cubic and non-cubic crystals and the theory of dislocation densities are linked to a non-linear constitutive algorithm that governs inhomogeneous deformation in crystalline materials at the continuum level.

This book is intended for beginning graduate students in programs in mechanical engineering, civil engineering, applied mathematics, and engineering mechanics. The book assumes some familiarity with the finite element method, such as a one-semester course or a four- to five-week section in a larger course. The student should be familiar with shape functions, stiffness, and force assembly; it is also helpful to have some background in variational or energy methods. In addition, students should have had some exposure to strength of materials and continuum mechanics. Familiarity with indicial notation and matrix notation is essential.

Most instructors will choose not to cover this entire book. To do so would require a one-year course. Our aim has been to include a wide selection of material to suit the needs and preferences of many instructors. Moreover, the additional material provides the interested student with a source of background reading before embarking into the literature.

Shorter courses, such as a 10-week quarter or a 16-week semester, require a judicious selection of material which reflects the aims and taste of the instructor. The book presents most material in both the total and the updated Lagrangian format. Thus, an introductory course can focus on the updated Lagrangian viewpoint from Chapter 2 to Chapter 4, with selected topics from Chapters 5 and 6 to familiarize the student with material models and solution procedures. Some instructors may opt to skip the one-dimensional treatment in Chapter 2, leaving it as perhaps required reading. The total Lagrangian formulation can then be introduced by simply showing the transformation in Chapter 4. A similar course can be designed with an emphasis on the total Lagrangian formulation.

We have endeavored to use a unified style and notation throughout this book. This is important because, for students, drastic changes in notation and formalism often impede learning. This, at times, causes divergence from notation customary in the literature of a particular area, but we hope that the consistency of presentation will help the student.

For the second edition of this book a solution manual is available, which includes solutions to all exercises in the book, including MATLAB® and/or FORTRAN codes for the prescribed computer problems.

We would like to thank our many friends, colleagues, and former students who read preliminary versions of this book and who provided numerous suggestions, feedback and corrections, especially

Zhanli Liu, Tsinghua University
Zhuo Zhuang, Tsinghua University
Danial Faghihi, University of Texas at Austin
J. S. Chen, University of Iowa
John Dolbow, Duke University
Thomas J. R. Hughes, Stanford University
Shaofan Li, Northwestern University
Arif Masud, University of Illinois at Chicago
Nicolas Moës, Northwestern University
Katerina Papoulia, Cornell University
Patrick Smolinski, University of Pittsburgh
Natarajan Sukumar, Northwestern University
Henry Stolarski, University of Minnesota
Ala Tabiel, University of Cincinnati

We would also like to thank our students Sheng Peng, Jifeng Zhao, Miguel Bessa, John Moore, Patrick Lea, Zulfiqar Ali, Debbie Burton, Hao Chen, Yong Guo, Dong Qian, Michael Singer, Pritpal Singh, Gregory Wagner, Shaoping Xiao and Lucy Zhang for help with preparation of figures, typing, and extensive proofreading. Any remaining errors are, of course, the responsibility of the authors.

A special thanks goes to Shaofan Li and Yong Guo for their contributions to several of the exercises and worked examples.

Ted Belytschko, Wing Kam Liu
Northwestern University, USA
Brian Moran
King Abdullah University of Science and Technology, KSA
Khalil I. Elkhodary
The American University in Cairo, Egypt

List of Boxes

1

Introduction

1.1 Nonlinear Finite Elements in Design

Nonlinear finite element analysis is an essential component of computer-aided design. Testing of prototypes is increasingly being replaced by simulation with nonlinear finite element methods because this provides a more rapid and less expensive way to evaluate design concepts and design details. For example, in the field of automotive design, simulation of crashes is replacing full-scale tests, for both the evaluation of early design concepts and details of the final design, such as accelerometer placement for airbag deployment, padding of the interior, and selection of materials and component cross-sections for meeting crashworthiness criteria. In many fields of manufacturing, simulation is speeding the design process by allowing simulation of processes such as sheet-metal forming, extrusion, and casting. In the electronics industries, simulation is replacing drop-tests for the evaluation of product durability.

Both analysts and developers of nonlinear finite element programs should understand the fundamental concepts of nonlinear finite element analysis. Without an understanding of the fundamentals, a finite element program is a black box that provides simulations. However, nonlinear finite element analysis confronts the analyst with many choices and pitfalls. Without an understanding of the implication and meaning of these choices and difficulties, an analyst is at a severe disadvantage.

The purpose of this book is to describe the methods of nonlinear finite element analysis for solid mechanics. The intent is to provide an integrated treatment so that the reader can gain an understanding of the fundamental methods, a feeling for the comparative usefulness of different approaches and an appreciation of the difficulties which lurk in the nonlinear world.

Nonlinear Finite Elements for Continua and Structures, Second Edition.
Ted Belytschko, Wing Kam Liu, Brian Moran, and Khalil I. Elkhodary.

Companion Website: www.wiley.com/go/belytschko

At the same time, enough detail about the implementation of various techniques is given so that they can be programmed.

Nonlinear analysis consists of the following steps:

1. Development of a model
2. Formulation of the governing equations
3. Discretization of the equations
4. Solution of the equations
5. Interpretation of the results.

Items 2 to 4 typically are within the analysis code, while the analyst is responsible for items 1 and 5.

Model development has changed markedly in the past decade. Until the 1990s, model development emphasized the extraction of the essential elements of mechanical behavior. The objective was to identify the simplest model which could replicate the behavior of interest.

It is now becoming common in industry to develop a single, detailed model of a design and to use it to examine all of the engineering criteria which are of interest. The impetus for this approach to modeling is that it costs far more to make several meshes for an engineering product than can be saved by specializing meshes for each application. For example, the same finite element model of a laptop computer can be used for a drop-test simulation, a linear static analysis and a thermal analysis. By using the same model for all of these analyses, a significant amount of engineering time can be saved. While this approach is not recommended in all situations, it is becoming commonplace in industry.

In the near future the finite element model may become a 'virtual' prototype that can be used for checking many aspects of a design's performance. The decreasing cost of computer time and the increasing speed of computers make this approach highly cost-effective. However, the user of finite element software must still be able to evaluate the suitability of a model for a particular analysis and understand its limitations.

The formulation of the governing equations and their discretization is today largely in the hands of the software developers. However, an analyst who does not understand the fundamentals of the software faces many perils, because some approaches and software may be unsuitable. Furthermore, to convert experimental data to input, the analyst must be aware of the stress and strain measures used in the program and by the experimentalist who provided material data. The analyst must understand the sensitivity of response to the data and how to assess it. An effective analyst must be aware of the likely sources of error, how to check for these errors and estimate their magnitudes, and the limitations and strengths of various algorithms.

The solution of the discrete equations also presents many choices. An inappropriate choice will result in very long run times which can prevent the analyst from obtaining the results within the time schedule. An understanding of the advantages and disadvantages and the approximate computer times required for various solution procedures is invaluable in the selection of a good strategy for developing a reasonable model and selecting the best solution procedure.

The analyst's role is most crucial in the interpretation of results. In addition to the approximations inherent even in linear finite element models, nonlinear analyses are often sensitive to many factors that can make a single simulation quite misleading. Nonlinear solids can undergo instabilities, their response can be sensitive to imperfections, and the results can depend dramatically on material parameters. Unless the analyst is aware of these phenomena, misinterpretation of simulation results is quite possible.

Despite these pitfalls, our views on the usefulness and potential of nonlinear finite element analyses are very sanguine. In many industries, nonlinear finite element analyses have shortened design cycles and dramatically reduced the need for prototype tests. Simulations, because of the wide variety of output they produce and the ease of doing 'what-ifs,' can lead to tremendous improvements of the engineer's understanding of the basic physics of a product's behavior under various environments. While tests give the gross but important result of whether the product withstands a certain environment, they usually provide little of the detail of the behavior of the product on which a redesign can be based if the product does not meet a test. Computer simulations, on the other hand, give detailed histories of stress and strain and other state variables, which in the hands of a good engineer give valuable insight into how to redesign the product.

Like many finite element books, this book presents a large variety of methods and recipes for the solution of engineering and scientific problems by the finite element method. However, to preserve a pedagogic character, we have interwoven several themes into the book which we feel are of central importance in nonlinear analysis. These include the following:

1. The selection of appropriate methods for the problem at hand
2. The selection of a suitable mesh description and kinematic and kinetic descriptions for a given problem
3. The examination of stability of the solution and the solution procedure
4. An awareness of the smoothness of the response of the model and its implication on the quality and difficulty of the solution
5. The role of major assumptions and the likely sources of error.

The selection of an appropriate mesh description, i.e. whether a Lagrangian, Eulerian or arbitrary Lagrangian Eulerian mesh is used, is important for many of the large deformation problems encountered in process simulation and failure analysis. The effects of mesh distortion need to be understood, and the advantages of different types of mesh descriptions should be borne in mind in the selection.

Stability is a ubiquitous issue in the simulation of nonlinear processes. In numerical simulations, it is possible to obtain solutions which are not physically stable and therefore quite meaningless. Many solutions are sensitive to imperfections of material and load parameters; in some cases, there is even sensitivity to the mesh employed in the solution. A knowledgeable user of nonlinear finite element software must be aware of these characteristics and the associated pitfalls, otherwise the results obtained by elaborate computer simulations can be quite misleading and lead to incorrect design decisions.

The issue of smoothness is also ubiquitous in nonlinear finite element analysis. Lack of smoothness degrades the robustness of most algorithms and can introduce undesirable noise into the solution. Techniques have been developed which improve the smoothness of the response; these are called regularization procedures. However, regularization procedures are often not based on physical phenomena and in many cases the constants associated with the regularization are difficult to determine. Therefore, an analyst is often confronted with the dilemma of whether to choose a method which leads to smoother solutions or to deal with a discontinuous response. An understanding of the effects of regularization parameters and of the presence of hidden regularizations, such as penalty methods in contact-impact, and an appreciation of the benefits of these methods, is highly desirable.

The accuracy and stability of solutions are important issues in nonlinear analysis. These issues manifest themselves in many ways. For example, in the selection of an element, the analyst must be aware of stability and locking characteristics of various elements. A judicious selection of an element involves factors such as the stability of the element for the problem at hand, the expected smoothness of the solution and the magnitude of deformations expected. In addition, the analyst must be aware of the complexity of nonlinear solutions. The possibility of both physical and numerical instabilities must be kept in mind and checked in a solution.

Thus the informed use of nonlinear software in both industry and research requires considerable understanding of nonlinear finite element methods. It is the objective of this book to provide this understanding and to make the reader aware of the many interesting challenges and opportunities in nonlinear finite element analysis.

1.2 Related Books and a Brief History of Nonlinear Finite Elements

Several excellent texts and monographs devoted either entirely or partially to nonlinear finite element analysis have already been published. Books dealing only with nonlinear finite element analysis include Oden (1972), Crisfield (1991), Kleiber (1989), and Zhong (1993). Oden's work is particularly noteworthy since it pioneered the field of nonlinear finite element analysis of solids and structures. Recent books are Simo and Hughes (1998) and Bonet and Wood (1997). Some of the books which are partially devoted to nonlinear analysis are Belytschko and Hughes (1983), Zienkiewicz and Taylor (1991), Bathe (1996) and Cook, Malkus and Plesha (1989). These books provide useful introductions to nonlinear finite element analysis. As a companion book, a treatment of linear finite element analysis is also useful. The most comprehensive are Hughes (1987) and Zienkiewicz and Taylor (1991).

In the following, we recount a brief history of nonlinear finite element methods. This account differs somewhat from those in many other books in that it focuses more on the software than published works. In nonlinear finite element analysis, as in many endeavors in this information-computer age, the software often represents a better guide to the state-of-the-art than the literature.

Nonlinear finite element methods have many roots. Not long after the linear finite element method became known through the work of the Boeing group and the famous paper of Turner, *et al.* (1956), engineers in many universities and research laboratories began extensions of the method to nonlinear, small-displacement static problems. It is difficult to convey the excitement of the early finite element community and the disdain of classical researchers for the method. For example, for many years the *Journal of Applied Mechanics* shunned papers on the finite element method because it was considered of no scientific substance. But to many, particularly engineers who had to deal with engineering problems, the promise of the finite element method was clear: it offered the possibility of dealing with the complex shapes of real designs.

The excitement in the 1960s was fueled by Ed Wilson's liberal distribution of his first programs. The first generation of these programs had no name. In many laboratories throughout the world, engineers developed new applications by modifying and extending these early codes developed at Berkeley; they had a tremendous impact on engineering and the subsequent development of finite element software. The second generation of linear programs developed at Berkeley were called SAP (Structural Analysis Program). The first nonlinear program which evolved from this work at Berkeley was NONSAP, which had capabilities for equilibrium solutions and the solution of transient problems by implicit integration.

Among the first papers on nonlinear finite element methods were Argyris (1965) and Marcal and King (1967). The number of papers soon proliferated, and software soon followed. Pedro Marcal taught at Brown University for a time, but he set up a firm to market the first nonlinear commercial finite element program in 1969; the program was called MARC and it is still a major player. At about the same time, John Swanson was developing a nonlinear finite element program at Westinghouse for nuclear applications. He left Westinghouse in 1969 to market the program ANSYS, which for many years dominated the commercial nonlinear finite element scene, although it focused more on nonlinear materials than the complete nonlinear problem.

Two other major players in the early commercial software scene were David Hibbitt and Klaus-Jürgen Bathe. Hibbitt worked with Pedro Marcal until 1972, and then co-founded HKS, which markets ABAQUS. This program has had substantial impact because it was one of the first finite element programs to introduce gateways for researchers to add elements and material models. Jürgen Bathe launched his program shortly after obtaining his PhD at Berkeley under the tutelage of Ed Wilson when he began teaching at MIT. It was an outgrowth of the NONSAP codes, and was called ADINA.

The commercial finite element programs marketed until about 1990 focused on static solutions and dynamic solutions by implicit methods. There were terrific advances in these methods in the 1970s, generated mainly by the Berkeley researchers and those with Berkeley roots: Thomas JR Hughes, Robert Taylor, Juan Simo, Jürgen Bathe, Carlos Felippa, Pal Bergan, Kaspar Willam, Ekerhard Ramm and Michael Ortiz are some of the prominent researchers who have been at Berkeley; it was undoubtedly the main incubator in the early years of finite elements.

Another lineage of modern nonlinear software is the explicit finite element codes. Explicit finite element methods in their early years were strongly influenced by the work in the DOE laboratories, particularly the so-called hydro-codes, Wilkins (1964).

In 1964, Costantino developed what was probably the first explicit finite element program, at the IIT Research Institute in Chicago (Costantino, 1967). It was limited to linear materials and small deformations, and computed the internal nodal forces by multiplying a banded form of the stiffness matrix by the nodal displacements. It was first run on an IBM 7040 series computer, which cost millions of dollars and had a speed of far less than 1 megaflop (million floating point operations per second) and 32 000 words of RAM. The stiffness matrix was stored on a tape and the progress of a calculation could be gauged by watching the tape drive; after every step, the tape drive would reverse to permit a read of the stiffness matrix. These and the later Control Data machines with similar specifications, the CDC 6400 and 6600, were the machines on which finite element codes were run in the 1960s. A CDC 6400 cost almost $10 million, had 32k words of memory (for storing everything including the operating system and compiler) and a real speed of about one megaflop.

In 1969, in order to sell a proposal to the Air Force, the senior author developed what has come to be known as the element-by-element technique: the computation of the nodal forces without use of a stiffness matrix. The resulting program, SAMSON, was a two-dimensional finite element program which was used for a decade by weapons laboratories in the US. In

1972, the program was extended to fully nonlinear three-dimensional transient analysis of structures and called WRECKER. The funding was provided by a visionary program manager, Lee Ovenshire, of the US Department of Transportation, who foresaw in the early 1970s that crash testing of automobiles could be replaced by simulation.

However, it was a little ahead of its time, for at that time a simulation of a 300-element model for a 20 ms simulation took about 30 hours of computer time, which cost about $30 000, the equivalent of three years' salary of an Assistant Professor. Lee Ovenshire's program funded several pioneering efforts: Hughes's work on contact-impact, Ivor McIvor's work on crush, and the research by Ted Shugar and Carly Ward on the modeling of the human head at Port Hueneme. But the Department of Transportation decided around 1975 that simulation was too expensive and all funding was redirected to testing, bringing this research effort to a screeching halt. WRECKER remained barely alive for the next decade at Ford, and the development of explicit codes by Belytschko was shifted to the nuclear safety industry at Argonne, where the code was called SADCAT and WHAMS.

Parallel work was initiated at the DOE national laboratories. In 1975, Sam Key, working at Sandia, completed HONDO, which also featured an element-by-element explicit method. The program treated both material nonlinearities and geometric nonlinearities and was carefully documented. However, this program suffered from the restrictive dissemination policies of Sandia, which did not permit codes to be released for security reasons. These programs evolved further under the work of Dennis Flanagan, a graduate of Northwestern, who named them PRONTO.

A milestone in the advancement of explicit finite element codes was John Hallquist's work at Lawrence Livermore Laboratories. John began his work in 1975, and the first release of the DYNA code was in 1976. He drew on the work which preceded his with discernment and interacted closely with many researchers from Berkeley, including Jerry Goudreau, Bob Taylor, Tom Hughes, and Juan Simo. Some of the key elements of his success were the development of contact-impact interfaces with Dave Benson, his awesome programming productivity, and the wide dissemination of the resulting codes, DYNA-2D and DYNA-3D. In contrast to Sandia, Livermore placed almost no impediments on the distribution of the program, and like Wilson's codes, John's codes were soon found in universities and government and industrial laboratories throughout the world. They were not as easy to modify, but many new ideas were developed with the DYNA codes as a testbed.

Hallquist's development of effective contact-impact algorithms (the first ones were crude compared to what is available today, but they often worked), the use of one-point quadrature elements and the high degree of vectorization made possible striking breakthroughs in engineering simulation. Vectorization has become somewhat irrelevant with the new generation of computers, but it was crucial for running large problems on the Cray machines which dominated the 1980s. The one-point quadrature elements with consistent hourglass control, to be discussed in Chapter 8, increased the speed of three-dimensional analysis by almost an order of magnitude over fully integrated three-dimensional elements.

The DYNA codes were first commercialized by a French firm, ESI, in the 1980s and called PAMCRASH, which also incorporated many routines from WHAMS. In 1989 John Hallquist left Livermore and started his own firm to distribute LSDYNA, a commercial version of DYNA.

The rapidly decreasing cost of computers and the robustness of explicit codes has revolutionized design in the past decade. The first major area of application was automotive

crashworthiness, but it proliferated rapidly. In more and more industries, prototype tests are being replaced by nonlinear finite element simulations. Products such as cellphones, laptops, washing machines, chain saws, and many others are designed with the help of simulations of normal operations, drop-tests and other extreme loadings. Manufacturing processes, such as forging, sheet-metal forming, and extrusion are also simulated by finite elements. For some of these simulations, implicit methods are becoming increasingly powerful, and it is clear that both capabilities are necessary. For example, while the explicit method is probably best suited for simulating sheet metal forming operations, in the springback simulation implicit methods are more suitable.

Today, the power of implicit methods is increasing more rapidly than that of explicit methods, perhaps because they still have such a long way to go. Implicit methods for the treatment of nonlinear constraints, such as contact and friction, have been improved tremendously. Sparse iterative solvers have also become much more effective. A robust capability today requires the availability of both classes of methods.

1.3 Notation

Nonlinear finite element analysis represents a nexus of three fields: (1) linear finite element methods, which evolved out of matrix methods of structural analysis; (2) nonlinear continuum mechanics; and (3) mathematics, including numerical analysis, linear algebra and functional analysis (Hughes, 1996). In each of these fields a standard notation has evolved. Unfortunately, the notations are quite different, and at times contradictory or overlapping. We have tried to keep the variety of notation to a minimum and consistent within the book and with the relevant literature. To aid readers who have some familiarity with the literature on continuum mechanics or finite elements, many equations are given in matrix, tensor and indicial notation.

Three types of notation are used in this book: indicial notation, tensor notation and matrix notation. Equations relating to continuum mechanics are written in tensor and indicial notation. Equations pertaining to the finite element implementation are given in indicial or matrix notation.

1.3.1 Indicial Notation

In indicial notation, the components of tensors or matrices are explicitly specified. Thus a vector, which is a first-order tensor, is denoted in indicial notation by x_i, where the range of the index is the number of dimensions n_{SD}. *Indices repeated twice in a term are summed*, in conformance with the rules of Einstein notation. For example in three dimensions, if x_i is the position vector with magnitude r,

$$r^2 = x_i x_i = x_1 x_1 + x_2 x_2 + x_3 x_3 = x^2 + y^2 + z^2 \tag{1.3.1}$$

where the second equation indicates that $x_1 = x$, $x_2 = y$, $x_3 = z$; we will usually write out the coordinates as x, y and z rather than using subscripts to avoid confusion with nodal values. For a vector such as the velocity v_i in three dimensions, $v_1 = v_x$, $v_2 = v_y$, $v_3 = v_z$; numerical subscripts are avoided in writing out expressions to avoid confusing components with node numbers. *Indices which refer to components of tensors are always lower case.*

Nodal indices are indicated by upper case Latin letters, for example, v_{iI} is the i-component of the velocity at node I. *Upper case indices repeated twice are summed over their range*, which depends on the context. When dealing with an element, the range is over the nodes of the element, whereas when dealing with a mesh, the range is over the nodes of the mesh.

Indicial notation at times leads to spaghetti-like equations, and the resulting equations are often only applicable to Cartesian coordinates. For those who dislike indicial notation, it should be pointed out that it is almost unavoidable in the implementation of finite element methods, for in programming the finite element equations the indices must be specified.

1.3.2 Tensor Notation

In tensor notation, the indices are not shown. While Cartesian indicial equations only apply to Cartesian coordinates, expressions in tensor notation are independent of the coordinate system and apply to other coordinates such as cylindrical coordinates, curvilinear coordinates, etc. Furthermore, equations in tensor notation are much easier to memorize. A large part of the continuum mechanics and finite element literature employs tensor notation, so a serious student should become familiar with it.

In tensor notation, we indicate tensors of order one or greater in boldface. Lower case boldface letters are almost always used for first-order tensors, while upper case boldface letters are used for higher-order tensors. For example, the velocity vector is $\mathbf{v}$ in tensor notation, while a second-order tensor, such as $\mathbf{E}$, is written in upper case. The major exception is the Cauchy stress tensor $\boldsymbol{\sigma}$, which is denoted by a lower case symbol. Equation (1.3.1) is written in tensor notation as $r^2 = \mathbf{x} \cdot \mathbf{x}$ where a dot denotes a contraction of the inner indices; in this case, the tensors on the RHS have only one index so the contraction applies to those indices.

Tensor expressions are distinguished from matrix expressions by using dots and colons between terms, as in $\mathbf{a} \cdot \mathbf{b}$, and $\mathbf{A} \cdot \mathbf{B}$. The symbol ':' denotes the contraction of a pair of repeated indices which appear in the same order, so $\mathbf{A} : \mathbf{B} \equiv A_{ij}B_{ij}$. As another example, a linear constitutive equation is given below in tensor notation and indicial notations:

$$\sigma_{ij} = C_{ijkl}\varepsilon_{kl} \quad \boldsymbol{\sigma} = \mathbf{C} : \boldsymbol{\varepsilon} \tag{1.3.2}$$

1.3.3 Functions

The functional dependence of a variable will be indicated wherever it first appears by listing the independent variables. For example, $\mathbf{v}(\mathbf{x}, t)$ indicates that the velocity v is a function of the space coordinates $\mathbf{x}$ and the time t. In subsequent appearances of $\mathbf{v}$, these independent variables are usually omitted. We will attach short words to some of the symbols. This is intended to help a reader who delves into the middle of the book. It is not intended that such complex symbols be used in working through derivations.

1.3.4 Matrix Notation

In implementing finite element methods, we will often use matrix notation. We will use the same notation for matrices as for tensors but we will not use connective symbols. Thus (1.3.1) in matrix notation is written as $r^2 = \mathbf{x}^T\mathbf{x}$. All first-order matrices will be denoted by lower case boldface letters, such as $\mathbf{v}$, *and will be considered column matrices*. Examples of column matrices are

$$\mathbf{x} = \begin{Bmatrix} x \\ y \\ z \end{Bmatrix}, \quad \mathbf{v} = \begin{Bmatrix} v_1 \\ v_2 \\ v_3 \end{Bmatrix} \tag{1.3.3}$$

Usually rectangular matrices will be denoted by upper case boldface, such as $\mathbf{A}$. The transpose of a matrix is denoted by a superscript 'T', and the first index always refers to a row number, the second to a column number. Thus a 2×2 matrix $\mathbf{A}$ and a 2×3 matrix $\mathbf{B}$ are written as follows (the order of a matrix is given with the number of rows first):

$$\mathbf{A} = \begin{bmatrix} A_{11} & A_{12} \\ A_{21} & A_{22} \end{bmatrix} \quad \mathbf{B} = \begin{bmatrix} B_{11} & B_{12} & B_{13} \\ B_{21} & B_{22} & B_{23} \end{bmatrix} \tag{1.3.4}$$

To illustrate the various notations, the quadratic form associated with $\mathbf{A}$ and the strain energy in the four notations is given next

$$\underbrace{\mathbf{x}\cdot\mathbf{A}\cdot\mathbf{x}}_{\text{tensor}} = \underbrace{\mathbf{x}^T\mathbf{A}\mathbf{x}}_{\text{matrix}} = \underbrace{x_i A_{ij} x_j}_{\text{indicial}} \quad \frac{1}{2}\underbrace{\boldsymbol{\varepsilon}:\mathbf{C}:\boldsymbol{\varepsilon}}_{\text{tensor}} = \frac{1}{2}\underbrace{\varepsilon_{ij} C_{ijkl} \varepsilon_{kl}}_{\text{indicial}} = \frac{1}{2}\underbrace{\{\varepsilon\}^T[\mathrm{C}]\{\varepsilon\}}_{\text{Voigt}} \tag{1.3.5}$$

Note that in converting a scalar product with a vector (column matrix) to matrix notation, the transpose of the column matrix is taken if it premultiplies the term. Second-order tensors are often converted to matrices in *Voigt notation*, which is described in Appendix 1.

1.4 Mesh Descriptions

One of the themes of this book is the different descriptions for the governing equations and their discretization. We will classify three aspects of the description (Belytschko, 1977):

1. The mesh description
2. The kinetic description, which is determined by the choice of the stress tensor and the form of the momentum equation
3. The kinematic description, which is determined by the choice of the strain measure.

In this section, we introduce the mesh descriptions. For this purpose, it is useful to introduce some definitions and concepts which will be used throughout this book.

Spatial coordinates are denoted by $\mathbf{x}$ and are also called Eulerian coordinates. A spatial coordinate specifies the location of a point in space. Material coordinates, also called Lagrangian coordinates, are denoted by $\mathbf{X}$. The material coordinate labels a material point: each material point has a unique material coordinate, which is usually taken to be its spatial coordinate in the initial configuration of the body, so at $t = 0$, $\mathbf{X} = \mathbf{x}$.

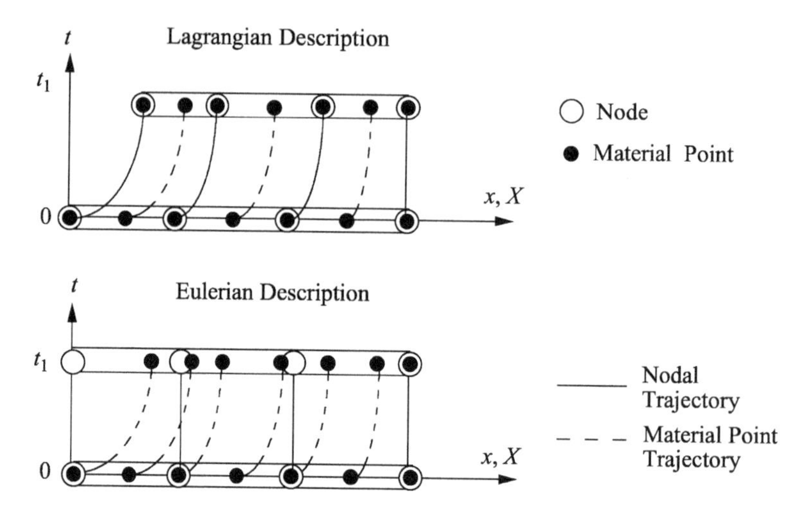

Figure 1.1 Space–time depiction of one-dimensional Lagrangian and Eulerian elements

The *motion* or deformation of a body is described by a function $\boldsymbol{\phi}(\mathbf{X}, t)$, with the material coordinates $\mathbf{X}$ and the time t as the independent variables. This function gives the spatial positions of the material points as a function of time through

$$\mathbf{x} = \boldsymbol{\phi}(\mathbf{X}, t) \tag{1.4.1}$$

This is also called a map between the initial and current configurations. The displacement $\mathbf{u}$ of a material point is the difference between its current position and its original position:

$$\mathbf{u}(\mathbf{X}, t) = \boldsymbol{\phi}(\mathbf{X}, t) - \mathbf{X} \tag{1.4.2}$$

To illustrate these definitions, consider the following motion in one dimension:

$$x = \phi(X,\ t) = (1 - X)t + \frac{1}{2}Xt^2 + X \tag{1.4.3}$$

In these equations, the material and spatial coordinates have been changed to scalars since the motion is one-dimensional. A motion is shown in Figure 1.1 (it differs from (1.4.3)); the motions of several material points are plotted in space-time to exhibit their trajectories. The velocity of a material point is the time derivative of the motion with the material coordinate fixed, that is, the velocity is given by

$$v = (X,\ t) = \frac{\partial \phi(X,\ t)}{\partial t} = 1 + X(t - 1) \tag{1.4.4}$$

The mesh description depends on the choice of independent variables. For purposes of illustration, let us consider the velocity field. We can describe the velocity field as a function of the Lagrangian (material) coordinates, as in (1.4.4), or we can describe the velocity as a function of the Eulerian (spatial) coordinates:

$$\bar{v}(x,\ t) = v(\phi^{-1}(x,\ t),\ t) \tag{1.4.5}$$

In these expressions we have placed a bar over the velocity symbol to indicate that the velocity field, when expressed in terms of the spatial coordinate x and the time t, will not be the same function as that given in (1.4.4). We have also used an inverse map to express the material coordinates in terms of the spatial coordinates:

$$X = \phi^{-1}(x, t) \tag{1.4.6}$$

Such inverse mappings can generally not be expressed in closed form for arbitrary motions, but they are an important conceptual device. For the simple motion given in (1.4.3), the inverse map is given by

$$X = \frac{x - t}{\frac{1}{2}t^2 - t + 1} \tag{1.4.7}$$

Substituting the (1.4.7) into (1.4.4) gives

$$\bar{v}(x,\ t) = 1 + \frac{(x-t)(t-1)}{\frac{1}{2}t^2 - t + 1} = \frac{1 - x + xt - \frac{1}{2}t^2}{\frac{1}{2}t^2 - t + 1} \tag{1.4.8}$$

Equations (1.4.4) and (1.4.8) give the same physical velocity fields, but express them in terms of different independent variables. Equation (1.4.4) is called a Lagrangian (material) description, for it expresses the dependent variable in terms of the Lagrangian (material) coordinates. Equation (1.4.8) is called an Eulerian (spatial) description, for it expresses the dependent variable as a function of the Eulerian (spatial) coordinates. Mathematically, the velocities in the two descriptions are different functions. Henceforth in this book, we will seldom use different symbols for different functions when they pertain to the same field, but keep in mind that if a field variable is expressed in terms of different independent variables, then the functions must be different. In this book, a symbol for a dependent variable is associated with the field, not the function.

The differences between Lagrangian and Eulerian meshes are most clearly seen in the behavior of the nodes. If the mesh is Eulerian, the Eulerian coordinates of nodes are fixed, that is, the nodes are coincident with spatial points. If the mesh is Lagrangian, the Lagrangian (material) coordinates of nodes are time invariant, that is, the nodes are coincident with material points. This is illustrated in Figure 1.1. In the Eulerian mesh, the nodal trajectories are vertical lines and material points pass across element interfaces. In the Lagrangian mesh, nodal trajectories are coincident with material point trajectories, and no material passes between elements. Furthermore, element quadrature points remain coincident with material points in Lagrangian meshes, whereas in Eulerian meshes the material point at a given quadrature point changes with time. We will see later that this complicates the treatment of materials for which the stress is history-dependent.

The comparative advantages of Eulerian and Lagrangian meshes can be seen even in this simple one-dimensional example. Since the nodes are coincident with material points in the Lagrangian mesh, boundary nodes remain on the boundary throughout the evolution of the problem. This simplifies the imposition of boundary conditions in Lagrangian meshes. In Eulerian meshes, on the other hand, boundary nodes do not remain coincident with the boundary. Therefore, boundary conditions must be imposed at points which are not nodes, and this engenders significant complications in multi-dimensional problems. Similarly, if a node is placed on an *interface between two materials*, it remains on the interface in a Lagrangian mesh, but not in an Eulerian mesh.

In Lagrangian meshes, since the material points remain coincident with mesh points, elements deform with the material. Therefore, elements in a Lagrangian mesh can become severely distorted. This effect is apparent in a one-dimensional problem only in the element lengths: in Eulerian meshes, element lengths are constant in time, whereas in Lagrangian meshes, element lengths change with time. In multi-dimensional problems, these effects are far more severe, and Lagrangian elements can get very distorted. Since element accuracy degrades with distortion, the magnitude of deformation that can be simulated with a Lagrangian mesh is limited. Eulerian elements, on the other hand, are unchanged by the deformation of the material, so no degradation in accuracy occurs because of material deformation.

To illustrate the differences between Eulerian and Lagrangian mesh descriptions, a two-dimensional example will be considered. The spatial coordinates are denoted by $\mathbf{x} = [x, y]^T$ and the material coordinates by $\mathbf{X} = [X, Y]^T$. The motion is given by

$$\mathrm{x} = \boldsymbol{\phi}(\mathbf{X}, t) \tag{1.4.9}$$

where $\boldsymbol{\phi}(\mathbf{X}, t)$ is a vector function, i.e. it gives a vector for every pair of the independent variables. Writing out the above expression gives

$$x = \phi_1(X, Y, t) \quad y = \phi_2(X, Y, t) \tag{1.4.10}$$

As an example of a motion, consider a pure shear

$$x = X + tY \quad y = Y \tag{1.4.11}$$

In a Lagrangian mesh, the nodes are coincident with material (Lagrangian) points, so for Lagrangian nodes, $\mathbf{X}_I$ = constant in time

For an Eulerian mesh, the nodes are coincident with spatial (Eulerian) points, so for Eulerian nodes, $\mathbf{x}_I$ = constant in time

Points on the edges of elements behave similarly to the nodes: in two-dimensional Lagrangian meshes, element edges remain coincident with material lines, whereas in Eulerian meshes, the element edges remain fixed in space.

To illustrate this statement, Figure 1.2 shows Lagrangian and Eulerian meshes for the shear deformation given by (1.4.11). As can be seen, a Lagrangian mesh is like an etching on the material: as the material is deformed, the etching (and the elements) deform with it. An Eulerian mesh is like an etching on a sheet of glass held in front of the material: as the material deforms, the etching is unchanged and the material passes across it.

The advantages and disadvantages of the two types of meshes in multi-dimensions are similar to those in one dimension. In Lagrangian meshes, element boundaries (lines in two dimensions, surfaces in three dimensions) remain coincident with boundaries and material interfaces. In Eulerian meshes, element sides do not remain coincident with boundaries or material interfaces. Hence tracking methods or approximate methods, such as volume of fluid approaches, have to be used for moving boundaries treated in Eulerian meshes. Furthermore, an Eulerian mesh must be large enough to enclose the material in its deformed state. On the

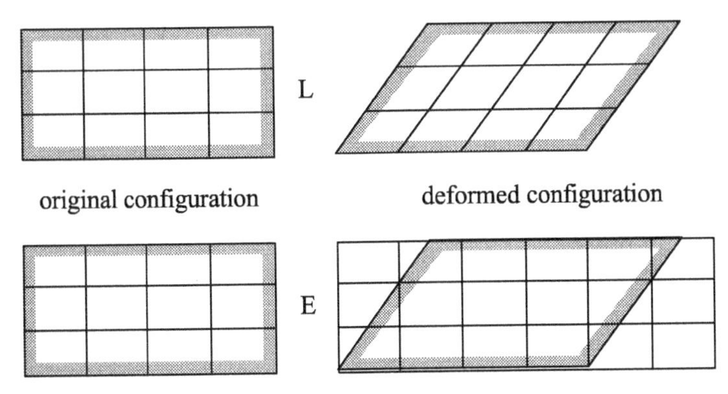

Figure 1.2 Two-dimensional shearing of a block showing Lagrangian (L) and Eulerian (E) elements

other hand, since Lagrangian meshes deform with the material, they become distorted in simulations with severe deformations. In Eulerian meshes, elements remain fixed in space, so their shapes never change.

A third type of mesh is an arbitrary Lagrangian Eulerian mesh, in which the nodes are programmed to move so that the advantages of both Lagrangian and Eulerian meshes can be exploited. In this type of mesh, the nodes can be programmed to move arbitrarily. Usually the nodes on the boundaries are moved to remain on the boundaries, while the interior nodes are moved to minimize mesh distortion. This type of mesh is described in Chapter 7.

1.5 Classification of Partial Differential Equations

For an understanding of the applicability of various finite element procedures, it is important to know the attributes of solutions to various types of partial differential equations (PDEs). The selection of an appropriate methodology depends on factors such as the smoothness of the solution, how information propagates, and the effects of initial conditions and boundary conditions; the latter are often collectively called the data for the problem. Considerable insight can be gained by knowing the type of partial differential equation one is dealing with, since the solution/attributes of different types of PDEs are markedly different.

PDEs are classified into three types:

1. Hyperbolic, which are typified by wave propagation problems
2. Parabolic, which are typified by diffusion equations, such as heat conduction
3. Elliptic; elasticity and the Laplace equations are examples.

We will shortly show why PDEs are classified in this manner. Before doing that, we briefly summarize the major characteristics of these different types of PDEs.

Hyperbolic PDEs arise from wave propagation phenomena. In hyperbolic PDEs, the smoothness of the solution depends on the smoothness of the data. If the data are rough, the solution will be rough; discontinuities in initial conditions and boundary condition propagate through the domain. Furthermore, in nonlinear hyperbolic PDEs, discontinuities may develop in the solution even for smooth data; examples are shocks in compressible flow. Information

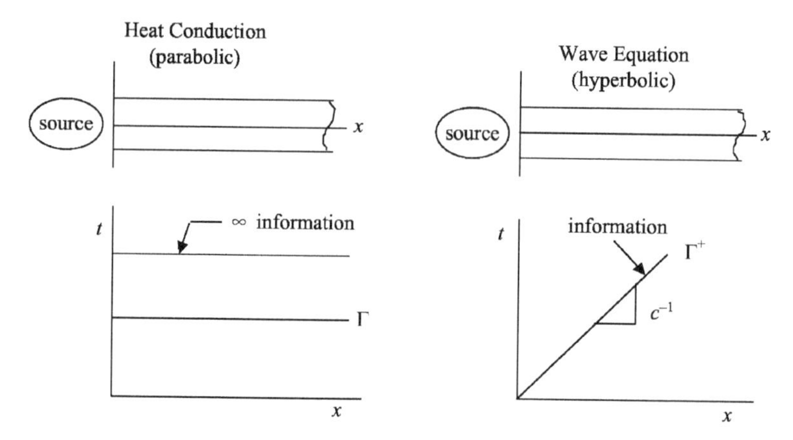

Figure 1.3 Flow of information in parabolic and hyperbolic systems of PDEs

in a hyperbolic model travels at a finite speed called the wavespeed. This is illustrated in Figure 1.3, which shows a rod with a force (source) applied at the left-hand end at time $t = 0$. An observer at a point x will not be aware of the source until the wave reaches him: the wave front is indicated by a line of slope c^{-1} in Figure 1.3; c is the wavespeed.

Elliptic PDEs are in a sense the opposite of hyperbolic PDEs. Examples of elliptic PDEs are the Laplace equation and the equations of elasticity. In elliptic PDEs the solutions are very smooth, that is, they are analytic, even if the data are rough. Furthermore, boundary data at any point tend to affect the entire solution, that is, the domain of influence of data is the entire domain. However, the effect of small irregularities in boundary data tends to be confined to the boundary: this is known as St Venant's principle. The major difficulty in the solution of elliptic PDEs is that acute corners in the boundary lead to singularities in the solution. For example, at a re-entrant corner such as a crack, the strains (derivatives of the displacements) in two-dimensional elastic solutions vary like $r^{-1/2}$, where r is the distance from the crack tip. This is the well-known crack tip singularity in fracture mechanics.

Parabolic PDEs are time-dependent PDEs with solutions that are smooth in space, but they may possess singularities at corners. Their attributes are intermediate between elliptic and hyperbolic equations. An example of a parabolic equation is the heat conduction equation. Information travels at an infinite speed in a parabolic system. For example, Figure 1.3 shows a heat source applied to a rod. The temperature rises instantaneously along the entire rod according to the heat conduction equation. Far from a source, the temperature increase may be very small. In hyperbolic systems, there is no response until the wave arrives.

The classification of PDEs rests on whether lines or surfaces exist across which the derivatives of the solution are discontinuous. This is equivalent to examining whether lines exist along which the PDEs can be reduced to ordinary differential equations.

The classification of PDEs is usually developed for first-order systems (any second-order system can be expressed as two first-order systems). Consider a quasilinear system in two unknowns:

$$A_1 u_{,x} + B_1 u_{,y} + C_1 v_{,x} + D_1 v_{,y} = E_1 \tag{1.5.1}$$

$$A_2 u_{,x} + B_2 u_{,y} + C_2 v_{,x} + D_2 v_{,y} = E_2 \tag{1.5.2}$$

In the above A_i, B_i, C_i and D_i are functions of the independent variables x and y and of the two dependent variables $u(x, y)$ and $v(x, y)$. The system is called quasilinear because it is linear in the derivatives.

Now let's examine whether u and v can have discontinuous derivatives in the x–y plane. Consider a curve Γ parametrized by s. Along Γ the derivatives are continuous but across Γ the derivatives may be discontinuous. By the chain rule, the derivatives of the dependent variables can be written as

$$u,_s = u,_x\, x,_s + u,_y\, y,_s \qquad v,_s = v,_x\, x,_s + v,_y\, y,_s \tag{1.5.3}$$

Writing (1.5.1–1.5.3) as a single matrix equation gives

$$\mathbf{Az} = \begin{bmatrix} A_1 & B_1 & C_1 & D_1 \\ A_2 & B_2 & C_2 & D_2 \\ x,_s & y,_s & 0 & 0 \\ 0 & 0 & x,_s & y,_s \end{bmatrix} \begin{Bmatrix} u,_x \\ u,_y \\ v,_x \\ v,_y \end{Bmatrix} = \begin{Bmatrix} E_1 \\ E_2 \\ u,_s \\ v,_s \end{Bmatrix} \tag{1.5.4}$$

If the derivatives are discontinuous, the solution of the above system of linear algebraic equations is indeterminate, that is, the solution is nonunique, which implies $\det(\mathbf{A}) = 0$. Enforcing this condition yields (after some algebra):

$$a y,_s^2 + 2b x,_s\, y,_s + c x,_s^2 = 0 \tag{1.5.5}$$

where

$$a = A_2 C_1 - A_1 C_2, \quad c = B_2 D_1 - B_1 D_2 \tag{1.5.6}$$

$$2b = B_1 C_2 - B_2 C_1 + A_1 D_2 - A_2 D_1$$

Dividing (1.5.5) by $x,_x^2$ and noting that $y,_s / x,_s = dy/dx \equiv y,_x$ we obtain

$$a y,_x^2 + 2b y,_x + c = 0 \tag{1.5.7}$$

The solution to (1.5.7) is given by the roots of the quadratic equation

$$y,x = \frac{-b \pm \sqrt{b^2 - ac}}{a} \tag{1.5.8}$$

The solution of the above gives the lines Γ along which the solutions may have discontinuous derivatives. If $b^2 - ac < 0$, then $y,_x$ is imaginary and such lines do not exist. If $b^2 - ac > 0$, these lines are real, so discontinuities can exist; such PDEs are called hyperbolic.

Since $y,_x$ by (1.5.8) is determined by the roots of a quadratic equation, there are two roots, which give two sets of lines Γ^+ and Γ^-, as shown in Figure 1.4. These lines are called characteristics. The classification of PDEs is summarized in Table 1.1. For time-dependent problems, the characteristics are lines along which information propagates in the x–t plane; the slope of these lines is the instantaneous wave speed c.

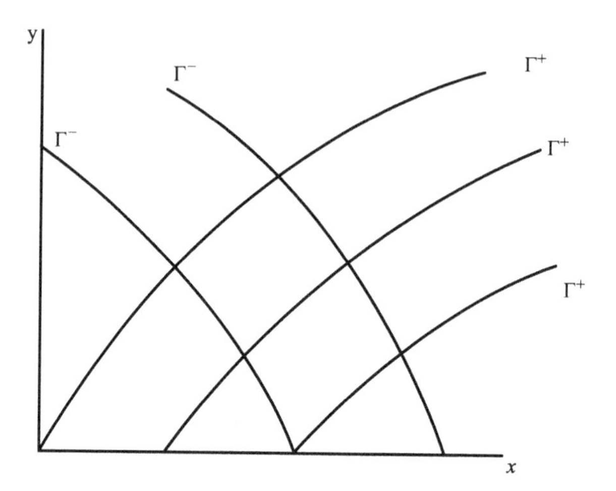

Figure 1.4 Characteristics in a hyperbolic system

Table 1.1 Classification of PDEs

$b^2 - ac$	PDE	Classification	Solution smoothness
>0	has two families of characteristics	hyperbolic	discontinuous derivatives
$=0$	has one family of characteristics	parabolic	smooth
<0	no real characteristic	elliptic	smooth

As an example, we consider the one-dimensional wave equation

$$u_{,tt} = c^2 u_{,xx} \tag{1.5.9}$$

To reduce this equation to first-order form (1.5.1–2) we let $f = u_{,x}$, $g = u_{,t}$. The wave equation then becomes a set of two first-order equations:

$$g_{,t} = c^2 f_{,x} \quad f_{,t} = g_{,x} \tag{1.5.10}$$

where the second equation is just the statement $u_{,xt} = u_{,tx}$. Writing the above system in matrix form with (1.5.4) gives

$$\mathbf{A} = \begin{bmatrix} 0 & 1 & -1 & 0 \\ c^2 & 0 & 0 & -1 \\ x_{,s} & y_{,s} & 0 & 0 \\ 0 & 0 & x_{,s} & y_{,s} \end{bmatrix}, \quad \mathbf{z}^T = [f_{,x} \; f_{,t} \; g_{,x} \; g_{,t}] \tag{1.5.11}$$

The characteristics are then found by setting det $(\mathbf{A}) = 0$, which gives

$$x_{,s}^2 - c^2 t_{,s}^2 = 0 \quad \text{or} \quad x_{,t}^2 = c^2 \tag{1.5.12}$$

From this it can be seen that the PDE is hyperbolic. The two sets of characteristic lines are given by

$$x_{,t} = \pm c \tag{1.5.13}$$

The characteristics are thus lines with slope $\pm c^{-1}$ in the x–t plane. In other words, in the wave equation information travels to the left or right by the wave speed. Across the characteristic lines, the derivatives of $f = u_{,x} = \varepsilon_x$ (ε_x is the linear strain) and of $g = u_{,t}$ (the velocity) can be discontinuous.

Consider next the Laplace equation $G_1 u_{,xx} + G_2 u_{,yy} = 0$. This is the governing equation for the elastic antiplane problem; $u(x, y)$ is the displacement in the z direction and G_α are the shear moduli. The procedure for examining the character of this equation is identical to that given before. The steps are sketched in the following:

$$f = u_{,x}, \quad g = u_{,y}, \quad \mathbf{A} = \begin{bmatrix} 0 & 1 & -1 & 0 \\ G_1 & 0 & 0 & G_2 \\ x_{,s} & t_{,s} & 0 & 0 \\ 0 & 0 & x_{,s} & t_{,s} \end{bmatrix}, \quad \mathbf{z}^T = [f_{,x}\ f_{,y}\ g_{,x}\ g_{,y}] \tag{1.5.14}$$

$$\det(\mathbf{A}) = 0 \quad \text{implies} \quad G_1 x_{,s}^2 + G_2 y_{,s}^2 = 0 \quad \text{or} \quad y_{,x}^2 = -\frac{G_1}{G_2} \tag{1.5.15}$$

If $G_1 > 0$ and $G_2 > 0$ (which is the case for stable elastic materials), the characteristic lines are then imaginary and the system is *elliptic*. No discontinuities are possible in the derivatives $f \equiv u_{,x}$ or $g \equiv u_{,y}$. Discontinuities in derivatives are possible when the material constants G_α are not homogeneous, that is, when the coefficients of the PDE G_α are discontinuous. However, discontinuities in derivatives of u coincide with the discontinuities in G_α. This equation differs from the wave equation in that both independent variables are spatial coordinates; it is difficult to give simple examples of PDEs in space-time which are elliptic.

It is left as an exercise to show that the equation $u_{,t} = \alpha u_{,xx}$ is parabolic. In a parabolic system, only one set of characteristics exists. These are parallel to the time axis, so information travels at infinite speed. In parabolic systems, discontinuities occur in space only if there are discontinuities in the data.

In a hyperbolic system, the governing equations become ordinary differential equations along the characteristics. By integrating these ODEs along the characteristics, very accurate solutions to hyperbolic PDEs can be obtained. This method is called the method of characteristics. Such methods are very appealing because of their high accuracy. However, they are quite difficult to program for more than one space dimension for arbitrary constitutive laws, so the method of characteristics is used only in special-purpose software.

1.6 Exercises

1.1. Show that the diffusion equation (heat conduction is one example) $u_{,xx} = \alpha u_{,t}$, where a is a positive constant, is parabolic.

1.2. Determine the classification of the equation for the dynamics of beams, $u_{,xxxx} = \alpha u_{,tt}$.

2

Lagrangian and Eulerian Finite Elements in One Dimension

2.1 Introduction

In this chapter, one-dimensional models of nonlinear continua are described and the corresponding finite element equations are developed. The development is restricted to one dimension so that the salient features of Lagrangian and Eulerian formulations can be demonstrated easily. These developments are applicable to nonlinear rods and one-dimensional phenomena in continua, including fluid flow. Both Lagrangian and Eulerian meshes will be considered.

This chapter also reviews some of the concepts of finite element discretization and procedures. These include the concepts of weak and strong forms, the operations of assembly, gather and scatter, and the imposition of essential boundary conditions and initial conditions. Continuity requirements for solutions and finite element approximations are presented. While this material may be familiar to those who have studied linear finite elements, it is worthwhile to skim this chapter to refresh your understanding and learn the notation.

In solid mechanics, Lagrangian meshes are most popular. Their attractiveness stems from the ease with which they handle complicated boundaries and their ability to follow material points, so that history-dependent materials can be treated accurately. In the development of Lagrangian finite elements, two approaches are commonly taken:

1. Formulations in terms of the Lagrangian measures of stress and strain in which derivatives and integrals are taken with respect to the Lagrangian (material) coordinates X, called *total Lagrangian formulations*.

Nonlinear Finite Elements for Continua and Structures, Second Edition.
Ted Belytschko, Wing Kam Liu, Brian Moran, and Khalil I. Elkhodary.

Companion Website: www.wiley.com/go/belytschko

2. Formulations expressed in terms of Eulerian measures of stress and strain in which derivatives and integrals are taken with respect to the Eulerian (spatial) coordinates x, called *updated Lagrangian formulations*.

Although the total and updated Lagrangian formulations are superficially quite different, it will be shown that the underlying mechanics of the two formulations are identical; furthermore, expressions in the total Lagrangian formulation can be transformed to updated Lagrangian expressions and vice versa. The major difference between the two formulations is in the point of view: in the total Lagrangian formulation variables are described in the original configuration, in the updated Lagrangian formulation in the current configuration. Different stress and deformation measures are typically used in these two formulations. For example, the total Lagrangian formulation customarily uses a total measure of strain, whereas the updated Lagrangian formulation often uses a rate measure of strain. However, these are not inherent characteristics of the formulations, for it is possible to use total measures of strain in updated Lagrangian formulations, and rate measures in total Lagrangian formulations. These attributes of the two Lagrangian formulations are discussed further in Chapter 4.

Until recently, Eulerian meshes have not been used much in solid mechanics. Eulerian meshes are most appealing in problems with very large deformations. Their advantage in these problems is a consequence of the fact that Eulerian elements do not deform with the material. Therefore, regardless of the magnitudes of the deformation in a process, Eulerian elements retain their original shape. Eulerian elements are particularly useful in modeling many manufacturing processes, where very large deformations are often encountered.

For each of the formulations, a weak form of the momentum equation, which is known as the principle of virtual work (or virtual power), will be developed. The weak form is developed by integrating the product of a test function with the momentum equation. The integration is performed over the material coordinates for the total Lagrangian formulation, or over the spatial coordinates for the Eulerian and updated Lagrangian formulations. It will also be shown how the traction boundary conditions are treated so that the approximate (trial) solutions need not satisfy traction boundary conditions. This procedure is identical to that in linear finite element analysis. The major difference in nonlinear formulations is the need to define the coordinates over which the integrals are evaluated and to specify the choice of stress and strain measures.

The discrete equations for a finite element approximation will then be derived. For problems in which the accelerations are important (often called dynamic problems) or those involving rate-dependent materials, the resulting discrete finite element equations are ordinary differential equations (ODEs). The process of discretizing in space is called a semidiscretization since the finite element procedure only converts the spatial differential operators to discrete form; the derivatives in time are not discretized. For static problems with rate-independent materials, the discrete equations are independent of time, and the finite element discretization results in a set of nonlinear algebraic equations.

Examples of the total and updated Lagrangian formulations are given for the two-node linear displacement and three-node quadratic displacement elements. Finally, to enable the student to solve some nonlinear problems, a central difference, explicit time-integration algorithm is described.

2.2 Governing Equations for Total Lagrangian Formulation

2.2.1 Nomenclature

Consider the rod shown in Figure 2.1. The *initial configuration*, also called the *undeformed configuration*, is shown in the lower part of the figure. This configuration plays an important role in the large deformation analysis of solids. An important configuration in the formulation of the equations is the reference configuration. As the name suggests, the equations are referred to this configuration. The reference configuration can be chosen to be the initial (undeformed) configuration or any configuration the body assumes during the motion. The *current* or *deformed configuration* is shown in the upper part of the figure. The spatial (Eulerian) coordinate is denoted by x and the material (Lagrangian) coordinates are denoted by X. The current cross-sectional area is denoted by $A(X, t)$ and the current density by $\rho(X, t)$. The initial cross-sectional area of the rod is denoted by $A_0(X)$ and its initial density by $\rho_0(X)$; variables pertaining to the reference (initial, undeformed) configuration will always be identified by a subscript or superscript zero. In this convention, we could denote the material coordinates by x_0 since they correspond to the initial coordinates, but this is not consistent with most of the continuum mechanics literature, so we will always use X for the material coordinates.

The cross-sectional area in the deformed state is denoted by $A(X, t)$; as indicated, it is a function of space and time. The spatial dependence of this variable and all others is expressed in terms of the material coordinates. The density is denoted by $\rho(X, t)$ and the displacement by $u(X, t)$. The boundary points in the reference configuration are X_a and X_b.

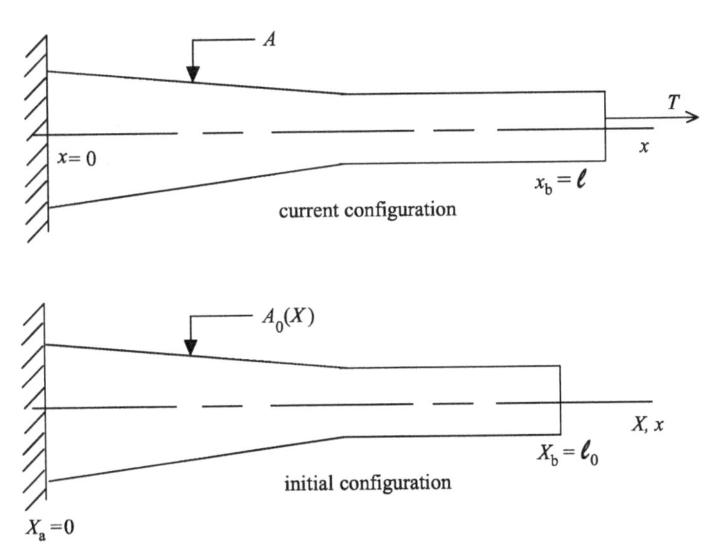

Figure 2.1 The undeformed initial configuration and deformed (current) configuration for a one-dimensional rod loaded at the right end; this is the model problem for Sections 2.2 to 2.8

2.2.2 Motion and Strain Measure

The motion of the body is described by a function of the Lagrangian coordinates and time:

$$x = \phi(X, t) \quad X \in [X_a, X_b] \tag{2.2.1}$$

where $\phi(X, t)$ is called a map between the initial and current domains. The material coordinates are the initial positions, so

$$X = \phi(X, 0) \tag{2.2.2}$$

The displacement $u(X, t)$ is given by the difference between the current position and the original position of a material point:

$$u(X, t) = \phi(X, t) - X \quad \text{or} \quad u = x - X \tag{2.2.3}$$

The deformation gradient is defined by

$$F = \frac{\partial \phi}{\partial X} = \frac{\partial x}{\partial X} \tag{2.2.4}$$

The second definition in (2.2.4) can at times be ambiguous, for it appears to involve the partial derivative of an independent variable x with respect to another independent variable X, which is meaningless. Therefore, it should be understood that whenever x appears in a context that implies it is a function, we are referring to $x = \phi(X, t)$.

Let J be the Jacobian between the current and reference configurations. The Jacobian is usually defined by $J(x(X)) = \partial x/\partial X$ for one-dimensional maps; however, to maintain consistency with multi-dimensional formulations, we will define the Jacobian as the ratio of an infinitesimal volume in the deformed body, $A\Delta x$, to the corresponding volume of the segment in the undeformed body $A_0\Delta X$:

$$J = \frac{\partial x}{\partial X}\frac{A}{A_0} = \frac{FA}{A_0} \tag{2.2.5}$$

The deformation gradient F is an unusual measure of strain since its value is unity when the body is undeformed. We will therefore define the measure of strain by

$$\varepsilon(X, t) = F(X, t) - 1 \equiv \frac{\partial x}{\partial X} - 1 = \frac{\partial u}{\partial X} \tag{2.2.6}$$

The above vanishes in the undeformed configuration. It is equivalent to the engineering strain. There are many other measures of strain, but this is the most convenient for this presentation.

2.2.3 Stress Measure

The stress measure which is used in total Lagrangian formulations does not correspond to the well-known physical stress. We will first define the physical stress, which is also known as the Cauchy stress. Let the total force across a given section be denoted by T and assume that the stress is constant in the cross-section. The Cauchy stress is given by

$$\sigma = \frac{T}{A} \tag{2.2.7}$$

This measure of stress refers to the *current* area A. In the total Lagrangian formulation, we will use the nominal stress. The nominal stress will be denoted by P and is given by

$$P = \frac{T}{A_0} \tag{2.2.8}$$

It can be seen that it differs from the physical stress in that the force is divided by the *initial*, or *undeformed, area* A_0. This is equivalent to the definition of engineering stress; however, no definition of the engineering stress is available in multi-dimensions.

Comparing (2.2.7) and (2.2.8), it can be seen that the physical and nominal stresses are related by

$$\sigma = \frac{A_0}{A} P \quad P = \frac{A}{A_0} \sigma \tag{2.2.9}$$

Therefore, if one of the stresses is known, the other can always be computed if the current and initial cross-sectional areas are known.

2.2.4 *Governing Equations*

The nonlinear rod is governed by the following equations:

1. Conservation of mass
2. Conservation of momentum
3. Conservation of energy
4. A measure of deformation, often called a strain–displacement equation
5. A constitutive equation, which describes material behavior and relates stress to a measure of deformation.

In addition, we require the deformation to be continuous, which is often called a compatibility requirement.

In the following we will simply state these equations. The equations are derived for multi-dimensional problems in Chapter 3. In processes with no heat or energy transfer, the energy equation need not be considered.

2.2.4.1 Conservation of Mass

The equation for conservation of mass for a Lagrangian formulation can be written as:

$$\rho J = \rho_0 J_0 \quad \text{or} \quad \rho(X,t)J(X,t) = \rho_0(X)J_0(X) \tag{2.2.10}$$

where J_0 is the initial Jacobian, which is taken to be unity. It is emphasized that the variables are functions of the Lagrangian coordinates. Conservation of matter can be expressed as an algebraic equation only when expressed in terms of material coordinates. Otherwise, it is a partial differential equation. For the rod, we can use (2.2.5) to write (2.2.10) as

$$\rho F A = \rho_0 A_0 \tag{2.2.11}$$

where we have used $J_0 = 1$.

2.2.4.2 Conservation of Momentum

Conservation of momentum in terms of the nominal stress and the Lagrangian coordinates gives:

$$(A_0 P)_{,X} + \rho_0 A_0 b = \rho_0 A_0 \ddot{u} \tag{2.2.12}$$

where the superposed dots denote the material time derivative, $\partial^2 u(X, t)/\partial t^2$ and b is the body force. The subscript following a comma denotes partial differentiation with respect to that variable, that is

$$P(X, t)_{,X} \equiv \frac{\partial P(X, t)}{\partial X} \tag{2.2.13}$$

Equation (2.2.12) is called the *momentum equation*. If the initial cross-sectional area is constant in space, the momentum equation becomes

$$P_{,X} + \rho_0 b = \rho_0 \ddot{u} \tag{2.2.14}$$

2.2.4.3 Equilibrium Equation

When the inertial term $\rho_0 \ddot{u}$ vanishes or can be neglected, that is, when the problem is static, the momentum equation becomes the *equilibrium equation*:

$$(A_0 P)_{,X} + \rho_0 A_0 b = 0 \tag{2.2.15}$$

Solutions of the equilibrium equations are called equilibrium solutions. Some authors call the momentum equation an equilibrium equation regardless of whether the inertial term is negligible; since equilibrium usually connotes a body at rest or moving with constant velocity, this nomenclature is avoided here.

2.2.4.4 Energy Conservation

The energy conservation equation for a rod of constant area in the absence of heat conduction or heat sources is

$$\rho_0 \dot{w}^{\text{int}} = \dot{F} P \tag{2.2.16}$$

which shows that the rate of internal work is given by the product of the rate-of-deformation gradient $\dot{F}$ and the nominal stress P. The energy conservation equation is not needed for the treatment of isothermal or adiabatic processes; its only purpose for this class of problems is to define the internal power by (2.2.16).

2.2.4.5 Constitutive Equations

The constitutive equations give the stresses resulting from the deformation. They relate the stress or stress rates to the measures of strain and/or strain rate at a material point. The constitutive equation can be written either in total form, which relates the current stress to the current deformation:

$$P(X,t) = S^{PF}(F(X,\bar{t}), \quad \dot{F}(X,\bar{t}), \quad \text{etc.}, \quad \bar{t} \le t) \tag{2.2.17}$$

or in rate form:

$$\dot{P}(X,t) = S_t^{PF}(\dot{F}(X,\bar{t}), \quad F(x,\bar{t}),\ P(X,\bar{t}), \quad \text{etc.}, \quad \bar{t} \le t) \tag{2.2.18}$$

Here S^{PF} and S_t^{PF} are functionals of the deformation history. The superscripts are appended to the constitutive functionals to indicate which measures of stress and strain they relate. The stress is assumed to be a continuous function of the strain.

As indicated in (2.2.17), the stress can depend on both F and $\dot{F}$ and on other state variables, such as temperature, void fraction; 'etc.' refers to these additional variables which can influence the stress. The stress can also depend on the history of deformation, as in an elastic–plastic material; this is indicated in (2.2.17) and (2.2.18) by letting the constitutive functions depend on deformations for all time prior to t. The constitutive equation for a solid is customarily expressed in material coordinates because it depends on the history of deformation at a material point. When a constitutive equation for a history-dependent material is written as a function of Eulerian coordinates, transport terms must be included in the stress rate, as seen in Chapter 7.

Examples of constitutive equations are

1. Linear elastic material:

$$\text{total form: } P(X,t) = E^{PF}\varepsilon(X,t) = E^{PF}(F(X,t)-1) \tag{2.2.19}$$

$$\text{rate form: } \dot{P}(X,t) = E^{PF}\dot{\varepsilon}(X,t) = E^{PF}\dot{F}(X,t) \tag{2.2.20}$$

2. Linear viscoelastic material:

$$P(X,t) = E^{PF}[(F(X,t)-1) + \alpha\dot{F}(X,t)] \quad \text{or} \quad P = E^{PF}(\varepsilon + \alpha\dot{\varepsilon}) \tag{2.2.21}$$

For small deformations the material parameter E^{PF} corresponds to Young's modulus; the constant α determines the magnitude of damping.

2.2.5 *Momentum Equation in Terms of Displacements*

A single governing equation can be obtained by substituting the relevant constitutive equation, (2.2.17) or (2.2.18), into the momentum equation (2.2.12) and expressing the strain measure in terms of the displacement by (2.2.6). For the total form of the constitutive equation (2.2.17), the resulting equation is

$$(A_0 P(u_{,X}, \dot{u}_{,X}, \cdots))_{,X} + \rho_0 A_0 b = \rho_0 A_0 \ddot{u} \tag{2.2.22}$$

which is a nonlinear partial differential equation (PDE) in the displacement $u(X, t)$. The character of this partial differential equation is not readily apparent from (2.2.22) and depends on the constitutive equation. For a linear elastic material (2.2.19), (2.2.22) becomes

$$\left(A_0 E^{PF} u_{,X}\right)_{,X} + \rho_0 A_0 b = \rho_0 A_0 \ddot{u} \tag{2.2.23}$$

For a rod of constant cross-section and modulus with zero body force, the above yields the well-known linear wave equation

$$u_{,XX} = \frac{1}{c_0^2}\ddot{u} \quad \text{where} \quad c_0^2 = \frac{E^{PF}}{\rho_0} \tag{2.2.24}$$

where the wave speed c_0 is relative to the undeformed configuration. This equation is hyperbolic when $E^{PF} > 0$ (see Section 1.5). If in addition the inertia is neglected, the governing equation becomes the equilibrium equation $E^{PF} u_{,XX} = 0$. This equation is elliptic. Thus the equations of solid mechanics change from hyperbolic to elliptic if the time dependence is removed.

2.2.5.1 Boundary Conditions

To complete the description of the problem, the boundary conditions and initial conditions must be given. The boundary in a one-dimensional problem consists of the two points at the ends of the domain, which in the model problem are the points X_a and X_b. We denote the boundary (points) by Γ.

A boundary is called a displacement boundary and denoted by Γ_u if the displacement is prescribed; it is called a traction boundary and denoted by Γ_t if the traction is prescribed. The prescribed values are designated by a superposed bar. The boundary conditions are

$$u = \bar{u} \quad \text{on} \quad \Gamma_u \tag{2.2.25}$$

$$n^0 P = \bar{t}_x^0 \quad \text{on} \quad \Gamma_t \tag{2.2.26}$$

where n^0 is the unit normal to the body, so $n^0 = 1$ at X_b, $n^0 = -1$ at X_a. The superscript zero on t_x^0 indicates that the traction is defined over the undeformed area; the subscript is always explicitly included on the traction t_x^0 to distinguish it from the time t.

From the linear form of the momentum equation, (2.2.23), it can be seen that it is second order in X. Therefore, at each end, either u or $u_{,X}$ must be prescribed as a boundary condition. In mechanics, instead of $u_{,X}$, the traction $t_x^0 = n^0 P$ is prescribed. Since the stress is a function of the measure of strain, which in turn depends on the derivative of the displacement by (2.2.6), prescribing t_x^0 is equivalent to prescribing $u_{,X}$.

For the rod shown in Figure 2.1, the boundary conditions are

$$u(X_a, t) = 0 \quad \text{and} \quad n^0(X_b)P(X_b, t) = P(X_b, t) = \frac{T(t)}{A_0(X_b)} \tag{2.2.27}$$

The traction and displacement cannot be prescribed at the same point, but one of these must be prescribed at each boundary point; this is indicated by

$$\Gamma_u \cap \Gamma_t = 0 \quad \Gamma_u \cup \Gamma_t = \Gamma \tag{2.2.28}$$

Thus in a one-dimensional mechanics problem any boundary is either a traction boundary or a displacement boundary, but no boundary is both a prescribed traction and a prescribed displacement boundary.

2.2.5.2 Initial Conditions

Since the governing equation for the rod is second order in time, two sets of initial conditions are needed. We will express the initial conditions in terms of the displacements and velocities:

$$u(X, 0) = u_0(X) \quad \text{for} \quad X \in [X_a, X_b] \tag{2.2.29a}$$

$$\dot{u}(X, 0) = v_0(X) \quad \text{for} \quad X \in [X_a, X_b] \tag{2.2.29b}$$

If the body is initially undeformed and at rest, the initial conditions can be written as

$$u(X, 0) = 0 \quad \dot{u}(X, 0) = 0 \tag{2.2.30}$$

2.2.5.3 Interior Continuity Conditions

Momentum balance requires that

$$[\![A_0 P]\!] = 0 \tag{2.2.31}$$

where $[\![f]\!]$ designates the jump in $f(X)$, i.e.

$$[\![f(X)]\!] = f(X + \varepsilon) - f(X - \varepsilon) \quad \varepsilon \to 0 \tag{2.2.32}$$

The above are also called jump conditions.

2.2.6 *Continuity of Functions*

In the discretization of the previous equations, the continuity of the dependent variables must be considered. We will describe the continuity of a function as follows: a function is C^n if the

nth derivative is a continuous function. Thus a C^1 function is continuously differentiable (its first derivative exists and is continuous everywhere). In a C^0 function, the derivative is only piecewise differentiable; discontinuities in the derivative occur at points for a one dimensional function. For a two dimensional C^0 function, discontinuities occur on lines, for a C^0 three dimensional function on surfaces. A C^{-1} function is itself discontinuous, but we assume that between the points of discontinuity the function is continuously differentiable as many times as we like. The derivative of a C^n function is C^{n-1}. The integral of a C^n function is C^{n+1}.

2.2.7 Fundamental Theorem of Calculus

The fundamental theorem of calculus states that for any C^0 function $f(x)$, the integral of the derivative gives the function, so for a definite integral

$$\int_a^b f_{,x}(x)dx = f(b) - f(a) \tag{2.2.33}$$

If the function is C^{-1}, then

$$\int_a^b f_{,x}(x)dx = \sum_i^k \int_{x_{i-1}}^{x_i} f_{,x}(x)dx = f(b) - f(a) - \sum_i^{k-1}[f(x_i)] \tag{2.2.34}$$

where x_i are the points of discontinuity and k is the number of piecewise segments of the function. The first equality is included to highlight that the integral is evaluated in a piecewise manner over open intervals (x_{i-1}, x_i), with $x_0 = a$ and $x_k = b$. Therefore, the integrals are obtained by the successive application of the fundamental theorem of calculus.

2.3 Weak Form for Total Lagrangian Formulation

The momentum equation cannot be discretized directly by the finite element method. In order to discretize this equation, a weak form, often called a variational form, is needed. The principle of virtual work, or weak form, which will be developed next, is equivalent to the momentum equation and the traction boundary conditions. Collectively, the latter are called the classical *strong form*.

2.3.1 Strong Form to Weak Form

A weak form will now be developed for the momentum equation (2.2.22) and the traction boundary conditions. For this purpose we require the trial functions $u(X, t)$ to satisfy all displacement boundary conditions and to be smooth enough so that all derivatives in the momentum equation are well-defined. The test functions $\delta u(X)$ are assumed to be smooth enough so that all of the following steps are well defined and to vanish on the prescribed displacement boundary. This is the standard, classical way of developing a weak form. Although it leads to continuity requirements that are more restrictive than those met by finite element approximations, we will go through this procedure before looking into the consequences of less restrictive continuity requirements.

The weak form is obtained by taking the product of the momentum equation with the test function and integrating over the domain. This gives

$$\int_{X_a}^{X_b} \delta u[(A_0 P)_{,X} + \rho_0 A_0 b - \rho_0 A_0 \ddot{u}]dX = 0 \tag{2.3.1}$$

In this, the nominal stress P is a function of the trial displacements. Expanding the derivative of the product in the first term in (2.3.1) and rearranging gives

$$\int_{X_a}^{X_b} \delta u(A_0 P)_{,X}\, dX = \int_{X_a}^{X_b} [(\delta u A_0 P)_{,X} - \delta u_{,X}\, A_0 P]dX \tag{2.3.2}$$

Applying the fundamental theorem of calculus to the above gives

$$\begin{aligned}\int_{X_a}^{X_b} \delta u(A_0 P)_{,X}\, dX &= \left(\delta u A_0 n^0 P\right)\Big|_{\Gamma} - \int_{X_a}^{X_b} \delta u_{,X}\,(A_0 P)dX \\ &= \left(\delta u A_0 \bar{t}_x^{\,0}\right)\Big|_{\Gamma_t} - \int_{X_a}^{X_b} \delta u_{,X}\,(A_0 P)dX\end{aligned} \tag{2.3.3}$$

where the second line follows because the test function δu vanishes on the prescribed displacement boundary, the complementarity conditions on the boundaries (2.2.28) and the traction boundary conditions. Substituting (2.3.3) into the first term of (2.3.1) gives (with a change of sign)

$$\int_{X_a}^{X_b} [\delta u_{,X}\, A_0 P - \delta u(\rho_0 A_0 b - \rho_0 A_0 \ddot{u})]dX - \left(\delta u A_0 \bar{t}_x^{\,0}\right)\Big|_{\Gamma_t} = 0 \tag{2.3.4}$$

The above is the weak form of the momentum equation and the traction boundary condition for the total Lagrangian formulation.

2.3.1.1 Smoothness of Test and Trial Functions; Kinematic Admissibility

We shall now investigate the smoothness requirement more closely. In classical derivations of the weak form, all functions in the strong form are assumed to be continuous. For the momentum equation (2.2.12) to be well defined in a classical sense, the product of the nominal stress and the initial area must be continuously differentiable, that is, C^1; otherwise the first derivative would have discontinuities. If the stress is a smooth function of the derivative of the displacement as in (2.2.17), then for the stress to be C^1, the trial functions must be C^2. For the functions in (2.3.2) to be smooth, the test function $\delta u(X)$ must be C^1.

However, the weak form is well defined for test and trial functions which are less smooth, and indeed the test and trial functions used in finite element methods do not meet these smoothness requirements. The weak form (2.3.4) involves only the first derivative of the test function, and only the first derivative of the trial function appears in the weak form if the nominal stress is only a function of the deformation gradient F. Thus the weak form (2.3.4) is integrable if the test and trial functions are C^0. The weak form can be developed with these less restrictive smoothness conditions if we add the interior continuity conditions (2.2.31) to the strong form.

We will now define these less restrictive continuity conditions on the test and trial function more precisely. We let the trial functions $u(X, t)$ be continuous functions with piecewise continuous derivatives, which is stated symbolically by $u(X, t) \in C^0(X)$, where the X in the parentheses following C^0 indicates that it pertains to the continuity in X; note that this definition permits discontinuities of the derivatives of $u(X, t)$ at discrete points.

In addition, the trial function $u(X, t)$ must satisfy all displacement boundary conditions. These conditions on the trial displacements are indicated symbolically by

$$u(X,t) \in \mathcal{u} \quad \text{where} \quad \mathcal{u} = \left\{ u(X,t) \middle| u(X,t) \in C^0(X),\ u = \bar{u} \text{ on } \Gamma_u \right\} \tag{2.3.5}$$

Displacement fields which satisfy the above conditions, i.e. displacement fields which are in $\mathcal{u}$, are called *kinematically admissible*.

The test functions are denoted by $\delta u(X)$; they are *not* functions of time. The test functions are required to be C^0 and to vanish on displacement boundaries, that is,

$$\delta u(X) \in \mathcal{u}_0 \quad \text{where} \quad \mathcal{u}_0 = \left\{ \delta u(X) \middle| \delta u(X) \in C^0(X),\ \delta u = 0 \text{ on } \Gamma_u \right\} \tag{2.3.6}$$

We will use the prefix δ for all variables which are test functions and for variables which are functions of the test functions. This notation originates in variational methods, where the test function emerges naturally as the difference between admissible functions. Although it is not necessary to know variational methods to understand weak forms, it provides an elegant framework. For example, in variational methods, any test function is a variation and defined as the difference between two trial functions, i.e. the variation $\delta u(X) = u^a(X) - u^b(X)$, where $u^a(X)$ and $u^b(X)$ are any two functions in $\mathcal{u}$. Since any function in $\mathcal{u}$ satisfies the displacement boundary conditions, the requirement in (2.3.6) that $\delta u(X) = 0$ on Γ_u follows immediately.

2.3.2 Weak Form to Strong Form

We will now develop the equations that emanate from the weak form with the less smooth test and trial functions given by (2.3.6) and (2.3.5), respectively. The weak form is given by

$$\int_{X_a}^{X_b} [\delta u_{,X} A_0 P - \delta u(\rho_0 A_0 b - \rho_0 A_0 \ddot{u})] dX - \left(\delta u A_0 \bar{t}_x^0\right)\Big|_{\Gamma_t} = 0 \quad \forall \delta u(X) \in \mathcal{u}_0 \tag{2.3.7}$$

The trial displacement fields are assumed to be kinematically admissible, that is, $u(X, t) \in \mathcal{u}$. The previous weak form is expressed in terms of the nominal stress P, but it is assumed that this stress can be expressed in terms of the first derivative of the displacement field through the strain measure and constitutive equation. Since $u(X, t)$ is C^0 and the strain measure involves first derivatives of $u(X, t)$ with respect to X, we expect $P(X, t)$ to be C^{-1} in X if the constitutive equation is continuous. The stress $P(X, t)$ will be discontinuous wherever the derivative of $u(X, t)$ is discontinuous.

To extract the strong form, the derivative of $\delta u(X)$ must be eliminated from the integrand. This is accomplished through integration by parts and the fundamental theorem of calculus. Taking the derivative of the product $\delta u A_0 P$ we have (after rearranging the terms)

$$\int_{X_a}^{X_b} \delta u_{,X} A_0 P \, dX = \int_{X_a}^{X_b} (\delta u A_0 P)_{,X} \, dX - \int_{X_a}^{X_b} \delta u (A_0 P)_{,X} \, dX \tag{2.3.8}$$

The first term on the RHS can be converted to point values by the fundamental theorem of calculus. Let the piecewise continuous function $(A_0P)_{,X}$ be continuous on intervals $\left[X_1^k, X_2^k\right]$, $k = 1$ to n. Then by the fundamental theorem of calculus

$$\int_{X_1^k}^{X_2^k} (\delta u A_0 P)_{,X} \, dX = (\delta u A_0 P)\big|_{X_2^k} - (\delta u A_0 P)\big|_{X_1^k} \tag{2.3.9}$$

Let $(X_a, X_b) = \bigcup_k \left(X_1^k, X_2^k\right)$; then applying (2.3.9) over the entire domain gives

$$\int_{X_a}^{X_b} (\delta u A_0 P)_{,X} \, dX = \sum_k \int_{X_1^k}^{X_2^k} (\delta u A_0 P)_{,X} \, dX = \left(\delta u A_0 n^0 P\right)\big|_{\Gamma_t} - \sum_i \delta u \left[\!\left[A_0 P_{\Gamma_i} \right]\!\right] \tag{2.3.10}$$

where n^0 are the normals to the segments: $n^0\left(X_1^k\right) = -1, n^0\left(X_2^k\right) = +1; \Gamma_i$ are the points of discontinuity. Only the traction boundary Γ_t appears in the above since $\delta u = 0$ on Γ_u and $\Gamma_u = \Gamma - \Gamma_t$ (see (2.3.6) and (2.2.28)). Substituting (2.3.10) into (2.3.8) gives

$$\int_{X_a}^{X_b} \delta u_{,X} (A_0 P) dX = -\int_{X_a}^{X_b} \delta u (A_0 P)_{,X} \, dX + \left(\delta u A_0 n^0 P\right)\big|_{\Gamma_t} - \sum_i \delta u [\![A_0 P]\!]_{\Gamma i} \tag{2.3.11}$$

Substituting the (2.3.11) into (2.3.7) gives (after changing signs)

$$\begin{aligned} &\int_{X_a}^{X_b} \delta u [(A_0 P)_{,X} + \rho_0 A_0 b - \rho_0 A_0 \ddot{u}] dX - \delta u A_0 \left(n^0 P - \bar{t}_x^0\right)\big|_{\Gamma_t} \\ &\quad + \sum_i \delta u [A_0 P]_{\Gamma_i} = 0 \quad \forall \delta u(X) \in u_0 \end{aligned} \tag{2.3.12}$$

From the arbitrariness of the virtual displacement $\delta u(X)$ it follows that:

$$(A_0 P)_{,X} + \rho_0 A_0 b - \rho_0 A_0 \ddot{u} = 0 \quad \text{for} \quad X \in (X_a, X_b) \setminus \Gamma_i \tag{2.3.13a}$$

$$n^0 P - \bar{t}_x^0 = 0 \quad \text{on} \quad \Gamma_t \tag{2.3.13b}$$

$$[\![A_0 P]\!] = 0 \quad \text{on} \quad \Gamma_i \tag{2.3.13c}$$

(a detailed derivation of this step is given in Section 4.3.2). Note that if the integrand $(A_0P)_{,X}$ is a piecewise continuous function, the domain of equation (2.3.13a) is not the entire domain $[X_a, X_b]$ but the union of open intervals $\bigcup_k \left(X_1^k, X_2^k\right)$, which can be written as $X \in (X_a, X_b) \setminus \Gamma_i$. Equations (2.3.13) are, respectively, the momentum equation, the traction boundary conditions, and the interior continuity conditions. These are called the *strong form*. Thus when we admit the less smooth test and trial functions, we have an additional equation in the strong form, the interior continuity condition (2.3.13c). If the test functions and trial functions satisfy the classical smoothness conditions, the interior continuity conditions are not part of the strong form. For smooth test and trial functions, the weak form implies only the momentum equation and the traction boundary conditions.

The less smooth test and trial functions are relevant to finite elements, where the test and trial functions are only C^0. They are also needed to deal with discontinuities in the cross-sectional area and material properties. At material interfaces, the classical strong form is not applicable, since it assumes that the second derivative is uniquely defined everywhere. However, at material interfaces the strains, and hence the derivatives of the displacement fields, are discontinuous. With the rougher test and trial functions, the conditions which hold at these interfaces, (2.3.13c), emerge naturally.

In the weak form for the total Lagrangian formulation, all integrals are over material domains, that is, the reference configuration. Since in total Lagrangian formulations, derivatives are taken with respect to the material coordinates X, integration by parts is most conveniently performed over the material domain.

2.3.3 *Physical Names of Virtual Work Terms*

For the purpose of obtaining a methodical procedure for developing the finite element equations, the virtual energies will be defined according to the type of work they represent; the corresponding nodal forces will subsequently carry identical names.

Each of the terms in the weak form represents a virtual work due to the virtual displacement δu. The test displacement $\delta u(X)$ is often called a 'virtual' displacement to indicate that it is not the actual displacement; according to Webster's dictionary, virtual means 'being in essence or effect, not in fact'. This is rather hazy and we prefer the name test displacement, but we use both names.

The virtual work of the body forces $b(X, t)$ and the prescribed tractions $\bar{t}_x^0$, which corresponds to the second and fourth terms in (2.3.4), is called the virtual external work since it results from the external loads. The external work is designated by the superscript 'ext' and given by

$$\delta W^{\text{ext}} = \int_{X_a}^{X_b} \delta u \rho_0 b A_0 \, dX + \left(\delta u A_0 \bar{t}_x^0\right)\Big|_{\Gamma_t} \tag{2.3.14}$$

The first term in (2.3.4) is the called the virtual internal work, for it arises from the stresses in the material. It can be written in two equivalent forms:

$$\delta W^{\text{int}} = \int_{X_a}^{X_b} \delta u_{,X} \, P A_0 \, dX = \int_{X_a}^{X_b} \delta F P A_0 dx \tag{2.3.15}$$

where the last form follows from (2.2.3) because:

$$\delta u_{,X}(X,t) = \delta(\phi(X,t) - X)_{,X} = \frac{\partial(\delta x)}{\partial X} = \delta F \tag{2.3.16}$$

Note that $\delta X = 0$ because X is an independent variable.

This definition of internal work in (2.3.15) is consistent with the internal work expression in the energy conservation equation, (2.2.16): if we change the rates in (2.2.16) to virtual increments, then $\rho_0 \delta w^{\text{int}} = \delta FP$. The virtual internal work δW^{int} is defined over the entire domain, so we have

$$\delta W^{\text{int}} = \int_{X_a}^{X_b} \delta w^{\text{int}} \rho_0 A_0 \, dX = \int_{X_a}^{X_b} \delta FPA_0 \, dX \tag{2.3.17}$$

which is the same term that appears in the weak form in (2.3.4).

The term $\rho_0 A_0 \ddot{u}$ can be considered as a body force which acts in the direction opposite to the acceleration, that is, a d'Alembert force. We will designate the corresponding virtual work by δW^{kin} and call it the virtual inertial work or virtual kinetic work, so

$$\delta W^{\text{kin}} = \int_{X_a}^{X_b} \delta u \rho_0 A_0 \ddot{u} \, dX \tag{2.3.18}$$

2.3.4 *Principle of Virtual Work*

The principle of virtual work is now given in terms of these physically motivated names. By using (2.3.14–2.3.18), (2.3.4) can be written as

$$\delta W(\delta u, u) \equiv \delta W^{\text{int}} - \delta W^{\text{ext}} + \delta W^{\text{kin}} = 0 \quad \forall \delta u \in u_0 \tag{2.3.19}$$

The above equation is the weak form of the momentum equation, the traction boundary conditions and the stress jump conditions. The weak form implies the strong form and the strong form implies the weak form. Thus the weak form and the strong form are equivalent. This equivalence of the strong and weak forms for the momentum equation is called the *principle of virtual work*. The principle of virtual work for the one-dimensional total Lagrangian formulation is summarized in Box 2.1.

Box 2.1 Principle of virtual work for one-dimensional total Lagrangian formulation

Weak Form: If the trial functions $u(X, t) \in \mathrm{U}$ then $\delta W = 0 \ \forall \ \delta u \in U_0$ (B2.1.1)

is equivalent to

Strong Form:

momentum equation (2.2.12): $(A_0 P)_{,X} + \rho_0 A_0 b = \rho_0 A_0 \ddot{u}$ (B2.1.2)

traction boundary conditions (2.2.26): $n^0 P = \bar{t}_x^0$ on Γ_t (B2.1.3)

interior continuity conditions (2.2.31): $[\![A_0 P]\!] = 0$ (B2.1.4)

Definitions:

$$\delta W \equiv \delta W^{\text{int}} - \delta W^{\text{ext}} + \delta W^{\text{kin}} \tag{B2.1.5}$$

$$\delta W^{\text{int}} = \int_{X_a}^{X_b} \delta u_{,X} \, P A_0 \; dX = \int_{X_a}^{X_b} \delta F \; P A_0 \; dX \tag{B2.1.6}$$

$$\delta W^{\text{kin}} = \int_{X_a}^{X_b} \delta u \rho_0 A_0 \ddot{u} \; dX \tag{B2.1.7}$$

$$\delta W^{\text{ext}} = \int_{X_a}^{X_b} \delta u \rho_0 b A_0 \; dX + \left(\delta u A_0 \bar{t}_x^{\,0} \right)\Big|_{\Gamma_t} \tag{B2.1.8}$$

All of the terms in the principle of virtual work δW are virtual energies. That the terms are energies is immediately apparent from δW^{ext}: since $\rho_0 b$ is a force per unit volume, its product with the virtual displacement δu gives a virtual work per unit volume, and the integral over the domain gives the total virtual work of the body force. Since the other terms in the weak form must be dimensionally consistent with the external work term, they must also be virtual energies.

This view of the weak form as a virtual work expression provides a unifying perspective which is useful for constructing weak forms for other coordinate systems and other types of problems: it is only necessary to write an equation for the virtual energies to obtain the weak form. Thus the steps we have just gone through of multiplying the equation by the test function and performing various manipulations can be avoided. The virtual work schema is also useful for memorizing the weak form. However, from a mathematical viewpoint, it is not necessary to think of the test functions $\delta u(X)$ as virtual displacements: they are simply test functions which satisfy continuity conditions and vanish on the displacement boundaries as specified by (2.3.6). This second viewpoint becomes useful for finite element discretizations of equations where the product with a test function does not have a physical meaning.

A key step in the development of the weak form is the integration by parts, (2.3.2–2.3.3). This eliminates the derivative on the stress P. Without this step, P would have to be C^0 and u would have to be C^1. Furthermore, the traction boundary conditions would have to be imposed on the trial functions. Equation (2.3.1) is perfectly acceptable as a weak form, but it is more convenient to integrate by parts and reduce the smoothness requirements on the stress and hence the trial displacements.

2.4 Finite Element Discretization in Total Lagrangian Formulation

2.4.1 *Finite Element Approximations*

The discrete equations for a finite element model are obtained from the principle of virtual work by using finite element interpolants for the test and trial functions. The domain $[X_a, X_b]$ is subdivided into elements $e = 1$ to n_e with n_N nodes. The nodes are denoted by X_I, $I = 1$ to n_N, and the nodes of a generic element by X_I^e, $I = 1$ to m, where m is the number of nodes per element. The domain of each element is $\left[X_1^e, X_m^e\right]$, which is denoted by Ω_e. For simplicity, we consider a model problem in which node 1 is a prescribed displacement boundary and node n_N a prescribed traction boundary. However, to derive the governing equations we first treat the model as if there were no prescribed displacement boundaries. We impose the displacement boundary conditions in the last step.

The finite element trial function $u(X, t)$ is

$$u(X,t) = \sum_{I=1}^{n_N} N_I(X) u_I(t) \tag{2.4.1}$$

In this, $N_I(X)$ are C^0 interpolants, often called shape functions in the finite element literature; $u_I(t)$, $I = 1$ to n_N, are the nodal displacements, which are functions of time, and are to be determined. The nodal displacements are considered functions of time even in static equilibrium problems, since in nonlinear problems we must follow the evolution of the problem; in many cases, t may simply be a monotonically increasing parameter. The shape functions, like all interpolants, satisfy the condition

$$N_I(X_J) = \delta_{IJ} \tag{2.4.2}$$

where δ_{IJ} is the Kronecker delta or unit matrix: $\delta_{IJ} = 1$ if $I = J$, $\delta_{IJ} = 0$ if $I \neq J$. We note here that if we set $u_1(t) = \bar{u}(0, t)$ for our model problem in Figure 2.1, then the trial function $u(X, t) \in u$, i.e. it is kinematically admissible since it has the requisite continuity and satisfies the essential boundary conditions. Equation (2.4.1) represents a separation of variables: the spatial dependence of the solution is entirely represented by the shape functions, whereas the time dependence is ascribed to the nodal variables.

The test functions (or virtual displacements) are

$$\delta u(X) = \sum_{I=1}^{n_N} N_I(X) \delta u_I \tag{2.4.3}$$

where δu_I are the nodal values of the test function; they are not functions of time.

2.4.2 Nodal Forces

To provide a systematic procedure for developing the finite element equations, nodal forces are defined for each of the virtual work terms. These nodal forces are given names which correspond to the names of the virtual energies. Thus

$$\delta W^{\text{int}} = \sum_{I=1}^{n_N} \delta u_I f_I^{\text{int}} = \delta \mathbf{u}^T \mathbf{f}^{\text{int}} \tag{2.4.4a}$$

$$\delta W^{\text{ext}} = \sum_{I=1}^{n_N} \delta u_I f_I^{\text{ext}} = \delta \mathbf{u}^T \mathbf{f}^{\text{ext}} \tag{2.4.4b}$$

$$\delta W^{\text{kin}} = \sum_{I=1}^{n_N} \delta u_I f_I^{\text{kin}} = \delta \mathbf{u}^T \mathbf{f}^{\text{kin}} \tag{2.4.4c}$$

$$\delta \mathbf{u}^T = \begin{bmatrix} \delta u_1 & \delta u_2 & \cdots & \delta u_{n_N} \end{bmatrix} \quad \mathbf{f}^T = \begin{bmatrix} f_1 & f_2 & \cdots & f_{n_N} \end{bmatrix} \tag{2.4.4d}$$

where $\mathbf{f}^{\text{int}}$ are the internal nodal forces, $\mathbf{f}^{\text{ext}}$ are the external nodal forces, and $\mathbf{f}^{\text{kin}}$ are the inertial, or kinetic, nodal forces. These names give a physical meaning to the nodal forces: the internal nodal forces correspond to the stresses 'in' the material, the external nodal forces correspond to the externally applied loads, while the kinetic or inertial nodal forces correspond to the inertia.

Nodal forces are always defined so that they are *conjugate* to the nodal displacements in the sense of work, that is, so that the scalar product of an increment of nodal displacements with the nodal forces gives an increment of work. This rule should be observed in the construction of the discrete equations, for when it is violated many of the important symmetries, such as that of the mass and stiffness matrices, are lost.

Next we develop expressions for the various nodal forces. In developing the nodal force expressions, we continue to ignore the displacement boundary conditions and consider δu_I arbitrary at all nodes. The expressions for the nodal forces are obtained by combining (2.3.14–2.3.18) with the definitions given in (2.4.4a–d) and the finite element approximations for the trial and test functions. Thus to define the internal nodal forces, we use (2.4.4a), (2.3.15) and the finite element approximation of the test function (2.4.3), giving

$$\delta W^{\text{int}} \equiv \sum_I \delta u_I f_I^{\text{int}} = \int_{X_a}^{X_b} \delta u_{,X} \, PA_0 \, dX = \sum_I \delta u_I \int_{X_a}^{X_b} N_{I,X} \, PA_0 \, dX \tag{2.4.5}$$

Since the above holds for arbitrary δu_I, it follows that

$$f_I^{\text{int}} = \int_{X_a}^{X_b} N_{I,X} \, PA_0 \, dX \tag{2.4.6}$$

The forces f_I^{int} are called the internal nodal forces. They are the nodal forces arising from the resistance of the solid to deformation.

The external and internal nodal forces are developed similarly. The external nodal forces are obtained by (2.4.4b) and (2.3.14):

$$\begin{aligned} \delta W^{\text{ext}} &= \sum_I \delta u_I f_I^{\text{ext}} = \int_{X_a}^{X_b} \delta u \rho_0 b A_0 \, dX + \left(\delta u A_0 \bar{t}_x^0\right)\Big|_{\Gamma_t} \\ &= \sum_I \delta u_I \left\{ \int_{X_a}^{X_b} N_I \rho_0 b A_0 \, dX + \left(N_I A_0 \bar{t}_x^0\right)\Big|_{\Gamma_t} \right\} \end{aligned} \tag{2.4.7}$$

where in the last step (2.4.3) has been used. The above gives

$$f_I^{\text{ext}} = \int_{X_a}^{X_b} \rho_0 N_I b A_0 \, dX + \left(N_I A_0 \bar{t}_x^0\right)\Big|_{\Gamma_t} \tag{2.4.8}$$

Since $N_I(X_J) = \delta_{IJ}$ the last term contributes only to those nodes which are on the prescribed traction boundary.

The kinetic nodal forces are obtained from the kinetic virtual work (2.4.4c) and (2.3.18):

$$\delta W^{\text{kin}} = \sum_I \delta u_I f_I^{\text{kin}} = \int_{X_a}^{X_b} \delta u \rho_0 \ddot{u} A_0 \, dX \tag{2.4.9}$$

Using the finite element approximation for the test functions (2.4.3) and the trial functions (2.4.1) gives

$$\sum_I \delta u_I f_I^{\text{kin}} = \sum_I \delta u_I \int_{X_a}^{X_b} \rho_0 N_I \sum_J N_J A_0 \, dX \, \ddot{u}_J \tag{2.4.10}$$

The inertial nodal force is usually expressed as a product of a mass matrix and the nodal accelerations. Therefore we define a mass matrix by

$$M_{IJ} = \int_{X_a}^{X_b} \rho_0 N_I N_J A_0 \, dX \quad \text{or} \quad \mathbf{M} = \int_{X_a}^{X_b} \rho_0 \mathbf{N}^T \mathbf{N} A_0 \, dX \tag{2.4.11}$$

Letting $\ddot{u}_I \equiv a_I$ the virtual kinetic work is

$$\delta W^{\text{kin}} = \sum_I \delta u_I f_I^{\text{kin}} = \sum_I \sum_J \delta u_I M_{IJ} a_J = \delta \mathbf{u}^T \mathbf{M} \mathbf{a}, \quad \mathbf{a} \equiv \ddot{\mathbf{u}} \tag{2.4.12}$$

The definition of the inertial nodal forces then gives the following expression

$$f_I^{\text{kin}} = \sum_J M_{IJ} a_J \quad \text{or} \quad \mathbf{f}^{\text{kin}} = \mathbf{M}\mathbf{a} \tag{2.4.13}$$

Note that the mass matrix as given by (2.4.11) will not change with time, so it needs to be computed only at the beginning of the calculation.

2.4.3 Semidiscrete Equations

We now develop the finite element equations for the model. At this point we will also consider the effect of the displacement boundary conditions. These can be satisfied for the model problem by letting

$$u_1(t) = \bar{u}_1(t) \quad \text{and} \quad \delta u_1 = 0 \tag{2.4.14}$$

It should be noted that definitions (2.4.4a–c) are made for convenience, and do not constitute the finite element equations. Substituting the definitions (2.4.4a–c) into (2.3.19) gives

$$\sum_{I=1}^{n_N} \delta u_I \left(f_I^{\text{int}} - f_I^{\text{ext}} + f_I^{\text{kin}} \right) = 0 \tag{2.4.15}$$

Since δu_I is arbitrary at all nodes except the displacement boundary node, node 1, it follows that

$$f_I^{\text{int}} - f_I^{\text{ext}} + f_I^{\text{kin}} = 0, \quad I = 2 \text{ to } n_N \tag{2.4.16}$$

Substituting (2.4.13) into (2.4.16) gives

$$\sum_{J=1}^{n_N} M_{IJ} \frac{d^2 u_J}{dt^2} + f_I^{\text{int}} - f_I^{\text{ext}} = 0, \quad I = 2 \text{ to } n_N \tag{2.4.17}$$

The acceleration of node 1 is not an unknown in this model problem, since node 1 is a prescribed displacement node. The acceleration of the prescribed displacement node can be obtained by differentiating the prescribed nodal displacement twice in time. Obviously, the prescribed displacement must be sufficiently smooth so that the second derivative can be taken; this requires it to be a C^1 function of time.

If the mass matrix is not diagonal, then the acceleration on the prescribed displacement node, node 1, will contribute to (2.4.17). The finite element equations are then

$$\sum_{J=1}^{n_N} M_{IJ} \frac{d^2 u_J}{dt^2} + f_I^{\text{int}} - f_I^{\text{ext}} = -M_{I1} \frac{d^2 \bar{u}_1}{dt^2}, \quad I = 2 \text{ to } n_N \tag{2.4.18}$$

Thus, when the mass matrix is not diagonal, prescribed displacements contribute to nodes which are not on the boundary. For a diagonal mass matrix, the above RHS terms do not appear.

In matrix form, (2.4.17) can be written as

$$\mathbf{Ma} = \mathbf{f}^{\text{ext}} - \mathbf{f}^{\text{int}} \quad \text{or} \quad \mathbf{f} = \mathbf{Ma}, \quad \text{where} \quad \mathbf{f} = \mathbf{f}^{\text{ext}} - \mathbf{f}^{\text{int}} \tag{2.4.19}$$

In the matrix form, modifications due to prescribed displacement boundary conditions cannot be indicated with simplicity, so the indicial forms (2.4.17–2.4.18) should be referred to for these purposes.

Equation (2.4.19) is the *semidiscrete momentum equation*, which is called the *equation of motion*. These equations are called semidiscrete because they are discrete in space but continuous in time. Sometimes they are simply called the discrete equations, but it should be remembered that they are only discrete in space. The equations of motion are systems of $n_N - 1$ second-order *ordinary differential equations* (*ODEs*); the independent variable is the time t. These equations can easily be remembered by the second form in (2.4.19), $\mathbf{f}=\mathbf{Ma}$, Newton's second law of motion. The mass matrix in finite element discretizations is often not diagonal, so the equations of motion differ from Newton's second law in that a force at node I can generate accelerations at node J if $M_{IJ} \neq 0$. A diagonal approximation to the mass matrix is often used. In that case, the discrete equations of motion are identical to Newton's equations for a system of particles interconnected by deformable elements. The force $f_I = f_I^{\text{ext}} - f_I^{\text{int}}$ is the net force on particle I. The negative sign appears on the internal nodal forces because these nodal forces are defined as acting on the elements; by Newton's third law, the forces on the nodes are equal and opposite, so a negative sign is needed. Viewing the semidiscrete equations of motion in terms of Newton's second law provides an intuitive feel for these equations and is useful in remembering them.

2.4.4 *Initial Conditions*

Since the equations of motion are second order in time, initial conditions on the displacements and velocities are needed. The continuous form of the initial conditions is given by (2.2.29). In many cases, the initial conditions can be applied by simply setting the nodal values of the variables to the initial values:

$$\begin{aligned} u_I(0) &= u_0(X_I) \quad \forall\ I \\ \dot{u}_I(0) &= \dot{u}_0(X_I) \quad \forall\ I \end{aligned} \tag{2.4.20}$$

Thus the initial conditions for a body which is initially at rest and undeformed are

$$u_I(0) = 0 \quad \text{and} \quad \dot{u}_I(0) = 0 \quad \forall \boldsymbol{I} \tag{2.4.21}$$

2.4.5 *Least-Square Fit to Initial Conditions*

For more complex initial conditions, the initial values of the nodal displacements and nodal velocities can be obtained by a least-square fit to the initial data. The least-square fit for the initial displacements results from minimizing the square of the difference between the finite element interpolate $\Sigma N_I(X)u_I(0)$ and the initial data $\bar{u}(X)$. Let

$$M = \frac{1}{2}\int_{X_a}^{X_b}\left(\sum_I u_I(0)N_I(X) - u_0(X)\right)^2 \rho_0 A_0\, dX \tag{2.4.22}$$

The density is not necessary in this expression but it leads to equations in terms of the mass matrix, which is quite convenient. To find the minimum of the above, set its derivative with respect to the initial nodal displacements to zero:

$$0 = \frac{\partial M}{\partial u K^{(0)}} = \int_{X_a}^{X_b} N_K(X)\left[\sum_I u_I(0)N_I(X) - u_0(X)\right]\rho_0 A_0\, dX \tag{2.4.23}$$

Using the definition of the mass matrix, (2.4.11), it can be seen that the above can be written as

$$\mathbf{Mu}(0) = \mathbf{g} \quad \text{where} \quad g_K = \int_{X_a}^{X_b} N_K(X)u_0(X)\rho_0 A_0\, dX \tag{2.4.24}$$

The least-square fit to the initial velocity data is obtained similarly. This method of fitting finite element approximations to functions is often called an L_2 projection, since it minimizes an L_2 norm.

2.4.6 *Diagonal Mass Matrix*

The mass matrix which results from a consistent derivation from the weak form is called a *consistent mass matrix*. In many applications, it is advantageous to use a diagonal mass matrix, also called a *lumped mass matrix*. Procedures for diagonalizing the mass matrix are quite *ad hoc*, and there is little theory for these procedures. One of the most common procedures is the row-sum technique, in which the diagonal elements of the mass matrix are obtained by

$$M_{II}^D = \sum_J M_{IJ}^C \tag{2.4.25}$$

where the sum is over the entire row of the matrix, M_{IJ}^C is the consistent mass matrix and M_{IJ}^D is the diagonal, or lumped, mass matrix.

The diagonal mass matrix can also be evaluated by

$$M_{II}^{D}=\sum_{J}M_{IJ}^{C}=\int_{X_a}^{X_b}\rho_0 N_I\left(\sum_{j}N_j\right)A_0\,dX=\int_{X_a}^{X_b}\rho_0 N_I A_0\,dX \tag{2.4.26}$$

where we have used the fact that the sum of the shape functions must equal 1; this is a reproducing condition discussed in Section 8.2.2. This diagonalization procedure conserves the total momentum of a body, that is, the momentum of the system with the diagonal mass is equivalent to that of the consistent mass:

$$\sum_{I,J}M_{IJ}^{C}\dot{u}_J=\sum_{I}M_{II}^{D}\dot{u}_I$$

for any nodal velocities.

2.5 Element and Global Matrices

In the previous section, we have developed the semidiscrete equations in terms of global shape functions. The global shape functions are nonzero over more than one element, so expressions such as the internal nodal force expression (2.4.6) involve computations over more than one element. In finite element programs, the nodal forces and the mass matrix are usually computed on an element level. The element nodal forces are combined into the global matrix by an operation called *scatter* or *vector assembly*. The mass matrix and other square matrices are similarly combined from the element level to the global level by an operation called *matrix assembly*. The element nodal displacements are extracted from the global matrix by an operation called *gather*. These operations are described in the following. In addition, we will show that there is no need to distinguish element and global shape functions or nodal forces in deriving finite element expressions: the expressions are identical and the element-related expressions can always be obtained by limiting the integration to the domain of the element.

The nodal displacements and nodal forces of element e are denoted by $\mathbf{u}_e$ and $\mathbf{f}_e$, respectively, and are column matrices of order m, where m is the number of nodes per element. Thus for a two-node element, the element nodal displacement matrix is $\mathbf{u}_e^T=[u_1, u_2]_e$. The corresponding element nodal force matrix is $\mathbf{f}_e^T=[f_1, f_2]_e$. We will place the element identifier 'e' as either a subscript or a superscript, but will *always use the letter* 'e' for identifying element-related quantities.

The element and global nodal force vectors must be defined so that their scalar products with the corresponding nodal displacement increments give an increment of work. This notion was used in defining the nodal forces in Section 2.4. In most cases, meeting this requirement entails little beyond care to arrange the nodal displacements and nodal forces in the correct order in the corresponding matrices. This feature of the nodal force and displacement matrices is crucial in the assembly procedure and for the symmetry of linear and linearized equations.

The element nodal displacements are related to the global nodal displacements by

$$\mathbf{u}_e=\mathbf{L}_e\mathbf{u}\quad \delta\mathbf{u}_e=\mathbf{L}e\delta\mathbf{u} \tag{2.5.1}$$

The matrix $\mathbf{L}_e$ is called the *connectivity matrix*. It is a Boolean matrix, that is, it consists of the integers 0 and 1. An example of the $\mathbf{L}_e$ matrix for a specific mesh is given later in this section. The operation of extracting $\mathbf{u}_e$ from $\mathbf{u}$ is called a *gather* because in this operation the small element vectors are *gathered* from the global vector.

The element nodal forces are defined analogously to (2.4.4), for example,

$$\delta W_e^{\text{int}} = \delta \mathbf{u}_e^T \mathbf{f}_e^{\text{int}} = \int_{X_1^e}^{X_m^e} \delta u_{,X} P A_0 \, dX \tag{2.5.2}$$

To obtain the relations between global and local nodal forces, we use the fact that the total virtual internal energy is the sum of the element internal energies:

$$\delta W^{\text{int}} = \sum_e \delta W_e^{\text{int}} \quad \text{or} \quad \delta \mathbf{u}^T \mathbf{f}^{\text{int}} = \sum_e \delta \mathbf{u}_e^T \mathbf{f}_e^{\text{int}} \tag{2.5.3}$$

Substituting (2.5.1) into (2.5.3) yields

$$\delta \mathbf{u}^T \mathbf{f}^{\text{int}} = \delta \mathbf{u}^T \sum_e \mathbf{L}_e^T \mathbf{f}_e^{\text{int}} \tag{2.5.4}$$

Since this must hold for arbitrary $\delta \mathbf{u}$, it follows that

$$\mathbf{f}^{\text{int}} = \sum_e \mathbf{L}_e^T \mathbf{f}_e^{\text{int}} \tag{2.5.5}$$

which is the relationship between element nodal forces and global nodal forces. The above operation is called a *scatter*, for each element vector is *scattered* into the global array according to the node numbers. Similar expressions can be derived for the external nodal forces and the inertial forces

$$\mathbf{f}^{\text{ext}} = \sum_e \mathbf{L}_e^T \mathbf{f}_e^{\text{ext}}, \quad \mathbf{f}^{\text{kin}} = \sum_e \mathbf{L}_e^T \mathbf{f}_e^{\text{kin}} \tag{2.5.6}$$

The gather and scatter operations are illustrated in Figure 2.2 for a one-dimensional mesh of two-node elements. The sequence of gather, compute and scatter is illustrated for two elements in the mesh. As can be seen, the displacements are gathered according to the node numbers of the element. Other nodal variables, such as nodal velocities and temperatures, can be gathered similarly. In the scatter, the nodal forces are then returned to the global force matrix according to the node numbers. The scatter operation is identical for the other nodal forces.

In order to describe the assembly of the global mass matrix from the element mass matrices, the element inertial nodal forces are defined as a product of an element mass matrix and the element acceleration:

$$\mathbf{f}_e^{\text{kin}} = \mathbf{M}_e \mathbf{a}_e \tag{2.5.7}$$

By taking time derivatives of (2.5.1), we can relate the element and global accelerations by $\mathbf{a}_e = \mathbf{L}_e \mathbf{a}$ (the connectivity matrix does not change with time). Inserting this into (2.5.7) and using (2.5.6) yields

$$\mathbf{f}^{\text{kin}} = \sum_e \mathbf{L}_e^T \mathbf{M}_e \mathbf{L}_e \mathbf{a} \tag{2.5.8}$$

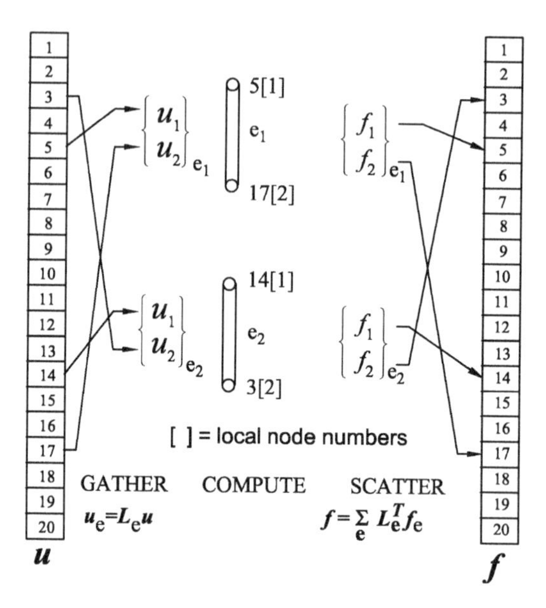

Figure 2.2 Illustration of gather and scatter for a one-dimensional mesh of two-node elements, showing the gather of two sets of element nodal displacements and the scatter of the computed nodal forces

Comparing (2.5.8) to (2.4.13), it can be seen that the global mass matrix is given in terms of the element matrices by

$$\mathbf{M} = \sum_{e} \mathbf{L}_e^T \mathbf{M}_e \mathbf{L}_e \tag{2.5.9}$$

This operation is the well-known procedure of *matrix assembly*. This is the same operation which is used to assemble the stiffness matrix in linear finite element methods.

Relations between element shape functions and global shape functions can also be developed by using the connectivity matrices. However, in most cases there is no need to distinguish them. The element shape functions $N_I^e(X)$ are nonzero only over element e. If we arrange the element shape functions $N_I^e(X)$ in a row matrix $\mathbf{N}^e(X)$, the displacement field in element e is given by

$$u^e(X) = \mathbf{N}^e(X)\mathbf{u}_e = \sum_{I=1}^{m} N_I^e(X) u_I^e \tag{2.5.10}$$

The global displacement field is obtained by summing the displacement fields for all elements, which gives

$$u(X) = \sum_{e=1}^{n_e} \mathbf{N}^e(X)\mathbf{L}_e \mathbf{u} = \sum_{e=1}^{n_e}\sum_{I=1}^{m}\sum_{J=1}^{n_N} N_I^e(X) L_{IJ}^e u_J \tag{2.5.11}$$

where (2.5.1) has been used in the above. Comparing the above with (2.4.1), we see that

$$\mathbf{N}(X) = \sum_{e=1}^{n_e} \mathbf{N}^e(X)\mathbf{L}_e \quad \text{or} \quad N_J(X) = \sum_{e=1}^{n_e}\sum_{I=1}^{m} N_I^e(X) L_{IJ}^e \tag{2.5.12}$$

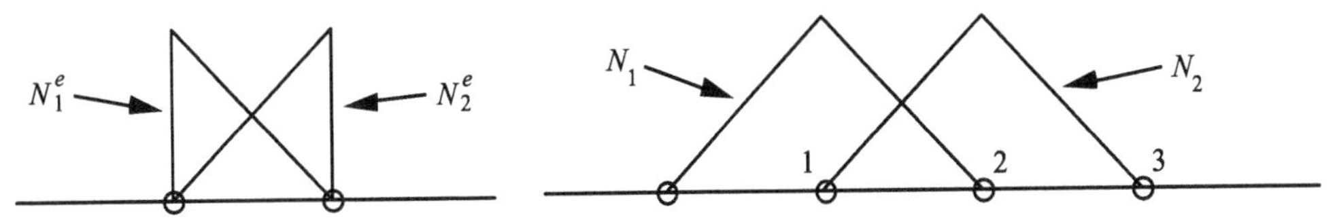

Figure 2.3 Illustration of element $N^e(X)$ and global shape functions $N(X)$ for a one-dimensional mesh of linear displacement, two-node elements

Thus the global shape functions are obtained from the element shape functions by summing according to the node numbers of the elements. This relationship is illustrated graphically for a two-node linear displacement element in Figure 2.3. Note we put the 'e' on $\mathbf{L}$ wherever it is convenient, sometimes as a subscript and sometimes as a superscript; we use this convention for 'universal' identifiers, such as '0', 'e', 'int', e and so on, throughout.

We will now show that the expressions for the element nodal forces are equivalent to the global nodal forces, except that the integrals are restricted to the elements. Using (2.5.2) and the element form of the displacement field, we obtain

$$\delta W_e^{\text{int}} = \delta \mathbf{u}_e^T \mathbf{f}_e^{\text{int}} = \delta \mathbf{u}_e^T \int_{X_1^e}^{X_m^e} \left(\mathbf{N}_{,X}^e\right)^T PA_0\, dX \tag{2.5.13}$$

Invoking the arbitrariness of the virtual nodal displacements, we obtain

$$\mathbf{f}_e^{\text{int}} = \int_{X_1^e}^{X_m^e} N_{,X}^T PA_0\, dX \quad \text{or} \quad f_{I,e}^{\text{int}} = \int_{X_1^e}^{X_m^e} N_{I,X} PA_0\, dX \tag{2.5.14}$$

where the superscript e has been removed from the above expressions since in element e, $\mathbf{N}(X)=\mathbf{N}^e(X)\mathbf{L}_e$.

Comparing this with (2.4.6), we can see that (2.5.14) is identical to the global expression (2.4.6) except that integrals here are limited to an element. Identical results can be obtained for the mass matrix and the external force matrix. Therefore, in subsequent derivations we will usually not distinguish element and global forms of the matrices: the element forms are identical to the global forms except that element matrices correspond to integrals over the element domain, whereas global force matrices correspond to integrals over the entire domain. The global matrices are almost always computed by assembly of the element matrices. Moreover, we will omit the superscript 'e' on the shape functions: the shape function to be used should be understood from the context.

In finite element programs, global nodal forces are not computed directly but are obtained from element nodal forces by assembly, that is, the scatter operation. Furthermore, the essential boundary conditions need not be considered until the final steps of the procedure. Therefore we will usually concern ourselves only with obtaining the element equations. The assembly of the element equations for the complete model and the imposition of boundary conditions is a standard procedure. Box 2.2 gives the discrete equations in the total Lagrangian formulation.

Box 2.2 Discrete equations in total Lagrangian formulation

$$u(X,t) = \mathbf{N}(X)\mathbf{u}_e(t) = \sum_I N_I(X)u_I^e(t) \tag{B2.2.1}$$

$$\varepsilon = \sum_I \frac{\partial N_I}{\partial X} u_I^e = \mathbf{B}_0\mathbf{u}^e, \quad F = \sum_I \frac{\partial N_I}{\partial X} x_I^e = \mathbf{B}_0\mathbf{x}^e \tag{B2.2.2}$$

$$\mathbf{f}_e^{\text{int}} = \int_{\Omega_0^e} \frac{\partial \mathbf{N}^T}{\partial X} P d\Omega_0 = \int_{\Omega_0^e} \mathbf{B}_0^T P d\Omega_0 \quad \text{or} \quad f_{I,e}^{\text{int}} = \int_{\Omega_0^e} \frac{\partial N_I}{\partial X} P d\Omega_0 \tag{B2.2.3}$$

$$\mathbf{f}_e^{\text{ext}} = \int_{\Omega_0^e} \rho_0 \mathbf{N}^T b d\Omega_0 + \left(\mathbf{N}^T A_0 \bar{t}_x^0\right)\Big|_{\Gamma_t^e} \tag{B2.2.4}$$

$$\mathbf{M}_e = \int_{\Omega_0^e} \rho_0 \mathbf{N}^T \mathbf{N} d\Omega_0 \tag{B2.2.5}$$

$$\mathbf{M}\ddot{\mathbf{u}} + \mathbf{f}^{\text{int}} = \mathbf{f}^{\text{ext}} \tag{B2.2.6}$$

We will often write the internal nodal force expressions for the total Lagrangian formulation in terms of a $\mathbf{B}_0$ matrix, where $\mathbf{B}_0$ in the one-dimensional case is a row matrix defined by

$$\mathbf{B}_0 = [B_{0I}] \quad \text{where} \quad B_{0I} = N_{I,X} \tag{2.5.15}$$

The subscript zero is included to indicate that the derivatives are with respect to the initial, or material, coordinates. The internal nodal forces (2.5.14) are then given by

$$\mathbf{f}_e^{\text{int}} = \int_{\Omega_0^e} \mathbf{B}_0^T P d\Omega_0 \quad \text{or} \quad f_{I,e}^{\text{int}} = \int_{\Omega_0^e} B_{0I} P d\Omega_0 \tag{2.5.16}$$

where we have used $d\Omega_0 = A_0\, dX$ and Ω_0^e is the initial domain of the element. In this notation the deformation gradient F and the one-dimensional strain are given by

$$F = \mathbf{B}_0\mathbf{x}^e, \quad \varepsilon = \mathbf{B}_0\mathbf{u}^e \tag{2.5.17}$$

Example 2.1 Two-node linear displacement element Consider a two-node element as shown in Figure 2.4. The element shown is initially of length ℓ_0 and constant cross-sectional area A_0. At any subsequent time t, the length is $\ell(t)$ and the cross-sectional area is $A(t)$; the dependence of ℓ and A on time t will not be explicitly noted henceforth. The initial cross-sectional area of the element is taken to be constant, that is, independent of X.

Displacement field, strain, and $\mathbf{B}_0$ *matrix* The displacement field is given by the linear Lagrange interpolant expressed in terms of the material coordinates:

$$u(X,t) = \frac{1}{\ell_0}\begin{bmatrix} X_2 - X & X - X_1 \end{bmatrix}\begin{Bmatrix} u_1(t) \\ u_2(t) \end{Bmatrix} \tag{E2.1.1}$$

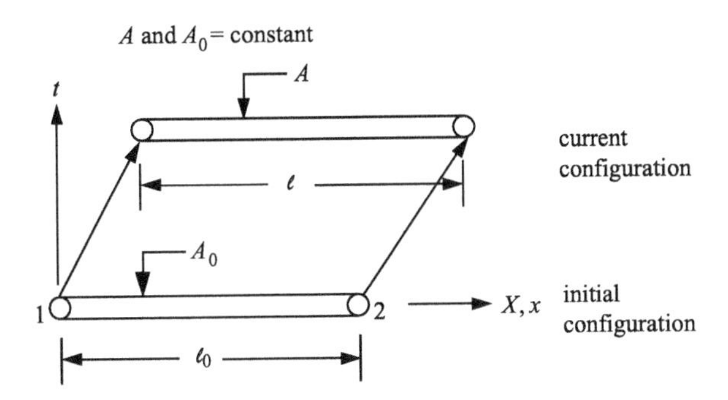

Figure 2.4 Two-node element in one dimension for total Lagrangian formulation showing the under-formed initial configuration and the deformed current configuration

where $\ell_0 = X_2 - X_1$. The strain measure is evaluated in terms of the nodal displacements by using (B2.2.2):

$$\varepsilon(X,t) = u_{,X} = \frac{1}{\ell_0}[-1 \quad +1]\begin{Bmatrix} u_1(t) \\ u_2(t) \end{Bmatrix} \tag{E2.1.2}$$

This defines the $\mathbf{B}_0$ matrix to be

$$B_0 = \frac{1}{\ell_0}[-1 \quad +1] \tag{E2.1.3}$$

Nodal internal forces The internal nodal forces are given by (2.5.16):

$$\mathbf{f}_e^{\text{int}} = \int_{\Omega_0^e} \mathbf{B}_0^T P \, d\Omega_0 = \int_{X_1}^{X_2} \frac{1}{\ell_0}\begin{Bmatrix} -1 \\ +1 \end{Bmatrix} P A_0 \, dX \tag{E2.1.4}$$

If we assume that the cross-sectional area and the nominal stress P are constant, the integrand in (E2.1.4) is then constant, so the integral can be evaluated by taking the product of the integrand and the initial length of the element ℓ_0, which gives

$$\mathbf{f}_e^{\text{int}} = \begin{Bmatrix} f_1 \\ f_2 \end{Bmatrix}_e^{\text{int}} = A_0 P \begin{Bmatrix} -1 \\ +1 \end{Bmatrix} \tag{E2.1.5}$$

From this, we can see that the nodal internal forces are equal and opposite, so the element internal nodal forces are in equilibrium, even in a dynamic problem. This characteristic of element nodal forces will apply to all elements which translate without deformation; it does not apply to axisymmetric elements. Since $P = T/A_0$ (see (2.2.8)) the nodal forces are equal to the load T carried by the element.

Nodal external forces The external nodal forces arising from the body force are given by (B2.2.4):

$$\mathbf{f}_e^{\text{ext}} = \int_{\Omega_0^e} \rho_0 \mathbf{N}^T b A_0 \, dX = \int_{X_1}^{X_2} \frac{\rho_0}{\ell_0} \begin{Bmatrix} X_2 - X \\ X - X_1 \end{Bmatrix} b A_0 \, dX \tag{E2.1.6}$$

If we approximate the body forces $b(X, t)$ by a linear Lagrange interpolant

$$b(X,t) = b_1(t)\left(\frac{X_2 - X}{\ell_0}\right) + b_2(t)\left(\frac{X - X_1}{\ell_0}\right) \tag{E2.1.7}$$

and take A_0 to be constant, the evaluation of the integral in (E2.1.6) gives

$$\mathbf{f}_e^{\text{ext}} = \frac{\rho_0 A_0 \ell_0}{6} \begin{Bmatrix} 2b_1 + b_2 \\ b_1 + 2b_2 \end{Bmatrix} \tag{E2.1.8}$$

The evaluation of the external nodal forces is facilitated by expressing the integral in terms of a parent element coordinate

$$\xi = (X - X_1)/\ell_0, \quad \xi \in [0, 1] \tag{E2.1.9}$$

Element mass matrix The element mass matrix is given by (B2.2.5):

$$\begin{aligned} \mathrm{M}_e &= \int_{\Omega_0^e} \rho_0 \mathbf{N}^T \mathbf{N} d\Omega_0 = \int_0^1 \rho_0 \mathbf{N}^T \mathbf{N} A_0 \ell_0 d\xi \\ &= \int_0^1 \rho_0 \begin{Bmatrix} 1-\xi \\ \xi \end{Bmatrix} [1-\xi \quad \xi] A_0 \ell_0 d\xi = \frac{\rho_0 A_0 \ell_0}{6} \begin{bmatrix} 2 & 1 \\ 1 & 2 \end{bmatrix} \end{aligned} \tag{E2.1.10}$$

It can be seen from (E2.1.10) that the mass matrix is independent of time, since it depends only on the initial density, initial cross-sectional area and initial length. The diagonal mass matrix as obtained by the row-sum technique (2.4.25) is

$$\mathbf{M}_e = \frac{\rho_0 A_0 \ell_0}{2} \begin{bmatrix} 1 & 0 \\ 0 & 1 \end{bmatrix} = \frac{\rho_0 A_0 \ell_0}{2} \mathbf{I} \tag{E2.1.11}$$

As can be seen from (E2.1.11), in the diagonal mass matrix for this element, half of the mass of the element is ascribed to each of the nodes. For this reason, it is often called the lumped mass matrix: half the mass has been 'lumped' at each node.

Example 2.2 Example of assembled equations Consider a mesh of two elements as shown in Figure 2.5. The body force $b(x)$ is constant, b. We will develop the governing equations for this mesh; the equation for the center node is of particular interest since it represents the typical equation for the interior node of any one-dimensional mesh.

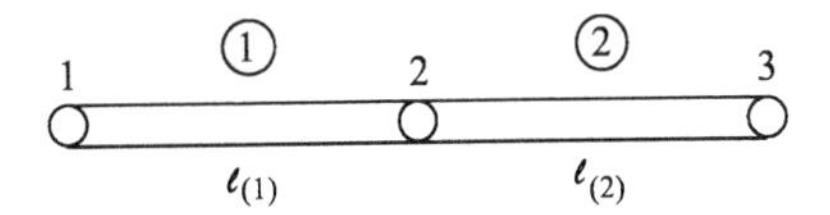

Figure 2.5 A mesh of two elements

The connectivity matrices $\mathbf{L}_e$ for this mesh are

$$\mathbf{L}_{(1)} = \begin{bmatrix} 1 & 0 & 0 \\ 0 & 1 & 0 \end{bmatrix} \tag{E2.2.1}$$

$$\mathbf{L}_{(2)} = \begin{bmatrix} 0 & 1 & 0 \\ 0 & 0 & 1 \end{bmatrix} \tag{E2.2.2}$$

The global internal force matrix by (2.5.5) is given in terms of the element internal forces by

$$\mathbf{f}^{\text{int}} = \mathbf{L}_{(1)}^T \mathbf{f}_{(1)}^{\text{int}} + \mathbf{L}_{(2)}^T \mathbf{f}_{(2)}^{\text{int}} = \begin{Bmatrix} f_1 \\ f_2 \\ 0 \end{Bmatrix}_{(1)}^{\text{int}} + \begin{Bmatrix} 0 \\ f_1 \\ f_2 \end{Bmatrix}_{(2)}^{\text{int}} \tag{E2.2.3}$$

which from (E2.1.4) gives

$$\mathbf{f}^{\text{int}} = A_0^{(1)} P^{(1)} \begin{Bmatrix} -1 \\ +1 \\ 0 \end{Bmatrix} + A_0^{(2)} P^{(2)} \begin{Bmatrix} 0 \\ -1 \\ +1 \end{Bmatrix} \tag{E2.2.4}$$

Similarly

$$\mathbf{f}^{\text{ext}} = \mathbf{L}_{(1)}^T \mathbf{f}_{(1)}^{\text{ext}} + \mathbf{L}_{(2)}^T \mathbf{f}_{(2)}^{\text{ext}} = \begin{Bmatrix} f_1 \\ f_2 \\ 0 \end{Bmatrix}_{(1)}^{\text{ext}} + \begin{Bmatrix} 0 \\ f_1 \\ f_2 \end{Bmatrix}_{(2)}^{\text{ext}} \tag{E2.2.5}$$

and using (E2.1.8) with constant body force gives

$$\mathbf{f}^{\text{ext}} = \frac{\rho_0^{(1)} A_0^{(1)} \ell_0^{(1)}}{2} \begin{Bmatrix} b \\ b \\ 0 \end{Bmatrix} + \frac{\rho_0^{(2)} A_0^{(2)} \ell_0^{(2)}}{2} \begin{Bmatrix} 0 \\ b \\ b \end{Bmatrix} \tag{E2.2.6}$$

The global, assembled mass matrix is given by (2.5.9):

$$\mathbf{M} = \mathbf{L}_{(1)}^T \mathbf{M}_{(1)} \mathbf{L}_{(1)} + \mathbf{L}_{(2)}^T \mathbf{M}_{(2)} \mathbf{L}_{(2)} \tag{E2.2.7}$$

and by (E2.1.10):

$$\mathbf{M} = \mathbf{L}_{(1)}^T \frac{\rho_0^{(1)} A_0^{(1)} \ell_0^{(1)}}{6} \begin{bmatrix} 2 & 1 \\ 1 & 2 \end{bmatrix} \mathbf{L}_{(1)} + \mathbf{L}_{(2)}^T \frac{\left(\rho_0^{(2)} A_0^{(2)} \ell_0^{(2)}\right)}{6} \begin{bmatrix} 2 & 1 \\ 1 & 2 \end{bmatrix} \mathbf{L}_{(2)} \tag{E2.2.8}$$

If we define $m_1 = \left(\rho_0^{(1)} A_0^{(1)} \ell_0^{(1)}\right)/6$, $m_2 = \left(\rho_0^{(2)} A_0^{(2)} \ell_0^{(2)}\right)/6$, the assembled mass matrix is

$$\mathbf{M} = \begin{bmatrix} 2m_1 & m_1 & 0 \\ m_1 & 2(m_1 + m_2) & m_2 \\ 0 & m_2 & 2m_2 \end{bmatrix} \tag{E2.2.9}$$

Writing out the second equation of motion for this system (which is obtained from the second row of **M**, $\mathbf{f}^{\text{ext}}$ and $\mathbf{f}^{\text{int}}$) gives

$$\begin{aligned} &\frac{1}{6}\rho_0^{(1)} A_0^{(1)} \ell_0^{(1)} \ddot{u}_1 + \frac{1}{3}\left(\rho_0^{(1)} A_0^{(1)} \ell_0^{(1)} + \rho_0^{(2)} A_0^{(2)} \ell_0^{(2)}\right) \ddot{u}_2 + \frac{1}{6}\rho_0^{(2)} A_0^{(2)} \ell_0^{(2)} \ddot{u}_3 \\ &+ A_0^{(1)} P^{(1)} - A_0^{(2)} P^{(2)} = \frac{b}{2}\left(\rho_0^{(1)} A_0^{(1)} \ell_0^{(1)} + \rho_0^{(2)} A_0^{(2)} \ell_0^{(2)}\right) \end{aligned} \tag{E2.2.10}$$

To simplify the form of the assembled equations, we now consider a uniform mesh with constant initial properties, so $\rho_0^{(1)} = \rho_0^{(2)} = \rho_0, A_0^{(1)} = A_0^{(2)} = A_0, \ell_0^{(1)} = \ell_0^{(2)} = \ell_0$. Dividing by $-A_0\ell_0$, we obtain the following equation of motion at node 2:

$$\frac{P^{(2)} - P^{(1)}}{\ell_0} + \rho_0 b = \rho_0 \left(\frac{1}{6}\ddot{u}_1 + \frac{2}{3}\ddot{u}_2 + \frac{1}{6}\ddot{u}_3\right) \tag{E2.2.11}$$

If the mass matrix is lumped, the corresponding expression is

$$\frac{P^{(2)} - P^{(1)}}{\ell_0} + \rho_0 b = \rho_0 \ddot{u}_2 \tag{E2.2.12}$$

Equation (E2.2.12) is equivalent to a finite difference expression for the momentum equation (2.2.14) with A_0 constant: it is only necessary to use the central difference expression $P_{,X}(X_2) = (P^{(2)} - P^{(1)})/\ell_0$ to reveal the identity. Thus the finite element procedure appears to be a circuitous way of obtaining what follows simply and directly from a finite difference approximation. The advantage of a finite element approach is that it gives a consistent procedure for obtaining semidiscrete equations when the element lengths, cross-sectional area, and density vary; try to obtain (E2.2.10) by finite differences! Furthermore, for linear problems, a finite element solution can be shown to provide the best approximation in the sense that the error is minimized in the energy norm (see Strang and Fix (1973)). The finite element method also gives the means of obtaining consistent mass matrices and higher-order elements, which are more accurate. But the main advantage of the finite element method, which undoubtedly has been the driving force behind its popularity, is the ease with which it can model complex shapes. This of course is not apparent in one-dimensional problems.

Example 2.3 Three-node quadratic displacement element Consider the three-node element of length L_0 and cross-sectional area A_0 shown later in this chapter in Figure 2.7. Although in this example we do not assume node 2 to be midway between the other nodes, this is recommended. The mapping between the material coordinates X and the element coordinate ξ is given by

$$X(\xi)=\mathbf{N}(\xi)\mathbf{X}_e=\begin{bmatrix}\frac{1}{2}\xi(\xi-1) & 1-\xi^2 & \frac{1}{2}\xi(\xi+1)\end{bmatrix}\begin{Bmatrix}X_1\\X_2\\X_3\end{Bmatrix} \tag{E2.3.1}$$

where $\mathbf{N}(\xi)$ is the matrix of Lagrange interpolants, or shape functions, and $\xi\in[-1,1]$ is the element coordinate. The displacement field is given by the same interpolants

$$u(\xi,t)=\mathbf{N}(\xi)\mathbf{u}_e(t)=\begin{bmatrix}\frac{1}{2}\xi(\xi-1) & 1-\xi^2 & \frac{1}{2}\xi(\xi+1)\end{bmatrix}\begin{Bmatrix}u_1(t)\\u_2(t)\\u_3(t)\end{Bmatrix} \tag{E2.3.2}$$

By the chain rule

$$\varepsilon=F-1=u_{,X}=u_{,\xi}\,\xi_{,X}=u_{,\xi}(X_{,\xi})^{-1}=\frac{1}{2X_{,\xi}}\begin{bmatrix}2\xi-1 & -4\xi & 2\xi+1\end{bmatrix}\mathbf{u}_e \tag{E2.3.3}$$

We have used the fact that in one dimension, $\xi_{,X}=(X_{,\xi})^{-1}$. We can write this as

$$\varepsilon=\mathbf{B}_0\mathbf{u}_e \quad \text{where} \quad \mathbf{B}_0=\frac{1}{2X_{,\xi}}\begin{bmatrix}2\xi-1 & -4\xi & 2\xi+1\end{bmatrix} \tag{E2.3.4}$$

The internal nodal forces are given by (2.5.16):

$$\mathbf{f}_e^{\text{int}}=\int_{\Omega_o^e}\mathbf{B}_0^T P\,d\Omega_0=\int_{-1}^{1}\frac{1}{2X_{,\xi}}\begin{Bmatrix}2\xi-1\\-4\xi\\2\xi+1\end{Bmatrix}PA_0X_{,\xi}\,d\xi=\int_{-1}^{1}\frac{1}{2}\begin{Bmatrix}2\xi-1\\-4\xi\\2\xi+1\end{Bmatrix}PA_0\,d\xi \tag{E2.3.5}$$

This integral is generally evaluated by numerical integration. For the purpose of examining this element further, let $P(\xi)$ be linear in ξ:

$$P(\xi)=P_1\frac{(1-\xi)}{2}+P_3\frac{(1+\xi)}{2} \tag{E2.3.6}$$

where P_1 and P_3 are the values of P at nodes 1 and 3, respectively. If $X_{,\xi}$ is constant, this is exact if the material is linear, since F is then linear in ξ by (E2.3.3). The internal forces are then given by

$$\mathbf{f}_e^{\text{int}}=\begin{Bmatrix}f_1\\f_2\\f_3\end{Bmatrix}_e^{\text{int}}=\frac{A_0}{6}\begin{Bmatrix}-5P_1-P_3\\4P_1-4P_3\\P_1+5P_3\end{Bmatrix} \tag{E2.3.7}$$

When P is constant, the nodal force at the center node vanishes and the nodal forces at the end nodes are equal and opposite with magnitude A_0P, as in the two-node element. In addition, for any values of P_1 and P_3, the sum of the internal nodal forces vanishes. Thus this element is also in equilibrium.

The external nodal forces are

$$\mathbf{f}_e^{\text{ext}} = \int_{-1}^{+1} \begin{Bmatrix} \frac{1}{2}\xi(\xi-1) \\ 1-\xi^2 \\ \frac{1}{2}\xi(\xi+1) \end{Bmatrix} \rho_0 b A_0 X_{,\xi}\, d\xi + \begin{Bmatrix} \frac{1}{2}\xi(\xi-1) \\ 1-\xi^2 \\ \frac{1}{2}\xi(\xi+1) \end{Bmatrix} A_0 \bar{t}_x^0 \Big|_{\Gamma_t^e} \tag{E2.3.8}$$

where the shape functions in the last term are either 1 or 0 at a traction boundary. Since $X_{,\xi} = \xi(X_1 + X_3 - 2X_2) + \frac{1}{2}(X_3 - X_1)$ by (E2.3.1), then

$$\mathbf{f}_e^{\text{ext}} = \frac{\rho_0 b A_0}{6} \begin{Bmatrix} L_0 - 2(X_1 + X_3 - 2X_2) \\ 4L_0 \\ L_0 + 2(X_1 + X_3 - 2X_2) \end{Bmatrix} + \begin{Bmatrix} \frac{1}{2}\xi(\xi-1) \\ 1-\xi^2 \\ \frac{1}{2}\xi(\xi+1) \end{Bmatrix} A_0 \bar{t}_x^0 \Big|_{\Gamma_t^e} \tag{E2.3.9}$$

Element mass matrix The element mass matrix is

$$\begin{aligned} \mathbf{M}_e &= \int_{-1}^{+1} \begin{Bmatrix} \frac{1}{2}\xi(\xi-1) \\ 1-\xi^2 \\ \frac{1}{2}\xi(\xi+1) \end{Bmatrix} \begin{bmatrix} \frac{1}{2}\xi(\xi-1) & 1-\xi^2 & \frac{1}{2}\xi(\xi+1) \end{bmatrix} \rho_0 A_0 X_{,\xi}\, d\xi \\ &= \frac{\rho_0 A_0}{30} \begin{bmatrix} 4L_0 - 6a & 2L_0 - 4a & -L_0 \\ & 16L_0 & 2L_0 + 4a \\ \text{sym} & & 4L_0 + 6a \end{bmatrix} \end{aligned} \tag{E2.3.10}$$

where $a = X_1 + X_3 - 2X_2$ If the node 2 is at the midpoint of the element, i.e., $X_1 + X_3 = 2X_2$, we have

$$\mathbf{M}_e = \frac{\rho_0 A_0 L_0}{30} \begin{bmatrix} 4 & 2 & -1 \\ 2 & 16 & 2 \\ -1 & 2 & 4 \end{bmatrix}, \quad \mathbf{M}_e^{\text{diag}} = \frac{\rho_0 A_0 L_0}{6} \begin{bmatrix} 1 & 0 & 0 \\ 0 & 4 & 0 \\ 0 & 0 & 1 \end{bmatrix} \tag{E2.3.11}$$

where the mass matrix has been diagonalized by the row-sum technique on the right.

This result displays one of the shortcomings of diagonal masses for higher-order elements: most of the mass is lumped at the center node. This causes rather strange behavior when high-order modes are excited. Therefore, higher-order elements are usually avoided when a lumped mass matrix is necessary for efficiency.

2.6 Governing Equations for Updated Lagrangian Formulation

In the updated Lagrangian formulation, the discrete equations are formulated in the current configuration. The stress is measured by the Cauchy (physical) stress σ given by (2.2.7). The dependent variables are chosen to be the stress $\sigma(X, t)$ and the velocity $v(X, t)$. In the total Lagrangian formulation, we have used the displacement $u(X, t)$ as the independent variable; this is only a formal difference since both the displacement and velocity are computed in a numerical implementation.

In developing the updated Lagrangian formulation, we will sometimes need the dependent variables to be expressed in terms of the Eulerian coordinates. Conceptually this is a simple matter, for we can invert (2.2.1) to obtain

$$X = \phi^{-1}(x, t) \equiv X(x, t) \tag{2.6.1}$$

Any variable can then be expressed in terms of the Eulerian coordinates; for example $\sigma(X, t)$ can be expressed as $\sigma(X(x, t), t)$. While the inverse of a function can easily be written in symbolic form, in practice the construction of an inverse function in closed form is difficult, if not impossible. Therefore the standard technique in finite elements is to express variables in terms of element coordinates, which are sometimes called parent coordinates or natural coordinates. By using element coordinates, we can always express a function, at least implicitly, in terms of either the Eulerian or Lagrangian coordinates.

In the updated Lagrangian formulations, the strain measure is the rate-of-deformation given by

$$D_x = \frac{\partial v}{\partial x} \tag{2.6.2}$$

This is also called the velocity-strain. It is a rate measure of deformation, as indicated by two of the names. It is shown in Chapter 5 that

$$\int_0^t D_x(X, \bar{t})d\bar{t} = \ln F(X, t)$$

in one dimension, so the time integral of the rate-of-deformation corresponds to the 'natural' or 'logarithmic' strain; as discussed in Chapter 5, this does not hold for multi-dimensional states of strain.

The governing equations for the nonlinear continuum in one dimension are:

1. Conservation of mass (continuity equation):

$$\rho J = \rho_0 \quad \text{or} \quad \rho F A = \rho_0 A_0 \tag{2.6.3}$$

2. Conservation of momentum:

$$\frac{\partial}{\partial x}(A\sigma) + \rho A b = \rho A \dot{v} \quad \text{or} \quad (A\sigma)_{,x} + \rho A b = \rho A \dot{v} \tag{2.6.4}$$

3. Measure of deformation:

$$D_x = \frac{\partial v}{\partial x} \quad \text{or} \quad D_x = v_{,x} \tag{2.6.5}$$

4. Constitutive equation:
   in total form:

$$\sigma(X,t) = S^{\sigma D}\left(D_x(X,t)\ldots, \quad \int_0^t D_x(X,\bar{t})d\bar{t}, \sigma(X,\bar{t}), \ \bar{t} \le t, \text{ etc.} \right) \tag{2.6.6a}$$

   in rate form:

$$\sigma_{,t}(X,t) = S_t^{\sigma D}(D_x(X,\bar{t}), \quad \sigma(X,\bar{t}), \ \bar{t} \le t, \text{ etc.}) \tag{2.6.6b}$$

5. Energy conservation:

$$\rho\dot{w}^{\text{int}} = \sigma D_x - q_{x,x} + \rho s, \quad \textit{where} \quad q_x = \text{heat flux}, s = \text{heat source} \tag{2.6.7}$$

The mass conservation equation in the updated Lagrangian formulation is the same as in the total Lagrangian formulation. The momentum equation in the updated formulation involves derivatives with respect to the Eulerian coordinates, whereas in the total Lagrangian formulation, derivatives were with respect to Lagrangian coordinates. In addition, the nominal stress is replaced by the Cauchy stress, and current values of the cross-sectional area A and density ρ are used. The constitutive equations as written here relate the rate-of-deformation $D_x(X, t)$ or its integral, the logarithmic strain, to the Cauchy stress or its rate. Note that all variables in the constitutive equation are functions of material coordinates. The subscript 't' on (2.6.6b) indicates that the constitutive equation is a rate equation. We can also use a constitutive equation expressed in terms of the nominal stress and s. It is then necessary to transform the nominal stress to the Cauchy stress before using the momentum equation.

2.6.1 *Boundary and Interior Continuity Conditions*

The boundary conditions are

$$v(X,t) = \bar{v}(X,t) \quad \text{on} \quad \Gamma_v \tag{2.6.8}$$

$$n\sigma(X,t) = \bar{t}_x(X,t) \quad \text{on} \quad \Gamma_t \tag{2.6.9}$$

where $\bar{v}(t)$ and $\bar{t}_x(t)$ are the prescribed velocity and traction, respectively, and n is the normal to the domain. Although the boundary condition is applied to the velocity, a velocity boundary condition is equivalent to a displacement boundary condition, since the velocity is a time derivative of the displacement. The tractions in (2.6.9) are in units of force per current area. They are related to the tractions on the undeformed area by

$$\bar{t}_x A = \bar{t}_x^0 A_0 \tag{2.6.10}$$

Note that we always retain the subscript on the traction to distinguish it from the time t. The relation between the traction and velocity boundaries is the same as in (2.2.28):

$$\Gamma_v \cup \Gamma_t = \Gamma \quad \Gamma_v \cup \Gamma_t = 0 \tag{2.6.11}$$

In addition we have the interior continuity conditions: $[\![\sigma A]\!] = 0$.

2.6.2 *Initial Conditions*

Since we have chosen the velocity and the stresses as the dependent variables, the initial conditions are imposed on these variables:

$$\sigma(X,0) = \sigma_0(X) \quad v(X,0) = v_0(X) \tag{2.6.12}$$

The initial displacements are assumed to vanish initially. In most practical problems, this choice of initial conditions is more suitable than conditions on velocities and displacements: this is discussed in Chapter 4.

2.7 Weak Form for Updated Lagrangian Formulation

In this section, the weak form for the momentum equation is developed. Recall that the dependent variables are the velocity $v(X, t)$ and the stress $\sigma(X, t)$. The conditions on the trial functions $v(X, t)$ and the test functions $\delta v(X)$ are:

$$v(X,t) \in u \quad u = \left\{ v(X,t) \mid v \in C^0(X), \quad v = \bar{v} \text{ on } \Gamma_v \right\} \tag{2.7.1}$$

$$\delta v(X) \in u_0 \quad u_0 = \left\{ \delta v(X) \mid \delta v \in C^0(X), \quad \delta v = 0 \text{ on } \Gamma_v \right\} \tag{2.7.2}$$

These admissibility conditions are identical to those for the test and trial displacements in the total Lagrangian formulation. As in the total Lagrangian formulation, the stress $\sigma(X, t)$ is assumed to be a C^{-1} function in space. The current domain is $[x_a(t), x_b(t)]$, where $x_a = \phi(X_a, t)$, $x_b = \phi(X_b, t)$.

The strong form consists of the momentum equation, the traction boundary conditions and the interior continuity conditions. The weak form is developed by multiplying the momentum equation (2.6.4) by the test function $\delta v(X)$ and integrating over the current domain. The current domain of the body is appropriate, since the momentum equation involves derivatives with respect to the spatial (Eulerian) coordinates. This gives

$$\int_{x_a}^{x_b} \delta v \left[(A\sigma)_{,x} + \rho A b - \rho A \frac{Dv}{Dt} \right] dx = 0 \tag{2.7.3}$$

Integration by parts is performed as in Section 2.3 (see (2.3.2)–(2.3.4)) and gives

$$\begin{aligned} \int_{x_a}^{x_b} \delta v (A\sigma)_{,x}\, dx &= \int_{x_a}^{x_b} \left[(\delta v A \sigma)_{,x} - \delta v_{,x} A\sigma \right] dx \\ &= (\delta v A n \sigma)\big|_{\Gamma_t} - \sum_i \delta v [\![A\sigma]\!]_{\Gamma_i} - \int_{x_a}^{x_b} \delta v_{,x} A\sigma\, dx \end{aligned} \tag{2.7.4}$$

where Γ_i are the points of discontinuity of $A\sigma$. We have used the fundamental theorem of calculus to convert a line (domain) integral to a sum of point (boundary and jump) values, with Γ changed to Γ_t because $\delta v\,(X)=0$ on Γ_v; see (2.7.2). Since the strong form holds, the traction boundary condition (2.6.9) gives $n\sigma = \bar{t}_x$ and the interior continuity condition gives $[A\sigma]=0$. When these are substituted into the above, we obtain

$$\int_{x_a}^{x_b}\left[\delta v_{,x}\, A\sigma - \delta v\left(\rho A b - \rho A \frac{Dv}{Dt}\right)\right]dx - \left(\delta v A \bar{t}_x\right)\Big|_{\Gamma_t} = 0 \tag{2.7.5}$$

This weak form is often called the *principle of virtual power* (or principle of virtual velocities: see Malvern, 1969, p. 241). If the test function is considered a velocity, then each term in the above corresponds to a virtual power; for example $\rho Ab\,dx$ is a force, and when multiplied by $\delta v(X)$ gives a virtual power. Therefore, the above weak form will be distinguished from the principle of virtual work in Section 2.3 by designating each term by P. However, it should be stressed that this physical interpretation of the weak form is entirely a matter of convenience; the test function $\delta v(X)$ need not have any of the attributes of a velocity; it can be any function which satisfies (2.7.2).

We define the virtual internal power by

$$\delta P^{\text{int}} = \int_{x_a}^{x_b} \delta v_{,x}\, \sigma A\, dx = \int_{x_a}^{x_b} \delta D_x \sigma A\, dx = \int_{\Omega} \delta D_x \sigma\, d\Omega \tag{2.7.6}$$

where the second equality is obtained by taking a variation of (2.6.5), i.e., $\delta D_x = \delta v_{,x}$, while the third equality results from the relation $d\Omega = A\,dx$. The integral in (2.7.6) corresponds to the internal energy rate in the energy conservation equation (2.6.7) except that the rate-of-deformation D_x is replaced by δD_x, so calling this a virtual internal power is consistent with the energy equation.

The virtual powers due to external and inertial forces are defined similarly:

$$\delta P^{\text{ext}} = \int_{x_a}^{x_b} \delta v\, \rho b A\, dx + \left(\delta v A \bar{t}_x\right)\Big|_{\Gamma_t} = \int_{\Omega} \delta v\, \rho b\, d\Omega + \left(\delta v A \bar{t}_x\right)\Big|_{\Gamma_t} \tag{2.7.7}$$

$$\delta P^{\text{kin}} = \int_{x_a}^{x_b} \delta v\, \rho \dot{v} A\, dx = \int_{\Omega} \delta v\ \rho \dot{v}\, d\Omega \tag{2.7.8}$$

Using (2.7.6–8), the weak form (2.7.5) can be written as

$$\delta P = \delta P^{\text{int}} - \delta P^{\text{ext}} + \delta P^{\text{kin}} = 0 \tag{2.7.9}$$

where the terms are defined previously. The principle of virtual power states that

$$\text{if } v(X,t) \in u \text{ and } \delta p = 0 \quad \forall \delta v(X) \in u_0 \tag{2.7.10}$$

then the momentum equation (2.6.4), the traction boundary conditions (2.6.9) and the interior continuity conditions are satisfied. The validity of this principle can be established by simply reversing the steps used to obtain (2.7.5). All of the steps are reversible so we can deduce the strong form from the weak form.

The key difference of this weak form, as compared to the weak form for the total Lagrangian formulation, is that all integrals are over the current domain. However, the two weak forms are just different forms of the same principle; it is left as an exercise to show that the principle of virtual work can be transformed to the principle of virtual power.

2.8 Element Equations for Updated Lagrangian Formulation

We will now develop the updated Lagrangian formulation. The updated Lagrangian formulation is simply a transformation of the total Lagrangian formulation. Numerically, the discrete equations are identical, and in fact, we can use the total Lagrangian formulation for some of the nodal forces and the updated Lagrangian form for others in the same program. Students often ask why both methods are presented when they are basically identical. We must confess that the major reason for presenting both formulations is that both are widely used, so to understand the software and literature, a familiarity with both formulations is essential. However, in a first course, it is not unwise to skip one of these Lagrangian formulations.

2.8.1 *Finite Element Approximation*

The domain is subdivided into elements Ω_e, so that $\Omega = \cup \Omega_e$. The coordinates of the nodes in the initial configuration are given by $X_1, X_2, \ldots X_{n_N}$ and the current positions of the nodes are given by $x_1(t), x_2(t), \ldots x_{n_N}(t)$. The positions of the m nodes of element e in the initial configuration are denoted by $X_1^e, X_2^e, \ldots X_m^e$, and the positions of these nodes in the current configuration are given by $x_1^e(t), x_2^e(t), \ldots x_m^e(t)$. These are given by the finite element approximation to the motion

$$x_I(t) = x(X_I, t) \tag{2.8.1}$$

Thus each node of the mesh remains coincident with a material point.

We will develop the equations on an element level and then obtain the global equations by assembly using the scheme described in Section 2.5. As before, the physically motivated names will be employed to systematize the procedure.

The dependent variables are the velocity and the stress. The constitutive equation and the mass conservation equation are treated in strong form, the momentum equation in weak form. The mass conservation equation can be used to easily compute the density at any point since it is an algebraic equation. We develop the semidiscrete equations as if there were no essential boundary conditions and then impose these subsequently.

The velocity field in each element is approximated by

$$v(X, t) = \sum_{I=1}^{m} N_I(X) v_I(t) = \mathbf{N}(X)\mathbf{v}(t) \tag{2.8.2}$$

Although the shape functions are functions of the material coordinates X, they can be expressed in terms of spatial coordinates. For this purpose, the mapping $x = \phi(X, t)$ is inverted to give $X = \phi^{-1}(x, t)$ so the velocity field is

$$v(X,t) = \mathbf{N}(\phi^{-1}(x,t))\mathbf{v}(t) \tag{2.8.3}$$

Although developing the inverse mapping is often impossible, partial derivatives with respect to the spatial coordinates can be obtained by implicit differentiation, so the inverse mapping need never be calculated.

The acceleration field is given by taking the material time derivative of (2.8.2), which gives

$$\dot{v}(X,t) = \mathbf{N}(X)\dot{\mathbf{v}}(t) \equiv \mathbf{N}(X)\mathbf{a}(t) \tag{2.8.4}$$

It can be seen from this step that it is crucial that the shape functions be expressed as functions of the *material coordinates*. If the shape functions are expressed in terms of the Eulerian coordinates by

$$v(X,t) = \mathbf{N}(x)\mathbf{v}(t) = \mathbf{N}(\phi(X,t))\mathbf{v}(t) \tag{2.8.5}$$

then the material time derivative of the shape functions does not vanish and the accelerations cannot be expressed as a product of the same shape functions and nodal accelerations.

2.8.2 *Element Coordinates*

Finite element calculations are usually performed with the parent element coordinates ξ, called element coordinates for brevity. Some authors call them natural coordinates. Element coordinates, such as triangular coordinates and isoparametric coordinates, are particularly convenient for multi-dimensional elements.

Figure 2.6 shows a two-node element in the initial (reference) and current configurations. The parent domain is the interval $0 \le \xi \le 1$. The parent domain can be mapped onto the initial and current configurations. For example, in the two-node element, the mapping between the Eulerian coordinates and the element coordinates is given by

$$x(\xi,t) = x_1(t)(1-\xi) + x_2(t)\xi \tag{2.8.6}$$

For a general one-dimensional element this map is expressed in terms of the shape functions by

$$x(\xi,t) = \mathbf{N}(\xi)\mathbf{x}^e(t) \tag{2.8.7}$$

Specializing this to the initial time gives the map between the initial configuration and the parent domain

$$X(\xi) = \sum_{I=1}^{m} N_I(\xi)X_I^e = \mathbf{N}(\xi)\mathbf{X}^e \tag{2.8.8}$$

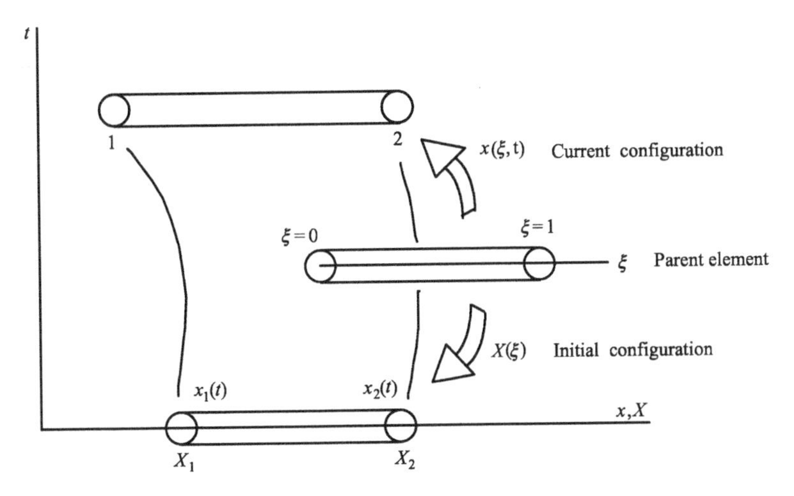

Figure 2.6 Role of parent configuration, showing mappings to the reference undeformed initial configuration and the current, deformed configuration in a Lagrangian mesh

which for the two-node element is

$$X(\xi) = X_1(1-\xi) + X_2\xi \tag{2.8.9}$$

The map between the Eulerian coordinates and the element coordinates, (2.8.6), changes with time, while the map between the initial configuration and the element domain is time invariant in a Lagrangian mesh. Therefore shape functions expressed in terms of the element coordinates as in (2.8.7) will be independent of time. If the initial map is such that every point in the parent element ξ maps onto a unique point of the initial configuration, and for every point X there exists a point ξ, then the parent element coordinates can serve as material labels. Such a map is called one-to-one. The map between the current configuration and the parent domain must also be one-to-one; this is discussed further in Chapter 3.

It follows from (2.8.7) and (2.8.8) that the displacements can also be interpolated by the same shape functions, since

$$u(\xi, t) = x(\xi, t) - X(\xi) = \mathbf{N}(\xi)(\mathbf{x}^e(t) - \mathbf{X}^e) = \mathbf{N}(\xi)\mathbf{u}^e(t) \tag{2.8.10}$$

The velocities and accelerations are also given by material derivatives of the displacement, while the test function is given by the same shape functions, so

$$v(\xi, 1) = \mathbf{N}(\xi)\mathbf{v}^e(t) \quad a(\xi, t) = \mathbf{N}(\xi)\ddot{\mathbf{u}}^e(t) \quad \delta v(\xi, t) = \mathbf{N}(\xi)\delta\mathbf{v}^e \tag{2.8.11}$$

since the shape functions are independent of time.

Using (2.8.2) and (2.6.5) and noting (2.8.3), the rate-of-deformation can be expressed in terms of the shape functions by

$$D_x(x, t) = v_{,x}(x, t) = \mathbf{N}_{,x}(X(x, t))\mathbf{v}^e(t) \tag{2.8.12}$$

where we have indicated the dependence of the shape functions on the Eulerian coordinates. The rate-of-deformation will be expressed in terms of nodal velocities via a **B** matrix by

$$D_x = v_{,x} = \mathbf{B}\mathbf{v}^e = \sum_{I=1}^{m} B_I v_I^e \tag{2.8.13}$$

where

$$\mathbf{B} = \mathbf{N}_{,x} \quad \text{or} \quad B_I = N_{I,x} \tag{2.8.14}$$

This **B** matrix differs from the $\mathbf{B}_0$ matrix used in the total Lagrangian formulation in that the derivatives are taken with respect to the Eulerian coordinates. The spatial derivatives of the shape function are obtained by the chain rule

$$\mathbf{N}_{,\xi} = \mathbf{N}_{,x}\, x_{,\xi} \quad \text{so} \quad \mathbf{N}_{,x} = \mathbf{N}_{,\xi}\, x_{,\xi}^{-1} \tag{2.8.15}$$

From the above, it follows that

$$D_x(\xi, t) = x_{,\xi}^{-1}\, \mathbf{N}_{,\xi}(\xi)\mathbf{v}^e(t) = \mathbf{B}(\xi)\mathbf{v}^e(t) \quad \mathbf{B}(\xi) = \mathbf{N}_{,\xi}\, x_{,\xi}^{-1} \tag{2.8.16}$$

2.8.3 *Internal and External Nodal Forces*

We now use the procedure given in Sections 2.4 and 2.5 to determine nodal forces corresponding to each term of the weak form on an element level. The assembly of the global equations and treatment of essential boundary conditions are identical to the procedure in the total Lagrangian formulation.

The internal nodal forces will be developed from the virtual internal power. Defining the element internal nodal forces so that the scalar product with the virtual velocities gives the internal virtual power, then from (2.7.6) and (2.8.13) we can write

$$\delta P_e^{\text{int}} \equiv \delta \mathbf{v}_e^T \mathbf{f}_e^{\text{int}} = \int_{x_1^e(t)}^{x_m^e(t)} \delta v_{,x}^T\, \sigma A\, dx = \delta \mathbf{v}_e^T \int_{x_1^e(t)}^{x_m^e(t)} \mathbf{N}_{,x}^T\, \sigma A\, dx \tag{2.8.17}$$

The transpose is taken of the first term in the integrand even though it is a scalar so that the expression remains consistent when δv is replaced by a matrix product. From the arbitrariness of $\delta \mathbf{v}_e$, it follows that

$$\mathbf{f}_e^{\text{int}} = \int_{x_1^e(t)}^{x_m^e(t)} \mathbf{N}_{,x}^T\, \sigma A\ dx \equiv \int_{x_1^e(t)}^{x_m^e(t)} \mathbf{B}^T \sigma A dx \quad \text{or} \quad \mathbf{f}_e^{\text{int}} = \int_{\Omega^e} \mathbf{B}^T \sigma d\Omega \tag{2.8.18}$$

The internal nodal forces can then be evaluated in terms of element coordinates by transforming (2.8.18) to the parent domain and with $dx = x_{,\xi}\, d\xi$, giving

$$\mathbf{f}_e^{\text{int}} = \int_{x_1^e(t)}^{x_m^e(t)} \mathbf{N}_{,x}^T\, \sigma A dx = \int_{\xi_1}^{\xi_m} \mathbf{N}_{,\xi}^T\, x_{,\xi}^{-1}\, \sigma A\, x_{,\xi}\, d\xi = \int_{\xi_1}^{\xi_m} \mathbf{N}_{,\xi}^T\, \sigma A d\xi \tag{2.8.19}$$

The last form in (2.8.19) is nice, but this simplification can be made only in one dimension.

The external nodal forces are obtained from the virtual external power (2.7.7):

$$\delta P_e^{\text{ext}} = \delta \mathbf{v}_e^T \mathbf{f}_e^{\text{ext}} = \int_{\Omega^e} \delta v^T \rho b d\Omega + \left(\delta v^T A \bar{t}_x\right)\Big|_{\Gamma_t} \tag{2.8.20}$$

Substituting (2.8.11) into the right-hand side of the above and using the arbitrariness of $\delta \mathbf{v}_e$ gives

$$\mathbf{f}_e^{\text{ext}} = \int_{x_1^e}^{x_m^e} \mathbf{N}^T \rho b A dx + \left(\mathbf{N}^T A \bar{t}_x\right)\Big|_{\Gamma_t^e} = \int_{\Omega^e} \mathbf{N}^T \rho b d\Omega + \left(\mathbf{N}^T A \bar{t}_x\right)\Big|_{\Gamma_t^e} \tag{2.8.21}$$

where the second term contributes only when the boundary coincides with a node of the element.

2.8.4 *Mass Matrix*

The inertial nodal forces and mass matrix are obtained from the virtual inertial power (2.7.8):

$$\delta P^{\text{kin}} = \delta \mathbf{v}_e^T \mathbf{f}_e^{\text{kin}} = \int_{x_1^e(t)}^{x_m^e(t)} \delta v^T \rho \frac{Dv}{Dt} A\, dx \tag{2.8.22}$$

Substituting (2.8.11) into this yields

$$\mathbf{f}_e^{\text{kin}} = \int_{x_1^e(t)}^{x_m^e(t)} \rho \mathbf{N}^T \mathbf{N} A \ dx \dot{\mathbf{v}}^e = \mathbf{M}^e \dot{\mathbf{v}}^e \tag{2.8.23}$$

where the inertial force has been written as the product of a mass matrix $\mathbf{M}$ and the nodal accelerations. The mass matrix is given by

$$\mathbf{M}^e = \int_{x_1^e(t)}^{x_m^e(t)} \rho \mathbf{N}^T \mathbf{N} A \ dx = \int_{\Omega_e} \rho \mathbf{N}^T \mathbf{N} \ d\Omega \tag{2.8.24}$$

The above form suggests that the mass matrix is a function of time, since the limits of integration and the cross-sectional area are functions of time. However, if we use the mass conservation equation (2.2.11) in the form $\rho_0 A_0 \ dX = \rho A \ dx$, we can obtain a time-invariant form:

$$\mathbf{M}^e = \int_{X_1^e}^{X_m^e} \rho_0 \mathbf{N}^T \mathbf{N} A_0 \, dX \tag{2.8.25}$$

This formula for the mass matrix is identical to the expression developed for the total Lagrangian formulation, (2.4.11). The advantage of this expression is that it clearly shows that the mass matrix in the updated Lagrangian formulation does not change with time and therefore need not be recomputed during the simulation, which is not clear from (2.8.24).

2.8.5 *Equivalence of Updated and Total Lagrangian Formulations*

The internal and external nodal forces in the updated and total Lagrangian formulations can be shown to be identical. To show the equivalence for the nodal internal forces, we express the spatial derivative of the shape function in terms of the material derivative by the chain rule:

$$\mathbf{N}_{,x}(X) = \mathbf{N}_{,X}\frac{\partial X}{\partial x} \tag{2.8.26}$$

From (2.8.26) we have $\mathrm{N}_{,x}\,dx = \mathrm{N}_{,X}\,dX$, and substituting this into (2.8.18) gives

$$\mathbf{f}_e^{\text{int}} = \int_{x_1^e(t)}^{x_m^e(t)} \mathbf{N}_{,x}^T \sigma A\,dx = \int_{X_1^e}^{X_m^e} \mathbf{N}_{,X}^T \sigma A\,dX \tag{2.8.27}$$

where the limits of integration in the third expression have been changed to the material coordinates of the nodes since the integral has been changed to the initial configuration. If we now use the identity $\sigma A = PA_0$ (2.2.9), we obtain from the previous that

$$\mathbf{f}_e^{\text{int}} = \int_{X_1^e}^{X_m^e} \mathbf{N}_{,X}^T PA_0\,dX \tag{2.8.28}$$

This expression is identical to the expression for the internal nodal forces in the total Lagrangian formulation, (2.5.14). Thus the expressions for the internal nodal forces in the updated and total Lagrangian formulations are simply two ways of expressing the same thing.

The equivalence of the external nodal forces is shown by using the conservation of mass (2.2.11). Starting with (2.8.21) and using (2.2.11) gives

$$\mathbf{f}_e^{\text{ext}} = \int_{x_1^e}^{x_m^e} \mathbf{N}^T \rho bA\;dx + \left(\mathbf{N}^T A\bar{t}_x\right)\Big|_{\Gamma_t^e} = \int_{X_1^e}^{X_m^e} \mathbf{N}^T \rho_0 bA_0\;dX + \left(\mathbf{N}^T A_0\bar{t}_x^{\,0}\right)\Big|_{\Gamma_t^e} \tag{2.8.29}$$

where we have used the identity $t_x A = t_x^0 A_0$ (2.6.10), in the last term. Equation (2.8.29) is identical to (2.4.8), the expression in the total Lagrangian formulation.

Thus, the updated Lagrangian formulation (Box 2.3) and the total Lagrangian formulation simply provide alternative expressions for the same discrete equations. The formulation which is chosen is simply a matter of convenience. Moreover, either of these formulations may be used for different nodal forces in the same calculation. For example, the internal nodal forces can be evaluated by an updated Lagrangian approach and the external nodal forces by a total Lagrangian approach in the same calculation. The total and updated Lagrangian formalisms simply reflect different ways of describing the stress and strain measures and different ways of evaluating derivatives and integrals. We have also used different dependent variables in the two formulations, the velocity and stress in the updated formulations, the nominal stress and the displacement in the total formulation. However, this difference is not intrinsic to the formulation.

Box 2.3 Discrete equations updated Lagrangian formulation

$$D_x = \sum_{I=1}^{m} \frac{\partial N_I}{\partial x} v_I^e = \mathbf{B}\mathbf{v}^e \tag{B2.3.1}$$

$$\mathbf{f}^{\text{int}} = \int_{\Omega} \frac{\partial \mathbf{N}^T}{\partial x} \sigma d\Omega \quad \text{or} \quad \mathbf{f}^{\text{int}} = \int_{\Omega} \mathbf{B}^T \sigma d\Omega \tag{B2.3.2}$$

$$\mathbf{f}^{\text{ext}} = \int_{\Omega} \rho \mathbf{N}^T b d\Omega + \left(\mathbf{N}^T A \bar{t}_x\right)\Big|_{\Gamma_t} \tag{B2.3.3}$$

$$\mathbf{M} = \int_{\Omega_0} \rho_0 \mathbf{N}^T \mathbf{N} d\Omega_0 \text{ (same as total Lagrangian)} \tag{B2.3.4}$$

Global equations:

$$\mathbf{M}\ddot{\mathbf{u}} + \mathbf{f}^{\text{int}} = \mathbf{f}^{\text{ext}}$$

2.8.6 *Assembly, Boundary Conditions and Initial Conditions*

The assembly of the element matrices to obtain the global equations is identical to the procedure described for the total Lagrangian formulation in Section 2.5. The operations of gather are used to obtain the nodal velocities of each element, from which the strain measure, in this case the rate-of-deformation, can be computed in each element. The constitutive equation is then used to evaluate the stresses, from which the nodal internal forces can be computed by (2.8.19). The internal and external nodal forces are assembled into the global arrays by the scatter operation. Similarly, the imposition of essential boundary conditions and initial conditions is identical to that described in Section 2.4. The resulting global equations are identical to (2.4.17) and (2.4.15). Initial conditions are now needed on the velocities and stresses. For an unstressed body at rest, the initial conditions are given by

$$v_I = 0, \quad I = 1 \text{ to } n_N \quad \sigma_Q = 0, \quad Q = 1 \text{ to } n_Q \tag{2.8.30}$$

where Q refers to the n_Q quadrature points. Initial conditions in terms of the stresses and velocities are more appropriate for engineering problems, as discussed in Section 4.2. Nonzero initial values can be fitted by an L_2 projection described in Section 2.4.5.

Example 2.4 Updated Lagrangian two-node element This element is the same as in Example 2.1, except that the updated Lagrangian treatment is now used; See Figure 2.4, (p.). Recall that A_0 and ρ_0 are assumed to be constant in each element.

The velocity field is:

$$v(X,t) = \underbrace{\frac{1}{\ell_0}\left[X_2 - X \quad X - X_1\right]}_{\mathrm{N}(X)} \begin{Bmatrix} v_1(t) \\ v_2(t) \end{Bmatrix} \tag{E2.4.1}$$

In terms of element coordinates, the velocity field is

$$v(\xi, t) = \underbrace{[1-\xi, \xi]}_{\mathbf{N}(\xi)} \begin{Bmatrix} v_1(t) \\ v_2(t) \end{Bmatrix} \quad \xi = \frac{X - X_1}{\ell_0} \tag{E2.4.2}$$

The displacement is the time integral of the velocity, and since ξ is independent of time:

$$u(\xi, t) = \mathbf{N}(\xi)\mathbf{u}_e(t) \tag{E2.4.3}$$

Therefore, since $x = X + u$

$$x(\xi, t) = \mathbf{N}(\xi)\mathbf{x}_e(t) = [1-\xi \quad \xi] \begin{Bmatrix} x_1(t) \\ x_2(t) \end{Bmatrix}, \quad x_{,\xi} = x_2 - x_1 = \ell \tag{E2.4.4}$$

where ℓ is the current length of the element. We can express ξ in terms of the Eulerian coordinates by

$$\xi = \frac{x - x_1}{x_2 - x_1} = \frac{x - x_1}{\ell}, \quad \xi_{,x} = \frac{1}{\ell} \tag{E2.4.5}$$

So $\xi_{,x}$ can be obtained directly, instead of through the inverse of $x_{,\xi}$. This is not the case for higher-order elements.

The **B** matrix is obtained by the chain rule:

$$\mathbf{B} = \mathbf{N}_{,x} = \mathbf{N}_{,\xi}\, \xi_{,x} = \frac{1}{\ell}[-1 \quad +1] \tag{E2.4.6}$$

so the rate-of-deformation is given by

$$D_x = \mathbf{B}\mathbf{v}^e = \frac{1}{\ell}(v_2 - v_1) \tag{E2.4.7}$$

Equation (2.8.18) then gives

$$\mathbf{f}_e^{\text{int}} = \int_{x_1}^{x_2} \mathbf{B}^T \sigma A\, dx = \int_{x_1}^{x_2} \frac{1}{\ell} \begin{Bmatrix} -1 \\ +1 \end{Bmatrix} \sigma A\, dx \quad \text{or} \quad \mathbf{f}_e^{\text{int}} = A\sigma \begin{Bmatrix} -1 \\ +1 \end{Bmatrix} \tag{E2.4.8}$$

if the integrand is constant. Thus the internal nodal forces for the element correspond to the forces resulting from the stress σ. Note that the internal nodal forces are in equilibrium.

The external nodal forces are evaluated by (2.8.21)

$$\mathbf{f}_e^{\text{ext}} = \int_{x_1}^{x_2} \begin{Bmatrix} 1-\xi \\ \xi \end{Bmatrix} \rho b A\, dx + \left(\begin{Bmatrix} 1-\xi \\ \xi \end{Bmatrix} A \bar{t}_x \right) \Big|_{\Gamma_t} \tag{E2.4.9}$$

where the last term makes a contribution only if a node of the element is on the traction boundary.

The data for $b(x, t)$ are customarily fitted by linear interpolants for linear displacement elements (the information in higher-order interpolations will be beyond the resolution of a two-node element). So we let $b(\xi, t) = b_1(1 - \xi) + b_2\xi$. Substituting this into (E2.4.9) and integrating gives

$$\mathbf{f}_e^{\text{ext}} = \frac{\rho A \ell}{6} \begin{Bmatrix} 2b_1 + b_2 \\ b_1 + 2b_2 \end{Bmatrix} \tag{E2.4.10}$$

Comparison to total Lagrangian We will now compare the nodal forces to those obtained by the total Lagrangian formulation. Replacing σ in (E2.4.8) by the nominal stress using (2.2.9), we see that (E2.4.8) and (E2.1.4–E2.1.5) are equivalent.

To compare the external nodal forces, we note that by the conservation of matter, $\rho A \ell = \rho_0 A_0 \ell_0$. Using this in (E2.4.10) gives (E2.1.8), the total Lagrangian form of the nodal external forces. In the updated Lagrangian formulation, the mass from the total Lagrangian formulation is used, so the equivalence is obvious.

Example 2.5 Updated Lagrangian form of three-node quadratic displacement element The three-node element is shown in Figure 2.7. Node 2 can be placed anywhere between the end-nodes, but we shall see there are restrictions on the placement of this node if the one-to-one condition is to be met. We will also examine the effects of *mesh distortion*.

The displacement and velocity fields will be written in terms of the element coordinates

$$u(\xi, t) = \mathbf{N}(\xi)\mathbf{u}_e(t), \quad v(\xi, t) = \mathbf{N}(\xi)\mathbf{v}_e(t), \quad x(\xi, t) = \mathbf{N}(\xi)\mathbf{x}_e(t) \tag{E2.5.1}$$

where

$$\mathbf{N}(\xi) = \left[\frac{1}{2}(\xi^2 - \xi) \quad 1 - \xi^2 \quad \frac{1}{2}(\xi^2 + \xi) \right] \tag{E2.5.2}$$

and

$$\mathbf{u}_e^T = [u_1,\ u_2,\ u_3] \quad \mathbf{v}_e^T = [v_1,\ v_2,\ v_3] \quad \mathbf{x}_e^T = [x_1,\ x_2,\ x_3] \tag{E2.5.3}$$

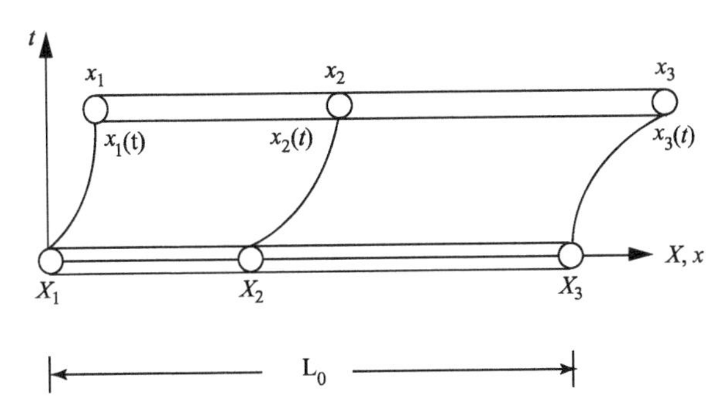

Figure 2.7 Three-node, quadratic displacement element in original and current configurations

The **B** matrix is given by

$$\mathbf{B} = \mathbf{N}_{,x} = x_{,\xi}^{-1}\,\mathbf{N}_{,\xi} = \frac{1}{2\mathbf{x}_{,\xi}}\begin{bmatrix} 2\xi - 1 & -4\xi & 2\xi + 1 \end{bmatrix} \tag{E2.5.4}$$

$$x_{,\xi} = \mathbf{N}_{,\xi}\,\mathbf{x}_e = \left(\xi - \frac{1}{2}\right)x_1 - 2\xi x_2 + \left(\xi + \frac{1}{2}\right)x_3$$

where the rate-of-deformation is given by

$$D_x = \mathbf{N}_{,x}\,\mathbf{v}_e = \mathbf{B}\mathbf{v}_e = \frac{1}{2x_{,\xi}}\begin{bmatrix} 2\xi - 1 & -4\xi & 2\xi + 1 \end{bmatrix}\mathbf{v}_e \tag{E2.5.5}$$

This rate-of-deformation varies linearly in the element if $x_{,\xi}$ is constant, which is the case when node 2 is midway between the other two nodes. However, when node 2 moves away from the midpoint due to element distortion, $x_{,\xi}$ becomes a linear function of ξ and the rate-of-deformation becomes a rational function. Furthermore, as node 2 moves from the center, it becomes possible for $x_{,\xi}$ to be negative or vanish. In that case, the mapping between the current spatial coordinates and the element coordinates is no longer one-to-one.

The internal forces are given by (2.8.18):

$$\mathbf{f}_e^{\text{int}} = \int_{x_1}^{x_3} \mathbf{B}^T \sigma A\, dx = \int_{-1}^{+1} \frac{1}{x_{,\xi}} \begin{Bmatrix} \xi - \frac{1}{2} \\ -2\xi \\ \xi + \frac{1}{2} \end{Bmatrix} \sigma A x_{,\xi}\, d\xi = \int_{-1}^{+1} \sigma A \begin{Bmatrix} \xi - \frac{1}{2} \\ -2\xi \\ \xi + \frac{1}{2} \end{Bmatrix} d\xi \tag{E2.5.6}$$

where we have used $dx = x_{,\xi}\, d\xi$. Using (2.2.9), we can see that this expression is identical to the internal force expression for the total Lagrangian formulation, E2.3.5).

2.8.7 *Mesh Distortion*

We will now examine the effects of mesh distortion on the quadratic three-node element. When $x_2 = \frac{1}{4}(x_3 + 3x_1)$, i.e. when node 2 of the element is one quarter of the element length from node 1, then $x_{,\xi} = \frac{1}{2}\,(x_3 - x_1)\,(\xi + 1) = 0$ at $\xi = -1$. From (2.2.5), the Jacobian is given by

$$J = \frac{A}{A_0} x_{,X} = \frac{A}{A_0} x_{,\xi}\, X_{,\xi}^{-1} \tag{2.8.31}$$

so it will also vanish. By (2.2.10) this implies that the current density becomes infinite at that point. As node 2 moves closer to node 1, the Jacobian becomes negative in part of the element, which implies a negative density and a violation of the one-to-one condition. This violates mass conservation. These situations are often masked by numerical quadrature, because the distortion must be more severe for the Jacobian to be negative at Gauss quadrature points.

The failure to meet the one-to-one condition can also lead to singularities in the rate-of-deformation, $D_x = \mathbf{B}\mathbf{v}_e$. From (E2.5.5) we can see the potential for difficulties when the denominator $x_{,\xi}$ vanishes or becomes negative. When $x_2 = \frac{1}{4}(x_3 + 3x_1)$, then $x_{,\xi}=0$ at $\xi=-1$, so the rate-of-deformation becomes infinite at node 1. This property of quadratic displacement elements has been exploited in fracture mechanics to develop elements with singular crack-tip stresses called quarter-point elements, but in large displacement analysis this behavior can be troublesome.

In one-dimensional elements the effects of mesh distortion are not as severe as in multi-dimensional problems. In fact, the effects of mesh distortion can be alleviated somewhat in this element by using F as a measure of deformation: see (E2.3.3). The deformation gradient F never becomes singular in the three-node element if the initial position of X_2 is at the midpoint.

Example 2.6 Axisymmetric two-node element As an example where the concept of the principle of virtual power or work becomes quite useful, we consider the analysis of an axisymmetric two-dimensional disk of constant thickness, a, which is thin compared to its dimensions, so $\sigma_z=0$ (Figure 2.8). The only nonzero velocity is $v_r(r)$ which, as shown, is only a function of the radial coordinate in an axisymmetric problem. The nonzero Cauchy stresses and rates of deformation are written in cylindrical coordinates using Voigt notation:

$$\{\mathbf{D}\} = \begin{Bmatrix} D_r \\ D_\theta \end{Bmatrix} \quad \{\sigma\} = \begin{Bmatrix} \sigma_r \\ \sigma_\theta \end{Bmatrix} \tag{E2.6.1}$$

The rate-of-deformation components are given by

$$D_r = v_{r,r} \quad D_\theta = \frac{v_r}{r} \tag{E2.6.2}$$

and the momentum equation is

$$\frac{\partial \sigma_r}{\partial r} + \frac{\sigma_r - \sigma_\theta}{r} + \rho b_r = \rho \dot{v}_r \tag{E2.6.3}$$

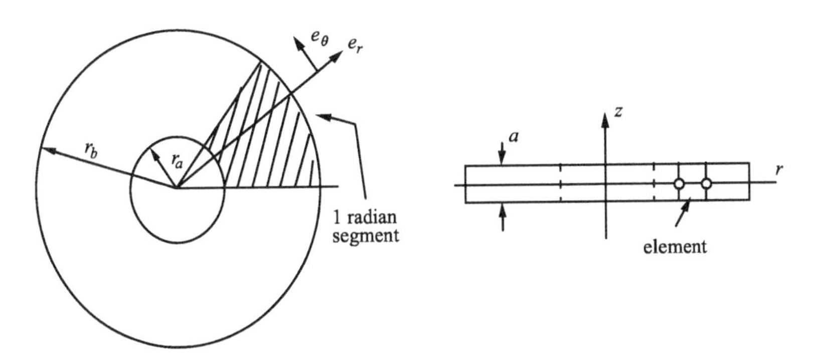

Figure 2.8 Schematic of axisymmetric disk: the shaded area is considered in work terms

It is not necessary to integrate the momentum equation to obtain its weak form. By the principle of virtual power the weak form is $\delta P=0\,\forall\,\delta v_r \in u_0$. The internal virtual power is obtained from the rate-of-deformation and stress:

$$\delta P_e^{\text{int}} = \int_{r_1^e}^{r_2^e} (\delta D_r\sigma_r + \delta D_\theta\sigma_\theta) ar\ dr = \int_{\Omega_e} \{\delta \mathbf{D}\}^T \{\sigma\} d\Omega \tag{E2.6.4}$$

where $d\Omega = ardr$ because a one radian segment in the circumferential direction has been chosen to avoid the factor 2π in all terms. The external virtual and kinetic power are given by

$$\delta P_e^{\text{ext}} = \int_{\Omega_e} \delta v_r \rho b_r\ d\Omega + \left(ar\bar{t}_r\right)\Big|_{\Gamma_t} \qquad \delta P_e^{\text{kin}} = \int_{\Omega_e} \delta v_r \rho \dot{v}_r\, d\Omega \tag{E2.6.5}$$

where ar in the last term is the area of a one-radian segment. Consider a two-node finite element with a linear velocity field written in terms of element coordinates

$$v(\xi, t) = \begin{bmatrix} 1-\xi & \xi \end{bmatrix} \begin{Bmatrix} v_1(t) \\ v_2(t) \end{Bmatrix} \tag{E2.6.6}$$

The rate-of-deformation is evaluated by (E2.6.2) using the above velocity field and immediately put into matrix form

$$\mathbf{D} = \begin{Bmatrix} D_r \\ D_\theta \end{Bmatrix} = \begin{bmatrix} -\dfrac{1}{r_{21}} & \dfrac{1}{r_{21}} \\ \dfrac{1-\xi}{r} & \dfrac{\xi}{r} \end{bmatrix} \begin{Bmatrix} v_1(t) \\ v_2(t) \end{Bmatrix} = \mathbf{B}\mathbf{v}_e \tag{E2.6.7}$$

where $r_{21} \equiv r_2 - r_1$. The internal nodal forces are given by an expression identical to (2.8.18) except that the stress is replaced by the column matrix

$$\mathbf{f}_e^{\text{int}} = \int_{\Omega_e} \mathbf{B}^T \{\boldsymbol{\sigma}\} d\Omega = \int_{r_1}^{r_2} \begin{bmatrix} -\dfrac{1}{r_{21}} & \dfrac{1}{r_{21}} \\ \dfrac{1-\xi}{r} & \dfrac{\xi}{r} \end{bmatrix} \begin{Bmatrix} \sigma_r \\ \sigma_\theta \end{Bmatrix} ar\ dr \tag{E2.6.8}$$

The element mass matrix is given by

$$\mathbf{M}_e = \int_{r_1}^{r_2} \begin{Bmatrix} 1-\xi \\ \xi \end{Bmatrix} [1-\xi \quad \xi] \rho ar\ dr = \frac{\rho a r_{21}}{12} \begin{bmatrix} 3r_1 + r_2 & r_1 + r_2 \\ r_1 + r_2 & r_1 + 3r_2 \end{bmatrix} \tag{E2.6.9}$$

A diagonal mass matrix can be computed by the row-sum technique or by lumping half the mass at each node, which give, respectively

$$\mathbf{M}_e = \frac{\rho a r_{21}}{6} \begin{bmatrix} 2r_1 + r_2 & 0 \\ 0 & r_1 + 2r_2 \end{bmatrix}_{\text{row-sum}} \qquad \mathbf{M}_e = \frac{\rho a r_{21}(r_1 + r_2)}{4} \begin{bmatrix} 1 & 0 \\ 0 & 1 \end{bmatrix}_{\text{lump}} \tag{E2.6.10}$$

As can be seen the two diagonalization procedures give slightly different results.

2.9 Governing Equations for Eulerian Formulation

In an Eulerian formulation, the nodes are fixed in space and the dependent variables are functions of the Eulerian spatial coordinate x and the time t. The stress measure is the Cauchy (physical) stress $\sigma(x, t)$, the measure of deformation is the rate-of-deformation $D_x(x, t)$, and the motion will be described by the velocity $v(x, t)$. In Eulerian formulations, the motion is not expressed as a function of the reference coordinates since an undeformed, initial configuration cannot be established, and no counterpart of (2.2.1) is available.

The governing equations are summarized in Box 2.4 for a constant area problem. In comparison with the updated Lagrangian formulation we have just discussed, four points are noteworthy:

1. The mass conservation equation is now written as a partial differential equation; the algebraic form used with Lagrangian meshes is not applicable because it applies only to material points.
2. The material time derivative for the velocity in the momentum equation has been written out in terms of the spatial time derivative and a transport term.
3. The constitutive equation is expressed in rate form.
4. The boundary conditions are imposed on fixed spatial points.

Box 2.4 Governing equations for Eulerian formulation

Continuity equation (mass conservation):

$$\frac{\partial \rho}{\partial t} + \frac{\partial (\rho v)}{\partial x} = 0 \tag{B2.4.1}$$

Momentum equation:

$$\rho\left(\frac{\partial v}{\partial t} + v\frac{\partial v}{\partial x}\right) = \frac{\partial \sigma}{\partial x} + \rho b \tag{B2.4.2}$$

Strain measure (rate-of-deformation):

$$D_x = v_{,x} \tag{B2.4.3}$$

Constitutive equation in rate form:

$$\frac{D\sigma}{Dt} = \sigma_{,t}(x,t) + \sigma_{,x}(x,t)v(x,t) = S_t^{\sigma D}(D_x, \sigma, \quad \text{etc.}, \bar{t} \le t) \tag{B2.4.4}$$

Energy conservation equation: same as before

In the general case, boundary conditions are required for the density, velocity and stress. As will be seen in Chapter 7, the boundary conditions for the density and stress in an Eulerian mesh depend on whether the material is flowing in or out at the boundary. In this introductory exposition, we consider only boundaries where there is no flow. The boundary points are then

Lagrangian, and the density and stress can be determined at these points by the Lagrangian mass conservation equation, (2.2.10) and the constitutive equation, respectively. Therefore, there is no need for boundary conditions for these variables.

2.10 Weak Forms for Eulerian Mesh Equations

In the Eulerian formulation, we have three dependent variables: the density $\rho(x, t)$, the velocity $v(x, t)$ and the stress $\sigma(x, t)$. The rate-of-deformation can easily be eliminated from the momentum equations by substituting (B2.4.3) into the constitutive equation (B2.4.4). Therefore, we will need three sets of discrete equations. Weak forms of the momentum equation, the mass conservation equation and the constitutive equation will be developed.

We will construct continuous solutions to the mass conservation equation. The trial functions for the density are $\rho(x, t)$, and the test functions are $\delta\rho(x)$, so

$$\rho(x,t) \in D, \quad D = \{\rho(x,t) | \rho(x,t) \in C^0(x)\} \tag{2.10.1}$$

$$\delta\rho(x) \in D_0 \quad D_0 = \{\delta\rho(x) | \delta\rho(x) \in C^0(x)\} \tag{2.10.2}$$

The weak form of the continuity equation is obtained by multiplying it by the test function $\delta\rho(x)$ and integrating over the domain. This gives

$$\int_{x_a}^{x_b} \delta\rho(\rho_{,t} + (\rho v)_{,x})dx = 0 \quad \forall \delta\rho \in D_0 \tag{2.10.3}$$

Only first derivatives with respect to the spatial variable of the density and velocity appear in the weak form, so there is no need for integration by parts.

The weak form of the constitutive equation is obtained in the same way. We express the material derivative in terms of a spatial derivative and a transport term, giving

$$\sigma_{,t} + \sigma_{,x} v - S_t^{\sigma D}(v_{,x}, \text{ etc.}) = 0 \tag{2.10.4}$$

where $S^{\sigma D}$ is defined in (2.6.6a–b). The test and trial functions, $\delta\sigma(x)$ and $\sigma(x)$, respectively, are subject to the same continuity as for the density in the continuity equation, that is, we let $\sigma \in D$, $\delta\sigma \in D_0$. The weak form of the constitutive equation is then obtained by multiplying it by the test function and integrating over the domain:

$$\int_{x_a}^{x_b} \delta\sigma\left(\sigma_{,t} + \sigma_{,x} v - S_t^{\sigma D}(v_{,x}, \text{ etc.})\right) dx = 0 \quad \forall \delta\sigma \in D_0 \tag{2.10.5}$$

As in the continuity equation, there is no benefit in integrating by parts.

The weak form of the momentum equation is obtained by integrating the test function $\delta v(x)$ over the spatial domain. The procedure is identical to that in the updated Lagrangian formulation in Section 2.7. The test and trial functions are defined by (2.7.1) and (2.7.2). The resulting weak form is

$$\int_{x_a}^{x_b} \left[\delta v_{,x} A\sigma - \delta v \left(\rho A b - \rho A \frac{Dv}{Dt} \right) \right] dx - \left(\delta v A \bar{t}_x \right) \Big|_{\Gamma_t} = 0 \tag{2.10.6}$$

after integration by parts. This gives

$$\int_{x_a}^{x_b}\left[\delta v_{,x} A\sigma + \delta v \rho A\left(\frac{\partial v}{\partial t} + v_{,x} v - b\right)\right] dx - \left(\delta v A\bar{t}_x\right)\Big|_{\Gamma_t} = 0 \tag{2.10.7}$$

when the total time derivative is written out. Note that the limits of the integration are fixed in space.

The weak form is identical to the principle of virtual power for the updated Lagrangian formulation except that the domain is fixed in space and the material time derivative is expressed in its Eulerian form. Thus the weak form of the momentum equation can be written

$$\delta P = \delta P^{\text{int}} - \delta P^{\text{ext}} + \delta P^{\text{kin}} = 0 \quad \forall \delta v \in u_0 \tag{2.10.8}$$

$$\delta P^{\text{int}} = \int_{x_a}^{x_b} \delta v_{,x} \sigma A dx = \int_{x_a}^{x_b} \delta D_x \sigma A \; dx = \int_{\Omega} \delta D_x \sigma \Omega \tag{2.10.9}$$

$$\delta P^{\text{ext}} = \int_{x_a}^{x_b} \delta v \rho b A \; dx + \left(\delta v A \bar{t}_x\right)\Big|_{\Gamma_t} \tag{2.10.10}$$

$$\delta P^{\text{kin}} = \int_{x_a}^{x_b} \delta v \rho \left(\frac{\partial v}{\partial t} + v_{,x} v\right) A dx = \int_{\Omega} \delta v \rho \left(\frac{\partial v}{\partial t} + v_{,x} v\right) d\Omega \tag{2.10.11}$$

All of the terms are identical to the corresponding terms in the principle of virtual power for the updated Lagrangian formulation, except that the limits of integration are fixed in space and the material time derivative in the inertial virtual power has been expressed in terms of the spatial time derivative and the transport term. Similar expressions for the virtual powers also hold on the element level.

2.11 Finite Element Equations

In a general Eulerian finite element formulation, approximations are needed for the density, stress and velocity. For each dependent variable, test and trial functions are needed. We will develop the equations for the entire mesh. For simplicity, we consider the case where the domain is $0 \le x \le L$. As mentioned before, the end points are fixed in space and the velocities on these points vanish. There are then no boundary conditions on the density or stress and the boundary conditions on the velocity are $v(0, t) = 0$, $v(L, t) = 0$.

The mapping between spatial and element parent coordinates is given by

$$x = \sum_{I=1}^{n_N} N_I(\xi) x_I \tag{2.11.1}$$

In contrast to the Lagrangian formulations, this mapping is constant in time since the nodal coordinates x_I are not functions of time. The trial and test functions are given by

$$\rho(x, t) = \sum_{I=1}^{n_N} N_I^{\rho}(x) \rho_I(t) \quad \delta\rho(x) = \sum_{I=1}^{n_N} N_I^{\rho}(x) \delta\rho_I \tag{2.11.2}$$

$$\sigma(x,t)=\sum_{I=1}^{n_N} N_I^\sigma(x)\sigma_I(t) \quad \delta\sigma(x)=\sum_{I=1}^{n_N} N_I^\sigma(x)\delta\sigma_I \tag{2.11.3}$$

$$v(x,t)=\sum_{I=2}^{n_N-1} N_I(x)v_I(t) \quad \delta v(x)=\sum_{I=2}^{n_N-1} N_I(x)\delta v_I \tag{2.11.4}$$

The velocity trial functions have been constructed so the velocity boundary condition is automatically satisfied.

Substituting the test and trial functions for the density into the weak continuity equation gives

$$\sum_{I=1}^{n_N}\sum_{J=1}^{n_N}\delta\rho_J\int_0^L\left(N_J^\rho N_I^\rho\rho_{I,t}+N_j^\rho(\rho v),x\right)dx=0 \tag{2.11.5}$$

Since this holds for arbitrary $\delta\rho_J$ at interior nodes, we obtain

$$\rho_{J,t}\int_0^L N_I^\rho N_J^\rho dx+\int_0^L N_I^\rho(\rho v)_{,x}\,dx=0 \quad I=1 \text{ to } n_N \tag{2.11.6}$$

We define the following matrices:

$$M_{IJ}^\rho=\int_0^L N_I^\rho N_J^\rho dx, \quad \mathbf{M}_e^\rho=\int_{\Omega_e}\left(\mathbf{N}^\rho\right)^T\mathbf{N}^\rho d\Omega \tag{2.11.7}$$

$$g_I^\rho=\int_0^L N_I^\rho(\rho v)_{,x}\,dx, \quad \mathbf{g}_e^\rho=\int_{\Omega_e}(\mathbf{N}^\rho)^T(\rho v)_{,x}\,d\Omega \tag{2.11.8}$$

The discrete continuity equation can then be written as

$$\sum_J M_{IJ}^\rho\dot{\rho}_J+g_I^\rho=0 \quad \text{for} \quad I=1 \text{ to } n_N, \quad \text{or} \quad \mathbf{M}^\rho\dot{\rho}+\mathbf{g}^\rho=0 \tag{2.11.9}$$

The matrices $\mathbf{M}^\rho$ can be assembled from element matrices just like the mass matrix. The column matrix $\mathbf{g}^\rho$ is obtained by a scatter. The matrix $\mathbf{M}^\rho$ is time invariant and closely resembles the mass matrix. However, the column matrix $\mathbf{g}^\rho$ varies with time and must be computed in every time step.

The discrete form of the constitutive equation is obtained similarly. The result is

$$\sum_J M_{IJ}^\sigma\dot{\sigma}_J+g_I^\sigma=h_I^\sigma \quad \text{for} \quad I=1 \text{ to } n_N, \quad \text{or} \quad \mathbf{M}^\sigma\dot{\sigma}+\mathbf{g}^\sigma=\mathbf{h}^\sigma \tag{2.11.10}$$

where

$$M_{IJ}^\sigma=\int_0^L N_I^\sigma N_J^\sigma dx, \quad \mathbf{M}_e^\sigma=\int_{\Omega_e}(\mathbf{N}^\sigma)^T\mathbf{N}^\sigma d\Omega \tag{2.11.11}$$

$$g_I^\sigma=\int_0^L N_I^\sigma v\sigma_{,x}\,dx, \quad \mathbf{g}_e^\sigma=\int_{\Omega_e}(\mathbf{N}^\sigma)^T v\sigma_{,x}\,d\Omega \tag{2.11.12}$$

where the matrix relations on the right have been obtained from the indicial forms.

2.11.1 *Momentum Equation*

The discrete momentum equation is identical to that for the updated Lagrangian formulation except for the kinetic term. The kinetic nodal forces for the Eulerian formulation are obtained in the following on an element level. We define the kinetic nodal forces by (2.10.11), which gives

$$\delta P_e^{\text{kin}} = \delta \mathbf{v}_e^T \mathbf{f}_e^{\text{kin}} = \delta \mathbf{v}_e^T \int_{\Omega_e} \rho \mathbf{N}^T (\mathbf{N}\dot{\mathbf{v}} + v_{,x}\, v) A \, dx \tag{2.11.13}$$

From this, it follows that the kinetic nodal forces are given by

$$\mathbf{f}_e^{\text{kin}} = \mathbf{M}_e \dot{\mathbf{v}}_e + \mathbf{f}_e^{\text{tran}} \tag{2.11.14}$$

where

$$\mathbf{M}_e = \int_{\Omega_e} \rho \mathbf{N}^T \mathbf{N} A \, dx, \qquad \mathbf{f}_e^{\text{tran}} = \int_{\Omega_e} \mathbf{N}^T \rho v_{,x}\, v A dx \tag{2.11.15}$$

The transport nodal forces have not been written out in terms of shape functions. This term is needed in the Eulerian formulation because the nodes are fixed in space, so the time derivatives of the nodal velocities correspond to spatial derivatives. The mass matrix is a *function of time*: as the density in the element changes, the mass matrix will change correspondingly.

Example 2.7 Two-node Eulerian finite element The finite element equations are developed for a one-dimensional, two-node element with linear velocity, density and stress fields. The element, shown in Figure 2.9, is of length $\ell = x_2 - x_1$ and unit cross-sectional area. As can be seen, the spatial configuration does not change with time since it is an Eulerian element. The map between element and spatial coordinates is given by

$$x(\xi) = [1-\xi \quad \xi] \begin{Bmatrix} x_1 \\ x_2 \end{Bmatrix} \equiv \mathbf{N}(\xi)\mathbf{x}_e \tag{E2.7.1}$$

The density, velocity and stress are also interpolated by the same linear shape functions

$$\rho(\xi) = \mathbf{N}(\xi)\rho_e \quad v(\xi) = \mathbf{N}(\xi)\mathbf{v}_e \quad \sigma(\xi) = \mathbf{N}(\xi)\sigma_e \tag{E2.7.2}$$

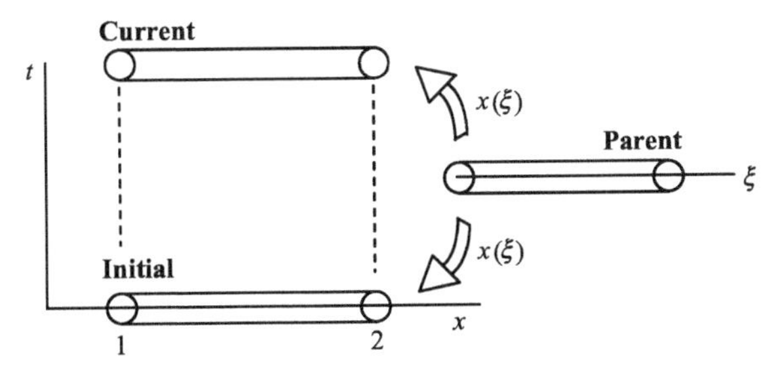

Figure 2.9 Eulerian element in current and original configurations, which are the same, and the mapping to the parent element

Superscripts are not appended to the shape functions because all variables are interpolated by the same shape functions.

Density equation The element matrices for the discrete continuity equation are given by (2.11.7) and (2.11.8):

$$\mathbf{M}_e^\rho = \int_{x_1}^{x_2} \mathbf{N}^T \mathbf{N} \, dx = \int_0^1 \begin{bmatrix} 1-\xi \\ \xi \end{bmatrix} \begin{bmatrix} 1-\xi & \xi \end{bmatrix} \ell d\xi = \frac{\ell}{6} \begin{bmatrix} 2 & 1 \\ 1 & 2 \end{bmatrix} \tag{E2.7.3}$$

$$\mathbf{g}_e^\rho = \int_{x_1}^{x_2} \mathbf{N}^T (\rho v)_{,x} \, dx = \int_0^1 \begin{bmatrix} 1-\xi \\ \xi \end{bmatrix} (\rho v)_{,x} \, \ell d\xi \tag{E2.7.4}$$

The $\mathbf{g}_e^\rho$ vector is usually evaluated by numerical quadrature. For linear interpolants it is given by

$$\mathbf{g}_e^\rho = \frac{1}{6}(\rho_2 - \rho_1) \begin{Bmatrix} 2v_1 + v_2 \\ v_1 + 2v_2 \end{Bmatrix} + \frac{1}{6}(v_2 - v_1) \begin{Bmatrix} 2\rho_1 + \rho_2 \\ \rho_1 + 2\rho_2 \end{Bmatrix} \tag{E2.7.5}$$

This matrix vanishes when the density and velocity are constant in the element.

Stress equation The element matrix for the stresses is given by (E2.7.3) since $\mathbf{M}_e^\sigma = \mathbf{M}_e^\rho$. The vector $\mathbf{g}_e^\rho$ is given by

$$\begin{aligned} \mathbf{g}_e^\sigma &= \int_{x_1}^{x_2} \mathbf{N}^T v \sigma_{,x} \, dx = \int_0^1 \begin{bmatrix} 1-\xi \\ \xi \end{bmatrix} (v_1(1-\xi) + v_2 \xi)(\sigma_2 - \sigma_1) d\xi \\ &= \frac{1}{6}(\sigma_2 - \sigma_1) \begin{Bmatrix} 2v_1 + v_2 \\ v_1 + 2v_2 \end{Bmatrix} \end{aligned} \tag{E2.7.6}$$

In summary, the finite element equations for the Eulerian formulation consist of three sets of discrete equations: the continuity equation, the constitutive equation, and the momentum equation, or equation of motion. The momentum equation is similar to the updated Lagrangian form, except that the inertial term includes a transport term and varies with time. The semidiscrete forms of the continuity and constitutive equations are first-order ordinary differential equations. We have only developed the discrete equations for the case where the endpoints are fixed.

2.12 Solution Methods

We have seen that the momentum equation can be discretized for a Lagrangian mesh in the form

$$\mathbf{M}\ddot{\mathbf{u}} = \mathbf{f}^{\mathrm{ext}} - \mathbf{f}^{\mathrm{int}} = \mathbf{f} \tag{2.12.1}$$

These are ordinary differential equations in time.

To solve some nonlinear problems at this point, we describe the simplest solution method, explicit time integration. The most widely used explicit method is the central difference method with a *diagonal* or *lumped mass* matrix.

We start at time $t=0$ using time steps Δt, so that at time step n, $t=n\Delta t$. The value of a function at $n\Delta t$ *is* denoted by a superscript n, i.e., $\mathbf{u}^n \equiv \mathbf{u}(n\Delta t)$. In the central difference method, the velocities are approximated by

$$\dot{\mathbf{u}}^n = \mathbf{v}^{n+\frac{1}{2}} = \frac{\mathbf{u}^{n+\frac{1}{2}} - \mathbf{u}^{n-\frac{1}{2}}}{\Delta t} = \frac{\mathbf{u}(t+\Delta t/2) - \mathbf{u}(t-\Delta t/2)}{\Delta t} \tag{2.12.2}$$

where the second equality is included to clarify the notation. Half-time-step values are used for the velocities. The accelerations are given by

$$\ddot{\mathbf{u}}^n = \mathbf{a}^n = \frac{\mathbf{v}^{n+\frac{1}{2}} - \mathbf{v}^{n-\frac{1}{2}}}{\Delta t} \quad \text{or} \quad \mathbf{v}^{n+\frac{1}{2}} = \Delta t \mathbf{M}^{-1}\mathbf{f}^n + \mathbf{v}^{n-\frac{1}{2}} \tag{2.12.3}$$

where the equation on the right is obtained by combining the equation on the left with (2.12.1). The value of the derivative at the center of a time interval is obtained from the difference of the function values at the ends of the interval, hence the name *central difference* formulas. The flowchart for an explicit program is given by Box 2.5.

Box 2.5 Flowchart for explicit time integration of Lagrangian mesh

1. Initial conditions and initialization: set $\mathbf{v}^0$, σ_e^0; $n=0$, $t=0$; compute $\mathbf{M}$
2. Get $\mathbf{f}^n$ (see following)
3. Compute accelerations $\mathbf{a}^n = \mathbf{M}^{-1}\mathbf{f}^n$
4. Update nodal velocities: $\mathbf{v}^{n+\frac{1}{2}} = \mathbf{v}^{n+\frac{1}{2}-\alpha} + \alpha\Delta t\mathbf{a}^n \,:\, \alpha = \begin{cases} 1/2 & \text{if } n=0 \\ 1 & \text{if } n>0 \end{cases}$
5. Enforce essential boundary conditions: if node I on Γ_v : $v_I^{n+\frac{1}{2}} = \bar{v}\left(x_I, t^{n+\frac{1}{2}}\right)$
6. Update nodal displacements: $\mathbf{u}^{n+1} = \mathbf{u}^n + \Delta t\mathbf{v}^{n+\frac{1}{2}}$
7. Update counter and time: $n \leftarrow n+1$, $t \leftarrow t+\Delta t$
8. Output; if simulation not complete, go to 2.

Module: get **f**

1. GATHER element nodal displacements $\mathbf{u}_e^n$ and velocities $\mathbf{v}_e^{n+\frac{1}{2}}$
2. if $n=0$, go to 5
3. Compute measure of deformation
4. Compute stress by constitutive equation
5. Compute internal nodal forces by relevant equation
6. Compute external nodal forces on element and $\mathbf{f}_e = \mathbf{f}_e^{\text{ext}} - \mathbf{f}_e^{\text{int}}$
7. SCATTER element nodal forces to global matrices.

Updating for the displacements by (2.12.3) does not require *any solution* of algebraic equations. Thus, in a sense, explicit integration is simpler than static linear stress analysis. As can be seen from the flowchart, most of the explicit program is a straightforward evaluation of the governing equations and the time integration formulas. The program begins with the enforcement of the initial conditions; procedures for fitting different initial conditions have already been described. The first time step is somewhat different from the others because only a half-step is taken. This enables the program to correctly account for the initial conditions on the stresses and velocities.

Most of the programming and computation time is in computing the element nodal forces, particularly the internal nodal forces. The nodal forces are computed element by element. Prior to starting the element computations, the element nodal velocities and displacements are gathered from the global arrays. As can be seen from the flowchart, the computation of the internal nodal forces involves the application of the equations which are left in strong form, the strain equation and the constitutive equation. This is followed by the evaluation of the internal nodal forces from the stress by a relationship which emanates from the weak form of the momentum equation. When the computation of the element nodal forces is completed, they are scattered to the global array according to their node numbers.

The essential boundary conditions are enforced quite easily as shown. By setting the nodal velocities equal to the prescribed nodal velocities at all nodes on prescribed velocity boundaries, the correct displacements result, since the velocities are subsequently integrated in time. The placement of this step in the flowchart insures that the correct velocities are available in the nodal force computation. The initial velocities must be compatible with the boundary conditions; this is not checked in this flowchart but would be checked in a production program. The reaction forces can be obtained by outputting the total nodal forces at the prescribed velocity nodes.

It can be seen from the flowchart that the effects of traction boundary conditions appear through the external nodal forces. For a traction-free boundary, nothing need be done: the homogeneous traction boundary condition is enforced naturally in a weak sense by the finite element solution. However, the traction boundary condition is satisfied only approximately.

Stability criterion The disadvantage of explicit integration is that the time step must be below a critical value or the solution 'blows up' due to a numerical instability. This is described in detail in Chapter 6. Here we limit ourselves to pointing out that the critical time step for the two-node elements with diagonal mass is given by

$$\Delta t_{\text{crit}} = \frac{\ell_0}{c_0} \tag{2.12.4}$$

where ℓ_0 is the *initial* length of the element and c_0 is the wave speed given by $c_0^2 = E^{PF} / \rho_0$, where E^{PF} is defined in (2.2.20).

2.13 Summary

The finite element equations have been developed for one-dimensional continua of varying cross-section. Two mesh descriptions have been described:

1. Lagrangian meshes, where the nodes and elements move with the material.
2. Eulerian meshes, in which nodes and elements are fixed in space.

Two formulations have been developed for Lagrangian meshes:

1. An updated Lagrangian formulation, in which the strong form is expressed in spatial coordinates, that is, the Eulerian coordinates.
2. A total Lagrangian formulation, where the strong form is expressed in the material, that is, the Lagrangian coordinates.

It has been shown that the updated and total Lagrangian formulations are two representations of the same discretization, and each can be transformed to the other. Thus the internal and external forces obtained by the total Lagrangian formulations are identical to those obtained by the updated formulation, and the choice of formulation is a matter of convenience.

The equations of motion correspond to the momentum equation and are obtained from its weak form. For of explicit time integration, the other equations, the measure of deformation and the constitutive equation, are used in their strong forms. The weak form and discrete equations have been structured so that their relationship to the corresponding terms in the momentum equation is readily apparent: the internal forces correspond to the stress; the external forces correspond to the body forces and tractions; **Ma** corresponds to the kinetic or inertial terms (d'Alembert forces).

If the inertial forces can be neglected, **Ma** is omitted from the discrete equations. The resulting equations are called equilibrium equations. They are either nonlinear algebraic equations or ordinary differential equations, depending on the character of the constitutive equation.

2.14 Exercises

2.1. Transform the principle of virtual work to the principle of virtual power by letting $\delta u = \delta v$ and using the conservation of mass and the transformations for the stresses. (Note that this is possible since the admissibility conditions on the two sets of test and trial function spaces are identical).

2.2. Consider a tapered two-node element with a linear displacement field as in Example 2.1 where the cross-sectional area $A_0 = A_{01}(1-\xi) + A_{02}\xi$, where A_{01} and A_{02} are the initial cross-sectional areas at nodes 1 and 2. Assume that the nominal stress P is also linear in the element, i.e. $P = P_1(1-\xi) + P_2\xi$.

(a) Using the total Lagrangian formulation, develop expressions for the internal nodal forces. For a constant body force, develop the external nodal forces. Compare the internal and external nodal forces for the case when $A_{01} = A_{02} = A_0$ and $P_1 = P_2$ to the results in Example 2.1.

(b) Develop the consistent mass matrix. Then obtain a diagonal form of the mass matrix by the row-sum technique. Find the frequencies of a single element with consistent mass and the diagonal mass by solving the eigenvalue problem

$$\mathbf{Ky} = \omega^2 \mathbf{My} \quad \text{where } \mathbf{K} = \frac{E^{PF}(A_{01} + A_{02})}{2\ell_0}\begin{bmatrix} 1 & -1 \\ -1 & 1 \end{bmatrix}$$

2.3. Consider a tapered two-node element with a linear displacement field in the updated Lagrangian formulation as in Example 2.4. Let the current cross-sectional area be given by $A = A_1(1 - \xi) + A_2\xi$, where A_1 and A_2 are the current cross-sectional areas at nodes 1 and 2. Develop the internal nodal forces in terms of the Cauchy stress for the updated Lagrangian formulation assuming $\sigma = \sigma_1(1 - \xi) + \sigma_2\xi$ where σ_1 and σ_2 are the Cauchy stresses at the two nodes. Develop the nodal external forces for a constant body force.

2.4. Consider a two-element mesh consisting of elements of length ℓ with constant cross-sectional area A. Assemble a consistent mass matrix and a stiffness matrix and obtain the frequency for the two element mesh with all nodes free (the eigenvalue problem is 3×3). The frequency analysis assumes a linear response so the initial and current geometry are identical. Repeat the same problem with a lumped mass. Compare the frequencies for the lumped and consistent mass matrices to the exact frequency for a free-free rod, $\omega = n\frac{\pi c}{L}$, where $n = 0, 1, \ldots$ Observe that the consistent mass frequencies are above the exact, whereas the diagonal mass frequencies are below the exact.

2.5. Repeat Example 2.6 for spherical symmetry, where

$$\mathbf{D} = \begin{Bmatrix} D_{rr} \\ D_{\theta\theta} \\ D_{\phi\phi} \end{Bmatrix}, \quad \boldsymbol{\sigma} = \begin{Bmatrix} \sigma_{rr} \\ \sigma_{\theta\theta} \\ \sigma_{\phi\phi} \end{Bmatrix}, \quad D_{rr} = v_{r,r}, \quad D_{\theta\theta} = D_{\phi\phi} = \frac{1}{r} v r$$

(a) Develop an expression for the principle of virtual power and derive the corresponding strong form.

(b) For a two node element with a linear velocity field, develop B, the internal nodal forces $\mathbf{f}_e^{\text{int}}$ in terms of the stresses, and the consistent mass matrix $\mathbf{M}_e$. For constant body force, develop an expression for the nodal external forces $\mathbf{f}_e^{\text{ext}}$.

3

Continuum Mechanics

3.1 Introduction

Continuum mechanics is a building block for nonlinear finite element analysis, and a mastery of continuum mechanics is essential for a good understanding of nonlinear finite elements. This chapter summarizes the nonlinear continuum mechanics that is needed for nonlinear finite element methods. It is, however, insufficient for thoroughly learning continuum mechanics. Instead, it provides a review of the topics that are needed for the remainder of the book.

Readers who have little or no familiarity with continuum mechanics should consult texts such as Hodge (1970), Mase and Mase (1992), Fung (1994), Malvern (1969), or Chandrasekharaiah and Debnath (1994). The first three are the most elementary. Hodge (1970) is particularly useful for learning indicial notation and the fundamental topics. Mase and Mase (1992) gives a concise introduction with notation almost identical to that used here. Fung (1994) is an interesting book with many discussions of how continuum mechanics is applied. The text by Malvern (1969) has become a classic for it provides a very lucid and comprehensive description of the field. Chandrasekharaiah and Debnath (1994) give a thorough introduction with an emphasis on tensor notation. The only topic treated here which is not presented in greater depth in all of these texts is the topic of objective stress rates, which is covered only in Malvern. Monographs of a more advanced character are Marsden and Hughes (1983) and Ogden (1984). Prager (1961), while an older book, provides a useful description of continuum mechanics for the reader with an intermediate background. The classic treatise on continuum mechanics is Truesdell and Noll (1965) which discusses the fundamental issues from a very general viewpoint.

Nonlinear Finite Elements for Continua and Structures, Second Edition.
Ted Belytschko, Wing Kam Liu, Brian Moran, and Khalil I. Elkhodary.

Companion Website: www.wiley.com/go/belytschko

This chapter begins with a description of deformation and motion. Rigid body motion is described with an emphasis on rotation. Rotation plays a central role in nonlinear continuum mechanics, and many of the more difficult and complicated aspects of nonlinear continuum mechanics stem from rotation.

Next, the concepts of stress and strain in nonlinear continuum mechanics are described. Stress and strain can be defined in many ways in nonlinear continuum mechanics. We will confine our attention to the strain and stress measures that are most frequently employed in nonlinear finite element programs: the Green strain tensor and the rate-of-deformation. The stress measures that are treated are the physical (Cauchy) stress, the nominal stress, and the second Piola–Kirchhoff stress, which we call PK2 for brevity. There are many others, but frankly even these are too many for most beginning students. The profusion of stress and strain measures is one of the obstacles to understanding nonlinear continuum mechanics. Once one understands the field, one realizes that the many measures add nothing fundamental, and are perhaps just a manifestation of academic excess. Nonlinear continuum mechanics could be taught with just one measure of stress and strain, but additional ones need to be covered so that the literature and software can be understood.

The conservation equations, which are often called balance equations, are derived next. These equations are identical in solid and fluid mechanics. They consist of the conservation of mass, momentum and energy. The equilibrium equation is a special case of the momentum equation which applies when the accelerations vanish. The conservation equations are derived in both the spatial and the material domains. In a first reading or introductory course, the derivations can be skipped, but the equations should be thoroughly understood in at least one form.

The chapter concludes with further study of the role of rotations in continuum mechanics. The polar decomposition theorem is derived and explained. Then objective rates, also called frame-invariant rates, of the Cauchy stress tensor are examined. It is shown why rate-type constitutive equations require objective rates, and several objective rates frequently used in nonlinear finite elements are presented.

3.2 Deformation and Motion

3.2.1 Definitions

Continuum mechanics is concerned with models of solids and fluids in which the properties and response can be characterized by smooth functions of spatial variables, with at most a limited number of discontinuities. It ignores inhomogeneities such as molecular, grain or crystal structures. Features such as crystal structure sometimes appear in continuum models through the constitutive equations, but the response and properties are assumed to be smooth with a finite number of discontinuities. The objective of continuum mechanics is to provide models for the macroscopic behavior of fluids, solids and structures.

Consider a body in an initial state at a time $t=0$ as shown in Figure 3.1; the domain of the body in the initial state is denoted by Ω_0 and called the *initial configuration*. In describing the motion of the body and deformation, we also need a configuration to which various equations are referred; this is called the *reference configuration*. Unless we specify otherwise, the initial configuration is used as the reference configuration. However, other configurations can also be used as the reference configuration and we will do so in some derivations. The significance

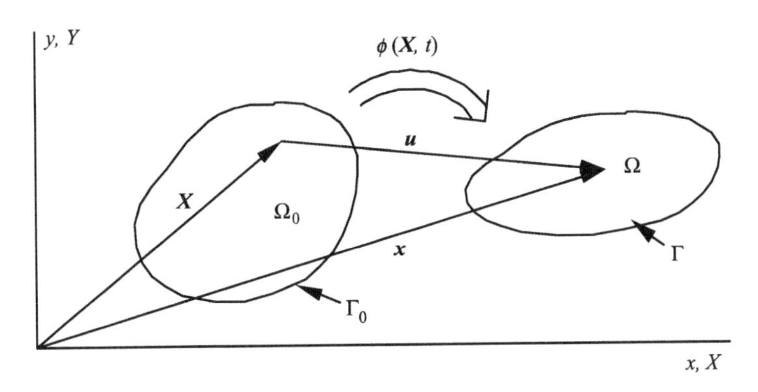

Figure 3.1 Undeformed (initial) and deformed (current) configurations of a body

of the reference configuration lies in the fact that motion is defined with respect to this configuration.

In many cases, we will also need to specify a configuration which is considered to be an *undeformed configuration*, which occupies domain Ω_0. Unless specified otherwise, the undeformed configuration is identical to the initial configuration. The notion of an 'undeformed' configuration should be viewed as an idealization, since undeformed objects seldom exist in reality. Most objects previously had a different configuration and were changed by deformations: a metal pipe was once a steel ingot, a cellular telephone housing was once a vat of liquid plastic, an airport runway was once many truckloads of concrete. So the term *undeformed configuration* is only relative and designates the configuration with respect to which we measure deformation.

The domain of the *current configuration* of the body is denoted by Ω; this will often also be called the *deformed configuration*. The domain can be one-, two- or three-dimensional; Ω then refers to a line, an area, or a volume, respectively. The boundary of the domain is denoted by Γ, and corresponds to the two end-points of a segment in one dimension, a curve in two dimensions, and a surface in three dimensions. The developments which follow hold for a model of any dimension from one to three. The dimension of a model is denoted by n_{SD}, where 'SD' denotes the number of space dimensions.

3.2.2 Eulerian and Lagrangian Coordinates

The position vector of a material point in the initial configuration is given by $\mathbf{X}$, where

$$\mathbf{X} = X_i \mathbf{e}_i \equiv \sum_{i=1}^{n_{SD}} X_i \mathbf{e}_i \qquad (3.2.1)$$

where X_i are the components of the position vector in the initial configuration and $\mathbf{e}_i$ are the unit base vectors of a rectangular Cartesian coordinate system; indicial notation as described in Section 1.3 has been used in the second expression and will be used throughout this book. Some authors, such as Malvern (1969), also define material particles and carefully distinguish between material points and particles in a continuum. The notion of particles in a continuum is somewhat confusing, for the concept of particles to most of us is discrete rather than continuous. Therefore we will refer only to material points of the continuum.

The vector variable **X** for a given material point does not change with time; the variables **X** are called *material coordinates* or *Lagrangian coordinates* and provide labels for material points. Thus if we want to track the function $f(\mathbf{X}, t)$ at a given material point, we simply track that function at a constant value of **X**. The position of a point in the current configuration is given by

$$\mathbf{x} = x_i \mathbf{e}_i \equiv \sum_{i=1}^{n_{SD}} x_i \mathbf{e}_i \tag{3.2.2}$$

where x_i are the components of the position vector in the current configuration.

3.2.3 Motion

The *motion of the body* is described by

$$\mathbf{x} = \boldsymbol{\phi}(\mathbf{X}, t) \text{ or } x_i = \phi_i(\mathbf{X}, t) \tag{3.2.3}$$

where **x** is the position of the material point **X** at time t. The coordinates x_i give the spatial position, and are called *spatial* or *Eulerian coordinates*. The function $\boldsymbol{\phi}(\mathbf{X}, t)$ maps the initial configuration into the current configuration at time t, and is called a mapping or map from the initial configuration to the current configuration.

When the reference configuration is identical to the initial configuration, the position vector **x** of any point at time $t=0$ coincides with the material coordinates, so

$$\mathbf{X} = \mathbf{x}(\mathbf{X}, 0) \equiv \boldsymbol{\phi}(\mathbf{X}, 0) \text{ or } X_i = x_i(\mathbf{X}, 0) = \phi_i(\mathbf{X}, 0) \tag{3.2.4}$$

Thus the mapping $\boldsymbol{\phi}(\mathbf{X}, 0)$ is the identity mapping.

Lines of constant values of the material coordinate, X_i, when etched into the material, behave just like a Lagrangian mesh; they deform with the material and, when viewed in the deformed configuration, these lines are no longer Cartesian. Viewed in this way, the material coordinates are often called convected coordinates. In pure shear, for example, they become skewed coordinates, just like a Lagrangian mesh becomes skewed (see Figure 1.2). In the equations to be developed here, the material coordinates are viewed in the initial configuration, so they are treated as a fixed Cartesian coordinate system. The spatial coordinate system, on the other hand, does not change with time regardless of how it is viewed.

3.2.4 Eulerian and Lagrangian Descriptions

Two approaches are used to describe the deformation and response of a continuum. In the first approach, the independent variables are the material coordinates X_i and the time t, as in (3.2.3); this description is called a *material description* or *Lagrangian description*. In the second approach, the independent variables are the spatial coordinates **x** and the time t. This is called a *spatial* or *Eulerian description*. The duality is similar to that in mesh descriptions.

In fluid mechanics, it is often impossible and unnecessary to describe the motion with respect to a reference configuration. For example, if we consider the flow around an airfoil, an undeformed configuration is not needed because the stresses and behavior of a Newtonian fluid are independent of its history. On the other hand, in solids, the stresses generally depend on the deformation and its history so an undeformed configuration must be specified. Because of the history dependence of most solids, Lagrangian descriptions are prevalent in solid mechanics.

In the mathematics and continuum mechanics literature (*cf.* Marsden and Hughes, 1983), different symbols are often used for the same field when it is expressed in terms of different independent variables, that is, when the description is Eulerian or Lagrangian. In this convention, the function which in an Eulerian description is $f(\mathbf{x}, t)$ is denoted by $F(\mathbf{X}, t)$ in a Lagrangian description. The two functions are related by

$$F(\mathbf{X}, t) = f(\boldsymbol{\phi}(\mathbf{X}, t), t) \quad \text{or} \quad F = f \circ \boldsymbol{\phi} \tag{3.2.5}$$

This is called a *composition of functions;* the notation on the right is frequently used in the mathematics literature – see, for example, Spivak (1965: 11). The notation for the composition of functions will be used infrequently in this book because it is unfamiliar to most engineers.

The convention of referring to different functions by different symbols is attractive and often adds clarity. However, in finite element methods, because of the need to refer to three or more sets of independent variables, this convention becomes quite awkward. Therefore in this book, we associate a symbol with a field, and the specific function is defined by specifying the independent variables. Thus $f(\mathbf{x}, t)$ is the function which describes the field f for the independent variables $\mathbf{x}$ and t, whereas $f(\mathbf{X}, t)$ is a different function which describes the same field in terms of the material coordinates. The independent variables are indicated near the beginning of a section or chapter, and if a change of independent variables is made, the new independent variables are noted.

3.2.5 Displacement, Velocity and Acceleration

The displacement of a material point is given by the difference between its current position and its original position (see Figure 3.1), so

$$\mathbf{u}(\mathbf{X}, t) = \boldsymbol{\phi}(\mathbf{X}, t) - \boldsymbol{\phi}(\mathbf{X}, 0) = \boldsymbol{\phi}(\mathbf{X}, t) - \mathbf{X}, \quad u_i = \phi_i(X_j, t) - X_i \tag{3.2.6}$$

where $\mathbf{u}(\mathbf{X}, t) = u_i \mathbf{e}_i$ and we have used (3.2.4). The displacement is often written as

$$\mathbf{u} = \mathbf{x} - \mathbf{X}, \quad u_i = x_i - X_i \tag{3.2.7}$$

where (3.2.3) has been used in (3.2.6) to replace $\boldsymbol{\phi}(\mathbf{X}, t)$ by $\mathbf{x}$. Equation (3.2.7) is somewhat ambiguous since it expresses the displacement as the difference of two variables, $\mathbf{x}$ and $\mathbf{X}$, both of which can be independent variables. The reader must keep in mind that in expressions such as (3.2.7) the symbol $\mathbf{x}$ represents the motion $\mathbf{x}(\mathbf{x}, t) \equiv \boldsymbol{\phi}(\mathbf{X}, t)$.

The velocity $\mathbf{v}(\mathbf{X}, t)$ is the rate of change of the position vector for a material point, i.e. the time derivative with $\mathbf{X}$ held constant. Time derivatives with $\mathbf{X}$ held constant are called *material time derivatives*, or sometimes *material derivatives*. Material time derivatives are also called *total derivatives*. The velocity can be written in various forms:

$$\mathbf{v}(\mathbf{X}, t) = \frac{\partial \boldsymbol{\phi}(\mathbf{X}, t)}{\partial t} = \frac{\partial \mathbf{u}(\mathbf{X}, t)}{\partial t} \equiv \dot{\mathbf{u}} \tag{3.2.8}$$

In the above, the motion is replaced by the displacement $\mathbf{u}$ in the third term by using (3.2.7) and the fact that $\mathbf{X}$ is independent of time. The superposed dot denotes a material time derivative, though it is also used for ordinary time derivatives when the variable is only a function of time.

The acceleration $\mathbf{a}(\mathbf{X}, t)$ is the rate of change of velocity of a material point, or in other words the material time derivative of the velocity, and can be written in the forms

$$\mathbf{a}(\mathbf{X}, t) = \frac{\partial \mathbf{v}(\mathbf{X}, t)}{\partial t} = \frac{\partial^2 \mathbf{u}(\mathbf{X}, t)}{\partial t^2} \equiv \dot{\mathbf{v}} \tag{3.2.9}$$

The above expression is called the material form of the acceleration.

When the velocity is expressed in terms of the spatial coordinates and the time, that is, in an Eulerian description as in $\mathbf{v}(\mathbf{x}, t)$, the material time derivative is obtained as follows. The spatial coordinates in $\mathbf{v}(\mathbf{x}, t)$ are first expressed as a function of the material coordinates and time by using (3.2.3), giving $\mathbf{v}(\boldsymbol{\phi}(\mathbf{X}, t), t)$. The material time derivative is then obtained by the chain rule:

$$\frac{Dv_i(\mathbf{x}, t)}{Dt} = \frac{\partial v_i(\mathbf{x}, t)}{\partial t} + \frac{\partial v_i(\mathbf{x}, t)}{\partial x_j}\frac{\partial \phi_j(\mathbf{X}, t)}{\partial t} = \frac{\partial v_i}{\partial t} + \frac{\partial v_i}{\partial x_j} v_j \tag{3.2.10}$$

where the second equality follows from (3.2.8). The second term on the RHS of (3.2.10) is the convective term, which is also called the transport term; $\partial v_i / \partial t$ is called the *spatial time derivative*. It is tacitly assumed throughout this book that when neither the independent variables nor the fixed variable are explicitly indicated in a partial derivative with respect to time, then the spatial coordinate is fixed and we are referring to the spatial time derivative. On the other hand, when the independent variables are specified as in (3.2.8–3.2.9), a partial derivative with respect to time is a material time derivative. Equation (3.2.10) is written in tensor notation as

$$\frac{D\mathbf{v}(\mathbf{x}, t)}{Dt} = \frac{\partial \mathbf{v}(\mathbf{x}, t)}{\partial t} + \mathbf{v} \cdot \nabla \mathbf{v} = \frac{\partial \mathbf{v}}{\partial t} + \mathbf{v} \cdot \text{grad } \mathbf{v} \tag{3.2.11}$$

where $\nabla \mathbf{v}$ and grad $\mathbf{v}$ are the left gradients of a vector field as defined in Malvern (1969: 58). The matrix of the *left gradient* is given by

$$[\nabla \mathbf{v}] \equiv [\text{grad } \mathbf{v}] = \begin{bmatrix} v_x, x & v_y, x \\ v_x, y & v_y, y \end{bmatrix} \tag{3.2.12}$$

The gradient index in the left gradient of a vector is the row number. This can be remembered by writing the left gradient $\nabla \mathbf{v}$ in indicial notation as $\partial_i v_j$. We will use only the left gradient in this book, but to maintain consistency with the notation of others such as Malvern (1969), we follow these conventions. Note that

$$\frac{D\mathbf{v}(\mathbf{x}, t)}{Dt} = \frac{\partial \mathbf{v}(\mathbf{X}, t)}{\partial t} \tag{3.2.13}$$

The material time derivative of any function of the spatial variables **x** and time t can similarly be obtained by the chain rule. Thus for a scalar function $f(\mathbf{x}, t)$ and a tensor function $\sigma_{ij}(\mathbf{x}, t)$, the material time derivatives are given by

$$\frac{Df}{Dt} = \frac{\partial f}{\partial t} + v_i \frac{\partial f}{\partial x_i} = \frac{\partial f}{\partial t} + \mathbf{v} \cdot \nabla f = \frac{\partial f}{\partial t} + \mathbf{v} \cdot \text{grad } f \tag{3.2.14}$$

$$\frac{D\sigma_{ij}}{Dt} = \frac{\partial \sigma_{ij}}{\partial t} + v_k \frac{\partial \sigma_{ij}}{\partial x_k} = \frac{\partial \boldsymbol{\sigma}}{\partial t} + \mathbf{v} \cdot \nabla \boldsymbol{\sigma} = \frac{\partial \boldsymbol{\sigma}}{\partial t} + \mathbf{v} \cdot \text{grad } \boldsymbol{\sigma} \tag{3.2.15}$$

where the first term on the RHS of each equation is the spatial time derivative and the second term is the convective term.

A complete description of the motion is not needed to develop the material time derivative in an Eulerian description. The motion can instead be described at each instant with respect to a reference configuration that coincides with the configuration at a fixed time t. For this purpose, let the configuration at a fixed time $t = \tau$ be the reference configuration and the position vectors of the material points at that time, denoted by $\mathbf{X}^\tau$, be the reference coordinates. These reference coordinates are given by

$$\mathbf{X}^\tau = \boldsymbol{\phi}(\mathbf{X}, \tau) \tag{3.2.16}$$

We use an upper case $\mathbf{X}^\tau$ for the position vectors at time τ since we wish to clearly identify it as another reference coordinate; the superscript τ distinguishes these reference coordinates from the initial reference coordinates. The motion can be described in terms of these reference coordinates by

$$\mathbf{x} = \boldsymbol{\phi}^\tau(\mathbf{X}^\tau, t) \qquad \text{for } t \geq \tau \tag{3.2.17}$$

Now the arguments used to develop (3.2.10) can be repeated; noting that $\mathbf{v}(\mathbf{x}, t) = \mathbf{v}(\boldsymbol{\phi}^\tau(\mathbf{X}^\tau, t), t)$, we can obtain an expression for the acceleration with the current configuration as the reference configuration:

$$\frac{Dv_i}{Dt} = \frac{\partial v_i(\mathbf{x}, t)}{\partial t} + \frac{\partial v_i(\mathbf{x}, t)}{\partial x_j} \frac{\partial \phi_j^\tau}{\partial t} = \frac{\partial v_i}{\partial t} + \frac{\partial v_i}{\partial x_j} v_j \tag{3.2.18}$$

Reference configurations which are not the initial configuration will also be employed in the development of finite element equations.

3.2.6 *Deformation Gradient*

The description of deformation and the measure of strain are essential parts of nonlinear continuum mechanics. An important variable in the characterization of deformation is the *deformation gradient*. The deformation gradient is defined by

$$F_{ij} = \frac{\partial \phi_i}{\partial X_j} \equiv \frac{\partial x_i}{\partial X_j} \quad \text{or} \quad \mathbf{F} = \frac{\partial \boldsymbol{\phi}}{\partial \mathbf{X}} \equiv \frac{\partial \mathbf{x}}{\partial \mathbf{X}} \equiv (\nabla_0 \boldsymbol{\phi})^T \tag{3.2.19}$$

In the terminology of mathematics, the deformation gradient $\mathbf{F}$ is the *Jacobian matrix* of the motion $\boldsymbol{\phi}(\mathbf{X}, t)$. Note in the above that the first index of F_{ij} refers to the motion, the second to the partial derivative. The operator ∇_0 is the *left gradient with respect to the material coordinates.*

If we consider an infinitesimal line segment $d\mathbf{X}$ in the reference configuration, then it follows from (3.2.19) that the corresponding line segment $d\mathbf{x}$ in the current configuration is given by

$$d\mathbf{x} = \mathbf{F} \cdot d\mathbf{X} \quad \text{or} \quad dx_i = F_{ij} dX_j \tag{3.2.20}$$

In the above expression, the dot could have been omitted between the $\mathbf{F}$ and $d\mathbf{X}$, since the expression is also valid as a matrix expression. We have retained it to conform to our convention of always explicitly indicating contractions in tensor expressions.

In two dimensions, the deformation gradient in a rectangular coordinate system is given by

$$\mathbf{F} = \begin{bmatrix} \dfrac{\partial x_1}{\partial X_1} & \dfrac{\partial x_1}{\partial X_2} \\ \dfrac{\partial x_2}{\partial X_1} & \dfrac{\partial x_2}{\partial X_2} \end{bmatrix} = \begin{bmatrix} \dfrac{\partial x}{\partial X} & \dfrac{\partial x}{\partial Y} \\ \dfrac{\partial y}{\partial X} & \dfrac{\partial y}{\partial Y} \end{bmatrix} \tag{3.2.21}$$

As can be seen in the above, in writing a second-order tensor in matrix form, we use the first index for the row number, and the second index for the column number. Note that $\mathbf{F}$ is the transpose of the left-gradient.

The determinant of $\mathbf{F}$ is denoted by J and called the *Jacobian determinant* or the determinant of the deformation gradient

$$J = \det(\mathbf{F}) \tag{3.2.22}$$

The Jacobian determinant can be used to relate integrals in the current and reference configurations by

$$\int_{\Omega} f(\mathbf{x}, t) d\Omega = \int_{\Omega_0} f(\boldsymbol{\phi}(\mathbf{X}, t), t) J \, d\Omega_0 \quad \text{or} \quad \int_{\Omega} f \, d\Omega = \int_{\Omega_0} f J \, d\Omega_0 \tag{3.2.23}$$

or in two dimensions

$$\int_{\Omega} f(x, y) dx dy = \int_{\Omega_0} f(X, Y) J \, dX dY \tag{3.2.24}$$

The material derivative of the Jacobian determinant is given by

$$\frac{DJ}{Dt} \equiv \dot{J} = J \operatorname{div} \mathbf{v} \equiv J \frac{\partial v_i}{\partial x_i} \tag{3.2.25}$$

The derivation of this formula is left as an exercise.

3.2.7 Conditions on Motion

The mapping $\boldsymbol{\phi}(\mathbf{X}, t)$ which describes the motion and deformation of the body is assumed to satisfy the following conditions except on a finite number of sets of measure zero:

1. The function $\boldsymbol{\phi}(\mathbf{X}, t)$ is continuously differentiable.
2. The function $\boldsymbol{\phi}(\mathbf{X}, t)$ is one-to-one.
3. The Jacobian determinant satisfies the condition $J > 0$.

These conditions ensure that $\boldsymbol{\phi}(\mathbf{X}, t)$ is sufficiently smooth so that compatibility is satisfied, i.e. there are no gaps or overlaps in the deformed body. The motion and its derivatives *can be discontinuous or possess discontinuous derivatives on sets of measure zero* (see Section 1.5), so it is piecewise continuously differentiable. The proviso excluding sets of measure zero is added to account for the possibility of crack formation. On surfaces which evolve into cracks, these conditions are not met. Sets of measure zero are points in one dimension, lines in two dimensions and planes in three dimensions, because a point has zero length, a line has zero area, and a surface has zero volume.

The deformation gradient is generally discontinuous on interfaces between materials. In some phenomena, such as a growing crack, the motion itself is discontinuous. We require the number of discontinuities in a motion and its derivatives to be finite. In fact, it has been found that some nonlinear solutions may possess an infinite number of discontinuities; see, for example, Belytschko *et al.* (1986). However, these solutions are quite unusual and cannot be treated effectively by finite elements, so we will not concern ourselves with them.

The second condition in the above list, that the motion be one-to-one, requires that for each point in the reference configuration Ω_0, there is a unique point in Ω and vice versa. This is a necessary and sufficient condition for the regularity of $\mathbf{F}$, i.e. that $\mathbf{F}$ be in vertible. When the deformation gradient $\mathbf{F}$ is regular, the Jacobian determinant J must be nonzero, since the inverse of $\mathbf{F}$ exists if and only if its determinant $J \neq 0$. Thus the second and third conditions are related. We have stated the stronger condition that J be positive rather than just nonzero, which will be seen in Section 3.5.4 to follow from mass conservation. This condition can also be violated on sets of measure zero. For example, on a surface which becomes a crack, each point becomes two points.

3.2.8 Rigid Body Rotation and Coordinate Transformations

Rigid body rotation plays a crucial role in the theory of nonlinear continuum mechanics. Many of the complexities which permeate the field arise from rigid body rotation. Furthermore, the decision as to whether linear or nonlinear software is appropriate for a particular linear material problem hinges on the magnitude of rotations. When the rotations are large enough to render a linear strain measure invalid, nonlinear software must be used.

A rigid body motion consisting of a translation $\mathbf{X}_T(t)$ and a rotation about the origin is written as

$$\mathbf{x}(\mathbf{X}, t) = \mathbf{R}(t) \cdot \mathbf{X} + \mathbf{x}_T(t) \quad x_i(\mathbf{X}, t) = R_{ij}(t) X_j + x_{Ti}(t) \tag{3.2.26}$$

where $\mathbf{R}(t)$ is the rotation tensor, also called a rotation matrix. Any rigid body motion can be expressed in this form.

The rotation matrix $\mathbf{R}$ is an orthogonal matrix, which means that its inverse is given by its transpose. This can be shown by noting that rigid body rotation preserves length, and since $d\mathbf{x}_T = \mathbf{0}$ we have

$$d\mathbf{x} \cdot d\mathbf{x} = d\mathbf{X} \cdot (\mathbf{R}^T \cdot \mathbf{R}) \cdot d\mathbf{X}, \quad dx_i dx_i = R_{ij} dX_j \; R_{ik} dX_k = dX_j (R_{ji}^T R_{ik}) dX_k$$

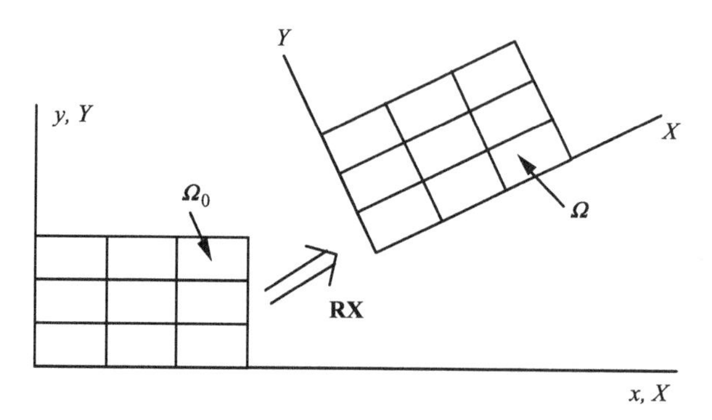

Figure 3.2 A rigid body rotation of a Lagrangian mesh showing the material coordinates when viewed in the reference (initial, undeformed) configuration and the current configuration

Since the length is unchanged in rigid body motion, it follows that $d\mathbf{x}\cdot d\mathbf{x}=d\mathbf{X}\cdot d\mathbf{X}$, for arbitrary $d\mathbf{X}$, so

$$\mathbf{R}^T \cdot \mathbf{R} = \mathbf{I} \tag{3.2.27}$$

This shows that the inverse of $\mathbf{R}$ is given by its transpose:

$$\mathbf{R}^{-1} = \mathbf{R}^T, \qquad R_{ij}^{-1} = R_{ij}^T = R_{ji} \tag{3.2.28}$$

The rotation tensor $\mathbf{R}$ is therefore said to be an orthogonal matrix. Any transformation by this matrix, such as $\mathbf{x}=\mathbf{R}\mathbf{X}$, is called an orthogonal transformation. Rotation is an example of an orthogonal transformation.

A rigid body rotation of a Lagrangian mesh of rectangular elements is shown in Figure 3.2. As can be seen, in the rigid body rotation, the element edges are rotated but the angles between the edges remain right angles. The element edges are lines of constant X and Y, so when viewed in the deformed configuration, the material coordinates are rotated when the body is rotated as shown in Figure 3.2.

Before obtaining the rotation matrix, we derive the formula relating the components of the vector $\mathbf{r}$ in two different coordinate systems. The two coordinate systems are described by orthonormal base vectors $\mathbf{e}_i$ and $\hat{\mathbf{e}}_i$, respectively. The orthonormality of the base vectors is expressed as

$$\mathbf{e}_i \cdot \mathbf{e}_j = \delta_{ij} \quad \hat{\mathbf{e}}_i \cdot \hat{\mathbf{e}}_j = \delta_{ij} \tag{3.2.29}$$

Figure 3.3 shows the vector $\mathbf{r}$ and the base vectors in the rotated and unrotated coordinate systems. Since the vector $\mathbf{r}$ is independent of the coordinate system,

$$\mathbf{r} = r_i \mathbf{e}_i = \hat{r}_i \hat{\mathbf{e}}_i \tag{3.2.30}$$

Taking the scalar product of (3.2.30) with $\mathbf{e}_j$ gives

$$r_i \mathbf{e}_i \cdot \mathbf{e}_j = \hat{r}_i \hat{\mathbf{e}}_i \cdot \mathbf{e}_j \quad \text{so} \quad r_i \delta_{ij} = \hat{r}_i \hat{\mathbf{e}}_i \cdot \mathbf{e}_j \quad \text{so} \quad r_j = R_{ji} \hat{r}_i \ \text{ where } R_{ji} = \mathbf{e}_j \cdot \hat{\mathbf{e}}_i \tag{3.2.31}$$

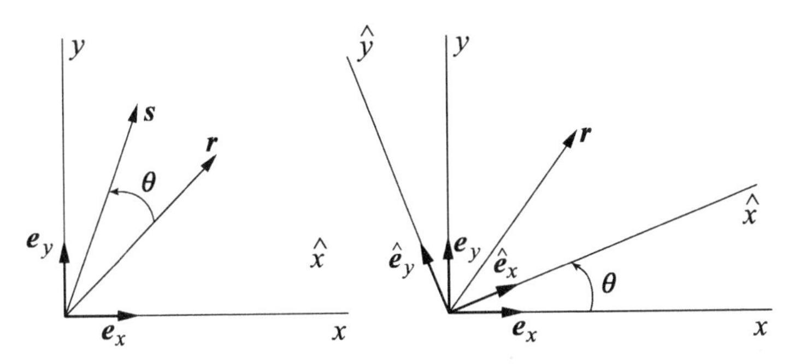

Figure 3.3 Nomenclature for rotation transformation in two dimensions

The second equation follows from the orthonormality of the base vectors, (3.2.29). We can write the RHS of (3.2.31) as

$$\mathbf{r} = \mathbf{R}\hat{\mathbf{r}} \quad \text{or} \quad r_i = R_{ij}\hat{r}_j \equiv R_{i\hat{j}}r_{\hat{j}} \tag{3.2.32}$$

On the far RHS we have put the 'hat' on the index; we will sometimes use that notation. The second index always pertains to the 'hatted' coordinate system of the rotation matrix; this convention helps in remembering the form of the transformation equation.

Premultiplying both sides of the above by $\mathbf{R}^T$ and using the orthogonality condition (3.2.27) gives

$$\hat{\mathbf{r}} = \mathbf{R}^T\mathbf{r} \quad \text{or} \quad \hat{r}_j \equiv r_{\hat{j}} = R^T_{ji}r_i = R_{i\hat{j}}r_i \tag{3.2.33}$$

The above are *matrix expressions*, as indicated by the absence of dots between the terms. The components of $\mathbf{r}$ and $\hat{\mathbf{r}}$ differ, but $\mathbf{r}$ and $\hat{\mathbf{r}}$ refer to the same vector. This distinction between the components of a vector and the vector itself is sometimes clarified by using different symbols for matrices and tensors, but the notation we have chosen does not permit this distinction.

In two dimensions (3.2.32) is

$$\begin{Bmatrix} r_x \\ r_y \end{Bmatrix} = \begin{bmatrix} R_{x\hat{x}} & R_{x\hat{y}} \\ R_{y\hat{x}} & R_{y\hat{y}} \end{bmatrix} \begin{Bmatrix} \hat{r}_x \\ \hat{r}_y \end{Bmatrix} = \begin{bmatrix} \mathbf{e}_x \cdot \hat{\mathbf{e}}_x & \mathbf{e}_x \cdot \hat{\mathbf{e}}_y \\ \mathbf{e}_y \cdot \hat{\mathbf{e}}_x & \mathbf{e}_y \cdot \hat{\mathbf{e}}_y \end{bmatrix} \begin{Bmatrix} \hat{r}_x \\ \hat{r}_y \end{Bmatrix} = \begin{bmatrix} \cos\theta & -\sin\theta \\ \sin\theta & \cos\theta \end{bmatrix} \begin{Bmatrix} \hat{r}_x \\ \hat{r}_y \end{Bmatrix} \tag{3.2.34}$$

In this, it can be seen that the subscripts on $\mathbf{R}$ correspond to the vector components which are related; for example, in the expression for the x component in row 1, the $R_{x\hat{y}}$ is the coefficient of the $\hat{y}$ component of $\hat{\mathbf{r}}$. The last form of the transformation in the above is obtained by evaluating the scalar products from Figure 3.3.

The equation for rotating the vector $\mathbf{r}$ is quite similar. Let the rotated vector be denoted by $\mathbf{s}$. The components of $\mathbf{s}$ in the rotated system, $\hat{s}_i$, equal the corresponding components of the vector $\mathbf{r}$ in the unrotated coordinates, r_i:

$$\hat{s}_i = r_i \tag{3.2.35}$$

For an example of the rotation of a vector, see Figure 3.3. Applying (3.2.33) to $\hat{s}_i$ gives

$$\mathbf{r} = \mathbf{R}^T\mathbf{s} \quad \text{or} \quad r_i = R_{ij}^T s_j \tag{3.2.36}$$

Premultiplying (3.2.36) by **R** and using the orthogonality condition (3.2.27) yields

$$\mathbf{s} = \mathbf{R}\mathbf{r} \quad \text{or} \quad s_i = R_{ij} r_j \tag{3.2.37}$$

This is the standard expression for the rotation of a vector without translation. Equation (3.2.26) is obtained by combining (3.2.37) with a translation. Note the difference between (3.2.33) and (3.2.37): the components of the rotated vector are obtained by multiplying by **R**, whereas the components of the same vector in a rotated coordinate system are obtained by multiplying by $\mathbf{R}^T$. Frankly, this is very difficult to keep straight, so in working with rotations it is often helpful to keep a picture like Figure 3.3 handy.

The components of a second-order tensor **D** are transformed by

$$\mathbf{D} = \mathbf{R}\hat{\mathbf{D}}\mathbf{R}^T, \quad D_{ij} = R_{ik}\hat{D}_{kl}R_{lj}^T \tag{3.2.38}$$

The inverse of the above is obtained by premultiplying by $\mathbf{R}^T$, postmultiplying by **R** and using the orthogonality of **R** (3.2.27):

$$\hat{\mathbf{D}} = \mathbf{R}^T\mathbf{D}\mathbf{R}, \quad \hat{D}_{ij} = R_{ik}^T D_{kl} R_{lj} \tag{3.2.39}$$

3.2.8.1 Angular Velocity

The velocity for a rigid body motion can be obtained by taking the time derivative of (3.2.26). This gives

$$\dot{\mathbf{x}}(\mathbf{X},t) = \dot{\mathbf{R}}(t)\cdot\mathbf{X} + \dot{\mathbf{x}}_T(t) \quad \text{or} \quad \dot{x}_i(\mathbf{X},t) = \dot{R}_{ij}(t)X_j + \dot{x}_{T_i}(t) \tag{3.2.40}$$

An Eulerian description of rigid body rotation can be obtained by expressing the material coordinates in (3.2.40) in terms of the spatial coordinates via (3.2.26), giving

$$\mathbf{v} \equiv \dot{\mathbf{x}} = \dot{\mathbf{R}}\cdot\mathbf{R}^T\cdot(\mathbf{x}-\mathbf{x}_T) + \dot{\mathbf{x}}_T = \Omega\cdot(\mathbf{x}-\mathbf{x}_T) + \dot{\mathbf{x}}_T \tag{3.2.41}$$

where

$$\Omega = \dot{\mathbf{R}}\cdot\mathbf{R}^T \tag{3.2.42}$$

The tensor $\boldsymbol{\Omega}$ is called the *angular velocity tensor* or *angular velocity matrix* (Dienes, 1979: 221). It is a skew-symmetric tensor; skew-symmetric tensors are also called *antisymmetric tensors*. To demonstrate the skew-symmetry of the angular velocity tensor, we take the time derivative of (3.2.27) which gives

$$\frac{D}{Dt}(\mathbf{R}\cdot\mathbf{R}^T) = \frac{D\mathbf{I}}{Dt} = 0 \rightarrow \dot{\mathbf{R}}\cdot\mathbf{R}^T + \mathbf{R}\cdot\dot{\mathbf{R}}^T = 0 \rightarrow \Omega = -\boldsymbol{\Omega}^T \tag{3.2.43}$$

Any skew-symmetric second-order tensor can be expressed in terms of the components of a vector, called the *axial vector*, and the matrix product $\mathbf{\Omega r}$ can be replicated by the cross product with the axial vector, $\boldsymbol{\omega}\times\mathbf{r}$. Thus

$$\mathbf{\Omega r}=\boldsymbol{\omega}\times\mathbf{r} \quad \text{or} \quad \Omega_{ij}r_j = e_{ijk}\omega_j r_k \tag{3.2.44}$$

for any $\mathbf{r}$. In the above e_{ijk} is the *alternator matrix* or *permutation symbol*, defined by

$$e_{ijk}=\begin{cases}1 \text{ for an even permutation of } ijk\\ -1 \text{ for an odd permutation of } ijk\\ 0 \text{ if any index is repeated}\end{cases} \tag{3.2.45}$$

The relations between the skew-symmetric tensor Ω and its axial vector ω are

$$\Omega_{ik}=e_{ijk}\omega_j=-e_{ikj}\omega_j, \qquad \omega_i=-\frac{1}{2}e_{ijk}\Omega_{jk} \tag{3.2.46}$$

The first is obtained by inspection from (3.2.44), the second follows from the first by premultiplying by e_{rij} and using the identity $e_{rij}e_{rkl}=\delta_{ik}\delta_{jl}-\delta_{il}\delta_{kj}$ (see Malvern, 1969: 23). In two dimensions, a skew-symmetric tensor has a single independent component and its axial vector is perpendicular to the two-dimensional plane of the model, so

$$\mathbf{\Omega}=\begin{bmatrix}0 & \Omega_{12}\\ -\Omega_{12} & 0\end{bmatrix}=\begin{bmatrix}0 & -\omega_3\\ \omega_3 & 0\end{bmatrix} \tag{3.2.47}$$

In three dimensions, a skew-symmetric tensor has three independent components which are related to the three components of its axial vector by (3.2.46), giving

$$\mathbf{\Omega}=\begin{bmatrix}0 & \Omega_{12} & \Omega_{13}\\ -\Omega_{12} & 0 & \Omega_{23}\\ -\Omega_{13} & -\Omega_{23} & 0\end{bmatrix}=\begin{bmatrix}0 & -\omega_3 & \omega_2\\ \omega_3 & 0 & -\omega_1\\ -\omega_2 & \omega_1 & 0\end{bmatrix} \tag{3.2.48}$$

The angular velocity matrix is sometimes defined to be the negative of the above.

When (3.2.41) is expressed in terms of the angular velocity vector, we have

$$v_i \equiv \dot{x}_i=\Omega_{ij}(x_j-x_{Tj})+v_{Ti}=e_{ijk}\omega_j(x_k-x_{Tk})+v_{Ti}$$

or

$$\mathbf{v}\equiv\dot{\mathbf{x}}=\boldsymbol{\omega}\times(\mathbf{x}-\mathbf{x}_T)+\mathbf{v}_T \tag{3.2.49}$$

where we have exchanged k and j in the second term and used $e_{kij}=e_{ijk}$. This is the equation for rigid body motion as given in dynamics texts. The first term on the RHS is the velocity due to the rotation about the point $\mathbf{x}_T$ and the second term is the translational velocity. The velocity in any rigid body motion can be expressed by (3.2.49).

This concludes the formal discussion of rotation in this chapter. However, the topic of rotation will reappear in many other parts of this chapter and this book. Rotation, especially when combined with deformation, is fundamental to nonlinear continuum mechanics, and it should be thoroughly understood by a student of this field.

Example 3.1 Rotation and stretch of a triangular element Consider the three-node triangular finite element shown in Figure 3.4. Let the motion of the nodes be given by

$$\begin{aligned} & x_1(t) = y_1(t) = 0 \\ & x_2(t) = 2(1+at)\cos\frac{\pi t}{2}, \quad y_2(t) = 2(1+at)\sin\frac{\pi t}{2} \\ & x_3(t) = -(1+bt)\sin\frac{\pi t}{2}, \quad y_3(t) = (1+bt)\cos\frac{\pi t}{2} \end{aligned} \tag{E3.1.1}$$

Find the deformation gradient and the Jacobian determinant as a function of time and find the values of a and b such that the Jacobian determinant remains constant.

In terms of the triangular element coordinates ξ_I, the configuration of a triangular three-node, linear displacement element at any time can be written as follows (see Appendix 3 if you are not familiar with triangular coordinates):

$$\begin{aligned} x(\boldsymbol{\xi}, t) = x_I(t)\xi_I = x_1(t)\xi_1 + x_2(t)\xi_2 + x_3(t)\xi_3 \\ y(\boldsymbol{\xi}, t) = y_I(t)\xi_I = y_1(t)\xi_1 + y_2(t)\xi_2 + y_3(t)\xi_3 \end{aligned} \tag{E3.1.2}$$

In the initial configuration, that is, at $t=0$:

$$\begin{aligned} X = x(\boldsymbol{\xi}, 0) = X_1\xi_1 + X_2\xi_2 + X_3\xi_3 \\ Y = y(\boldsymbol{\xi}, 0) = Y_1\xi_1 + Y_2\xi_2 + Y_3\xi_3 \end{aligned} \tag{E3.1.3}$$

Substituting the coordinates of the nodes in the undeformed configuration into this, $X_1 = X_3 = 0$, $X_2 = 2$, $Y_1 = Y_2 = 0$, $Y_3 = 1$ yields

$$X = 2\xi_2, \quad Y = \xi_3 \tag{E3.1.4}$$

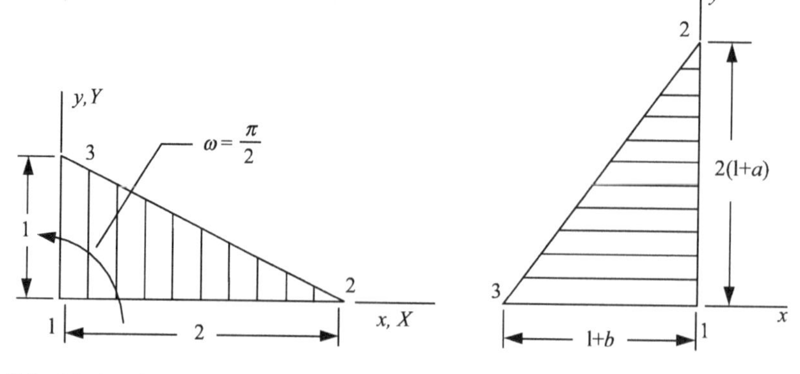

Figure 3.4 Motion described by (E3.1.1) with the initial configuration on the left and the deformed configuration at $t = 1$ shown on the right

In this case, the relations between the triangular coordinates and the material coordinates can be inverted by inspection to give

$$\xi_2 = \frac{1}{2}X, \quad \xi_3 = Y \tag{E3.1.5}$$

Substituting (E3.1.1) and (E3.1.5) into (E3.1.2) gives the following expressions for the motion:

$$\begin{aligned} x(\mathbf{X}, t) &= X(1+at)\cos\frac{\pi t}{2} - Y(1+bt)\sin\frac{\pi t}{2} \\ y(\mathbf{X}, t) &= X(1+at)\sin\frac{\pi t}{2} + Y(1+bt)\cos\frac{\pi t}{2} \end{aligned} \tag{E3.1.6}$$

The deformation gradient is given by (3.2.21):

$$\mathbf{F} = \begin{bmatrix} \dfrac{\partial x}{\partial X} & \dfrac{\partial x}{\partial Y} \\ \dfrac{\partial y}{\partial X} & \dfrac{\partial y}{\partial Y} \end{bmatrix} = \begin{bmatrix} (1+at)\cos\dfrac{\pi t}{2} & -(1+bt)\sin\dfrac{\pi t}{2} \\ (1+at)\sin\dfrac{\pi t}{2} & (1+bt)\cos\dfrac{\pi t}{2} \end{bmatrix} \tag{E3.1.7}$$

The deformation gradient is a function of time only, and at any time it is constant in the element because the displacement in this element is a linear function of the material coordinates. The Jacobian determinant is given by

$$J = \det(\mathbf{F}) = (1+at)(1+bt)\left(\cos^2\frac{\pi t}{2} + \sin^2\frac{\pi t}{2}\right) = (1+at)(1+bt) \tag{E3.1.8}$$

When $a=b=0$ the Jacobian determinant remains constant, $J=1$. This motion is a rotation without deformation. The Jacobian determinant J also remains constant when $b=-a/(1+at)$, which corresponds to a shearing deformation and a rotation in which the area of the element remains constant. This type of deformation is called *isochoric*; the deformation of incompressible materials is isochoric.

Example 3.2 Consider an element which is rotating at a constant angular velocity ω about the origin. Obtain the accelerations using both the material and spatial descriptions. Find the deformation gradient **F** and its rate.

The motion for a pure rotation about the origin is obtained from (3.2.26) using the rotation matrix in two dimensions (3.2.34):

$$\mathbf{x}(t) = \mathbf{R}(t)\mathbf{X} \Rightarrow \begin{Bmatrix} x \\ y \end{Bmatrix} = \begin{bmatrix} \cos\omega t & -\sin\omega t \\ \sin\omega t & \cos\omega t \end{bmatrix} \begin{Bmatrix} X \\ Y \end{Bmatrix} \tag{E3.2.1}$$

where we have used $\theta = \omega t$ to express the motion as a function of time; ω is the angular velocity of the body. The velocity is obtained by taking the derivative of this motion with respect to time, which gives

$$\begin{Bmatrix} v_x \\ v_y \end{Bmatrix} = \begin{Bmatrix} \dot{x} \\ \dot{y} \end{Bmatrix} = \omega \begin{bmatrix} -\sin\omega t & -\cos\omega t \\ \cos\omega t & -\sin\omega t \end{bmatrix} \begin{Bmatrix} X \\ Y \end{Bmatrix} \tag{E3.2.2}$$

The acceleration in the material description is obtained by taking time derivatives of the velocities

$$\begin{Bmatrix} a_x \\ a_y \end{Bmatrix} = \begin{Bmatrix} \dot{v}_x \\ \dot{v}_y \end{Bmatrix} = \omega^2 \begin{bmatrix} -\cos\omega t & \sin\omega t \\ -\sin\omega t & -\cos\omega t \end{bmatrix} \begin{Bmatrix} X \\ Y \end{Bmatrix} \tag{E3.2.3}$$

To obtain a spatial description for the velocity, the material coordinates X and Y in (E3.2.2) are first expressed in terms of the spatial coordinates x and y by inverting (E3.2.1):

$$\begin{aligned} \begin{Bmatrix} v_x \\ v_y \end{Bmatrix} &= \omega \begin{bmatrix} -\sin\omega t & -\cos\omega t \\ \cos\omega t & -\sin\omega t \end{bmatrix} \begin{bmatrix} \cos\omega t & \sin\omega t \\ -\sin\omega t & \cos\omega t \end{bmatrix} \begin{Bmatrix} x \\ y \end{Bmatrix} \\ &= \omega \begin{bmatrix} 0 & -1 \\ 1 & 0 \end{bmatrix} \begin{Bmatrix} x \\ y \end{Bmatrix} = \omega \begin{Bmatrix} -y \\ x \end{Bmatrix} \end{aligned} \tag{E3.2.4}$$

The material time derivative of the velocity field in the spatial description, (E3.2.4), is obtained via (3.2.11):

$$\begin{aligned} \frac{D\mathbf{v}}{Dt} &= \frac{\partial \mathbf{v}}{\partial t} + \mathbf{v}\cdot\nabla\mathbf{v} = \begin{Bmatrix} \partial v_x/\partial t \\ \partial v_y/\partial t \end{Bmatrix}^T + \begin{bmatrix} v_x & v_y \end{bmatrix} \begin{bmatrix} \partial v_x/\partial x & \partial v_y/\partial x \\ \partial v_x/\partial y & \partial v_y/\partial y \end{bmatrix} \\ &= 0 + \begin{bmatrix} v_x & v_y \end{bmatrix} \begin{bmatrix} 0 & \omega \\ -\omega & 0 \end{bmatrix} = \omega[-v_y \quad v_x] \end{aligned} \tag{E3.2.5}$$

If we then express the velocity field in (E3.2.5) in terms of the spatial coordinates x and y by (E3.2.4), we have

$$\begin{Bmatrix} a_x \\ a_y \end{Bmatrix} = -\omega^2 \begin{Bmatrix} x \\ y \end{Bmatrix} \tag{E3.2.6}$$

This is the well-known result for centripetal acceleration: the acceleration vector points toward the center of rotation and its magnitude is $\omega^2 (x^2 + y^2)^{1/2}$.

To compare the above with the material form of the acceleration (E3.2.3), we use (E3.2.1) to express the spatial coordinates in (E3.2.6) in terms of the material coordinates, which gives

$$\begin{Bmatrix} \dot{v}_x \\ \dot{v}_y \end{Bmatrix} = \omega^2 \begin{bmatrix} -1 & 0 \\ 0 & -1 \end{bmatrix} \begin{bmatrix} \cos\omega t & -\sin\omega t \\ \sin\omega t & \cos\omega t \end{bmatrix} \begin{Bmatrix} X \\ Y \end{Bmatrix} = \omega^2 \begin{bmatrix} -\cos\omega t & \sin\omega t \\ -\sin\omega t & -\cos\omega t \end{bmatrix} \begin{Bmatrix} X \\ Y \end{Bmatrix}$$

which agrees with (E3.2.3).

The deformation gradient is obtained from its definition (3.2.19) and (E3.2.1):

$$\mathbf{F} = \frac{\partial \mathbf{x}}{\partial \mathbf{X}} = \mathbf{R} = \begin{bmatrix} \cos\omega t & -\sin\omega t \\ \sin\omega t & \cos\omega t \end{bmatrix}, \quad \mathbf{F}^{-1} = \begin{bmatrix} \cos\omega t & \sin\omega t \\ -\sin\omega t & \cos\omega t \end{bmatrix} \tag{E3.2.7}$$

Example 3.3 Consider a unit square four-node element, with three of the nodes fixed as shown in Figure 3.5. Find the locus of positions of node 3 which results in a vanishing Jacobian.

The displacement field for the rectangular element with all nodes but node 3 fixed is given by the bilinear field

$$u_x(X,Y) = u_{x3}XY, \quad u_y(X,Y)v = u_{y3}XY \tag{E3.3.1}$$

Since this element is a square, an isoparametric mapping is not needed. This displacement field vanishes along edges defined by nodes 1 and 2 and nodes 1 and 4. The motion is given by

$$\begin{aligned} x &= X + u_x = X + u_{x3}XY \\ y &= Y + u_y = Y + u_{y3}XY \end{aligned} \tag{E3.3.2}$$

The deformation gradient is obtained from this and (3.2.19):

$$\mathbf{F} = \begin{bmatrix} 1 + u_{x3}Y & u_{x3}X \\ u_{y3}Y & 1 + u_{y3}X \end{bmatrix} \tag{E3.3.3}$$

The Jacobian determinant is then

$$J = \det(\mathbf{F}) = 1 + u_{x3}Y + u_{y3}X \tag{E3.3.4}$$

We now examine when the Jacobian determinant will vanish. We need only consider the Jacobian determinant for material particles in the undeformed configuration of the element, that is, the unit square $X \in [0, 1]$, $Y \in [0, 1]$. From (E3.3.4), it is apparent that J is minimum when $u_{x3} < 0$ and $u_{y3} < 0$. Then the minimum value of J occurs at $X = Y = 1$, so

$$J \geq 0 \Rightarrow 1 + u_{x3}Y + u_{y3}X \geq 0 \Rightarrow 1 + u_{x3} + u_{y3} \geq 0 \tag{E3.3.5}$$

The locus of points along which $J = 0$ is given by a linear function of the nodal displacements as shown on the right in Figure 3.5. The right figure also shows a deformed configuration of the element for which $J < 0$. As can be seen, the Jacobian becomes negative when node 3 crosses the diagonal of the undeformed element.

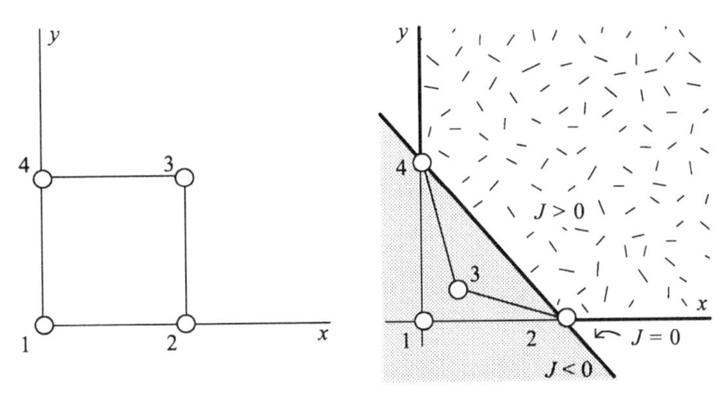

Figure 3.5 Original configuration of a square element and the locus of points for which $J = 0$; a deformed configuration with $J < 0$ is also shown

Example 3.4 The displacement field around a growing crack for small deformations is given by

$$u_x = kf(r)\left(a + 2\sin^2\frac{\theta}{2}\right)\cos\frac{\theta}{2}$$
$$u_y = kf(r)\left(b - 2\cos^2\frac{\theta}{2}\right)\sin\frac{\theta}{2} \tag{E3.4.1}$$

$$r^2 = (X - ct)^2 + Y^2, \quad \theta = \tan^{-1}(Y/(X - ct)), \theta \in (-\pi, \pi) \quad \text{for } X \neq ct \tag{E3.4.2}$$

where a, b, c and k are parameters determined by the solution of the governing equations. This displacement field corresponds to a crack opening along the X-axis with a cracktip velocity c; the initial configuration of the body and two subsequent configurations are shown in Figure 3.6.

Find the discontinuity in the displacement along the line $Y=0, X<ct$. Does this displacement field conform with the continuity requirements on motion given in Section 3.2.7?

The motion is $x=X+u_x$, $y=Y+u_y$. The discontinuity in the displacement field is the difference in (E3.4.1) for $\theta=\pi^-$ and $\theta=\pi^+$

$$\theta = -\pi \Rightarrow u_x = 0, \quad u_y = -kf(r)b, \quad \theta = \pi \Rightarrow u_x = 0, \quad u_y = kf(r)b \tag{E3.4.3}$$

so the jumps, or discontinuities, in the displacement are

$$[\![u_x]\!] = u_x(\pi, r) - u_x(-\pi, r) = 0, \quad [\![u_y]\!] = u_y(\pi, r) - u_y(-\pi, r) = 2kf(r)b \tag{E3.4.4}$$

Everywhere else the displacement field is continuous.

This motion meets the criteria given in Section 3.2.7 because the discontinuity occurs along only a line, which is a set of measure zero in two dimensions. From Figure 3.6 it can be seen that in this motion, the line behind the crack tip splits into two lines. It is also possible to devise motions where the line does not separate but a discontinuity occurs in the tangential displacement field. Both motions are now common in nonlinear finite element analysis.

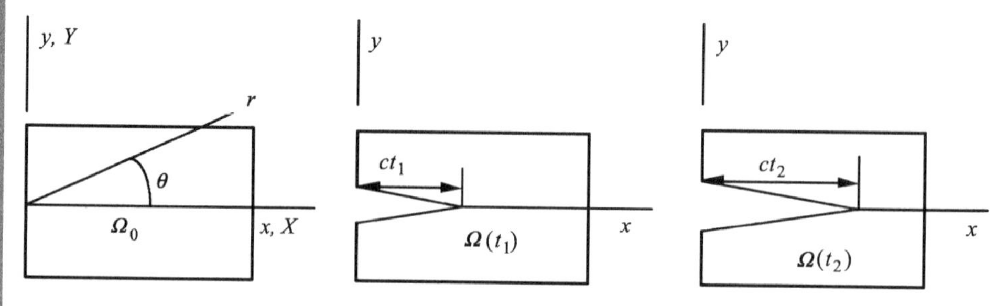

Figure 3.6 The initial uncracked configuration and two subsequent configurations for a crack growing along the x-axis

3.3 Strain Measures

In contrast to linear elasticity, many different measures of strain and strain rate are used in nonlinear continuum mechanics. Only two of these measures are considered here:

1. The Green (Green–Lagrange) strain, **E**.
2. The rate-of-deformation tensor, **D**.

In the following, these measures are defined and some key properties are given. Many other measures of strain and strain rate appear in the continuum mechanics literature; however, the above are the most widely used in finite element methods. It is sometimes advantageous to use other measures in describing constitutive equations as discussed in Chapter 5, and these other strain measures will be introduced as needed.

A strain measure must vanish for any rigid body motion, and in particular for rigid body rotation. If a strain measure fails to meet this requirement, it will predict nonzero strains, and in turn nonzero stresses, in rigid body rotation. This is the key reason why the usual linear strain displacement equations are abandoned in nonlinear theory; see Example 3.6. It will be shown in the following that **E** and **D** vanish in rigid body motion. A strain measure should satisfy other criteria, that is, it should increase as the deformation increases, and so on (Hill, 1978). However, the ability to represent rigid body motion is crucial and indicates when a geometrically nonlinear theory must be used.

3.3.1 Green Strain Tensor

The Green strain tensor **E** is defined by

$$ds^2 - dS^2 = 2d\mathbf{X}\cdot\mathbf{E}\cdot d\mathbf{X} \quad \text{or} \quad dx_i dx_i - dX_i dX_i = 2dX_i E_{ij} dX_j \tag{3.3.1}$$

so it gives the change in the square of the length of the material vector $d\mathbf{X}$. Recall the vector $d\mathbf{X}$ pertains to the undeformed configuration. Therefore, the Green strain measures the difference of the square of the length of an infinitesimal segment in the current (deformed) configuration and the initial (undeformed) configuration. To evaluate the Green strain tensor, we use (3.2.20) to rewrite the LHS of (3.3.1) as

$$\begin{aligned} d\mathbf{x}\cdot d\mathbf{x} &= (\mathbf{F}\cdot d\mathbf{X})\cdot(\mathbf{F}\cdot d\mathbf{X}) \\ &= (\mathbf{F}d\mathbf{X})^T(\mathbf{F}d\mathbf{X}) = d\mathbf{X}^T\mathbf{F}^T\mathbf{F}d\mathbf{X} = d\mathbf{X}\cdot(\mathbf{F}^T\cdot\mathbf{F})\cdot d\mathbf{X} \end{aligned} \tag{3.3.2}$$

where in the second line we have switched to matrix notation until the last step. This is clearer in indicial notation:

$$d\mathbf{x}\cdot d\mathbf{x} = dx_i dx_i = F_{ij} dX_j F_{ik} dX_k = dX_j F_{ji}^T F_{ik} dX_k = d\mathbf{X}\cdot(\mathbf{F}^T\cdot\mathbf{F})\cdot d\mathbf{X}$$

Using this with (3.3.1) and $d\mathbf{X}\cdot d\mathbf{X}=d\mathbf{X}\cdot\mathbf{I}\cdot d\mathbf{X}$ gives

$$d\mathbf{X}\cdot\mathbf{F}^T\cdot\mathbf{F}\cdot d\mathbf{X}-d\mathbf{X}\cdot\mathbf{I}\cdot d\mathbf{X}-d\mathbf{X}\cdot 2\mathbf{E}\cdot d\mathbf{X}=0 \tag{3.3.3}$$

Factoring out the common terms then yields

$$d\mathbf{X}\cdot(\mathbf{F}^T\cdot\mathbf{F}-\mathbf{I}-2\mathbf{E})\cdot d\mathbf{X}=0 \tag{3.3.4}$$

Since (3.3.4) must hold for all $d\mathbf{X}$, it follows that

$$\mathbf{E}=\frac{1}{2}(\mathbf{F}^T\cdot\mathbf{F}-\mathbf{I}) \quad \text{or} \quad E_{ij}=\frac{1}{2}(F_{ik}^T F_{kj}-\delta_{ij}) \tag{3.3.5}$$

The Green strain tensor can also be expressed in terms of displacement gradients by

$$\mathbf{E}=\frac{1}{2}((\nabla_0\mathbf{u})^T+\nabla_0\mathbf{u}+\nabla_0\mathbf{u}\cdot(\nabla_0\mathbf{u})^T),\quad E_{ij}=\frac{1}{2}\left(\frac{\partial u_i}{\partial X_j}+\frac{\partial u_j}{\partial X_i}+\frac{\partial u_k}{\partial X_i}\frac{\partial u_k}{\partial X_j}\right) \tag{3.3.6}$$

This expression is derived as follows. We first evaluate $\mathbf{F}^T\cdot\mathbf{F}$ in terms of the displacements using indicial notation:

$$
\begin{aligned}
F_{ik}^T F_{kj} = F_{ki}F_{kj} &= \frac{\partial x_k}{\partial X_i}\frac{\partial x_k}{\partial X_j} \qquad \text{(definition of transpose, and (3.2.19))}\\
&=\left(\frac{\partial u_k}{\partial X_i}+\frac{\partial X_k}{\partial X_i}\right)\left(\frac{\partial u_k}{\partial X_j}+\frac{\partial X_k}{\partial X_j}\right)\\
&=\left(\frac{\partial u_k}{\partial X_i}+\delta_{ki}\right)\left(\frac{\partial u_k}{\partial X_j}+\delta_{kj}\right)\\
&=\left(\frac{\partial u_i}{\partial X_j}+\frac{\partial u_j}{\partial X_i}+\frac{\partial u_k}{\partial X_j}\frac{\partial u_k}{\partial X_j}+\delta_{ij}\right)
\end{aligned}
$$

Substituting the above into (3.3.5) gives (3.3.6).

To show that the Green strain vanishes in rigid body motion, we consider the rigid body motion (3.2.26): $\mathbf{x}=\mathbf{R}\cdot\mathbf{X}+\mathbf{x}_T$. The deformation gradient $\mathbf{F}$ according to (3.2.19) is then given by $\mathbf{F}=\mathbf{R}$. Using the expression for the Green strain, (3.3.5) gives

$$\mathbf{E}=\frac{1}{2}(\mathbf{R}^T\cdot\mathbf{R}-\mathbf{I})=\frac{1}{2}(\mathbf{I}-\mathbf{I})=0$$

where the second equality follows from the orthogonality of the rotation tensor, (3.2.27). This demonstrates that the Green strain will vanish in any rigid body motion, so it meets an important requirement of a strain measure.

3.3.2 Rate-of-Deformation

The second kinematic measure to be considered here is the *rate-of-deformation* **D**. It is also called the *velocity strain*. In contrast to the Green strain tensor, it is a rate measure of deformation.

To develop an expression for the rate-of-deformation, we first define the velocity gradient **L** by

$$\mathbf{L} = \frac{\partial \mathbf{v}}{\partial \mathbf{x}} = (\nabla \mathbf{v})^T = (\text{grad}\, \mathbf{v})^T \quad \text{or} \quad L_{ij} = \frac{\partial v_i}{\partial x_j} \tag{3.3.7a}$$

$$d\mathbf{v} = \mathbf{L} \cdot d\mathbf{x} \quad \text{or} \quad dv_i = L_{ij} dx_j \tag{3.3.7b}$$

We have shown several tensor forms of the definition, but we will primarily use the indicial form. In (3.3.7), the symbol ∇ or the abbreviation 'grad' preceding the function denotes the left spatial gradient of the function: in a spatial gradient, the derivatives are taken with respect to the spatial (Eulerian) coordinates. The symbol ∇ specifies the spatial gradient; ∇_0 is the material gradient.

The velocity gradient tensor can be decomposed into symmetric and skew-symmetric parts by

$$\mathbf{L} = \frac{1}{2}(\mathbf{L} + \mathbf{L}^T) + \frac{1}{2}(\mathbf{L} - \mathbf{L}^T) \quad \text{or} \quad L_{ij} = \frac{1}{2}(L_{ij} + L_{ji}) + \frac{1}{2}(L_{ij} - L_{ji}) \tag{3.3.8}$$

This is a standard decomposition of a second-order tensor or square matrix: any second-order tensor can be expressed as the sum of its symmetric and skew-symmetric parts in this manner.

The rate-of-deformation D is defined as the symmetric part of **L**, that is, the first term on the RHS of (3.3.8), and the spin **W** is the skew-symmetric part of **L**, that is, the second term on the RHS of (3.3.8). Using these definitions, we can write

$$\mathbf{L} = (\nabla \mathbf{v})^T = \mathbf{D} + \mathbf{W} \quad \text{or} \quad L_{ij} = v_{i,j} = D_{ij} + W_{ij} \tag{3.3.9}$$

$$\mathbf{D} = \frac{1}{2}(\mathbf{L} + \mathbf{L}^T) \quad \text{or} \quad D_{ij} = \frac{1}{2}\left(\frac{\partial v_i}{\partial x_j} + \frac{\partial v_j}{\partial x_i}\right) \tag{3.3.10}$$

$$\mathbf{W} = \frac{1}{2}(\mathbf{L} - \mathbf{L}^T) \quad \text{or} \quad W_{ij} = \frac{1}{2}\left(\frac{\partial v_i}{\partial x_j} - \frac{\partial v_j}{\partial x_i}\right) \tag{3.3.11}$$

The rate-of-deformation is a measure of the rate of change of the square of the length of infinitesimal material line segments:

$$\frac{\partial}{\partial t}(ds^2) = \frac{\partial}{\partial t}(d\mathbf{x}(\mathbf{X}, t) \cdot d\mathbf{x}(\mathbf{X}, t)) = 2d\mathbf{x} \cdot \mathbf{D} \cdot d\mathbf{x} \quad \forall d\mathbf{x} \tag{3.3.12}$$

The equivalence of (3.3.10) and (3.3.12) is now shown. The expression for the rate-of-deformation is obtained from (3.3.12) as follows:

$$\begin{aligned} 2d\mathbf{x}\cdot\mathbf{D}\cdot d\mathbf{x} &= \frac{\partial}{\partial t}(d\mathbf{x}(\mathbf{X},t)\cdot d\mathbf{x}(\mathbf{X},t)) = 2d\mathbf{x}\cdot d\mathbf{v} \qquad (\text{using}\,(3.2.8)) \\ &= 2d\mathbf{x}\cdot\frac{\partial \mathbf{v}}{\partial \mathbf{x}}\cdot d\mathbf{x} \qquad (\text{by chain rule}) \\ &= 2d\mathbf{x}\cdot\mathbf{L}\cdot d\mathbf{x} \qquad (\text{using}\,(3.37)) \\ &= d\mathbf{x}\cdot(\mathbf{L}+\mathbf{L}^T+\mathbf{L}-\mathbf{L}^T)\cdot d\mathbf{x} \\ &= d\mathbf{x}\cdot(\mathbf{L}+\mathbf{L}^T)\cdot d\mathbf{x} \end{aligned} \tag{3.3.13}$$

where the last step follows from the antisymmetry of $\mathbf{L}-\mathbf{L}^T$; (3.3.10) follows from the last line in (3.3.13) due to the arbitrariness of $d\mathbf{x}$.

In the absence of deformation, the spin tensor and angular velocity tensor are equal: $\mathbf{W}=\Omega$. This is shown as follows. In rigid body motion $\mathbf{D}=\mathbf{0}$, so $\mathbf{L}=\mathbf{W}$ and by integrating (3.3.7b) we have

$$\mathbf{v}=\mathbf{W}\cdot(\mathbf{x}-\mathbf{x}_T)+\mathbf{v}_T \tag{3.3.14}$$

where $\mathbf{x}_T$ and $\mathbf{v}_T$ are constants of integration. Comparison with (3.2.41) then shows that the spin and angular velocity tensors are identical in rigid body rotation. When the body undergoes deformation in addition to rotation, the spin tensor generally differs from the angular velocity tensor.

3.3.3 Rate-of-Deformation in Terms of Rate of Green Strain

The rate-of-deformation can be related to the rate of the Green strain tensor. To obtain this relation, we first obtain the material gradient of the velocity field in terms of the spatial gradient by the chain rule:

$$\mathbf{L}=\frac{\partial \mathbf{v}}{\partial \mathbf{x}}=\frac{\partial \mathbf{v}}{\partial \mathbf{X}}\cdot\frac{\partial \mathbf{X}}{\partial \mathbf{x}}, \qquad L_{ij}=\frac{\partial v_i}{\partial x_j}=\frac{\partial v_i}{\partial X_k}\frac{\partial X_k}{\partial x_j} \tag{3.3.15}$$

Recall the definition of the deformation gradient, (3.2.19), $F_{ij}=\partial x_i/\partial X_j$. Taking the material time derivative of the deformation gradient gives

$$\dot{\mathbf{F}}=\frac{\partial}{\partial t}\left(\frac{\partial \boldsymbol{\phi}(\mathbf{X},t)}{\partial \mathbf{X}}\right)=\frac{\partial \mathbf{v}}{\partial \mathbf{X}}, \qquad \dot{F}_{ij}=\frac{\partial}{\partial t}\left(\frac{\partial \phi_i(\mathbf{X},t)}{\partial X_j}\right)=\frac{\partial v_i}{\partial X_j} \tag{3.3.16}$$

where the last step follows from (3.2.8). Using the chain rule to expand the identity $\partial x_i/\partial x_j=\delta_{ij}$ gives

$$\frac{\partial x_i}{\partial X_k}\frac{\partial X_k}{\partial x_j}=\delta_{ij}\to F_{ik}\frac{\partial X_k}{\partial x_j}=\delta_{ij}\to F_{kj}^{-1}=\frac{\partial X_k}{\partial x_j} \quad \text{or} \quad \mathbf{F}^{-1}=\frac{\partial \mathbf{X}}{\partial \mathbf{x}} \tag{3.3.17}$$

Using this, (3.3.15) can be rewritten as

$$\mathbf{L}=\dot{\mathbf{F}}\cdot\mathbf{F}^{-1}, \qquad L_{ij}=\dot{F}_{ik}F_{kj}^{-1} \tag{3.3.18}$$

To obtain a single expression relating these two measures of strain rate, we note that from (3.3.10) and (3.3.18) we have

$$\mathbf{D} = \frac{1}{2}(\mathbf{L} + \mathbf{L}^T) = \frac{1}{2}(\dot{\mathbf{F}} \cdot \mathbf{F}^{-1} + \mathbf{F}^{-T} \cdot \dot{\mathbf{F}}^T) \tag{3.3.19}$$

Taking the time derivative of the Green strain, (3.3.5) gives

$$\dot{\mathbf{E}} = \frac{1}{2}\frac{D}{Dt}(\mathbf{F}^T \cdot \mathbf{F} - \mathbf{I}) = \frac{1}{2}(\mathbf{F}^T \cdot \dot{\mathbf{F}} + \dot{\mathbf{F}}^T \cdot \mathbf{F}) \tag{3.3.20}$$

Premultiplying (3.3.19) by $\mathbf{F}^T$ and postmultiplying by F gives

$$\mathbf{F}^T \cdot \mathbf{D} \cdot \mathbf{F} = \frac{1}{2}(\mathbf{F}^T \cdot \dot{\mathbf{F}} + \dot{\mathbf{F}}^T \cdot \mathbf{F}), \quad \text{so} \quad \dot{\mathbf{E}} = \mathbf{F}^T \cdot \mathbf{D} \cdot \mathbf{F} \quad \text{or} \quad \dot{E}_{ij} = F_{ik}^T D_{kl} F_{lj} \tag{3.3.21}$$

where the last equality follows from (3.3.20). The above can easily be inverted to yield

$$\mathbf{D} = \mathbf{F}^{-T} \cdot \dot{\mathbf{E}} \cdot \mathbf{F}^{-1} \quad \text{or } D_{ij} = F_{ik}^{-T} \dot{E}_{kl} F_{lj}^{-1} \tag{3.3.22}$$

As we shall see in Chapter 5, (3.3.22) is an example of a *push forward operation*, (3.3.21) a *pull-back operation*. The two measures are two ways of viewing the same process: the rate of Green strain expresses in the initial configuration what the rate-of-deformation expresses in the current configuration. However, the properties of the two forms are somewhat different. For instance, in Example 3.7 we shall see that the integral of the Green strain rate in time is path independent, whereas the integral of the rate-of-deformation is not path independent.

Example 3.5 Strain measures in combined stretch and rotation Consider the motion

$$x(\mathbf{X}, t) = (1 + at)X \cos\frac{\pi}{2}t - (1 + bt)Y \sin\frac{\pi}{2}t \tag{E3.5.1}$$

$$y(\mathbf{X}, t) = (1 + at)X \sin\frac{\pi}{2}t + (1 + bt)Y \cos\frac{\pi}{2}t \tag{E3.5.2}$$

where a and b are positive constants. Evaluate the deformation gradient F, the Green strain **E** and rate-of-deformation tensor as functions of time and examine for $t=0$ and $t=1$.

For convenience, we define

$$A(t) \equiv (1 + at), \; B(t) \equiv (1 + bt), \quad c \equiv \cos\frac{\pi}{2}t, \quad s \equiv \sin\frac{\pi}{2}t \tag{E3.5.3}$$

The deformation gradient **F** is evaluated by (3.2.21) using (E3.5.1):

$$\mathbf{F} = \begin{bmatrix} \dfrac{\partial x}{\partial X} & \dfrac{\partial x}{\partial Y} \\ \dfrac{\partial y}{\partial X} & \dfrac{\partial y}{\partial Y} \end{bmatrix} = \begin{bmatrix} Ac & -Bs \\ As & Bc \end{bmatrix} \tag{E3.5.4}$$

This deformation consists of the simultaneous stretching of the material lines along the X and Y axes and the rotation of the element. The deformation gradient is constant in the element at any time, and the other measures of strain will also be constant at any time. The Green strain tensor is obtained from (3.3.5), with $\mathbf{F}$ given by (E3.5.4), which gives

$$\begin{aligned}\mathbf{E} &= \frac{1}{2}(\mathbf{F}^T \cdot \mathbf{F} - \mathbf{I}) = \frac{1}{2}\left(\begin{bmatrix} Ac & As \\ -Bs & Bc \end{bmatrix}\begin{bmatrix} Ac & -Bs \\ As & Bc \end{bmatrix} - \begin{bmatrix} 1 & 0 \\ 0 & 1 \end{bmatrix}\right) \\ &= \frac{1}{2}\left(\begin{bmatrix} A^2 & 0 \\ 0 & B^2 \end{bmatrix} - \begin{bmatrix} 1 & 0 \\ 0 & 1 \end{bmatrix}\right) = \frac{1}{2}\begin{bmatrix} 2at + a^2t^2 & 0 \\ 0 & 2bt + b^2t^2 \end{bmatrix}\end{aligned} \tag{E3.5.5}$$

It can be seen that the components of the Green strain tensor correspond to what would be expected from its definition: the line segments which are in the $\mathbf{X}$ and $\mathbf{Y}$ directions are extended by at and bt, respectively. The constants are restricted by $at > -1$ and $bt > -1$, for otherwise the Jacobian determinant becomes negative. When $t = 0$, $\mathbf{x} = \mathbf{X}$ and $\mathbf{E} = \mathbf{0}$.

For the purpose of evaluating the rate-of-deformation, we first obtain the velocity, which is the material time derivative of (E3.5.1):

$$v_x = \left(ac - \frac{\pi}{2}As\right)X - \left(bs + \frac{\pi}{2}Bc\right)Y \tag{E3.5.6}$$

$$v_y = \left(ac + \frac{\pi}{2}Ac\right)X + \left(bc - \frac{\pi}{2}Bs\right)Y \tag{E3.5.7}$$

Since at $t = 0$, $x = X$, $y = Y$, $c = 1$, $s = 0$, $A = B = 1$, the velocity gradient at $t = 0$ is given by

$$\mathbf{L} = (\nabla\mathbf{v})^T = \begin{bmatrix} a & -\frac{\pi}{2} \\ \frac{\pi}{2} & b \end{bmatrix} \rightarrow \mathbf{D} = \begin{bmatrix} a & 0 \\ 0 & b \end{bmatrix}, \quad \mathbf{W} = \frac{\pi}{2}\begin{bmatrix} 0 & -1 \\ 1 & 0 \end{bmatrix} \tag{E3.5.8}$$

To determine the time history of the rate-of-deformation, we first evaluate the time derivative of the deformation gradient and the inverse of the deformation gradient. Recall that $\mathbf{F}$ is given in (E3.5.4), from which we obtain

$$\dot{\mathbf{F}} = \begin{bmatrix} A_{,t}\,c - \frac{\pi}{2}As & -B_{,t}\,s - \frac{\pi}{2}Bc \\ A_{,t}\,s + \frac{\pi}{2}Ac & B_{,t}\,c - \frac{\pi}{2}Bs \end{bmatrix}, \quad \mathbf{F}^{-1} = \frac{1}{AB}\begin{bmatrix} Bc & Bs \\ -As & Ac \end{bmatrix} \tag{E3.5.9}$$

$$\mathbf{L} = \dot{\mathbf{F}} \cdot \mathbf{F}^{-1} = \frac{1}{AB}\begin{bmatrix} Bac^2 + Abs^2 & cs(Ba - Ab) \\ cs(Ba - Ab) & Bas^2 + Abc^2 \end{bmatrix} + \frac{\pi}{2}\begin{bmatrix} 0 & -1 \\ 1 & 0 \end{bmatrix} \tag{E3.5.10}$$

The first term on the RHS is the rate-of-deformation since it is the symmetric part of the velocity gradient, while the second term is the spin, which is skew-symmetric. The rate-of-deformation at $t = 1$ is given by

$$\mathbf{D} = \frac{1}{AB}\begin{bmatrix} Ab & 0 \\ 0 & Ba \end{bmatrix} = \frac{1}{1+a+b+ab}\begin{bmatrix} b+ab & 0 \\ 0 & a+ab \end{bmatrix} \tag{E3.5.11}$$

Thus, while in the intermediate stages, the shear velocity-strains are nonzero, in the configuration at $t=1$ only the elongational velocity-strains are nonzero. For comparison, from (E3.5.5) the rate of the Green strain at $t=1$ is given by

$$\dot{\mathbf{E}} = \begin{bmatrix} Aa & 0 \\ 0 & Bb \end{bmatrix} = \begin{bmatrix} a+a^2 & 0 \\ 0 & b+b^2 \end{bmatrix} \tag{E3.5.12}$$

Example 3.6 An element is rotated by an angle θ about the origin. Evaluate the linear strain.

For a pure rotation, the motion is given by (3.2.26), $\mathbf{x}=\mathbf{R}\cdot\mathbf{X}$, where the translation has been dropped and $\mathbf{R}$ is given in (3.2.34), so

$$\begin{Bmatrix} x \\ y \end{Bmatrix} = \begin{bmatrix} \cos\theta & -\sin\theta \\ \sin\theta & \cos\theta \end{bmatrix}\begin{Bmatrix} X \\ Y \end{Bmatrix} \quad \begin{Bmatrix} u_x \\ u_y \end{Bmatrix} = \begin{bmatrix} \cos\theta-1 & -\sin\theta \\ \sin\theta & \cos\theta-1 \end{bmatrix}\begin{Bmatrix} X \\ Y \end{Bmatrix} \tag{E3.6.1}$$

In the definition of the linear strain tensor, the spatial coordinates with respect to which the derivatives are taken are not specified. We take them with respect to the material coordinates (the conclusion is unchanged if we choose the spatial coordinates). The linear strain is then given by

$$\varepsilon_x = \frac{\partial u_x}{\partial X} = \cos\theta - 1, \quad \varepsilon_y = \frac{\partial u_y}{\partial Y} = \cos\theta - 1, \quad 2\varepsilon_{xy} = \frac{\partial u_x}{\partial Y} + \frac{\partial u_y}{\partial X} = 0 \tag{E3.6.2}$$

Thus, if θ is large, the extensional strains do not vanish. Therefore, the linear strain tensor cannot be used for large deformation problems, that is, in geometrically nonlinear problems.

A question that often arises is: 'how large can the rotations be before a nonlinear analysis is required?'. The previous example provides some guidance to this decision. The magnitudes of the linear strains in (E3.6.2) are an indication of the error due to the small strain assumption. To get a more convenient handle on this error, we expand $\cos\theta$ in a Taylor series and substitute into (E3.6.2), which gives

$$\varepsilon_x = \cos\theta - 1 = 1 - \frac{\theta^2}{2} + O(\theta^4) - 1 \approx -\frac{\theta^2}{2} \tag{3.3.23}$$

This shows that the error in the linear strain is second order in the rotation. The adequacy of a linear analysis then hinges on what magnitude of error can be tolerated, which depends in turn on the magnitudes of the strains of interest. If the strains of interest are of order 10^{-2}, and 1% error is acceptable (it almost always is), then the rotations can be of order 10^{-2} radians, since the error due to the small strain assumption is of order 10^{-4}. If the strains of interest are smaller, the acceptable rotations are smaller: for strains of order 10^{-4}, the rotations should be of order 10^{-3} radians for 1% error. These guidelines assume that the equilibrium solution is stable, that is, that buckling is not possible. Buckling is possible even with very small strains, so when buckling is a possibility, measures which can properly account for large deformations should be used.

Example 3.7 An element is deformed through the stages shown in Figure 3.7. The motions between these stages are linear functions of time. Evaluate the rate-of-deformation tensor $\mathbf{D}$ in each of these stages and obtain the time integral of the rate-of-deformation for the complete cycle of deformation ending in the undeformed configuration.

Each stage of the deformation is assumed to occur over a unit time interval. The time scaling is irrelevant to the results, and we adopt this particular scaling to simplify the algebra. The results would be identical with any other scaling. The motion that takes state 1 to state 2 is

$$x(\mathbf{X},t)=X+atY, \quad y(\mathbf{X},t)=Y \quad 0\le t\le 1 \tag{E3.7.1}$$

To determine the rate-of-deformation, we will use (3.3.18), $\mathbf{L}=\dot{\mathbf{F}}\cdot\mathbf{F}^{-1}$, so we first have to determine $\mathbf{F}$, $\dot{\mathbf{F}}$ and $\mathbf{F}^{-1}$. These are

$$\mathbf{F}=\begin{bmatrix}1 & at\\ 0 & 1\end{bmatrix}, \quad \dot{\mathbf{F}}=\begin{bmatrix}0 & a\\ 0 & 0\end{bmatrix}, \quad \mathbf{F}^{-1}=\begin{bmatrix}1 & -at\\ 0 & 1\end{bmatrix} \tag{E3.7.2}$$

The velocity gradient and rate-of-deformation are then given by (3.3.10):

$$\mathbf{L}=\dot{\mathbf{F}}\cdot\mathbf{F}^{-1}=\begin{bmatrix}0 & a\\ 0 & 0\end{bmatrix}\begin{bmatrix}1 & -at\\ 0 & 1\end{bmatrix}=\begin{bmatrix}0 & a\\ 0 & 0\end{bmatrix}, \quad \mathbf{D}=\frac{1}{2}(\mathbf{L}+\mathbf{L}^T)=\frac{1}{2}\begin{bmatrix}0 & a\\ a & 0\end{bmatrix} \tag{E3.7.3}$$

Thus the rate-of-deformation is a pure shear, that is both elongational components vanish. The Green strain is obtained by (3.3.5):

$$\mathbf{E}=\frac{1}{2}(\mathbf{F}^T\cdot\mathbf{F}-\mathbf{I})=\frac{1}{2}\begin{bmatrix}0 & at\\ at & a^2t^2\end{bmatrix}, \quad \dot{\mathbf{E}}=\frac{1}{2}\begin{bmatrix}0 & a\\ a & 2a^2t\end{bmatrix} \tag{E3.7.4}$$

Note that $\dot{E}_{22}$ is nonzero whereas $D_{22}=0$. However, $\dot{E}_{22}$ is small when the constant a is small.

The following gives the motion, the deformation gradient, its inverse and rate, and the rate-of-deformation and Green strain tensors for the subsequent stages.

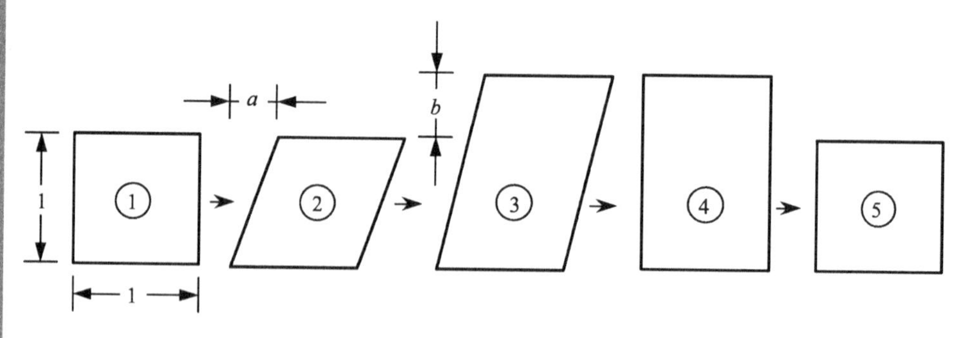

Figure 3.7 An element sheared in the x-direction followed by an extension in the y-direction and then subjected to deformations so that it returns to its initial configuration

Configuration 2 to configuration 3:

$$x(\mathbf{X},t) = X + aY, \quad y(\mathbf{X},t) = (1+bt)Y, \quad 1 \le \bar{t} \le 2, \quad t = \bar{t} - 1 \tag{E3.7.5a}$$

$$\mathbf{F} = \begin{bmatrix} 1 & a \\ 0 & 1+bt \end{bmatrix}, \quad \dot{\mathbf{F}} = \begin{bmatrix} 0 & 0 \\ 0 & b \end{bmatrix}, \quad \mathbf{F}^{-1} = \frac{1}{1+bt}\begin{bmatrix} 1+bt & -a \\ 0 & 1 \end{bmatrix} \tag{E3.7.5b}$$

$$\mathbf{L} = \dot{\mathbf{F}} \cdot \mathbf{F}^{-1} = \frac{1}{1+bt}\begin{bmatrix} 0 & 0 \\ 0 & b \end{bmatrix}, \quad \mathbf{D} = \frac{1}{2}(\mathbf{L} + \mathbf{L}^T) = \frac{1}{1+bt}\begin{bmatrix} 0 & 0 \\ 0 & b \end{bmatrix} \tag{E3.7.5c}$$

$$\mathbf{E} = \frac{1}{2}(\mathbf{F}^T \cdot \mathbf{F} - \mathbf{I}) = \frac{1}{2}\begin{bmatrix} 0 & a \\ a & a^2 + bt(bt+2) \end{bmatrix}, \quad \dot{\mathbf{E}} = \frac{1}{2}\begin{bmatrix} 0 & 0 \\ 0 & 2b(bt+1) \end{bmatrix} \tag{E3.7.5d}$$

Configuration 3 to configuration 4:

$$x(\mathbf{X},t) = X + a(1-t)Y, \quad y(\mathbf{X},t) = (1+b)Y, \quad 2 \le \bar{t} \le 3, \quad t = \bar{t} - 2 \tag{E3.7.6a}$$

$$\mathbf{F} = \begin{bmatrix} 1 & a(1-t) \\ 0 & 1+b \end{bmatrix}, \quad \dot{\mathbf{F}} = \begin{bmatrix} 0 & -a \\ 0 & 0 \end{bmatrix}, \quad \mathbf{F}^{-1} = \frac{1}{1+b}\begin{bmatrix} 1+b & a(t-1) \\ 0 & 1 \end{bmatrix} \tag{E3.7.6b}$$

$$\mathbf{L} = \dot{\mathbf{F}} \cdot \mathbf{F}^{-1} = \frac{1}{1+b}\begin{bmatrix} 0 & -a \\ 0 & 0 \end{bmatrix}, \quad \mathbf{D} = \frac{1}{2}(\mathbf{L} + \mathbf{L}^T) = \frac{1}{2(1+b)}\begin{bmatrix} 0 & -a \\ -a & 0 \end{bmatrix} \tag{E3.7.6c}$$

Configuration 4 to configuration 5:

$$x(\mathbf{X},t) = X, \quad y(\mathbf{X},t) = (1+b-bt)Y, \quad 3 \le \bar{t} \le 4, \quad t = \bar{t} - 3 \tag{E3.7.7a}$$

$$\mathbf{F} = \begin{bmatrix} 1 & 0 \\ 0 & 1+b-bt \end{bmatrix}, \quad \dot{\mathbf{F}} = \begin{bmatrix} 0 & 0 \\ 0 & -b \end{bmatrix}, \quad \mathbf{F}^{-1} = \frac{1}{1+b-bt}\begin{bmatrix} 1+b-bt & 0 \\ 0 & 1 \end{bmatrix} \tag{E3.7.7b}$$

$$\mathbf{L} = \dot{\mathbf{F}} \cdot \mathbf{F}^{-1} = \frac{1}{1+b-bt}\begin{bmatrix} 0 & 0 \\ 0 & -b \end{bmatrix}, \quad \mathbf{D} = \mathbf{L} \tag{E3.7.7c}$$

The Green strain in configuration 5 vanishes, since at $\bar{t} = 4$ the deformation gradient is the unit tensor, $\mathbf{F} = \mathbf{I}$. The time integral of the rate-of-deformation is given by

$$\begin{aligned} \int_0^4 \mathbf{D}(t)dt &= \frac{1}{2}\begin{bmatrix} 0 & a \\ a & 0 \end{bmatrix} + \begin{bmatrix} 0 & 0 \\ 0 & \ln(1+b) \end{bmatrix} + \frac{1}{2(1+b)}\begin{bmatrix} 0 & -a \\ -a & 0 \end{bmatrix} + \begin{bmatrix} 0 & 0 \\ 0 & -\ln(1+b) \end{bmatrix} \\ &= \frac{ab}{2(1+b)}\begin{bmatrix} 0 & 1 \\ 1 & 0 \end{bmatrix} \end{aligned} \tag{E3.7.8}$$

Thus the integral of the rate-of-deformation over a cycle ending in the initial configuration does not vanish. While the final configuration in this problem corresponds to the undeformed configuration so that a measure of strain should vanish, the integral of the rate-of-deformation is nonzero. This has significant repercussions for hypoelastic materials described in Chapter 5. It also implies that the integral of the rate-of-deformation is not a good measure of total strain. It should be noted that the integral of **D** over the cycle is second order in the constants that characterize the deformation, so the error is negligible as long as these constants are small. The integral of the Green strain rate vanishes in any closed cycle, since it is the time derivative of the Green strain **E**. In other words, the integral of the Green strain rate is path independent.

3.4 Stress Measures

3.4.1 Definitions of Stresses

In nonlinear problems, various stress measures can be defined. We will consider three measures of stress:

1. The Cauchy stress, $\boldsymbol{\sigma}$
2. The nominal stress tensor, **P**, which as described later, is closely related to the first Piola–Kirchhoff stress
3. The second Piola–Kirchhoff (PK2) stress tensor, **S**.

The definitions of the first three stress tensors are given in Box 3.1.

Box 3.1 Definition of stress measures

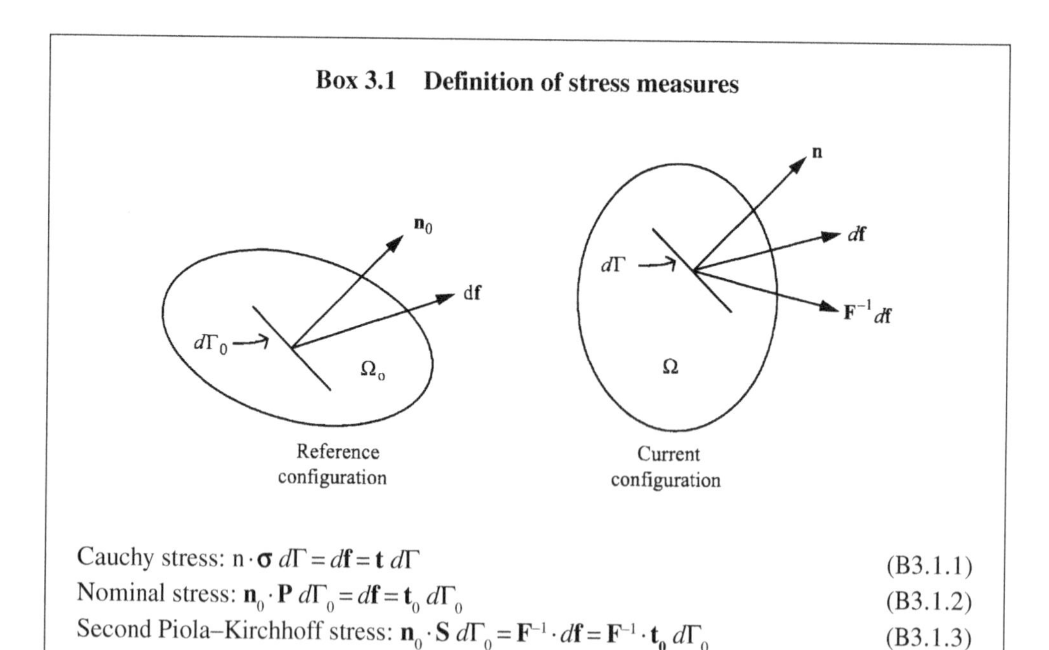

Cauchy stress: $\mathbf{n}\cdot\boldsymbol{\sigma}\,d\Gamma = d\mathbf{f} = \mathbf{t}\,d\Gamma$ (B3.1.1)

Nominal stress: $\mathbf{n}_0\cdot\mathbf{P}\,d\Gamma_0 = d\mathbf{f} = \mathbf{t}_0\,d\Gamma_0$ (B3.1.2)

Second Piola–Kirchhoff stress: $\mathbf{n}_0\cdot\mathbf{S}\,d\Gamma_0 = \mathbf{F}^{-1}\cdot d\mathbf{f} = \mathbf{F}^{-1}\cdot\mathbf{t}_0\,d\Gamma_0$ (B3.1.3)

$$d\mathbf{f} = \mathbf{t}\,d\Gamma = \mathbf{t}_0\,d\Gamma_0 \qquad \text{(B3.1.4)}$$

The stresses are defined by Cauchy's law

$$\mathbf{n}\cdot\boldsymbol{\sigma}d\Gamma = t d\Gamma \tag{3.4.1}$$

where **t** is the traction. In the initial configuration the counterpart of (3.4.1) is

$$\mathbf{n}_0\cdot\mathbf{P}d\Gamma_0 = \mathbf{t}_0 d\Gamma_0 \tag{3.4.2}$$

Note that the normal is always to the left. The PK2 stress is defined by

$$\mathbf{n}_0\cdot\mathbf{S}d\Gamma_0 = \mathbf{F}^{-1}\cdot\mathbf{t}_0 d\Gamma_0 \tag{3.4.3}$$

The expression for the traction in terms of the Cauchy stress, (B3.1.1), is called Cauchy's law or sometimes the Cauchy hypothesis. It involves the normal to the current surface and the traction (force per unit area) on the current surface. For this reason, the Cauchy stress is often called the physical stress or true stress. For example, the trace of the Cauchy stress,

$$\frac{1}{3}\text{trace }(\boldsymbol{\sigma}) = \frac{1}{3}\sigma_{ii} = -p \tag{3.4.4}$$

gives the true pressure p commonly used in fluid mechanics. The traces of the stress measures **P** and **S** do not give the true pressure because they are referred to the undeformed area. We will use the convention that the normal components of the Cauchy stress are positive in tension, and by (3.4.4) the pressure is positive in compression. The Cauchy stress tensor is symmetric, that is, $\boldsymbol{\sigma}^T=\boldsymbol{\sigma}$, which we shall see follows from the conservation of angular momentum.

The definition of the nominal stress **P** is similar to that of the Cauchy stress except that it is expressed in terms of the area and normal of the reference surface, that is, the undeformed surface. It will be shown in Section 3.6.3 that the nominal stress is not symmetric. The transpose of the nominal stress is called the first Piola–Kirchhoff stress. (The nomenclature used by different authors for nominal stress and first Piola–Kirchhoff stress is contradictory; Truesdell and Noll (1965), Ogden (1984), and Marsden and Hughes (1983) use the definition given here, whereas Malvern (1969) calls **P** the first Piola–Kirchhoff stress.) Since **P** is not symmetric, it is important to note that in the definition given in (3.4.2), the normal is to the left of the tensor **P**.

The second Piola–Kirchhoff stress is defined by (B3.1.3). It differs from **P** in that the force is transformed by $\mathbf{F}^{-1}$. This transformation has a definite purpose: it makes the second Piola–Kirchhoff stress symmetric and, as we shall see, conjugate to the rate of Green strain in the sense of power. The second Piola–Kirchhoff stress is widely used for path-independent materials such as rubber. We will use the acronyms PK1 and PK2 stress for the first and second Piola–Kirchhoff stresses, respectively.

3.4.2 *Transformation between Stresses*

The different stress tensors are interrelated by functions of the deformation. The relations between the stresses are given in Box 3.2. These relations can be obtained by using (3.4.1)–(3.4.3) along with Nanson's relation (Malvern, 1969: 169) which relates the current normal to the reference normal by

$$\mathbf{n}d\Gamma = J\mathbf{n}_0\cdot\mathbf{F}^{-1}d\Gamma_0,\quad n_i d\Gamma = Jn_j^0 F_{ji}^{-1}d\Gamma_0 \tag{3.4.5}$$

Note that the zero is placed wherever it is convenient: '0' and 'e' have a special, invariant meaning in this book and can appear as subscripts or superscripts!

Box 3.2 Transformations of stresses

Cauchy stress $\boldsymbol{\sigma}$	Nominal stress $\mathbf{P}$	2nd Piola–Kirchhoff stress $\mathbf{S}$	Corotational Cauchy stress $\hat{\boldsymbol{\sigma}}$
$\boldsymbol{\sigma} =$	$J^{-1}\mathbf{F}\cdot\mathbf{P}$	$J^{-1}\mathbf{F}\cdot\mathbf{S}\cdot\mathbf{F}^T$	$\mathbf{R}\cdot\hat{\boldsymbol{\sigma}}\cdot\mathbf{R}^T$
$\mathbf{P} = J\mathbf{F}^{-1}\cdot\boldsymbol{\sigma}$		$\mathbf{S}\cdot\mathbf{F}^T$	$J\mathbf{U}^{-1}\cdot\hat{\boldsymbol{\sigma}}\cdot\mathbf{R}^T$
$\mathbf{S} = J\mathbf{F}^{-1}\cdot\boldsymbol{\sigma}\cdot\mathbf{F}^{-T}$	$\mathbf{P}\cdot\mathbf{F}^{-T}$		$J\mathbf{U}^{-1}\cdot\hat{\boldsymbol{\sigma}}\cdot\mathbf{U}^{-1}$
$\hat{\boldsymbol{\sigma}} = \mathbf{R}^T\cdot\boldsymbol{\sigma}\cdot\mathbf{R}$	$J^{-1}\mathbf{U}\cdot\mathbf{P}\cdot\mathbf{R}$	$J^{-1}\mathbf{U}\cdot\mathbf{S}\cdot\mathbf{U}$	
$\boldsymbol{\tau} = J\boldsymbol{\sigma}$	$\mathbf{F}\cdot\mathbf{P}$	$\mathbf{F}\cdot\mathbf{S}\cdot\mathbf{F}^T$	$J\mathbf{R}\cdot\hat{\boldsymbol{\sigma}}\cdot\mathbf{R}^T$

Notes: $d\mathbf{x} = \mathbf{F}\cdot d\mathbf{X} = \mathbf{R}\cdot\mathbf{U}\cdot d\mathbf{X}$
$\mathbf{U}$ is the stretch tensor; see Section 3.7.1
$d\mathbf{x} = \mathbf{R}\cdot d\mathbf{X} = \mathbf{R}\cdot d\hat{\mathbf{x}}$ in rotation
$\boldsymbol{\tau}$ = Kirchhoff stress

To illustrate how the transformations between different stress measures are obtained, we will develop an expression for the nominal stress in terms of the Cauchy stress. To begin, we equate $d\mathbf{f}$ written in terms of the Cauchy stress and the nominal stress, (B3.1.1) and (B3.1.2), giving

$$d\mathbf{f} = \mathbf{n}\cdot\boldsymbol{\sigma}d\Gamma = \mathbf{n}_0\cdot\mathbf{P}d\Gamma_0 \tag{3.4.6}$$

Substituting the expression for normal $\mathbf{n}$ given by Nanson's relation, (3.4.5), into (3.4.6) gives

$$J\mathbf{n}_0\cdot\mathbf{F}^{-1}\cdot\boldsymbol{\sigma}d\Gamma_0 = \mathbf{n}_0\cdot\mathbf{P}\cdot d\Gamma_0 \tag{3.4.7}$$

Since this holds for all $\mathbf{n}_0$, it follows that

$$\mathbf{P} = J\mathbf{F}^{-1}\cdot\boldsymbol{\sigma} \quad \text{or} \quad P_{ij} = JF_{ik}^{-1}\sigma_{kj} \quad \text{or} \quad P_{ij} = J\frac{\partial X_i}{\partial x_k}\sigma_{kj} \tag{3.4.8}$$

$$J\boldsymbol{\sigma} = \mathbf{F}\cdot\mathbf{P} \quad \text{or} \quad J\sigma_{ij} = F_{ik}P_{kj} \tag{3.4.9}$$

It can be seen immediately from (3.4.8) that $\mathbf{P}\neq\mathbf{P}^T$, i.e. the nominal stress tensor is not symmetric. The nominal stress can be related to the PK2 stress by multiplying (3.4.3) by $\mathbf{F}$, giving

$$d\mathbf{f} = \mathbf{F}\cdot(\mathbf{n}_0\cdot\mathbf{S})d\Gamma_0 = \mathbf{F}\cdot(\mathbf{S}^T\cdot\mathbf{n}_0)d\Gamma_0 = \mathbf{F}\cdot\mathbf{S}^T\cdot\mathbf{n}_0 d\Gamma_0 \tag{3.4.10}$$

This is somewhat confusing in tensor notation, so it is rewritten next in indicial notation:

$$df_i = F_{ik}(n_j^0 S_{jk})d\Gamma_0 = F_{ik}S_{kj}^T n_j^0 d\Gamma_0 \tag{3.4.11}$$

The force $d\mathbf{f}$ in (3.4.11) is now written in terms of the nominal stress using (3.4.2):

$$d\mathbf{f} = \mathbf{n}_0 \cdot \mathbf{P} d\Gamma_0 = \mathbf{P}^T \cdot \mathbf{n}_0 d\Gamma_0 = \mathbf{F} \cdot \mathbf{S}^T \cdot \mathbf{n}_0 d\Gamma_0 \tag{3.4.12}$$

In (3.4.12) the last equality in (3.4.10) is repeated. Since this holds for all $\mathbf{n}_0$, we have

$$\mathbf{P} = \mathbf{S} \cdot \mathbf{F}^T \quad \text{or} \quad P_{ij} = S_{ik} F_{kj}^T = S_{ik} F_{jk} \tag{3.4.13}$$

Taking the inverse transformation of (3.4.8) and substituting into (3.4.13) gives

$$\boldsymbol{\sigma} = J^{-1} \mathbf{F} \cdot \mathbf{S} \cdot \mathbf{F}^T \quad \text{or} \quad \sigma_{ij} = J^{-1} F_{ik} S_{kl} F_{lj}^T \tag{3.4.14a}$$

The relation (3.4.14a) can be inverted to express the PK2 stress in terms of the Cauchy stress:

$$\mathbf{S} = J\mathbf{F}^{-1} \cdot \boldsymbol{\sigma} \cdot \mathbf{F}^{-T} \quad \text{or} \quad S_{ij} = J F_{ik}^{-1} \sigma_{kl} F_{lj}^{-T} \tag{3.4.14b}$$

These relations between the PK2 stress and the Cauchy stress, like (3.4.8), depend only on the deformation gradient $\mathbf{F}$ and the Jacobian determinant $J = \det(\mathbf{F})$. Thus, if the deformation is known, the state of stress can always be expressed in terms of either the Cauchy stress $\boldsymbol{\sigma}$, the nominal stress $\mathbf{P}$ or the PK2 stress $\mathbf{S}$. It can be seen from (3.4.14b) that if the Cauchy stress is symmetric, then $\mathbf{S}$ is also symmetric: $\mathbf{S} = \mathbf{S}^T$.

3.4.3 Corotational Stress and Rate-of-Deformation

In the corotational approach, a coordinate system is constructed for each point in the body with base vectors $\hat{\mathbf{e}}_i$. This coordinate system is rotated with the material or the element. By expressing these tensors in a coordinate system that rotates with the material, it is easier to deal with structural elements and anisotropic materials. The rate-of-deformation is then expressed in terms of its corotational components $\hat{D}_{ij}$, which can be obtained from the global components by (3.2.39). These components can be obtained directly from the velocity field by

$$\hat{D}_{ij} = \frac{1}{2}\left(\frac{\partial \hat{v}_i}{\partial \hat{x}_j} + \frac{\partial \hat{v}_j}{\partial \hat{x}_i}\right) \equiv \text{sym}\left(\frac{\partial \hat{v}_i}{\partial \hat{x}_j}\right) \equiv v_{\hat{i},\hat{j}} \tag{3.4.15}$$

where $\hat{v}_i \equiv v_{\hat{i}}$ are the components of the velocity field in the corotational system. The corotational system can be obtained from the polar decomposition theorem to be described later or by other techniques; see Section 4.6.

The corotational approach is often confusing to experienced mechanicians because they interpret it as a curvilinear coordinate system with base vectors $\hat{\mathbf{e}}_i$ which are functions of $\mathbf{x}$. Then the gradient of a vector $\hat{v}_i \hat{\mathbf{e}}_i$ would be given by $\hat{v}_{i,j} \hat{\mathbf{e}}_i + \hat{v}_i \hat{\mathbf{e}}_{i,j}$. However, this interpretation is not correct. The corotational system is a rotated global system and all vectors are expressed in this basis, so the correct gradient of the velocity $\mathbf{v}$ is $\hat{v}_{i,j} \hat{\mathbf{e}}_i$. Each point may have a different corotational system. Nevertheless, this approach provides the correct physical components of strains in a curved member or element; see Exercise 3.3. This is obvious in hindsight, since the Cartesian equations are applicable to any orientation. We will defer the details of how the rotation and the rotation matrix $\mathbf{R}$ is defined until we consider specific elements in Chapters 4 and 9. For the present, we assume that we can find a coordinate system that rotates with the material.

The corotational Cauchy stress and the corotational rate-of-deformation are defined by

$$\hat{\boldsymbol{\sigma}} = \mathbf{R}^T \cdot \boldsymbol{\sigma} \cdot \mathbf{R} \quad \text{or} \quad \hat{\sigma}_{ij} = R_{ik}^T \sigma_{kl} R_{lj} \tag{3.4.16a}$$

$$\hat{\mathbf{D}} = \mathbf{R}^T \cdot \mathbf{D} \cdot \mathbf{R} \quad \text{or} \quad \hat{D}_{ij} = R_{ik}^T D_{kl} R_{lj} \tag{3.4.16b}$$

The corotational Cauchy stress tensor is the same tensor as the Cauchy stress, but it is expressed in terms of components in a coordinate system that rotates with the material. Strictly speaking, a tensor does not depend on the coordinate system in which its components are expressed.

The corotational Cauchy stress, $\hat{\boldsymbol{\sigma}}$, is also called the rotated-stress tensor. The corotational stress is sometimes called the *unrotated stress*, which seems like a contradictory name: the difference arises as to whether you consider the 'hatted' coordinate system to be moving with the material (or element) or whether you consider it to be a fixed, independent entity. Both viewpoints are valid and the choice is just a matter of preference. We prefer the corotational viewpoint because it is easier to picture: see Example 4.6.

Example 3.8 Consider the motion given in Example 3.2, (E3.2.1). Let the Cauchy stress in the initial state be given by

$$\boldsymbol{\sigma}_{(t=0)} = \begin{bmatrix} \sigma_x^0 & 0 \\ 0 & \sigma_y^0 \end{bmatrix} \tag{E3.8.1}$$

Consider the stress to be frozen into the material, so as the body rotates, the initial stress rotates also, as shown in Figure 3.8.

This corresponds to the behavior of an initial state of stress in a rotating solid, which will be explored further in Section 3.6. Evaluate the PK2 stress, the nominal stress and the corotational stress in the initial configuration and the configuration at $t = \pi/2\omega$.

In the initial state, $\mathbf{F} = \mathbf{I}$, so

$$\mathbf{S} = \mathbf{P} = \hat{\boldsymbol{\sigma}} = \boldsymbol{\sigma} = \begin{bmatrix} \sigma_x^0 & 0 \\ 0 & \sigma_y^0 \end{bmatrix} \tag{E3.8.2}$$

In the deformed configuration at $t = \pi/2\omega$, the deformation gradient is given by

$$\mathbf{F} = \begin{bmatrix} \cos\pi/2 & -\sin\pi/2 \\ \sin\pi/2 & \cos\pi/2 \end{bmatrix} = \begin{bmatrix} 0 & -1 \\ 1 & 0 \end{bmatrix}, \quad J = \det(\mathbf{F}) = 1 \tag{E3.8.3}$$

Since the stress is considered frozen in the material, the stress state in the rotated configuration is given by

$$\boldsymbol{\sigma} = \begin{bmatrix} \sigma_y^0 & 0 \\ 0 & \sigma_x^0 \end{bmatrix} \tag{E3.8.4}$$

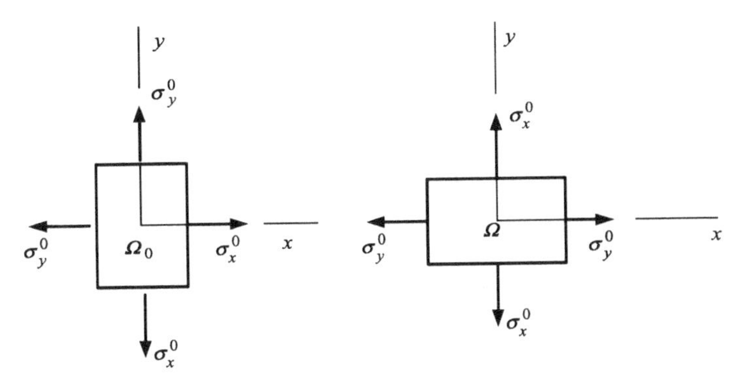

Figure 3.8 Prestressed body rotated by 90°

The nominal stress in the configuration is given by Box 3.2:

$$\mathbf{P} = J\mathbf{F}^{-1}\boldsymbol{\sigma} = \begin{bmatrix} 0 & 1 \\ -1 & 0 \end{bmatrix}\begin{bmatrix} \sigma_y^0 & 0 \\ 0 & \sigma_x^0 \end{bmatrix} = \begin{bmatrix} 0 & \sigma_x^0 \\ -\sigma_y^0 & 0 \end{bmatrix} \tag{E3.8.5}$$

Note that the nominal stress is not symmetric. The second Piola–Kirchhoff stress can be expressed in terms of the nominal stress **P** by Box 3.2 as follows:

$$\mathbf{S} = \mathbf{P}\cdot\mathbf{F}^{-T} = \begin{bmatrix} 0 & \sigma_x^0 \\ -\sigma_y^0 & 0 \end{bmatrix}\begin{bmatrix} 0 & -1 \\ 1 & 0 \end{bmatrix} = \begin{bmatrix} \sigma_x^0 & 0 \\ 0 & \sigma_y^0 \end{bmatrix} \tag{E3.8.6}$$

Since the mapping in this case is a pure rotation, $\mathbf{R} = \mathbf{F}$, so when $t = \pi/2\omega$, $\hat{\boldsymbol{\sigma}} = \mathbf{S}$.

Example 3.8 used the notion that an initial state of stress can be considered frozen into the material and rotated with the solid. It showed that in a pure rotation, the PK2 stress is unchanged; thus the PK2 stress behaves as if it were frozen into the material. This can also be explained by noting that the material coordinates rotate with the material and the components of the PK2 stress are related to the orientation of the material coordinates. Thus in the previous example, the component S_{11} corresponds to σ_{22} in the final configuration and σ_{11} in the initial configuration. The corotational components of the Cauchy stress $\hat{\boldsymbol{\sigma}}$ are also unchanged by the rotation of the material, and in the absence of deformation, equal the components of the PK2 stress. If the motion is a pure rotation, the corotational Cauchy stress components differ from the components of the PK2 stress in the final configuration.

The nominal stress at $t = 1$ is more difficult to interpret physically. From (E3.8.5) it can be seen that the PK1 stress is not constant and therefore, unlike the PK2 stress, does not rotate with the material; in fact, the nonzero stresses become the shear stresses. The nominal stress is a kind of expatriate, living partially in the current configuration and partially in the reference configuration. For this reason, it is often described as a two-point tensor, with a leg or index in each configuration, the reference configuration and the current configuration. The left leg is associated with the normal in the reference configuration, the right leg with a force on a surface element in the current configuration, as seen from in its definition, (B3.1.2). For this reason and because of the absence of symmetry in the nominal stress **P**, it is seldom used in constitutive equations. Its attractiveness lies in the simplicity of the momentum and finite element equations when expressed in terms of **P**.

Example 3.9 Uniaxial stress and associated strain Consider a bar in a state of uniaxial stress as shown in Figure 3.9. Relate the nominal stress and the PK2 stress to the uniaxial Cauchy stress.

The initial dimensions (the dimensions of the bar in the reference configuration) are ℓ_0, a_0 and b_0, and the current dimensions are ℓ, a and b, so

$$x=\frac{\ell}{\ell_0}X,\quad y=\frac{a}{a_0}Y,\quad z=\frac{b}{b_0}Z \tag{E3.9.1}$$

$$\mathbf{F}=\begin{bmatrix}\partial x/\partial X & \partial x/\partial Y & \partial x/\partial Z\\ \partial y/\partial X & \partial y/\partial Y & \partial y/\partial Z\\ \partial z/\partial X & \partial z/\partial Y & \partial z/\partial Z\end{bmatrix}=\begin{bmatrix}\ell/\ell_0 & 0 & 0\\ 0 & a/a_0 & 0\\ 0 & 0 & b/b_0\end{bmatrix} \tag{E3.9.2}$$

$$J=\det(\mathbf{F})=\frac{ab\ell}{a_0b_0\ell_0} \tag{E3.9.3}$$

$$\mathbf{F}^{-1}=\begin{bmatrix}\ell_0/\ell & 0 & 0\\ 0 & a_0/a & 0\\ 0 & 0 & b_0/b\end{bmatrix} \tag{E3.9.4}$$

The state of stress is uniaxial with the x-component the only nonzero component, so

$$\boldsymbol{\sigma}=\begin{bmatrix}\sigma_x & 0 & 0\\ 0 & 0 & 0\\ 0 & 0 & 0\end{bmatrix} \tag{E3.9.5}$$

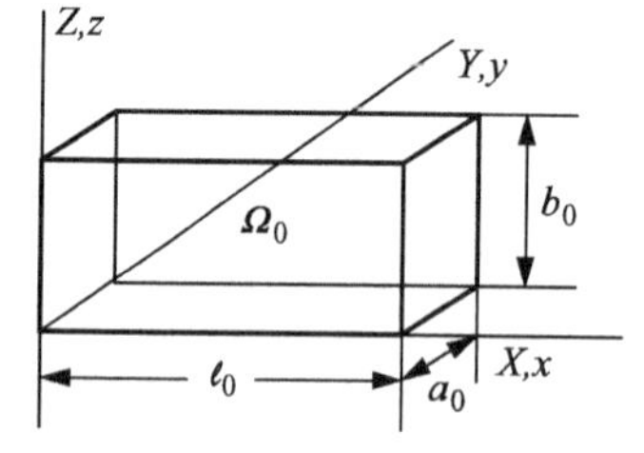

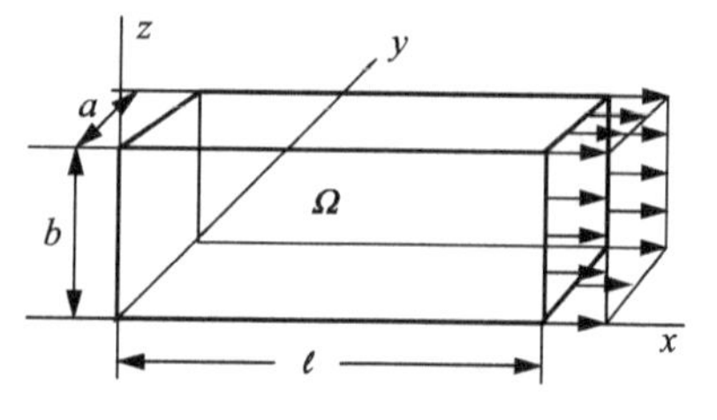

Figure 3.9 Undeformed and current configurations of a body in a uniaxial state of stress

Evaluating **P** as given by Box 3.2 using (E3.9.3)–(E3.9.5) then gives

$$\mathbf{P} = \frac{ab\ell}{a_0 b_0 \ell_0}\begin{bmatrix} \ell_0/\ell & 0 & 0 \\ 0 & a_0/a & 0 \\ 0 & 0 & b_0/b \end{bmatrix}\begin{bmatrix} \sigma_x & 0 & 0 \\ 0 & 0 & 0 \\ 0 & 0 & 0 \end{bmatrix} = \begin{bmatrix} \dfrac{ab\sigma_x}{a_0 b_0} & 0 & 0 \\ 0 & 0 & 0 \\ 0 & 0 & 0 \end{bmatrix} \tag{E3.9.6}$$

Thus the only nonzero component of the nominal stress is

$$P_{11} = \frac{ab}{a_0 b_0}\sigma_x = \frac{A\sigma_x}{A_0} \tag{E3.9.7}$$

where the last equality is based on the formulas for the cross-sectional area, $A = ab$ and $A_0 = a_0 b_0$; (E3.9.7) agrees with (2.2.8) and (2.2.9). Thus, in a state of uniaxial stress, P_{11} corresponds to the engineering stress.

The relationship between the PK2 stress and Cauchy stress for a uniaxial state of stress is obtained by using (E3.9.3)–(E3.9.5) with (3.4.14), which gives

$$S_{11} = \frac{\ell_0}{\ell}\left(\frac{A\sigma_x}{A_0}\right) \tag{E3.9.8}$$

where the quantity in the parentheses can be recognized as the nominal stress. It can be seen from this that it is difficult to ascribe a physical meaning to the PK2 stress. This, as will be seen in Chapter 5, influences the selection of stress measures for plasticity theories, since yield functions must be described in terms of physical stresses. Because of the non-physical nature of the nominal and PK2 stresses, it is awkward to formulate plasticity in terms of these stresses.

The Green strain is evaluated by (3.3.5):

$$E_{11} = \frac{\ell^2 - \ell_0^2}{2\ell_0^2}, \quad E_{22} = \frac{a^2 - a_0^2}{2a_0^2}, \quad E_{33} = \frac{b^2 - b_0^2}{2b_0^2} \tag{E3.9.9}$$

The remaining strain components vanish.

3.5 Conservation Equations

3.5.1 Conservation Laws

One group of the fundamental equations of continuum mechanics arises from the conservation laws. These equations must always be satisfied by physical systems. Four conservation laws relevant to thermomechanical systems are considered here:

1. Conservation of mass
2. Conservation of linear momentum, often called conservation of momentum

3. Conservation of energy
4. Conservation of angular momentum.

The conservation laws are also known as balance laws, for example, the conservation of energy is often called the balance of energy.

The conservation laws are usually expressed as partial differential equations (PDEs). These PDEs are derived by applying the conservation laws to a domain of the body, which leads to an integral relation. The following relationship is used to extract the PDEs from the integral relation:

if $f(\mathbf{x}, t)$ is C^{-1} and $\int_\Omega f(\mathbf{x}, t)d\Omega = 0$ for any subdomain Ω of $\bar{\Omega}$ and time $t \in [0, \bar{t}]$, then

$$f(\mathbf{x}, t) = 0 \text{ in } \Omega \quad \text{for} \quad t \in [0, \bar{t}] \tag{3.5.1}$$

In the following, Ω is an *arbitrary subdomain* of the body $\bar{\Omega}$ under consideration; we also refer to Ω as simply a domain. Prior to deriving the balance equations, several theorems useful for this purpose are derived.

3.5.2 Gauss's Theorem

In the derivation of the conservation equations, Gauss's theorem is frequently used. This theorem relates integrals over a domain to an integral over the boundary of this domain. It can be used to relate a volume integral to a surface integral or an area integral to a contour integral. The one-dimensional form of Gauss's theorem is the fundamental theorem of calculus, given in Chapter 2.

Gauss's theorem states that when $f(\mathbf{x})$ is piecewise continuously differentiable, that is, a C^0 function, then

$$\int_\Omega \frac{\partial f(\mathbf{x})}{\partial x_i} d\Omega = \int_\Gamma n_i f(\mathbf{x}) d\Gamma \quad \text{or} \quad \int_\Omega \nabla f(\mathbf{x}) d\Omega = \int_\Gamma \mathbf{n} f(\mathbf{x}) d\Gamma \tag{3.5.2a}$$

The theorem holds for any domain, including the reference domain, where for a C^0 function $f(\mathbf{X})$, we have

$$\int_{\Omega_0} \frac{\partial f(\mathbf{X})}{\partial X_i} d\Omega_0 = \int_{\Gamma_0} n_i^0 f(\mathbf{X}) d\Gamma_0 \quad \text{or} \quad \int_{\Omega_0} \nabla_0 f(\mathbf{X}) d\Omega_0 = \int_{\Gamma_0} \mathbf{n}_0 f(\mathbf{X}) d\Gamma_0 \tag{3.5.2b}$$

This theorem holds for a tensor of any order; for example if $f(\mathbf{x})$ is replaced by a tensor of first order, then

$$\int_\Omega \frac{\partial g_i(\mathbf{x})}{\partial x_i} d\Omega = \int_\Gamma n_i g_i(\mathbf{x}) d\Gamma \quad \text{or} \quad \int_\Omega \nabla \cdot \mathbf{g}(\mathbf{x}) d\Omega = \int_\Gamma \mathbf{n} \cdot \mathbf{g}(\mathbf{x}) d\Gamma \tag{3.5.3}$$

which is often known as the divergence theorem. The theorem also holds for gradients of the vector field:

$$\int_{\Omega}\frac{\partial g_i(\mathbf{x})}{\partial x_j}d\Omega=\int_{\Gamma}n_j g_i(\mathbf{x})d\Gamma \quad \text{or} \quad \int_{\Omega}\nabla\mathbf{g}(\mathbf{x})d\Omega=\int_{\Gamma}\mathbf{n}\otimes\mathbf{g}(\mathbf{x})d\Gamma \tag{3.5.4}$$

and to tensors of arbitrary order. Note that with the left gradient in the domain integral, the normal appears on the left in the surface integral.

If the function $f(\mathbf{x})$ is not continuously differentiable, that is, if its derivatives are discontinuous along a finite number of lines in two dimensions or on surfaces in three dimensions, then Ω must be subdivided into subdomains so that the function is C^0 within each subdomain. Discontinuities in the derivatives of the function will then occur on the interfaces between the subdomains. Gauss's theorem is applied to each of the subdomains, and summing the results yields the following counterparts of (3.5.2) and (3.5.3):

$$\int_{\Omega}\frac{\partial f}{\partial x_i}d\Omega=\int_{\Gamma}fn_i d\Gamma+\int_{\Gamma_{\text{int}}}[\![fn_i]\!]d\Gamma,\quad \int_{\Omega}\frac{\partial g_i}{\partial x_i}d\Omega=\int_{\Gamma}n_i g_i d\Gamma+\int_{\Gamma_{\text{int}}}[\![n_i g_i]\!]d\Gamma \tag{3.5.5}$$

where Γ_{int} is the set of all interfaces between these subdomains and $[\![f\mathbf{n}]\!]$ and $[\![\mathbf{n}\cdot\mathbf{g}]\!]$ are the jumps defined by

$$[\![f\mathbf{n}]\!]=f^A\mathbf{n}^A+f^B\mathbf{n}^B \quad [\![f\mathbf{n}]\!]\cdot\mathbf{n}^A=f^A-f^B \tag{3.5.6a}$$

$$[\![\mathbf{n}\cdot\mathbf{g}]\!]=[\![g_i n_i]\!]=g_i^A n_i^A+g_i^B n_i^B=(g_i^A-g_i^B)n_i^A=(g_i^B-g_i^A)n_i^B \tag{3.5.6b}$$

where A and B are a pair of subdomains which border on the interface Γ_{int}, $\mathbf{n}^A$ and $\mathbf{n}^B$ are the outward normals for the two subdomains, and f^A and f^B are the function values at the points adjacent to the interface in subdomains A and B, respectively. All of the forms in (3.5.6b) are equivalent and make use of the fact that on the interface, $\mathbf{n}^A=-\mathbf{n}^B$. The first of the formulas is the easiest to remember because of its symmetry with respect to A and B.

3.5.3 *Material Time Derivative of an Integral and Reynolds' Transport Theorem*

The material time derivative of an integral is the rate of change of an integral on a material domain. A material domain moves with the material, so that the material points on the boundary remain on the boundary and no mass flux occurs across the boundaries. A material domain is analogous to a Lagrangian mesh; a Lagrangian element or group of Lagrangian elements is a nice example of a material domain. The various forms for material time derivatives of integrals are called Reynolds' transport theorem.

The material time derivative of an integral is defined by

$$\frac{D}{Dt}\int_{\Omega}f\,d\Omega=\lim_{\Delta t\to 0}\frac{1}{\Delta t}\left(\int_{\Omega_{\tau+\Delta t}}f(\mathbf{x},\tau+\Delta t)d\Omega-\int_{\Omega_\tau}f(\mathbf{x},\tau)d\Omega\right) \tag{3.5.7}$$

where Ω_τ is the spatial domain at time τ and $\Omega_{\tau+\Delta t}$ is the spatial domain occupied by the same material points at time $\tau+\Delta t$. The notation on the left-hand side is a little confusing because it appears to refer to a single spatial domain. However, in this notation, which is standard, the material derivative on the integral implies that the domain Ω is a material domain. We now transform both integrals on the RHS to the reference domain using (3.2.23):

$$\frac{D}{Dt}\int_{\Omega} f\,d\Omega = \lim_{\Delta t \to 0} \frac{1}{\Delta t}\left(\int_{\Omega_0} f(\mathbf{X},\ \tau + \Delta t)J(\mathbf{X},\ \tau + \Delta t)d\Omega_0 - \int_{\Omega_0} f(\mathbf{X},\ \tau)J(\mathbf{X},\ \tau)d\Omega_0 \right) \tag{3.5.8}$$

With this change in the domain of integration, f becomes a function of the material coordinates, that is, $f(\Phi(\mathbf{X}, t), t) \equiv f \circ \Phi$.

Since the domain of integration is now independent of time, we can pull the limit operation inside the integral and take the limit, which yields

$$\frac{D}{Dt}\int_{\Omega} f\,d\Omega = \int_{\Omega_0} \frac{\partial}{\partial t}(f(\mathbf{X}, t)J(\mathbf{X}, t))d\Omega_0 \tag{3.5.9}$$

The partial derivative with respect to time in the integrand is a material time derivative since the independent space variables are the material coordinates. We next use the product rule for derivatives on the previous:

$$\frac{D}{Dt}\int_{\Omega} f\,d\Omega = \int_{\Omega_0} \frac{\partial}{\partial t}(f(\mathbf{X}, t)J(\mathbf{X}, t))d\Omega_0 = \int_{\Omega_0}\left(\frac{\partial f}{\partial t}J + f\frac{\partial J}{\partial t} \right)d\Omega_0 \tag{3.5.10}$$

Bearing in mind that the partial time derivatives are material time derivatives, we can use (3.2.25) to obtain

$$\frac{D}{Dt}\int_{\Omega} f\,d\Omega = \int_{\Omega_0}\left(\frac{\partial f}{\partial t}J + fJ\frac{\partial v_i}{\partial x_i} \right)d\Omega_0 \tag{3.5.11}$$

We can now transform the RHS integral to the current domain by (3.2.23) and change the independent variables to an Eulerian description, which gives

$$\frac{D}{Dt}\int_{\Omega} f(\mathbf{x}, t)d\Omega = \int_{\Omega}\left(\frac{Df(\mathbf{x}, t)}{Dt} + f\frac{\partial v_i}{\partial x_i} \right)d\Omega \tag{3.5.12}$$

where we have used $Df(\mathbf{x}, t)/Dt \equiv \partial f(\mathbf{X}, t)/\partial t$ as indicated in (3.2.13). This is one form of *Reynolds' transport theorem*.

An alternative form of Reynolds' transport theorem can be obtained by the definition of the material time derivative, (3.2.14), in (3.5.12). This gives

$$\frac{D}{Dt}\int_{\Omega} f\,d\Omega = \int_{\Omega}\left(\frac{\partial f}{\partial t} + v_i\frac{\partial f}{\partial x_i} + \frac{\partial v_i}{\partial x_i}f \right)d\Omega = \int_{\Omega}\left(\frac{\partial f}{\partial t} + \frac{\partial (v_i f)}{\partial x_i} \right)d\Omega \tag{3.5.13}$$

which can be written in tensor form as

$$\frac{D}{Dt}\int_{\Omega} f\,d\Omega = \int_{\Omega}\left(\frac{\partial f}{\partial t} + \text{div}\ (\mathbf{v}f) \right)d\Omega \tag{3.5.14}$$

Equation (3.5.14) can be put into another form by using Gauss's theorem on the second term of the RHS, which gives

$$\frac{D}{Dt}\int_{\Omega} f\,d\Omega = \int_{\Omega}\frac{\partial f}{\partial t}d\Omega + \int_{\Gamma} f v_i n_i d\Gamma \quad \text{or} \quad \frac{D}{Dt}\int_{\Omega} f\,d\Omega = \int_{\Omega}\frac{\partial f}{\partial t}d\Omega + \int_{\Gamma} f\mathbf{v}\cdot\mathbf{n}d\Gamma \tag{3.5.15}$$

where the product $f\mathbf{v}$ is assumed to be C^0 in Ω. Reynolds' transport theorem, which in the above has been given for a scalar, applies to a tensor of any order. Thus to apply it to a first-order tensor (vector) g_k, replace f by g_k in (3.5.13), which gives

$$\frac{D}{Dt}\int_{\Omega} g_k d\Omega = \int_{\Omega}\left(\frac{\partial g_k}{\partial t} + \frac{\partial (v_i g_k)}{\partial x_i}\right)d\Omega \tag{3.5.16}$$

3.5.4 *Mass Conservation*

The mass $m(\Omega)$ of a material domain Ω is given by

$$m(\Omega) = \int_{\Omega}\rho(\mathbf{x}, t)d\Omega \tag{3.5.17}$$

where $\rho(\mathrm{x}, t)$ is the density. Mass conservation requires that the mass of any material domain be constant, since no material flows through the boundaries of a material domain and we are not considering mass to energy conversion. Therefore, according to the principle of mass conservation, the material time derivative of $m(\Omega)$ vanishes, that is,

$$\frac{Dm}{Dt} = \frac{D}{Dt}\int_{\Omega}\rho\, d\Omega = 0 \tag{3.5.18}$$

Applying Reynolds' theorem, (3.5.12), to (3.5.18) yields

$$\int_{\Omega}\left(\frac{D\rho}{Dt} + \rho\operatorname{div}(\mathbf{v})\right)d\Omega = 0 \tag{3.5.19}$$

Since this holds for any subdomain Ω, it follows from (3.5.1) that

$$\frac{D\rho}{Dt} + \rho\operatorname{div}(\mathbf{v}) = 0 \quad \text{or} \quad \frac{D\rho}{Dt} + \rho v_{i,i} = 0 \quad \text{or} \quad \dot{\rho} + \rho v_{i,i} = 0 \tag{3.5.20}$$

The above is the *equation of mass conservation*, often called the *continuity equation*. It is a first-order partial differential equation.

Several special forms of the mass conservation equation are of interest. When a material is incompressible, the material time derivative of the density vanishes, and it can be seen from (3.5.20) that the mass conservation equation becomes

$$\operatorname{div}(\mathbf{v}) = 0 \quad \text{or} \quad v_{i,i} = 0 \tag{3.5.21}$$

In other words, mass conservation requires the divergence of the velocity field of an incompressible material to vanish.

If the definition of a material time derivative, (3.2.14) is invoked in (3.5.20), then the continuity equation can be written in the form

$$\frac{\partial \rho}{\partial t} + \rho,_i v_i + \rho v_{i,i} = \frac{\partial \rho}{\partial t} + (\rho v_i),_i = 0 \tag{3.5.22}$$

This is called the *conservative form* of the mass conservation equation. It is often preferred in computational fluid dynamics because discretizations of (3.5.22) are thought to enforce mass conservation more accurately.

For Lagrangian descriptions, the mass conservation equation, (3.5.18), can be integrated in time to obtain an algebraic equation for the density:

$$\int_{\Omega} \rho d\Omega = \text{constant} = \int_{\Omega_0} \rho_0 d\Omega_0 \tag{3.5.23}$$

Transforming the left-hand integral in (3.5.23) to the reference domain by (3.2.23) gives

$$\int_{\Omega_0} (\rho J - \rho_0) d\Omega_0 = 0 \tag{3.5.24}$$

Then invoking the smoothness of the integrand and (3.5.1) gives the following equation for mass conservation:

$$\rho(\mathbf{X}, t) J(\mathbf{X}, t) = \rho_0(\mathbf{X}) \quad \text{or} \quad \rho J = \rho_0 \tag{3.5.25}$$

We have explicitly indicated the independent variables on the left to emphasize that this equation holds only for material points; this follows from the fact that the domain of integration in (3.5.24) must be a material domain.

The algebraic equation (3.5.25) is usually used to enforce mass conservation in Lagrangian meshes. In Eulerian meshes, the algebraic form of mass conservation, (3.5.25), cannot be used, and mass conservation is imposed by the partial differential equation (3.5.20) or (3.5.22), that is, the continuity equation.

3.5.5 *Conservation of Linear Momentum*

The equation emanating from the principle of linear momentum conservation is a key equation in nonlinear finite element procedures. Linear momentum conservation is equivalent to Newton's second law of motion, which relates the forces acting on a body to its acceleration. The principle is often called the momentum conservation principle, or the balance of momentum principle.

We will here state the principle in integral form and then derive an equivalent partial differential equation. We consider an arbitrary domain Ω with boundary Γ subjected to body forces $\rho\mathbf{b}$ and to surface tractions $\mathbf{t}$, where $\mathbf{b}$ is a force per unit mass and $\mathbf{t}$ is a force per unit area. The total force is given by

$$\mathbf{f}(t)=\int_{\Omega}\rho\mathbf{b}(\mathbf{x},t)d\Omega+\int_{\Gamma}\mathbf{t}(\mathbf{x},t)d\Gamma \tag{3.5.26}$$

The linear momentum is given by

$$\mathbf{p}(t)=\int_{\Omega}\rho\mathbf{v}(\mathbf{x},t)d\Omega \tag{3.5.27}$$

where $\rho\mathbf{v}$ is the linear momentum per unit volume.

Newton's second law of motion for a continuum, the momentum conservation principle, states that the material time derivative of the linear momentum equals the net force. Using (3.5.26) and (3.5.27), this gives

$$\frac{D\mathbf{p}}{Dt}=\mathbf{f}\Rightarrow\frac{D}{Dt}\int_{\Omega}\rho\mathbf{v}\,d\Omega=\int_{\Omega}\rho\mathbf{b}\;d\Omega+\int_{\Gamma}\mathbf{t}\;d\Gamma \tag{3.5.28}$$

We now convert the first and third integrals in the above to obtain a single domain integral so (3.5.1) can be applied. Reynolds' transport theorem applied to the LHS integral in (3.5.28) gives

$$\begin{aligned}\frac{D}{Dt}\int_{\Omega}\rho\mathbf{v}\,d\Omega&=\int_{\Omega}\left(\frac{D}{Dt}(\rho\mathbf{v})+\operatorname{div}(\mathbf{v})\rho\mathbf{v}\right)d\Omega\\&=\int_{\Omega}\left[\rho\frac{D\mathbf{v}}{Dt}+\mathbf{v}\left(\frac{D\rho}{Dt}+\rho\operatorname{div}(\mathbf{v})\right)\right]d\Omega\end{aligned} \tag{3.5.29}$$

where the second equality is obtained by using the product rule of derivatives for the first term of the integrand and rearranging terms.

The term multiplying the velocity in the RHS of the above can be recognized as the continuity equation (3.5.20), which vanishes, giving

$$\frac{D}{Dt}\int_{\Omega}\rho\mathbf{v}\,d\Omega=\int_{\Omega}\rho\frac{D\mathbf{v}}{Dt}d\Omega \tag{3.5.30}$$

To convert the second term on the RHS of (3.5.28) to a domain integral, we invoke Cauchy's relation and Gauss's theorem in sequence, giving

$$\int_{\Gamma}\mathbf{t}\,d\Gamma=\int_{\Gamma}\mathbf{n}\cdot\boldsymbol{\sigma}\,d\Gamma=\int_{\Omega}\nabla\cdot\boldsymbol{\sigma}\,d\Omega\quad\text{or}\quad\int_{\Gamma}t_j\,d\Gamma=\int_{\Gamma}n_i\sigma_{ij}d\Gamma=\int_{\Omega}\frac{\partial\sigma_{ij}}{\partial x_i}d\Omega \tag{3.5.31}$$

Note that since the normal is *to the left* on the boundary integral, the divergence is *to the left* and contracts with the first index on the stress tensor. When the divergence operator acts on the first index of the stress tensor it is called the left divergence operator and is placed to the left. When it acts on the second index, it is placed to the right and called the right divergence. Since the Cauchy stress is symmetric, the left and right divergence operators have the same effect. However, in contrast to linear continuum mechanics, in nonlinear continuum mechanics it is important to become accustomed to placing the divergence operator where it belongs because some stress tensors, such as the nominal stress, are not symmetric. When the stress is not symmetric, the left and right divergence operators lead to different results. In this book we use the convention that *the divergence and gradient operators are placed on the left and the normal appears on the left in the surface integrals.*

Substituting (3.5.30) and (3.5.31) into (3.5.28) gives

$$\int_{\Omega}\left(\rho\frac{D\mathbf{v}}{Dt}-\rho\mathbf{b}-\nabla\cdot\boldsymbol{\sigma}\right)d\Omega=0 \tag{3.5.32}$$

Therefore, if the integrand is C^{-1}, since (3.5.32) holds for an arbitrary domain, applying (3.5.1) yields

$$\rho\frac{D\mathbf{v}}{Dt}=\nabla\cdot\boldsymbol{\sigma}+\rho\mathbf{b}\equiv\operatorname{div}\boldsymbol{\sigma}+\rho\mathbf{b}\quad\text{or}\quad\rho\frac{Dv_i}{Dt}=\frac{\partial\sigma_{ji}}{\partial x_j}+\rho b_i \tag{3.5.33}$$

This is called the *momentum equation*; it is also called the balance of linear momentum equation. The LHS term represents the change in momentum, since it is a product of the acceleration and the density; it is also called the inertial or kinetic term. The first term on the RHS is the net resultant internal force per unit volume due to the divergence of the stress field.

This form of the momentum equation is applicable to both Lagrangian and Eulerian descriptions. In a Lagrangian description, the dependent variables are assumed to be functions of the Lagrangian coordinates **X** and time t, so the momentum equation is

$$\rho(\mathbf{X},t)\frac{\partial\mathbf{v}(\mathbf{X},t)}{\partial t}=\operatorname{div}\boldsymbol{\sigma}(\Phi^{-1}(\mathbf{x},t),t)+\rho(\mathbf{X},t)\mathbf{b}(\mathbf{X},t) \tag{3.5.34}$$

Note that the stress must be expressed as a function of the Eulerian coordinates through the inverse of the motion $\Phi^{-1}(\mathbf{x},t)$ so that the spatial divergence of the stress field can be evaluated, but it is considered a function of **X** and time t, $\boldsymbol{\sigma}(\mathbf{X},t)$. The material derivative of the velocity with respect to time in (3.5.33) becomes a partial derivative with respect to time when the independent variables are changed from **x** to **X**.

This would not be considered a true Lagrangian description in classical texts on continuum mechanics because of the appearance of the derivative with respect to Eulerian coordinates. However, the essential feature of a Lagrangian description is that the independent variables are the Lagrangian (material) coordinates. This requirement is met by this, and we will see in the development of the updated Lagrangian finite element method that this form of the momentum equation can be discretized with a Lagrangian mesh.

In an Eulerian description, the material derivative of the velocity is written out by (3.2.10) and all variables are considered functions of the Eulerian coordinates. Equation (3.5.33) becomes

$$\rho(\mathbf{x},t)\left(\frac{\partial\mathbf{v}(\mathbf{x},t)}{\partial t}+(\mathbf{v}(\mathbf{x},t)\cdot\operatorname{grad})\mathbf{v}(\mathbf{x},t)\right)=\operatorname{div}\boldsymbol{\sigma}(\mathbf{x},t)+\rho(\mathbf{x},t)\mathbf{b}(\mathbf{x},t) \tag{3.5.35}$$

or

$$\rho\left(\frac{\partial v_i}{\partial t}+v_{i,j}\,v_j\right)=\frac{\partial \sigma_{ji}}{\partial x_j}+\rho b_i$$

As can be seen from (3.5.33), when the independent variables are all explicitly written out, the equations are quite awkward, so we will usually omit them.

In computational fluid dynamics, the momentum equation is sometimes used without the changes made by (3.5.29–3.5.30). The resulting equation is

$$\frac{D(\rho\mathbf{v})}{Dt}\equiv\frac{\partial(\rho\mathbf{v})}{\partial t}+\mathbf{v}\cdot\operatorname{grad}(\rho\mathbf{v})=\operatorname{div}\boldsymbol{\sigma}+\rho\mathbf{b} \tag{3.5.36}$$

This is called the *conservative form of the momentum equation*. In the conservative form, the momentum per unit volume $\rho\mathbf{v}$ is a dependent variable. This variant of the momentum equation is said to observe momentum conservation more accurately.

3.5.6 Equilibrium Equation

In many problems, the loads are applied slowly and the inertial forces are very small and can be neglected. In that case, the acceleration in the momentum equation (3.5.35) can be dropped and we have

$$\nabla\cdot\boldsymbol{\sigma}+\rho\mathbf{b}=0 \quad \text{or} \quad \frac{\partial \sigma_{ji}}{\partial x_j}+\rho b_i=0 \tag{3.5.37}$$

The above equation is called the *equilibrium equation*. Problems to which the equilibrium equation is applicable are often called static problems. The equilibrium equation should be carefully distinguished from the momentum equation: equilibrium processes are static and do not include acceleration. The momentum and equilibrium equations are tensor equations, and (3.5.33) and (3.5.37) each represent n_{SD} scalar equations.

3.5.7 Reynolds' Theorem for a Density-Weighted Integrand

Equation (3.5.30) is a special case of a general result: the material time derivative of an integral in which the integrand is a product of the density and a function f is given by

$$\frac{D}{Dt}\int_{\Omega}\rho f\,d\Omega=\int_{\Omega}\rho\frac{Df}{Dt}\,d\Omega \tag{3.5.38}$$

This holds for a tensor of any order and is a consequence of Reynolds' theorem and mass conservation; it is another variant of Reynolds' theorem. It can be verified by repeating the steps in (3.5.29) and (3.5.30).

3.5.8 *Conservation of Angular Momentum*

The integral form of the conservation of angular momentum is obtained by taking the cross-product of each term in the corresponding linear momentum principle with the position vector **x**, giving

$$\frac{D}{Dt}\int_{\Omega}\mathbf{x}\times\rho\mathbf{v}\,d\Omega=\int_{\Omega}\mathbf{x}\times\rho\mathbf{b}\,d\Omega+\int_{\Gamma}\mathbf{x}\times\mathbf{t}\,d\Gamma \tag{3.5.39}$$

We will leave the derivation of the conditions which follow from (3.5.39) as an exercise and only state them:

$$\boldsymbol{\sigma}=\boldsymbol{\sigma}^{T}\quad\text{or}\quad\sigma_{ij}=\sigma_{ji} \tag{3.5.40}$$

In other words, conservation of angular momentum requires that the Cauchy stress be a symmetric tensor. Therefore, the Cauchy stress tensor represents three distinct dependent variables in two-dimensional problems, and six in three-dimensional problems. The conservation of angular momentum does not result in any additional equations when the Cauchy stress is used.

3.5.9 *Conservation of Energy*

We consider thermomechanical processes where the only sources of energy are mechanical work and heat. The principle of conservation of energy, that is, the energy balance principle, states that the rate of change of total energy is equal to the work done by the body forces and surface tractions plus the heat energy delivered to the body by the heat flux and other sources of heat. The internal energy per unit volume is denoted by ρw^{int} where w^{int} is the internal energy per unit mass. The heat flux per unit area is denoted by a vector **q**, in units of power per area, and the heat source per unit volume is denoted by ρs. The conservation of energy then requires that the rate of change of the total energy in the body, which includes both internal energy and kinetic energy, equals the power of the applied forces and the energy added to the body by heat conduction and any heat sources.

The rate of change of the total energy in the body is given by

$$P^{\text{tot}}=P^{\text{int}}+P^{\text{kin}},\quad P^{\text{int}}=\frac{D}{Dt}\int_{\Omega}\rho w^{\text{int}}\,d\Omega,\quad P^{\text{kin}}=\frac{D}{Dt}\int_{\Omega}\frac{1}{2}\rho\mathbf{v}\cdot\mathbf{v}\,d\Omega \tag{3.5.41}$$

where P^{int} denotes the rate of change of internal energy and P^{kin} the rate of change of the kinetic energy. The rate of work by the body forces in the domain and the tractions on the surface is

$$P^{\text{ext}}=\int_{\Omega}\mathbf{v}\cdot\rho\mathbf{b}\,d\Omega+\int_{\Gamma}\mathbf{v}\cdot\mathbf{t}\,d\Gamma=\int_{\Omega}v_i\rho b_i\,d\Omega+\int_{\Gamma}v_i t_i\,d\Gamma \tag{3.5.42}$$

The power supplied by heat sources s and the heat flux **q** is

$$P^{\text{heat}}=\int_{\Omega}\rho s\,d\Omega-\int_{\Gamma}\mathbf{n}\cdot\mathbf{q}\,d\Gamma=\int_{\Omega}\rho s\,d\Omega-\int_{\Gamma}n_i q_i\,d\Gamma \tag{3.5.43}$$

where the sign of the heat flux term is negative since positive heat flow is out of the body.

The statement of the conservation of energy is

$$P^{\text{tot}} = P^{\text{ext}} + P^{\text{heat}} \tag{3.5.44}$$

that is, the rate of change of the total energy in the body (consisting of the internal and kinetic energies) is equal to the rate of work by the external forces and rate of work provided by heat flux and energy sources. This is known as the *first law of thermodynamics*. The disposition of the internal energy depends on the material. In an elastic material, it is stored as elastic internal energy and is fully recoverable upon unloading. In an elastic–plastic material, some of the internal energy is converted to heat and some is dissipated in changes of the internal structure of the material.

Substituting (3.5.41)–(3.5.43) into (3.5.44) gives the full statement of the conservation of energy:

$$\frac{D}{Dt}\int_{\Omega}\left(\rho w^{\text{int}} + \frac{1}{2}\rho\mathbf{v}\cdot\mathbf{v}\right)d\Omega = \int_{\Omega}\mathbf{v}\cdot\rho\mathbf{b}\,d\Omega + \int_{\Gamma}\mathbf{v}\cdot\mathbf{t}\,d\Gamma + \int_{\Omega}\rho s\,d\Omega - \int_{\Gamma}\mathbf{n}\cdot\mathbf{q}\,d\Gamma \tag{3.5.45}$$

We will now derive the equation which emerges from the above integral statement using the same procedure as before: we use Reynolds' theorem to bring the total derivative inside the integral and convert all surface integrals to domain integrals. Using Reynolds' theorem, (3.5.38), on the first integral in (3.5.45) gives

$$\begin{aligned}\frac{D}{Dt}\int_{\Omega}\left(\rho w^{\text{int}} + \frac{1}{2}\rho\mathbf{v}\cdot\mathbf{v}\right)d\Omega &= \int_{\Omega}\left(\rho\frac{Dw^{\text{int}}}{Dt} + \frac{1}{2}\rho\frac{D(\mathbf{v}\cdot\mathbf{v})}{Dt}\right)d\Omega \\ &= \int_{\Omega}\left(\rho\frac{Dw^{\text{int}}}{Dt} + \rho\mathbf{v}\cdot\frac{D\mathbf{v}}{Dt}\right)d\Omega\end{aligned} \tag{3.5.46}$$

Applying Cauchy's law (3.4.1) and Gauss's theorem (3.5.2) to the traction boundary integrals on the RHS of (3.5.45) yields:

$$\begin{aligned}\int_{\Gamma}\mathbf{v}\cdot\mathbf{t}\,d\Gamma &= \int_{\Gamma}\mathbf{n}\cdot\boldsymbol{\sigma}\cdot\mathbf{v}\,d\Gamma = \int_{\Gamma}n_j\sigma_{ji}v_i\,d\Gamma \\ &= \int_{\Omega}(\sigma_{ji}v_i)_{,j}\,d\Omega = \int_{\Omega}(v_{i,j}\sigma_{ji} + v_i\sigma_{ji,j})\,d\Omega \\ &= \int_{\Omega}(D_{ji}\sigma_{ji} + W_{ji}\sigma_{ji} + v_i\sigma_{ji,j})\,d\Omega \quad \text{(using (3.3.9))} \\ &= \int_{\Omega}(D_{ji}\sigma_{ji} + v_i\sigma_{ji,j})\,d\Omega \quad \text{(by sym. of } \boldsymbol{\sigma} \text{ and skew sym. of } \mathbf{W}) \\ &= \int_{\Omega}(\mathbf{D}:\boldsymbol{\sigma} + (\nabla\cdot\boldsymbol{\sigma})\cdot\mathbf{v})\,d\Omega\end{aligned} \tag{3.5.47}$$

Inserting (3.5.47) into (3.5.45), application of Gauss's theorem to the heat flux integral and rearrangement of terms yields

$$\int_{\Omega}\left(\rho\frac{Dw^{\text{int}}}{Dt} - \mathbf{D}:\boldsymbol{\sigma} + \nabla\cdot\mathbf{q} - \rho s + \mathbf{v}\cdot\left(\rho\frac{D\mathbf{v}}{Dt} - \nabla\cdot\boldsymbol{\sigma} - \rho\mathbf{b}\right)\right)d\Omega = 0 \tag{3.5.48}$$

The last term in the integral can be recognized as the momentum equation, (3.5.33), so it vanishes. Then invoking the arbitrariness of the domain gives

$$\rho\frac{Dw^{\text{int}}}{Dt} = \mathbf{D}:\boldsymbol{\sigma} - \nabla\cdot\mathbf{q} + \rho s \tag{3.5.49}$$

which is the partial differential equation of energy conservation.

When the heat flux and heat sources vanish, that is, in a purely mechanical process, the energy equation becomes

$$\rho\frac{Dw^{\text{int}}}{Dt} = \mathbf{D}:\boldsymbol{\sigma} = \boldsymbol{\sigma}:\mathbf{D} = \sigma_{ij}D_{ij} \tag{3.5.50}$$

which is no longer a partial differential equation. The above defines the rate of energy imparted to a unit volume of the body in terms of the measures of stress and strain; this is called the internal energy rate or internal power. It can be seen from this that the internal power is given by the contraction of the rate-of-deformation and the Cauchy stress. We therefore say that the rate-of-deformation and the Cauchy stress are *conjugate in power*. As we shall see, conjugacy in power is helpful in the development of weak forms: measures of stress and strain rate which are conjugate in power can be used to construct principles of virtual work or power, that is, weak forms of the momentum equation. Variables which are conjugate in power are also said to be *conjugate in work or energy*, but we will often use the phrase conjugate in power because it is more accurate.

The rate of change of the internal energy of the system is obtained by integrating (3.5.50) over the domain of the body, which gives

$$\frac{DW^{\text{int}}}{Dt} = \int_{\Omega}\rho\frac{Dw^{\text{int}}}{Dt}d\Omega = \int_{\Omega}\mathbf{D}:\boldsymbol{\sigma}d\Omega = \int_{\Omega}D_{ij}\sigma_{ij}\,d\Omega = \int_{\Omega}\frac{\partial v_i}{\partial x_j}\sigma_{ij}\,d\Omega \tag{3.5.51}$$

where the last expression follows from the symmetry of the Cauchy stress tensor.

The conservation equations are summarized in Box 3.3 in both tensor and indicial form. The equations are written without specifying the independent variables; they can be expressed in terms of either the spatial coordinates or the material coordinates. The equations are not expressed in conservative form because this does not seem to be as useful in solid mechanics as it is in fluid mechanics. The reasons for this are not explored in the literature, but it appears to be related to the much smaller changes in density which occur in solid mechanics problems.

Box 3.3 Conservation equations

Eulerian description

Mass conservation

$$\frac{D\rho}{Dt} + \rho\,\text{div}(\mathbf{v}) = 0 \quad \text{or} \quad \frac{D\rho}{Dt} + \rho v_{i,i} = 0 \quad \text{or} \quad \dot{\rho} + \rho v_{i,i} = 0 \tag{B3.3.1}$$

Linear momentum conservation

$$\rho\frac{D\mathbf{v}}{Dt}=\nabla\cdot\sigma+\rho\mathbf{b}\equiv\operatorname{div}\sigma+\rho\mathbf{b}\quad\text{or}\quad\rho\frac{Dv_i}{Dt}=\frac{\partial\sigma_{ji}}{\partial x_j}+\rho b_i \tag{B3.3.2}$$

Angular momentum conservation

$$\sigma==\sigma^T\quad\text{or}\quad\sigma_{ij}=\sigma_{ji} \tag{B3.3.3}$$

Energy conservation

$$\rho\frac{Dw^{\text{int}}}{Dt}=\mathbf{D}:\sigma-\nabla\cdot\mathbf{q}+\rho s \tag{B3.3.4}$$

Lagrangian description

Mass conservation

$$\rho(\mathbf{X},t)J(\mathbf{X},t)=\rho_0(\mathbf{X})\quad\text{or}\quad\rho J=\rho_0 \tag{B3.3.5}$$

Linear momentum conservation

$$\rho_0\frac{\partial\mathbf{v}(\mathbf{X},t)}{\partial t}=\nabla_0\cdot\mathbf{P}+\rho_0\mathbf{b}\quad\text{or}\quad\rho_0\frac{\partial v_i(\mathbf{X},t)}{\partial t}=\frac{\partial P_{ji}}{\partial X_j}+\rho_0 b_i \tag{B3.3.6}$$

Angular momentum conservation

$$\mathbf{F}\cdot\mathbf{P}=\mathbf{P}^T\cdot\mathbf{F}^T,\qquad F_{ik}P_{kj}=P_{ik}^T F_{kj}^T=F_{jk}P_{ki},\quad \mathbf{S}=\mathbf{S}^T \tag{B3.3.7}$$

Energy conservation

$$\rho_0\dot{w}^{\text{int}}=\rho_0\frac{\partial w^{\text{int}}(\mathbf{X},t)}{\partial t}=\dot{\mathbf{F}}^T:\mathbf{P}-\nabla_0\cdot\tilde{\mathbf{q}}+\rho_0 s \tag{B3.3.8}$$

3.6 Lagrangian Conservation Equations

3.6.1 Introduction and Definitions

It is instructive to directly develop the conservation equations in terms of the Lagrangian measures of stress and strain in the reference configuration. In the continuum mechanics literature such formulations are called Lagrangian, whereas in the finite element literature these formulations are called *total Lagrangian formulations*. For a total Lagrangian formulation, a Lagrangian mesh is always used. The conservation equations in a Lagrangian framework are fundamentally identical to those which have just been developed; they are just expressed in terms of different variables. In fact, as we shall show, they can be obtained by the transformations in Box 3.2 and the chain rule. This section can be skipped in a first reading. It is included here because much of the finite element literature for nonlinear mechanics employs total Lagrangian formulations, so it is essential for a serious student of the field.

The independent variables in the total Lagrangian formulation are the Lagrangian (material) coordinates $\mathbf{X}$ and the time t. The major dependent variables are the initial density $\rho_0(\mathbf{X})$, the displacement $\mathbf{u}(\mathbf{X}, t)$ and the Lagrangian measures of stress and strain. We will use the nominal stress $\mathbf{P}(\mathbf{X}, t)$ as the measure of stress. This leads to a momentum equation which is strikingly similar to the momentum equation in the Eulerian description, (3.5.33), so it is easy to remember. The deformation will be described by the deformation gradient $\mathbf{F}(\mathbf{X}, t)$. The pair $\mathbf{P}$ and $\mathbf{F}$ is not especially useful for constructing constitutive equations, since $\mathbf{F}$ does not vanish in rigid body motion and $\mathbf{P}$ is not symmetric. Therefore constitutive equations are usually formulated in terms of the PK2 stress $\mathbf{S}$ and the Green strain $\mathbf{E}$. However, keep in mind that relations between $\mathbf{S}$ and $\mathbf{E}$ can easily be transformed to relations between $\mathbf{P}$ and $\mathbf{E}$ by the relations in Box 3.2.

The applied loads are defined on the reference configuration. The traction $\mathbf{t}_0$ is defined in (3.4.2); $\mathbf{t}_0$ is in units of force per unit initial area. As mentioned in Chapter 1, the zeros that indicate that the variables pertain to the reference configuration, are either subscripts or superscripts, whichever is more convenient. The body force is denoted by $\mathbf{b}$, which is in units of force per unit mass; the body force per initial unit volume is given by $\rho_0\mathbf{b}$, which is equivalent to $\rho\mathbf{b}$. This equivalence is shown by

$$d\mathbf{f} = \rho\mathbf{b}\, d\Omega = \rho\mathbf{b}J\, d\Omega_0 = \rho_0\mathbf{b}d\Omega_0 \tag{3.6.1}$$

where the last equality follows from the conservation of mass, (3.5.25). Many authors, including Malvern (1969), use different symbols for the body forces in the two formulations, but this is not necessary with our convention of associating symbols with fields.

The conservation of mass has already been developed in a form that applies to the total Lagrangian formulation, (3.5.25). Therefore we develop only the conservation of momentum and energy.

3.6.2 *Conservation of Linear Momentum*

In a Lagrangian description, the linear momentum of a body is given in terms of an integral over the reference configuration by

$$\mathbf{p}(t) = \int_{\Omega_0} \rho_0 \mathbf{v}(\mathbf{X}, t) d\Omega_0 \tag{3.6.2}$$

The total force on the body is given by integrating the body forces over the reference domain and the traction over the reference boundaries:

$$\mathbf{f}(t) = \int_{\Omega_0} \rho_0 \mathbf{b}(\mathbf{X}, t) d\Omega_0 + \int_{\Gamma_0} \mathbf{t}_0(\mathbf{X}, t) d\Gamma_0 \tag{3.6.3}$$

Newton's second law then gives

$$\frac{d\mathbf{p}}{dt} = \mathbf{f} \tag{3.6.4}$$

Substituting (3.6.2) and (3.6.3) into this gives

$$\frac{d}{dt}\int_{\Omega_0}\rho_0\mathbf{v}\,d\Omega_0 = \int_{\Omega_0}\rho_0\mathbf{b}\,d\Omega_0 + \int_{\Gamma_0}\mathbf{t}_0\,d\Gamma_0 \tag{3.6.5}$$

On the LHS, the material derivative can be taken inside the integral because the reference domain is constant in time, so

$$\frac{d}{dt}\int_{\Omega_0}\rho_0\mathbf{v}\,d\Omega_0 = \int_{\Omega_0}\rho_0\frac{\partial\mathbf{v}(\mathbf{X},t)}{\partial t}d\Omega_0 \tag{3.6.6}$$

Using Cauchy's law (3.4.2) and Gauss's theorem in sequence gives

$$\int_{\Gamma_0}\mathbf{t}_0\,d\Gamma_0 = \int_{\Gamma_0}\mathbf{n}_0\cdot\mathbf{P}\,d\Gamma_0 = \int_{\Omega_0}\nabla_0\cdot\mathbf{P}\,d\Omega_0$$

or

$$\int_{\Gamma_0}t_i^0\,d\Gamma_0 = \int_{\Gamma_0}n_j^0P_{ji}\;d\Gamma_0 = \int_{\Omega_0}\frac{\partial P_{ji}}{\partial X_j}d\Omega_0 \tag{3.6.7}$$

Note that in tensor notation, the left gradient appears in the domain integral because the nominal stress is defined with the normal on the left side. The definition of the material gradient, which is distinguished with the subscript '0', should be clear from the indicial expression. The index on the material coordinate is the same as the first index on the nominal stress: the order is important because the nominal stress is not symmetric.

Substituting (3.6.6) and (3.6.7) into (3.6.5) gives

$$\int_{\Omega_0}\left(\rho_0\frac{\partial\mathbf{v}(\mathbf{X},t)}{\partial t} - \rho_0\mathbf{b} - \nabla_0\cdot\mathbf{P}\right)d\Omega_0 = 0 \tag{3.6.8}$$

which, because of the arbitrariness of Ω_0, gives

$$\rho_0\frac{\partial\mathbf{v}(\mathbf{X},t)}{\partial t} = \nabla_0\cdot\mathbf{P} + \rho_0\mathbf{b} \quad \text{or} \quad \rho_0\frac{\partial v_i(\mathbf{X},t)}{\partial t} = \frac{\partial P_{ji}}{\partial X_j} + \rho_0 b_i \tag{3.6.9}$$

The above is often called the Lagrangian form of the momentum equation. Comparing the above with the Eulerian form, (3.5.33), we can see that they are quite similar: the Cauchy stress is replaced by the nominal stress and the density is replaced by the initial density.

The equilibrium equation for the Lagrangian description is obtained by neglecting the accelerations, so

$$\nabla_0 \cdot \mathbf{P} + \rho_0 \mathbf{b} = 0 \quad \text{or} \quad \frac{\partial P_{ji}}{\partial X_j} + \rho_0 b_i = 0 \tag{3.6.10}$$

The equilibrium equations are often given in terms of the PK2 stress, by substituting the transformation given in Box 3.2. However, this form is much easier to remember.

This form of the momentum equation can also be obtained directly by transforming all of the terms in (3.5.33) using the chain rule and Box 3.2. Actually, this is somewhat difficult, particularly for the gradient term. Using the transformation from Box 3.2 and the chain rule gives

$$\begin{aligned} \frac{\partial \sigma_{ji}}{\partial x_j} &= \frac{\partial (J^{-1} F_{jk} P_{ki})}{\partial x_j} = P_{ki} \frac{\partial}{\partial x_j}(J^{-1} F_{jk}) + J^{-1} F_{jk} \frac{\partial P_{ki}}{\partial x_j} \\ &= J^{-1} \frac{\partial x_j}{\partial X_k} \frac{\partial P_{ki}}{\partial x_j} \end{aligned} \tag{3.6.11}$$

In (3.6.11) we have used the definition of the deformation gradient F, (3.2.19), and the interesting relation $\partial(J^{-1}F_{jk})/\partial x_j=0$ (see Ogden, 1984: 89). Thus (3.5.33) becomes

$$\rho \frac{\partial v_i}{\partial t} = J^{-1} \frac{\partial x_j}{\partial X_k} \frac{\partial P_{ki}}{\partial x_j} + \rho b_i \tag{3.6.12}$$

By the chain rule, the first term on the RHS is $J^{-1}\partial P_{ki}/\partial X_k$. Multiplying the equation by J and using mass conservation, $\rho J = \rho_0$ then gives (3.6.9).

3.6.3 *Conservation of Angular Momentum*

The balance equations for angular momentum will not be rederived in the total Lagrangian framework. We will use (3.5.40) in conjunction with the stress transformations in Box 3.2 to derive the consequences for the Lagrangian measures of stress. This gives

$$J^{-1} \mathbf{F} \cdot \mathbf{P} = (J^{-1} \mathbf{F} \cdot \mathbf{P})^T \tag{3.6.13}$$

Multiplying both sides of the above by J and taking the transpose inside the parentheses then gives

$$\mathbf{F} \cdot \mathbf{P} = \mathbf{P}^T \cdot \mathbf{F}^T \quad \text{or} \quad F_{ik} P_{kj} = P_{ik}^T F_{kj}^T = F_{jk} P_{ki} \tag{3.6.14}$$

These equations are nontrivial only when $i \neq j$. Thus the above gives one nontrivial equation in two dimensions, three nontrivial equations in three dimensions. So, while the nominal stress is not symmetric, the number of conditions imposed by angular momentum balance equals the number of symmetry conditions on the Cauchy stress, (3.5.40). In two dimensions, the angular momentum equation is

$$F_{11}P_{12} + F_{12}P_{22} = F_{21}P_{11} + F_{22}P_{21} \tag{3.6.15}$$

These conditions are usually imposed directly on the constitutive equation, as will be seen in Chapter 5.

For the PK2 stress, the conditions emanating from conservation of angular momentum can be obtained by expressing $\mathbf{P}$ in terms of $\mathbf{S}$ in (3.6.13) (the same equations are obtained if $\boldsymbol{\sigma}$ is replaced by $\mathbf{S}$ in the symmetry conditions (3.5.40)), which gives

$$\mathbf{F} \cdot \mathbf{S} \cdot \mathbf{F}^T = \mathbf{F} \cdot \mathbf{S}^T \cdot \mathbf{F}^T \tag{3.6.16}$$

Since $\mathbf{F}$ must be a regular (nonsingular) matrix, its inverse exists and we can premultiply the above by $\mathbf{F}^{-1}$ and postmultiply it by $\mathbf{F}^{-T} \equiv (\mathbf{F}^{-1})^T$ to obtain

$$\mathbf{S} = \mathbf{S}^T \tag{3.6.17}$$

So the conservation of angular momentum requires the PK2 stress to be symmetric.

3.6.4 *Conservation of Energy in Lagrangian Description*

The counterpart of (3.5.45) in the reference configuration can be written as

$$\begin{aligned} &\frac{d}{dt}\int_{\Omega_0}\left(\rho_0 w^{\text{int}} + \frac{1}{2}\rho_0 \mathbf{v}\cdot\mathbf{v}\right)d\Omega_0 \\ &= \int_{\Omega_0} \mathbf{v}\cdot\rho_0\mathbf{b}\,d\Omega_0 + \int_{\Gamma_0}\mathbf{v}\cdot\mathbf{t}_0\,d\Gamma_0 + \int_{\Omega_0}\rho_0 s\,d\Omega_0 - \int_{\Gamma_0}\mathbf{n}_0\cdot\tilde{\mathbf{q}}\,d\Gamma_0 \end{aligned} \tag{3.6.18}$$

The heat flux in a total Lagrangian formulation is defined as energy per unit reference area and is denoted by $\tilde{\mathbf{q}}$ to distinguish it from the heat flux per unit current area $\mathbf{q}$; they are related by

$$\tilde{\mathbf{q}} = J\mathbf{F}^{-1}\cdot\mathbf{q} \tag{3.6.19}$$

The above follows from Nanson's law (3.4.5) and the equivalence

$$\int_{\Gamma}\mathbf{n}\cdot\mathbf{q}\,d\Gamma = \int_{\Gamma_0}\mathbf{n}_0\cdot\tilde{\mathbf{q}}\,d\Gamma_0$$

Substituting (3.4.5) for $\mathbf{n}$ into the previous gives (3.6.19).

The internal energy per unit initial volume in the above is related to the internal energy per unit current volume in (3.5.45) as follows:

$$\rho_0 w^{\text{int}} d\Omega_0 = \rho_0 w^{\text{int}} J^{-1} d\Omega = \rho w^{\text{int}} d\Omega \tag{3.6.20}$$

where the last step follows from the mass conservation equation (3.5.25). On the LHS of (3.6.18), the time derivative can be taken inside the integral since the domain is fixed, giving

$$\frac{d}{dt}\int_{\Omega_0}\left(\rho_0 w^{\text{int}}+\frac{1}{2}\rho_0\mathbf{v}\cdot\mathbf{v}\right)d\Omega = \int_{\Omega_0}\left(\rho_0\frac{\partial w^{\text{int}}(\mathbf{X},t)}{\partial t}+\rho_0\mathbf{v}\cdot\frac{\partial \mathbf{v}(\mathbf{X},t)}{\partial t}\right)d\Omega_0 \tag{3.6.21}$$

The second term on the RHS of (3.6.18) can be modified as follows by using (B3.1.2) and Gauss's theorem:

$$\begin{aligned}\int_{\Gamma_0}\mathbf{v}\cdot\mathbf{t}_0 d\Gamma_0 &= \int_{\Gamma_0}v_j t_j^0 d\Gamma_0 = \int_{\Gamma_0}v_j n_i^0 P_{ji}\,d\Gamma_0\\ &= \int_{\Omega_0}\frac{\partial}{\partial X_i}(v_j P_{ij})d\Omega_0 = \int_{\Omega_0}\left(\frac{\partial v_j}{\partial X_i}P_{ij}+v_j\frac{\partial P_{ij}}{\partial X_i}\right)d\Omega_0\\ &= \int_{\Omega_0}\left(\frac{\partial F_{ji}}{\partial t}P_{ij}+\frac{\partial P_{ij}}{\partial X_i}v_j\right)d\Omega_0 = \int_{\Omega_0}\left(\frac{\partial \mathbf{F}^T}{\partial t}:\mathbf{P}+(\nabla_0\cdot\mathbf{P})\cdot\mathbf{v}\right)d\Omega_0\end{aligned} \tag{3.6.22}$$

Applying Gauss's theorem to the heat flux term of the RHS of (3.6.18) and some manipulation gives

$$\int_{\Omega_0}\left(\rho_0\frac{\partial w^{\text{int}}}{\partial t}-\frac{\partial \mathbf{F}^T}{\partial t}:\mathbf{P}+\nabla_0\cdot\tilde{\mathbf{q}}-\rho_0 s+\left(\rho_0\frac{\partial \mathbf{v}(\mathbf{X},t)}{\partial t}-\nabla_0\cdot\mathbf{P}-\rho_0\mathbf{b}\right)\cdot\mathbf{v}\right)d\Omega_0 = 0 \tag{3.6.23}$$

The term inside the inner parentheses of the integrand is the Lagrangian form of the momentum equation, (3.6.9), so it vanishes. Then because of the arbitrariness of the domain, the rest of the integrand vanishes, giving

$$\rho_0\dot{w}^{\text{int}} = \rho_0 = \frac{\partial w^{\text{int}}(\mathbf{X},t)}{\partial t} = \dot{\mathbf{F}}^T:\mathbf{P}-\nabla_0\cdot\tilde{\mathbf{q}}+\rho_0 s \tag{3.6.24}$$

In the absence of heat conduction or heat sources, (3.6.24) gives

$$\rho_0\dot{w}^{\text{int}} = \dot{F}_{ji}P_{ij} = \dot{\mathbf{F}}^T:\mathbf{P} = \mathbf{P}:\dot{\mathbf{F}}^T \tag{3.6.25}$$

This is the Lagrangian counterpart of (3.5.50). It shows that the *nominal stress is conjugate in power to the material time derivative of the deformation gradient.*

These energy conservation equations could also be obtained directly from (3.5.50) by transformations. This is most easily done in indicial notation.

$$
\begin{aligned}
D_{ij}\sigma_{ij}J &= \frac{\partial v_i}{\partial x_j}\sigma_{ij}J \quad \text{by definition of } \mathbf{D} \text{ and symmetry of stress } \boldsymbol{\sigma} \\
&= \frac{\partial v_i}{\partial X_k}\frac{\partial X_k}{\partial x_j}\sigma_{ij}J \quad \text{by chain rule} \\
&= \dot{F}_{ik}\frac{\partial X_k}{\partial x_j}\sigma_{ij}J \quad \text{by definition of } \mathbf{F}, (3.2.19) \\
&= \dot{F}_{ik}P_{ki} \quad \text{by Box 3.2 and mass conservation}
\end{aligned}
\tag{3.6.26}
$$

The factor J is added because $\mathbf{D}{:}\boldsymbol{\sigma}$ is the work per current unit volume whereas $\mathbf{P}{:}\dot{\mathbf{F}}^T$ is work per initial unit volume.

3.6.5 *Power of PK2 Stress*

The stress transformations in Box 3.2 can also be used to express the internal energy in terms of the PK2 stress.

$$
\begin{aligned}
\dot{\mathbf{F}}^T : \mathrm{P} &\equiv \dot{F}_{ik}P_{ki} = \dot{F}_{ik}S_{kr}F_{ri}^T \quad \text{by Box 3.2} \\
&= F_{ri}^T\dot{F}_{ik}S_{kr} = (\mathbf{F}^{\mathrm{T}}\cdot\dot{\mathbf{F}}) : \mathrm{S} \quad \text{by symmetry of } \mathbf{S} \\
&= \left(\frac{1}{2}(\mathbf{F}^{\mathrm{T}}\cdot\dot{\mathbf{F}}+\dot{\mathbf{F}}^{\mathrm{T}}\cdot\mathbf{F})+\frac{1}{2}(\mathbf{F}^{\mathrm{T}}\cdot\dot{\mathbf{F}}-\dot{\mathbf{F}}^{\mathrm{T}}\cdot\mathbf{F})\right) : \mathbf{S} \quad \text{decomposing tensor into symmetric and antisymmetric parts} \\
&= \frac{1}{2}(\mathbf{F}^{\mathrm{T}}\cdot\dot{\mathbf{F}}+\dot{\mathbf{F}}^{\mathrm{T}}\cdot\mathbf{F}) : \mathbf{S} \quad \text{since contraction of symmetric and antisymmetrictensors vanishes}
\end{aligned}
\tag{3.6.27}
$$

Then, using the time derivative of $\mathbf{E}$ as defined in (3.3.20) gives

$$
\rho_0\dot{w}^{\mathrm{int}} = \dot{\mathbf{E}} : \mathbf{S} = \mathbf{S} : \dot{\mathbf{E}} = \dot{E}_{ij}S_{ij} \tag{3.6.28}
$$

This shows that *the rate of the Green strain tensor is conjugate in power (energy) to the PK2 stress*.

Thus we have identified three stress and strain rate measures which are conjugate in power. These conjugate measures are listed in Box 3.4 along with a fourth conjugate pair, the corotational Cauchy stress and corotational rate-of-deformation.

Box 3.4 Stress-deformation (strain) rate pairs conjugate in power

Cauchy stress/rate of deformation: $\rho\dot{w}^{\mathrm{int}} = \mathbf{D} : \boldsymbol{\sigma} = \boldsymbol{\sigma} : \mathbf{D} = D_{ij}\sigma_{ij}$

Nominal stress/rate of deformation gradient: $\rho_0\dot{w}^{\mathrm{int}} = \dot{\mathbf{F}}^T : \mathbf{P} = \mathbf{P}^T : \dot{\mathbf{F}} = \dot{F}_{ij}P_{ji}$

PK2 stress/rate of Green strain: $\rho_0\dot{w}^{\mathrm{int}} = \dot{\mathbf{E}} : \mathbf{S} = \mathbf{S} : \dot{\mathbf{E}} = \dot{E}_{ij}S_{ij}$

Corotational Cauchy stress/rate-of-deformation: $\rho\dot{w}^{\mathrm{int}} = \hat{\mathbf{D}} : \hat{\boldsymbol{\sigma}} = \hat{\boldsymbol{\sigma}} : \hat{\mathbf{D}} = \hat{D}_{ij}\hat{\sigma}_{ij}$

Conjugate stress and strain rate measures are useful in developing weak forms of the momentum equation, that is, the principles of virtual work and power. The conjugate pairs presented here just scratch the surface: many other conjugate pairs have been developed in continuum mechanics (Ogden, 1984; Hill, 1978). However, those presented here are the most frequently used in nonlinear finite element methods.

3.7 Polar Decomposition and Frame-Invariance

In this section, the role of rigid body rotation is explored further. First, a theorem known as the polar decomposition theorem is presented. This theorem enables the rigid body rotation to be obtained for any motion. Next, we consider the effect of rigid body rotations on constitutive equations. We show that for the Cauchy stress, a modification of the time derivatives is needed to formulate rate-constitutive equations. This is known as a *frame-invariant or objective rate of stress*. Three frame-invariant rates are presented: the Jaumann rate, the Truesdell rate and the Green–Naghdi rate. Some startling differences in results obtained by misuse of hypoelastic constitutive equations with these various rates are then demonstrated.

3.7.1 Polar Decomposition Theorem

A fundamental theorem which elucidates the role of rotation in large-deformation problems is the polar decomposition theorem. This theorem states that any deformation gradient tensor **F** can be multiplicatively decomposed into the product of an orthogonal matrix **R** and a symmetric tensor **U**, called the right stretch tensor (the adjective 'right' is often omitted):

$$\mathbf{F} = \mathbf{R} \cdot \mathbf{U} \qquad \text{or} \qquad F_{ij} = \frac{\partial x_i}{\partial X_j} = R_{ik} U_{kj} \tag{3.7.1}$$

where

$$\mathbf{R}^{-1} = \mathbf{R}^T \quad \text{and} \quad \mathbf{U} = \mathbf{U}^T \tag{3.7.2}$$

Rewriting these with (3.2.20) gives

$$d\mathbf{x} = \mathbf{R} \cdot \mathbf{U} \cdot d\mathbf{X} \tag{3.7.3}$$

This shows that any motion of a body consists of a deformation, which is represented by the symmetric mapping **U**, and a rigid body rotation **R**; **R** can be recognized as a rigid-body rotation because all orthogonal transformations are rotations. Rigid-body translation does not appear in this equation because $d\mathbf{x}$ and $d\mathbf{X}$ are differential line segments in the current and reference configurations, respectively, and the maps of differential line segments are not affected by translation. If (3.7.3) were integrated to obtain the notion $\mathbf{x} = \phi(\mathbf{X}, t)$, then the rigid body translation would appear as a constant of integration. In a translation, $\mathbf{F} = \mathbf{I}$, and $d\mathbf{x} = d\mathbf{X}$.

The polar decomposition theorem is proven in the following. To simplify the notation, we treat the tensors as matrices. Premultiplying both sides of (3.7.1) by its transpose gives

$$\mathbf{F}^T\mathbf{F} = (\mathbf{RU})^T(\mathbf{RU}) = \mathbf{U}^T\mathbf{R}^T\mathbf{RU} = \mathbf{U}^T\mathbf{U} = \mathbf{UU} \tag{3.7.4}$$

where (3.7.2) is used to obtain the third and fourth equalities. The last term is the square of the U matrix. It follows that

$$\mathbf{U} = (\mathbf{F}^T \cdot \mathbf{F})^{\frac{1}{2}} \tag{3.7.5}$$

The fractional power of a matrix is defined in terms of its spectral representation: see, for example, Chandrasekharaiah and Debnath (1994: 96). It is computed by first transforming the matrix to its principal coordinates, where the matrix is diagonal with the eigenvalues on the diagonal. The fractional power is then applied to all of the diagonal terms, and the matrix is transformed back. This is illustrated in the following examples. The matrix $\mathbf{F}^T\mathbf{F}$ is positive definite, so all of its eigenvalues are positive. Consequently the matrix U is always real.

The rotation **R**, can then be found by applying (3.7.1), which gives

$$\mathbf{R} = \mathbf{F} \cdot \mathbf{U}^{-1} \tag{3.7.6}$$

The existence of the inverse of U follows from the fact that all of its eigenvalues are always positive, since the right-hand side of (3.7.5) is always a positive matrix.

The matrix **U** is closely related to engineering strain. Its principal values are the elongations of line segments in the principal directions of **U**. Therefore, many researchers have found this tensor to be appealing for developing constitutive equations. The tensor $\mathbf{U} - \mathbf{I}$ is called the *Biot strain tensor.*

A motion can also be decomposed in terms of a left stretch tensor and a rotation according to

$$\mathbf{F} = \mathbf{V} \cdot \mathbf{R} \tag{3.7.7}$$

This form of the polar decomposition is used less frequently and we note it only in passing. It will play a role in discussions of material symmetry for elastic materials at finite strain. The polar decomposition theorem applies to any invertible square matrix. Any square matrix can be multiplicatively decomposed into a rotation matrix and a symmetric matrix.

It is emphasized that the rotations of different line segments at the same point depend on the orientation of the line segment. In a three-dimensional body, only three line segments are rotated exactly by $\mathbf{R}(\mathbf{X}, t)$ at any point **X**. These are the line segments corresponding to the principal directions of the stretch tensor **U**. It can be shown that these are also the principal directions of the Green strain tensor. The rotations of line segments which are oriented in directions other than the principal directions of **E** are not given by **R**.

Example 3.10 Consider the motion of the triangular element shown in Figure 3.10 in which the nodal coordinates $x_I(t)$ and $y_I(t)$ are given by

$$\begin{aligned} x_1(t) &= a + 2at & y_1(t) &= 2at \\ x_2(t) &= 2at & y_2(t) &= 2a - 2at \\ x_3(t) &= 3at; & y_3(t) &= 0 \end{aligned} \tag{E3.10.1}$$

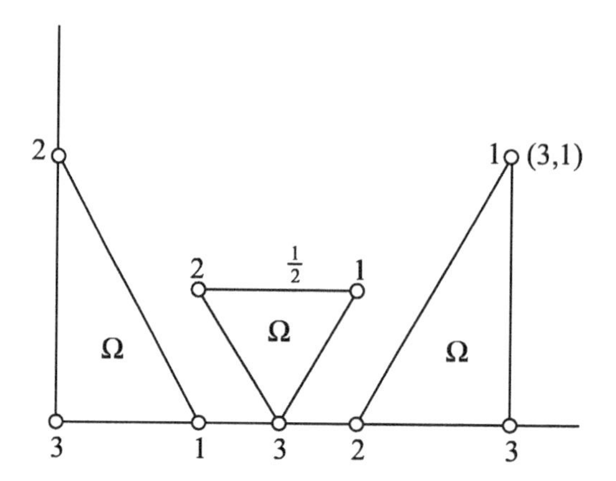

Figure 3.10 Motion (E3.10.1) with $a = 1$

Find the rigid body rotation and the stretch tensors by the polar decomposition theorem at $t = 1.0$ and at $t = 0.5$.

The motion of a triangular domain can most easily be expressed by using the shape functions for triangular elements, that is, the area coordinates. In terms of the area coordinates, the motion is given by

$$x(\xi, t) = x_1(t)\xi_1 + x_2(t)\xi_2 + x_3(t)\xi_3 \quad \text{(E3.10.2)}$$

$$y(\xi, t) = y_1(t)\xi_1 + y_2(t)\xi_2 + y_3(t)\xi_3 \quad \text{(E3.10.3)}$$

where are the area coordinates; see Appendix 3. The material coordinates appear implicitly in the RHS of these through the relationship between the area coordinates and the coordinates at time $t = 0$. To extract those relationships we write (E3.10.2–3) at time $t = 0$, which gives

$$x(\xi, 0) = X = X_1\xi_1 + X_2\xi_2 + X_3\xi_3 = a\xi_1 \quad \text{(E3.10.4)}$$

$$y(\xi, 0) = Y = Y_1\xi_1 + Y_2\xi_2 + Y_3\xi_3 = 2a\xi_2 \quad \text{(E3.10.5)}$$

In this case, the relations between the area and material coordinates are particularly simple, so the relations developed here could be obtained by inspection.

Using (E3.10.4–5) to express the area coordinates in terms of the material coordinates, (E3.10.2–3) at $t = 1$ can be written

$$x(\mathbf{X}, 1) = 3a\xi_1 + 2a\xi_2 + 3a\xi_3 = 3X + Y + 3a\left(1 - \frac{X}{a} - \frac{Y}{2a}\right) = 3a - \frac{Y}{2} \quad \text{(E3.10.6)}$$

$$y(\mathbf{X}, 1) = 2a\xi_1 + 0\xi_2 + 0\xi_3 = 2X \quad \text{(E3.10.7)}$$

The deformation gradient is then obtained by (3.2.21):

$$\mathbf{F} = \begin{bmatrix} \dfrac{\partial x}{\partial X} & \dfrac{\partial x}{\partial Y} \\ \dfrac{\partial y}{\partial X} & \dfrac{\partial y}{\partial Y} \end{bmatrix} = \begin{bmatrix} 0 & -0.5 \\ 2 & 0 \end{bmatrix} \tag{E3.10.8}$$

The stretch tensor $\mathbf{U}$ is then evaluated by (3.7.5):

$$\mathbf{U} = (\mathbf{F}^T\mathbf{F})^{\frac{1}{2}} = \begin{bmatrix} 4 & 0 \\ 0 & 0.25 \end{bmatrix}^{\frac{1}{2}} = \begin{bmatrix} 2 & 0 \\ 0 & 0.5 \end{bmatrix} \tag{E3.10.9}$$

In this case the $\mathbf{U}$ matrix is diagonal, so the principal values are simply the diagonal terms. The positive square roots are chosen in evaluating the square root of the matrix because the principal stretches must be positive. The rotation matrix $\mathbf{R}$ is then given by (3.7.6):

$$\mathbf{R} = \mathbf{F}\mathbf{U}^{-1} = \begin{bmatrix} 0 & -0.5 \\ 2 & 0 \end{bmatrix}\begin{bmatrix} 0.5 & 0 \\ 0 & 2 \end{bmatrix} = \begin{bmatrix} 0 & -1 \\ 1 & 0 \end{bmatrix} \tag{E3.10.10}$$

Comparing the above rotation matrix $\mathbf{R}$ and (3.2.34), it can be seen that the rotation is a counterclockwise 90° rotation. This is also readily apparent from Figure 3.10. The deformation consists of an elongation of the line segment between nodes 1 and 3, by a factor of 2 (see U_{11} in (E3.10.9)), and a contraction of the line segment between nodes 3 and 2, by a factor of 0.5 (see U_{22} in (E3.10.9)), followed by a translation of $3a$ in the x-direction and a 90° rotation. Since the original line segments along the x- and y-directions correspond to the principal directions, or eigenvectors of $\mathbf{U}$, the rotations of these line segments correspond to the rotation of the body in the polar decomposition theorem.

The configuration at $t=0.5$ is obtained by (E3.10.2–3):

$$x(\mathbf{X}, 0.5) = 2a\xi_1 + a\xi_2 + 1.5a\xi_3$$
$$= 2a\frac{X}{a} + a\frac{Y}{2a} + 1.5a\left(1 - \frac{X}{a} - \frac{Y}{2a}\right) = 1.5a + 0.5X - 0.25Y \tag{E3.10.11a}$$

$$y(\mathbf{X}, 0.5) = a\xi_1 + a\xi_2 + 0\xi_3 = a\frac{X}{a} + a\frac{Y}{2a} = X + 0.5Y \tag{E3.10.11b}$$

The deformation gradient $\mathbf{F}$ is then given by

$$\mathbf{F} = \begin{bmatrix} \dfrac{\partial x}{\partial X} & \dfrac{\partial x}{\partial Y} \\ \dfrac{\partial y}{\partial X} & \dfrac{\partial y}{\partial Y} \end{bmatrix} = \begin{bmatrix} 0.5 & -0.25 \\ 1 & 0.5 \end{bmatrix} \tag{E3.10.12}$$

and the stretch tensor **U** is given by (3.7.5):

$$\mathbf{U} = (\mathbf{F}^T\mathbf{F})^{\frac{1}{2}} = \begin{bmatrix} 1.25 & 0.375 \\ 0.375 & 0.3125 \end{bmatrix}^{\frac{1}{2}} = \begin{bmatrix} 1.0932 & 0.2343 \\ 0.2343 & 0.5076 \end{bmatrix} \qquad \text{E3.10.13}$$

The last matrix in the above is obtained by finding the eigenvalues λ_i of $\mathbf{F}^T\mathbf{F}$, taking their positive square roots, and placing them on a diagonal matrix $\mathbf{H}=\text{diag}\left(\sqrt{\lambda_1}, \sqrt{\lambda_2}\right)$. The matrix **H** is transformed back to the global components by $\mathbf{U}=\mathbf{AHA}^T$ where **A** is the matrix whose columns are the eigenvectors of $\mathbf{F}^T\mathbf{F}$. These matrices are:

$$\mathbf{A} = \begin{bmatrix} -0.9436 & 0.3310 \\ -0.3310 & -0.9436 \end{bmatrix}, \quad \mathbf{H} = \begin{bmatrix} 1.3815 & 0 \\ 0 & 0.1810 \end{bmatrix} \qquad \text{(E3.10.14)}$$

The rotation matrix **R** is then found by

$$\mathbf{R} = \mathbf{F}\mathbf{U}^{-1} = \begin{bmatrix} 0.5 & -0.25 \\ 1 & 0.5 \end{bmatrix} \begin{bmatrix} 1.0932 & 0.2343 \\ 0.2343 & 0.5076 \end{bmatrix}^{-1} = \begin{bmatrix} 0.6247 & -0.7809 \\ 0.7809 & 0.6247 \end{bmatrix} \qquad \text{(E3.10.15)}$$

Example 3.11 Consider the deformation gradient

$$\mathbf{F} = \begin{bmatrix} c - as & ac - s \\ s + ac & as + c \end{bmatrix} \qquad \text{(E3.11.1)}$$

where $c=\cos\theta$, $s=\sin\theta$ and a is a constant. Find the stretch tensor and the rotation matrix when $a = \frac{1}{2}$, $\theta = \pi/2$.

For the particular values given

$$\mathbf{F} = \begin{bmatrix} -\frac{1}{2} & -1 \\ 1 & \frac{1}{2} \end{bmatrix}, \quad \mathbf{C} = \mathbf{F}^T \cdot \mathbf{F} = \begin{bmatrix} 1.25 & 1 \\ 1 & 1.25 \end{bmatrix} \qquad \text{(E3.11.2)}$$

The eigenvalues and corresponding eigenvectors of **C** are

$$\begin{aligned} \lambda_1 &= 0.25 \qquad \mathbf{y}_1^T = \frac{1}{\sqrt{2}}[1 \quad -1] \\ \lambda_2 &= 2.25 \qquad \mathbf{y}_2^T = \frac{1}{\sqrt{2}}[1 \quad 1] \end{aligned} \qquad \text{(E3.11.3)}$$

The diagonal form of **C**, diag(**C**), consists of these eigenvalues, and the square root of diag(**C**) is obtained by taking the positive square roots of these eigenvalues:

$$\text{diag}(\mathbf{C}) = \begin{bmatrix} \frac{1}{4} & 0 \\ 0 & \frac{9}{4} \end{bmatrix} \Rightarrow \text{diag}(\mathbf{C}^{\frac{1}{2}}) = \begin{bmatrix} \frac{1}{2} & 0 \\ 0 & \frac{3}{2} \end{bmatrix} \tag{E3.11.4}$$

The $\mathbf{U}$ matrix is then obtained by transforming diag($\mathbf{C}$) back to the x–y coordinate system:

$$\mathbf{U} = \mathbf{Y} \cdot \text{diag}(\mathbf{C}^{\frac{1}{2}}) \cdot \mathbf{Y}^T = \frac{1}{\sqrt{2}} \begin{bmatrix} 1 & 1 \\ -1 & 1 \end{bmatrix} \begin{bmatrix} \frac{1}{2} & 0 \\ 0 & \frac{3}{2} \end{bmatrix} \frac{1}{\sqrt{2}} \begin{bmatrix} 1 & -1 \\ 1 & 1 \end{bmatrix} = \frac{1}{2} \begin{bmatrix} 2 & 1 \\ 1 & 2 \end{bmatrix} \tag{E3.11.5}$$

The rotation matrix is obtained by (3.7.6):

$$\mathbf{R} = \mathbf{F}\mathbf{U}^{-1} = \begin{bmatrix} -\frac{1}{2} & -1 \\ 1 & \frac{1}{2} \end{bmatrix} \frac{2}{3} \begin{bmatrix} 2 & -1 \\ -1 & 2 \end{bmatrix} = \begin{bmatrix} 0 & -1 \\ 1 & 0 \end{bmatrix} \tag{E3.11.6}$$

3.7.2 *Objective Rates in Constitutive Equations*

To explain why objective rates are needed for the Cauchy stress tensor, we consider the rod shown in Figure 3.11. Consider the simplest example of a rate-constitutive equation, the hypoelastic law where the stress rate is linearly related to the rate-of-deformation:

$$\frac{D\sigma_{ij}}{Dt} = C^{\sigma D}_{ijkl} D_{kl} \quad \text{or} \quad \frac{D\boldsymbol{\sigma}}{Dt} = \mathbf{C}^{\sigma D} : \mathbf{D} \tag{3.7.8}$$

We pose the following question: is the above a valid constitutive equation?

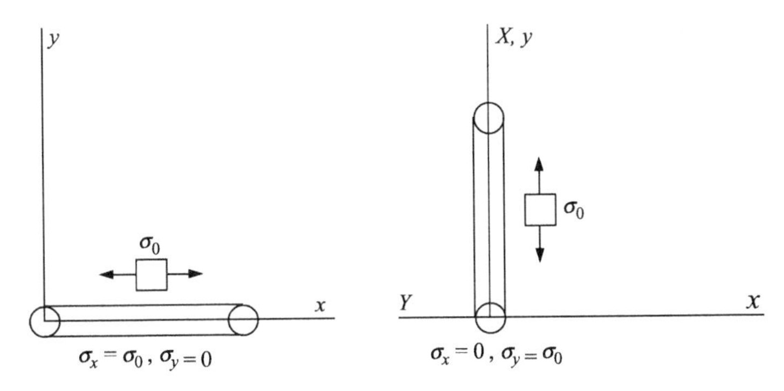

Figure 3.11 Rotation of a bar under initial stress showing the change of Cauchy stress which occurs without any deformation

The answer is negative, and can be explained as follows. Consider a solid, such as the bar in Figure 3.11, which is stressed in its initial configuration with $\sigma_x = \sigma_0$. Now assume that the bar rotates as shown at constant length, so there is no deformation, that is, $\mathbf{D}=\mathbf{0}$. Recall that in rigid body motion a state of initial stress (or prestress) is frozen in a solid, that is, since the deformation does not change in a rigid body rotation, the stress as viewed by an observer riding with the body should not change. Therefore the components of Cauchy stress in a fixed coordinate system will change during the rotation, so the material derivative of the stress must be nonzero. However, in a pure rigid body rotation, the right-hand side of (3.7.8) will vanish throughout the motion, for we have already shown that the rate-of-deformation vanishes in rigid body motion. Therefore, something must be missing in (3.7.8): $\mathbf{D}=\mathbf{0}$ but $D\sigma/Dt$ should not be zero!

The situation in the previous paragraph is not just hypothetical; it is representative of what happens in real situations and simulations. Large rotations of stressed bodies occur often. A body may be in a state of initial stress due to thermal stresses or prestressing, for example, prestressed reinforcement bars. An element may undergo large rotations in rigid body motions of the body, as in a space vehicle or a moving car, or large local rotations, as in beam buckling. The rotation need not be as large as 90° to exhibit this effect; we have chosen 90° to simplify the presentation.

The fallacy in (3.7.8) is that it does not account for the rotation of the material. The material rotation can be accounted for by an *objective rate of the stress tensor*; it is also called a *frame-invariant rate*. We will consider three objective rates: the Jaumann rate, the Truesdell rate and the Green–Naghdi rate. All of these are frequently used in current finite element software. There are many other objective rates, some of which will be discussed in Chapter 5.

3.7.3 Jaumann Rate

The Jaumann rate of the Cauchy stress is given by

$$\sigma^{\nabla J} = \frac{D\sigma}{Dt} - \mathbf{W}\cdot\sigma - \sigma\cdot\mathbf{W}^T \quad \text{or} \quad \sigma_{ij}^{\nabla J} = \frac{D\sigma_{ij}}{Dt} - W_{ik}\sigma_{kj} - \sigma_{ik}W_{kj}^T \tag{3.7.9}$$

where $\mathbf{W}$ is the spin tensor given by (3.3.11). The superscript '∇' here designates an objective rate; the Jaumann rate is designated by the subsequent superscript 'J'. One appropriate hypoelastic constitutive equation is given by

$$\sigma^{\nabla J} = \mathbf{C}^{\sigma J} : \mathbf{D} \quad \text{or} \quad \sigma_{ij}^{\nabla J} = C_{ijkl}^{\sigma J} D_{kl} \tag{3.7.10}$$

The material rate for the Cauchy stress tensor, that is, the correct equation corresponding to (3.7.8), is then

$$\frac{D\sigma}{Dt} = \sigma^{\nabla J} + \mathbf{W}\cdot\sigma + \sigma\cdot\mathbf{W}^T = \underbrace{\mathbf{C}^{\sigma J} : \mathbf{D}}_{\text{material}} + \underbrace{\mathbf{W}\cdot\sigma + \sigma\cdot\mathbf{W}^T}_{\text{rotation}} \tag{3.7.11}$$

where the first equality is just a rearrangement of (3.7.9) and the second equality follows from (3.7.10). We see in this that the material response is specified in terms of an objective stress rate, here the Jaumann rate. The material derivative of the Cauchy stress then consists of two parts: the rate of change due to material response, which is reflected in the objective rate, and the change of stress due to rotation, which corresponds to the last two terms in (3.7.11).

3.7.4 Truesdell Rate and Green–Naghdi Rate

Two other frequently used rates are the Truesdell rate and Green–Naghdi rate, given in Box 3.5. The Green–Naghdi rate differs from the Jaumann rate only in using a different measure of the rotation of the material: the Green–Naghdi rate employs the angular velocity $\Omega = \dot{\mathbf{R}}\mathbf{R}^T$ from (3.2.42). This markedly changes the behavior of the material model as we shall see.

Box 3.5 Objective rates

Jaumann rate

$$\sigma^{\nabla J} = \frac{D\sigma}{Dt} - \mathbf{W}\cdot\sigma - \sigma\cdot\mathbf{W}^T, \quad \sigma_{ij}^{\nabla J} = \frac{D\sigma_{ij}}{Dt} - W_{ik}\sigma_{kj} - \sigma_{ik}W_{kj}^T \tag{B3.5.1}$$

Truesdell rate

$$\sigma^{\nabla J} = \frac{D\sigma}{Dt} + \text{div }(\mathbf{v})\sigma - \mathbf{L}\cdot\sigma - \sigma\cdot\mathbf{L}^T \tag{B3.5.2}$$

$$\sigma_{ij}^{\nabla T} = \frac{D\sigma_{ij}}{Dt} + \frac{\partial v_k}{\partial x_k}\sigma_{ij} - \frac{\partial v_i}{\partial x_k}\sigma_{kj} - \sigma_{ik}\frac{\partial v_j}{\partial x_k} \tag{B3.5.3}$$

Green–Naghdi rate

$$\sigma^{\nabla G} = \frac{D\sigma}{Dt} - \Omega\cdot\sigma - \sigma\cdot\Omega^T, \quad \sigma_{ij}^{\nabla G} = \frac{D\sigma_{ij}}{Dt} - \Omega_{ik}\sigma_{kj} - \sigma_{ik}\Omega_{kj}^T \tag{B3.5.4}$$

$$\Omega = \dot{\mathbf{R}}\cdot\mathbf{R}^T, \quad \mathbf{L} = \frac{\partial \mathbf{v}}{\partial \mathbf{x}} = \mathbf{D} + \mathbf{W}, \quad L_{ij} = \frac{\partial v_i}{\partial x_j} = D_{ij} + W_{ij} \tag{B3.5.5}$$

The relationship between the Truesdell rate and the Jaumann rate can be examined by replacing the velocity gradient by its symmetric and antisymmetric parts, that is, using (3.3.9):

$$\boldsymbol{\sigma}^{\nabla T} = \frac{D\boldsymbol{\sigma}}{Dt} + \text{div}(\mathbf{v})\boldsymbol{\sigma} - (\mathbf{D}+\mathbf{W})\cdot\boldsymbol{\sigma} - \boldsymbol{\sigma}\cdot(\mathbf{D}+\mathbf{W})^T \tag{3.7.12}$$

A comparison of (3.7.9) and (3.7.12) then shows that the Truesdell rate includes the same spin-related terms as the Jaumann rate, but also includes additional terms which depend on the rate-of-deformation. Consider a rigid body rotation; the Truesdell rate when $\mathbf{D}=\mathbf{0}$ is then

$$\boldsymbol{\sigma}^{\nabla T} = \frac{D\boldsymbol{\sigma}}{Dt} - \mathbf{W}\cdot\boldsymbol{\sigma} - \boldsymbol{\sigma}\cdot\mathbf{W}^T \tag{3.7.13}$$

Comparison of this with (3.7.9) shows that the Truesdell rate is equivalent to the Jaumann rate in the absence of deformation. However, when the body deforms, the two rates differ, so a constitutive equation relating the Jaumann rate to **D** will give a different material rate of stress than one in terms of the Truesdell rate unless the constitutive equation is changed appropriately.

In other words, if the two laws

$$\boldsymbol{\sigma}^{\nabla T} = \mathbf{C}^{\sigma T}:\mathbf{D}, \quad \boldsymbol{\sigma}^{\nabla J} = \mathbf{C}^{\sigma J}:\mathbf{D} \tag{3.7.14}$$

are to model the same material response, $\mathbf{C}_{\sigma}^{J}$ will differ from $\mathbf{C}_{\sigma}^{J}$. For this reason, we will add superscripts to specify the objective rate associated with the material response tensor.

In addition to (3.7.14), the following forms are valid for large rotations:

$$\text{(a) } \boldsymbol{\sigma}^{\nabla G} = \mathbf{C}^{\sigma G}:\mathbf{D} \quad \text{(b) } \dot{\hat{\boldsymbol{\sigma}}} = \hat{\mathbf{C}}^{\hat{\sigma}\hat{D}}:\hat{\mathbf{D}} \quad \text{(c) } \dot{\mathbf{S}} = \mathbf{C}^{SE}:\dot{\mathbf{E}} \tag{3.7.15}$$

The second and third are applicable to arbitrary anisotropic materials. Form (a) and (3.7.14) with constant **C** are applicable only to isotropic materials or when the constitutive response matrix **C** is only a function of isotropic tensors: see Chapter 5.

Example 3.12 Consider a body rotating in the x–y plane about the origin with an angular velocity ω; the original configuration is prestressed as shown in Figure 3.12. The motion is rigid body rotation and the related tensors are given in Example 3.2. Evaluate the material time derivative of the Cauchy stress using the Jaumann rate and integrate it to obtain the Cauchy stress as a function of time.

From Example 3.2, (E3.2.7), we note that

$$\mathbf{F} = \mathbf{R} = \begin{bmatrix} c & -s \\ s & c \end{bmatrix}, \quad \dot{\mathbf{F}} = \omega \begin{bmatrix} -s & -c \\ c & -s \end{bmatrix}, \quad \mathbf{F}^{-1} = \begin{bmatrix} c & s \\ -s & c \end{bmatrix} \tag{E3.12.1a}$$

where $s = \sin \omega t$, $c = \cos \omega t$. The spin is evaluated in terms of the velocity gradient **L**, which is given for this case by (3.3.18) and then using (E3.12.1a):

$$\begin{gathered} \mathbf{L} = \dot{\mathbf{F}} \cdot \mathbf{F}^{-1} = \omega \begin{bmatrix} -s & -c \\ c & -s \end{bmatrix} \begin{bmatrix} c & s \\ -s & c \end{bmatrix} = \omega \begin{bmatrix} 0 & -1 \\ 1 & 0 \end{bmatrix} \Rightarrow \\ \mathbf{W} = \frac{1}{2}(\mathbf{L} - \mathbf{L}^{T}) = \omega \begin{bmatrix} 0 & -1 \\ 1 & 0 \end{bmatrix} \end{gathered} \tag{E3.12.1b}$$

The material time derivative based on the Jaumann rate is then given by

$$\frac{D\boldsymbol{\sigma}}{Dt} = \mathbf{W} \cdot \sigma + \sigma \cdot \mathbf{W}^{T} \tag{E3.12.1c}$$

(since $\mathbf{D} = 0$, the material part of the stress rate vanishes). We now change the material time derivative to an ordinary derivative, since the stress is constant in space, and write out the matrices in (E3.12.1c):

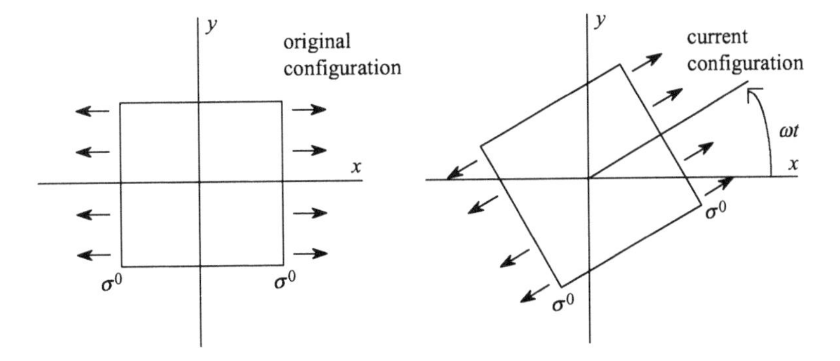

Figure 3.12 Rotation of a prestressed element with no deformation

$$\frac{d\sigma}{dt} = \omega \begin{bmatrix} 0 & -1 \\ 1 & 0 \end{bmatrix} \begin{bmatrix} \sigma_x & \sigma_{xy} \\ \sigma_{xy} & \sigma_y \end{bmatrix} + \begin{bmatrix} \sigma_x & \sigma_{xy} \\ \sigma_{xy} & \sigma_y \end{bmatrix} \omega \begin{bmatrix} 0 & 1 \\ -1 & 0 \end{bmatrix} \tag{E3.12.2}$$

$$\frac{d\sigma}{dt} = \omega \begin{bmatrix} -2\sigma_{xy} & \sigma_x - \sigma_y \\ \sigma_x - \sigma_y & 2\sigma_{xy} \end{bmatrix} \tag{E3.12.3}$$

It can be seen that the material time derivative of the Cauchy stress is symmetric. We now write out the three ordinary differential equations in the three unknowns, σ_x, σ_y, and σ_{xy}, corresponding to (E3.12.3) (the fourth scalar equation of the previous tensor equation is omitted because of symmetry):

$$\frac{d\sigma_x}{dt} = -2\omega\sigma_{xy}, \quad \frac{d\sigma_y}{dt} = 2\omega\sigma_{xy}, \quad \frac{d\sigma_{xy}}{dt} = \omega(\sigma_x - \sigma_y) \tag{E3.12.4}$$

The initial conditions are

$$\sigma_x(0) = \sigma_x^0, \quad \sigma_y(0) = 0, \quad \sigma_{xy}(0) = 0 \tag{E3.12.5}$$

It can be shown that the solution to the above differential equations is

$$\sigma = \sigma_x^0 \begin{bmatrix} c^2 & cs \\ cs & s^2 \end{bmatrix} \tag{E3.12.6}$$

We verify the solution only for $\sigma_x(t)$:

$$\frac{d\sigma_x}{dt} = \sigma_x^0 \frac{d(\cos^2 \omega t)}{dt} = \sigma_x^0 \omega(-2\cos\omega t \ \sin\omega t) = -2\omega\sigma_{xy} \tag{E3.12.7}$$

where the last step follows from the solution for $\sigma_{xy}(t)$ as given in (E3.12.6); comparing with (E3.12.4) we see that the differential equation is satisfied.

Examining (E3.12.6) we can see that the solution corresponds to a constant state of the corotational stress $\hat{\sigma}$, that is, if we let the corotational stress be given by

$$\hat{\sigma} = \begin{bmatrix} \sigma_x^0 & 0 \\ 0 & 0 \end{bmatrix}$$

then the Cauchy stress components in the global coordinate system are given by (E3.12.6). This follows from $\boldsymbol{\sigma} = \mathbf{R} \cdot \hat{\boldsymbol{\sigma}} \cdot \mathbf{R}^T$ (Box 3.2) with $\mathbf{R}$ specified in (E3.12.1a).

We leave it as an exercise to show that when all of the initial stresses are nonzero, then the solution to (E3.12.4) is

$$\sigma = \begin{bmatrix} c & -s \\ s & c \end{bmatrix} \begin{bmatrix} \sigma_x^0 & \sigma_{xy}^0 \\ \sigma_{xy}^0 & \sigma_y^0 \end{bmatrix} \begin{bmatrix} c & s \\ -s & c \end{bmatrix} \tag{E3.12.8}$$

Thus in rigid body rotation, the Jaumann rate changes the Cauchy stress so that the corotational stress is constant. Therefore, the Jaumann rate is often called the corotational rate of the Cauchy stress. The Truesdell, Jaumann, Green–Naghdi and corotational stress rates are identical in rigid body rotation.

Example 3.13 Consider an element in shear as shown in Figure 3.13. Find the shear stress using the Jaumann, Truesdell and Green–Naghdi rates for a hypoelastic, isotropic material.

The motion of the element is given by

$$x = X + tY, \quad y = Y \tag{E3.13.1}$$

The deformation gradient is given by (3.2.19), so

$$\mathbf{F} = \begin{bmatrix} 1 & t \\ 0 & 1 \end{bmatrix}, \quad \dot{\mathbf{F}} = \begin{bmatrix} 0 & 1 \\ 0 & 0 \end{bmatrix}, \quad \mathbf{F}^{-1} = \begin{bmatrix} 1 & -t \\ 0 & 1 \end{bmatrix} \tag{E3.13.2}$$

The velocity gradient is given by (E3.12.1b), and the rate-of-deformation and spin are its symmetric and skew-symmetric parts, so

$$\mathbf{L} = \dot{\mathbf{F}}\mathbf{F}^{-1} = \begin{bmatrix} 0 & 1 \\ 0 & 0 \end{bmatrix}, \quad \mathbf{D} = \frac{1}{2}\begin{bmatrix} 0 & 1 \\ 1 & 0 \end{bmatrix}, \quad \mathbf{W} = \frac{1}{2}\begin{bmatrix} 0 & 1 \\ -1 & 0 \end{bmatrix} \tag{E3.13.3}$$

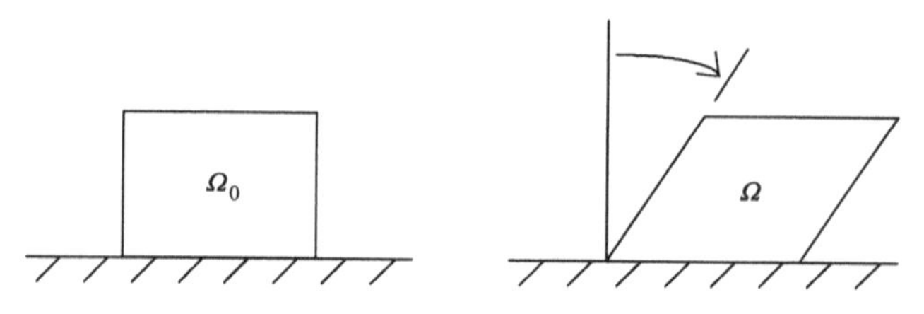

Figure 3.13 Shearing of a block

The hypoelastic, isotropic constitutive equation in terms of the Jaumann rate is given by

$$\dot{\boldsymbol{\sigma}} = (\lambda^J \text{trace } \mathbf{D})\mathbf{I} + 2\mu^J \mathbf{D} + \mathbf{W}\cdot\boldsymbol{\sigma} + \boldsymbol{\sigma}\cdot\mathbf{W}^T \quad \text{(E3.13.4)}$$

We have placed the superscripts on the material constants to distinguish the material constants which are used with different objective rates. Writing out the matrices in the previous gives

$$\begin{bmatrix} \dot{\sigma}_x & \dot{\sigma}_{xy} \\ \dot{\sigma}_{xy} & \dot{\sigma}_y \end{bmatrix} = \mu^J \begin{bmatrix} 0 & 1 \\ 1 & 0 \end{bmatrix} + \frac{1}{2}\begin{bmatrix} 0 & 1 \\ -1 & 0 \end{bmatrix}\begin{bmatrix} \sigma_x & \sigma_{xy} \\ \sigma_{xy} & \sigma_y \end{bmatrix} + \frac{1}{2}\begin{bmatrix} \sigma_x & \sigma_{xy} \\ \sigma_{xy} & \sigma_y \end{bmatrix}\begin{bmatrix} 0 & -1 \\ 1 & 0 \end{bmatrix} \quad \text{(E3.13.5)}$$

$$\dot{\sigma}_x = \sigma_{xy}, \quad \dot{\sigma}_y = -\sigma_{xy}, \quad \dot{\sigma}_{xy} = \mu^J + \frac{1}{2}(\sigma_y - \sigma_x) \quad \text{(E3.13.6)}$$

The solution to these differential equations is

$$\sigma_x = -\sigma_y = \mu^J(1-\cos t), \quad \sigma_{xy} = \mu^J \sin t \quad \text{(E3.13.7)}$$

For the Truesdell rate, the constitutive equation is

$$\dot{\sigma} = \lambda^T \text{trace } \mathbf{D} + 2\mu^T \mathbf{D} + \mathbf{L}\cdot\boldsymbol{\sigma} + \boldsymbol{\sigma}\cdot\mathbf{L}^T - (\text{trace}\,\mathbf{D})\boldsymbol{\sigma} \quad \text{(E3.13.8)}$$

This gives

$$\begin{bmatrix} \dot{\sigma}_x & \dot{\sigma}_{xy} \\ \dot{\sigma}_{xy} & \dot{\sigma}_y \end{bmatrix} = \mu^T \begin{bmatrix} 0 & 1 \\ 1 & 0 \end{bmatrix} + \begin{bmatrix} 0 & 1 \\ 0 & 0 \end{bmatrix}\begin{bmatrix} \sigma_x & \sigma_{xy} \\ \sigma_{xy} & \sigma_y \end{bmatrix} + \begin{bmatrix} \sigma_x & {}_{xy} \\ \sigma_{xy} & {}_{y} \end{bmatrix}\begin{bmatrix} 0 & 0 \\ 1 & 0 \end{bmatrix} \quad \text{(E3.13.9)}$$

where we have used the results trace $\mathbf{D}=0$: see (E3.13.3). The differential equations for the stresses are

$$\dot{\sigma}_x = 2\sigma_{xy}, \quad \dot{\sigma}_y = 0, \quad \dot{\sigma}_{xy} = \mu^T + \sigma_y \quad \text{(E3.13.10)}$$

and the solution is

$$\sigma_x = \mu^T t^2, \quad \sigma_y = 0, \quad \sigma_{xy} = \mu^T t \quad \text{(E3.13.11)}$$

To obtain the solution for the Cauchy stress by means of the Green–Naghdi rate, we need to find the rotation matrix $\mathbf{R}$ by the polar decomposition theorem. To obtain the rotation, we diagonalize $\mathbf{F}^T\mathbf{F}$

$$\mathbf{F}^T\mathbf{F} = \begin{bmatrix} 1 & t \\ t & 1+t^2 \end{bmatrix}, \quad \text{eigenvalues } \bar{\lambda}_i = \frac{2+t^2 \pm t\sqrt{4+t^2}}{2} \quad \text{(E3.13.12)}$$

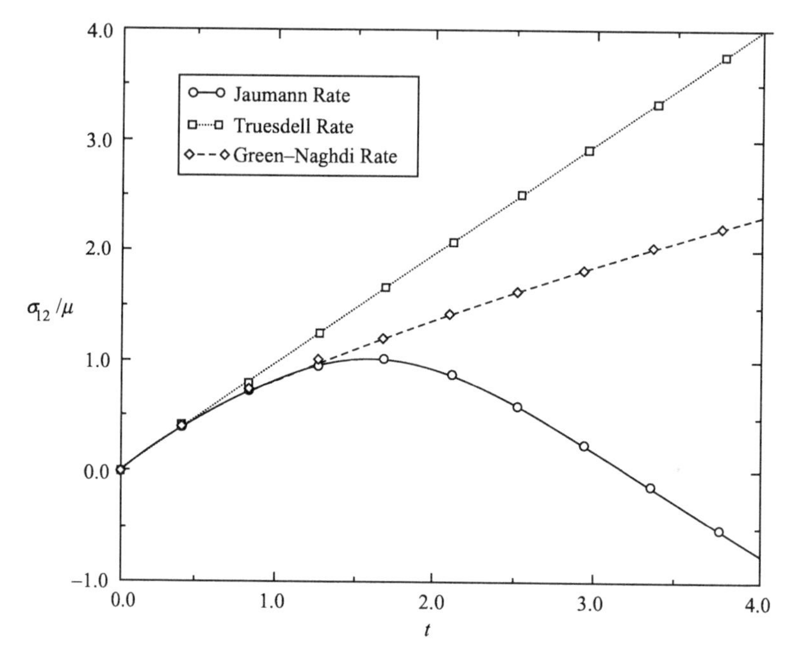

Figure 3.14 Comparison of stresses for various objective stress rates for simple shear problem with same material constants

The closed-form solution by hand is quite involved and we recommend a computer solution. A closed-form solution has been given by Dienes (1979):

$$\sigma_x = -\sigma_y = 4\mu^G(\cos 2\beta\ \ln\ \cos\beta + \beta\sin 2\beta - \sin^2\beta) \qquad \text{(E3.13.13)}$$

$$\sigma_{xy} = 2\mu^G \cos 2\beta(2\beta - 2\tan 2\beta\ \ln\cos\beta - \tan\beta), \quad \tan\beta = \frac{t}{2} \qquad \text{(E3.13.14)}$$

The results are shown in Figure 3.14, and are strikingly different. In fact, it is a misuse of material models to employ the same constants with different objective rates.

3.7.5 Explanation of Objective Rates

One underlying characteristic of objective rates can be gleaned from the previous example: an objective rate of the Cauchy stress instantaneously coincides with the rate of a stress field whose material rate already accounts for rotation. Therefore, if we take a stress measure which rotates with the material, such as the corotational stress or the PK2 stress then we can obtain an objective stress rate. This is not the most general framework for developing objective rates. A general framework is provided by using objectivity in the sense that the stress rate should be invariant for observers who are rotating with respect to each other. A derivation based on these principles may be found in Chapter 5.

To illustrate the first approach, we develop an objective rate from the corotational Cauchy stress $\hat{\boldsymbol{\sigma}}$. Its material rate is given by

$$\frac{D\hat{\boldsymbol{\sigma}}}{Dt} = \frac{D(\mathbf{R}^T\boldsymbol{\sigma}\mathbf{R})}{Dt} = \frac{D\mathbf{R}^T}{Dt}\boldsymbol{\sigma}\mathbf{R} + \mathbf{R}^T\frac{D\boldsymbol{\sigma}}{Dt}\mathbf{R} + \mathbf{R}^T\boldsymbol{\sigma}\frac{D\mathbf{R}}{Dt} \tag{3.7.16}$$

where the first equality follows from the stress transformation in Box 3.2 and the second equality is based on the derivative of a product. If we now consider the corotational coordinate system coincident with the reference coordinates but rotating with a spin $\mathbf{W}$ then

$$\mathbf{R} = \mathbf{I}, \quad \frac{D\mathbf{R}}{Dt} = \mathbf{W} \tag{3.7.17}$$

Inserting (3.7.17) into (3.7.16), it follows that, at the instant when the corotational coordinate system coincides with the global system, the rate of the Cauchy stress in rigid body rotation is given by

$$\frac{D\hat{\boldsymbol{\sigma}}}{Dt} = \mathbf{W}^T\cdot\boldsymbol{\sigma} + \frac{D\boldsymbol{\sigma}}{Dt} + \boldsymbol{\sigma}\cdot\mathbf{W} \tag{3.7.18}$$

The first and last terms of the RHS of this expression can be seen to be identical to the rotation correction in the Jaumann rate.

The Truesdell rate is derived similarly by considering the time derivative of the PK2 stress when the reference configuration instantaneously coincides with the current configuration. This is left as an exercise. The result is that if $\mathbf{x}$ coincides with $\mathbf{X}$,

$$\boldsymbol{\sigma}^{\nabla T} = \dot{\mathbf{S}} \tag{3.7.19}$$

Readers familiar with fluid mechanics may wonder why frame-invariant rates are rarely discussed in courses in fluids, since the Cauchy stress is widely used in fluid mechanics. The reason for this lies in the structure of constitutive equations which are used in fluid mechanics. For a Newtonian fluid, for example, $\boldsymbol{\sigma} = 2\mu\mathbf{D}^{\mathrm{dev}} - p\mathbf{I}$, where μ is the viscosity and $\mathbf{D}^{\mathrm{dev}}$ is the deviatoric part of the rate-of-deformation tensor. A major difference between this constitutive equation and the hypoelastic law (3.7.14) can be seen immediately: the hypoelastic law gives the stress rate in terms of $\mathbf{D}$, whereas the Newtonian fluid constitutive equation gives the stress in terms of $\mathbf{D}$. The stress transforms in a rigid body rotation exactly like $\mathbf{D}$, so the relation for a Newtonian fluid behaves properly in a rigid body rotation.

3.8 Exercises

3.1. Consider the element shown in Figure 3.4. Let the motion be given by

$$x = X + Yt, \quad y = Y + \frac{1}{2}Xt$$

(a) Sketch the element at time $t = 1$. Evaluate the deformation gradient and the Green strain tensor at this time.
(b) Evaluate the velocity and acceleration of the element at $t = 1$.

(c) Evaluate the rate-of-deformation and the spin tensor of the element at $t=1$.
(d) Repeat this at $t=0.5$.
(e) Evaluate the Jacobian determinant as a function of time and determine for how long it remains positive. Sketch the element at the time that the Jacobian changes sign. What can you say about the motion at that time?

3.2. Consider the motion given in Example 3.13, (E3.13.1). Find the velocity gradient L, the rate-of-deformation D, the spin tensor W and the angular velocity Ω as functions of time. Plot the spin and the angular velocity as function of time on the interval $t \in [0,4]$. Does this shed any light on the difference between the Green–Naghdi and Jaumann material shown in Figure 3.13?

3.3. Consider the three-node rod element shown in Figure 3.15. Use the standard 3-node shape functions for $\hat{v}_x$ and $\hat{v}_y$. The nodal coordinates are given by

$$x_1 = -r\sin\theta, \quad x_2 = 0, \quad x_3 = r\sin\theta \qquad y_1 = 0, \quad y_2 = r(1-\cos\theta), \quad y_3 = 0$$

The nodal velocities at each node are in the radial direction as shown. Evaluate the corotational rate-of-deformation at node 2 in terms of the nodal velocities. For this point, the corotational coordinate system is coincident with the global system. Compare the result with the result obtained by using cylindrical coordinates, $D_{\theta\theta} = \frac{v_r}{r}$. Repeat the procedure at the Gauss quadrature point $\xi = -3^{-1/2}$ for $\theta = 0.1$ rad and $\theta = 0.05$ rad and compare to $D_{\theta\theta} = \frac{v_r}{r}$; the corotational system for the quadrature point is shown on the RHS of Figure 3.15.

3.4. Use Nanson's relation (3.4.5) to show that the material time derivative of a surface integral is given by

$$\frac{d}{dt}\int_S g\mathbf{n}\, dS = \int_S \left[(\dot{g} + g\nabla\cdot\mathbf{v})\mathbf{I} - g\mathbf{L}^T\right]\cdot\mathbf{n}\, dS$$

This result is used in Chapter 6 in the derivation of load stiffness.

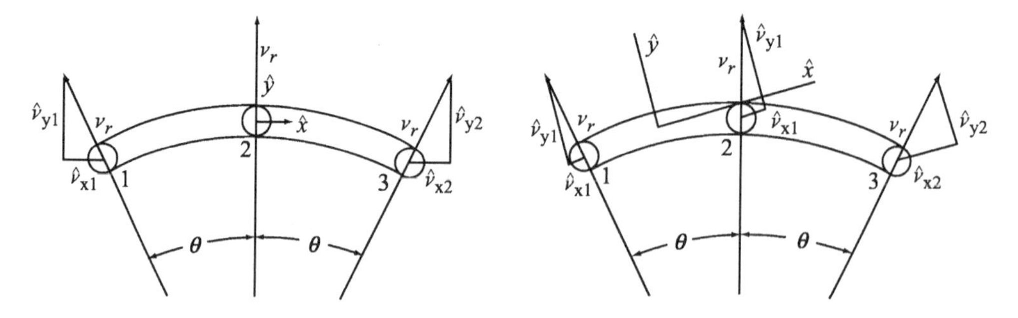

Figure 3.15 A 3 node element showing the corotational coordinates at node 2 and at a Gauss quadrature point for Exercise 3.3

3.5.

(a) Show that for any two second order tensors A and B, the Jaumann rate has the property that

$$\frac{D}{Dt}(\mathbf{A}:\mathbf{B}) = \dot{\mathbf{A}}:\mathbf{B}+\mathbf{A}:\dot{\mathbf{B}} = \mathbf{A}^{\nabla J}:\mathbf{B}+\mathbf{A}:\mathbf{B}^{\nabla J}$$

(b) Show that for symmetric tensors A and B, the additional results

$$\mathbf{A}:\mathbf{B} = \mathbf{A}:\mathbf{B}^{\nabla J} \quad \text{or} \quad \dot{\mathbf{A}}:\mathbf{B} = \mathbf{A}^{\nabla J}:\mathbf{B}$$

hold if A and B commute (i.e., are coaxial or have the same principal directions).

(c) Finally, show that the results in (a) and (b) hold for any spin-based rate, that is,

$$\mathbf{A}^{\nabla} = \dot{\mathbf{A}} - \boldsymbol{\Omega}\cdot\mathbf{A} - \mathbf{A}\cdot\boldsymbol{\Omega}^T$$

where $\Omega = -\Omega^T$ is a spin tensor.

These results, due to Prager, are used in Chapter 5 in developing the elasto-plastic tangent modulus.

3.6.

(a) Use the results in Exercise 3.5 and the expressions for the principal invariants of a tensor in Box 5.2 to show that the material time derivatives of the principal invariants of the Cauchy stress can be written as

$$\dot{I}_1 = \dot{\boldsymbol{\sigma}}:\mathbf{I} = \boldsymbol{\sigma}^{\nabla J}:\mathbf{I}$$
$$\dot{I}_2 = I_1\dot{\boldsymbol{\sigma}}:\mathbf{I} - (\dot{\boldsymbol{\sigma}}\cdot\dot{\boldsymbol{\sigma}}):\mathbf{I} = \boldsymbol{\sigma}^{\nabla J}:\mathbf{I} - (\boldsymbol{\sigma}^{\nabla J}\cdot\boldsymbol{\sigma}):\mathbf{I}$$
$$\dot{I}_3 = I_3 \text{ Trace }(\dot{\boldsymbol{\sigma}}\cdot\boldsymbol{\sigma}^{-1}) = I_3 \text{ trace}(\boldsymbol{\sigma}^{\nabla J}\cdot\boldsymbol{\sigma}^{-1})$$

It follows that if the Jaumann rate of Cauchy stress vanishes, that is, $\boldsymbol{\sigma}^{\nabla J}=0$, then the principal invariants of the Cauchy stress are stationary.

(b) Show that if the material time-derivative of the Cauchy stress is deviatoric then the Jaumann rate of Cauchy stress is deviatoric.

From Exercise 3.5(c), it follows that these results also hold for any symmetric tensor and for any spin-based rate.

3.7. Starting from Equations. (3.3.4) and (3.3.12), show that

$$2d\mathbf{x}\cdot\mathbf{D}\cdot d\mathbf{x} = 2d\mathbf{x}\cdot\mathbf{F}^{-T}\cdot\dot{\mathbf{E}}\cdot\mathbf{F}^{-1}\cdot d\mathbf{x}$$

and hence that Equation (3.3.22) holds.

3.8. Using the statement of the conservation of angular momentum in the Lagrangian description in the initial configuration, show that it implies

$$\mathbf{P}^T\mathbf{F}^T = \mathbf{FP}$$

3.9. Extend Example 3.3 by finding the conditions at which the Jacobian becomes negative at the Gauss quadrature points for 2×2 quadrature when the initial element is rectangular with dimension $a\times b$. Repeat for one-point quadrature, with the quadrature point at the center of the element.

3.10. Derive (3.2.19).

4

Lagrangian Meshes

4.1 Introduction

In Lagrangian meshes, the nodes and elements move with the material. Boundaries and interfaces remain coincident with element edges, so that their treatment is simplified. Quadrature points also move with the material, so constitutive equations are always evaluated at the same material points, which is advantageous for history-dependent materials. For these reasons, Lagrangian meshes are widely used for solid mechanics.

The formulations described in this chapter apply to large deformations and nonlinear materials, that is, they consider both geometric and material nonlinearities. They are only limited by the element's capabilities to deal with large distortions. The limited distortions most elements can sustain without degradation in performance or failure is an important factor in nonlinear analysis with Lagrangian meshes.

Finite element discretizations with Lagrangian meshes are commonly classified as updated Lagrangian formulations and total Lagrangian formulations. Both formulations use Lagrangian descriptions, that is, the dependent variables are functions of the material (Lagrangian) coordinates and time. In the updated Lagrangian formulation, the derivatives are with respect to the spatial (Eulerian) coordinates; the weak form involves integrals over the deformed (or current) configuration. In the total Lagrangian formulation, the weak form involves integrals over the initial configuration (considered henceforth to be the reference configuration) and derivatives are taken with respect to the material coordinates.

This chapter begins with the development of the updated Lagrangian formulation. The key equation to be discretized is the momentum equation, which is expressed in terms of the

Nonlinear Finite Elements for Continua and Structures, Second Edition.
Ted Belytschko, Wing Kam Liu, Brian Moran, and Khalil I. Elkhodary.

Companion Website: www.wiley.com/go/belytschko

Eulerian (spatial) coordinates and the Cauchy (physical) stress. A weak form for the momentum equation is then developed, which is known as the principle of virtual power. The momentum equation in the updated Lagrangian formulation employs derivatives with respect to the spatial coordinates, so it is natural that the weak form involves integrals taken with respect to the spatial coordinates, that is, on the current configuration. It is common practice to use the rate-of-deformation as a measure of strain rate, but other measures of strain or strain rate can be used in an updated Lagrangian formulation. For many applications, the updated Lagrangian formulation provides the most efficient formulation.

The total Lagrangian formulation is developed next. In the total Lagrangian formulation, we will use the nominal stress, although the second Piola–Kirchhoff stress (PK2) is also used in the formulations presented here. As a measure of strain we will use the Green strain tensor in the total Lagrangian formulation. A weak form of the momentum equation is developed, which is known as the principle of virtual work. The development of the total Lagrangian formulation closely parallels that of the updated Lagrangian formulation, and it is stressed that the two are basically identical. Any of the expressions in the updated Lagrangian formulation can be transformed to the total Lagrangian formulation by transformations of tensors and mappings of configurations. However, the total Lagrangian formulation is often used in practice, so to understand the literature, an advanced student must be familiar with it. In introductory courses one of the formulations can be skipped.

Implementations of the updated and total Lagrangian formulations are given for several elements. In this chapter, only the expressions for the nodal forces are developed. It is stressed that the nodal forces represent the discretization of the momentum equation. The tangential stiffness matrices, which are emphasized in many texts, are simply a means to solving the equations for certain solution procedures. They are not central to the finite element discretization. Stiffness matrices are developed in Chapter 6.

For the total Lagrangian formulation, a variational principle is presented. This principle is applicable only to static problems with conservative loads and hyperelastic materials, that is, materials which are described by a path-independent, rate-independent elastic constitutive law. The variational principle is of value in interpreting and understanding numerical solutions and the stability of nonlinear solutions. It can also sometimes be used to develop numerical procedures.

4.2 Governing Equations

We consider a body which occupies a domain Ω with a boundary Γ (Figure 4.1). The governing equations for the mechanical behavior of a continuous body are:

1. Conservation of mass (or matter)
2. Conservation of linear momentum and angular momentum
3. Conservation of energy, often called the first law of thermodynamics
4. Constitutive equations
5. Strain–displacement equations.

We will first develop the updated Lagrangian formulation. The conservation equations have been developed in Chapter 3 and are given in both tensor form and indicial form in Box 4.1. As can be seen, the dependent variables in the conservation equations are written in terms of material coordinates but are expressed in terms of what are classically Eulerian variables, such as the Cauchy stress and the rate-of-deformation.

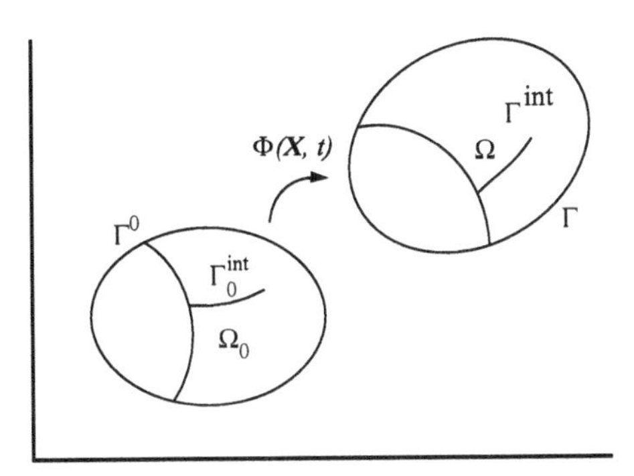

Figure 4.1 Deformed and undeformed body showing a set of discontinuities Γ^{int}

Box 4.1 Governing equations for updated Lagrangian formulation

Conservation of mass:

$$\rho(\mathbf{X},t)J(\mathbf{X},t)=\rho_0(\mathbf{X})J_0(\mathbf{X})=\rho_0(\mathbf{X}) \tag{B4.1.1}$$

Conservation of linear momentum:

$$\nabla\cdot\boldsymbol{\sigma}+\rho\mathbf{b}=\rho\dot{\mathbf{v}}\equiv\rho\frac{D\mathbf{v}}{Dt}\quad\text{or}\quad\frac{\partial\sigma_{ji}}{\partial x_j}+\rho b_i=\rho\dot{v}_i\equiv\rho\frac{Dv_i}{Dt} \tag{B4.1.2}$$

Conservation of angular momentum:

$$\boldsymbol{\sigma}=\boldsymbol{\sigma}^T\quad\text{or}\quad\sigma_{ij}=\sigma_{ji} \tag{B4.1.3}$$

Conservation of energy:

$$\rho\dot{w}^{\text{int}}=\mathbf{D}:\boldsymbol{\sigma}-\nabla\cdot\mathbf{q}+\rho s\quad\text{or}\quad\rho\dot{w}^{\text{int}}=D_{ij}\sigma_{ij}-\frac{\partial qi}{\partial x_i}+\rho s \tag{B4.1.4}$$

Constitutive equation:

$$\boldsymbol{\sigma}^{\nabla}=S_t^{\sigma D}(\mathbf{D},\boldsymbol{\sigma},\text{etc.}) \tag{B4.1.5}$$

Rate of deformation:

$$\mathbf{D}=\text{sym}(\nabla\mathbf{v}),\quad D_{ij}=\frac{1}{2}\left(\frac{\partial v_i}{\partial x_j}+\frac{\partial v_j}{\partial x_i}\right) \tag{B4.1.6}$$

Boundary conditions:

$$n_j\sigma_{ji}=\bar{t}_i\text{ on }\Gamma_{t_i}\qquad v_i=\bar{v}_i\text{ on }\Gamma_{v_i} \tag{B4.1.7}$$

$$\Gamma_{t_i}\cap\Gamma_{v_i}=0,\Gamma_{t_i}\cup\Gamma_{v_i}=\Gamma\quad i=1\text{ to }n_{\text{SD}} \tag{B4.1.8}$$

Initial conditions:

$$\mathbf{v}(\mathbf{X},0)=\mathbf{v}_0(\mathbf{X}),\quad \boldsymbol{\sigma}(\mathbf{X},0)=\boldsymbol{\sigma}_0(\mathbf{X}) \tag{B4.1.9}$$

or

$$\mathbf{v}(\mathbf{X},0)=\mathbf{v}_0(\mathbf{X}),\quad \mathbf{u}(\mathbf{X},0)=\mathbf{u}_0(\mathbf{X}) \tag{B4.1.10}$$

Interior continuity conditions (stationary):

$$\text{on } \Gamma_{\text{int}}: [\![\mathbf{n}\cdot\boldsymbol{\sigma}]\!]=0 \quad \text{or} \quad [\![n_i\sigma_{ij}]\!]\equiv n_i^A\sigma_{ij}^A+n_i^B\sigma_{ij}^B=0 \tag{B4.1.11}$$

We next give a count of the number of equations and unknowns. The conservation of mass and conservation of energy equations are scalar equations. The equation for the conservation of linear momentum, or momentum equation for short, is a tensor equation which consists of n_{SD} partial differential equations, where n_{SD} is the number of space dimensions. The constitutive equation relates the stress to the strain or strain-rate measure. Both the strain measure and the stress are symmetric tensors, so this provides n_σ equations where

$$n_\sigma \equiv n_{\text{SD}}\left(n_{\text{SD}}+1\right)/2 \tag{4.2.1}$$

In addition, we have the n_σ equations which express the rate-of-deformation $\mathbf{D}$ in terms of the velocities or displacements. Thus we have a total of $2n_\sigma + n_{\text{SD}} + 1$ equations and unknowns. For example, in two-dimensional problems without energy transfer, $n_{\text{SD}} = 2$, so we have nine partial differential equations in nine unknowns: the two momentum equations, the three constitutive equations, the three equations relating $\mathbf{D}$ to the velocity and the mass conservation equation. The unknowns are the three stress components (symmetry of the stress follows from angular momentum conservation), the three components of $\mathbf{D}$, the two velocity components, and the density ρ, for a total of nine unknowns. Additional unknown stresses (plane strain) and strains (plane stress) are evaluated using the plane strain and plane stress conditions, respectively. In three dimensions ($n_{\text{SD}} = 3$, $n_\sigma = 6$), we have 16 equations in 16 unknowns.

When a process is neither adiabatic nor isothermal, the energy equation must be appended to the system. This adds one equation and n_{SD} unknowns, the heat flux vector q_i. However, the heat flux vector can be determined from a single scalar, the temperature, so only one unknown is added; the heat flux is related to the temperature by a type of constitutive law which depends on the material. Usually a simple linear relation, Fourier's law, is used. This then completes the system of equations, although often a law is needed for conversion of some of the mechanical energy to thermal energy.

The dependent variables are the velocity $\mathbf{v}(\mathbf{X}, t)$, the Cauchy stress $\boldsymbol{\sigma}(\mathbf{X}, t)$, the rate-of-deformation $\mathbf{D}(\mathbf{X}, t)$, and the density $\rho(\mathbf{X}, t)$. As seen from the preceding the dependent variables are functions of the material (Lagrangian) coordinates. The expression of all functions in terms of material coordinates is intrinsic in any treatment by a Lagrangian mesh. In principle, the functions can be expressed in terms of the spatial coordinates at any time t by using the inverse of the map $\mathbf{x} = \boldsymbol{\phi}(\mathbf{X}, t)$. However, inverting this map is quite difficult. We shall see that it is only necessary to obtain derivatives with respect to the spatial coordinates. This is accomplished by implicit differentiation, so the map corresponding to the motion is never explicitly inverted.

In Lagrangian meshes, the mass conservation equation is used in its integrated form (B4.1.1) rather than as a partial differential equation. This eliminates the need to include the continuity equation, (3.5.20). Although the continuity equation can be used to obtain the density in a Lagrangian mesh, it is simpler and more accurate to use the integrated form (B4.1.1).

The constitutive equation (B4.1.5), when expressed in rate form in terms of a rate of Cauchy stress, requires a frame-invariant rate. For this purpose, any of the frame-invariant rates, such as the Jaumann rate or the Truesdell rate, can be used as described in Chapter 3. It is not necessary for the constitutive equation in the updated Lagrangian formulation to be expressed in terms of the Cauchy stress or its frame-invariant rate. It is also possible to use constitutive equations expressed in terms of the PK2 stress and then to convert the PK2 stress to a Cauchy stress by the transformations developed in Chapter 3, prior to computing the internal forces.

The rate-of-deformation is used as the measure of strain rate in (B4.1.6). However, other measures of strain or strain rate, for example the Green strain, can also be used in an updated Lagrangian formulation. As indicated in Chapter 3, simple hypoelastic laws in terms of the rate-of-deformation can cause difficulties in the simulation of cyclic loading because its integral is not path independent. However, for many simulations, such as the single application of a large load, the errors due to the path dependence of the integral of the rate-of-deformation are insignificant compared to other sources of error, such as inaccuracies and uncertainties in the material data and material model. The appropriate selection of stress and strain measures depends on the constitutive equation, that is, whether the material response is reversible or not, time dependence, and the load history under consideration.

The boundary conditions are summarized in (B4.1.7). In two-dimensional problems, each component of the traction or velocity must be prescribed on the entire boundary; however, the same component of the traction and velocity cannot be prescribed at the same point of the boundary, as indicated by (B4.1.8). Traction and velocity components can also be specified in local coordinate systems which differ from the global system.

A velocity boundary condition is equivalent to a displacement boundary condition: if a displacement is specified as a function of time, then the prescribed velocity can be obtained by time differentiation; if a velocity is specified, then the displacement can be obtained by time integration.

Initial conditions can be applied either to the velocities and the stresses or to the displacements and velocities. The first set of initial conditions is more suitable for most engineering problems, since it is usually difficult to determine the initial displacement of a body. On the other hand, initial stresses, often known as residual stresses, can sometimes be measured or estimated by equilibrium solutions. For example, it is almost impossible to determine the displacements of a steel part after it has been formed from an ingot. On the other hand, good estimates of the residual stress field in the engineering component can often be made. Similarly, in a buried tunnel, the notion of initial displacements of the soil or rock enclosing the tunnel is quite meaningless, whereas the initial stress field can be estimated by equilibrium analysis. Therefore, initial conditions in terms of the stresses are more useful.

We have also included the interior continuity conditions on the stresses, (B4.1.11). In this equation, superscripts A and B refer to the stresses and normal on two sides of the discontinuity.

These continuity conditions must be met by the stresses wherever stationary discontinuities in certain stress components are possible, such as at material interfaces. They must hold for bodies in equilibrium and in transient problems.

4.3 Weak Form: Principle of Virtual Power

In this section, the principle of virtual power is developed for the updated Lagrangian formulation. The principle of virtual power is the weak form of the momentum equation, the traction boundary conditions and the interior traction continuity conditions. These three are collectively called *generalized momentum balance*. The relationship of the principle of virtual power to the momentum equations will be described in two parts:

1. The principle of virtual power (weak form) will be developed from the generalized momentum balance (strong form), that is, strong form to weak form.
2. The principle of virtual power (weak form) will be shown to imply the generalized momentum balance (strong form), that is, weak form to strong form.

We first define the spaces for the test functions and trial functions. We will consider the minimum smoothness required for the functions to be defined in the sense of distributions, that is, we allow Dirac delta functions to be derivatives of functions. Thus, the derivatives will not be defined according to classical definitions of derivatives: this was discussed at the end of Section 2.3.3.

The space of test functions is defined by:

$$\delta v_i(\mathbf{X}) \in U_0, \quad U_0 = \left\{ \delta v_i \mid \delta v_i \in C^0(\mathbf{X}),\ \delta v_i = 0 \text{ on } \Gamma_{v_i} \right\} \tag{4.3.1}$$

This selection of the space for the test functions $\delta\mathbf{v}$ is dictated by foresight from what will ensue in the development of the weak form; with this construction of the test function, the integral over the kinematic boundary vanishes and the only boundary integral in the weak form is over the traction boundary. The test functions $\delta\mathbf{v}$ are sometimes called the virtual velocities.

The velocity trial functions live in the space given by

$$v_i(\mathbf{X}, t) \in U, \quad U = \left\{ v_i \mid v_i \in C^0(\mathbf{X}),\ v_i = \bar{v}_i \ 0 \text{ on } \Gamma_{v_i} \right\} \tag{4.3.2}$$

The space of velocities in U is often called kinematically admissible velocities or compatible velocities; they satisfy the continuity conditions required for compatibility and the velocity boundary conditions. Note that the space of test functions is identical to the space of trial functions except that the test velocities vanish wherever the trial velocities are prescribed. We have selected a specific class of test and trial spaces that are applicable to finite elements; the weak form holds also for more general spaces, which is the space of functions with square integrable derivatives, called a Hilbert space.

Since the displacement $u_i(\mathbf{X}, t)$ is the time integral of the velocity, the displacement field can also be considered to be the trial function. Whether the displacements or the velocities are considered to be the trial functions is a matter of taste.

4.3.1 Strong Form to Weak Form

As we have already noted, the strong form, or generalized momentum balance, consists of the momentum equation, the traction boundary conditions and the traction continuity conditions, which are respectively:

$$\frac{\partial \sigma_{ji}}{\partial x_j} + \rho b_i = \rho \dot{v}_i \text{ in } \Omega \tag{4.3.3}$$

$$n_j \sigma_{ji} = \bar{t}_i \text{ on } \Gamma_{t_i} \tag{4.3.4}$$

$$[\![n_j \sigma_{ji}]\!] = 0 \text{ on } \Gamma_{\text{int}} \tag{4.3.5}$$

where Γ_{int} is the union of all surfaces (lines in two dimensions) on which the stresses are discontinuous in the body. These are usually material interfaces.

Since the velocities are $C^0(\mathbf{X})$, the displacements are similarly $C^0(\mathbf{X})$; the rate-of-deformation and the rate of Green strain will then be $C^{-1}(\mathbf{X})$ since they are related to spatial derivatives of the velocity. The stress $\boldsymbol{\sigma}$ is a function of the velocities via the constitutive equation; it can also be expressed as a function of the Green strain tensor. It is assumed that the constitutive equation leads to a *stress that is a well-behaved function of the Green strain tensor*, so that the stresses will also be $C^{-1}(\mathbf{X})$.

The first step in the development of the weak form, as in the one-dimensional case in Chapter 2, consists of taking the product of a test function δv_i with the momentum equation and integrating over the *current configuration*:

$$\int_\Omega \delta v_i \left(\frac{\partial \sigma_{ji}}{\partial x_j} + \rho b_i - \rho \dot{v}_i \right) d\Omega = 0 \tag{4.3.6}$$

In the above integral, the independent variables are the Eulerian coordinates. However, the dependent variables in the integrand need never be expressed as explicit functions of the Eulerian coordinates in the implementation.

The first term in (4.3.6) is next expanded by the product rule, which gives

$$\int_\Omega \delta v_i \frac{\partial \sigma_{ji}}{\partial x_j} d\Omega = \int_\Omega \left[\frac{\partial}{\partial x_j} (\delta v_i \sigma_{ji}) - \frac{\partial (\delta v_i)}{\partial x_j} \sigma_{ji} \right] d\Omega \tag{4.3.7}$$

Since the velocities are C^0 and the stresses are C^{-1}, the term $\delta v_i \sigma_{ji}$ on the RHS of the above is C^{-1}. We assume that the discontinuities occur over a finite set of surfaces Γ_{int}, so Gauss's theorem, (3.5.5), gives

$$\int_\Omega \frac{\partial}{\partial x_j} (\delta v_i \sigma_{ji}) d\Omega = \int_{\Gamma_{\text{int}}} \delta v_i (n_j \sigma_{ji}) d\Gamma + \int_\Gamma \delta v_i n_j \sigma_{ji} d\Gamma \tag{4.3.8}$$

From traction continuity (4.3.5), the first integral on the RHS vanishes. For the second integral on the RHS we use the traction boundary conditions (4.3.4). Since the test function vanishes on the complement of the traction boundaries, (4.3.8) becomes

$$\int_\Omega \frac{\partial}{\partial x_j} (\delta v_i \sigma_{ji}) d\Omega = \sum_{i=1}^{n_{SD}} \int_{\Gamma_{t_i}} \delta v_i \bar{t}_i d\Gamma \tag{4.3.9}$$

The summation sign is included to avoid any confusion because the index i appears three times on the RHS of (4.3.9).

When (4.3.9) is substituted into (4.3.7) we obtain

$$\int_{\Omega} \delta v_i \frac{\partial \sigma_{ji}}{\partial x_j} d\Omega = \sum_{i=1}^{n_{SD}} \int_{\Gamma_{t_i}} \delta v_i \bar{t}_i d\Gamma - \int_{\Omega} \frac{\partial(\delta v_i)}{\partial x_j} \sigma_{ji} d\Omega \tag{4.3.10}$$

The process of obtaining (4.3.10) is called integration by parts. If (4.3.10) is then substituted into (4.3.6), we obtain

$$\int_{\Omega} \frac{\partial(\delta v_i)}{\partial x_j} \sigma_{ji} d\Omega - \int_{\Omega} \delta v_i \rho b_i d\Omega - \sum_{i=1}^{n_{SD}} \int_{\Gamma_{t_i}} \delta v_i \bar{t}_i d\Gamma + \int_{\Omega} \delta v_i \rho \dot{v}_i d\Omega = 0 \tag{4.3.11}$$

This is the weak form for the momentum equation, the traction boundary conditions and the interior continuity conditions. It is known as the *principle of virtual power* (Malvern, 1969), for each of the terms in the weak form is a virtual power; see Section 2.7.

4.3.2 *Weak Form to Strong Form*

It will now be shown that the weak form (4.3.11) implies the strong form or generalized momentum balance: the momentum equation, the traction boundary conditions and the interior continuity conditions, (4.3.3–5). To obtain the strong form, the derivative of the test function must be eliminated from (4.3.11). This is accomplished by using the derivative product rule, which gives

$$\int_{\Omega} \frac{\partial(\delta v_i)}{\partial x_j} \sigma_{ji} d\Omega = \int_{\Omega} \frac{\partial(\delta v_i \sigma_{ji})}{\partial x_j} d\Omega - \int_{\Omega} \delta v_i \frac{\partial \sigma_{ji}}{\partial x_j} d\Omega \tag{4.3.12}$$

We now apply Gauss's theorem (see Section 3.5.2) to the first term on the RHS of (4.3.12):

$$\begin{aligned} \int_{\Omega} \frac{\partial(\delta v_i \sigma_{ji})}{\partial x_j} d\Omega &= \int_{\Gamma} \delta v_i n_j \sigma_{ji} d\Gamma + \int_{\Gamma_{\text{int}}} \delta v_i [\![n_j \sigma_{ji}]\!] d\Gamma \\ &= \sum_{i=1}^{n_{SD}} \int_{\Gamma_{t_i}} \delta v_i n_j \sigma_{ji} d\Gamma + \int_{\Gamma_{\text{int}}} \delta v_i [\![n_j \sigma_{ji}]\!] d\Gamma \end{aligned} \tag{4.3.13}$$

where the second equality follows because $\delta v_i = 0$ on Γ_{v_i} (see (4.3.1) and (B4.1.8)).

Substituting (4.3.13) into (4.3.12) and in turn into (4.3.11), we obtain

$$\int_{\Omega} \delta v_i \left(\frac{\partial \sigma_{ji}}{\partial x_j} + \rho b_i - \rho \dot{v}_i \right) d\Omega - \sum_{i=1}^{n_{SD}} \int_{\Gamma_{t_i}} \delta v_i \left(n_j \sigma_{ji} - \bar{t}_i \right) d\Gamma - \int_{\Gamma_{\text{int}}} \delta v_i [\![n_j \sigma_{ji}]\!] d\Gamma = 0 \tag{4.3.14}$$

We will now prove that the coefficients of the test functions in the above integrals must vanish. For this purpose, we prove the following theorem:

$$\text{if} \quad \alpha_i(\mathbf{x}), \beta_i(\mathbf{x}), \gamma_i(\mathbf{x}) \in C^{-1} \quad \text{and} \quad \delta v_i(\mathbf{x}) \in u_0$$

$$\text{and} \int_\Omega \delta v_i \alpha_i \, d\Omega + \sum_{i=1}^{n_{SD}} \int_{\Gamma_{t_i}} \delta v_i \beta_i \, d\Gamma + \int_{\Gamma_{int}} \delta v_i \gamma_i \, d\Gamma = 0 \quad \forall \delta v_i(\mathbf{x}) \tag{4.3.15}$$

$$\text{than } \alpha_i(\mathbf{x}) = 0 \text{ in } \Omega, \beta_i(\mathbf{x}) = 0 \text{ on } \Gamma_{t_i}, \gamma_i(\mathbf{x}) = 0 \text{ on } \Gamma_{int}$$

In functional analysis, the statement in (4.3.15) is called the *density theorem* (Oden and Reddy, 1976: 19). It is also called the *fundamental theorem of variational calculus*; sometimes we call it the *function scalar product theorem*. We follow Hughes (1987: 80) in proving (4.3.15). As a first step we show that $\alpha_i(\mathbf{x}) = 0$ in Ω. For this purpose, we assume that

$$\delta v_i(\mathbf{x}) = \alpha_i(\mathbf{x}) f(\mathbf{x}) \tag{4.3.16}$$

where

1. $f(\mathbf{x}) > 0$ on Ω but $f(\mathbf{x}) = 0$ on Γ_{int} and $f(\mathbf{x}) = 0$ on Γ_{t_i}
2. $f(\mathbf{x})$ is C^{-1}.

Substituting the previous expression for δv_i into (4.3.15) gives

$$\int_\Omega \alpha_i(\mathbf{x}) \alpha_i(\mathbf{x}) f(\mathbf{x}) d\Omega = 0 \tag{4.3.17}$$

The integrals over the boundary and interior surfaces of discontinuity vanish because the arbitrary function $f(\mathbf{x})$ has been chosen to vanish on these surfaces. Since $f(\mathbf{x}) > 0$, and the functions $f(\mathbf{x})$ and $\alpha_i(\mathbf{x})$ are sufficiently smooth, (4.3.17) implies $\alpha_i(\mathbf{x}) = 0$ in Ω for $i = 1$ to n_{SD}.

To show that the $\gamma_i(\mathbf{x}) = 0$, let

$$\delta v_i(\mathbf{x}) = \gamma_i(\mathbf{x}) f(\mathbf{x}) \tag{4.3.18}$$

where

1. $f(\mathbf{x}) > 0$ on Γ_{int}; $f(\mathbf{x}) = 0$ on Γ_{t_i}
2. $f(\mathbf{x})$ is C^{-1}.

Substituting (4.3.18) into (4.3.15) gives

$$\int_{\Gamma_{int}} \gamma_i(\mathbf{x}) \gamma_i(\mathbf{x}) f(\mathbf{x}) d\Gamma = 0 \tag{4.3.19}$$

which implies $\gamma_i(\mathbf{x}) = 0$ on Γ_{int} (since $f(\mathbf{x}) > 0$).

The final step in the proof, showing that $\beta_i(\mathbf{x}) = 0$, is accomplished by using a function $f(\mathbf{x}) > 0$ on Γ_{t_i}. The steps are exactly as before. Thus $\alpha_i(\mathbf{x})$, $\beta_i(\mathbf{x})$, and $\gamma_i(\mathbf{x})$ must vanish on the relevant domain or surface. Thus (4.3.11) implies the strong form: the momentum equation, the traction boundary conditions, and the interior continuity conditions, (4.3.3–5).

Let us now recapitulate what has been accomplished so far in this section. We first developed a weak form, called the principle of virtual power, from the strong form. The strong form consists of the momentum equation, the traction boundary conditions and interior continuity conditions. The weak form was obtained by multiplying the momentum equation by a test function and integrating over the *current configuration*. A key step in obtaining the weak form is the elimination of the derivatives of the stresses, (4.3.7–8). This step is crucial since, as a result, the stresses can be C^{-1} functions. As a consequence, if the constitutive equation is smooth, the velocities need only be C^0.

Equation (4.3.6) could also be used as the weak form. However, since the derivatives of the stresses would appear in this alternative weak form, the displacements and velocities would have to be C^1 functions (see Chapter 2); C^1 functions are difficult to construct in more than one dimension. Furthermore, the trial functions would then have to be constructed so as to satisfy the traction boundary conditions, which would be difficult. The removal of the derivative of the stresses through integration by parts also leads to certain symmetries in the linearized equations, as will be seen in Chapter 6. Thus the integration by parts is a key step in the development of the weak form.

Next, we started with the weak form and showed that it implies the strong form. This, combined with the development of the weak form from the strong form, shows that the weak and strong forms are equivalent. Therefore, if the space of test functions is infinite-dimensional, a solution to the weak form is a solution of the strong form. However, the test functions used in computational procedures must be finite-dimensional. Therefore, satisfying the weak form in a computation leads only to an approximate solution of the strong form. In linear finite element analysis, it has been shown that the solution of the weak form is the best solution in the sense that it minimizes the error in energy (Strang and Fix, 1973). In nonlinear problems, such optimality results are not available in general.

4.3.3 *Physical Names of Virtual Power Terms*

We will next ascribe a physical name to each of the terms in the virtual power equation. This will be useful in systematizing the development of finite element equations. The nodal forces in the finite element discretization will be identified according to the same physical names.

To identify the first integrand in (4.3.11), note that it can be written as

$$\frac{\partial(\delta v_i)}{\partial x_j}\sigma_{ij} = \delta L_{ij}\sigma_{ij} = (\delta D_{ij} + \delta W_{ij})\sigma_{ij} = \delta D_{ij}\sigma_{ij} = \delta\mathbf{D}:\boldsymbol{\sigma} \tag{4.3.20}$$

Here we have used the decomposition of the velocity gradient into its symmetric and skew-symmetric parts and the fact that $\delta W_{ij}\sigma_{ij} = 0$ since δW_{ij} is skew-symmetric while σ_{ij} is symmetric. Comparison with (B4.1.4) then indicates that we can interpret $\delta D_{ij}\sigma_{ij}$ as the rate of virtual internal work, or the *virtual internal power*, per unit volume. Observe that $\dot{w}^{\text{int}}$ in (B4.1.4) is power per unit mass, so $\rho\dot{w}^{\text{int}} = \mathbf{D}:\boldsymbol{\sigma}$ is the power per unit volume. The total virtual internal power δP^{int} is defined by the integral of $\delta D_{ij}\sigma_{ij}$ over the domain

$$\delta P^{\text{int}} = \int_\Omega \delta D_{ij}\sigma_{ij}d\Omega = \int_\Omega \frac{\partial(\delta v_i)}{\partial x_j}\sigma_{ij}\,d\Omega \equiv \int_\Omega \delta L_{ij}\sigma_{ij}\,d\Omega = \int_\Omega \delta\mathbf{D}:\boldsymbol{\sigma}\,d\Omega \tag{4.3.21}$$

where the third and fourth terms have been added to remind us that they are equivalent to the second term because of the symmetry of the Cauchy stress tensor.

The second and third terms in (4.3.11) are the *virtual external power*.

$$\delta P^{\text{ext}} = \int_{\Omega} \delta v_i \rho b_i d\Omega + \sum_{j=1}^{n_{SD}} \int_{\Gamma_{t_j}} \delta v_j \bar{t}_i d\Gamma = \int_{\Omega} \sigma \mathbf{v} \cdot \rho \mathbf{b} d\Omega + \sum_{j=1}^{n_{SD}} \int_{\Gamma_{t_j}} \delta v_j \mathbf{e}_j \cdot \bar{\mathbf{t}}_i d\Gamma \tag{4.3.22}$$

This name is selected because the virtual external power arises from the external body forces $\mathbf{b}(\mathbf{X}, t)$ and prescribed tractions $\bar{\mathbf{t}}(\mathbf{X}, t)$.

The last term in (4.3.11) is the *virtual inertial (or kinetic) power*:

$$\delta P^{\text{kin}} = \int_{\Omega} \delta v_i \rho \dot{v}_i \, d\Omega \tag{4.3.23}$$

which is the power corresponding to the inertial force. The inertial force can be considered a body force in the d'Alembert sense.

Inserting (4.3.21–23) into (4.3.11), we can write the principle of virtual power as

$$\delta P = \delta P^{\text{int}} - \delta P^{\text{ext}} + \delta P^{\text{kin}} = 0 \quad \forall \delta v_i \in U_0 \tag{4.3.24}$$

which is the weak form for the momentum equation. The physical meanings help in remembering the weak form and in the derivation of the finite element equations. The weak form is summarized in Box 4.2.

Box 4.2 Weak form for updated Lagrangian formulation: principle of virtual power

If σ_{ij} is a smooth function of the displacements and velocities and $v_i \in u$, then if

$$\delta P = \delta P^{\text{int}} - \delta P^{\text{ext}} + \delta P^{\text{kin}} = 0 \quad \forall \delta v_i \in u_0 \tag{B4.2.1}$$

then

$$\frac{\partial \sigma_{ji}}{\partial x_j} + \rho b_i = \rho \dot{v}_i \text{ in } \Omega \tag{B4.2.2}$$

$$n_j \sigma_{ji} = \bar{t}_i \text{ on } \Gamma_{t_i} \tag{B4.2.3}$$

$$[\![n_j \sigma_{ji}]\!] = 0 \text{ on } \Gamma_{t_i} \tag{B4.2.4}$$

where

$$\delta P^{\text{int}} = \int_{\Omega} \partial \mathbf{D} : \boldsymbol{\sigma} d\Omega = \int_{\Omega} \delta D_{ij} \sigma_{ij} \, d\Omega = \int_{\Omega} \frac{\partial(\delta v_i)}{\partial x_j} \sigma_{ij} \, d\Omega \tag{B4.2.5}$$

$$\delta P^{\text{ext}} = \int_{\Omega} \delta \mathbf{v} \cdot \rho \mathbf{b} d\Omega + \sum_{j=1}^{n_{SD}} \int_{\Gamma_{t_j}} (\delta \mathbf{v} \cdot \mathbf{e}_j) \bar{\mathbf{t}} \cdot \mathbf{e}_j \, d\Gamma = \int_{\Omega} \delta v_i \rho b_i \, d\Omega + \sum_{j=1}^{n_{SD}} \int_{\Gamma_{t_j}} \delta v_j \bar{t}_j d\Gamma \tag{B4.2.6}$$

$$\delta P^{\text{kin}} = \int_{\Omega} \delta \mathbf{v} \cdot \rho \dot{\mathbf{v}} d\Omega = \int_{\Omega} \delta v_i \rho \dot{v}_i \, d\Omega \tag{B4.2.7}$$

4.4 Updated Lagrangian Finite Element Discretization

4.4.1 Finite Element Approximation

In this section, the finite element equations for the updated Lagrangian formulation are developed by means of the principle of virtual power. For this purpose the current domain Ω is subdivided into elements Ω_e so that the union of the elements comprises the total domain, $\Omega = \bigcup_e \Omega_e$. The nodal coordinates in the current configuration are denoted by x_{iI}, $I = 1$ to n_N. Lower case subscripts are used for components, upper case subscripts for nodal values. In two dimensions, $x_{iI}=[x_I, y_I]$; three dimensions $x_{iI} = [x_I, y_I, z_I]$. The nodal coordinates in the undeformed configuration are X_{iI}.

In the finite element method, the motion $\mathbf{x}(\mathbf{X}, t)$ is approximated by

$$x_i(\mathbf{X},t) = N_I(\mathbf{X})x_{iI}(t) \quad \text{or} \quad \mathbf{x}(\mathbf{X},t) = N_I(\mathbf{X})\mathbf{x}_I(t) \tag{4.4.1}$$

where $N_I(\mathbf{X})$ are the interpolation (shape) functions and $\mathbf{x}_I$ is the position vector of node I. Summation over repeated indices is implied; in the case of lower case indices, the sum is over the number of space dimensions, while for upper case indices the sum is over the number of nodes. The nodes in the sum depend on the entity considered: when the total domain is considered, the sum is over all nodes in the domain, whereas when an element is considered, the sum is over the nodes of the element.

Writing (4.4.1) at a node with initial position $\mathbf{X}_J$ we have

$$\mathbf{x}(\mathbf{X}_J,t) = \mathbf{x}_I(t)N_I(\mathbf{X}_J) = \mathbf{x}_I(t)\delta_{IJ} = \mathbf{x}_J(t) \tag{4.4.2}$$

where we have used the interpolation property $N_I(\mathbf{X}_J) = \delta_{IJ}$ of the shape functions in the third term. Interpreting this equation, we see that node J always corresponds to the same material point $\mathbf{X}_J$: in a Lagrangian mesh, nodes remain coincident with material points.

We define the nodal displacements by using (3.2.7) at the nodes:

$$u_{iI}(t) = x_{iI}(t) - X_{iI} \quad \text{or} \quad \mathbf{u}_I(t) = \mathbf{x}_I(t) - \mathbf{X}_I \tag{4.4.3}$$

The displacement field is

$$u_i(\mathbf{X}_J, t) = x_i(\mathbf{X}, t) - X_i = u_{iI}(t)N_I(\mathbf{X}) \quad \text{or} \quad \mathbf{u}(\mathbf{X}, t) = \mathbf{u}_I(t)N_I(\mathbf{X}) \tag{4.4.4}$$

which follows from (4.4.1) and (4.4.3).

The velocities are obtained by taking the material time derivative of the displacements, giving

$$v_i(\mathbf{X},t) = \frac{\partial u_i(\mathbf{X},t)}{\partial t} = \dot{u}_{iI}(t)N_I(\mathbf{X}) = v_{iI}(t)N_I(\mathbf{X}) \quad \text{or} \quad \mathbf{v}(\mathbf{X},t) = \dot{\mathbf{u}}_I(t)N_I(\mathbf{X}) \tag{4.4.5}$$

where we have indicated that the velocity is a material time derivative of the displacement, that is, the partial derivative with respect to time with the material coordinate fixed. Note the velocities are given by the same shape function since the shape functions are constant in time.

The superposed dot on the nodal displacements is an ordinary derivative, since the nodal displacements are only functions of time.

The accelerations are similarly given by the material time derivative of the velocities:

$$\ddot{u}_i(\mathbf{X},t) = \ddot{u}_{iI}(t)N_I(\mathbf{X}) \quad \text{or} \quad \ddot{\mathbf{u}}(\mathbf{X},t) = \ddot{\mathbf{u}}_I(t)N_I(\mathbf{X}) = \dot{\mathbf{v}}_I(t)N_I(\mathbf{X}) \tag{4.4.6}$$

It is emphasized that the shape functions are expressed in terms of the material coordinates in the updated Lagrangian formulation even though we will use the weak form in the current configuration. As pointed out in Section 2.8, it is crucial to express the shape functions in terms of material coordinates for a Lagrangian mesh, because we want the time dependence in the finite element approximation of the motion to reside entirely in the nodal variables.

The velocity gradient is obtained by substituting (4.4.5) into (3.3.7), which yields

$$L_{ij} = v_{i,j} = v_{iI}\frac{\partial N_I}{\partial x_j} = v_{iI}N_{I,j} \quad \text{or} \quad \mathbf{L} = \mathbf{v}_I \nabla N_I = \mathbf{v}_I N_{I,\mathbf{x}} \tag{4.4.7}$$

and the rate-of-deformation is given by

$$D_{ij} = \frac{1}{2}(L_{ij} + L_{ji}) = \frac{1}{2}(v_{iI}N_{I,j} + v_{jI}N_{I,i}) \tag{4.4.8}$$

In the construction of the finite element approximation to the motion, (4.4.1), we have ignored the velocity boundary conditions, that is, the velocities given by (4.4.5) are not in the space defined by (4.3.2). We will first develop the equations for an unconstrained body with no velocity boundary conditions, and then modify the discrete equations to account for the velocity boundary conditions.

In (4.4.1), all components of the motion are approximated by the same shape functions. This construction of the motion facilitates the representation of rigid body rotation, which is an essential requirement for convergence. This is discussed further in Chapter 8.

The test function, or variation, is not a function of time, so we approximate the test function as

$$\delta v_i(\mathbf{X}) = \delta v_{iI}N_I(\mathbf{X}) \quad \text{or} \quad \delta\mathbf{v}(\mathbf{X}) = \delta\mathbf{v}_I N_I(\mathbf{X}) \tag{4.4.9}$$

where δv_{iI} are the virtual nodal velocities.

As a first step in the construction of the discrete finite element equations, the test function is substituted into the principle of virtual power, giving

$$\delta v_{iI}\int_\Omega \frac{\partial N_I}{\partial x_j}\sigma_{ji}d\Omega - \delta v_{iI}\int_\Omega N_I\rho b_i d\Omega - \sum_{i=1}^{n_{SD}}\delta v_{iI}\int_{\Gamma_{t_i}} N_I\bar{t}_i d\Gamma + \delta v_{iI}\int_\Omega N_I\rho\dot{v}_i d\Omega = 0 \tag{4.4.10}$$

The stresses in (4.4.10) are functions of the trial velocities and trial displacements. From the definition of the test space, (4.3.1), the virtual velocities must vanish wherever the velocities are prescribed, that is, $\delta v_i = 0$ on Γ_{v_i}, and therefore only the virtual nodal velocities for nodes not on Γ_{v_i} are arbitrary. Using the arbitrariness of the virtual nodal velocities everywhere except on Γ_{v_i}, it then follows that the weak form of the momentum equation is

$$\int_{\Omega}\frac{\partial N_I}{\partial x_j}\sigma_{ji}d\Omega-\int_{\Omega}N_I\rho b_i d\Omega-\sum_{j=1}^{n_{SD}}\int_{\Gamma_{t_i}}N_I\bar{t}_i d\Gamma+\int_{\Omega}N_I\rho\dot{v}_i d\Omega=0\qquad\forall(I,\ i)\notin\Gamma_{v_i}\tag{4.4.11}$$

Where, as shown, the degrees of freedom which are prescribed are excluded from the above. The above form is difficult to remember. For a better physical interpretation, it is worthwhile to ascribe physical names to each of the terms in this equation.

4.4.2 *Internal and External Nodal Forces*

We define the nodal forces corresponding to each term in the virtual power equation. This helps in remembering the equation and also provides a systematic procedure which is found in most finite element software. The internal nodal forces are defined by

$$\delta P^{\text{int}}=\delta v_{iI}f_{iI}^{\text{int}}=\int_{\Omega}\frac{\partial(\delta v_i)}{\partial x_j}\sigma_{ji}d\Omega=\delta v_{iI}\int_{\Omega}\frac{\partial N_I}{\partial x_j}\sigma_{ji}d\Omega\tag{4.4.12}$$

where the third term is the definition of internal virtual power as given in (B4.2.5), and (4.4.7) has been used in the last term. From (4.4.12) it can be seen that the internal nodal forces are given by

$$f_{iI}^{\text{int}}=\int_{\Omega}\frac{\partial N_I}{\partial x_j}\sigma_{ji}d\Omega\tag{4.4.13}$$

These nodal forces are called internal because they represent the *stresses in the body*. These expressions apply both to a complete mesh and to any element or group of elements, as has been described in Chapter 2. Note that this expression involves derivatives of the shape functions with respect to spatial coordinates and integration over the current configuration. Equation (4.4.13) is a key equation in nonlinear finite element methods for updated Lagrangian meshes; it applies also to Eulerian and ALE meshes.

The external nodal forces are defined similarly in terms of the virtual external power:

$$\begin{aligned}\delta P^{\text{ext}}&=\delta v_{iI}f_{iI}^{\text{ext}}=\int_{\Omega}\delta v_i\rho b_i d\Omega+\sum_{i=1}^{n_{SD}}\int_{\Gamma_{t_i}}\delta v_i\bar{t}_i d\Gamma\\&=\delta v_{iI}\int_{\Omega}N_I\rho b_i d\Omega+\sum_{i=1}^{n_{SD}}\delta v_{iI}\int_{\Gamma_{t_i}}N_I\bar{t}_i d\Gamma\end{aligned}\tag{4.4.14}$$

so the external nodal forces are given by

$$f_{iI}^{\text{ext}}=\int_{\Omega}N_I\rho b_i d\Omega+\int_{\Gamma_{t_i}}N_I\bar{t}_i d\Gamma\quad\text{or}\quad\mathbf{f}_I^{\text{ext}}=\int_{\Omega}N_I\rho\mathbf{b}d\Omega+\int_{\Gamma_{t_i}}N_I\mathbf{e}_i\cdot\bar{\mathbf{t}}d\Gamma\tag{4.4.15}$$

4.4.3 *Mass Matrix and Inertial Forces*

The inertial (or kinetic) nodal forces are defined by

$$\delta P^{\mathrm{kin}} = \delta v_{iI} f_{iI}^{\mathrm{kin}} = \int_{\Omega} \delta v_i \rho \dot{v}_i d\Omega = \delta v_{iI} \int_{\Omega} N_I \rho \dot{v}_i d\Omega \tag{4.4.16}$$

so

$$f_{iI}^{\mathrm{kin}} = \int_{\Omega} \rho N_I \dot{v}_i d\Omega \quad \text{or} \quad \mathbf{f}_I^{\mathrm{kin}} = \int_{\Omega} \rho N_I \dot{\mathbf{v}} d\Omega \tag{4.4.17}$$

Using the expression (4.4.6) for the accelerations in the previous gives

$$f_{iI}^{\mathrm{kin}} = \int_{\Omega} \rho N_I N_J d\Omega \dot{v}_{iJ} \tag{4.4.18}$$

It is convenient to define these nodal forces as a product of a mass matrix and the nodal accelerations. Defining the mass matrix by

$$M_{ijIJ} = \delta_{ij} \int_{\Omega} \rho N_I N_J d\Omega \tag{4.4.19}$$

it follows from (4.4.16) and (4.4.17) that the inertial forces are given by

$$f_{iI}^{\mathrm{kin}} = M_{ijIJ} \dot{v}_{jJ} \quad \text{or} \quad \mathbf{f}_I^{\mathrm{kin}} = \mathbf{M}_{IJ} \dot{\mathbf{v}}_J \tag{4.4.20}$$

4.4.4 *Discrete Equations*

With the definitions of the internal, external and inertial nodal forces, (4.4.13), (4.4.15) and (4.4.20), we can concisely write the discrete approximation to the weak form (4.4.11) as

$$\delta v_{iI} (f_{iI}^{\mathrm{int}} - f_{iI}^{\mathrm{ext}} + M_{ijIJ} \dot{v}_{jJ}) = 0 \quad \forall \delta v_{iI} \notin \Gamma_{v_i} \tag{4.4.21}$$

We can also write (4.4.21) as

$$\delta \mathbf{v}^T (\mathbf{f}^{\mathrm{int}} - \mathbf{f}^{\mathrm{ext}} + \mathbf{Ma}) = 0 \tag{4.4.22}$$

where **v, a** and **f** are column matrices of the unconstrained virtual velocities, accelerations and nodal forces, and **M** is the mass matrix for the unconstrained degrees of freedom. Invoking the arbitrariness of the unconstrained virtual nodal velocities in (4.4.21) and (4.4.22), respectively, gives

$$M_{ijIJ} \dot{v}_{jJ} + f_{iI}^{\mathrm{int}} = f_{iI}^{\mathrm{ext}} \qquad \forall (I,i) \notin \Gamma_{v_i} \tag{4.4.23}$$

or

$$\mathbf{Ma} + \mathbf{f}^{\text{int}} = \mathbf{f}^{\text{ext}} \tag{4.4.24}$$

These are the *discrete momentum equations* or the *equations of motion*; they are also called the *semidiscrete momentum equations* since they have not been discretized in time. The implicit sums are over all components and all nodes of the mesh; any prescribed velocity component that appears in the above is not an unknown. Equation (4.4.24) can also be written in the form of Newton's second law

$$\mathbf{f} = \mathbf{Ma} \quad \text{where} \quad \mathbf{f} = \mathbf{f}^{\text{ext}} - \mathbf{f}^{\text{int}} \tag{4.4.25}$$

The semidiscrete momentum equations are a system of n_{DOF} ordinary differential equations in the nodal velocities, where n_{DOF} is the number of nodal velocity components which are unconstrained; n_{DOF} is often called the number of degrees of freedom. To complete the system of equations, we append the constitutive equations at the element quadrature points and the expression for the rate-of-deformation in terms of the nodal velocities. Let the n_Q quadrature points in the mesh be denoted by

$$\mathbf{x}_Q(t) = N_I(\mathbf{X}_Q)\mathbf{x}_I(t) \tag{4.4.26}$$

Note that the quadrature points are coincident with material points. Let n_σ be the number of independent components of the stress tensor: in a two-dimensional plane stress problem, $n_\sigma = 3$, since the stress tensor σ is symmetric; in three-dimensional problems, $n_\sigma = 6$.

The semidiscrete equations for the finite element approximation then consist of the following ordinary differential equations in time:

$$M_{ijIJ}\dot{v}_{jJ} + f_{iI}^{\text{int}} = f_{iI}^{\text{ext}} \quad \text{for} \quad (I,i) \notin \Gamma_{v_i} \tag{4.4.27}$$

$$\sigma_{ij}^{\nabla}(\mathbf{X}_Q) = \sigma_{ij}(D_{kl}(\mathbf{X}_Q), \text{ etc.}) \quad \forall \mathbf{X}_Q \tag{4.4.28}$$

where

$$D_{ij}(\mathbf{X}_Q) = \frac{1}{2}(L_{ij} + L_{ji}) \quad \text{and} \quad L_{ij} = N_{I,j}(\mathbf{X}_Q)v_{iI} \tag{4.4.29}$$

This is a standard initial value problem, consisting of first-order ordinary differential equations in the velocities $v_{iI}(t)$ and the stresses $\sigma_{ij}(\mathbf{X}_q, t)$. If we substitute (4.4.29) into (4.4.28) to eliminate the rate-of-deformation from the equations, the total number of unknowns is $n_{\text{DOF}} + n_\sigma n_Q$. This system of ordinary differential equations can be integrated in time by any of the methods for integrating ordinary differential equations, such as Runge–Kutta methods or the central difference method; this is discussed in Chapter 6.

The nodal velocities on prescribed velocity boundaries, v_{iI}, $(I, i) \in \Gamma_{v_i}$, are obtained from velocity boundary conditions (B4.1.7). The initial conditions (B4.1.9) are applied at the nodes and quadrature points

$$v_{iI}(0) = v_{iI}^0 \tag{4.4.30}$$

$$\sigma_{ij}(\mathbf{X}_Q, 0) = \sigma_{ij}^0(\mathbf{X}_Q) \tag{4.4.31}$$

where v_{iI}^0 and σ_{ij}^0 are initial data. If data for the initial conditions are given at a different set of points, the values at the nodes and quadrature points can be estimated by least-square fits, as in Section 2.4.5.

For an equilibrium problem, the accelerations vanish and the governing equations are

$$f_{iI}^{\text{int}} = f_{iI}^{\text{ext}} \quad \text{for} \quad (I,i) \notin \Gamma_{v_i} \quad \text{or} \quad \mathbf{f}^{\text{int}} = \mathbf{f}^{\text{ext}} \tag{4.4.32}$$

along with (4.4.28) and (4.4.29). These are called the discrete *equilibrium equations*. If the constitutive equations are rate-independent, then the discrete equilibrium equations are a set of nonlinear algebraic equations in the stresses and nodal displacements. For rate-dependent materials, any rate terms must be discretized in time to obtain a set of nonlinear algebraic equations.

4.4.5 Element Coordinates

Finite elements are usually developed with shape functions expressed in terms of parent element coordinates, which we will often call element coordinates for brevity. Examples of element coordinates are triangular coordinates and isoparametric coordinates. We will next describe the use of shape functions expressed in terms of element coordinates. As part of this description, we will show that the element coordinates can be considered an alternative set of material coordinates in a Lagrangian mesh. Therefore, expressing the shape functions in terms of element coordinates is intrinsically equivalent to expressing them in terms of material coordinates in a Lagrangian mesh. We denote the parent element coordinates by ξ_i^e, or $\boldsymbol{\xi}^e$ in tensor notation, and the parent domain by □; the superscript e will be carried only in the beginning of this description. The shape of the parent domain depends on the type of element and the dimension of the problem; it may be a biunit square, a triangle, or a cube, for example. Specific parent domains are given in the examples which follow.

When a Lagrangian element is treated in terms of element coordinates, we are concerned with three domains that correspond to an element:

1. The parent element domain □
2. The current element domain $\Omega^e = \Omega^e(t)$
3. The initial (reference) element domain Ω_0^e.

The following maps are pertinent:

1. Parent domain to current configuration: $\mathbf{x} = \mathbf{x}(\boldsymbol{\xi}^e, t)$
2. Parent domain to initial configuration: $\mathbf{X} = \mathbf{X}(\boldsymbol{\xi}^e)$
3. Initial configuration to current configuration, that is, the motion $\mathbf{x} = \mathbf{x}(\mathbf{X}, t) = \boldsymbol{\phi}(\mathbf{X}, t)$.

The map $\mathbf{X} = \mathbf{X}(\boldsymbol{\xi}^e)$ corresponds to $\mathbf{x} = \mathbf{x}(\boldsymbol{\xi}^e, 0)$. These maps are illustrated in Figure 4.2 for a triangular element where a space–time plot of a two-dimensional triangular element is shown.

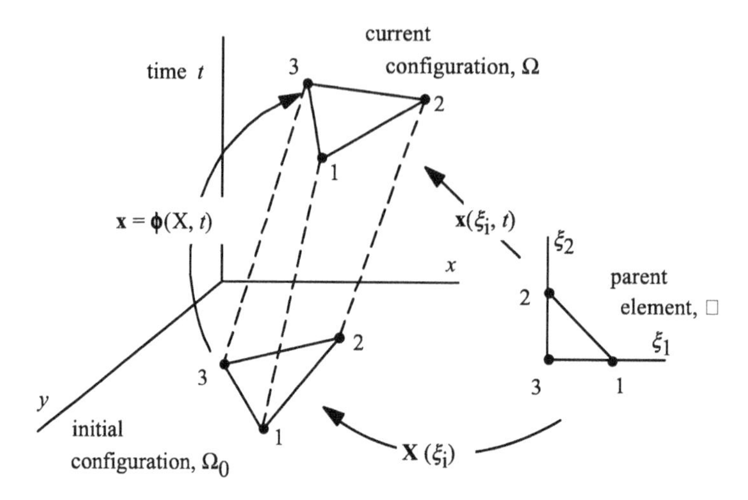

Figure 4.2 Initial and current configurations of a Lagrangian element and their relationships to the parent element

The motion in each element is described by a composition of the maps

$$\mathbf{x} = \mathbf{x}(\mathbf{X}, t) = \mathbf{x}(\boldsymbol{\xi}^e(\mathbf{X}), t) \quad \mathbf{x}(\mathbf{X}, t) = \mathbf{x}(\boldsymbol{\xi}^e, t) \circ \boldsymbol{\xi}^e(\mathbf{X}) \quad \text{in } \Omega_e \tag{4.4.33}$$

where $\boldsymbol{\xi}^e(\mathbf{X}) = \mathbf{X}^{-1}(\boldsymbol{\xi}^e)$. For the motion to be well-defined and smooth, the inverse map $\mathbf{X}^{-1}(\boldsymbol{\xi}^e)$ must exist and the function $\mathbf{x} = \mathbf{x}(\boldsymbol{\xi}^e, t)$ must be sufficiently smooth and meet certain conditions of regularity so that $\mathbf{x}^{-1}(\boldsymbol{\xi}^e, t)$ exists; these conditions are given in Section 4.4.8. The inverse map $\mathbf{x}^{-1}(\boldsymbol{\xi}^e, t)$ is usually not constructed because in most cases it cannot be obtained explicitly, so instead the derivatives with respect to the spatial coordinates are obtained in terms of the derivatives with respect to the parent coordinates by implicit differentiation.

The motion is approximated by

$$x_i(\boldsymbol{\xi}, t) = x_{iI}(t) N_I(\boldsymbol{\xi}) \quad \text{or} \quad x(\boldsymbol{\xi}, t) = \mathbf{x}_I(t) N_I(\boldsymbol{\xi}) \tag{4.4.34}$$

where we have dropped the superscript e on the element coordinates. As can be seen in the above, the shape functions $N_I(\boldsymbol{\xi})$ are only functions of the parent element coordinates; the time dependence of the motion resides entirely in the nodal coordinates. This represents a time-dependent mapping between the parent domain and the current configuration of the element.

Writing this map at time $t = 0$ we obtain

$$X_i(\boldsymbol{\xi}) = x_i(\boldsymbol{\xi}, 0) = x_{iI}(0) N_I(\boldsymbol{\xi}) = X_{iI} N_I(\boldsymbol{\xi}) \quad \text{or} \quad \mathbf{X}(\boldsymbol{\xi}) = \mathbf{X}_I N_I(\boldsymbol{\xi}) \tag{4.4.35}$$

It can be seen from (4.4.35) that the map between the material coordinates and the element coordinates is time invariant in a Lagrangian element. If this map is one-to-one, then *the element coordinates can be considered surrogate material coordinates in a Lagrangian mesh*, since each material point in an element then has a unique element coordinate label. To establish a unique correspondence between element coordinates and the material coordinates in Ω_0, the element number must be part of the label. Element coordinates are not surrogate material coordinates for meshes which are not Lagrangian, as will be seen in Chapter 7. The use of the initial coordinates $\mathbf{X}$ as material coordinates in fact originates mainly in analysis; in finite element methods, the use of element coordinates as material labels is more natural.

As before, since the element coordinates are time invariant, we can express the displacements, velocities and accelerations in terms of the shape functions:

$$u_i(\boldsymbol{\xi}, t) = u_{iI}(t)N_I(\boldsymbol{\xi}) \quad \mathbf{u}(\boldsymbol{\xi}, t) = \mathbf{u}_I(t)N_I(\boldsymbol{\xi}) \tag{4.4.36}$$

$$\dot{u}_i(\boldsymbol{\xi}, t) = v_i(\boldsymbol{\xi}, t) = v_{iI}(t)N_I(\boldsymbol{\xi}), \quad \dot{\mathbf{u}}(\boldsymbol{\xi}, t) = \mathbf{v}(\boldsymbol{\xi}, t) = \mathbf{v}_I(t)N_I(\boldsymbol{\xi}) \tag{4.4.37}$$

$$\dot{v}_i(\boldsymbol{\xi}, t) = \dot{v}_{iI}(t)N_I(\boldsymbol{\xi}), \quad \dot{\mathbf{v}}(\boldsymbol{\xi}, t) = \dot{\mathbf{v}}_I(t)N_I(\boldsymbol{\xi}) \tag{4.4.38}$$

where we have obtained (4.4.37) by taking the material time derivative of (4.4.36) and we have obtained (4.4.38) by taking the material time derivative of (4.4.37). The time dependence, as before, resides entirely in the nodal variables, since the element coordinates are independent of time.

4.4.6 Derivatives of Functions

The spatial derivatives of the velocity field are obtained by implicit differentiation because the function $\mathbf{x}(\boldsymbol{\xi}, t)$ is generally not explicitly invertible; that is, it is not possible to write closed-form expressions for $\boldsymbol{\xi}$ in terms of $\mathbf{x}$. By the chain rule,

$$\frac{\partial v_i}{\partial \xi_j} = \frac{\partial v_i}{\partial x_k}\frac{\partial x_k}{\partial \xi_j} \quad \text{or} \quad \mathbf{v}_{,\boldsymbol{\xi}} = \mathbf{v}_{,\mathbf{x}}\, \mathbf{x}_{,\boldsymbol{\xi}} \tag{4.4.39}$$

The matrix $\partial x_k/\partial \xi_j$ is the Jacobian of the map between the current configuration of the element and the parent element configuration. We will use two symbols for this matrix: $\mathbf{x}_{,\boldsymbol{\xi}}$ and $\mathbf{F}_{\xi}$, where $F_{ij}^{\xi} = \partial x_i/\partial \xi_j$. The second symbol is used to convey the notion that the Jacobian with respect to the element coordinates can be viewed as a deformation gradient with respect to the parent element configuration. In two dimensions

$$\mathbf{x}_{,\boldsymbol{\xi}}(\boldsymbol{\xi}, t) \equiv \mathbf{F}_{\xi}(\boldsymbol{\xi}, t) = \begin{bmatrix} x_{,\xi_1} & x_{,\xi_2} \\ y_{,\xi_1} & y_{,\xi_2} \end{bmatrix} = \begin{bmatrix} x_{,\xi} & x_{,\eta} \\ y_{,\xi} & y_{,\eta} \end{bmatrix} \tag{4.4.40}$$

As indicated in (4.4.40), the Jacobian of the map between the current and parent configurations is a function of time.

Inverting (4.4.39), we obtain

$$L_{ij} = \frac{\partial v_i}{\partial \xi_k}\left(F_{kj}^{\xi}\right)^{-1} = \left(\frac{\partial v_i}{\partial \xi_k}\right)\left(\frac{\partial \xi_k}{\partial x_j}\right) \quad \text{or} \quad \mathbf{L} = \mathbf{v}_{,\mathbf{x}} = \mathbf{v}_{,\boldsymbol{\xi}}\, \mathbf{x}_{,\boldsymbol{\xi}}^{-1} = \mathbf{v}_{,\boldsymbol{\xi}}\, \mathbf{F}_{\xi}^{-1} \tag{4.4.41}$$

Thus computing the derivatives with respect to ξ involves finding the inverse of the Jacobian between the current and parent element coordinates; the matrix to be inverted in the two-dimensional case is given in (4.4.40). Similarly for the shape functions N_I, we have

$$N_{I,\mathbf{x}}^T = N_{I,\xi,}^T \mathbf{x}_{,\xi}^{-1} = N_{I,\xi,}^T \mathbf{F}_\xi^{-1} \tag{4.4.42}$$

where the transpose appears in the matrix expressions because we consider $N_{I,x}$ and $N_{I,\xi}$ to be column matrices and the matrix on the RHS of the above must be a row matrix. The determinant of the element Jacobian $\mathbf{F}_\xi$,

$$J_\xi = \det(\mathbf{x}_{,\xi}) \tag{4.4.43}$$

is called the *element Jacobian determinant*; we append the subscript to distinguish it from the determinant of the deformation gradient, J. Substituting (4.4.42) into (4.4.41) gives

$$L_{ij} = v_{iI} \frac{\partial N_I}{\partial \xi_k} \left(F_{kj}^\xi\right)^{-1} \quad \text{or} \quad \mathbf{L} = \mathbf{v}_I N_{I,\xi}^T \mathbf{x}_{,\xi}^{-1} \tag{4.4.44}$$

The rate-of-deformation is obtained from the velocity gradient by (3.3.10).

4.4.7 Integration and Nodal Forces

Integrals on the current configuration are related to integrals over the reference domain and the parent domain by

$$\int_{\Omega^e} g(\mathbf{x}) d\Omega = \int_{\Omega_0^e} g(\mathbf{x}(\mathbf{X})) J \, d\Omega_0 = \int_{\Box} g(\xi) J_\xi \, d\Box \quad \text{and} \quad \int_{\Omega_0^e} g(\mathbf{X}) d\Omega_0 = \int_{\Box} g(\mathbf{X}(\xi)) J_\xi^0 d\Box \tag{4.4.45}$$

where J and J_ξ are the determinants of the Jacobians between the current configuration and the reference and parent element configurations, respectively; J_ξ^0 is the determinant of the Jacobian between the reference configurations and the parent element.

When the internal nodal forces are computed by integration over the parent domain, (4.4.13) is transformed to the parent element domain by (4.4.45), giving

$$f_{iI}^{\text{int}} = \int_{\Omega^e} \frac{\partial N_I}{\partial x_j} \sigma_{ji} d\Omega = \int_{\Box} \frac{\partial N_I}{\partial x_j} \sigma_{ji} J_\xi d\Box \tag{4.4.46}$$

The external nodal forces and the mass matrix can similarly be integrated over the parent domain.

4.4.8 Conditions on Parent to Current Map

The finite element approximation to the motion $\mathbf{x}(\xi, t)$, which maps the parent domain of an element onto the current domain of the element, is subject to the same conditions as $\phi(\mathbf{X}, t)$, as given in Section 3.2.6, except that no discontinuities are allowed. These conditions are:

1. $\mathbf{x}(\boldsymbol{\xi}, t)$ must be one-to-one
2. $\mathbf{x}(\boldsymbol{\xi}, t)$ must be at least C^0 in space
3. The element Jacobian determinant must be positive, that is,

$$J_\xi \equiv \det(\mathbf{x}_{,\xi}) > 0 \tag{4.4.47}$$

These conditions insure that $\mathbf{x}(\boldsymbol{\xi}, t)$ is invertible.

We now explain why the condition $\det(\mathbf{x}_{,\xi}) > 0$ is necessary. We first use the chain rule to express $\mathbf{x}_{,\xi}$ in terms of $\mathbf{F}$ and $\mathbf{X}_{,\xi}$:

$$\frac{\partial x_i}{\partial \xi_j} = \frac{\partial x_i}{\partial X_k}\frac{\partial X_k}{\partial \xi_j} = F_{ik}\frac{\partial X_k}{\partial \xi_j} \quad \text{or} \quad \mathbf{x}_{,\xi} = \mathbf{x}_{,X}\mathbf{X}_{,\xi} = \mathbf{F}\mathbf{X}_{,\xi} \tag{4.4.48}$$

We can also write this as

$$\mathbf{F}_\xi = \mathbf{F}\cdot\mathbf{F}_\xi^0 \tag{4.4.49}$$

which highlights the fact that the deformation gradient with respect to the parent element coordinates is the product of the standard deformation gradient and the initial deformation gradient with respect to the parent element coordinates. The determinant of the product of two matrices is equal to the product of the determinants, so

$$J_\xi = \det(\mathbf{x}_{,\xi}) = \det(\mathbf{F})\det(\mathbf{X}_{,\xi}) \equiv JJ_\xi^0 \tag{4.4.50}$$

We assume that the elements in the initial mesh are properly constructed so that $J_\xi^0 = J_\xi(0) > 0$ for all elements; otherwise the initial mapping would not be one-to-one. If $J_\xi(t) \le 0$ at any time then by (4.4.50), $J \le 0$. By the conservation of matter $\rho = \rho_0/J$, so $J \le 0$ implies $\rho \le 0$, which is physically impossible. Therefore it is necessary that $J_\xi(t) > 0$. In some calculations, excessive distortion can result in severely deformed meshes in which $J_\xi \le 0$. This implies a negative density, so such calculations violate the physical principle that mass is always positive.

4.4.9 Simplifications of Mass Matrix

When the same shape functions are used for all components, it is convenient to note that (4.4.19) can be written as

$$M_{ijIJ} = \delta_{ij}\tilde{M}_{IJ} \tag{4.4.51}$$

where

$$\tilde{M}_{IJ} = \int_\Omega \rho N_I N_J d\Omega, \quad \text{or} \quad \tilde{\mathrm{M}} = \int_\Omega \rho \mathbf{N}^T \mathbf{N} d\Omega \tag{4.4.52}$$

Then the equations of motion (4.4.27) become

$$\tilde{M}_{IJ}\dot{v}_{iJ} + f_{iI}^{\text{int}} = f_{iI}^{\text{ext}} \tag{4.4.53}$$

This form is advantageous when the consistent mass matrix is used with explicit time integration, since the order of the matrix which needs to be inverted is reduced by a factor of n_{SD}.

We next show that the mass matrix for a Lagrangian mesh is constant in time. If the shape functions are expressed in terms of parent element coordinates, then

$$M_{ijIJ} = \delta_{ij}\int_{\square}\rho N_I N_J \det(\mathbf{x}_{,\xi})d\square = \delta_{ij}\int_{\Omega}\rho N_I N_J d\Omega \tag{4.4.54}$$

Since $\det(\mathbf{x}_{,\xi})$ and the density are time dependent, this mass matrix appears to be time dependent. To show that the matrix is in fact time independent, we transform the above integral to the undeformed configuration by (3.2.23), giving

$$M_{ijIJ} = \delta_{ij}\int_{\Omega}\rho N_I N_J J d\Omega_0 \tag{4.4.55}$$

From mass conservation, (B4.1.1), it follows that $\rho J = \rho_0$. Hence (4.4.55) becomes

$$M_{ijIJ} = \delta_{ij}\int_{\Omega_0}\rho_0 N_I N_J \, d\Omega_0 \quad \text{or} \quad M_{ijIJ} = \delta_{ij}\int_{\square}\rho_0 N_I N_J \, J_{\xi}^0 \, d\square \tag{4.4.56}$$

The compact form of the mass matrix, (4.4.52), can similarly be written as

$$\tilde{M}_{IJ} = \int_{\Omega_0}\rho_0 N_I N_J d\Omega_0 \quad \text{and} \quad \mathbf{M}_{IJ} = \mathbf{I}\tilde{M}_{IJ} = \mathbf{I}\int_{\Omega_0}\rho_0 N_I N_J d\Omega_0 \tag{4.4.57}$$

In these integrals, the integrand is independent of time, so the mass matrix is constant in time. It needs to be evaluated only at the beginning of a computation. The same result could be obtained by computing the mass matrix by (4.4.54) at the initial time, that is, in the initial configuration. The mass matrix in (4.4.55–7) can be called *total Lagrangian* since it is evaluated in the reference (undeformed) configuration. We take the view here and subsequently that the each term in the discrete equations should be evaluated in whatever configuration is most convenient.

4.5 Implementation

In the implementation of the finite element equations, two approaches are popular:

1. The indicial expressions are directly treated as matrix equations.
2. Voigt notation is used, as in linear finite element methods, and the square stress and strain matrices are converted to column matrices.

Each of these methods has advantages, so both methods will be described. In Box 4.3 the discrete equations are summarized in both forms.

Box 4.3 Discrete equations and internal nodal force algorithm for updated Lagrangian formulation

Equations of motion (discrete momentum equation):

$$M_{ijIJ}\dot{v}_{jJ} + f_{iI}^{\text{int}} = f_{iI}^{\text{ext}} \text{ for } (I,i) \notin \Gamma_{v_i} \tag{B4.3.1}$$

Internal nodal forces:

$$f_{iI}^{\text{int}} = \int_\Omega B_{Ij}\sigma_{ji}\,d\Omega = \int_\Omega \frac{\partial N_I}{\partial x_j}\sigma_{ji}\,d\Omega \quad \text{or} \quad (\mathbf{f}_I^{\text{int}})^T = \int_\Omega \boldsymbol{B}_I^T \boldsymbol{\sigma}\,d\Omega \tag{B4.3.2}$$

$$\mathbf{f}_I^{\text{int}} = \int_\Omega \mathbf{B}_I^T\{\boldsymbol{\sigma}\}d\Omega \text{ in Voigt notation}$$

External nodal forces:

$$f_{iI}^{\text{ext}} = \int_\Omega N_I \rho b_i\,d\Omega + \int_{\Gamma_{t_i}} N_I \bar{t}_i d\Gamma \quad \text{or} \quad \mathbf{f}_I^{\text{ext}} = \int_\Omega N_{I\rho}\mathbf{b}d\Omega + \int_{\Gamma_{t_i}} N_I \mathbf{e}_i \cdot \bar{t}\,d\Gamma \tag{B4.3.3}$$

Mass matrix (updated Lagrangian):

$$M_{ijIJ} = \delta_{ij}\int_{\Omega_0}\rho_0 N_I N_J\,d\Omega_0 = \delta_{ij}\int_{\square}\rho_0 N_I N_J J_\xi^0\,d\square \tag{B4.3.4}$$

$$\mathbf{M}_{IJ} = \mathbf{I}\tilde{M}_{IJ} = \mathbf{I}\int_{\Omega_0}\rho_0 N_I N_J\,d\Omega_0 \tag{B4.3.5}$$

Internal nodal force computation for element

1. $\mathbf{f}^{\text{int}} = 0$
2. For all quadrature points $\boldsymbol{\xi}_Q$
   i. compute $[B_{Ij}] = [\partial N_I(\boldsymbol{\xi}_Q)/\partial x_j]$ for all I
   ii. $\mathbf{L} = [L_{ij}] = [v_{iI}B_{Ij}] = \mathbf{v}_I \boldsymbol{B}_I^T; L_{ij} = \dfrac{\partial N_I}{\partial x_j}v_{iI}$
   iii. $\mathbf{D} = \frac{1}{2}(\mathbf{L}^T + \mathbf{L})$
   iv. if needed compute $\mathbf{F}$ and $\mathbf{E}$ by procedures in Box 4.6
   v. compute Cauchy stress $\boldsymbol{\sigma}$ or PK2 stress $\mathbf{S}$ by constitutive equation
   vi. if S computed, compute $\boldsymbol{\sigma}$ by $\boldsymbol{\sigma} = J^{-1}\,\mathbf{FSF}^T$
   vii. $\mathbf{f}_I^{\text{int}} \leftarrow \mathbf{f}_I^{\text{int}} + \boldsymbol{B}_I^T \sigma J_\xi \bar{w}_Q$ for all nodes I

   end loop

   ($\bar{w}_Q$ are quadrature weights)

4.5.1 *Indicial to Matrix Notation Translation*

The conversion of indicial expressions to matrix form is somewhat arbitrary and depends on individual preferences. In this book, we interpret single-index variables as column matrices in most cases; the procedures are somewhat different when there is a preference for row matrices.

To illustrate the conversion of indicial expressions to matrix form, consider the expression for the velocity gradient (4.4.7):

$$L_{ij} = \frac{\partial v_i}{\partial x_j} = v_{iI} \frac{\partial N_I}{\partial x_j} \tag{4.5.1}$$

This expression can be put into the form of a matrix product if we associate the index I with a column number in v_{iI} and a row number in $\partial N_I / \partial x_j$. To simplify the resulting matrix expression, we define a matrix $\boldsymbol{B}$ by

$$B_{jI} = \frac{\partial N_I}{\partial x_j} \quad \text{or} \quad \boldsymbol{B} = [\boldsymbol{B}_{jI}] = [\partial N_I / \partial x_j] \tag{4.5.2}$$

where j is the row number in the $\boldsymbol{B}$ matrix. The velocity gradient can then be expressed in terms of the nodal displacements through (4.5.1) and (4.5.2) by

$$[L_{ij}] = [v_{iI}][B_{Ij}] = [v_{iI}][B_{jI}]^T \quad \text{or} \quad \mathbf{L} = \mathbf{v}\boldsymbol{B}^T \tag{4.5.3}$$

so, because of the implicit sum on I, the indicial expression corresponds to a matrix product.

We can also write (4.5.1). The $\boldsymbol{B}$ matrix is then subdivided into $\boldsymbol{B}_I$ matrices, each associated with node I:

$$\boldsymbol{B} = \left[\boldsymbol{B}_1, \boldsymbol{B}_2, \boldsymbol{B}_3, \ldots, \boldsymbol{B}_m\right] \text{ where } \boldsymbol{B}_I^T = \{B_j\}_I = N_{I,\mathrm{x}} \tag{4.5.4}$$

For each node I, the $\boldsymbol{B}_I$ matrix is a column matrix. Then the expression for the velocity gradient can be written as:

$$\mathbf{L} = \mathbf{v}_I \boldsymbol{B}_I^T = \begin{Bmatrix} v_{xI} \\ v_{yI} \end{Bmatrix} \begin{bmatrix} N_{I,x} & N_{I,y} \end{bmatrix} = \begin{bmatrix} v_{xI} N_{I,x} & v_{xI} N_{I,y} \\ v_{yI} N_{I,x} & v_{yI} N_{I,y} \end{bmatrix} \tag{4.5.5}$$

To put the internal force expression (4.4.13) in matrix form, we first rearrange the terms so that adjacent terms correspond to matrix products. This entails interchanging the row and column number on the internal forces as shown here:

$$\left(f_{iI}^{\mathrm{int}}\right)^T = f_{Ii}^{\mathrm{int}} = \int_\Omega \frac{\partial N_I}{\partial x_j} \sigma_{ji} d\Omega = \int_\Omega B_{Ij}^T \sigma_{ji} d\Omega \tag{4.5.6}$$

This can be put in the following matrix form:

$$\left[f_{iI}^{\mathrm{int}}\right]^T = \left[f_{Ii}^{\mathrm{int}}\right] = \int_\Omega [\partial N_I / \partial x_j][\sigma_{ji}] \; d\Omega = \int_\Omega [B_{jI}]^T [\sigma_{ji}] \; d\Omega, \quad (\mathbf{f}_I^{\mathrm{int}})^T = \int_\Omega \mathbf{B}_I^T \boldsymbol{\sigma} d\Omega \tag{4.5.7}$$

For example, in two dimensions this gives

$$[f_{xI},\ f_{yI}]^{\text{int}} = \int_{\Omega}[N_{I,x}\ N_{I,y}]\begin{bmatrix}\sigma_{xx} & \sigma_{xy}\\ \sigma_{xy} & \sigma_{yy}\end{bmatrix}d\Omega \tag{4.5.8}$$

There are many other ways of converting indicial expressions to matrix form, but the above is convenient. An expression for the complete matrix of internal nodal forces can be obtained with the $\boldsymbol{B}$ matrix defined in (4.5.4), which gives

$$(\mathbf{f}^{\text{int}})^T = \int_{\Omega}\boldsymbol{B}^T\boldsymbol{\sigma}\,d\Omega$$

4.5.2 *Voigt Notation*

An alternative implementation which is widely used in linear and nonlinear finite element programs is based on Voigt notation: see Appendix 1. Voigt notation is useful for computing tangent stiffness matrices in Newton methods: see Chapter 6. In Voigt notation the stresses and rate-of-deformation are expressed in column vectors, so in two dimensions

$$\{\mathbf{D}\}^T = [D_x\ D_y\ 2D_{xy}] \quad \{\boldsymbol{\sigma}\}^T = [\sigma_x\ \sigma_y\ \sigma_{xy}] \tag{4.5.9}$$

We define the $\boldsymbol{B}_I$ matrix so it relates the rate-of-deformation to the nodal velocities by

$$\{\mathbf{D}\} = \mathbf{B}_I\mathbf{v}_I \quad \{\delta\mathbf{D}\} = \mathbf{B}_I\delta\mathbf{v}_I \tag{4.5.10}$$

where the summation convention as usual applies to repeated indices. The elements of the $\boldsymbol{B}_I$ matrix are constructed so as to meet the definition (4.5.10); this is illustrated in the following examples. Note that a variable is enclosed in brackets only when this is needed to distinguish it from its usual form as a square matrix.

The expression for the internal force vector can be derived in this notation by the definition of the virtual internal power in (4.3.21). Since $\{\mathbf{D}\}^T\{\boldsymbol{\sigma}\}$ gives the internal power per unit volume (the column matrices are constructed to meet this definition), it follows that

$$\delta P^{\text{int}} = \delta\mathbf{v}_I^T\mathbf{f}_I^{\text{int}} = \int_{\Omega}\{\delta\mathbf{D}\}^T\{\boldsymbol{\sigma}\}d\Omega \tag{4.5.11}$$

Substituting (4.5.10) into (4.5.11) and invoking the arbitrariness of $\{\delta\mathrm{v}\}$ gives

$$\mathbf{f}_I^{\text{int}} = \int_{\Omega}\mathbf{B}_I^T\{\boldsymbol{\sigma}\}d\Omega \tag{4.5.12}$$

As will be shown in the examples, (4.5.12) gives the same expression for the internal nodal forces as (4.5.7): (4.5.12) uses the symmetric part of the velocity gradient, whereas the complete velocity gradient has been used in (4.5.7). Since the Cauchy stress is symmetric, the two expressions are equivalent.

It is sometimes convenient to place the displacement, velocities and nodal forces for an element or a complete mesh in a single column matrix. We will use the symbol $\mathbf{d}$ for the column matrix of all nodal displacements, $\dot{\mathbf{d}}$ for the column matrix of nodal velocities and $\{\mathbf{f}\}$ for the column matrix of nodal forces, that is,

$$\mathbf{d} = \begin{Bmatrix} \mathbf{u}_1 \\ \mathbf{u}_2 \\ \vdots \\ \mathbf{u}_m \end{Bmatrix} \quad \dot{\mathbf{d}} = \begin{Bmatrix} \mathbf{v}_1 \\ \mathbf{v}_2 \\ \vdots \\ \mathbf{v}_m \end{Bmatrix} \quad \mathbf{f} = \begin{Bmatrix} \mathbf{f}_1 \\ \mathbf{f}_2 \\ \vdots \\ \mathbf{f}_m \end{Bmatrix} \tag{4.5.13}$$

where m is the number of nodes. The correspondence between the two matrices is given by

$$d_a = u_{iI} \quad where \quad a = (I-1)n_{\mathrm{SD}} + i \tag{4.5.14}$$

Note that we use a different symbol for the column matrix of all nodal displacements and nodal velocities because the symbols $\mathbf{u}$ and $\mathbf{v}$ refer to the displacement and velocity vector fields in the continuum mechanics description.

In this notation, we can write the counterpart of (4.5.10) as

$$\{\mathbf{D}\} = \mathbf{B}\dot{\mathbf{d}} \quad \text{where} \quad \mathbf{B} = [\mathbf{B}_1, \mathbf{B}_2, \cdots, \mathbf{B}_m] \tag{4.5.15}$$

where the brackets around $\mathbf{D}$ indictate that the tensor is in column matrix form; we do not put brackets around $\mathbf{B}$ because this is always a rectangular matrix. The nodal forces are given by the counterpart of (4.5.12):

$$\{\mathbf{f}\}^{\text{int}} = \int_{\Omega} \mathbf{B}^T \{\boldsymbol{\sigma}\} d\Omega \tag{4.5.16}$$

Often we omit the brackets on the nodal force, since the presence of a single term in Voigt notation indicates that the entire equation is in Voigt notation. The Voigt form can also be obtained by rewriting (4.5.6) as

$$f_{rI}^{\text{int}} = \int_{\Omega} \frac{\partial N_I}{\partial x_j} \delta_{ri} \sigma_{ji} d\Omega \tag{4.5.17}$$

Then defining the $\mathbf{B}$ matrix by

$$B_{ijrI} = \underset{(i,j)}{\text{sym}} \left(\frac{\partial N_I}{\partial x_j} \delta_{ri} \right) \tag{4.5.18}$$

and converting the indices (i, j) by the kinematic Voigt rule to b and the indices (I, r) by the matrix-column vector rule to a gives

$$f_a^{\text{int}} = \int_\Omega B_{da}\sigma_b d\Omega \quad \text{or} \quad \mathbf{f}^{\text{int}} = \int_\Omega \mathbf{B}^T\{\boldsymbol{\sigma}\}d\Omega \tag{4.5.19}$$

More detail on translating indicial notation to Voigt notation can be found in Appendix 1.

4.5.3 *Numerical Quadrature*

The integrals for the nodal forces, mass matrix and other element matrices are not evaluated in closed form, and are instead integrated numerically (often called numerical quadrature). The most widely used procedure for numerical integration in finite elements is Gauss quadrature. The Gauss quadrature formulas are (e.g., Dhatt and Touzot, 1984: 240; Hughes, 1987: 137)

$$\int_{-1}^{1} f(\xi)d\xi = \sum_{Q=1}^{n_Q} w_Q f(\xi_Q) \tag{4.5.20}$$

where the weights w_Q and coordinates ξ_Q of the n_Q quadrature points are available in tables; a short table is given in Appendix 3. Equation (4.5.20) integrates $f(\xi)$ exactly if it is a polynomial of order $m \le 2n_Q - 1$. Equation (4.5.20) is tailored to quadrature over the parent element domains, since it is over the interval $[-1, 1]$.

To integrate over a two-dimensional element, the procedure is repeated in the second direction, yielding

$$\int_\square f(\boldsymbol{\xi})d\square = \int_{-1}^{1}\int_{-1}^{1} f(\xi,\eta)d\xi d\eta = \sum_{Q_1=1}^{n_{Q_1}}\sum_{Q_2=1}^{n_{Q_2}} w_{Q_1} w_{Q_2} f\left(\xi_{Q_1}, \eta_{Q_2}\right) \tag{4.5.21}$$

In three dimensions, the Gauss quadrature formula is

$$\int_\square f(\boldsymbol{\xi})d\square = \int_{-1}^{1}\int_{-1}^{1}\int_{-1}^{1} f(\xi)d\xi d\eta d\zeta = \sum_{Q_1=1}^{n_{Q_1}}\sum_{Q_2=1}^{m_{Q_2}}\sum_{Q_3=1}^{n_{Q_3}} w_{Q_1} w_{Q_2} w_{Q_3} f\left(\xi_{Q_1}, \eta_{Q_2}, \zeta_{Q_3}\right) \tag{4.5.22}$$

For example, the internal nodal forces for a biunit square parent are

$$\begin{aligned} \mathbf{f}^{\text{int}} &= \int_\square \mathbf{B}^T\{\boldsymbol{\sigma}\}J_\xi \, d\square = \int_{-1}^{1}\int_{-1}^{1} \mathbf{B}^T\{\boldsymbol{\sigma}\}J_\xi \, d\xi d\eta \\ &= \sum_{Q_1=1}^{n_{Q_1}}\sum_{Q_2=1}^{n_{Q_2}} w_{Q_1} w_{Q_2} \mathbf{B}^T\left(\xi_{Q_1}, \eta_{Q_2}\right)\left\{\boldsymbol{\sigma}\left(\xi_{Q_1}, \eta_{Q_2}\right)\right\} J_\xi\left(\xi_{Q_1}, \eta_{Q_2}\right) \end{aligned} \tag{4.5.23}$$

To simplify the notation for multi-dimensional quadrature, we often combine the weights into a single weight

$$\int_{\square} f(\boldsymbol{\xi})\, d\square = \sum_{Q} \bar{w}_Q f(\boldsymbol{\xi}_Q) \tag{4.5.24}$$

where $\bar{w}_Q$ is a product of the weights for one-dimensional quadrature w_Q.

The number of quadrature points used in nonlinear analysis is generally based on the same rules as for linear analysis; the number of quadrature points is chosen to exactly integrate the nodal internal forces for a regular element. A regular form of an element is one that can be obtained by only stretching but not shearing the parent element; for example, a rectangle for two-dimensional isoparametric elements. To choose the number of quadrature points for the internal nodal forces for a four-node quadrilateral, we use the following argument. The rate-of-deformation and the **B** matrix are linear in this element since the velocities are bilinear. If the stress is linearly related to the rate-of-deformation, it will vary linearly within the element. The integrand for the internal nodal forces is approximately quadratic, since it is a product of the **B** matrix and the stresses. In Gauss quadrature, two quadrature points are needed in each direction for exact quadrature of a quadratic function, so 2×2 quadrature integrates the internal nodal forces exactly for a linear material. Quadrature formulas which integrate the nodal internal forces almost exactly for a linear constitutive equation are called *full quadrature.*

Gauss quadrature is very powerful for smooth functions which are polynomials or nearly polynomials. In linear finite element analysis, the integrand in the expression for the stiffness matrix consists of polynomials for rectangular elements and is smooth and nearly a polynomial for isoparametric elements. In nonlinear analysis, the integrand is not always smooth. For example, for an elastic–plastic material, the stress may have discontinuous derivatives at the surface separating elastic and plastic material. Therefore, the errors in Gauss quadrature of an element that contains an elastic–plastic interface are likely to be large. However, higher-order quadrature is not recommended for circumventing these errors, since it often leads to stiff behavior or locking.

4.5.4 Selective-Reduced Integration

For incompressible or nearly incompressible materials, full quadrature of the nodal internal forces may cause an element to lock, that is, the displacements are very small and do not converge, or converge very slowly. The easiest way to circumvent this difficulty is to use selective-reduced integration.

In selective-reduced integration, the pressure is underintegrated, whereas the remainder of the stress matrix is fully integrated. For this purpose, the stress tensor is decomposed into the hydrostatic and deviatoric components

$$\sigma_{ij} = \sigma_{ij}^{\text{dev}} + \sigma^{\text{hyd}} \delta_{ij} \tag{4.5.25}$$

where

$$\sigma^{\text{hyd}} = \frac{1}{3}\sigma_{kk} = -p, \quad \sigma_{ij}^{\text{dev}} = \sigma_{ij} - \sigma^{\text{hyd}} \delta_{ij} \tag{4.5.26}$$

where p is the pressure. The rate-of-deformation is similarly split into dilatational (volumetric) and deviatoric components which are defined by

$$D_{ij}^{\text{dev}} = D_{ij} - \frac{1}{3} D_{kk} \delta_{ij}, \quad D_{ij}^{\text{vol}} = \frac{1}{3} D_{kk} \delta_{ij} \tag{4.5.27}$$

It is noted that the dilatational and deviatoric components are orthogonal to each other so that the virtual internal power as defined in (4.3.21) becomes

$$\delta P^{\text{int}} = \int_{\Omega} \delta D_{ij} \sigma_{ij} \, d\Omega = -\int_{\Omega} \delta D_{ii} p \, d\Omega + \int_{\Omega} \delta D_{ij}^{\text{dev}} \sigma_{ij}^{\text{dev}} \, d\Omega \tag{4.5.28}$$

After expressing the rate-of-deformation in terms of the shape functions by (4.4.8) and (4.5.27), the dilatational and deviatoric integrands become

$$\delta D_{ii} p = \delta v_{iI} N_{I,i} p \tag{4.5.29}$$

and

$$\delta D_{ij}^{\text{dev}} \sigma_{ji}^{\text{dev}} = \frac{1}{2} (N_{I,j} \delta v_{iI} + N_{I,i} \delta v_{jI}) \sigma_{ji}^{\text{dev}} \tag{4.5.30}$$

Using the symmetry of σ_{ji}^{dev}, the deviatoric integrand simplifies to

$$\delta D_{ij}^{\text{dev}} \sigma_{ji}^{\text{dev}} = \delta v_{iI} N_{I,j} \sigma_{ji}^{\text{dev}} \tag{4.5.31}$$

Selective-reduced integration consists of full integration on the deviatoric power and reduced integration on the dilatational power in δP^{int}. Thus, for a four-node quadrilateral, selective-reduced integration gives

$$\delta P^{\text{int}} = \delta v_{iI} \left(-J_{\xi}(\mathbf{0}) N_{I,i}(\mathbf{0}) p(\mathbf{0}) + \sum_{Q=1}^{4} \bar{w}_Q J_{\xi}(\boldsymbol{\xi}_Q) N_{I,j}(\boldsymbol{\xi}_Q) \sigma_{ji}^{\text{dev}}(\boldsymbol{\xi}_Q) \right) \tag{4.5.32}$$

Hence the selective-reduced integration expression for the internal forces is

$$(f_{iI}^{\text{int}})^T = f_{Ii}^{\text{int}} = -J_{\xi}(\mathbf{0}) N_{I,i}(\mathbf{0}) p(\mathbf{0}) + \sum_{Q=1}^{4} \bar{w}_Q J_{\xi}(\boldsymbol{\xi}_Q) N_{I,j}(\boldsymbol{\xi}_Q) \sigma_{ji}^{\text{dev}}(\boldsymbol{\xi}_Q) \tag{4.5.33}$$

where, as indicated in the above, the single quadrature point for the reduced quadrature is the centroid of the parent element. The deviatoric part is integrated by full quadrature. This scheme is similar to the scheme used in linear analysis of incompressible materials. Selective-reduced schemes for other elements can be developed by similarly modifying the schemes for linear finite elements; see, for example, Hughes (1987).

4.5.5 Element Force and Matrix Transformations

Often element nodal forces and element matrices must be expressed in terms of alternative degrees of freedom, that is, for a different set of nodal displacements. In the following, transformations are developed for nodal forces and element matrices.

Consider an element or assemblage of elements with nodal displacements $\hat{\mathbf{d}}$. We wish to express the nodal forces for the nodal displacements $\mathbf{d}$ which are related to $\hat{\mathbf{d}}$ by

$$\frac{d\hat{\mathbf{d}}}{dt} = \mathbf{T}\frac{d\mathbf{d}}{dt} \quad \text{and} \quad \delta\hat{\mathbf{d}} = \mathbf{T}\delta\mathbf{d} \tag{4.5.34}$$

The nodal forces associated with $\mathbf{d}$ are then given by

$$\mathbf{f} = \mathbf{T}^T\hat{\mathbf{f}} \tag{4.5.35}$$

This transformation holds because the nodal forces and velocities are assumed to be conjugate in power: see Section 2.4.2. It is proven as follows. An increment of work is given by

$$\delta W = \delta\mathbf{d}^T\mathbf{f} = \delta\hat{\mathbf{d}}^T\hat{\mathbf{f}} \quad \forall\delta\mathbf{d} \tag{4.5.36}$$

Either set of nodal forces and virtual displacements must give an increment in work, since work is a scalar independent of the coordinate system or choice of generalized displacements. Substituting (4.5.34) into (4.5.36) gives

$$\delta\mathbf{d}^T\mathbf{f} = \delta\hat{\mathbf{d}}^T\mathbf{T}^T\hat{\mathbf{f}} \qquad \forall\delta\mathbf{d} \tag{4.5.37}$$

Since (4.5.37) holds for all $\delta\mathbf{d}$, (4.5.35) follows.

When $\mathbf{T}$ is independent of time, the mass matrices for the two sets of degrees of freedom are related by

$$\mathbf{M} = \mathbf{T}^T\hat{\mathbf{M}}\mathbf{T} \tag{4.5.38}$$

This is shown as follows. By (4.5.35)

$$\mathbf{f}^{\text{kin}} = \mathbf{T}^T\hat{\mathbf{f}}^{\text{kin}} \tag{4.5.39}$$

and using (4.4.20), we have

$$\mathbf{M}\dot{\mathbf{v}} = \mathbf{T}^T\hat{\mathbf{M}}\dot{\hat{\mathbf{v}}} \tag{4.5.40}$$

If $\mathbf{T}$ is independent of time, from (4.5.34), $\dot{\hat{\mathbf{v}}} = \mathbf{T}\dot{\mathbf{v}}$, and substituting this into the above and since this must hold for the arbitrary nodal accelerations, we obtain (4.5.38). If the $\mathbf{T}$ matrix is time dependent, then $\dot{\hat{\mathbf{v}}} = \mathbf{T}\dot{\mathbf{v}} + \dot{\mathbf{T}}\mathbf{v}$, so

$$\mathbf{f}^{\text{kin}} = \mathbf{T}^T\hat{\mathbf{M}}\mathbf{T}\ddot{\mathbf{d}} + \mathbf{T}^T\hat{\mathbf{M}}\dot{\mathbf{T}}\dot{\mathbf{d}} \tag{4.5.41}$$

A transformation similar to (4.5.38) holds for the linear stiffness matrix and the tangent stiffness discussed in Chapter 6:

$$\mathbf{K} = \mathbf{T}^T\hat{\mathbf{K}}\mathbf{T}, \quad \mathbf{K}^{\tan} = \mathbf{T}^T\hat{\mathbf{K}}^{\tan}\mathbf{T} \tag{4.5.42}$$

These transformations enable us to evaluate element matrices in other coordinate systems to simplify the derivation as in Example 4.6. They are also useful for treating slave nodes: see Example 4.5.

Example 4.1 Triangular three-node element The triangular element will be developed using triangular coordinates (also called area coordinates and barycentric coordinates). The element is shown in Figure 4.3. It is a three-node element with a linear displacement field; the thickness of the element is a. The nodes are numbered counterclockwise in the parent element, and they must be numbered counterclockwise in the initial configuration, otherwise the determinant of the map between the initial and parent domains will be negative.

The shape functions for the linear displacement triangle are the triangular coordinates, so $N_I = \xi_I$. The spatial coordinates are expressed in terms of the triangular coordinates ξ_I by

$$\begin{Bmatrix} x \\ y \\ 1 \end{Bmatrix} = \begin{bmatrix} x_1 & x_2 & x_3 \\ y_1 & y_2 & y_3 \\ 1 & 1 & 1 \end{bmatrix} \begin{Bmatrix} \xi_1 \\ \xi_2 \\ \xi_3 \end{Bmatrix} \tag{E4.1.1}$$

where we have appended the condition that the sum of the triangular element coordinates is 1. The inverse of (E4.1.1) is given by

$$\begin{Bmatrix} \xi_1 \\ \xi_2 \\ \xi_3 \end{Bmatrix} = \frac{1}{2A} \begin{bmatrix} y_{23} & x_{32} & x_2 y_3 & -x_3 y_2 \\ y_{31} & x_{13} & x_3 y_1 & -x_1 y_3 \\ y_{12} & x_{21} & x_1 y_2 & -x_2 y_1 \end{bmatrix} \begin{Bmatrix} x \\ y \\ 1 \end{Bmatrix} \tag{E4.1.2}$$

where we have used the notation

$$x_{IJ} = x_I - x_J \qquad y_{IJ} = y_I - y_J \tag{E4.1.3}$$

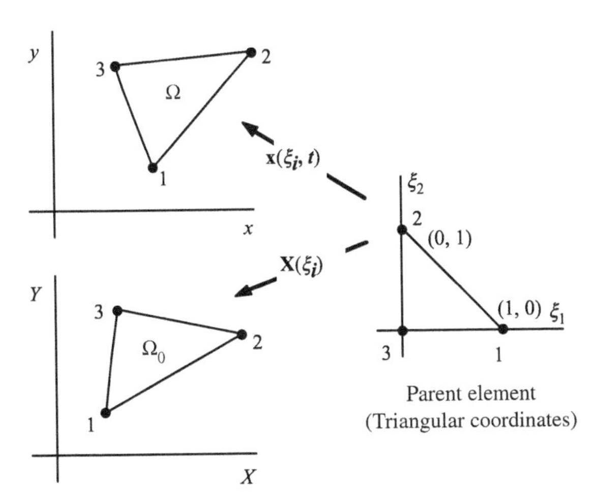

Figure 4.3 Triangular element showing node numbers and the mappings of the parent element to the initial and current configurations

and

$$2A = x_{32}y_{12} - x_{12}y_{32} \tag{E4.1.4}$$

where A is the current area of the element. As can be seen from the previous, in the triangular three-node element, the parent to current map (E4.1.1) can be inverted explicitly. This unusual circumstance is due to the fact that the map for this element is linear. However, the parent to current map is nonlinear for most other elements, so for most elements it cannot be inverted.

The derivatives of the shape functions can be determined directly from (E4.1.2) by inspection:

$$\left[N_{I,j}\right] = \left[\xi_{I,j}\right] = \begin{bmatrix} \xi_{1,x} & \xi_{1,y} \\ \xi_{2,x} & \xi_{2,y} \\ \xi_{3,x} & \xi_{3,y} \end{bmatrix} = \frac{1}{2A}\begin{bmatrix} y_{23} & x_{32} \\ y_{31} & x_{13} \\ y_{12} & x_{21} \end{bmatrix} \tag{E4.1.5}$$

We can obtain the map between the parent element and the initial configuration by writing (E4.1.1) at time $t = 0$, which gives

$$\begin{Bmatrix} X \\ Y \\ 1 \end{Bmatrix} = \begin{bmatrix} X_1 & X_2 & X_3 \\ Y_1 & Y_2 & Y_3 \\ 1 & 1 & 1 \end{bmatrix} \begin{Bmatrix} \xi_1 \\ \xi_2 \\ \xi_3 \end{Bmatrix} \tag{E4.1.6}$$

The inverse of this relation is identical to (E4.1.2) except that it is in terms of the initial coordinates

$$\begin{Bmatrix} \xi_1 \\ \xi_2 \\ \xi_3 \end{Bmatrix} = \frac{1}{2A_0}\begin{bmatrix} Y_{23} & X_{32} & X_2Y_3 - X_3Y_2 \\ Y_{31} & X_{13} & X_3Y_1 - X_1Y_3 \\ Y_{12} & X_{21} & X_1Y_2 - X_2Y_1 \end{bmatrix} \begin{Bmatrix} X \\ Y \\ 1 \end{Bmatrix} \tag{E4.1.7}$$

$$2A_0 = X_{32}Y_{12} - X_{12}Y_{32} \tag{E4.1.8}$$

where A_0 is the initial area of the element.

4.5.5.1 Voigt Notation

We first develop the element equations in Voigt notation, which should be familiar to those who have studied linear finite elements. Those who like the more condensed matrix notation can skip directly to that form. In Voigt notation, the displacement field is written in terms of triangular coordinates as

$$\begin{Bmatrix} u_x \\ u_y \end{Bmatrix} = \begin{bmatrix} \xi_1 & 0 & \xi_2 & 0 & \xi_1 & 0 \\ 0 & \xi_1 & 0 & \xi_2 & 0 & \xi_3 \end{bmatrix} \mathbf{d} = \mathbf{N}\mathbf{d} \tag{E4.1.9}$$

where $\mathbf{d}$ is the column matrix of nodal displacements, which is given by

$$\mathbf{d}^T = \left[u_{x1}, u_{y1}, u_{x2}, u_{y2}, u_{x3}, u_{y3}\right] \tag{E4.1.10}$$

The velocities are obtained by taking the material time derivatives of the displacements, giving

$$\begin{Bmatrix} v_x \\ v_y \end{Bmatrix} = \begin{bmatrix} \xi_1 & 0 & \xi_2 & 0 & \xi_3 & 0 \\ 0 & \xi_1 & 0 & \xi_2 & 0 & \xi_3 \end{bmatrix} \dot{\mathbf{d}} \tag{E4.1.11}$$

$$\dot{\mathbf{d}}^T = \left[v_{x1}, v_{y1}, v_{x2}, v_{y2}, v_{x3}, v_{y3}\right] \tag{E4.1.12}$$

Nodal velocities and nodal forces are shown in Figure 4.4.

The rate-of-deformation and stress column matrices in Voigt form are

$$\{\mathbf{D}\} = \begin{Bmatrix} D_{xx} \\ D_{yy} \\ 2D_{xy} \end{Bmatrix} \qquad \{\boldsymbol{\sigma}\} = \begin{Bmatrix} \sigma_{xx} \\ \sigma_{yy} \\ \sigma_{xy} \end{Bmatrix} \tag{E4.1.13}$$

where the factor of 2 on the shear velocity strain is needed in Voigt notation; see Appendix 1. Only the in-plane stresses are needed for the internal nodal forces in either plane stress or plain strain (σ_{zz} is needed in plane strain plasticity in the constitutive update). Since $\sigma_{zz} = 0$ in plane stress whereas $D_{zz} = 0$ in plane strain, $D_{zz}\sigma_{zz}$ makes no contribution to the power in either case. The transverse shear stresses, σ_{xz} and σ_{yz}, and the corresponding components of the rate-of-deformation, D_{xz} and D_{yz}, also vanish in both plane stress and plane strain.

By the definition of the rate-of-deformation, (3.3.10), and the velocity approximation, we have

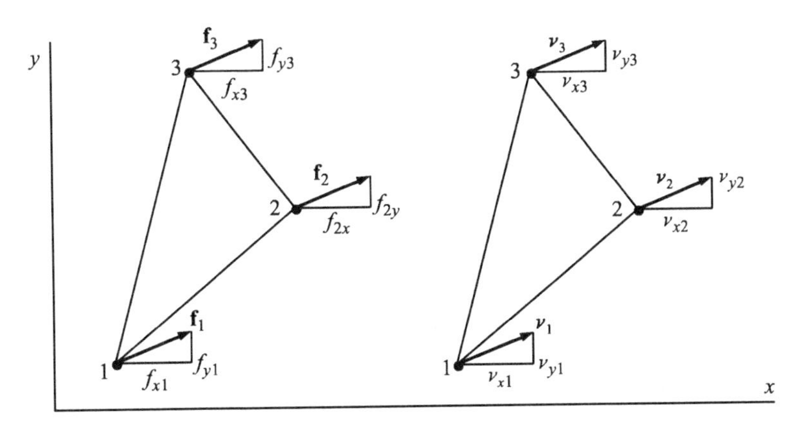

Figure 4.4 Triangular element showing the nodal force and velocity components

$$D_{xx} = \frac{\partial v_x}{\partial x} = \frac{\partial N_I}{\partial x} v_{xI}$$
$$D_{yy} = \frac{\partial v_y}{\partial y} = \frac{\partial N_I}{\partial y} v_{yI} \qquad \text{(E4.1.14)}$$
$$2D_{xy} = \frac{\partial v_x}{\partial y} + \frac{\partial v_y}{\partial x} = \frac{\partial N_I}{\partial x} v_{xI} + \frac{\partial N_I}{\partial_x} v_{yI}$$

In Voigt notation, the **B** matrix is constructed so it relates the rate-of-deformation to the nodal velocities by $\{D\} = \mathbf{B}\dot{\mathbf{d}}$, so using (E4.1.14) and the formulas for the derivatives of the triangular coordinates (E4.1.5), we have

$$\mathbf{B}_I = \begin{bmatrix} N_{I,x} & 0 \\ 0 & N_{I,}y \\ N_{I,y} & N_{I,x} \end{bmatrix} \quad [\mathbf{B}] = [\mathbf{B}_1 \;\; \mathbf{B}_2 \;\; \mathbf{B}_3] = \frac{1}{2A} \begin{bmatrix} y_{23} & 0 & y_{31} & 0 & y_{12} & 0 \\ 0 & x_{32} & 0 & x_{13} & 0 & x_{21} \\ x_{32} & y_{23} & x_{13} & y_{31} & x_{21} & y_{12} \end{bmatrix} \qquad \text{(E4.1.15)}$$

The internal nodal forces are then given by (4.5.12):

$$\begin{Bmatrix} f_{x1} \\ f_{y1} \\ f_{x2} \\ f_{x2} \\ f_{x3} \\ f_{y3} \end{Bmatrix} = \int_{\Omega} \mathbf{B}^T \{\boldsymbol{\sigma}\} d\Omega = \int_{\Omega} \frac{a}{2A} \begin{bmatrix} y_{23} & 0 & x_{32} \\ 0 & x_{32} & y_{23} \\ y_{31} & 0 & x_{13} \\ 0 & x_{13} & y_{31} \\ y_{12} & 0 & x_{21} \\ 0 & x_{21} & y_{12} \end{bmatrix} \begin{Bmatrix} \sigma_{xx} \\ \sigma_{yy} \\ \sigma_{xy} \end{Bmatrix} dA \qquad \text{(E4.1.16)}$$

where a is the thickness and we have used $d\Omega = adA$; if we assume that the stresses and thickness a are constant in the element, we obtain

$$\begin{Bmatrix} f_{x1} \\ f_{y1} \\ f_{x2} \\ f_{y_2} \\ f_{x3} \\ f_{y3} \end{Bmatrix}^{\text{int}} = \frac{a}{2} \begin{bmatrix} y_{23} & 0 & x_{32} \\ 0 & x_{32} & y_{23} \\ y_{31} & 0 & x_{13} \\ 0 & x_{13} & y_{31} \\ y_{12} & 0 & x_{21} \\ 0 & x_{21} & y_{12} \end{bmatrix} \begin{Bmatrix} \sigma_{xx} \\ \sigma_{yy} \\ \sigma_{xy} \end{Bmatrix} \qquad \text{(E4.1.17)}$$

In the three-node triangle, the stresses are sometimes not constant within the element; for example, when thermal stresses are included for a linear temperature field, the stresses are linear. In this case, or when the thickness a varies in the element, one-point quadrature is usually adequate. One-point quadrature is equivalent to (E4.1.17) with the stresses and thickness evaluated at the centroid of the element.

4.5.5.2 Matrix Form based on Indicial Notation

In the following, the expressions for the element are developed using a direct translation of the indicial expression to matrix form. The equations are more compact but not in the form commonly seen in linear finite element analysis.

Rate-of-deformation The velocity gradient is given by a matrix form of (4.4.7):

$$\mathbf{L}=[L_{ij}]=[v_{il}][N_{I,j}]=\begin{bmatrix} v_{x1} & v_{x2} & v_{x3} \\ v_{y1} & v_{y2} & v_{y3} \end{bmatrix}\frac{1}{2A}\begin{bmatrix} y_{23} & x_{32} \\ y_{31} & x_{13} \\ y_{12} & x_{21} \end{bmatrix}$$
$$=\frac{1}{2A}\begin{bmatrix} y_{23}v_{x1}+y_{31}v_{x2}+y_{12}v_{x3} & x_{32}v_{x1}+x_{13}v_{x2}+x_{21}v_{x3} \\ y_{23}v_{y1}+y_{31}v_{y2}+y_{12}v_{y3} & x_{32}v_{y1}+x_{13}v_{y2}+x_{21}v_{y3} \end{bmatrix} \tag{E4.1.18}$$

The rate-of-deformation is obtained from t(E4.1.18) by (3.3.10):

$$\mathbf{D}=\frac{1}{2}(\mathbf{L}+\mathbf{L}^T) \tag{E4.1.19}$$

As can be seen from (E4.1.18) and (E4.1.19), the rate-of-deformation is constant in the element. *Internal nodal forces* The internal forces are obtained by (4.5.16):

$$\mathbf{f}_{\text{int}}^T=\left|f_{Ii}\right|^{\text{int}}=\begin{bmatrix} f_{1x} & f_{1y} \\ f_{2x} & f_{2y} \\ f_{3x} & f_{3y} \end{bmatrix}^{\text{int}}=\int_\Omega [N_{I,j}]\,[\sigma_{ji}]\,d\Omega$$
$$=\int_A \frac{1}{2A}\begin{bmatrix} y_{23} & x_{32} \\ y_{31} & x_{13} \\ y_{12} & x_{21} \end{bmatrix}\begin{bmatrix} \sigma_{xx} & \sigma_{xy} \\ \sigma_{xy} & \sigma_{yy} \end{bmatrix} a\,dA \tag{E4.1.20}$$

where a is the thickness. If the stresses and thickness are constant within the element, the integrand is constant and the integral can be evaluated by multiplying the integrand by the volume aA, giving

$$\mathbf{f}_{\text{int}}^T=\frac{a}{2}\begin{bmatrix} y_{23} & x_{32} \\ y_{31} & x_{13} \\ y_{12} & x_{21} \end{bmatrix}\begin{bmatrix} \sigma_{xx} & \sigma_{xy} \\ \sigma_{xy} & \sigma_{yy} \end{bmatrix}=\frac{a}{2}\begin{bmatrix} y_{23}\sigma_{xx}+x_{32}\sigma_{xy} & y_{23}\sigma_{xy}+x_{32}\sigma_{yy} \\ y_{31}\sigma_{xx}+x_{13}\sigma_{xy} & y_{31}\sigma_{xy}+x_{13}\sigma_{yy} \\ y_{12}\sigma_{xx}+x_{21}\sigma_{xy} & y_{12}\sigma_{xy}+x_{21}\sigma_{yy} \end{bmatrix} \tag{E4.1.21}$$

This expression gives the same result as (E4.1.17). It is easy to show that the sums of each component of the nodal forces vanish, that is, the element is in equilibrium. Comparing (E4.1.20) with (E4.1.17), we see that the former involves fewer multiplications. Evaluating the Voigt form (E4.1.17) involves many multiplications by zero, which slows computations, particularly in the three-dimensional counterparts of these equations. However, the matrix indicial form is difficult to extend to the computation of stiffness matrices, so as will be seen in Chapter 6, the Voigt form is indispensable when stiffness matrices are needed.

Mass matrix The mass matrix is evaluated in the undeformed configuration by (4.4.57). The mass matrix is given by

$$\tilde{M}_{IJ} = \int_{\Omega} \rho_0 N_I N_J \, d\Omega_0 = \int_{\square} a_0 \rho_0 \xi_I \xi_J J_\xi^0 \, d\square \tag{E4.1.22}$$

where we have used; $d\Omega_0 = a_0 J_\xi^0 d\square$; the quadrature in the far right expression is over the parent element domain. Putting this in matrix form gives

$$\tilde{\mathbf{M}} = \int_{\square} a_0 \rho_0 \begin{bmatrix} \xi_1 \\ \xi_2 \\ \xi_3 \end{bmatrix} [\xi_1 \;\; \xi_2 \;\; \xi_3] J_\xi^0 \, d\square \tag{E4.1.23}$$

where the element Jacobian determinant for the initial configuration is given by $J_\xi^0 = 2A_0$, where A_0 is the initial area. Using the quadrature rule for triangular coordinates, the consistent mass matrix is

$$\tilde{\mathbf{M}} = \frac{\rho_0 A_0 a_0}{12} \begin{bmatrix} 2 & 1 & 1 \\ 1 & 2 & 1 \\ 1 & 1 & 2 \end{bmatrix} \tag{E4.1.24}$$

The mass matrix can be expanded to full size by using (4.4.51), $M_{iIjJ} = \delta_{ij} \tilde{M}_{IJ}$, and then using the rule of (2.4.25), which gives

$$\mathbf{M} = \frac{\rho_0 A_0 a_0}{12} \begin{bmatrix} 2 & 0 & 1 & 0 & 1 & 0 \\ 0 & 2 & 0 & 1 & 0 & 1 \\ 1 & 0 & 2 & 0 & 1 & 0 \\ 0 & 1 & 0 & 2 & 0 & 1 \\ 1 & 0 & 1 & 0 & 2 & 0 \\ 0 & 1 & 0 & 1 & 0 & 2 \end{bmatrix} \tag{E4.1.25}$$

The diagonal or lumped mass matrix can be obtained by the row-sum technique, giving

$$\tilde{\mathbf{M}} = \frac{\rho_0 A_0 a_0}{3} \begin{bmatrix} 1 & 0 & 0 \\ 0 & 1 & 0 \\ 0 & 0 & 1 \end{bmatrix} \tag{E4.1.26}$$

This matrix could also be obtained by simply assigning one-third of the mass of the element to each of the nodes.

External nodal forces To evaluate the external nodal forces, an interpolation of the body forces is needed. Let the body forces be approximated by linear interpolants expressed in terms of the triangular coordinates as

$$\begin{Bmatrix} b_x \\ b_y \end{Bmatrix} = \begin{bmatrix} b_{x1} & b_{x2} & b_{x3} \\ b_{y1} & b_{y2} & b_{y3} \end{bmatrix} \begin{Bmatrix} \xi_1 \\ \xi_2 \\ \xi_3 \end{Bmatrix} \tag{E4.1.27}$$

The matrix form of (4.4.15) gives

$$\begin{aligned} [\mathbf{f}_{iI}]^{\text{ext}} &= \begin{bmatrix} f_{x1} & f_{x2} & f_{x3} \\ f_{y1} & f_{y2} & f_{y3} \end{bmatrix}^{\text{ext}} = \int_\Omega \{b_i\}\{N_I\}^T \rho a\, dA \\ &= \begin{bmatrix} b_{x1} & b_{x2} & b_{x3} \\ b_{y1} & b_{y2} & b_{y3} \end{bmatrix} \int_\Omega \begin{bmatrix} \xi_1 \\ \xi_2 \\ \xi_3 \end{bmatrix} [\xi_1\, \xi_2\, \xi_3] \rho a\, dA \end{aligned} \tag{E4.1.28}$$

Using the integration rule for triangular coordinates with the thickness and density considered constant then gives

$$\mathbf{f}_{\text{ext}}^T = \frac{\rho A a}{12} \begin{bmatrix} b_{x1} & b_{x2} & b_{x3} \\ b_{y1} & b_{y2} & b_{y3} \end{bmatrix} \begin{bmatrix} 2 & 1 & 1 \\ 1 & 2 & 1 \\ 1 & 1 & 2 \end{bmatrix} \tag{E4.1.29}$$

To illustrate the computation of the external forces due to a prescribed traction, consider component i of the traction to be prescribed between nodes 1 and 2. If we approximate the traction by a linear interpolation, then

$$\bar{t}_i = \bar{t}_{i1}\xi_1 + \bar{t}_{i2}\xi_2 \tag{E4.1.30}$$

The external nodal forces are given by (4.4.15). We develop a row of the matrix:

$$[f_{i1} \quad f_{i2} \quad f_{13}]^{\text{ext}} = \int_{\Gamma_{12}} \bar{t}_i N_I d\Gamma \quad \int_0^1 (\bar{t}_{i1}\xi_1 + \bar{t}_{i2}\xi_2)[\xi_1\, \xi_2\, \xi_3] a\ell_{12} d\xi_1 \tag{E4.1.31}$$

where we have used $ds = \ell_{12}\, d\xi_1$; ℓ_{12} is the current length of the side connecting nodes 1 and 2. Along this side, $\xi_2 = 1 - \xi_1$, $\xi_3 = 0$, and evaluation of the integral in (E4.1.31) gives

$$[f_{i1} \quad f_{i2} \quad f_{i3}]^{\text{ext}} = \frac{a\ell_{12}}{6}[2\bar{t}_{i1} + \bar{t}_{i2} \quad \bar{t}_{i1} + 2\bar{t}_{i2} \quad 0] \tag{E4.1.32}$$

The external nodal forces are nonzero only on the nodes belonging to the side to which the traction is applied. This equation holds for an arbitrary local coordinate system. For an applied pressure, (E4.1.32) would be evaluated in a local coordinate system.

Example 4.2 Quadrilateral Element and Other Isoparametric 2D Elements Develop the expressions for the deformation gradient, the rate-of-deformation, the nodal forces and the mass matrix for two-dimensional isoparametric elements. Detailed expressions are given for the four-node quadrilateral. Expressions for the nodal internal forces are given in matrix form.

Shape functions and nodal variables The element shape functions are expressed in terms of the element coordinates (ξ, η). The spatial coordinates are expressed in terms of the shape functions and nodal coordinates by

$$\begin{Bmatrix} x(\boldsymbol{\xi}, t) \\ x(\boldsymbol{\xi}, t) \end{Bmatrix} = N_I(\boldsymbol{\xi}) \begin{Bmatrix} x_I(t) \\ y_I(t) \end{Bmatrix}, \quad \boldsymbol{\xi} = \begin{Bmatrix} \xi \\ \eta \end{Bmatrix} \tag{E4.2.1}$$

For the quadrilateral, the isoparametric shape functions are

$$N_I(\xi) = \frac{1}{4}(1 + \xi_I \xi)(1 + \eta_I \eta) \tag{E4.2.2}$$

where (ξ_I, η_I), $I = 1$ to 4, are the nodal coordinates of the parent element shown in Figure 4.5. They are given by

$$[\xi_{iI}] = \begin{Bmatrix} \xi_I \\ \eta_I \end{Bmatrix} = \begin{bmatrix} -1 & 1 & 1 & -1 \\ -1 & -1 & 1 & 1 \end{bmatrix} \tag{E4.2.3}$$

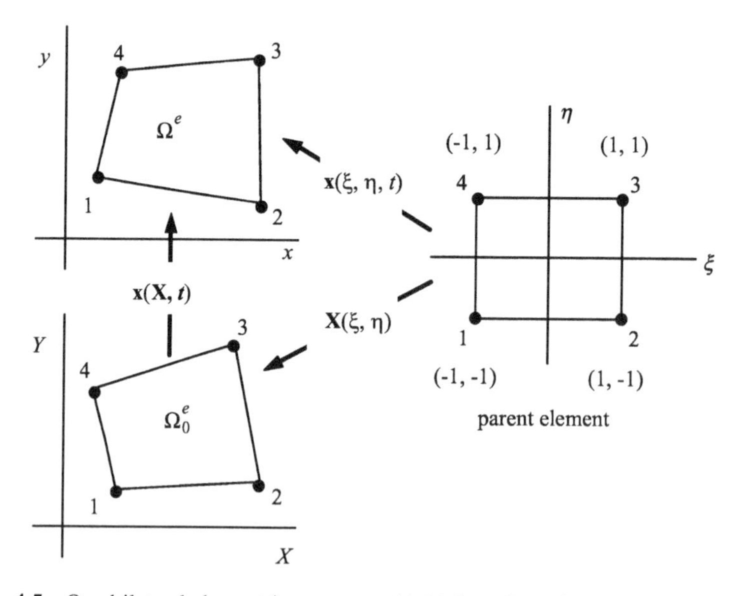

Figure 4.5 Quadrilateral element in current and initial configurations and the parent domain

Since (E4.2.1) also holds for $t = 0$, we can write

$$\begin{Bmatrix} X(\xi) \\ Y(\xi) \end{Bmatrix} = \begin{Bmatrix} X_I \\ Y_I \end{Bmatrix} N_I(\xi) \tag{E4.2.4}$$

where X_I, Y_I are the coordinates in the undeformed configuration. The nodal velocities are given by

$$\begin{Bmatrix} v_x(\xi, t) \\ v_y(\xi, t) \end{Bmatrix} = \begin{Bmatrix} v_{xI}(t) \\ v_{yI}(t) \end{Bmatrix} N_I(\xi) \tag{E4.2.5}$$

which is the material time derivative of the displacement.

Rate-of-deformation and internal nodal forces The map (E4.2.1) is not invertible for the shape functions (E4.2.2). Therefore it is impossible to write explicit expressions for ξ_i in terms of x and y, so the derivatives of the shape functions are evaluated by implicit differentiation. Referring to (4.4.42) and (4.4.40), we have

$$N_{I,\mathbf{x}}^T = [N_{I,x} \ N_{I,y}] = N_{I,\xi}^T \mathbf{x}_{,\xi}^{-1} = [N_{I,\xi} \ N_{I,\eta}] \begin{bmatrix} x_{,\xi} & x_{,\eta} \\ y_{,\xi} & y_{,\eta} \end{bmatrix}^{-1} \tag{E4.2.6}$$

The Jacobian of the current configuration with respect to the element coordinates is given by

$$\mathbf{x}_{,\xi} = \begin{bmatrix} x_{,\xi} & x_{,\eta} \\ y_{,\xi} & y_{,\eta} \end{bmatrix} = [x_{iI}][\partial N_I / \partial \xi_j] = \begin{Bmatrix} x_I \\ y_I \end{Bmatrix} [N_{I,\xi} \ N_{I,\eta}] = \begin{bmatrix} x_I N_{I,\xi} & x_I N_{I,\eta} \\ y_I N_{I,\xi} & y_I N_{I,\eta} \end{bmatrix} \tag{E4.2.7}$$

For the four-node quadrilateral this is

$$\mathbf{x}_{,\xi} = \frac{1}{4} \sum_{I=1}^{4} \begin{bmatrix} x_I(t)\xi_I(1+\eta_I\eta) & x_I(t)\eta_I(1+\xi_I\xi) \\ y_I(t)\xi_I(1+\eta_I\eta) & y_I(t)\eta_I(1+\xi_I\xi) \end{bmatrix} \tag{E4.2.8}$$

In (E4.2.8), the summation has been indicated explicitly because the index I appears three times. As can be seen from the RHS, the Jacobian matrix is a function of time. The inverse of $\mathbf{x}_{,\xi}$ is given by

$$\mathbf{x}_{,\xi}^{-1} = \frac{1}{J_\xi} \begin{bmatrix} y_{,\eta} & -x_{,\eta} \\ -y_{,\xi} & x_{,\xi} \end{bmatrix}, \quad J_\xi = x_{,\xi}\, y_{,\eta} - x_{,\eta} - x_{,\eta}\, y_{,\xi} \tag{E4.2.9}$$

The gradients of the shape functions for the four-node quadrilateral with respect to the element coordinates are given by

$$\mathbf{N},_{\xi}^{T}=\left[\partial N_I/\partial \xi_i\right]=\begin{bmatrix}\partial N_1/\partial\xi & \partial N_1/\partial\eta\\ \partial N_2/\partial\xi & \partial N_2/\partial\eta\\ \partial N_1/\partial\xi & \partial N_3/\partial\eta\\ \partial N_4/\partial\xi & \partial N_4/\partial\eta\end{bmatrix}=\frac{1}{4}\begin{bmatrix}\xi_1(1+\eta_1\eta) & \eta_1(1+\xi_1\xi)\\ \xi_2(1+\eta_2\eta) & \eta_2(1+\xi_2\xi)\\ \xi_3(1+\eta_3\eta) & \eta_3(1+\xi_3\xi)\\ \xi_4(1+\eta_4\eta) & \eta_4(1+\xi_4\xi)\end{bmatrix}$$

The gradients of the shape functions with respect to the spatial coordinates can then be computed by

$$\boldsymbol{B}_I=N_{I,\mathbf{x}}^{T}=N_{I,\xi}^{T}\mathbf{x},_{\xi}^{-1}=\frac{1}{4}\begin{bmatrix}\xi_1(1+\eta_1\eta) & \eta_1(1+\xi_1\xi)\\ \xi_2(1+\eta_2\eta) & \eta_2(1+\xi_2\xi)\\ \xi_3(1+\eta_3\eta) & \eta_3(1+\xi_3\xi)\\ \xi_4(1+\eta_4\eta) & \eta_4(1+\xi_4\xi)\end{bmatrix}\frac{1}{J_{\xi}}\begin{bmatrix}y,_{\eta} & -x,_{\eta}\\ -y,_{\xi} & x,_{\xi}\end{bmatrix}\qquad\text{(E4.2.10)}$$

and the velocity gradient is given by (4.5.3):

$$\mathbf{L}=\mathbf{v}_I\boldsymbol{B}_I^{T}=\mathbf{v}_I N_{I,\mathbf{x}}^{T}\qquad\text{(E4.2.11)}$$

For a four-node quadrilateral which is not rectangular, the velocity gradient, and hence the rate-of-deformation, is a rational function because $J_{\xi}=\det(\mathbf{x},_{\xi})$ appears in the denominator of $\mathbf{x},_{\xi}$. The determinant J_{ξ} is a linear function in (ξ,η).

The internal nodal forces are obtained by (4.5.7), which gives

$$\left(\mathbf{f}_I^{\text{int}}\right)^T=[f_{xI}\quad f_{yI}]^{\text{int}}=\int_{\Omega}\boldsymbol{B}_I^T\boldsymbol{\sigma}d\Omega=\int_{\Omega}[N_{I,x}\quad N_{I,y}]\begin{bmatrix}\sigma_{xx} & \sigma_{xy}\\ \sigma_{xy} & \sigma_{yy}\end{bmatrix}d\Omega\qquad\text{(E4.2.12)}$$

The integration is performed over the parent domain. For this purpose, we use

$$d\Omega=J_{\xi}a\,d\xi\,d\eta\qquad\text{(E4.2.13)}$$

where a is the thickness, so

$$\left(\mathbf{f}_I^{\text{int}}\right)^T=[f_{xI}\quad f_{yI}]^{\text{int}}=\int_{\square}[N_{I,x}\quad N_{I,y}]\begin{bmatrix}\sigma_{xx} & \sigma_{xy}\\ \sigma_{xy} & \sigma_{yy}\end{bmatrix}aJ_{\xi}\,d\square\qquad\text{(E4.2.14)}$$

This applies to any isoparametric element in two dimensions. The integrand is a rational function of the element coordinates, since J_{ξ} appears in the denominator (see (E4.2.9)), so analytic quadrature of the above is not feasible. Therefore numerical quadrature is generally used. For the four-node quadrilateral, 2×2 Gauss quadrature is full quadrature. However, for full quadrature, the element locks for incompressible and nearly incompressible materials in plane strain problems, so selective-reduced quadrature as described in Section 4.5.4 must be used. The displacement for a four-node quadrilateral is linear along each edge. Therefore, the external nodal forces are identical to those for the three-node triangle: see (E4.1.28–32).

Mass matrix The consistent mass matrix is obtained by (4.4.57), which gives

$$\tilde{\mathbf{M}} = \int_{\Omega_0} \begin{bmatrix} N_1 \\ N_2 \\ N_3 \\ N_4 \end{bmatrix} [N_1\; N_2\; N_3\; N_4]\, \rho_0 \, d\Omega_0 \qquad \text{(E4.2.15)}$$

We use

$$d\Omega_0 = J_\xi^0(\xi,\eta)\, a_0 \, d\xi \, d\eta \qquad \text{(E4.2.16)}$$

where $J_\xi^0(\xi,\eta)$ is the determinant of the Jacobian of the transformation of the parent element to the initial configuration and a_0 is the thickness of the undeformed element. The expression for $\tilde{\mathrm{M}}$ when evaluated on the parent domain is given by

$$\tilde{\mathbf{M}} = \int_{-1}^{+1}\int_{-1}^{+1} \begin{bmatrix} N_1^2 & N_1 N_2 & N_1 N_3 & N_1 N_4 \\ & N_2^2 & N_2 N_3 & N_2 N_4 \\ \text{sym} & & N_3^2 & N_3 N_4 \\ & & & N_4^2 \end{bmatrix} \rho_0 a_0 J_\xi^0(\xi,\eta)\, d\xi \, d\eta \qquad \text{(E4.2.17)}$$

The matrix is evaluated by numerical quadrature. This mass matrix can be expanded to an 8×8 matrix by the same procedure as described for the triangle in the previous example.

A lumped, diagonal mass matrix can be obtained by Lobatto quadrature with the quadrature points coincident with the nodes. If we denote the integrand of (E4.2.17) by $\mathbf{m}(\xi_I, \eta_I)$, then Lobatto quadrature gives

$$\tilde{\mathbf{M}} = \sum_{I=1}^{4} \mathbf{m}(\xi_I, \eta_I) \qquad \text{(E4.2.18)}$$

Alternatively, a diagonal mass matrix can be obtained by apportioning the total mass of the element equally among the four nodes. The total mass is $\rho_0 A_0 A_0$ when A_0 is constant, so dividing it among the four nodes gives

$$\tilde{\mathbf{M}} = \frac{1}{4}\rho_0 A_0 a_0 \mathbf{I}_4 \qquad \text{(E4.2.19)}$$

where $\mathbf{I}_4$ is the unit matrix of order 4.

Example 4.3 Three-Dimensional Isoparametric Element Develop the expressions for the rate-of-deformation, the nodal forces and the mass matrix for three-dimensional isoparametric elements. An example of this class of elements, the eight-node hexahedron, is shown in Figure 4.6.

Motion and strain measures The motion of the element is given by

$$\begin{Bmatrix} x \\ y \\ z \end{Bmatrix} = N_I(\boldsymbol{\xi}) \begin{Bmatrix} x_I(t) \\ y_I(t) \\ z_I(t) \end{Bmatrix}, \quad \boldsymbol{\xi} = (\xi, \eta, \zeta) \tag{E4.3.1}$$

where the shape functions for particular elements are given in Appendix 3. Equation (E4.3.1) also holds at time $t = 0$, so

$$\begin{Bmatrix} X \\ Y \\ Z \end{Bmatrix} = N_I(\boldsymbol{\xi}) \begin{Bmatrix} X_I \\ Y_I \\ Z_I \end{Bmatrix} \tag{E4.3.2}$$

The velocity field is given by

$$\begin{Bmatrix} v_x \\ v_y \\ v_z \end{Bmatrix} = N_I(\boldsymbol{\xi}) \begin{Bmatrix} v_{xI} \\ v_{yI} \\ v_{zI} \end{Bmatrix} \tag{E4.3.3}$$

The velocity gradient is obtained from (4.5.3), giving

$$\mathbf{B}_I^T = [N_{I,x} \quad N_{I,y} \quad N_{I,z}] \tag{E4.3.4}$$

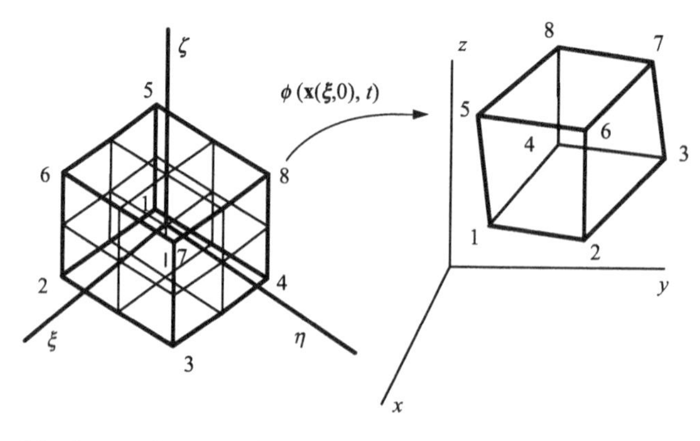

Figure 4.6 Parent element and current configuration for an eight-node hexahedron

$$\mathbf{L} = \mathbf{v}_I B_I^T = \begin{Bmatrix} v_{xI} \\ v_{yI} \\ v_{zI} \end{Bmatrix} [N_{I,x} \quad N_{I,y} \quad N_{I,z}] \tag{E4.3.5}$$

$$= \begin{bmatrix} v_{xI} N_{I,x} & v_{xI} N_{I,y} & v_{xI} N_{I,z} \\ v_{yI} N_{I,x} & v_{yI} N_{I,y} & v_{yI} N_{I,z} \\ v_{zI} N_{I,x} & v_{zI} N_{I,y} & v_{zI} N_{I,z} \end{bmatrix} \tag{E4.3.6}$$

The derivatives with respect to spatial coordinates are obtained in terms of derivatives with respect to the element coordinates by (4.4.42):

$$N_{I,\mathbf{x}}^T = N_{I,\xi}^T \mathbf{x}_{,\xi}^{-1} \tag{E4.3.7}$$

$$\mathbf{x}_{,\xi} \equiv \mathbf{F}_\xi = \mathbf{x}_I N_{I,\xi}^T = \begin{Bmatrix} x_I \\ y_I \\ z_I \end{Bmatrix} [N_{I,\xi} \quad N_{I,\eta} \quad N_{I,\zeta}] \tag{E4.3.8}$$

The deformation gradient can be computed by (3.2.19), (E4.3.1) and (E4.3.7):

$$\mathbf{F} = \frac{\partial \mathbf{x}}{\partial \mathbf{X}} = \mathbf{x}_I N_{I,\mathbf{x}} = \mathbf{x}_I N_{I,\xi}^T \mathbf{X}_{,\xi}^{-1} \equiv \mathbf{x}_I N_{I,\xi}^T \left(\mathbf{F}_\xi^0\right)^{-1} \tag{E4.3.9}$$

where

$$\mathbf{X}_{,\xi} \equiv \mathbf{F}_\xi^0 = \mathbf{X}_I N_{I,\xi}^T \tag{E4.3.10}$$

The Green strain is then computed by (3.3.5); a more accurate procedure is described in Example 4.11.

Internal nodal forces The internal nodal forces are obtained by (4.5.7):

$$\left(\mathbf{f}_I^{\text{int}}\right)^T = \left[f_{xI}, \quad f_{yI}, \quad f_{zI}\right]^{\text{int}} \int_\Omega \mathbf{B}_I^T \boldsymbol{\sigma}\, d\Omega = \int_\square \left[N_{I,x} \quad N_{I,y} \quad N_{I,z}\right] \begin{bmatrix} \sigma_{xx} & \sigma_{xy} & \sigma_{xz} \\ \sigma_{xy} & \sigma_{yy} & \sigma_{yz} \\ \sigma_{xz} & \sigma_{yz} & \sigma_{zz} \end{bmatrix} J_\xi d_\square \tag{E4.3.11}$$

The integral is evaluated by numerical quadrature (4.5.22).

External nodal forces We consider first the nodal forces due to the body force. By (4.4.15), we have

$$f_{iI}^{\text{ext}} = \int_{\Omega} N_I \rho b_i \, d\Omega = \int_{\square} N_I(\boldsymbol{\xi})\rho(\boldsymbol{\xi})b_i(\boldsymbol{\xi})J_{\xi} \, d\square \tag{E4.3.12}$$

where we have transformed the integral to the parent domain. The integral over the parent domain is evaluated by numerical quadrature.

Next we obtain the external nodal forces due to an applied pressure $\mathbf{t} = -p\mathbf{n}$ on a surface of the element. For example, consider the external surface corresponding to the parent element surface $\zeta = -1$.

On any surface, any dependent variable can be expressed as a function of two parent coordinates; in this case they are ζ and η. The vectors $\mathbf{x},_{\xi}$ and $\mathbf{x},_{\eta}$ are tangent to the surface. The vector $\mathbf{x},_{\xi} \times x,_{\eta}$ is in the direction of the normal $\mathbf{n}$ and, as shown in any advanced calculus text, its magnitude is the surface Jacobian, so we can write

$$p\mathbf{n} \, d\Gamma = -p\mathbf{x},_{\xi} \times \mathbf{x},_{\eta} \, d\xi \, d\eta \tag{E4.3.13}$$

The nodal external forces are then given by

$$f_{iI}^{\text{ext}} = \int_{\Gamma} t_i N_I \, d\Gamma = -\int_{\Gamma} pn_i N_I \, d\Gamma = \int_{-1}^{1}\int_{-1}^{1} p \, e_{ijk} x_{j,\xi} x_{k,\eta} N_I \, d\xi d\eta \tag{E4.3.14}$$

where we have used (E4.3.13) in indicial form in the last step. Another form of the above is

$$\mathbf{f}_I^{\text{ext}} = -\int_{-1}^{1}\int_{-1}^{1} pN_I \mathbf{x},_{\xi} \times \mathbf{x},_{\eta} \, d\xi \, d\eta \tag{E4.3.15}$$

We can expand (E4.3.15) by using (4.4.1) to express the tangent vectors in terms of the shape functions and by writing the cross product in determinant form, giving

$$\mathbf{f}_I^{\text{ext}} = f_{xI}\mathbf{e}_x + f_{yI}\mathbf{e}_y + f_{zI}\mathbf{e}_z = -\int_{-1}^{1}\int_{-1}^{1} pN_I \det \begin{bmatrix} \mathbf{e}_x & \mathbf{e}_y & \mathbf{e}_z \\ x_J N_{J,\xi} & y_J N_{J,\xi} & z_J N_{J,\xi} \\ x_K N_{K,\eta} & y_K N_{K,\eta} & z_K N_{K,\eta} \end{bmatrix} d\xi \, d\eta \tag{E4.3.16}$$

This integral can readily be evaluated by numerical quadrature over the loaded surfaces of the parent element.

Example 4.4 Axisymmetric quadrilateral The expressions for the rate-of-deformation and the nodal forces for the axisymmetric quadrilateral element are developed. The element is shown in Figure 4.7. The domain of the element is the volume swept out by rotating the quadrilateral 2π radians about the axis of symmetry, the z-axis. The expressions in indicial notation, (4.5.3) and (4.5.7), are not directly applicable since they do not apply to curvilinear coordinates.

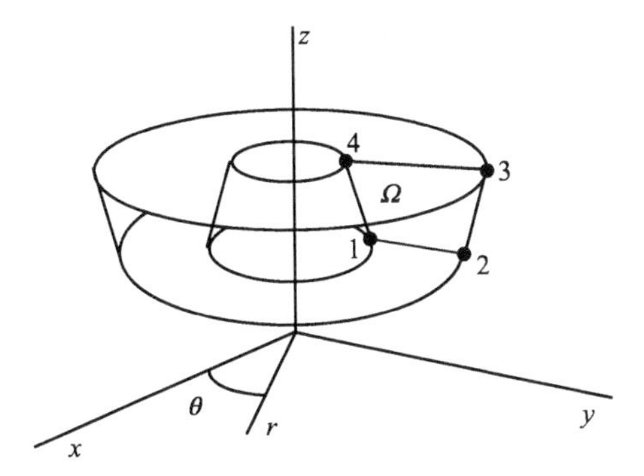

Figure 4.7 Current configuration of quadrilateral axisymmetric element; the element consists of the volume generated by rotating the quadrilateral 2π radians about the z-axis

In this case, the isoparametric map relates the cylindrical coordinates $[r, z]$ to the parent element coordinates $[\xi, \eta]$:

$$\begin{Bmatrix} r(\xi, \eta, t) \\ z(\xi, \eta, t) \end{Bmatrix} = \begin{Bmatrix} r_I(t) \\ z_I(t) \end{Bmatrix} N_I(\xi, \eta) \tag{E4.4.1}$$

where the shape functions N_I are given in (E4.2.2). The expression for the rate-of-deformation is based on standard expressions for the gradient in cylindrical coordinates (the expressions are identical to those for the linear strain):

$$\begin{Bmatrix} D_r \\ D_z \\ D_\theta \\ 2D_{rz} \end{Bmatrix} = \begin{bmatrix} \frac{\partial}{\partial r} & 0 \\ 0 & \frac{\partial}{\partial z} \\ \frac{1}{r} & 0 \\ \frac{\partial}{\partial z} & \frac{\partial}{\partial r} \end{bmatrix} \begin{Bmatrix} v_r \\ v_z \end{Bmatrix} = \begin{Bmatrix} \frac{\partial v_r}{\partial r} \\ \frac{\partial v_z}{\partial z} \\ \frac{v_r}{r} \\ \frac{\partial v_r}{\partial z} + \frac{\partial v_z}{\partial r} \end{Bmatrix} \tag{E4.4.2}$$

The conjugate stress is

$$\{\sigma\}^T = [\sigma_r, \sigma_z, \sigma_\theta, \sigma_{rz}] \tag{E4.4.3}$$

The velocity field is given by

$$\begin{Bmatrix} v_r \\ v_z \end{Bmatrix} = N_I(\xi, \eta) \begin{Bmatrix} v_{rI} \\ v_{zI} \end{Bmatrix} = \begin{bmatrix} N_1 & 0 & N_2 & 0 & N_3 & 0 & N_4 & 0 \\ 0 & N_1 & 0 & N_2 & 0 & N_3 & 0 & N_4 \end{bmatrix} \dot{\mathbf{d}} \tag{E4.4.4}$$

$$\dot{\mathbf{d}}^T = [v_{r1}, v_{z1}, v_{r2}, v_{z2}, v_{r3}, v_{z3}, v_{r4}, v_{z4}] \tag{E4.4.5}$$

The submatrices of the **B** matrix are given from (E4.4.2) by

$$[\mathbf{B}]_I = \begin{bmatrix} \dfrac{\partial N_I}{\partial r} & 0 \\ 0 & \dfrac{\partial N_I}{\partial z} \\ \dfrac{N_I}{r} & 0 \\ \dfrac{\partial N_I}{\partial z} & \dfrac{\partial N_I}{\partial r} \end{bmatrix} \tag{E4.4.6}$$

The derivatives in (E4.4.6) now have to be expressed in terms of derivatives with respect to the parent element coordinates. Rather than obtaining these with a matrix product, we just write out the expressions using (E4.2.9) with x, y replaced by r, z, which gives

$$\frac{\partial N_I}{\partial r} = \frac{1}{J_\xi}\left(\frac{\partial z}{\partial \eta}\frac{\partial N_I}{\partial \xi} - \frac{\partial z}{\partial \xi}\frac{\partial N_I}{\partial \eta}\right) \tag{E4.4.7}$$

$$\frac{\partial N_I}{\partial z} = \frac{1}{J_\xi}\left(\frac{\partial r}{\partial \xi}\frac{\partial N_I}{\partial \eta} - \frac{\partial r}{\partial \eta}\frac{\partial N_I}{\partial \xi}\right) \tag{E4.4.8}$$

where

$$\frac{\partial z}{\partial \eta} = z_I \frac{1}{J_\xi}\frac{\partial N_I}{\partial \eta} \quad \frac{\partial z}{\partial \xi} = z_I \frac{\partial N_I}{\partial \xi} \tag{E4.4.9}$$

$$\frac{\partial r}{\partial \eta} = r_I \frac{\partial N_I}{\partial \eta} \quad \frac{\partial r}{\partial \xi} = r_I \frac{\partial N_I}{\partial \xi} \tag{E4.4.10}$$

The nodal forces are obtained from (4.5.16), which yields

$$\mathbf{f}_I^{\text{int}} = \int_\Omega \mathbf{B}_I^T \{\boldsymbol{\sigma}\} d\Omega = 2\pi \int_\Box \mathbf{B}_I^T \{\boldsymbol{\sigma}\} J_\xi r \, d\Box \tag{E4.4.11}$$

where $\mathbf{B}_I$ is given by (E4.4.6) and we have used $d\Omega = 2\pi r J_\xi d\Box$ where r is given by (E4.4.1). The factor 2π is often omitted from all nodal forces.

Example 4.5 Master–slave tieline A master–slave tieline is shown in Figure 4.8. Tielines are frequently used to connect parts of the mesh which use different element sizes, for they are more convenient than connecting the elements of different sizes by triangles or tetrahedra. Continuity of the motion across the tieline is enforced by constraining the motion of the slave nodes to the field of the adjacent edge connecting the master nodes. In the following, the resulting nodal forces and mass matrix are developed by the transformation rules of Section 4.5.5.

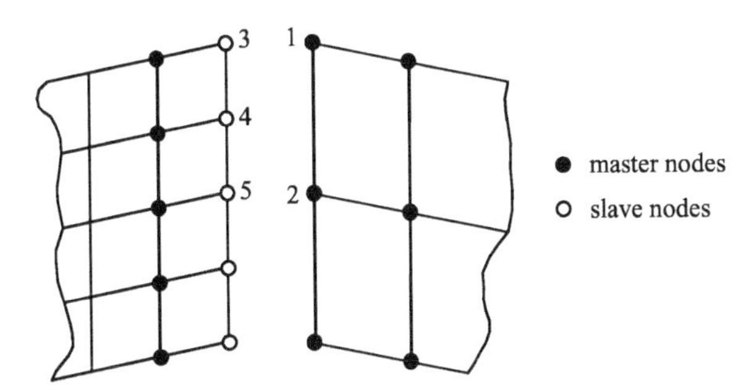

Figure 4.8 Exploded view of a tieline; when joined together, the velocities of nodes 3 and 5 equal the nodal velocities of nodes 1 and 2 respectively, and the velocity of node 4 is given in terms of nodes 1 and 2 by a linear constraint

The slave node velocities are given by the kinematic constraint that the velocities along the two sides of the tieline must remain compatible, that is, C^0. This constraint can be expressed as a linear relation in the nodal velocities, so the relation corresponding to (4.5.34) can be written as

$$\begin{Bmatrix} \hat{\mathbf{v}}_M \\ \hat{\mathbf{v}}_S \end{Bmatrix} = \begin{bmatrix} \mathbf{I} \\ \mathbf{A} \end{bmatrix} \{\mathbf{v}_M\} \quad \text{so} \quad \mathbf{T} = \begin{bmatrix} \mathbf{I} \\ \mathbf{A} \end{bmatrix} \tag{E4.5.1}$$

where the matrix $\mathbf{A}$ is obtained from the linear constraint and the superposed 'hats' indicate the velocities of the disjoint model before the two sides are tied together. We denote the nodal forces of the disjoint model by $\hat{\mathbf{f}}_S$ and $\hat{\mathbf{f}}_M$, respectively, where $\hat{\mathbf{f}}_S$ is the matrix of nodal forces assembled from the elements on the slave side of the tieline and $\hat{\mathbf{f}}_M$ is the matrix of nodal forces assembled from the elements on the master side of the tieline. The nodal forces for the joined model are then given by (4.5.35):

$$\{\mathbf{f}_M\} = \mathbf{T}^T \begin{Bmatrix} \hat{\mathbf{f}}_M \\ \hat{\mathbf{f}}_S \end{Bmatrix} = \begin{bmatrix} \mathbf{I} \ \mathbf{A}^T \end{bmatrix} \begin{Bmatrix} \hat{\mathbf{f}}_M \\ \hat{\mathbf{f}}_S \end{Bmatrix} \tag{E4.5.2}$$

where $\mathbf{T}$ is given by (E4.5.1). As can be seen from the above, the master nodal forces are the sum of the master nodal forces for the disjoint model and the transformed slave node forces. These formulas apply to both the external and internal nodal forces.

The consistent mass matrix is given by (4.5.40):

$$\mathbf{M} = \mathbf{T}^T \hat{\mathbf{M}} \mathbf{T} = \begin{bmatrix} \mathbf{I} \ \mathbf{A}^T \end{bmatrix} \begin{bmatrix} \hat{\mathbf{M}}_M & 0 \\ 0 & \hat{\mathbf{M}}_S \end{bmatrix} \begin{bmatrix} \mathbf{I} \\ \mathbf{A} \end{bmatrix} = \hat{\mathbf{M}}_M + \mathbf{A}^T \hat{\mathbf{M}}_S \mathbf{A} \tag{E4.5.3}$$

We illustrate these transformations in more detail for the five nodes which are numbered in Figure 4.8. The elements are four-node quadrilaterals, so the velocity along any edge is linear. Slave nodes 3 and 5 are coincident with master nodes 1 and 2, and slave node 4 is at a distance $\xi\ell$ from node 1, where $\ell = \|\mathbf{x}_2 - \mathbf{x}_1\|$. Therefore,

$$\mathbf{v}_3 = \mathbf{v}_1, \quad \mathbf{v}_5 = \mathbf{v}_2, \quad \mathbf{v}_4 = \xi\mathbf{v}_2 + (1-\xi)\mathbf{v}_1 \tag{E4.5.4}$$

and (E4.5.1) can be written as

$$\begin{Bmatrix} \mathbf{v}_1 \\ \mathbf{v}_2 \\ \mathbf{v}_3 \\ \mathbf{v}_4 \\ \mathbf{v}_5 \end{Bmatrix} = \begin{bmatrix} \mathbf{I} & \mathbf{0} \\ \mathbf{0} & \mathbf{I} \\ \mathbf{I} & 0 \\ (1-\xi)\mathbf{I} & \xi\mathbf{I} \\ 0 & \mathbf{I} \end{bmatrix} \begin{Bmatrix} \mathbf{v}_1 \\ \mathbf{v}_2 \end{Bmatrix}, \quad \mathbf{T} = \begin{bmatrix} \mathbf{I} & \mathbf{0} \\ \mathbf{0} & \mathbf{I} \\ \mathbf{I} & 0 \\ (1-\xi)\mathbf{I} & \xi\mathbf{I} \\ 0 & \mathbf{I} \end{bmatrix} \tag{E4.5.5}$$

The nodal forces are then given by

$$\begin{bmatrix} \mathbf{f}_1 \\ \mathbf{f}_2 \end{bmatrix} = \begin{bmatrix} \mathbf{I} & \mathbf{0} & \mathbf{I} & (1-\xi)\mathbf{I} & \mathbf{0} \\ \mathbf{0} & \mathbf{I} & \mathbf{0} & \xi\mathbf{I} & \mathbf{I} \end{bmatrix} \begin{Bmatrix} \hat{\mathbf{f}}_1 \\ \hat{\mathbf{f}}_2 \\ \hat{\mathbf{f}}_3 \\ \hat{\mathbf{f}}_4 \\ \hat{\mathbf{f}}_5 \end{Bmatrix} \tag{E4.5.6}$$

The force for master node 1 is

$$\mathbf{f}_1 = \hat{\mathbf{f}}_1 + \hat{\mathbf{f}}_3 + (1-\xi)\hat{\mathbf{f}}_4 \tag{E4.5.7}$$

Both components of the nodal force transform identically; the transformation applies to both internal and external nodal forces.

If the two lines are only tied in the normal direction, a local coordinate system needs to be set up at the nodes. The normal components of the nodal forces are then related by a relation similar to (E4.5.6), whereas the tangential components remain independent.

4.6 Corotational Formulations

In structural elements such as bars, beams and shells, it is awkward to deal with fixed coordinate systems. Consider, for example, a rotating rod such as shown in Figure 3.10. Initially, the only nonzero stress is σ_x, whereas σ_y vanishes. Subsequently, as the rod rotates it becomes awkward to express the state of uniaxial stress in terms of the global components of the stress tensor.

A natural approach to overcoming this difficulty is to embed a coordinate system in the bar and rotate it with the rod. Such coordinate systems are known as *corotational coordinates*. For example, consider a coordinate system, $\hat{\mathbf{x}} = [\hat{x}, \hat{y}]$ so that $\hat{x}$ always connects nodes 1 and 2, as

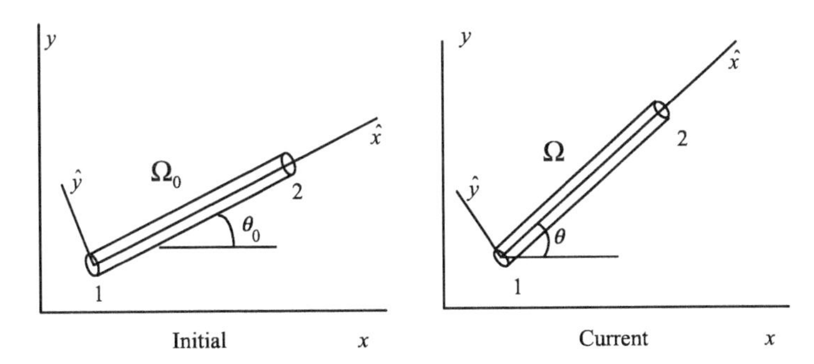

Figure 4.9 Two-node rod element showing initial configuration and current configuration and the corotational coordinate

shown in Figure 4.9. In a uniaxial state of stress $\hat{\sigma}_y = \hat{\sigma}_{xy} = 0$ and $\hat{\sigma}_x$ is nonzero. Similarly the rate-of-deformation of the rod is described by the component $\hat{D}_x$.

There are two approaches to corotational finite element formulations:

1. A coordinate system is embedded at each quadrature point and rotated with material in some sense.
2. A coordinate system is embedded in an element and rotated with the element.

The first procedure is valid for arbitrarily large strains and large rotations. A key consideration in corotational formulations lies in defining the rotation of the material. In most cases, the polar decomposition theorem is used to define a rotation. However, when particular directions of the material have a large stiffness which must be represented accurately, the rotation provided by a polar decomposition does not necessarily provide the best rotation for a Cartesian coordinate system; this is discussed in Chapter 9.

For some elements, such as a rod or the constant-strain triangle, the rigid body rotation is the same throughout the element. It is then sufficient to embed a single coordinate system in the element. For higher-order elements, if the strains are small, the coordinate system does not need to rotate exactly with the material. For example, the corotational coordinate system can be defined to be coincident to one side of the element. If the rotations relative to the embedded coordinate system are of order 9, then the error in the strains is of order θ^2. Therefore, as long as θ^2 is small compared to the strains, a single embedded coordinate system is adequate. These applications are often known as *small-strain, large-rotation* problems; see Wempner (1969) and Belytschko and Hsieh (1973)).

The components of a vector $\mathbf{v}$ in the corotational system are related to the global components by

$$\hat{v}_i = R_{ji} v_j \quad \text{or} \quad \hat{\mathbf{v}} = \mathbf{R}^T \mathbf{v} \quad \text{and} \quad \mathbf{v} = \mathbf{R}\hat{\mathbf{v}} \tag{4.6.1}$$

where $\mathbf{R}$ is an orthogonal transformation matrix defined in (3.2.33–34) and the superposed 'hat' indicates the corotational components.

The corotational components of the finite element approximation to the velocity field can be written as

$$\hat{v}_i(\xi, t) = N_I(\xi)\hat{v}_{iI}(t) \tag{4.6.2}$$

This expression is identical to (4.4.37) except that it pertains to the corotational components. Equation (4.6.2) can be obtained from (4.4.37) by multiplying both sides by $\mathbf{R}^T$.

The corotational components of the velocity gradient tensor are given by

$$\hat{L}_{ij} = \frac{\partial \hat{v}_i}{\partial \hat{x}_j} = \frac{\partial N_I(\xi)}{\partial \hat{x}_j}\hat{v}_{iI}(t) = \hat{B}_{jI}\hat{v}_{iI} \quad \text{or} \quad \hat{\mathbf{L}} = \hat{\mathbf{v}}_I \frac{\partial N_I}{\partial \hat{\mathbf{x}}} = \hat{\mathbf{v}}_I N_{I,\hat{x}}^T = \hat{\mathbf{v}}_I \hat{\boldsymbol{B}}_I^T \tag{4.6.3}$$

where

$$\hat{B}_{jI} = \frac{\partial N_I}{\partial \hat{x}_j} \tag{4.6.4}$$

The corotational rate-of-deformation tensor is then given by

$$\hat{D}_{ij} = \frac{1}{2}\left(\hat{L}_{ij} + \hat{L}_{ji}\right) = \frac{1}{2}\left(\frac{\partial \hat{v}_i}{\partial \hat{x}_j} + \frac{\partial \hat{v}_j}{\partial \hat{x}_i}\right) \tag{4.6.5}$$

The corotational formulation is used only for the evaluation of internal nodal forces. The external nodal forces and the mass matrix are usually evaluated in the global system. The semidiscrete equations of motion are treated in terms of global components. We therefore concern ourselves only with the evaluation of the internal nodal forces in the corotational formulation.

The expression for $\hat{\mathbf{f}}_I^{\text{int}}$ in terms of corotational components is developed as follows. We start with the standard expression for the nodal internal forces, (4.5.6):

$$f_{iI}^{\text{int}} = \int_\Omega \frac{\partial N_I}{\partial x_j}\sigma_{ji} d\Omega \quad \text{or} \quad (\mathbf{f}_{iI}^{\text{int}})^T \int_\Omega N_I^T{}_{,\mathbf{x}} \boldsymbol{\sigma} \, d\Omega \tag{4.6.6}$$

By the chain rule and (3.2.31)

$$\frac{\partial N_I}{\partial x_j} = \frac{\partial N_I}{\partial \hat{x}_k}\frac{\partial \hat{x}_k}{\partial x_j} = \frac{\partial N_I}{\partial \hat{x}_k}R_{jk} \quad \text{or} \quad N_{I,\mathbf{x}} = \mathbf{R}N_{I,\hat{\mathbf{x}}} \tag{4.6.7}$$

Substituting the transformation for the Cauchy stress into the corotational stress, Box 3.2, and (4.6.7) into (4.6.6), we obtain

$$\left(\mathrm{f}_I^{\text{int}}\right)^T = \int_\Omega N_{I,\hat{\mathbf{x}}}^T \mathbf{R}^T \mathbf{R}\hat{\boldsymbol{\sigma}}\mathbf{R}^T d\Omega \tag{4.6.8}$$

and using the orthogonality of $\mathbf{R}$, we have

$$\left(\mathbf{f}_I^{\text{int}}\right)^T = \int_{\Omega} N_{I,\hat{x}}^T \hat{\boldsymbol{\sigma}} \mathbf{R}^T d\Omega \quad \text{or} \quad \left[\mathbf{f}_{iI}^{\text{int}}\right]^T = \left[\mathbf{f}_{Ii}^{\text{int}}\right] = \int_{\Omega} \frac{\partial N_I}{\partial \hat{x}_j} \hat{\sigma}_{jk} R_{ki}^T \, d\Omega \tag{4.6.9}$$

Comparing this to the standard expression for the nodal internal forces, (4.6.6), we can see that the expressions are similar, but the stress is expressed in the corotational system and the rotation matrix $\mathbf{R}$ now appears. In the expression on the right, the indices on $\mathbf{f}^{\text{int}}$ have been exchanged so that the expression can be converted to matrix form.

If we use the $\hat{\mathbf{B}}$ matrix defined by (4.6.4) we can write

$$\left(\mathbf{f}_I^{\text{int}}\right)^T = \int_{\Omega} \hat{B}_I^T \hat{\boldsymbol{\sigma}} \mathbf{R}^T d\Omega, \quad \mathbf{f}_{\text{int}}^T = \int_{\Omega} \hat{B}^T \hat{\boldsymbol{\sigma}} \mathbf{R}^T d\Omega \tag{4.6.10}$$

Corresponding relations for the internal nodal forces can be developed in Voigt notation:

$$\mathbf{f}_I^{\text{int}} = \int_{\Omega} \mathbf{R}^T \hat{\mathbf{B}}_I^T \{\hat{\boldsymbol{\sigma}}\} d\Omega \quad \text{where} \quad \{\hat{\mathbf{D}}\} = \hat{\mathbf{B}}_I \hat{\mathbf{v}}_I \tag{4.6.11}$$

and $\hat{\mathbf{B}}_I$ is obtained from $\hat{\mathbf{B}}_I$ by the Voigt rule.

The rate of the corotational Cauchy stress is objective (frame-invariant), so the constitutive equation can be expressed directly as a relationship between the rate of the corotational Cauchy stress and the corotational rate-of-deformation:

$$\frac{D\hat{\boldsymbol{\sigma}}}{Dt} = S^{\hat{\sigma}\hat{D}}(\hat{\mathbf{D}}, \hat{\boldsymbol{\sigma}}, \text{etc.}) \tag{4.6.12}$$

In particular, for a hypoelastic material,

$$\frac{D\hat{\boldsymbol{\sigma}}}{Dt} = \hat{\mathbf{C}} : \hat{\mathbf{D}} \quad \text{or} \quad \frac{D\hat{\sigma}_{ij}}{Dt} = \hat{C}_{ijkl} \hat{D}_{kl} \tag{4.6.13}$$

where the elastic response matrix is also expressed in terms of the corotational components. An attractive feature of the above is that the $\hat{\mathbf{C}}$ matrix for anisotropic materials need not be changed to reflect rotations. Since the coordinate system rotates with the material, material rotation has no effect on $\hat{\mathbf{C}}$. On the other hand, for an anisotropic material, the $\mathbf{C}$ matrix changes as the material rotates since the components of the $\mathbf{C}$ matrix are expressed in a fixed coordinate system.

Example 4.6 Rods in two dimensions A two-node element is shown in Figure 4.9. The element uses linear displacement and velocity fields. The corotational coordinate $\hat{x}$ is chosen to coincide with the axis of the element at all times as shown. Obtain an expression for the corotational rate-of-deformation and the internal nodal forces. Then extend the methodology to a three-node rod.

The displacement and velocity fields are linear in $\hat{x}$ and given by

$$\begin{Bmatrix} x \\ y \end{Bmatrix} = \begin{Bmatrix} x_1 & x_2 \\ y_1 & y_2 \end{Bmatrix} \begin{Bmatrix} 1-\xi \\ \xi \end{Bmatrix}$$

$$\begin{Bmatrix} \hat{v}_x \\ \hat{v}_y \end{Bmatrix} = \begin{Bmatrix} \hat{v}x_1 & \hat{v}x_2 \\ \hat{v}y_1 & \hat{v}y_2 \end{Bmatrix} \begin{Bmatrix} 1-\xi \\ \xi \end{Bmatrix} \quad \xi = \frac{\hat{x}}{\ell} \tag{E4.6.1}$$

where ℓ is the current length of the element. The corotational velocities are related to the global components by (4.6.1):

$$\begin{Bmatrix} v_{xI} \\ v_{yI} \end{Bmatrix} = \mathbf{R} \begin{Bmatrix} \hat{v}_{xI} \\ \hat{v}_{yI} \end{Bmatrix}, \mathbf{R} = \begin{bmatrix} R_{x\hat{x}} & R_{x\hat{y}} \\ R_{y\hat{x}} & R_{y\hat{y}} \end{bmatrix} = \begin{bmatrix} \cos\theta & -\sin\theta \\ \sin\theta & \cos\theta \end{bmatrix} = \frac{1}{\ell} \begin{bmatrix} x_{21} & -y_{21} \\ y_{21} & x_{21} \end{bmatrix} \tag{E4.6.2}$$

A state of uniaxial stress is assumed; the only nonzero stress is $\hat{\sigma}_x$, the stress component along the axis of the bar. Only the axial component of the rate-of-deformation tensor, $\hat{D}_x$, contributes to the internal power. It is given by the derivative of the velocity field (E4.6.1):

$$\hat{D}_x = \frac{\partial \hat{v}_x}{\partial \hat{x}} = \left[N_{I,\hat{x}}\right] \begin{Bmatrix} \hat{v}_{x1} \\ \hat{v}_{x2} \end{Bmatrix} = \frac{1}{\ell}[-1,\ 1] \begin{Bmatrix} \hat{v}_{x1} \\ \hat{v}_{x2} \end{Bmatrix} = \hat{\mathbf{B}}\hat{\mathbf{v}} \quad \hat{\mathbf{B}} = \left[N_{I,\hat{x}}\right] = \frac{1}{\ell}[-1,\ 1] \tag{E4.6.3}$$

Internal nodal forces The internal nodal forces are obtained from (4.6.8), which can be rewritten as

$$[\mathbf{f}_{Ii}]^{\text{int}} = \int_{\Omega} \frac{\partial N_I}{\partial \hat{x}_j} \hat{\sigma}_{jk} R_{ki}^T d\Omega = \int_{\Omega} \frac{\partial N_I}{\partial \hat{x}} \hat{\sigma}_{xx} R_{\hat{x}i}^T d\Omega = \int_{\Omega} \hat{\mathbf{B}}_I^T \hat{\sigma}_{xx} R_{\hat{x}i}^T \, d\Omega \tag{E4.6.4}$$

where the second expression omits the zeros which appear in the more general expression; the subscripts on the internal nodal forces have been interchanged. Substituting (E4.6.2) and (E4.6.3) into (E4.6.4) gives

$$[f_{Ii}]^{\text{int}} = \int \frac{1}{\ell} \begin{bmatrix} -1 \\ +1 \end{bmatrix} [\hat{\sigma}_x][\cos\theta \quad \sin\theta] d\Omega \tag{E4.6.5}$$

If we assume the stress is constant in the element, we can evaluate the integral by multiplying the integral by the volume of the element, $V = A\ell$, which gives

$$[f_{Ii}]^{\text{int}} = \begin{bmatrix} f_{1x} & f_{1y} \\ f_{2x} & f_{2y} \end{bmatrix} = A\hat{\sigma}_x \begin{bmatrix} -\cos\theta & -\sin\theta \\ \cos\theta & \sin\theta \end{bmatrix} \tag{E4.6.6}$$

The result shows that the nodal forces are along the axis of the rod and equal and opposite at the two nodes.

The stress–strain law in this element is computed in the corotational system. Thus, the rate form of the hypoelastic law is

$$\frac{D\hat{\sigma}_x}{Dt} = E\hat{D}_x \tag{E4.6.7}$$

where E is a tangent modulus in uniaxial stress. The rotation terms which appear in the objective rates are not needed, since this is objective.

To evaluate the nodal forces, the current cross-sectional area A must be known. The change in area can then be expressed in terms of the transverse strains; the exact formula depends on the shape of the cross-section. For a rectangular cross-section

$$\dot{A} = A(\hat{D}_y + \hat{D}_z) \tag{E4.6.8}$$

The internal nodal forces can also be obtained from Example 2.4, (E2.4.8), and then transforming by (4.5.39). In the corotational system, (E2.4.8) gives:

$$\hat{\mathbf{f}}^{\text{int}} = \begin{Bmatrix} \hat{f}_{x1} \\ \hat{f}_{x2} \end{Bmatrix}^{\text{int}} = \int_0^{\ell} \frac{1}{\ell}\begin{bmatrix} -1 \\ +1 \end{bmatrix} \hat{\sigma}_x A \, d\hat{x} \tag{E4.6.9}$$

Since we are considering a slender rod with no stiffness normal to its axis, the transverse nodal forces vanish, that is, $\hat{f}_{y1} = \hat{f}_{y2} = 0$.

The global components of the nodal forces are obtained by the transformation equations, (4.5.35). We first construct $\mathbf{T}$ by relating the local degrees of freedom (which are conjugate to $\hat{\mathbf{f}}^{\text{int}}$) to the global degrees of freedom of the element:

$$\begin{Bmatrix} \hat{v}_{x1} \\ \hat{v}_{x2} \end{Bmatrix} = \begin{bmatrix} \cos\theta & \sin\theta & 0 & 0 \\ 0 & 0 & \cos\theta & \sin\theta \end{bmatrix} \begin{Bmatrix} v_{x1} \\ v_{y1} \\ v_{x2} \\ v_{y2} \end{Bmatrix} \quad \text{so} \quad \mathbf{T} = \begin{bmatrix} \cos\theta & \sin\theta & 0 & 0 \\ 0 & 0 & \cos\theta & \sin\theta \end{bmatrix} \tag{E4.6.10}$$

Using (4.5.35), $\mathbf{f} = \mathbf{T}^T\hat{\mathbf{f}}$, and assuming the stress is constant in the element then gives

$$\mathbf{f}^{\text{int}} = \begin{Bmatrix} f_{x1} \\ f_{y1} \\ f_{x2} \\ f_{y2} \end{Bmatrix} = \mathbf{T}^T\hat{\mathbf{f}}^{\text{int}} = \begin{bmatrix} \cos\theta & 0 \\ \sin\theta & 0 \\ 0 & \cos\theta \\ 0 & \sin\theta \end{bmatrix} A\hat{\sigma}_x \begin{Bmatrix} -1 \\ 1 \end{Bmatrix} = A\hat{\sigma}_x \begin{Bmatrix} -\cos\theta \\ -\sin\theta \\ \cos\theta \\ \sin\theta \end{Bmatrix} \tag{E4.6.11}$$

which is identical to (E4.6.6).

Three-node element We consider the three-node curved rod element shown in Figure 4.10. The configurations, displacement, and velocity are given by quadratic fields. The expression for the internal nodal forces will be developed by the corotational approach.

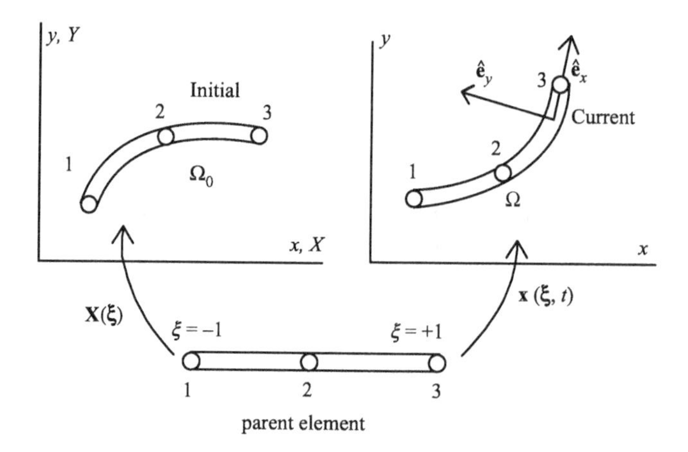

Figure 4.10 Initial, current, and parent elements for a three-node rod; the corotational base vector $\hat{e}_x$ is tangent to the current configuration

The initial and current configurations are given by

$$\mathbf{X}(\xi) = \mathbf{X}_I N_I(\xi) \quad \mathbf{x}(\xi,t) = x_I(t)N_I(\xi) \tag{E4.6.12}$$

$$[N_I] = \left[\frac{1}{2}\xi(\xi-1) \quad 1-\xi^2 \quad \frac{1}{2}\xi(\xi+1)\right] \tag{E4.6.13}$$

The displacement and velocity are given by

$$\mathbf{u}(\xi,t) = \mathbf{u}_I(t)N_I(\xi) \quad \mathbf{v}(\xi,t) = \mathbf{v}_I(t)N_I(\xi) \tag{E4.6.14}$$

The corotational system is defined at each point of the rod (in practice it is needed only at the quadrature points). Let $\hat{\mathbf{e}}_x$ be tangent to the rod, so

$$\hat{\mathbf{e}}_x = \frac{\mathbf{x}_{,\xi}}{\|\mathbf{x}_{,\xi}\|} \quad \text{where} \quad \mathbf{x}_{,\xi} = \mathbf{x}_I N_{I,\xi}(\xi) \tag{E4.6.15}$$

The normal to the element is given by

$$\hat{\mathbf{e}}_y = \mathbf{e}_z \times \hat{\mathbf{e}}_x \quad \text{where} \quad \mathbf{e}_z = [0,0,1] \tag{E4.6.16}$$

The rate-of-deformation is given by

$$\hat{D}_x = \frac{\partial \hat{v}_x}{\partial \hat{x}} = \frac{\partial \hat{v}_x}{\partial \xi}\frac{\partial \xi}{\partial \hat{x}} = \frac{1}{\|\mathbf{x}_{,\xi}\|}\frac{\partial \hat{v}_x}{\partial \xi} \tag{E4.6.17}$$

From (E4.6.14) and (4.6.1)

$$\hat{v}_x = N_I(\xi)(R_{x\hat{x}}v_{xI} + R_{y\hat{x}}v_{yI}) \tag{E4.6.18}$$

The rate-of-deformation is given by

$$\hat{D}_x = \frac{1}{\|\mathbf{x}_{,\xi}\|} N_{I,\xi}(\xi) \begin{Bmatrix} v_{xI} \\ v_{yI} \end{Bmatrix} \tag{E4.6.19}$$

The above shows the $\hat{\boldsymbol{B}}_I$ matrix to be

$$\hat{\boldsymbol{B}}_I = \frac{1}{\|\mathbf{x}_{,\xi}\|} N_{I,\xi} \tag{E4.6.20}$$

The internal nodal forces are then given by

$$\left(\mathbf{f}_I^{\text{int}}\right)^T = [f_{xI} \quad f_{yI}]^{\text{int}} = \int_{-1}^{1} A\hat{\boldsymbol{B}}_I \hat{\sigma}_x \|\mathbf{x}_{,\xi}\| [R_{x\hat{x}} \quad R_{y\hat{x}}] d\xi \tag{E4.6.21}$$

An interesting feature of the above development is that it avoids curvilinear tensors completely.

Example 4.7 Triangular element Develop the expression for the velocity strain and the internal nodal forces for a three-node triangle using the corotational approach.

The element in its initial and current configurations is shown in Figure 4.11. The corotational system is initially at an angle of θ_0 with the global coordinate system; θ_0 is often chosen to vanish, but for an anisotropic material it may be desirable to orient the initial $\hat{x}$-axis in a direction of anisotropy; for example, in a composite material it may be useful to orient x in a fiber direction. The current angle of the corotational coordinate system is θ.

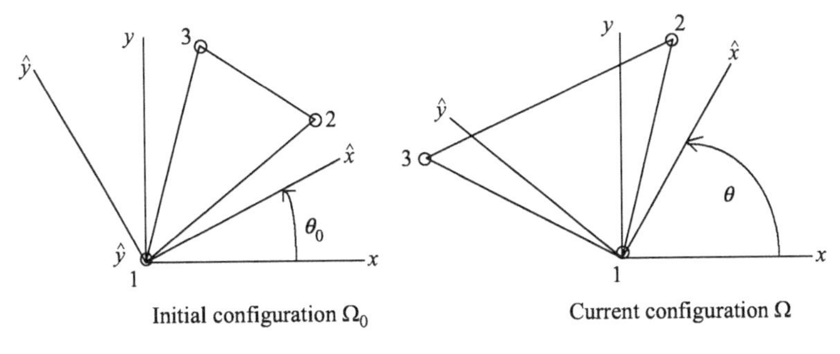

Figure 4.11 Triangular three-node element treated by corotational coordinate system

The motion can be expressed in terms of the triangular coordinates, as in Example 4.1:

$$\begin{Bmatrix} x \\ y \end{Bmatrix} = \begin{bmatrix} x_1 & x_2 & x_3 \\ y_1 & y_2 & y_3 \end{bmatrix} \begin{Bmatrix} \xi_1 \\ \xi_2 \\ \xi_3 \end{Bmatrix} \tag{E4.7.1}$$

The displacement and velocity fields in the element are then given by

$$\begin{Bmatrix} \hat{u}_x \\ \hat{u}_y \end{Bmatrix} = \begin{bmatrix} \hat{u}_{x1} & \hat{u}_{x2} & \hat{u}_{x3} \\ \hat{u}_{y1} & \hat{u}_{y2} & \hat{u}_{y3} \end{bmatrix} \begin{Bmatrix} \xi_1 \\ \xi_2 \\ \xi_3 \end{Bmatrix} \tag{E4.7.2}$$

$$\begin{Bmatrix} \hat{v}_x \\ \hat{v}_y \end{Bmatrix} = \begin{bmatrix} \hat{v}_{x1} & \hat{v}_{x2} & \hat{v}_{x3} \\ \hat{v}_{y1} & \hat{v}_{y2} & \hat{v}_{y3} \end{bmatrix} \begin{Bmatrix} \xi_1 \\ \xi_2 \\ \xi_3 \end{Bmatrix} \tag{E4.7.3}$$

The derivatives of the shape functions with respect to the corotational coordinate system are given by the counterpart of (E4.1.5):

$$[\partial N_I / \partial \hat{x}_j] \equiv [\partial \xi_I / \partial \hat{x}_j] = \frac{1}{2A} \begin{bmatrix} \hat{y}_{23} & \hat{x}_{32} \\ \hat{y}_{31} & \hat{x}_{13} \\ \hat{y}_{12} & \hat{x}_{21} \end{bmatrix} \equiv \hat{\boldsymbol{B}} \tag{E4.7.4}$$

The corotational components of the rate-of-deformation are given by

$$\hat{\mathbf{D}} = \frac{1}{2}\left(\hat{\boldsymbol{B}}_I \hat{\mathbf{v}}_I^T + \hat{\mathbf{v}}_I \hat{\boldsymbol{B}}_I^T\right) \tag{E4.7.5}$$

The internal nodal forces are given by (4.6.10):

$$[f_{Ii}]^{\text{int}} = \int_{\Omega} \hat{B}_{Ij} \hat{\sigma}_{jk} R_{ki}^T d\Omega = \int_{\Omega} \frac{\partial \xi_I}{\partial \hat{x}_j} \hat{\sigma}_{jk} R_{ki}^T d\Omega \tag{E4.7.6}$$

Writing out the matrices in the above using (E4.6.2) and (E4.7.4) gives

$$\begin{bmatrix} f_{1x} & f_{1y} \\ f_{2x} & f_{2y} \\ f_{3x} & f_{3y} \end{bmatrix}^{\text{int}} = \int_A \frac{1}{2A} \begin{bmatrix} \hat{y}_{23} & \hat{x}_{32} \\ \hat{y}_{31} & \hat{x}_{13} \\ \hat{y}_{12} & \hat{x}_{21} \end{bmatrix} \begin{bmatrix} \hat{\sigma}_x & \hat{\sigma}_{xy} \\ \hat{\sigma}_{xy} & \hat{\sigma}_y \end{bmatrix} \begin{bmatrix} \cos\theta & \sin\theta \\ -\sin\theta & \cos\theta \end{bmatrix} a dA \tag{E4.7.7}$$

The rotation of the coordinate system can be obtained in several ways:

1. By polar decomposition
2. By rotating the corotational coordinate system at each quadrature point with a material line in the element, for example, a preferred direction in a composite
3. By rotating the corotational coordinate system with a side of the element (this is only correct for small strain problems).

Readers are referred to Wempner (1969) and Belytschko and Hsieh (1973) for further details.

4.7 Total Lagrangian Formulation

4.7.1 Governing Equations

The physical principles which govern the total Lagrangian formulation are the same as those for the updated Lagrangian formulation given in Section 4.2. The form of the governing equations is different, but as has been seen in Chapter 3, they express the same physical principles and can be obtained by transforming the associated conservation equations from Eulerian to Lagrangian form.

Similarly, the finite element equations for the total Lagrangian formulation can be obtained by transforming the equations for the updated Lagrangian formulation. It is only necessary to transform the integrals to the reference (undeformed) domain and transform the stress and strain measures to the Lagrangian type. This approach is used in Section 4.7.2, and for most readers will suffice as an introduction to the total Lagrangian formulation. However, for readers who would like to see the entire structure of the total Lagrangian formulation or prefer to learn it first, Section 4.8 gives a development of the weak form in the total Lagrangian description, followed by a derivation of the finite element equations from this weak form.

The governing equations are given in both tensor form and indicial form in Box 4.4. We have chosen to use the nominal stress $\mathbf{P}$ in the momentum equation, because the resulting momentum equation and its weak form are simpler than for the PK2 stress. However, the nominal stress is awkward in constitutive equations because of its lack of symmetry, so we have used the PK2 stress for constitutive equations. Once the PK2 stress has been evaluated by the constitutive equations, the nominal stress can then easily be obtained by a transformation given in Box 3.2, (B4.4.5). The constitutive equation can relate the Cauchy stress $\boldsymbol{\sigma}$ to the rate-of-deformation $\mathbf{D}$. The stress would then be converted to the nominal stress $\mathbf{P}$ prior to evaluation of the nodal forces. However, this entails additional transformations and hence additional computational expense, so when the constitutive equations are expressed in terms of $\boldsymbol{\sigma}$ it is advantageous to use the updated Lagrangian formulation.

Box 4.4 Governing equations for total Lagrangian formulation

Conservation of mass:

$$\rho J = \rho_0 J_0 = \rho_0 \tag{B4.4.1}$$

Conservation of linear momentum:

$$\nabla_0 \cdot \mathbf{P} + \rho_0 \mathbf{b} = \rho_0 \ddot{\mathbf{u}} \quad \text{or} \quad \frac{\partial P_{ji}}{\partial X_j} + \rho_0 b_i = \rho_0 \ddot{u}_i \tag{B4.4.2}$$

Conservation of angular momentum:

$$\mathbf{F}\cdot\mathbf{P}=\mathbf{P}^{T}\cdot\mathbf{F}^{T} \quad \text{or} \quad F_{ij}P_{jk}=F_{kj}P_{ji} \tag{B4.4.3}$$

Conservation of energy:

$$\rho_0\dot{w}^{\text{int}}=\dot{\mathbf{F}}^{T}:\mathbf{P}-\nabla_0\cdot\overline{\mathbf{q}}+\rho_0 s \quad \text{or} \quad \rho_0\dot{w}^{\text{int}}=\dot{F}_{ij}P_{ji}-\frac{\partial\overline{q}_i}{\partial X_i}+\rho_0 s \tag{B4.4.4}$$

where $\overline{\mathbf{q}}=J\mathbf{F}^{-1}\cdot\mathbf{q}$

Constitutive equation:

$$\mathbf{S}=\mathbf{S}(\mathbf{E},\ldots,\text{etc.}),\quad \mathbf{P}=\mathbf{S}\cdot\mathbf{F}^{T} \tag{B4.4.5}$$

Measure of strain:

$$\mathbf{E}=\frac{1}{2}(\mathbf{F}^{T}\cdot\mathbf{F}-\mathbf{I}) \quad \text{or} \quad E_{ij}=\frac{1}{2}(F_{ki}F_{kj}-\delta_{ij}) \tag{B4.4.6}$$

Boundary conditions:

$$n_j^0 P_{ji}=\overline{t}_i^0 \quad \text{or} \quad \mathbf{e}_i\cdot\mathbf{n}^0\cdot\mathbf{P}=\mathbf{e}_i\cdot\overline{\mathbf{t}}_0 \quad \text{on} \quad \Gamma_{t_i}^0, \tag{B4.4.7}$$

$$u_i=\overline{u}_i \quad \text{on} \quad \Gamma_{u_i}^0,\ \Gamma_{t_i}^0\cup\Gamma_{u_i}^0=\Gamma^0,\ \Gamma_{t_i}^0\cap\Gamma_{u_i}^0=0 \quad \text{for} \quad i=1 \quad \text{to} \quad n_{\text{SD}} \tag{B4.4.8}$$

Initial conditions:

$$\mathbf{P}(\mathbf{X},0)=\mathbf{P}_0(\mathbf{X}) \tag{B4.4.9}$$

$$\dot{\mathbf{u}}(\mathbf{X},0)=\dot{\mathbf{u}}_0(\mathbf{X}) \tag{B4.4.10}$$

Internal continuity conditions:

$$[\![n_j^0 P_{ji}]\!]=0 \quad \text{on} \quad \Gamma_{\text{int}}^0 \tag{B4.4.11}$$

The nominal stress is conjugate to the material time derivative of the deformation tensor, $\dot{\mathbf{F}}$: see Box 3.4. Thus in (B4.4.4) the internal work is expressed in terms of these two tensors. Note that $\mathbf{n}^0$ appears before $\mathbf{P}$ in the traction expression in (B4.4.7); if the order is reversed the resulting matrix corresponds to the transpose of $\mathbf{P}$, which is the PK1 stress; see Section 3.4.1.

The deformation tensor $\mathbf{F}$ is not suitable as a measure of strain in constitutive equations since it does not vanish in rigid body rotation. Therefore constitutive equations in total Lagrangian formulations are usually formulated in terms of the Green strain tensor $\mathbf{E}$, which can be obtained from $\mathbf{F}$. In the continuum mechanics literature, one often sees constitutive equations expressed as $\mathbf{P}=\mathbf{P}(\mathbf{F})$, which gives the impression that the constitutive equation uses $\mathbf{F}$ as a measure of strain. In fact, when writing $\mathbf{P}(\mathbf{F})$, it is implicit that the constitutive stress depends on $\mathbf{F}^T\mathbf{F}$ (i.e., $\mathbf{E}+\mathbf{I}$, where the unit matrix $\mathbf{I}$ makes no difference) or some other measure of deformation which is independent of rigid body rotation. Similarly, the nominal stress $\mathbf{P}$ in constitutive equations is generated so it satisfies conservation of angular momentum, (B4.4.3).

As in any mechanical system, the same component of traction and displacement cannot be prescribed at any point of a boundary, but one of these must be prescribed; see (B4.4.7–8). In the Lagrangian formulation, tractions are prescribed in units of force per undeformed area.

The total Lagrangian formulation can be obtained in two ways:

1. Transforming the finite element equations for the updated Lagrangian formulation to the initial (reference) configuration and expressing it in terms of Lagrangian variables.
2. Developing the weak form in terms of the initial configuration and Lagrangian variables and then using this weak form to obtain discrete equations.

We will begin with the first approach since it is quicker and more convenient.

4.7.2 *Total Lagrangian Finite Element Equations by Transformation*

To obtain the discrete finite element equations for total Lagrangian formulation, we will transform each of the nodal force expressions in the updated Lagrangian formulation. The mass conservation equation (B4.4.1), $\rho J = \rho_0$, and the relation

$$d\Omega = J\,d\Omega_0 \tag{4.7.1}$$

will be used. The internal nodal forces are given in the updated Lagrangian formulation by (4.4.13):

$$f_{iI}^{\text{int}} = \int_\Omega \frac{\partial N_I}{\partial x_j}\sigma_{ji}\,d\Omega \tag{4.7.2}$$

Using the transformation from Box 3.2, $J\sigma_{ji} = F_{jk}P_{ki} = \frac{\partial x_j}{\partial X_k}P_{ki}$, we convert (4.7.2) to:

$$f_{iI}^{\text{int}} = \int_\Omega \frac{\partial N_I}{\partial x_j}\frac{\partial x_j}{\partial X_k}P_{ki}J^{-1}\,d\Omega \tag{4.7.3}$$

Recognizing that the product of the first two terms is a chain rule expression of $\partial N_I/\partial X_k$ and using (4.7.1), we get

$$f_{iI}^{\text{int}} = \int_{\Omega_0} \frac{\partial N_I}{\partial X_k}P_{ki}\,d\Omega_0 = \int_{\Omega_0} B_{0kI}P_{ki}\,d\Omega_0 \tag{4.7.4}$$

where

$$B_{0\,kI} = \frac{\partial N_I}{\partial X_k} \tag{4.7.5}$$

In matrix form, this can be written as

$$\left(\mathbf{f}_I^{\text{int}}\right)^T = \int_\Omega \boldsymbol{B}_0^T\mathbf{P}\,d\Omega_0 \tag{4.7.6}$$

The expression has been written in the above form to stress the analogy to the updated Lagrangian form: if we consider the current configuration to be the reference configuration, $\boldsymbol{B}_0$ is replaced by $\boldsymbol{B}$, Ω_0 by Ω, and $\mathbf{P}$ by $\boldsymbol{\sigma}$, so we obtain the updated Lagrangian form from the previous.

The external nodal forces are next obtained by transforming the updated Lagrangian expression to the total Lagrangian form. We start with (4.4.15):

$$f_{iI}^{\text{ext}} = \int_{\Omega} N_I \rho b_i \, d\Omega + \int_{\Gamma_{t_i}} N_I \bar{t}_i d\Gamma \tag{4.7.7}$$

Substituting (3.6.1), $\rho\mathbf{b}\, d\Omega = \rho_0 \mathbf{b} d\Omega_0$, and (B3.1.4), $\bar{\mathbf{t}}\, d\Gamma = \mathbf{t}_0 \, d\Gamma_0$, into (4.7.7) gives

$$f_{iI}^{\text{ext}} = \int_{\Omega_0} N_I \rho_0 b_i \, d\Omega_0 + \int_{\Gamma_{t_i}^0} N_I \bar{t}_i^0 \, d\Gamma_0 \tag{4.7.8}$$

which is the total Lagrangian form of the external nodal forces. The two integrals are over the initial (reference) domain and boundary; note that $\rho_0 \mathbf{b}$ is the body force per unit reference volume: see (3.6.1). The previous can be written in matrix form as

$$\mathbf{f}_I^{\text{ext}} = \int_{\Omega_0} N_I \rho_0 \mathbf{b}_i \, d\Omega_0 + \int_{\Gamma_{t_i}^0} N_I \mathbf{e}_i \cdot \bar{\mathbf{t}}_0 \, d\Gamma \tag{4.7.9}$$

The inertial nodal forces and the mass matrix were expressed in terms of the initial configuration in the development of the updated Lagrangian form, (4.4.55). Thus, all of the nodal forces have been expressed in terms of Lagrangian variables on the initial (reference) configuration. The equations of motion for the total Lagrangian discretization are identical to those of the updated Lagrangian discretization, (4.4.53).

4.8 Total Lagrangian Weak Form

In this section, we develop the weak form from the strong form in a total Lagrangian format. Subsequently, we will show that the weak form implies the strong form. The strong form consists of the momentum equation, (B4.4.2), the traction boundary condition, (B4.4.7), and the internal continuity conditions, (B4.4.11). We define the spaces for the test and trial functions as in Section 4.3:

$$\delta\mathbf{u}(\mathbf{X}) \in \mathcal{U}_0, \quad \mathbf{u}(\mathbf{X}, t) \in \mathcal{U} \tag{4.8.1}$$

where $\mathcal{U}$ is the space of kinematically admissible displacements and $\mathcal{U}_0$ is the same space with the additional requirement that the displacements vanish on displacement boundaries.

4.8.1 *Strong Form to Weak Form*

To develop the weak form, we multiply the momentum equation (B4.4.2) by the test function and integrate over the *initial (reference) configuration*:

$$\int_{\Omega_0} \delta u_i \left(\frac{\partial P_{ji}}{\partial X_j} + \rho_0 b_i - \rho_0 \ddot{u}_i \right) d\Omega_0 = 0 \tag{4.8.2}$$

In (4.8.2), the nominal stress is a function of the trial displacements via the constitutive equation and the strain–displacement equation. This weak form is not useful because it requires the trial displacements to be C^1, since a derivative of the nominal stress appears in (4.8.2); see Sections 4.3.1 and 4.3.2.

To eliminate the derivative of the nominal stress from (4.8.2), the derivative product formula is used:

$$\int_{\Omega_0} \delta u_i \frac{\partial P_{ji}}{\partial X_j} d\Omega_0 = \int_{\Omega_0} \frac{\partial}{\partial X_j} (\delta u_i P_{ji}) d\Omega_0 - \int_{\Omega_0} \frac{\partial (\delta u_i)}{\partial X_j} P_{ji}\, d\Omega_0 \tag{4.8.3}$$

The first term of the RHS of (4.8.3) can be expressed as a boundary integral by Gauss's theorem (3.5.5):

$$\int_{\Omega_0} \frac{\partial}{\partial X_j} (\delta u_i P_{ji}) d\Omega_0 = \int_{\Gamma_0} \delta u_i n_j^0 P_{ji}\, d\Gamma_0 + \int_{\Gamma_{\text{int}}^0} \delta u_i \left[\!\left[n_j^0 P_{ji} \right]\!\right] d\Gamma_0 \tag{4.8.4}$$

From the traction continuity condition, (B4.4.11), the last term vanishes. The first term on the RHS can be reduced to the traction boundary since $\delta u_i = 0$ on $\Gamma_{u_i}^0$ and $\Gamma_{t_i}^0 = \Gamma_0 - \Gamma_{u_i}^0$, so

$$\int_{\Omega_0} \frac{\partial}{\partial X_j} (\delta u_i P_{ji}) d\Omega_0 = \int_{\Gamma_0} \delta u_i n_j^0 P_{ji}\, d\Gamma_0 = \sum_{i=1}^{n_{\text{SD}}} \int_{\Gamma_{t_i}^0} \delta u_i \bar{t}_i^0\, d\Gamma_0 \tag{4.8.5}$$

where the last equality follows from the strong form (B4.4.7). From (3.2.19) we note that

$$\delta F_{ij} = \delta \left(\frac{\partial u_i}{\partial X_j} \right) = \frac{\partial (\delta u_i)}{\partial X_j} \tag{4.8.6}$$

Substituting (4.8.5) into (4.8.3) and the result into (4.8.2) gives, after a change of sign and using (4.8.6):

$$\int_{\Omega_0} (\delta F_{ij} P_{ji} - \delta u_i \rho_0 b_i + \delta u_i \rho_0 \ddot{u}_i) d\Omega_0 - \sum_{i=1}^{n_{\text{SD}}} \int_{\Gamma_{t_i}^0} \delta u_i \bar{t}_i^0\, d\Gamma_0 = 0 \tag{4.8.7}$$

or

$$\int_{\Omega_0} (\delta \mathbf{F}^T : \mathbf{P} - \rho_0 \delta \mathbf{u} \cdot \mathbf{b} + \rho_0 \delta \mathbf{u} \cdot \ddot{\mathbf{u}}) d\Omega_0 - \sum_{i=1}^{n_{\text{SD}}} \int_{\Gamma_{t_i}^0} (\delta \mathbf{u} \cdot \mathbf{e}_i)(\mathbf{e}_i \cdot \bar{\mathbf{t}}_i^0) d\Gamma_0 = 0 \tag{4.8.8}$$

This is the weak form of the momentum equation, traction boundary conditions, and interior continuity conditions. It is called the *principle of virtual work*, since each of the terms in (4.8.7) is a virtual work increment. The weak form is summarized in Box 4.5, in which physical names are ascribed to each of the terms.

Box 4.5 Weak form for total Lagrangian formulation: principle of virtual work

If $u \in U$ and

$$\delta W^{\text{int}}(\delta\mathbf{u},\mathbf{u}) - \delta W^{\text{ext}}(\delta\mathbf{u},\mathbf{u}) + \delta W^{\text{kin}}(\delta\mathbf{u},\mathbf{u}) = 0 \quad \forall \delta\mathbf{u} \in u_0 \tag{B4.5.1}$$

then equilibrium, the traction boundary conditions and internal continuity conditions are satisfied. In the above,

$$\delta W^{\text{int}} = \int_{\Omega_0} \delta\mathbf{F}^T : \mathbf{P} d\Omega_0 = \int_{\Omega_0} \delta F_{ij} P_{ji} d\Omega_0 \tag{B4.5.2}$$

$$\begin{aligned} \delta W^{\text{ext}} &= \int_{\Omega_0} \rho_0 \delta\mathbf{u} \cdot \mathbf{b} d\Omega_0 + \sum_{i=1}^{n_{SD}} \int_{\Gamma_{t_i}^0} (\delta\mathbf{u}\cdot\mathbf{e}_i)(\mathbf{e}_i \cdot \bar{\mathbf{t}}_t^0) d\Gamma_0 \\ &= \int_{\Omega_0} \delta u_i \rho_0 b_i d\Omega_0 + \sum_{i=1}^{n_{SD}} \int_{\Gamma_{t_i}^0} \delta u_i \bar{t}_i^0 d\Gamma_0 \end{aligned} \tag{B4.5.3}$$

$$\delta W^{\text{kin}} = \int_{\Omega_0} \delta\mathbf{u} \cdot \rho_0 \ddot{\mathbf{u}} d\Omega_0 = \int_{\Omega_0} \delta u_i \rho_0 \ddot{u}_i d\Omega_0 \tag{B4.5.4}$$

This weak form can also be developed by replacing the test velocity by a test displacement in (4.3.11) and transforming each term to the reference configuration. The total Lagrangian weak form, (4.8.8), is thus simply a transformation of the updated Lagrangian weak form.

4.8.2 Weak Form to Strong Form

Next we deduce the strong form from the weak form. Substituting (4.8.6) into the first term of (4.8.7) and using the derivative product rule gives

$$\int_{\Omega_0} \frac{\partial(\delta u_i)}{\partial X_j} P_{ji} d\Omega_0 = \int_{\Omega_0} \left[\frac{\partial}{\partial X_j}(\delta u_i P_{ji}) - \delta u_i \frac{\partial P_{ji}}{\partial X_j} \right] d\Omega_0 \tag{4.8.9}$$

Gauss's theorem on the first term on the RHS then yields

$$\int_{\Omega_0} \frac{\partial(\delta u_i)}{\partial X_j} P_{ji} d\Omega_0 = \sum_{i=1}^{n_{SD}} \int_{\Gamma_{t_i}^0} \delta u_i n_j^0 P_{ji} d\Gamma_0 + \int_{\Gamma_{\text{int}}^0} \delta u_i [\![n_j^0 P_{ji}]\!] d\Gamma_0 - \int_{\Omega_0} \delta u_i \frac{\partial P_{ji}}{\partial X_j} d\Omega_0 \tag{4.8.10}$$

where the surface integral is changed to the traction boundary because $\delta u_i = 0$ on $\Gamma_{u_i}^0$ and $\Gamma_{t_i}^0 = \Gamma^0 - \Gamma_{u_i}^0$.

Substituting (4.8.10) into (4.8.7) and collecting terms gives

$$0=\int_{\Omega_0}\delta u_i\left(\frac{\partial P_{ji}}{\partial X_j}+\rho_0 b_i-\rho_0\ddot{u}_i\right)d\Omega_0+\sum_{i=1}^{n_{SD}}\int_{\Gamma^0_{t_i}}\delta u_i\left(n^0_j P_{ji}-\bar{t}^0_i\right)d\Gamma_0 \tag{4.8.11}$$

$$+\int_{\Gamma^0_{\text{int}}}\delta u_i\left[\!\left[n^0_j P_{ji}\right]\!\right]d\Gamma_0=0$$

Since the above holds for all $\delta\mathbf{u}\in u_0$, it follows by the density theorem given in Section 4.3.2 that the momentum equation (B4.4.2) holds on Ω_0, the traction boundary conditions (B4.4.7) hold on Γ^0_t, and the internal continuity conditions (B4.4.11) hold on Γ^0_{int}. Thus the weak form implies the momentum equation, the traction boundary conditions, and the interior continuity conditions.

4.9 Finite Element Semidiscretization

4.9.1 Discrete Equations

We consider a Lagrangian mesh with the same properties as described in Section 4.4.1. The finite element approximation to the motion is given by

$$x_i(\mathbf{X},t)=x_{iI}(t)N_I(\mathbf{X}) \tag{4.9.1}$$

where $N_I(\mathbf{X})$ are the shape functions; as in the updated Lagrangian formulation, the shape functions are functions of the material (Lagrangian) coordinates, or element coordinates. The trial displacement field is given by

$$u_i(\mathbf{X},t)=u_{iI}(t)N_I(\mathbf{X})\quad\text{or}\quad\mathbf{u}(X,t)=\mathbf{u}_I(t)N_I(\mathbf{X}) \tag{4.9.2}$$

The test functions, or variations, are not functions of time, so

$$\delta u_i(\mathbf{X})=\delta u_{iI}N_I(\mathbf{X})\quad\text{or}\quad\delta\mathbf{u}(\mathbf{X})=\delta\mathbf{u}_I N_I(\mathbf{X}) \tag{4.9.3}$$

As before, we will use indicial notation where *all* repeated indices are summed; upper case indices pertain to nodes and are summed over all relevant nodes, and lower case indices pertain to components and are summed over the number of dimensions.

Taking material time derivatives of (4.9.2) gives the velocity and acceleration

$$\dot{u}_i(\mathbf{X},\ t)=\dot{u}_{iI}(t)N_I(\mathbf{X}) \tag{4.9.4}$$

$$\ddot{u}_i(\mathbf{X},\ t)=\ddot{u}_{iI}(t)N_I(\mathbf{X}) \tag{4.9.5}$$

The deformation gradient is then given by

$$F_{ij} = \frac{\partial x_i}{\partial X_j} = \frac{\partial N_I}{\partial X_j} x_{iI} \tag{4.9.6}$$

It is sometimes convenient to write this as

$$F_{ij} = B^0_{jI} x_{iI} \quad \text{where} \quad B^0_{jI} = \frac{\partial N_I}{\partial X_j} \quad \text{so} \quad \mathbf{F} = \mathbf{x}\boldsymbol{B}_0^T \tag{4.9.7}$$

$$\delta F_{ij} = \frac{\partial N_I}{\partial X_j} \delta x_{iI} = \frac{\partial N_I}{\partial X_j} \delta u_{iI} \quad \text{so} \quad \delta \mathbf{F} = \delta \mathbf{u} \boldsymbol{B}_0^T \tag{4.9.8}$$

where we have used $\delta x_{iI} = \delta(X_{iI} + u_{iI}) = \delta u_{iI}$.

4.9.1.1 Internal Nodal Forces

The internal nodal forces are defined in terms of the internal virtual work using

$$\delta W^{\text{int}} = \delta u_{iI} f^{\text{int}}_{iI} = \int_{\Omega_0} \delta F_{ij} P_{ji}\, d\Omega_0 = \delta u_{iI} \int_{\Omega_0} \frac{\partial N_I}{\partial X_j} P_{ji}\, d\Omega_0 \tag{4.9.9}$$

where (4.9.8) has been used in the last step. Then the arbitrariness of δu_{iI} yields

$$f^{\text{int}}_{iI} = \int_{\Omega_0} \frac{\partial N_I}{\partial X_j} P_{ji}\, d\Omega_0 \quad \text{or} \quad f^{\text{int}}_{iI} = \int_{\Omega_0} B^0_{jI} P_{ji}\, d\Omega_0 \quad \text{or} \quad \mathbf{f}^{\text{int},T} = \int_{\Omega_0} \boldsymbol{B}_0^T \mathbf{P}\, d\Omega_0 \tag{4.9.10}$$

which is identical to (4.7.6), the expression developed by transformation.

4.9.1.2 External Nodal Forces

The external nodal forces are defined by equating the virtual external work (B4.5.3) to the virtual work of the external nodal forces:

$$\delta W^{\text{ext}} = \delta u_{iI} f^{\text{ext}}_{iI} = \int_{\Omega} \delta u_i \rho_0 b_i\, d\Omega_0 + \int_{\Gamma^0_{t_i}} \delta u_i \bar{t}_i^{\,0} d\Gamma_0 = \delta u_{iI} \left\{ \int_{\Omega_0} N_I \rho_0 b_i d\Omega_0 + \int_{\Gamma^0_{t_i}} N_I \bar{t}_i^{\,0} d\Gamma_0 \right\} \tag{4.9.11}$$

This gives

$$f^{\text{ext}}_{iI} = \int_{\Omega_0} N_I \rho_0 b_i d\Omega_0 + \int_{\Gamma^0_{t_i}} N_I \bar{t}_i^{\,0} d\Gamma_0 \tag{4.9.12}$$

4.9.1.3 Mass Matrix

Defining a nodal force equivalent to the inertial force (B4.5.4) gives

$$\delta W^{\text{kin}} = \delta u_{iI} f_{iI}^{\text{kin}} = \int_{\Omega_0} \delta u_i \rho_0 \ddot{u}_i \, d\Omega_0 \tag{4.9.13}$$

Substituting (4.9.3) and (4.9.5) in the right-hand side of (4.9.13) gives

$$\delta u_{iI} f_{iI}^{\text{kin}} = \delta u_{iI} \int_{\Omega_0} \rho_0 N_I N_J \, d\Omega_0 \ddot{u}_{jJ} = \delta u_{iI} M_{ijIJ} \ddot{u}_{jJ} \tag{4.9.14}$$

Since this holds for arbitrary $\delta\mathbf{u}$ and $\ddot{\mathbf{u}}$, it follows that the mass matrix is given by

$$M_{ijIJ} = \delta_{ij} \int_{\Omega_0} \rho_0 N_I N_J d\Omega_0 \tag{4.9.15}$$

Comparing this mass matrix $\mathbf{M}$ to that for the updated Lagrangian formulation, (4.4.56), we see that they are identical, which is expected since we transformed the mass to the reference configuration to highlight its time invariance for a Lagrangian mesh.

Substituting these expressions into the weak form, (B4.5.1), we have

$$\delta u_{iI} (f_{iI}^{\text{int}} - f_{iI}^{\text{ext}} + M_{ijIJ} \ddot{u}_{jJ}) = 0 \quad \forall I, i \notin \Gamma_{u_i} \tag{4.9.16}$$

Since this applies for arbitrary values of all nodal displacement components that are not constrained by displacement boundary conditions, it follows that

$$M_{ijIJ} \ddot{u}_{jJ} + f_{iI}^{\text{int}} = f_{iI}^{\text{ext}} \quad \forall I, i \notin \Gamma_{u_i} \tag{4.9.17}$$

These equations are identical to the governing equations for the updated Lagrangian formulation, as given in Box 4.3. The nodal forces in the updated and total Lagrangian formulations are expressed in terms of different variables and integrated over different domains, but the discrete equations for the updated Lagrangian and total Lagrangian formulations are identical. Each of these formulations can be advantageous for certain constitutive equations or loadings by reducing the number of transformations which are needed.

4.9.2 Implementation

The procedure for the computation of the internal nodal forces is given in Box 4.6. In the procedure shown, the nodal forces are evaluated by numerical quadrature.

Box 4.6 Discrete equations and internal nodal force algorithm for total Lagrangian formulation

Equations of motion (discrete momentum equation):

$$M_{ijIJ} \ddot{u}_{jJ} + f_{iI}^{\text{int}} = f_{iI}^{\text{ext}} \quad \text{for} \quad (I, i) \notin \Gamma_{v_i} \tag{B4.6.1}$$

Internal nodal forces:

$$f_{iI}^{\text{int}} = \int_{\Omega_0} \left(B_{Ij}^{0}\right)^{T} P_{ji}\, d\Omega_0 = \int_{\Omega_0} \frac{\partial N_I}{\partial X_j} P_{ji}\, d\Omega_0 \quad \text{or} \quad \left(\mathbf{f}_I^{\text{int}}\right)^{T} = \int_{\Omega_0} \boldsymbol{B}_{0I}^{T} \mathbf{P}\, d\Omega_0 \tag{B4.6.2}$$

$$\mathbf{f}_I^{\text{int}} = \int_{\Omega_0} \mathbf{B}_{0I}^{T} \{\mathbf{S}\}\, d\Omega_0 \text{ in Voigt notation}$$

External nodal forces:

$$f_{iI}^{\text{ext}} = \int_{\Omega_0} N_I \rho_0 b_i\, d\Omega_0 + \int_{\Gamma_{t_i}^0} N_I \bar{t}_i^{0} d\Gamma_0 \quad \text{or} \quad f_I^{\text{ext}} = \int_{\Omega_0} N_I \rho_0 \mathbf{b} d\Omega_0 + \int_{\Gamma_{t_i}^0} N_I \mathbf{e}_i \cdot \bar{\mathbf{t}}^{0} d\Gamma_0 \tag{B4.6.3}$$

Mass matrix:

$$M_{ijIJ} = \delta_{ij} \int_{\Omega_0} \rho N_I N_J\, d\Omega = \delta_{ij} \int_{\Omega_0} \rho_0 N_I N_J J_{\xi}^{0} d\square \tag{B4.6.4}$$

$$\mathbf{M}_{IJ} = \mathbf{I}\tilde{M}_{IJ} = \mathbf{I} \int_{\Omega_0} \rho_0 N_I N_J\, d\Omega_0 \tag{B4.6.5}$$

Internal nodal force computation for element

1. $\mathbf{f}^{\text{int}} = 0$
2. For all quadrature points ξ_Q
   i. compute $\left[B_{Ij}^{0}\right] = [\partial N_I(\xi_Q)/\partial X_j]$ for all I
   ii. $\mathbf{H} = \boldsymbol{B}_{0I}\mathbf{u}I;\ H_{ij} = \partial N_I/\partial X_j u_{iI}$
   iii. $\mathbf{F} = \mathbf{I} + \mathbf{H}$, $J = \det(\mathbf{F})$
   iv. $\mathbf{E} = \frac{1}{2}(\mathbf{H} + \mathbf{H}^T + \mathbf{H}^T\mathbf{H})$
   v. if needed, compute $\dot{\mathbf{E}} = \Delta\mathbf{E}/\Delta t$, $\dot{\mathbf{F}} = \Delta\mathbf{F}/\Delta t, \mathbf{D} = \text{sym}(\dot{\mathbf{F}}\mathbf{F}^{-1})$
   vi. compute PK2 stress $\mathbf{S}$ or Cauchy stress $\boldsymbol{\sigma}$ by constitutive equation
   vii. $\mathbf{P} = \mathbf{S}\mathbf{F}^{\mathrm{T}}$ or $\mathbf{P} = J\mathbf{F}^{-1}\boldsymbol{\sigma}$
   viii. $\mathbf{f}_I^{\text{int}} \leftarrow \mathbf{f}_I^{\text{int}} + \boldsymbol{B}_{0I}^{T}\mathbf{P}J_{\xi}^{0}\bar{w}_Q$ for all nodes I

   end loop
   ($\bar{w}_Q$ are quadrature weights)

Usually the shape functions are expressed in terms of element coordinates ξ, such as the area coordinates in triangular elements or reference coordinates in isoparametric elements. The derivatives with respect to the material coordinates are then found by

$$\mathbf{N}_{\mathbf{X}} \equiv \boldsymbol{B}^{0} = \mathbf{N}_{,\xi}\, \mathbf{X}_{,\xi}^{-1} = \mathbf{N}_{,\xi} \left(\mathbf{F}_{\xi}^{0}\right)^{-1} \tag{4.9.18}$$

where $\mathbf{F}_{\xi}^{0}$ is the Jacobian between the material and element coordinates. As shown in Box 4.6, the Green strain tensor is usually not computed in terms of the deformation gradient $\mathbf{F}$, because the resulting computation is susceptible to round-off errors for small strains.

Voigt form It is of little use to write the nodal forces in terms of **P** using Voigt notation since **P** is not symmetric. Therefore, we will write the Voigt form in terms of the PK2 stress **S**. Using the transformation $\mathbf{P} = \mathbf{S} \cdot \mathbf{F}^T$, the expression for the internal nodal forces becomes

$$f_{jI}^{\text{int}} = \int_{\Omega_0} \frac{\partial N_I}{\partial X_i} F_{jk} S_{ik} \, d\Omega_0 \quad \text{or} \quad \left(\mathbf{f}_I^{\text{int}}\right)^T = \int_{\Omega_0} \frac{\partial N_I}{\partial \mathbf{X}} \mathbf{S}\mathbf{F}^T \, d\Omega_0 \tag{4.9.19}$$

We define a $\mathbf{B}_0$ matrix by

$$B_{ikjI}^0 = \underset{(i,k)}{\text{sym}} \left(\frac{\partial N_I}{\partial X_i} F_{jk} \right) \tag{4.9.20}$$

Note that the above specializes to the updated form (4.5.18) where the current configuration and the reference configuration coincide, so that $F_{ij} \rightarrow \delta_{ij}$. The Voigt form of this matrix (see Appendix 1) is

$$B_{ikjI}^0 \rightarrow B_{ab}^0 \begin{array}{l} (i,k) \rightarrow a \text{ by the Voigt kinematic rule} \\ (j,I) \rightarrow b \text{ by the rectangular to column matrix rule} \end{array} \tag{4.9.21}$$

Similarly, S_{ik} is converted to S_a by the kinetic Voigt rule. Then

$$f_b^{\text{int}} = \int_{\Omega_0} \left(B_{ab}^0\right)^T S_a \, d\Omega_0 \quad \text{or} \quad \mathbf{f} = \int_{\Omega_0} \mathbf{B}_0^T \{\mathbf{S}\} \, d\Omega_0 \quad \text{or} \quad \mathbf{f}_I = \int_{\Omega_0} \mathbf{B}_{0I}^T \{\mathbf{S}\} \, d\Omega_0 \tag{4.9.22}$$

The construction of the $\mathbf{B}_0$ matrix hinges on the correspondence between the index o and the j indices given in Table A1.1. Using this correspondence for a two-dimensional element, we obtain:

$$B_{ijkI}^0 \rightarrow B_{akI}^0$$

$$\begin{aligned}
i=1,\ k=1 \rightarrow a=1 \quad & \left[B_{1j}^0\right]_I = \frac{\partial N_I}{\partial X} F_{j1} = \frac{\partial N_I}{\partial X}\frac{\partial x_j}{\partial X} \\
i=2,\ k=2 \rightarrow a=2 \quad & \left[B_{2j}^0\right]_I = \frac{\partial N_I}{\partial Y} F_{j2} = \frac{\partial N_I}{\partial Y}\frac{\partial x_j}{\partial Y} \\
i=1,\ k=2 \rightarrow a=3 \quad & \left[B_{3j}^0\right]_I = \frac{\partial N_I}{\partial X} F_{j2} = \frac{\partial N_I}{\partial Y} F_{j1} = \frac{\partial N_I}{\partial X}\frac{\partial x_j}{\partial Y} + \frac{\partial N_I}{\partial Y}\frac{\partial x_j}{\partial X}
\end{aligned} \tag{4.9.23}$$

The $\mathbf{B}_I^0$ matrix is then written out by letting $k = 1$ and 2 correspond to columns 1 and 2 of the matrix, respectively:

$$\mathbf{B}_I^0 = \begin{bmatrix} \dfrac{\partial N_I}{\partial X}\dfrac{\partial x}{\partial X} & \dfrac{\partial N_I}{\partial X}\dfrac{\partial y}{\partial X} \\ \dfrac{\partial N_I}{\partial Y}\dfrac{\partial x}{\partial Y} & \dfrac{\partial N_I}{\partial Y}\dfrac{\partial y}{\partial Y} \\ \dfrac{\partial N_I}{\partial X}\dfrac{\partial x}{\partial Y} + \dfrac{\partial N_I}{\partial Y}\dfrac{\partial x}{\partial X} & \dfrac{\partial N_I}{\partial X}\dfrac{\partial y}{\partial Y} + \dfrac{\partial N_I}{\partial Y}\dfrac{\partial y}{\partial X} \end{bmatrix} \tag{4.9.24}$$

In three dimensions, a similar procedure yields

$$\mathbf{B}_I^0 = \begin{bmatrix} \frac{\partial N_I}{\partial X}\frac{\partial x}{\partial X} & \frac{\partial N_I}{\partial X}\frac{\partial y}{\partial X} & \frac{\partial N_I}{\partial X}\frac{\partial z}{\partial X} \\ \frac{\partial N_I}{\partial Y}\frac{\partial x}{\partial Y} & \frac{\partial N_I}{\partial Y}\frac{\partial y}{\partial Y} & \frac{\partial N_I}{\partial Y}\frac{\partial z}{\partial Y} \\ \frac{\partial N_I}{\partial Z}\frac{\partial x}{\partial Z} & \frac{\partial N_I}{\partial Z}\frac{\partial y}{\partial Z} & \frac{\partial N_I}{\partial Z}\frac{\partial z}{\partial Z} \\ \frac{\partial N_I}{\partial Y}\frac{\partial x}{\partial Z}+\frac{\partial N_I}{\partial Z}\frac{\partial x}{\partial Y} & \frac{\partial N_I}{\partial Y}\frac{\partial y}{\partial Z}+\frac{\partial N_I}{\partial Z}\frac{\partial y}{\partial Y} & \frac{\partial N_I}{\partial Y}\frac{\partial z}{\partial Z}+\frac{\partial N_I}{\partial Z}\frac{\partial z}{\partial Y} \\ \frac{\partial N_I}{\partial X}\frac{\partial x}{\partial Z}+\frac{\partial N_I}{\partial Z}\frac{\partial x}{\partial X} & \frac{\partial N_I}{\partial X}\frac{\partial y}{\partial Z}+\frac{\partial N_I}{\partial Z}\frac{\partial y}{\partial X} & \frac{\partial N_I}{\partial X}\frac{\partial z}{\partial Z}+\frac{\partial N_I}{\partial Z}\frac{\partial z}{\partial X} \\ \frac{\partial N_I}{\partial X}\frac{\partial x}{\partial Y}+\frac{\partial N_I}{\partial Y}\frac{\partial x}{\partial X} & \frac{\partial N_I}{\partial X}\frac{\partial y}{\partial Y}+\frac{\partial N_I}{\partial Y}\frac{\partial y}{\partial X} & \frac{\partial N_I}{\partial X}\frac{\partial z}{\partial Y}+\frac{\partial N_I}{\partial Y}\frac{\partial z}{\partial X} \end{bmatrix} \tag{4.9.25}$$

Many writers construct the $\mathbf{B}_0$ matrix through a sequence of multiplications by Boolean matrices. The procedure shown here can easily be coded and is much faster.

It can be easily shown that $\mathbf{B}_0$ relates the rate of Green strain $\dot{\mathbf{E}}$ to the node velocities by

$$\{\dot{\mathbf{E}}\} = \mathbf{B}_I^0 \mathbf{v}_I = \mathbf{B}_0 \dot{\mathbf{d}} \tag{4.9.26}$$

The reader should be cautioned about one characteristic of the $\mathbf{B}_0$ matrix: although it carries a subscript zero, the matrix $\mathbf{B}_0$ is not time invariant. This can easily be seen from (4.9.20) or (4.9.24–25), which show that the $\mathbf{B}_0$ matrix depends on $\mathbf{F}$, which varies with time.

The total Lagrangian equation for internal nodal forces, (4.9.22), can easily be reduced to the updated Lagrangian form, (4.5.12) without any transformations. This is accomplished by letting the configuration at a fixed time t be the reference configuration. When the current configuration is the reference configuration

$$\mathbf{F} = \mathbf{I} \quad \text{or} \quad F_{ij} = \frac{\partial x_i}{\partial X_j} = \delta_{ij} \tag{4.9.27}$$

since the two coordinate systems are now coincident at time t. In addition, when the current configuration becomes the reference configuration:

$$\mathbf{B}_0 = \mathbf{B} \quad \mathbf{S} = \boldsymbol{\sigma} \quad \Omega_0 = \Omega \quad J = 1 \quad d\Omega_0 = d\Omega \tag{4.9.28}$$

To verify the first of these, substitute (4.9.27) into (4.9.20) and compare to (4.5.18); since $\mathbf{F} = \mathbf{I}$, from Box 3.2, $\mathbf{S} = \boldsymbol{\sigma}$. The expression for the nodal forces (4.9.22) becomes

$$\mathbf{f}_I = \int_\Omega \mathbf{B}_I^T \{\boldsymbol{\sigma}\} d\Omega \tag{4.9.29}$$

which agrees with the updated Lagrangian form (4.5.12). This process of making the current configuration the reference configuration is a helpful trick which we will use again later.

Example 4.8 Rod in Two Dimensions Develop the internal nodal forces for a two-node rod element in two-dimensions. The bar element is shown in Figure 4.12. It is in a uniaxial state of stress with the only nonzero stress along the axis of the bar.

To simplify the formulation, we place the material coordinate system so that the X-axis coincides with the axis of the rod, as shown in Figure 4.12, with the origin of the material coordinates at node 1. The parent element coordinate is $\xi \in [0, 1]$. The material coordinates are related to the element coordinates by

$$X = X_2\xi = \ell_0\xi \tag{E4.8.1}$$

where ℓ_0 is the initial length of the element. In this example, the coordinates X, Y are used in a somewhat different sense than before: it is no longer true that $\mathbf{x}(t=0) = \mathbf{X}$. However, the definition used here corresponds to a rotation and translation of $\mathbf{x}(t=0)$. Since neither rotation nor translation affects $\mathbf{E}$ or any strain measure, this choice of an X, Y coordinate system is perfectly acceptable. We could have used the element coordinates ξ as material coordinates, but this complicates the definition of physical strain components.

The motion is given in terms of the element coordinates by

$$\begin{aligned} x(\xi,t) &= x_1(t)(1-\xi) + x_2(t)\xi \\ y(\xi,t) &= y_1(t)(1-\xi) + y_2(t)\xi \end{aligned} \quad \text{or} \quad \begin{Bmatrix} x \\ y \end{Bmatrix} = \begin{bmatrix} x_1 & x_2 \\ y_1 & y_2 \end{bmatrix} \begin{Bmatrix} 1-\xi \\ \xi \end{Bmatrix} \tag{E4.8.2}$$

or

$$\mathbf{x}(\xi,t) = \mathbf{x}_I(t)N_I(\xi) \tag{E4.8.3}$$

where

$$\{N_I(\xi)\}^T = [(1-\xi)\ \xi] = \left[1-\frac{X}{\ell_0},\ \frac{X}{\ell_0}\right] \tag{E4.8.4}$$

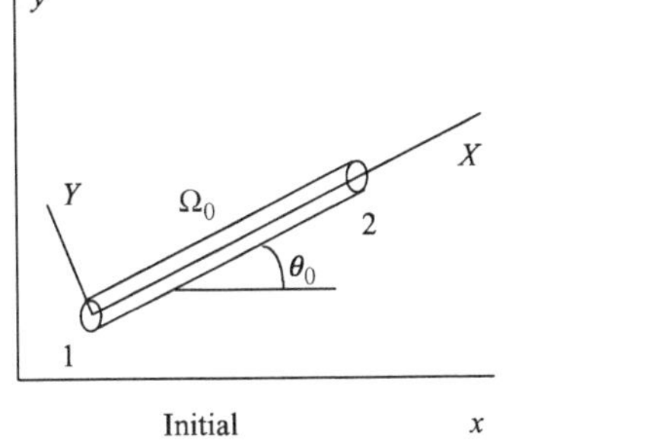

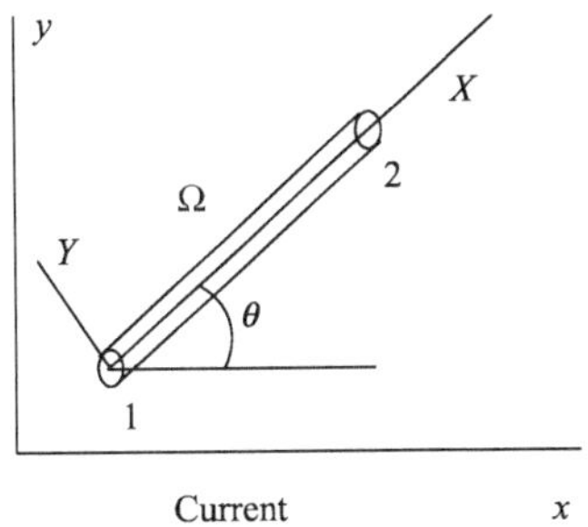

Figure 4.12 Rod element in two dimensions in total Lagrangian formulation

The $\boldsymbol{B}_0$ matrix as defined in (4.9.7) is given by

$$[\boldsymbol{B}_{0iI}] \equiv [\partial N_I / \partial X_i]^T = \begin{bmatrix} \frac{\partial N_1}{\partial X} & \frac{\partial N_2}{\partial X} \end{bmatrix} = \frac{1}{\ell_0}[-1 \quad +1] \tag{E4.8.5}$$

where (E4.8.1) has been used to give $\frac{\partial N_I}{\partial X} = \frac{1}{\ell_0}\frac{\partial N_I}{\partial \xi}$. The deformation gradient is given by (4.9.7):

$$\mathbf{F} = \mathbf{x}_I \left(\boldsymbol{B}_I^0\right)^T = \begin{bmatrix} x_1 & x_2 \\ y_1 & y_2 \end{bmatrix} \frac{1}{\ell_0} \begin{Bmatrix} -1 \\ 1 \end{Bmatrix} = \frac{1}{\ell_0}[x_2 - x_1 \quad y_2 - y_1] \equiv \frac{1}{\ell_0}[x_{21} \quad y_{21}] \tag{E4.8.6}$$

The deformation gradient $\mathbf{F}$ is not a square matrix for the rod since there are two space dimensions but only one independent variable.

The only nonzero stress is along the axis of the rod. To take advantage of this, we use the nodal force formula in terms of the PK2 stress, since S_{11} is the only nonzero component of this stress. For the nominal stress, P_{11} is not the only nonzero component. The X axis as defined here is corotational with the axis of the rod, so S_{11} is always the stress component along the axis of the rod. Substituting (E4.8.5) and (E4.8.6) into (4.9.19) then gives the following expression for the internal nodal forces:

$$\mathbf{f}_{\text{int}}^T = \int_{\Omega_0} \boldsymbol{B}_0^T \mathbf{S}\mathbf{F}^T \, d\Omega_0 = \int_{\Omega_0} N,_{\mathbf{x}} \, \mathbf{S}\mathbf{F}^T \, d\Omega_0 = \int_{\Omega_0} \frac{1}{\ell_0} \begin{bmatrix} -1 \\ +1 \end{bmatrix} [S_{11}] \frac{1}{\ell_0} [x_{21}, \; y_{21}] \, d\Omega_0 \tag{E4.8.7}$$

We assume the integrand is constant, so multiplying the integrand by the volume $A_0\ell_0$ we have

$$\begin{bmatrix} f_{1x} & f_{1y} \\ f_{2x} & f_{2y} \end{bmatrix}^{\text{int}} = \frac{A_0 S_{11}}{\ell_0} \begin{bmatrix} -x_{21} & -y_{21} \\ x_{21} & y_{21} \end{bmatrix} \tag{E4.8.8}$$

This result can be transformed to the result for the corotational formulation if we use (E3.9.8) and note that $\cos\theta = x_{21}/\ell$ and $\sin\theta = y_{21}/\ell$.

In Voigt notation, only the first row of the $\mathbf{B}_0$ matrix need be considered since the only nonzero stress is S_{11}. From (4.9.24), and using (E4.8.3–4) to obtain $x,_X = x_{21}/\ell_0 = \ell/\ell_0 \cos\theta, y,_X = y_{21}/\ell_0 = \ell/\ell_0 \sin\theta$ the first row is given by

$$\mathbf{B}_I^0 = [N_{IX} x,_X \quad N_{IX} y,_X] = \frac{1}{\ell_0}[N_{I,X} x_{21} \quad N_{I,X} y_{21}] \tag{E4.8.9}$$

From (E4.8.4), $N_{1,X} = -1/\ell_0, N_2,_X = 1/\ell_0$ and so

$$\mathbf{B}_0 = [\mathbf{B}_1^0 \quad \mathbf{B}_2^0] = \frac{\ell}{\ell_0^2}[-\cos\theta \quad -\sin\theta \qquad \cos\theta \qquad \sin\theta] \tag{E4.8.10}$$

The expression for the nodal forces, (4.9.22) then becomes

$$\mathbf{f}^{\text{int}} \equiv \begin{Bmatrix} f_{x1} \\ f_{y1} \\ f_{x2} \\ f_{y2} \end{Bmatrix}^{\text{int}} = \int_{\Omega_0} \mathbf{B}_0^T \{\mathbf{S}\} d\Omega_0 = \int_{\Omega_0} \frac{\ell}{\ell_0^2} \begin{Bmatrix} -\cos\theta \\ -\sin\theta \\ \cos\theta \\ \sin\theta \end{Bmatrix} \{S_{11}\} d\Omega_0 = \frac{A_0 \ell S_{11}}{\ell_0} \begin{Bmatrix} -\cos\theta \\ -\sin\theta \\ \cos\theta \\ \sin\theta \end{Bmatrix} \tag{E4.8.11}$$

where we have used the fact that the integrand is constant and the volume is $A_0\ell_0$ in the last step.

Example 4.9 Triangular element Develop expressions for the deformation gradient, internal nodal forces and external nodal forces for the three-node, linear displacement triangle. The element was developed in the updated Lagrangian formulation in Example 4.1; the element is shown in Figure 4.3.

The motion of the element is given by the same linear map as in Example 4.1, (E4.1.7) in terms of the triangular coordinates ξ_I. The $\boldsymbol{B}_0$ matrix is given by (4.9.7):

$$\boldsymbol{B}_I^0 = \left[\boldsymbol{B}_{jI}^0\right] = \left[\partial N_I / \partial N_j\right]$$

$$\boldsymbol{B}_0 = \left[\boldsymbol{B}_1^0 \quad \boldsymbol{B}_2^0 \quad \boldsymbol{B}_3^0\right] = \begin{bmatrix} \dfrac{\partial N_1}{\partial X} & \dfrac{\partial N_2}{\partial X} & \dfrac{\partial N_3}{\partial X} \\ \dfrac{\partial N_1}{\partial Y} & \dfrac{\partial N_2}{\partial Y} & \dfrac{\partial N_3}{\partial X} \end{bmatrix} = \frac{1}{2A_0} \begin{bmatrix} Y_{23} & Y_{31} & Y_{12} \\ X_{32} & X_{13} & X_{21} \end{bmatrix} \tag{E4.9.1}$$

where A_0 is the area of the undeformed element, $X_{IJ} = X_I - X_J$ and $Y_{IJ} = Y_I - Y_J$. These equations are identical to those given in the updated Lagrangian formulation except that the initial nodal coordinates and initial area are used. The internal forces are then given by (4.9.10):

$$\mathbf{f}_{\text{int}}^T = [f_{iI}] = \begin{bmatrix} f_{1x} & f_{1y} \\ f_{2x} & f_{2y} \\ f_{3x} & f_{3y} \end{bmatrix}^{\text{int}} = \int_{\Omega_0} \boldsymbol{B}_0^T \mathbf{P}\, d\Omega_0$$

$$= \int_{A_0} \frac{1}{2A_0} = \begin{bmatrix} Y_{23} & X_{32} \\ Y_{31} & X_{13} \\ Y_{12} & X_{21} \end{bmatrix} \begin{bmatrix} P_{11} & P_{12} \\ P_{21} & P_{22} \end{bmatrix} a_0\, dA_0 = \frac{a_0}{2} \begin{bmatrix} Y_{23} & X_{32} \\ Y_{31} & X_{13} \\ Y_{12} & X_{21} \end{bmatrix} \begin{bmatrix} P_{11} & P_{12} \\ P_{21} & P_{22} \end{bmatrix} \tag{E4.9.2}$$

where a_0 is the initial thickness.

Voigt notation The expression for the internal nodal forces in Voigt notation requires the $\mathbf{B}_0$ matrix. Using (4.9.24) and the derivatives of the shape functions in (E4.9.1) gives

$$\mathbf{B}_0 = \begin{bmatrix} Y_{23}x,_X & Y_{23}y,_X & Y_{31}x,_X & Y_{31}y,_X & Y_{12}x,_X & Y_{12}y,_X \\ X_{32}x,_Y & X_{32}y,_Y & X_{13}x,_Y & X_{13}y,_Y & X_{21}x,_Y & X_{21}y,_Y \\ Y_{23}x,_Y + X_{32}x,_X & Y_{23}y,_Y + X_{32}y,_X & Y_{31}x,_Y + X_{13}x,_X & Y_{31}y,_Y + X_{13}y,_X & Y_{12}x,_Y + X_{21}x,_X & Y_{12}y,_Y + X_{21}y,_X \end{bmatrix} \tag{E4.9.3}$$

The terms of the **F** matrix are evaluated by (4.9.6); for example:

$$x,_X = N_{I,X}x_I = \frac{1}{2A_0}(Y_{23}x_1 + Y_{31}x_2 + Y_{12}x_3) \tag{E4.9.4}$$

Note that the **F** matrix is constant in the element, and so is $\mathbf{B}_0$. The nodal forces are then given by (4.9.22):

$$\mathbf{f}^{\text{int}} = \{f_a\} = \begin{Bmatrix} f_{x1} \\ f_{y1} \\ f_{x2} \\ f_{y2} \\ f_{x3} \\ f_{y3} \end{Bmatrix}^{\text{int}} = \int_{\Omega_0} \mathbf{B}_0^T \begin{Bmatrix} S_{11} \\ S_{22} \\ S_{12} \end{Bmatrix} d\Omega_0 \tag{E4.9.5}$$

Example 4.10 Two-dimensional isoparametric element Construct the discrete equations for two-dimensional isoparametric elements. The element is shown in Figure 4.5; the same element in the updated Lagrangian form was considered in Example 4.2.

The motion of the element is given in (E4.2.1). The key difference in the formulation in the total Lagrangian formulation is that the derivatives of the shape functions with respect to the material coordinates must be found. By implicit differentiation

$$\begin{Bmatrix} N_{I,X} \\ N_{I,Y} \end{Bmatrix} = \mathbf{X},_\xi^{-T} \begin{Bmatrix} N_{I,\xi} \\ N_{I,\eta} \end{Bmatrix} \tag{E4.10.1}$$

where

$$\mathbf{X},_\xi = \mathbf{X}_I N_{I,\xi} \quad \text{or} \quad \frac{\partial X_i}{\partial \xi_j} = X_{iI}\frac{\partial N_I}{\partial \xi_j} \tag{E4.10.2}$$

Writing out these gives

$$\begin{bmatrix} X_{,\xi} & X_{,\eta} \\ Y_{,\xi} & Y_{,\eta} \end{bmatrix} = \begin{Bmatrix} X_I \\ Y_I \end{Bmatrix} [N_{I,\xi} \quad N_{I,\eta}] \tag{E4.10.3}$$

which can be evaluated from the shape functions and nodal coordinates; details are given for the four-node quadrilateral in (E4.2.7–9) in terms of the updated coordinates; the formulas for the material coordinates can be obtained by replacing (x_I, y_I) by (X_I, Y_I). The inverse of $\mathbf{X}_{,\xi}$ is then given by

$$\mathbf{X}_{,\xi}^{-1} = \begin{bmatrix} X_{,\xi} & X_{,\eta} \\ Y_{,\xi} & Y_{,\eta} \end{bmatrix}^{-1} = \frac{1}{J_\xi^0} \begin{bmatrix} Y_{,\eta} & X_{,\eta} \\ Y_{,\xi} & Y_{,\xi} \end{bmatrix}$$

where the last step is an alternate expression for $\mathbf{X}_{,\xi}^{-1}$. The determinant of the Jacobian between the parent and reference configurations is given by

$$J_\xi^0 = X_{,\xi} Y_{,\eta} - Y_{,\xi} X_{,\eta}$$

The $\boldsymbol{B}_{0I}$ matrices are given by

$$\boldsymbol{B}_{0I}^T = [N_{I,X} N_{I,Y}] = [N_{I,\xi} N_{I,\eta}] \mathbf{X}_{,\xi}^{-1} \tag{E4.10.4}$$

The gradient of the displacement field $\mathbf{H}$ is given by

$$\mathbf{H} = \mathbf{u}_I \boldsymbol{B}_{0I}^T = \begin{Bmatrix} u_{xI} \\ u_{yI} \end{Bmatrix} [N_{I,X} N_{I,Y}] \tag{E4.10.5}$$

The deformation gradient is then given by

$$\mathbf{F} = \mathbf{I} + \mathbf{H} \tag{E4.10.6}$$

The Green strain $\mathbf{E}$ is obtained as shown in Box 4.6 and the stress $\mathbf{S}$ is evaluated by the constitutive equation; the nominal stress $\mathbf{P}$ can then be computed by $\mathbf{P} = \mathbf{S}\mathbf{F}^T$; see Box 3.2.

The internal nodal forces are given by (4.9.10):

$$\left(\mathbf{f}_I^{\text{int}}\right)^T = \int_{\Omega 0} \boldsymbol{B}_{0I}^T \mathbf{P}\, d\Omega_0 = \int_{-1}^{1}\int_{-1}^{1} \begin{bmatrix} N_{I,X} & N_{I,Y} \end{bmatrix} \begin{bmatrix} P_{11} & P_{12} \\ P_{21} & P_{22} \end{bmatrix} J_\xi^0 d\xi d\eta \tag{E4.10.7}$$

where

$$J_\xi^0 = \det(\mathbf{X}_{,\xi}) = \det\left(\mathbf{F}_\xi^0\right) \tag{E4.10.8}$$

The external nodal forces, particularly those due to pressure, are usually best computed in the updated form. The mass matrix was given in Example 4.2.

Example 4.11 Three-dimensional element Develop the strain and nodal force equations for a general three-dimensional element in the total Lagrangian format. The element is shown in Figure 4.6. The parent element coordinates are $\boldsymbol{\xi} = (\xi_1, \xi_2, \xi_3) \equiv (\xi, \eta, \zeta)$ for an isoparametric element, $\boldsymbol{\xi} = (\xi_1, \xi_2, \xi_3)$ for a tetrahedral element, where for the latter ξ_i are the volume (barycentric) coordinates.

Matrix form The standard expressions for the motion, (4.9.1–5), are used. The deformation gradient is given by (4.9.6). The Jacobian matrix relating the reference configuration to the parent is

$$\mathbf{X}_{,\xi} = \begin{bmatrix} X_{,\xi} & X_{,\eta} & X_{,\zeta} \\ Y_{,\xi} & Y_{,\eta} & Y_{,\zeta} \\ Z_{,\xi} & Z_{,\eta} & Z_{,\zeta} \end{bmatrix} = \{X_{iI}\}[\partial N_I / \partial \xi_j] = \begin{Bmatrix} X_I \\ Y_I \\ Z_I \end{Bmatrix} [N_{I,\xi} \quad N_{I,\eta} \quad N_{I,\zeta}] \quad \text{(E4.11.1)}$$

The deformation gradient is given by

$$[F_{ij}] = \{x_{iI}\}\left[\frac{\partial N_I}{\partial X_j}\right] = \begin{Bmatrix} x_I \\ y_I \\ z_I \end{Bmatrix} [N_{I,X} \quad N_{I,Y} \quad N_{I,Z}] \quad \text{(E4.11.2)}$$

where

$$\left[\frac{\partial N_I}{\partial X_j}\right] = [N_{I,X} \quad N_{I,Y} \quad N_{I,Z}] = \left[\frac{\partial N_I}{\partial \xi_k}\right]\left[\frac{\partial \xi_k}{\partial X_j}\right] = N_{I,\xi}\mathbf{X}_{,\xi}^{-1} \quad \text{(E4.11.3)}$$

where $\mathbf{X}_{,\xi}^{-1}$ is evaluated numerically from (E4.11.1). The Green strain tensor should not be computed directly from $\mathbf{F}$, because the round-off errors can be huge; it is better to compute

$$[H_{ij}] = \{u_{iI}\}\left[\frac{\partial N_I}{\partial X_j}\right] = \begin{Bmatrix} u_{xI} \\ u_{yI} \\ u_{zI} \end{Bmatrix} [N_{I,\xi} \quad N_{I,\eta} \quad N_{I,\zeta}] \quad \text{(E4.11.4)}$$

The Green strain tensor is then obtained as in Box 4.6. If the constitutive law relates the PK2 stress $\mathbf{S}$ to $\mathbf{E}$, the nominal stress is then computed by $\mathbf{P} = \mathbf{S}\mathbf{F}^{\mathrm{T}}$, using $\mathbf{F}$ from (E4.11.2). The internal nodal forces are then given by

$$\begin{Bmatrix} f_{xI} \\ f_{yI} \\ f_{zI} \end{Bmatrix}^{\text{int}} = \int_{\square} \boldsymbol{B}_{0I}^T \mathbf{P} J_\xi^0 \, d\square = \int_{\square} [N_{I,X} \quad N_{I,Y} \quad N_{I,Z}] \begin{bmatrix} P_{11} & P_{12} & P_{13} \\ P_{21} & P_{22} & P_{23} \\ P_{31} & P_{32} & P_{33} \end{bmatrix} J_\xi^0 \, d\square \quad \text{(E4.11.5)}$$

where $J_0^\xi = \det(\mathbf{X}_{,\xi})$.

Voigt form All of the variables needed for the evaluation of the $\mathbf{B}_0$ matrix given in (4.9.25) can be obtained from the previous. In Voigt form

$$\begin{aligned}\{\mathbf{E}\}^T &= [E_{11}, E_{22}, E_{33}, 2E_{23}, 2E_{13}, 2E_{12}] \\ \{\mathbf{S}\}^T &= [S_{11}, S_{22}, S_{33}, S_{23}, S_{13}, S_{12}]\end{aligned} \tag{E4.11.6}$$

The rate of Green strain can be computed by (4.9.26):

$$\begin{aligned}\{\dot{\mathbf{E}}\} &= \mathbf{B}_0\dot{\mathbf{d}} \\ \dot{\mathbf{d}} &= [u_{x1}, u_{y1}, u_{z1}, \ldots u_{xN}, u_{yN}, u_{zN}]\end{aligned} \tag{E4.11.7}$$

The Green strain is computed by the procedure in Box 4.6. The nodal forces are given by

$$\mathrm{f}_I^{\text{int}} = \int_{\square} \mathbf{B}_{0I}^T \{\mathbf{S}\} J_\xi^0 \, d\square \tag{E4.11.8}$$

4.9.3 *Variational Principle for Large Deformation Statics*

For static problems, weak forms for nonlinear analysis with path-independent materials can be obtained from variational principles. For many nonlinear problems, variational principles cannot be formulated. However, when the problem is conservative, it is possible to develop a variational principle. In a conservative problem, the constitutive equations and loads are path independent and nondissipative, so that the stresses and external loads can be obtained from potentials. Such external loads are called conservative loads, for they are path independent and return to their original values when the body returns to its original configuration. The materials for which stress is derivable from a potential are called hyperelastic materials: see Section 5.4.

In a hyperelastic material, the nominal stress is given in terms of the potential by

$$\mathbf{P}^T = \frac{\partial w(\mathbf{F})}{\partial \mathbf{F}} \quad \text{or} \quad P_{ji} = \frac{\partial w}{\partial F_{ij}}, \quad \text{where} \quad w = \rho_0 w^{\text{int}} \tag{4.9.30}$$

Note the order of the subscripts on the stress and rate-of-deformation, which follows from the definition of the conjugate pairing of the nominal stress, Box 3.4. The total internal work is given by

$$W^{\text{int}} = \int_{\Omega_0} w \, d\Omega_0 \tag{4.9.31}$$

It follows from (4.9.30–31) that

$$\delta W^{\text{int}} = \int_{\Omega_0} \delta w \, d\Omega_0 = \int_{\Omega_0} \frac{\partial w}{\partial F_{ij}} \delta F_{ij} \, d\Omega_0 = \int_{\Omega_0} P_{ji} \delta F_{ij} \, d\Omega_0 = \int_{\Omega_0} \mathbf{P}^T : \delta\mathbf{F} \, d\Omega_0 \tag{4.9.32}$$

The potential also relates any other stress and strain measures which are conjugate in energy, so it also follows that

$$\mathbf{S} = \frac{\partial w(\mathbf{E})}{\partial \mathbf{E}} \quad \text{or} \quad S_{ij} = \frac{\partial w}{\partial E_{ij}} \tag{4.9.33}$$

These relationships do not hold for stress and strain rate measures where the strain rate is not integrable, so the relationship does not hold for the Cauchy stress and rate-of-deformation pair.

A conservative load is also derivable from a potential, that is, the loads must be related to a potential so that

$$\rho_0 b_i = \frac{\partial w_b^{\text{ext}}}{\partial u_i}, \quad \bar{t}_i^0 = \frac{\partial w_t^{\text{ext}}}{\partial u_i} \tag{4.9.34}$$

$$W^{\text{ext}}(\mathbf{u}) = \int_{\Omega_0} w_b^{\text{ext}}(\mathbf{u}) d\Omega_0 + \int_{\Gamma_t^0} w_t^{\text{ext}}(\mathbf{u}) d\Gamma_0 \tag{4.9.35}$$

Theorem of stationary potential energy We consider only boundaries that are strictly displacement or traction boundaries; the results are easily extended to the more general case. The theorem states that for a static process with conservative loads and constitutive equations, the stationary points of

$$W(\mathbf{u}) = W^{\text{int}}(\mathbf{F}(\mathbf{u})) - W^{\text{ext}}(\mathbf{u}), \quad \mathbf{u}(\mathbf{X}, t) \in u \tag{4.9.36}$$

satisfy the strong form of the equilibrium equation ((B4.4.2) with zero accelerations), the traction boundary conditions (B4.4.7), and the internal continuity conditions (B4.4.11). The equilibrium equation which emanates from this stationary principle is in terms of the displacements.

The theorem is proven by showing the equivalence of the stationary principle to the weak form for equilibrium, traction boundary conditions and interior continuity conditions. The stationary condition of (4.9.36) gives

$$0 = \delta W(\mathbf{u}) = \int_{\Omega_0} \left(\frac{\partial w}{\partial F_{ij}} \delta F_{ij} \, d\Omega - \frac{\partial w_b^{\text{ext}}}{\partial u_i} \delta u_i \right) d\Omega_0 - \int_{\Gamma_0} \frac{\partial w_t^{\text{ext}}}{\partial u_i} \delta u_i \, d\Gamma_0 \tag{4.9.37}$$

Substituting (4.9.30) and (4.9.34) into (4.9.37) gives

$$0 = \int_{\Omega_0} (P_{ji} \delta F_{ij} - \rho_0 b_i \delta u_i) \, d\Omega_0 - \int_{\Gamma_0} t_i^0 \delta u_i d\Gamma_0 \tag{4.9.38}$$

which is the weak form given in (4.8.7) for the case when the accelerations vanish. The same steps given in Section 4.3.2 can then be used to establish the equivalence of (4.9.38) to the strong form of the equilibrium equation, the traction boundary conditions and the interior continuity conditions.

Stationary principles are in a sense more restrictive weak forms: they apply only to conservative, static problems. However, they can improve our understanding of stability problems and the solution of equilibrium equations; they are also used in the study of the existence and uniqueness of solutions.

The discrete equations are obtained from the stationary principle by using the usual finite element approximation to motion with a Lagrangian mesh, (4.9.2) which we write in the form

$$\mathbf{u}(\mathbf{X},t)=\mathbf{N}(\mathbf{X})\mathbf{d}(t) \tag{4.9.39}$$

The potential energy can then be expressed in terms of the nodal displacements, giving

$$W(\mathbf{d})=W^{\text{int}}(\mathbf{d})-W^{\text{ext}}(\mathbf{d}) \tag{4.9.40}$$

The solutions to (4.9.10) correspond to the stationary points of this function, so the discrete equations are

$$0=\frac{\partial W(\mathbf{d})}{\partial \mathbf{d}}=\frac{\partial W^{\text{int}}(\mathbf{d})}{\partial \mathbf{d}}-\frac{\partial W^{\text{ext}}(\mathbf{d})}{\partial \mathbf{d}} \tag{4.9.41}$$

We now define

$$\mathbf{f}^{\text{int}}=\frac{\partial W^{\text{int}}(\mathbf{d})}{\partial \mathbf{d}},\quad \mathbf{f}^{\text{ext}}=\frac{\partial W^{\text{ext}}(\mathbf{d})}{\partial \mathbf{d}} \tag{4.9.42}$$

Then it follows from (4.9.41–42) that the discrete equations of equilibrium are

$$\mathbf{f}^{\text{int}}=\mathbf{f}^{\text{ext}} \tag{4.9.43}$$

The internal and external nodal forces defined above are equivalent to the internal and external nodal forces defined in Box 4.6. It will be shown in Chapter 6 that when the equilibrium point is stable, it corresponds to a local minimum of the potential energy.

Example 4.12 Rod element by stationary principle Consider a structural model consisting of two-node rod elements in three dimensions. Let the internal potential energy be given by

$$w=\frac{1}{2}C^{\text{SE}}E_{11}^{2} \tag{E4.12.1}$$

where the E_{11} component is along the axis of the rod and $C^{SE}=E^{SE}$, which is given in Box 5.1. Let the only load on the structure be gravity, for which the external potential is

$$w^{\text{ext}}=-\rho_0 gz \tag{E4.12.2}$$

where g is the acceleration due to gravity. Find expressions for the internal and external nodal forces of an element.

From (4.9.31) and (E4.12.1), the total internal potential is given by

$$W^{\text{int}} = \sum_e W_e^{\text{int}}, \quad W_e^{\text{int}} = \frac{1}{2}\int_{\Omega_0^e} C^{SE} E_{11}^2 \, d\Omega_0 \qquad \text{(E4.12.3)}$$

For the two-node element, the displacement field is linear and the Green strain is constant, so (E4.12.3) can be simplified by multiplying the integrand by the initial volume of the element $A_0\ell_0$:

$$W_e^{\text{int}} = \frac{1}{2} A_0 \ell_0 C^{SE} E_{11}^2 \qquad \text{(E4.12.4)}$$

To develop the internal nodal forces, we will need the derivatives of the Green strain with respect to the nodal displacements. Since the strain is constant in the element (see E3.9.9)

$$E_{11} = \frac{\ell^2 - \ell_0^2}{2\ell_0^2} = \frac{\mathbf{x}_{21} \cdot \mathbf{x}_{21} - \mathbf{X}_{21} \cdot \mathbf{X}_{21}}{2\ell_0^2} \qquad \text{(E4.12.5)}$$

where $\mathbf{x}_{IJ} \equiv \mathbf{x}_I - \mathbf{x}_J$ and $\mathbf{X}_{IJ} \equiv \mathbf{X}_I - \mathbf{X}_J$. Noting that

$$\mathbf{x}_{IJ} \equiv \mathbf{X}_{IJ} + \mathbf{u}_{IJ} \qquad \text{(E4.12.6)}$$

where $\mathbf{u}_{IJ} \equiv \mathbf{u}_I - \mathbf{u}_J$ are the nodal displacements, and substituting (E4.12.6) into (E4.12.5) gives, after some algebra,

$$E_{11} = \frac{2\mathbf{X}_{21} \cdot \mathbf{u}_{21} + \mathbf{u}_{21} \cdot \mathbf{u}_{21}}{2\ell_0^2} \qquad \text{(E4.12.7)}$$

The derivatives of E_{11}^2 with respect to the nodal displacements are then given by

$$\frac{\partial (E_{11}^2)}{\partial \mathbf{u}_2} = \frac{\mathbf{X}_{21} + \mathbf{u}_{21}}{\ell_0^2} = \frac{\mathbf{x}_{21}}{\ell_0^2}, \qquad \frac{\partial (E_{11}^2)}{\partial \mathbf{u}_1} = -\frac{\mathbf{X}_{21} + \mathbf{u}_{21}}{\ell_0^2} = -\frac{\mathbf{x}_{21}}{\ell_0^2} \qquad \text{(E4.12.8)}$$

Using the definition for internal nodal forces (4.9.42) in conjunction with (E4.12.4) and (E4.12.8) gives

$$\mathbf{f}_2^{\text{int}} = -\mathbf{f}_1^{\text{int}} = \frac{A_0 C^{SE} E_{11} \mathbf{x}_{21}}{\ell_0} \qquad \text{(E4.12.9)}$$

By using the constitutive equation emanating from (E4.12.1) and (4.9.33), it follows that $S_{11} = C^{SE} E_{11}$, so

$$\left(\mathbf{f}_2^{\text{int}}\right)^T\left[f_{x2}, f_{y2}, f_{z2}\right] = -\left(\mathbf{f}_1^{\text{int}}\right)^T = \frac{A_0 S_{11}}{\ell_0}[x_{21} \quad y_{21} \quad z_{21}] \tag{E4.12.10}$$

This result, as expected, is identical to the result obtained for the bar by the principle of virtual work, (E4.8.8). The external potential for a gravity load is given by

$$W^{\text{ext}} = -\int_{\Omega_0} \rho_0 g z \, d\Omega_0 \tag{E4.12.11}$$

If we make the finite element approximation $z = z_I N_I$, where N_I are the shape functions given in (E4.8.4), then

$$W^{\text{ext}} = -\int_{\Omega_0} \rho_0 g z_I N_I \, d\Omega_0 \tag{E4.12.12}$$

and

$$f_{zI}^{\text{ext}} = \frac{\partial W^{\text{ext}}}{\partial u_{zI}} = -\int_0^1 \rho_0 g(Z_I + u_{zI}) N_I(\xi) \ell_0 A_0 \, d\xi = -\frac{1}{2} A_0 \ell_0 \rho_0 g \begin{Bmatrix} 1 \\ 1 \end{Bmatrix} \tag{E4.12.13}$$

so the external nodal force on each node is half the force on the rod element due to gravity.

4.10 Exercises

4.1. Consider the element shown in Figure 3.4 with the motion

$$x(\mathbf{X}, t) = (1 + at) X \cos\frac{\pi}{2}t - (1 + bt) Y \sin\frac{\pi}{2}t$$

$$y(\mathbf{X}, t) = (1 + at) X \sin\frac{\pi}{2}t + (1 + bt) Y \cos\frac{\pi}{2}t$$

Sketch the element in the deformed configuration at $t = 1$ (this was already done in Exercise 3.1).

(a) Let the only nonzero PK2 stress component in the deformed configuration be S_{11}. Find the nodal internal forces, using the updated Lagrangian formulation.

(b) For the same state of stress, find the nodal internal forces using the total Lagrangian configuration. What is the effect of rotating the body on the nodal internal forces?

(c) Repeat parts (a) and (b) with the only non zero components being S_{22} and S_{12}. Explain the nodal internal forces in the undeformed and deformed configurations.

4.2. Consider the block under shear shown in Figure 3.13 with the motion given in (E3.13.1). Evaluate the Green strain as a function of time. Plot E_{12} and E_{22} for $t \in [0, 4]$; explain why E_{22} is nonzero. Evaluate the PK2 stress for a Kirchhoff material, using a $[C^{SE}]$ given by (5.4.58) (the matrix given in (5.4.58) is $[C^{\tau}]$, but use the same matrix).

4.3. (a) Use Nanson's relation (3.4.5)

$$\mathbf{n}\, d\Gamma = J\mathbf{n}_0 \cdot \mathbf{F}^{-1}\, d\Gamma_0 \quad n_i\, d\Gamma = Jn_j^0 F_{ji}^{-1}\, d\Gamma_0$$

to show that the external nodal forces for an applied pressure p acting on the plane $\varsigma = -1$ are given by

$$\mathbf{f}_I^{\text{ext}} = -\int_{-1}^{1}\int_{-1}^{1} pN_I J_\xi \mathbf{F}_\xi^{-T} \cdot \mathbf{n}_{0\xi}\, d\xi d\eta,$$

where $\mathbf{n}_{0\xi} = -\hat{\mathbf{e}}_3$ is the unit normal vector in parent coordinates to the plane $\varsigma = -1$ in the parent element,

(b) By using the definition of the inverse of a tensor in terms of Cramer's rule, that is,

$$\mathbf{F}_\xi^{-1} = \frac{1}{\det(\mathbf{F}_\xi)}\left[\mathbf{F}_\xi^*\right]$$

where $\mathbf{F}_\xi^*$ is the adjoint (transpose of the matrix of co-factors) of $\mathbf{F}_\xi$, show that the above expression for the external nodal forces reduces to (E4.3.16).

4.4. To illustrate the flexibility in choice of reference configuration for the formulation of the finite element equations, consider the tensor quantity, $\mathbf{P}_\xi = J_\xi \mathbf{F}_\xi^{-1} \cdot \boldsymbol{\sigma}$, which can be thought of as the nominal stress tensor on the parent element domain. Show that the equilibrium equation and boundary conditions can be written as

$$\begin{aligned} &\nabla_\xi . \mathbf{P}_\xi = 0 \quad \text{in } \cup\Delta_e \text{ (union of parent element domains)} \\ &\mathbf{n}_{0\xi} . \mathbf{P}_\xi = \mathrm{t}_\xi \quad \text{on } \Gamma_t \text{ (traction boundary)} \end{aligned}$$

and derive the corresponding weak form. Introduce parent element shape functions $N_I(\xi)$ and show that the element internal force vector can be written directly in terms of the parent element domain as

$$f_{iI}^{\text{int}} = \int_{\square} (P_\xi)_{ji} \frac{\partial N_I}{\partial \xi_j}\, d\square$$

5

Constitutive Models

5.1 Introduction

In the mathematical description of material behavior, the response of the material is characterized by a constitutive equation which gives the stress as a function of the deformation history of the body. Different constitutive relations allow us to distinguish between a viscous fluid and rubber or concrete, for example. In one-dimensional solid mechanics, the constitutive relation is often referred to as the stress–strain law for the material. There is an extensive body of literature on the thermodynamic foundations of constitutive equations at finite strains and the interested reader is referred to Truesdell and Noll (1965). Mathematical foundations of elasticity are presented in the monograph by Marsden and Hughes (1983) and computational aspects of plasticity are covered in Simo and Hughes (1998). In the present discussion, emphasis is on the mechanical response, although thermodynamic restrictions on material behavior and some additional stability postulates are considered. Computational aspects of constitutive modeling including stress update algorithms and algorithmic moduli are also presented.

In this chapter, some of the most common constitutive models used in solid mechanics applications are described. Constitutive equations for different classes of materials are first presented for the one-dimensional case and are then generalized to multiaxial stress states. Special emphasis is placed on elastic–plastic constitutive equations for both small and large strains. Some fundamental properties such as reversibility, stability and smoothness are also discussed.

In the following section, the tensile test is introduced to motivate different classes of material behavior. One-dimensional constitutive relations for elastic materials are then discussed in Section 5.3. Multiaxial constitutive equations for large deformation elasticity are

Nonlinear Finite Elements for Continua and Structures, Second Edition.
Ted Belytschko, Wing Kam Liu, Brian Moran, and Khalil I. Elkhodary.

Companion Website: www.wiley.com/go/belytschko

described in Section 5.4. The special cases of hypoelasticity (which often plays an important role in large deformation elastic–plastic constitutive relations) and hyperelasticity are considered. Well-known constitutive models such as Neo–Hookean, Saint Venant–Kirchhoff and Mooney–Rivlin models are given as examples of hyperelastic materials.

In Section 5.5 constitutive relations for elastic–plastic materials in one dimension are presented, and in Section 5.6, the extensions to multiaxial stress states are given. The commonly used von Mises J_2 flow theory plasticity models (representative of the behavior of metals) for rate-independent and rate-dependent plastic deformation and the Drucker–Prager relation (for the deformation of soils and rock) are presented. The Gurson constitutive model which accounts for void growth and coalescence is given as an illustration of a constitutive relation which models material deformation together with damage and failure.

Hyperelastic–plastic constitutive equations are considered in Section 5.7. In these models, the elastic response is modeled as hyperelastic (rather than hypoelastic) as a means of circumventing some of the difficulties associated with rotations in problems involving geometric nonlinearity and to incorporate general anisotropic elastic and plastic behavior.

Constitutive models for the viscoelastic response of polymeric materials are described in Section 5.8. Straightforward generalizations of one-dimensional viscoelastic models to multiaxial stress states are presented for small and large deformations.

The implementation of the constitutive relation in a finite element code requires a procedure for the evaluation of the stress given the deformation (or an increment of deformation from a previous state). This may be a straightforward function evaluation as in hyperelasticity or it may require the integration of the rate form of the constitutive equations. The algorithm for the integration of the rate form of the constitutive relation is called a *stress update algorithm* (also called a constitutive update algorithm). Stress update algorithms for the integration of constitutive relations are presented in Section 5.9. A class of return mapping algorithms, including the radial return method, is presented. The concept of algorithmic moduli consistent with the underlying stress update scheme is also presented. Incrementally objective stress update schemes for large deformation problems are discussed. Finally, stress update schemes based on the formulation of the elastic response in terms of a hyperelastic potential that automatically satisfy objectivity are described.

Some additional continuum mechanics concepts which are helpful in the description of constitutive models are given in Section 5.10. The concepts of Eulerian, Lagrangian and two-point tensors are presented, and the operations of pull-back, push-forward and the Lie derivative are described. These concepts are then used in a more general treatment of objectivity (or material frame indifference) than that given in Chapter 3. Material symmetry is also briefly presented and some aspects of invariance in the tensorial representation of constitutive behavior are discussed. Restrictions on material behavior due to the second law of thermodynamics and some additional stability postulates are also discussed.

5.2 The Stress–Strain Curve

To the analyst, the choice of material model is important and may not always be obvious. Often the only information available is general knowledge and experience about the material's behavior along with perhaps a few stress–strain curves. It is the analyst's task to choose the appropriate constitutive model from available libraries in the finite element code or to develop a user-supplied constitutive routine if no suitable constitutive model is available. It is important for the analyst to understand the key features of the constitutive model, the assumptions in the development of the model, how suitable the model is for the material in question, how appropriate the model is for the expected load and deformation regime and the numerical issues in the implementation.

Many of the essential features of the stress–strain behavior of a material can be obtained from a set of stress–strain curves for the material response in a state of one-dimensional stress (uniaxial stress or shear). For this reason, we begin with a discussion of the tensile test. As will be seen, constitutive equations for multiaxial states are often based on simple generalizations of the one-dimensional behavior observed in tensile tests.

5.2.1 The Tensile Test

The stress–strain behavior of a material in a state of uniaxial (one-dimensional) stress can be obtained by a tensile test (Figure 5.1). In the tensile test, a specimen is gripped at each end in a testing machine and elongated at a prescribed rate. The elongation δ of the gage section and the force T are measured. A plot of load versus elongation (for a typical metal) is shown in Figure 5.1. This plot represents the response of the specimen as a structure. In order to extract meaningful information about the material behavior from this plot, the effects of the specimen geometry must be removed. To do this, we plot load per unit cross-sectional area (stress) versus elongation per unit length (strain). Even at this stage, decisions need to be made. Do we use the original area and length or the current ones? Another way of stating this question is: what stress and strain measures should we use? If the deformations are sufficiently small so that distinctions between original and current geometries are negligible, a small-strain theory suffices. Otherwise, full nonlinear theory is needed. From Chapter 3 (Box 3.2), it can be seen that we can always transform from one stress or strain measure to another, but it is important to know precisely how the stress–strain data are measured.

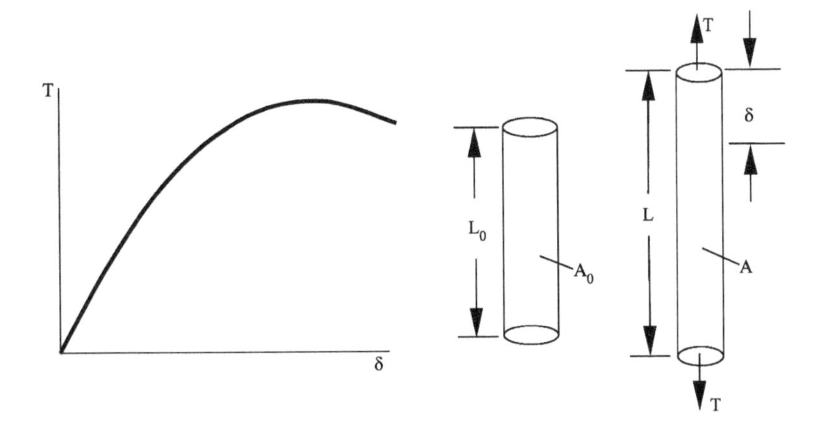

Figure 5.1 Load-elongation curve

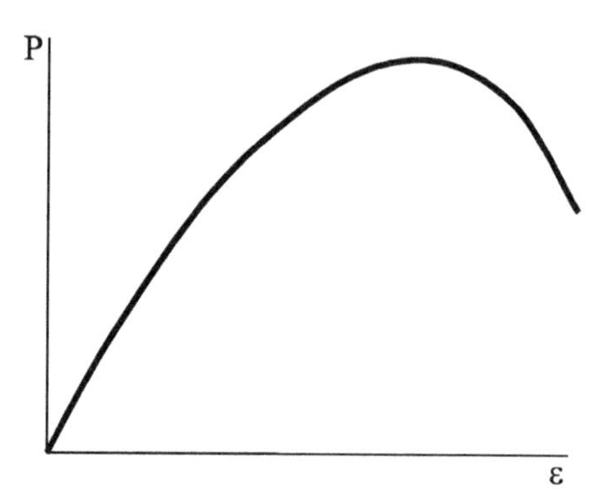

Figure 5.2 Engineering stress-engineering strain curve

A typical procedure is as follows. Define the stretch $\lambda_x = L/L_0$ where $L = L_0 + \delta$ is the length of the gage section associated with elongation δ. The nominal (or engineering stress) is given by

$$P_x = \frac{T}{A_0} \tag{5.2.1}$$

where A_0 is the original cross-sectional area. The engineering strain is defined as

$$\varepsilon_x = \frac{\delta}{L_0} = \lambda_x - 1 \tag{5.2.2}$$

A plot of engineering stress versus engineering strain for a typical metal is given in Figure 5.2.

Alternatively, the stress–strain response may be given in terms of true stress. The Cauchy (or true) stress is given by

$$\sigma_x = \frac{T}{A} \tag{5.2.3}$$

where A is the current (instantaneous) area of the cross-section. An alternative measure of strain is derived by considering an increment of strain as the change in length per unit current length, that is, $de_x = dL/L$. Integrating this relation from the initial length L_0 to the current length L gives

$$e_x = \int_{L_0}^{L} \frac{dL}{L} = \ln(L/L_0) = \ln \lambda_x \tag{5.2.4}$$

and e_x is called the logarithmic strain (also called the true strain). Taking the material time derivative of this expression gives

$$\dot{e}_x = \frac{\dot{\lambda}_x}{\lambda_x} = D_x \tag{5.2.5}$$

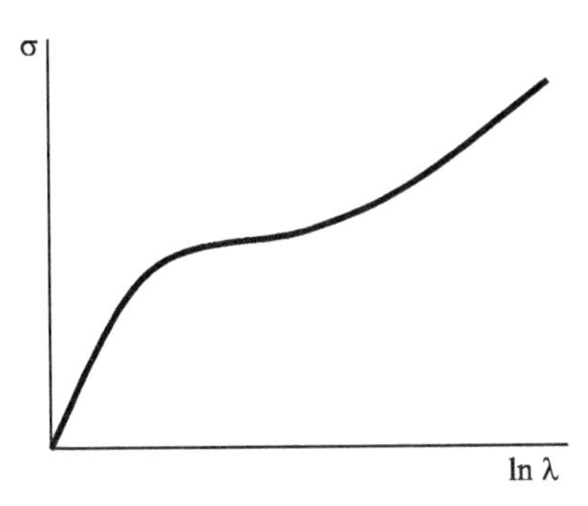

Figure 5.3 True stress strain curve

that is, in the one-dimensional case, the time derivative of the logarithmic strain is equal to the rate-of-deformation given by (3.3.19). This is not true for general multiaxial states of deformation but does hold if the principal axes of the deformation are fixed.

To plot true stress versus logarithmic strain, the cross-sectional area A is needed as a function of the deformation. This can be measured during the test. From (3.2.23), the Jacobian determinant relates reference to current volumes. Prior to the onset of necking or other instabilities in the bar, the deformation is homogeneous and the relation between the original and current volumes of the gage length is given by $JA_0L_0 = AL$ where J is the determinant of the deformation gradient. An expression for the current area A is given by

$$A = JA_0 L_0 / L = JA_0 / \lambda_x \tag{5.2.6}$$

The Cauchy stress is therefore

$$\sigma_x = \frac{T}{A} = \lambda_x \frac{T}{JA_0} = \lambda_x J^{-1} P_x \tag{5.2.7}$$

(compare with the tensor expression in Box 3.2). An example plot of true stress versus logarithmic strain is given in Figure 5.3.

As an example, consider an incompressible material ($J=1$). The relationship between nominal stress and engineering strain is

$$P_x = s_0(\varepsilon_x) \tag{5.2.8}$$

where $\varepsilon_x = \lambda_x - 1$ is the engineering strain. We can regard (5.2.8) as a stress–strain equation for the material under uniaxial stress at a given rate of deformation. From (5.2.7), the true stress (for an incompressible material) can be written as

$$\sigma_x = \lambda_x s_0(\varepsilon_x) = s(\lambda_x) \tag{5.2.9}$$

where the relation between the functions is $s(\lambda_x) = \lambda_z s_0(\lambda_x - 1)$. This illustrates how different functional representations of the constitutive behavior are obtained for the same material

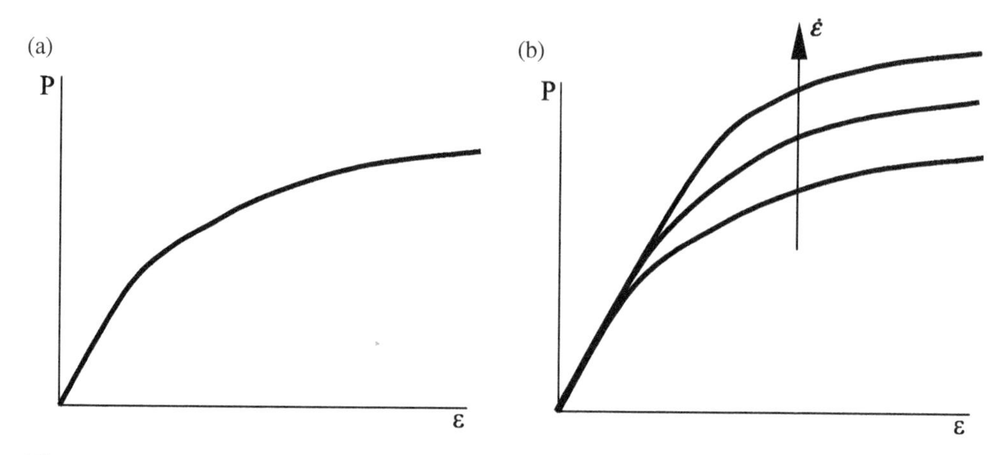

Figure 5.4 Engineering stress–strain curve (a) rate-independent material (b) rate-dependent material

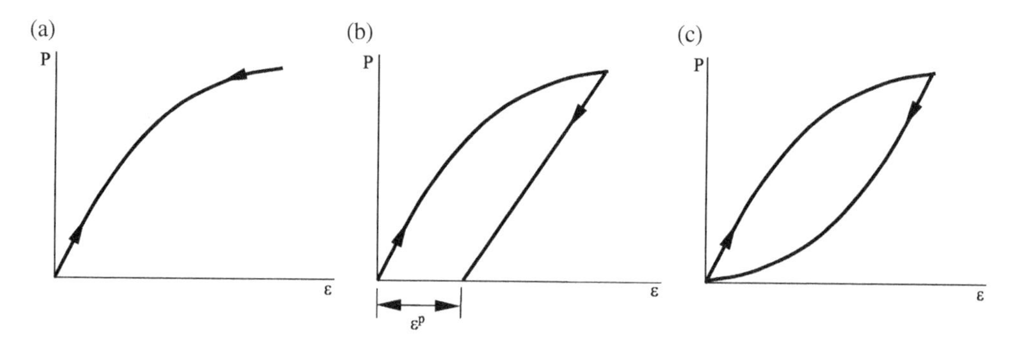

Figure 5.5 Unloading behavior (a) elastic (b) elastic–plastic (c) elastic with microcracking

depending on what measures of stress and deformation are used. It is especially important to keep this in mind when dealing with multiaxial constitutive relations at large strains.

A material for which the stress–strain response is independent of the rate of deformation is said to be *rate-independent*; otherwise it is *rate-dependent*. The one-dimensional responses of rate-independent and rate-dependent materials are illustrated in Figures 5.4(a) and (b), respectively, for different nominal strain rates. The nominal strain rate is defined as $\dot{\varepsilon}_x = \dot{\delta}/L_0$. Because $\dot{\delta} = \dot{L}$ and $\dot{\delta}/L_0 = \dot{L}/L_0 = \dot{\lambda}_x$ it follows that the nominal strain rate is equivalent to the rate of stretching, that is, $\dot{\varepsilon}_x = \dot{\lambda}_x$. As can be seen, the stress–strain curve for the rate-independent material is independent of the strain rate, while for the rate-dependent material the stress–strain curve is elevated at higher rates.

In the description of the tensile test given here, no unloading was considered. In Figure 5.5, unloading for different types of material is illustrated. For elastic materials, the unloading stress–strain curve simply retraces the loading curve. Upon complete unloading, the material returns to its initial unstretched state. For elastic–plastic materials, however, the unloading curve differs from the loading curve. The slope of the unloading curve is typically that of the elastic (initial) portion of the stress–strain curve. This results in permanent strains upon unloading, as shown in Figure 5.5(b). Other materials behave between these two extremes. For example, the unloading behavior for a brittle material which is damaged due to the formation of microcracks during loading is shown in Figure 5.5(c). In this case the elastic strains are recovered when the microcracks close upon removal of the load. The initial slope of the unloading curve gives information about the extent of damage due to microcracking. For the development of constitutive models in damage mechanics, see Krajcinovic (1996).

5.3 One-Dimensional Elasticity

A fundamental property of elastic materials is that the stress depends only on the current level of the strain. This implies that the loading and unloading stress–strain curves are identical and that the original shape is recovered upon unloading. In this case the strains are said to be *reversible*. Furthermore, an elastic material is rate-independent (no dependence on strain rate). It follows that, for an elastic material, there is a one-to-one correspondence between stress and strain.

5.3.1 *Small Strains*

We first consider elastic behavior in the small-strain regime. When strains and rotations are small, a small-strain theory (kinematics, equations of motion and constitutive equation) is often used. In this case we make no distinction between the various measures of stress and strain. We also confine our attention to a purely mechanical theory in which thermodynamic effects (such as heat conduction) are not considered. For a nonlinear elastic material under uniaxial stress, the stress–strain law can be written as

$$\sigma_x = s(\varepsilon_x) \tag{5.3.1}$$

where σ_x is the Cauchy stress and $\varepsilon_x = \delta/L_0$ is the linear strain, often known as the engineering strain. Thus, the stress is given as a function of the current strain and is independent of the deformation history or path. Here $s(\varepsilon_x)$ is assumed to be a monotonically increasing function. The assumption that the function $s(\varepsilon_x)$ is monotonically increasing is crucial to the stability of the material: if at any strain ε_x, the slope of the stress strain curve is negative, that is, $ds/d\varepsilon_x < 0$, then the material is said to exhibit strain-softening and the response is unstable (material stability is discussed in Section 6.5). Such behavior can occur in constitutive models for materials which exhibit phase transformations (for example, see Abeyaratne and Knowles, 1988). Note that reversibility and path-independence are implied by the structure of (5.3.1): the stress σ_x for any strain ε_x is uniquely given by (5.3.1). It does not matter how the strain reaches the value ε_x.

The generalization of (5.3.1) to multiaxial large strains is a formidable mathematical problem which has been addressed by some of the keenest minds in the twentieth century and still encompasses open questions (see Ogden, 1984, and references therein). The extension of (5.3.1) to large-strain uniaxial behavior is presented later in this section. Some of the most common multiaxial generalizations to large strain are discussed in Section 5.4.

In a purely mechanical theory, reversibility and path-independence also imply the absence of energy dissipation in deformation. In other words, in an elastic material, deformation is not accompanied by any dissipation of energy and all energy expended in deformation is stored in

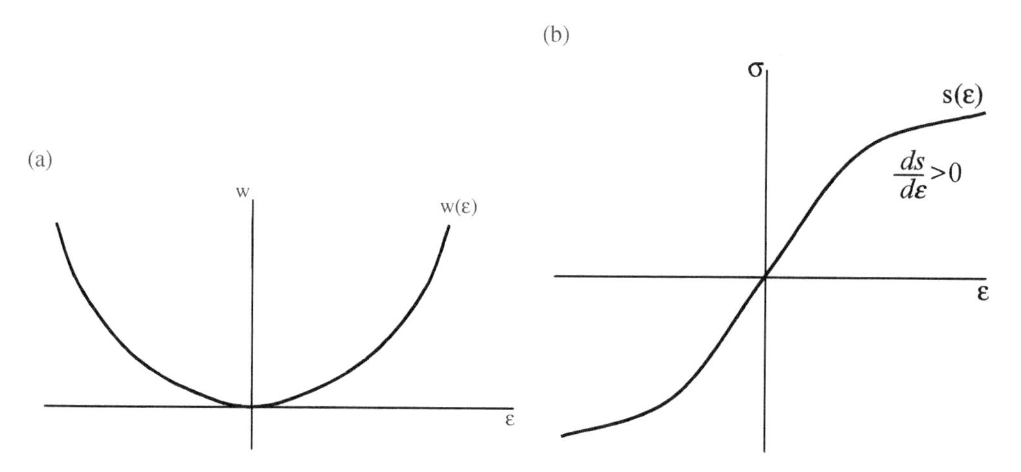

Figure 5.6 (a) Convex strain energy function (b) stress–strain curve

the body and can be recovered upon unloading. This implies that there exists a *potential* function $w(\varepsilon_x)$ such that

$$\sigma_x = s(\varepsilon_x) = \frac{dw(\varepsilon_x)}{d\varepsilon_x} \tag{5.3.2}$$

where $w(\varepsilon_x)$ is the strain energy density per unit volume. The strain energy density can be identified with the internal energy density ρw^{int} (see Section 3.5.9) under adiabatic conditions and with the Helmholtz free energy under isothermal conditions (Malvern, 1969: 265). We omit the superscript 'int' in this chapter. From (5.3.2) it follows that

$$dw(\varepsilon_x) = \sigma_x d\varepsilon_x \tag{5.3.3}$$

which when integrated gives

$$w = \int_0^{\varepsilon_x} \sigma_x d\varepsilon_x \tag{5.3.4}$$

This can also be seen by noting that $\sigma_x d\varepsilon_x = \sigma_x \dot{\varepsilon}_x dt$ is the one-dimensional, small-strain version of the multiaxial expression (3.5.50).

The strain energy w is usually a convex function of strain, that is, $(w'(\varepsilon_x^1) - w'(\varepsilon_x^2))(\varepsilon_x^1 - \varepsilon_x^2) \geq 0$,with the equality if $\varepsilon_x^1 = \varepsilon_x^2$. An example of a convex strain energy function is shown in Figure 5.6(a). In this case the function $s(\varepsilon_x)$ is monotonically increasing as shown in Figure 5.6(b). If w is a non-convex function, $s(\varepsilon_x)$ first increases and then decreases and the material is said to exhibit *strain-softening*. This is an unstable material response (i.e., $ds/d\varepsilon_x < 0$). A schematic of a non-convex strain energy function and the corresponding stress–strain curve are shown in Figures 5.7(a) and 5.7(b), respectively.

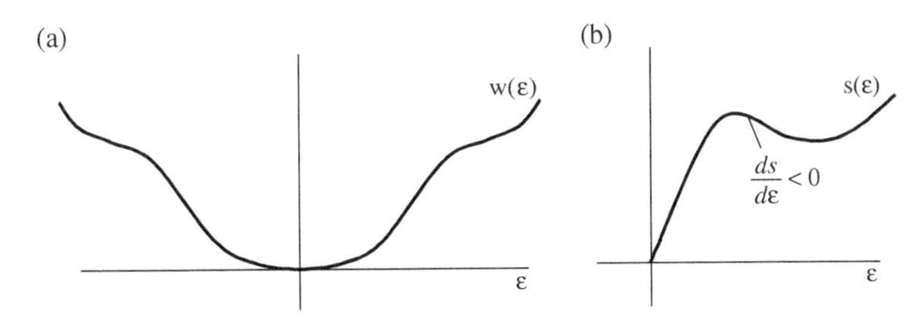

Figure 5.7 (a) Non-convex strain energy function (b) Corresponding stress–strain curve

In summary, the one-dimensional behavior of an elastic material is characterized by three properties which are all interrelated:

$$\text{path-independence} \Leftrightarrow \text{reversible} \Leftrightarrow \text{nondissipative}$$

In two and three dimensions, the same properties hold for hyperelastic materials. There are other types of multiaxial elastic material models such as hypoelasticity for which these properties are not observed exactly.

One of the salient characteristics of a stress–strain curve is its degree of nonlinearity. For many materials, the stress–strain curve consists of an initial elastic portion which can be modeled as linear. The regime of linear elastic behavior is typically confined to strains of no more than a few percent and, consequently, small-strain theory is used to describe linearly elastic materials.

For a linear elastic material, the stress–strain law can be written as

$$\sigma_x = E\varepsilon_x \tag{5.3.5}$$

where the constant of proportionality is Young's modulus, E. This relation is often referred to as Hooke's law. From (5.3.4) the strain energy density is therefore given by

$$w = \frac{1}{2}E\varepsilon_x^2 \tag{5.3.6}$$

which is a quadratic function of strains.

5.3.2 *Large Strains*

The extension of elasticity to large strains in one dimension is rather straightforward: it is only necessary to choose a measure of strain and define an elastic potential for the (work conjugate) stress. Keep in mind that the existence of a potential implies reversibility, path-independence and absence of dissipation in the deformation process. We choose the Green strain E_x as the strain measure and write the work conjugate second Piola–Kirchhoff stress as

$$S_x = \frac{\partial w}{\partial E_x} \tag{5.3.7}$$

To avoid confusion of the Green strain and Young's modulus, the Green strain is always in boldface or subscripted. The fact that the corresponding stress is the second Piola–Kirchhoff stress follows from the work (power) conjugacy of the second Piola–Kirchhoff stress and the Green strain, that is,, recalling Box 3.4, and specializing to one dimension, the stress power per unit reference volume is given by $\dot{w} = S_x \dot{E}_x$.

The potential w in (5.3.7) reduces to the potential (5.3.2) as the strains become small. Elastic stress–strain relationships in which the stress can be obtained from a potential function of the strains are called *hyperelastic*. The simplest hyperelastic relation (for large deformation problems in one dimension) results from a potential which is quadratic in the Green strain:

$$w = \frac{1}{2} E^{SE} E_x^2 \tag{5.3.8}$$

where the modulus, E^{SE}, is constant. Then,

$$S_x = E^{SE} E_x \tag{5.3.9}$$

by (5.3.7), so the relation between these stress and strain measures is linear. At small strains, the above reduces to Hooke's law (5.3.5) with Young's modulus, $E = E^{SE}$.

We could also express the elastic potential in terms of any other conjugate stress and strain measures. For example, it was pointed out in Chapter 3 that the tensor quantity $\bar{\mathbf{U}} = \mathbf{U} - \mathbf{I}$ is a valid strain measure (called the Biot strain), and that in *one dimension* the conjugate stress is the nominal stress P_x, so

$$P_x = \frac{\partial w}{\partial \bar{U}_x} = \frac{\partial w}{\partial U_x} \tag{5.3.10}$$

We can write the second form in (5.3.10) because the unit tensor I is constant and hence $d\bar{U}_x = dU_x$. It is interesting to observe that linearity in the relationship between a certain pair of stress and strain measures does not imply linearity in other conjugate pairs. For example, if $S_x = E^{SE} E_x$ it follows that $P_x = E^{SE}(1 + \bar{U}_x)(\bar{U}_x^2 + 2\bar{U}x)/2$.

A material for which the rate of Cauchy stress is related to the rate-of-deformation is said to be *hypoelastic*. The relation is generally nonlinear and is given by

$$\dot{\sigma}_x = f(\sigma_x, D_x) \tag{5.3.11}$$

where a superposed dot denotes the material time derivative and D_x is the rate-of-deformation. A particular linear hypoelastic relation is given by

$$\dot{\sigma}_x = ED_x = E\frac{\dot{\lambda}_x}{\lambda_x} \tag{5.3.12}$$

where E is Young's modulus and λ_x is the stretch. The relationship of E to other uniaxial stress measures is examined in Example 5.1. Integrating the relation in (5.3.12), we obtain

$$\sigma_x = E \ln \lambda_x \tag{5.3.13}$$

or

$$\sigma_x = \frac{d}{x\lambda_x} \int_1^{\lambda_x} E \ln \lambda_x \, d\lambda_x \tag{5.3.14}$$

which is a *hyperelastic* relation and thus path-independent. However, for multiaxial problems, hypoelastic relations can not, in general, be transformed to hyperelastic.

A hypoelastic material is, in general, strictly path-independent only in the one-dimensional case. However, if the elastic strains are small, the behavior is close enough to path-independent to model elastic behavior. Because of the simplicity of hypoelastic laws, a multiaxial generalization of (5.3.11) is often used in finite element software to model elastic response of materials in large-strain elastic–plastic problems (see Section 5.6).

5.4 Nonlinear Elasticity

In this section, more general elastic constitutive relations for finite strain are presented. As will be seen, in finite strain elasticity, many different constitutive relations can be developed for multiaxial elasticity. In addition, because of the many different stress and deformation measures for finite strain, the same constitutive relation can be written in several different ways. It is important to distinguish between these two situations. The first case gives different material models, while in the second the same material model is represented by different mathematical expressions. In the latter, it is always possible to transform from one form of the constitutive relation to the other. The constitutive models for large-strain elasticity are first illustrated for the special case of a Kirchhoff material, which is a straightforward generalization of linear elasticity to large deformations. As we saw in Section 5.3, path-independence, reversibility, and nondissipative behavior are closely related. Thus the degree of path-independence may be viewed as a measure of the elasticity of the material model. Hypoelastic materials, which are most weakly path-independent, are then presented, followed by Cauchy elasticity where the stress is path-independent but the energy is not. Finally we present hyperelasticity or Green elasticity, which is path-independent and fully reversible and where the stress is derived from a strain (or stored) energy potential.

5.4.1 Kirchhoff Material

Many engineering applications involve small strains and large rotations. In these problems the effects of large deformation are primarily due to rotations (such as in the bending of a marine riser or a fishing rod). The response of the material may then be modeled by a simple extension of the linear elastic laws by replacing the stress by the PK2 stress and the linear strain by the Green strain. This is called a *Saint Venant–Kirchhoff material*, or a *Kirchhoff material* for brevity. The most general Kirchhoff model is

$$S_{ij} = C_{ijkl} E_{kl} \quad \mathbf{S} = \mathbf{C} : \mathbf{E} \tag{5.4.1}$$

where $\mathbf{C}$ is the fourth-order tensor of elastic moduli which are constant for the Kirchhoff material. This represents the generalization of (5.3.5) to multiaxial states of stress and strain. It can incorporate fully anisotropic material response. We do not append a superscript to this material response matrix $\mathbf{C}$ because it relates stresses to strains; note that the corresponding rate relationship is $\dot{\mathbf{S}} = \mathbf{C}^{SE} : \dot{\mathbf{E}}$, where $\mathbf{C}^{SE} = \mathbf{C}$ is called the *tangent modulus tensor*; its elements are called the *tangent moduli*.

In the following, we will describe the properties of $\mathbf{C}$, particularly the symmetry properties which enable the number of material constants to be reduced to a manageable number. Many of these properties also apply to material tangent moduli. From the symmetries of the stress and strain tensors, the material coefficients in (5.4.1) have the *minor symmetries*

$$C_{ijkl} = C_{jikl} = C_{ijlk} \tag{5.4.2}$$

The Saint Venant–Kirchhoff material is path-independent and possesses an elastic strain energy potential. The strain energy per unit volume, which in one dimension is given by (5.3.4), is generalized to multiaxial states by

$$w = \int S_{ij}\, dE_{ij} = \int C_{ijkl} E_{kl}\, dE_{ij} = \frac{1}{2} C_{ijkl} E_{ij} E_{kl} = \frac{1}{2} \mathbf{E}:\mathbf{C}:\mathbf{E} \tag{5.4.3}$$

The stress is then given by

$$S_{ij} = \frac{\partial w}{\partial E_{ij}} \quad \text{or} \quad \mathbf{S} = \frac{\partial w}{\partial \mathbf{E}} \tag{5.4.4}$$

which is the tensor equivalent of (5.3.7). The strain energy is assumed to be positive-definite, that is,

$$w = \frac{1}{2} C_{ijkl} E_{ij} E_{kl} = \frac{1}{2} \mathbf{E}:\mathbf{C}:\mathbf{E} \geq 0 \quad \forall\, \mathbf{E} \tag{5.4.5}$$

The equality holds if and only if $\mathbf{E} = \mathbf{0}$, which implies that $\mathbf{C}$ is a positive-definite fourth-order tensor. From the existence of a strain energy potential (5.4.3) it follows that

$$C_{ijkl} = \frac{\partial^2 w}{\partial E_{ij} \partial E_{kl}}, \quad \mathbf{C} = \frac{\partial^2 w}{\partial \mathbf{E} \partial \mathbf{E}} \tag{5.4.6}$$

Smoothness of the potential (i.e., w is a C^1 function of $\mathbf{E}$) implies that

$$\frac{\partial^2 w}{\partial E_{ij} \partial E_{kl}} = \frac{\partial^2 w}{\partial E_{kl} \partial E_{ij}} \tag{5.4.7}$$

Therefore if the material possesses a smooth potential, w, $\mathbf{C}$ has *major symmetry*:

$$C_{ijkl} = C_{klij} \tag{5.4.8}$$

Major symmetry of the material response matrix is a necessary condition for the symmetry of the tangent stiffness matrix to be developed in Section 6.4. If the tangent moduli do not exhibit major symmetry, the tangent stiffness is not symmetric. Major symmetry of the tangent moduli is not a sufficient condition for symmetry of the tangent stiffness. For some objective rates, the tangent stiffness is not symmetric even when the tangent moduli possess major symmetry. We will examine the major symmetry of the tangent moduli for the material laws that follow as they are developed.

The general fourth-order tensor C_{ijkl} has $3^4=81$ independent constants. These 81 constants relate nine components of the complete stress tensor to nine components of the complete strain tensor, that is, $81=9\times9$. The symmetries of the stress and strain tensors require only that six independent components of stress be related to six independent components of strain. The resulting minor symmetries of the elastic moduli reduce the number of independent constants to $6\times6=36$. Major symmetry of the moduli (5.4.8) reduces the number of independent elastic constants to $n(n+1)/2=21$, for $n=6$, that is, the number of independent components of a symmetric 6×6 matrix. Considerations of material symmetry further reduce the number of independent material constants.

The matrix of elastic constants and tangent moduli matrices are usually implemented in Voigt form, since it is difficult to work with fourth-order matrices. The Voigt form is obtained from the tensor components by mapping the first and second pairs of indices according to the Voigt rule (described in Appendix 1)

$$\{\mathbf{S}\} = [\mathbf{C}]\{\mathbf{E}\}, \quad S_a = C_{ab}E_b \tag{5.4.9}$$

Major symmetry (5.4.8) implies that the matrix $[\mathbf{C}]$ is symmetric, so

$$\begin{Bmatrix} S_{11} \\ S_{22} \\ S_{33} \\ S_{23} \\ S_{13} \\ S_{12} \end{Bmatrix} = \begin{bmatrix} C_{11} & C_{12} & C_{13} & C_{14} & C_{15} & C_{16} \\ & C_{22} & C_{23} & C_{24} & C_{25} & C_{26} \\ & & C_{33} & C_{34} & C_{35} & C_{36} \\ & & & C_{44} & C_{45} & C_{46} \\ & \text{sym} & & & C_{55} & C_{56} \\ & & & & & C_{66} \end{bmatrix} \begin{Bmatrix} E_{11} \\ E_{22} \\ E_{33} \\ 2E_{23} \\ 2E_{13} \\ 2E_{12} \end{Bmatrix} \tag{5.4.10}$$

It can be seen from the above that the $[\mathbf{C}]$ matrix in three dimensions has 21 independent entries and that major symmetry of the fourth-order tensor manifests itself as symmetry of the Voigt matrix form. This holds for fully anisotropic Kirchhoff materials.

The number of independent material constants can be further reduced by consideration of material symmetry (Nye, 1985). This theory has been developed for linear elasticity, but it applies also to the Kirchhoff material. For example, if the material has a plane of symmetry, say the X_1-plane, the elastic moduli must remain unchanged when the X_1-axis is reflected through the X_1-plane. This reflection introduces a factor of -1 on the C_{ijkl} for each appearance of the index 1. Therefore any term in which the index 1 appears an odd number of times must vanish. This occurs for $C_{\alpha 5}$ and $C_{\alpha 6}$ for $\alpha=1, 2, 3, 4$ and reduces the number of constants from 21 to 13. For an orthotropic material (e.g., wood or composites with aligned fibers) with three mutually orthogonal planes of symmetry, this procedure can be repeated for all three planes to show that there are only nine independent elastic constants and the Kirchhoff stress–strain relation is

$$\begin{Bmatrix} S_{11} \\ S_{22} \\ S_{33} \\ S_{23} \\ S_{13} \\ S_{12} \end{Bmatrix} = \begin{bmatrix} C_{11} & C_{12} & C_{13} & 0 & 0 & 0 \\ & C_{22} & C_{23} & 0 & 0 & 0 \\ & & C_{33} & 0 & 0 & 0 \\ & & & C_{44} & 0 & 0 \\ & \text{sym} & & & C_{55} & 0 \\ & & & & & C_{66} \end{bmatrix} \begin{Bmatrix} E_{11} \\ E_{22} \\ E_{33} \\ 2E_{23} \\ 2E_{13} \\ 2E_{12} \end{Bmatrix} \tag{5.4.11}$$

An important example of material symmetry is isotropy. An isotropic material is one which has no preferred orientations or directions, so that the stress–strain relation is identical when expressed in any rectangular Cartesian coordinate system. Many materials (such as metals and ceramics) can be modeled as isotropic for small strains. In an isotropic Kirchhoff material, the tensor **C** is isotropic. An isotropic tensor has the same components in any (rectangular Cartesian) coordinate system. The most general isotropic fourth-order tensor can be shown to be a linear combination of terms comprised of Kronecker deltas:

$$C_{ijkl} = \lambda\delta_{ij}\delta_{kl} + \mu(\delta_{ik}\delta_{jl} + \delta_{il}\delta_{jk}) + \mu'(\delta_{ik}\delta_{jl} - \delta_{il}\delta_{jk}) \tag{5.4.12}$$

Because of the symmetry of the strain, the product of the third term in (5.4.12) with the strain vanishes, so it follows that we can set $\mu'=0$. Thus (5.4.12) reduces to

$$C_{ijkl} = \lambda\delta_{ij}\delta_{kl} + \mu(\delta_{ik}\delta_{jl} + \delta_{il}\delta_{jk}),\quad \mathbf{C} = \lambda\mathbf{I}\otimes\mathbf{I} + 2\mu\boldsymbol{I} \tag{5.4.13}$$

The two independent material constants λ and μ are called the Lamé constants. The fourth-order symmetric identity tensor $\boldsymbol{I}$ has components $I_{ijkl} = \frac{1}{2}(\delta_{ik}\delta_{jl} + \delta_{il}\delta_{jk})$. The stress–strain relation for an isotropic Kirchhoff material may therefore be written as

$$S_{ij} = \lambda E_{kk}\delta_{ij} + 2\mu E_{ij} = C_{ijkl}E_{kl},\quad \mathbf{S} = \lambda\,\text{trace}(\mathbf{E})\mathbf{I} + 2\mu\mathbf{E} = \mathbf{C}:\mathbf{E} \tag{5.4.14}$$

The Lamé constants can be expressed in terms of other constants which are more closely related to physical measurements, the bulk modulus K, Young's modulus E and Poisson's ratio ν, by

$$\mu = \frac{E}{2(1+\nu)},\ \lambda = \frac{\nu E}{(1+\nu)(1-2\nu)},\ K = \lambda + \frac{2\mu}{3} \tag{5.4.15}$$

For two-dimensional problems the stress–strain relations depend on whether the problem is plane stress or plane strain (Malvern, 1969: 512).

5.4.2 *Incompressibility*

In an incompressible material, the volume is unchanged during deformation and the density remains constant. A motion of an incompressible material is called *isochoric*. From the conservation of matter, (3.5.25) or (3.5.21), it can be deduced that for an isochoric motion the following conditions hold:

$$J = \det \mathbf{F} = 1, \quad \text{and} \quad \operatorname{div} \mathbf{v} = \operatorname{trace}(\mathbf{D}) = 0 \tag{5.4.16}$$

The first expression is in terms of the total deformation; the second is a rate form of the isochoric constraint. For situations with large incremental steps, the first is more accurate.

It is often useful, particularly for incompressible materials, to write the stress and strain rate measures as the sum of deviatone and hydrostatic (volumetric) parts. The latter are also called the spherical part of a tensor. The decompositions are:

$$\boldsymbol{\sigma} = \boldsymbol{\sigma}^{\text{dev}} + \boldsymbol{\sigma}^{\text{hyd}},\ \boldsymbol{\sigma}^{\text{hyd}} = -p\mathbf{I},\ \mathbf{S} = \mathbf{S}^{\text{dev}} + \mathbf{S}^{\text{hyd}},\ \mathbf{S}^{\text{hyd}} = \frac{1}{3}(\mathbf{S}:\mathbf{C})\mathbf{C}^{-1} \tag{5.4.17a}$$

$$\mathbf{D} = \mathbf{D}^{\text{dev}} + \mathbf{D}^{\text{vol}},\ \mathbf{D}^{\text{vol}} = \frac{1}{3}\operatorname{trace}(\mathbf{D})\mathbf{I},\ \dot{\mathbf{E}} = \dot{\mathbf{E}}^{\text{dev}} + \dot{\mathbf{E}}^{\text{vol}},\ \dot{\mathbf{E}}^{\text{vol}} = \frac{1}{3}(\dot{\mathbf{E}}:\mathbf{C}^{-1})\mathbf{C} \tag{5.4.17b}$$

In these, $p = \frac{1}{3}\sigma_{kk}$ is the pressure and $\mathbf{C}=\mathbf{F}^T\cdot\mathbf{F}$ is the right Cauchy–Green deformation tensor, often called the deformation tensor for brevity (not to be confused with the tensor of elastic moduli for the Kirchhoff model). The origin of the expressions in the total Lagrangian form is not easy to explain. In Section 5.10 it is shown that they are the pull-backs of the corresponding Eulerian variables.

The Green strain can not be additively split into volumetric and deviatoric parts. The spherical part of the Green strain does not vanish in isochoric motion. Therefore, a multiplicative decomposition must be used to separate the volumetric parts of the deformation. The multiplicative split is given by

$$\mathbf{F} = \mathbf{F}^{\text{vol}} \cdot \mathbf{F}^{\text{dev}}, \quad \text{where} \quad \mathbf{F}^{\text{vol}} = J^{\frac{1}{3}}\mathbf{I},\ \mathbf{F}^{\text{dev}} = J^{-\frac{1}{3}}\mathbf{F} \tag{5.4.18}$$

The determinant of the deviatoric deformation gradient is always unity, that is, $\det(\mathbf{F}^{\text{dev}})=1$, so any function of $\mathbf{F}^{\text{dev}}$ alone is independent of the volumetric deformation.

The deviatoric and spherical components of a tensor are orthogonal in the sense that $\boldsymbol{\sigma}^{\text{hyd}}:\mathbf{D}^{\text{dev}} = \boldsymbol{\sigma}^{\text{dev}}:\mathbf{D}^{\text{vol}} = 0$ and $\mathbf{S}^{\text{hyd}}:\dot{\mathbf{E}}^{\text{dev}} = \mathbf{S}^{\text{dev}}:\dot{\mathbf{E}}^{\text{vol}} = 0$; verifying these is left as an exercise. As a consequence, the power (3.5.50) can be split into deviatoric and hydrostatic parts:

$$\rho\dot{w} = \boldsymbol{\sigma}:\mathbf{D} = \boldsymbol{\sigma}^{\text{dev}}:\mathbf{D}^{\text{dev}} + \boldsymbol{\sigma}^{\text{hyd}}:\mathbf{D}^{\text{vol}} \quad \text{or} \quad \rho_0\dot{w} = \mathbf{S}:\dot{\mathbf{E}} = \mathbf{S}^{\text{dev}}:\dot{\mathbf{E}}^{\text{dev}} + \mathbf{S}^{\text{hyd}}:\dot{\mathbf{E}}^{\text{vol}} \tag{5.4.19}$$

where the different forms of power are defined in Box 3.4.

Positive-definiteness of the strain energy (5.4.5) imposes restrictions on the elastic moduli (see Malvern, 1969: 293). For an isotropic Kirchhoff material undergoing large rotations but small strains, positive-definiteness of w requires

$$K > 0 \quad \text{and} \quad \mu > 0, \quad \text{or} \quad E > 0 \quad \text{and} \quad -1 < \nu \leq \frac{1}{2} \tag{5.4.20}$$

When $\nu = \frac{1}{2}(K = \infty)$ the material is incompressible. These restrictions are identical to those for an isotropic linear-elastic material. For an incompressible material, the pressure can not be determined from the constitutive equation, but instead must be determined from the momentum equation; see Section 8.5.5.

5.4.3 Kirchhoff Stress

The Kirchhoff stress is defined by

$$\boldsymbol{\tau} = J\boldsymbol{\sigma} \quad \text{or} \quad \boldsymbol{\tau} = \mathbf{F} \cdot \mathbf{S} \cdot \mathbf{F}^T \tag{5.4.21}$$

The second relation follows from the first by Box 3.2. The Kirchhoff stress is almost identical to the Cauchy stress, but is scaled by the determinant of the Jacobian. Therefore it is also called the weighted Cauchy stress. For isochoric motion, it is identical to the Cauchy stress. It arises naturally in hyperelastic constitutive relations (as will be seen in Section 5.4.7) and is useful in hypoelastic–plastic models (Section 5.6.1) because it leads to symmetric tangent moduli.

The material time derivative of the Kirchhoff stress, like the material time derivative of the Cauchy stress, is not objective. The most appealing objective time derivative of the Kirchhoff stress is denoted by $\boldsymbol{\tau}^{\nabla c}$ and called the convected rate. It is given by

$$\boldsymbol{\tau}^{\nabla c} = \dot{\boldsymbol{\tau}} - \mathbf{L} \cdot \boldsymbol{\tau} - \boldsymbol{\tau} \cdot \mathbf{L}^T = J(\dot{\boldsymbol{\sigma}} - \mathbf{L} \cdot \boldsymbol{\sigma} - \boldsymbol{\sigma} \cdot \mathbf{L}^T + \text{trace}\,(\mathbf{L})\boldsymbol{\sigma}) = J\boldsymbol{\sigma}^{\nabla T} \tag{5.4.22}$$

As indicated in the above, the convected rate of the Kirchhoff stress is equivalent to $J\boldsymbol{\sigma}^{\nabla T}$, the weighted Truesdell rate of the Cauchy stress. It is shown in Section 5.10 that the convected rate of the Kirchhoff stress is the *Lie derivative* of the Kirchhoff stress. This endows it with a certain naturalness which is manifested in the simplicity of constitutive relations based on the Kirchhoff stress.

5.4.4 Hypoelasticity

Hypoelastic material laws relate the rate of stress to the rate-of-deformation. As discussed in Section 3.7.2, in order to satisfy the principle of material frame indifference, the stress rate should be objective and should be related to an objective measure of the rate-of-deformation. A more detailed treatment of material frame indifference is given Section 5.10 and we will draw on results from that section as needed. A general form of the hypoelastic relation is given by

$$\boldsymbol{\sigma}^{\nabla} = \mathbf{f}(\boldsymbol{\sigma}, \mathbf{D}) \tag{5.4.23}$$

where $\boldsymbol{\sigma}^{\nabla}$ represents any objective rate of the Cauchy stress and $\mathbf{D}$ is the rate-of-deformation, which is also objective. The function $\mathbf{f}$ must also be an objective function of the stress and rate-of-deformation.

A large class of hypoelastic constitutive relations can be written in the form of a linear relation between the objective measure of stress rate and the rate-of-deformation, that is,

$$\boldsymbol{\sigma}^{\nabla} = \mathbf{C}:\mathbf{D} \tag{5.4.24}$$

The fourth-order tensor of elastic moduli $\mathbf{C}$ may depend on stress, in which case it must be an objective function of the stress state. As noted by Prager (1961), the relation (5.4.24) is rate-independent and incrementally linear and reversible. This means that for small increments about a finitely deformed state, the increments in stress and strain are linearly related and are recovered upon unloading. However, for large deformations, energy is not necessarily conserved and the work done in a closed deformation path is not necessarily zero. Hypoelastic laws are used primarily for representing the elastic response in phenomenological elastic–plastic laws where the elastic deformations are small, and the dissipative effects are also small.

Some commonly used forms of hypoelastic constitutive relations are

$$\boldsymbol{\sigma}^{\nabla J} = \mathbf{C}^{\sigma J}:\mathbf{D} \quad \boldsymbol{\sigma}^{\nabla T} = \mathbf{C}^{\sigma T}:\mathbf{D} \quad \boldsymbol{\sigma}^{\nabla G} = \mathbf{C}^{\sigma G}:\mathbf{D} \tag{5.4.25}$$

where $\boldsymbol{\sigma}^{\nabla J}$ is the Jaumann rate of Cauchy stress given in (3.7.9), $\boldsymbol{\sigma}^{\nabla T}$ is the Truesdell rate given by (3.7.12), and $\boldsymbol{\sigma}^{\nabla G}$, given by (B3.5.4), is the Green–Naghdi rate (also called the corotational or the Dienes rate) of Cauchy stress. The superscripts on the moduli identify the objective rate of stress given by the moduli.

These tangent moduli possess the minor symmetries since the rate-of-deformation and the objective stress rates are symmetric. In hypoelastic models, the tangent moduli are generally assumed to also have the major symmetries. The symmetry properties are identical to those for the Kirchhoff material described in Section 5.3. For an isotropic material, the tangent modulus tensor is a fourth-order isotropic tensor as in (5.4.13).

For example, the tangent moduli for the Jaumann rate for an isotropic material are

$$C^{\sigma J}_{ijkl} = \lambda\delta_{ij}\delta_{kl} + \mu(\delta_{ik}\delta_{jl} + \delta_{il}\delta_{jk}), \; \mathbf{C}^{\sigma J} = \lambda\mathbf{I}\otimes\mathbf{I} + 2\mu\boldsymbol{I}$$

For an orthotropic material, the symmetries yield a tangent modulus of the form given in Voigt notation in (5.4.11), whereas for a general anisotropic material the Voigt form of the tangent moduli is (5.4.10). It is important to observe that, for anisotropic materials, the tangent moduli depend on the deformation. Therefore, they have to be updated as the material deforms. One such update formula is given below in (5.4.50).

5.4.5 Relations between Tangent Moduli

The tangent moduli $\mathbf{C}^{\sigma T}$, $\mathbf{C}^{\sigma J}$ and $\mathbf{C}^{\sigma G}$ will differ for the same material. However, as we have seen in Example 3.13, when $\mathbf{C}^{\sigma T}=\mathbf{C}^{\sigma J}=\mathbf{C}^{\sigma G}$, the material response differs. To illustrate how the same constitutive relation can be written in terms of different stress rates, consider the constitutive equation $(5.4.25)_1$. Then, using the definitions of the stress rates (3.7.9) and (3.7.12), the constitutive relation (5.4.25) i in terms of the Truesdell rate is

$$\begin{aligned}\boldsymbol{\sigma}^{\nabla T} &= \mathbf{C}^{\sigma J}:\mathbf{D} - \mathbf{D}\cdot\boldsymbol{\sigma} - \boldsymbol{\sigma}\cdot\mathbf{D}^{T} + \text{trace}(\mathbf{D})\boldsymbol{\sigma} \\ &= (\mathbf{C}^{\sigma J} - \mathbf{C}' + \boldsymbol{\sigma}\otimes\mathbf{I}):\mathbf{D} = \mathbf{C}^{\sigma T}:\mathbf{D}\end{aligned} \tag{5.4.26}$$

So the relation between the Jaumann and Truesdell moduli is

$$\mathbf{C}^{\sigma T} = \mathbf{C}^{\sigma J} - \mathbf{C}' + \boldsymbol{\sigma} \otimes \mathbf{I} = \mathbf{C}^{\sigma J} - \mathbf{C}^{*} \quad \text{where} \quad \mathbf{C}':\mathbf{D} = \mathbf{D} \cdot \boldsymbol{\sigma} + \boldsymbol{\sigma} \cdot \mathbf{D} \tag{5.4.27}$$

and

$$\mathbf{C}^{*} = \mathbf{C}' - \boldsymbol{\sigma} \otimes \mathbf{I} \; C^{*}_{ijkl} = \frac{1}{2}(\delta_{ik}\sigma_{jl} + \delta_{il}\sigma_{jk} + \delta_{jk}\sigma_{il} + \delta_{jl}\sigma_{ik}) - \sigma_{ij}\delta_{kl} \tag{5.4.28}$$

Note that if $\mathbf{C}^{\sigma J}$ are constant, the tangent moduli $\mathbf{C}^{\sigma T}$ are not constant. Also, $\mathbf{C}'$ has major symmetry, but $\mathbf{C}^{*}$ does not ($\boldsymbol{\sigma} \otimes \mathbf{I} \neq \mathbf{I} \otimes \boldsymbol{\sigma}$).

The relation between the Green–Naghdi moduli $\mathbf{C}^{\sigma G}$ and the Truesdell moduli $\mathbf{C}^{\sigma T}$ is obtained as follows. Note that the Green–Naghdi rate of Cauchy stress can be written as

$$\boldsymbol{\sigma}^{\nabla G} = \mathbf{R} \cdot \dot{\hat{\boldsymbol{\sigma}}} \cdot \mathbf{R}^{T} = \mathbf{R} \cdot \frac{D}{Dt}(\mathbf{R}^{T} \cdot \boldsymbol{\sigma} \cdot \mathbf{R}) \cdot \mathbf{R}^{T} = \dot{\boldsymbol{\sigma}} - \boldsymbol{\Omega} \cdot \boldsymbol{\sigma} - \boldsymbol{\sigma} \cdot \boldsymbol{\Omega}^{T} \tag{5.4.29}$$

where $\boldsymbol{\Omega} = \dot{\mathbf{R}} \cdot \mathbf{R}^{T}$ is the spin tensor (skew-symmetric: $\Omega = -\Omega^{T}$) associated with the rotation $\mathbf{R}$. From (3.7.11), (3.7.12) and (5.4.29)

$$\begin{aligned} \boldsymbol{\sigma}^{\nabla T} &= \boldsymbol{\sigma}^{\nabla G} - (\mathbf{L} - \boldsymbol{\Omega}) \cdot \boldsymbol{\sigma} - \boldsymbol{\sigma} \cdot (\mathrm{L} - \Omega)^{T} + \operatorname{trace}(\mathbf{D})\boldsymbol{\sigma} \\ &= \mathbf{C}^{\sigma G}:\mathbf{D} - (\mathbf{L} - \boldsymbol{\Omega}) \cdot \boldsymbol{\sigma} - \boldsymbol{\sigma} \cdot (\mathrm{L} - \Omega)^{T} + \operatorname{trace}(\mathbf{D})\boldsymbol{\sigma} \\ &= \mathbf{C}^{\sigma G}:\mathbf{D} - \mathbf{D} \cdot \boldsymbol{\sigma} - \boldsymbol{\sigma} \cdot \mathbf{D} - (\mathbf{W} - \Omega) \cdot \boldsymbol{\sigma} - \boldsymbol{\sigma} \cdot (\mathrm{W} - \Omega)^{T} + \operatorname{trace}(\mathbf{D})\boldsymbol{\sigma} \end{aligned} \tag{5.4.30}$$

where the constitutive relation is used in the second of the above and $\mathbf{L} = \mathbf{D} + \mathbf{W}$ in the third. A complex derivation (see Simo and Hughes, 1998: 273; Mehrabadi and Nemat-Nasser, 1987) leads to the final expression

$$\boldsymbol{\sigma}^{\nabla T} = \mathbf{C}^{\sigma T}:\mathbf{D}, \; \mathbf{C}^{\sigma T} = \mathbf{C}^{\sigma G} - \mathbf{C}^{*} - \mathbf{C}^{\text{spin}} \tag{5.4.31}$$

The fourth-order tensor $\mathbf{C}^{\text{spin}}$ is a function of the left stretch tensor $\mathbf{V}$ and the Cauchy stress and accounts for the terms involving the difference in spins in (5.4.30). It is defined in Box 5.1 and $\mathbf{C}^{*}$ is as defined in (5.4.28). The tensor $\mathbf{C}^{\text{spin}}$ does not have major symmetry and, consequently, $\mathbf{C}^{\sigma T}$ is not symmetric (Simo and Hughes, 1998: 273). A summary of some key stress rates and associated moduli is given in Box 5.1.

The Jaumann rate of Kirchhoff stress is included because it is often used in plasticity and can give rise to a symmetric tangent modulus. To see this, note that if $\mathbf{C}^{\tau J}$ has major symmetry, then the tangent modulus $\mathbf{C}^{\sigma T}$ for the Jaumann rate of Kirchhoff stress in Box 5.1 is symmetric because $\mathbf{C}'$ is symmetric. In contrast, if $\mathbf{C}^{\sigma J}$ has major symmetry, then the tangent modulus for the Jaumann rate of Cauchy stress is not symmetric because $\mathbf{C}^{*}$ does not have major symmetry.

Box 5.1 Relations between tangent moduli

Stress rate	Constitutive relation	Tangent modulus $\boldsymbol{\sigma}^{\nabla T}=\mathbf{C}^{\sigma T}:\mathbf{D}$
Jaumann (Cauchy) $\boldsymbol{\sigma}^{\nabla J}=\dot{\boldsymbol{\sigma}}-\mathbf{W}\cdot\boldsymbol{\sigma}-\boldsymbol{\sigma}\cdot\mathbf{W}^T$	$\boldsymbol{\sigma}^{\nabla J}=\mathbf{C}^{\sigma J}:\mathbf{D}$	$\mathbf{C}^{\sigma T}=\mathbf{C}^{\sigma J}-\mathbf{C}^*$ $\mathbf{C}^*=\mathbf{C}'-\boldsymbol{\sigma}\otimes\mathbf{I}$, (See 5.4.28) $\mathbf{C}':\mathbf{D}=\mathbf{D}\cdot\boldsymbol{\sigma}+\boldsymbol{\sigma}\cdot\mathbf{D}$
Jaumann (Kirchhoff) $\boldsymbol{\tau}^{\nabla J}=\dot{\boldsymbol{\tau}}-\mathbf{W}\cdot\boldsymbol{\tau}-\boldsymbol{\tau}\cdot\mathbf{W}^T$	$\boldsymbol{\tau}^{\nabla J}=\mathbf{C}^{\tau J}:\mathbf{D}$	$\mathbf{C}^{\sigma T}=J^{-1}\mathbf{C}^{\tau J}-\mathbf{C}'$
Green–Naghdi (Cauchy) $\boldsymbol{\sigma}^{\nabla G}=\dot{\boldsymbol{\sigma}}-\boldsymbol{\Omega}\cdot\boldsymbol{\sigma}-\boldsymbol{\sigma}\cdot\boldsymbol{\Omega}^T$	$\boldsymbol{\sigma}^{\nabla G}=\mathbf{C}^{\sigma G}:\mathbf{D}$	$\mathbf{C}^{\sigma T}=\mathbf{C}^{\sigma G}-\mathbf{C}^*-\mathbf{C}^{\text{spin}}$ $\mathbf{C}^{\text{spin}}:\mathbf{D}=(\mathbf{W}-\boldsymbol{\Omega})\cdot\boldsymbol{\sigma}+\boldsymbol{\sigma}\cdot(\mathbf{W}-\boldsymbol{\Omega})^T$

1D Uniaxial Stress

$$\sigma_{11}^{\nabla T}=E^{\sigma T}D_{11},\ E^{\sigma T}=C_{1111}^{\sigma T}-2\hat{\nu}C_{1122}^{\sigma T}$$
$$\sigma_{11}^{\nabla J}=E^{\sigma J}D_{11},\ E^{\sigma J}=C_{1111}^{\sigma J}-2\hat{\nu}C_{1122}^{\sigma J}=E^{\sigma T}+(1+2\hat{\nu})\sigma_{11}$$
$$\dot{S}_{11}=E^{SE}\dot{E}_{11},\ E^{SE}=C_{1111}^{SE}-2\hat{\nu}C_{1122}^{SE}=J\lambda_1^{-4}E^{\sigma T}$$
$$\dot{\sigma}_{11}=E^{\sigma}D_{11},\ \dot{\sigma}_{11}=\sigma_{11}^{\nabla J},\ E^{\sigma}=E^{\sigma J}$$

for constant $\mathbf{C}^{SE}$, $S_{11}=E^{SE}E_{11}$ where E^{SE} is defined above.

1D Uniaxial Strain

$$\sigma_{11}^{\nabla T}=C_{1111}^{\sigma T}D_{11}$$
$$\sigma_{11}^{\nabla J}=C_{1111}^{\sigma J}D_{11}\quad C_{1111}^{\sigma J}=C_{1111}^{\sigma T}+\sigma_{11}$$
$$\dot{S}_{11}=C_{1111}^{SE}\dot{E}_{11},\ C_{1111}^{SE}=J\lambda_1^{-4}C_{1111}^{\sigma T}\quad J=\lambda_1$$
$$\dot{\sigma}_{11}=\sigma_{11}^{\nabla J}=C_{1111}^{\sigma J}D_{11}$$

An expression for the material time derivative of the Cauchy stress is sometimes required in stress update algorithms or in the enforcement of uniaxial or plane stress conditions, for example. This expression can be obtained from the expression for the Truesdell rate:

$$\begin{aligned}\dot{\boldsymbol{\sigma}}&=\boldsymbol{\sigma}^{\nabla T}+\mathbf{L}\cdot\boldsymbol{\sigma}+\boldsymbol{\sigma}\cdot\mathbf{L}^T-(\text{trace}\,\mathbf{D})\boldsymbol{\sigma}\\&=(\mathbf{C}^{\sigma T}+\mathbf{C}''-\boldsymbol{\sigma}\otimes\mathbf{I}):\mathbf{L},\ C''_{ijkl}=\delta_{ik}\sigma_{jl}+\sigma_{il}\delta_{jk}\end{aligned}\qquad(5.4.32)$$

where the appropriate expression for $\mathbf{C}^{\sigma T}$ is used.

Example 5.1 Tangent moduli for uniaxial strain and stress Consider a transversely isotropic material with the axis of isotropy in the x_1-direction. The material is under a state of (a) uniaxial strain in the x_1-direction and (b) uniaxial stress in the x_1-direction. Derive expressions for the instantaneous slope (tangent modulus) of the true stress vs. log stretch curve (σ_{11} vs. In λ_1) and relate this modulus to the moduli in Box 5.1.

First note that the instantaneous slope of the Cauchy stress-log stretch curve is given by the relation between $\dot{\sigma}_{11}$, the material time derivative of the Cauchy stress, and $D_{11}=D(\ln\lambda_1)/Dt$, the material time derivative of the log stretch. This relation will be obtained from $(5.4.32)_1$.

(1) *Uniaxial strain* In uniaxial strain, the spin is zero and the only nonzero components of $\mathbf{L}$ and $\mathbf{D}$ are $L_{11}=D_{11}$. Thus trace $(\mathbf{D})=D_{11}$. From (5.4.32), then,

$$\begin{aligned}\dot{\sigma}_{11} &= C^{\sigma T}_{1111}D_{11} + D_{11}\sigma_{11} + \sigma_{11}D_{11} - D_{11}\sigma_{11} \\ &= (C^{\sigma T}_{1111} + \sigma_{11})D_{11}\end{aligned} \tag{E5.1.1}$$

Note that, because the spin is zero, the Jaumann rate is equal to the material time derivative of the Cauchy stress and therefore $C^{\sigma J}_{1111} = C^{\sigma T}_{1111} + \sigma_{11}$. The material time derivative of the second Piola–Kirchhoff stress is given by

$$\dot{S}_{11} = C^{SE}_{1111}\dot{E}_{11} \quad \text{where} \quad C^{SE}_{1111} = J\lambda_1^{-4}C^{\sigma T}_{1111} \quad \text{and} \quad J = \lambda_1 \tag{E5.1.2}$$

(2) *Uniaxial stress* In uniaxial stress, the spin is again zero and $\mathbf{L}=\mathbf{D}$. The only nonzero component of stress is σ_{11} and the only nonzero components of the rate-of-deformation tensor are D_{11}, D_{22} and D_{33}. From $(5.4.32)_1$

$$\dot{\sigma}_{11} = C^{\sigma T}_{1111}D_{11} + C^{\sigma T}_{1122}D_{22} + C^{\sigma T}_{1133}D_{33} + \sigma_{11}D_{11} + D_{11}\sigma_{11} - \text{trace}(\mathbf{D})\sigma_{11} \tag{E5.1.3}$$

Because of the uniaxial stress condition, the transverse stresses and their material rates vanish:

$$\dot{\sigma}_{22} = \dot{\sigma}_{33} = 0 \tag{E5.1.4}$$

which, from (5.4.32) can be written

$$\begin{aligned}\dot{\sigma}_{22} &= C^{\sigma T}_{2211}D_{11} + C^{\sigma T}_{2222}D_{22} + C^{\sigma T}_{2233}D_{33} = 0 \\ \dot{\sigma}_{33} &= C^{\sigma T}_{3311}D_{11} + C^{\sigma T}_{3322}D_{22} + C^{\sigma T}_{3333}D_{33} = 0\end{aligned} \tag{E5.1.5}$$

For uniaxial stressing in the x_1-direction, the terms involving the stress in (5.4.31) make no contribution to the expressions in (E5.1.5). For transverse isotropy, the tangent moduli are related by $C^{\sigma T}_{1133} = C^{\sigma T}_{1122}$ and $C^{\sigma T}_{2222} = C^{\sigma T}_{3333}$. Furthermore, uniaxial stressing in the direction of the axis of isotropy preserves the transverse isotropy and these relations hold throughout the deformation. Solving (E5.1.5) for D_{22} and D_{33} yields

$$D_{22} = D_{33},\ D_{22} = -\hat{\nu}D_{11} \quad \text{where} \quad \hat{\nu} = \frac{C^{\sigma T}_{2211}}{C^{\sigma T}_{2222} + C^{\sigma T}_{2233}} \tag{E5.1.6}$$

From (E5.1.6)

$$\text{trace}\,\mathbf{D} = D_{11} + D_{22} + D_{33} = (1+2\hat{v})D_{11} \tag{E5.1.7}$$

Substituting (E5.1.6) and (E5.1.7) into (E5.1.3) gives the uniaxial stress relation

$$\dot{\sigma}_{11} = E^{\sigma} D_{11} \quad \text{where} \quad E^{\sigma} = C^{\sigma T}_{1111} - 2\hat{v}C^{\sigma T}_{1122} + (1+2\hat{v})\sigma_{11} \tag{E5.1.8}$$

and E^{σ} is the tangent modulus for uniaxial stress. Because the spin is zero, $E^{\sigma}=E^{\sigma J}$ where the latter relates the Jaumann rate of stress to the rate-of-deformation. An expression for the Truesdell rate of Cauchy stress follows from (5.4.26) and is given by

$$\sigma^{\nabla T}_{11} = E^{\sigma T} D_{11} \quad \text{where} \quad E^{\sigma T} = C^{\sigma T}_{1111} - 2\hat{v}C^{\sigma T}_{1122} \tag{E5.1.9}$$

The second Piola–Kirchhoff stress is given by

$$\dot{S}_{11} = E^{SE}\dot{E}_{11} \quad \text{where} \quad E^{SE} = J\lambda_1^{-4}(C^{\sigma T}_{1111} - 2\hat{v}C^{\sigma T}_{1122}) = C^{SE}_{1111} - 2\hat{v}C^{SE}_{1122} \tag{E5.1.10}$$

where $C^{SE}_{1111} = J\lambda_1^{-4}C^{\sigma T}_{1111}$, $C^{SE}_{1122} = J\lambda_1^{-4}C^{\sigma T}_{1122}$ and $J=\det\mathbf{F}$. For a rod, the Jacobian is given by (E3.9.3), that is, $J=A\lambda_1/A_0$ where A and A_0 are the current and original cross-sectional areas, respectively.

5.4.6 Cauchy Elastic Material

A Cauchy elastic material may be characterized as a material which has no dependence on the history of the motion. The constitutive relation for a Cauchy elastic material is given by

$$\boldsymbol{\sigma} = \boldsymbol{G}(\mathbf{F}) \tag{5.4.33}$$

where $\boldsymbol{G}$ is called the material response function and the explicit dependence on position $\mathbf{X}$ and time t has been suppressed for notational convenience. The response function depends only on the current value of the deformation gradient and not on its history. Restrictions on the material response imposed by frame invariance are discussed in Section 5.10. For a Cauchy elastic material, the restriction is given by (5.10.39), that is,

$$\boldsymbol{\sigma} = \mathbf{R}\cdot\boldsymbol{G}(\mathbf{U})\cdot\mathbf{R}^T \tag{5.4.34}$$

Thus as indicated in Section 5.10, a constitutive equation can not be expressed in terms of $\mathbf{F}$ unless the $\mathbf{F}$ dependence has a special form. The dependence can be on $\mathbf{F}^T\cdot\mathbf{F}=2\mathbf{E}+\mathbf{I}$ or on $\mathbf{U}=(\mathbf{F}^T\cdot\mathbf{F})^{\frac{1}{2}}$ see Section 3.7.1. Alternative forms of the same constitutive relation for other representations of stress and strain follow from the stress transformation relations in Box 3.2; for example, the nominal stress for a Cauchy elastic material is given by $\mathbf{P}=J\mathbf{U}^{-1}\cdot\boldsymbol{G}(\mathbf{U})\cdot\boldsymbol{R}^T$. The relationship for at the second Piola–Kirchhoff stress takes the form $\mathbf{S}=J\mathbf{U}^{-1}\cdot\boldsymbol{G}(\mathbf{U})\cdot\mathbf{U}^{-1}=\mathbf{h}(\mathbf{U})=\tilde{\mathbf{h}}(\mathbf{C})$ where $\mathbf{C}$ is the right Cauchy–Green deformation tensor. The stress can be computed for a Cauchy elastic material according to (5.4.33) and is independent of the deformation history.

However, the work done may depend on the deformation history or load path. Thus the Cauchy elastic material has some of the elastic properties but not all: the stress is path-independent, but the energy may not be.

5.4.7 Hyperelastic Materials

Elastic materials for which the work is independent of the load path are said to be *hyperelastic* (or Green elastic) materials. In this section, some general features of hyperelastic materials are described and then examples of hyperelastic constitutive models that are used in practice are given. Hyperelastic materials are characterized by the existence of a stored (or strain) energy function that is a potential for the stress:

$$\mathbf{S} = 2\frac{\partial \psi(\mathbf{C})}{\partial \mathbf{C}} = \frac{\partial w(\mathbf{E})}{\partial \mathbf{E}} \tag{5.4.35}$$

where ψ is the stored energy potential. When the potential is written as a function of the Green strain $\mathbf{E}$, we use the notation w where the relation between the two scalar functions is given by $w(\mathbf{E})=\psi(2\mathbf{E}+\mathbf{I})$. Hyperelastic materials provide a natural framework for frame-invariant formulation of anisotropic material response by simply embodying the anisotropy in the potential w. Expressions for different stress measures are obtained through the appropriate transformations (given in Box 3.2), for example, the Kirchhoff stress is given by

$$\boldsymbol{\tau} = J\boldsymbol{\sigma} = \mathbf{F}\cdot\mathbf{S}\cdot\mathbf{F}^T = 2\mathbf{F}\cdot\frac{\partial \psi(\mathbf{C})}{\partial \mathbf{C}}\cdot\mathbf{F}^T = \mathbf{F}\cdot\frac{\partial w(\mathbf{E})}{\partial \mathbf{E}}\cdot\mathbf{F}^T \tag{5.4.36}$$

A consequence of the existence of a stored energy function is that the work done on a hyperelastic material is independent of the deformation path. This behavior is approximately observed in many rubber-like materials. To illustrate the independence of work on deformation path, consider the stored energy per unit reference volume in going from deformation state $\mathbf{C}_1$ to $\mathbf{C}_2$. Because the second Piola–Kirchhoff stress tensor $\mathbf{S}$ and the Green strain $\mathbf{E}=(\mathbf{C}-\mathbf{I})/2$ are work conjugates, we have

$$\int_{\mathbf{E}_1}^{\mathbf{E}_2}\mathbf{S}{:}d\mathbf{E} = w(\mathbf{E}_2)-w(\mathbf{E}_1),\ \text{or}\ \frac{1}{2}\int_{\mathbf{C}_1}^{\mathbf{C}_2}\mathbf{S}{:}d\mathbf{C} = \psi(\mathbf{C}_2)-\psi(\mathbf{C}_1) \tag{5.4.37}$$

and the energy stored in the material depends only on the initial and final states of deformation and is independent of the deformation (or load) path.

To obtain an expression for the nominal stress tensor $\mathbf{P}$ as a derivative of the potential, we recall that $\mathbf{P}$ is conjugate in power to $\dot{\mathbf{F}}^T$. The nominal stress is then given in terms of the potential by

$$\frac{\partial w}{\partial \mathbf{F}^T} = \frac{\partial \psi}{\partial \mathbf{C}}:\frac{\partial \mathbf{C}}{\partial \mathbf{F}^T} = \mathbf{S}\cdot\mathbf{F}^{\mathrm{T}} = \mathbf{P},\ \text{or}\ P_{ij} = \frac{\partial w}{\partial F_{ji}} \tag{5.4.38}$$

As the deformation gradient tensor $\mathbf{F}$ is not symmetric, the nine components of the nominal stress tensor $\mathbf{P}$ are not symmetric.

5.4.8 Elasticity Tensors

Expressions for the rate of stress are required in the linearization of the weak form in Chapter 6. These are often expressed through what are known as the four elasticity tensors which are given next. Taking the time derivative of the nominal stress and using (5.4.38),

$$\dot{\mathbf{P}} = \frac{\partial \mathbf{P}}{\partial \mathbf{F}^T}:\dot{\mathbf{F}}^T = \frac{\partial^2 w}{\partial \mathbf{F}^T \partial \mathbf{F}^T}:\dot{\mathbf{F}}^T = \boldsymbol{A}^{(1)}:\dot{\mathbf{F}}^T \tag{5.4.39}$$

where

$$\boldsymbol{A}^{(1)} = \frac{\partial^2 w}{\partial \mathbf{F}^T \partial \mathbf{F}^T}, \ \boldsymbol{A}^{(1)}_{ijkl} = \frac{\partial^2 w}{\partial F_{ji} \partial F_{lk}} \tag{5.4.40}$$

is called the *first elasticity tensor*. The first elasticity tensor has major symmetry, $A^{(1)}_{ijkl} = A^{(1)}_{klij}$, *but* does not have minor symmetries. Sometimes the first elasticity tensor is defined in terms of the first Piola–Kirchhoff stress which is the transpose of the nominal stress:

$$\dot{\mathbf{P}}^T = \hat{\boldsymbol{A}}^{(1)}:\dot{\mathbf{F}}, \ \hat{\boldsymbol{A}}^{(1)}_{ijkl} = A^{(1)}_{jilk} \tag{5.4.41}$$

The rate forms of constitutive equations for hyperelastic materials, required in linearization of the weak form in Chapter 6, can be obtained by taking the material time derivative of (5.4.35):

$$\dot{\mathbf{S}} = 4\frac{\partial^2 \psi(\mathbf{C})}{\partial \mathbf{C} \partial \mathbf{C}}:\frac{\dot{\mathbf{C}}}{2} = \frac{\partial^2 w(\mathbf{E})}{\partial \mathbf{E} \partial \mathbf{E}}:\dot{\mathbf{E}} = \mathbf{C}^{SE}:\frac{\dot{\mathbf{C}}}{2} = \mathbf{C}^{SE}:\dot{\mathbf{E}} \tag{5.4.42}$$

where

$$\mathbf{C}^{SE} = 4\frac{\partial^2 \psi(\mathbf{C})}{\partial \mathbf{C} \partial \mathbf{C}} = \frac{\partial^2 w(\mathbf{E})}{\partial \mathbf{E} \partial \mathbf{E}} \tag{5.4.43}$$

is the tangent modulus, also called the *second elasticity tensor*, denoted $\boldsymbol{A}^{(2)} \equiv \boldsymbol{C}^{SE}$. It follows that the tangent modulus for a hyperelastic material has the major symmetry $C^{SE}_{ijkl} = C^{SE}_{klij}$. Since it relates symmetric measures of stress rate and strain rate, it also has the minor symmetries.

The relation between the first and second elasticity tensors can be derived by noting that since $\mathbf{P} = \mathbf{S} \cdot \mathbf{F}^T$, then $\dot{\mathbf{P}} = \dot{\mathbf{S}} \cdot \mathbf{F}^T + \mathbf{S} \cdot \dot{\mathbf{F}}^T$. Substituting (5.4.42) for $\dot{\mathbf{S}}$ and from (3.3.20) and the minor symmetry of $\mathbf{C}^{SE}$ we obtain (after some manipulation of the indices)

$$\begin{aligned} \dot{P}_{ij} &= A^{(1)}_{ijkl}\dot{F}_{lk} = (C^{SE}_{inpk}F_{jn}F_{lp} + S_{ik}\delta_{lj})\dot{F}_{lk} \\ A^{(1)}_{ijkl} &= C^{SE}_{inpk}F_{jn}F_{lp} + S_{ik}\delta_{lj} = A^{(2)}_{ijkl}F_{jn}F_{lp} + S_{ik}\delta_{lj} \end{aligned} \tag{5.4.44}$$

The *third elasticity tensor* $\mathbf{A}^{(3)}$ is defined through the push-forward of $\dot{\mathbf{P}}$, that is, $\mathbf{F}\cdot\dot{\mathbf{P}}$ which appears in the linearization of the weak form (Chapter 6). From (5.4.44)

$$F_{ir}\dot{P}_{rj} = A^{(3)}_{ijkl}L^T_{kl} = (F_{im}F_{jn}F_{kp}F_{lq}C^{SE}_{mnpq} + F_{im}F_{kn}S_{mn}\delta_{lj})L^T_{kl} \quad (5.4.45)$$

where we have used the relation $\dot{\mathbf{F}}^T = \mathbf{F}^T\cdot\mathbf{L}^T$. The first term in parentheses in the previous expression is the spatial form of the second elasticity tensor and is called the *fourth elasticity tensor*

$$A^{(4)}_{ijkl} \equiv C^{\tau}_{ijkl} = F_{im}F_{jn}F_{kp}F_{lq}C^{SE}_{mnpq} \equiv F_{im}F_{jn}F_{kp}F_{lq}A^{(2)}_{mnpq} \quad (5.4.46)$$

Finally, the second term in parentheses in (5.4.45) is the Kirchhoff stress tensor, and therefore

$$A^{(3)}_{ijkl} = A^{(4)}_{ijkl} + \tau_{ik}\delta_{jl} \quad (5.4.47)$$

In a finite element implementation, the minor symmetry of $A^{(4)}_{ijkl}$ is taken into account and

$$A^{(3)}_{ijkl}L^T_{kl} = A^{(4)}_{ijkl}D_{kl} + \tau_{ik}\delta_{jl}L^T_{kl} \quad (5.4.48)$$

where the term involving $A^{(4)}_{ijkl}$ gives rise to the material tangent stiffness matrix and the term involving $\tau_{ik}\delta_{jl}$ gives rise to the geometric stiffness (see Chapter 6).

In the linearization of the updated Lagrangian discretization, the relation between the convected rate of the Kirchhoff stress and the rate-of-deformation is useful. To this end we recall the convected rate of Kirchhoff stress (5.4.22) and note that it can be written (for a derivation see (5.10.3))

$$\boldsymbol{\tau}^{\nabla c} = \dot{\boldsymbol{\tau}} - \mathbf{L}\cdot\boldsymbol{\tau} - \boldsymbol{\tau}\cdot\mathbf{L}^T = \mathbf{F}\cdot\frac{D}{Dt}(\mathbf{F}^{-1}\cdot\boldsymbol{\tau}\cdot\mathbf{F}^{-T})\cdot\mathbf{F}^T = \mathbf{F}\cdot\dot{\mathbf{S}}\cdot\mathbf{F}^T \equiv L_v\boldsymbol{\tau} \quad (5.4.49)$$

In the second step, we have used the relation between the Kirchhoff stress and the PK2 stress from Box 3.2. Substituting (5.4.42) into the last form of (5.4.49) and using (3.3.21), $\dot{\mathbf{E}} = \mathbf{F}^T\cdot\mathbf{D}\cdot\mathbf{F}$, gives

$$L_v\boldsymbol{\tau} \equiv \boldsymbol{\tau}^{\nabla c} = \mathbf{C}^{\tau}:\mathbf{D}, \quad \text{where} \quad C^{\tau}_{ijkl} = F_{im}F_{jn}F_{kp}F_{lq}C^{SE}_{mnpq} \equiv A^{(4)}_{ijkl} \quad (5.4.50)$$

where $\mathbf{C}^{\tau}$ are referred to as the spatial tangent moduli (or spatial form of the second elasticity tensor, i.e., the fourth elasticity tensor). The connection between (5.4.50) and (5.4.48) is made by noting that $\mathbf{F}\cdot\dot{\mathbf{P}} = \mathbf{F}\cdot\dot{\mathbf{S}}\cdot\mathbf{F}^T + \mathbf{F}\cdot\mathbf{S}\cdot\dot{\mathbf{F}}^T = L_v\boldsymbol{\tau} + \boldsymbol{\tau}\cdot\mathbf{L}^T$.

It can be seen from the above that the Lie derivative (or convected rate) of Kirchhoff stress arises naturally as a stress rate in hyperelasticity. The relation (5.4.50) can be expressed in terms of the Truesdell rate of Cauchy stress as follows:

$$\boldsymbol{\sigma}^{\nabla T} = J^{-1}\boldsymbol{\tau}^{\nabla c} = J^{-1}\mathbf{C}^{\tau}:\mathbf{D} = \mathbf{C}^{\sigma T}:\mathbf{D},\ \mathbf{C}^{\sigma T} = J^{-1}\mathbf{C}^{\tau} \quad (5.4.51)$$

The elasticity tensors defined above also play an important role in stability and uniqueness of solutions in finite strain elasticity; see Section 6.7, Ogden (1984) and Marsden and Hughes (1983).

5.4.9 *Isotropic Hyperelastic Materials*

It can be shown (Malvern, 1969: 409) that the stored strain energy (potential) for a hyperelastic material which is isotropic with respect to the initial, unstressed configuration can be written as a function of the principal invariants (I_1, I_2, I_3) of the right Cauchy–Green deformation tensor, that is, $\psi = \psi(I_1, I_2, I_3)$. The principal invariants of a second-order tensor and their derivatives figure prominently in elastic and elastic–plastic constitutive relations. For reference, Box 5.2 summarizes some key relations involving principal invariants.

Box 5.2 Principal invariants of a second-order tensor

The principal invariants of a second-order tensor **A** are given by

$$I_1(\mathbf{A}) = \mathrm{trace}(\mathbf{A}) = A_{ii} \tag{B5.2.1a}$$

$$I_2(\mathbf{A}) = \frac{1}{2}\left\{(\mathrm{trace}(\mathbf{A}))^2 - \mathrm{trace}(\mathbf{A}^2)\right\} = \frac{1}{2}\left\{(A_{ii})^2 - A_{ij}A_{ji}\right\} \tag{B5.2.1b}$$

$$I_3(\mathbf{A}) = \det \mathbf{A} = \varepsilon_{ijk} A_{i1} A_{j2} A_{k3} \tag{B5.2.1c}$$

When the tensor in question is clear from the context, the argument **A** is omitted and the principal invariants are denoted simply as I_1, I_2, and I_3.

If **A** is symmetric, $\mathbf{A}=\mathbf{A}^T$, and **A** possesses three real eigenvalues (or principal values)λ_1, λ_2, λ_3; then

$$\begin{aligned} I_1(\mathbf{A}) &= \lambda_1 + \lambda_2 + \lambda_3 \\ I_1(\mathbf{A}) &= \lambda_1\lambda_2 + \lambda_2\lambda_3 + \lambda_3\lambda_1 \\ I_1(\mathbf{A}) &= \lambda_1\lambda_2\lambda_3 \end{aligned} \tag{B5.2.2}$$

The derivatives of the principal invariants of a second-order tensor with respect to the tensor itself are often required in constitutive equations. For reference:

$$\frac{\partial I_1}{\partial \mathbf{A}} = \mathbf{I}; \frac{\partial I_1}{\partial A_{ij}} = \delta_{ij} \tag{B5.2.3}$$

$$\frac{\partial I_1}{\partial \mathbf{A}} = I_1\mathbf{I} - \mathbf{A}^T; \frac{\partial I_2}{\partial A_{ij}} = A_{kk}\delta_{ij} - A_{ji} \tag{B5.2.4}$$

$$\frac{\partial I_3}{\partial \mathbf{A}} = I_3\mathbf{A}^{-T}; \frac{\partial I_3}{\partial A_{ij}} = I_3 A_{ji}^{-1} \tag{B5.2.5}$$

The second Piola–Kirchhoff stress tensor for a hyperelastic material is given by (5.4.35). Thus, for an isotropic material we have

$$\begin{aligned}\mathbf{S} &= 2\frac{\partial\psi}{\partial\mathbf{C}} = 2\left(\frac{\partial\psi}{\partial I_1}\frac{\partial I_1}{\partial\mathbf{C}} + \frac{\partial\psi}{\partial I_2}\frac{\partial I_2}{\partial\mathbf{C}} + \frac{\partial\psi}{\partial I_3}\frac{\partial I_3}{\partial\mathbf{C}}\right)\\ &= 2\left(\frac{\partial\psi}{\partial I_1} + I_1\frac{\partial\psi}{\partial I_2}\right)\mathbf{I} - 2\frac{\partial\psi}{\partial I_2}\mathbf{C} + 2I_3\frac{\partial\psi}{\partial I_3}\mathbf{C}^{-1}\end{aligned} \tag{5.4.52}$$

where we have used the results in Box 5.2. The Kirchhoff stress tensor is given by

$$\boldsymbol{\tau} = \mathbf{F}\cdot\mathbf{S}\cdot\mathbf{F}^T = 2\left(\frac{\partial\psi}{\partial I_1} + I_1\frac{\partial\psi}{\partial I_2}\right)\mathbf{B} - 2\frac{\partial\psi}{\partial I_2}\mathbf{B}^2 + 2I_3\frac{\partial\psi}{\partial I_3}\mathbf{I} \tag{5.4.53}$$

where $\mathbf{B}=\mathbf{F}\cdot\mathbf{F}^T$ is the left Cauchy–Green deformation tensor and the expression for $\mathbf{S}$ in (5.4.52) has been used. Note that $\mathbf{S}$ is coaxial (has the same principal directions) with $\mathbf{C}$ while $\boldsymbol{\tau}$ is coaxial with $\mathbf{B}$.

5.4.10 Neo-Hookean Material

The Neo-Hookean material model is an extension of the isotropic linear law (Hooke's law) to large deformations. The stored energy function for a compressible Neo-Hookean material (isotropic with respect to the initial, unstressed configuration) is

$$\psi(\mathbf{C}) = \frac{1}{2}\lambda_0(\ln J)^2 - \mu_0\ln J + \frac{1}{2}\mu_0(\operatorname{trace}\mathbf{C} - 3) \tag{5.4.54}$$

Here λ_0 and μ_0 are the Lamé constants of the linearized theory and $J=\det\mathbf{F}$. From (5.4.52) and (5.4.53), the stresses are given by

$$\mathbf{S} = \lambda_0\ln J\mathbf{C}^{-1} + \mu_0(\mathbf{I} - \mathbf{C}^{-1}),\quad \boldsymbol{\tau} = \lambda_0\ln J\mathbf{I} + \mu_0(\mathbf{B} - \mathbf{I}) \tag{5.4.55}$$

Letting $\lambda=\lambda_0$, $\mu=\mu_0 - \lambda\ln J$ and using (5.4.43) and (5.4.50), the elasticity tensors (tangent moduli) in component form are

$$C^{SE}_{ijkl} = \lambda C^{-1}_{ij}C^{-1}_{kl} + \mu(C^{-1}_{ik}C^{-1}_{jl} + C^{-1}_{il}C^{-1}_{kj}) \tag{5.4.56}$$

$$C^{\tau}_{ijkl} = \lambda\delta_{ij}\delta_{kl} + \mu(\delta_{ik}\delta_{jl} + \delta_{il}\delta_{kj}) \tag{5.4.57}$$

The tangent moduli in (5.4.57) have the same form as Hooke's law for small-strain elasticity, except for the dependence of the shear modulus, μ, on the deformation. Nearly incompressible behavior is obtained for $\lambda_0 \gg \mu_0$. In Voigt matrix notation the spatial elastic moduli for a Neo-Hookean material for plane strain are

$$\left[C^{\tau}_{ab}\right]=\begin{bmatrix}\lambda+2\mu & \lambda & 0\\ \lambda & \lambda+2\mu & 0\\ 0 & 0 & \mu\end{bmatrix} \tag{5.4.58}$$

5.4.11 *Modified Mooney–Rivlin Material*

Rivlin and Saunders (1951) developed a hyperelastic constitutive model for large deformations of rubber. The model is incompressible and initially isotropic. The stored energy function is therefore of the form given in Section 5.4.9, where $\psi=\psi(I_1, I_2, I_3)$. For an incompressible material, $J=\det \mathbf{F}=1$ and therefore $I_3=\det \mathbf{C}=J^2=1$. The potential can be written as a series expansion in I_1 and I_2:

$$\psi=\psi(I_1,I_2)=\sum_{i=0}^{\infty}\sum_{j=0}^{\infty}\bar{c}_{ij}(I_1-3)^i(I_2-3)^j,\ \bar{c}_{00}=0 \tag{5.4.59}$$

where $\bar{c}_{ij}$ are constants. Mooney and Rivlin showed that the simple form

$$\psi=\psi(I_1,I_2)=c_1(I_1-3)+c_2(I_2-3) \tag{5.4.60}$$

closely matches experimental results. The Mooney–Rivlin material is an example of a Neo-Hookean material.

The components of the second Piola–Kirchhoff stress are obtained by differentiating (5.4.60) with respect to the right Cauchy–Green deformation tensor (as in (5.4.52). However, since the material is incompressible the deformation is constrained such that $I_3=1$ which by (B5.2.1c) is equivalent to $J=1$. One way in which this constraint can be enforced is through the use of a constrained potential. Alternatively, a penalty function formulation can be used. The corresponding penalty function is expressed as $\ln I_3=0$. In this case, the modified strain energy function and the constitutive equation are:

$$\hat{\psi}=\psi+p_0\ln I_3+\frac{1}{2}\beta(\ln I_3)^2 \tag{5.4.61}$$

$$\mathbf{S}=2\frac{\partial\hat{\psi}}{\partial\mathbf{C}}=2\frac{\partial\psi}{\partial\mathbf{C}}+2(p_0+\beta(\ln I_3))\mathbf{C}^{-1} \tag{5.4.62}$$

respectively. The penalty parameter β must be large enough so that the compressibility error is negligible (i.e., I_3 is approximately equal to 1), yet not so large that numerical ill-conditioning occurs. Numerical experiments reveal that $\beta=10^3\times\max(c_1,\ c_2)$ to $\beta=10^7\times\max(c_1,\ c_2)$ is adequate for a floating-point word length of 64 bits. The constant p_0 is chosen so that the components of $\mathbf{S}$ are all zero in the initial configuration, that is, $p_0=-(c_1+2c_2)$. See Section 6.3 for a description of Lagrange multipliers and penalty methods.

5.5 One-Dimensional Plasticity

Materials for which permanent strains are developed upon unloading are called plastic materials. Many materials such as metals, soils and concrete exhibit elastic behavior up to a stress called the yield strength. Once loaded beyond the initial yield strength, plastic strains are developed. Elastic–plastic materials are further classified as rate-independent materials, where the stress is independent of the strain rate, and rate-dependent materials in which the stress depends on the strain rate. The latter are also referred to as rate-sensitive materials.

The major ingredients of the theory of plasticity are:

1. A decomposition of each increment of strain into an elastic, reversible part $d\varepsilon^e$, and an irreversible plastic part $d\varepsilon^p$.
2. A yield function $f(\sigma, q_\alpha)$ which governs the onset and continuance of plastic deformation; q_α are a set of internal variables.
3. A flow rule which governs the plastic flow, that is, determines the plastic strain increments.
4. Evolution equations for internal variables, including a strain-hardening relation which governs the evolution of the yield function.

Elastic–plastic laws are path-dependent and dissipative. A large part of the work expended in plastically deforming the material is irreversibly converted to other forms of energy, particularly heat. The stress depends on the entire history of the deformation, and can not be written as a single-valued function of the strain; rather it can only be specified as a relation between rates of stress and strain.

5.5.1 Rate-Independent Plasticity in One Dimension

A typical stress–strain curve for a metal under uniaxial stress is shown in Figure 5.8. Upon initial loading, the material behaves elastically (usually assumed linear) until the initial yield stress is attained. The elastic regime is followed by an elastic–plastic regime where permanent irreversible plastic strains are induced upon further loading. Reversal of the stress is called unloading. In unloading, the stress–strain response is assumed to be governed by the

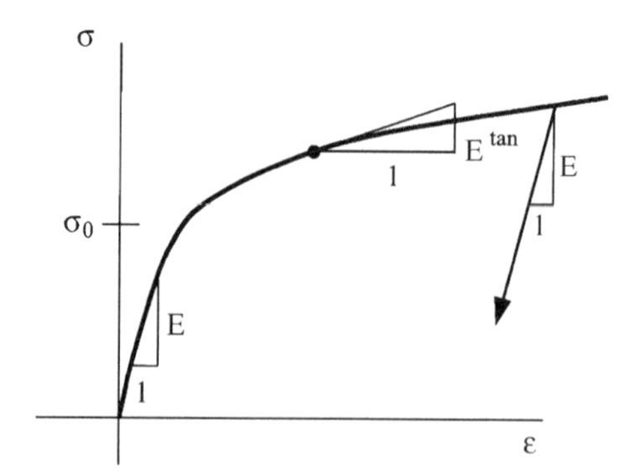

Figure 5.8 Stress–strain curve for typical elastic–plastic material

elastic law. The increments in strain are assumed to be *additively decomposed into elastic and plastic parts*. Thus we write

$$d\varepsilon = d\varepsilon^e + d\varepsilon^p \tag{5.5.1}$$

Dividing both sides of this equation by a differential time increment dt gives the rate form

$$\dot{\varepsilon} = \dot{\varepsilon}^e + \dot{\varepsilon}^p \tag{5.5.2}$$

The stress increment (rate) is always related by the elastic modulus to the increment (rate) of elastic strain:

$$d\sigma = E d\varepsilon^e, \; \dot{\sigma} = E \dot{\varepsilon}^e \tag{5.5.3}$$

In the nonlinear elastic–plastic regime, the stress–strain relation is given by

$$d\sigma = E d\varepsilon^e, \; E^{\tan} d\varepsilon, \; \dot{\sigma} = E \dot{\varepsilon}^e = E^{\tan} \dot{\varepsilon} \tag{5.5.4}$$

where the elastic–plastic tangent modulus, $E^{\tan}$, is the slope of the stress–strain curve (Figure 5.8).

These relations are homogeneous in the rates of stress and strain, so if time is scaled by an arbitrary factor, the constitutive relation remains unchanged. Therefore the material response is *rate-independent* even though it is expressed in terms of a strain rate. In the sequel, the rate form of the constitutive relations will be used, as the notation for the incremental relations can get cumbersome, especially for large-strain formulations.

The plastic strain rate is given by a flow rule which is often specified in terms of a plastic flow potential Ψ:

$$\dot{\varepsilon}^p = \dot{\lambda} \frac{\partial \Psi}{\partial \sigma} \tag{5.5.5}$$

where $\dot{\lambda}$ is called the plastic rate parameter. An example of a flow potential is

$$\Psi = |\sigma| = \bar{\sigma} = \sigma \operatorname{sign}(\sigma), \; \frac{\partial \Psi}{\partial \sigma} = \operatorname{sign}(\sigma) \tag{5.5.6}$$

where $\bar{\sigma}$ is called the effective stress and the sign function is defined in the Glossary.

The yield condition is

$$f = \bar{\sigma} - \sigma_Y(\bar{\varepsilon}) = 0 \tag{5.5.7}$$

where $\sigma_Y(\bar{\varepsilon})$ is the yield strength in uniaxial tension and $\bar{\varepsilon}$ is the effective plastic strain. The increase of the yield strength after initial yield is called work hardening or strain hardening. The hardening behavior of the material is generally a function of the prior history of plastic

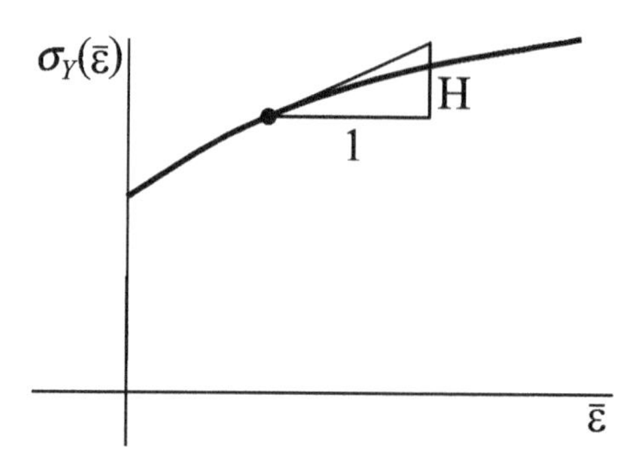

Figure 5.9 Typical hardening curve, $\sigma_Y(\bar{\varepsilon})$. The plastic modulus is $\mathrm{H} = \mathrm{d}\sigma_Y / \mathrm{d}\bar{\varepsilon}$

deformation. In metal plasticity, the history of plastic deformation is often characterized by the effective plastic strain which is given by

$$\bar{\varepsilon} = \int \dot{\bar{\varepsilon}}\, dt, \ \dot{\bar{\varepsilon}} = \sqrt{\dot{\varepsilon}^P \dot{\varepsilon}^P} \tag{5.5.8}$$

where $\dot{\bar{\varepsilon}}$ is the effective plastic strain rate. The effective plastic strain, $\bar{\varepsilon}$, is an example of an internal variable which is used to characterize the inelastic response of the material. An alternative internal variable for the representation of hardening is the plastic work (Hill, 1950) which is given by $W^P = \int \sigma \dot{\varepsilon}^P dt$.

The yield behavior given by (5.5.7) is called isotropic hardening: the yield strengths in tension and compression are always equal and are given by $\sigma_Y(\bar{\varepsilon})$. A typical hardening curve is shown in Figure 5.9. The slope of this curve is the plastic modulus, H, that is, $H = d\sigma_Y(\bar{\varepsilon}) / d\bar{\varepsilon}$. An extension of the model to the case of kinematic hardening is given next. More general constitutive relations use additional internal variables.

For this particular model, combining (5.5.6), (5.5.5) and (5.5.8) yields $\dot{\lambda} = \dot{\bar{\varepsilon}}$. Therefore the plastic strain rate (5.5.5) can be written as

$$\dot{\varepsilon}^P = \dot{\bar{\varepsilon}} \operatorname{sign}(\sigma) = \dot{\bar{\varepsilon}} \frac{\partial f}{\partial \sigma} \tag{5.5.9}$$

where the result $\partial f / \partial \sigma = \partial \bar{\sigma} / \partial \sigma = \operatorname{sign}(\sigma)$ has been used. Here we see that $\partial f / \partial \sigma = \partial \Psi / \partial \sigma$. Plastic models for which $\partial f / \partial \sigma \sim \partial \Psi / \partial \sigma$ are called *associative*; otherwise plastic flow is said to be *non-associative*. For associative plasticity, plastic flow is in the direction normal to the yield surface. These distinctions are important in multiaxial plasticity models and will be elaborated upon in Section 5.6.

Plastic deformation occurs only when the yield condition $f=0$ is met. During plastic loading, the stress must remain on the yield surface, so $\dot{f} = 0$ Enforcement of this leads to the *consistency* condition:

$$\dot{f} = \dot{\bar{\sigma}} - \dot{\sigma}_Y(\bar{\varepsilon}) = 0 \tag{5.5.10}$$

which gives

$$\dot{\bar{\sigma}} = \frac{d\sigma_Y(\bar{\varepsilon})}{d\bar{\varepsilon}}\dot{\bar{\varepsilon}} = H\dot{\bar{\varepsilon}} \tag{5.5.11}$$

where $H = d\sigma_Y(\bar{\varepsilon})/d\bar{\varepsilon}$ is called the plastic modulus. Using (5.5.2), (5.5.4) (5.5.11) and (5.5.5) it follows that

$$\frac{1}{E^{\tan}} = \frac{1}{E} + \frac{1}{H}, \quad \text{or} \quad E^{\tan} = \frac{EH}{E+H} = E - \frac{E^2}{E+H} \tag{5.5.12}$$

Consider a plastic switch parameter β with $\beta=1$ corresponding to plastic loading and $\beta=0$ corresponding to purely elastic response (loading or unloading). Then the tangent modulus is

$$E^{\tan} = E - \beta\frac{E^2}{E+H} \tag{5.5.13}$$

The loading-unloading conditions can also be stated as

$$\dot{\lambda} \geq 0, \quad f \leq 0, \quad \dot{\lambda}f = 0 \tag{5.5.14}$$

which are sometimes referred to as Kuhn–Tucker conditions. The first of these indicates that the plastic rate parameter is non-negative while the second indicates that the stress state must lie on or within the yield surface. The last condition assures that the stress lies on the yield surface during plastic loading, $\dot{\lambda} > 0$. This latter condition can also be stated in rate form through what is known as the *consistency* condition, $\dot{f} = 0$. For plastic loading ($\dot{\lambda} > 0$) the stress state must remain on the yield surface, f=0, therefore $\dot{f} = 0$. For elastic loading or unloading $\dot{\lambda} = 0$, that is, there is no plastic flow.

5.5.2 *Extension to Kinematic Hardening*

In cyclic loading, an isotropic hardening model provides a poor representation of the stress–strain response for many metals. For example, Figure 5.10 illustrates a phenomenon observed in cyclic plasticity known as the Bauschinger effect, in which the yield strength in compression is reduced following an initial yield in tension. A way to conceptualize this behavior is to observe that the center of the yield surface moves in the direction of the plastic flow. The schematic in Figure 5.10(b) is actually for multiaxial stress states – expansion of the circular yield surface corresponds to isotropic hardening and translation of its center corresponds to kinematic hardening. Prager (1945) and Ziegler (1950) introduced a simple kinematic

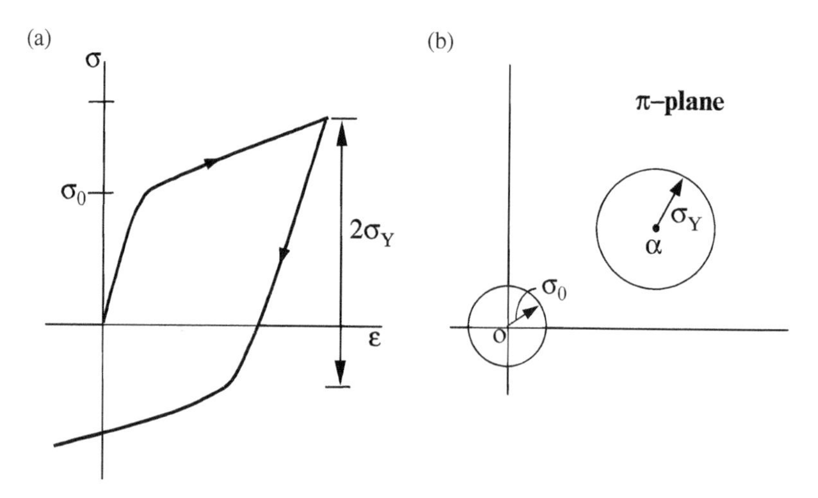

Figure 5.10 Combined isotropic-kinematic hardening illustrating (a) Bauschinger effect (b) translation and expansion of yield surface

hardening plasticity model to account for this phenomenon. In the kinematic hardening model another internal variable α, called the backstress, is introduced into both the plastic flow relation and the yield condition as follows:

Plastic flow law

$$\dot{\varepsilon}^p = \dot{\lambda}\frac{\partial \Psi}{\partial \sigma}, \Psi = |\sigma - \alpha| \tag{5.5.15}$$

Yield condition

$$f = |\sigma - \alpha| - \sigma_Y(\bar{\varepsilon}) \tag{5.5.16}$$

Note that $\partial \Psi/\partial \sigma = \partial f/\partial \sigma = \text{sign}(\sigma - \alpha)$ and also, from (5.5.8), $\dot{\bar{\varepsilon}} = \dot{\lambda}$. An evolution equation for the internal variable α (backstress) is required. The simplest form is called linear kinematic hardening and is specified as $\dot{\alpha} = \kappa\dot{\varepsilon}^p$.

The consistency condition is given by differentiating (5.5.16):

$$\dot{f} = (\dot{\sigma} - \dot{\alpha})\text{sign}(\sigma - \alpha) - H\dot{\bar{\varepsilon}} = 0 \tag{5.5.17}$$

and thus

$$\dot{\bar{\varepsilon}} = \frac{1}{H}(\dot{\sigma} - \dot{\alpha})\text{sign}(\sigma - \alpha) \tag{5.5.18}$$

From (5.5.3), (5.5.15) and (5.5.18)

$$\dot{\sigma} = E(\dot{\varepsilon} - \dot{\varepsilon}^p) = E(\dot{\varepsilon} - \dot{\bar{\varepsilon}}\,\text{sign}(\sigma - \alpha)) \tag{5.5.19}$$

Subtracting $\dot{\alpha}$

$$\begin{aligned}(\dot{\sigma} - \dot{\alpha}) &= E(\dot{\varepsilon} - \dot{\bar{\varepsilon}}\,\text{sign}(\sigma - \alpha)) - \dot{\alpha} \\ &= E(\dot{\varepsilon} - \dot{\bar{\varepsilon}}\,\text{sign}(\sigma - \alpha)) - \kappa\dot{\bar{\varepsilon}}\,\text{sign}(\sigma - \alpha)\end{aligned} \tag{5.5.20}$$

where (5.5.15) and the expression for $\dot{\alpha}$ have been used. From (5.5.18) and (5.5.20), an expression for the effective plastic strain rate is obtained:

$$\dot{\bar{\varepsilon}} = \frac{1}{H}[E\dot{\varepsilon}\,\text{sign}(\sigma - \alpha) - E\dot{\bar{\varepsilon}} - \kappa\dot{\bar{\varepsilon}}] = \frac{E\dot{\varepsilon}\,\text{sign}(\sigma - \alpha)}{E + (H + \kappa)} \tag{5.5.21}$$

Substituting this expression into (5.5.19) gives $\dot{\sigma} = E^{\tan}\dot{\varepsilon}$ where

$$E^{\tan} = \frac{E(H + \kappa)}{(H + \kappa) + E} = E - \frac{E^2}{E + (H + \kappa)} \tag{5.5.22}$$

during plastic loading. For elastic loading or unloading, the tangent modulus is simply the elastic modulus, $E^{tan} = E$. The loading unloading conditions can alternatively be written in terms of the Kuhn–Tucker conditions (5.5.14). The constitutive equations for rate-independent plasticity in one dimension with combined isotropic and (linear) kinematic hardening are summarized in Box 5.3.

Box 5.3 Constitutive relations for one-dimensional rate-independent plasticity with combined isotropic and kinematic hardening

1. Strain rate:

$$\dot{\varepsilon} = \dot{\varepsilon}^e + \dot{\varepsilon}^p \tag{B5.3.1}$$

2. Stress rate:

$$\dot{\sigma} = E\dot{\varepsilon}^e = E(\dot{\varepsilon} - \dot{\varepsilon}^p) \tag{B5.3.2}$$

3. Plastic flow rule:

$$\dot{\varepsilon}^p = \dot{\lambda}\frac{\partial \Psi}{\partial \sigma}, \quad \dot{\bar{\varepsilon}} = \dot{\lambda}, \quad \sigma' = \sigma - \alpha, \quad \Psi = |\sigma'| \tag{B5.3.3}$$

4. Evolution equation for backstress:

$$\dot{\alpha} = \kappa\dot{\varepsilon}^{p} \tag{B5.3.4}$$

5. Yield condition:

$$f = |\sigma - \alpha| - \sigma_Y(\bar{\varepsilon}) = 0 \tag{B5.3.5}$$

6. Loading–unloading conditions:

$$\dot{\lambda} \geq 0, \quad f \leq 0, \quad \dot{\lambda} f = 0 \tag{B5.3.6}$$

7. Consistency condition:

$$\dot{f} = 0 \Rightarrow \dot{\bar{\varepsilon}} \equiv \dot{\lambda} = \frac{E\dot{\varepsilon}\,\text{sign}\,\sigma'}{E + H + \kappa} \tag{B5.3.7}$$

8. Tangent modulus:

$$\dot{\sigma} = E^{\tan}\dot{\varepsilon}; \quad E^{\tan} = E - \beta\frac{E^2}{E + (H + \kappa)} \tag{B5.3.8}$$

($\beta = 1$ for plastic loading, $\beta = 0$ for elastic loading or unloading.)

5.5.3 Rate-Dependent Plasticity in One Dimension

In rate-dependent plasticity, the plastic response of the material depends on the rate of loading. The elastic response is given as before (in rate form) by

$$\dot{\sigma} = E\dot{\varepsilon}^{e} = E(\dot{\varepsilon} - \dot{\varepsilon}^{p}) \tag{5.5.23}$$

For plastic deformation to occur, the yield condition must be met or exceeded, in contrast to rate-independent plasticity where the yield condition can not be exceeded. The plastic strain rate is given by (combined isotropic and kinematic hardening)

$$\dot{\varepsilon}^{p} = \dot{\lambda}\frac{\partial \Psi}{\partial \sigma}, \dot{\bar{\varepsilon}} = \dot{\lambda}, \sigma' = \sigma - \alpha, \Psi = |\sigma'| \tag{5.5.24}$$

For rate-dependent materials, one way to describe the plastic response is by an overstress model

$$\dot{\bar{\varepsilon}} = \frac{\phi(\sigma, \bar{\varepsilon}, \alpha)}{\eta} \tag{5.5.25}$$

where ϕ is the overstress and η is the viscosity. In an overstress model, the rate of effective plastic strain depends on by how much the yield stress is exceeded. Instead of the consistency condition (5.5.17–18) to obtain $\dot{\varepsilon}$, the effective plastic strain rate is given by an empirical law. For example, the overstress model by Perzyna (1971) is

$$\phi = \sigma_Y \left\langle \frac{|\sigma - \alpha|}{\sigma_Y} - 1 \right\rangle^n \tag{5.5.26}$$

where n is called the rate-sensitivity exponent. Plastic straining occurs when the yield condition, $|\sigma - \alpha| - \sigma_Y(\bar{\varepsilon}) = 0$ is exceeded. This is incorporated through the use of the Macaulay bracket $\langle f \rangle = f$ if $f > 0$, $\langle f \rangle = 0$ if $f \leq 0$. Using (5.5.2–3) and (5.5.24–25) the expression for the stress rate is given by

$$\dot{\sigma} = E\left(\dot{\varepsilon} - \frac{\phi(\sigma, \bar{\varepsilon}, \alpha)}{\eta}\text{sign}(\sigma')\right) \tag{5.5.27}$$

which is a differential equation for the evolution of the stress. Comparing this expression to (5.5.5), it can be seen that (5.5.27) is inhomogeneous in the rates and therefore the material response is *rate-dependent*. More elaborate models with additional internal variables are given in Lubliner (1990) and Khan and Huang (1995).

An alternative form of the effective plastic strain rate in rate-dependent plasticity, given by Peirce, Shih and Needleman (1984), is

$$\dot{\bar{\varepsilon}} = \dot{\varepsilon}_0 \left(\frac{|\sigma - \alpha|}{\sigma_Y(\bar{\varepsilon})}\right)^{1/m} \tag{5.5.28}$$

This model does not include an explicit yield surface. For plastic straining at the rate $\dot{\varepsilon}_0$, the reference response $|\sigma - \alpha| = \sigma_Y$ is obtained. For rates which exceed $\dot{\varepsilon}_0$, the stress is elevated above the reference stress, σ_Y, while for lower rates the stress falls below the reference stress. A case of particular interest is the near rate-independent limit when the rate-sensitivity exponent $m \to 0$. It can be seen from (5.5.28) (with $m \to 0$) that, for $|\sigma - \alpha| < \sigma_Y$, the effective plastic strain rate is negligible, while for a finite plastic strain rate $|\sigma - \alpha|$ is approximately equal to the reference stress, σ_Y. In this way, the model exhibits yielding together with near-elastic unloading and rate-independent response. The constitutive relations for rate-dependent plasticity in one dimension are summarized in Box 5.4.

Box 5.4 Constitutive equations for rate-dependent plasticity in one dimension with combined isotropic and (linear) kinematic hardening

1. Strain rate:

$$\dot{\varepsilon} = \dot{\varepsilon}^e + \dot{\varepsilon}^{ep} \tag{B5.4.1}$$

2. Stress rate:

$$\dot{\sigma} = E\dot{\varepsilon}^e = E(\dot{\varepsilon} - \dot{\varepsilon}^p) \tag{B5.4.2}$$

3. Plastic flow rule:

$$\sigma' = \sigma - \alpha, \quad \Psi = |\sigma'| \tag{B5.4.3}$$

$$\dot{\varepsilon}^p = \dot{\lambda}\frac{\partial \Psi}{\partial \sigma'} = \dot{\bar{\varepsilon}}\,\text{sign}(\sigma') \tag{B5.4.4}$$

4. Overstress function:

$$\dot{\bar{\varepsilon}} = \frac{\phi}{\eta}(\sigma, \bar{\varepsilon}, \alpha) \quad \text{e.g. } \phi = \sigma_Y \left\langle \frac{|\sigma - \alpha|}{\sigma_Y} - 1 \right\rangle^n \tag{B5.4.5}$$

5. Evolution equations for stress:

$$\dot{\sigma} = E\left(\dot{\varepsilon} - \frac{\phi}{\eta}(\sigma, \bar{\varepsilon}, \alpha)\,\text{sign}(\sigma - \alpha)\right) \tag{B5.4.6}$$

6. Effective plastic strain rate:

$$\dot{\bar{\varepsilon}} = \frac{\phi}{\eta}(\sigma, \bar{\varepsilon}, \alpha) \tag{B5.4.7}$$

7. Backstress:

$$\dot{\alpha} = \kappa \dot{\bar{\varepsilon}}\,\text{sign}(\sigma') \tag{B5.4.8}$$

where, in steps 5–7, the representation of the constitutive relation as a set of differential (evolution) equations for the stress and internal variables is emphasized.

5.6 Multiaxial Plasticity

The one-dimensional plastic constitutive relations presented in Section 5.5 will now be generalized to the multiaxial case. We begin with a general treatment of hypoelastic-plastic constitutive relations for large strains. These formulations are typically based on an additive decomposition of the rate-of-deformation tensor into elastic and plastic parts and the elastic response is taken as hypoelastic. Special forms such as the J_2 flow theory of metal plasticity, the Drucker-Prager model for soil plasticity and the Gurson constitutive model for porous plastic solids are then given. The reduction of the general large-strain formulation to the case of small strains is given as a special case. Modifications of the rate-independent results for the case of rate-dependent plasticity (viscoplasticity) are given. Large-deformation plasticity formulations based on a multiplicative decomposition of the deformation gradient into elastic and plastic parts are then discussed. The elastic-plastic behavior is based on a hyperelastic representation of the elastic response. The special case of single-crystal plasticity is also considered.

5.6.1 Hypoelastic-Plastic Materials

Hypoelastic-plastic models are typically used when elastic strains are small compared to the plastic strains. As discussed in Section 5.5, energy is not conserved in a closed deformation cycle for hypoelastic materials. However, for small elastic strains, the energy error is insignificant and hypoelastic descriptions of the elastic response are often adequate. In these constitutive models, the additive decomposition of the rate-of-deformation tensor, $\mathbf{D}$, into elastic and plastic parts is assumed:

$$\mathbf{D} = \mathbf{D}^e + \mathbf{D}^p \tag{5.6.1}$$

The elastic response is hypoelastic: a suitable objective rate of stress is related to the *elastic* part of the rate-of-deformation tensor (in pure hypoelasticity, considered in Section 5.5.1, the stress rate is related to the *total* rate-of-deformation tensor – see (5.4.24)). The choice of objective stress rate in the constitutive response depends on several factors. As seen in Box 5.1, the Jaumann rate of the Cauchy (true) stress leads to nonsymmetric tangent stiffness matrices, while the Jaumann rate of the Kirchhoff stress $\boldsymbol{\tau}=J\boldsymbol{\sigma}$ can lead to symmetric stiffness matrices. We will identify the circumstances under which these choices are appropriate and what advantages are to be gained by the choice of one rate over another. The choice of objective stress rate should not be confused with the expression of a *given* constitutive relation in terms of different stress rates. The latter is accomplished by simply using the appropriate transformation between rates.

We first present a model based on the Cauchy stress with elastic response specified in terms of the Jaumann rate. The elastic response is specified by applying the hypoelastic law (5.4.24) to the elastic part of the rate-of-deformation,

$$\boldsymbol{\sigma}^{\nabla J} = \mathbf{C}^{\sigma J}_{el}:\mathbf{D}^e = \mathbf{C}^{\sigma J}_{el}:(\mathbf{D} - \mathbf{D}^p) \tag{5.6.2}$$

If the elastic moduli, $\mathbf{C}^{\sigma J}_{el}$, are taken to be constant, they must be isotropic in order to satisfy the principle of material frame indifference (Section 5.10).

The rate of plastic flow is given by

$$\mathbf{D}^p = \dot{\lambda}\mathbf{r}(\boldsymbol{\sigma},\mathbf{q}), \quad D^p_{ij} = \dot{\lambda}r_{ij}(\boldsymbol{\sigma},\mathbf{q}) \tag{5.6.3}$$

where $\dot{\lambda}$ is a scalar plastic flow rate and $\mathbf{r}(\boldsymbol{\sigma}, \mathbf{q})$ is the plastic flow direction. The plastic flow direction is often specified as $\mathbf{r}=\partial\psi/\partial\boldsymbol{\sigma}$ where ψ is called the *plastic flow potential.* To avoid confusion of the plasticity parameter with the Lamé constant, the Lamé constant will subsequently be denoted as λ^e. The plastic flow direction depends on the Cauchy stress, $\boldsymbol{\sigma}$, and on a set of internal variables denoted collectively as $\mathbf{q}$. Examples of scalar internal variables are the accumulated effective plastic strain and the void volume fraction. The backstress in kinematic hardening models is an example of an internal variable which is a second-order tensor.

Evolution equations for the internal variables are required and, for most plasticity models, can be specified as

$$\dot{\mathbf{q}} = \dot{\lambda}\mathbf{h}(\boldsymbol{\sigma},\mathbf{q}), \quad \dot{q}_\alpha = \dot{\lambda}h_\alpha(\boldsymbol{\alpha},\mathbf{q}) \tag{5.6.4}$$

where α ranges over the number of internal variables. Here, the internal variables are a collection of scalars and the material time derivative is an objective rate. Note that the plastic parameter λ or some function of it may be one of the internal variables. The evolution equation for the plastic parameter is obtained through the consistency condition below. The yield condition is

$$f(\boldsymbol{\sigma},\mathbf{q}) = 0 \tag{5.6.5}$$

As in the one-dimensional case, the loading-unloading conditions are given by

$$\dot{\lambda} \geq 0, \quad f \leq 0, \quad \dot{\lambda} f = 0 \tag{5.6.6}$$

During plastic loading ($\dot{\lambda} > 0$) the stress is required to remain on the yield surface $f=0$. This can be also be stated in terms of the *consistency* condition $\dot{f} = 0$ which can be expanded by the chain rule to give

$$\dot{f} = f_{\sigma} : \dot{\boldsymbol{\sigma}} + f_{\mathbf{q}} \cdot \dot{\mathbf{q}} = 0 \quad \dot{f} = (f_{\sigma})_{ij} \dot{\sigma}_{ij} + (f_{q})_{\alpha} \dot{q}_{\alpha} = 0 \tag{5.6.7}$$

where we have adopted the notation $f_{\sigma} = \partial f / \partial \boldsymbol{\sigma}$ and $f_{q} = \partial f / \partial \mathbf{q}$.

The consistency condition involves the normal, f_{σ}, to the yield surface. If the plastic flow direction is proportional to the normal to the yield surface that is, $\mathbf{r} \sim f_{\sigma}$, plastic flow is said to be *associative*, otherwise it is said to be *non-associative*. When the flow direction is given by the derivative of a plastic flow potential, the condition for associative plasticity is $\psi_{\sigma} \sim f_{\sigma}$. For many materials, an appropriate choice for the plastic potential is $\psi = f$ which gives rise to an associative flow rule. Drucker has shown that, when the yield surface is convex, associative plasticity models are stable for small strains if the strain hardening is positive.

Several useful results regarding the Jaumann rate are given in Chapter 3, Exercise 3.6. From these results it follows that if f_{σ} and $\boldsymbol{\sigma}$ commute, that is,

$$f_{\sigma} \cdot \boldsymbol{\sigma} = \boldsymbol{\sigma} \cdot f_{\sigma} \tag{5.6.8}$$

then

$$f_{\sigma} : \dot{\boldsymbol{\sigma}} = f_{\sigma} : \boldsymbol{\sigma}^{\nabla J} \tag{5.6.9}$$

(Prager, 1961). Referring to Box 5.2 for the derivatives of the principal invariants of a second-order tensor, it can be seen that when f is a function of the invariants of the stress, f_{σ} and $\boldsymbol{\sigma}$ commute and thus (5.6.8) and (5.6.9) hold. It is shown in Section 5.10 that objectivity requires yield functions of the form (5.6.5) to be isotropic functions of the stress and hence functions of the principal invariants of stress. For example, the von Mises yield function depends on the second invariant of the deviatoric stress, $I_2(\boldsymbol{\sigma}^{\text{dev}}) \equiv -J_2 = -\frac{1}{2}\boldsymbol{\sigma}^{\text{dev}} : \boldsymbol{\sigma}^{\text{dev}}$. Substituting (5.6.9) into (5.6.7) gives

$$\dot{f} = f_{\sigma} : \boldsymbol{\sigma}^{\nabla J} + f_{\mathbf{q}} \cdot \dot{\mathbf{q}} = 0 \tag{5.6.10}$$

Using the hypoelastic relation (5.6.2), the plastic flow relation (5.6.3) and the evolution equations (5.6.4) in (5.6.10) gives

$$0 = f_{\sigma}:\mathbf{C}_{el}^{\sigma J}:(\mathbf{D}-\mathbf{D}^{p}) + f_{\mathbf{q}}\cdot\dot{\mathbf{q}} = f_{\sigma}:\mathbf{C}_{el}^{\sigma J}:(\mathbf{D}-\dot{\lambda}\mathbf{r}) + f_{\mathbf{q}}\cdot\dot{\lambda}\mathbf{h} \tag{5.6.11}$$

which can be solved for $\dot{\lambda}$ to give

$$\dot{\lambda} = \frac{f_{\sigma}:\mathbf{C}_{el}^{\sigma J}:\mathbf{D}}{-f_{\mathbf{q}}\cdot\mathbf{h} + f_{\sigma}:\mathbf{C}_{el}^{\sigma J}:\mathbf{r}} \tag{5.6.12}$$

Equation (5.6.9) also holds for other spin-based rates of stress (see Exercise 3.6) but does not hold for the Truesdell rate, for example. When the elastic response is specified in terms of a spin-based rate, the simplification (5.6.11) occurs. With the Truesdell rate, additional terms need to be accounted for.

Substituting (5.6.12) together with the plastic flow rule (5.6.3) into (5.6.2), we obtain a relation between the Jaumann rate of Cauchy stress and the total rate-of-deformation tensor:

$$\boldsymbol{\sigma}^{\nabla J} = \mathbf{C}_{el}^{\sigma J}:(\mathbf{D}-\dot{\lambda}\mathbf{r}) = \mathbf{C}_{el}^{\sigma J}:\left(\mathbf{D} - \frac{f_{\sigma}:\mathbf{C}_{el}^{\sigma J}:\mathbf{D}}{-f_{\mathbf{q}}\cdot\mathbf{h} + f_{\sigma}:\mathbf{C}_{el}^{\sigma J}:\mathbf{r}}\mathbf{r}\right) = \mathbf{C}^{\sigma J}:\mathbf{D} \tag{5.6.13}$$

The fourth-order tensor $\mathbf{C}^{\sigma J}$ is called the *continuum* elasto-plastic tangent modulus and is obtained by rearrangement of the expression in (5.6.13):

$$\mathbf{C}^{\sigma J} = \mathbf{C}_{el}^{\sigma J} - \frac{(\mathbf{C}_{el}^{\sigma J}:\mathbf{r})\otimes(f_{\sigma}:\mathbf{C}_{el}^{\sigma J})}{-f_{\mathbf{q}}\cdot\mathbf{h} + f_{\sigma}:\mathbf{C}_{el}^{\sigma J}:\mathbf{r}},\quad C_{ijkl}^{\sigma J} = \left(C_{el}^{\sigma J}\right)_{ijkl} - \frac{\left(C_{el}^{\sigma J}\right)_{ijmn} r_{mn}(f_{\sigma})_{pq}\left(C_{el}^{\sigma J}\right)_{pqkl}}{-(f_{q})_{\alpha}h_{\alpha} + (f_{\sigma})_{rs}\left(C_{el}^{\sigma J}\right)_{rstu} r_{tu}} \tag{5.6.14}$$

The symbol ⊗ denotes the tensor or open product and is defined in the Glossary. The elasto-plastic tangent modulus consists of the elastic tangent modulus and a term due to plastic flow. When written in Voigt matrix form, the plastic flow contribution is a matrix of rank 1 and is often referred to as a rank 1 correction (to the elastic moduli). From symmetry of the stress rate and the rate-of-deformation, the elasto-plastic tangent modulus $\mathbf{C}^{\sigma J}$ has both minor symmetries. It has major symmetry, $C_{ijkl}^{\sigma J} = C_{klij}^{\sigma J}$, when $\mathbf{C}_{el}^{\sigma J}:\mathbf{r} \sim f_{\sigma}:\mathbf{C}_{el}^{\sigma J}$, or, alternatively, when plastic flow is associative, that is, $\mathbf{r} \sim f_{\sigma}$ (major symmetry of the elastic moduli is assumed). The above equations are summarized in Box 5.5.

Box 5.5 Hypoelastic–plastic constitutive model (Cauchy stress formulation)

1. Rate-of-deformation tensor:

$$\mathbf{D} = \mathbf{D}^{e} + \mathbf{D}^{p} \tag{B5.5.1}$$

2. Stress rate:

$$\boldsymbol{\sigma}^{\nabla J} = \mathbf{C}_{el}^{\sigma J}:\mathbf{D}^{e} = \mathbf{C}_{el}^{\sigma J}:(\mathbf{D}-\mathbf{D}^{p}) \tag{B5.5.2}$$

3. Plastic flow rule and evolution equations:

$$\mathbf{D}^p = \dot{\lambda}\mathbf{r}(\boldsymbol{\sigma},\mathbf{q}) \quad \dot{\mathbf{q}} = \dot{\lambda}\mathbf{h}(\boldsymbol{\sigma},\mathbf{q}) \tag{B5.5.3}$$

4. Yield condition:

$$f(\boldsymbol{\sigma},\mathbf{q}) = 0 \tag{B5.5.4}$$

5. Loading–unloading conditions:

$$\dot{\lambda} \geq 0, \quad f \leq 0, \quad \dot{\lambda} f = 0 \tag{B5.5.5}$$

6. Plastic rate parameter (consistency condition):

$$\dot{\lambda} = \frac{f_{\sigma} : \mathbf{C}_{el}^{\sigma J} : \mathbf{D}}{-f_{\mathbf{q}} \cdot \mathbf{h} + f_{\mathbf{q}} : \mathbf{C}_{el}^{\sigma J} : \mathbf{r}} \tag{B5.5.6}$$

7. Stress rate – total rate-of-deformation relation:

$$\boldsymbol{\sigma}^{\nabla J} = \mathbf{C}^{\sigma J} : \mathbf{D} \; \sigma_{ij}^{\nabla J} = C_{ijkl}^{\sigma J} D_{kl} \tag{B5.5.7}$$

During elastic loading or unloading, $\mathbf{C}^{\sigma J} = \mathbf{C}_{el}^{\sigma J}$; for plastic loading $\mathbf{C}^{\sigma J}$ is given by the continuum elasto-plastic tangent modulus

$$\mathbf{C}^{\sigma J} = \mathbf{C}_{el}^{\sigma J} - \frac{\left(\mathbf{C}_{el}^{\sigma J} : \mathbf{r}\right) \otimes \left(f_{\sigma} : \mathbf{C}_{el}^{\sigma J}\right)}{-f_{\mathbf{q}} \cdot \mathbf{h} + f_{\sigma} : \mathbf{C}_{el}^{\sigma J} : \mathbf{r}}$$

$$C_{ijkl}^{\sigma J} = \left(C_{el}^{\sigma J}\right)_{ijkl} - \frac{\left(C_{el}^{\sigma J}\right)_{ijmn} r_{mn} (f_{\sigma})_{pq} \left(C_{el}^{\sigma J}\right)_{pqkl}}{-(f_q)_{\alpha} h_{\alpha} + (f_{\sigma})_{rs} \left(C_{el}^{\sigma J}\right)_{rstu} r_{tu}} \tag{B5.5.8}$$

The elasto-plastic tangent modulus $\mathbf{C}^{\sigma J}$ in (B5.5.8) has major symmetry when plastic flow is associative. However, from Box 5.1, it can be seen that the corresponding modulus $\mathbf{C}^{\sigma T}$ for the Truesdell rate of Cauchy stress (which appears in the linearized weak form in Chapter 6) is not symmetric. This lack of symmetry is a result of basing the plastic flow equations on the Cauchy stress. If the plasticity equations are formulated in terms of the Kirchhoff stress and if plastic flow is associative, $\mathbf{C}^{\sigma T}$ will have major symmetry (Box 5.1).

The Cauchy stress is preferred in the plastic yield function and flow rule because it is the true stress. For plastic constitutive relations where plastic flow is essentially isochoric (volume-preserving) we have $\mathbf{J} \approx 1$ (elastic strains are small) and the Kirchhoff stress is virtually indistinguishable from the Cauchy stress. This is the case for a broad class of metals described by classical J_2 flow theory, where experiments show that plastic strains produce little or no volume change.

For dilatant materials and porous plastic solids, as in the Gurson model (next), large dilations accompany plastic deformations and the assumption $J \approx 1$ is no longer valid. In this case, the yield function is preferably expressed in terms of the Cauchy stress and the resulting tangent stiffness is not symmetric. The Kirchhoff stress formulation is analogous to the Cauchy stress formulation and may be obtained from Box 5.5 by replacing the Cauchy stress with the Kirchhoff stress everywhere. The specific case of J_2 flow theory plasticity is illustrated in the following section.

The restriction to isotropy of both the elastic moduli and the yield function is discussed further in Section 5.10. This restriction is a limitation of hypoelastic–plastic constitutive models based on the Cauchy (or Kirchhoff) stresses. It will be seen later that the hypoelastic–plastic model based on the corotational stress and the hyperelastic–plastic model formulated on the intermediate configuration are not limited to isotropic response in this way.

5.6.2 *J_2 Flow Theory Plasticity*

A special case of the general model presented above is the J_2 flow model based on a von Mises yield surface. This model is especially useful for metal plasticity, for which it was developed. For a thorough discussion of J_2 flow theory plasticity see Lubliner (1990). The key assumption of the model is that plastic flow in metals is unaffected by pressure; this was experimentally demonstrated by Bridgman (1949). The yield condition and plastic flow direction are based on the deviatoric part of the stress tensor. The von Mises effective stress is used to generalize the observed behavior in uniaxial stress (alternative treatments generalize the behavior in shear) to multiaxial stress states.

The Kirchhoff stress formulation of J_2 flow theory plasticity is given in Box 5.6.

Box 5.6 J_2 flow theory hypoelastic–plastic constitutive model

1. Rate-of-deformation tensor:

$$\mathbf{D} = \mathbf{D}^e + \mathbf{D}^p \tag{B5.6.1}$$

2. Stress rate relation:

$$\boldsymbol{\tau}^{\nabla J} = \mathbf{C}^{\tau J}_{el} : \mathbf{D}^e = \mathbf{C}^{\tau J}_{el} : (\mathbf{D} - \mathbf{D}^p) \tag{B5.6.2}$$

3. Plastic flow rule and evolution equations:

$$\mathbf{D}^p = \dot{\lambda}\mathbf{r}(\boldsymbol{\tau},\mathbf{q}), \quad \mathbf{r} = \frac{3}{2\bar{\sigma}}\boldsymbol{\tau}^{\text{dev}}, \quad \boldsymbol{\tau}^{\text{dev}} = \boldsymbol{\tau} - \frac{1}{3}\text{trace}(\boldsymbol{\tau})\mathbf{I}, \quad \bar{\sigma} = \left[\frac{3}{2}\boldsymbol{\tau}^{\text{dev}} : \boldsymbol{\tau}^{\text{dev}}\right]^{3} \tag{B5.6.3}$$

$$\dot{q}_1 = \dot{\lambda} h_1 \quad q_1 = \bar{\varepsilon} = \int \dot{\bar{\varepsilon}}\, dt \quad \dot{\lambda} = \dot{\bar{\varepsilon}} \quad h_1 = 1$$

where the only internal variable is $q_1 \equiv \bar{\varepsilon}$, the accumulated effective plastic strain, $\boldsymbol{\tau}'$ is the deviatoric part of the Kirchhoff stress, $\bar{\sigma}$ is the von Mises effective stress. Note that $\bar{\sigma}$ and $\bar{\varepsilon}$ are plastic work conjugates: $\boldsymbol{\sigma}:\mathbf{D}^p = \bar{\sigma}\dot{\bar{\varepsilon}}$. Also, for the case of uniaxial stress, $\bar{\sigma} = \sigma$.

4. Yield condition:

$$f(\boldsymbol{\tau},\mathbf{q}) = \bar{\sigma} - \sigma_Y(\bar{\varepsilon}) = 0 \tag{B5.6.4}$$

$$\frac{\partial f}{\partial \boldsymbol{\tau}} = \frac{3}{2\bar{\sigma}}\boldsymbol{\tau}^{\mathrm{dev}} = \mathbf{r}(\text{associative plasticity}), \quad \frac{\partial f}{\partial q_1} = -\frac{d}{d\bar{\varepsilon}}\sigma_Y(\bar{\varepsilon}) = -H(\bar{\varepsilon}) \tag{B5.6.5}$$

where $\sigma_Y(\bar{\varepsilon})$ is the yield stress in uniaxial tension and $H(\bar{\varepsilon})$ is the plastic modulus.

5. Loading–unloading conditions:

$$\dot{\lambda} \geq 0, \quad f \leq 0, \quad \dot{\lambda} f = 0 \tag{B5.6.6}$$

6. Plastic rate parameter (consistency condition):

$$\dot{\lambda} = \frac{f_{\boldsymbol{\tau}}:\mathbf{C}_{el}^{\tau J}:\mathbf{D}}{-f_{\mathbf{q}}\cdot\mathbf{h} + f_{\boldsymbol{\tau}}:\mathbf{C}_{el}^{\tau J}:\mathbf{r}} = \dot{\bar{\varepsilon}} = \frac{\mathbf{r}:\mathbf{C}_{el}^{\tau J}:\mathbf{D}}{H + \mathbf{r}:\mathbf{C}_{el}^{\tau J}:\mathbf{r}} \tag{B5.6.7}$$

7. Stress rate – total rate-of-deformation relation:

$$\boldsymbol{\tau}^{\nabla J} = \mathbf{C}^{\tau J}:\mathbf{D} \quad \tau_{ij}^{\nabla J} = C_{ijkl}^{\tau J} D_{kl} \tag{B5.6.8}$$

8. Continuum elasto-plastic tangent modulus:

$$\mathbf{C}^{\tau J} = \mathbf{C}_{el}^{\tau J} - \frac{\left(\mathbf{C}_{el}^{\tau J}:\mathbf{r}\right)\otimes\left(f_{\tau}:\mathbf{C}_{el}^{\tau J}\right)}{-f_{\mathbf{q}}\cdot\mathbf{h} + f_{\tau}:\mathbf{C}_{el}^{\tau J}:\mathbf{r}} = \mathbf{C}_{el}^{\tau J} - \frac{\left(\mathbf{C}_{el}^{\tau J}:\mathbf{r}\right)\otimes\left(\mathbf{r}:\mathbf{C}_{el}^{\tau J}\right)}{H + \mathbf{r}:\mathbf{C}_{el}^{\tau J}:\mathbf{r}} \tag{B5.6.9}$$

Writing the elastic moduli in terms of bulk and deviatoric parts,

$$\mathbf{C}_{el}^{\tau J} = K\mathbf{I}\otimes\mathbf{I} + 2\mu \boldsymbol{I}^{\mathrm{dev}}, \quad \boldsymbol{I}^{\mathrm{dev}} = \boldsymbol{I} - \frac{1}{3}\mathbf{I}\otimes\mathbf{I} \tag{B5.6.10}$$

and noting that $\mathbf{r}$ is deviatoric, it follows that

$$\mathbf{C}_{el}^{\tau J}:\mathbf{r} = 2\mu\mathbf{r}, \quad \mathbf{r}:\mathbf{C}_{el}^{\tau J}:\mathbf{r} = 3\mu \tag{B5.6.11}$$

The elasto-plastic modulus is given by

$$\mathbf{C}^{\tau J} = K\mathbf{I}\otimes\mathbf{I} + 2\mu(\boldsymbol{I}^{\mathrm{dev}} - \gamma\hat{\mathbf{n}}\otimes\hat{\mathbf{n}}) = \lambda^{e}\mathbf{I}\otimes\mathbf{I} + 2\mu\boldsymbol{I} - 2\mu\gamma\hat{\mathbf{n}}\otimes\hat{\mathbf{n}}$$
$$\gamma = \frac{1}{1+(H/3\mu)}, \quad \hat{\mathbf{n}} = \sqrt{\frac{2}{3}}\mathbf{r} \tag{B5.6.12}$$

Here the Lamé constant is denoted by λ^e to avoid confusion with the plasticity parameter λ. For elastic loading or unloading, $\mathbf{C}^{\tau J} = \mathbf{C}^{\tau J}_{el}$.

9. Overall tangent modulus. From Box 5.1, the overall tangent modulus which relates the Truesdell rate of Cauchy stress and the rate-of-deformation tensor, $\boldsymbol{\sigma}^{\nabla T} = \mathbf{C}^{\sigma T}:\mathbf{D}$, is given by

$$\mathbf{C}^{\sigma T} = J^{-1}\mathbf{C}^{\tau J} - \mathbf{C}' \tag{B5.6.13}$$

which has major and both minor symmetries. For plane strain, the tangent modulus is written in matrix form using Voigt notation as

$$\left[C^{\sigma T}_{ab}\right] = J^{-1}\begin{bmatrix} \lambda^e + 2\mu & \lambda^e & 0 \\ \lambda^e & \lambda^e + 2\mu & 0 \\ 0 & 0 & \mu \end{bmatrix} - 2\mu\gamma J^{-1}\begin{bmatrix} \hat{n}_1\hat{n}_1 & \hat{n}_1\hat{n}_2 & \hat{n}_1\hat{n}_3 \\ \hat{n}_2\hat{n}_1 & \hat{n}_2\hat{n}_2 & \hat{n}_2\hat{n}_3 \\ \hat{n}_3\hat{n}_1 & \hat{n}_3\hat{n}_2 & \hat{n}_3\hat{n}_3 \end{bmatrix} - \frac{1}{2}\begin{bmatrix} 4\sigma_1 & 0 & 2\sigma_3 \\ 0 & 4\sigma_2 & 2\sigma_3 \\ 2\sigma_3 & 2\sigma_3 & \sigma_1 + \sigma_2 \end{bmatrix} \tag{B5.6.14}$$

where $\hat{n}_1 = \hat{n}_{11}, \hat{n}_2 = \hat{n}_{22}, \hat{n}_3 = \hat{n}_{12}$ and $\sigma_1 = \sigma_{11}, \sigma_2 = \sigma_{22}, \sigma_3 = \sigma_{12}$.

5.6.3 *Extension to Kinematic Hardening*

The isotropic hardening formulation presented above can be extended to combined kinematic and isotropic hardening following the same procedure outlined in Section 5.5. In multiaxial large-strain kinematic hardening models, an objective rate of the backstress tensor $\boldsymbol{\alpha}$ is required. To generalize the one-dimensional kinematic hardening model presented in Section 5.5, the overstress tensor $\boldsymbol{\Sigma} = \boldsymbol{\tau} - \boldsymbol{\alpha}$ is introduced, where $\boldsymbol{\alpha}$ is the center of the yield surface. The evolution of the backstress tensor is given in terms of the Jaumann rate, that is, $\boldsymbol{\alpha}^{\nabla J} = \kappa\mathbf{D}^p$, where κ is the kinematic hardening modulus. Nagtegaal and DeJong (1981) showed that the Jaumann rate in the backstress evolution law gives rise to nonphysical stress oscillations in simple shear at large deformations; these are related to the oscillations in the elastic response shown in Example 3.13. The model is acceptable when strains are smaller than about 0.4 and we present it here with that caveat. The plastic flow equations and tangent modulus are summarized in Box 5.7.

Box 5.7 J_2 flow theory hypoelastic–plastic constitutive model with combined isotropic kinematic hardening

1. Plastic flow rule and evolution equations:

$$\mathbf{D}^p = \dot{\lambda}\mathbf{r}\left(\Sigma, \mathbf{q}\right), \mathbf{r} = \frac{3}{2\bar{\sigma}}\overset{\text{dev}}{\Sigma}$$

$$\Sigma = \boldsymbol{\tau} - \boldsymbol{\alpha}, \overset{\text{dev}}{\Sigma} = \boldsymbol{\tau}^{\text{dev}} - \boldsymbol{\alpha}, \boldsymbol{\tau}^{\text{dev}} = \boldsymbol{\tau} - \frac{1}{3}\text{trace}(\boldsymbol{\tau})\mathbf{I}, \quad \bar{\sigma} = \left[\frac{3}{2}\overset{\text{dev}}{\Sigma} : \overset{\text{dev}}{\Sigma}\right]^{\frac{1}{2}} \tag{B5.7.1}$$

$$\dot{q}_1 = \dot{\lambda}h_1, \quad q_1 = \bar{\varepsilon} = \int \dot{\bar{\varepsilon}}\, dt, \quad \dot{\lambda} = \dot{\bar{\varepsilon}}, \quad h_1 = 1$$

$$\boldsymbol{\alpha}^{\nabla J} = \kappa \mathbf{D}^p = \kappa \dot{\lambda}\mathbf{r}$$

where κ is the kinematic hardening modulus and the internal variables are the accumulated effective plastic strain, $\bar{\varepsilon}$, and the backstress tensor, $\boldsymbol{\alpha}$.

2. Yield condition:

$$f\left(\Sigma, \mathbf{q}\right) = \bar{\sigma} - \sigma_Y(\bar{\varepsilon}) = 0 \tag{B5.7.2}$$

$$\frac{\partial f}{\partial \Sigma} = \frac{3}{2\bar{\sigma}}\overset{\text{dev}}{\Sigma} = \mathbf{r}(\text{associative plasticity}), \quad \frac{\partial f}{\partial q_1} = -\frac{d}{d\bar{\varepsilon}}\sigma_Y(\bar{\varepsilon}) = -H(\bar{\varepsilon}) \tag{B5.7.3}$$

where $H(\bar{\varepsilon})$ is the plastic modulus and $\sigma_Y(\bar{\varepsilon})$ is the yield stress in uniaxial tension.

3. Loading–unloading conditions:

$$\dot{\lambda} \geq 0, \quad f \leq 0, \quad \dot{\lambda} f = 0, \tag{B5.7.4}$$

4. Plastic rate parameter (from consistency condition):

$$\dot{\lambda} = \frac{f_\Sigma : \mathbf{C}_{el}^{\tau J} : \mathbf{D}}{-f_{\mathbf{q}} \cdot \mathbf{h} + f_\Sigma : \kappa \mathbf{r} + f_\Sigma : \mathbf{C}_{el}^{\tau J} : \mathbf{r}} = \dot{\bar{\varepsilon}} = \frac{\mathbf{r} : \mathbf{C}_{el}^{\tau J} : \mathbf{D}}{H + \kappa' + \mathbf{r} : \mathbf{C}_{el}^{\tau J} : \mathbf{r}} \tag{B5.7.5}$$

where $\kappa' = \frac{3}{2}\kappa$.

5. Stress rate – total rate-of-deformation relation:

$$\boldsymbol{\tau}^{\nabla J} = \mathbf{C}^{\sigma J} : \mathbf{D} \quad \tau_{ij}^{\nabla J} = C_{ijkl}^{\tau J} D_{kl} \tag{B5.7.6}$$

6. Continuum elasto-plastic tangent modulus:

$$\mathbf{C}^{\tau J} = \mathbf{C}_{el}^{\tau J} - \frac{(\mathbf{C}_{el}^{\tau J}:\mathbf{r}) \otimes (f_{\Sigma}:\mathbf{C}_{el}^{\tau J})}{-f_{\mathbf{q}} \cdot \mathbf{h} + f_{\Sigma}:\kappa\mathbf{r} + f_{\Sigma}:\mathbf{C}_{el}^{\tau J}:\mathbf{r}} = \mathbf{C}_{el}^{\tau J} - \frac{(\mathbf{C}_{el}^{\tau J}:\mathbf{r}) \otimes (\mathbf{r}:\mathbf{C}_{el}^{\tau J})}{H + \kappa' + \mathbf{r}:\mathbf{C}_{el}^{\tau J}:\mathbf{r}} \quad \text{(B5.7.7)}$$

The elasto-plastic tangent modulus is also

$$\mathbf{C}^{\tau J} = K\mathbf{I} \otimes \mathbf{I} + 2\mu(\boldsymbol{I}^{\text{dev}} - \gamma\hat{\mathbf{n}} \otimes \hat{\mathbf{n}}) = \lambda^{e}\mathbf{I} \otimes \mathbf{I} + 2\mu\boldsymbol{I} - 2\mu\gamma\hat{\mathbf{n}} \otimes \hat{\mathbf{n}}$$

$$\gamma = \frac{1}{1 + ((H + \kappa')/3\mu)}, \quad \hat{\mathbf{n}} = \sqrt{\frac{2}{3}}\mathbf{r} \quad \text{(B5.7.8)}$$

For elastic loading or unloading, $\mathbf{C}^{\tau J} = \mathbf{C}_{el}^{\tau J}$. The overall tangent modulus is obtained in an analogous manner to (B5.6.13) and (B5.6.14).

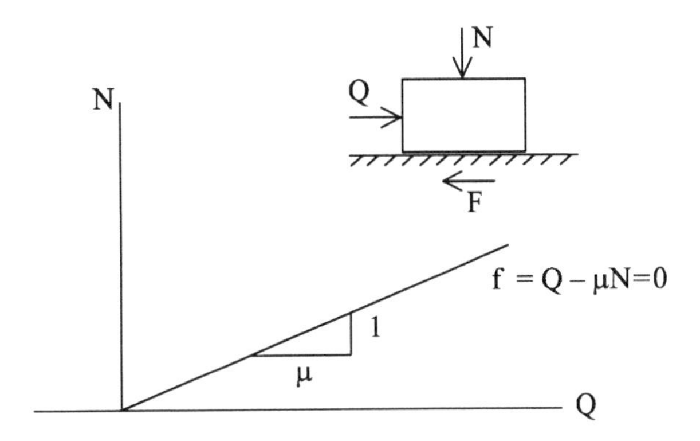

Figure 5.11 Yield surface for frictional sliding

Johnson and Bammann (1984) show that the Green–Naghdi rate of stress and backstress eliminate nonphysical oscillations. A hypoelastic–plastic formulation based on the Green–Naghdi rate is given later.

5.6.4 *Mohr–Coulomb Constitutive Model*

For materials such as soils and rock, frictional and dilatational effects are significant. The J_2 flow models presented previously are not appropriate for these materials. Instead, yield functions have been developed which represent the behavior of frictional materials. In these materials, the plastic behavior depends on the pressure, in contrast to von Mises plasticity which is independent of pressure. Furthermore, for frinctional materials, associative plasticity laws are often inappropriate. To illustrate frictional behavior, consider the block in Figure 5.11 loaded as shown with a normal load N and a tangential load Q. The block rests on a rough

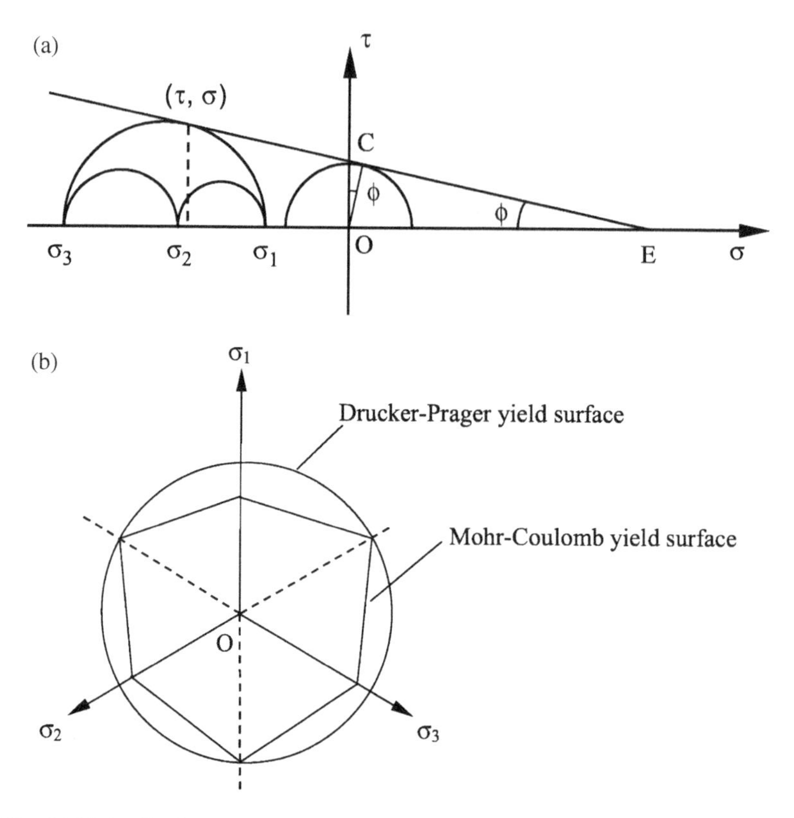

Figure 5.12 (a) Mohr–Coulomb Yield behavior; (b) Drucker–Prager and Mohr–Coulomb yield surfaces

surface with coefficient of static friction μ. If Coulomb's law is assumed to hold, the maximum frictional resistance is given by $F_{max} = \mu N$. The onset of sliding occurs when the 'yield' condition

$$f = Q - \mu N = 0 \tag{5.6.15}$$

is satisfied. The yield surface (5.6.15) is shown in Figure 5.11. Note that the direction of sliding (plastic flow) is horizontal (in the direction of Q) and not normal to the yield surface. This is a simple example of non-associative plastic behavior. The Mohr–Coulomb criterion is a generalization of this behavior to continua and to multiaxial states of stress and strain. It is widely used in modeling the behavior of granular materials (soils) and rock.

The Mohr–Coulomb criterion is based on the concept that yielding in the material occurs when a critical combination of shear stress and mean normal stress is reached on any plane. The criterion is stated as

$$\tau = c - \mu\sigma \tag{5.6.16}$$

where τ is the magnitude of the shear stress, σ is the normal stress on the plane and c is the cohesion. The angle of internal friction ϕ is defined through $\mu = \tan\phi$. Equation (5.6.16) is represented by two straight lines in the Mohr plane which are envelopes of the Mohr circles and are called the Mohr failure or rupture envelopes. In more general forms, these lines are curves (Khan and Huang, 1995). If all three Mohr's circles associated with the principal stresses lie between the failure envelopes, no yielding occurs. When the yield surface is tangential to one of the Mohr's circles, yield occurs. For example, the stress state at yield is depicted in Figure 5.12(a) where it is assumed that the principal stresses $\sigma_1 > \sigma_2 > \sigma_3$. The stress state is given by $\tau = \frac{1}{2}(\sigma_1 - \sigma_3)\cos\phi$ and $\sigma = \frac{1}{2}(\sigma_1 - \sigma_3) + \frac{1}{2}(\sigma_1 - \sigma_3)\sin\phi$. The yield criterion (5.6.16) is therefore given by

$$f(\boldsymbol{\sigma}) = \sigma_1 - \sigma_3 + (\sigma_1 + \sigma_3)\sin\phi - 2c\cos\phi = 0 \tag{5.6.17}$$

This is an equation of a conical surface in principal stress space. The intersection of the yield surface with the π-plane ($\sigma_{kk}=0$) is shown in Figure 5.12(b) and is seen to be an irregular hexagon. Considering the special case $\phi = 0$ and letting $c=k$ denote the yield strength in shear, (5.6.17) reduces to $\sigma_1 - \sigma_3 - 2k=0$, which is the Tresca criterion (Hill, 1950).

5.6.5 *Drucker–Prager Constitutive Model*

The straight line segments on the Tresca and Mohr–Coulomb yield surfaces make these surfaces convenient for analytical treatments of plasticity problems. However, the corners make the constitutive equations difficult from a computational standpoint (in computing the normal to the yield surface, for example). The Drucker–Prager yield criterion avoids the problems associated with corners by modifying the von Mises yield criterion (B5.4.4) to incorporate the effects of pressure:

$$f = \bar{\sigma} - \alpha\boldsymbol{\sigma}:\mathbf{I} - Y = 0 \tag{5.6.18}$$

which is the equation of a smooth circular cone. In (5.6.18), $\bar{\sigma}$ is the effective Cauchy stress. By choosing the constants α and Y as

$$\alpha = \frac{2\sin\phi}{3 \pm \sin\phi}, \ Y = \frac{6c\cos\phi}{3 \pm \sin\phi} \tag{5.6.19}$$

the Drucker–Prager yield surface passes through the inner or outer apexes of the Mohr–Coulomb yield surface (with the plus sign corresponding to the inner apexes and the minus sign corresponding to the outer).

The elastic response is given by a hypoelastic relation for the Jaumann rate of Cauchy stress. Both associative and non-associative models can be developed. The associative plastic flow rule is $\mathbf{D}^p = \dot{\lambda}\mathbf{r}(\boldsymbol{\sigma},\mathbf{q})$ where

$$\mathbf{r} = \frac{\partial f}{\partial \boldsymbol{\sigma}} = \frac{3}{2\bar{\sigma}}\boldsymbol{\sigma}^{\text{dev}} - \alpha\mathbf{I} \tag{5.6.20}$$

Many non-associative rules can be developed. One example is

$$\mathbf{r} = \frac{\partial \psi}{\partial \boldsymbol{\sigma}} = \frac{3}{2\bar{\sigma}}\boldsymbol{\sigma}^{\text{dev}}, \ \psi = \bar{\sigma} \tag{5.6.21}$$

It can be seen from this that, for the associative law (5.6.20), the volumetric plastic flow is nonzero and the material dilates under compression. This contradicts the observed behavior of granular materials. In the non-associative law (5.6.21), plastic flow is isochoric.

The complete formulation of the model is given by Box 5.5 with the definitions (5.6.18) and (5.6.20) or (5.6.21) of the yield condition and flow rule, respectively. Because the model is based on the Cauchy stress, the overall tangent modulus is not symmetric (Box 5.1).

5.6.6 *Porous Elastic–Plastic Solids: Gurson Model*

The Gurson constitutive model (Gurson, 1977) was developed to model progressive microrupture through void nucleation and growth. It is used extensively in modeling ductile rupture of metals (e.g., Tvergaard and Needleman, 1984).

Different versions of the Gurson model can be formulated. For example, Narasimhan, Rosakis and Moran (1992) considered fracture initiation in a ductile steel using a small-strain rate-independent plasticity version of the model. Here we present a large-deformation, hypoelastic, rate-independent plasticity version. A rate-dependent formulation is given in Pan, Saje and Needleman (1983).

The material consists of a matrix and voids with volume fraction f (in this section, the yield function will be denoted by Φ, and f denotes the void volume fraction). The void volume fraction and the accumulated plastic strain in the matrix material are internal variables in the model. The starting point for the constitutive model is the additive decomposition of the rate-of-deformation tensor into elastic and plastic parts. The Jaumann rate of Cauchy stress is used in the hypoelastic stress rate relation (the moduli are usually taken to be constant and isotropic) and the plastic flow equations are based on the Cauchy (true) stress. A von Mises-type yield condition is used.

The yield function Φ also acts as a potential for the plastic flow, so this theory is associative. The yield condition is given by

$$\Phi = \frac{\sigma_e^2}{\bar{\sigma}^2} + 2f^*\beta_1 \cosh\left(\frac{\beta_2 \boldsymbol{\sigma}:\mathbf{I}}{2\bar{\sigma}}\right) - 1 - (\beta_1 f^*)^2 = 0 \tag{5.6.22}$$

where $\sigma_e = \left(\frac{3}{2}\boldsymbol{\sigma}^{\text{dev}}:\boldsymbol{\sigma}^{\text{dev}}\right)^{\frac{1}{2}}$ is the effective macroscopic Cauchy stress ($\boldsymbol{\sigma}^{\text{dev}} = \boldsymbol{\sigma} - \frac{1}{3}\text{trace}(\boldsymbol{\sigma})\mathbf{I}$ is the deviatoric Cauchy stress), $\bar{\sigma}$ is the effective stress in the matrix material, and f^* is a function of the void volume fraction given below. The model was originally introduced by Gurson (1977) for rate-independent plasticity with the values of β_1 and β_2 set to unity. The parameters β_1 and β_2 were introduced by Tvergaard (1981) to model behavior more accurately at low void volume fractions. The parameter f^* was introduced by Tvergaard and Needleman (1984) to simulate the rapid loss of strength in the final stages of void coalescence. In Gurson's original model, the parameter f^* is simply the void volume fraction f. In the Tvergaard–Needleman approach, a modification is invoked when the void volume fraction reaches a critical value f_c and is given by

$$f^* = \begin{cases} f & \text{for } f \le f_c \\ f_c + (f_u - f_c)(f - f_c)/(f_f - f_c) & \text{for } f > f_c \end{cases} \tag{5.6.23}$$

where $f_u = 1/\beta_1$ and $f^*(f_f) = fu$. Note that f_f is the void volume fraction at which the stress carrying capacity of the material is completely lost.

The plastic flow direction is given by the associative rule $\mathbf{D}^p = \dot{\lambda}\mathbf{r}$ where

$$\mathbf{r} = \frac{\partial \Phi}{\partial \boldsymbol{\sigma}} = \frac{3}{\bar{\sigma}^2}\boldsymbol{\sigma}^{\text{dev}} + (f^* \beta_1 \beta_2 / \bar{\sigma}) \sinh\left(\frac{\beta_2 \boldsymbol{\sigma}:\mathbf{I}}{2\bar{\sigma}}\right)\mathbf{I} \tag{5.6.24}$$

Evolution equations for the internal variables $q_1 = f$ and $q_2 = \bar{\varepsilon}$ are also required. The increase of voids in the material is due to the growth of existing voids and the nucleation of new ones and can be written as

$$\dot{f} = \dot{f}_{\text{growth}} + \dot{f}_{\text{nucleation}} \tag{5.6.25}$$

The expression for void growth

$$\dot{f}_{\text{growth}} = (1-f)\,\text{trace}\,(\mathbf{D}^p) = \dot{\lambda}(1-f)\,\text{trace}\,(\mathbf{r}) \tag{5.6.26}$$

is obtained from the kinematics of void growth in an incompressible matrix (the small contribution from elastic strain is neglected) and the macroscopic plastic flow rule has been used. Nucleation is typically regarded as strain controlled or stress controlled. For simplicity, we neglect nucleation here.

During plastic loading, the effective stress in the matrix material must lie on the matrix yield surface $\bar{\sigma} - \sigma_Y(\bar{\varepsilon}) = 0$. The consistency condition in the matrix material is obtained by differentiating this expression to give

$$\dot{\bar{\sigma}} = H(\bar{\varepsilon})\dot{\bar{\varepsilon}} \tag{5.6.27}$$

where $H(\bar{\varepsilon}) = d\sigma_Y(\bar{\varepsilon})/d\bar{\varepsilon}$ is the matrix plastic modulus. It follows from (5.6.27) that

$$\frac{\partial}{\partial q_2} \equiv \frac{\partial}{\partial \bar{\varepsilon}} = H\frac{\partial}{\partial \bar{\sigma}} \tag{5.6.28}$$

which is used below in obtaining the derivatives of the yield function.

An expression for the evolution of the accumulated effective plastic strain is obtained by equating macroscopic and microscopic rates of plastic work, that is,

$$\boldsymbol{\sigma}:\mathbf{D}^p = (1-f)\bar{\sigma}\dot{\bar{\varepsilon}} \tag{5.6.29}$$

from which we obtain

$$\dot{\bar{\varepsilon}} = \frac{\boldsymbol{\sigma}:\mathbf{D}^p}{(1-f)\bar{\sigma}} = \dot{\lambda}\frac{\boldsymbol{\sigma}:\mathbf{r}}{(1-f)\bar{\sigma}} \tag{5.6.30}$$

The equations for the Gurson model are summarized in Box 5.8.

When $f=0$, the Gurson model in Box 5.8 reduces to rate-independent J_2 flow theory plasticity.

Box 5.8 Rate-independent Gurson model

1. Rate-of-deformation tensor:

$$\mathbf{D} = \mathbf{D}^e + \mathbf{D}^p \tag{B5.8.1}$$

2. Stress rate relation:

$$\boldsymbol{\sigma}^{\nabla J} = \mathbf{C}_{el}^{\sigma J} : \mathbf{D}^e = \mathbf{C}_{el}^{\sigma J} : (\mathbf{D} - \mathbf{D}^p) \tag{B5.8.2}$$

3. Plastic flow rule and evolution equations:

$$\begin{aligned} &\mathbf{D}^p = \dot{\lambda}\mathbf{r}(\boldsymbol{\sigma},\mathbf{q}), \quad \mathbf{r} = \frac{\partial \Phi}{\partial \boldsymbol{\sigma}} \\ &\dot{\mathbf{q}} = \dot{\lambda}\mathbf{h}, \quad q_1 = f, \quad q_2 = \bar{\varepsilon} \\ &h_1 = (1-f)tr(\mathbf{r}), \quad h_2 = \frac{\boldsymbol{\sigma}:\mathbf{r}}{(1-f)\bar{\sigma}} \\ &\dot{\bar{\sigma}} = H\dot{\bar{\varepsilon}} \end{aligned} \tag{B5.8.3}$$

4. Yield condition (see (5.6.15)):

$$\Phi(\boldsymbol{\sigma},\mathbf{q}) = 0 \tag{B5.8.4}$$

5. Loading–unloading conditions:

$$\dot{\lambda} \geq 0, \quad \Phi \leq 0, \quad \dot{\lambda}\Phi = 0 \tag{B5.8.5}$$

6. Plastic rate parameter (from consistency condition $\dot{\Phi} = 0$):

$$\dot{\lambda} = \frac{\Phi_{\boldsymbol{\sigma}} : \mathbf{C}_{el}^{\sigma J} : \mathbf{D}}{-\Phi_{\mathbf{q}} \cdot \mathbf{h} + \Phi_{\mathbf{q}} : \mathbf{C}_{el}^{\sigma J} : \mathbf{r}} = \frac{\mathbf{r} : \mathbf{C}_{el}^{\sigma J} : \mathbf{D}}{-\Phi_{\mathbf{q}} \cdot \mathbf{h} + \mathbf{r} : \mathbf{C}_{el}^{\sigma J} : \mathbf{r}} \tag{B5.8.6}$$

Note that from (5.6.28) $\partial\Phi/\partial q_2 = H\partial\Phi/\partial\bar{\sigma}$.

7. Stress rate – rate-of-deformation relation:

$$\boldsymbol{\sigma}^{\nabla J} = \mathbf{C}^{\sigma J} : \mathbf{D}, \quad \sigma_{ij}^{\nabla J} = \mathbf{C}_{ijkl}^{\sigma J} D_{kl} \tag{B5.8.7}$$

8. Continuum elasto-plastic tangent modulus

$$\mathbf{C}^{\sigma J} = \mathbf{C}_{el}^{\sigma J} - \frac{(\mathbf{C}_{el}^{\sigma J} : \mathbf{r}) \otimes (\Phi_{\boldsymbol{\sigma}} : \mathbf{C}_{el}^{\sigma J})}{-\Phi_{\mathbf{q}} \cdot \mathbf{h} + \Phi_{\boldsymbol{\sigma}} : \mathbf{C}_{el}^{\sigma J} : \mathbf{r}} = \mathbf{C}_{el}^{\sigma J} - \frac{(\mathbf{C}_{el}^{\sigma J} : \mathbf{r}) \otimes (\mathbf{r} : \mathbf{C}_{el}^{\sigma J})}{-\Phi_{\mathbf{q}} \cdot \mathbf{h} + \mathbf{r} : \mathbf{C}_{el}^{\sigma J} : \mathbf{r}} \tag{B5.8.8}$$

which has major symmetry. For elastic loading or unloading, $\mathbf{C}^{\sigma J} = \mathbf{C}_{el}^{\sigma J}$. The overall tangent modulus is given by $\mathbf{C}^{\sigma T} = \mathbf{C}^{\sigma J} - \mathbf{C}^*$ which does not possess major symmetry because $\mathbf{C}^*$ is not symmetric (Box 5.1).

5.6.7 *Corotational Stress Formulation*

The previously described hypoelastic-plastic formulations are typically used with constant elastic moduli. As will be seen in Section 5.9, objectivity requires these moduli to be isotropic and, as discussed above, the yield function is restricted to be an isotropic function of the stress. The corotational stress formulation presented here is not restricted to isotropic response, although, as we will see, the tangent modulus is not symmetric even when the model is based on the Kirchhoff stress.

The corotational Kirchhoff stress tensor, $\hat{\boldsymbol{\tau}}$, is defined by

$$\hat{\boldsymbol{\tau}} = \mathbf{R}^T \cdot \boldsymbol{\tau} \cdot \mathbf{R} = J\hat{\boldsymbol{\sigma}} \tag{5.6.31}$$

where $\hat{\boldsymbol{\sigma}}$ is the corotational Cauchy stress given in Box 3.2. The relation between stress rate and elastic strain rate is written as

$$\dot{\hat{\boldsymbol{\tau}}} = \hat{\mathbf{C}}_{el}^{\tau} : \hat{\mathbf{D}}^e \tag{5.6.32}$$

where $\hat{\mathbf{D}}^e$ is the elastic part of the corotational rate-of-deformation tensor $\hat{\mathbf{D}} = \mathbf{R}^T \cdot \mathbf{D} \cdot \mathbf{R}$ (see 3.4.16b). The plasticity equations and the elasto-plastic tangent modulus are analogous to those in Section 5.6.1 and are given in Box 5.9.

Box 5.9 Hypoelastic–plastic constitutive model: rotated Kirchhoff stress formulation

1. Rate-of-deformation tensor:

$$\hat{\mathbf{D}} = \hat{\mathbf{D}}^e + \hat{\mathbf{D}}^p \tag{B5.9.1}$$

2. Stress rate relation:

$$\dot{\hat{\boldsymbol{\tau}}} = \hat{\mathbf{C}}_{el}^{\tau} : \hat{\mathbf{D}}^e = \hat{\mathbf{C}}_{el}^{\tau} : (\hat{\mathbf{D}} - \hat{\mathbf{D}}^p) \tag{B5.9.2}$$

3. Plastic flow rule and evolution equations:

$$\hat{\mathbf{D}}^p = \dot{\lambda}\hat{\mathbf{r}}(\hat{\boldsymbol{\tau}}, \hat{\mathbf{q}}), \ \dot{\hat{\mathbf{q}}} = \dot{\lambda}\hat{\mathbf{h}}(\hat{\boldsymbol{\tau}}, \hat{\mathbf{q}}) \tag{B5.9.3}$$

4. Yield condition:

$$\hat{f}(\hat{\boldsymbol{\tau}}, \hat{\mathbf{q}}) = 0 \tag{B5.9.4}$$

5. Loading–unloading conditions:

$$\dot{\lambda} \geq 0, \ \hat{f} \leq 0, \ \dot{\lambda}\hat{f} = 0 \tag{B5.9.5}$$

6. Plastic rate parameter (consistency condition):

$$\dot{\lambda} = \frac{\hat{f}_{\hat{\tau}} : \hat{\mathbf{C}}_{el}^{\tau J} : \hat{\mathbf{D}}}{-\hat{f}_{\hat{\mathbf{q}}} \cdot \hat{\mathbf{h}} + \hat{f}_{\hat{\tau}} : \hat{\mathbf{C}}_{el}^{\tau} : \hat{\mathbf{r}}} \tag{B5.9.6}$$

7. Relation between stress rate and rate-of-deformation:

$$\dot{\hat{\boldsymbol{\tau}}} = \hat{\mathbf{C}}^{\tau}:\hat{\mathbf{D}},\ \dot{\hat{\tau}}_{ij} = \hat{C}^{\tau}_{ijkl}\hat{D}_{kl} \tag{B5.9.7}$$

Elastic loading or unloading:

$$\hat{\mathbf{C}}^{\tau} = \hat{\mathbf{C}}^{\tau}_{el}$$

8. Plastic loading (continuum elasto-plastic tangent modulus):

$$\hat{\mathbf{C}}^{\tau} = \hat{\mathbf{C}}^{\tau}_{el} - \frac{\left(\hat{\mathbf{C}}^{\tau}_{el}:\hat{\mathbf{r}}\right)\otimes\left(\hat{f}_{\hat{\tau}}:\hat{\mathbf{C}}^{\tau}_{el}\right)}{-\hat{f}_{\hat{\mathbf{q}}}\cdot\hat{\mathbf{h}} + \hat{f}_{\hat{\tau}}:\hat{\mathbf{C}}^{\tau}_{el}:\hat{\mathbf{r}}}$$

$$\hat{C}^{\tau}_{ijkl} = \left(\hat{C}^{\tau}_{el}\right)_{ijkl} - \frac{\left(\hat{C}^{\tau}_{el}\right)_{ijmn}\hat{r}_{mn}\left(\hat{f}_{\hat{\tau}}\right)_{pq}\left(\hat{C}^{\tau}_{el}\right)_{pqkl}}{-\left(\hat{f}_{\hat{q}}\right)_{\alpha}\hat{h}_{\alpha} + \left(\hat{f}_{\hat{\tau}}\right)_{rs}\left(\hat{C}^{\tau}_{el}\right)_{rstu}\hat{r}_{tu}} \tag{B5.9.8}$$

The relation between the material time derivative of the Kirchhoff stress and the corotational rate-of-deformation tensor is denoted by

$$\dot{\hat{\boldsymbol{\tau}}} = \hat{\mathbf{C}}^{\tau}:\hat{\mathbf{D}} \tag{5.6.33}$$

Now noting that the Green–Naghdi rate of Kirchhoff stress is given by

$$\boldsymbol{\tau}^{\nabla G} = \dot{\boldsymbol{\tau}} - \Omega\cdot\boldsymbol{\tau} - \boldsymbol{\tau}\cdot\Omega^{T} = \mathbf{R}\cdot\dot{\hat{\boldsymbol{\tau}}}\cdot\mathbf{R}^{T} \tag{5.6.34}$$

then from (5.6.33) we obtain

$$\boldsymbol{\tau}^{\nabla G} = \mathbf{C}^{\tau G}:\mathbf{D}\quad C^{\tau G}_{ijkl} = R_{im}R_{jn}R_{kp}R_{lp}\hat{C}^{\tau}_{mnpq} \tag{5.6.35}$$

where $\mathbf{D} = \mathbf{R}\cdot\hat{\mathbf{D}}\cdot\mathbf{R}^{T}$. The overall tangent modulus is given by $\mathbf{C}^{\sigma T} = J^{-1}\mathbf{C}^{\tau G} - \mathbf{C}' - \mathbf{C}^{\text{spin}}$ (see Box 5.1) which is not symmetric due to $\mathbf{C}^{\text{spin}}$.

An advantage of the corotational stress formulation is that frame invariance requirements do not limit the model to isotropic elastic moduli or isotropic yield behavior, as was the case for the previously described Jaumann rate of Cauchy or Kirchhoff stress formulations. This can be seen by noting that the rotated stress is insensitive to rigid rotations of the current configuration (see also Section 5.10.3):

$$\hat{\boldsymbol{\tau}}^{*} = \mathbf{R}^{*T}\cdot\boldsymbol{\tau}^{*}\cdot\mathbf{R}^{*} = \mathbf{R}^{T}\cdot\mathbf{Q}^{T}\cdot\mathbf{Q}\cdot\boldsymbol{\tau}\cdot\mathbf{Q}^{T}\cdot\mathbf{Q}\cdot\mathbf{R} = \mathbf{R}^{T}\cdot\boldsymbol{\tau}\cdot\mathbf{R} = \hat{\boldsymbol{\tau}} \tag{5.6.36}$$

where $\mathbf{R}^{*} = \mathbf{Q}\cdot\mathbf{R}$ and $\boldsymbol{\tau}^{*} = \mathbf{Q}\cdot\boldsymbol{\tau}\cdot\mathbf{Q}^{T}$ (Section 5.10). Thus the elastic moduli $\hat{\mathbf{C}}^{\tau}$ may be anisotropic and the yield function f may be an arbitrary function of the corotational stress $\hat{\boldsymbol{\tau}}$.

5.6.8 *Small-Strain Formulation*

The general formulation for rate-independent large-deformation plasticity presented in Box 5.5 above can be readily reduced to the small-strain case as follows. No distinction between stress measures is required and we use the Cauchy stress $\boldsymbol{\sigma}$ because objectivity requirements are not relevant in the small-strain setting, the material time derivative is the relevant stress rate, and strain rate $\dot{\varepsilon}$ replaces the rate-of-deformation. Also, the small-strain formulation is valid for anisotropic elastic moduli, $\mathbf{C}$, and yield function f. The small-strain formulation is summarized in Box 5.10.

Box 5.10 Elasto-plastic constitutive model – small strains

1. Additive decomposition of rate of strain into elastic and plastic parts:

$$\dot{\varepsilon} = \dot{\varepsilon}^e + \varepsilon^p \qquad \text{(B.5.10.1)}$$

2. Relation between stress rate and elastic strain rate:

$$\dot{\boldsymbol{\sigma}} = \mathbf{C}:\dot{\varepsilon}^e = \mathbf{C}:(\dot{\varepsilon} - \dot{\varepsilon}^p) \qquad \text{(B.5.10.2)}$$

3. Plastic flow rule and evolution equations:

$$\begin{aligned} \dot{\varepsilon}^p &= \dot{\lambda}\mathbf{r}(\boldsymbol{\sigma}, \mathbf{q}) \\ \dot{\mathbf{q}} &= \dot{\lambda}\mathbf{h} \end{aligned} \qquad \text{(B.5.10.3)}$$

4. Yield condition:

$$f(\boldsymbol{\sigma}, \mathbf{q}) = 0 \qquad \text{(B.5.10.4)}$$

5. Loading–unloading conditions:

$$\dot{\lambda} \geq 0, \quad f \leq 0, \quad \dot{\lambda} f = 0 \qquad \text{(B.5.10.5)}$$

6. Plastic rate parameter (from consistency condition):

$$\dot{\lambda} = \frac{f_{\sigma}:\mathbf{C}:\dot{\varepsilon}}{-f_{\mathbf{q}} \cdot \mathbf{h} + f_{\sigma}:\mathbf{C}:\mathbf{r}} \qquad \text{(B.5.10.6)}$$

7. Relation between stress rate and strain rate:

$$\dot{\boldsymbol{\sigma}} = \mathbf{C}^{ep}:\dot{\varepsilon} \qquad \text{(B.5.10.7)}$$

8. Continuum elasto-plastic tangent modulus:

$$\mathbf{C}^{ep} = \mathbf{C} - \frac{(\mathbf{C}:\mathbf{r}) \otimes (f_{\sigma}:\mathbf{C})}{-f_{\mathbf{q}} \cdot \mathbf{h} + f_{\sigma}:\mathbf{C}:\mathbf{r}} \qquad \text{(B.5.10.8)}$$

which is symmetric if plastic flow is associative ($\mathbf{C}:\mathbf{r} \sim f_{\sigma}:\mathbf{C}$).

5.6.9 Large-Strain Viscoplasticity

The rate-dependent plasticity (viscoplasticity) constitutive relations can be extended to multi-dimensions by generalizing the one-dimensional rate-dependent plasticity equations of Section 5.4 in a similar manner to the generalization of the rate-independent equations described previously. While in rate-independent plasticity, the plastic rate parameter is obtained from the consistency condition, in rate-dependent plasticity, this parameter is given as an empirical function of stress and internal variables. It is typically given by an overstress function, as will be seen below. We therefore have the same form of the plastic flow rule and evolution equations for the internal variables, namely

$$\mathbf{D}^p = \dot{\lambda}\mathbf{r}(\boldsymbol{\sigma}, \mathbf{q}), \ \dot{\mathbf{q}} = \dot{\lambda}\mathbf{h} \tag{5.6.37}$$

where the plastic rate parameter is given by

$$\dot{\lambda} = \frac{\phi(\boldsymbol{\sigma}, \mathbf{q})}{\eta} \tag{5.6.38}$$

where ϕ is an overstress function and η is the viscosity. Note that ϕ has dimensions of stress and can be thought of as the driving force for the plastic strain rate. The viscosity η has dimensions of stress$\times$time.

A typical example of (5.6.38) of the overstress function for J_2 flow plasticity, due to Perzyna (1971), is

$$\phi = \sigma_Y(\bar{\varepsilon})\left\langle \frac{\bar{\sigma}}{\sigma_Y(\bar{\varepsilon})} - 1 \right\rangle^n \tag{5.6.39}$$

where <•> are Macaulay brackets, $\bar{\sigma}$ is the von Mises effective stress (B5.6.3), n is the rate-sensitivity exponent and $\sigma_Y(\bar{\varepsilon})$ is the yield stress in uniaxial tension. An alternative viscoplastic model is given by Peirce, Shih and Needleman (1984) for J_2 flow theory as

$$\dot{\lambda} = \dot{\bar{\varepsilon}} = \dot{\varepsilon}_0\left(\frac{\bar{\sigma}}{\sigma_Y(\bar{\varepsilon})} \right)^{1/m} \tag{5.6.40}$$

where m is the rate-sensitivity exponent and $\dot{\varepsilon}_0$ is a reference strain rate. This model does not use an explicit yield function. However, as $m \to 0$, rate-independent plasticity with yield stress $\sigma_Y(\bar{\varepsilon})$ is approached. The constitutive equations for rate-dependent plasticity (viscoplasticity) are summarized in Box 5.11.

Box 5.11 Large strain rate-dependent plasticity

1. Additive decomposition of rate-of-deformation tensor into elastic and plastic parts:

$$\mathbf{D} = \mathbf{D}^e + \mathbf{D}^p \tag{B5.11.1}$$

2. Stress rate relation:

$$\boldsymbol{\sigma}^{\nabla J} = \mathbf{C}_{el}^{\sigma J} : \mathbf{D}^e = \mathbf{C}_{el}^{\sigma J} : (\mathbf{D} - \mathbf{D}^p) \tag{B5.11.2}$$

3. Plastic flow rule and evolution equations:

$$\mathbf{D}^p = \dot{\lambda} \mathbf{r}(\boldsymbol{\sigma}, \mathbf{q}),\ \dot{\lambda} = \frac{\phi(\boldsymbol{\sigma}, \mathbf{q})}{\eta},\ \dot{\mathbf{q}} = \dot{\lambda} \mathbf{h}(\boldsymbol{\sigma}, \mathbf{q}) \tag{B5.11.3}$$

4. Stress rate – total rate-of-deformation relation:

$$\boldsymbol{\sigma}^{\nabla J} = \mathbf{C}_{el}^{\sigma J} : \mathbf{D} - \frac{\phi}{\eta} \mathbf{C}_{el}^{\sigma J} : \mathbf{r} \tag{B5.11.4}$$

5.7 Hyperelastic–Plastic Models

There are several drawbacks to the hypoelastic–plastic constitutive relations described in Section 5.6:

1. The elastic response is hypoelastic and therefore the work done in a closed cycle of deformation is not exactly zero.
2. If the elastic moduli are assumed to be constant, frame invariance restricts the moduli to be isotropic.
3. The yield function is required to be an isotropic function of stress.
4. The hypoelastic relation must be integrated in time to compute the stress. Hypoelastic formulations require incrementally objective stress update schemes (Section 5.9) to ensure that finite rotations do not induce unacceptably large erroneous stresses.

For the corotational formulation, items 2 and 3 do not apply, that is, the elastic and plastic response is not restricted to be isotropic.

Hyperelastic–plastic constitutive models were developed to eliminate the above drawbacks of the hypoelastic–plastic formulations (Simo and Ortiz, 1985; Moran, Ortiz and Shih, 1990; Miehe, 1994). Since the elastic response is derived from a hyperelastic potential, the work done in a closed elastic deformation path vanishes exactly. Furthermore, no stress rate equations are integrated to compute the stress, and the need for incrementally objective stress update algorithms is eliminated. In addition, the hyperelastic–plastic formulation provides a natural framework for frame-invariant formulations of anisotropic elasticity *and* anisotropic plastic yield.

For a thorough understanding of this material, the reader should be familiar with pull-back and push-forward operations and the Lie derivative. These are described in Section 5.10.2 along with several other useful results and derivations for the material in this section. However, the outline of this development can be followed without this background. Two key concepts distinguish hyperelastic–plastic materials from hypoelastic–plastic materials:

1. a multiplicative decomposition of the deformation into elastic and plastic parts by

$$\mathbf{F} = \mathbf{F}^{e} \cdot \mathbf{F}^{p} \tag{5.7.1}$$

   where $\mathbf{F}^e$ and $\mathbf{F}^P$ are the elastic and plastic parts of the deformation gradient, respectively;
2. the calculation of the stress in terms of the elastic strain through a hyperelastic potential

$$\bar{\mathbf{S}} = \frac{\partial \bar{w}(\bar{\mathbf{E}}^{e})}{\partial \bar{\mathbf{E}}^{e}} \tag{5.7.2}$$

The stress $\bar{\mathbf{S}}$ here is not in the reference configuration and so it differs from $\mathbf{S}$ and is defined later along with $\bar{\mathbf{E}}^e$.

5.7.1 *Multiplicative Decomposition of Deformation Gradient*

The multiplicative decomposition $\mathbf{F}=\mathbf{F}^e \cdot \mathbf{F}^P$ of the deformation gradient into elastic and plastic parts introduces three configurations shown in Figure 5.13 (Lee, 1969; Asaro and Rice, 1977). As can be seen from Figure 5.13, the deformation gradient maps a point $\mathbf{X}$ in the reference configuration Ω_0 to $\bar{\mathbf{X}}$ in the intermediate configuration $\bar{\Omega}$ by $\mathbf{F}^P$, and then to $\mathbf{x}$ in the spatial configuration Ω by $\mathbf{F}^e$. Any rigid rotations of the body are incorporated in $\mathbf{F}^e$. The intermediate configuration is actually a misnomer in that this configuration does not exist as a continuous map however $\mathbf{F}^e$ is used for pull-backs and we refer to these as pull-backs to the intermediate configuration $\bar{\Omega}$. The decomposition (5.7.1) is used only to represent the constitutive response at a material point.

To formulate the elastic and plastic constitutive relations on the intermediate configuration, we work with the pull-back by the elastic part of the deformation gradient $\mathbf{F}^e$ of the various spatial kinematic and stress measures to $\bar{\Omega}$. For example, the elastic Green strain, $\bar{\mathbf{E}}^e$ is defined by

$$\bar{\mathbf{E}}^{e} = \frac{1}{2}(\bar{\mathbf{C}}^{e} - \mathbf{I}), \quad \bar{\mathbf{C}}^{e} = \mathbf{F}^{eT} \cdot \mathbf{F}^{e} = \mathbf{F}^{eT} \cdot \mathbf{g} \cdot \mathbf{F}^{e} \equiv \phi_{e}^{*}\mathbf{g} \tag{5.7.3}$$

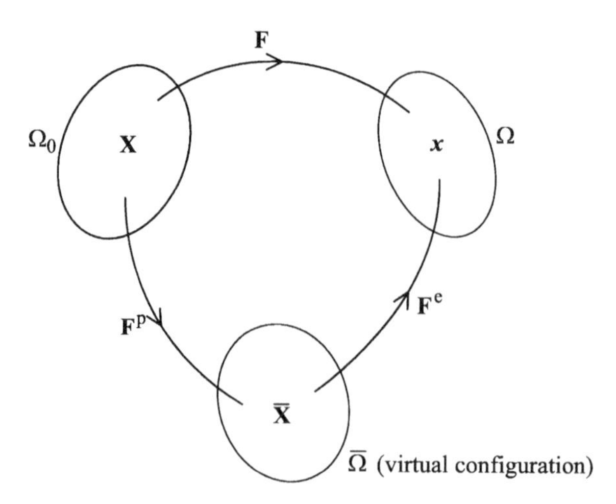

Figure 5.13 Decomposition of deformation gradient and definition of intermediate configuration $\bar{\Omega}$

The second equation interprets $\bar{\mathbf{C}}^e$ as the pull-back of the spatial metric tensor $\mathbf{g}$ by $\mathbf{F}^e$. In Euclidean space which is flat, $\mathbf{g}=\mathbf{I}$.

5.7.2 *Hyperelastic Potential and Stress*

The second Piola–Kirchhoff stress $\bar{\mathbf{S}}$ on $\bar{\Omega}$ is defined by the pull-back of the Kirchhoff stress with the elastic part of the deformation gradient:

$$\bar{\mathbf{S}} = (\mathbf{F}^e)^{-1} \cdot \boldsymbol{\tau} \cdot (\mathbf{F}^e)^{-T} \equiv \phi_e^* \boldsymbol{\tau} \tag{5.7.4}$$

The elastic response is then given by a hyperelastic potential

$$\bar{\mathbf{S}} = 2\frac{\partial \bar{\psi}(\bar{\mathbf{C}}^e)}{\partial \bar{\mathbf{C}}^e} = \frac{\partial \bar{w}(\bar{\mathbf{E}}^e)}{\partial \bar{\mathbf{E}}^e} \tag{5.7.5}$$

where $\bar{\psi}(\bar{\mathbf{C}}^e) = \bar{w}(\bar{\mathbf{E}}^e)$ is the elastic stored energy potential. This relation is identical to (5.4.35) except that the potential is a function of the elastic strain (see (5.7.3)).

An expression for the stress rate is required in the development of the elasto-plastic tangent modulus next. Taking the material time derivative of $\bar{\mathbf{S}}$ in (5.7.5) we obtain

$$\dot{\bar{\mathbf{S}}} = \frac{\partial^2 \bar{w}}{\partial \bar{\mathbf{E}}^e \partial \bar{\mathbf{E}}^e} : \dot{\bar{\mathbf{E}}}^e = \mathbf{C}_{el}^{\bar{S}} : \dot{\bar{\mathbf{E}}}^e \tag{5.7.6}$$

where $\mathbf{C}_{el}^{\bar{S}} = \partial^2 \bar{w} / \partial \bar{\mathbf{E}}^e \partial \bar{\mathbf{E}}^e$ is the elasticity tensor which relates the material time derivative of the second Piola–Kirchhoff stress on $\bar{\Omega}$ to the material time derivative of the Green strain on $\bar{\Omega}$.

Note that, because $\bar{\mathbf{E}}^e$ is invariant under an observer transformation (Section 5.10), then from (5.7.5) $\bar{\mathbf{S}}$ is also invariant. Therefore defining the elastic response in terms of a potential ensures objectivity. Furthermore, from (5.7.6) $\mathbf{C}_{el}^{\bar{S}}$ is also unaffected by rotations, so the elastic moduli may be anisotropic.

5.7.3 *Decomposition of Rates of Deformation*

The formulation of the plastic flow equations on the intermediate configuration requires the elastic and plastic parts of rates of deformation. The material time-derivative of the elastic Green strain rate is given by

$$\dot{\bar{\mathbf{E}}}^e = \frac{1}{2}\dot{\bar{\mathbf{C}}}^e = (\mathbf{F}^e)^T \cdot \mathbf{D}^e \cdot \mathbf{F}^e \equiv \bar{\mathbf{D}}^e \tag{5.7.7}$$

The first of these is obtained by differentiating (5.7.3). The second equality identifies the material time-derivative of the Green strain as the pull-back (by $\mathbf{F}^e$) of the elastic part of the rate-of-deformation tensor, to give $\bar{\mathbf{D}}^e$. This can be interpreted as the elastic part of the rate-of-deformation tensor on $\bar{\Omega}$. This relation is developed below along with other relations required for a description of plastic flow.

From (3.3.18) the spatial velocity gradient is given by

$$\mathbf{L} = \dot{\mathbf{F}} \cdot \mathbf{F}^{-1} = \frac{D}{Dt}(\mathbf{F}^e \cdot \mathbf{F}^p) \cdot (\mathbf{F}^e \cdot \mathbf{F}^p)^{-1} = \dot{\mathbf{F}}^e \cdot (\mathbf{F}^e)^{-1} + \mathbf{F}^e \cdot \dot{\mathbf{F}}^p \cdot (\mathbf{F}^p)^{-1} \cdot (\mathbf{F}^e)^{-1} \quad (5.7.8)$$

where we have used (5.7.1), $\mathbf{F} = \mathbf{F}^e \cdot \mathbf{F}^p$. $\mathbf{L}$ is next split into elastic and plastic parts, that is, $\mathbf{L} = \mathbf{L}^e + \mathbf{L}^p$, where

$$\mathbf{L}^e = \dot{\mathbf{F}}^e \cdot (\mathbf{F}^e)^{-1}, \mathbf{L}^p = \dot{\mathbf{F}}^e \cdot \dot{\mathbf{F}}^p \cdot (\dot{\mathbf{F}}^p)^{-1} \cdot (\dot{\mathbf{F}}^e)^{-1} \quad (5.7.9)$$

It can be seen that the elastic part has the usual structure of a velocity gradient but is defined by $\mathbf{F}^e$ rather than $\mathbf{F}$. The plastic part defined by $\mathbf{F}^p$ is mapped or pushed forward by $\mathbf{F}^e$. The symmetric and antisymmetric parts of $\mathbf{L}^e$ and $\mathbf{L}^p$ are given by

$$\mathbf{D}^e = \frac{1}{2}(\mathbf{L}^e + \mathbf{L}^{eT}), \mathbf{W}^e = \frac{1}{2}(\mathbf{L}^e + \mathbf{L}^{eT}), \mathbf{D}^p = \frac{1}{2}(\mathbf{L}^p + \mathbf{L}^{pT}), \mathbf{W}^p = \frac{1}{2}(\mathbf{L}^p + \mathbf{L}^{pT}) \quad (5.7.10)$$

The velocity gradient $\bar{\mathbf{L}}$ on $\bar{\Omega}$ is defined as the pull-back of $\mathbf{L}$ by $\mathbf{F}^e$:

$$\bar{\mathbf{L}} = \phi_e^* \mathbf{L} = (\mathbf{F}^e)^{-1} \cdot \mathbf{L} \cdot \mathbf{F}^e = (\mathbf{F}^e)^{-1} \cdot \dot{\mathbf{F}}^e + \dot{\mathbf{F}}^p \cdot (\mathbf{F}^p)^{-1} = \bar{\mathbf{L}}^e + \bar{\mathbf{L}}^p \quad (5.7.11)$$

As indicated previously, $\bar{\mathbf{L}}^e = (\mathbf{F}^e)^{-1} \cdot \dot{\mathbf{F}}^e$ and $\bar{\mathbf{L}}^p = \dot{\mathbf{F}}^p \cdot (\mathbf{F}^p)^{-1}$ are defined as the elastic and plastic parts of $\bar{\mathbf{L}}$. The particular form of the pull-back in (5.7.11) is discussed in Section 5.10.2. The plastic part $\bar{\mathbf{L}}^p$ is used in the formulation of the plastic flow equation next. The symmetric parts of $\bar{\mathbf{L}}, \bar{\mathbf{L}}^e$ and $\bar{\mathbf{L}}^p$ are defined as

$$\begin{aligned}
\bar{\mathbf{D}} &= \operatorname{sym} \bar{\mathbf{L}} = \frac{1}{2}(\bar{\mathbf{C}}^e \cdot \bar{\mathbf{L}} + \bar{\mathbf{L}}^T \cdot \bar{\mathbf{C}}^e) \\
\bar{\mathbf{D}}^e &= \operatorname{sym} \bar{\mathbf{L}}^e = \frac{1}{2}(\bar{\mathbf{C}}^e \cdot \bar{\mathbf{L}}^e + \bar{\mathbf{L}}^{eT} \cdot \bar{\mathbf{C}}^e), \\
\bar{\mathbf{D}}^p &= \operatorname{sym} \bar{\mathbf{L}}^p = \frac{1}{2}(\bar{\mathbf{C}}^e \cdot \bar{\mathbf{L}}^p + \bar{\mathbf{L}}^{pT} \cdot \bar{\mathbf{C}}^e)
\end{aligned} \quad (5.7.12)$$

The use of $\bar{\mathbf{C}}^e$ in forming the symmetric parts in (5.7.12) is justified by noting that $\bar{\mathbf{D}}, \bar{\mathbf{D}}^e$, and $\bar{\mathbf{D}}^p$ are the pull-backs of their spatial counterparts in (5.7.10), that is,

$$\bar{\mathbf{D}} = \mathbf{F}^{eT} \cdot \mathbf{D} \cdot \mathbf{F}^e = \bar{\mathbf{D}}^e + \bar{\mathbf{D}}^p, \bar{\mathbf{D}}^e = \mathbf{F}^{eT} \cdot \mathbf{D}^e \cdot \mathbf{F}^e, \bar{\mathbf{D}}^p = \mathbf{F}^{eT} \cdot \mathbf{D}^p \cdot \mathbf{F}^e \quad (5.7.13)$$

which can be verified from (5.7.10), (5.7.11) and the relation $\bar{\mathbf{C}}^e = (\mathbf{F}^e)^T \cdot \mathbf{F}^e$. Substituting this expression for $\bar{\mathbf{C}}^e$ into the second of (5.7.13) gives (5.7.7), $\dot{\bar{\mathbf{E}}}^e = \bar{\mathbf{D}}^e$. The expression for $\bar{\mathbf{D}}^p$ in (5.7.12) will be used later in the formulation of the J_2 flow rule.

From (5.7.13) $\bar{\mathbf{D}}^e = \bar{\mathbf{D}} - \bar{\mathbf{D}}^p$, and noting that $\dot{\bar{\mathbf{E}}}^e = \bar{\mathbf{D}}^e$, (5.7.6) can be written as

$$\dot{\bar{\mathbf{S}}} = \mathbf{C}^{\bar{S}}_{el}:\bar{\mathbf{D}}^e = \mathbf{C}^{\bar{S}}_{el}:(\bar{\mathbf{D}} - \bar{\mathbf{D}}^p) \tag{5.7.14}$$

In the following, we describe the plastic flow equations and derive the tangent modulus in the usual manner from (5.7.14) in conjunction with the flow rule and the consistency condition.

5.7.4 *Flow Rule*

The plastic flow rule which determines $\dot{\mathbf{F}}^p$ is

$$\bar{\mathbf{L}}^p = \dot{\mathbf{F}}^p \cdot (\mathbf{F}^p)^{-1} = \dot{\lambda}\bar{\mathbf{r}}(\bar{\mathbf{S}}, \bar{\mathbf{q}}) \tag{5.7.15}$$

where $\bar{\mathbf{r}}$ is the plastic flow direction, $\dot{\lambda}$ is the plastic rate parameter and $\bar{\mathbf{q}}$ is a set of internal variables defined on $\bar{\Omega}$. This differs from the hypoelastic–plastic materials in Section 5.6 where plastic flow is specified in terms of $\mathbf{D}^p$, the plastic part of the rate-of-deformation tensor. The evolution of the internal variables is assumed to be governed by a hardening (softening) law of the form

$$\dot{\bar{\mathbf{q}}} = \dot{\lambda}\bar{\mathbf{h}}(\bar{\mathbf{S}}, \bar{\mathbf{q}}) \tag{5.7.16}$$

where $\bar{\mathbf{h}}$ are the plastic moduli. The yield condition is expressed in terms of $\bar{\mathbf{S}}$ as

$$\bar{f}(\bar{\mathbf{S}}, \bar{\mathbf{q}}) = 0 \tag{5.7.17}$$

Because $\bar{\mathbf{S}}$ is invariant under rotations, objectivity imposes no restrictions on the functional dependence on $\bar{\mathbf{S}}$ in (5.7.17) and anisotropic plastic yield behavior can be incorporated.

The loading–unloading conditions are given by $\dot{\lambda} \geq 0, \bar{f} \leq 0, \dot{\lambda}\bar{f} = 0$. From the consistency condition $\dot{\bar{f}} = 0$, it follows that

$$\dot{\lambda} = \frac{\bar{f}_{\bar{\mathbf{S}}}:\mathbf{C}^{\bar{S}}_{el}:\bar{\mathbf{D}}}{-\bar{f}_{\bar{\mathbf{q}}} \cdot \bar{\mathbf{h}} + \bar{f}_{\bar{\mathbf{S}}}:\mathbf{C}^{\bar{S}}_{el}:\mathrm{sym}\bar{\mathbf{r}}}, \quad \bar{f}_{\bar{\mathbf{S}}} = \frac{\partial \bar{f}}{\partial \bar{\mathbf{S}}}, \quad \bar{f}_{\bar{\mathbf{q}}} = \frac{\partial \bar{f}}{\partial \bar{\mathbf{q}}} \tag{5.7.18}$$

where (5.7.15) and ((5.7.10) have been used to obtain an expression for $\bar{\mathbf{D}}^p$ as

$$\bar{\mathbf{D}}^p = \dot{\lambda}\,\mathrm{sym}\,\bar{\mathbf{r}} = \frac{1}{2}\dot{\lambda}(\bar{\mathbf{C}}^e \cdot \bar{\mathbf{r}} + \bar{\mathbf{r}}^T \cdot \bar{\mathbf{C}}^e) \tag{5.7.19}$$

Substituting the result into (5.7.14) leads to the following expression for the elasto-plastic tangent modulus, denoted by $\mathbf{C}^{\bar{S}}$:

$$\dot{\bar{\mathbf{S}}} = \mathbf{C}^{\bar{S}}_{el}:(\bar{\mathbf{D}} - \bar{\mathbf{D}}^p) = \mathbf{C}^{\bar{S}}:\bar{\mathbf{D}}, \; \bar{\mathbf{C}}^{\bar{S}} = \mathbf{C}^{\bar{S}}_{el} - \frac{\left(\mathbf{C}^{\bar{S}}_{el}:\mathrm{sym}\,\bar{\mathbf{r}}\right) \otimes \left(\bar{f}_{\bar{\mathbf{S}}}:\mathbf{C}^{\bar{S}}_{el}\right)}{-\bar{f}_{\bar{\mathbf{q}}} \cdot \bar{\mathbf{h}} + \bar{f}_{\bar{\mathbf{S}}}:\mathbf{C}^{\bar{S}}_{el}:\mathrm{sym}\,\bar{\mathbf{r}}} \tag{5.7.20}$$

Symmetry of the elasto-plastic tangent modulus is obtained for associative plasticity where $\text{sym}\,\bar{\mathbf{r}} \sim f_{\bar{\mathbf{S}}}$.

Alternatively, for rate-dependent plasticity we use (5.7.14) and write the relation between stress rate and elastic strain rate as

$$\dot{\lambda} = \frac{\bar{\phi}(\bar{\mathbf{S}}, \bar{\mathbf{q}})}{\eta}, \quad \dot{\bar{\mathbf{S}}} = \mathbf{C}_{el}^{\bar{S}} : \left(\bar{\mathbf{D}} - \frac{\bar{\phi}}{\eta} \text{sym}\,\bar{\mathbf{r}} \right) \tag{5.7.21}$$

where ϕ is an overstress function and η is the viscosity.

5.7.5 Tangent Moduli

The preceding suffices for a description of the hyperelastic material model. For linearization compatible with Section 6.4, the tangent moduli are needed in terms of the Truesdell rate. To obtain this, we first write $(5.7.20)_1$ in terms of the *elastic* Lie derivative of Kirchhoff stress which we then relate to the Truesdell rate of Cauchy stress. This introduces the *plastic* Lie derivative of the second Piola–Kirchhoff stress on $\bar{\Omega}$. We will make frequent use of the Lie derivative in this section. For a more detailed discussion see Section 5.10.

The elastic Lie derivative of the Kirchhoff stress is given by the usual form of the Lie derivative of a kinetic quantity (see Box 5.17 later); the elastic part of the deformation gradient is used for the pull-back and push-forward:

$$L_v^e \boldsymbol{\tau} \equiv \phi_*^e \left(\frac{D}{Dt} (\phi_e^* \boldsymbol{\tau}) \right) = \mathbf{F}^e \cdot \frac{D}{Dt} ((\mathbf{F}^e)^{-1} \cdot \boldsymbol{\tau} \cdot (\mathbf{F}^e)^{-T}) \cdot \mathbf{F}^{eT} \tag{5.7.22}$$

The push-forward and pull-back operations are performed with $\mathbf{F}^e$ as indicated. The last expression in (5.7.22) can be written as

$$L_v^e \boldsymbol{\tau} = \mathbf{F}^e \cdot \dot{\bar{\mathbf{S}}} \cdot \mathbf{F}^{eT} \tag{5.7.23}$$

which shows that $L_v^e \boldsymbol{\tau}$ is the push-forward of $\dot{\bar{\mathbf{S}}}$ by $\mathbf{F}^e$. Carrying out the differentiation in (5.7.22) and using (5.7.9) yields

$$L_v^e \boldsymbol{\tau} = \dot{\boldsymbol{\tau}} - \mathbf{L}^e \cdot \boldsymbol{\tau} - \boldsymbol{\tau} \cdot \mathbf{L}^{eT} \equiv \boldsymbol{\tau}^{\nabla_{ce}} \tag{5.7.24}$$

that is, the elastic Lie derivative of the Kirchhoff stress is equivalent to the elastic convected rate of stress, $\boldsymbol{\tau}^{\nabla_{ce}}$. The Truesdell rate of Cauchy stress is related to the elastic Lie derivative as follows by a modification of (5.4.22):

$$J\boldsymbol{\sigma}^{\nabla T} = L_v \boldsymbol{\tau} = \dot{\boldsymbol{\tau}} - \mathbf{L} \cdot \boldsymbol{\tau} - \boldsymbol{\tau} \cdot \mathbf{L}^T \equiv L_v^e \boldsymbol{\tau} - \mathbf{L}^p \cdot \boldsymbol{\tau} - \boldsymbol{\tau} \cdot \mathbf{L}^{pT} \tag{5.7.25}$$

where (5.7.24) is used. Pulling the last expression in (5.7.25) back to the intermediate configuration gives

$$\phi_e^*(L_v\boldsymbol{\tau}) = \phi_e^*\left(L_v^e\boldsymbol{\tau} - \mathbf{L}^p\cdot\boldsymbol{\tau} - \boldsymbol{\tau}\cdot\mathbf{L}^{pT}\right)$$
$$= \dot{\overline{\mathbf{S}}} - \overline{\mathbf{L}}^p\cdot\overline{\mathbf{S}} - \overline{\mathbf{S}}\cdot\mathbf{L}^{pT} \tag{5.7.26}$$

The last of these can be thought of as the *plastic* Lie derivative of $\overline{\mathbf{S}}$, that is,

$$L_v^p(\overline{\mathbf{S}}) = \phi_*^p\left(\frac{D}{Dt}(\phi_p^*(\overline{\mathbf{S}}))\right)$$
$$= \mathbf{F}^p\cdot\frac{D}{Dt}((\mathbf{F}^p)^{-1}\cdot\overline{\mathbf{S}}\cdot(\mathbf{F}^p)^{-T})\cdot\mathbf{F}^{pT} = \dot{\overline{\mathbf{S}}} - \overline{\mathbf{L}}^p\cdot\overline{\mathbf{S}} - \overline{\mathbf{S}}\cdot\mathbf{L}^{pT} \tag{5.7.27}$$

where the pull-back and push-forward operations are now from $\overline{\Omega}$ to Ω_0 with the plastic part of the deformation gradient. Comparison of (5.7.26) and (5.7.27) then yields

$$\phi_e^*(L_v\,\boldsymbol{\tau}) = L_v^p\overline{\mathbf{S}},\ \phi_*^e(L_v^p\overline{\mathbf{S}}) = L_v\boldsymbol{\tau} \tag{5.7.28}$$

An alternative way of viewing this is:

$$L_v\boldsymbol{\tau} = \mathbf{F}\cdot\dot{\mathbf{S}}\cdot\mathbf{F}^T = \mathbf{F}^e\cdot(\mathbf{F}^p\cdot\dot{\mathbf{S}}\cdot\mathbf{F}^{pT})\cdot\mathbf{F}^{eT} = \mathbf{F}^e\cdot L_v^p\overline{\mathbf{S}}\cdot\mathbf{F}^{eT} = \phi_*^e(L_v^p\overline{\mathbf{S}}) \tag{5.7.29}$$

To obtain the required tangent modulus we substitute the first equation in (5.7.20) into the last expression in (5.7.27) to give

$$L_v^p(\overline{\mathbf{S}}) = \mathbf{C}^{\overline{S}}:\overline{\mathbf{D}} - \overline{\mathbf{L}}^p\cdot\overline{\mathbf{S}} - \overline{\mathbf{S}}\cdot\mathbf{L}^{pT} \tag{5.7.30}$$

Now using (5.7.15) and (5.7.19) and rearranging gives

$$L_v^p(\overline{\mathbf{S}}) = \left(\mathbf{C}^{\overline{S}} - \frac{(\overline{\mathbf{r}}\cdot\overline{\mathbf{S}} + \overline{\mathbf{S}}\cdot\overline{\mathbf{r}}^T)\otimes\left(\overline{f}_{\overline{\mathbf{S}}}:\mathbf{C}_{el}^{\overline{S}}\right)}{-\overline{f}_{\overline{\mathbf{q}}}\cdot\overline{\mathbf{h}} + \overline{f}_{\overline{\mathbf{S}}}:\mathbf{C}_{el}^{\overline{S}}:\operatorname{sym}\overline{\mathbf{r}}}\right):\overline{\mathbf{D}} = \tilde{\mathbf{C}}^{\overline{S}}:\overline{\mathbf{D}} \tag{5.7.31}$$

which defines the plastic convected modulus $\tilde{\mathbf{C}}^{\overline{S}}$. The final expression is obtained from (5.7.28):

$$L_v\boldsymbol{\tau} = \phi_*^e(\tilde{\mathbf{C}}^{\overline{S}}:\overline{\mathbf{D}}) = \tilde{\mathbf{C}}^{\tau}:\mathbf{D} \tag{5.7.32}$$

where $\mathbf{D} = \phi_*^e(\overline{\mathbf{D}}) = (\mathbf{F}^e)^{-T}\cdot\overline{\mathbf{D}}\cdot(\mathbf{F}^e)^T$ and the spatial moduli are given by

$$\tilde{\mathbf{C}}^{\tau} = \phi_*^e\tilde{\mathbf{C}}^{\overline{S}},\ \tilde{\mathbf{C}}_{ijkl}^{\tau} = F_{im}^eF_{jn}^eF_{kp}^eF_{lq}^e\tilde{\mathbf{C}}_{mnpq}^{\overline{S}} \tag{5.7.33}$$

The spatial moduli $\tilde{\mathbf{C}}^{\tau}$ can also be obtained by pushing every term in the expression for $\tilde{\mathbf{C}}^{\overline{S}}$ in (5.7.31) forward to the spatial configuration:

$$\tilde{\mathbf{C}}^{\tau} = \mathbf{C}^{\tau} - \frac{(\mathbf{r}\cdot\boldsymbol{\tau} + \boldsymbol{\tau}\cdot\mathbf{r}^T)\otimes\left(f_\tau:\mathbf{C}_{el}^{\tau}\right)}{-f_{\mathbf{q}}\cdot\mathbf{h} + f_\tau:\mathbf{C}_{el}^{\tau}:\operatorname{sym}\mathbf{r}} \tag{5.7.34}$$

where

$$\begin{aligned}
&\mathbf{r} = \phi_*^e \bar{\mathbf{r}} = \mathbf{F}^e \cdot \bar{\mathbf{r}} \cdot (\mathbf{F}^e)^{-1}, \ \mathrm{sym}\,\mathbf{r} = \phi_*^e \mathrm{sym}\,\bar{\mathbf{r}} = (\mathbf{F}^e)^{-T} \cdot \mathrm{sym}\,\bar{\mathbf{r}} \cdot (\mathbf{F}^e)^{-1}, \\
&f = \phi_*^e \bar{f} = \bar{f}, \mathbf{q} = \phi_*^e \bar{\mathbf{q}} = \bar{\mathbf{q}}, \mathbf{h} = \phi_*^e \bar{\mathbf{h}} = \bar{\mathbf{h}}, \ f_\tau = \phi_*^e f_{\bar{S}} = (\mathbf{F}^e)^{-T} \cdot f_{\bar{S}} \cdot (\mathbf{F}^e)^{-1} \\
&\mathbf{C}_{el}^{\tau} = \phi_*^e \mathbf{C}_{el}^{\bar{S}}, (C_{el}^{\tau})_{ijkl} = F_{im}^e F_{jn}^e F_{kp}^e F_{lq}^e (C_{el}^{\bar{S}})_{mnpq}, \ \mathbf{C}^{\tau} = \phi_*^e \mathbf{C}^{\bar{S}}, \ C_{ijkl}^{\tau} = F_{im}^e F_{jn}^e F_{kp}^e F_{lp}^e C_{mnpq}^{\bar{S}}
\end{aligned} \tag{5.7.35}$$

5.7.6 J_2 Flow Theory

We illustrate the hyperelastic–plastic formulation by an elastic–plastic J_2 flow model with isotropic hardening and neo-Hookean elasticity. We begin with the hyperelastic potential for neo-Hookean elasticity (see (5.4.54–58)) here specified on the intermediate configuration $\bar{\Omega}$:

$$\bar{w} = \frac{1}{2}\lambda_0^e (\ln J_e)^2 - \mu_0 \ln J_e + \frac{1}{2}\mu_0 (\mathrm{trace}\,\bar{\mathbf{C}}^e - 3) \tag{5.7.36}$$

where $J^e = \det \mathbf{F}^e$ and λ_0^e and μ_0 are the Lamé constants. The stress is derived from the elastic potential through (5.7.5) and is given by

$$\bar{\mathbf{S}} = \lambda_0^e \ln J_e (\bar{\mathbf{C}}^e)^{-1} + \mu_0 (\mathbf{I} - (\bar{\mathbf{C}}^e)^{-1}), \ \boldsymbol{\tau} = \lambda_0^e \ln J_e \mathbf{g}^{-1} + \mu_0 (\mathbf{B}^e - \mathbf{g}^{-1}) \tag{5.7.37}$$

where $\mathbf{B}^e = \mathbf{F}^e \cdot \mathbf{F}^{eT}$ (recall that $\mathbf{g} = \mathbf{I} = \mathbf{g}^{-1}$). Let $\lambda^e = \lambda_0$ and $\mu = \mu_0 - \lambda^e \ln J_e$; from (5.7.6) the elasticity tensor in component form is

$$\begin{aligned}
\left(C_{el}^{\bar{S}}\right)_{ijkl} &= \lambda^e (\bar{C}^e)_{ij}^{-1} C_{kl}^{-1} + \mu\left((\bar{C}^e)_{ik}^{-1} (\bar{C}^e)_{jl}^{-1} + (\bar{C}^e)_{il}^{-1} (\bar{C}^e)_{kj}^{-1}\right) \text{ on } \bar{\Omega} \\
\left(C_{el}^{\tau}\right)_{ijkl} &= \lambda^e \delta_{ij}\delta_{kl} + \mu(\delta_{ik}\delta_{jl} + \delta_{il}\delta_{kj}) \text{ on } \Omega
\end{aligned} \tag{5.7.38}$$

To specify the flow rule, we introduce $\bar{\mathbf{S}}^{\mathrm{dev}}$, the deviatoric part of $\bar{\mathbf{S}}$:

$$\bar{\mathbf{S}}^{\mathrm{dev}} = \bar{\mathbf{S}} - \frac{1}{3}(\bar{\mathbf{S}}{:}\bar{\mathbf{C}}^e)\bar{\mathbf{C}}^{e-1}, \ \boldsymbol{\tau}^{\mathrm{dev}} = \boldsymbol{\tau} - \frac{1}{3}(\boldsymbol{\tau}{:}\mathbf{g})\mathbf{g}^{-1} = \mathbf{F}^e \cdot \bar{\mathbf{S}}^{\mathrm{dev}} \cdot \mathbf{F}^{eT} = \phi_*^e \bar{\mathbf{S}}^{\mathrm{dev}} \tag{5.7.39}$$

where the last relation indicates that $\boldsymbol{\tau}^{\mathrm{dev}}$ is the push-forward of $\bar{\mathbf{S}}^{\mathrm{dev}}$ and illustrates that the role of $\bar{\mathbf{C}}^e$ in forming the deviatoric part of $\bar{\mathbf{S}}$ is analogous to that of $\mathbf{g} = \mathbf{I}$ in forming $\boldsymbol{\tau}^{\mathrm{dev}}$. In the phenomenological J_2 flow theory given here, it is assumed that the plastic spin vanishes, that is, $\bar{\mathbf{W}}^p = 0$. It therefore suffices to specify the flow rule through the symmetric part of $\bar{\mathbf{L}}^p$, that is,

$$\bar{\mathbf{D}}^p = \dot{\lambda}\,\mathrm{sym}\,\bar{\mathbf{r}} = \dot{\lambda}\frac{3}{2\bar{\sigma}}\bar{\mathbf{C}}^e \cdot \bar{\mathbf{S}}^{\mathrm{dev}} \cdot \bar{\mathbf{C}}^e, \ \ \mathbf{D}^p = \dot{\lambda}\,\mathrm{sym}\,\mathbf{r} = \dot{\lambda}\frac{3}{2\bar{\sigma}}\mathbf{g} \cdot \boldsymbol{\tau}^{\mathrm{dev}} \cdot \mathbf{g} \tag{5.7.40}$$

which is deviatoric in the sense that $\bar{\mathbf{C}}^{e-1}{:}\,\bar{\mathbf{D}}^p = \mathbf{g}^{-1}{:}\mathbf{D}^p \equiv \mathbf{I}{:}\mathbf{D}^p = 0$. Note that (5.7.40) is of form (5.7.15) because of the one-to-one correspondence of $\bar{\mathbf{C}}^e$ and $\bar{\mathbf{S}}$ through the hyperelastic potential (5.7.5). In (5.7.40), $\bar{\sigma}$ is the von Mises effective stress given by

$$\bar{\sigma}^2 = \frac{3}{2}(\bar{\mathbf{S}}^{\text{dev}} \cdot \mathbf{C}^e) : (\bar{\mathbf{S}}^{\text{dev}} \cdot \bar{\mathbf{C}}^e)^T = \frac{3}{2}(\boldsymbol{\tau}^{\text{dev}} \cdot \mathbf{g}) : (\boldsymbol{\tau}^{\text{dev}} \cdot \mathbf{g})^T \tag{5.7.41}$$

With the above specifications of the elastic and plastic response, the elasto-plastic tangent modulus can be derived from (5.7.31). The equations for hyperelastic–plastic J_2 flow theory with von Mises yield surface are summarized in Box 5.12.

Box 5.12 Hyperelastic–plastic J_2 flow theory constitutive model

Multiplicative decomposition

$$\mathbf{F} = \mathbf{F}^e \cdot \mathbf{F}^p \tag{B5.12.1}$$

Hyperelastic response

$$\bar{\mathbf{S}} = 2\frac{\partial \bar{\psi}(\bar{\mathbf{C}}^e)}{\partial \bar{\mathbf{C}}^e} = \frac{\partial \bar{w}(\bar{\mathbf{E}}^e)}{\partial \bar{\mathbf{E}}^e} \tag{B5.12.2}$$

$$\mathbf{C}^{\bar{S}}_{el} = 2\frac{\partial \bar{\mathbf{S}}}{\partial \bar{\mathbf{C}}^e} = 4\frac{\partial^2 \bar{\psi}(\bar{\mathbf{C}}^e)}{\partial \bar{\mathbf{C}}^e \partial \bar{\mathbf{C}}^e} = \frac{\partial^2 \bar{w}(\mathbf{E}^e)}{\partial \mathbf{E}^e \partial \mathbf{E}^e} \tag{B5.12.3}$$

Rate form of hyperelasticity:

$$\dot{\bar{\mathbf{S}}} = \mathbf{C}^{\bar{S}}_{el} : \bar{\mathbf{D}}^e \tag{B5.12.4}$$

Plastic response

Flow rule:

$$\bar{\mathbf{D}}^p = \dot{\lambda} \ \text{sym}\,\bar{\mathbf{r}}(\bar{\mathbf{S}}, \bar{\mathbf{q}}), \dot{\bar{\varepsilon}} = \dot{\lambda} \tag{B5.12.5}$$

Plastic flow direction:

$$\text{sym}\,\bar{\mathbf{r}} = \frac{3}{2\bar{\sigma}}\bar{\mathbf{C}}^e \cdot \bar{\mathbf{S}}^{\text{dev}} \cdot \bar{\mathbf{C}}^e, \text{sym}\,\mathbf{r} = \frac{3}{2\bar{\sigma}}\mathbf{g} \cdot \boldsymbol{\tau}^{\text{dev}} \cdot \mathbf{g} \tag{B5.12.6}$$

Effective stress:

$$\bar{\sigma}^2 = \frac{3}{2}(\bar{\mathbf{S}}^{\text{dev}} \cdot \bar{\mathbf{C}}^e) : (\bar{\mathbf{S}}^{\text{dev}} \cdot \bar{\mathbf{C}}^e)^T = \frac{3}{2}(\boldsymbol{\tau}^{dev} \cdot \mathbf{g}) : (\boldsymbol{\tau}^{dev} \cdot \mathbf{g})^T \tag{B5.12.7}$$

Yield condition:

$$\bar{f}(\bar{\mathbf{S}}, \bar{\mathbf{q}}) = \bar{\sigma} - \sigma_Y(\bar{\varepsilon}) = 0 \tag{B5.12.8}$$

Loading-unloading conditions:

$$\dot{\lambda} \geq 0, \bar{f} \leq 0, \dot{\lambda}\bar{f} = 0 \tag{B5.12.9}$$

Plastic flow rate – rate-independent:

$$\dot{\lambda} = \frac{\bar{f}_{\bar{\mathbf{S}}} : \mathbf{C}_{el}^{\bar{S}} : \bar{\mathbf{D}}}{-\bar{f}_{\bar{\mathbf{q}}} \cdot \bar{\mathbf{h}} + \bar{f}_{\bar{\mathbf{S}}} : \mathbf{C}_{el}^{\bar{S}} : \operatorname{sym} \bar{\mathbf{r}}} \tag{B5.12.10}$$

Plastic flow rate – rate-dependent:

$$\dot{\lambda} = \dot{\bar{\varepsilon}} = \frac{\bar{\phi}}{\eta}(\bar{\mathbf{S}}, \bar{\mathbf{q}}) \tag{B5.12.11}$$

Stress rate:

$$\dot{\bar{\mathbf{S}}} = \mathbf{C}^{\bar{S}} : \bar{\mathbf{D}}$$

$$\mathbf{C}^{\bar{S}} = \mathbf{C}_{el}^{\bar{S}} - \frac{\left(\mathbf{C}_{el}^{\bar{S}} : \operatorname{sym} \bar{\mathbf{r}}\right) \otimes \left(\bar{f}_{\bar{\mathbf{S}}} : \mathbf{C}_{el}^{\bar{S}}\right)}{-\bar{f}_{\bar{\mathbf{q}}} \cdot \bar{\mathbf{h}} + \bar{f}_{\bar{\mathbf{S}}} : \operatorname{sym} \bar{\mathbf{r}}} \tag{B5.12.12}$$

Plastic Lie derivative:

$$L_v^p(\bar{\mathbf{S}}) = \tilde{\mathbf{C}}^{\bar{S}} : \bar{\mathbf{D}}$$

$$\tilde{\mathbf{C}}^{\bar{S}} = \mathbf{C}^{\bar{S}} - \frac{(\bar{\mathbf{r}} \cdot \bar{\mathbf{S}} + \bar{\mathbf{S}} \cdot \bar{\mathbf{r}}^T) \otimes \left(\bar{f}_{\bar{\mathbf{S}}} : \mathbf{C}_{el}^{\bar{S}}\right)}{-\bar{f}_{\bar{\mathbf{q}}} \cdot \bar{\mathbf{h}} + \bar{f}_{\bar{\mathbf{S}}} : \mathbf{C}_{el}^{\bar{S}} : \operatorname{sym} \bar{\mathbf{r}}} \tag{B5.12.13}$$

where it is understood that $\tilde{\mathbf{C}}^{\bar{S}} = \mathbf{C}_{el}^{\bar{S}}$ for elastic loading or unloading. The spatial tangent moduli can be obtained from (5.7.33) or (5.7.34).

$$\mathbf{C}^{\sigma T} = J^{-1}\left(\mathbf{C}^{\tau} - \frac{(\mathbf{r} \cdot \boldsymbol{\tau} + \boldsymbol{\tau} \cdot \mathbf{r}) \otimes \left(f_{\tau} : \mathbf{C}_{el}^{\tau}\right)}{-f_{\mathbf{q}} \cdot \mathbf{h} + f_{\tau} : \mathbf{C}_{el}^{\tau} : \operatorname{sym} \mathbf{r}}\right) \tag{B5.12.14}$$

For Neo-Hookean hyperelastic response and letting $\hat{\mathbf{n}} = \sqrt{\frac{2}{3}}\mathbf{r}$, the elasto-plastic modulus is given by

$$\mathbf{C}^{\tau} = \lambda^e \mathbf{I} \otimes \mathbf{I} + 2\mu \mathbf{I} - 2\mu\gamma \hat{\mathbf{n}} \otimes \hat{\mathbf{n}}, \ \gamma = \frac{1}{1 + (H/3\mu)} \tag{B5.12.15}$$

where $\lambda^e = \lambda_0^e, \mu = \mu_0 - \lambda^e \ln J_e$ and $J_e = \det \mathbf{F}^e$, and the Truesdell modulus is therefore

$$\mathbf{C}^{\sigma T} = J^{-1}(\mathbf{C}^{\tau} - 2\mu\gamma(\hat{\mathbf{n}} \cdot \boldsymbol{\tau} + \boldsymbol{\tau} \cdot \hat{\mathbf{n}}) \otimes \hat{\mathbf{n}}) \tag{B5.12.16}$$

5.7.7 Implications for Numerical Treatment of Large Rotations

The advantages of the hyperelastic representation (5.7.5) over hypoelastic treatments for problems involving large rotations arise from the issue of material objectivity or frame invariance. See Section 5.10.3 for a more extensive discussion of material frame indifference and Section 5.10.8 for applications to the hyperelastic–plastic model. To satisfy frame invariance, the material response must be independent of the frame of reference of the observer. This requires that various kinematic and stress measures be objective, that is,, they must transform appropriately to preserve the correct material relations in the rotating frame. An alternative approach to frame invariance is to formulate the constitutive response in terms of tensors which are not affected by rotation. In the following, we show that the hyperelastic-plastic formulation is frame invariant. Let $\mathbf{Q}$ be a time-dependent rotation of the current configuration leaving the reference and intermediate configurations unrotated. The deformation gradient after rotation is given by $\mathbf{F}^*=\mathbf{Q}\cdot\mathbf{F}$ and from (5.7.1) the elastic and plastic parts are given by $\mathbf{F}^{*e}=\mathbf{Q}\cdot\mathbf{F}^e$, $\mathbf{F}^{*P}=\mathbf{F}^P$. It follows therefore from (5.7.3) that $\bar{\mathbf{E}}^{e*}=\bar{\mathbf{E}}^e$, so the Lagrange strain on $\bar{\Omega}$ is unaffected by rotation. The corresponding stress is given by (5.7.5), that is, $\bar{\mathbf{S}}^*=\partial\bar{w}/\partial\bar{\mathbf{E}}^{e*}=\partial\bar{w}/\partial\bar{\mathbf{E}}^e=\bar{\mathbf{S}}$, and is therefore also unaffected by rotation. The stress on the intermediate configuration is thus completely independent of rotations. This eliminates the need for incrementally objective integration schemes which are required in the hypoelastic-plastic formulations.

5.7.8 Single-Crystal Plasticity

In crystal plasticity models (Asaro and Rice, 1977; Asaro, 1983; Harren et *al.*, 1989) plastic slip takes place on a set of crystallographic planes and is specified by the flow rule

$$\bar{\mathbf{L}}^P=\sum_{\alpha}\dot{\gamma}^{(\alpha)}\mathbf{m}^{(\alpha)}\otimes\mathbf{n}^{(\alpha)} \tag{5.7.42}$$

where $\mathbf{m}^{(\alpha)}$ is the plastic slip direction and $\mathbf{n}^{(\alpha)}$ is the normal to slip plane α. It remains to specify the scalar plastic strain rate $\dot{\gamma}^{(\alpha)}$ on the slip planes. Both rate-independent and rate-dependent formulations have been developed (see Havner (1992) for a thorough treatment). For a rate-dependent model (Asaro, 1983; Harren *et al.*, 1989) which takes the form

$$\dot{\gamma}^{(\alpha)}=\dot{\gamma}_0\left(\frac{\left|\tau^{(\alpha)}\right|}{g(\gamma^{(\alpha)})}\right)^{1/m} \tag{5.7.43}$$

where $\tau^{(\alpha)}=\mathbf{m}^{(\alpha)}\cdot(\bar{\mathbf{C}}^e\cdot\bar{\mathbf{S}})\cdot\mathbf{n}^{(\alpha)}$ is the resolved shear stress and $\gamma^{(\alpha)}$ is the accumulated plastic strain on the slip system. The plastic strain rate (5.7.43) is similar to the strain rate in the empirical model (5.6.40). The quantity $g(\gamma^{(\alpha)})$ acts as a yield strength. With these expressions for the flow rule and the specification of the elastic response in terms of a suitable hyperelastic potential which accounts for the crystal elasticity, the formulation of the constitutive relations parallels that presented in Box 5.12.

5.8 Viscoelasticity

5.8.1 Small Strains

Many materials, such as polymers, exhibit a rate- and time-dependent behavior which is known as viscoelastic. The characteristic feature of viscoelastic material response is a fading memory. A simple schematic of linear viscoelastic material behavior is given by the Maxwell model shown in Figure 5.14. It consists of a linear spring in series with a dashpot. As we will see, the Maxwell model exhibits behaviors which are fluid-like and solid-like. The spring models the elastic response and is assigned a stiffness E. The dashpot models the viscous response and is assigned a viscosity η. The total strain of the spring and the dashpot combined is taken to be the sum of the elastic and viscous strains

$$\varepsilon = \varepsilon^{e} + \varepsilon^{v} \tag{5.8.1}$$

Taking the material time derivative of this expression gives

$$\dot{\varepsilon} = \dot{\varepsilon}^{e} + \dot{\varepsilon}^{v} \tag{5.8.2}$$

Noting that $\dot{\varepsilon}^{e} = \dot{\sigma} / E$ and $\dot{\varepsilon}^{v} = \sigma/\eta$, (5.8.2) can be written as

$$\dot{\sigma} + \frac{\sigma}{\tau} = E\dot{\varepsilon} \tag{5.8.3}$$

where $\tau = \eta/E$ is called the relaxation time; (5.8.3) is an ordinary differential equation for stress σ with constant coefficients. The right-hand side term $E\dot{\varepsilon}$ can be interpreted as the forcing function. A solution is given by the convolution integral

$$\sigma(t) = \int_{-\infty}^{t} E \exp\left[-(t-t')/\tau\right] \frac{d\varepsilon(t')}{dt'} dt' \tag{5.8.4}$$

For more general one-dimensional models, the stress is given by

$$\sigma(t) = \int_{-\infty}^{t} R(t-t') \frac{d\varepsilon(t')}{dt'} dt' \tag{5.8.5}$$

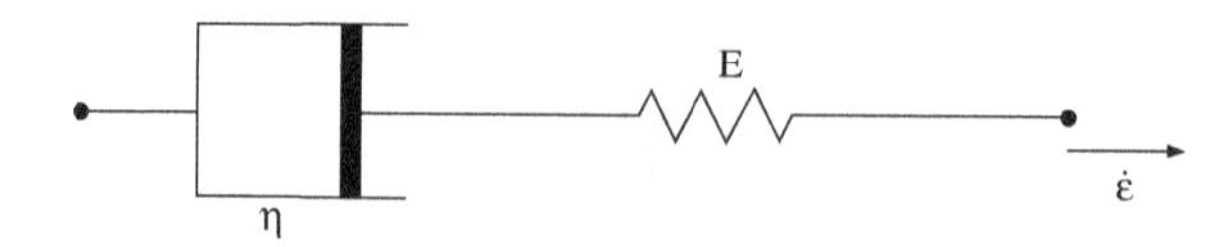

Figure 5.14 Maxwell Element: spring stiffness E, dashpot viscosity η

where the kernel $R(t)$ is called the relaxation modulus. For the special case of the Maxwell model the relaxation modulus is given by $R(t)=E\exp(-t/\tau)$.

The convolution integral (5.8.5) can be extended to the multiaxial case by

$$\sigma_{ij}=\int_{-\infty}^{t}\hat{C}_{ijkl}(t-t')\frac{d\varepsilon_{ij}(t')}{\partial t'}dt' \tag{5.8.6}$$

where $\hat{C}_{ijkl}(t)$ are the relaxation moduli. The relaxation moduli have the same minor symmetries as the linear elastic moduli. Major symmetry is assumed.

An as example, for an isotropic material (5.8.6) can be written in terms of two relaxation functions relating the deviatoric and hydrostatic parts of stress and strain:

$$\sigma_{ij}^{\text{dev}}=2\int_{-\infty}^{t}\hat{\mu}(t-t')\frac{\partial\varepsilon_{ij}^{\text{dev}}(t')}{\partial t'}dt',\ \sigma_{kk}=3\int_{-\infty}^{t}\hat{K}(t-t')\frac{\partial\varepsilon_{kk}(t')}{\partial t'}dt' \tag{5.8.7}$$

To fit polymer behavior, the relaxation moduli may be represented by a finite Dirichlet series of exponential relaxation functions (generalized Maxwell model), that is,

$$K(t)=\sum_{i=1}^{N_b}K^i\exp\left(-t/\tau_i^b\right),\ \mu(t)=\sum_{i=1}^{N_s}\mu^i\exp\left(-t/\tau_i^s\right) \tag{5.8.8}$$

where K^i and μ^i are the bulk and shear moduli of the N_b and N_s elements in the series and τ_i^b and τ_i^s are the corresponding relaxation times. For a crosslinked polymer (viscoelastic *solid*) one of the relaxation times will be infinite in each of the previous expressions (so that the functions $K(t)$ and $\mu(t)$ relax to constant values, called the long-term or rubbery moduli, and not to zero). For extensions of the above models into the nonlinear regime, see Losi and Knauss (1992), for example.

5.8.2 *Finite Strain Viscoelasticity*

The generalization of small-strain viscoelastic constitutive relations to finite strain can be carried out in several different ways. As with elasticity, many different stress and strain measures can be used. Care must also be taken to assure that the constitutive equation is frame-indifferent. Here, for illustrative purposes, we develop the finite strain generalization of (5.8.6). We consider the case where the model reduces to a hyperelastic material in the absence of viscosity and to a Newtonian viscous fluid in the absence of elasticity.

To satisfy frame invariance, we write a straightforward generalization of (5.8.6) in terms of the second Piola–Kirchhoff stress as follows:

$$\mathbf{S}=\int_{-\infty}^{t}R\left(t,t',\mathbf{E}\right):\frac{\partial\mathbf{E}(t')}{\partial t'}dt' \tag{5.8.9}$$

where R is the relaxation function. Here we consider relaxation functions which are written as a Prony series

$$R(t, t', \mathbf{E}) = \sum_{\alpha=1}^{N} \mathbf{C}_{\alpha}^{SE}(\mathbf{E}(t')) \exp[-(t-t')/t_{\alpha}] \tag{5.8.10}$$

To recover the response of a purely hyperelastic material, we set the relaxation times t_{α} to infinity in (5.8.10) to obtain

$$\mathbf{S} = \int_{-\infty}^{t} \mathbf{C}^{SE} : \frac{\partial \mathbf{E}}{\partial t'} dt' = \frac{\partial w}{\partial \mathbf{E}}, \; \mathbf{C}^{SE} = \sum_{\alpha=1}^{N} \mathbf{C}_{\alpha}^{SE} = \frac{\partial^2 w}{\partial \mathbf{E} \partial \mathbf{E}} \tag{5.8.11}$$

where $\mathbf{C}^{SE}$ is the elasticity tensor and we have indicated that it is derivable from a potential.
Differentiating (5.8.9) using the relaxation function (5.8.10) gives

$$\dot{\mathbf{S}}^{\alpha} + \frac{\mathbf{S}^{\alpha}}{t_{\alpha}} = \mathbf{C}_{\alpha}^{SE} : \dot{\mathbf{E}}, \; \mathbf{S} = \sum_{\alpha=1}^{N} \mathbf{S}^{\alpha} \tag{5.8.12}$$

which is a (parallel) series of Maxwell elements; $\mathbf{S}^{\alpha}$ are called the partial stresses. Pushing the expression (5.8.12) forward to the spatial configuration gives

$$L_v \boldsymbol{\tau}^{\alpha} + \frac{\boldsymbol{\tau}^{\alpha}}{t^{\alpha}} = \mathbf{C}_{\alpha}^{\tau} : \mathbf{D}, \; \boldsymbol{\tau} = \sum_{\alpha=1}^{N} \boldsymbol{\tau}^{\alpha} = \sum_{\alpha=1}^{N} \phi_* \mathbf{S}^{\alpha} = \sum_{\alpha=1}^{N} \mathbf{F} \cdot \mathbf{S}^{\alpha} \cdot \mathbf{F}^T \tag{5.8.13}$$

where $L_v \boldsymbol{\tau}^{\alpha} = \phi_* \dot{\mathbf{S}}^{\alpha} \equiv \mathbf{F} \cdot \dot{\mathbf{S}}^{\alpha} \cdot \mathbf{F}^T$ $\mathbf{F}^T$ is the Lie derivative of the partial Kirchhoff stress and $\mathbf{C}_{\alpha}^{\tau} = \phi_* \mathbf{C}_{\alpha}^{SE}$ are the spatial elastic moduli: $\left(C_{\alpha}^{\tau}\right)_{ijkl} = F_{im} F_{jn} F_{kp} F_{lq} \left(C_{\alpha}^{SE}\right)_{mnpq}$.

For constitutive modeling of viscoelastic materials at large strains, see Green and Rivlin (1957) and Coleman and Noll (1961). Constitutive developments and applications to large deformation of polymers are given in Boyce, Parks and Argon (1988). For an extension of the model (5.8.9) to account for nonlinear thermorheological effects based on the free-volume concept, see O'Dowd and Knauss (1995).

5.9 Stress Update Algorithms

The numerical algorithm for integrating the rate constitutive equations is called a *constitutive integration algorithm* or a *stress update algorithm*. Constitutive integration algorithms are presented for rate-independent and rate-dependent materials. For simplicity we start with small strain plasticity. The extension of the small strain algorithms to large deformations is then discussed. It is desirable that integration algorithms for large deformation analysis preserve the underlying objectivity of the constitutive equations. Incrementally objective integration algorithms for large deformation plasticity are presented. A stress update algorithm for large deformation hyperelastic–plastic materials which avoids the integration of stress rate equations is also discussed. The *algorithmic moduli* that are associated with the constitutive integration algorithms, and that are used in the development of the material tangent stiffness matrix (Chapter 6) in implicit solution schemes, are also presented.

5.9.1 Return Mapping Algorithms for Rate-Independent Plasticity

Consider the constitutive equations for small strain, rate-independent elasto-plasticity given in Box 5.10:

$$\begin{aligned}
&\dot{\boldsymbol{\sigma}} = \mathbf{C} : \dot{\varepsilon}^e = \mathbf{C} : (\dot{\varepsilon} - \dot{\varepsilon}^p) \\
&\dot{\varepsilon}^p = \dot{\lambda}\mathbf{r} \\
&\dot{\mathbf{q}} = \dot{\lambda}\mathbf{h} \\
&\dot{f} = f_\sigma : \dot{\boldsymbol{\sigma}} + f_\mathbf{q} \cdot \dot{\mathbf{q}} = 0 \\
&\dot{\lambda} \geq 0, \quad f \leq 0, \quad \dot{\lambda} f \leq 0
\end{aligned} \tag{5.9.1}$$

The purpose of a constitutive integration algorithm is, given the set $(\varepsilon_n, \varepsilon_n^p, \mathbf{q}_n)$ at time n and the strain increment $\Delta\varepsilon = \Delta t\dot{\varepsilon}$, to compute $(\varepsilon_{n+1}, \varepsilon_{n+1}^p, \mathbf{q}_{n+1})$ and satisfy the loading-unloading conditions. Note that the stress at time $n+1$ is given by $\boldsymbol{\sigma}_{n+1} = \mathbf{C}:(\varepsilon_{n+1} - \varepsilon_{n+1}^p)$. From Box 5.10, the consistency condition is solved for $\dot{\lambda}$ to give

$$\dot{\lambda} = \frac{f_\sigma : \mathbf{C} : \dot{\varepsilon}}{-f_\mathbf{q} \cdot \mathbf{h} + f_\sigma : \mathbf{C} : \mathbf{r}} \tag{5.9.2}$$

It might be supposed that we could now use this value of the plasticity parameter to provide rates of stress, plastic strain and internal variables for updating and write the simple forward Euler integration scheme

$$\begin{aligned}
\varepsilon_{n+1} &= \varepsilon_n + \Delta\varepsilon \\
\varepsilon_{n+1}^p &= \varepsilon_n^p + \Delta\lambda_n \mathbf{r}_n \\
\mathbf{q}_{n+1} &= \mathbf{q}_n + \Delta\lambda_n \mathbf{h}_n \\
\boldsymbol{\sigma}_{n+1} &= \mathbf{C} : \left(\varepsilon_{n+1} - \varepsilon_{n+1}^p\right) = \boldsymbol{\sigma}_n + \mathbf{C}^{ep} : \Delta\varepsilon
\end{aligned} \tag{5.9.3}$$

where $\Delta\lambda_n = \Delta t \dot{\lambda}_n$. However, these updated values of stress and internal variables do not satisfy the yield condition at the next step so $f_{n+1} = f(\boldsymbol{\sigma}_{n+1}, \mathbf{q}_{n+1}) \neq 0$ and the solution drifts from the yield surface, often resulting in inaccurate solutions. The integration algorithm (5.9.3) is sometimes called a tangent modulus update scheme. This approach formed the basis for early work in computational rate-independent plasticity but because of the inaccuracies of the method it is no longer favored.

This leads us to consider alternative methods for integrating the rate constitutive equations. One of the objectives of these alternative methods is to enforce consistency at the *end* of the time step, that is, $f_{n+1}=0$, to avoid drift from the yield surface. There are many different approaches to integrating the constitutive equations. A summary of the principal methods is given by Simo and Hughes (1998). Some key issues in the numerical implementation of constitutive models are addressed by Hughes (1984). Here we focus mainly on a class of methods called *return mapping algorithms* which are robust and accurate and which are widely used in practice. The popular *radial return method* for von Mises plasticity is a special case of the return mapping algorithms.

Return mapping schemes consist of an initial elastic-predictor step, involving an excursion (in stress space) away from the yield surface, and a plastic-corrector step which returns the stress to the updated yield surface. Two ingredients of the method are an integration scheme which transforms the set of constitutive equations into a set of nonlinear algebraic equations and a solution scheme for the nonlinear algebraic equations. The method can be based on different integration schemes such as the generalized trapezoidal rule, generalized mid-point rule or Runge–Kutta methods, for example. Here we consider a fully implicit method and a semi-implicit method based on a backward Euler scheme.

5.9.2 Fully Implicit Backward Euler Scheme

In the fully implicit backward Euler method, the increments in plastic strain and internal variables are calculated at the end of the step and the yield condition is enforced at the end of the step. Thus the integration scheme is written as

$$\begin{aligned} \varepsilon_{n+1} &= \varepsilon_n + \Delta\varepsilon \\ \varepsilon^p_{n+1} &= \varepsilon^p_n + \Delta\lambda_{n+1}\mathbf{r}_{n+1} \\ \mathbf{q}_{n+1} &= \mathbf{q}_n + \Delta\lambda_{n+1}\mathbf{h}_{n+1} \\ \boldsymbol{\sigma}_{n+1} &= \mathbf{C}:\left(\varepsilon_{n+1} - \varepsilon^p_{n+1}\right) \\ f_{n+1} &= f\left(\boldsymbol{\sigma}_{n+1}, \mathbf{q}_{n+1}\right) = 0 \end{aligned} \tag{5.9.4}$$

Given the set $\left(\varepsilon_n, \varepsilon^p_n, \mathbf{q}_n\right)$ at time n and the strain increment $\Delta\varepsilon$, (5.9.4) are a set of nonlinear algebraic equations for $\left(\varepsilon_{n+1}, \varepsilon^p_{n+1}, \mathbf{q}_{n+1}\right)$. It is noted that the variables are updated from the converged values at the end of the previous time-step. This avoids nonphysical effects such as spurious unloading which can occur when path-dependent plasticity equations are driven by nonconverged values of the plastic strain and internal variables. The strain ε_{n+1} is the strain obtained from the solution of the system of equations through (6.2.4) or (6.2.31) at time $n+1$. If the solution procedure is implicit, it is understood that ε_{n+1} is the total strain after the last iteration of the implicit solution scheme.

A geometric interpretation of the algorithm is given as follows. First note that from $(5.9.4)_2$, the plastic strain increment is given by

$$\Delta\varepsilon^p_{n+1} \equiv \varepsilon^p_{n+1} - \varepsilon^p_n = \Delta\lambda_{n+1}\mathbf{r}_{n+1} \tag{5.9.5}$$

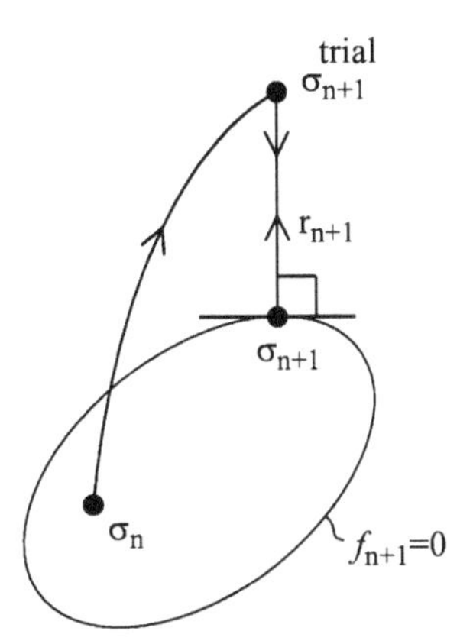

Figure 5.15 Closest point projection scheme for associative plasticity: $\mathbf{r}_{n+1} \sim \partial f/\boldsymbol{\sigma}_{n+1}$

Substituting this expression into $(5.9.4)_4$ gives

$$
\begin{aligned}
\boldsymbol{\sigma}_{n+1} &= \mathbf{C}:\left(\varepsilon_{n+1} - \varepsilon_n^p - \Delta\varepsilon_{n+1}^p\right) \\
&= \mathbf{C}:\left(\varepsilon_n + \Delta\varepsilon - \varepsilon_n^p - \Delta\varepsilon_{n+1}^p\right) = \mathbf{C}:\left(\varepsilon_n - \varepsilon_n^p\right) + \mathbf{C}:\Delta\varepsilon - \mathbf{C}:\Delta\varepsilon_{n+1}^p \\
&= (\boldsymbol{\sigma}_n + \mathbf{C}:\Delta\varepsilon) - \mathbf{C}:\Delta\varepsilon_{n+1}^p \\
&= \boldsymbol{\sigma}_{n+1}^{\text{trial}} - \mathbf{C}:\Delta\varepsilon_{n+1}^p = \boldsymbol{\sigma}_{n+1}^{\text{trial}} - \Delta\lambda_{n+1}\mathbf{C}:\mathbf{r}_{n+1}
\end{aligned}
\tag{5.9.6}
$$

where $\boldsymbol{\sigma}_{n+1}^{\text{trial}} = \boldsymbol{\sigma}_n + \mathbf{C}:\Delta\varepsilon$ is the *trial* stress of *elastic predictor* and the quantity $-\Delta\lambda_{n+1}\mathbf{C}:\mathbf{r}_{n+1}$ is the *plastic corrector* which returns or projects the trial stress onto the suitably updated (accounting for hardening) yield surface along a direction specified by the plastic flow direction at the end-point (see Figure 5.15). The elastic-predictor phase is driven by the increment in total strain while the plastic-corrector phase is driven by the increment $\Delta\lambda_{n+1}$ in the plasticity parameter. Thus, during the elastic-predictor stage, the plastic strain and internal variables remain fixed, and during the plastic-corrector stage, the total strain is fixed. A consequence of this is that, from (5.9.4) during the plastic-corrector phase,

$$
\Delta\boldsymbol{\sigma}_{n+1} = -\mathbf{C}:\Delta\varepsilon_{n+1}^p = -\Delta\lambda_{n+1}\mathbf{C}:\mathbf{r}_{n+1}
\tag{5.9.7}
$$

which is a result we will use next in the solution of (5.9.4).

The solution of the set of nonlinear algebraic equations (5.9.4) is typically obtained by a Newton procedure. As discussed by Simo and Hughes (1998), a Newton procedure based on the systematic linearization of equations (5.9.4) gives rise to a plastic-corrector return to the yield surface based on the concept of *closest point projection*. During the plastic-corrector stage of the algorithm, the total strain is constant and linearization is with respect to the increment $\Delta\lambda$ in the plasticity parameter. We use the following notation in the Newton procedure: for the linearization of an equation $g\,(\Delta\lambda)=0$, with $\Delta\lambda^{(0)}=0$, at the kth iteration we write

$$
g^{(k)} + \left(\frac{dg}{d\Delta\lambda}\right)^{(k)} \delta\lambda^{(k)} = 0, \quad \Delta\lambda^{(k+1)} = \Delta\lambda^{(k)} + \delta\lambda^{(k)}
\tag{5.9.8}
$$

where $\delta\lambda^{(k)}$ is the increment in $\Delta\lambda$ at the kth iteration. For the most part, we will omit the load or time increment subscript $n+1$ on quantities in the remainder of this chapter. Thus, unless indicated otherwise, all quantities are evaluated at time $n+1$.

We write the plastic updates and yield condition in (5.9.4) in the form (5.9.8), suitable for Newton iteration:

$$\begin{aligned} \mathbf{a} &= -\varepsilon^{p} + \varepsilon_{n}^{p} + \Delta\lambda\mathbf{r} = \mathbf{0} \\ \mathbf{b} &= -\mathbf{q} + \mathbf{q}_{n} + \Delta\lambda\mathbf{h} = \mathbf{0} \\ f &= f(\sigma, \mathbf{q}) = 0 \end{aligned} \tag{5.9.9}$$

Linearization of these equations gives (using (5.9.7) in the form $\Delta\varepsilon^{p^{(k)}} = -\mathbf{C}^{-1} : \Delta\boldsymbol{\sigma}^{(k)}$)

$$\begin{aligned} &\mathbf{a}^{(k)} + \mathbf{C} : \Delta\boldsymbol{\sigma}^{(k)} + \Delta\lambda^{(k)}\Delta\mathbf{r}^{(k)} + \delta\lambda^{(k)}\mathbf{r}^{(k)} = \mathbf{0} \\ &\mathbf{b}^{(k)} - \Delta\mathbf{q}^{(k)} + \Delta\lambda^{(k)}\Delta\mathbf{h}^{(k)} + \delta\lambda^{(k)}\mathbf{h}^{(k)} = \mathbf{0} \\ &f^{(k)} + f_{\sigma}^{(k)} : \Delta\boldsymbol{\sigma}^{(k)} + f_{\mathbf{q}}^{(k)} \cdot \Delta_{\mathbf{q}}^{(k)} = 0 \end{aligned} \tag{5.9.10}$$

where

$$\begin{aligned} \Delta\mathbf{r}^{(k)} &= \mathbf{r}_{\sigma}^{(k)} : \Delta\boldsymbol{\sigma}^{(k)} + \mathbf{r}_{\mathbf{q}}^{(k)} \cdot \Delta\mathbf{q}^{(k)} \\ \Delta\mathbf{h}^{(k)} &= \mathbf{h}_{\sigma}^{(k)} : \Delta\boldsymbol{\sigma}^{(k)} + \mathbf{h}_{\mathbf{q}}^{(k)} \cdot \Delta\mathbf{q}^{(k)} \end{aligned} \tag{5.9.11}$$

and where a subscript $\boldsymbol{\sigma}$ or $\mathbf{q}$ denotes a partial derivative. Equations (5.9.10) are a set of three equations which can be solved for $\Delta\boldsymbol{\sigma}^{(k)}$, $\Delta\mathbf{q}^{k}$ and $\delta\lambda^{(k)}$. Substituting (5.9.11) into the first two of (5.9.10) and writing the resulting pair of equations in matrix form gives

$$[\mathbf{A}^{(k)}]^{-1} \begin{Bmatrix} \Delta\boldsymbol{\sigma}^{(k)} \\ \Delta\boldsymbol{\sigma}^{(k)} \end{Bmatrix} = -\{\tilde{\mathbf{a}}^{(k)}\} - \delta\lambda^{(k)}\{\tilde{\mathbf{r}}\} \tag{5.9.12}$$

where

$$\left[\mathbf{A}^{(k)}\right]^{-1} = \begin{bmatrix} \mathbf{C}^{-1} + \Delta\lambda\mathbf{r}_{\sigma} & \Delta\lambda\mathbf{r}_{\mathbf{q}} \\ \Delta\lambda\mathbf{h}_{\sigma} & -\mathbf{I} + \Delta\lambda\mathbf{h}_{\mathbf{q}} \end{bmatrix}^{(k)}, \left\{\tilde{\mathbf{a}}^{(k)}\right\} = \begin{Bmatrix} \mathbf{a}(k) \\ \mathbf{b}(k) \end{Bmatrix}, \left\{\tilde{\mathbf{r}}^{(k)}\right\} = \begin{Bmatrix} \mathbf{r}^{(k)} \\ \mathbf{h}^{(k)} \end{Bmatrix} \tag{5.9.13}$$

Solving (5.9.12) for the stress and internal variable increments gives

$$\begin{Bmatrix} \Delta\boldsymbol{\sigma}^{(k)} \\ \Delta\mathbf{q}^{(k)} \end{Bmatrix} = -\left[\mathbf{A}^{(k)}\right]\left\{\tilde{\mathbf{a}}^{(k)}\right\} - \delta\lambda^{(k)}\left[\mathbf{A}^{(k)}\right]\left\{\tilde{\mathbf{r}}^{(k)}\right\} \tag{5.9.14}$$

Substituting this result into $(5.9.10)_3$ and solving for $\delta\lambda^{(k)}$ we get

$$\delta\lambda^{(k)} = \frac{f^{(k)} - \partial\mathbf{f}^{(k)}\mathbf{A}^{(k)}\tilde{\mathbf{a}}^{(k)}}{\partial\mathbf{f}^{(k)}\mathbf{A}^{(k)}\tilde{\mathbf{r}}^{(k)}} \tag{5.9.15}$$

where we have used the notation $\partial\mathbf{f}=[f_{\sigma} f_{\mathbf{q}}]$.

Thus, the update of the plastic strain, internal variables and the plasticity parameter is

$$\begin{aligned} \varepsilon^{p(k+1)} &= \varepsilon^{p(k)} + \Delta\varepsilon^{p(k)} = \varepsilon^{p(k)} - \mathbf{C}^{-1} : \Delta\sigma^{(k)} \\ \mathbf{q}^{(k+1)} &= \mathbf{q}^{(k)} + \Delta\mathbf{q}^{(k)} \\ \Delta\lambda^{(k+1)} &= \Delta\lambda^{(k)} + \delta\lambda^{(k)} \end{aligned} \tag{5.9.16}$$

with the increments as given in (5.9.14) and (5.9.15). The Newton procedure is continued until convergence to the updated yield surface is achieved to within a sufficient tolerance. As noted by Simo and Hughes (1998), this procedure is implicit and involves the solution of a local (at the level of the element integration points) system of equations (5.9.12). A complicating feature of the method is that the gradients $\mathbf{r}_{\sigma}$, $\mathbf{r}_{\mathbf{q}}$, $\mathbf{h}_{\sigma}$ and $\mathbf{h}_{\mathbf{q}}$ of the plastic flow direction and the plastic moduli are required. These expressions may be difficult to obtain for complex constitutive models. The complete stress update algorithm is given in Box 5.13.

Box 5.13 Backward Euler return mapping scheme

1. Initialization: set initial values of plastic strain and internal variables to converged values at end of previous time-step, zero the increment in plasticity parameter and evaluate the elastic trial stress:

$$k = 0; \quad \varepsilon^{p^{(0)}} = \varepsilon^{p}_{n}, \quad \mathbf{q}^{(0)} = \mathbf{q}_{n}, \quad \Delta\lambda^{(0)} = 0, \quad \boldsymbol{\sigma}^{(0)} = \mathbf{C} : \left(\varepsilon_{n+1} - \varepsilon^{p(0)}\right)$$

2. Check yield condition and convergence at kth iteration:

$$f^{(k)} = f(\boldsymbol{\sigma}^{(k)}, \mathbf{q}^{(k)}), \quad \left\{\tilde{\mathbf{a}}^{(k)}\right\} = \left\{\begin{matrix}\mathbf{a}^{(k)} \\ \mathbf{b}^{(k)}\end{matrix}\right\}$$

If: $f^{(k)} < TOL_1$ and $\|\tilde{\mathbf{a}}^{(k)}\| < TOL_2$, converged

Else: go to 3.

3. Compute increment in plasticity parameter:

$$\left[\mathbf{A}^{(k)}\right]^{-1} = \begin{bmatrix} \mathbf{C}^{-1} + \Delta\lambda\mathbf{r}_{\sigma} & \Delta\lambda\mathbf{r}_{\mathbf{q}} \\ \Delta\lambda\mathbf{h}_{\sigma} & -\mathbf{I} + \Delta\lambda\mathbf{h}_{\mathbf{q}} \end{bmatrix}^{(k)}, \quad \left\{\tilde{\mathbf{r}}^{(k)}\right\} = \left\{\begin{matrix}\mathbf{r}^{(k)} \\ \mathbf{h}^{(k)}\end{matrix}\right\}, \quad \left[\partial\mathbf{f}^{(k)}\right] = \left[f_{\sigma}^{(k)} \quad f_{\mathbf{q}}^{(k)}\right]$$

$$\delta\lambda^{(k)} = \frac{f^{(k)} - \partial\mathbf{f}^{(k)}\mathbf{A}^{(k)}\tilde{\mathbf{a}}^{(k)}}{\partial\mathbf{f}^{(k)}\mathbf{A}^{(k)}\tilde{\mathbf{r}}^{(k)}}$$

4. Obtain increments in stress and internal variables:

$$\begin{Bmatrix} \Delta\boldsymbol{\sigma}^{(k)} \\ \Delta\mathbf{q}^{(k)} \end{Bmatrix} = -\left[\mathbf{A}^{(k)}\right]\left\{\tilde{\mathbf{a}}^{(k)}\right\} - \delta\lambda^{(k)}\left[\mathbf{A}^{(k)}\right]\left\{\tilde{\mathbf{r}}^{(k)}\right\}$$

5. Update plastic strain and internal variables:

$$\begin{aligned} \varepsilon^{p(k+1)} &= \varepsilon^{p(k)} + \Delta\varepsilon^{p(k)} = \varepsilon^{p(k)} - \mathbf{C}^{-1}:\Delta\boldsymbol{\sigma}^{(k)} \\ \mathbf{q}^{(k+1)} &= \mathbf{q}^{(k)} + \Delta\mathbf{q}^{(k)} \\ \Delta\lambda^{(k+1)} &= \Delta\lambda^{(k)} + \delta\lambda^{(k)} \\ \boldsymbol{\sigma}^{(k+1)} &= \boldsymbol{\sigma}^{(k)} + \Delta\boldsymbol{\sigma}^{(k)} = \mathbf{C}:\left(\varepsilon_{n+1} - \varepsilon^{p(k+1)}\right) \\ & k \leftarrow k+1, \text{go to } 2 \end{aligned}$$

5.9.3 *Application to J_2 Flow Theory – Radial Return Algorithm*

For the special case of J_2 flow theory plasticity, the general return mapping algorithm previously reduces to the well-known radial return method (Krieg and Key, 1976; Simo and Taylor, 1985). To facilitate the description some key results for radial return are first derived. These results will also be used in the determination of the consistent algorithmic modulus.

Recall from (5.9.6) that the trial stress, denoted here by $\boldsymbol{\sigma}^{(0)}$, is given by the elastic predictor, that is,

$$\boldsymbol{\sigma}^{(0)} = \mathbf{C}:\left(\varepsilon_{n+1} - \varepsilon^{p(0)}\right) \tag{5.9.17}$$

The stress at iteration k is given by

$$\boldsymbol{\sigma}^{(k)} = \boldsymbol{\sigma}^{(0)} - \Delta\lambda^{(k)}\mathbf{C}:\mathbf{r}^{(k)} \tag{5.9.18}$$

Referring to Box 5.10 for the elastic–plastic constitutive relation at small strains and to Box 5.6 for details of J_2 flow theory, we note that the plastic flow direction is in the direction of the deviatoric stress and is given by $\mathbf{r}=3\boldsymbol{\sigma}^{\text{dev}}/2\bar{\sigma}$ which is also the normal to the yield surface, that is, $\mathbf{r}=f_{\boldsymbol{\sigma}}$. In deviatoric stress space, the von Mises yield surface is circular and therefore the normal to the yield surface is radial (see Figure 5.16). We define a unit normal vector in the (radial) direction of plastic flow as

$$\hat{\mathbf{n}} = \mathbf{r}^{(0)} / \left\|\mathbf{r}^{0}\right\| = \boldsymbol{\sigma}^{(0)}_{\text{dev}} / \left\|\boldsymbol{\sigma}^{(0)}_{\text{dev}}\right\|, \quad \mathbf{r}^{(0)} = \sqrt{3/2}\hat{\mathbf{n}} \tag{5.9.19}$$

The key feature of the radial return method is that $\hat{\mathbf{n}}$ remains radial and unchanged throughout the plastic-corrector phase of the algorithm. Referring to (5.9.9), the update of the plastic strain is therefore a linear function of $\Delta\lambda$ and the plastic flow residual is identically zero: $\mathbf{a}^{(k)}=\mathbf{0}$. The only internal variable (isotropic hardening) is the accumulated plastic strain, given by $q_1 \equiv \bar{\varepsilon} = \lambda$ with $h=1$. Therefore the update of the internal variable is also a linear function of $\Delta\lambda$ and the corresponding residual is zero, that is, $\mathbf{b}^{(k)}=\mathbf{0}$.

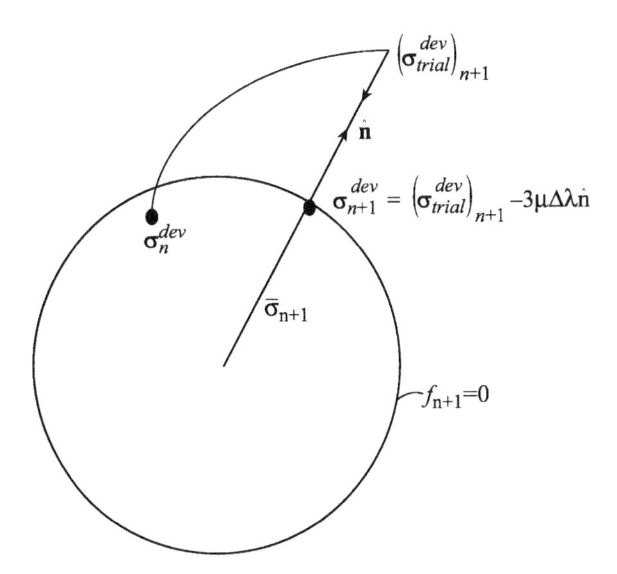

Figure 5.16 Radial return method for J_2 plasticity–shown at the converged state. The von Mises yield surface is a circle in deviatoric stress space

By differentiating the expression for the plastic flow direction with respect to stress we obtain the result

$$\mathbf{r}_\sigma = \frac{3}{2\bar{\sigma}}\hat{\boldsymbol{I}}, \quad \hat{\boldsymbol{I}} = \boldsymbol{I}^{\text{dev}} - \hat{\mathbf{n}} \otimes \hat{\mathbf{n}}, \quad \boldsymbol{I}^{\text{dev}} = \boldsymbol{I} - \frac{1}{3}\mathbf{I} \otimes \mathbf{I} \tag{5.9.20}$$

where $\boldsymbol{I}^{\text{dev}}$ is the fourth-order symmetric deviatoric tensor and the projection tensor $\hat{\boldsymbol{I}}$ has the properties

$$\hat{\boldsymbol{I}}^n = \hat{\boldsymbol{I}} \forall n, \quad \hat{\boldsymbol{I}} : \hat{\mathbf{n}} = 0, \quad \hat{\boldsymbol{I}} : \mathbf{I} = 0, \quad \hat{\boldsymbol{I}} : \boldsymbol{I}^{\text{dev}} = \hat{\boldsymbol{I}} \tag{5.9.21}$$

The plastic flow direction is independent of the accumulated plastic strain and therefore $\mathbf{r}_q = \mathbf{0}$. Also, since $h = 1$, $h_\sigma = \mathbf{0}$ and $h_q = 0$. The yield condition is given by $f = \bar{\sigma} - \sigma_Y(\bar{\varepsilon}) = 0$ and the derivatives of f are $f_\sigma = \mathbf{r}$ and $f_q = -d\sigma_Y/d\bar{\varepsilon}, = -H$. The matrix $\mathbf{A}$ is therefore written as

$$[\mathbf{A}^{(k)}] = \begin{bmatrix} \left(\mathbf{C}^{-1}\Delta\lambda\mathbf{r}_\sigma\right)^{-1} & \mathbf{0} \\ \mathbf{0} & -\mathbf{I} \end{bmatrix}^{(k)} \tag{5.9.22}$$

Now note that

$$(\mathbf{C}^{-1} + \Delta\lambda\mathbf{r}_\sigma) = (\mathbf{C}^{-1} + a\hat{\boldsymbol{I}}), \quad a = 3\Delta\lambda / 2\bar{\sigma} \tag{5.9.23}$$

For isotropic elastic moduli and using (5.9.21), the inverse can be written as

$$(\mathbf{C}^{-1}+\Delta\lambda\mathbf{r}_{\sigma})^{-1}=(\mathbf{C}-2\mu b\hat{\boldsymbol{I}}),\quad b=\frac{2\mu a}{1+2\mu a} \tag{5.9.24}$$

and **A** is

$$\left[\mathbf{A}^{(k)}\right]=\begin{bmatrix}(\mathbf{C}-2\mu b\hat{\boldsymbol{I}}) & \mathbf{0}\\ \mathbf{0} & -\mathbf{I}\end{bmatrix}^{(k)} \tag{5.9.25}$$

For isotropic elastic moduli, we have the identities $\mathbf{C}:\mathbf{r}=2\mu\mathbf{r}=2\mu\sqrt{3/2}\hat{\mathbf{n}}, \hat{\boldsymbol{I}}:(\mathbf{C}:\mathbf{r})=\mathbf{0}$ and $\mathbf{r}:\mathbf{C}:\mathbf{r}=3\mu$. Using these identities and the expression (5.9.25) for **A**, and recalling that $\tilde{\mathbf{a}}^{(k)}=\mathbf{0}$ (because $\mathbf{a}^{(k)}=\mathbf{b}^{(k)}=\mathbf{0}$), the increment in the plasticity parameter (5.9.15) is given by

$$\delta\lambda^{(k)}=\frac{f^{(k)}}{3\mu+H^{(k)}} \tag{5.9.26}$$

To obtain an alternative expression, we note that the deviatoric stress can be written as $\boldsymbol{\sigma}^{\text{dev}}=\sqrt{\frac{2}{3}}\,\bar{\sigma}\hat{\mathbf{n}}$ and therefore, from (5.9.18),

$$\boldsymbol{\sigma}_{\text{dev}}^{(k)}=\boldsymbol{\sigma}_{\text{dev}}^{(0)}-2\mu\Delta\lambda^{(k)}\mathbf{r}^{(k)}=\left(\sqrt{\frac{2}{3}}\bar{\sigma}^{(0)}-2\mu\Delta\lambda^{(k)}\sqrt{\frac{3}{2}}\right)\hat{\mathbf{n}} \tag{5.9.27}$$

Using this expression, the effective stress is

$$\bar{\sigma}^{(k)}=\bar{\sigma}^{(0)}-3\mu\Delta\lambda^{(k)} \tag{5.9.28}$$

which is substituted into the yield function $f^{(k)}$ in (5.9.26) to give the following expression for the increment in plasticity parameter:

$$\delta\lambda^{(k)}=\frac{(\bar{\sigma}^{(0)}-3\mu\Delta\lambda^{(k)})-\sigma_Y(\bar{\varepsilon}^{(k)})}{3\mu+H^{(k)}} \tag{5.9.29}$$

The radial return method is summarized in Box 5.14.

To illustrate the backward Euler return mapping scheme, we have considered the case of J_2 flow plasticity and shown how the general scheme reduces to the popular radial return algorithm. The radial return algorithm is usually coded directly along the lines of Box 5.14. For more complex constitutive models, the general scheme in Box 5.13 is used.

5.9.4 Algorithmic Moduli

In implicit methods, an appropriate tangent modulus is needed. Due to the abrupt transition to plastic behavior at yield, the continuum elasto-plastic tangent modulus can cause spurious loading and unloading. To avoid this, an *algorithmic modulus* (also called *the consistent*

Box 5.14 Radial return method

1. Initialization:

$$k=0: \quad \varepsilon^{p^{(0)}} = \varepsilon_n^p, \quad \bar{\varepsilon}^{(0)} = \bar{\varepsilon}_n, \quad \Delta\lambda^{(0)} = 0, \quad \boldsymbol{\sigma}^{(0)} = \mathbf{C} : (\varepsilon_{n+1} - \varepsilon^{p(0)})$$

2. Check yield condition at kth iteration:

$$f^{(k)} = \bar{\sigma}^{(k)} - \sigma_Y(\bar{\varepsilon}^{(k)}) = (\bar{\sigma}^{(0)} - 3\mu\Delta\lambda^{(k)}) - \sigma_Y(\bar{\varepsilon}^{(k)})$$

   If: $f^{(k)} < TOL_1$ then converged

   Else: go to 3

3. Compute increment in plasticity parameter:

$$\delta\lambda^{(k)} = \frac{(\bar{\sigma}^{(0)} - 3\mu\Delta\lambda^{(k)}) - \sigma_Y(\bar{\varepsilon}^{(k)})}{3\mu + H^{(k)}}$$

4. Update plastic strain and internal variables:

$$\hat{\mathbf{n}} = \boldsymbol{\sigma}_{\text{dev}}^{(0)} / \left\| \boldsymbol{\sigma}_{\text{dev}}^{(0)} \right\|, \quad \Delta\varepsilon^{p^{(k)}} = -\delta\lambda^{(k)} \sqrt{\frac{3}{2}} \hat{\mathbf{n}}, \quad \Delta\bar{\varepsilon}^{(k)} = \delta\lambda^{(k)}$$

$$\varepsilon^{p(k+1)} = \varepsilon^{p(k)} + \Delta\varepsilon^{p(k)}$$

$$\boldsymbol{\sigma}^{(k+1)} = \mathbf{C} : \left(\varepsilon_{n+1} - \varepsilon^{p(k+1)} \right) = \boldsymbol{\sigma}^{(k)} + \Delta\boldsymbol{\sigma}^{(k)} = \boldsymbol{\sigma}^{(k)} - 2\mu\delta\lambda^{(k)} \sqrt{\frac{3}{2}} \hat{\mathbf{n}}$$

$$\bar{\varepsilon}^{(k+1)} = \bar{\varepsilon}^{(k)} + \delta\lambda^{(k)}$$

$$\Delta\lambda^{(k+1)} = \Delta\lambda^{(k)} + \delta\lambda^{(k)}$$

   $k \leftarrow k+1$, go to 2

tangent modulus) based on a systematic linearization of the constitutive integration algorithm is used instead of the continuum elasto-plastic tangent modulus. The development of the algorithmic modulus for the fully implicit backward Euler scheme is given below.

The algorithmic tangent modulus for the backward Euler update is defined as

$$\mathbf{C}^{\text{alg}} = \left(\frac{d\boldsymbol{\sigma}}{d\boldsymbol{\varepsilon}} \right)_{n+1} \tag{5.9.30}$$

To derive an expression for the algorithmic tangent modulus, we write (5.9.4) in incremental form (again dropping the subscripts $n+1$) as

$$
\begin{aligned}
d\boldsymbol{\sigma} &= \mathbf{C}:(d\boldsymbol{\varepsilon} - d\boldsymbol{\varepsilon}^p) \\
d\boldsymbol{\varepsilon}^p &= d(\Delta\lambda)\mathbf{r} + \Delta\lambda d\mathbf{r} \\
d\mathbf{q} &= d(\Delta\lambda)\mathbf{h} + \Delta\lambda d\mathbf{h} \\
df &= f_\sigma : d\boldsymbol{\sigma} + f_\mathbf{q} \cdot d\mathbf{q} = 0
\end{aligned}
\tag{5.9.31}
$$

where

$$
d\mathbf{r} = \mathbf{r}_\sigma : d\boldsymbol{\sigma} + \mathbf{r}_\mathbf{q} \cdot d\mathbf{q} \quad d\mathbf{h} = \mathbf{h}_\sigma : d\boldsymbol{\sigma} + \mathbf{h}_\mathbf{q} \cdot d\mathbf{q}
\tag{5.9.32}
$$

Substituting $(5.9.31)_2$ into $(5.9.31)_1$, using (5.9.32) and solving for $d\boldsymbol{\sigma}$ and $d\mathbf{q}$, we obtain

$$
\begin{Bmatrix} d\boldsymbol{\sigma} \\ d\mathbf{q} \end{Bmatrix} = [\mathbf{A}] \begin{Bmatrix} d\boldsymbol{\varepsilon} \\ \mathbf{0} \end{Bmatrix} - d(\Delta\lambda)\mathbf{A} : \tilde{\mathbf{r}}
\tag{5.9.33}
$$

where

$$
[\mathbf{A}] = \begin{bmatrix} \mathbf{C}^{-1} + \Delta\lambda \mathbf{r}_\sigma & \Delta\lambda \mathbf{r}_\mathbf{q} \\ \Delta\lambda \mathbf{h}_\sigma & -\mathbf{I} + \Delta\lambda \mathbf{h}_\mathbf{q} \end{bmatrix}^{-1}
\tag{5.9.34}
$$

For notational convenience, let $\partial\mathbf{f} = [f_\sigma \;\; f_q]$. Substituting (5.9.33) into the incremental consistency condition $(5.9.31)_4$ and solving for $d\,(\Delta\lambda)$ gives

$$
d\Delta\lambda = \frac{-\partial\mathbf{f} : \mathbf{A} : \begin{Bmatrix} d\boldsymbol{\varepsilon} \\ \mathbf{0} \end{Bmatrix}}{\partial\mathbf{f} : \mathbf{A} : \tilde{\mathbf{r}}}
\tag{5.9.35}
$$

Substituting this result into (5.9.33) we obtain

$$
\begin{Bmatrix} d\boldsymbol{\sigma} \\ d\mathbf{q} \end{Bmatrix} = \left[\frac{(\mathbf{A}:\tilde{\mathbf{r}}) \otimes (\partial\mathbf{f}:\mathbf{A})}{\partial\mathbf{f}:\mathbf{A}:\tilde{\mathbf{r}}} \right] : \begin{Bmatrix} d\boldsymbol{\varepsilon} \\ \mathbf{0} \end{Bmatrix}
\tag{5.9.36}
$$

which is an expression for the algorithmic moduli for stress and internal variable increments.

Ortiz and Martin (1989) examined the conditions for which constitutive symmetries are preserved by the algorithmic modulus in (5.9.36). They noted that underlying symmetries are preserved if the plastic flow direction and the plastic moduli are derivable from a common potential, that is, $\mathbf{r} = \partial\psi/\partial\boldsymbol{\sigma}$ and $\mathbf{h} = \partial\psi/\partial\mathbf{q}$, from which it follows that $\mathbf{A}$ is symmetric. A simple closed-form expression for $\mathbf{A}$ may be obtained if the coupling terms in (5.9.34) vanish, that is, if $\partial\mathbf{r}/\partial\mathbf{q} = 0$ and $\partial\mathbf{h}/\partial\boldsymbol{\sigma} = 0$, which corresponds to the uncoupling of the plastic flow direction and plastic moduli from the internal variables and the stress respectively. In several widely used constitutive relations such as J_2 flow theory plasticity these coupling terms are zero. Under these conditions, **A** is given by

$$[\mathbf{A}] = \begin{bmatrix} (\mathbf{C}^{-1} + \Delta\lambda \mathbf{r}_\sigma)^{-1} & \mathbf{0} \\ \mathbf{0} & (-\mathbf{I} + \Delta\lambda \mathbf{h}_q)^{-1} \end{bmatrix} = \begin{bmatrix} \tilde{\mathbf{C}} & \mathbf{0} \\ \mathbf{0} & \mathbf{Y} \end{bmatrix} \tag{5.9.37}$$

where $\tilde{\mathbf{C}} = (\mathbf{C}^{-1} + \Delta\lambda \mathbf{r}_\sigma)^{-1}$ and $\mathbf{Y} = (-\mathbf{I} + \Delta\lambda \mathbf{h}_q)^{-1}$. Using this result in (5.9.36), an expression for the algorithmic modulus is obtained as

$$\mathbf{C}^{\text{alg}} = \left(\tilde{\mathbf{C}} - \frac{(\tilde{\mathbf{C}} : \mathbf{r}) \otimes (f_\sigma : \tilde{\mathbf{C}})}{f_\sigma : \tilde{\mathbf{C}} : \mathbf{r} + f_q \cdot \mathbf{Y} \cdot \mathbf{h}} \right) \tag{5.9.38}$$

which has the same form as the continuum elasto-plastic tangent modulus (B5.10.8) except that the elastic modulus is replaced by $\tilde{\mathbf{C}}$ and the term $-fq \cdot \mathbf{h}$ in the denominator is replaced by $fq \cdot \mathbf{Y} \cdot \mathbf{h}$.

5.9.5 *Algorithmic Moduli: J_2 Flow and Radial Return*

For the case of J_2 flow theory, the algorithmic modulus is consistent with the radial return stress update. For isotropic hardening, $f_\sigma = \mathbf{r}$, $\mathbf{h}_q = \mathbf{0}$, $\mathbf{Y} = -\mathbf{I}$. From (5.9.24), $\tilde{\mathbf{C}} = \mathbf{C} - 2\mu b\hat{\boldsymbol{I}}$ where $b = 3\mu\Delta\lambda/(\bar{\sigma} + 3\mu\Delta\lambda)$. It follows from (5.9.21) that $\mathbf{C}:\mathbf{r} = \mathbf{C}:\mathbf{r}$, and the algorithmic modulus can be written as

$$\mathbf{C}^{\text{alg}} = \mathbf{C}^{ep} - 2\mu b\hat{\boldsymbol{I}} \tag{5.9.39}$$

where $\mathbf{C}^{ep}$ is given by the small strain version of (B5.6.12) with isotropic moduli

$$\mathbf{C}^{ep} = K\mathbf{I} \otimes \mathbf{I} + 2\mu \boldsymbol{I}^{\text{dev}} - 2\mu\gamma \hat{\mathbf{n}} \otimes \hat{\mathbf{n}}, \quad \gamma = \frac{1}{1 + (H/3\mu)} \tag{5.9.40}$$

Substituting this expression into (5.9.39) we obtain the consistent algorithmic modulus for the radial return algorithm for J_2 flow theory:

$$\begin{aligned} \mathbf{C}^{\text{alg}} &= K\mathbf{I} \otimes \mathbf{I} + 2\mu\beta \boldsymbol{I}^{\text{dev}} - 2\mu\bar{\gamma} \hat{\mathbf{n}} \otimes \hat{\mathbf{n}}, \quad \bar{\gamma} = \gamma - (1 - \beta) \\ \beta &= (1 - b) = \bar{\sigma} / (\bar{\sigma} + 3\mu\Delta\lambda) = \sigma_Y / \bar{\sigma}^{(0)} \end{aligned} \tag{5.9.41}$$

In deriving this expression for β we have used (5.9.29) with $\delta\lambda^{(k)} = 0$, i.e., at the converged state of the last stress-update iteration, to write $\bar{\sigma} = \sigma_Y, \bar{\sigma} + 3\mu\Delta\lambda = \bar{\sigma}^{(0)}$. The radial return algorithm and consistent algorithmic moduli for combined isotropic–kinematic hardening are given by Simo and Taylor (1985).

5.9.6 Semi-Implicit Backward Euler Scheme

The semi-implicit backward Euler method (Moran, Ortiz and Shih, 1990) is implicit in the plasticity parameter and explicit in the plastic flow direction and plastic moduli, that is, the increments in the plasticity parameter are calculated at the end of the step but the plastic flow direction and plastic moduli are calculated at the beginning of the step. To avoid drift from the yield surface, the yield condition is enforced at the end of the step (see Figure 5.17). The integration scheme is

$$\begin{aligned} &\varepsilon_{n+1} = \varepsilon_n + \Delta\varepsilon, \quad \varepsilon^p_{n+1} = \varepsilon^p_n + \Delta\lambda_{n+1}\mathbf{r}_n \\ &\mathbf{q}_{n+1} = \mathbf{q}_n + \Delta\lambda_{n+1}\mathbf{h}_n, \quad \boldsymbol{\sigma}_{n+1} = \mathbf{C}:\left(\varepsilon_{n+1} - \varepsilon^p_{n+1}\right) \\ &f_{n+1} = f(\boldsymbol{\sigma}_{n+1}, \mathbf{q}_{n+1}) = 0 \end{aligned} \tag{5.9.42}$$

Following an analogous procedure to that of the fully implicit method, we write the plasticity equations in (5.9.42) as

$$\begin{aligned} &\mathbf{a} = -\varepsilon^p + \varepsilon^p_n + \Delta\lambda\mathbf{r}_n = \mathbf{0} \\ &\mathbf{b} = -\mathbf{q} + \mathbf{q}_n + \Delta\lambda\mathbf{h}_n = \mathbf{0} \\ &f = f(\boldsymbol{\sigma}, \mathbf{q}) = 0 \end{aligned} \tag{5.9.43}$$

Linearization of these equations gives (the subscript $n+1$ is dropped for convenience)

$$\begin{aligned} &\mathbf{a}^{(k)} + \mathbf{C}^{-1} : \Delta\sigma^{(k)} + \delta\lambda^{(k)}\mathbf{r}_n = \mathbf{0} \\ &\mathbf{b}^{(k)} - \Delta\mathbf{q}^{(k)} + \delta\lambda^{(k)}\mathbf{h}_n = \mathbf{0} \\ &f^{(k)} + f^{(k)}_{\boldsymbol{\sigma}} : \Delta\boldsymbol{\sigma}^{(k)} + f^{(k)}_{\mathbf{q}} \cdot \Delta\mathbf{q}^{(k)} = 0 \end{aligned} \tag{5.9.44}$$

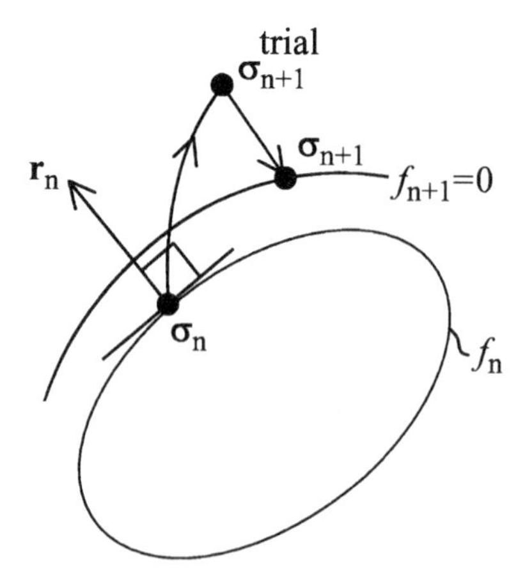

Figure 5.17 Return mapping scheme for semi-implicit constitutive integration algorithm: associative plasticity: $\mathbf{r}_n \sim \partial f/\partial\boldsymbol{\sigma}_n$

These are solved for $\Delta\boldsymbol{\sigma}^{(k)}$, $\Delta_{\mathrm{q}}{}^{k}$ and $\delta\lambda^{(k)}$. Note that because the plastic flow direction and plastic moduli are evaluated at the beginning of the time step, their gradients do not appear in (5.9.44). In addition, the updates of the plastic strain and the internal variables are linear functions of $\Delta\lambda$ and therefore $\tilde{\mathbf{a}}^{(k)} = \mathbf{0}$. Solving the first two of (5.9.44) for the stress and internal variable increments gives

$$\begin{Bmatrix} \Delta\boldsymbol{\sigma}^{(k)} \\ \Delta\mathbf{q}^{(k)} \end{Bmatrix} = -[\mathbf{A}^{(k)}] - \delta\lambda^{(k)}[\mathbf{A}^{(k)}]\{\tilde{\mathbf{r}}_n\} = -\delta\lambda^{(k)}[\mathbf{A}^{(k)}]\{\tilde{\mathbf{r}}_n\} \tag{5.9.45}$$

where

$$[\mathbf{A}^{(k)}] = \begin{bmatrix} \mathbf{C} & \mathbf{0} \\ \mathbf{0} & -\mathbf{I} \end{bmatrix}^{(k)}, \quad \{\tilde{\mathbf{a}}^{(k)}\} = \begin{Bmatrix} \mathbf{a}^{(k)} \\ \mathbf{b}^{(k)} \end{Bmatrix}, \quad \{\tilde{\mathbf{r}}_n\} = \begin{Bmatrix} \mathbf{r}_n \\ \mathbf{h}_n \end{Bmatrix} \tag{5.9.46}$$

Due to this explicit treatment of the plastic flow direction and plastic moduli, a simple closed-form expression is obtained for $\mathbf{A}$ which involves only the instantaneous elastic moduli. Substituting the result (5.9.45) into $(5.9.44)_3$ and solving for $\delta\lambda^{(k)}$ we get

$$\delta\lambda^{(k)} = \frac{f^{(k)}}{\partial\mathbf{f}^{(k)}:\mathbf{A}^{(k)}:\tilde{\mathbf{r}}_n} \tag{5.9.47}$$

Thus the variables are updated as

$$\begin{aligned} \varepsilon^{p(k+1)} &= \varepsilon^{p(k)} + \Delta\varepsilon^{p(k)} = \varepsilon^{p(k)} - \mathbf{C}^{-1} : \Delta\sigma^{(k)} \\ \mathbf{q}^{(k+1)} &= \mathbf{q}^{(k)} + \Delta\mathbf{q}^{(k)} \\ \Delta\lambda^{(k+1)} &= \Delta\lambda^{(k)} + \delta\lambda^{(k)} \end{aligned} \tag{5.9.48}$$

with the increments given by (5.9.47) and (5.9.45).

5.9.7 *Algorithmic Moduli – Semi-Implicit Scheme*

Because a simple closed-form expression for $\mathbf{A}$ is obtained, the algorithmic modulus for the semi-implicit scheme can be obtained in closed form for general elastic–plastic materials. Following an identical treatment to that for the fully implicit scheme (and noting that the plastic flow direction and the plastic moduli are evaluated at the beginning of the time step) we obtain, analogous to (5.9.36):

$$\begin{Bmatrix} d\boldsymbol{\sigma} \\ d\mathbf{q} \end{Bmatrix} = \left[\mathbf{A} - \frac{(\mathbf{A}:\tilde{\mathbf{r}}) \otimes (\partial\mathbf{f}:\mathbf{A})}{\partial\mathbf{f}:\mathbf{A}:\tilde{\mathbf{r}}} \right] : \begin{Bmatrix} d\varepsilon \\ \mathbf{0} \end{Bmatrix} \tag{5.9.49}$$

Using (5.9.46), the algorithmic modulus is

$$\mathbf{C}^{\text{alg}} = \left(\frac{d\boldsymbol{\sigma}}{d\boldsymbol{\varepsilon}}\right)_{n+1} = \left(\mathbf{C} - \frac{(\mathbf{C}:\mathbf{r}_n)\otimes f_{\boldsymbol{\sigma}}:\mathbf{C}}{-f_{\mathbf{q}}\cdot\mathbf{h}_n + f_{\boldsymbol{\sigma}}:\mathbf{C}:\mathbf{r}_n}\right)_{n+1} \tag{5.9.50}$$

In this expression, all quantities except for the plastic flow direction and moduli are evaluated at time $n+1$. Note that because of the appearance of $\boldsymbol{r}_n$, the plastic flow direction at the *beginning* of the step, and f_{σ}, the normal to the yield surface at the *end* of the step, this algorithmic modulus is in general not symmetric even when plastic flow is associative. The lack of symmetry due to the algorithmic properties of the semi-implicit scheme is not a drawback when lack of symmetry arises from non-associative plastic flow or formulations based on the Cauchy stress or multiplicative decompositions of the deformation gradient. We do interject a word of caution, however: because the stability of deformation processes is closely associated with the symmetry properties of the tangent moduli, care should be taken in the use of algorithmic moduli which do not preserve symmetries. Along the same lines, for constitutive relations which lead to nonsymmetric moduli, the practice of using the symmetric part of the moduli (on the supposed basis that only the rate of convergence is affected) should only be undertaken if it is clear that stability is not of concern.

5.9.8 Return Mapping Algorithms for Rate-Dependent Plasticity

The return mapping constitutive integration algorithms and algorithmic tangent moduli for rate-independent plasticity can be easily modified to account for rate-dependence. For rate-dependent materials, other integration schemes such as Runge–Kutta methods could be used in place of the return mapping schemes presented here.

The rate-dependent model is summarized in Box 5.11. The rate relation for the plasticity parameter can be expressed as an increment by

$$\Delta\lambda = \Delta t \frac{\phi(\boldsymbol{\sigma}, \mathbf{q})}{\eta} \tag{5.9.51}$$

where ϕ is the overstress function and η is the viscosity. This variable is now a known function of stress and internal variables. For a fully implicit scheme, the update can be written in incremental form as

$$\begin{gathered}\varepsilon_{n+1} = \varepsilon_n + \Delta\varepsilon, \quad \varepsilon^p_{n+1} = \varepsilon^p_n + \Delta\lambda_{n+1}\mathbf{r}_{n+1} \\ \mathbf{q}_{n+1} = \mathbf{q}_n + \Delta\lambda_{n+1}\mathbf{h}_{n+1}, \quad \boldsymbol{\sigma}_{n+1} = \mathbf{C}:\left(\varepsilon_{n+1} - \varepsilon^p_{n+1}\right), \quad \Delta\lambda_{n+1} = \frac{\Delta t}{\eta}\phi_{n+1}\end{gathered} \tag{5.9.52}$$

The steps leading to expression for the stress and internal variable increments parallel those leading to (5.9.14) for rate-independent plasticity. The expressions are rewritten here for convenience:

$$\begin{Bmatrix} \Delta\boldsymbol{\sigma}^{(k)} \\ \Delta\mathbf{q}^{(k)} \end{Bmatrix} = -\left[\mathbf{A}^{(k)}\right]\left\{\tilde{\mathbf{a}}^{(k)}\right\} - \delta\lambda^{(k)}\left[\mathbf{A}^{(k)}\right]\left\{\tilde{\mathbf{r}}^{(k)}\right\} \tag{5.9.53}$$

where $\mathbf{A}^k$ is given by (5.9.13) as in the rate-independent case.

The quantity $\delta\lambda^{(k)}$ is now obtained from the incremental form of $(5.9.52)_5$ as opposed to the consistency condition. The incremental form of $(5.9.52)_5$ is given by

$$\begin{aligned} \delta\lambda^{(k)} &= \frac{\Delta t}{\eta}\phi_{\boldsymbol{\sigma}}^{(k)} : \Delta\boldsymbol{\sigma}^{(k)} + \frac{\Delta t}{\eta}\phi_{\mathbf{q}}^{(k)} \cdot \Delta\mathbf{q}^{(k)} \\ &= \frac{\Delta t}{\eta}\left[\phi_{\boldsymbol{\sigma}}^{(k)}\ \phi_{\mathbf{q}}^{(k)}\right]\begin{Bmatrix} \Delta\boldsymbol{\sigma}^{(k)} \\ \Delta\mathbf{q}^{(k)} \end{Bmatrix} \\ &= \frac{\Delta t}{\eta}\left[\partial\Phi^{(k)}\right]\begin{Bmatrix} \Delta\boldsymbol{\sigma}^{(k)} \\ \Delta\mathbf{q}^{(k)} \end{Bmatrix} \end{aligned} \tag{5.9.54}$$

where the last equation is introduced for notational convenience. Substituting (5.9.53) into the last of these expressions and solving for $\delta\lambda^{(k)}$ gives

$$\delta\lambda^{(k)} = -\frac{\partial\Phi^{(k)}: \mathbf{A}^{(k)}:\tilde{\mathbf{a}}^{(k)}}{\dfrac{\eta}{\Delta t} + \partial\Phi^{(k)} : \mathbf{A}^{(k)} : \tilde{\mathbf{r}}^{(k)}} \tag{5.9.55}$$

The update of the constitutive equations is now given by

$$\begin{aligned} \varepsilon^{p(k+1)} &= \varepsilon^{p^{(k)}} + \Delta\varepsilon^{p(k)} = \varepsilon^{p(k)} - \mathbf{C}^{-1} : \Delta\boldsymbol{\sigma}^{(k)} \\ \mathbf{q}^{(k+1)} &= \mathbf{q}^{(k)} + \Delta_{\mathbf{q}}^{(k)} \\ \boldsymbol{\sigma}^{(k+1)} &= \mathbf{C} : \left(\varepsilon_{n+1} - \varepsilon^{p(k+1)}\right) = \boldsymbol{\sigma}^{(k)} + \Delta\boldsymbol{\sigma}^{(k)} \\ \Delta\lambda^{(k+1)} &= \Delta\lambda^{(k)} + \delta\lambda^{(k)} \end{aligned} \tag{5.9.56}$$

The algorithmic tangent modulus for the backward Euler scheme is derived similarly to that for the rate-independent case. The result is given by

$$\begin{Bmatrix} d\boldsymbol{\sigma} \\ d\mathbf{q} \end{Bmatrix} = \left[\mathbf{A} - \frac{(\mathbf{A} : \tilde{\mathbf{r}}) \otimes (\partial\boldsymbol{\Phi} : \mathbf{A})}{\dfrac{\eta}{\Delta t} + \partial\boldsymbol{\Phi} : \mathbf{A} : \tilde{\mathbf{r}}}\right] : \begin{Bmatrix} d\varepsilon \\ \mathbf{0} \end{Bmatrix} \tag{5.9.57}$$

Referring to (5.9.13), the parameter $\Delta\lambda$ which appears in the expression for $\mathbf{A}$ is given by $\Delta\lambda = \Delta t\phi/\eta$. Under suitable conditions (as discussed for the rate-independent case above), the coupling terms in $\mathbf{A}$ vanish and we obtain the closed-form expression

$$\mathbf{C}^{\text{alg}} = \left(\frac{d\boldsymbol{\sigma}}{d\varepsilon}\right)_{n+1} = \left(\tilde{\mathbf{C}} - \frac{(\tilde{\mathbf{C}}:\mathbf{r})\otimes(\phi_{\sigma}:\tilde{\mathbf{C}})}{\frac{\eta}{\Delta t} + \phi_{\sigma}:\tilde{\mathbf{C}}\mathbf{r} + \phi_{\mathbf{q}}\cdot\mathbf{Y}\cdot\mathbf{h}}\right)_{n+1} \tag{5.9.58}$$

for the algorithmic tangent modulus where $\tilde{\mathbf{C}}$ is defined in (5.9.37). Compare this result with the algorithmic modulus in (5.9.38).

5.9.8.1 Semi-Implicit Scheme

Similar results are obtained for the semi-implicit scheme. In particular, a closed-form expression for the algorithmic modulus is obtained for arbitrary constitutive response and is given by

$$\mathbf{C}^{\text{alg}} = \left(\frac{d\boldsymbol{\sigma}}{d\varepsilon}\right)_{n+1} = \left(\mathbf{C} - \frac{(\mathbf{C}:\mathbf{r}_n)\otimes(\phi_{\sigma}:\mathbf{C})}{\frac{\eta}{\Delta t} + \phi_{\sigma}:\mathbf{C}:\mathbf{r}_n - \phi_{\mathbf{q}}\cdot\mathbf{h}_n}\right)_{n+1} \tag{5.9.59}$$

5.9.9 *Rate Tangent Modulus Method*

Another popular approach to the integration of rate-dependent constitutive relations is the rate tangent modulus scheme introduced by Peirce, Shih and Needleman (1984) and discussed further by Moran (1987). The integration scheme for this method is forward Euler on all variables except $\Delta\lambda$ which is integrated by a generalized trapezoidal rule

$$\begin{aligned}
\varepsilon_{n+1} &= \varepsilon_n + \Delta\varepsilon \\
\varepsilon^p_{n+1} &= \varepsilon^p_n + \Delta\lambda\mathbf{r}_n \\
\mathbf{q}_{n+1} &= \mathbf{q}_n + \Delta\lambda\mathbf{h}_n \\
\boldsymbol{\sigma}_{n+1} &= \mathbf{C}:\left(\varepsilon_{n+1} - \varepsilon^p_{n+1}\right) \\
\Delta\lambda &= \frac{\Delta t}{\eta}\left[(1-\theta)\phi_n + \theta\phi_{n+1}\right]
\end{aligned} \tag{5.9.60}$$

The overstress function ϕ_{n+1} is expanded about the state at time n using a forward gradient approximation to give an effective increment of the plasticity parameter $\Delta\lambda$ as follows:

$$\phi_{n+1} = \phi_n + (\phi_{\sigma})_n : \Delta\boldsymbol{\sigma} + (\phi_{\mathbf{q}})_n \cdot \Delta_{\mathbf{q}} \tag{5.9.61}$$

This expression is substituted into (5.9.60) which gives

$$\Delta\lambda = \frac{\Delta t}{\eta}\phi_n + \frac{\theta\Delta t}{\eta}((\phi_{\sigma})_n : \Delta\sigma + (\phi_{\mathbf{q}})_n \cdot \Delta_{\mathbf{q}}) \tag{5.9.62}$$

Increments in stress and internal variables for use in this expression are obtained by writing (5.9.60) in incremental form as

$$\Delta\boldsymbol{\sigma} = \mathbf{C}:(\Delta\varepsilon - \Delta\lambda\mathbf{r}_n), \quad \Delta\mathbf{q} = \Delta\lambda\mathbf{h}_n \tag{5.9.63}$$

Using these expressions in (5.9.62) and solving for $\Delta\lambda$ gives

$$\Delta\lambda = \frac{\phi_n + \theta(\phi_\sigma)_n\mathbf{:C:}\Delta\varepsilon}{\dfrac{\eta}{\Delta t} + \theta\left((\phi_\sigma)_n\mathbf{:C:r}_n - (\phi_{\mathbf{q}})_n \cdot \mathbf{h}_n\right)} \tag{5.9.64}$$

which when substituted into (5.9.63) gives expression (5.9.65) (all quantities evaluated at time n):

$$\begin{aligned} &\Delta\boldsymbol{\sigma} = \mathbf{C}^{\text{alg}}\mathbf{:}\Delta\varepsilon - \mathbf{p} \\ &\mathbf{C}^{\text{alg}} = \mathbf{C} - \frac{\theta(\mathbf{C:r})\otimes(\phi_\sigma\mathbf{:C})}{(\eta/\Delta t) + 0(\phi_\sigma\mathbf{:C:r} - \phi_{\mathbf{q}}\mathbf{:h})}, \quad \mathbf{p} = \frac{\phi_n\mathbf{C:r}}{(\eta/\Delta t) + 0(\phi_\sigma\mathbf{:C:r} - \phi_{\mathbf{q}}\mathbf{:h})} \end{aligned} \tag{5.9.65}$$

Here $\mathbf{C}^{\text{alg}}$ is sometimes called the rate tangent modulus and $\mathbf{p}$ is a pseudo-load which is added into the external nodal force vector in a one-step solution of the incremental weak form. Note that, for $\theta = 1$, this algorithmic modulus is identical to the tangent modulus (5.9.59) of the semi-implicit backward Euler scheme on the first iteration. The pseudo load $\mathbf{p}$, which appears in the present one-step algorithm, does not appear in the iterative semi-implicit scheme.

5.9.10 Incrementally Objective Integration Schemes for Large Deformations

An important issue in constitutive algorithms for large deformations is the observance of material frame indifference. Many researchers advocate that the constitutive algorithm should exactly preserve the underlying objectivity of the constitutive relation; that is, the algorithm should exactly account for the proper rotation of the stress in a rigid body rotation. Hughes and Winget (1980) introduced the concept of *incremental objectivity*: an update algorithm is incrementally objective if, in a rigid rotation where $\mathbf{F}_{n+1} = \mathbf{Q}(t)\cdot\mathbf{F}_n$ (where det $\mathbf{Q} = 1$) the Cauchy stress is given by $\sigma_{n+1} = \mathbf{Q}(t)\cdot\boldsymbol{\sigma}_n\cdot\mathbf{Q}^{\mathrm{T}}(t)$. Rashid (1993) further distinguishes between *weak* objectivity, in which the update scheme properly rotates the stress tensor under pure rotations, and *strong* objectivity, in which the stress is properly rotated for motions comprised of stretching and rotation. An extensive discussion of incrementally objective stress update algorithms based on the concept of Lie derivatives is given in Simo and Hughes (1998).

As an example of incremental objectivity, we consider a simple update scheme based on the Jaumann rate of Kirchhoff stress. In what follows, rates of deformation are effective rates for the time increment and are defined next. The stress update is given by

$$\boldsymbol{\tau}_{n+1} = \mathbf{Q}_{n+1}\cdot\boldsymbol{\tau}_n\cdot\mathbf{Q}_{n+1}^T + \Delta t\boldsymbol{\tau}^{\nabla J} \tag{5.9.66}$$

where $\mathbf{Q}_{n+1} = \exp[\mathbf{W}\Delta t]$ is the incremental rotation tensor associated with the effective spin $\mathbf{W}$. Substituting the constitutive response in terms of the Jaumann rate gives

$$\boldsymbol{\tau}_{n+1} = \mathbf{Q}_{n+1}\cdot\boldsymbol{\tau}_n\cdot\mathbf{Q}_{n+1}^T + \Delta t\mathbf{C}^{\tau J}\mathbf{:D} \tag{5.9.67}$$

where $\mathbf{D}$ is the effective rate-of-deformation. Different algorithms are used to compute the effective rates of deformation. Here we take a straightforward approach based on the incremental deformation gradient:

$$\begin{aligned}\mathbf{F}_{n+1} &= \mathbf{F}_n + (\nabla_0 \Delta\mathbf{u})^T = \Delta\mathbf{F}_n \cdot \mathbf{F}_n \\ \Delta\mathbf{F}_n &= \mathbf{I} + (\nabla_0 \Delta\mathbf{u})^T \cdot \mathbf{F}_n^{-1} = \mathbf{I} + (\nabla_n \Delta\mathbf{u})^T\end{aligned} \tag{5.9.68}$$

where $\nabla_n = \partial/\partial\mathbf{x}_n$, $\Delta\mathbf{u} = \Delta t\mathbf{v}$ is the incremental displacement and $\mathbf{v}$ is the effective velocity for the increment. The effective rate-of-deformation is defined through the push-forward of the increment in Green strain:

$$\begin{aligned}\Delta\mathbf{E} &= \frac{1}{2}\left(\mathbf{F}_{n+1}^T \cdot \mathbf{F}_{n+1} - \mathbf{F}_n^T \cdot \mathbf{F}_n\right) \\ \Delta t\mathbf{D} &= \mathbf{F}_{n+1}^{-T} \cdot \Delta\mathbf{E} \cdot \mathbf{F}_{n+1}^{-1} = \frac{1}{2}\left(\mathbf{I} - (\Delta\mathbf{F}_n)^{-T} \cdot \Delta\mathbf{F}_n^{-1}\right)\end{aligned} \tag{5.9.69}$$

The effective rate-of-deformation $\mathbf{D}$ vanishes in a rigid rotation and incremental objectivity is achieved. The effective spin is defined through

$$\Delta t\mathbf{W} = \frac{1}{2}\left(\left(\nabla_0 \Delta\mathbf{u}\right)^T \cdot \mathbf{F}_{n+1}^{-1} - \mathbf{F}_{n+1}^{-T} \cdot \left(\nabla_0 \Delta\mathbf{u}\right)\right) \tag{5.9.70}$$

Approximations to the exponential map for computing $\mathbf{Q}_{n+1}$ are given in Chapter 9.

For a hypoelastic–plastic material (5.9.67) takes the form

$$\begin{aligned}\boldsymbol{\tau}_{n+1} &= \mathbf{Q}_{n+1} \cdot \boldsymbol{\tau}_n \cdot \mathbf{Q}_{n+1}^T + \Delta t\mathbf{C}_{el}^{\tau J} : (\mathbf{D} - \mathbf{D}^p) \\ &= \left(\mathbf{Q}_{n+1} \cdot \boldsymbol{\tau}_n \cdot \mathbf{Q}_{n+1}^T + \Delta t\mathbf{C}_{el}^{\tau J} : \mathbf{D}\right) - \Delta t\mathbf{C}_{el}^{\tau J} : \mathbf{D}^p\end{aligned} \tag{5.9.71}$$

The elastic moduli are taken to be constant and isotropic. The term in parentheses defines the trial stress $\boldsymbol{\tau}_{n+1}^{\text{trial}} = \mathbf{Q}_{n+1} \cdot \boldsymbol{\tau}_n \cdot \mathbf{Q}_{n+1}^T + \Delta t\mathbf{C}_{el}^{\tau J} : \mathbf{D}$. The radial return algorithm for von Mises plasticity then parallels the small strain formulation in Section 5.9.3 with, for example,

$$\hat{\mathbf{n}} = \frac{\left(\tau_{\text{dev}}^{\text{trial}}\right)_{n+1}}{\left\|\left(\tau_{\text{dev}}^{\text{trial}},\right)\right\|_{n+1}} \qquad \delta\lambda^{(k)} = \frac{f^{(k)}}{3\mu + H^{(k)}} \tag{5.9.72}$$

5.9.11 Semi-Implicit Scheme for Hyperelastic–Plastic Constitutive Models

In hyperelastic–plastic constitutive relations the need for incrementally objective algorithms can be avoided as discussed is Section 5.7.7. In this section, a stress update scheme for the hyperelastic–plastic model (rate-dependent case) described in Section 5.7 is presented. The procedure (Moran, Ortiz and Shih, 1990) is shown in Box 5.15. Because the stress-update scheme is formulated on the intermediate configuration, it is independent of rigid body rotations and thus automatically satisfies the requirements of material objectivity. The consistent tangent modulus for this stress update scheme can be derived in closed form (Moran, Ortiz and Shih, 1990) because it is explicit in the plastic flow direction, and plastic moduli, that is, $\bar{\mathbf{r}}_n$ and $\bar{\mathbf{h}}_n$, are evaluated at the beginning of the time step.

Box 5.15 Stress update scheme for hyperelastic–viscoplastic model

1. Given $\mathbf{F}_{n+1}$, $\mathbf{F}_n$, $\mathbf{F}_n^p$, $\bar{\mathbf{S}}_n$, $\bar{\mathbf{q}}_n$ and time increment Δt, compute $\mathbf{F}_{n+1}^p$, $\bar{\mathbf{S}}_{n+1}$, $\bar{\mathbf{q}}_{n+1}$

$$\begin{aligned}
\mathbf{F}_{n+1}^p &= (\mathbf{I} + \Delta\lambda_{n+1}\bar{\mathbf{r}}_n)\cdot\mathbf{F}_n^p \\
\mathbf{F}_{n+1}^e &= \mathbf{F}_{n+1}\cdot\left(\mathbf{F}_{n+1}^p\right)^{-1} \\
\bar{\mathbf{C}}_{n+1}^e &= \mathbf{F}_{n+1}^{eT}\cdot\mathbf{F}_{n+1}^e \\
\bar{\mathbf{S}}_{n+1} &= \bar{\mathbf{S}}\left(\bar{\mathbf{C}}_{n+1}^e\right) = 2\partial\bar{\psi}/\partial\bar{\mathbf{C}}_{n+1}^e \\
\bar{\mathbf{q}}_{n+1} &= \bar{\mathbf{q}}_n + \Delta\lambda_{n+1}\bar{\mathbf{h}}_n
\end{aligned} \qquad \text{(B5.15.1)}$$

Note that from this, the stress can be written as an implicit function of $\Delta\lambda_{n+1}$. First express the elastic Cauchy deformation tensor as a function of $\Delta\lambda_{n+1}$:

$$\begin{aligned}
\bar{\mathbf{C}}_{n+1}^e(\Delta\lambda_{n+1}) = \mathbf{F}_{n+1}^{eT}\cdot\mathbf{F}_{n+1}^e &= \left(\left(\mathbf{F}_{n+1}^p\right)^{-T}\cdot\mathbf{F}_{n+1}^T\right)\cdot\left(\mathbf{F}_{n+1}\cdot\left(\mathbf{F}_{n+1}^p\right)^{-1}\right) \\
&= (\mathbf{I} + \Delta\lambda_{n+1}\bar{\mathbf{r}}_n)^{-T}\cdot\left(\mathbf{F}_n^p\right)^{-T}\cdot\left(\mathbf{F}_{n+1}^T\cdot\mathbf{F}_{n+1}\right)\cdot\left(\mathbf{F}_n^p\right)^{-1}\cdot(\mathbf{I} + \Delta\lambda_{n+1}\bar{\mathbf{r}}_n)^{-1}
\end{aligned} \qquad \text{(B5.15.2)}$$

2. Use this to write the increment in plasticity parameter as

$$\begin{aligned}
\Delta\lambda_{n+1} &= \Delta t\frac{\bar{\phi}\left(\bar{\mathbf{S}}_{n+1}, \bar{\mathbf{q}}_{n+1}\right)}{\eta} \\
&= \frac{\Delta t}{\eta}\bar{\phi}\left(\bar{\mathbf{S}}\left(\bar{\mathbf{C}}_{n+1}^e\,\Delta\lambda_{n+1}\right)\right), \bar{\mathbf{q}}_n + \Delta\lambda_{n+1}\bar{\mathbf{h}}_n\right)
\end{aligned} \qquad \text{(B5.15.3)}$$

which can be solved for $\Delta\lambda_{n+1}$ using a Newton method.

3. The algorithmic modulus for the above update scheme can be obtained in closed form and is given by

$$\begin{aligned}
\boldsymbol{\sigma}^{\nabla T} &= \mathbf{C}_{\mathrm{alg}}^{\sigma T} : \mathbf{D} \\
\mathbf{C}_{\mathrm{alg}}^{\sigma T} &= J^{-1}\mathbf{C}_{\mathrm{alg}}^{\tau} - \frac{(\hat{\mathbf{r}}\cdot\boldsymbol{\sigma} + \boldsymbol{\sigma}\cdot\hat{\mathbf{r}}^T)\otimes\left(\phi_\tau:\mathbf{C}_{el}^\tau\right)}{\frac{\eta}{\Delta t} - \phi_{\mathbf{q}}\cdot\mathbf{h}_n + \phi_\tau:\mathbf{C}_{el}^\tau:\mathrm{sym}\,\hat{\mathbf{r}}} \\
\mathbf{C}_{\mathrm{alg}}^{\tau} &= \mathbf{C}_{el}^\tau - \frac{\left(\mathbf{C}_{el}^\tau : \mathrm{sym}\ \hat{\mathbf{r}}\right)\otimes\left(\phi_\tau:\mathbf{C}_{el}^\tau\right)}{\frac{\eta}{\Delta t} - \phi_{\mathbf{q}}\cdot\mathbf{h}_n + \phi_\tau : \mathbf{C}_{el}^\tau : \mathrm{sym}\,\hat{\mathbf{r}}} \\
\hat{\mathbf{r}} &= (\mathbf{F}^e)^{-T}\cdot\left(\bar{\mathbf{C}}^e\cdot\bar{\mathbf{r}}_n\cdot\mathbf{F}_n^p\cdot(\mathbf{F}^p)^{-1}\right)\cdot(\mathbf{F}^e)^{-1} \\
\phi_\tau &= \partial\phi/\partial\tau, \quad \phi_{\mathbf{q}} = \partial\phi/\mathbf{q}
\end{aligned} \qquad \text{(B5.15.4)}$$

where quantities are evaluated at $n+1$ unless noted otherwise. By setting $\eta=0$, the algorithmic modulus for the rate-independent case is obtained.

5.10 Continuum Mechanics and Constitutive Models

Considerable insight into the relations and mappings for various tensors in large deformation elasticity and plasticity can be gained through concepts from analysis on manifolds. The interested reader is referred to Marsden and Hughes (1983) for a through treatment of these topics. In this section, a brief introduction to some of the salient features of analysis on manifolds is given and connection is made with some of the key relations developed earlier in this chapter. In particular, issues associated with frame invariance, stress rates and the formulation of constitutive models are more clearly seen in the manifold setting.

5.10.1 Eulerian, Lagrangian and Two-Point Tensors

For the treatment of material frame indifference, it is useful to introduce the concepts of Lagrangian, Eulerian and two-point tensors. The line element $d\mathbf{X}$ is a Lagrangian line element and is called a Lagrangian vector. Second-order tensors defined through contractions with Lagrangian vectors are termed Lagrangian second-order tensors. Thus, from (3.3.1) the Green strain tensor, $\mathbf{E}$, is a Lagrangian tensor, as is the right Cauchy–Green deformation tensor, $\mathbf{C}$. The second Piola–Kirchhoff stress which is work conjugate to the Green strain tensor is a Lagrangian stress tensor. Material time-derivatives of Lagrangian tensors are also Lagrangian tensors.

The line element $d\mathbf{x}$ in the current configuration is an Eulerian line element and dx is called an Eulerian vector. The material time-derivative of $d\mathbf{x}$ is $d\mathbf{v}$; and is also an Eulerian vector. Second-order tensors defined through contractions with Eulerian vectors are termed Eulerian second-order tensors. Thus from (3.3.13) the velocity gradient $\mathbf{L}$ and its symmetric and antisymmetric parts, $\mathbf{D}$ and $\mathbf{W}$, are Eulerian tensors. It follows from work conjugacy that the Cauchy stress $\boldsymbol{\sigma}$ and the Kirchhoff (weighted Cauchy) stress $\boldsymbol{\tau}$ are Eulerian tensors.

Second-order tensors defined through contractions with a Lagrangian vector and an Eulerian vector belong to the class of two-point tensors and are called Eulerian–Lagrangian if the contraction is with an Eulerian vector on the left and a Lagrangian vector on the right and Lagrangian–Eulerian if vice versa. Thus the deformation gradient $\mathbf{F}$ is an Eulerian–Lagrangian two-point tensor and the nominal stress $\mathbf{P}$ is a Lagrangian–Eulerian two-point tensor.

5.10.2 Pull-Back, Push-Forward and the Lie Derivative

5.10.2.1 Fundamental Concepts

A unified description of the maps between Eulerian and Lagrangian tensors can be given by the pull-back and push-forward operations. For example, the push-forward by $\mathbf{F}$ of the Lagrangian vector $d\mathbf{X}$ to the current configuration gives the Eulerian vector $d\mathbf{x}$, that is,

$$d\mathbf{x} = \mathbf{F} \cdot d\mathbf{X} \equiv \phi_* \, d\mathbf{X} \tag{5.10.1}$$

The pull-back by $\mathbf{F}^{-1}$ of the Eulerian vector $d\mathbf{x}$ to the reference configuration gives $d\mathbf{X}$, that is,

$$d\mathbf{X} = \mathbf{F}^{-1} \cdot d\mathbf{x} \equiv \phi^* \, d\mathbf{x} \tag{5.10.2}$$

In (5.10.2) ϕ_* and ϕ^* represent the indicated push-forward and pull-back operations respectively.

The pull-back and push-forward operations on second-order tensors give the relationships between these tensors in the deformed and undeformed configurations. The pull-backs and push-forwards of some key second-order tensors are given in Box 5.16. These definitions depend on whether a tensor is kinetic or kinematic. The difference arises from the conjugacy in power which is observed by these tensors: as work conjugate kinematic and kinetic tensors are pulled-back or pushed-forward, the power must remain invariant (this issue is explained further in the next subsection). Many of the relations follow from Box 3.2, but these concepts enable us to develop relationships which are not readily apparent otherwise.

Box 5.16 Summary of pull-back and push-forward operations (note that the metric tensor $\mathbf{g}=\mathbf{I}$)

Push-forward ϕ_*	**Pull-back ϕ^***

Kinematic (covariant-covariant tensors)

Push-forward:

$$\phi_*(\bullet)=\mathbf{F}^{-T}\cdot(\bullet)\cdot\mathbf{F}^{-1}$$
$$\phi_*\mathbf{C}=\mathbf{F}^{-T}\cdot\mathbf{C}\cdot\mathbf{F}^{-1}=\mathbf{g}$$
$$\phi_*\dot{\mathbf{E}}=\mathbf{F}^{-T}\cdot\dot{\mathbf{E}}\cdot\mathbf{F}^{-1}=\mathbf{D}$$

$$\begin{aligned}\phi_*\dot{\mathbf{E}}^{dev}&=\mathbf{F}^{-T}\cdot\dot{\mathbf{E}}^{\mathrm{dev}}\cdot\mathbf{F}^{-1}\\&=\mathbf{F}^{-T}\cdot\left(\dot{\mathbf{E}}-\left(\frac{1}{3}\dot{\mathbf{E}}:\mathbf{C}^{-1}\right)\mathbf{C}\right)\cdot\mathbf{F}^{-1}\\&=\mathbf{D}-\left(\frac{1}{3}\mathbf{D}:\mathbf{g}^{-1}\right)\mathbf{g}\equiv\mathbf{D}-\left(\frac{1}{3}\mathbf{D}:\mathbf{I}\right)\mathbf{I}=\mathbf{D}^{\mathrm{dev}}\end{aligned}$$

Pull-back:

$$\phi^*(\bullet)=\mathbf{F}^{T}\cdot(\bullet)\cdot\mathbf{F}$$
$$\phi^*\mathbf{g}=\mathbf{F}^{T}\cdot\mathbf{g}\cdot\mathbf{F}=\mathbf{C}$$
$$\phi^*\mathbf{D}=\mathbf{F}^{T}\cdot\mathbf{D}\cdot\mathbf{F}=\dot{\mathbf{E}}$$

$$\begin{aligned}\phi^*\mathbf{D}^{\mathrm{dev}}&=\mathbf{F}^{T}\cdot\mathbf{D}^{\mathrm{dev}}\cdot\mathbf{F}\\&=\mathbf{F}^{T}\cdot\left(\mathbf{D}-\left(\frac{1}{3}\mathbf{D}:\mathbf{g}^{-1}\right)\mathbf{g}\right)\cdot\mathbf{F}\\&=\dot{\mathbf{E}}-\left(\frac{1}{3}\dot{\mathbf{E}}:\mathbf{C}^{-1}\right)\mathbf{C}=\dot{\mathbf{E}}^{\mathrm{dev}}\end{aligned}$$

Kinetic (contravariant–contravariant tensors)

Push-forward:

$$\phi_*(\bullet)=\mathbf{F}\cdot(\bullet)\cdot\mathbf{F}^{T}$$
$$\phi_*\mathbf{S}=\mathbf{F}\cdot\mathbf{S}\cdot\mathbf{F}^{T}=\tau$$
$$\phi_*\dot{\mathbf{S}}=\mathbf{F}\cdot\dot{\mathbf{S}}\cdot\mathbf{F}^{T}=L_v\tau\equiv\tau^{\nabla c}$$

$$\begin{aligned}\phi_*\mathbf{S}^{\mathrm{dev}}&=\mathbf{F}\cdot\mathbf{S}^{\mathrm{dev}}\cdot\mathbf{F}^{T}\\&=\mathbf{F}\cdot\left(\mathbf{S}-\left(\frac{1}{3}\mathbf{S}:\mathbf{C}\right)\mathbf{C}^{-1}\right)\cdot\mathbf{F}^{T}\\&=\tau-\left(\frac{1}{3}\tau:\mathbf{g}\right)\mathbf{g}^{-1}\equiv\tau-\left(\frac{1}{3}\tau:\mathbf{I}\right)\mathbf{I}=\tau^{\mathrm{dev}}\end{aligned}$$

Pull-back:

$$\phi^*(\bullet)=\mathbf{F}^{-1}\cdot(\bullet)\cdot\mathbf{F}^{-T}$$
$$\phi^*\tau=\mathbf{F}^{-1}\cdot\tau\cdot\mathbf{F}^{-T}=\mathbf{S}$$
$$\phi^*\tau^{\nabla c}=\mathbf{F}^{-1}\cdot\tau^{\nabla c}\cdot\mathbf{F}^{-T}=\dot{\mathbf{S}}$$

$$\begin{aligned}\phi^*\tau^{\mathrm{dev}}&=\mathbf{F}^{-1}\cdot\tau^{\mathrm{dev}}\cdot\mathbf{F}^{-T}\\&=\mathbf{F}^{-1}\cdot\left(\tau-\left(\frac{1}{3}\tau:\mathbf{g}\right)\mathbf{g}^{-1}\right)\cdot\mathbf{F}^{-T}\\&=\mathbf{S}-\left(\frac{1}{3}\mathbf{S}:\mathbf{C}\right)\mathbf{C}^{-1}=\mathbf{S}^{\mathrm{dev}}\end{aligned}$$

We encountered the metric tensor **g** in Section 5.7 and noted there that for Euclidean space the metric tensor is the identity tensor **I**. However, we retain the metric tensor in some formulas in order to illustrate the fundamental role it and its various pull-backs (e.g., $\mathbf{C} = \phi^* \mathbf{g}$) play in tensor operations such as the formation of deviatoric or symmetric parts of tensors (see Box 5.16).

The concepts of pull-back and push-forward provided a mathematically consistent method for defining time derivatives of tensors, called Lie derivatives. As shown in Box 5.17, the Lie derivative of the Kirchhoff stress is the push-forward of the time derivative of the pull-back of the Kirchhoff stress. Loosely speaking, in the Lie derivative, we take the time derivative in the reference configuration, which is fixed, and then push forward to the current configuration. The Lie derivative of a kinematic tensor is defined in a power-conjugate manner as given in Box 5.17 (i.e., using the kinematic pull-back and push-forward defined in Box 5.16).

Box 5.17 Lie derivatives

Lie derivative $L_v(\bullet) = \phi_* \left(\dfrac{D}{Dt} \phi^*(\bullet) \right)$

$$L_v \mathbf{g} = \phi_* \left(\frac{D}{Dt}(\phi^* \mathbf{g}) \right) = \mathbf{F}^{-T} \cdot \frac{D}{Dt}(\mathbf{F}^T \cdot \mathbf{g} \cdot \mathbf{F}) \cdot \mathbf{F}^{-1} = \mathbf{F}^{-T} \cdot \dot{\mathbf{C}} \cdot \mathbf{F}^{-1} = 2\mathbf{D}$$

$$L_v \boldsymbol{\tau} = \phi_* \left(\frac{D}{Dt}(\phi^* \boldsymbol{\tau}) \right) = \mathbf{F} \cdot \frac{D}{Dt}(\mathbf{F}^{-1} \cdot \boldsymbol{\tau} \cdot \mathbf{F}^{-T}) \cdot \mathbf{F}^T = \mathbf{F} \cdot \dot{\mathbf{S}} \cdot \mathbf{F}^T = \boldsymbol{\tau}^{\nabla c}$$

The convected rate of the Kirchhoff stress corresponds to its Lie derivative. This is shown as follows. We start with the definition as given in Box 5.17 and then write out push-forwards and pull-backs.

$$L_v \boldsymbol{\tau} = \phi_* \left(\frac{D}{Dt}(\phi^* \boldsymbol{\tau}) \right) = \mathbf{F} \cdot \frac{D\mathbf{S}}{Dt} \cdot \mathbf{F}^T = \mathbf{F} \cdot \frac{D}{Dt}(\mathbf{F}^{-1} \cdot \boldsymbol{\tau} \cdot \mathbf{F}^{-T}) \cdot \mathbf{F}^T \tag{5.10.3}$$

Next we take the material time derivative of the product

$$\begin{aligned} L_v \boldsymbol{\tau} = \phi_* \left(\frac{D}{Dt}(\phi^* \boldsymbol{\tau}) \right) &= \mathbf{F} \cdot (\dot{\mathbf{F}}^{-1} \cdot \boldsymbol{\tau} \cdot \mathbf{F}^{-T} + \mathbf{F}^{-1} \cdot \dot{\boldsymbol{\tau}} \cdot \mathbf{F}^{-T} + \mathbf{F}^{-1} \cdot \boldsymbol{\tau} \cdot \dot{\mathbf{F}}^{-T}) \cdot \mathbf{F}^T \\ &= \mathbf{F} \cdot (-\mathbf{F}^{-1} \cdot \dot{\mathbf{F}} \cdot \mathbf{F}^{-1} \cdot \boldsymbol{\tau} \cdot \mathbf{F}^{-T} + \mathbf{F}^{-1} \cdot \dot{\boldsymbol{\tau}} \cdot \mathbf{F}^{-T} - \mathbf{F}^{-1} \cdot \boldsymbol{\tau} \cdot \mathbf{F}^{-T} \cdot \dot{\mathbf{F}}^T \cdot \mathbf{F}^{-T}) \cdot \mathbf{F}^T \end{aligned} \tag{5.10.4}$$

where we have substituted $\dot{\mathbf{F}}^{-1} = -\mathbf{F}^{-1} \cdot \dot{\mathbf{F}} \cdot \mathbf{F}^{-1}$. Recalling $\mathbf{L} = \dot{\mathbf{F}} \cdot \mathbf{F}^{-1}$ we obtain

$$L_v \boldsymbol{\tau} = \boldsymbol{\tau}^{\nabla c} = \dot{\boldsymbol{\tau}} - \mathbf{L} \cdot \boldsymbol{\tau} - \boldsymbol{\tau} \cdot \mathbf{L}^T \tag{5.10.5}$$

Thus the Lie derivative is equivalent to the convected rate of the Truesdell stress defined in (5.4.22).

Pull-back of tensors to the intermediate configuration $\bar{\Omega}$ (Section 5.7) is accomplished by the elastic part of the deformation gradient, for example, $\bar{\mathbf{S}} = (\mathbf{F}^e)^{-1} \cdot \boldsymbol{\tau} \cdot (\mathbf{F}^e)^{-T} = \phi_e^* \boldsymbol{\tau}$. Similarly, pull-back of tensors from $\bar{\Omega}$ to the reference configuration Ω is with the plastic part of the deformation gradient, for example, $\mathbf{S} = (\mathbf{F}^p)^{-1} \cdot \bar{\mathbf{S}} \cdot (\mathbf{F}^p)^{-T} = \phi_p^* \bar{\mathbf{S}}$.

Elastic and plastic Lie derivatives can be defined using the elastic and plastic parts, respectively, in the pull-back and push-forward operations. For example,

$$L_v^e \boldsymbol{\tau} = \phi_*^e \left(\frac{D}{Dt} (\phi_e^* \tau) \right) = \mathbf{F}^e \cdot \frac{D}{Dt} ((\mathbf{F}^e)^{-1} \cdot \boldsymbol{\tau} \cdot (\mathbf{F}^e)^{-T}) \cdot \mathbf{F}^{eT} \tag{5.10.6}$$

is the elastic Lie derivative of the Kirchhoff stress and

$$L_v^p \bar{\mathbf{S}} = \phi_*^p \left(\frac{D}{Dt} (\phi_p^* \bar{\mathbf{S}}) \right) = \mathbf{F}^p \cdot \frac{D}{Dt} ((\mathbf{F}^p)^{-1} \cdot \bar{\mathbf{S}} \cdot (\mathbf{F}^p)^{-T}) \cdot \mathbf{F}^{pT} \tag{5.10.7}$$

is the plastic Lie derivative of $\bar{\mathbf{S}}$.

5.10.2.2 Pull-Back and Push-Forward in Hyperelastic–Plastic Models

In this subsection we elaborate further on the pull-back and push-forward operations discussed previously. We also provide some motivation for some of the expressions that arise in the formulation of hyperelastic–plastic constitutive models based on the multiplicative decomposition of the deformation gradient (Section 5.7). To begin, recall that the second Piola-Kirchhoff stress on $\bar{\Omega}$ can be interpreted as the pull-back (via the elastic part of the deformation gradient) to $\bar{\Omega}$ of the Kirchhoff stress tensor, that is,

$$\bar{\mathbf{S}} = (\mathbf{F}^e)^{-1} \cdot \boldsymbol{\tau} (\cdot \mathbf{F}^e)^{-T} \equiv \phi_e^* (\boldsymbol{\tau}) \tag{5.10.8}$$

The particular form of the pull-back or push-forward operation is based on the character of the tensor within the framework of analysis on general manifolds. In this approach, $\boldsymbol{\tau}$ is a contravariant second-order tensor and the pull-back operation is performed with $(\mathbf{F}^e)^{-1}$ and $(\mathbf{F}^e)^{-T}$ on the left and right legs as indicated. However, $\mathbf{g}$ and $\mathbf{D}^e$, for example, are covariant second-order tensors and are pulled-back using $\mathbf{F}^{eT}$ on the left and $\mathbf{F}^e$ on the right, respectively. The interested reader is referred to Marsden and Hughes (1983) for a detailed treatment of these concepts. Note that while these concepts can be substantially avoided by working within the framework of Euclidean spaces, even there they re-emerge, in part, when general curvilinear coordinates are used. In this book, the particular pull-back or push-forward operation to be employed is explicitly given.

The push-forward operation can be used to obtain a useful expression for the Kirchhoff stress by replacing each term in the expression $(5.7.5)_1$ for $\bar{\mathbf{S}}$ by its push-forward:

$$\boldsymbol{\tau} = \phi_*^e \bar{\mathbf{S}} = 2 \frac{\partial \psi}{\partial \mathbf{g}} \tag{5.10.9}$$

where $\psi(\mathbf{g}, \mathbf{F}^e) = \bar{\psi}(\phi_e^*(\mathbf{g})) = \bar{\psi}(\mathbf{F}^{eT} \cdot \mathbf{g} \cdot \mathbf{F}^e)$ is the push-forward of the hyperelastic potential $\bar{\psi}$. An expression for the elasticity tensor on Ω is obtained through (5.7.6) and is given by

$$\mathbf{C}_{el}^{\tau} = 2\frac{\partial \boldsymbol{\tau}}{\partial \mathbf{g}} = 4\frac{\partial^2 \psi}{\partial \mathbf{g} \partial \mathbf{g}} \tag{5.10.10}$$

The expressions (5.10.9) and (5.10.10) are known as the Doyle–Ericksen formulae (Doyle and Ericksen, 1956).

The time derivative of the elastic right Cauchy–Green deformation tensor can be related to the pull-back of the elastic part of the rate-of-deformation tensor to $\bar{\Omega}$, that is,

$$\dot{\bar{\mathbf{C}}}^e = 2\mathbf{F}^{eT} \cdot \mathbf{D}^e \cdot \mathbf{F}^e = \phi_e^*(\mathbf{D}^e) \tag{5.10.11}$$

but

$$\dot{\bar{\mathbf{C}}}^e = \frac{D}{Dt}(\mathbf{F}^{eT} \cdot \mathbf{g} \cdot \mathbf{F}^e) \tag{5.10.12}$$

and thus from (5.10.11)

$$2\mathbf{D}^e = (\mathbf{F}^e)^{-T} \cdot \frac{D}{Dt}(\mathbf{F}^{eT} \cdot \mathbf{g} \cdot \mathbf{F}^e) \cdot \mathbf{F}^e = \phi_*^e\left(\frac{D}{Dt}\phi_e^*(\mathbf{g})\right) = L_v^e \mathbf{g} \tag{5.10.13}$$

that is, an expression for twice the elastic part of the rate-of-deformation tensor is obtained by pulling the metric tensor back to the intermediate configuration, taking the material time-derivative and pushing the result forward to the current configuration which is the elastic Lie derivative of the metric tensor. Using the concept of Lie derivatives, the time-derivative of the hyperelastic potential $\bar{\psi}$ can be written as

$$\dot{\bar{\psi}} = \frac{1}{2}\bar{\mathbf{S}} : \dot{\bar{\mathbf{C}}}^e = \frac{1}{2}(\mathbf{F}^{e-1} \cdot \boldsymbol{\tau} \cdot \mathbf{F}^{e-T}) : 2(\mathbf{F}^{eT} \cdot \mathbf{D}^e \cdot \mathbf{F}^e) = \boldsymbol{\tau} : \mathbf{D}^e = \frac{1}{2}\boldsymbol{\tau} : L_v^e \mathbf{g} \tag{5.10.14}$$

and $\boldsymbol{\tau}$ and $\mathbf{g}$ are regarded as work conjugates. The last of these results can be obtained from the first by replacing each term with its push-forward (i.e., $\bar{\mathbf{S}}$ by $\boldsymbol{\tau}$ and $\bar{\mathbf{C}}^e$ by $L_v^e \mathbf{g}$).

Now note that the symmetric and antisymmetric parts of $\mathbf{L}^e$ and $\mathbf{L}^p$ (in 5.7.10) can be written as

$$\begin{aligned}
\mathbf{D}^e &= \frac{1}{2}(\mathbf{g} \cdot \mathbf{L}^e + \mathbf{L}^{eT} \cdot \mathbf{g}), \quad \mathbf{W}^e = \frac{1}{2}(\mathbf{g} \cdot \mathbf{L}^e - \mathbf{L}^{eT} \cdot \mathbf{g}) \\
\mathbf{D}^p &= \frac{1}{2}(\mathbf{g} \cdot \mathbf{L}^p + \mathbf{L}^{pT} \cdot \mathbf{g}), \quad \mathbf{W}^p = \frac{1}{2}(\mathbf{g} \cdot \mathbf{L}^p - \mathbf{L}^{pT} \cdot \mathbf{g})
\end{aligned} \tag{5.10.15}$$

The explicit retention of the metric tensor, $\mathbf{g}=\mathbf{I}$ on Ω, in these expressions is to facilitate the interpretation of the corresponding quantities mapped back to the intermediate configuration. The pull-back of the elastic and plastic parts of the spatial velocity gradient is given by

$$\begin{aligned} \bar{\mathbf{L}}^e &= \phi_e^*(\mathbf{L}^e) = (\mathbf{F}^e)^{-1}\cdot\mathbf{L}^e\cdot\mathbf{F}^e = (\mathbf{F}^e)^{-1}\cdot\dot{\mathbf{F}}^e \\ \bar{\mathbf{L}}^p &= \phi_e^*(\mathbf{L}^p) = (\mathbf{F}^e)^{-1}\cdot\mathbf{L}^p\cdot\mathbf{F}^e = \dot{\mathbf{F}}^p\cdot(\mathbf{F}^p)^{-1} \end{aligned} \tag{5.10.16}$$

where (5.7.9) has been used in obtaining the final expressions. Thus, $\mathbf{L}^e$ is the elastic part of the velocity gradient on Ω and $\bar{\mathbf{L}}^e$ is its pull-back, while $\bar{\mathbf{L}}^p$ is the plastic part of the velocity gradient on $\bar{\Omega}$ and $\mathbf{L}^p$ is its push-forward. In tensor analysis on general manifolds, the spatial velocity gradient $\mathbf{L}$ and its elastic and plastic parts ($\mathbf{L}^e$ and $\mathbf{L}^p$, respectively) have mixed contravariant–covariant character. Hence the pull-back operations in (5.10.16) are also mixed and use $(\mathbf{F}^e)^{-1}$ on the left (contravariant) leg and $\mathbf{F}^e$ on the right (covariant) leg.

The symmetric and antisymmetric parts of the elastic and plastic velocity gradients on Ω are given by

$$\begin{aligned} \bar{\mathbf{D}}^e &= \phi_e^*(\mathbf{D}^e) = \frac{1}{2}(\bar{\mathbf{C}}^e\cdot\bar{\mathbf{L}}^e + \bar{\mathbf{L}}^{eT}\cdot\bar{\mathbf{C}}^e), \quad \bar{\mathbf{W}}^e = \phi_e^*(\mathbf{W}^e) = \frac{1}{2}(\bar{\mathbf{C}}^e\cdot\bar{\mathbf{L}}^e - \bar{\mathbf{L}}^{eT}\cdot\bar{\mathbf{C}}^e) \\ \bar{\mathbf{D}}^p &= \phi_e^*(\mathbf{D}^p) = \frac{1}{2}(\bar{\mathbf{C}}^e\cdot\bar{\mathbf{L}}^p + \bar{\mathbf{L}}^{pT}\cdot\bar{\mathbf{C}}^e), \quad \bar{\mathbf{W}}^p = \phi_e^*(\mathbf{W}^p) = \frac{1}{2}(\bar{\mathbf{C}}^e\cdot\bar{\mathbf{L}}^p - \bar{\mathbf{L}}^{pT}\cdot\bar{\mathbf{C}}^e) \end{aligned} \tag{5.10.17}$$

where $\bar{\mathbf{C}}^e = \phi_e^*(\mathbf{g})$ plays the role of the metric on the intermediate configuration. The role of the metric in these expressions is akin to lowering the contravariant indexes so that the tensors are purely covariant and the symmetric and antisymmetric parts can be formed.

5.10.3 Material Frame Indifference

In addition to appropriate stress and deformation measures, the extension of a small strain constitutive relation to finite strains requires the consideration of finite rotations. A constitutive relation should be independent of any rigid body motions. Alternatively stated, the constitutive relation should be the same for two observers in relative motion (translation plus rotation). This is referred to as the *principle of material objectivity*. Here we consider its implications for the formulation of constitutive relations at finite strain.

Let $\mathbf{x}(\mathbf{X}, t)$ and $\mathbf{x}^*(\mathbf{X}, t)$ be the motion of the body as described by two observers in relative motion where

$$\mathbf{x}^* = \mathbf{Q}(t)\cdot\mathbf{x} + \mathbf{c}(t), \quad \mathbf{Q}^{-1} = \mathbf{Q}^T \tag{5.10.18}$$

with $\mathbf{Q}(0)=\mathbf{I}$ and $\mathbf{c}(0)=0$. Equations (5.10.18) represent the transformation between two observers whose frames of reference are rotating by $\mathbf{Q}$ and translating by $\mathbf{c}$ with respect to each other. Equations (5.10.18) are also mathematically equivalent to the superposition of a rigid body rotation and translation on the current configuration of the body. It is often convenient to consider objectivity requirements from this second point of view (see Malvern (1969) and Ogden (1984) for additional details).

It follows from (5.10.18) that

$$d\mathbf{x}^* = \mathbf{Q} \cdot d\mathbf{x} = \mathbf{Q} \cdot \mathbf{F} \cdot d\mathbf{X} = \mathbf{F}^* \cdot d\mathbf{X} \tag{5.10.19}$$

Since this holds for arbitrary $d\mathbf{X}$, the deformation gradient transforms like the Eulerian vector $d\mathbf{x}$ under an observer transformation:

$$\mathbf{F}^* = \mathbf{Q} \cdot \mathbf{F} \tag{5.10.20}$$

Taking the material time-derivative of the expression (5.10.18), we obtain

$$\mathbf{v}^* = \mathbf{Q} \cdot \mathbf{v} + \dot{\mathbf{Q}} \cdot \mathbf{x} + \dot{\mathbf{c}} \tag{5.10.21}$$

It can be seen from (5.10.19) and (5.10.21) that the vector $d\mathbf{x}$ and the velocity vector field $\mathbf{v}$ do not transform in the same manner under the rigid body motion. To see how other kinematic quantities transform under an observer transformation (or superposed rigid body rotation) we can simply derive the transformations from (5.10.18–21).

For example, the transformation of the right Cauchy–Green deformation tensor follows immediately from (5.10.20) as

$$\mathbf{C}^* = \mathbf{F}^{*T} \cdot \mathbf{F}^* = \mathbf{F}^T \cdot \mathbf{Q}^T \cdot \mathbf{Q} \cdot \mathbf{F} = \mathbf{F}^T \cdot \mathbf{F} = \mathbf{C} \tag{5.10.22}$$

Using (3.3.18) and (5.10.20), the spatial velocity gradient transforms according to

$$\mathbf{L}^* = \dot{\mathbf{F}}^* \cdot (\mathbf{F}^*)^{-1} = \mathbf{Q} \cdot \mathbf{L} \cdot \mathbf{Q}^T + \dot{\mathbf{Q}} \cdot \mathbf{Q}^T \tag{5.10.23}$$

The transformations of the rate-of-deformation and spin tensors follow immediately as

$$\mathbf{D}^* = \mathbf{Q} \cdot \mathbf{D} \cdot \mathbf{Q}^T, \quad \mathbf{W}^* = \mathbf{Q} \cdot \mathbf{W} \cdot \mathbf{Q}^T + \dot{\mathbf{Q}} \cdot \mathbf{Q}^T \tag{5.10.24}$$

Equations (5.10.22–24) indicate how various kinematic quantities change under an observer transformation or the superposition of a rigid body rotation on the current configuration.

Now consider a rectangular Cartesian coordinate system which rotates with the observer, that is, $\mathbf{e}_i^* = \mathbf{Q} \cdot \mathbf{e}_i$. An Eulerian tensor field is said to be objective if the transformed tensor has the same components in the starred coordinate system as the untransformed tensor has in the unstarred system. It is easily shown that the Eulerian tensors $d\mathbf{x}$ and $\mathbf{D}$ are objective, that is,

$$dx_i^* = \mathbf{e}_i^* \cdot d\mathbf{x}^* = \mathbf{e}_i \cdot \mathbf{Q}^T \cdot \mathbf{Q} \cdot d\mathbf{x} = dx_i \tag{5.10.25}$$

and

$$D_{ij}^* = \mathbf{e}_i^* \cdot \mathbf{D}^* \cdot \mathbf{e}_j^* = \mathbf{e}_i \cdot \mathbf{Q}^T \cdot \mathbf{Q} \cdot \mathbf{D} \cdot \mathbf{Q}^T \cdot \mathbf{Q} \cdot \mathbf{e}_j = D_{ij} \tag{5.10.26}$$

This same approach can be used to show that the quantities **v, L** and **W** are not objective due to the presence of the rate of rotation terms in the transformation relations (5.10.23) and (5.10.24). More generally, an Eulerian vector **a** and an Eulerian second-order tensor **A** are said to be objective if

$$\mathbf{a}^* = \mathbf{Q}\cdot\mathbf{a}, \quad \mathbf{A}^* = \mathbf{Q}\cdot\mathbf{A}\cdot\mathbf{Q}^T \tag{5.10.27}$$

A Lagrangian tensor field is said to be objective if it remains unchanged under an observer transformation. From (5.10.18) at time $t = 0$,

$$d\mathbf{X}^* = d\mathbf{X} \tag{5.10.28}$$

and is an objective Lagrangian vector field, while from (5.10.22) **C** is an objective Lagrangian second-order tensor. A Lagrangian vector $\mathbf{a}_o$ and second-order tensor $\mathbf{A}_o$ are said to be objective if

$$\mathbf{a}_o^* = \mathbf{a}_o, \quad \mathbf{A}_o^* = \mathbf{A}_o \tag{5.10.29}$$

This requirement can be written in terms of components as

$$a_{oi}^* = \mathbf{e}_i\cdot\mathbf{a}_o^* = \mathbf{e}_i\cdot\mathbf{a}_o = a_{oi} \quad A_{oij}^* = \mathbf{e}_i\cdot\mathbf{A}_o^* \cdot\mathbf{e}_j = \mathbf{e}_i\cdot\mathbf{A}_o\cdot\mathbf{e}_j = A_{oij} \tag{5.10.30}$$

The definition of objectivity for an Eulerian–Lagrangian tensor field incorporates both of the above definitions. For example, the deformation gradient is a two-point tensor because it maps quantities from the reference to the current configuration. An Eulerian–Lagrangian second-order tensor **B** is said to be objective if

$$B_{ij}^* = \mathbf{e}_i^* \cdot\mathbf{B}^*\cdot\mathbf{e}_j = B_{ij} = \mathbf{e}_i\cdot\mathbf{B}\cdot\mathbf{e}_j \tag{5.10.31}$$

or

$$\mathbf{B}^* = \mathbf{Q}\cdot\mathbf{B} \tag{5.10.32}$$

Thus, from (5.10.20), the deformation gradient **F** is an objective Eulerian–Lagrangian second-order tensor.

5.10.4 *Implications for Constitutive Relations*

Having established how various kinematic quantities transform under an observer transformation, we now consider the concept of frame invariance or material objectivity and the restrictions it imposes on constitutive relations. Consider a Cauchy elastic material with response function $\boldsymbol{G}$:

$$\boldsymbol{\sigma} = G(\mathbf{FX}, t) \tag{5.10.33}$$

Here, the Cauchy stress, as seen by an observer O, is given by the response function G of the deformation gradient at $\mathbf{X}$ at time t. For an observer, O^* moving with respect to O according to (5.10.18), the Cauchy stress is written as

$$\boldsymbol{\sigma}^* = G^*(\mathbf{F}^*) \tag{5.10.34}$$

where the arguments $\mathbf{X}$ and t are suppressed for notational convenience.

The principle of *material objectivity* or *material frame indifference* states that the material response is independent of the observer. A mathematical statement of the principle is written as

$$\boldsymbol{G}^*(\mathbf{F}^*) = \boldsymbol{G}(\mathbf{F}^*) \tag{5.10.35}$$

that is, and $\boldsymbol{G}^*$ are $\boldsymbol{G}$ the same function. Moreover, material objectivity implies that, to determine the Cauchy stress, observer O^* treats $\mathbf{F}^*$ the same way as observer O treats $\mathbf{F}$.

The Cauchy stress is an objective (Eulerian) tensor and thus the components of the Cauchy stress seen by observer O^* in the rotated coordinate system are the same as those seen by observer O in the unrotated system, that is,

$$\boldsymbol{\sigma}^* = \mathbf{Q}\cdot\boldsymbol{\sigma}\cdot\mathbf{Q}^T, \quad \sigma_{ij}^* = \mathbf{e}_i^* \cdot\mathbf{Q}\cdot\boldsymbol{\sigma}\cdot\mathbf{Q}^T\cdot\mathbf{e}_j^* = \mathbf{e}_i\cdot\boldsymbol{\sigma}\cdot\mathbf{e}_j = \sigma_{ij} \tag{5.10.36}$$

Using (5.10.36) and (5.10.34) gives

$$\mathbf{Q}\cdot\boldsymbol{\sigma}\cdot\mathbf{Q}^T = \boldsymbol{\sigma}^* = G(\mathbf{Q}\cdot\mathbf{F}) \tag{5.10.37}$$

or

$$\boldsymbol{\sigma} = \mathbf{Q}^T\cdot G(\mathbf{Q}\cdot\mathbf{F})\cdot\mathbf{Q} \tag{5.10.38}$$

Recall that from the polar decomposition theorem, the deformation gradient can be written as $\mathbf{F}=\mathbf{R}\cdot\mathbf{U}$ where $\mathbf{R}$ is the rotation tensor and $\mathbf{U}$ is the right stretch tensor. The previous must be true for all rotations $\mathbf{Q}(t)$ and for $\mathbf{Q}=\mathbf{R}^T$ in particular. For this choice, (5.10.38) reduces to

$$\boldsymbol{\sigma} = \mathbf{R}\cdot G(\mathbf{U})\cdot\mathbf{R}^T \tag{5.10.39}$$

Equation (5.10.39) represents the restriction, due to material objectivity, on the constitutive relation (5.10.33): the dependence of the constitutive relation on the rotation $\mathbf{R}$ can only take the form (5.10.39), that is, the constitutive relation can only depend on the current value of the rotation and on the right stretch tensor (or related measures such as the Green strain). Note that the response function for the rotated Cauchy stress $\hat{\sigma} = \mathbf{R}^T\cdot\boldsymbol{\sigma}\cdot\mathbf{R}$ must therefore take the simple form $\hat{\sigma} = \boldsymbol{G}(\mathbf{U})$.

5.10.5 Objective Scalar Functions

Scalar functions of tensor variables arise frequently in constitutive relations. An example is the yield function in elasto-plastic problems. An objective scalar function f satisfies the condition $f^*=f$ under an observer transformation. Consider a scalar function $f(\boldsymbol{\sigma})$, where $\boldsymbol{\sigma}$ is the Cauchy stress, an objective Eulerian second-order tensor. Material objectivity then states that

$$f^*(\boldsymbol{\sigma}^*) = f(\boldsymbol{\sigma}), \quad f^* = f, \quad \text{or} \quad f(\mathbf{Q}\cdot\boldsymbol{\sigma}\cdot\mathbf{Q}^T) = f(\boldsymbol{\sigma}) \tag{5.10.40}$$

for arbitrary rotations $\mathbf{Q}$. This requires that f be only a function of the principal invariants of $\boldsymbol{\sigma}$ or, in other words, that f is an isotropic function of $\boldsymbol{\sigma}$. This means that if anisotropic yield behavior is required, it can not be represented by a function of the form $f(\boldsymbol{\sigma})$. To represent anisotropic yield, it is expedient to define the yield function in terms of a stress measure which is invariant under superposed rotations of Ω. For example, the second Piola–Kirchhoff stress on $\bar{\Omega}$ is invariant under rotation. Material objectivity of the yield function requires that

$$\bar{f}(\bar{\mathbf{S}}^*) = \bar{f}(\bar{\mathbf{S}}) \tag{5.10.41}$$

From (5.10.61) $\bar{\mathbf{S}}^* = \bar{\mathbf{S}}$, and therefore (5.10.41) is automatically satisfied and material objectivity imposes no restrictions on $\bar{f}$. To illustrate, consider the anisotropic yield function

$$\bar{f}(\bar{\mathbf{S}}) = \bar{f}(\bar{\mathbf{S}}:\bar{\mathrm{H}}:\bar{\mathbf{S}}) \tag{5.10.42}$$

where $\bar{\mathrm{H}}$ is taken to be a fourth-order tensor of material constants. Note that this yield function satisfies (5.10.41) and thus there are no frame invariance restrictions on $\bar{f}$ or $\bar{\mathrm{H}}$. Pushing this expression forward to the spatial configuration (with the elastic part of the deformation gradient), we get

$$f = \phi_*^e \bar{f} = f(\boldsymbol{\tau}:\mathrm{H}:\boldsymbol{\tau}) \tag{5.10.43}$$

where

$$\begin{aligned} &\boldsymbol{\tau} = \phi_*^e \bar{\mathbf{S}} = \mathbf{F}^e \cdot \bar{\mathbf{S}} \cdot \mathbf{F}^{eT}, \\ &\mathrm{H} = \phi_*^e \bar{\mathrm{H}}, \quad H_{ijkl} = \left(F_{im}^e\right)^{-T} \left(F_{jn}^e\right)^{-T} \left(F_{kp}^e\right)^{-T} \left(F_{lq}^e\right)^{-T} \bar{H}_{mnpq} \end{aligned} \tag{5.10.44}$$

The particular form of the push-forward of $\bar{\mathrm{H}}$ is determined through the invariance of the scalar argument of f, that is, $\boldsymbol{\tau}:\mathrm{H}:\boldsymbol{\tau} = \bar{\mathbf{S}}:\bar{\mathrm{H}}:\bar{\mathbf{S}}$ and $\bar{\mathrm{H}}$ is pushed-forward in the manner of kinematic tensors (conjugate to the stresses). Invariance of the spatial form of f can be verified by noting that under an observer transformation $\boldsymbol{\tau}^* = \mathbf{Q}\cdot\boldsymbol{\tau}\cdot\mathbf{Q}^T$, $(\mathbf{F}^e)^* = \mathbf{Q}\cdot\mathbf{F}^e$ and therefore $(\mathbf{F}^{e*})^{-T} = \mathbf{Q}\cdot(\mathbf{F}^e)^{-T}$.

5.10.6 Restrictions on Elastic Moduli

In Sections 5.4 and 5.6 we noted that, if the spatial elastic moduli are assumed to be constant, then material objectivity requires that they be isotropic. To see this, consider the hypoelastic relation (B5.6.2) with constant moduli $\mathbf{C}_{el}^{\tau J}$. Frame invariance requires $(\boldsymbol{\tau}^{\nabla J})^* = \mathbf{C}_{el}^{\tau J}:(\mathbf{D}^e)^*$ which, upon using the transformation (5.10.27) for objective Eulerian second-order tensors, can be written in component form as:

$$Q_{im}Q_{jn}\tau_{mn}^{\nabla J} = \left(C_{el}^{\tau J}\right)_{ijkl}\left(Q_{kr}Q_{ls}Q_{rs}^e\right) \tag{5.10.45}$$

This expression is rearranged to give

$$\tau_{ij}^{\nabla J} = \left(Q_{mi}Q_{nj}Q_{pk}Q_{ql}\left(C_{el}^{\tau J}\right)_{mnpq}\right)D_{kl}^{e} \tag{5.10.46}$$

But $\boldsymbol{\tau}^{\nabla J} = \mathbf{C}_{el}^{\tau J} : \mathbf{D}^{e}$, therefore

$$\left(C_{el}^{\tau J}\right)_{ijkl} = Q_{mi}Q_{nj}Q_{pk}Q_{ql}\left(C_{el}^{\tau J}\right)_{mnpq} \quad \forall Q_{ij}, \quad \det Q_{ij} = 1 \tag{5.10.47}$$

which restricts the moduli to be isotropic.

If the elastic response is formulated on some intermediate configuration which is pulled-back from the spatial configuration, this restriction to isotropy is removed. To see this, consider the hypoelastic relation based on the corotational Kirchhoff stress (5.6.31). Pushing this expression forward to the spatial configuration using the rotation $\mathbf{R}$ we get

$$\boldsymbol{\tau}^{\nabla G} = \mathbf{C}_{el}^{\tau G} : \mathbf{D}^{e}, \quad \left(C_{el}^{\tau G}\right)_{ijkl} = R_{im}R_{jn}R_{kp}R_{lp}\hat{C}_{mnpq} \tag{5.10.48}$$

Frame invariance requires that the constitutive relation transforms as $(\boldsymbol{\tau}^{\nabla G})^{*} = (\mathbf{C}_{el}^{\tau G})^{*} : (\mathbf{D}^{e})^{*}$. Note that the moduli $\mathbf{C}_{el}^{\tau G}$ are now not constant and are given through the transformation of $\hat{\mathbf{C}}$ under the rotation $\mathbf{R}$. From the polar decomposition $\mathbf{F}=\mathbf{R}\cdot\mathbf{U}$ and the transformation (5.10.20) of the deformation gradient, the rotation tensor transforms as $\mathbf{R}^{*}=\mathbf{Q}\cdot\mathbf{R}$. Using this expression and transformation (5.10.48), the transformed constitutive relation (component form) is

$$\begin{aligned}
\left(\tau^{\nabla G}\right)_{ij}^{*} &= \left(C_{el}^{\tau G}\right)_{ijrs}^{*} \; (D^{e})_{rs}^{*} \\
Q_{im}Q_{jn}\tau_{mn}^{\nabla G} &= R_{im}^{*} \; R_{jn}^{*} \; R_{kp}^{*} \; R_{lq}^{*} \; \hat{C}_{mnpq}\left(Q_{kr}Q_{ls}D_{rs}^{e}\right) \\
&= Q_{it}R_{tm}Q_{ju}R_{un}Q_{kv}R_{vp}Q_{lw}R_{wq}\hat{C}_{mnpq}\left(Q_{kr}Q_{ls}Q_{rs}^{e}\right)
\end{aligned} \tag{5.10.49}$$

This expression can be rearranged to give

$$\tau_{ij}^{\nabla G} = R_{im}R_{jn}R_{kp}R_{lq}\hat{C}_{mnpq}D_{kl}^{e} = \left(C_{el}^{\tau G}\right)_{ijkl} D_{kl}^{e} \tag{5.10.50}$$

which is the component form of (5.10.48). The rigid rotation imposes no restrictions on the moduli $\mathbf{C}_{el}^{\tau G}$ which may therefore be anisotropic.

5.10.7 *Material Symmetry*

Restrictions on the material response due to material symmetry have been addressed by Noll (Malvern, 1969). For simplicity, we again consider a Cauchy elastic material. A tensor $\mathbf{Q}$ is said to be an element of the symmetry group of a material if the stress on a material element is independent of whether that element is first operated upon by $\mathbf{Q}$ or not. We consider only orthogonal tensors for which $\det \mathbf{Q}=\pm 1$. This allows for the symmetry operations of rotation, $\det \mathbf{Q}=+1$, and reflection, $\det \mathbf{Q}=-1$. If $\mathbf{Q}$ is an element of the symmetry group then

$$G(\mathbf{F}) = G(\mathbf{F}\cdot\mathbf{Q}) \tag{5.10.51}$$

where G is the response function of the material ($\boldsymbol{\sigma} = G(\mathbf{F})$). A material for which the symmetry group includes all rotations is said to be isotropic (with respect to the initial configuration). Using the polar decomposition in the form $\mathbf{F} = \mathbf{V}\cdot\mathbf{R}$ and setting $\mathbf{Q} = \mathbf{R}^T$, (5.10.51) is written as

$$G(\mathbf{F}) = G(\mathbf{V}) \tag{5.10.52}$$

For isotropic materials it can be further shown (Ogden, 1984) through the representation theorem that

$$G(\mathbf{V}) = G(I_1(\mathbf{V}), I_2(\mathbf{V}), I_3(\mathbf{V})) \tag{5.10.53}$$

where $I_I(\mathbf{V})$ denote the principal invariants of $\mathbf{V}$. The principal invariants of a second-order tensor are defined in Box 5.2. For an isotropic material, then,

$$\boldsymbol{\sigma} = G(\mathbf{V}) = \mathbf{R}\cdot G(\mathbf{U})\cdot\mathbf{R}^T \tag{5.10.54}$$

where we have used (5.10.39) to write the second equality previously.

5.10.8 Frame Invariance in Hyperelastic–Plastic Models

When the elastic response is given by a hyperelastic potential (5.7.5), the principle of material frame indifference is trivially satisfied as the stress is directly evaluated through (5.7.5) and no integration of stress rates is required. To illustrate this further, consider the rigid body motions superimposed on the spatial configuration. Let $\mathbf{Q}(t)$ be a time-dependent rotation of the spatial configuration Ω. We regard Ω_0 and $\bar{\Omega}$ as fixed, that is, unaffected by the rotation $\mathbf{Q}(t)$ (Dashner, 1986). Under the superimposed rotation $\mathbf{Q}(t)$, the deformation gradient is given by

$$\mathbf{F}^* = \mathbf{Q}\cdot\mathbf{F} \tag{5.10.55}$$

Thus the deformation gradient, which is a two-point tensor connecting the reference and current configurations, transforms like a vector under the superimposed rotation $\mathbf{Q}(t)$. Writing the deformation gradient $\mathbf{F}^*$ as the product of elastic and plastic parts we have

$$\mathbf{F}^* = \mathbf{F}^{*e}\cdot\mathbf{F}^{*p} \tag{5.10.56}$$

Because the intermediate configuration $\bar{\Omega}$ is unaffected by a rotation of the current configuration Ω, we have

$$\mathbf{F}^{*p} = \mathbf{F}^p \tag{5.10.57}$$

and using this result in (5.10.56) gives

$$\mathbf{F}^* = \mathbf{F}^{*e}\cdot\mathbf{F}^p \tag{5.10.58}$$

It then follows from (5.10.55) that

$$\mathbf{F}^* = \mathbf{Q}\cdot\mathbf{F} = \mathbf{Q}\cdot\mathbf{F}^e\cdot\mathbf{F}^p, \quad \mathbf{F}^{*e} = \mathbf{Q}\cdot\mathbf{F}^e \tag{5.10.59}$$

The transformed Cauchy deformation tensor is given by $(\bar{\mathbf{C}}^e)^* = \mathbf{F}^{*eT}\cdot\mathbf{g}^*\cdot\mathbf{F}^{*e}$. However, $\mathbf{g}^* = \mathbf{Q}\cdot\mathbf{g}\cdot\mathbf{Q}^T = \mathbf{g}$ (the observer transformation is known as an isometry; it leaves the metric unchanged) and from (5.10.59) we get the result $(\bar{\mathbf{C}}^e)^* = \bar{\mathbf{C}}^e$ which demonstrates that $(\bar{\mathbf{C}}^e)$ is invariant under rigid body rotation of Ω. It follows that the principle of material frame indifference, which requires the hyperelastic potential to be an objective scalar quantity on $\bar{\Omega}$, gives

$$\bar{\psi} = \bar{\psi}((\bar{\mathbf{C}}^e)^*) = \bar{\psi}(\bar{\mathbf{C}}^e) \tag{5.10.60}$$

The second Piola–Kirchhoff stress is obtained from the hyperelastic potential

$$\bar{\mathbf{S}}^* = 2\frac{\partial\bar{\psi}((\bar{\mathbf{C}}^e)^*)}{\partial(\bar{\mathbf{C}}^e)^*} = 2\frac{\partial\bar{\psi}((\bar{\mathbf{C}}^e))}{\partial\bar{\mathbf{C}}^e} = \bar{\mathbf{S}} \tag{5.10.61}$$

and is also invariant under rigid body rotation of Ω. Thus, by formulating the elastic response in terms of a hyperelastic potential on the intermediate configuration $\bar{\Omega}$, the principle of material frame indifference is automatically satisfied and issues associated with the integration of stress rates and the enforcement of incremental objectivity are circumvented.

5.10.9 Clausius–Duhem Inequality and Stability Postulates

In addition to objectivity, restrictions are imposed on constitutive relations by the second law of thermodynamics. These are expressed through the Clausius–Duhem inequality. We derive the inequality here in a variety of forms and examine its implications for associative and non-associative plasticity models. Let the specific entropy in the body be denoted as η. The second law of thermodynamics states that the total rate of increase of entropy in the body is greater than or equal to the entropy input:

$$\frac{D}{Dt}\int_{\Omega}\rho\eta\, d\Omega \geq -\int_{\Gamma}\mathbf{h}\cdot\mathbf{n}d\Gamma + \int_{\Omega}\frac{1}{\theta}\rho s\, d\Omega \tag{5.10.62}$$

where θ is the absolute temperature, $\mathbf{h} = \mathbf{q}/\theta$ is called the entropy flow vector, $\mathbf{q}$ is the heat flux vector (not to be confused with the collection of internal variables in the plasticity models), and s/θ is the specific rate of entropy production. Applying the divergence theorem to the surface integral and noting that the inequality is valid for arbitrary volumes gives

$$\rho\dot{\eta} \geq -\text{div}\left(\frac{\mathbf{q}}{\theta}\right) + \frac{1}{\theta}\rho s, \quad \text{or} \quad \rho\dot{\eta} \geq -\frac{1}{\theta}\text{div}\ \mathbf{q} + \frac{1}{\theta^2}\mathbf{q}\cdot\nabla\theta + \frac{1}{\theta}\rho s \tag{5.10.63}$$

Because heat flows from hot to cold, we have $-\left(\frac{1}{\theta^2}\right)\mathbf{q}\cdot\nabla\theta \geq 0$, and (5.10.63) is sometimes written under the stronger assumption

$$\rho\dot{\eta} \geq -\frac{1}{\theta}\operatorname{div}(\mathbf{q}) + \frac{1}{\theta}\rho s, \quad \text{or} \quad \rho\theta\dot{\eta} + \operatorname{div}(\mathbf{q}) - \rho s \geq 0 \tag{5.10.64}$$

Now define the specific free energy as $\psi = w^{\text{int}} - \theta\eta$. Differentiating this expression we get

$$\dot{\psi} = \dot{w}^{\text{int}} - \dot{\theta}\eta - \theta\dot{\eta} \tag{5.10.65}$$

From the energy equation (3.5.49), this can be written as

$$\rho\dot{\psi} = \boldsymbol{\sigma}:\mathbf{D} - \operatorname{div}\mathbf{q} + \rho s - \rho\dot{\theta}\eta - \rho\theta\dot{\eta} \tag{5.10.66}$$

Solving for $\rho\theta\dot{\eta}$ and substituting into (5.10.64) gives

$$\boldsymbol{\sigma}:\mathbf{D} - \rho\dot{\psi} - \rho\dot{\theta}\eta \geq 0, \quad \text{or} \quad \mathbf{S}:\dot{\mathbf{E}} - \rho_0\dot{\psi} - \rho_0\dot{\theta}\eta \geq 0 \tag{5.10.67}$$

where the second of these is obtained by pulling the first back to the reference configuration using the deformation gradient $\mathbf{F}$ and multiplying by $J = \rho_0/\rho$. Consider inelastic materials for which the free energy is $\rho_0\psi = \rho_0\psi(\mathbf{E}^e, \xi_\alpha, \theta)$ where $\mathbf{E}^e = \mathbf{E} - \mathbf{E}^p$ is the elastic part of the Green strain and $\xi\alpha$ are a set of internal variables (assumed to be scalars here). Taking the material time-derivative of the free energy gives

$$\rho_0\dot{\psi} = \mathbf{S}:\dot{\mathbf{E}}^e + \rho_0\frac{\partial\psi}{\partial\xi_\alpha}\dot{\xi}_\alpha - \rho_0\eta\dot{\theta}, \quad \text{where} \quad \mathbf{S} = \rho_0\frac{\partial\psi}{\partial\mathbf{E}^e}, \quad \eta = -\frac{\partial\psi}{\partial\theta}, \tag{5.10.68}$$

and substituting the right-hand side of the expression for the reference form of the inequality, (5.10.67) can be written as

$$\mathbf{S}:\dot{\mathbf{E}}^p - \rho_0\frac{\partial\psi}{\partial\xi_\alpha}\dot{\xi}_\alpha \geq 0 \tag{5.10.69}$$

Pushing this result forward to the current configuration and dividing by $J = \rho_0/\rho$ gives

$$\boldsymbol{\sigma}:\mathbf{D}^p - \rho\frac{\partial\psi}{\partial\xi_\alpha}\dot{\xi}_\alpha \geq 0 \tag{5.10.70}$$

If we further restrict attention to materials for which the free energy ψ has no *explicit* dependence on the internal variables ξ_α, then this reduces to

$$\boldsymbol{\sigma}:\mathbf{D}^p \geq 0 \tag{5.10.71}$$

which states that the plastic dissipation must be non-negative and indicates that plastic deformation is irreversible. Note that the same result is obtained if, instead, the free energy is defined as $\rho_0\psi = \rho_0\bar{\psi}(\bar{\mathbf{E}}^e, \xi_\alpha, \theta)$ which is appropriate for a hyperelastic–plastic model based on the multiplicative decomposition. In this case, (5.10.69) becomes $\bar{\mathbf{S}}:\bar{\mathbf{D}}^p - \rho_0 \frac{\partial \psi}{\partial \xi_\alpha}\dot{\xi}_\alpha \geq 0$. Dividing by J and pushing forward $\mathbf{F}^e$ gives (5.10.70) and, when there is no explicit dependence on the internal variables, (5.10.71).

We will now examine the implications of the plastic dissipation inequality for elastic–plastic constitutive relations. Consider the plastic flow rule given by $\mathbf{D}^p = \dot{\lambda}\mathbf{r}$. The plastic rate parameter is defined to be non-negative and therefore the dissipation inequality (5.10.71) requires $\boldsymbol{\sigma}:\mathbf{r} \geq 0$. Constitutive equations must be constructed to assure that this condition is not violated. For rate-independent problems with a convex yield surface which encloses the origin (in stress space), *the plastic dissipation inequality is always satisfied for associative flow rules*. For non-associative plastic flow rules, it is necessary to restrict the flow direction from drifting too far from the normal to the yield surface. The dissipation inequality applies to strain hardening *and* to strain softening plastic materials.

5.10.9.1 Principle of Maximum Plastic Dissipation

An additional postulate on material behavior is the principle of maximum plastic work

$$(\boldsymbol{\sigma} - \boldsymbol{\sigma}^*):\mathbf{D}^p \geq 0 \qquad (5.10.72)$$

where $\boldsymbol{\sigma}^*$ is any stress state which is inside or on the yield surface. Note that by taking $\boldsymbol{\sigma}^* = \mathbf{0}$, we recover the dissipation inequality (5.10.71). If (5.10.72) is satisfied it can be shown (Lubliner, 1990) that the yield surface must be convex and that plastic flow must be associated, that is, in the normal direction to the yield surface. The principle of maximum plastic dissipation can be satisfied for strain-hardening and strain-softening materials.

5.10.9.2 Drucker's Postulate

A definition of a plastically stable material is given for small strains by Drucker's postulate which can be stated in different ways, one of which is that the second-order plastic work is non-negative, $\dot{\boldsymbol{\sigma}}:\dot{\varepsilon}^p \geq 0$. It can be shown that the maximum plastic dissipation (5.10.72) is a necessary condition for a material to be stable in the Drucker sense (Lubliner, 1990). A discussion of stability for large deformations is given in Chapter 6. For additional discussions of material and structural stability see Bazant and Cedolin (1991).

5.11 Exercises

5.1. Show that if p is the pressure, the relations $3Jp = \boldsymbol{\tau}:\mathbf{g} = \bar{\mathbf{S}}:\bar{\mathbf{C}}^e$ hold.

5.2. Show that $(\text{sym}\ \bar{\mathbf{r}}):(\bar{\mathbf{C}}^e)^{-1} = 0$ and hence that $\bar{\mathbf{S}}^{\text{dev}}:\bar{\mathbf{D}}^p = \bar{\mathbf{S}}:\bar{\mathbf{D}}^p$. See (5.7.39) and (5.7.40).

5.3. Derive expressions for the Lie derivatives $L_v\boldsymbol{\tau}^{\text{dev}}$ and $L_v\boldsymbol{\tau}^{\text{hyd}}$ in terms of the material time derivative of the stress and the spatial velocity gradient $\mathbf{L}$.

6

Solution Methods and Stability

6.1 Introduction

This chapter describes solution procedures for nonlinear finite element discretizations. Explicit and implicit methods for transient problems and methods for the solution of equilibrium problems are described and their implementation and properties are examined. Stability is an important topic in nonlinear finite element methods. In this chapter, the stability of solutions, the stability of numerical procedures and material stability are presented in a unified manner.

We begin with a description of explicit time integration. We focus on the central difference method, but some remarks on other time integrators are given at the end of the section. The implementation of the method is described in detail. Some of the companion techniques, such as mass scaling, subcycling, and dynamic relaxation are also briefly considered.

We next describe implicit methods, with the Newmark β-method as a model. Methods for solution of static problems, that is, the solution of equilibrium problems, are developed simultaneously. Solution of the discrete equations by the Newton method is described, including techniques for convergence checks and line searches. A critical step in the solution of implicit systems and equilibrium problems is the linearization of the governing equations. Linearization procedures for the equations of motion, and as a special case the equilibrium equations, are described. Explicit and implicit methods are compared and their relative advantages are examined. In addition, special techniques for equilibrium solutions, such as continuation methods (parametrization and arc length methods), are described.

Section 6.5 deals with the stability of solutions to the discrete equations. We begin with a general description of stability and then focus on linear stability analysis of the discretized

Nonlinear Finite Elements for Continua and Structures, Second Edition.
Ted Belytschko, Wing Kam Liu, Brian Moran, and Khalil I. Elkhodary.

Companion Website: www.wiley.com/go/belytschko

system. In linear stability analysis, a perturbation is applied to a nonlinear state. Simplified methods for examining stability are described and the conditions under which the stability depends on the positive-definiteness of the Jacobian matrix are given.

Section 6.6 then examines the stability of time integration procedures. The fundamental concepts closely parallel those in the stability analysis of discrete solutions. Emphasis is placed on the stability analysis of linear systems, since nonlinear computations are usually guided by linear stability estimates. The stability of the Euler methods for first-order systems is first examined to provide a simple framework for introducing the ideas. Next the z-transform and Hurwitz matrices are introduced, so that the analysis of the more complex central difference method and the Newmark β-method can be performed. The section concludes with a proof of the stability in energy of the trapezoidal method nonlinear systems.

Section 6.7 then introduces the topic of material stability. Again the methodology is a linear stability analysis. The conditions for linear stability of a material are developed and it is shown that a material with a positive-definite response matrix is linear stable. A brief survey of methods for regularizing unstable material computations is also given.

6.2 Explicit Methods

6.2.1 Central Difference Method

The central difference method is among the most popular of the explicit methods in computational mechanics and physics. It has already been discussed in Chapter 2, where it was chosen to demonstrate some nonlinear solutions in one dimension. The central difference method is developed from central difference formulas for the velocity and acceleration. We consider here its application to Lagrangian meshes. Methods for Eulerian and ALE meshes are discussed in Chapter 7. Geometric and material nonlinearities are included, and in fact have little effect on the time integration algorithm.

For the purpose of developing this and other time integrators we will use the following notation. Let the time of the simulation $0 \leq t \leq t_E$ be subdivided into time steps Δt^n, $n=1$ to n_{TS} where n_{TS} is the number of time steps and t_E is the end of the simulation. The superscript indicates the time step: t^n and $\mathbf{d}^n \equiv \mathbf{d}(t^n)$ are the time and displacement, respectively, at time step n.

We consider here an algorithm with a variable time step. This is necessary in most practical calculations since the stable time step changes as the mesh deforms and the wave speed changes due to the stress. For this purpose, we define the time increments by

$$\Delta t^{n+\frac{1}{2}} = t^{n+1} - t^n, \quad t^{n+\frac{1}{2}} = \frac{1}{2}(t^{n+1} + t^n), \quad \Delta t^n = t^{n+\frac{1}{2}} - t^{n-\frac{1}{2}} \tag{6.2.1}$$

The central difference formula for the velocity is

$$\dot{\mathbf{d}}^{n+\frac{1}{2}} \equiv \mathbf{v}^{n+\frac{1}{2}} = \frac{\mathbf{d}^{n+1} - \mathbf{d}^n}{t^{n+1} - t^n} = \frac{1}{\Delta t^{n+\frac{1}{2}}}(\mathbf{d}^{n+1} - \mathbf{d}^n) \tag{6.2.2}$$

where the definition of $\Delta t^{n+\frac{1}{2}}$ from (6.2.1) has been used in the last step. This difference formula can be converted to an integration formula by rearranging the terms as follows:

$$\mathbf{d}^{n+1} = \mathbf{d}^n + \Delta t^{n+\frac{1}{2}} \mathbf{v}^{n+\frac{1}{2}} \tag{6.2.3}$$

The acceleration and the corresponding integration formula are

$$\ddot{\mathbf{d}}^n \equiv \mathbf{a}^n = \left(\frac{\mathbf{v}^{n+\frac{1}{2}} - \mathbf{v}^{n-\frac{1}{2}}}{t^{n+\frac{1}{2}} - t^{n-\frac{1}{2}}}\right), \quad \mathbf{v}^{n+\frac{1}{2}} = \mathbf{v}^{n-\frac{1}{2}} + \Delta t^n \mathbf{a}^n \tag{6.2.4}$$

As can be seen from the above, the velocities are defined at the midpoints of the time intervals, which are called half-steps or midpoint steps. By substituting (6.2.2) and its counterpart for the previous time step into (6.2.4), the acceleration can be expressed directly in terms of the displacements:

$$\ddot{\mathbf{d}}^n \equiv \mathbf{a}^n = \frac{\Delta t^{n-\frac{1}{2}}(\mathbf{d}^{n+1} - \mathbf{d}^n) - \Delta t^{n+\frac{1}{2}}(\mathbf{d}^n - \mathbf{d}^{n-1})}{\Delta t^{n+\frac{1}{2}} - \Delta t^n \Delta t^{n-\frac{1}{2}}} \tag{6.2.5}$$

For the case of equal time steps the above reduces to

$$\ddot{\mathbf{d}}^n \equiv \mathbf{a}^n = \frac{(\mathbf{d}^{n+1} - 2\mathbf{d}^n + \mathbf{d}^{n-1})}{(\Delta t^n)^2} \tag{6.2.6}$$

This is the well-known central difference formula for the second derivative of a function.

We now consider the time integration of the equations of motion, (4.4.24), which at time step n are given by

$$\mathbf{M}\mathbf{a}^n = \mathbf{f}^n = \mathbf{f}^{\text{ext}}(\mathbf{d}^n, t^n) - \mathbf{f}^{\text{int}}(\mathbf{d}^n, t^n) \tag{6.2.7}$$

subject to

$$g_I(\mathbf{d}^n) = 0, \quad I = 1 \text{ to } n_c \tag{6.2.8}$$

These equations in (6.2.7), are ordinary differential equations of second order in time. They are often called semidiscrete, since they have been discretized in space but not in time. Equation (6.2.8) is a generalized representation of the n_c displacement boundary conditions and other constraints on the model. These constraints are linear or nonlinear algebraic functions of the nodal displacements. If the constraint involves integral or differential relationships, it can be put in the above form by using difference equations or a numerical approximation of the integral. The mass matrix is constant for a Lagrangian mesh as noted in Section 4.4.9.

The internal and external nodal forces are functions of the nodal displacements and the time. The external loads are usually prescribed as functions of time; they may also be functions of the nodal displacements because they may depend on the configuration of the structure, as when pressure forces are applied to surfaces which undergo large deformations. The dependence of the internal nodal forces on displacements is obvious: the nodal internal forces depend on the stresses, which depend on the strain and strain rates by the constitutive equations, which in turn depend on the displacements and their derivatives. The internal nodal forces can also depend directly on time, for example, when the temperature is prescribed as a function of time, then the stresses and hence the internal nodal forces are directly functions of time.

The equations for updating the nodal velocities and displacements are obtained as follows. Substituting (6.2.7) into (6.2.4) gives

$$\mathbf{v}^{n+½} = \mathbf{v}^{n-½} + \Delta t^n \mathbf{M}^{-1}\mathbf{f}^n \tag{6.2.9}$$

At any time step n, the displacements $\mathbf{d}^n$ are known. The nodal forces $\mathbf{f}^n$ can be determined by sequentially evaluating the strain–displacement equations, the constitutive equation expressed in terms of $\mathbf{D}^{n-½}$ or $\mathbf{E}^n$ and the nodal external forces. Thus the entire right-hand side of (6.2.9) can be evaluated, and (6.2.9) can be used to obtain $\mathbf{v}^{n+½}$. The displacements $\mathbf{d}^{n+1}$ can then be determined by (6.2.3).

The update of the nodal velocities and nodal displacements can be accomplished without solving any equations provided that the mass matrix $\mathbf{M}$ is diagonal. This is the salient characteristic of an explicit method: *in an explicit method, the time integration of the discrete momentum equations does not require the solution of any equations*. The avoidance of solution of equations of course hinges critically on the use of a diagonal mass matrix.

In numerical analysis, integration methods are classified according to the structure of the time difference equation. The difference equations for first and second derivatives, respectively, can be written in the general expressions

$$\sum_{k=0}^{n_S}\left(\alpha_k \mathbf{d}^{n_S-k} - \Delta t \beta_k \dot{\mathbf{d}}^{n_S-k}\right) = \mathbf{0}, \quad \sum_{k=0}^{n_S}\left(\bar{\alpha}_k \mathbf{d}^{n_S-k} - \Delta t^2 \bar{\beta}_k \ddot{\mathbf{d}}^{n_S-k}\right) = \mathbf{0} \tag{6.2.10}$$

where n_S is the number of steps in the difference equation. The difference formulas for the first or second derivatives are called explicit if $\beta_0=0$ or $\bar{\beta}_0 = 0$, respectively. Thus a difference formula is called explicit if the equation for the function at time step n_S involves only the derivatives at previous time steps. In the central difference formula for the second derivative, (6.2.6), $\bar{\beta}_0=0$, $\bar{\beta}_1=1$, $\bar{\beta}_2=0$, so it is explicit. Difference equations which are explicit according to this classification generally lead to solution schemes which require no solution of equations. In most cases there is no benefit in using explicit schemes which involve the solution of equations, so such explicit schemes are rare. There are exceptions. For example, the consistent mass is sometimes used with the central difference method in wave propagation problems, and the update then involves the solution of equations.

6.2.2 *Implementation*

Box 6.1 gives a flowchart for explicit time integration. This flowchart generalizes the one given in Chapter 2 by including nonzero initial conditions, a variable time step, elements with more than one quadrature point, and damping. The damping is modeled by a linear viscous force $\mathbf{f}^{\text{damp}}=\mathbf{C}^{\text{damp}}\mathbf{v}$, so that the total force in (6.2.9) is $\mathbf{f} - \mathbf{C}^{\text{damp}}\mathbf{v}$. The implementation of the velocity update is broken into two substeps by

$$\mathbf{v}^n = \mathbf{v}^{n-½} + (t^n - t^{n-½})\mathbf{a}^n, \quad \mathbf{v}^{n+½} = \mathbf{v}^n + (t^{n+½} - t^n)\mathbf{a}^n \tag{6.2.11}$$

This enables energy balance to be checked at integer time steps.

Box 6.1 Flowchart for explicit time integration

1. Initial conditions and initialization:
   set $\mathbf{v}^0$, σ^0, and initial values of other material state variables;
   $\mathbf{d}^0=\mathbf{0}$, $n=0$, $t=0$; compute $\mathbf{M}$
2. *getforce*
3. Compute accelerations $\mathbf{a}^n=\mathbf{M}^{-1}(\mathbf{f}^n-\mathbf{C}^{\text{damp}}\mathbf{v}^{n-1/2})$
4. Time update: $t^{n+1}=t^n+\Delta t^{n+1/2}$, $t^{n+1/2}=\frac{1}{2}(t^n+t^{n+1})$
5. First partial update nodal velocities: $\mathbf{v}^{n+1/2}=\mathbf{v}^n+(t^{n+1/2}-t^n)\mathbf{a}^n$
6. Enforce velocity boundary conditions:
   if node I on Γ_{v_i} : $v_{iI}^{n+1/2}=\bar{v}_i\left(\mathbf{x}_I,t^{n+1/2}\right)$
7. Update nodal displacements: $\mathbf{d}^{n+1}=\mathbf{d}^n+\Delta t^{n+1/2}\mathbf{v}^{n+1/2}$
8. *getforce*
9. compute $\mathbf{a}^{n+1}$
10. Second partial update nodal velocities: $\mathbf{v}^{n+1}=\mathbf{v}^{n+1/2}+(t^{n+1}-t^{n+1/2})\mathbf{a}^{n+1}$
11. Check energy balance at time step $n+1$: see (6.2.14–18)
12. Update counter: $n\leftarrow n+1$
13. Output; if simulation not complete, go to 4.

Subroutine *getforce*

0. Initialization: $\mathbf{f}^n=0$, $\Delta t_{\text{crit}}=\infty$
1. Compute global external nodal forces $\mathbf{f}_{\text{ext}}^n$
2. Loop over elements e
   i. GATHER element nodal displacements and velocities
   ii. $\mathbf{f}_e^{\text{int},n}=\mathbf{0}$
   iii. Loop over quadrature points $\boldsymbol{\xi}_Q$
      1. if $n=0$, go to 4
      2. compute measures of deformation: $\mathbf{D}^{n-1/2}(\boldsymbol{\xi}_Q)$, $\mathbf{F}^n(\boldsymbol{\xi}_Q)$, $\mathbf{E}^n(\boldsymbol{\xi}_Q)$
      3. compute stress $\sigma^n(\boldsymbol{\xi}_Q)$ by constitutive equation
      4. $\mathbf{f}_e^{\text{int},n}\leftarrow\mathbf{f}_e^{\text{int},n}+\boldsymbol{B}^T\sigma^n\bar{w}_Q J\,|_{\xi_Q}$
      END quadrature point loop
   iv. Compute external nodal forces on element, $\mathbf{f}_e^{\text{ext},n}$
   v. $\mathbf{f}_e^n=\mathbf{f}_e^{\text{ext},n}-\mathbf{f}_e^{\text{int},n}$
   vi. Compute Δt_{crit}^e, if $\Delta t_{\text{crit}}^e<\Delta t_{\text{crit}}$ then $\Delta t_{\text{crit}}=\Delta t_{\text{crit}}^e$
   vii. SCATTER $\mathbf{f}_e^n$ to global $\mathbf{f}^n$
5. END loop over elements
6. $\Delta t=\alpha\Delta t_{\text{crit}}$

The cardinal dependent variables in this flowchart are the velocities and the Cauchy stress. Initial conditions must be given for the velocities, the Cauchy stress, and all material state variables. The initial displacements are assumed to vanish (initial displacements are meaningless in nonlinear analysis except for a hyperelastic material since the stress generally depends on the history of deformation).

The main part of the procedure is the calculation of the nodal forces, which is done in *getforce*. The major steps in this subroutine are:

1. Extract the nodal displacements and velocities of the element from the global arrays by the GATHER operation.
2. The strain measures are computed at each quadrature point of the element.
3. The stresses are computed by the constitutive equation at each quadrature point.
4. Evaluate the internal nodal forces by integrating the product of the $\boldsymbol{B}$ matrix and the Cauchy stress over the domain of the element.
5. The nodal forces of the element are SCATTERED into the global array.

In the first time step, the strain measures and the stresses are not computed. Instead, as shown in the flowchart, the internal nodal forces are computed directly from the initial stresses.

In the flowchart, the matrix form of the internal force computation, in which the stress tensor is stored as a square matrix, is shown. To change to Voigt form, replace $\boldsymbol{B}$ by $\mathbf{B}$ and the square matrix of stresses by the column matrix $\{\boldsymbol{\sigma}\}$. Similarly, the internal force computation can be changed to the total Lagrangian format by replacing step iii.4 by (B4.6.2).

Most essential boundary conditions are easily handled in explicit methods. For example, if the velocities or displacements are prescribed as functions of time along any boundary, then the velocity/displacement boundary conditions can be enforced by setting the nodal velocities according to the data:

$$v_{iI}^{n+1/2} = \bar{v}_i\left(\mathbf{x}_I, t^{n+1/2}\right) \tag{6.2.12}$$

If the data are not available on the nodes, they can be obtained by the least-square procedure given in Section 2.4.5. When the boundary conditions are posed in terms of displacements, imposition of (6.2.12) involves a numerical differentiation of the prescribed displacements to obtain the prescribed velocities; these velocities are then integrated in step 7. This roundabout procedure can be avoided by setting the prescribed boundary displacements after step 7.

The velocity boundary conditions can also be enforced in local coordinate systems. In that case, the equations of motion at these nodes must be expressed in the local coordinate components; the nodal forces must be transformed into global components before assembly. The orientation of the local coordinate system may vary with time but the time integration formulas must then be modified to account for the rotation of the coordinate system.

When essential boundary conditions are given as linear or nonlinear algebraic equations relating the displacements, the implementation is more complicated. The penalty or Lagrange multiplier methods described below in Section 6.3.8 are commonly utilized.

Any damping in the system lags by a half time step: see step 3 in Box 6.1. This also holds for any rate-dependent terms in the constitutive equation evaluation in step 2.iii. of *getforce*. The time lag is unavoidable if the implementation is to be fully explicit, that is, not require the solution of any equations. As we shall see in Section 6.6.6, this decreases the stable time step for the method.

As can be seen from the flowchart, an explicit method is easily implemented. Furthermore, explicit time integration is very robust, by which we mean that the explicit procedure seldom aborts due to failure of the numerical algorithm. The salient disadvantage of explicit integration, the price you pay for the simplicity of the method and its avoidance of the solution of equations, is *the conditional stability of explicit methods*. If the time step exceeds a critical value Δt_{crit}, the solution will grow unboundedly.

The *critical time step* is also called the *stable time step*. A stable time step for a mesh of constant strain elements with rate-independent materials is given by

$$\Delta t = \alpha \Delta t_{crit}, \quad \Delta t_{crit} = \frac{2}{\omega_{max}} \leq \min_{e,I} \frac{2}{\omega_I^e} = \min_e \frac{\ell_e}{c_e} \qquad (6.2.13)$$

where ω_{max} is the maximum frequency of the linearized system, ℓ_e is a characteristic length of element e, c_e the current wavespeed in element e, and α is a reduction factor that accounts for the destabilizing effects of nonlinearities; a good choice for a is $0.8 \leq \alpha \leq 0.98$. The development of this and additional discussion of time steps for explicit methods are given in Section 6.6.6.

The above is called the Courant condition in finite difference methods after one of its discoverers; the result was first published by Courant, Friedrichs and Lewy (1928). The ratio of the time step to the critical time step, α, is called the Courant number. From (6.2.13) it can be seen that the critical time step decreases with mesh refinement and increasing stiffness of the material. The cost of an explicit simulation is independent of the frequency range that is of interest and depends only on the size of the model and the number of time steps.

An interesting question for elastic–plastic materials is whether the slower wavespeed in the plastic response enables one to increase the time step. Based on our experience, the answer appears to be negative. An elastic–plastic material can unload at any moment, and in numerical solutions unloading often occurs due to numerical noise. During elastic unloading, the critical time step depends on the elastic wavespeed, and a time step which exceeds the critical time step results in instability.

The mesh time step is obtained from element time steps. For each element, an element time step is calculated and the minimum element time step is chosen as the mesh time step. The theoretical justification for setting the critical time step on an element basis is given in Section 6.6.8–9.

6.2.3 Energy Balance

These stability conditions emanate from a stability analysis of the integrator for the linear equations of motion. At this time, there are no stability theorems that cover the range of nonlinear phenomena encountered in engineering problems, such as contact-impact, tearing, etc.

It is possible for instabilities to develop even when (6.2.13) is observed. In contrast to linear problems, where an instability leads to an exponential growth of the solution and cannot be overlooked, unstable solutions of nonlinear problems are sometimes not readily discernible. For example, Belytschko (1983) describes a numerical phenomenon called an *arrested instability*. The scenario is as follows. An instability is triggered by nonlinearities, such as geometric stiffening, while the material is elastic. This instability causes local exponential growth of the solution, which in turn leads to plastic behavior. The plastic response softens the structure and decreases the wavespeed, so the integrator regains stability.

Such arrested instabilities can lead to a large overprediction of displacements, but they are not detectable by perusing the results. However, they can easily be detected by an energy balance check. Any instability results in the spurious generation of energy which leads to a violation of the conservation of energy. Therefore, whether stability was maintained during a nonlinear computation can be established by checking energy balance.

In low-order methods like the central difference method, the energy is usually integrated in time by a method of similar order, such as the trapezoidal rule. The internal and external energies are integrated as follows:

$$W_{\text{int}}^{n+1} = W_{\text{int}}^{n} + \frac{\Delta t^{n+1/2}}{2}(\mathbf{v}^{n+1/2})^T\left(\mathbf{f}_{\text{int}}^{n} + \mathbf{f}_{\text{int}}^{n+1}\right) = W_{\text{int}}^{n} + \frac{1}{2}\Delta\mathbf{d}^T\left(\mathbf{f}_{\text{int}}^{n} + \mathbf{f}_{\text{int}}^{n+1}\right) \tag{6.2.14}$$

$$W_{\text{ext}}^{n+1} = W_{\text{ext}}^{n} + \frac{\Delta t^{n+1/2}}{2}(\mathbf{v}^{n+1/2})^T\left(\mathbf{f}_{\text{ext}}^{n} + \mathbf{f}_{\text{ext}}^{n+1}\right) = W_{\text{ext}}^{n} + \frac{1}{2}\Delta\mathbf{d}^T\left(\mathbf{f}_{\text{ext}}^{n} + \mathbf{f}_{\text{ext}}^{n+1}\right) \tag{6.2.15}$$

where $\Delta\mathbf{d} = \mathbf{d}^{n+1} - \mathbf{d}^{n}$. The kinetic energy is given by

$$W_{\text{kin}}^{n} = \frac{1}{2}(\mathbf{v}^{n})^T\mathbf{M}\mathbf{v}^{n} \tag{6.2.16}$$

Note that integer time steps are used for the velocities, which is why the velocities at integer time steps are computed in Box 6.1.

The internal energies can also be computed on the element or quadrature point level by

$$\begin{aligned} W_{\text{int}}^{n+1} &= W_{\text{int}}^{n} + \frac{1}{2}\sum_{e}\Delta\mathbf{d}_e^T\left(\mathbf{f}_{e,\,\text{int}}^{n} + \mathbf{f}_{e,\,\text{int}}^{n+1}\right) \\ &= W_{\text{int}}^{n} + \frac{\Delta t^{n+1/2}}{2}\sum_{e}\sum_{n_Q}\bar{w}_Q\mathbf{D}_Q^{n+1/2} : \left(\boldsymbol{\sigma}_Q^{n} + \boldsymbol{\sigma}_Q^{n+1}\right)J_{\xi Q} \end{aligned} \tag{6.2.17}$$

where $\boldsymbol{\sigma}_Q^{n} = \boldsymbol{\sigma}^{n}(\xi_Q)$, etc. Energy conservation requires that

$$|\, W_{\text{kin}} + W_{\text{int}} - W_{\text{ext}} \,| \leq \varepsilon \max\left(W_{\text{ext}},\, W_{\text{int}},\, W_{\text{kin}}\right) \tag{6.2.18}$$

where ε is a small tolerance, generally on the order of 10^{-2}.

If the system is very large, on the order of 10^5 nodes or larger, the energy balance should be performed on subdomains of the model. The internal forces from adjacent subdomains are then treated as external forces for each subdomain.

6.2.4 Accuracy

The central difference method is second order in time, that is, the truncation error is of order Δt^2 in the displacements. We will see that the spatial error in the displacements in the L_2 norm for linear complete elements is of order h^2, where h is the element size. Although there are some technical differences between these two measures of error, the outcome is similar. Since the time step and the element size must be of the same order to meet the stability condition, (6.2.13), the time integration error and the spatial error are of the same order for central difference time integration. However, for materials with rapidly varying stiffness, such as viscoplastic materials, the accuracy of the central difference method is sometimes inadequate. In those cases, we suggest Runge–Kutta methods. The Runge–Kutta method need not be used on all equations: it can be applied to the constitutive equation while integrating the equations of motion by the central difference method.

6.2.5 *Mass Scaling, Subcycling and Dynamic Relaxation*

When a model contains a few very small or stiff elements, the efficiency of explicit integration is compromised severely, since the time step of the entire mesh is set by these very stiff elements. Several techniques are available for circumventing this difficulty:

1. *mass scaling*: the masses of stiffer elements is increased so that the time step is not decreased by these elements;
2. *subcycling*: a smaller time step is used for the stiffer elements,

Mass scaling should be used for problems where high frequency effects are not important. For example, in sheet-metal forming, which is essentially a static process, it causes no difficulties. On the other hand, if high frequency response is important, mass scaling is not recommended.

Subcycling was introduced by Belytschko, Yen and Mullen (1979). In this technique the model is split into subdomains and each is integrated with its own stable time step. The crucial issue in subcycling is the treatment of the interface between subdomains. The early methods used linear interpolation. These can be shown to be stable for first-order systems (Belytschko, Smolinski and Liu, 1985). But as shown by Daniel (1998), in second-order systems, linear interpolation leads to narrow bands of instability. These can be eliminated either by the addition of artificial viscosity, or by alternative subcycling methods, which are more complex. Stable subcycling methods for second-order systems are given by Smolinski, Sleith and Belytschko (1996) and Daniel (1997).

Dynamic relaxation is often used in explicit codes to obtain static solutions. The basic idea is to apply the load very slowly and solve the dynamic system equations with enough damping so that oscillations are minimized. In path-dependent materials, dynamic relaxation often yields poor solutions. Furthermore, it is very slow. Newton methods combined with effective iterative solvers, such as preconditioned conjugate gradient or multigrid methods, are much faster and more accurate.

6.3 Equilibrium Solutions and Implicit Time Integration

6.3.1 *Equilibrium and Transient Problems*

We will combine the description of the solution of the equilibrium equations with time integration by implicit methods since they share many common features. To begin, we write the discrete momentum equation at time step $n+1$ in a form applicable to both equilibrium and dynamic problems:

$$0 = \mathbf{r}(\mathbf{d}^{n+1}, t^{n+1}) = s_D\mathbf{M}\mathbf{a}^{n+1} + \mathbf{f}^{\text{int}}(\mathbf{d}^{n+1}, t^{n+1}) - \mathbf{f}^{\text{ext}}(\mathbf{d}^{n+1}, t^{n+1}) \tag{6.3.1}$$

where s_D is a switch which is set by:

$$s_D = \begin{cases} 0 & \text{for static (equilibrium) problem} \\ 1 & \text{for a dynamic (transient) problem} \end{cases} \tag{6.3.2}$$

The column matrix $\mathbf{r}(\mathbf{d}^{n+1}, t^{n+1})$ is called a residual. The discrete equations for both the implicit update of the equations of motion and the equilibrium equations are nonlinear algebraic equations in the nodal displacements, $\mathbf{d}^{n+1}$.

6.3.2 Equilibrium Solutions and Equilibrium Points

When the accelerations vanish or are negligible, a system is in equilibrium and the solution of (6.3.1) is called an *equilibrium solution.* The equilibrium equations are given by (6.3.1) with s_D=0:

$$0 = \mathbf{r}(\mathbf{d}^{n+1}, t^{n+1}) = \mathbf{f}^{\text{int}}(\mathbf{d}^{n+1}, t^{n+1}) - \mathbf{f}^{\text{ext}}(\mathbf{d}^{n+1}, t^{n+1}) \tag{6.3.3}$$

In equilibrium problems, the residuals correspond to the out-of-balance forces; problems in which the accelerations can be neglected are called *static problems.* A solution to the above is called an *equilibrium point* and a continuous locus of solutions is called an *equilibrium branch* or *equilibrium path.*

In equilibrium problems with rate-independent materials, t need not be the real time. Instead it can be any monotonically increasing parameter. If the constitutive equation is a differential or integral equation, it must also be discretized in time to obtain a set of algebraic equations for the system.

6.3.3 Newmark β-Equations

To illustrate the formulation of the discrete equations, we consider a popular class of time integrators called the Newmark β-method. For this time integrator, the updated displacements and velocities are

$$\mathbf{d}^{n+1} = \tilde{\mathbf{d}}^{n+1} + \beta\Delta t^2\mathbf{a}^{n+1} \quad \text{where} \quad \tilde{\mathbf{d}}^{n+1} = \mathbf{d}^n + \Delta t\mathbf{v}^n + \frac{\Delta t^2}{2}(1-2\beta)\mathbf{a}^n \tag{6.3.4}$$

$$\mathbf{v}^{n+1} = \tilde{\mathbf{v}}^{n+1} + \gamma\Delta t\mathbf{a}^{n+1} \quad \text{where} \quad \tilde{\mathbf{v}}^{n+1} = \mathbf{v}^n + (1-\gamma)\Delta t\mathbf{a}^n \tag{6.3.5}$$

where $\Delta t = t^{n+1} - t^n$ and β and γ are parameters whose useful values and stability properties are summarized in Box 6.2. The parameter γ controls artificial viscosity, a damping introduced by the numerical method. It is used to suppress noise in the solution. When $\gamma = \frac{1}{2}$, the Newmark integrator adds no damping; for $\gamma > \frac{1}{2}$ artificial damping proportional to $\gamma - \frac{1}{2}$ is added by the integrator. In the previous, we have segregated the historical values of the nodal variables, i.e. those pertaining to time step n, in $\tilde{\mathbf{v}}^{n+1}$ and $\tilde{\mathbf{d}}^{n+1}$; it would be more logical to use a superscript n on the preceding variables, but we bow to convention. The above correspond to the predictor-corrector form given by Hughes and Liu (1978).

Box 6.2 Newmark β-method

$\beta=0,\ \gamma=\frac{1}{2}$ explicit central difference method

$\beta=\frac{1}{4},\ \gamma=\frac{1}{2}$ undamped trapezoidal rule

$\gamma>\frac{1}{2}$ numerically damped integrator with damping proportional to $\gamma-\frac{1}{2}$

Stability

Unconditionally stable for $\beta\geq\frac{\gamma}{2}\geq\frac{1}{4}$

Conditional stability:

$$\omega_{\max}\Delta t=\frac{\xi\bar{\gamma}+\left[\bar{\gamma}+\frac{1}{4}-\beta+\xi^2\bar{\gamma}^2\right]^{\frac{1}{2}}}{\left(\frac{\gamma}{2}-\beta\right)},\quad \bar{\gamma}\equiv\gamma-\frac{1}{2}\geq 0$$

ξ=fraction of critical damping in frequency $\omega_{\max}$, see (6.6.34).

Equation (6.3.4) can be solved for the updated accelerations, giving

$$\mathbf{a}^{n+1}=\frac{1}{\beta\Delta t^2}(\mathbf{d}^{n+1}-\tilde{\mathbf{d}}^{n+1})\quad\text{when}\quad\beta>0 \tag{6.3.6}$$

Substituting (6.3.6) into (6.3.1) gives

$$\mathbf{0}=\mathbf{r}=\frac{s_D}{\beta\Delta t^2}\mathbf{M}(\mathbf{d}^{n+1}-\tilde{\mathbf{d}}^{n+1})-\mathbf{f}^{\text{ext}}(\mathbf{d}^{n+1},t^{n+1})+\mathbf{f}^{\text{int}}(\mathbf{d}^{n+1}) \tag{6.3.7}$$

which is a set of nonlinear algebraic equations in the nodal displacements $\mathbf{d}^{n+1}$. Equation (6.3.7) applies to both the static and dynamic problems. In both cases the discrete problem is

$$\text{find }\mathbf{d}^{n+1}\text{ so that }\mathbf{r}(\mathbf{d}^{n+1},t^{n+1})=\mathbf{0}\text{ subject to }\mathbf{g}(\mathbf{d}^{n+1},t^{n+1})=\mathbf{0} \tag{6.3.8}$$

where $\mathbf{r}(\mathbf{d}^{n+1})$ is given by (6.3.7).

6.3.4 Newton's Method

The most widely used and most robust method for the solution of the nonlinear algebraic equations (6.3.7) is Newton's method. The method is often called the Newton–Raphson method in computational mechanics. It is identical to the Newton method taught in introductory calculus courses.

We first illustrate the Newton method for one equation in one unknown d without a displacement boundary condition. It is then generalized to an arbitrary number of unknowns. For the case of one unknown, (6.3.7) for $\beta>0$ reduces to a single nonlinear algebraic equation

$$r(d^{n+1}, t^{n+1}) = \frac{s_D}{\beta \Delta t^2} M\left(d^{n+1} - \tilde{d}^{n+1}\right) - f(d^{n+1}, t^{n+1}) = 0 \tag{6.3.9}$$

The solution of (6.3.9) by Newton's method is an iterative procedure. The iteration number is indicated by Greek subscript: $d_\upsilon^{n+1} \equiv d_\upsilon$ is the displacement in iteration υ at time step $n+1$; the time step number $n+1$ will be omitted in the following.

To begin the iterative procedure, a starting value for the unknown must be chosen; usually the solution d^n from the last time step is selected, so $d_0 \equiv d^n$. In dynamic solutions by the Newmark β-method, a better starting value is $\tilde{\mathbf{d}}^{n+1}$. A Taylor expansion of the residual about the current value of the nodal displacement d_υ and setting the resulting residual equal to zero gives

$$0 = r(d_{\upsilon+1}, t^{n+1}) = r(d_\upsilon, t^{n+1}) + \frac{\partial r(d_\upsilon, t^{n+1})}{\partial d} \Delta d + O\left(\Delta d^2\right) \tag{6.3.10}$$

where

$$\Delta d = d_{\upsilon+1} - d_\upsilon \tag{6.3.11}$$

If the terms which are higher order than linear in Δd are dropped, then (6.3.10) gives a linear equation for Δd:

$$0 = r(d_\upsilon, t^{n+1}) + \frac{\partial r(d_\upsilon, t^{n+1})}{\partial d} \Delta d \tag{6.3.12}$$

The above is called a *linear model* or *linearized model of the nonlinear equations* (Dennis and Schnabel, 1983). The linear model is the tangent to the nonlinear residual function; the process of obtaining the linear model is called *linearization*.

Note that in the Taylor expansion, the residual is written in terms of the time t^{n+1}. The time-dependence of the residual is usually explicitly given. For example, the tractions and body forces are usually given as functions of time, and any change in the external nodal forces is due to changes in the nodal displacements. Therefore the residual is ordinarily computed using the load at time t^{n+1} and the latest value of the nodal displacements.

Solving this linear model for the incremental displacements gives

$$\Delta d = -\left(\frac{\partial r(d_\upsilon)}{\partial d}\right)^{-1} r(d_\upsilon) \tag{6.3.13}$$

In the Newton procedure, the solution to the nonlinear equation is obtained by iteratively solving a sequence of linear models (6.3.13). The new value for the unknown in each step of the iteration is obtained by rewriting (6.3.11) as

$$d_{\upsilon+1} = d_\upsilon + \Delta d \tag{6.3.14}$$

The procedure is illustrated in Figure 6.1. The process is continued until the solution is obtained with the desired level of accuracy.

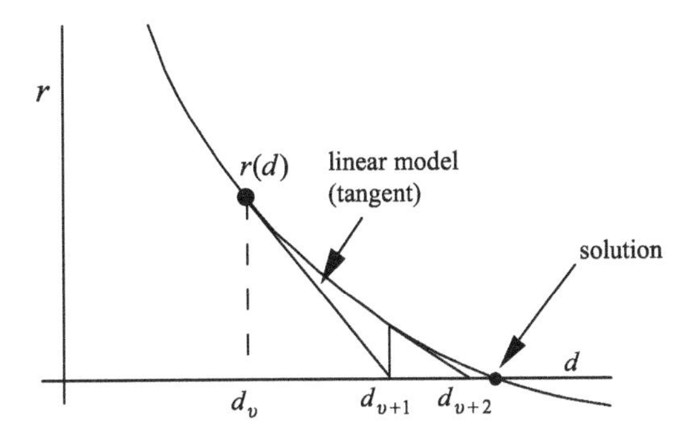

Figure 6.1 Linear models for a nonlinear equation $r(d)=0$

6.3.5 *Newton's Method for* n *Unknowns*

Newton's method is generalized to n_{dof} unknowns by replacing the above scalar equations by matrix equations. The counterpart of (6.3.10) is

$$\mathbf{r}(\mathbf{d}_\upsilon)+\frac{\partial\mathbf{r}(\mathbf{d}_\upsilon)}{\partial\mathbf{d}}\Delta\mathbf{d}+O(\Delta\mathbf{d}^2)=\mathbf{0} \quad \text{or} \quad r_a(\mathbf{d}_\upsilon)+\frac{\partial r_a(\mathbf{d}_\upsilon)}{\partial d_b}\Delta d_b+O(\Delta\mathbf{d})^2=\mathbf{0} \tag{6.3.15}$$

where the summation convention is retained and the range for indices *a, b* is the number of degrees of freedom, n_{dof}. The matrix $\partial\mathbf{r}/\partial\mathbf{d}$ is called the *system Jacobian matrix* and will be denoted by $\mathbf{A}$:

$$\mathbf{A}=\frac{\partial\mathbf{r}}{\partial\mathbf{d}}, \quad \text{or} \quad A_{ab}=\frac{\partial r_a}{\partial d_b} \tag{6.3.16}$$

Substituting (6.3.16) into (6.3.15) and dropping terms of higher order than linear in $\Delta\mathbf{d}$ gives

$$\mathbf{r}+\mathbf{A}\Delta\mathbf{d}=\mathbf{0} \tag{6.3.17}$$

which is the *linear model* of the nonlinear equations. The linear model is difficult to picture for problems with more than one unknown, since $\mathbf{r}(\mathbf{d})$ maps R^n to R^n. Figure 6.2 shows an example of the first component of the residual for a function of two unknowns. The linear model is a plane tangent to the nonlinear function $r_1(d_1, d_2)$. The other residual component is another nonlinear function $r_2(d_1, d_2)$, which is not drawn.

The increment in the nodal displacements in the Newton iterative procedure is obtained by solving (6.3.17), which gives a system of linear algebraic equations

$$\mathbf{A}\Delta\mathbf{d}=-\mathbf{r}\left(\mathbf{d}_\upsilon,t^{n+1}\right) \tag{6.3.18}$$

Once the increments in nodal displacements have been obtained, they are added to the previous iterate

$$\mathbf{d}_{\upsilon+1}=\mathbf{d}_\upsilon+\Delta\mathbf{d} \tag{6.3.19}$$

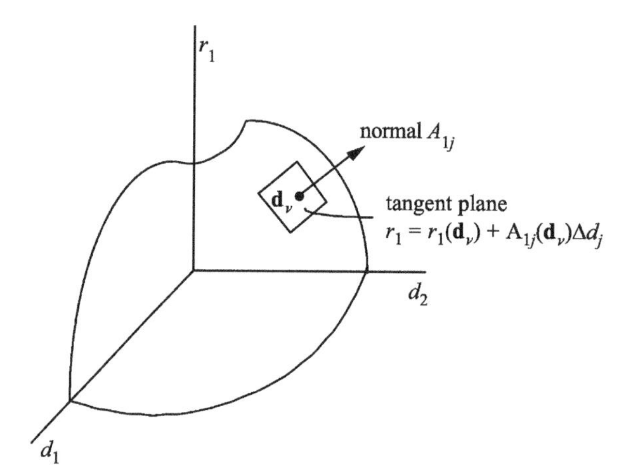

Figure 6.2 Depiction of a residual component r_1 as a function of d_1 and d_2 and the tangent plane at $\mathbf{d}_\nu$

The new displacement is checked for convergence: see Section 6.3.9. If the convergence criterion is not met, a new linear model is constructed and the process is repeated. The iterative procedure continues until the convergence criterion is met.

In computational mechanics, the Jacobian is called the *effective tangent stiffness matrix* and the contributions of the inertial, internal and external nodal forces are linearized separately. From (6.3.7) we can write the Jacobian for the Newmark integrator as

$$\mathbf{A} = \frac{\partial \mathbf{r}}{\partial \mathbf{d}} = \frac{s_D}{\beta \Delta t^2}\mathbf{M} + \frac{\partial \mathbf{f}^{\text{int}}}{\partial \mathbf{d}} - \frac{\partial \mathbf{f}^{\text{ext}}}{\partial \mathbf{d}} \quad \text{for} \quad \beta > 0 \tag{6.3.20}$$

where we have used the fact that the mass matrix in a Lagrangian mesh is constant in time. The Jacobian for other integrators is identical except for the coefficient of the mass matrix. For equilibrium problems, the coefficient of the mass matrix vanishes, as indicated by the switch s_D. The Jacobian of the internal nodal forces is called the *tangent stiffness matrix* and will be denoted by $\mathbf{K}^{\text{int}}$:

$$K_{iIjJ}^{\text{int}} = \frac{\partial f_{iI}^{\text{int}}}{\partial u_{jJ}}, \quad \mathbf{K}_{IJ}^{\text{int}} = \frac{\partial \mathbf{f}_I^{\text{int}}}{\partial \mathbf{u}_J}, \quad K_{ab}^{\text{int}} = \frac{\partial f_a^{\text{int}}}{\partial d_b}, \quad \mathbf{K}^{\text{int}} = \frac{\partial \mathbf{f}^{\text{int}}}{\partial \mathbf{d}} \tag{6.3.21}$$

We have shown this in four notations which are used in this book.

The Jacobian matrix of the external nodal forces is called the *load stiffness matrix* and given by

$$\mathbf{K}^{\text{ext}} = \frac{\partial \mathbf{f}^{\text{ext}}}{\partial \mathbf{d}} \tag{6.3.22}$$

Other notations as in (6.3.21) are also used. The name load stiffness is rather odd, since a load cannot have a stiffness, but the name has persisted.

The development of these matrices is called linearization and is treated in Section 6.4. From (6.3.21–22), the Jacobian matrix (6.3.20) can be written as

$$\mathbf{A} = \frac{s_D}{\beta \Delta t^2}\mathbf{M} + \mathbf{K}^{\text{int}} - \mathbf{K}^{\text{ext}} \tag{6.3.23}$$

This Jacobian matrix applies to both dynamic and equilibrium problems with the switch s_D, (6.3.2).

The Jacobians in (6.3.21–22) can be used to relate differentials of the nodal forces to differentials of the nodal displacements by

$$d\mathbf{f}^{\text{int}} = \mathbf{K}^{\text{int}} d\mathbf{d}, \quad d\mathbf{f}^{\text{ext}} = \mathbf{K}^{\text{ext}} d\mathbf{d}, \quad d\mathbf{r} = \mathbf{A} d\mathbf{d} \tag{6.3.24}$$

6.3.6 Conservative Problems

We next develop the discrete problem for the stationary principle described in Section 4.9.3. This stationary principle only applies to conservative equilibrium problems, but it provides insight into the character of nonlinear solutions, so it is important. Equilibrium solutions are found by setting the derivative of the potential to zero. From (4.9.41–42) and the definition of the residual (6.3.3) we then obtain

$$0 = \mathbf{r} = \frac{\partial W}{\partial \mathbf{d}} = \frac{\partial W^{\text{int}}}{\partial \mathbf{d}} - \frac{\partial W^{\text{ext}}}{\partial \mathbf{d}} = \mathbf{f}^{\text{int}} - \mathbf{f}^{\text{ext}} \tag{6.3.25}$$

An equilibrium point is stable if the potential energy is a local minimum, as shown in Section 6.5 (see also Figure 6.3). Thus stable equilibrium solutions can be found by minimizing the potential W.

The discrete problem for a stable equilibrium solution is:

$$\min W(\mathbf{d}) \text{ subject to } g_I(\mathbf{d}) = 0, \quad I = 1 \text{ to } n_C \tag{6.3.26}$$

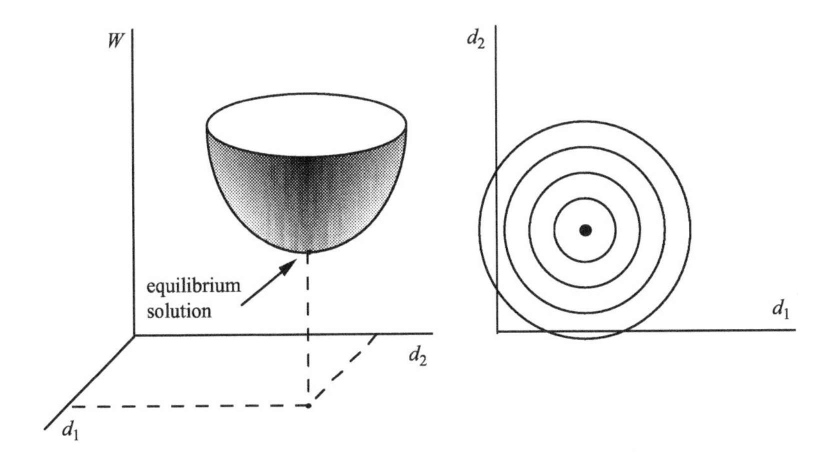

Figure 6.3 Schematic of potential energy, a stable equilibrium solution, and the contours for the potentia

where $g_i(\mathbf{d})=0$ are constraints on the system. In this form, the equilibrium solution can be obtained by mathematical programming and optimization techniques, such as the method of steepest descent.

If both stable and unstable solutions are desired, the stationary points of $W(\mathbf{d})$ must be found. The discrete problem is then

$$\text{find } \mathbf{d} \text{ so that } \frac{\partial W(\mathbf{d})}{\partial \mathbf{d}} = -\mathbf{f} = \mathbf{r} = \mathbf{0} \text{ subject to } g_I(\mathbf{d}) = 0 \tag{6.3.27}$$

To obtain the linear model for a conservative system, we write a first-order Taylor expansion of the residual (6.3.27) about the last iteration point, which gives

$$0 = \mathbf{r}(\mathbf{d}_\upsilon) + \frac{\partial \mathbf{r}(\mathbf{d}_\upsilon)}{\partial \mathbf{d}} \Delta\mathbf{d} = \mathbf{r}(\mathbf{d}_\upsilon) + \frac{\partial^2 W(\mathbf{d}_\upsilon)}{\partial \mathbf{d} \partial \mathbf{d}} \Delta\mathbf{d} \equiv \mathbf{r}(\mathbf{d}_\upsilon) + \mathbf{A}\Delta\mathbf{d} \tag{6.3.28}$$

(see 6.3.15–16) where

$$A_{ab} = \frac{\partial^2 W}{\partial d_a \partial d_b} \quad \text{or} \quad \mathbf{A} = \frac{\partial^2 W}{\partial \mathbf{d} \partial \mathbf{d}} \tag{6.3.29}$$

The matrix $\mathbf{A}$, when defined by the second derivatives of the potential, is called the *Hessian matrix*.

The Hessian matrix is identical to the Jacobian (the two different names stem from their different origins), and the Hessian matrix for a discrete continuum finite element model is

$$\mathbf{A} = \mathbf{K}^{\text{int}} - \mathbf{K}^{\text{ext}}, \quad K_{ab}^{\text{int}} = \frac{\partial^2 W^{\text{int}}}{\partial d_a \partial d_b}, \quad K_{ab}^{\text{ext}} = \frac{\partial^2 W^{\text{ext}}}{\partial d_a \partial d_b} \tag{6.3.30}$$

The linearized equations for a conservative system then follow from (6.3.29–30):

$$(\mathbf{K}^{\text{int}} - \mathbf{K}^{\text{ext}})\,\Delta\mathbf{d} = -\mathbf{r}_\upsilon \tag{6.3.31}$$

These are identical to (6.3.17) except that the mass matrix is omitted, since dynamic effects cannot be included in a conservative problem.

6.3.7 Implementation of Newton's Method

Flowcharts for implicit integration and equilibrium solutions are given in Boxes 6.3 and 6.4, respectively. Both the dynamic problem and the equilibrium problem are solved by time-stepping: the external loads and other conditions are described as functions of time, which is incremented over the range of interest. In equilibrium problems, the time is often replaced by a monotonically increasing parameter. Solutions of equilibrium processes obtained in this manner are called *incremental solutions*.

Box 6.3 Flowchart for implicit time integration

1. Initial conditions and initialization of parameters:

   set $\mathbf{v}^0$, $\boldsymbol{\sigma}^0$; $\mathbf{d}^0=\mathbf{0}$, $n=0$, $t=0$; compute $\mathbf{M}$

2. Get $\mathbf{f}^0=\mathbf{f}(\mathbf{d}^0, 0)$
3. Compute initial accelerations $\mathbf{a}^n=\mathbf{M}^{-1}\mathbf{f}^n$
4. Estimate next solution: $\mathbf{d}_{\text{new}}=\mathbf{d}^n$ or $\mathbf{d}_{\text{new}}=\tilde{\mathbf{d}}^{n+1}$
5. Newton iterations for time step $n+1$:
   a. *getforce* computes $\mathbf{f}(\mathbf{d}_{\text{new}}, t^{n+1})$, see Box 6.1
   b. $\mathbf{a}^{n+1}=1/\beta\Delta t^2\left(\mathbf{d}_{\text{new}}-\tilde{\mathbf{d}}^{n+1}\right), \mathbf{v}^{n+1}=\tilde{\mathbf{v}}^{n+1}+\gamma\Delta t\mathbf{a}^{n+1}$: see (6.3.4–5)
   c. $\mathbf{r}=\mathbf{M}\mathbf{a}^{n+1}-\mathbf{f}$
   d. compute Jacobian $\mathbf{A}(\mathbf{d})$
   e. modify $\mathbf{A}(\mathbf{d})$ for essential boundary conditions
   f. solve linear equations $\Delta\mathbf{d}=-\mathbf{A}^{-1}\mathrm{r}$
   g. $\mathbf{d}_{\text{new}}\leftarrow\mathbf{d}_{\text{old}}+\Delta\mathbf{d}$
   h. check convergence criterion; if not met, go to step 5a.
6. Update displacements, counter and time: $\mathbf{d}^{n+1}=\mathbf{d}_{\text{new}}$, $n\leftarrow n+1$, $t\leftarrow t+\Delta t$
7. Check energy balance
8. Output; if simulation not complete, go to 4.

Box 6.4 Flowchart for equilibrium solution

1. Initial conditions and initialization: set $\mathbf{d}^0=\mathbf{0}$; $\boldsymbol{\sigma}^0$; $n=0$; $t=0$; $\mathbf{d}_{\text{new}}=\mathbf{d}^0$
2. Newton iterations for load increment $n+1$:
   a. *getforce* computes $\mathbf{f}(\mathbf{d}_{\text{new}},t^{n+1})$; $\mathbf{r}=\mathbf{f}(\mathbf{d},t^{n+1})$
   b. compute $\mathbf{A}(\mathbf{d}_{\text{new}})$
   c. modify $\mathbf{A}(\mathbf{d}_{\text{new}})$ for essential boundary conditions
   d. solve linear equations $\Delta\mathbf{d}=-\mathbf{A}^{-1}\mathbf{r}$
   e. $\mathbf{d}_{\text{new}}\leftarrow\mathbf{d}_{\text{new}}+\Delta\mathbf{d}$
   f. check error criterion; if not met, go to 2a.
3. Update displacements, step count and time: $\mathbf{d}^{n+1}=\mathbf{d}_{\text{new}}$, $n\leftarrow n+1$, $t\leftarrow t+\Delta t$
4. Output; if simulation not complete, go to 2.

The flowchart shows a procedure often called a full Newton algorithm, where the Jacobian matrix is evaluated and inverted in every iteration of the procedure. Many programs use a modified Newton algorithm, in which the Jacobian is assembled and triangulated only at the beginning of a step or intermittently during a step. For example, the Jacobian may be triangulated only when the iterative procedure does not seem to be converging well or at the beginning of the iterations for a time step. These modified schemes are faster but less robust.

The implicit algorithm begins with the imposition of the initial conditions. The initial conditions can be handled exactly as in explicit methods. The initial displacements are considered to be zero. The initial accelerations are computed as shown in steps 2 and 3.

The displacements $\mathbf{d}^{n+1}$ are obtained by an iterative procedure. To begin the iterative procedure, a starting value of $\mathbf{d}$ is needed; usually the solution from the preceding time step is used. The residual is then calculated for this starting value. In an equilibrium solution, the residual depends only on the internal and external nodal forces, which is obtained in the module *getforce*. This module, *getforce*, is the same as in the explicit procedure, Box 6.1, except that the calculation of the stable time step is omitted. In transient implicit solutions, the residuals also depend on the accelerations.

The Jacobian matrix is calculated for the latest state of the body. Homogeneous displacement boundary conditions can be enforced by modifying the Jacobian matrix as follows: the equation corresponding to the vanishing displacement component is either omitted or replaced by a dummy equation which states that the component vanishes. This can be done by zeroing the corresponding row and column of the Jacobian and the RHS of the equations and placing a positive constant on the diagonal. For more complex constraints, Lagrange multipliers methods or penalty methods are employed; these are described in the next section.

6.3.8 *Constraints*

We will describe four methods for treating the constraints (6.2.8). They are:

1. Penalty
2. Lagrange multiplier
3. Augmented Lagrangian
4. Perturbed Lagrangian

These methods originate in optimization theory. As will be seen, they can readily be adapted to the solution of the discrete momentum and equilibrium equations.

To explain these techniques for constrained problems, it is useful to first consider the conservative problem, where the solution is found by minimization. This provides guidance in the formulation of the nonconservative problem. The unconstrained conservative problem has been described in Section 6.3.6.

6.3.8.1 Lagrange Multiplier Method

In this method, the constraints are appended to the objective function with Lagrange multipliers. In the conservative problem, the objective function, that is, the function to be minimized, is the potential. The minimization of a function subject to constraints can be posed as a Lagrange multiplier problem: the minimum of the function corresponds to the stationary points of the sum of the function and the constraints weighted by the Lagrange multipliers. The solution of (6.3.27) is equivalent to finding the stationary points of

$$W_L = W + \lambda_I g_I \equiv W + \lambda^T \mathbf{g} \tag{6.3.32}$$

where $\boldsymbol{\lambda} = \{\lambda_I\}$ are the Lagrange multipliers and the subscript L indicates the Lagrange multiplier modified potential. At the equilibrium points

$$0 = dW_L = dW + d(\lambda_I g_I) \equiv dW + d(\boldsymbol{\lambda}^T \mathbf{g}) \quad \forall d\mathbf{d} \quad \text{and} \quad \forall d\boldsymbol{\lambda} \tag{6.3.33}$$

Note that a stable equilibrium point is a minimum with respect to **d** and a maximum with respect to $\boldsymbol{\lambda}$, that is, a saddle point. The derivatives of (6.3.32) with respect to **d** and $\boldsymbol{\lambda}$ vanish at the stationary point, so:

$$\frac{\partial W_L}{\partial d_a} = \frac{\partial W}{\partial d_a} + \lambda_I \frac{\partial g_I}{\partial d_a} \equiv r_a + \lambda_I \frac{\partial g_I}{\partial d_a} = 0, \quad a = 1 \text{ to } n_{\text{dof}} \tag{6.3.34}$$

$$g_I = 0, \quad I = 1 \text{ to } n_c \tag{6.3.35}$$

The above is the system of $n_{\text{dof}}+n_c$ algebraic equations; note that repeated indices are summed. Equation (6.3.34) can be rewritten by (6.3.25) as

$$f_a^{\text{int}} - f_a^{\text{ext}} + \lambda_I \frac{\partial g_I}{\partial d_a} = 0 \quad \text{or} \quad \mathbf{f}^{\text{int}} - \mathbf{f}^{\text{ext}} + \boldsymbol{\lambda}^T \frac{\partial \mathbf{g}}{\partial \mathbf{d}} = 0 \tag{6.3.36}$$

As can be seen in the above, the constraints introduce additional forces $\lambda_I \partial g_I / \partial d_a$, which are linear combinations of the Lagrange multipliers. If the constraints are linear, the additional forces are independent of the nodal displacements.

To obtain a linear model for (6.3.34–35), we take a Taylor expansion of (6.3.34–35) and set the result to zero, giving

$$r_a + \lambda_I \frac{\partial g_I}{\partial d_a} + \frac{\partial r_a}{\partial d_b} \Delta d_b + \frac{\partial g_I}{\partial d_a} \Delta \lambda_I + \lambda_I \frac{\partial^2 g_I}{\partial d_a \partial d_b} + \Delta d_b = 0 \tag{6.3.37}$$

$$g_I \frac{\partial g_I}{\partial d_a} \Delta d_a = 0 \tag{6.3.38}$$

Note the summations on repeated indices. To put this into matrix notation, we define

$$\mathbf{G} = \left[G_{I_a} \right] = \left[\frac{\partial g_I}{\partial d_a} \right] \quad \mathbf{H}_I = \left[H_{ab} \right]_I = \left[\frac{\partial^2 g_I}{\partial d_a \partial d_b} \right] \tag{6.3.39}$$

In this notation the linear model (6.3.37–38) is

$$\begin{bmatrix} \mathbf{A} + \lambda_I \mathbf{H}_I & \mathbf{G}^T \\ \mathbf{G} & \mathbf{0} \end{bmatrix} \begin{Bmatrix} \Delta \mathbf{d} \\ \Delta \lambda \end{Bmatrix} = \begin{Bmatrix} -\mathbf{r} - \boldsymbol{\lambda}^T \mathbf{G} \\ -\mathbf{g} \end{Bmatrix} \tag{6.3.40}$$

where **A** is defined in (6.3.30). As can be seen in the above, the linear model has n_c additional equations due to the constraints. Even when the matrix **A** is positive-definite, the augmented system of equations will not be positive-definite because of the zeroes on the diagonal in the lower right-hand corner of the matrix.

For a linear statics problem with a linear constraints **Gd**=**a**, the equations are

$$\begin{bmatrix} \mathbf{K} & \mathbf{G}^T \\ \mathbf{G} & \mathbf{0} \end{bmatrix} \begin{Bmatrix} \mathbf{d} \\ \boldsymbol{\lambda} \end{Bmatrix} = \begin{Bmatrix} \mathbf{f}^{\text{ext}} \\ \mathbf{a} \end{Bmatrix} \tag{6.3.41}$$

To obtain (6.3.41) from (6.3.40), we have used the following properties of the linear static system:

1. $\mathbf{A}=\mathbf{K}$, where $\mathbf{K}$ is the linear stiffness.
2. $\mathbf{H}_I=0$ for linear constraints.
3. The starting value is zero, $\Delta\mathbf{d}=\mathbf{d}$, $\Delta\boldsymbol{\lambda}=\boldsymbol{\lambda}$.

For the general problem involving nonconservative materials, dynamics, etc., the Lagrange multiplier method is formulated as follows. Using (6.3.33) as a guide to the construction of the differential with the Lagrange multiplier, we have:

$$0 = dW_L = dW + d(\lambda_I g_I) = 0 \tag{6.3.42}$$

where from (B4.6.1) and (6.3.1)

$$\begin{aligned} dW &= dW^{\text{int}} - dW^{\text{ext}} + dW^{\text{kin}} \\ &= d\mathbf{d}^T(\mathbf{f}^{\text{int}} - \mathbf{f}^{\text{ext}} + s_D\mathbf{M}\ddot{\mathbf{d}}) = d\mathbf{d}^T\mathbf{r} = d\mathrm{d}_a r_a \end{aligned} \tag{6.3.43}$$

Substituting (6.3.43) into (6.3.42) and writing out the differentials in the second term gives

$$d\mathrm{d}_a r_a + d\lambda_I g_I + \lambda_I \frac{\partial g_I}{\partial \mathrm{d}_a} d\mathrm{d}_a = 0 \quad \forall d\mathrm{d}_a, d\lambda_I \tag{6.3.44}$$

Using the arbitrariness of the differentials $d\mathrm{d}_a$ and $d\lambda_I$, respectively, in this implies

$$r_a + \lambda_I \frac{\partial g_I}{\partial d_a} = 0, \quad g_I = 0 \tag{6.3.45}$$

The above appear to be identical to (6.3.34–35), but the residual may include kinetic forces and the system need not be conservative. The linearized equations are identical to those for the conservative system, (6.3.40). While the development has been given for a differential of work dW, it also applies to a differential of power.

6.3.8.2 Penalty Method

Again, we first consider a conservative problem where the solution is determined by minimization. In the penalty method, the constraint is enforced by adding the square of the constraints, $g_I g_I$, multiplied by a large number called the penalty parameter, to the potential. The modified potential is

$$W_P(\mathbf{d}) = W(\mathbf{d}) + \frac{1}{2}\beta g_I(\mathbf{d}) g_I(\mathbf{d}) \equiv W + \frac{1}{2}\beta \mathbf{g}^T\mathbf{g} \tag{6.3.46}$$

where β is the penalty parameter and the subscript P indicates the penalty form of the potential. The penalty parameter is chosen to be orders of magnitude greater than other parameters of the problem. The idea is that if β is large enough, the minimum of $W_P(\mathbf{d})$ cannot be attained without satisfying the constraints.

The stationary (or minimum) conditions give

$$\frac{\partial W_P}{\partial d_a} = \frac{\partial W}{\partial d_a} + \beta g_I \frac{\partial g_I}{\partial d_a} = 0 \quad \text{or} \quad \mathbf{r} + \beta \mathbf{g}^T \mathbf{G} = \mathbf{0} \tag{6.3.47}$$

The linear model is

$$\left(\frac{\partial r_a}{\partial d_b} + \beta \frac{\partial g_I}{\partial d_b} \frac{\partial g_I}{\partial d_a} + \beta g_I \frac{\partial^2 g_I}{\partial d_a \partial d_b} \right) \Delta d_b = \left(-r_a - \beta g_I \frac{\partial g_I}{\partial d_a} \right) \tag{6.3.48}$$

or in matrix form

$$\mathbf{A}_P \Delta \mathbf{d} = (\mathbf{A} + \beta \mathbf{G}^T \mathbf{G} + \beta g_I \mathbf{H}_I)\, \Delta \mathbf{d}_b = -\mathbf{r} - \beta \mathbf{g}^T \mathbf{G} \tag{6.3.49}$$

This system is the same size as the unconstrained system. For linear constraints, if $\mathbf{A}>0$, then $\mathbf{A}_P>0$, that is, if the original Jacobian matrix is positive-definite, then the augmented system is positive-definite. The major drawbacks of penalty methods is that they impair the conditioning of the equations and they require the selection of a penalty parameter. Often, it is not possible to make the penalty parameter large enough to satisfy the constraints accurately, and it is difficult to ascertain the error due to the approximate satisfaction of the constraints.

The discrete equations for nonconservative systems are obtained with the guidance of the differential of (6.3.46):

$$0 = dW_P = dW + \frac{1}{2} \beta d(g_I g_I) = dW + \beta g_I \; dg_I \tag{6.3.50}$$

Now apply (6.3.43) to replace dW in the above. The discrete equations and linear model are then given by the RHS equation in (6.3.47) and (6.3.48), respectively.

6.3.8.3 Augmented Lagrangian Method

The augmented Lagrangian method can be viewed as a combination of the Lagrange multiplier and penalty methods. In the penalty method, the conditioning of the linearized equations decreases with increasing β. In the Lagrange multiplier method, extra unknowns are introduced and the resulting equation system is not necessarily positive-definite. The augmented Lagrangian method improves the conditioning of the matrix and improves the penalty selection.

The mathematical programming problem for the augmented Lagrangian method can be stated as follows. For a given penalty parameter β, determine the displacement $\mathbf{d}$ and Lagrange multiplier $\boldsymbol{\lambda}$ so that

$$W_{AL}(\mathbf{d}, \boldsymbol{\lambda}, \beta) = W(\mathbf{d}) + \boldsymbol{\lambda}^T \mathbf{g}(\mathbf{d}) + \frac{1}{2} \beta \mathbf{g}^T(\mathbf{d}) \mathbf{g}(\mathbf{d}) \tag{6.3.51}$$

is a stationary point. It is noted that if we set $\boldsymbol{\lambda}=\mathbf{0}$, the method reduces to the penalty method (6.3.46), and if we set $\beta=0$, it reduces to the Lagrange multiplier method (6.3.32). Because of the introduction of the Lagrange multiplier, the parameter β need not be as large as in penalty methods to satisfy the constraints $\mathbf{g}(\mathbf{d})=\mathbf{0}$ adequately. This improves the conditioning of the equations.

The stationary points can be determined by setting the partial derivatives with respect to $\mathbf{d}$ and $\boldsymbol{\lambda}$ equal to zero:

$$\frac{\partial W_{AL}}{\partial d_a}=\frac{\partial W}{\partial d_a}+\lambda_I\frac{\partial g_I}{\partial d_a}+\beta g_I\frac{\partial g_I}{\partial d_a}=0 \tag{6.3.52}$$

$$\frac{\partial W_{AL}}{\partial \lambda_I}=g_I=0 \tag{6.3.53}$$

Rewriting these equations in terms of the residual $\mathbf{r}$ and the gradient of the matrix $\mathbf{G}$ yields

$$\mathbf{r}+\boldsymbol{\lambda}^T\mathbf{G}+\beta\mathbf{g}^T\mathbf{G}=\mathbf{0} \tag{6.3.54}$$

$$\mathbf{g}(\mathbf{d})=\mathbf{0} \tag{6.3.55}$$

The linearized model is:

$$\begin{aligned}&r_a+\lambda_I\frac{\partial g_I}{\partial d_a}+\beta g_I\frac{\partial g_I}{\partial d_a}+\frac{\partial r_a}{\partial d_b}\Delta d_b+\Delta\lambda_I\frac{\partial g_I}{\partial d_a}\\&+\lambda_I\frac{\partial^2 g_I}{\partial d_a\partial d_b}\Delta d_b+\beta\frac{\partial g_I}{\partial d_a}\frac{\partial g_I}{\partial d_b}\Delta d_b+\beta g_I\frac{\partial^2 g_I}{\partial d_a\partial d_b}\Delta d_b=0\end{aligned} \tag{6.3.56}$$

and

$$g_I+\frac{\partial g_I}{\partial d_b}\Delta d_b=0 \tag{6.3.57}$$

These can be put into the matrix form:

$$\begin{bmatrix}\mathbf{A}+\lambda_I\mathbf{H}_I+\beta(\mathbf{G}^T\mathbf{G}+g_I\mathbf{H}_I) & \mathbf{G}^T\\ \mathbf{G} & \mathbf{0}\end{bmatrix}\begin{Bmatrix}\Delta\mathbf{d}\\ \Delta\boldsymbol{\lambda}\end{Bmatrix}=\begin{bmatrix}-(\mathbf{r}-\boldsymbol{\lambda}^T\mathbf{G}+\beta\mathbf{g}^T\mathbf{G})\\ -\mathbf{g}\end{bmatrix} \tag{6.3.58}$$

This is simply a combination of (6.3.40) and (6.3.49) with the removal of one $\mathbf{r}$ and one $\mathbf{A}$. The discrete equations for the nonconservative system are identical.

6.3.8.4 Perturbed Lagrangian

In the perturbed Lagrangian, the product of a small constant with the sum of squares of the Lagrange multipliers is added to the Lagrange multiplier form. The potential for the perturbed Lagrangian is

$$W_{PL}(\mathbf{d}, \boldsymbol{\lambda}, \beta) = W(\mathbf{d}) + \boldsymbol{\lambda}^T \mathbf{g}(\mathbf{d}) - \frac{1}{2}\varepsilon \boldsymbol{\lambda}^T \boldsymbol{\lambda} \tag{6.3.59}$$

We leave it as an exercise to show that the resulting equations are

$$\mathbf{r} + \boldsymbol{\lambda}^T \mathbf{G} = \mathbf{0}, \quad \mathbf{g} - \varepsilon \boldsymbol{\lambda} = \mathbf{0} \tag{6.3.60}$$

The linearized equations are

$$\begin{bmatrix} \mathbf{A} + \lambda_I \mathbf{H}_I & \mathbf{G}^T \\ \mathbf{G} & -\varepsilon \mathbf{I} \end{bmatrix} \begin{Bmatrix} \Delta \mathbf{d} \\ \Delta \boldsymbol{\lambda} \end{Bmatrix} = \begin{bmatrix} -(\mathbf{r} - \boldsymbol{\lambda}^T \mathbf{G}) \\ -(\mathbf{g} - \varepsilon \boldsymbol{\lambda}) \end{bmatrix} \tag{6.3.61}$$

We leave it to the reader (Exercise 6.3) to show that the above is identical to a penalty formulation. Therefore the perturbed Lagrangian method is primarily of theoretical interest.

Example 6.1 Consider a nonlinear elastic rod OA of length $\ell_0 = a$ hinged at the origin O. Use the hyperelastic, Kirchhoff material model (5.4.39). The rod is loaded by a constant external force $\mathbf{f}^{\text{ext}}$ (conservative loading) and stretched from point A to A′ as shown in Figure 6.4. The right end of the rod is constrained to lie in the circular groove as shown.

(a) Use the definition (3.3.1), which for a uniaxial state of strain can be written as $2\ell_0 E_{11} \ell_0 = \ell^2 - \ell_0^2$, to evaluate E_{11}.
(b) Develop the potential for the unconstrained problem and its first and second derivatives.
(c) Formulate the Lagrange multiplier method.
(d) Repeat (c) with the penalty method.

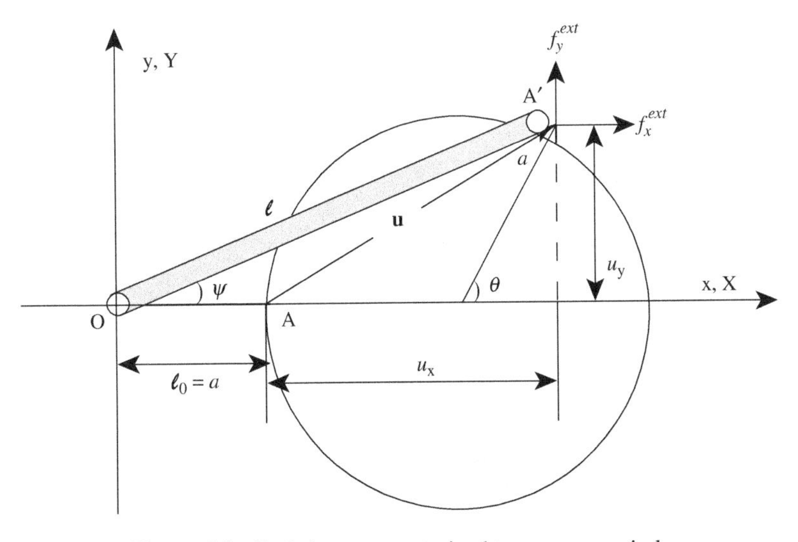

Figure 6.4 Rod element constrained to move on a circle

(a) From the definition of the Green strain (3.3.1) it follows that

$$E_{11} = \frac{\ell^2 - \ell_0^2}{2\ell_0^2} = \frac{(a+u_{x1})^2 + u_{y1}^2 - a^2}{2a^2} = \frac{u_{x1}}{a} + \frac{1}{2a^2}\left(u_{x1}^2 + u_{y1}^2\right) \tag{E6.1.1}$$

(b) For a Kirchhoff material, the internal energy for a rod element is given by $W^{\text{int}} = \frac{1}{2}\alpha E_{11}^2$, where $\alpha = A_0 a E^{SE}$ and E^{SE} is defined in Box 5.1.
For a conservative load, the external potential is

$$W^{\text{ext}} = f_{x1}^{\text{ext}} x_1 + f_{y1}^{\text{ext}} y_1 = f_{x1}^{\text{ext}}(X_1 + u_{x1}) + f_{y1}^{\text{ext}}(Y_1 + u_{y1}) \tag{E6.1.2}$$

where (X_1, Y_1) is the initial position of node 1. The total potential is given by

$$W(u_{x1}, u_{y1}) = \frac{1}{2}\alpha\left(\frac{u_{x1}}{a} + \frac{1}{2a^2}\left(u_{x1}^2 + u_{y1}^2\right)\right)^2 - f_{x1}^{\text{ext}}(X_1 + u_{x1}) - f_{y1}^{\text{ext}}(Y_1 + u_{y1}) \tag{E6.1.3}$$

Partial differentiation of W with respect to u_{x1} and u_{y1} gives the residual $\mathbf{r}$:

$$\begin{Bmatrix} r_{x1} \\ r_{y1} \end{Bmatrix} = \begin{Bmatrix} \dfrac{\partial W}{\partial u_{x1}} \\ \dfrac{\partial W}{\partial u_{y1}} \end{Bmatrix} = \begin{Bmatrix} \alpha\left(\dfrac{u_{x1}}{a^2} + \dfrac{3u_{x1}^2}{2a^3} + \dfrac{u_{y1}^2}{2a^3} + \dfrac{u_{x1}^3}{2a^4} + \dfrac{u_{x1}u_{y1}^2}{2a^4}\right) - f_{x1}^{\text{ext}} \\ \alpha\left(\dfrac{u_{x1}u_{y1}}{a^3} + \dfrac{u_{x1}^2 u_{y1}}{2a^4} + \dfrac{u_{y1}^3}{2a^4}\right) - f_{y1}^{\text{ext}} \end{Bmatrix} \tag{E6.1.4}$$

The tangent stiffness matrix $\mathbf{K}^{\text{int}}$ is

$$\mathbf{K}^{\text{int}} = \begin{bmatrix} \dfrac{\partial r_{x1}}{\partial u_{x1}} & \dfrac{\partial r_{x1}}{\partial u_{y1}} \\ \dfrac{\partial r_{y1}}{\partial u_{x1}} & \dfrac{\partial r_{y1}}{\partial u_{y1}} \end{bmatrix} = \alpha\begin{bmatrix} \left(\dfrac{1}{a^2} + \dfrac{3u_{x1}}{a^3} + \dfrac{3u_{x1}^2}{2a^4} + \dfrac{u_{y1}^2}{2a^4}\right) & \left(\dfrac{u_{y1}}{a^3} + \dfrac{u_{x1}u_{y1}}{a^4}\right) \\ \left(\dfrac{u_{y1}}{a^3} + \dfrac{u_{x1}u_{y1}}{a^4}\right) & \left(\dfrac{u_{x1}}{a^3} + \dfrac{u_{x1}^2}{2a^4} + \dfrac{3u_{y1}^2}{2a^4}\right) \end{bmatrix} \tag{E6.1.5}$$

For conservative loads, $\mathbf{K}^{\text{ext}} = 0$ and the Jacobian $\mathbf{A} = \mathbf{K}^{\text{int}}$.

(c) Next we impose the constraints by Lagrange multipliers. The modified potential is given by (6.3.42):

$$W_L(u_{x1}, u_{y1}, \lambda) = W(u_{x1}, u_{y1}) + \lambda g(u_{x1}, u_{y1}) \tag{E6.1.6}$$

where λ is the Lagrange multiplier and $g = 0$ is the constraint. The constraint requires node 1 to remain on the circle centered at $(2a, 0)$. In terms of the nodal coordinates, the constraint is

$$0 = g(u_{x1}, u_{y1}) = (x_1 - 2a)^2 + y_1^2 - a^2 = (X_1 + u_{x1} - 2a)^2 + (Y_1 + u_{y1})^2 - a^2 \tag{E6.1.7}$$
$$= u_{x1}^2 - 2au_{x1} + u_{y1}^2$$

since $Y_1 = 0$ and $X_1 = a$. The gradient of the constraint is

$$\mathbf{G}^T = \begin{Bmatrix} \dfrac{\partial g}{\partial u_{x1}} \\ \dfrac{\partial g}{\partial u_{y1}} \end{Bmatrix} = \begin{Bmatrix} 2(u_{x1} - a) \\ 2u_{y1} \end{Bmatrix} \tag{E6.1.8}$$

The linearized equations are given by (6.3.40) with the matrices **A** and **G** defined above and **H** given by

$$\mathbf{H} = \begin{bmatrix} \dfrac{\partial^2 g}{\partial u_{x1}^2} & \dfrac{\partial^2 g}{\partial u_{x1} \partial u_{y1}} \\ \dfrac{\partial^2 g}{\partial u_{y1} \partial u_{x1}} & \dfrac{\partial^2 g}{\partial u_{y1}^2} \end{bmatrix} = \begin{bmatrix} 2 & 0 \\ 0 & 2 \end{bmatrix} \tag{E6.1.9}$$

The Lagrange multiplier λ corresponds to the constraint $g = 0$.

(d) The modified potential for the penalty formulation is given by (6.3.46), $W_p = W + \frac{1}{2}\beta g^2$, where W is given in (E6.1.3) and g is given in (E6.1.7). The stationary conditions are given in (6.3.47) with **G** and **r** given in (E6.1.8) and (E6.1.4). The matrix form of the linear model is (6.3.49) with **H** given by (E6.1.9) and

$$\mathbf{G}^T\mathbf{G} = 4\begin{bmatrix} (u_{x1} - a)^2 & (u_{x1} - a)u_{y1} \\ u_{y1}(u_{x1} - a) & u_{y1}^2 \end{bmatrix} \tag{E6.1.10}$$

6.3.9 *Convergence Criteria*

The termination of the iterations in Newton algorithms for implicit and equilibrium solutions is determined by convergence criteria. These criteria pertain to the convergence of the iterate to the solution of $\mathbf{r}(\mathbf{d}^n, t^n) = \mathbf{0}$, not the convergence of the discrete solution to the solution of the partial differential equations. Three types of convergence criteria are used to terminate the iterations:

1. criteria based on the magnitude of the residual **r**
2. criteria based on the magnitude of the displacement increments $\Delta\mathbf{d}$
3. energy error criteria.

Customarily an ℓ_2 norm of the vectors (see Appendix 2 for definitions of norms) is used for the first two criteria. The criteria then are:

residual error criterion:

$$\|\mathbf{r}\|_{\ell_2} = \left(\sum_{a=1}^{n_{\rm dof}} r_a^2\right)^{\frac{1}{2}} \le \varepsilon \max\left(\|\mathbf{f}^{\rm ext}\|_{\ell_2}, \|\mathbf{f}^{\rm int}\|_{\ell_2}, \|\mathbf{Ma}\|_{\ell_2}\right) \quad (6.3.62)$$

displacement increment error criterion:

$$\|\Delta\mathbf{d}\|_{\ell_2} = \left(\sum_{a=1}^{n_{\rm dof}} \Delta d_a^2\right)^{\frac{1}{2}} \le \varepsilon \|\mathbf{d}\|_{\ell_2} \quad (6.3.63)$$

The ℓ_2 norm is appropriate when the mean error over all degrees of freedom is to be controlled. If this is replaced by the maximum norm $\|\bullet\|_{\ell\infty}$ we can limit the maximum error at any degree-of-freedom. The norms on the right-hand side of (6.3.62) and (6.3.63) are scaling factors. Without these, the criterion depends on the parameters of the problem. The error tolerance s determines the precision with which the displacements are calculated before terminating the iterative procedure; with $\varepsilon = 10^{-3}$ and the ℓ_2 norm, the mean error in the nodal displacements is in the third significant digit. The convergence tolerance determines the speed and accuracy of a calculation. If the criterion is too coarse, the solution may be quite inaccurate. On the other hand, a criterion which is too tight results in unnecessary computations.

The energy convergence criterion measures the energy flow to the system resulting from the residual, which is like an error in energy (Belytschko and Schoeberle, 1975). It is given by

$$\left|\Delta\mathbf{d}^T\mathbf{r}\right| = \left|\Delta d_a r_a\right| \le \varepsilon \max(W^{\rm ext}, W^{\rm int}, W^{\rm kin}) \quad (6.3.64)$$

The computation of the energies on the RHS which scale the error criterion is described in Section 6.2.3. The left-hand side in the above represents an error in the energy, that is, the energy associated with the error in the momentum equation.

6.3.10 Line Search

Line search increases the effectiveness of Newton methods when convergence is slow due to substantial deviations of the residual from the underlying linear model and roughness of the residual. The rationale behind the line search is that the direction $\Delta\mathbf{d}$ found by the Newton method is often a good direction, but the step size $\|\Delta\mathbf{d}\|$ is not optimal. It is cheaper to find the best point along the direction $\Delta\mathbf{d}$ by several computations of the residual than to get a new direction by using a new Jacobian. Therefore, before proceeding to the next direction, a measure of the residual is minimized along the line

$$\mathbf{d} = \mathbf{d}_{\rm old} + \xi\Delta\mathbf{d} \quad (6.3.65)$$

where $\mathbf{d}_{\text{old}}$ is the last iterate and ξ is a parameter which defines the position along the line. In other words, we find the parameter ξ so that $\mathbf{d}_{\text{old}}+\xi\Delta\mathbf{d}$ minimizes some measure of the residual along the line. We can use as a measure of the residual its ℓ_2 norm, the maximum norm, or some other measure.

An energy measure of the residual can be determined in terms of the potential energy when the system is conservative. If the potential $W(\mathbf{d})$ along the search direction is a minimum, the derivative with respect to the line parameter must vanish:

$$0=\frac{dW(\xi)}{d\xi}=\frac{\partial W}{\partial \mathbf{d}}\cdot\frac{d\mathbf{d}}{d\xi}=\mathbf{r}^T\Delta\mathbf{d} \tag{6.3.66}$$

where the last step follows from (6.3.25) and (6.3.65). The above can be interpreted as follows: at the minimum, the residual $\mathbf{r}$ is orthogonal to the direction $\Delta\mathbf{d}$. Note that this criterion is equivalent to minimizing the error in energy; compare to (6.3.64). It can also be applied to nonconservative systems. Equation (6.3.66) should be normalized similarly to (6.3.64) in an implementation.

The line search can use any method for minimizing a function of a single parameter: the method of bisection or searches based on interpolation or combinations thereof. A widely used technique is based on quadratic interpolation. Once the residual has been evaluated at two points, the residual measure is interpolated by a quadratic function of ξ (its value at $\xi=0$ is known, so we have three points). This quadratic interpolate is then used to estimate the position of the minimum, and the measure is interpolated again with the latest three points. The procedure is terminated when the measure has been minimized to the desired precision.

6.3.11 The α-Method

A major drawback of the Newmark integrator is the tendency for high frequency noise to persist in the solution. On the other hand, when linear damping or artificial viscosity is added via the parameter γ, the accuracy is markedly degraded. The α-method (Hilber Hughes, and Taylor, 1977) improves numerical dissipation for high frequencies without degrading the accuracy as much. A nice variant of the method is found in Chung and Hulbert (1993).

In the α-method, the Newmark-β formulas, (6.3.4–5), are used but the equation of motion (6.3.1) is modified as follows. Equation (6.3.1) is replaced by

$$0=\mathbf{r}(\mathbf{d}^{n+1},t^{n+1})=s_D\mathbf{M}\mathbf{a}^{n+1}-\mathbf{f}^{\text{ext}}(\mathbf{d}^{n+\alpha},t^{n+1})+\mathbf{f}^{\text{int}}(\mathbf{d}^{n+\alpha},t^{n+1}) \tag{6.3.67}$$

where the major change compared to the Newmark method is the displacement that drives the nodal force computation:

$$\mathbf{d}^{n+\alpha}=(1+\alpha)\mathbf{d}^{n+1}-\alpha\mathbf{d}^n \tag{6.3.68}$$

For a linear system, the above definition of the nodal internal force vector gives $\mathbf{f}^{\text{int}}=\mathbf{K}\mathbf{d}^{n+\alpha}=(1+\alpha)\,\mathbf{K}\mathbf{d}^{n+1}-\alpha\mathbf{K}\mathbf{d}^n$. Thus in going to the α method, the term $\alpha\mathbf{K}(\mathbf{d}^{n+1}-\mathbf{d}^n)$ is added; this can be seen to be like stiffness proportional damping.

The linearized equation for the residual corresponding to (6.3.15) is

$$\mathbf{r}\left(\mathbf{d}_{\nu}^{n+\alpha}\right)+\frac{\partial \mathbf{r}\left(\mathbf{d}_{\nu}^{n+\alpha}\right)}{\partial \mathbf{d}}\Delta\mathbf{d}+\mathbf{0}\,(\Delta\mathbf{d}^{2})=0 \tag{6.3.69}$$

The Jacobian matrix (or effective tangent stiffness matrix) corresponding to (6.3.20) can be shown to be

$$\mathbf{A}=\frac{\partial \mathbf{r}\left(\mathbf{d}_{\nu}^{n+\alpha}\right)}{\partial \mathbf{d}}=\frac{s_D}{\beta\Delta t^{2}}\mathbf{M}+(1+\alpha)\frac{\partial \mathbf{f}^{\text{int}}\left(\mathbf{d}_{\nu}^{n+\alpha}\right)}{\partial \mathbf{d}}-(1+\alpha)\frac{\partial \mathbf{f}^{\text{ext}}\left(\mathbf{d}_{\nu}^{n+\alpha}\right)}{\partial \mathbf{d}} \tag{6.3.70}$$

The rest of the formulation is the same as for the Newmark method. If $\alpha=0$, the method as given here corresponds to the trapezoidal rule. The method is unconditionally stable for a linear system when

$$\alpha\in\left[-\frac{1}{3},0\right],\ \gamma=\frac{(1-2\alpha)}{2},\ \text{and}\ \beta=\frac{(1-\alpha)^{2}}{4} \tag{6.3.71}$$

There are no general stability results for this method for nonlinear problems in the literature.

6.3.12 *Accuracy and Stability of Implicit Methods*

The advantage of an implicit method over an explicit method is that *for linear transient problems, suitable implicit integrators are unconditionally stable.* This is proven in Section 6.6 for some specific integrators. Although theoretical results for specific situations indicate that unconditional stability holds at least for certain nonlinear systems, no proof of unconditional stability exists which encompasses the wide range of conditions found in practice. However, experience indicates that the time steps for implicit integrators can be much larger than those for explicit integration without incurring instabilities.

The major restrictions on the time step in implicit methods arise from accuracy requirements and the decreasing robustness of the Newton procedure as the time step increases. The Newmark method is *second-order accurate*, that is, the truncation error is of order Δt^2, the same order as for the central difference method. Therefore, for large time steps, truncation error becomes a concern.

Large time steps also impair the convergence of the Newton method, particularly in problems with very rough response, such as contact-impact. With a large time step, the starting iterate may be far from the solution, so the possibility of failure of the Newton method to converge increases. Small time steps improve the robustness of the Newton algorithm.

In return for their enhanced stability, implicit methods exact a significant price: they require the solution of nonlinear algebraic equations in each time step. The construction of the linear models for the Newton procedure is often quite involved. Furthermore, the storage of these equations requires significant amounts of memory. The memory requirements can be reduced substantially by iterative linear equation solvers (an iterative method within an iterative Newton method). Recently, iterative solvers have improved dramatically, so implicit solutions are now feasible in many problems where they failed before. We are certain that further improvements are imminent. Nevertheless, high cost and insufficient robustness still plague Newton methods.

6.3.13 Convergence and Robustness of Newton Iteration

The rate of the convergence of the iterations in the Newton method is quadratic when the Jacobian matrix A satisfies certain conditions. These conditions may roughly be described as follows:

1. The Jacobian **A** should be a sufficiently smooth function of **d**.
2. The Jacobian **A** should be regular (invertable) and well-conditioned in the entire domain in the displacement space that the iterative procedure traverses.

Quadratic convergence means that the ℓ_2 norm of the difference between the solution and the iterate $\mathbf{d}_\nu$ decreases quadratically in each iteration:

$$\left\|\mathbf{d}_{\nu+1}-\mathbf{d}\right\| \leq c\left\|\mathbf{d}_\nu-\mathbf{d}\right\|^2 \tag{6.3.72}$$

where c is a constant that depends on the nonlinearity of the problem and **d** is the solution to the nonlinear algebraic equations. Thus the convergence of the Newton algorithm is quite rapid when **A** meets these conditions. The above gives the requirements for convergence only in broad terms and convergence has been proven for various specific conditions on **A**. One set of conditions for quadratic convergence are: the residual must be continuously differentiable and the inverse of the Jacobian matrix must exist and be uniformly bounded in the neighborhood of the solution (Dennis and Schnabel, 1983: 90).

These conditions are usually not satisfied by engineering problems. For example, in an elastic–plastic material, the residual is not a continuously differentiable function of the nodal displacements: when an element quadrature point changes from elastic to plastic or vice versa, the derivative is discontinuous. In contact-impact problems with Lagrange multiplier methods, the residual also lacks smoothness. Thus in many engineering problems, the conditions for quadratic convergence of the Newton method are not met. Yet, Newton's method is still remarkably effective, though the rate of convergence does indeed decrease. In some problems, the conditions for quadratic convergence are satisfied: for example, these conditions are satisfied in a smoothly loaded model with a Mooney–Rivlin material when the load is small enough so that the equilibrium solutions are stable.

Convergence difficulties occur most often in equilibrium problems. At points of instability, the Jacobian matrix is no longer regular and the proof of quadratic convergence does not apply. The Newton method often fails to converge in the vicinity of unstable states. In dynamic problems, these difficulties are ameliorated by the mass matrix, which makes the Jacobian more positive. As the time step increases, the beneficial effects of the mass matrix on the Jacobian decrease, since the coefficient of the mass matrix is inversely proportional to the square of the time step; see (6.3.9). For many problems, a straightforward application of the Newton method will fail completely, and enhancements of the Newton method such as the arc length method are needed.

6.3.14 Selection of Integration Method

The selection of an integration method depends on:

1. the type of partial differential equation
2. the smoothness of the data
3. the response of interest.

For parabolic PDEs, implicit methods are generally preferred. The solutions of parabolic systems are smooth, even when the data is rough. The stable time step in a parabolic system decreases by a factor of four each time the element size is halved, so that refinement in an explicit method becomes prohibitively expensive. In fact, Richtmeyer and Morton (1967) state unequivocally that parabolic systems should never be integrated by explicit methods.

However, in modern simulations, there are many situations where explicit methods are preferable even when the system is partially or fully parabolic. For example, in car crash simulation, the equations for the shells which model the sheet metal are parabolic (see Section 6.6), yet because of the noise introduced by contact-impact, explicit methods are preferred at this time. Similarly, in complex heat conduction problems, it is often impossible to take advantage of the large time steps permitted by implicit methods. So even in parabolic systems, the answer unfortunately is not clear cut.

In hyperbolic systems, the choice depends on the response of interest. For the purpose of making a selection, hyperbolic problems can be classified as wave propagation problems and inertial problems. The latter are also called structural dynamics problems. (Structural elements often have a parabolic character due to bending.) In structural dynamics problems, the frequency spectrum of the input is far below the resolution limits of the mesh; the refinement of the mesh is dictated by the need for very high accuracy in the frequency band of interest. In this class of problems, implicit methods are definitely preferred. Examples include the seismic response of structures and vibrations in structures and smaller entities, such as cars, tools, and electronic devices.

Wave propagation problems are those in which relatively high frequency parts of the spectrum are of interest. They encompass entities as large as the core of the earth for seismic studies to waves in a cellular phone in a droptest. These simulations require a small time step to track the high frequency portion of the response, and explicit methods are preferred. In general, for hyperbolic systems, particularly with rough data or where the interest lies in high frequencies, explicit methods are more efficient.

6.4 Linearization

6.4.1 Linearization of the Internal Nodal Forces

In the following, we derive expressions for the tangent stiffness matrix $\mathbf{K}^{\text{int}}$ for Lagrangian elements. As will be seen, part of the expression can be derived independently of the material response. Linearization of the constitutive equation is carried out in two ways:

1. with the continuum tangent moduli, which does not account for the actual constitutive update algorithm; the resulting material tangent stiffness matrix is called the *tangent stiffness matrix*.
2. with the algorithmic tangent moduli, which gives rise to the so-called *consistent tangent stiffness*.

The choice rests on practical considerations related to ease of implementation and on the smoothness of the problem. The continuum tangent modulus approach is straightforward to implement. However, it can run into convergence difficulties when the derivatives of the constitutive equation are discontinuous, as at the yield point of elastic–plastic materials. The consistent tangent stiffness, which is based on the algorithmic moduli, exhibits better convergence for rough constitutive equations. One drawback of the method is that it is not always possible to derive explicit forms for the algorithmic moduli for complex constitutive relations. Numerical differentiation is sometimes used to obtain the algorithmic moduli.

In the following, the tangent stiffness matrix will be expressed in terms of $\mathbf{C}^{SE}$, the tangent moduli relating the rate of the **PK2** stress to the rate of the Green strain in the total Lagrangian framework, and in terms of $C^{\sigma T}$, the tangent moduli relating the Truesdell rate of the Cauchy stress to the rate-of-deformation in the updated Lagrangian formulation. These tangent moduli are related: if the current configuration is chosen to be the reference configuration, $C^{\sigma T} = \mathbf{C}^{SE}$. We will consider these forms of the tangent stiffness as canonical. The tangent stiffnesses for any other rates of stress and strain can be obtained by transforming the tangent moduli. One of these transformations are given in Box 5.8, (B5.8.8).

We will develop the tangent stiffness matrix by relating rates of the internal nodal forces $\dot{\mathbf{f}}^{\text{int}}$ to the nodal velocities $\dot{\mathbf{d}}$. The procedure is identical to relating an infinitesimal increment of nodal forces $d\mathbf{f}^{\text{int}}$ to an infinitesimal increment of nodal displacements $d\mathbf{d}$, and we will occasionally recast the equations in that form; the superposed dot notation is chosen for convenience. The results are exact for any continuously differentiable residual; for rougher residuals, directional derivatives are needed and are described later.

By (4.9.10–11), the internal nodal forces in the total Lagrangian form are given by

$$\mathbf{f}^{\text{int}} = \int_{\Omega_0} \boldsymbol{B}_0^T \mathbf{P} d\Omega_0, \quad f_{iI}^{\text{int}} = \int_{\Omega_0} \frac{\partial N_I}{\partial X_j} P_{ji} \, d\Omega_0 = \int_{\Omega_0} B_{jI}^0 P_{ji} \, d\Omega_0 \tag{6.4.1}$$

where $\mathbf{P}$ is the nominal stress tensor with components P_{ji}, N_I are the nodal shape functions, and $B_{jI}^0 = \partial N_I / \partial X_j$. We have chosen the total Lagrangian form because this leads to the simplest derivation. In the total Lagrangian form, (6.4.1), the only variable which depends on the deformation is the nominal stress, that is, it is the only variable which varies with time. In the updated Lagrangian form, (B4.3.2) the domain of the element (or body), the spatial derivatives $\partial N_I / \partial x_j$ and the Cauchy stress depend on the deformation, and hence on time, which complicates the derivation of the tangent stiffness. Taking the material time derivative of (6.4.1) gives

$$\dot{\mathbf{f}}^{\text{int}} = \int_{\Omega_0} \boldsymbol{B}_0^T \dot{\mathbf{P}} d\Omega_0, \quad \text{or} \quad \dot{f}_{iI}^{\text{int}} = \int_{\Omega_0} \frac{\partial N_I}{\partial X_j} \dot{P}_{ji} \, d\Omega_0 \tag{6.4.2}$$

since $\boldsymbol{B}_0$ and $d\Omega_0$ are independent of the deformation or time. To obtain the stiffness matrix $\mathbf{K}^{\text{int}}$ it is now necessary to express the stress rate $\dot{\mathbf{P}}$ in terms of nodal velocities via the constitutive equation and the strain measure. However, constitutive equations are usually not expressed directly in terms of $\dot{\mathbf{P}}$ because this stress rate is not objective: see Sections 3.7 and 5.10. So we will use the material time derivative of the PK2 stress, which is objective.

The material time derivative of the PK2 stress is related to the material time derivative of the nominal stress by taking the time derivative of the transformation $\mathbf{P}=\mathbf{S}\cdot\mathbf{F}^T$ (Box 3.2):

$$\dot{\mathbf{P}}=\dot{\mathbf{S}}\cdot\mathbf{F}^T+\mathbf{S}\cdot\dot{\mathbf{F}}^T \quad \text{or} \quad \dot{P}_{ij}=\dot{S}_{ir}F_{rj}^T+S_{ir}\dot{F}_{rj}^T \tag{6.4.3}$$

Substituting (6.4.3) into (6.4.2) yields

$$\dot{f}_{iI}^{\text{int}}=\int_{\Omega_0}\frac{\partial N_I}{\partial X_j}(\dot{S}_{jr}F_{ir}+S_{jr}\dot{F}_{ir})\,d\Omega_0 \quad \text{or} \quad df_{iI}^{\text{int}}=\int_{\Omega_0}\frac{\partial N_I}{\partial X_j}(dS_{jr}F_{ir}+S_{jr}dF_{ir})d\Omega_0 \tag{6.4.4}$$

This shows that the rate (or increment) of the internal nodal forces consists of two parts:

1. The first part involves the rate of stress ($\dot{\mathbf{S}}$) and thus depends on the material response and leads to what is called the material tangent stiffness matrix which we denote by $\mathbf{K}^{\text{mat}}$.
2. The second part involves the current state of stress $\mathbf{S}$, and accounts for geometric effects of the deformation (including rotation and stretching). This term is called the geometric stiffness. It is also called the initial stress matrix to indicate the role of the existing state of stress. It is denoted by $\mathbf{K}^{\text{geo}}$.

The changes in the nodal forces due to these two effects are given analogous names, so (6.4.4) is written as

$$\dot{\mathbf{f}}^{\text{int}}=\dot{\mathbf{f}}^{\text{mat}}+\dot{\mathbf{f}}^{\text{geo}} \quad \text{or} \quad \dot{f}_{iI}^{\text{int}}=\dot{f}_{iI}^{\text{mat}}+\dot{f}_{iI}^{\text{geo}} \tag{6.4.5}$$

where

$$\dot{f}_{iI}^{\text{mat}}=\int_{\Omega_0}\frac{\partial N_I}{\partial X_j}F_{ir}\dot{S}_{jr}\,d\Omega_0, \quad \dot{f}_{iI}^{\text{geo}}=\int_{\Omega_0}\frac{\partial N_I}{\partial X_j}S_{jr}\dot{F}_{ir}\,d\Omega_0 \tag{6.4.6a, b}$$

6.4.2 Material Tangent Stiffness

To simplify the remaining development, we put the above expression into Voigt form; see Appendix 1 for details on this notation. Voigt form is convenient in developing the tangent stiffness matrices because the tensor of tangent moduli C_{ijkl} is fourth-order; this tensor cannot be handled readily by standard matrix operations. We can rewrite the material rate of the internal nodal forces, (6.4.6a), in Voigt notation as

$$\dot{\mathbf{f}}^{\text{mat}}=\int_{\Omega_0}\mathbf{B}_0^T\{\dot{\mathbf{S}}\}d\Omega_0 \tag{6.4.7}$$

where $\{\dot{\mathbf{S}}\}$ is the rate of the PK2 stress in Voigt column matrix form. It is stressed that (6.4.7) is identical to (6.4.6a); it is just a change in notation. The constitutive equation in rate form is

$$\dot{S}_{ij}=C_{ijkl}^{SE}\dot{E}_{kl} \quad \text{or} \quad \{\dot{\mathbf{S}}\}=[\mathbf{C}^{SE}]\{\dot{\mathbf{E}}\} \tag{6.4.8}$$

Recall (4.9.26), which relates the rate of the Green strain to the nodal velocities in Voigt notation, $\{\dot{\mathbf{E}}\} = \mathbf{B}_0\dot{\mathbf{d}}$. Substituting this into (6.4.8) and the result into (6.4.7) gives

$$\dot{\mathbf{f}}^{\text{int}}_{\text{mat}} = \int_{\Omega_0} \mathbf{B}_0^T[\mathbf{C}^{SE}]\,\mathbf{B}_0\,d\Omega_0\dot{\mathbf{d}} \quad \text{or} \quad d\mathbf{f}^{\text{int}}_{\text{mat}} = \int_{\Omega_0} \mathbf{B}_0^T[\mathbf{C}^{SE}]\,\mathbf{B}_0\,d\Omega_0 d\mathbf{d} \tag{6.4.9}$$

So the material tangent stiffness matrix is given by

$$\mathbf{K}^{\text{mat}} = \int_{\Omega_0} \mathbf{B}_0^T[\mathbf{C}^{SE}]\,\mathbf{B}_0\,d\Omega_0 \quad \text{or} \quad \mathbf{K}^{\text{mat}}_{IJ} = \int_{\Omega_0} \mathbf{B}_{0I}^T[\mathbf{C}^{SE}]\,\mathbf{B}_{0J}\,d\Omega_0 \tag{6.4.10}$$

The material tangent stiffness relates increments (or rates) of the internal nodal forces to the increment (or rate) of the displacements due to material response, which is reflected in the tangent moduli $\mathbf{C}^{SE}$. Its form is identical to the stiffness matrix in linear finite elements.

6.4.3 Geometric Stiffness

The geometric stiffness is obtained as follows. From the definition $B^0_{iI} = \partial N_I/\partial X_i$, and (6.4.4), we can write

$$\dot{f}^{\text{geo}}_{iI} = \int_{\Omega_0} B^0_{Ij}S_{jr}\dot{F}_{ir}\,d\Omega_0 = \int_{\Omega_0} B^0_{Ij}S_{jr}B^0_{rJ}\,d\Omega_0\dot{u}_{iJ} = \int_{\Omega_0} B^0_{Ij}S_{jr}B^0_{rJ}\,d\Omega_0\,\delta_{ik}\dot{u}_{kJ} \tag{6.4.11}$$

where in the second step we have used (4.9.7), $\dot{F}_{ir} = B^0_{rI}\dot{u}_{iI}$, and in the third step we have added a dummy unit matrix so that the component indices in $\dot{f}^{\text{geo}}_{iI}$ and $\dot{u}_{kJ}$ are not the same. Writing the above in matrix form gives

$$\dot{\mathbf{f}}^{\text{geo}}_I = \mathbf{K}^{\text{geo}}_{IJ}\dot{\mathbf{u}}_J \quad \text{where} \quad \mathbf{K}^{\text{geo}}_{IJ} = \mathbf{I}\int_{\Omega_0} \boldsymbol{B}_{0I}^T\mathbf{S}\boldsymbol{B}_{0J}\,d\Omega_0 \tag{6.4.12}$$

Note that the PK2 stress in the above is the tensor form, i.e. a square matrix. Each submatrix of the geometric stiffness matrix is a unit matrix. Therefore the geometric stiffness matrix is invariant with rotation, i.e. $\hat{\mathbf{K}}^{\text{geo}}_{IJ} = \mathbf{K}^{\text{geo}}_{IJ}$, where the superposed hat denotes the geometric stiffness matrix on rotated coordinate system.

These forms are easily converted to updated Lagrangian forms by letting the current configuration be the reference configuration, as in Section 4.9.2. From (4.9.29), we recall that taking the current configuration as the reference configuration gives $\mathbf{B}_0 = \mathbf{B}$, $\boldsymbol{B}_0 = \boldsymbol{B}$, $\mathbf{S} = \sigma$ and $d\Omega_0 = d\Omega$. Also, we note that when the current configuration is taken to be the reference configuration, $\mathbf{F} = \mathbf{I}$, $\mathbf{C}^{SE} = \mathbf{C}^{\sigma T}$. Expressions for $\mathbf{C}^{\sigma T}$ for different constitutive relations are given in Box 5.1. Thus, (6.4.10) and (6.4.12) become

$$\begin{aligned} \mathbf{K}^{\text{mat}}_{IJ} &= \int_{\Omega} \mathbf{B}_I^T[\mathbf{C}^{\sigma T}]\,\mathbf{B}_J\,d\Omega, \quad \mathbf{K}^{\text{mat}} = \int_{\Omega_0} \mathbf{B}^T[\mathbf{C}^{\sigma T}]\,\mathbf{B}\,d\Omega \\ \mathbf{K}^{\text{geo}}_{IJ} &= \mathbf{I}\int_{\Omega} \boldsymbol{B}_I^T\sigma\boldsymbol{B}_J\,d\Omega \end{aligned} \tag{6.4.13}$$

The integrand in the geometric stiffness is a scalar for given values of I and J, so (6.4.13) can be written as

$$\mathbf{K}_{IJ}^{\text{geo}} = \mathbf{I}H_{IJ} \quad \text{where} \quad H_{IJ} = \int_{\Omega} \boldsymbol{B}_I^T \sigma \boldsymbol{B}_J \, d\Omega \tag{6.4.14}$$

The tangent moduli for a large variety of materials are given Chapter 5. Tangent material stiffness matrices for specific finite elements are given in Examples 6.2 and 6.3.

The updated Lagrangian forms (6.4.13) are generally easier to use than the total Lagrangian forms, since $\mathbf{B}$ is more easily constructed than $\mathbf{B}_0$ and many material laws are developed in terms of rates of Cauchy stress. The material stiffness in total Lagrangian form can be combined with the geometric stiffness in updated Lagrangian form or vice versa. The numerical values of these matrices in the total and updated Lagrangian forms are identical, and the choice of which to use is a matter of convenience.

We next address the issue of symmetry of the tangent stiffness matrix. Symmetry is important because it speeds the solution of the equations, reduces storage requirements and simplifies stability analyses. From (6.4.10) it can be seen that the material tangent stiffness is symmetric when the Voigt form of the tangent modulus matrix $[\mathbf{C}^{SE}]$ is symmetric. The Voigt form is symmetric when the tensor tangent moduli C_{ijkl}^{SE} have major symmetry. Therefore, the material part of the tangent stiffness is symmetric when the tangent moduli possess major symmetry. Similar arguments hold for the updated Lagrangian form (6.4.13): the material tangent stiffness is symmetric when the tangent moduli $C_{ijkl}^{\sigma T}$ possess major symmetry.

The geometric stiffness given previously is always symmetric. Therefore, the tangent stiffness matrix $\mathbf{K}^{\text{int}}$ is symmetric whenever the tangent moduli possess major symmetry. Note that these conclusions pertain only to the specific rates of stress chosen in this derivation: $\dot{\mathbf{S}}$ and the Truesdell rate $\sigma^{\nabla T}$. For other objective rates, major symmetry of the tangent moduli is not necessarily sufficient for symmetry of the tangent stiffness matrix. For example, it is shown in Example 6.1 that the tangent stiffness for the material described by $\sigma^{\nabla J}=\mathbf{C}^{\sigma J}:\mathbf{D}$ is not symmetric when $\mathbf{C}^{\sigma J}$ has major symmetry.

6.4.4 *Alternative Derivations of Tangent Stiffness*

In this section the tangent stiffness matrix is derived in terms of the convected rate of the Kirchhoff stress. Many of the relations in nonlinear mechanics take on a particular elegance and simplicity when expressed in terms of the Kirchhoff stress.

The Kirchhoff stress is related to the nominal stress by $\boldsymbol{\tau}=\mathbf{F}\cdot\mathbf{P}$ (Box 3.2). Taking the time derivative of the preceding expression gives

$$\dot{\boldsymbol{\tau}} = \dot{\mathbf{F}}\cdot\mathbf{P}+\mathbf{F}\cdot\dot{\mathbf{P}} \tag{6.4.15}$$

Solving (6.4.15) for $\dot{\mathbf{P}}$ gives

$$\dot{\mathbf{P}} = \mathbf{F}^{-1}(\dot{\boldsymbol{\tau}}-\dot{\mathbf{F}}\cdot\mathbf{P}) = \mathbf{F}^{-1}(\dot{\boldsymbol{\tau}}-\mathbf{L}\cdot\mathbf{F}\cdot\mathbf{P}) \tag{6.4.16}$$

where the second step follows from (3.3.18), $\dot{\mathbf{F}} = \mathbf{L}\cdot\mathbf{F}$. Recognizing that $\boldsymbol{\tau}=\mathbf{F}\cdot\mathbf{P}$, the above can be simplified to:

$$\dot{\mathbf{P}} = \mathbf{F}^{-1}\cdot\,(\dot{\boldsymbol{\tau}}-\mathbf{L}\cdot\boldsymbol{\tau}) \tag{6.4.17}$$

Using (5.4.49) to relate the material rate of the Kirchhoff stress to its convected rate (or Lie derivative), $\boldsymbol{\tau}^{\nabla c}=\dot{\boldsymbol{\tau}}-\mathbf{L}\cdot\boldsymbol{\tau}-\boldsymbol{\tau}\cdot\mathbf{L}^T$, (6.4.17) becomes

$$\dot{\mathbf{P}}=\mathbf{F}^{-1}(\boldsymbol{\tau}^{\nabla c}+\boldsymbol{\tau}\cdot\mathbf{L}^T) \quad \text{or} \quad \dot{P}_{ji}=F_{jk}^{-1}\left(\tau_{ki}^{\nabla c}+\tau_{kl}L_{il}\right) \tag{6.4.18}$$

This is a canonical relationship that will often be useful. It cleanly separates the rate of the nominal stress into material and geometric parts.

Substituting (6.4.18) into (6.4.2) gives

$$\dot{f}_{iI}^{\text{int}}=\int_{\Omega_o}\frac{\partial N_I}{\partial X_j}\frac{\partial X_j}{\partial x_k}\left(\tau_{ki}^{\nabla c}+\tau_{kl}L_{il}\right)d\Omega_0=\int_{\Omega_o}\frac{\partial N_I}{\partial x_k}\left(\tau_{ki}^{\nabla c}+\tau_{kl}L_{il}\right)d\Omega_0 \tag{6.4.19a}$$

where the second expression follows from the first by the chain rule. This is the counterpart of (6.4.4) in terms of the PK2 stress. Subdividing into material and geometric parts as in (6.4.5) gives

$$\dot{f}_{iI}^{\text{mat}}=\int_{\Omega_o}N_{I,k}\tau_{ki}^{\nabla c}\,d\Omega_0,\quad \dot{f}_{iI}^{\text{geo}}=\int_{\Omega_o}N_{I,k}\tau_{kl}L_{il}\,d\Omega_0 \tag{6.4.19b}$$

This result can easily be transformed to an updated Lagrangian format with the integral over the current domain. From (3.2.23), $d\Omega=J\,d\Omega_0$, and the relation (5.4.22) between the convected rate of Kirchhoff stress and the Truesdell rate of Cauchy stress, ($\boldsymbol{\tau}^{\nabla c}=J\sigma^{\nabla T}$), (6.4.19a) yields

$$\dot{f}_{iI}=\int_{\Omega}N_{I,k}\left(\sigma_{ki}^{\nabla T}+\sigma_{kl}L_{il}\right)d\Omega \tag{6.4.20}$$

This is the updated Lagrangian counterpart of (6.4.4); (6.4.20) could also be obtained by making the current configuration the reference configuration (Exercise 6.2).

To complete the derivation of the material tangent stiffness matrix, it is necessary to introduce the constitutive relation to relate the convected stress rate to the nodal velocities. We write the constitutive relation (rate-independent material response) in the form (see Box 6.5)

$$\tau_{ij}^{\nabla c}=C_{ijkl}^{\tau}D_{kl} \tag{6.4.21}$$

Substituting this into the first part of (6.4.19b) gives

$$K_{ijIJ}^{\text{mat}}\dot{u}_{jJ}=\int_{\Omega_o}\frac{\partial N_I}{\partial x_k}\,C_{kijl}^{\tau}D_{jl}\,d\Omega_o \tag{6.4.22}$$

Recall that the rate-of-deformation tensor is the symmetric part of the spatial velocity gradient, $D_{kl}=v_{(k,\ell)}=\text{sym}\,(v_{kI}N_{I,\ell})$. Substituting this into (6.4.22) we obtain

$$\begin{aligned}K_{ijIJ}^{\text{mat}}\dot{u}_{jJ}&=\int_{\Omega_o}N_{I,k}C_{kijl}^{\tau}v_{j,l}\,d\Omega_0=\int_{\Omega_o}N_{I,k}C_{kijl}^{\tau}N_{J,l}\dot{u}_{jJ}\,d\Omega_o\\&=\int_{\Omega}N_{I,k}C_{kijl}^{\tau}N_{J,l}J\;\;d\Omega\,\dot{u}_{jJ}=\int_{\Omega}N_{I,k}C_{kijl}^{\sigma T}N_{J,l}\,d\Omega\dot{u}_{jJ}\end{aligned} \tag{6.4.23}$$

where in the second expression we have used $C^{\tau}_{kijl}D_{jl} = C^{\tau}_{kijl}v_{j,l}$ which follows from the minor symmetry of the tangent modulus matrix; in the third expression we have transformed to the current configuration, and the last follows from (5.4.51).

We now convert the previous to Voigt notation. Using the minor symmetries, (6.4.23) becomes

$$K^{\text{mat}}_{rsIJ} = \int_{\Omega} N_{I,k}\delta_{ri}C^{\sigma T}_{kijl}N_{J,l}\delta_{sj}\,d\Omega = \int_{\Omega} B_{ikrI}J^{-1}C^{\tau}_{kijl}B_{jlsJ}\,d\Omega \tag{6.4.24}$$

where (4.5.18) was used in the second step. The above in Voigt matrix form is

$$\mathbf{K}^{\text{mat}}_{IJ} = \int_{\Omega} \mathbf{B}^T_I[\mathbf{C}^{\sigma T}]\,\mathbf{B}_J\,d\Omega = \int_{\Omega} J^{-1}\mathbf{B}^T_I[\mathbf{C}^{\tau}]\,\mathbf{B}_J\,d\Omega \tag{6.4.25}$$

which is identical to the material tangent stiffness in (6.4.13).

Box 6.5 Jacobian of internal nodal forces (tangent stiffness matrix)

Material, $\mathbf{K}^{\text{mat}}_{IJ}$	Geometric, $\mathbf{K}^{\text{geo}}_{IJ}$	
$\mathbf{K}^{\text{mat}}_{IJ} = \int_{\Omega} \mathbf{B}^T_I[\mathbf{C}^{\sigma T}]\mathbf{B}_J\,d\Omega$	$\mathbf{K}^{\text{geo}}_{IJ} = \mathbf{I}\int_{\Omega} \boldsymbol{B}^T_I\boldsymbol{\sigma}\boldsymbol{B}_J\,d\Omega,$	updated
$= \int_{\Omega_0} \mathbf{B}^T_{0I}[\mathbf{C}^{SE}]\mathbf{B}_{0J}\,d\Omega$	$= \mathbf{I}\int_{\Omega_0} \boldsymbol{B}^T_{0I}\mathbf{S}\boldsymbol{B}_{0J}\,d\Omega_0,$	total
$= \int_{\Omega_0} \mathbf{B}^T_I[\mathbf{C}^{\tau}]\mathbf{B}_J\,d\Omega_0$	$= \mathbf{I}\int_{\Omega_0} \boldsymbol{B}^T_I\tau\boldsymbol{B}_J\,d\Omega_0$	
$= \int_{\Omega} \mathbf{B}^T_I[\mathbf{C}^{\sigma J} - \mathbf{C}^*]\mathbf{B}_J\,d\Omega$		
$K^{\text{mat}}_{rsIJ} = \int_{\Omega} B_{ikrI}C^{\sigma T}_{kijl}B_{jlsJ}\,d\Omega$	$K^{\text{geo}}_{rsIJ} = \int_{\Omega} B_{Ij}\sigma_{jk}B_{kJ}\,d\Omega\,\delta_{rs},$	updated
$= \int_{\Omega_0} B^{0T}_{ikrI}C^{SE}_{kijl}B^0_{jlsJ}\,d\Omega,$	$= \int_{\Omega_0} B^0_{Ij}S_{jk}B^0_{kJ}\,d\Omega_0\delta_{rs},$	total

6.4.5 *External Load Stiffness*

Follower loads are loads that change with the configuration of the body. They appear in many geometrically nonlinear problems. Pressure is a common example of a follower load. A pressure load is always normal to the surface, so as the surface moves the nodal external forces change even if the pressure is constant. These effects are accounted for in the Jacobian matrix of the nodal external forces, $\mathbf{K}^{\text{ext}}$, which is also called the load stiffness.

The load stiffness $\mathbf{K}^{\text{ext}}$ relates the rate of the external nodal forces to the nodal velocities. Consider a pressure field, $p(\mathbf{x}, t)$. The external nodal forces on a surface of element e are given by letting $\bar{\mathbf{t}} = -p\mathbf{n}$ in (B4.3.3):

$$\mathbf{f}^{\text{ext}}_I = -\int_{\Gamma} N_I p\mathbf{n}\,d\Gamma \tag{6.4.26}$$

Let the surface Γ be parametrized by two variables ξ and η. For a quadrilateral surface element, these variables can be the parent element coordinates on the biunit square. Since $\mathbf{n}\, d\Gamma = \mathbf{x},_{\xi} \times \mathbf{x},_{\eta} d\xi\, d\eta$ (see (E4.3.13)), the above becomes

$$\mathbf{f}_I^{\text{ext}} = -\int_{-1}^{1}\int_{-1}^{1} p(\xi,\eta) N_I(\xi,\eta)\, \mathbf{x},_{\xi} \times \mathbf{x},_{\eta}\, d\xi\, d\eta \tag{6.4.27}$$

Taking the time derivative of the previous gives

$$\dot{\mathbf{f}}_I^{\text{ext}} = -\int_{-1}^{1}\int_{-1}^{1} N_I(\dot{p}\mathbf{x},_{\xi} \times \mathbf{x},_{\eta} + p\mathbf{v},_{\xi} \times \mathbf{x},_{\eta} + p\mathbf{x},_{\xi} \times \mathbf{v},_{\eta})\, d\xi\, d\eta \tag{6.4.28}$$

The first term is the rate of change of the external forces due to the rate of change of the pressure. In many problems the rate of change of pressure is prescribed as part of the problem definition. In other problems, such as in fluid–structure interaction problems, the pressure may arise from changes of the geometry; these effects must then be linearized and added to the load stiffness. We omit this term in the following.

The last two terms represent the changes in the external nodal forces due to the changes in the direction of the surface and the area of the surface. These are related by the external load stiffness, so the RHS of (6.4.28) becomes

$$\mathbf{K}_{IK}^{\text{ext}} \mathbf{v}_K = -\int_{-1}^{1}\int_{-1}^{1} pN_I(\mathbf{v},_{\xi} \times \mathbf{x},_{\eta} + \mathbf{x},_{\xi} \times \mathbf{v},_{\eta})\, d\xi\, d\eta \tag{6.4.29}$$

At this point, it is convenient to switch to indicial notation. Taking the dot product of the above with the unit vector $\mathbf{e}_i$ gives

$$\begin{aligned} \mathbf{e}_i \cdot \mathbf{K}_{IK}^{\text{ext}} \mathbf{v}_K \equiv \mathbf{K}_{ikIJ}^{\text{ext}} v_{kJ} &= -\int_{-1}^{1}\int_{-1}^{1} pN_I(\mathbf{e}_i \cdot \mathbf{v},_{\xi} \times \mathbf{x},_{\eta} + \mathbf{e}_i \cdot \mathbf{x},_{\xi} \times \mathbf{v},_{\eta})\, d\xi\, d\eta \\ &= -\int_{-1}^{1}\int_{-1}^{1} pN_I(e_{ik\ell} v_{k,\xi} x_{\ell,\eta} + e_{ik\ell} x_{k,\xi} v_{\ell,\eta})\, d\xi\, d\eta \end{aligned} \tag{6.4.30}$$

Next we expand the velocity field in terms of the shape functions by $v_{i,\xi} = v_{iJ} N_{J,\xi}$ and interchange the indices on the second term to obtain

$$\mathbf{K}_{ikIJ}^{\text{ext}} v_{kJ} = -\int_{-1}^{1}\int_{-1}^{1} pN_I(e_{ik\ell} N_{J,\xi} x_{\ell,\eta} - e_{ik\ell} x_{\ell,\xi} N_{J,\eta})\, d\xi\, d\eta\, v_{kJ}$$

We define

$$H_{ik}^{\eta} \equiv e_{ik\ell} x_{\ell,\eta} \qquad H_{ik}^{\xi} = e_{ik\ell} x_{\ell,\xi} \tag{6.4.31}$$

Using these definitions in (6.4.30), we obtain

$$K^{\text{ext}}_{ijIJ} = -\int_{-1}^{1}\int_{-1}^{1} pN_I\left(N_{J,\xi}H^{\eta}_{ij} - N_{J,\eta}H^{\xi}_{ij}\right)d\xi d\eta \quad \text{or}$$
$$\mathbf{K}^{\text{ext}}_{IJ} = -\int_{-1}^{1}\int_{-1}^{1} pN_I\left(N_{J,\xi}\mathbf{H}^{\eta} - N_{J,\eta}\mathbf{H}^{\xi}\right)d\xi d\eta \tag{6.4.32}$$

Writing out the matrices $\mathbf{H}^{\xi}$ and $\mathbf{H}^{\eta}$:

$$\mathbf{K}^{\text{ext}}_{IJ} = -\int_{-1}^{1}\int_{-1}^{1} pN_I\left(N_{J,\xi}\begin{bmatrix} 0 & z,_{\eta} & -y,_{\eta} \\ -z,_{\eta} & 0 & x,_{\eta} \\ y,_{\eta} & -x,_{\eta} & 0 \end{bmatrix} - N_{J,\eta}\begin{bmatrix} 0 & z,_{\xi} & -y,_{\xi} \\ -z,_{\xi} & 0 & x,_{\xi} \\ y,_{\xi} & -x,_{\xi} & 0 \end{bmatrix}\right)d\xi d\eta \tag{6.4.33}$$

The above applies to any surface which is generated from a biunit parent surface loaded by a pressure p. For a surface with a triangular parent element, this is expressed in terms of the area coordinates and the limits of integration are changed. The load stiffness reflects the effect of the change in geometry on the external nodal forces: alterations both in the direction of the loaded surfaces and in the size of the surface will change the nodal forces. The load stiffness can also be derived by using Nanson's formula and the derivative of a surface integral (see Exercise 6.1).

It is immediately apparent from (6.4.33) that the submatrices of the load stiffness matrix are not symmetric, so the total Jacobian matrix generally is not symmetric in the presence of follower forces. However, it can be shown that for a closed structure in a constant pressure field, the assembled external load stiffness is symmetric.

Example 6.2 Two-node rod element Consider the two-node rod element shown in Figure 4.9, Example 4.6. The rod is in a state of uniaxial stress. The $\hat{x}$-axis lies along the axis of the rod and rotates with the rod, that is, it is a corotational coordinate. The only nonzero Cauchy stress component is $\hat{\sigma}_{11} \equiv \hat{\sigma}_{xx}$. The tangent and load stiffness are derived in the updated Lagrangian form, that is, in the current configuration.

Material tangent stiffness matrix The tangent stiffness matrix for a rate-independent material in the current configuration is given by (6.4.13), which we write in the local coordinate system as

$$\hat{\mathbf{K}}^{\text{mat}} = \int_{\Omega}\hat{\mathbf{B}}^T[\hat{\mathbf{C}}^{\sigma T}]\hat{\mathbf{B}}d\Omega \tag{E6.2.1}$$

The **B** matrix from (E4.6.4) is expanded to a 4×1 matrix by adding zeros to reflect that the rate-of-deformation is independent of the $\hat{y}$ components of the nodal velocities and $[C^{\sigma T}]=[E^{\sigma T}]$, which is given in Box 5.1 for uniaxial stress, so (E6.2.1) becomes

$$\hat{\mathbf{K}}^{\text{mat}} = \int_0^1 \frac{1}{\ell}\begin{Bmatrix} -1 \\ 0 \\ 1 \\ 0 \end{Bmatrix}[E^{\sigma T}]\frac{1}{\ell}[-1 \quad 0 \quad +1 \quad 0]\; A\ell d\xi \tag{E6.2.2}$$

If we assume $E^{\sigma T}$ is constant in the element, then

$$\hat{\mathbf{K}}^{\text{mat}} = \frac{AE^{\sigma T}}{\ell}\begin{bmatrix} +1 & 0 & -1 & 0 \\ 0 & 0 & 0 & 0 \\ -1 & 0 & +1 & 0 \\ 0 & 0 & 0 & 0 \end{bmatrix} \tag{E6.2.3}$$

This is identical to the linear stiffness matrix for a rod if $E^{\sigma T}$ is replaced by Young's modulus E. The material tangent stiffness relating global components of nodal internal forces and velocities is given by (4.5.42):

$$\mathbf{K}^{\text{mat}} = \mathbf{T}^T\hat{\mathbf{K}}^{\text{mat}}\mathbf{T} \tag{E6.2.4a}$$

where $\mathbf{T}$ is given by

$$\mathbf{T} = \begin{bmatrix} \cos\theta & \sin\theta & 0 & 0 \\ -\sin\theta & \cos\theta & 0 & 0 \\ 0 & 0 & \cos\theta & \sin\theta \\ 0 & 0 & -\sin\theta & \cos\theta \end{bmatrix} \tag{E6.2.4b}$$

$$\mathbf{K}^{\text{mat}} = \frac{AE^{\sigma T}}{\ell}\begin{bmatrix} \cos^2\theta & \cos\theta\sin\theta & -\cos^2\theta & -\cos\theta\sin\theta \\ & \sin^2\theta & -\cos\theta\sin\theta & -\sin^2\theta \\ & & \cos^2\theta & \cos\theta\sin\theta \\ \text{symmetric} & & & \sin^2\theta \end{bmatrix} \tag{E6.2.5}$$

Geometric stiffness matrix Consider a coordinate system that at time t coincides with the axis of the bar but is fixed in time. Note that since the coordinate system is fixed in the orientation shown in Figure 4.9, the rotation corrections of an objective rate must be considered. We will use the Truesdell rate. We could also consider the $\hat{x}, \hat{y}$ coordinate system corotational and derive the geometric stiffness by accounting for the change of the transformation matrix $\mathbf{T}$. Such derivations are given in Crisfield (1991). The result is identical, since the same mechanical effect is linearized, but the latter derivation is generally more difficult. The geometric stiffness matrix is given by (6.4.13):

$$\hat{\mathbf{K}}_{IJ} = \hat{H}_{IJ}\mathbf{I}, \quad \hat{\mathbf{H}} = \int_{\Omega} \hat{\boldsymbol{B}}^T \boldsymbol{\sigma} \hat{\boldsymbol{B}} d\Omega \tag{E6.2.6}$$

where the geometric stiffness has been expressed in the local coordinate system for simplicity. Since $\boldsymbol{B} = \mathbf{B}$ from (E4.6.4), it follows that

$$\hat{\mathbf{H}} = \int_{\Omega} \frac{1}{\ell}\begin{bmatrix} -1 \\ +1 \end{bmatrix}\left[\hat{\sigma}_{xx}\right]\frac{1}{\ell}\left[-1 \quad +1\right] d\Omega \tag{E6.2.7}$$

Assuming that the stress is constant gives

$$\hat{\mathbf{H}} = \frac{\hat{\sigma}_{xx}A}{\ell}\begin{bmatrix} +1 & -1 \\ - & +1 \end{bmatrix} \tag{E6.2.8}$$

Expanding this by (E6.2.6), we obtain

$$\hat{\mathbf{K}}^{\text{geo}} = \frac{A\hat{\sigma}_{xx}}{\ell}\begin{bmatrix} +1 & 0 & -1 & 0 \\ 0 & +1 & 0 & -1 \\ -1 & 0 & +1 & 0 \\ 0 & -1 & 0 & +1 \end{bmatrix} \tag{E6.2.9}$$

By the transformation formula, (4.5.42), it can be show that the geometric stiffness is independent of the orientation of the rod: $\mathbf{K}^{\text{geo}} = \mathbf{T}^T\hat{\mathbf{K}}^{\text{geo}}\mathbf{T} = \hat{\mathbf{K}}^{\text{geo}}$. The total tangent stiffness is given by the sum of the material and geometric stiffnesses:

$$\mathbf{K}^{\text{int}} = \mathbf{K}^{\text{mat}} + \mathbf{K}^{\text{geo}} \tag{E6.2.10}$$

As can be seen from the above, the tangent stiffness matrix is symmetric.

Load stiffness The load stiffness for the rod is developed from (6.4.33). We write only the nonzero terms, noting that $N_{I,\eta}=0$ and $x,_{\eta}=y,_{\eta}=0$, since the shape function is only a function of ξ, $\xi \in [0, 1]$. For simplicity, we first evaluate (6.4.33) in the corotational system, which gives

$$\hat{\mathbf{K}}_{IJ}^{\text{ext}} = -\int_0^1 pN_I N_{J,\xi}\begin{bmatrix} 0 & z,_{\eta} \\ -z,_{\eta} & 0 \end{bmatrix} d\xi \tag{E6.2.11}$$

In this, $z,_{\eta}$ is the width of the element a, i.e. $z,_{\eta}=a$.

$$\hat{\mathbf{K}}_{IJ}^{\text{ext}} = -\int_0^1 pN_I N_{J,\xi}\begin{bmatrix} 0 & 1 \\ -1 & 0 \end{bmatrix} a\, d\xi \tag{E6.2.12}$$

Let

$$H_{IJ} = \int_0^1 N_I N_{J,\xi}\, d\xi = \int_0^1 \begin{bmatrix} 1-\xi \\ \xi \end{bmatrix}[-1 \quad +1]\, d\xi = \frac{1}{2}\begin{bmatrix} -1 & 1 \\ -1 & 1 \end{bmatrix} \tag{E6.2.13}$$

Then if the pressure is constant, from (E6.2.13) and (E6.2.12) we have

$$\hat{\mathbf{K}}_{IJ}^{\text{ext}} = -paH_{IJ}\begin{bmatrix} 0 & 1 \\ -1 & 0 \end{bmatrix} \tag{E6.2.14}$$

Writing out (E6.2.14) gives

$$\hat{\mathbf{K}}^{\text{ext}} = -\frac{pa}{2}\begin{bmatrix} 0 & -1 & 0 & 1 \\ 1 & 0 & -1 & 0 \\ 0 & -1 & 0 & 1 \\ 1 & 0 & -1 & 0 \end{bmatrix} \tag{E6.2.15}$$

This matrix is also invariant with rotation, i.e. $\mathbf{K}^{\text{ext}} = \mathbf{T}^T\hat{\mathbf{K}}^{\text{ext}}\mathbf{T} = \hat{\mathbf{K}}^{\text{ext}}$.

Material tangent stiffness matrix in total Lagrangian form The material tangent stiffness matrix for a rate-independent material is given in total Lagrangian form by (6.4.10):

$$\mathbf{K}^{\text{mat}} = \int_{\Omega} \mathbf{B}_0^T \left[\mathbf{C}^{SE}\right] \mathbf{B}_0 \, d\Omega_0 \tag{E6.2.16}$$

Using the **B** matrix from (E4.8.10) and $\mathbf{C}^{SE}$ as given by Box 5.1, we obtain

$$\mathbf{K}^{\text{mat}} = \int_0^1 \frac{\ell}{\ell_0^2}\begin{Bmatrix} -\cos\theta \\ -\sin\theta \\ \cos\theta \\ \sin\theta \end{Bmatrix}[E^{SE}]\frac{\ell}{\ell_0^2}[-\cos\theta \quad -\sin\theta \quad \cos\theta \quad \sin\theta]\, A_0\ell_0\, d\xi \tag{E6.2.17}$$

where the material constant E^{SE} relates the rate of the PK2 stress to the rate of the Green strain in a uniaxial state of stress. If we assume E^{SE} is constant in the element, then

$$\mathbf{K}^{\text{mat}} = \frac{A_0 E^{SE}}{\ell_0}\left(\frac{\ell}{\ell_0}\right)^2 \begin{bmatrix} \cos^2\theta & \cos\theta\sin\theta & -\cos^2\theta & -\cos\theta\sin\theta \\ & \sin^2\theta & -\cos\theta\sin\theta & -\sin^2\theta \\ & & \cos^2\theta & \cos\theta\sin\theta \\ \text{symmetric} & & & \sin^2\theta \end{bmatrix} \tag{E6.2.18}$$

By transforming the material modulus, we show that the above is identical to (E6.2.5). Referring to the equation following (E5.1.10) the relation between the moduli is

$$E^{SE} = \frac{J}{\lambda^4}E^{\sigma T} = \frac{(A\ell/A_0\ell_0)}{(\ell/\ell_0)^4}E^{\sigma T} = \frac{A\ell_0^3}{A_0\ell^3}E^{\sigma J}$$

Substituting this into (E6.2.18) gives (E6.2.5).

Geometric stiffness matrix in total Lagrangian form The geometric stiffness is developed from (6.4.12):

$$\mathbf{K}_{IJ}^{\text{geo}} = H_{IJ}\mathbf{I}, \quad \mathbf{H} = \int_{\Omega} \boldsymbol{B}_0^T \mathbf{S} \boldsymbol{B}_0 \, d\Omega_0 \tag{E6.2.19}$$

where the $\boldsymbol{B}_0$ matrix is given in (E4.8.5), so

$$\mathbf{H} = \int_{\Omega_0} \frac{1}{\ell_0}\begin{bmatrix}-1\\+1\end{bmatrix}[S_{11}]\frac{1}{\ell_0}[-1 \quad +1]d\Omega_0 \tag{E6.2.20}$$

Assuming that the stress is constant gives

$$\mathbf{H} = \frac{S_{11}A_0}{\ell_0}\begin{bmatrix}+1 & -1\\-1 & +1\end{bmatrix} \tag{E6.2.21}$$

Substituting (E6.2.21) into (E6.2.19), we obtain the geometric stiffness

$$\mathbf{K}^{\text{geo}} = \frac{A_0 S_{11}}{\ell_0}\begin{bmatrix}+1 & 0 & -1 & 0\\0 & +1 & 0 & -1\\-1 & 0 & +1 & 0\\0 & -1 & 0 & +1\end{bmatrix} \tag{E6.2.22}$$

The above holds for any orientation of the element. The total tangent stiffness is then given by the sum of the material and geometric stiffnesses.

Example 6.3 Three-node triangular element We consider the three-node triangle in two dimensions, as in Example 4.1. The element is in a state of plane strain. The only non-zero velocity components are v_x and v_y and their derivatives with respect to z vanish. The material tangent stiffness matrix is derived. The geometric tangent stiffness matrix, which is independent of material response, is then developed.

Material tangent stiffness matrix The material tangent stiffness matrix for a rate-independent material is given by (6.4.25):

$$\mathbf{K}^{\text{mat}} = \int_A \mathbf{B}^T[\mathbf{C}^{\sigma T}]\mathbf{B}\,dA \tag{E6.3.1}$$

where A is the current area of the element; we have set the thickness $a=1$. The tangent modulus matrix in Voigt matrix form is

$$[\mathbf{C}^{\sigma T}] = \begin{bmatrix}C^{\sigma T}_{1111} & C^{\sigma T}_{1122} & C^{\sigma T}_{1112}\\C^{\sigma T}_{2211} & C^{\sigma T}_{2222} & C^{\sigma T}_{2212}\\C^{\sigma T}_{1211} & C^{\sigma T}_{1222} & C^{\sigma T}_{1212}\end{bmatrix} \tag{E6.3.2}$$

Substituting (E6.3.2) and the **B** matrix given in (E4.1.15) into (E6.3.1) gives

$$\mathbf{K}^{\text{mat}} = \int_A \left(\frac{1}{2A}\right)^2 \begin{bmatrix} y_{23} & 0 & x_{32} \\ 0 & x_{32} & y_{23} \\ y_{31} & 0 & x_{13} \\ 0 & x_{13} & y_{31} \\ y_{12} & 0 & x_{21} \\ 0 & x_{21} & y_{12} \end{bmatrix} \begin{bmatrix} C_{1111}^{\sigma T} & C_{1122}^{\sigma T} & C_{1112}^{\sigma T} \\ C_{2211}^{\sigma T} & C_{2222}^{\sigma T} & C_{2212}^{\sigma T} \\ C_{1211}^{\sigma T} & C_{1222}^{\sigma T} & C_{1212}^{\sigma T} \end{bmatrix} \times \begin{bmatrix} y_{23} & 0 & y_{31} & 0 & y_{12} & 0 \\ 0 & x_{32} & 0 & x_{13} & 0 & x_{21} \\ x_{32} & y_{23} & x_{13} & y_{31} & x_{21} & y_{12} \end{bmatrix} dA \tag{E6.3.3}$$

This integrand is frequently constant. In that case, the tangent stiffness is the product of the integrand with the area.

Geometric stiffness matrix The geometric stiffness matrix is given by (6.4.13):

$$\mathbf{K}_{IJ}^{\text{geo}} = \mathbf{I}_{2\times 2} \int_A \boldsymbol{B}_I^T \boldsymbol{\sigma} \boldsymbol{B}_J \, dA = \mathbf{I}_{2\times 2} H_{IJ} \tag{E6.3.4}$$

From Example 4.1,

$$\boldsymbol{B} = \frac{1}{2A} \begin{bmatrix} y_{23} & y_{31} & y_{12} \\ x_{32} & x_{13} & x_{21} \end{bmatrix} \tag{E6.3.5}$$

Substituting (E6.3.5) into (E6.3.4) and assuming the integrand to be constant gives

$$\mathbf{H} = \frac{1}{4A} \begin{bmatrix} y_{23} & x_{32} \\ y_{31} & x_{13} \\ y_{12} & x_{21} \end{bmatrix} \begin{bmatrix} \boldsymbol{\sigma}_{xx} & \boldsymbol{\sigma}_{xy} \\ \boldsymbol{\sigma}_{xy} & \boldsymbol{\sigma}_{yy} \end{bmatrix} \begin{bmatrix} y_{23} & y_{31} & y_{12} \\ x_{32} & x_{13} & x_{21} \end{bmatrix} \tag{E6.3.6}$$

The geometric stiffness matrix is

$$\mathbf{K}^{\text{geo}} = \frac{1}{4A} \begin{bmatrix} H_{11} & 0 & H_{12} & 0 & H_{13} & 0 \\ 0 & H_{11} & 0 & H_{12} & 0 & H_{13} \\ H_{21} & 0 & H_{22} & 0 & H_{23} & 0 \\ 0 & H_{21} & 0 & H_{22} & 0 & H_{23} \\ H_{31} & 0 & H_{32} & 0 & H_{33} & 0 \\ 0 & H_{31} & 0 & H_{32} & 0 & H_{33} \end{bmatrix} \tag{E6.3.7}$$

The geometric stiffness matrix is independent of material response. As can be seen from (E6.3.6–7), it depends only on the current stress state and the current geometry of the element.

Jaumann rate When the constitutive equation is expressed in terms of the Jaumann rate, the tangent stiffness matrix changes due to the alteration in the tangent moduli as given in Box 5.1. This change may be incorporated in two ways:

1. change the material tangent stiffness by replacing $\mathbf{C}^{\sigma T}$ by (5.4.27);
2. incorporate the additional stress-dependent terms in the geometric stiffness.

The first approach is easier, but is inappropriate when the tangent stiffness is used for critical point estimates as described in Section 6.5.8–9. We apply the first approach here. Writing the matrix $\mathbf{C}^*$ given in (5.4.28) in Voigt notation:

$$[\mathbf{C}^*] = \begin{bmatrix} \sigma_{xx} & -\sigma_{xx} & \sigma_{xy} \\ -\sigma_{yy} & \sigma_{yy} & \sigma_{xy} \\ 0 & 0 & \frac{1}{2}(\sigma_{xx}+\sigma_{yy}) \end{bmatrix} \tag{E6.3.8}$$

This matrix is subtracted from $\mathbf{C}^{\sigma T}$ in (E6.3.3) to obtain the material tangent stiffness for the Jaumann rate. The geometric stiffness is unchanged.

In the second approach, the material tangent stiffness is unchanged and the geometric stiffness is given by

$$\mathbf{K}_{IJ}^{\text{geo}} = \int_A (\mathbf{I}_{2\times2}\boldsymbol{B}_I^T \sigma \boldsymbol{B}_J - \mathbf{B}_I^T[\mathbf{C}^*]\mathbf{B}_J)\,dA \tag{E6.3.9}$$

In both cases, the tangent stiffness matrix is not symmetric if the tangent moduli for the Jaumann rate have major symmetry.

6.4.6 *Directional Derivatives*

Four difficulties arise in applying the traditional Newton method to solid mechanics problems:

1. The nodal forces are not continuously differentiable functions of the nodal displacements for materials such as elastic–plastic materials.
2. For path-dependent materials, the classical Newton method pollutes the constitutive models since the intermediate solutions to the linear problem in the iterative procedure, $\mathbf{d}_\nu$, are not part of the actual load path.
3. For large incremental rotations and deformation, the linearized increments introduce a substantial error.
4. Derivatives and integrals need to be discretized; the tangent for the discrete forms is step-size dependent.

In order to overcome these difficulties, the Newton method is often modified as follows:

1. Directional derivatives, also called Gateaux derivatives, are used to develop the tangent stiffness.
2. A secant method is used instead of a tangent method and the last converged solution is used as the iteration point.
3. Formulas depending on increment size are used to relate the increments of forces and displacements.

To illustrate the need for directional derivatives in constructing the Jacobian for elastic–plastic materials, consider the following example. The two-bar truss shown in Figure 6.5 has been loaded so that the stresses in both bars are compressive and equal, and both bars are at the compressive yield stress. For simplicity, we consider only material nonlinearities and neglect geometric nonlinearities. If an arbitrary load increment $\Delta\mathbf{f}_1^{\text{ext}}$ is applied to node 1, the tangent stiffness matrix will depend on the incremental displacement $\Delta\mathbf{u}_1$ since the change of the internal nodal forces depends on the displacement increment. The residual is not a continuously differentiable function of the incremental nodal displacements. In this case, there are four lines of discontinuity in the Jacobian, as shown in Figure 6.5. These result from the fact that if the displacement increment results in a tensile strain increment, then the bar unloads elastically, so the tangent modulus changes from the elastic modulus E to the plastic modulus H_p.

The internal nodal forces in the current configuration are

$$\begin{Bmatrix} f_{x1} \\ f_{y1} \end{Bmatrix}^{\text{int}} = A\sigma_0 \begin{Bmatrix} 0 \\ 2\sin\theta \end{Bmatrix} \tag{6.4.34}$$

where σ_0 is the current yield stress; (6.4.34) is obtained by assembling the internal nodal forces for rod elements as given by (E4.6.6). For each rod element, there are two possibilities

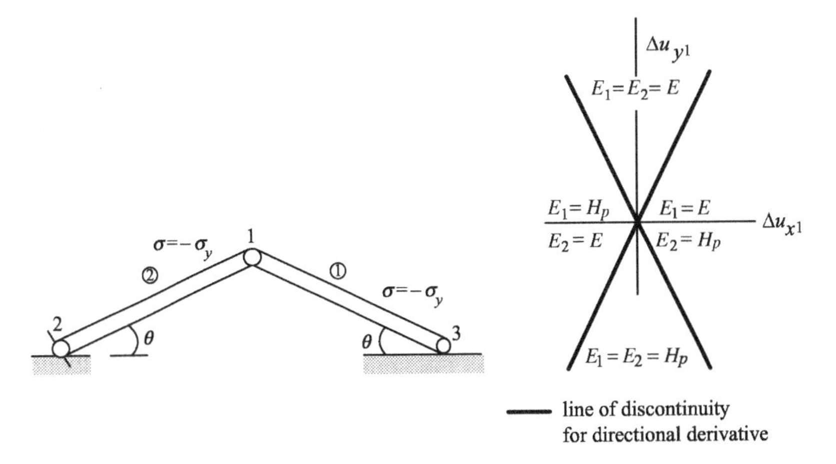

Figure 6.5 A two-bar truss with both bars in compressive yield and the four quadrants of directional derivatives

depending on the incremental displacement: either the rod continues to load with a plastic modulus, or it unloads with an elastic modulus. As a result, the tangent stiffness in this configuration can take on four different values.

The regimes of different nodal force behavior are shown in Figure 6.5. Obviously, a standard derivative cannot be evaluated since it has four different values. The tangent stiffness for displacement increments in the four quadrants is given by the following:

in quadrant 1:

$$\mathbf{K}^{\text{int}} = \frac{AE}{\ell}\begin{bmatrix} 2\cos^2\theta & 0 \\ 0 & 2\sin^2\theta \end{bmatrix} \tag{6.4.35}$$

in quadrant 2:

$$\mathbf{K}^{\text{int}} = \frac{A}{\ell}\begin{bmatrix} (E+H_p)\cos^2\theta & (E-H_p)\sin\theta\cos\theta \\ (E-H_p)\sin\theta\cos\theta & (E+H_p)\sin^2\theta \end{bmatrix} \tag{6.4.36}$$

in quadrant 3:

$$\mathbf{K}^{\text{int}} = \frac{AH_p}{\ell}\begin{bmatrix} 2\cos^2\theta & 0 \\ 0 & 2\sin^2\theta \end{bmatrix} \tag{6.4.37}$$

in quadrant 4:

$$\mathbf{K}^{\text{int}} = \frac{A}{\ell}\begin{bmatrix} (E+H_p)\cos^2\theta & (H_p-E)\sin\theta\cos\theta \\ (H_p-E)\sin\theta\cos\theta & (E+H_p)\sin^2\theta \end{bmatrix} \tag{6.4.38}$$

To deal with this type of behavior, a directional derivative must be used. The directional derivative is usually defined as a differential:

$$d_{\mathbf{g}}f(\mathbf{d}) \equiv d_{\mathbf{g}}f = \lim_{\varepsilon\to 0}\frac{f(\mathbf{d}+\varepsilon\mathbf{g}) - f(\mathbf{d})}{\varepsilon} = \frac{d}{d_\varepsilon}\bigg|_{\varepsilon=0} f(\mathbf{d}+\varepsilon\mathbf{g}) \tag{6.4.39}$$

The subscript $\mathbf{g}$ gives the direction. The notations $Df(\mathbf{d})[\mathbf{g}]$ and $Df(\mathbf{d})\cdot\mathbf{g}$ are often used for the directional derivative in the finite element literature.

6.4.7 *Algorithmically Consistent Tangent Stiffness*

The continuum tangent modulus relates rates, or infinitesimal increments, of stress and strain. In contrast, an algorithmic modulus relates *finite* increments of stress and strain. When the incremental stress–strain relation is obtained through consistent linearization of the stress update algorithm, the algorithmic modulus is called the *consistent* algorithmic modulus.

The algorithmic moduli should be used when the solution is not smooth. Examples of algorithmic moduli for implicit and semi-implicit backward Euler stress updates and for rate-independent and rate-dependent materials are given in Chapter 5 (see (5.9.38), (5.9.41), and (5.9.58–59), for example).

The standard Newton procedure is not suitable for algorithmic moduli. Instead, a secant Newton method must be used. The essential difference between the secant and STANDARD Newton methods is that, during the secant Newton iteration, all variables are updated from their values at the end of the previous time-step, i.e., the last converged point, as opposed to the last iteration. This avoids non-converged values of stress and internal variables from erroneously driving the constitutive equation in path-dependent materials. The secant Newton method is given in Box 6.6 for an equilibrium solution method. Implicit time integration schemes are similarly modified. See Simo and Hughes (1998) and Hughes and Pister (1978) for more details.

Box 6.6 Flowchart for equilibrium solution: Newton method with algorithmic moduli

1. Initial conditions and initialization: set $\mathbf{d}^0=0$; σ^0; $n=0$; $t=0$
2. Newton iterations for load increment $n+1$:
   (a) Set $\mathbf{d}_{\text{new}}=\mathbf{d}^n$
   (b) *getforce* computes $\mathbf{f}(\mathbf{d}_{\text{new}}, t^{n+1})$; $\mathbf{r}(\mathbf{d}_{\text{new}}, t^{n+1})$
      i. Compute new stresses and internal variables by stress update algorithm: $\boldsymbol{\sigma}_{\text{new}}=\boldsymbol{\sigma}(\mathbf{d}_{\text{new}}-\mathbf{d}^n, \boldsymbol{\sigma}^n, \mathbf{q}^n)$, $\mathbf{q}_{\text{new}}=\mathbf{q}(\mathbf{d}_{\text{new}}-\mathbf{d}^n, \boldsymbol{\sigma}^n, \mathbf{q}^n)$; update is from the converged values at time n
      ii. compute $\mathbf{f}$ and $\mathbf{r}$
   (c) Compute $\mathbf{A}(\mathbf{d}_{\text{new}})$; use $\boldsymbol{\sigma}_{\text{new}}$ and $\mathbf{q}_{\text{new}}$ in forming $\mathbf{K}^{\text{int}}=\mathbf{K}^{\text{mat}}+\mathbf{K}^{\text{geo}}$; compute $\mathbf{K}^{\text{mat}}$ with the algorithmic modulus $\mathbf{C}^{\text{alg}}$ (see Chapter 5)
   (d) Modify $\mathbf{A}(\mathbf{d})$ for essential boundary conditions
   (e) Solve linear equations $\Delta\mathbf{d}=-\mathbf{A}^{-1}\mathbf{r}$
   (f) $\mathbf{d}_{\text{new}} \leftarrow \mathbf{d}_{\text{new}}+\Delta\mathbf{d}$
   (g) Check convergence criterion; if not met, go to 2(a).
3. (a) Update displacements, stress and internal variables:
   (b) Update step count and time:

$$\mathbf{d}^{n+1}=\mathbf{d}_{\text{new}}, \boldsymbol{\sigma}^{n+1}=\boldsymbol{\sigma}_{\text{new}}, \mathbf{q}^{n+1}=\mathbf{q}_{\text{new}}$$

4. Output; if simulation not complete, go to 2.

6.5 Stability and Continuation Methods

6.5.1 Stability

In nonlinear problems, stability of solutions is of considerable interest. There are many definitions of stability: stability is a concept that depends on the observer and his objectives. However, some general definitions are widely accepted. We will here describe a theory of

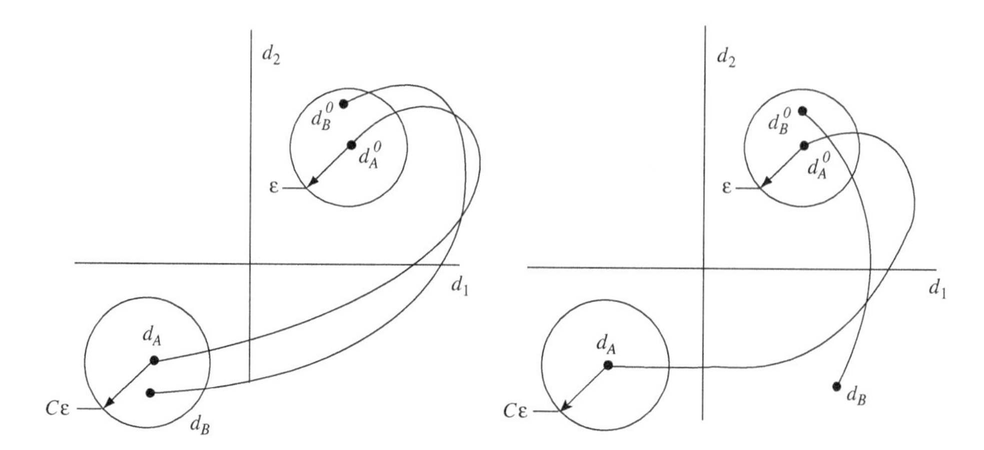

Figure 6.6 Solution trajectories for a stable system (left) and an unstable system (right)

stability that originates from Liapunov and is widely used throughout mathematical analysis; see Seydel (1994) for a lucid account of its application to a variety of computational problems and Bazant and Cedolin (1991) for a comprehensive treatment. We will focus on its application to finite element methods.

We will first give a definition of stability and explore its implications. Consider a process that is governed by an evolution equation such as the equations of motion or heat conduction. Let the solution for the initial conditions $\mathbf{d}_A(0) = \mathbf{d}_A^0$ be denoted by $\mathbf{d}_A(t)$. Now consider solutions for initial conditions $\mathbf{d}_B(0) = \mathbf{d}_B^0$, where $\mathbf{d}_B^0$ are small perturbations of $\mathbf{d}_A^0$. This means that $\mathbf{d}_B^0$ is close to $\mathbf{d}_A^0$ in some norm (to be specific we will use the ℓ_2 vector norm):

$$\left\|\mathbf{d}_A^0 - \mathbf{d}_B^0\right\|_{\ell_2} \leq \varepsilon \tag{6.5.1}$$

A solution is stable if for all initial conditions that satisfy (6.5.1), the solutions satisfy

$$\left\|\mathbf{d}_A(t) - \mathbf{d}_B(t)\right\|_{\ell_2} \leq C\varepsilon \quad \forall t > 0 \tag{6.5.2}$$

Note that all initial conditions which satisfy (6.5.1) lie in a hypersphere (a circle in two dimensions) centered at $\mathbf{d}_A^0$; a simpler way to say this is: 'the initial conditions lie in a ball around $\mathbf{d}_A^0$. According to this definition, the solution is stable if all solutions $\mathbf{d}_B(t)$ lie in a ball around the solution $\mathbf{d}_A(t)$ for any time whenever $\mathbf{d}_B^0$ is a small perturbation of $\mathbf{d}_A^0$. This definition is illustrated for a system with two dependent variables in Figure 6.6. The left-hand side shows the behavior of a stable system. Here we have only shown two solutions resulting from perturbations of the initial data, since it is impossible to show an infinite number of solutions, but for the system to be stable, any solution with initial conditions in the ball about $\mathbf{d}_A^0$ must be in ball about $\mathbf{d}_A(t)$. The right-hand side shows an unstable system.

It suffices for a single solution starting in the ball about $\mathbf{d}_A^0$ to diverge from $\mathbf{d}_A(t)$ to indicate an unstable solution.

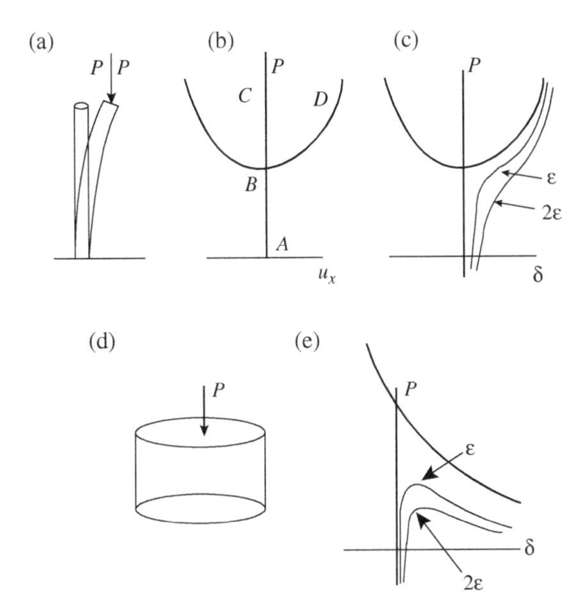

Figure 6.7 Examples of instability: (a) beam stability; (b) equilibrium branches of perfect beam; (c) equilibrium branches of imperfect beam with two imperfections; (d) cylindrical shell in compression; (e) equilibrium branches of cylindrical shell with two imperfections

The relation of this definition with intuitive notions of stability is illustrated in the following examples. Consider the process of a beam loaded axially as shown in Figure 6.7. If we perturb the location of the load by a distance ε or 2ε, then the equilibrium paths are as shown in Figure 6.7(c). It can be seen that when the load is below the buckling load, the paths for different initial conditions remain close to *AC*. However, when the load exceeds the buckling load, the solutions for different initial conditions diverges. Therefore any process in which the load exceeds the buckling load is unstable. Although it is not shown in the figure, the direction of the unstable branch depends on the sign of the initial imperfection; only one direction is shown in the figure.

As an aside, we remark that when the beam is perfectly straight, the numerical solution usually remains on the path *AC* shown in Figure 6.7(b). The lateral displacement is zero even when the load exceeds the buckling load. If you do not believe this, try it. In a simulation, a straight beam will usually not buckle in an incremental static solution or a dynamic solution, regardless of whether explicit or implicit integration is used. Only when roundoff error introduces a 'numerical imperfection' or when an imperfection is introduced in the data will the straight beam buckle in a simulation that is, an imperfection is needed to break the symmetry.

In the stability analysis of systems, the *stability of an equilibrium state* is usually considered. The notion of stability of a state differs somewhat from the stability of a process. The stability of an equilibrium state is determined by examining whether perturbations applied to that equilibrium state grow. The outcomes are less intuitive than the notion of the equilibrium of a process. For example, if we consider the beam in Figure 6.7, then it can be shown that any perturbation of an equilibrium state on the branches *AB* and *BD* will not grow. In other words,

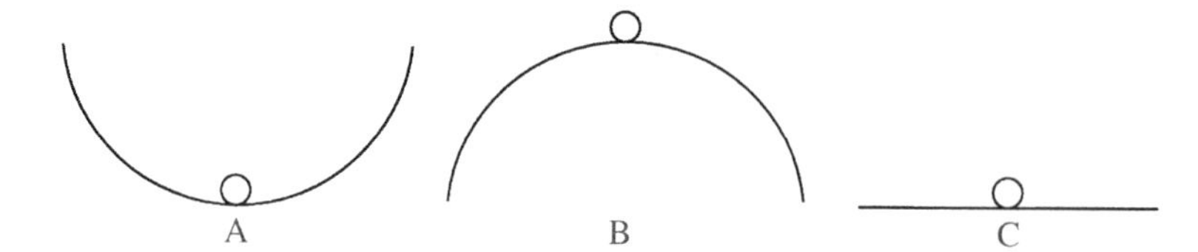

Figure 6.8 Stability conditions for a ball on a surface: (a) stable, (b) unstable, (c) neutral stability

any equilibrium state on these branches is stable; this is shown Example 6.5. On the other hand, any perturbation of an equilibrium solution on the branch BC can be shown grow, that is, to be unstable. The branch AC changes from stable to unstable at the point B. This point is called a critical point, or buckling point.

This approach is also used in studying the stability of fluid flow of a liquid in a pipe. When the flow velocity is below a critical Reynolds number, the flow is stable. A perturbation of the flow leads to small changes. On the other hand, when the flow is above the critical Reynolds number, a small perturbation leads to large changes because the flow changes from laminar to turbulent. Thus the flow above the critical Reynolds number is unstable.

An example of stable and unstable states often given in introductory dynamics texts is shown in Figure 6.8. It is clear that state A is stable, since small perturbations of the position of the ball will not significantly change the evolution of the system. State B is unstable, since small perturbations will lead to large changes: the ball will roll either to the right or to the left. State C is often called neutral stability in introductory texts. According to the definition of (6.5.1a), state C is an unstable state, since small changes in the velocity will lead to large changes in the position at large times, though they are not as large as in B. Thus the definition of stability given in introductory texts does not completely conform to the one given here.

6.5.2 Branches of Equilibrium Solutions

To obtain a good understanding of a system, its equilibrium paths, or branches, and their stability must be determined. It is widely believed among structural mechanicians that the difficulties associated with unstable behavior can be circumvented by simply obtaining a dynamic solution. When a structure is loaded above its limit point or a bifurcation point in a dynamic simulation, the structure passes dynamically to the nearest stable branch. However, the instability is not readily apparent, and the possibility of imperfection sensitivity is not clear. Therefore, to understand the behavior of a structure or process thoroughly, its equilibrium behavior should be carefully examined. Many vagaries of structural behavior may be hidden by dynamic simulations. For example, in a cylindrical shell under axial load the intersecting branch is asymmetric as shown in Figure 6.7(e). A system with an asymmetric branch is very sensitive to imperfections, as can be seen by the large change in the maximum load with change in the imperfection in Figure 6.7(e). The theoretical bifurcation point of a perfect structure is then not a realistic measure of the strength; the actual structure may buckle at a much lower load than the theoretical value because imperfections are unavoidable in a real structure. A single numerical simulation could miss this sensitivity completely. This sensitivity to imperfections for cylindrical shells was analyzed by Koiter and is a classical example of imperfection sensitivity. To ascertain this type of behavior, the equilibrium branches must be known.

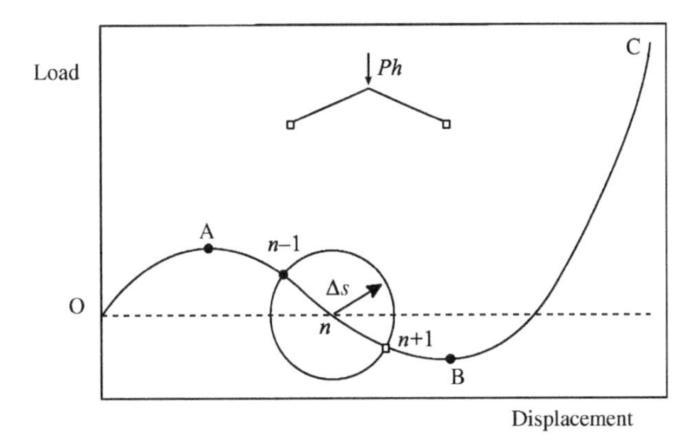

Figure 6.9 Arc length method: next solution is the intersection of the equilibrium branch with the circle about the last solution

As a first step in studying the equilibrium behavior of a system, the load and any other parameters of interest, such as the temperature, must be parametrized. Up to now we have parametrized the load by the time *t*, which is convenient in many practical problems. However, a single parameter does not always suffice in the study of equilibrium problems. We will now parametrize the load by n_γ parameters γ_a, so the load is then given by $\gamma_a q_a$, where q_a denotes a load distribution. We continue the convention that repeated indices are summed over the range, in this case n_γ.

For purposes of characterizing the nonlinear system, the equilibrium solutions are usually grouped into branches, which are continuous lines describing the response as one parameter is changed. These branches are called *equilibrium branches*. Our intention is to trace the equilibrium branches of the model as a function of the parameters γ_α. The problem then is to find $\mathbf{d}(\gamma_a)$ such that the residual vanishes (i.e., the system is in equilibrium):

$$\mathbf{r}(\mathbf{d}(\gamma_a)) = \mathbf{0} \tag{6.5.3}$$

Nonlinear systems exhibit three types of branching behavior:

1. Turning points, usually called limit points in structural analysis, in which the slope of the branch changes sign.
2. Stationary bifurcations, often called simply bifurcations, in which two equilibrium branches intersect.
3. Hopf bifurcations, in which an equilibrium branch intersects with a branch on which there is periodic motion.

The behavior of the shallow truss exhibits a turning (or limit) point, as can be seen from Figure 6.9; points *A* and *B* are turning points. Subsequent to a turning point, a branch can be either stable or unstable. In this case, as shown later in the analysis in Example 6.4, the branch after the first turning point, point *A*, is unstable, while the branch after the second turning point, point *B*, is stable.

The beam problem shown in Figure 6.7(a) is a classic example of a bifurcation. The point *B* where the two branches intersect is the point of bifurcation. Subsequent to the bifurcation point, the continuation *BC* of the fundamental branch becomes unstable. Point *B*, the bifurcation point, corresponds to the buckling load of the Euler beam. This type of branching is often called a pitchfork.

Hopf bifurcations are quite uncommon in passive rate-independent structures. They occur in problems with rate-dependent materials or active control. In a Hopf bifurcation, stable equilibrium solutions become impossible at the end of a branch. Instead, the equilibrium branch splits into two branches on which solutions are periodic in time.

6.5.3 Methods of Continuation and Arc Length Methods

Methods for tracing equilibrium branches are called *continuation methods*. The tracing of equilibrium branches is often quite difficult; robust, automatic procedures for continuation are not yet available. In the following, we describe continuation methods based on parametrization, such as the arc length method. To motivate this discussion, we first describe how branches are traced with a load parameter.

In tracing the branches, the load parameter is usually started at zero and incremented. For each increment of the parameter γ, an equilibrium solution is computed, that is, we find a solution $\mathbf{d}^{n+1}$ for

$$\mathbf{r}(\mathbf{d}^{n+1}, \gamma^{n+1}) = 0 \quad \text{or} \quad \mathbf{f}^{\text{int}}(\mathbf{d}^{n+1}) - \gamma^{n+1}\mathbf{f}^{\text{ext}} = 0 \tag{6.5.4}$$

where n is the step number and $\mathbf{f}^{\text{ext}}$ is the load distribution.

In the arc length method, instead of incrementing the load parameter γ, a measure of the arc length in the displacement–load parameter space is incremented. This is accomplished by adding a parametrization equation to the equilibrium equations:

$$p(\mathbf{d}^{n+1}, \gamma^{n+1}) = 0 \tag{6.5.5}$$

In the arc length method the parametrization equation is

$$p(\mathbf{d}, \gamma) = (\mathbf{d}^{n+1} - \mathbf{d}^{n})^T(\mathbf{d}^{n+1} - \mathbf{d}^{n}) + \alpha\Delta\gamma^2 - \Delta s^2 = 0, \quad \Delta\gamma = \gamma^{n+1} - \gamma^{n} \tag{6.5.6}$$

where Δs is an approximate value of the arc length to be traversed during the step, and α is a scaling factor. One possible scaling factor is related to the diagonal elements of the material stiffness matrix, that is,

$$\alpha^{-1} = \frac{1}{n_{\text{dof}}}\sum_{a}(K_{aa}^{\text{mat}})$$

Many other types of parametrization equations can be devised. The total system of equations then consists of the equilibrium equations and the parametrization equation

$$\begin{Bmatrix} \mathbf{r}(\mathbf{d}^{n+1}, \gamma^{n+1}) \\ p(\mathbf{d}^{n+1}, \gamma^{n+1}) \end{Bmatrix} = \begin{Bmatrix} \mathbf{f}^{\text{int}} - \gamma\mathbf{f}^{\text{ext}} \\ p(\mathbf{d}^{n+1}, \gamma^{n+1}) \end{Bmatrix} = \begin{Bmatrix} \mathbf{0} \\ 0 \end{Bmatrix} \tag{6.5.7}$$

Thus we now have an additional equation and an additional unknown γ^{n+1}; the arc length s is incremented instead of the load parameter γ^{n+1}. We assume that the distribution of nodal external forces does not change with the deformation of the model; when the load multiplier is the coefficient of a follower load such as a pressure, the methods must be altered to account for changes in the load distribution due to load geometric effects. The inertial term is not included in the above because continuation methods are applicable only to equilibrium problems. The arc length method is often called the Riks method in structural mechanics (Riks, 1972).

This procedure is most easily explained for a one degree-of-freedom problem such as the shallow truss shown in Figure 6.9. We assume that an equilibrium solution has been obtained at point n. The arc length equation (6.5.6) in the load-displacement plane is a circle about point n. The next solution of the parametrized equations (6.179) is the intersection of the equilibrium branch with the circle. Incrementing the load parameter would be fruitless at point n, since it would bring us back up the branch. In the arc length method, the problem has been restated in terms of the arc length along the branch, so that a branch with a decreasing load can be followed. The load need not increase in the step, and may in fact decrease. It is only necessary for the arc length parameter to increase, which is a perfectly natural way of tracing the branch. In a problem with two degrees of freedom, the arc length equation would define a sphere or ball about the equilibrium point, and the next solution the intersection of the branch with the ball.

The parametrized equations for the symmetric truss can then be posed as follows: find a solution to

$$r(d^{n+1}, \gamma^{n+1}) = 0 \tag{6.5.8}$$

subject to

$$\alpha(\gamma^{n+1}(s) - \gamma^{n})^2 + (d^{n+1}(s) - d^{n})^2 - \Delta s^2 = 0 \tag{6.5.9}$$

In this, d is the vertical displacement, r the corresponding residual; we assume the horizontal displacement vanishes. Alternatively, we can write the above in terms of increments in the displacements and the load parameters: find a solution to

$$r = 0 \text{ subject to } \alpha\Delta\gamma^2 + (\Delta d)^2 = \Delta s^2 \tag{6.5.10}$$

Thus the original set of discrete equations in one unknown is augmented by one equation, and one unknown γ is added. The resulting equations can be solved by the standard Newton methods we have described. Let

$$\begin{aligned}\delta\mathbf{d} &= \mathbf{d}_{\upsilon+1}^{n+1} - \mathbf{d}_{\upsilon}^{n+1} = \mathbf{d}_{\text{new}} - \mathbf{d}_{\text{old}}\\ \delta\gamma &= \gamma_{\upsilon+1}^{n+1} - \gamma_{\upsilon}^{n+1} = \gamma_{\text{new}} - \gamma_{\text{old}}\end{aligned}$$

Note that the δ is used to denote the change in a variable during an iteration; it does not refer to a variation here. This symbol has been introduced so that we can use the notation $\Delta\mathbf{d}$ in the arc length equation. The linearized equations for the Newton method are given by

$$\begin{bmatrix} \partial\mathbf{r}/\partial\mathbf{d} & \partial\mathbf{r}/\partial\gamma \\ \partial p/\partial\mathbf{d} & \partial p/\partial\gamma \end{bmatrix} \begin{Bmatrix} \delta\mathbf{d} \\ \delta\gamma \end{Bmatrix} = \begin{Bmatrix} -\mathbf{r}_{\upsilon} \\ -p_{\upsilon} \end{Bmatrix} \tag{6.5.11}$$

Using the definitions of the derivatives of the nodal forces (6.3.15–16), the fact that $\mathbf{r},_g = -\mathbf{f}^{\text{ext}}$ from (6.176), and the arc length equation (6.178) gives

$$\begin{bmatrix} \mathbf{K}^{\text{int}} - \gamma \mathbf{K}^{\text{ext}} & -\mathbf{f}^{\text{ext}} \\ 2\Delta \mathbf{d}^T & 2\alpha\Delta\gamma \end{bmatrix} \begin{Bmatrix} \delta \mathbf{d} \\ \delta\gamma \end{Bmatrix} = \begin{Bmatrix} -\mathbf{r}_\upsilon \\ -p_\upsilon \end{Bmatrix} \tag{6.5.12}$$

As can be seen, the above equations are not symmetric, so in large problems considerable advantage can be gained by solving the parametrization equation separately. Difficulties can also arise because the augmented system has two solutions (points $n-1$ and $n+1$ in Figure 6.9), and the solution which retraces the path is not wanted. The correct solution is usually the one which maximizes $\delta\mathbf{d}^T(\mathbf{d}^n - \mathbf{d}^{n-1})$, that is, the one in the same direction as the previous step. The above procedures can be extended to multiple load parameters by replacing scalar parameter γ in the above equations by a matrix $\boldsymbol{\gamma}$.

6.5.4 *Linear Stability*

The most widely used technique for examining the stability of an equilibrium solution is linear stability analysis. In this procedure, a dynamic solution is obtained for a perturbation of the equilibrium state. The perturbations are assumed to be small, so the dynamic equations are linear, that is, *linearized model* is used. If the dynamic solution grows, then the equilibrium solution is called *linear unstable*, otherwise it is *linear stable*. In most cases, it is not necessary to actually integrate the solution in time to ascertain linear stability; instead it can be ascertained from eigenvalues of the linearized system as shown in the following.

Consider an equilibrium point $\mathbf{d}^{\text{eq}}$ of a rate-independent system associated with a parametrized load, $\gamma\mathbf{f}^{\text{ext}}$. A Taylor series expansion of $\mathbf{f} = \mathbf{f}^{\text{int}} - \mathbf{f}^{\text{ext}}$ about the equilibrium solution gives

$$\mathbf{f}(\mathbf{d}^{\text{eq}} + \tilde{\mathbf{d}}) = \mathbf{f}(\mathbf{d}^{\text{eq}}) + \frac{\partial \mathbf{f}(\mathbf{d}^{\text{eq}})}{\partial \mathbf{d}}\tilde{\mathbf{d}} + \text{higher order terms} \tag{6.5.13}$$

where $\tilde{\mathbf{d}}$ is a perturbation of the equilibrium solution. The first term on the RHS vanishes because $\mathbf{d}^{\text{eq}}$ is an equilibrium solution. From (6.3.17–19) we can see that the second term can be linearized as follows:

$$\frac{\partial \mathbf{f}(\mathbf{d}^{\text{eq}})}{\partial \mathbf{d}} = \mathbf{K}^{\text{ext}}(\mathbf{d}^{\text{eq}}) - \mathbf{K}^{\text{int}}(\mathbf{d}^{\text{eq}}) \equiv -\tilde{\mathbf{A}}(\mathbf{d}^{\text{eq}}) \tag{6.5.14}$$

where $\tilde{\mathbf{A}}$ is defined above as the Jacobian of the nodal forces. Note that the mass matrix is not included in the Jacobian matrix $\tilde{\mathbf{A}}$. We now add the inertial forces to the system. Since the mass matrix does not change with displacements, we can then write the equations of motion for a small perturbations about the equilibrium point as

$$\mathbf{M}\frac{d^2\tilde{\mathbf{d}}}{dt^2} + \tilde{\mathbf{A}}\tilde{\mathbf{d}} = 0 \tag{6.5.15}$$

The above is a set of linear ordinary differential equations in $\tilde{\mathbf{d}}$. Since the solutions to such linear ordinary differential equations are exponential, we assume solutions of the form

$$\tilde{\mathbf{d}} = \mathbf{y}e^{\mu t} \quad \text{or} \quad \tilde{d}_a = y_a e^{\mu t} \tag{6.5.16}$$

Substituting (6.5.16) into (6.5.15) gives

$$(\tilde{\mathbf{A}} + \mu^2 \mathbf{M})\mathbf{y}e^{\mu t} = 0 \tag{6.5.17}$$

The characteristic values μ_i of the system can be obtained from the eigenvalue problem

$$\tilde{\mathbf{A}}\mathbf{y}_i = \lambda_i \mathbf{M}\mathbf{y}_i, \quad \text{where} \quad \lambda_i = -\mu_i^2 \tag{6.5.18}$$

and λ_i, $i = 1$ to n, are the eigenvalues and $\mathbf{y}_i$ are the corresponding eigenvectors.

The linear stability of the system is determined by the character of the square roots of the eigenvalues, $\mu_i = \sqrt{-\lambda_i}$, which are generally complex. If the real part of the μ_i is positive, the solution will grow, that is,

$$\text{if for any } i, \text{Real}(\mu_i) > 0, \text{ the equilibrium point is linear unstable} \tag{6.5.19}$$

On the other hand, if the real parts of all eigenvalues are negative, then the linearized solutions about the equilibrium point do not grow. So we can say that

$$\text{if for all } i, \text{Real}(\mu_i) \le 0, \text{ the equilibrium point is linear stable} \tag{6.5.20}$$

The equality in the above corresponds to neutral stability.

The equilibrium point at which a branch changes from stable to unstable or vice versa is called a critical point. At a critical point at least one of the eigenvalues must vanish, so the determinant of $\tilde{\mathbf{A}}$ vanishes.

6.5.5 *Symmetric Systems*

For systems with symmetric Jacobians, the stability of an equilibrium point can be ascertained by checking the Jacobian matrix for positive-definiteness. In the absence of follower loads the linearized equations (6.4.15) are often symmetric. The mass matrix $\mathbf{M}$ is always symmetric for a Lagrangian continuum mesh; conditions under which the tangent stiffness is symmetric were discussed in Section 6.4.4.

When the linearized equations are symmetric, the eigenvalues must be real. Since the mass matrix $\mathbf{M}$ is positive-definite, if the matrix $\tilde{\mathbf{A}}$ is also positive-definite, the eigenvalues λ_i of (6.5.18) must be positive for a symmetric system. The characteristic values μ_i are then strictly imaginary with no real parts, so by (6.5.20) the system is stable.

On the other hand, when $\tilde{\mathbf{A}}$ is not positive definite, then at least one of the eigenvalues λ_i is negative and one of the corresponding μ_i will be real and positive. Hence by (6.5.19) the system is unstable. Therefore, for a system with a symmetric Jacobian, *the equilibrium point is linear stable if and only if the Jacobian of the nodal forces is positive-definite*. So in systems with a symmetric Jacobian, stability can be checked by simply examining the eigenvalues of the Jacobian. Positive-definiteness can be ascertained from the minimum eigenvalue of the Jacobian: if it is positive, the Jacobian is positive-definite and the system is stable. At a critical point, where the equilibrium branch changes from stable to unstable, one of the eigenvalues must vanish (since at least one eigenvalue has changed from positive to negative). The determinant of $\tilde{\mathbf{A}}$ will also vanish at the critical point, since it is the product of the eigenvalues.

6.5.6 *Conservative Systems*

If the system is conservative, that is, if the stresses and loads are derivable from a potential, the stability of equilibrium points can be ascertained from properties of the potential function. Note that for a conservative system, the matrix $\tilde{\mathbf{A}}$ is symmetric and corresponds to the Hessian of the potential energy, that is, $\tilde{A}_{ab} = \partial^2 W / \partial d_a \partial d_b$ by (6.3.29). Recall that an equilibrium solution is a stationary point of the potential. The positive-definiteness of $\tilde{\mathbf{A}}$ implies that

$$\Delta d_a \frac{\partial^2 W(\mathbf{d}^{\text{eq}})}{\partial d_a \partial d_b} \Delta d_b = \tilde{A}_{ab}(\mathbf{d}^{\text{eq}}) \Delta d_a \Delta d_b = \Delta \mathbf{d}^T \tilde{\mathbf{A}} \Delta \mathbf{d} > 0 \quad \forall \quad \Delta \mathbf{d} \neq \mathbf{0} \tag{6.5.21}$$

Note that the condition for a local minimum is equivalent to the definition of positive-definiteness. Therefore, at any stable equilibrium solution, the Hessian matrix is positive definite.

Any equilibrium solution $\mathbf{d}^{\text{eq}}$ that satisfies the above must be linear stable. On the other hand, if there exists a $\Delta\mathbf{d}$ for which the above inequality is violated, then the equilibrium solution must be a saddle point or a local maximum, and the equilibrium solution is not linear stable. Thus any equilibrium solution that does not correspond to a local minimum of the potential energy is unstable.

6.5.7 *Remarks on Linear Stability Analysis*

The information provided by a linear stability analysis is not conclusive from an engineering viewpoint. Since linear stability analysis assumes the linearity of the response in the vicinity of the equilibrium solution, perturbations must be small enough so that the response can be predicted by a linear model. Linear stability of an equilibrium solution does not preclude the possibility that a physically realistic perturbation will grow. If the system is highly nonlinear in the neighborhood of the equilibrium solution, moderate perturbations of the system may lead to unstable growth. A linear stability analysis only reveals how a system with properties obtained by a linearization of the system behaves. Nevertheless, it gives information which is useful in engineering and scientific analysis of systems.

Linear stability analyses of path-dependent materials present special difficulties since the tangent matrix does not describe the behavior of the system upon unloading. This has led to the concept of *elastic comparison* materials with identical behavior in loading and unloading. In these materials, the tangent matrix is based on the plastic moduli, and unloading is neglected. These models sometimes underestimate the bifurcation loads, but generally provide very good estimates.

6.5.8 Estimates of Critical Points

It is often desirable to estimate bifurcation points as the equilibrium path is generated by continuation methods. Bifurcation points which have been passed or which are upcoming are of interest. Whether a bifurcation point has been passed can be determined by checking when the determinant of the Jacobian changes sign, although the test is not conclusive, since the Jacobian sometimes does not change sign at a bifurcation point. A change of sign in the determinant of the Jacobian is an indication of the change of sign of an eigenvalue. The determinant of the Jacobian vanishes at a critical point and will usually change sign at a critical point. It does not always change sign: when two eigenvalues change sign simultaneously at a bifurcation point, the determinant of the Jacobian does not change sign. Therefore the determinant test is not conclusive.

Critical points can also be estimated by eigenvalue problems. For this purpose, we assume that the Jacobian, $\tilde{\mathbf{A}}$, is a linear function between the current state n and the previous state $n-1$:

$$\tilde{\mathbf{A}}(\mathbf{d}, \gamma) = (1-\xi)\tilde{\mathbf{A}}(\mathbf{d}^{n-1}, \gamma^{n-1}) + \xi\tilde{\mathbf{A}}(\mathbf{d}^{n}, \gamma^{n}) \equiv (1-\xi)\,\tilde{\mathbf{A}}^{n-1} + \xi\tilde{\mathbf{A}}^{n} \tag{6.5.22}$$

The load factor is interpolated similarly:

$$\gamma = (1-\xi)\gamma^{n-1} + \xi\gamma^{n} \tag{6.5.23}$$

At a critical point, the determinant of the Jacobian $\tilde{\mathbf{A}}$ vanishes: det $\tilde{\mathbf{A}}(\mathbf{d}, \gamma_{\text{crit}})=0$. Since a system with a zero determinant has a nontrivial homogeneous solution, it follows that there exist a ξ and a $\mathbf{y}$ such that

$$\tilde{\mathbf{A}}(\mathbf{d}, \gamma_{\text{crit}})\mathbf{y} = (1-\xi)\tilde{\mathbf{A}}^{n-1}\mathbf{y} + \xi\tilde{\mathbf{A}}^{n}\mathbf{y} = 0 \tag{6.5.24}$$

This can be put in the standard form of the generalized eigenvalue problem by the following rearrangement of terms:

$$\tilde{\mathbf{A}}^{n-1}\mathbf{y} = \xi(\tilde{\mathbf{A}}^{n-1} - \tilde{\mathbf{A}}^{n})\mathbf{y} \tag{6.5.25}$$

An alternative form of the above eigenvalue problem is

$$\tilde{\mathbf{A}}^{n}\mathbf{y} = \mu\tilde{\mathbf{A}}^{n-1}\mathbf{y} \quad \text{where} \quad \mu = \frac{\xi-1}{\xi}, \quad \xi = \frac{1}{1-\mu} \tag{6.5.26}$$

which is more robust numerically since it does not involve differences of numbers which may be nearly equal.

The procedure of determining the location of a nearby critical point then consists of the following. The Jacobian is saved from the last step, and with the current Jacobian, the eigenvalues ξ are obtained by (6.5.25) or (6.5.26). The eigenvalues which are the smallest in absolute value are the ones of interest. When $\xi > 1$, the critical point estimate is further along the branch. When $0 \leq \xi \leq 1$, the critical point is estimated to be between points $n-1$ and n. When $\xi < 0$ the critical point estimate is before equilibrium point $n-1$. In the latter cases, the current equilibrium points $n-1$ and/or n may be unstable.

The above are estimates of the critical points. To obtain accurate values of the critical loads, iterative procedures are used. For example, the critical point estimated by (6.5.25) can be used as one of the states and the process repeated. At the same time, a higher-order interpolation of the determinant of the Jacobian should be made to guide the search, using a procedure similar to that in line search.

6.5.9 *Initial Estimates of Critical Points*

For problems with linear materials without follower loads, the critical point, often called the buckling load, can often be estimated accurately after a single load step. Such estimates are based on the following assumptions and arguments:

1. The material tangent stiffness matrix for an elastic material will not change significantly if the displacements from the initial configuration to the configuration at the critical point are small.
2. The stresses are proportional to the loads, so that geometric stiffness depends linearly on the load parameter (recall that the geometric stiffness is linear in the stresses if the displacements are small, as can be seen from the geometric stiffnesses in Examples 6.2 and 6.3).
3. In the absence of follower loads, the load is independent of the displacements, so the load stiffness vanishes.

In this procedure, a single load step is taken with the load $\gamma^1 \mathbf{f}^{\text{ext}}$, where the superscript is the step number. The Jacobian at that point is given by

$$\mathbf{A}^1 = \mathbf{K}^0_{\text{mat}} + \mathbf{K}_{\text{geo}}(\gamma^1) \tag{6.5.27}$$

where $\mathbf{K}^{\text{geo}}(\gamma^1)$ is the geometric stiffness associated with the load $\gamma^1 \mathbf{f}^{\text{ext}}$. The Jacobian in the initial configuration is given by

$$\mathbf{A}^0 = \mathbf{K}^0_{\text{mat}} \tag{6.5.28}$$

since the initial stresses are assumed to vanish. Substituting the above into (6.5.25) with $n = 1$ gives

$$\mathbf{K}^{\text{mat}} \mathbf{y} = -\xi \mathbf{K}^{\text{geo}}(\gamma^1) \mathbf{y} \tag{6.5.29}$$

The critical load is then given by

$$\gamma_{\text{crit}} = \xi \gamma^1 \tag{6.5.30}$$

This formula is commonly given in matrix structural mechanics texts for the buckling load of a structure. Note that it assumes that the geometry of the structure changes so little with increasing load that the first estimate of the geometric stiffness suffices for extrapolating the Jacobian to the critical point. It is much more effective for bifurcation points than for limit points, because prior to reaching a limit point, the geometric stiffness usually changes significantly.

Stability is a rich field with many exciting intellectual developments in the past four decades. The most noteworthy are catastrophe theory and dynamical systems theory. In catastrophe theory, it has been shown that in a four degree-of-freedom system with a potential, only seven types of turning point are possible. Dynamical systems includes topics such as chaos, fractals, attractors, and repellers. These topics are beyond the scope of this book.

Example 6.4 A simple example of a problem with stable and unstable paths connected by turning points is the shallow truss shown in Figure 6.10. The initial cross-sectional areas of the elements are A_0 and the initial lengths of the two elements are ℓ_0, which is given by $\ell_0^2 = a^2 + b^2$. A vertical load p is applied as shown at node 1. Since this is the only load, we let p be the load parameter. The material is governed by a Kirchhoff law, $S_{11} = E^{SE} E_{11}$, where E^{SE} is given in Box 5.1. The system is conservative, so we will determine the equilibrium path by finding the stationary points of the potential. The stability of the branches will then be examined by linear stability analysis.

The deformation of the truss is described by the current vertical coordinate of the center node, denoted y_1. This leads to simpler equations than those in terms of the displacement. The potential energy, (E4.12.3), is

$$W = W^{\text{int}} - W^{\text{ext}},\; W^{\text{int}} = \frac{1}{2}\sum_{e=1}^{2}\int_{\Omega_0^e} E^{SE}\hat{E}_{11}^2\, d\Omega, \quad W^{\text{ext}} = p(b + y_1) \tag{E6.4.1}$$

The Green strain for both elements is most easily evaluated by (E3.9.9), which gives

$$\hat{E}_{11} = \frac{1}{2}\frac{(\ell^2 - \ell_0^2)}{\ell_0^2} = \frac{(a^2 + y_1^2 - a^2 - b^2)}{2(a^2 + b^2)} = \frac{y_1^2 - b^2}{2(a^2 + b^2)} \tag{E6.4.2}$$

so the internal energy for the two elements is given by

$$W^{\text{int}} = k(y_1^2 - b^2)^2 \quad \text{where} \quad k = \frac{E^{SE} A_0}{4\ell_0^3} \tag{E6.4.3}$$

Combining the above with the potential of the external force gives the total potential

$$W = W^{\text{int}} - W^{\text{ext}} = k(y_1^2 - b^2)^2 - p(b + y_1) \tag{E6.4.4}$$

The equilibrium equation is obtained by applying the theorem of stationary potential energy. The equilibrium equation is obtained by setting the derivative of the above to zero, giving

$$0 = \frac{dW}{dy_1} 4ky_1(y_1^2 - b^2) - p \tag{E6.4.5}$$

The nodal force is a cubic function of the vertical position of the centerpoint. The equilibrium branches are shown in Figure 6.10. There are two turning points, called limit points in structural mechanics, B and C, and three branches, denoted by AB, BC and CD.

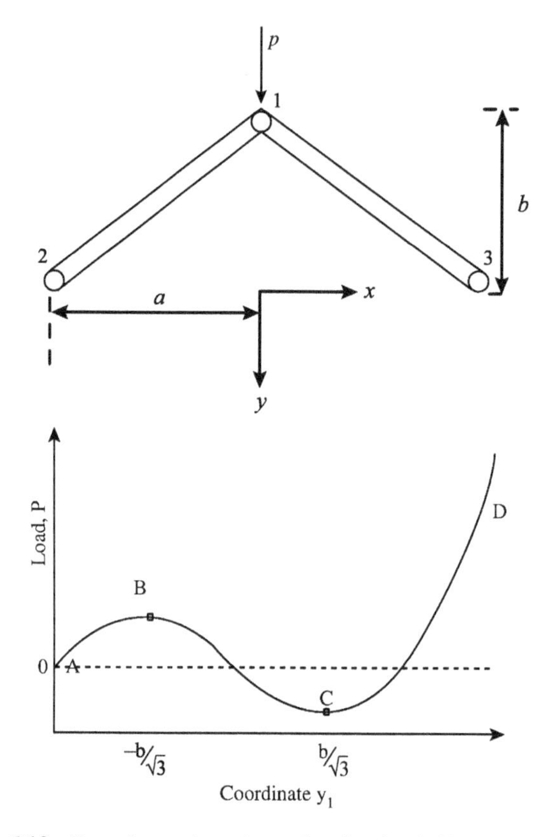

Figure 6.10 Branches and turning points for the shallow truss example

We will now examine the stability of the branches by solving the perturbed, linearized equations of motion about the state y_1. Usually, the tangent stiffness matrix would be used to obtain the linearized equations, but in this case it is simpler to just work with the derivatives of the nodal forces. Let the position of node 1 be y_1; the applied load $f_{y1}^{\text{ext}} = p$ is given by (E6.4.5) in terms of y_1. The perturbed solution is

$$y_1(t) = \bar{y}_1 + \tilde{y}_1(t) \quad \text{where} \quad \tilde{y}_1(t) = \varepsilon e^{\mu t} \tag{E6.4.6}$$

where ε is a small parameter. The equations of motion are given by

$$0 = M\frac{d^2 y_1}{dt^2} + f_{y1}^{\text{int}} - f_{y1}^{ext} = M\frac{d^2 \tilde{y}_1}{dt^2} + 4ky_1(y_1^2 - b^2) - p \tag{E6.4.7}$$

where M is the mass of the node; we have replaced d^2y_1/dt^2 by $d^2\tilde{y}_1/dt^2$ using (E6.4.6).

Substituting (E6.4.6) into (E6.4.7) gives

$$M\frac{d^2 \tilde{y}_1}{dt^2} + 4k(\bar{y}_1 + \tilde{y}_1)((\bar{y}_1 + \tilde{y}_1)^2 - b^2) - p = 0 \tag{E6.4.8}$$

Expanding the above and dropping all terms which are higher order than linear in $\tilde{y}_1$ gives the perturbed equations of motion:

$$M\frac{d^2\tilde{y}_1}{dt^2}+4k\left[\bar{y}_1\left(\bar{y}_1^2-b^2\right)+\tilde{y}_1(3\bar{y}_1^2-b^2)\right]-p=0 \tag{E6.4.9}$$

The load p cancels the first term in the brackets of the above since y_1 is an equilibrium state (see (E6.4.5)) so the equations of motion become

$$M\frac{d^2\tilde{y}_1}{dt^2}+4k\tilde{y}_1\left(3\bar{y}_1^2-b^2\right)=0 \tag{E6.4.10}$$

Substituting (E6.4.6) into the above yields $\mu=\pm i\left(3\bar{y}_1^2-b^2\right)^{1/2}$. For $3\bar{y}_1^2-b^2<0$, one of the parameters μ in (E6.4.6) is real and positive so the perturbation solution will grow. So the branch BC defined by

$$-b/\sqrt{3}<\bar{y}_1<b/\sqrt{3} \tag{E6.4.11}$$

is unstable. For any other values of $\bar{y}_1$, the coefficient μ is imaginary, so the perturbation solution is harmonic with constant amplitude ε and the equilibrium point is linear stable.

The results of the above stability analysis can be obtained directly by examining the second derivative of the potential energy function, which from (E6.4.5) is given by $d^2W/dy_1^2+4k\left(3y_1^2-b^2\right)$. Examining whether the above Jacobian is positive or not leads to

$$\frac{d^2W}{dy_1^2}+<0 \text{ when } -b<\sqrt{3}y_1<b,\quad \frac{d^2W}{dy_1^2}\geq 0 \text{ otherwise} \tag{E6.4.12}$$

so the stability conditions are identical to those obtained by the perturbation analysis (E6.4.11). In summary, the equilibrium branches AB and CD are stable, whereas branch BC is unstable.

Example 6.5 Consider a linear stability analysis of the beam element shown in Figure 6.11. Node 2 is clamped; node 1 is free to rotate and move in the x-direction. Find the equilibrium equation and the equilibrium branches of the system.

The problem parameters are Young's modulus E, the moment of the cross-section I, and the original length of the beam ℓ_0. The beam is modeled by a single element with a linear axial displacement field and a cubic transverse displacement field. The unknowns are $\mathbf{d}^T=[u_x\ u_y\ \theta]$, where θ is the rotation of the node; nodal subscripts have been dropped because they all refer to node 1. The displacement boundary conditions are $u_{x2}=u_{y2}=\theta_2=u_{y1}=0$. Therefore, the only nonzero degrees of freedom are $u_{x1}\equiv u_1$ and θ_1. The potential energy of the beam is given by

$$W=\frac{EA}{2\ell}u_1^2-\frac{EA}{15}u_1\theta_1^2+\frac{EA\ell}{140}\theta_1^4+\frac{2EI}{\ell}\theta_1^2-Pu_1 \tag{E6.5.1}$$

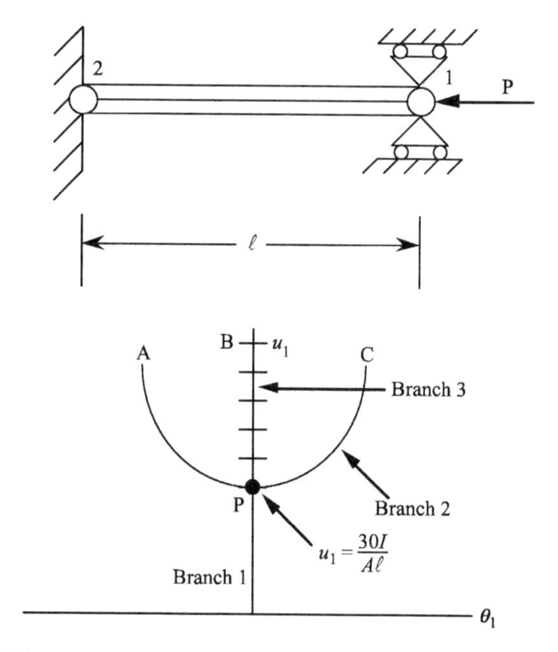

Figure 6.11 Beam model used for stability analysis and equilibrium paths

The equations of equilibrium are obtained by taking the derivatives of the potential energy with respect to u_1 and θ_1, which yields, respectively:

$$\frac{EA}{\ell}u_1 - \frac{EA}{15}\theta_1^2 = P \tag{E6.5.2}$$

$$\left(\frac{4EI}{\ell} - \frac{2EA}{15}u_1 + \frac{EA\ell}{35}\theta_1^2\right)\theta_1 = 0 \tag{E6.5.3}$$

This system of two nonlinear algebraic equations in two unknowns possesses two branches:

$$\text{Branch 1}: \ \theta_1 = 0, \quad u_1 = \frac{P\ell}{EA} \tag{E6.5.4}$$

$$\text{Branch 2}: \ u_1 = \frac{3\ell}{14}\theta_1^2 + \frac{30I}{A\ell} \tag{E6.5.5}$$

These two curves are plotted in Figure 6.11. It can be seen that a pitchfork bifurcation occurs at $u_1 = 30I/A\ell$. The corresponding load can be found by substituting this displacement and $\theta_1 = 0$ into (E6.5.2), which gives $P_{crit} = 30EI/\ell^2$.

Since the system equations are symmetric, the stability of any of the equilibrium paths can be determined from the eigenvalues of the tangent stiffness. The tangent stiffness is given by the second derivatives of the potential energy

$$\mathbf{K}^{\tan} = \frac{\partial^2 W}{\partial \mathbf{d} \partial \mathbf{d}} = \begin{bmatrix} \frac{\partial^2 W}{\partial u_1^2} & \frac{\partial^2 W}{\partial u_1 \partial \theta_1} \\ \frac{\partial^2 W}{\partial \theta_1 \partial u_1} & \frac{\partial^2 W}{\partial \theta_1^2} \end{bmatrix} = \frac{EA}{\ell} \begin{bmatrix} 1 & \frac{2\theta_1 \ell}{15} \\ -\frac{2\theta_1 \ell}{15} & 4r - \frac{2}{15}\ell u_1 + \frac{3}{15}\ell^2 \theta_1^2 \end{bmatrix} \quad \text{(E6.5.6)}$$

where $r = I/A$. If $\theta_1 = 0$, then

$$\mathbf{K}^{\tan} = \frac{EA}{\ell} = \begin{bmatrix} 1 & 0 \\ 0 & 4r - \frac{2}{15}\ell u_1 \end{bmatrix} \quad \text{(E6.5.7)}$$

This matrix is positive-definite when $u_1 \leq 30I/A\ell$. When $\theta_1 \neq 0$, $\det(\mathbf{K}^{\tan}) = 62E^2 A\theta_1^2 / 1575 > 0$. Therefore, when $\theta_1 \neq 0$, the structure is stable. On branch 3, the solutions are unstable.

From Example 6.4, it can be seen that a linearized stability analysis is not conclusive from an engineering viewpoint. For example, if $y_1 = -0.99b$, the test indicates that the equilibrium point is unstable. However, the structure when perturbed at that equilibrium point will displace by a distance 0.002, which to most engineers would not be an instability. Linear stability analysis checks whether any perturbation will grow based on linearized properties of the system. This does not conform exactly to an engineer's intuitive notion of instability.

Restricting the perturbations to the initial conditions also appears unrepresentative of actual physical processes, where imperfections are found in the geometry and material properties and perturbations in the loading occur throughout the process. However, as Thompson and Hunt (1984: 7) say: 'the only reason for this approach is that it makes the problem mathematically tractable and it leads to useful results.' We cannot think of a better reason for a mathematical method.

6.6 Numerical Stability

6.6.1 *Definition and Discussion*

The definition of numerical stability is similar to that of stability of solutions of systems (6.5.1–2). A numerical procedure is stable if small perturbations of initial data result in small changes in the numerical solution. More formally, the numerical solution $\mathbf{u}_A^n$ is stable if

$$\left\| \mathbf{u}_A^n - \mathbf{u}_B^n \right\| \leq C\varepsilon \quad \forall n > 0 \quad \text{for all } \mathbf{u}_A^0 \quad \text{such that } \left\| \mathbf{u}_A^0 - \mathbf{u}_B^0 \right\| \leq \varepsilon \quad (6.6.1)$$

In this C is an arbitrary positive constant. An algorithm that yields stable numerical solutions is said to be stable.

General results for numerical stability of time integrators are largely based on the analysis of linear systems. These results are extrapolated to nonlinear systems by examining linearized models of nonlinear systems. Therefore, we will first develop stability theory for linear systems. Next we describe the procedures for applying these results to nonlinear systems. In conclusion, we will describe some stability analyses of time integrators which apply directly to nonlinear systems. However, we stress that at the present time there is no stability theory which encompasses the nonlinear problems which are routinely solved by nonlinear finite element methods, and most of our understanding of stability stems from the analysis of linear models.

At this point it is worthwhile to comment on the differences between physical stability and numerical stability. Physical stability pertains to the stability of a solution of a model, whereas numerical stability pertains to the stability of the numerical method. Numerical instabilities arise from the discretization of the model equations, whereas physical instabilities are instabilities in the solutions of the model equations independent of the numerical discretization. Numerical stability is usually only examined for processes which are physically stable. Very little is known about how 'stable' numerical procedures behave in physically unstable processes. This shortcoming has important practical ramifications, because many computations today simulate physical instabilities, and if we cannot guarantee that our methods track these instabilities accurately, then these simulations may be suspect.

Numerical stability of a process that is physically unstable cannot be examined by the definition (6.6.1), that is, we cannot say anything about the stability of a numerical procedure when applied to an unstable system. The reason can be seen as follows. If a system is unstable, then the solution to the system will not satisfy the stability conditions (6.5.1), so even if the numerical solution procedure is stable, it will not satisfy (6.6.1). The philosophy today is to study numerical stability on stable systems, and then to hope that any algorithm that is stable for a stable system behaves well on an unstable system.

6.6.2 *Stability of a Model Linear System: Heat Conduction*

Most of the theory of stability of numerical methods is concerned with linear and linearized systems. The prevailing notion is that if a numerical method is unstable for linear systems, it will of course be unstable for nonlinear systems, since linear systems are a subset of nonlinear systems. Luckily, the converse has also turned out to be true: numerical methods which are stable for linear systems in almost all cases turn out to be stable for nonlinear systems. Therefore, the stability of numerical procedures for linear systems provides a useful guide to their behavior in both linear and nonlinear systems.

To begin our exploration of stability of numerical procedures, and in particular the stability of time integrators, we first consider the equations of heat conduction:

$$\mathbf{M}\dot{\mathbf{u}} + \mathbf{K}\mathbf{u} = \mathbf{f} \tag{6.6.2}$$

where $\mathbf{M}$ is the capacitance matrix, $\mathbf{K}$ is the conductance matrix, $\mathbf{f}$ is the forcing term and $\mathbf{u}$ is a matrix of nodal temperatures. This system is chosen as a starting point because it is a first-order system of ordinary differential equations. The equations of motion are second-order in time, which complicates their analysis.

To apply the definition of stability (6.6.1), we consider two solutions for the same system with the same discrete forcing function but slightly different initial data. We will consider the stability of the *process* of time integration. The two solutions satisfy the same equation with the same inhomogeneous term $\mathbf{f}$, i.e.

$$\mathbf{M}\dot{\mathbf{u}}_A + \mathbf{K}\mathbf{u}_A = \mathbf{f} \quad \mathbf{M}\dot{\mathbf{u}}_B + \mathbf{K}\mathbf{u}_B = \mathbf{f} \tag{6.6.3}$$

If we now take the difference of the two equations, we obtain

$$\mathbf{M}\dot{\mathbf{d}} + \mathbf{K}\mathbf{d} = \mathbf{0} \quad \text{where} \quad \mathbf{d} + \mathbf{u}_A = \mathbf{u}_B \tag{6.6.4}$$

Stability according to (6.6.1) then requires that $\mathbf{d}(t)$ does not grow.

We next recast the semidiscrete equations, (6.6.2), as uncoupled equations. For this purpose we need the eigenvectors of the associated system. The matrix $\mathbf{M}$ is positive-definite and symmetric, whereas the matrix $\mathbf{K}$ is positive-semidefinite and symmetric. Because of the symmetry of the matrices, the eigenvectors $\mathbf{y}_I$ of the associated eigenproblem

$$\mathbf{K}\mathbf{y}_I = \lambda_I \mathbf{M}\mathbf{y}_I \tag{6.6.5}$$

are orthogonal with respect to $\mathbf{M}$ and $\mathbf{K}$ and the eigenvalues λ_I are real. The orthogonality conditions can be written as

$$\mathbf{y}_J^T \mathbf{M}\mathbf{y}_I = \delta_{IJ}, \quad \mathbf{y}_J \mathbf{K}\mathbf{y}_I = \lambda_I \delta_{IJ} \text{ (no sum on } I) \tag{6.6.6}$$

Because of the positive-semidefiniteness of the matrices, the eigenvalues are nonnegative. The eigenvectors span the space $R^{n_{\text{dof}}}$, so any vector $\mathbf{d} \in R^{n_{\text{dof}}}$ is a linear combination of the eigenvectors:

$$\mathbf{d}(t) = \eta_J(t)\mathbf{y}_J \tag{6.6.7}$$

Substituting (6.6.7) into (6.6.4) gives

$$\mathbf{M}\dot{\eta}_J \mathbf{y}_J + \mathbf{K}\eta_J \mathbf{y}_J = \mathbf{0} \tag{6.6.8}$$

Premultiplying by $\mathbf{y}_k$ and using the orthogonality conditions (6.6.6) then gives a set of uncoupled equations

$$\dot{\eta}_K + \lambda_K \eta_K = 0, \quad K = 1 \text{ to } n_{\text{dof}} \text{ (no sum on } K) \tag{6.6.9}$$

This procedure of obtaining uncoupled equations by exploiting the eigenstructure is often called a spectral decomposition or modal decomposition. The stability of the time integration of (6.6.9) is identical to the stability of integrating (6.6.4): since $\mathbf{d}$ is linearly related to η_J by (6.6.7), if one blows up, so does the other. A nice explanation of the relationship of the spectral problem to the discrete equations is given in Hughes (1997: 494).

We now consider a two-step family of time integrators for the discrete equations (6.6.2):

$$\mathbf{d}_{n+1} = \mathbf{d}_n + (1-\alpha)\Delta t \dot{\mathbf{d}}_n + \alpha \Delta t \dot{\mathbf{d}}_{n+1} \tag{6.6.10}$$

with $\alpha \geq 0$; time step numbers are denoted by subscripts. This is called the generalized trapezoidal rule. For $\alpha = 0$ this gives the forward Euler method, an explicit method for first-order equations. For $\alpha = 1$ it gives the backward Euler method, an implicit method. The most useful of the implicit methods encompassed in the above is the standard trapezoidal rule, given by $\alpha = \frac{1}{2}$.

The (6.6.10) integration formula, when applied to the spectral coefficients $\eta \equiv \eta_K$ at time step n, is

$$\eta_{n+1} = \eta_n + (1-\alpha)\Delta t\dot{\eta}_n + \alpha\Delta t\dot{\eta}_{n+1} \tag{6.6.11}$$

We have dropped the uppercase subscripts which give the mode number, but keep in mind that we are dealing with n_{dof} uncoupled equations. This is a linear difference equation called the modal equation. The solutions of linear difference equations, like those of linear ODEs, are exponentials. We write this solution in the form

$$\eta_n = \mu^n \quad \text{where} \quad \mu = e^{\gamma\Delta t} \tag{6.6.12}$$

Compare the above to the solution of an ordinary linear differential equation $\eta = e^{\bar{\gamma}t}$, where $\bar{\gamma}$ is the characteristic value of the ODE. Note that the solutions to the difference equation and the ODE are almost identical: in the discrete solution, time t has been replaced by $n\Delta t$ and γ has been replaced by $\bar{\gamma}$. The difference between γ and $\bar{\gamma}$ is a measure of the accuracy of the discrete solution, but we will not pursue this here. From the exponential character of the discrete solution, it can be seen that if μ is a complex number, for the solution to be stable, μ must lie in the unit circle in the complex plane. We write this condition as $|\mu| \le 1$. Although we have included the equality in this stability condition, we caution the reader that when $|\mu| = 1$, the solution can sometimes be unstable; this is discussed later.

Substituting (6.6.9), $\dot{\eta} = -\lambda\eta$, into (6.6.11) wherever the derivative appears, and using (6.6.12), gives

$$\mu^{n+1} = \mu^n - (1-\alpha)\Delta t\lambda\mu^n - \alpha\Delta t\lambda\mu^{n+1} \tag{6.6.13}$$

Factoring out μ^n yields a linear equation in μ:

$$(1+\alpha\Delta t\lambda)\mu - 1 + (1-\alpha)\Delta t\lambda = 0 \tag{6.6.14}$$

This is called the characteristic equation of the integrator. The solution is

$$\mu = \frac{1-(1-\alpha)\Delta t\lambda}{1+\alpha\Delta t\lambda} \tag{6.6.15}$$

We deduce the conditions on the time step necessary for numerical stability as follows (we reattach the subscripts to the spectral coefficients and the eigenvalues to remind the reader that we are dealing with n_{dof} uncoupled equations in (6.6.13–15)). Since μ is real in this case, the condition for stability, $|\mu_J| \le 1$, implies the following:

$$\mu_J \le 1 \text{ so } \frac{1-(1-\alpha)\Delta t\lambda_J}{1+\alpha\Delta t\lambda_J} \le 1 \tag{6.6.16a}$$

$$\mu_J \ge -1 \text{ so } \frac{1-(1-\alpha)\Delta t\lambda_J}{1+\alpha\Delta t\lambda_J} \ge -1 \tag{6.6.16b}$$

Condition (6.6.16a) is always met since $\lambda_J \Delta t \geq 0$. From (6.6.16b) a little algebra gives that

$$(1-2\alpha)\,\Delta t \lambda_J \leq 2 \text{ for stability} \tag{6.6.17}$$

There are two consequences of (6.6.17). If $1-2\alpha \leq 0$, i.e. $\alpha \geq 0.5$, then the method is stable regardless of the size of the time step. The method is then called *unconditionally stable*. When $1-2\alpha > 0$, that is, $\alpha < 0.5$, (6.6.16b) yields the requirement that

$$\Delta t \leq \frac{2}{(1-2\alpha)\lambda_J} \quad \forall\, J \tag{6.6.18}$$

where, as indicated, the condition must be met for all J. The maximum eigenvalue is the most restrictive on the size of the time step, and yields the critical time step. From (6.6.18), we can deduce that the critical time step is given by

$$\Delta t_{\text{crit}} = \max_K \frac{2}{(1-2\alpha)\lambda_K} \quad \text{or} \quad \Delta t_{\text{crit}} = \frac{2}{(1-2\alpha)\lambda_{\max}},\ \lambda_{\max} = \max_K \lambda_K \tag{6.6.19}$$

An integrator that is stable only when the time step is below a critical value is called *conditionally stable*.

If we consider the explicit form of the generalized trapezoidal rule, the forward Euler method, given by $\alpha = 0$, (6.6.19) gives

$$\Delta t_{\text{crit}} = \frac{2}{\lambda_{\max}} \tag{6.6.20}$$

As can be seen from the above, the stable time step is inversely proportional to the maximum eigenvalue of the system. The stiffer a system, the greater the maximum eigenvalue and the smaller the stable time step.

To gain an appreciation of the explosive growth of an exponential instability, note that when $\mu = 1.0001$, $\mu^n = 2.2\cdot 10^4$ for $n = 10^5$. This number of time steps is not unusual in an explicit computation, and obviously the unstable mode will completely obliterate the rest of the solution if the other spectral coefficients are of order 1. Exponential growth is truly startling. It is also the reason why compound interest can make you very rich if you live long enough and start saving early.

In summary, we have shown that the determination of the stability of an integration formula for the semidiscrete initial value problem (6.6.2) can be reduced to examining the roots of the characteristic equation (6.6.14). The roots are complex numbers, and the stability condition is $|\mu| \leq 1$. The stable domain corresponds to the unit circle: to see this, let $\mu = a + ib$, where a and b are the real and imaginary parts of μ, respectively. Then $|\mu|^2 = (a+ib)\,(a-ib) = a^2 + b^2$. Thus the stable part of the μ domain corresponds to the unit circle, as shown in Figure 6.12. If any root lies outside the unit circle, the perturbation grows exponentially, so the method is unstable. Otherwise, the method is stable.

The stability conditions are identical to those in the analysis of the stability of discrete systems in Section 6.5. At first glance, they appear different since the condition for the discrete

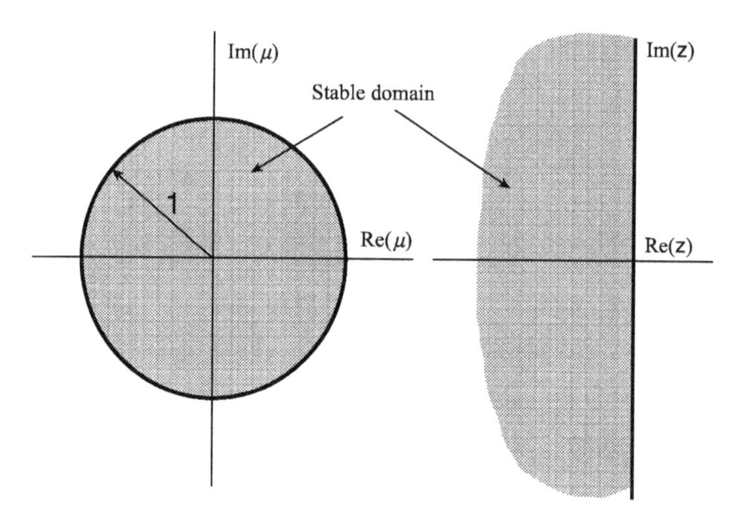

Figure 6.12 Stable domains in the complex μ-plane and complex z-plane

system is that the real part be nonpositive whereas the stable characteristic value for the time integrator lies in the unit circle. However, the discrete system condition is on the exponent, whereas the numerical stability condition concerns the value of the exponential.

6.6.3 *Amplification Matrices*

The stability of an integration method is often examined in terms of the eigenvalues of an amplification matrix. The results are identical to the preceding but the method provides a broader perspective on stability and is often needed for the stability analysis of higher-order ODEs.

An amplification matrix gives the solution at the next time step in terms of the solutions at the preceding time steps. We can express this in terms of a generalized amplification matrix or a standard amplification matrix, depending on the system:

$$\mathbf{Bd}_{n+1} = \mathbf{Ad}_n \text{ (general)}, \quad \mathbf{d}_{n+1} = \mathbf{Ad}_n \text{ (standard)} \tag{6.6.21}$$

The standard amplification matrix is a special case of the generalized amplification matrix with $\mathbf{B}=\mathbf{I}$. We shall show that the time integrator is stable if the eigenvalues of the amplification matrix lie within the unit circle in the complex plane. The generalized and standard amplification equations are associated with the generalized and standard eigenvalue problems:

$$\mathbf{Ay}_I = \mu_I \mathbf{By}_I \text{ (general)}, \quad \mathbf{Ay}_I = \mu_I \mathbf{y}_I \text{ (standard)} \quad \text{no sum on } I \tag{6.6.22}$$

The eigenvectors of the above span the space $R^{n_{\mathrm{dof}}}$, so any vector $\mathbf{d} \in R^{n_{\mathrm{dof}}}$ can be written as a linear combination of the eigenvectors. We now ascertain the conditions under which the eigenvalues μ_i fall within the unit circle, which corresponds to a stable numerical integrator. Expanding the initial condition in terms of the eigenvectors, we obtain

$$\mathbf{d}_0 = \sum_{I=1}^{n_{\rm dof}} \alpha_I \mathbf{y}_I \tag{6.6.23}$$

where α_I is determined by the initial conditions. Substituting the previous into (6.6.21) and using the fact that $\mathbf{y}_I$ are eigenvectors of (6.6.22) with eigenvalues μ_I, we obtain that

$$\mathbf{d}_1 = \sum_{I=1}^{n_{\rm dof}} \mu_I \alpha_I \mathbf{y}_I, \quad \mathbf{d}_2 = \sum_{I=1}^{n_{\rm dof}} (\mu_I)^2 \alpha_I \mathbf{y}_I, \quad \mathbf{d}_n = \sum_{I=1}^{n_{\rm dof}} (\mu_I)^n \alpha_I \mathbf{y}_I \tag{6.6.24}$$

where the second equation in the above is obtained by repeating the process; the third equation follows by induction.

We can see immediately from the above that if the modulus of any of the eigenvalues of the generalized amplification matrix μ_i is greater than 1, the solution will grow exponentially. Since we are examining the behavior of the difference of two solutions, this indicates that the process is unstable. Some readers may argue that this unstable growth will occur only if the initial data contains the eigenvector associated with μ_I. In fact, due to roundoff error, the constant α_I will initially be nonzero or will become nonzero later in almost any computation. No matter how small the constant, the exponential growth of the unstable mode will eventually dominate the solution.

The stability condition is as follows: the update (6.6.21) is stable provided that the moduli of all eigenvalues of the associated generalized eigenproblem (6.6.22) are less than or equal to 1, that is, $|\mu_I| \le 1$, and the multiplicity of any eigenvalue with $|\mu_I| = 1$ is not greater than 1. The last point is explained next.

For a root μ_I of multiplicity 2, in addition to (6.6.24), $\mathbf{d}_n = \Sigma_{I=1}^{n_{\rm dof}} n(\mu_I)^n \alpha_I \mathbf{y}_I$ is also a solution. This parallels the results for ODEs, where te^{at} is a solution when a is a multiple root of the characteristic equation. When $|\mu_I| = 1$, the discrete solution will then grow like n. This is called a *weak instability* to differentiate it from the exponential instability associated with $|\mu_I| > 1$. See Hughes (1987: 497) for a more detailed explanation.

6.6.4 Amplification Matrix for Generalized Trapezoidal Rule

We will now express the trapezoidal rule ((6.6.10) with $\alpha=0.5$) for the semidiscrete heat conduction equation in amplification matrix form. We first multiply (6.6.10) by $\mathbf{M}$. We then eliminate from the resulting equation $\mathbf{M}\dot{\mathbf{d}}$ by (6.6.4), yielding

$$(\mathbf{M} + \alpha \Delta t \mathbf{K}) \mathbf{d}_{n+1} = (\mathbf{M} - (1-\alpha) \Delta t \mathbf{K}) \mathbf{d}_n \tag{6.6.25}$$

The associated eigenvalue problem by (6.6.22) is

$$(\mathbf{M} - (1-\alpha) \Delta t \mathbf{K}) \mathbf{z} = \mu (\mathbf{M} + \alpha \Delta t \mathbf{K}) \mathbf{z} \tag{6.6.26}$$

The eigenvalues of this system control the stability of the time integrator. The simplest way to find the eigenvalue of (6.6.26) is to diagonalize the system: expand $\mathbf{z}$ in the eigenvectors $\mathbf{y}_I$ and premultiply by $\mathbf{y}_J$ and invoke orthogonality. This yields (6.6.14), and the remainder of the analysis is identical to that in Section 6.6.2.

6.6.5 The z-Transform

The analysis of stability in the μ-plane is often difficult because it is necessary to determine the roots of an mth-order algebraic equation. The procedure can be simplified by using the z-transform and Hurwitz matrices, which are closely related to the Routh–Hurwitz criterion. Stability is examined by determining whether a characteristic equation has any nonnegative real roots, which is easily done.

The z-transform is given by

$$\mu = \frac{1+z}{1-z} \tag{6.6.27}$$

where both μ and z may be complex. This transformation maps the unit circle in the complex μ-plane into the left-hand side of the z-plane, that is, the open set $|\mu| \equiv \mu\bar{\mu} < 1$ maps to $Re(z) < 0$. This map and the stable domains are shown in Figure 6.12. As can be seen, the stable domain in the μ-plane is the interior of the unit circle, while the stable part of the z-plane is the left-hand plane. The lines $|\mu| = 1$ and $z=0$ are stable only when these roots of the characteristic equation are of multiplicity 1.

To verify the map, let $z = a + ib$ and substitute in (6.6.27), so

$$|\mu|^2 = (\mu\bar{\mu}) = \left(\frac{1+a+ib}{1-a-ib}\right)\left(\frac{1+a-ib}{1-a+ib}\right) = \frac{1+2a+a^2+b^2}{1-2a+a^2-b^2} \leq 1 \tag{6.6.28}$$

Since the denominator is the sum of the squares, it must be positive. So we can multiply both sides of (6.6.28) by the denominator without changing the inequality; a little algebra then yields $a \leq 0$. This shows that the domain $|\mu| < 1$ corresponds to the domain where the real part of z is negative, that is, the left-hand plane.

The z-transform makes it possible to determine the stability of a numerical integrator by checking the real parts of the roots. This is easily done by Hurwitz matrices, which are described next. The Hurwitz matrix of the polynomial equation of order p,

$$\sum_{i=0}^{p} c_i z^{p-i} = 0 \quad \text{with} \quad c_0 > 0 \tag{6.6.29}$$

is given by

$$H_{ij} = \begin{cases} c_{2j-i} & \text{if } 0 \leq 2j-i \leq p \\ 0 & \text{otherwise} \end{cases} \tag{6.6.30}$$

The real parts of the roots of (6.6.29) are negative if and only if the leading principal minors of the Hurwitz matrix are positive. The ith principal minor is the determinant of the matrix after deleting all columns and rows with column number or row number greater than i.

Example 6.6 Find the conditions on the coefficients of the quadratic characteristic equation

$$a\mu^2 + b\mu + c = 0 \tag{E6.6.1}$$

so that $|\mu| \le 1$. The z-transform (6.6.27) applied to (E6.6.1) gives

$$a\left(\frac{1+z}{1-z}\right)^2 + b\left(\frac{1+z}{1-z}\right) + c = 0 \tag{E6.6.2}$$

Multiply this by $(1-z)^2$ (it is nonzero, since $z=1$ is not of interest) and rearrange the terms:

$$z^2(a-b+c) + 2z(a-c) + (a+b+c) = 0 \tag{E6.6.3}$$

Comparing to (6.6.29) we see that

$$c_0 = a-b+c, \quad c_1 = 2(a-c), \quad c_2 = a+b+c$$

The Hurwitz matrix is given by

$$\mathbf{H} = \begin{bmatrix} c_1 & 0 \\ c_0 & c_2 \end{bmatrix} \tag{E6.6.4}$$

The leading principal minors are

$$\begin{aligned} \Delta_1 &= c_1 \ge 0 \\ \Delta_2 &= c_1 c_2 \ge 0 \Rightarrow c_2 \ge 0 \end{aligned} \tag{E6.6.5}$$

Combining this with the requirement that $c_0 > 0$ gives

$$\begin{aligned} c_0 &= a-b+c \ge 0 \\ c_1 &= 2(a-c) \ge 0 \\ c_2 &= a+b+c \ge 0 \end{aligned} \tag{E6.6.6}$$

6.6.6 *Stability of Damped Central Difference Method*

The linear equations of motion for a damped system are

$$\mathbf{M}\ddot{\mathbf{d}} + \mathbf{C}^v\dot{\mathbf{d}} + \mathbf{K}\mathbf{d} = \mathbf{f}^{\text{ext}} \tag{6.6.31}$$

where $\mathbf{C}^v$ is the damping matrix. For the purpose of a linear stability analysis, it is convenient to diagonalize (uncouple) these equations by a spectral method. In the absence of damping the above can be diagonalized with the eigenvectors of (6.6.5). In general, the damping matrix cannot be diagonalized along with the **K** and **M** matrices unless it is a linear combination of these matrices. One example of this is Rayleigh damping:

$$\mathbf{C}^{\nu} = a_1\mathbf{M} + a_2\mathbf{K} \tag{6.6.32}$$

where a_1 and a_2 are arbitrary parameters. We let $\mathbf{d}=\sum_J \alpha_J(t)\mathbf{y}_J$, where the $\mathbf{y}_J$ are the eigenvectors of (6.6.5) and premultiply by $\mathbf{y}_K$. Then the orthogonality of the eigenvectors with respect to **M**, (6.6.6), gives

$$\ddot{\alpha} + (a_1 + a_2\omega_J^2)\dot{\alpha} + \omega_J^2\alpha = 0 \quad \text{where} \quad \omega_J^2 = \lambda_J, J = 1 \text{ to } n_{\text{dof}} \tag{6.6.33}$$

The above are the modal equations for the damped system. Note that they are simply the uncoupled form of (6.6.31) with the damping given by (6.6.32). The damping is usually given as a fraction of critical damping ξ:

$$\ddot{\alpha} + 2\xi\omega\dot{\alpha} + \omega^2\alpha = 0 \tag{6.6.34}$$

In (6.6.34) the modal indices have been dropped. The fraction of critical damping for Raleigh damping is given by

$$\xi = \frac{a_1}{2\omega} + \frac{a_2\omega}{2} \tag{6.6.35}$$

In integrating the equations of motion by the explicit central difference method, any velocity-dependent terms lag the equation by a half time step, that is, the discrete equations are

$$\mathbf{M}\ddot{\mathbf{d}}_n + \mathbf{C}^{\nu}\dot{\mathbf{d}}_{n-1/2} + \mathbf{K}\mathbf{d}_n = \mathbf{f}_n^{\text{ext}} \tag{6.6.36}$$

The reason for this lag is explained in Box 6.1 and the accompanying text. The uncoupled form of (6.6.34) for the central difference method with the lag is

$$\frac{\alpha_{n+1} - 2\alpha_n + \alpha_{n-1}}{\Delta t^2} + 2\xi\omega\frac{\alpha_n - \alpha_{n-1}}{\Delta t} + \omega^2\alpha_n = 0 \tag{6.6.37}$$

We will now perform a stability analysis of the above. Since the difference equation is linear, the solution is an exponential $\alpha_n = \mu^n$, which when substituted into (6.6.37) gives (after factoring out μ^n, multiplying by Δt^2 and rearranging)

$$\mu^2 + \mu(g + h - 2) + (1 - g) = 0 \quad \text{where} \quad g = 2\xi\omega\Delta t, \quad h = \omega^2\Delta t^2 \tag{6.6.38}$$

Applying the z-transform yields a quadratic equation in z. We can directly use (E6.6.6) to obtain the conditions for stability:

$$a - b + c = 4 - 2g - h = 4 - 4\xi\omega\Delta t - \omega^2\Delta t^2 \geq 0 \tag{6.6.39}$$

$$a - c = g = 2\xi\omega\Delta t \geq 0 \tag{6.6.40}$$

$$a+b+c=h=\omega^2\Delta t^2 \geq 0 \tag{6.6.41}$$

The third condition (6.6.41), is automatically satisfied; (6.6.40) is satisfied provided that the damping is not negative, that is, $\xi \geq 0$; (6.6.39) yields a quadratic equation in the factor $\omega\Delta t$ with the solution

$$\omega\Delta t = -2\xi \pm 2\sqrt{\xi^2+1} \tag{6.6.42}$$

The negative root of the above equation is irrelevant since it yields a negative time step. Between the points where (6.6.39) vanishes, the inequality is satisfied. So the positive root gives the critical time step:

$$\Delta t_{\text{crit}} = \max_I \frac{2}{\omega_I}\left(\sqrt{\xi_I^2+1}-\xi_I\right) \equiv \max_I \frac{2}{\sqrt{\lambda_I}}\left(\sqrt{\xi_I^2+1}-\xi_I\right) \tag{6.6.43}$$

where we have included the last term to stress that by (6.6.33), $\omega_I^2 = \lambda_I$. The factor in the parentheses equals 1 when $\xi_I=0$ and is less than 1 when $\xi_I>0$. So the lag in the velocity decreases the stable time step when the system is damped. The critical time step decreases both for explicit damping by a linear law such as $\mathbf{C}^{\nu}\mathbf{v}$ and for any damping that arises in a material law.

The issue of negative damping is a fascinating one. Although negative damping probably never occurs in nature, it arises in simplified models. For example, flutter is often modeled by negative damping. According to this analysis, the central difference method is always unstable for negative damping: see (6.6.40). However, this is an artifact of our definition of stability, which we discussed earlier in comparing physical and numerical stability. According to (6.6.1), the numerical integrator is unstable if any disturbance grows. However, the response of any model with negative damping will grow and violate (6.5.1). If you apply the central difference method to a problem with negative damping, you will find that the method tracks the exact solution quite well (provided Δt is small enough to keep truncation errors reasonable). The dilemma arises from the definition of stability: these methods cannot analyze the stability of a numerical method in an unstable process. This is examined further in Kulkarni, Belytschko, and Bayliss (1995).

6.6.7 *Linearized Stability Analysis of Newmark β-Method*

We will now perform a linear stability analysis of the Newmark β-method. The linearized stability analysis of a numerical method closely parallels the linear stability analysis of discrete systems described in Section 6.5. The integrator is applied to the linearized equations of motion and then its stability under small perturbations is examined. If the perturbation does not grow, the integrator is considered stable.

The linearization of the discrete equations (6.3.1) with Rayleigh damping gives

$$\mathbf{M}\mathbf{a}_n + \mathbf{C}^{\nu}\mathbf{v}_n + \mathbf{K}^{\text{int}}\mathbf{d}_n = 0 \tag{6.6.44}$$

where $\mathbf{d}_n$, $\mathbf{v}_n$ and $\mathbf{a}_n$ are the nodal displacements, velocities and accelerations, respectively, at time step n, and $\mathbf{C}^{\nu}=\mathbf{C}^{\text{damp}}=a_1\mathbf{M}+a_2\mathbf{K}$ is the Rayleigh damping matrix, (6.6.32). For tractability we restrict the analysis to symmetric Jacobians and neglect the load stiffness. Note that $\mathbf{d}_n$ here denotes a perturbation in the nodal displacements. The updates (6.3.4–6) can be written as

$$\mathbf{d}_{n+1} = \mathbf{d}_n + \Delta t \mathbf{v}_n + \Delta t^2 (\bar{\beta}\mathbf{a}_n + \beta \mathbf{a}_{n+1}) \tag{6.6.45}$$

$$\mathbf{v}_{n+1} = \mathbf{v}_n + \Delta t (\bar{\gamma}\mathbf{a}_n + \gamma \mathbf{a}_{n+1}) \tag{6.6.46}$$

where $\bar{\gamma} = 1 - \gamma$ and $\bar{\beta} = (1 - 2\beta)/2$. Multiplying (6.6.45) and (6.6.46) by M and substituting the equation of motion (6.6.44) wherever $\mathbf{Ma}_n$ or $\mathbf{Ma}_{n+1}$ occurs, we obtain

$$\mathbf{Md}_{n+1} = \mathbf{Md}_n + \Delta t \mathbf{Mv}_n - \Delta t^2 [\bar{\beta}(\mathbf{C}^v \mathbf{v}_n + \mathbf{K}^{\text{int}} \mathbf{d}_n) + \beta(\mathbf{C}^v \mathbf{v}_{n+1} + \mathbf{K}^{\text{int}} \mathbf{d}_{n+1})] \tag{6.6.47}$$

$$\mathbf{Mv}_{n+1} = \mathbf{Mv}_n - \bar{\gamma}\Delta t (\mathbf{C}^v \mathbf{v}_n + \mathbf{K}^{\text{int}} \mathbf{d}_n) - \gamma \Delta t (\mathbf{C}^v \mathbf{v}_{n+1} + \mathbf{K}^{\text{int}} \mathbf{d}_{n+1}) \tag{6.6.48}$$

These are now put in generalized amplification matrix form by rearranging the terms:

$$\begin{bmatrix} \mathbf{M} + \beta \Delta t^2 \mathbf{K}^{\text{int}} & \beta \Delta t^2 \mathbf{C}^v \\ \gamma \Delta t \mathbf{K}^{\text{int}} & \mathbf{M} + \gamma \Delta t \mathbf{C}^v \end{bmatrix} \begin{Bmatrix} \mathbf{d}_{n+1} \\ \mathbf{v}_{n+1} \end{Bmatrix} = \begin{bmatrix} \mathbf{M} - \bar{\beta}\Delta t^2 \mathbf{K}^{\text{int}} & \Delta t \mathbf{M} - \bar{\beta}\Delta t^2 \mathbf{C}^v \\ -\bar{\gamma}\Delta t \mathbf{K}^{\text{int}} & \mathbf{M} - \bar{\gamma}\Delta t \mathbf{C}^v \end{bmatrix} \begin{Bmatrix} \mathbf{d}_n \\ \mathbf{v}_n \end{Bmatrix} \tag{6.6.49}$$

The corresponding eigenvalue problem is (see (6.6.21) and (6.6.22)):

$$\begin{bmatrix} \mathbf{M} - \bar{\beta}\Delta t^2 \mathbf{K}^{\text{int}} & \Delta t \mathbf{M} - \bar{\beta}\Delta t^2 \mathbf{C}^v \\ -\bar{\gamma}\Delta t \mathbf{K}^{\text{int}} & \mathbf{M} - \bar{\gamma}\Delta t \mathbf{C}^v \end{bmatrix} \{\mathbf{z}\} = \mu \begin{bmatrix} \mathbf{M} + \beta \Delta t^2 \mathbf{K}^{\text{int}} & \beta \Delta t^2 \mathbf{C}^v \\ \gamma \Delta t \mathbf{K}^{\text{int}} & \mathbf{M} + \gamma \Delta t \mathbf{C}^v \end{bmatrix} \{\mathbf{z}\} \tag{6.6.50}$$

where μ is an eigenvalue of the above generalized amplification problem. It can be shown that the eigenvectors of the above are linear combinations of the eigenvectors of $\mathbf{K}^{\text{int}}\mathbf{y}_I = \mu \mathbf{My}_I$, i.e. $\{\mathbf{z}\}^T = \{a_K \mathbf{y}_K^T,\ b_K \mathbf{y}_K^T\}$. Substituting this expression for $\{\mathbf{z}\}$ into (6.6.50) and premultiplying by $\{\mathbf{y}_J, \mathbf{y}_J\}$ enables us to diagonalize each of the submatrices:

$$\begin{bmatrix} \mathbf{I} - \bar{\beta}\Delta t^2 \mathbf{L} & \Delta t \mathbf{I} - \bar{\beta}\Delta t^2 \mathbf{G} \\ -\bar{\gamma}\Delta t \mathbf{L} & \mathbf{I} - \bar{\gamma}\Delta t \mathbf{G} \end{bmatrix} \begin{Bmatrix} \mathbf{a} \\ \mathbf{b} \end{Bmatrix} = \mu \begin{bmatrix} \mathbf{I} + \beta \Delta t^2 \mathbf{L} & \beta \Delta t^2 \mathbf{G} \\ \gamma \Delta t \mathbf{L} & \mathbf{I} + \gamma \Delta t \mathbf{G} \end{bmatrix} \begin{Bmatrix} \mathbf{a} \\ \mathbf{b} \end{Bmatrix} \tag{6.6.51}$$

$$\mathbf{L} = \begin{bmatrix} \omega_1^2 & 0 & \cdot & 0 \\ 0 & \cdot & \cdot & 0 \\ \cdot & \cdot & \cdot & \cdot \\ 0 & 0 & 0 & \omega_{n_{\text{dof}}}^2 \end{bmatrix}, \quad \mathbf{G} = \begin{bmatrix} 2\xi_1\omega_1 & 0 & \cdot & 0 \\ 0 & 2\xi_2\omega_2 & \cdot & 0 \\ \cdot & \cdot & \cdot & \cdot \\ 0 & 0 & 0 & 2\xi_{n_{\text{dof}}}\omega_{n_{\text{dof}}} \end{bmatrix} \tag{6.6.52}$$

In previous above, ξ_I is the fraction of critical damping in mode I: see (6.6.35). For each mode, (6.6.51) yields two equations which are uncoupled from the other modal equations. The two equations for mode I can be written as

$$\mathbf{H}\begin{Bmatrix} a_I \\ b_I \end{Bmatrix} = 0 \quad \text{where} \quad \mathbf{H} = \begin{bmatrix} A - a\mu & B - b\mu \\ C - c\mu & D - d\mu \end{bmatrix} \tag{6.6.53}$$

$$A = 1 - \bar{\beta}\Delta t^2 \omega^2, \quad B = \Delta t - 2\Delta t^2 \bar{\beta}\xi\omega, \quad C = -\Delta t \bar{\gamma}\omega^2, \quad D = 1 - 2\bar{\gamma}\Delta t \xi \omega$$
$$a = 1 + \beta \Delta t^2 \omega^2, \quad b = 2\beta \Delta t^2 \xi \omega, \quad c = \Delta t \gamma \omega^2, \quad d = 1 + 2\gamma \Delta t \xi \omega$$

The critical time step is obtained by setting det(**H**)=0 and is given in Box 6.2. The time step for the central difference method (β=0, ξ>0) is independent of damping whereas according to (6.6.43) the critical time step of the central difference method decreases with damping. This disagreement arises because the above analysis is for an implicit treatment of damping. This form of the central difference method entails the solution of equations when β=0 and is not truly explicit for a damped system.

6.6.8 *Eigenvalue Inequality and Time Step Estimates*

The critical time steps in the preceding are given in terms of the maximum eigenvalue of the system $\mathbf{Ky} = \lambda \mathbf{My}$. For large systems, the computation of even a single eigenvalue requires a substantial amount of computer time. In nonlinear systems, the stiffness changes with time, so the maximum eigenvalue needs to be recomputed frequently. Therefore, estimates on the maximum eigenvalue that are easy to compute are useful.

Such estimates are provided by the element eigenvalue inequality. The element eigenvalue inequality relates the eigenvalues of the symmetric matrices **A**, $\boldsymbol{A}_e$, **B** and $\boldsymbol{B}_e$, where

$$\mathbf{A} = \sum_e \mathbf{L}_e^T \mathbf{A}_e \mathbf{L}_e, \quad \mathbf{B} = \sum_e \mathbf{L}_e^T \mathbf{B}_e \mathbf{L}_e \tag{6.6.54}$$

and $\mathbf{L}_e$ are the connectivity matrices (see (2.5.1)). The element and system eigenvalue problems are

$$\mathbf{A}_e \mathbf{y}_i^e = \lambda_i^e \mathbf{B}_e \mathbf{y}_i^e, \quad e = 1 \text{ to } n_e, \quad \mathbf{Ay}_i = \lambda_i \mathbf{By}_i \tag{6.6.55}$$

Then the element eigenvalue inequality states that

$$\left|\lambda^{\max}\right| \le \left|\lambda_E^{\max}\right| \quad \text{where} \quad \lambda_E^{\max} = \max_{i,\,e} \lambda_i^e \tag{6.6.56}$$

The above is an extension of a theorem first given by Rayleigh; a proof is given in Belytschko, Smolinski and Liu (1985). Although the theorem is usually given for finite element systems, the matrices $\mathbf{L}_e$ may be unit matrices, so the theorem applies to any sum of matrices. The eigenvalue inequality also applies to the integrand of the stiffness at each quadrature point, as pointed out by Lin (1991). This can be readily seen by noting that the stiffness and mass matrices are the sums of the integrands at the quadrature points.

The eigenvalue inequality is a special case of the Rayleigh nesting theorem, which states: if λ_i are the eigenvalues of $\mathbf{Ay} = \lambda \mathbf{By}$, if the system is constrained by $\mathbf{g}^T\mathbf{y} = a$, where **g** is a column matrix of constants, then the eigenvalues $\bar{\lambda}_i$ of the constrained system are nested by those of the unconstrained system, that is,

$$\lambda_1 \le \bar{\lambda}_1 \le \lambda_2 \le \bar{\lambda}_2 \le \cdots \lambda_{n_{\text{dof}}} \tag{6.6.57}$$

To illustrate the nesting theorem, consider two disconnected elements with no boundary conditions. The stiffness of the two unconnected rod elements of equal length and material properties is

$$\mathbf{K}=\frac{AE}{\ell}\begin{bmatrix}1 & -1 & 0 & 0\\ -1 & 1 & 0 & 0\\ 0 & 0 & 1 & -1\\ 0 & 0 & -1 & 1\end{bmatrix}\quad \mathbf{d}=\left[d_1^{e=1}\;\; d_2^{e=1}\;\; d_2^{e=2}\;\; d_3^{e=2}\right] \tag{6.6.58}$$

The assembly of the element stiffness then corresponds to the imposition of the constraint $d_2^{e=1}=d_1^{e=2}$. Therefore, if the eigenvalues of (6.6.58) are λ_i, $i=1$ to 4, the eigenvalues after the constraint will be $\bar{\lambda}_i$, $i=1$ to 3, and the latter are nested by λ_i. Similarly, the imposition of each essential boundary condition nests the next set of eigenvalues, and reduces the maximum eigenvalue. By the nesting theorem, the maximum eigenvalue of the unconstrained, disjoint set of elements bounds the eigenvalue of the final assembled system, so $\bar{\lambda}^{\max}\le\lambda_4$.

6.6.9 *Element Eigenvalues*

For speed of computation, the element eigenvalues are usually obtained by simple formulas. In the following, such formulas are given for one-and multi-dimensional elements. As part of this, we develop the Courant condition for a one-dimensional mesh.

To illustrate the application of the element eigenvalue inequality, consider a two-node element in a state of uniaxial strain with a diagonal mass matrix. The element represents a section of an infinite slab. We will use the updated Lagrangian formulation and write the uniaxial constitutive equation in terms of the Truesdell rate, $\sigma_{xx}^{\nabla T}=C^{\sigma T}D_{xx}$, where $C^{\sigma T}=C_{1111}^{\sigma T}$ for uniaxial strain as shown in Example 5.1. The element eigenvalue problem $\mathbf{K}_e^{\text{int}}\mathbf{y}=\lambda_e\mathbf{M}_e\mathbf{y}$ for the rod is obtained by combining the material and geometric stiffnesses, (E6.2.3) and (E6.2.9), respectively, and taking the diagonal mass matrix (E2.1.11) on the RHS:

$$\frac{A\left(C^{\sigma T}+\sigma_{xx}\right)}{\ell}\begin{bmatrix}1 & -1\\ -1 & 1\end{bmatrix}\begin{Bmatrix}y_1\\ y_2\end{Bmatrix}=\frac{\lambda\rho A\ell}{2}\begin{bmatrix}1 & 0\\ 0 & 1\end{bmatrix}\begin{Bmatrix}y_1\\ y_2\end{Bmatrix} \tag{6.6.59}$$

The subscripts e have been omitted. The eigenvalues of the previous are obtained by setting the determinant equal to zero, giving

$$\det\begin{bmatrix}1-\alpha & -1\\ -1 & 1-\alpha\end{bmatrix}=0,\quad \text{where}\quad \alpha=\frac{\lambda\ell^2}{2c^2},\quad c^2=\frac{C^{\sigma T}+\sigma_{xx}}{\rho} \tag{6.6.60}$$

where c is the instantaneous wave speed. Note that the instantaneous wave speed depends on the state of stress. The roots of this are $\alpha=0$ and $\alpha=2$, which gives $\lambda_{\max}=4c^2/\ell^2$. The element eigenvalue inequality (6.6.56) and (6.6.20) then gives the following critical time step for the central difference method in the absence of damping:

$$\Delta t_{\text{crit}}\le\min_e\frac{\ell_e}{c_e} \tag{6.6.61}$$

(the element identifiers are added for clarity). The previous is the same as the critical time step given in (6.2.13). This estimate is often used for two-and three-dimensional problems with ℓ_e the smallest distance between any two nodes of the element.

The above critical time step was first obtained in finite difference methods by Courant, Lewy and Friedrichs (1928). However, their analysis was limited to uniform meshes of infinite bodies. The theory that has evolved in finite element methods is applicable to arbitrary meshes with arbitrary linear boundary conditions.

When the same procedure is applied to a total Lagrangian formulation, the eigenvalue problem is

$$\frac{A_0(F^2C^{SE}+S_{11})}{\ell_0}\begin{bmatrix} 1 & -1 \\ -1 & 1 \end{bmatrix}\begin{Bmatrix} y_1 \\ y_2 \end{Bmatrix} = \frac{\lambda \rho_0 A_0 \ell_0}{2}\begin{bmatrix} 1 & 0 \\ 0 & 1 \end{bmatrix}\begin{Bmatrix} y_1 \\ y_2 \end{Bmatrix} \tag{6.6.62}$$

where $F=\ell/\ell_0$; the stiffness is taken from Example 6.2. The maximum eigenvalue is

$$\lambda_{\max} = \frac{4c_0^2}{\ell_0^2} \quad \text{where} \quad c_0^2 = \frac{F^2C^{SE}+S_{11}}{\rho_0}, \qquad \Delta t_{\text{crit}} \le \min_e \frac{\ell_0^e}{c_0^e} \tag{6.6.63}$$

The above analyses are also applicable to rods if C^{SE} is replaced by E^{SE}: see Example 5.1. We leave it as an exercise to show that the time step (6.6.63) for the reference configuration is identical to (6.6.61), which was developed in the current configuration. So the pull-back of the element eigenproblem leads to identical eigenvalues and an identical critical time step.

For continuum elements, eigenvalue estimates are given for the 4-node quadrilateral and 4-node to 8-node one-point quadrature elements for isotropic materials by Flanagan and Belytschko (1981). Both upper and lower bounds and the maximum eigenvalue are given:

$$\frac{1}{n_{SD}}\mathbf{b}_i^T(\xi_Q)\mathbf{b}_i(\xi_Q) \le \frac{\lambda_{\max}}{n_N c_{\tan}^2} \le \mathbf{b}_i^T(\xi_Q)\mathbf{b}_i(\xi_Q) \quad \text{(no sum on } Q) \tag{6.6.64}$$

where $1 \le n_{SD} \le 3$ is the dimension of the problem, n_N is the number of nodes in the element, ξ_Q is the quadrature point, and $\mathbf{b}_I = \mathrm{N}_{I,X}$ (or $b_{iI} = \partial N_I/\partial x_i$). These inequalities can also be applied to elements with more than one quadrature point as follows. Since the eigenvalue inequality holds for any sum of matrices, (6.6.64) holds for any quadrature point (Lin, 1991).

The time steps for other elements are given in Table 6.1. Note that the subscript e is dropped from ℓ and c, and that r_g is the radius of gyration. The following remarks are of interest:

1. Δt_{crit} for the consistent mass is smaller than for the diagonal mass.
2. Δt_{crit} is smaller for higher-order elements.
3. For the beam, Δt_{crit} varies either with ℓ^2 when $\ell/r_g < 4\sqrt{3}$, or with ℓ when it is greater. This results from an interplay of parabolic and hyperbolic behavior in the governing PDE: hyperbolic behavior governs the stable time step for long elements whereas parabolic behavior due to bending governs for short elements.

Critical time step estimates based on the eigenvalue inequality and linearization are very good for problems with C^1 constitutive laws and smooth response (no impact), but even then,

Table 6.1 Element eigenvalues and time steps

Element	**M**	$\omega^e_{\max}$	Δt^e_{crit}
2-node rod	diagonal by row-sum	$\dfrac{2c}{\ell}$	$\dfrac{\ell}{c}$
2-node rod	consistent	$\dfrac{2\sqrt{3}c}{\ell}$	$\dfrac{\ell}{\sqrt{3}c}$
3-node rod	diagonal by row-sum	$\dfrac{2\sqrt{6}c}{\ell}$	$\dfrac{\ell}{\sqrt{6}c}$
2-node beam: cubic transverse, linear axial $\mathbf{v}(\xi)$	diagonal by row-sum		$\min\begin{cases}\sqrt{3}\ell^2/12cr_g\\ \ell/c\end{cases}$

2–5% reductions are advisable for nonlinear problems. For rough problems, slightly bigger reductions should be made in the time step, say 7–20%. The critical time step estimates based on the element eigenvalue inequality are conservative, that is, the estimated time step is smaller than or equal to the critical time step. For uniform meshes, the difference is small. However, when large changes of element stiffness or element size occur between adjacent elements, the estimates become very conservative, that is, the estimated time step is much smaller than the critical time step for the mesh. LS-DYNA has the option to compute the maximum eigenvalue by an iterative algorithm. This is quite expensive, but in long calculations the increased time step more than compensates for the eigenvalue calculation. A better recourse is to compute the eigenvalues for small element groups where the element size varies quite much and to use the element eigenvalue inequality on the element groups.

6.6.10 Stability in Energy

For some classes of nonlinear problems it is possible to prove unconditional stability of a time integrator by showing that a positive-definite quantity, like the energy, is constant or decreases. We here describe a proof from Belytschko and Schoeberle (1975) for the trapezoidal rule (Newmark β-method with $\beta = \frac{1}{4}$, $\gamma = \frac{1}{2}$) for the equations of motion. We consider the energy as defined in (6.2.14–16) in terms of $\Delta\mathbf{d}$ and assume that the internal energy is a norm for the displacements. The initial conditions are nonzero and external forces are neglected. We will show that the sum of the kinetic and internal energy is bounded, that is, that

$$W^{n+1} \equiv W^{n+1}_{\text{kin}} + W^{n+1}_{\text{int}} \leq (1+\varepsilon)\,W^{0}_{\text{kin}} \tag{6.6.65}$$

where ε is a small number. The concept underlying this proof differs somewhat from the definition of stability (6.6.1) we have used so far. We do not consider a perturbation of a solution but instead we show that the energy is bounded for any solution with nonzero initial conditions. Since the kinetic energy is a positive-definite function of the velocities and the internal energy grows monotonically with the displacements, boundedness of the total energy implies that the response is bounded, and thus stable.

To obtain the previous energy inequality, we start with the energy as defined in Section 6.2.3:

$$W^{n+1} = W_{\text{kin}}^{n+1} + W_{\text{int}}^{n+1} = W_{\text{kin}}^{n+1} + W_{\text{int}}^{n} + \frac{1}{2}\Delta\mathbf{d}^T\left(\mathbf{f}_{\text{int}}^n + \mathbf{f}_{\text{int}}^{n+1}\right) \tag{6.6.66}$$

where $\Delta\mathbf{d} = \mathbf{d}^{n+1} - \mathbf{d}^n$. Using (6.2.16) and (6.3.4–5) with $\beta = \frac{1}{4}$, $\gamma = \frac{1}{2}$ gives

$$\begin{aligned} W_{\text{kin}}^{n+1} &= W_{\text{kin}}^{n} + \frac{\Delta t}{4}(\mathbf{v}^n)^T\mathbf{M}(\mathbf{a}^n + \mathbf{a}^{n+1}) + \frac{\Delta t^2}{8}(\mathbf{a}^n + \mathbf{a}^{n+1})^T\mathbf{M}(\mathbf{a}^n + \mathbf{a}^{n+1}) \\ &= W_{\text{kin}}^{n} + \frac{1}{2}\left(\frac{\Delta t}{2}\mathbf{v}^n + \frac{\Delta t^2}{4}(\mathbf{a}^n + \mathbf{a}^{n+1})\right)^T\mathbf{M}(\mathbf{a}^n + \mathbf{a}^{n+1}) \\ &= W_{\text{kin}}^{n} + \frac{1}{2}\Delta\mathbf{d}^T\mathbf{M}(\mathbf{a}^n + \mathbf{a}^{n+1}) \end{aligned} \tag{6.6.67}$$

where the last step follows from (6.3.4–5). Substituting (6.6.67) into (6.6.66) gives

$$W^{n+1} = W^n + \frac{1}{2}\Delta\mathbf{d}^T\left(\mathbf{M}\mathbf{a}^n + \mathbf{f}_{\text{int}}^n + \mathbf{M}\mathbf{a}^{n+1} + \mathbf{f}_{\text{int}}^{n+1}\right) \tag{6.6.68}$$

If we recall $\mathbf{r} = \mathbf{M}\mathbf{a} + \mathbf{f}_{\text{int}} - \mathbf{f}_{\text{ext}}$ and that $\mathbf{f}_{\text{ext}} = \mathbf{0}$, then (6.6.68) becomes

$$W^{n+1} = W^n + \frac{1}{2}\Delta\mathbf{d}^T(\mathbf{r}^n + \mathbf{r}^{n+1}) \tag{6.6.69}$$

From the previous it can be seen that the energy can only increase as much as the error in the solution of the nonlinear algebraic equations. In fact, if we hypothesize that the nonlinear equations are solved to infinite precision so that $\mathbf{r} = \mathbf{0}$, then the last term on the RHS of the above vanishes and the energy is constant. This implies that energy cannot grow. When the energy decreases, as in a damped system, it is said that the *energy is contractive*. The notion of the equivalence of contractivity and stability was introduced by Banach. Since the kinetic energy is a positive-definite function of the velocities and the internal energy grows with the displacements, the boundedness of the total energy implies that the solution cannot grow unboundedly, so it is stable. Since this result is independent of the time step, the integrator is unconditionally stable.

6.7 Material Stability

6.7.1 Description and Early Work

An important issue in nonlinear mechanics is the stability of the material model. In this section, we describe criteria for material stability. For this purpose, we consider an infinite slab of material in a homogeneous state and examine its response to a small perturbation. As before, the growth of the perturbation is taken to indicate an instability. Some remedies for the numerical difficulties incurred by material instabilities are also discussed.

The literature on material instability goes back at least as far as Hadamard (1903) who examined the question of what happens when the tangent modulus in a small deformation problem is negative. A material with negative tangent moduli is said to *strain soften*. Hadamard identified the conditions for a vanishing propagation speed of an acceleration wave, which he identified as a material instability. A milestone in the study of unstable materials is the work of Hill (1962). In his examination of material stability, he considered an infinite body of the material in a homogeneous state of stress and deformation. He then applied a small perturbation to the body and obtained an expression for its response. If the perturbation grows, the material is considered unstable; otherwise it is stable. Another milestone paper is that of Rudnicki and Rice (1975), who showed that instabilities or localization can occur even in the presence of strain hardening when plasticity is non-associative. In other words, when the tangent moduli lack major symmetry, the material may be unstable even if there is no strain softening.

Material instabilities are usually associated with a localized growth of the deformation. This is called *localization*. It corresponds to phenomena which are observed in nature: for certain stress states, metals, rocks and soils will exhibit narrow bands of intense deformation: these are often called *shear bands*, since the deformation mode in these bands is usually shear.

6.7.2 Material Stability Analysis

In the following we analyze a rate-independent material model for stability; the analysis is based on Rice (1976). An infinite body in a homogeneous state of stress is subjected to a perturbation. Consider the total Lagrangian form of the governing equations for the continuum: the momentum equation, the constitutive equation and the relationship of the Green strain rate to the rate of the deformation gradient. These are

$$\nabla_0 \cdot \mathbf{P} = \rho_0 \ddot{\mathbf{u}}, \quad \dot{\mathbf{S}} = \mathbf{C}^{SE} : \dot{\mathbf{E}}, \quad \dot{\mathbf{E}} = \frac{1}{2}(\dot{\mathbf{F}}^T \mathbf{F} + \mathbf{F}^T \dot{\mathbf{F}}) \tag{6.7.1}$$

(see Box 3.3 and Section 6.4.1, particularly (6.4.3), (6.4.9), and (6.4.10)). No body forces are considered and heat transfer is neglected so the energy equation is omitted. Since the body is infinite, there are no boundary conditions. The position of the body prior to the perturbation is $\bar{\mathbf{x}}(\mathbf{X})$ and the state of nominal stress and the deformation gradient are $\bar{\mathbf{P}}$ and $\bar{\mathbf{F}}$, respectively, which are both constant throughout the body.

The body is perturbed by $\tilde{\mathbf{u}}(\mathbf{X}, t)$, so the total perturbed motion is

$$\Phi(\mathbf{X}, t) = \bar{\mathbf{x}}(\mathbf{X}) + \tilde{\mathbf{u}}(\mathbf{X}, t) \tag{6.7.2}$$

In the above and the following, any variable associated with the perturbations is indicated by a superposed ($\tilde{\bullet}$). The perturbation is assumed to be a plane harmonic wave in an arbitrary direction defined by $\mathbf{n}^0$ in the reference configuration:

$$\tilde{\mathbf{u}} = \mathbf{g}e^{(\omega t + ik\mathbf{n}^0 \cdot \mathbf{X})} \equiv \mathbf{g}e^{\alpha(\mathbf{X}, t)} \quad \text{where} \quad \alpha(\mathbf{X}, t) = \omega t + ik\mathbf{n}^0 \cdot \mathbf{X} \tag{6.7.3}$$

where k is real, $\mathbf{g}$ is a constant vector and $i = \sqrt{-1}$. The perturbation in the deformation gradient is

$$\tilde{F}_{rs} = \frac{\partial \tilde{u}_r}{\partial X_s} = ikg_r n_s^0 e^{\alpha} \quad \text{or} \quad \tilde{\mathbf{F}} = ike^{\alpha}\mathbf{g} \otimes \mathbf{n}^0 \tag{6.7.4}$$

The perturbation in the stress can be obtained by starting with $\mathbf{P}=\mathbf{S}\cdot\mathbf{F}^T$ (Box 3.2):

$$\tilde{P}_{ij} = \tilde{S}_{ik}\bar{F}_{kj}^T + \bar{S}_{ik}\tilde{F}_{kj}^T = C_{ikas}^{SE}\bar{F}_{ar}^T\tilde{F}_{rs}\bar{F}_{kj}^T + \bar{S}_{ib}\tilde{F}_{bj}^T = A_{ijrs}\tilde{F}_{rs} \tag{6.7.5}$$

where we have used the minor symmetry of $\mathbf{C}^{SE}$ and defined $\boldsymbol{A}$ by

$$A_{ijrs} = \bar{F}_{jb}\bar{F}_{ra}C_{ibas}^{SE} + \bar{S}_{is}\delta_{rj} \tag{6.7.6}$$

The tensor $\boldsymbol{A}$ is related to the first elasticity tensor when the material is hyperelastic so it usually lacks minor symmetry; see Section 5.4.8. Note the similarity of the above to (6.4.3), which is also a linearization of the constitutive relation.

The perturbed (i.e., linearized) equation of motion is obtained by substituting $\mathbf{P}=\bar{\mathbf{P}}+\tilde{\mathbf{P}}$ into (6.7.1) and noting that $\bar{\mathbf{P}}$ is an equilibrium solution:

$$\frac{\partial \tilde{P}_{ji}}{\partial X_j} = \rho_0 \frac{\partial^2 \tilde{u}_i}{\partial t^2} \tag{6.7.7}$$

Substituting (6.7.4) into (6.7.5) and the result into (6.7.7) gives

$$\rho_0 \frac{\partial^2 \tilde{u}_i}{\partial t^2} = \frac{\partial}{\partial X_j}(A_{jisr}\tilde{F}_{rs}) = -k^2 e^{\alpha} A_{jisr} g_r n_s^0 n_j^0 \tag{6.7.8}$$

Substituting (6.7.3) into the LHS of above yields

$$\left(\rho_0 \omega^2 \delta_{ri} + k^2 A_{jisr} n_s^o n_j^o\right) g_r = 0 \quad \text{for} \quad i = 1 \text{ to } n_{SD} \tag{6.7.9}$$

This can be rewritten in the form

$$\left(\frac{\omega^2}{k^2}\delta_{ir} + \frac{1}{\rho_0}\bar{A}_{ir}\right) g_r = 0 \quad \text{where} \quad \bar{A}_{ir}(\mathbf{n}^0) = A_{jisr} n_j^0 n_s^0 \tag{6.7.10}$$

$\bar{A}(\mathbf{n}^0)$ is called the *acoustic tensor*. The above is a set of homogeneous linear algebraic equations. Nontrivial solutions exist only when the determinant vanishes. This yields the characteristic equation for the complex frequencies ω_I:

$$\det\left[\frac{\omega_I^2}{k^2}\delta_{ir} + \frac{1}{\rho_0}\bar{A}_{ir}\right] = 0 \tag{6.7.11}$$

The stability conditions can be ascertained from (6.7.3). If we let $\omega_I = a + ib$, then the perturbation (6.7.3) can be written as

$$\tilde{\mathbf{x}} = \tilde{\mathbf{u}} = \mathbf{g}e^{(at+ibt+ik\mathbf{n}^0\cdot\mathbf{X})} = \mathbf{g}\cdot \underbrace{e^{at}}_{\substack{\text{growth}\\ \text{or decay}}} \cdot \underbrace{e^{i(bt+k\mathbf{n}^0\cdot\mathbf{X})}}_{\substack{\text{constant}\\ \text{amplitude}\\ \text{wave}}} \tag{6.7.12}$$

From this we can see that the solution consists of the product of an exponential and a wave of constant amplitude. The real part the exponent ω_I, denoted by a, governs the growth or decay of the perturbation. If the real part of ω_I is negative, then the perturbation decays and the material is stable. The response must be stable for any direction of propagation, i.e. $\forall\, \mathbf{n}^o$.On the other hand, if the real part of ω_I for any direction of propagation is positive, then the response grows and the material is unstable. To summarize:

$$\textit{if}\ \mathrm{Re}(\omega_I) = a \le 0 \quad \text{for all } I \text{ and all } \mathbf{n}^0, \text{ the material is stable} \tag{6.7.13}$$

$$\textit{if}\ \mathrm{Re}(\omega_I) = a > 0 \quad \text{for any } I \text{ or any } \mathbf{n}^0, \text{ the material is unstable} \tag{6.7.14}$$

Note the similarity of this analysis to the linear stability analysis of a discrete system in Section 6.5. In both cases a perturbation was applied and an eigenvalue problem was derived for the exponent. The characteristic equation that determines stability is then obtained by setting the determinant of the characteristic matrix equal to zero. However, the stability conditions in this case are less tractable than in Section 6.5, since the real parts of the frequencies ω_I must be nonpositive for all directions $\mathbf{n}^o$.

As in the discrete system, a simple sufficient condition for stability can be deduced when the acoustic tensor $\bar{\boldsymbol{A}}$ is symmetric: the positive-semidefiniteness of the acoustic tensor for all $\mathbf{n}^0$ is sufficient for stability. The acoustic tensor $\bar{\boldsymbol{A}}$ is symmetric when $\boldsymbol{A}$ has major symmetry. The eigenvalue problem corresponding to (6.7.10) is $\bar{\mathbf{A}}\mathbf{g} = \lambda\mathbf{g}$ where $\lambda_I = -\rho_0\omega_I^2 / k^2$. f the acoustic tensor is symmetric and positive-definite for all $\mathbf{n}^0$, then the eigenvalues are real and positive, that is, $\lambda_I \ge 0$. Therefore, all ω_I will be imaginary with no real parts, and by (6.7.13) the response is stable.

The positive-definiteness of $\bar{A}$ for all $\mathbf{n}^0$ can also be expressed as

$$A_{ijsr} n_i^0 n_s^0 h_j h_r > 0 \quad \forall \mathbf{h} \text{ and } \mathbf{n}^0 \tag{6.7.15}$$

This is called the *strong ellipticity condition*. When the strong ellipticity holds, the PDE for equilibrium is elliptic.

The properties of $\boldsymbol{A}$ with respect to arbitrary second-order tensor ε_{ij} are also sometimes examined. The condition for stability is

$$\varepsilon_{ij} A_{ijsr} \varepsilon_{sr} > 0 \quad \forall \varepsilon \ne 0 \tag{6.7.16}$$

This is a stronger condition than the strong ellipticity condition (6.7.15), as can be seen by noting that in (6.7.16) $\varepsilon_{ij} = n_i^0 h_j$, so that ε_{ij} is restricted to rank 1 tensors: see Ogden (1984: 389). The set of rank 1 second-order tensors $\varepsilon_{ij} = n_i^0 h_j$ is a subset of the space of arbitrary second-order tensors ε_{ij} in (6.7.16). Note that (6.7.16) corresponds to positive definiteness of $\boldsymbol{A}$.

If we write out $\boldsymbol{A}$ by (6.7.6), the strong ellipticity condition (6.7.15) becomes

$$\left(\bar{F}_{ra}\bar{F}_{ib}C_{jbas}^{SE} + S_{js}\delta_{ir}\right) n_j^0 n_s^0 h_i h_r > 0 \quad \forall \mathbf{h} \text{ and } \mathbf{n}^0 \tag{6.7.17}$$

Note that the above stability condition depends on the state of stress. Material stability always depends the state of stress.

This condition can be expressed in the current configuration by taking the current configuration to be the reference configuration; using the procedure in Section 6.4.4, we note that when the current configuration is the reference configuration, $\mathbf{F}=\mathbf{I}$, $\mathbf{C}^{SE}=\mathbf{C}^{\sigma T}$, $\mathbf{S}=\sigma$, and $\mathbf{n}=\mathbf{n}^0$, giving that for material stability

$$\left(C^{\sigma T}_{jirb}+\sigma_{jb}\delta_{ir}\right)n_b n_j h_i h_r>0 \quad \forall\mathbf{h} \text{ and } \mathbf{n} \tag{6.7.18}$$

The assumptions in this stability analysis should be borne in mind. We have taken an infinite slab of a material in a specified state of deformation and stress and perturbed it by plane waves. Then we assumed that the material responds linearly according to the tangent modulus matrix $\mathbf{C}^{SE}$ or its spatial counterpart $\mathbf{C}^{\sigma T}$. We have also assumed that the material response does not change in loading or unloading, so the material tangent moduli are constant. This type of material when applied to elastoplasticity is called a *linear comparison solid*. In most cases it gives good stability estimates.

Although the model in this analysis is highly idealized, it is mathematically tractable and it works: materials which are unstable according to this idealized analysis will show unstable behavior in complicated inhomogeneous stress states. For inhomogeneous stress states, the material instability initiates in a narrow band. The instability is local, and it often grows in band-like structures, such as shear bands. The *system* does not become unstable until the material instability or localization has grown sufficiently to permit the failure of the system. Thus the system is not unstable until a shear band has grown across a body or a crack has traversed the structure.

The acoustic matrix $\hat{\mathbf{A}}$ is not symmetric when $C^{\sigma T}$ does not possess major symmetry, as in non-associative plasticity or associative plasticity based on the Jaumann rate of Cauchy stress (see Box 5.4). The rich and varied possibilities for localization in materials are examined in Rudnicki and Rice (1975) and Dobovsek and Moran (1996); in the latter is it shown that in some viscoplastic materials the equilibrium branches are split by a Hopf bifurcation.

6.7.3 *Material Instability and Change of Type of PDEs in 1D*

The loss of material stability changes the character of the PDE. For simplicity, we show this in one dimension for an infinite slab in a state of uniaxial stress. We begin with a perturbation analysis in one dimension. The one-dimensional counterpart of (6.7.8) is

$$\rho_0\frac{\partial^2\tilde{u}}{\partial t^2}=\frac{\partial(A\tilde{F}_{11})}{\partial X} \quad \text{where} \quad A\equiv A_{1111}=E^{SE}F_{11}^2+S_{11} \tag{6.7.19}$$

In the above, A is an element of the first elasticity tensor. If we apply the perturbation $\tilde{u}=e^{(\omega t+ikX)}$, we obtain the characteristic equation

$$\frac{\omega^2}{k^2}+\frac{A}{\rho_0}=0 \tag{6.7.20}$$

When $A\geq 0$, ω is imaginary with no real part so the response is stable. If $A<0$, ω is real and the response is unstable.

We next consider (6.7.20) in the current configuration. When the reference configuration corresponds to the configuration $\rho_0=\rho$ and $A=E^{\sigma T}+\sigma_{11}$, so (6.7.20) becomes

$$\frac{\omega^2}{k^2}+\frac{E^{\sigma T}+\sigma_{11}}{\rho}=0 \tag{6.7.21}$$

It can immediately be seen that when $E^{\sigma T}+\sigma_{11}>0$, the material response is stable, whereas when $E^{\sigma T}+\sigma_{11}<0$, the material response is unstable. As an example, consider the constitutive relation $\sigma_{11}^{\nabla J}=E^{\sigma J}D_{11}$ for an incompressible material in a state of uniaxial stress where $E^{\sigma J}$ is the material tangent stiffness for the Jaumann rate. From the relation in Box 5.1 between Jaumann and Truesdell rates of Cauchy stress, it follows that for an incompressible material in uniaxial stress, $E^{\sigma T}+2\sigma_{11}=E^{\sigma J}$. Thus the critical stress where the material changes from stable to unstable, $E^{\sigma T}+\sigma_{11}=0$ corresponds to $E^{\sigma J}=\sigma_{11}$. This is the famous Considere criterion for the onset of necking in a tensile bar. From (6.4.18) it can be seen that it corresponds to a maximum in the nominal stress $(\dot{P}_{11}=\lambda^{-1}(E^{\sigma T}+\sigma_{11})D_{11}=0)$.

When the reference configuration is the current configuration the perturbed wave equation (6.7.19) becomes

$$\rho\tilde{u}_{,tt}=(E^{\sigma T}+\sigma_{11})\tilde{u}_{,xx} \tag{6.7.22}$$

Comparing (6.7.22) with (1.5.9), we can see that when the material is stable (and $E^{\sigma T}+\sigma_{11}>0$), the PDE is hyperbolic. When the material is unstable (and $E^{\sigma T}+\sigma_{11}<0$), the system becomes elliptic (compare with (1.5.14)). Thus when $E^{\sigma T}+\sigma_{11}$ changes from positive to negative, the equations change type from hyperbolic to elliptic. This is called *loss of hyperbolicity* of the PDEs. If we consider the antiplane equilibrium equations in two dimensions, the perturbed equations are $G_1\tilde{u}_{,xx}+G_2\tilde{u}_{,yy}=0$. When one of the moduli becomes negative due to material instability, the PDE changes from elliptic to hyperbolic (i.e., it looks like the wave equation in space). This is called *loss of ellipticity*. Thus material instability is associated with a change in type of the PDE.

6.7.4 Regularization

Shortly after nonlinear finite element programs became available in the 1970s, computational analysts began to include unstable material models, both intentionally and inadvertently, and they discovered many difficulties. Numerical solutions often became unstable and it was discovered that results depended very much on the mesh. At that time it was argued by some mechanicians that material models that violate the stability postulates should never be used. Their arguments had some merit, for unless a constitutive law is carefully designed to be unstable only in the rare situations when the material is unstable, many difficulties are encountered. However, there is no way to replicate observed phenomena such as shear banding without a material model that exhibits strain softening.

In an effort to explain these difficulties, Bazant and Belytschko (1985) constructed a closed-form solution for a rate-independent material model with strain softening in one dimension. They showed that for rate-independent materials, when the material reaches a state of instability, the strain grows to infinity at a single point. Thus the strain localizes, as expected for a material instability, but it localizes to a set of measure zero. They also showed that the dissipation over a set of measure zero vanishes, so that these models cannot represent fracture; fracture always involves significant dissipation.

This led to the search for an effective regularization of the governing equations, which were also called localization limiters. At Northwestern, we soon discovered that both gradient models and nonlocal models regularize the solution (Bazant, Belytschko and Chang, 1984; Lasry and Belytschko, 1988). This remedy of the difficulties associated with negative moduli had already been made in another context, the heat equation, where Cahn and Hilliard circumvented the difficulty by a gradient theory. This has come to be known as the Cahn–Hilliard theory. Hilliard was incidentally also at Northwestern but we were unaware of his work until later. Aifantis (1984) was one of the first to study gradient regularization in solid mechanics; the work in Triantifyllides and Aifantis (1986) is also interesting.

Subsequently a plethora of work emerged in this area. It had two major goals: to obtain physical justifications for regularization procedures and to simplify the treatment of nonlocal and gradient models. Schreyer and Chen (1986) introduced a regularization based on the gradient of the plasticity parameter $\dot{\lambda}$ (see (5.5.5)). Pijaudier-Cabot and Bazant (1987) introduced the gradient on the damage parameter in material models with damage. These are important contributions because introducing nonlocality in the six strain components is awkward indeed. Mulhaus and Vardoulakis (1987) showed that a coupled stress theory also regularizes the equations, and Needleman (1988) showed that viscoplasticity provides a regularization of strain-softening. However, Bayliss *et al.* (1994) showed that viscous regularization still leads to exponential growth. deBorst *et al.* (1993) showed that the plasticity consistency requirement introduces another partial differential equation; the boundary conditions for these partial differential equations are still an enigma. Fleck *et al.* (1994) reported experiments in which metal plasticity depends on scale, and developed a gradient plasticity theory motivated by dislocation movement.

Four regularization techniques have been proposed for unstable materials:

1. Gradient regularization, in which a gradient of a field variable is introduced in the constitutive equation.
2. Integral, or nonlocal, regularization, in which the constitutive equation is a function of a nonlocal variable, such as nonlocal damage, a nonlocal invariant of a strain, or a nonlocal strain.
3. Coupled stress regularization.
4. Regularization by introducing rate-dependence in the material.

All of these except the last are still in early states of development. Little is known about the material constants for these models and the material length scales.

Regularization via viscoplastic material laws has achieved substantial robustness. However, viscoplastic regularization has some idiosyncrasies: there is no intrinsic length scale in the viscoplastic model, and material instability is associated with exponential growth of the response in the localization band. Therefore, although a discontinuity does not develop in the displacement, the gradient in the displacement becomes unbounded. Wright and Walter (1987) have shown that this anomaly can be rectified by coupling the momentum equation to heat conduction via the energy equation: the computed length scales then agree well with observed shear band widths in metals.

The computation of localization still poses substantial difficulties. For most materials, the characteristic widths of shear bands are much smaller than body dimensions. Therefore tremendous resolution is required to obtain a reasonably accurate solution; see Belytschko *et al.*

(1994) for what were high-resolution computations in the early 1990s. Solutions of localization problems converge very slowly with mesh refinement. This behavior is often called mesh sensitivity or lack of objectivity, though it has nothing to do with objectivity or its absence: it is simply a consequence of the inability of coarse meshes to resolve sharp gradients.

Several techniques have evolved to improve the coarse-mesh accuracy of finite element models for unstable materials. These involve the embedment of discontinuities or enriched fields in the element. Ortiz, Leroy and Needleman (1987) were the first to modify an element at the point of material instability: they embedded discontinuities in the strain field of the 4-node quadrilateral when the acoustic tensor indicated a material instability in the element. Belytschko, Fish and Englemann (1988) embedded a displacement discontinuity by enriching the strain field with a narrow band in which the material is unstable. In the band, the material behavior was considered homogeneous, which is ridiculous since an unstable material cannot remain in a homogeneous state of stress: any perturbation will trigger a growth on the scale of the perturbation. Such is hindsight. Nevertheless these models were able to capture the evolving discontinuity in displacement more effectively. Simo, Oliver and Armero (1993) invoked the theory of distributions to justify a similar method of enrichment. They also categorized discontinuities as strong (in the displacements) and weak (in the strains).

Just as shear bands can be viewed as the outcome of a material instability in the shear component, fracture can be considered the outcome of a material instability in the components normal (and tangential in the case of mode 2 fracture) to the discontinuity. The relationship between damage and fracture has long been noted (see Lemaitre and Chaboche, 1994, where a fracture is assumed to occur when the damage variable reaches 0.7). The origin of the number 0.7 is quite hazy in most works on damage mechanics, but the relation to the phase transition point based on percolation theory of 0.59275 is interesting. The modeling of fracture by constitutive models with damage poses some of the same difficulties encountered in shear band modeling, since the material law becomes unstable when the damage exceeds a threshold value. All of the peculiarities associated with a material instability occur: localization to a set of measure zero for rate-independent models (or exponential growth for simple rate-dependent local models), zero energy dissipation and absence of a length scale.

These difficulties were resolved in a novel manner early in the evolution of finite element crack modeling by Hillerborg *et al.* (1976). Their idea was to match the dissipation in fracture to the energy dissipated in the element which exceeds the stability threshold. This is accomplished by treating the fracture energy as a material parameter. The dissipation in the strain-softening element is equated to the fracture energy:

$$W^{\text{fract}} = A^e W^{\text{cont}}(\varepsilon^{\text{final}}, h) \tag{6.7.23}$$

where W^{fract} is the fracture energy associated with crack growth across the element, A^e is the area of the element, and $\varepsilon^{\text{final}}$ is the strain at which the stress goes to zero. The strain $\varepsilon^{\text{final}}$ is of course an artifice and has no relationship to the failure strain of the material. Instead it is chosen so that the energy in the discrete model matches the measured fracture energy. When the element size A^e is changed, the strain $\varepsilon^{\text{final}}$ must be altered so that the energy dissipated by the fracture is invariant with element size. Thus the constitutive equation depends on the element size! This is a strange but intriguing idea because the PDE now depends on a discretization parameter, the element size. Experience shows it works very well. It is a step in the direction of constitutive formulations directly in terms of finite elements rather than the PDE.

6.8 Exercises

6.1. Use Nanson's law (3.4.5) and the result obtained in Exercise 3.4 (Chapter 3) for the material time derivative of a surface integral,

$$\frac{d}{dt}\int_S g\mathbf{n}\,dS = \int_S \left[\left(\dot{g} + g\nabla\cdot\mathbf{v}\right)\mathbf{I} - g\mathbf{L}^T\right]\cdot\mathbf{n}\,dS,$$

to develop a linearization of the load stiffness $\mathbf{K}^{\text{ext}} = \partial\mathbf{f}^{\text{ext}}/\partial\mathbf{d}$ for a pressure load applied on a surface mapped from the biunit square in the parent element.

6.2. Show that (6.3.60) corresponds to the stationary points of (6.3.59).

6.3. Show that (6.3.61), the linearized perturbed Lagrangian equations, can be converted to the linearized penalty equations by eliminating the Lagrange multipliers.

6.4. Obtain (6.4.20) by letting the reference configuration in (6.4.4) be the current configuration.

6.5. Show that the critical time steps given by the updated and total Lagrangian formulations in (6.6.61) and (6.6.63) are identical. Use the relations between tangent moduli in Example 5.1 for uniaxial strain.

6.6. Develop the tangent stiffness for an axisymmetric 2-node membrane element.

6.7. Examine the stability of a solution of the two-dimensional heat conduction equation $(k_{ij}\theta_{,j})_{,i}=0$ in the following way. Consider an infinite slab under a uniform temperature and apply the perturbation $\tilde{\theta} = e^{\omega t + i\kappa\mathbf{n}\cdot\mathbf{x}}$ where κ is real. Using the transient equations of heat conduction, determine the conditions under which the solution is stable if k_{ij} is symmetric.

7

Arbitrary Lagrangian Eulerian Formulations

7.1 Introduction

Many problems cannot be treated effectively with Lagrangian meshes. When the material is severely deformed, Lagrangian elements become similarly distorted since they deform with the material. The approximation accuracy of the elements then deteriorates, particularly for higher order elements. Furthermore, the Jacobian determinants may become negative at quadrature points, aborting the calculations or causing severe local inaccuracies. In addition, the conditioning of the linearized Newton equations deteriorates and explicit stable time steps decrease markedly. In many simulations of with severe deformations, remeshing Lagrangian meshes becomes unavoidable. This is burdensome and introduces errors due to projections.

In some problems, Lagrangian methods are totally inappropriate. For example, in fluid mechanics problems with high velocity flows, interest is usually focused on a particular spatial subdomain, such as the domain around an airfoil. Similarly, the modeling of processes such as extrusion involve fixed spatial domains through which the material flows. These types of problems are more suited to Eulerian elements. In Eulerian finite elements, the elements are fixed in space and material convects through the elements. Eulerian finite elements thus undergo no distortion due to material motion; however, the treatment of constitutive equations and updates is complicated due to the convection of material through the elements.

Unfortunately, the treatment of moving boundaries and interfaces is difficult with Eulerian elements. Therefore, hybrid techniques which combine the advantages of Eulerian and Lagrangian methods have been developed. These are called ALE methods: arbitrary Lagrangian Eulerian. The aim of ALE finite element formulations is to capture the advantages of both

Nonlinear Finite Elements for Continua and Structures, Second Edition.
Ted Belytschko, Wing Kam Liu, Brian Moran, and Khalil I. Elkhodary.

Companion Website: www.wiley.com/go/belytschko

Lagrangian and Eulerian finite elements while minimizing the disadvantages. As the name suggests, ALE descriptions are arbitrary combinations of the Lagrangian and Eulerian descriptions. The word arbitrary here refers to the fact that the combination is specified by the user through the selection of a mesh motion. Of course, a judicious choice of the mesh motion is required if severe mesh distortions are to be eliminated, and this often imposes a substantial burden on the user.

As we shall see, the formulation of ALE methods closely parallels that of Eulerian methods. In fact, Eulerian methods are a subset of ALE methods. The literature is quite mature: readers should consult Belytschko and Kennedy (1978), Hughes, Liu and Zimmerman (1981), Liu (1981), Liu and Ma (1982), Liu and Chang (1984), Belytschko and Liu (1985), Liu and Chang (1985), Liu, Belytschko and Chang (1986), Huerta and Liu (1988), Liu, Chang, Chen and Belytschko (1988), Benson (1989), Liu, Chen, Belytschko and Zhang (1991), and Hu and Liu (1993).

The chapter begins with a discussion of motion in an ALE framework in Section 7.2. For this more general framework, another reference coordinate system known as a referential system (which we also call an ALE system) is needed. In an ALE system, mesh motion should be described. Velocities and accelerations of both the material and mesh are developed. In addition, we examine the reduction of ALE formulations to Lagrangian and Eulerian formulations and the special conditions on the mappings that arise in an ALE description.

Section 7.3 describes the conservation equations for an ALE description. We will take the convenient viewpoint that most of what we need is in the Eulerian conservation equations developed in Section 3.5. The ALE weak forms are very similar to those in the updated Lagrangian formulation. One major difference which arises in ALE descriptions is that mass conservation must be treated by the continuity equation, i.e. by a PDE, whereas in Lagrangian formulations it was enforced by an algebraic equation.

A summary of the ALE governing equations is given in Section 7.4, and the weak forms are presented in Section 7.5. The finite element approximations will be developed in Section 7.5.3. The approximations are written immediately in terms of element coordinates. Because of the similarity of the weak form to the updated Lagrangian formulation, the discrete momentum equation is also quite similar. The major difference is the inertial term, which involves a nonconstant mass matrix.

In ALE meshes, the element coordinates no longer serve as surrogate material coordinates. The procedures for updating the stress and the mesh must be carefully designed. These are described in Sections 7.8 and 7.10 respectively.

A major feature of ALE and Eulerian methods is the appearance of convection terms in the momentum and continuity equations. Straightforward treatment of these terms by the methods described in Chapter 6 leads to what are called spatial instabilities. These are similar to the temporal instabilities studied in Chapter 6, but they also appear in time-independent problems. They manifest themselves in spatial oscillations in the dependent variables. We focus on stabilization of these instabilities by Streamline Upwind Petrov–Galerkin (SUPG) methods which are described in Section 7.6. The momentum equation in the Petrov–Galerkin formulation is developed in Section 7.7.

Linearization of the internal force and the external force in the discrete momentum equation will be developed in Section 7.9. An updated ALE numerical example of an elastic–plastic wave propagation problem will be presented in Section 7.11. Finally, in Section 7.12, the total ALE formulation is derived.

7.2 ALE Continuum Mechanics

7.2.1 Material Motion, Mesh Displacement, Mesh Velocity, and Mesh Acceleration

In an ALE method, both the *motion of the mesh and the material must be described.* The motion of the material is described as before by:

$$\mathbf{x} = \boldsymbol{\phi}(\mathbf{X}, t) \tag{7.2.1}$$

where $\mathbf{X}$ are the material coordinates. The function $\boldsymbol{\phi}(\mathbf{X}, t)$ maps the body from the initial configuration Ω_0 to the current or spatial configuration Ω. Although it is called the motion throughout this book, in this chapter we will often call it the *material motion* to distinguish it from the mesh motion. It is identical to the map used to describe the motion of Lagrangian elements.

In the ALE formulation, we consider another reference domain $\hat{\Omega}$ as shown in Figure 7.1. This domain is called the *referential domain* or the *ALE domain*. The initial values of the position of particles are denoted by $\boldsymbol{\chi}$, so:

$$\boldsymbol{\chi} = \boldsymbol{\phi}(\mathbf{X}, 0) \tag{7.2.2}$$

The coordinates $\boldsymbol{\chi}$ are called the *referential* or *ALE coordinates*. In most cases $\boldsymbol{\phi}(\mathbf{X}, 0) = \mathbf{X}$, so $\boldsymbol{\chi}(\mathbf{X}, 0) = \mathbf{X}$. The referential domain $\hat{\Omega}$ is used to describe the motion of the mesh independent of the motion of the material. In the implementation, the domain $\hat{\Omega}$ is used to construct the initial mesh. It remains coincident with the mesh throughout the computation, so it can also be considered the *computational domain*.

The *motion of the mesh* is described by

$$\mathbf{x} = \hat{\boldsymbol{\phi}}(\boldsymbol{\chi}, t) \tag{7.2.3}$$

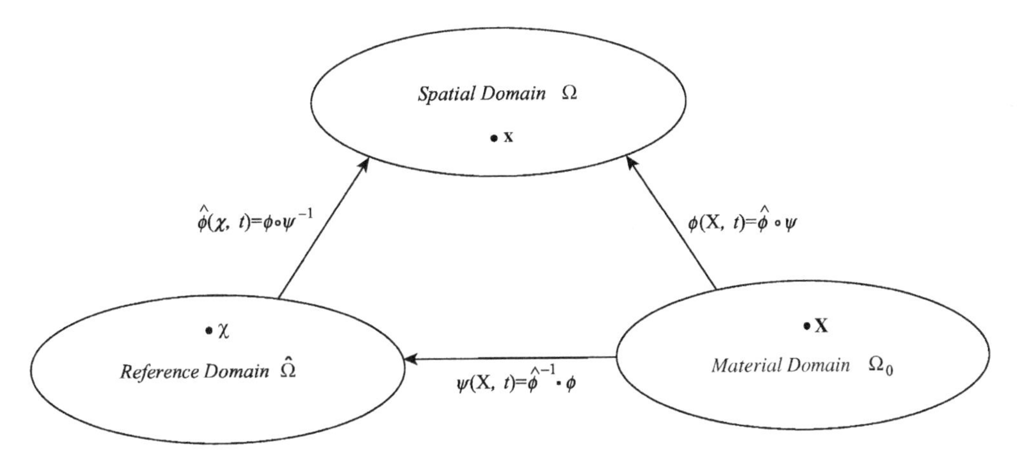

Figure 7.1 Maps between Lagrangian, Eulerian and ALE domains

This map $\hat{\phi}$ plays a crucial role in the ALE finite element formulation. Points χ in the ALE domain, $\hat{\Omega}$, are mapped to points $\mathbf{x}$ in the spatial domain, Ω, via this map.

As is apparent from Figure 7.1, (7.2.1) and (7.2.3), we can relate the ALE coordinates to the material coordinates by a composition of functions:

$$\chi = \hat{\phi}^{-1}(\mathbf{x}, t) = \hat{\phi}^{-1}(\phi(\mathbf{X}, t), t) = \psi(\mathbf{X}, t) \quad \text{or} \quad \psi = \hat{\phi}^{-1} \circ \phi \tag{7.2.4}$$

As can be seen from the previous, the relation between the material coordinates and the ALE coordinates is a function of time.

The material motion can be expressed as a composition of the mesh motion and the ψ map:

$$\mathbf{x} = \phi(\mathbf{X}, t) = \hat{\phi}(\psi(\mathbf{X}, t), t) \quad \text{or} \quad \phi = \hat{\phi} \circ \psi \tag{7.2.5}$$

As will be seen, in the ALE algorithm the mesh motion is prescribed or computed. The material motion can then be reconstructed through the above composition of functions if the map ψ is invertible.

We will now define the displacement, velocity and acceleration of the mesh motion, which will be called the mesh displacement, mesh velocity and mesh acceleration. The *mesh displacement*, $\hat{\mathbf{u}}$, is defined by

$$\hat{\mathbf{u}}(\chi, t) = \mathbf{x} - \chi = \hat{\phi}(\chi, t) - \chi \tag{7.2.6}$$

Note the similarity of the above definition to the definition of material displacement, which is $\mathbf{u} = \mathbf{x} - \mathbf{X}$: the material coordinate in the material description has been replaced by the ALE referential coordinate to obtain the mesh displacement. The *mesh velocity* is also defined analogously to the material velocity:

$$\hat{\mathbf{v}}(\chi, t) = \frac{\partial \hat{\phi}(\chi, t)}{\partial t} \equiv \left.\frac{\partial \hat{\phi}}{\partial t}\right|_{\chi} \equiv \hat{\phi}_{,t}\,[\chi] \tag{7.2.7}$$

In the above, the ALE coordinate χ is fixed; in the expression for the material velocity, the material coordinate $\mathbf{X}$ is fixed. Three notations to be used are shown in (7.2.7). When the independent variables are explicitly given, we simply use the partial derivative with respect to time to indicate the mesh velocity. If the independent variables are not explicitly given, we will designate the coordinate which is fixed either by a subscript following a bar or in brackets following the subscript '$,t$' as shown previously.

The *mesh acceleration* is given by

$$\hat{\mathbf{a}} = \frac{\partial \hat{\mathbf{v}}(\chi, t)}{\partial t} = \frac{\partial^2 \hat{\mathbf{u}}(\chi, t)}{\partial t^2} = \hat{\mathbf{u}}_{,tt[\chi]} \tag{7.2.8}$$

Neither the mesh acceleration nor the mesh velocity have any physical meaning in an ALE mesh which is not Lagrangian. When the mesh is Lagrangian, they correspond to the material velocity and acceleration.

7.2.2 Material Time Derivative and Convective Velocity

In ALE descriptions, fields are usually expressed as functions of the ALE coordinates $\boldsymbol{\chi}$ and time t. The material time derivative (or total derivative) must then be obtained by the chain rule, similar to the process used in Section 3.2.5 to obtain the material time derivative in an Eulerian description. Consider a specific function, $f(\boldsymbol{\chi}, t)$. Using the chain rule gives

$$\frac{Df}{Dt} \equiv \dot{f}(\boldsymbol{\chi}, t) = \frac{\partial f(\boldsymbol{\chi}, t)}{\partial t} + \frac{\partial f(\boldsymbol{\chi}, t)}{\partial \chi_i}\frac{\partial \psi_i(\mathbf{X}, t)}{\partial t} = f_{,t[\chi]} + \frac{\partial f}{\partial \chi_i}\frac{\partial \chi_i}{\partial t} \tag{7.2.9}$$

We now define the *referential particle velocity* w_i by

$$w_i = \frac{\partial \psi_i(\mathbf{X}, t)}{\partial t} = \left.\frac{\partial \chi_i}{\partial t}\right|_{[\mathbf{X}]} \tag{7.2.10}$$

Substituting (7.2.10) into (7.2.9) gives the following expression for the material time derivative (or total derivative):

$$\frac{Df}{Dt} \equiv \dot{f}(\boldsymbol{\chi}, t) = f_{,t[\chi]} + \frac{\partial f}{\partial \chi_i} w_i \tag{7.2.11}$$

In the formulations to be given later, the ALE field variables are often treated as functions of the material coordinates $\mathbf{X}$ and time. Hence, it is convenient to develop expressions for the material time derivative in terms of the spatial gradient. For this purpose, we first develop a relationship between material velocity, mesh velocity and referential velocity. We start with expression for material motion (7.2.1) and equate it to a composition of the functions $\hat{\boldsymbol{\phi}} \circ \boldsymbol{\psi}$, which from Figure 7.1 can easily be seen to be equivalent to the material motion. This yields as in (7.2.5):

$$\mathbf{x} = \boldsymbol{\phi}(\mathbf{X}, t) = \hat{\boldsymbol{\phi}}(\boldsymbol{\psi}(\mathbf{X}, t), t) = \hat{\boldsymbol{\phi}} \circ \boldsymbol{\Psi} \tag{7.2.12}$$

where the last expression shows the motion as a composition of mesh motion and the $\boldsymbol{\psi}$ map. Using the third term to develop a chain rule expression for the material velocity gives

$$v_j = \frac{\partial \phi_j(\mathbf{X}, t)}{\partial t} = \frac{\partial \hat{\phi}_j(\boldsymbol{\chi}, t)}{\partial t} + \frac{\partial \hat{\phi}_j(\boldsymbol{\chi}, t)}{\partial \chi_i}\frac{\partial \psi_i(\mathbf{X}, t)}{\partial t} = \hat{v}_j + \left.\frac{\partial x_j}{\partial \chi_i}\frac{\partial \chi_i}{\partial t}\right|_{[\mathbf{X}]} \tag{7.2.13}$$

where we have used the definition of the mesh velocity (7.2.7). From (7.2.10) we can rewrite the second term on the RHS of (7.2.13) as

$$\frac{\partial x_j(\boldsymbol{\chi}, t)}{\partial \chi_i}\frac{\partial \chi_i(\mathbf{X}, t)}{\partial t} = \frac{\partial x_j}{\partial \chi_i} w_i \tag{7.2.14}$$

Now we define the *convective velocity*, $\mathbf{c}$, as the difference between the material and mesh velocities:

$$c_i = v_i - \hat{v}_i \tag{7.2.15}$$

Using (7.2.13) to express $v_i - \hat{v}_i$ and then substituting in (7.2.14) gives

$$c_i = v_i - \hat{v}_i = \frac{\partial x_i(\boldsymbol{\chi},t)}{\partial \chi_j}\frac{\partial \chi_j(\mathbf{X},t)}{\partial t} = \frac{\partial x_i(\boldsymbol{\chi},t)}{\partial \chi_j} w_j \tag{7.2.16}$$

This relationship between the convected velocity $\mathbf{c}$, material velocity $\mathbf{v}$, mesh velocity $\hat{\mathbf{v}}$ and the referential velocity $\mathbf{w}$ will be used frequently in the ALE formulation.

To develop an expression for the material time derivative with a spatial gradient we note that from (7.2.3), the chain rule gives:

$$\left.\frac{\partial f}{\partial \chi_i}\right|_t = \left.\frac{\partial f}{\partial x_j}\right|_t \left.\frac{\partial x_j}{\partial \chi_i}\right|_t$$

where we have added the bars to emphasize the relations hold only with time fixed. Substituting the previous into (7.2.9) gives

$$\frac{Df}{Dt} = f_{,t[\chi]} + \left.\frac{\partial f}{\partial x_j}\frac{\partial x_j}{\partial \chi_i}\frac{\partial \chi_i}{\partial t}\right|_{[X]} = f_{,t[\chi]} + f_{,j}\frac{\partial x_j}{\partial \chi_i} w_i = f_{,t[\chi]} + f_{,j}\, c_j \tag{7.2.17}$$

where (7.2.10) has been used for the third equality and (7.2.16) for the last equality. The above gives the material time derivative for the function in terms of the partial time derivative with the ALE coordinates fixed and a spatial gradient. Note that a comma followed by an index represents the spatial derivative with respect to an Eulerian coordinate, as in the rest of this book. In vector notation, the above can be written as

$$\frac{Df}{Dt} = f_{,t[\chi]} + \mathbf{c}\cdot \operatorname{grad} f = f_{,t[\chi]} + \mathbf{c}\cdot\nabla f \tag{7.2.18}$$

7.2.3 Relationship of ALE Description to Eulerian and Lagrangian Descriptions

It is worthwhile at this point to relate Lagrangian and Eulerian descriptions to the ALE description. We begin by letting $\boldsymbol{\chi} = \mathbf{X}$, that is, by letting the ALE coordinates be coincident with the material coordinates. The mesh motion (7.2.3) is then given by

$$\mathbf{x} = \hat{\boldsymbol{\phi}}(\mathbf{X}, t)$$

Since the mesh motion is now identical to the material motion (7.2.1), this indicates that the mesh is now Lagrangian. This can also be seen by examining the $\boldsymbol{\psi}$ map, (7.2.4), which becomes

$$\mathbf{X} = \boldsymbol{\psi}(\mathbf{X}, t) = I(\mathbf{X})$$

and as indicated above, the $\boldsymbol{\psi}$ map becomes the identity map, i.e. in this case the ALE coordinates are identical to the material coordinates. This does not really say anything new, since this was our starting point. Nevertheless it is of interest because of the correspondence which will emerge when we examine the reduction of an ALE formulation to an Eulerian formulation.

When we let the ALE coordinates correspond to the Eulerian coordinate, i.e. $\boldsymbol{\chi} = \mathbf{x}$, then the mesh motion is given by

$$\mathbf{x} = \hat{\boldsymbol{\phi}}(\mathbf{x}, t) = I(\mathbf{x})$$

so the mesh motion is the identity map, i.e. the mesh is fixed in space. The motion for an Eulerian description is given by

$$\mathbf{x} = \boldsymbol{\phi}\left(\boldsymbol{\Psi}^{-1}(\mathbf{x}, t), t\right) = I(\mathbf{x}) \quad \text{or} \quad \boldsymbol{\phi} \circ \boldsymbol{\Psi}^{-1} = I(\mathbf{x})$$

So in the reduction of the ALE description to the Eulerian description,

$$\boldsymbol{\phi} = \boldsymbol{\Psi}$$

The reductions are thus duals of each other. In the reduction of ALE to the Lagrangian description, the $\boldsymbol{\psi}$ map becomes the identity and the mesh motion becomes the material motion. In the degeneration to the Eulerian description, the mesh motion becomes the identity map, and the $\boldsymbol{\psi}$ map becomes the material motion.

It is also interesting to examine the Eulerian and Lagrangian forms of the material time derivative which are embedded in the ALE form. Recall the material time derivative can be expressed for the different descriptions as follows:

$$\frac{Df}{Dt} = \dot{f} = \frac{\partial f(\mathbf{X}, t)}{\partial t} \quad \text{Lagrangian description}\,(\mathbf{X}, t) \tag{7.2.19}$$

$$= f_{,t[x]} + \left.\frac{\partial f}{\partial x_i}\frac{\partial x_i}{\partial t}\right|_{[X]} = f_{,t[x]} + f_{,i} v_i \quad \text{Eulerian description}\,(\mathbf{x}, t) \tag{7.2.20}$$

$$= f_{,t[\chi]} + \left.\frac{\partial f}{\partial \chi_i}\frac{\partial \chi_i}{\partial t}\right|_{[X]} = f_{,t[\chi]} + \frac{\partial f}{\partial \chi_i} w_i \quad \text{ALE description}\,(\boldsymbol{\chi}, t) \tag{7.2.21}$$

When the description becomes Lagrangian $\boldsymbol{\chi} = \mathbf{X}$ and the convected velocity $\mathbf{c} = \mathbf{0}$. The ALE form (7.2.21) then reduces to the Lagrangian form (7.2.19). When the description becomes Eulerian, $\boldsymbol{\chi} = \mathbf{x}$ and the convected velocity equals the material velocity, $\mathbf{c} = \mathbf{v}$ and the ALE form (7.2.21) reduces to the Eulerian form (7.2.20).

The descriptions of the relationships between ALE, Lagrangian and Eulerian are summarized in Table 7.1. This table gives the material motion, displacement and velocity and the mesh motion, displacement and velocity for each of the description. It can be seen from the table that material motion is independent of the description, but the mesh motion hinges on the description. The definitions of mesh motion are all analogous to the definitions of material motion, with the material coordinates replaced by the ALE coordinates.

Table 7.1 Comparison of the kinematics for an ALE formulation with Lagrangian and Eulerian descriptions

Description		General ALE	Lagrangian	Eulerian
Motion	Material	$\mathbf{x}=\boldsymbol{\phi}(\mathbf{X},t)$	$\mathbf{x}=\boldsymbol{\phi}(\mathbf{X},t)$	$\mathbf{x}=\boldsymbol{\phi}(\mathbf{X},t)$
	Mesh	$\mathbf{x}=\hat{\boldsymbol{\phi}}(\boldsymbol{\chi},t)$	$\mathbf{x}=\boldsymbol{\phi}(\mathbf{X},t)$ $(\boldsymbol{\chi}=\mathbf{X},\hat{\boldsymbol{\phi}}=\boldsymbol{\phi})$	$\mathbf{x}=\mathbf{I}(\mathbf{x})$ $(\boldsymbol{\chi}=\mathbf{x},\hat{\boldsymbol{\phi}}=\mathrm{I})$
Displacement	Material	$\mathbf{u}=\mathbf{x}-\mathbf{X}$	$\mathbf{u}=\mathbf{x}-\mathbf{X}$	$\mathbf{u}=\mathbf{x}-\mathbf{X}$
	Mesh	$\hat{\mathbf{u}}=\mathbf{x}-\boldsymbol{\chi}$	$\hat{\mathbf{u}}=\mathbf{x}-\mathbf{X}=\mathbf{u}$	$\hat{\mathbf{u}}=\mathbf{x}-\mathbf{x}=\mathbf{0}$
Velocity	Material	$\mathbf{v}=\mathbf{u},_{t[X]}$	$\mathbf{v}=\mathbf{u},_{t[X]}$	$\mathbf{v}=\mathbf{u},_{t[X]}$
	Mesh	$\hat{\mathbf{v}}=\hat{\mathbf{u}},_{t[\chi]}$	$\hat{\mathbf{v}}=\hat{\mathbf{u}},_{t[X]}=\mathbf{v}$	$\hat{\mathbf{v}}=\hat{\mathbf{u}},_{t[X]}=\mathbf{0}$
Acceleration	Material	$\mathbf{a}=\mathbf{v},_{t[X]}$	$\mathbf{a}=\mathbf{v},_{t[X]}$	$\mathbf{a}=\mathbf{v},_{t[X]}$
	Mesh	$\hat{\mathbf{a}}=\hat{\mathbf{v}},_{t[\chi]}$	$\hat{\mathbf{a}}=\hat{\mathbf{v}},_{t[X]}=\mathbf{a}$	$\hat{\mathbf{a}}=\hat{\mathbf{v}},_{t[X]}=\mathbf{0}$

Example 7.1 At $t = 0$, the material, reference, and spatial domains coincide. Suppose the bar is rotated with constant velocity ω_1, and the reference domain is rotated with ω_2 as shown in Figure 7.2. The angle between the material domain and the spatial domain, θ_1, is equal to $\omega_1 t$; the angle between the spatial domain and the referential domain, θ_2, is equal to $\omega_2 t$. The material, reference, and spatial coordinates are denoted by X_1, X_2, χ_1, χ_2 and x_1, x_2 respectively.

The relation between the spatial coordinates, $\mathbf{x}$, and material coordinates, $\mathbf{X}$, can be described as

$$\boldsymbol{\phi}(\mathbf{X})=\begin{Bmatrix} x_1 \\ x_2 \end{Bmatrix}=\begin{bmatrix} \cos\theta_1 & \sin\theta_1 \\ -\sin\theta_1 & \cos\theta_1 \end{bmatrix}\begin{bmatrix} X_1 \\ X_2 \end{bmatrix} \tag{E7.1.1}$$

Inversely, we can find $\boldsymbol{\phi}^{-1}(\mathbf{x})$:

$$\boldsymbol{\phi}^{-1}(\mathbf{x})=\begin{Bmatrix} X_1 \\ X_2 \end{Bmatrix}=\begin{bmatrix} \cos\theta_1 & -\sin\theta_1 \\ \sin\theta_1 & \cos\theta_1 \end{bmatrix}\begin{Bmatrix} x_1 \\ x_2 \end{Bmatrix} \tag{E7.1.2}$$

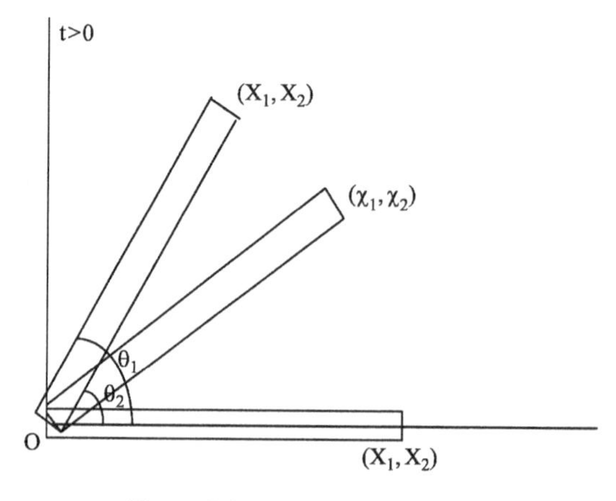

Figure 7.2 Rotating rod example

Similar with the relation that we defined above, except the angle is θ_2 instead θ_1, the referential coordinates, $\boldsymbol{\chi}$, and spatial coordinates, $\mathbf{x}$, relation can be expressed as

$$\hat{\boldsymbol{\phi}}(\boldsymbol{\chi}) = \begin{Bmatrix} x_1 \\ x_2 \end{Bmatrix} = \begin{bmatrix} \cos\theta_2 & \sin\theta_2 \\ -\sin\theta_2 & \cos\theta_2 \end{bmatrix} \begin{Bmatrix} \chi_1 \\ \chi_2 \end{Bmatrix} \tag{E7.1.3}$$

$$\hat{\boldsymbol{\phi}}^{-1}(\mathbf{x}) = \begin{Bmatrix} \chi_1 \\ \chi_2 \end{Bmatrix} = \begin{bmatrix} \cos\theta_2 & -\sin\theta_2 \\ \sin\theta_2 & \cos\theta_2 \end{bmatrix} \begin{Bmatrix} x_1 \\ x_2 \end{Bmatrix} \tag{E7.1.4}$$

The last mapping involves the relationship between the referential coordinates, $\boldsymbol{\chi}$, and the material coordinates $\mathbf{X}$:

$$\boldsymbol{\Psi}(\mathbf{X}) = \begin{Bmatrix} \chi_1 \\ \chi_2 \end{Bmatrix} = \begin{bmatrix} \cos(\theta_1-\theta_2) & \sin(\theta_1-\theta_2) \\ -\sin(\theta_1-\theta_2) & \cos(\theta_1-\theta_2) \end{bmatrix} \begin{bmatrix} X_1 \\ X_2 \end{bmatrix} \tag{E7.1.5}$$

$$\boldsymbol{\Psi}^{-1}(\boldsymbol{\chi}) = \begin{Bmatrix} X_1 \\ X_2 \end{Bmatrix} = \begin{bmatrix} \cos(\theta_1-\theta_2) & -\sin(\theta_1-\theta_2) \\ \sin(\theta_1-\theta_2) & \cos(\theta_1-\theta_2) \end{bmatrix} \begin{Bmatrix} \chi_1 \\ \chi_2 \end{Bmatrix} \tag{E7.1.6}$$

By now, all the domains are defined, we can prove the relations that are given in the previous section.

To prove that $\boldsymbol{\phi} = \hat{\boldsymbol{\phi}} \circ \boldsymbol{\Psi}$, we use (E7.1.3) and (E7.1.5):

$$\begin{aligned} \hat{\boldsymbol{\phi}} \circ \boldsymbol{\Psi} &= \begin{bmatrix} \cos\theta_2 & \sin\theta_2 \\ -\sin\theta_2 & \cos\theta_2 \end{bmatrix} \begin{bmatrix} \cos(\theta_1-\theta_2) & \sin(\theta_1-\theta_2) \\ -\sin(\theta_1-\theta_2) & \cos(\theta_1-\theta_2) \end{bmatrix} \begin{bmatrix} X_1 \\ X_2 \end{bmatrix} \\ &= \begin{bmatrix} \cos\theta_1 & \sin\theta_1 \\ -\sin\theta_1 & \cos\theta_1 \end{bmatrix} \begin{bmatrix} X_1 \\ X_2 \end{bmatrix} \\ &= \boldsymbol{\phi} \end{aligned}$$

By following the same routine, we are able to prove that $\boldsymbol{\Psi} = \hat{\boldsymbol{\phi}}^{-1} \circ \boldsymbol{\phi}$ and $\hat{\boldsymbol{\phi}} = \boldsymbol{\phi} \circ \boldsymbol{\Psi}^{-1}$.

Between the referential and the material domain, the displacement, $\mathbf{u}$, can be defined as

$$\mathbf{u} = \mathbf{x} - \mathbf{X} = \begin{bmatrix} \cos\theta_1 - 1 & \sin\theta_1 \\ -\sin\theta_1 & \cos\theta_1 - 1 \end{bmatrix} \mathbf{X} \tag{E7.1.7}$$

The velocity and acceleration can be found respectively:

$$\mathbf{v} = \omega_1 \begin{bmatrix} -\sin\theta_1 & \cos\theta_1 \\ -\cos\theta_1 & -\sin\theta_1 \end{bmatrix} \mathbf{X} \tag{E7.1.8}$$

Between the spatial and the referential domain, the displacement and velocity can be found to be

$$\hat{\mathbf{u}} = \mathbf{x} - \boldsymbol{\chi} = \begin{bmatrix} \cos\theta_2 - 1 & \sin\theta_2 \\ -\sin\theta_2 & \cos\theta_2 - 1 \end{bmatrix} \boldsymbol{\chi} \tag{E7.1.9}$$

$$\hat{\mathbf{v}} = \omega_2 \begin{bmatrix} -\sin\theta_2 & \cos\theta_2 \\ -\cos\theta_2 & -\sin\theta_2 \end{bmatrix} \boldsymbol{\chi} \tag{E7.1.10}$$

The referential velocity, **w**, is defined as

$$\mathbf{w} = \frac{\partial \boldsymbol{\Psi}(\mathbf{X}, t)}{\partial t} = (\omega_1 - \omega_2) \begin{bmatrix} -\sin(\theta_1 - \theta_2) & \cos(\theta_1 - \theta_2) \\ -\cos(\theta_1 - \theta_2) & -\sin(\theta_1 - \theta_2) \end{bmatrix} \mathbf{X} \tag{E7.1.11}$$

The convective velocity, $c_i = \dfrac{\partial x_i(\chi, t)}{\partial \chi_j} w_j$, is calculated using (E7.1.3) and (E7.1.11):

$$\mathbf{c} = (\omega_1 - \omega_2) \begin{bmatrix} -\sin\theta_1 & \cos\theta_1 \\ -\cos\theta_1 & -\sin\theta_1 \end{bmatrix} \mathbf{X} \tag{E7.1.12}$$

Finally, we can substitute (E7.1.12), (E7.1.8) and (E7.1.10) into $\mathbf{c} = \mathbf{v} - \hat{\mathbf{v}}$ to verify the relations between the velocities. This will be left as an exercise for the reader.

7.3 Conservation Laws in ALE Description

We will use conservation laws in a form almost identical to those of the Eulerian description given in Chapter 3. The only modification which will be made is that all material time derivatives will be replaced by the ALE form of material time derivatives (7.2.17). Consequently, the *only* difference between the Eulerian description in the updated Lagrangian formulation (Box 4.1) and the ALE description is in the material time derivative terms.

One major difference from the Lagrangian formulation developed in Chapter 4 is that we will now need to consider the mass conservation equation in the form of a partial differential equation, the continuity equation. Therefore, we will almost always deal with two systems of partial differential equation: the scalar continuity equation and the vector momentum equation. When this is coupling with heat transfer or other energy transfer, the energy equation must also be included.

7.3.1 Conservation of Mass (Equation of Continuity)

The continuity equation in an Eulerian description is given by

$$\dot{\rho} + \rho v_{j,j} = 0 \tag{7.3.1}$$

Replacing the material time derivative in the above by the ALE form (7.2.17), the continuity equation becomes

$$\rho_{,t[\chi]} + \rho_{,j} c_j + \rho v_{j,j} = 0 \quad \rho_{,t[\chi]} + \mathbf{c} \cdot \text{grad}\ \rho + \rho \nabla \cdot \mathbf{v} = 0 \tag{7.3.2}$$

where $\nabla \cdot \mathbf{v}$ is the divergence with respect to the spatial coordinates.

An alternative way of deriving the continuity equation is to employ the Reynolds transport theorem (3.5.14):

$$\int_{\Omega} \left[\left. \frac{\partial \rho}{\partial t} \right|_{x} + \frac{\partial (\rho v_i)}{\partial x_i} \right] d\Omega = 0 \tag{7.3.3a}$$

Assuming there are no discontinuities in the linear momentum, i.e. that the linear momentum is C^0, the product rule for derivatives yields

$$\int_{\Omega} \left[\left. \frac{\partial \rho}{\partial t} \right|_{x} + v_i \frac{\partial \rho}{\partial x_i} + \rho \frac{\partial v_i}{\partial x_i} \right] d\Omega = 0 \tag{7.3.3b}$$

Observing that the first two terms correspond to the material time derivative of ρ and using (7.2.17), this becomes

$$\int_{\Omega} \left[\left. \frac{\partial \rho}{\partial t} \right|_{\chi} + c_i \frac{\partial \rho}{\partial x_i} + \rho \frac{\partial v_i}{\partial x_i} \right] d\Omega = 0 \tag{7.3.3c}$$

Since Ω is arbitrary, it follows that

$$\left. \frac{\partial \rho}{\partial t} \right|_{\chi} + c_i \frac{\partial \rho}{\partial x_i} + \rho \frac{\partial v_i}{\partial x_i} = 0 \text{ in } \Omega \tag{7.3.3d}$$

which is identical to (7.3.2). If the linear momentum is discontinuous, we cannot apply the chain rule to the linear momentum since there is a jump in ρv_i. Hence in the presence of discontinuites in the momentum, we have to employ the conservative form, (7.3.3a), instead of (7.3.1).

7.3.2 *Conservation of Linear and Angular Momenta*

The momentum equation in an Eulerian description is given in (3.5.33):

$$\rho \dot{v}_i = \sigma_{ji,j} + \rho b_i \tag{7.3.4}$$

After applying the material time derivative operator (7.2.17) to (7.3.4), the momentum equation becomes

$$\rho\{v_{i,t[\chi]} + c_j v_{i,j}\} = \sigma_{ji,j} + \rho b_i \quad \text{or} \quad \rho\{\mathbf{v}_{,t[\chi]} + \mathbf{c} \cdot \text{grad } \mathbf{v}\} = \text{div}(\boldsymbol{\sigma}) + \rho \mathbf{b} \tag{7.3.5}$$

The conservation of angular momentum, as before, leads to the symmetry of the Cauchy (true) stress tensor $\boldsymbol{\sigma} = \boldsymbol{\sigma}^T$.

7.3.3 *Conservation of Energy*

Energy conservation is expressed as (see (3.5.45)):

$$\frac{D}{Dt}\int_{\Omega}\rho E d\Omega = \int_{\Gamma}\sigma_{ji}n_j v_i d\Gamma + \int_{\Omega}b_i v_i d\Omega - \int_{\Gamma_q}q_i n_i d\Gamma + \int_{\Omega}\rho s d\Omega \tag{7.3.6}$$

where $q_i n_i$ is the outward heat flux on the boundary Γ_q; w^{int} is the specific total energy density and is related to the specific internal energy, E, by

$$E = w^{\text{int}} + \frac{V^2}{2} \tag{7.3.7a}$$

where $E = E(\theta, \rho)$ with θ the temperature and ρs the specific heat source, that is, the heat source per unit spatial volume and $V^2 = v_i v_i$. The Fourier law of heat conduction and heat flux boundary conditions are

$$q_i = -k_{ij}\theta_{,j} \quad q_i = -k_{ij}\theta_{,j} + k_i(\theta - \theta_0) \tag{7.3.7b}$$

where k_{ij} and k_i are the components of the thermal conductivity matrix and convective heat transfer coefficients, respectively; θ_0 is a reference temperature.

Using the divergence theorem and (7.2.17), the energy equation can be shown to be

$$\rho\{E_{,t[\chi]} + E_{,j}c_j\} = (\sigma_{ij}v_i)_{,j} + b_j v_j + (k_{ij}\theta_{,j})_{,i} + \rho s \tag{7.3.8}$$

7.4 ALE Governing Equations

In the following we give the system equations and initial and boundary conditions for an ALE formulation. The equations are given in nonconservative form. Energy transfer is included, so the energy equation is also included. The objective of the initial/boundary value problem is to find the following dependent variables:

$\mathbf{u}(\boldsymbol{\chi}, t)$ material displacement
$\boldsymbol{\sigma}(\boldsymbol{\chi}, t)$ Cauchy stress tensor
$\theta(\boldsymbol{\chi}, t)$ thermodynamic temperature
$\hat{\mathbf{u}}(\boldsymbol{\chi}, t)$ mesh displacement
$\rho(\boldsymbol{\chi}, t)$ density

plus any internal variables in the constitutive model such that they satisfy the field equations and constitutive (state) equations shown in Box 7.1.

Box 7.1 ALE governing equations

Continuity equation

$$\dot{\rho}+\rho v_{k,k}=0 \quad \text{or} \quad \rho_{,t[\chi]}+\rho_{,i}c_i+\rho v_{k,k}=0 \tag{B7.1.1}$$

Momentum equations

$$\rho\dot{v}_i=\rho(v_{i,t[\chi]}+v_{i,j}c_j)=\sigma_{ji,j}+\rho b_i \tag{B7.1.2}$$

Energy equation

$$\rho(E_{,t[\chi]}+E_{,i}c_i)=\sigma_{ij}D_{ij}+b_iv_i+(k_{ij}\theta_{,j})_{,i}+\rho s \tag{B7.1.3}$$

Equations of state

supplemented by the constitutive equations given in Chapter 5

Natural boundary conditions

$$t_i(\boldsymbol{\chi},t)=n_j(\boldsymbol{\chi},t)\sigma_{ji}(\boldsymbol{\chi},t) \quad \text{on } \Gamma_{t_i} \tag{B7.1.4}$$

$$q_i(\boldsymbol{\chi},t)=-k_{ij}(\theta,\boldsymbol{\chi},t)\theta_{,j}(\boldsymbol{\chi},t)+\kappa_i(\theta,t)(\theta-\theta_0) \quad \text{on } \Gamma_q \tag{B7.1.5}$$

Essential boundary conditions

$$u_i(\boldsymbol{\chi},t)=\bar{u}_i(\boldsymbol{\chi},t) \quad \text{on } \Gamma_{u_i} \qquad \theta(\boldsymbol{\chi},t)=\bar{\theta}(\boldsymbol{\chi},t) \quad \text{on } \Gamma_{u\theta} \tag{B7.1.6}$$

Initial conditions

$$\sigma(\mathbf{X},0)=\sigma_0(\mathbf{X}) \qquad \theta(\mathbf{X},0)=\theta_0(\mathbf{X}) \tag{B7.1.7}$$

Mesh motion

$$\hat{\mathbf{u}}(\boldsymbol{\chi},t) \text{ given except, perhaps, on part of the boundary} \tag{B7.1.8}$$

7.5 Weak Forms

Since we have chosen to use the conservation equations with spatial derivatives in terms of the Eulerian coordinates, the difference between the updated Lagrangian field equations and the updated ALE field equations is the material time derivatives. Hence the weak form of the momentum equation, and subsequently the discrete finite element equations of motion, are identical to those derived in Chapter 4 except for the kinetic term. However, the spatial domain of each ALE element depends on how the mesh motion is updated.

7.5.1 Continuity Equation – Weak Form

Since mass conservation is enforced in the ALE method as a partial differential equation, a weak form needs to be developed. Let the trial solution be $\rho \in C^0$, the weak form of the continuity equation, is obtained by multiplying the strong form (7.3.2) over the current spatial domain by a test function $\delta\tilde{\rho} \in C^0$:

$$\int_{\Omega} \delta\tilde{\rho}\rho_{,t[\chi]} \, d\Omega + \int_{\Omega} \delta\tilde{\rho} c_i \rho_{,i} \, d\Omega + \int_{\Omega} \delta\tilde{\rho}\rho v_{i,i} \, d\Omega = 0 \tag{7.5.1}$$

7.5.2 Momentum Equation – Weak Form

The weak form of the momentum equation, is obtained by multiplying the strong form (B7.1.2), by the test function $\delta\tilde{v}_i \in u_0$ (see (4.3.1) and (4.3.2) for the definitions of $\delta\tilde{v}_i$ and v_i). Following the same procedures as in Chapter 4, we obtain the following weak form:

$$\begin{aligned}
\int_{\Omega} \delta\tilde{v}_i \rho \dot{v}_i \, d\Omega &= \int_{\Omega} \delta\tilde{v}_i \rho v_{i,t[\chi]} \, d\Omega + \int_{\Omega} \delta\tilde{v}_i \rho c_j v_{i,j} \, d\Omega = \int_{\Omega} \delta\tilde{v}_i (\sigma_{ji,j} + \rho b_i) d\Omega \\
&= -\int_{\Omega} \delta\tilde{v}_{i,j} \sigma_{ij} \, d\Omega + \int_{\Omega} \delta\tilde{v}_i \rho b_i \, d\Omega + \int_{\Gamma_t} \delta\tilde{v}_i \bar{t}_i \, d\Gamma
\end{aligned} \tag{7.5.2}$$

7.5.3 Finite Element Approximations

In the finite element approximation, we define all dependent variables as functions of element coordinates. The ALE domain is subdivided into elements and for element e the ALE coordinates are given by

$$\boldsymbol{\chi}(\boldsymbol{\xi}^e) = \boldsymbol{\varphi}^e(\boldsymbol{\xi}^e) = \boldsymbol{\chi}_I N_I(\boldsymbol{\xi}^e)$$

where $\boldsymbol{\xi}^e$ are the element coordinates of element e. The mesh motion is given by

$$\mathbf{x}(\boldsymbol{\xi}^e, t) = \hat{\boldsymbol{\phi}}^h(\boldsymbol{\chi}(\boldsymbol{\xi}^e), t) = \mathbf{x}_I(t) N_I(\boldsymbol{\xi}^e) \tag{7.5.3}$$

where $\mathbf{x}_I(t)$, are the motions of the nodes. Note that the above gives the mesh motion, which differs from the material motion unless the node becomes Lagrangian. The above represents a composition of two maps: the parent element to ALE map and the map of the mesh motion, so in terms of compositions we write that in element e: $\mathbf{x} = \hat{\boldsymbol{\phi}}^h \circ \varphi^e$.

The mesh velocity is given by applying (7.2.7) to the previous:

$$\hat{\mathbf{v}} = \frac{\partial \hat{\boldsymbol{\phi}}^h(\boldsymbol{\chi}, t)}{\partial t} = \dot{\mathbf{x}}_I(t) N_I(\boldsymbol{\xi}^e) \equiv \hat{\mathbf{v}}_I(t) N_I(\boldsymbol{\xi}^e) \tag{7.5.4}$$

In this, $\dot{\mathbf{x}}_I(t)$ is the mesh velocity of the node I. The maps between the element coordinates $\boldsymbol{\xi}$, mesh coordinates $\boldsymbol{\chi}$, and spatial coordinates $\mathbf{x}$ are illustrated in Figure 7.3.

In the ALE formulation, the density is also a dependent variable. The density is approximated by

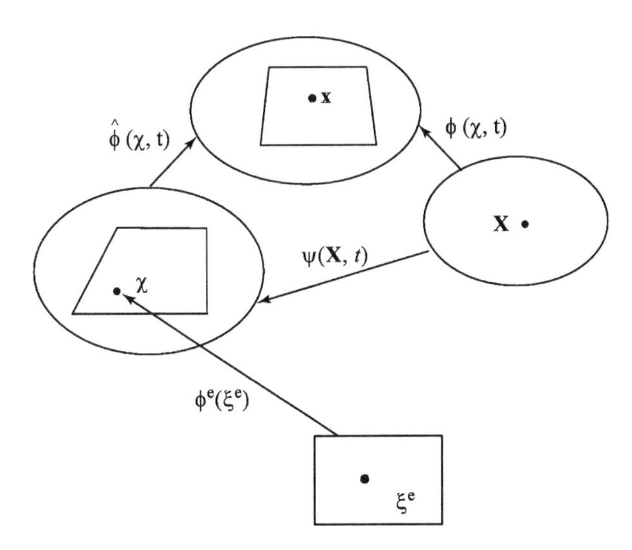

Figure 7.3 Maps between Lagrangian, Eulerian, ALE and natural coordinates domains

$$\rho(\xi^e, t) = \rho_I(t) N_I^\rho(\xi^e) \tag{7.5.5}$$

where $N_I^\rho(\boldsymbol{\xi}^e)$ are the shape functions for the density; these may differ from the shape functions for the mesh motion.

The gradients of the shape functions can be obtained by the usual implicit differentiation formulas described in Section 4.4.6, (4.4.42). It is noted that because convective terms ($\rho_{,i}c_i$ and $v_{i,j}c_j$) appear in the continuity equation and the momentum equations, a Galerkin finite element formulation will encounter numerical instabilities. One approach to overcoming these instabilities which will be emphasized in this chapter is the Petrov–Galerkin formulation. In a Petrov–Galerkin discretization, the test and trial functions differ. The trial shape functions for the velocity and density will be denoted by $\mathbf{N}$ and $\mathbf{N}^\rho$, respectively. The test shape functions for the velocity and density will be denoted by $\bar{\mathbf{N}}$ and $\bar{\mathbf{N}}^\rho$, respectively. If $\bar{\mathbf{N}} = \mathbf{N}$, the formulation reduces to a Galerkin formulation. The selection of $\bar{\mathbf{N}}$ and $\bar{\mathbf{N}}^\rho$ to stabilize the convective terms will be described in Section 7.7.

The material time derivative of the velocity $\dot{\mathbf{v}}$ is given by (7.2.17) with f replaced by $\mathbf{v}$.

$$\dot{\mathbf{v}} = \mathbf{v}_{,t[\chi]} + \mathbf{c} \cdot \nabla \mathbf{v} \tag{7.5.6}$$

The kinetic term in the discrete equations of motion will be expressed in terms of the mesh acceleration $\mathbf{v}_{,t[\chi]}$. Once $\mathbf{v}_{,t[\chi]}$ is known, the material acceleration can be obtained by (7.5.6), which is integrated in time to give the material velocity $\mathbf{v}$.

The material velocity is interpolated as

$$\mathbf{v}(\boldsymbol{\chi}(\boldsymbol{\xi}), t) = \mathbf{v}_I(t) N_I(\boldsymbol{\xi}) \tag{7.5.7}$$

A similar interpolation is used for the convective velocity:

$$\mathbf{c}(\boldsymbol{\chi}(\boldsymbol{\xi}), t) = \mathbf{c}_I(t) N_I(\boldsymbol{\xi}) \tag{7.5.8}$$

Applying (7.2.15) to this equation gives

$$\mathbf{c}(\boldsymbol{\chi}(\boldsymbol{\xi}),t)=(\mathbf{v}_I(t)-\hat{\mathbf{v}}_I(t))N_I(\boldsymbol{\xi}) \tag{7.5.9}$$

Substituting the interpolation (7.5.7) into (7.5.6), we obtain the ALE time derivative of the material velocity:

$$\mathbf{v}_{,t[\chi]}=\frac{d\mathbf{v}_I(t)}{dt}N_I(\boldsymbol{\xi}) \tag{7.5.10}$$

Consequently, the expression for the material time derivative of the material velocity, $\dot{\mathbf{v}}$, that appears in the weak form is

$$\dot{\mathbf{v}}=\frac{d\mathbf{v}_I(t)}{dt}N_I(\boldsymbol{\xi})+\mathbf{c}(\boldsymbol{\xi},t)\cdot\nabla N_I(\boldsymbol{\xi})\mathbf{v}_I(t) \tag{7.5.11}$$

Applying the same procedures to the material time derivatives of the density gives

$$\dot{\rho}=\frac{d\rho_I}{dt}(t)N_I^\rho(\boldsymbol{\xi})+\mathbf{c}(\boldsymbol{\xi},t)\cdot\nabla N_I^\rho(\boldsymbol{\xi})\rho_I(t) \tag{7.5.12}$$

where the density is given in (7.5.5).

7.5.4 *The Finite Element Matrix Equations*

The continuity equation is

$$\mathbf{M}^\rho\frac{d\boldsymbol{\rho}}{dt}+\mathbf{L}^\rho\boldsymbol{\rho}+\mathbf{K}^\rho\boldsymbol{\rho}=\mathbf{0} \tag{7.5.13}$$

where $\mathbf{M}^\rho$, $\mathbf{L}^\rho$, $\mathbf{K}^\rho$ are capacitance, transport, and divergence matrices, respectively:

$$\mathbf{M}^\rho=\left[M_{IJ}^\rho\right]=\int_\Omega \bar{N}_I^\rho N_J^\rho d\Omega \tag{7.5.14}$$

$$\mathbf{L}^\rho=\left[L_{IJ}^\rho\right]=\int_\Omega \bar{N}_I^\rho c_i N_{J,i}^\rho d\Omega \tag{7.5.15}$$

$$\mathbf{K}^\rho=\left[K_{IJ}^\rho\right]=\int_\Omega \bar{N}_I^\rho v_{i,i} N_J^\rho d\Omega \tag{7.5.16}$$

$$\boldsymbol{\rho}=[\rho_J];\quad \frac{d\boldsymbol{\rho}}{dt}=\left[\frac{d\rho_J}{dt}\right] \tag{7.5.17}$$

The momentum equation is

$$\mathbf{M}\frac{d\mathbf{v}}{dt}+\mathbf{L}\mathbf{v}+\mathbf{f}^{\text{int}}=\mathbf{f}^{\text{ext}} \tag{7.5.18}$$

where **M** and **L** are generalized mass and convective matrices, respectively, for velocity under a reference description; while $\mathbf{f}^{\text{int}}$ and $\mathbf{f}^{\text{ext}}$ are the internal and external force vectors respectively, such that

$$\mathbf{M} = \mathbf{I}[M_{IJ}] = \left(\int_{\Omega} \rho \bar{N}_I N_J \, d\Omega \right) \mathbf{I} \tag{7.5.19}$$

$$\mathbf{L} = \mathbf{I}[L_{IJ}] = \left(\int_{\Omega} \rho \bar{N}_I c_i N_{J,i} \, d\Omega \right) \mathbf{I} \tag{7.5.20}$$

$$\mathbf{f}^{\text{int}} = \left[f_{iI}^{\text{int}} \right] = \int_{\Omega} \bar{N}_{I,j} \sigma_{ij} \, d\Omega \tag{7.5.21}$$

$$\mathbf{f}^{\text{ext}} = \left[f_{iI}^{\text{ext}} \right] = \int_{\Omega} \rho \bar{N}_I b_i \, d\Omega + \int_{\Gamma_{t_i}} \bar{N}_I \bar{t}_i \, d\Gamma \tag{7.5.22}$$

$$\mathbf{v} = [\mathbf{v}_J]; \quad \frac{d\mathbf{v}}{dt} = \left[\frac{d\mathbf{v}_J}{dt} \right] \tag{7.5.23}$$

Note that the internal and external nodal forces are identical to those of the updated Lagrangian formulation (Box 4.3) except that they are defined in terms of the test shape functions. The mass matrix is not constant in time, since the density and domain vary with time. For a Petrov–Galerkin formulation, the mass matrix is not symmetric. All the matrices and vectors defined in (7.5.13–23) are integrated over the spatial domain Ω. The mesh update procedure will be described in Section 7.10.

7.6 Introduction to the Petrov–Galerkin Method

In this section, *streamline upwinding* by the Petrov–Galerkin method (SUPG) is formulated. Prior to the development of this method, the upwinding scheme is motivated by an examination of the classical advection–diffusion equation. The advection-diffusion equation is a useful model since it corresponds to a linearization of the momentum equation. A closed-form solution for the discrete steady-state advection-diffusion equation will be obtained. It will be shown that this solution is oscillatory in space when a mesh parameter known as the Peclet number exceeds a critical value. This is an example of a *spatial instability*. Next, a Petrov–Galerkin method will be developed which eliminates these oscillations, that is, corrects this instability.

The steady-state linear advection-diffusion equation is

$$\mathbf{u} \cdot \nabla \phi - \upsilon \nabla^2 \phi = 0 \tag{7.6.1}$$

where ϕ is the dependent variable, υ is the kinematic viscosity, and **u** is a given velocity. For the study of the numerical instabilities, we restrict (7.6.1) to one dimension so that

$$u \frac{d\phi}{dx} = \upsilon \frac{d^2 \phi}{dx^2} \tag{7.6.2}$$

with boundary conditions

$$\phi(0) = 0 \quad \text{and} \quad \phi(L) = 1 \tag{7.6.3}$$

is a two-point boundary value problem on the domain $0 \leq x \leq L$. It is easy to verify that the exact solution to (7.6.2) and (7.6.3) is

$$\phi(x) = \frac{1 - e^{ux/\upsilon}}{1 - e^{uL/\upsilon}} \tag{7.6.4}$$

7.6.1 Galerkin Discretization of the Advection–Diffusion Equation

In this section, we develop a Galerkin discretization with linear shape functions. Let the test function be $w(x)$. Multiplying (7.6.2) by w and integrating over the domain gives

$$\int_{\Omega} w(u\phi_{,x} - \upsilon\phi_{,xx})dx = 0, \quad u > 0 \tag{7.6.5}$$

Integrating by parts and making use of the divergence theorem, the weak form of the advection–diffusion equation, (7.6.2), is

$$\int_{\Omega} wu\phi_{,x}\, dx + \int_{\Omega} \upsilon w_{,x}\, \phi_{,x}\, dx = 0 \quad \forall w \in u_0 \tag{7.6.6}$$

The domain (0, L) is divided into equally sized finite elements, Ω^e, on which the discrete equations are given by

$$\left(\int_{\Omega^e} uN_a N_{b,x}\, dx\right)\phi_b + \left(\int_{\Omega^e} \upsilon N_{a,x} N_{b,x}\, dx\right)\phi_b = 0 \quad a,b = 1,2 \tag{7.6.7}$$

where N_a and N_b are the finite element shape functions. This can be written in indicial form as

$$L_{ab}\phi_b + K_{ab}\phi_b = 0 \quad a,b = 1,2 \tag{7.6.8}$$

where the convective matrix and the diffusion matrix are

$$L_{ab} = \int_{x_e}^{x_{e+1}} uN_a N_{b,x}\, dx; \quad K_{ab} = \int_{x_e}^{x_{e+1}} \upsilon N_{a,x} N_{b,x}\, dx \tag{7.6.9}$$

It is a simple exercise to show that for linear shape functions, for elements of length Δx,

$$\mathbf{L} = \frac{u}{2}\begin{bmatrix} -1 & 1 \\ -1 & 1 \end{bmatrix} \quad \mathbf{K} = \frac{\upsilon}{\Delta x}\begin{bmatrix} 1 & -1 \\ -1 & 1 \end{bmatrix} \tag{7.6.10}$$

After assembly, the equation for the jth node interior is

$$u\left(\frac{\phi_{j+1}-\phi_{j-1}}{2\Delta x}\right)-\upsilon\left(\frac{\phi_{j+1}-2\phi_j+\phi_{j-1}}{\Delta x^2}\right)=0 \quad (7.6.11)$$

which is exactly the central difference equation. It is convenient to rewrite the previous as

$$\frac{u\Delta x}{2\upsilon}(\phi_{j+1}-\phi_{j-1})-(\phi_{j+1}-2\phi_j+\phi_{j-1})=0 \quad (7.6.12)$$

The Peclet number, P_e, may then be defined as

$$P_e=\frac{u\Delta x}{2\upsilon} \quad (7.6.13)$$

In terms of the Peclet number, (7.6.12) becomes

$$(p_e-1)\phi_{j+1}+2\phi_j-(P_e+1)\phi_{j-1}=0 \quad (7.6.14)$$

Ignoring the boundary conditions, (7.6.14) can be written out as

$$P_e\underbrace{\begin{bmatrix} & & & & \\ -1 & 0 & 1 & & \\ & -1 & 0 & 1 & \\ & & -1 & 0 & 1 \\ & & & & \end{bmatrix}\begin{bmatrix}\vdots\\ \phi_{j-1}\\ \phi_j\\ \phi_{j+1}\\ \vdots\end{bmatrix}}_{\text{convective term}}+\underbrace{\begin{bmatrix} & & & & \\ -1 & 2 & -1 & & \\ & -1 & 2 & -1 & \\ & & -1 & 2 & -1 \\ & & & & \end{bmatrix}\begin{bmatrix}\vdots\\ \phi_{j-1}\\ \phi_j\\ \phi_{j+1}\\ \vdots\end{bmatrix}}_{\text{diffusion term}}=\begin{bmatrix}\vdots\\ 0\\ 0\\ 0\\ \vdots\end{bmatrix} \quad (7.6.15)$$

Since this discrete equation is linear, its solution is an exponential:

$$\phi(x_j)\equiv\phi_j=e^{ax_j}=e^{a(j\Delta x)}=e^{(a\Delta x)j}\equiv\mu^j \quad (7.6.16)$$

where $\mu=e^{a\Delta x}$ and a is an unknown coefficient to be determined. From (7.6.16), we can see that $\phi_{j+1}=\mu^{j+1}$ and $\phi_{j-1}=\mu^{j-1}$. Substituting (7.6.16) into (7.6.14) yields

$$(P_e-1)\mu^{j+1}+2\mu^j-(P_e+1)\mu^{j-1}=0 \quad (7.6.17)$$

Assuming that $\mu^{j-1}\neq0$ and dividing the above by μ^{j-1}, we obtain a quadratic characteristic equation for μ (note the similarity to Section 6.6 where we examined numerical stability of time integrators):

$$(P_e-1)\mu^2+2\mu-(P_e+1)=0 \quad (7.6.18)$$

The two roots for (7.6.18) are

$$\mu=1 \quad \text{and} \quad \mu=\frac{1+P_e}{1-P_e} \quad (7.6.19)$$

Recalling that $\phi_j = \mu^j$, the discrete solution to (7.6.14) takes the form

$$\phi_j = c_1 + c_2 \left(\frac{1+P_e}{1-p_e} \right)^j \tag{7.6.20}$$

where c_1 and c_2 are coefficients to be determined from the boundary conditions. The exact solution to (7.6.2) is given by (7.6.4), evaluated at $x = x_j$, which is

$$\phi(x_j) = \frac{1}{1-e^{uL/\upsilon}} \left[1 - e^{ux_j/\upsilon} \right] = c_1 + c_2 e^{uj\Delta x/v} \tag{7.6.21}$$

Comparing the finite difference solution (7.6.20) with the exact solution (7.6.21), it can be seen that:

1. If the Peclet number is less than one, i.e., $P_e < 1$, then the discrete solution is similar to the exact solution since $\left(\frac{1+P_e}{1-P_e}\right)^j > 0$.
2. If the Peclet number is greater than one, i.e., $P_e > 1$, then the discrete solution becomes:

$$\left(\frac{1+P_e}{1-P_e} \right)^j = (-m)^j \quad \text{with} \quad m > 0$$

Hence the solution is oscillatory with positive or negative depending whether ϕ_j is even or odd. In comparing this analysis to those presented in Chapter 6, we can see that we have performed a linearized stability analysis of the discrete steady-state equation. The instability for $P_e > 1$ is a spatial instability of the numerical discretization. Note the similarities and differences with the stability analysis of the time integrations the linearized equations in both cases have exponential solutions, but in this case we are examining spatial stability.

7.6.2 *Petrov–Galerkin Stabilization*

The Petrov–Galerkin (PG) formulation for (7.6.2) is obtained by replacing the test function w in (7.6.5) by $\tilde{w}$, which is defined as

$$\tilde{w} \equiv \underbrace{w}_{\substack{\text{Galerkin} \\ \text{test} \\ \text{function}}} + \underbrace{\gamma w_{,x}}_{\substack{\text{discontinuous} \\ \text{test function}}} \tag{7.6.22}$$

where $\gamma = \alpha \frac{\Delta x}{2}$. Replacing w by $\tilde{w}$, (7.6.5) becomes

$$\int_{\Omega} \tilde{w} \left(u\phi_{,x} - \upsilon\phi_{,xx} \right) dx = 0 \tag{7.6.23}$$

Note that while $w \in u_0$, the PG test function $\tilde{w}$ is not in this space, that is, $\tilde{w} \notin u_0$. The parameter α is chosen so as to eliminate oscillations for $P_e > 1$ and hopefully get accurate solutions.

In one dimension, it is possible to select α so as to match the exact solution at the nodes. Substituting the definition of $\tilde{w}$, (7.6.22), into (7.6.23) yields

$$0=\underbrace{\int_{\Omega} w(u\phi_{,x}-\upsilon\phi_{,xx})dx}_{\text{Galerkin term}}+\underbrace{\sum_{e=1}^{N_e}\int_{\Omega^e}\gamma w_{,x}(u\phi_{,x}-\upsilon\phi_{,xx})dx}_{\text{upwind Petrov-Galerkin term}} \tag{7.6.24}$$

The second term has been subdivided into element integrals because $w_{,x}$ and $\phi_{,x}\in C^{-1}$ (they are both discontinuous in their derivatives).

After integrating by parts (using $w(0)=w(L)=0$), (7.6.24) becomes

$$0=\int_{\Omega} wu\phi_{,x}\,dx+\int_{\Omega}\upsilon w_{,x}\,\phi_{,x}\,dx+\sum_{e=1}^{N_e}\int_{\Omega^e}u\gamma w_{,x}\,\phi_{,x}\,dx-\sum_{e=1}^{N_e}\int_{\Omega^e}u\gamma w_{,x}\,\phi_{,xx}\,dx \tag{7.6.25}$$

This is known as the upwind Petrov–Galerkin formulation (Brooks and Hughes, 1982). In this formulation the second derivative of ϕ is required. Guidelines for setting the free parameter α are given in the following section after the presentation of an alternative formulation which only requires the first derivative of ϕ. The discrete equations are similar to the standard upwind equations in one dimension. However, in multiple dimensions they provide a consistent theoretical framework for directing the upwind term along the streamlines.

7.6.3 *Alternative Derivation of the SUPG*

SUPG is widely used for stabilizing advection–diffusion equations. It is the acronym for streamline upwind Petrov–Galerkin. In one dimension, there are no streamlines; we only wish to present the basic features. This section describes an alternative derivation of SUPG. It is motivated by a desire to reduce the continuity of the trial functions. We begin with the weak form of the one-dimensional advection–diffusion equation (7.6.23).

The order of the trial function derivative is reduced by increasing the order of the test function derivative, $\tilde{w}$, thereby relaxing trial function continuity requirements. To this end, we integrate the second term of (7.6.23) by parts:

$$I\equiv\int_{\Omega}\tilde{w}\upsilon\phi_{,xx}\,dx=\int_{\Omega}(\tilde{w}\upsilon\phi_{,x})_{,x}\,dx-\int_{\Omega}\tilde{w}_{,x}\,\upsilon\phi_{,x}\,dx$$

Applying the divergence theorem and substituting in the definition of $\tilde{w}$ gives

$$\begin{aligned}I&=\tilde{w}\upsilon\phi_{,x}\big|_0^L-\int_{\Omega}\tilde{w}_{,x}\,\upsilon\phi_{,x}\,dx=\left[w+\gamma w_{,x}\right]\upsilon\phi_{,x}\big|_0^L-\int_{\Omega}\tilde{w}_{,x}\,\upsilon\phi_{,x}\,dx\\&=\gamma w_{,x}\,\upsilon\phi_{,x}\big|_0^L-\int_{\Omega}\tilde{w}_{,x}\,\upsilon\phi_{,x}\,dx\end{aligned}$$

since $w(0)=0$ and $w(L)=0$.

Combining this with the advection term yields the following alternative weak form:

$$\int_{\Omega}[\tilde{w}u\phi_{,x}+\tilde{w}_{,x}\,\upsilon\phi_{,x}]dx-\gamma w_{,x}\,\upsilon\phi_{,x}\Big|_0^L=0$$

It is apparent that reduction of the order of the trial function derivative gives rise to a boundary term which was not present in the Petrov–Galerkin formulation presented earlier. In the particular case when $\tilde{w}$ is defined as in (7.6.22), it is straightforward to show that this alternative formulation yields the same results as the formulation presented in the previous section. To show this, we substitute the explicit expression for $\tilde{w}$ into the equation to give

$$\boxed{\int_{\Omega}[w+\gamma w_{,x}]u\phi_{,x}\,dx+\int_{\Omega}[w_{,x}+\gamma w_{,xx}]\,\upsilon\phi_{,x}\,dx-\gamma w_{,x}\,\upsilon\phi_{,x}\Big|_0^L=0}$$

Integration by parts of the fourth term gives

$$\int_{\Omega}\gamma w_{,xx}\,\upsilon\phi_{,x}\,dx=\gamma\upsilon w_{,x}\,\phi_{,x}\Big|_0^L-\int_{\Omega}\gamma\upsilon w_{,x}\,\phi_{,xx}\,dx$$

which, upon rearrangement, yields an expression identical to the previously presented Petrov–Galerkin formulation, (7.6.25).

We may select either formulation depending on which is more convenient computationally for the problem at hand. Finally, note that when linear elements are used, $\phi_{,xx}=0$, so all terms involving second derivatives vanish. Consequently, (7.6.25) may be written as

$$0=\sum_e\int_{\Omega^e}[wu\phi_{,x}+\upsilon^*w_{,x}\,\phi_{,x}]dx \tag{7.6.26}$$

where υ^* is the sum of two viscosities defined next.

7.6.4 Parameter Determination

In this section, we examine a physical interpretation of (7.6.26). For this purpose we make the following definitions:

$$\upsilon^*=\upsilon+\bar{\upsilon}=\text{total viscosity}\quad\text{and}\quad\bar{\upsilon}=\alpha u\frac{\Delta x}{2},\quad\alpha\geq 0 \tag{7.6.27a, b}$$

Clearly $\bar{\upsilon}$ may be thought of as an artificial viscosity which is added to the 'normal' flow viscosity, υ, to ensure stability. Without this artificial damping, our numerical solution is spatially unstable.

To define these viscosities in terms of the Peclet numbers, consider the following:

$$P_e=\frac{u\Delta x}{2\upsilon};\quad\bar{P}_e=\frac{u\Delta x}{2\bar{\upsilon}};\quad P_e^*=\frac{u\Delta x}{2\upsilon^*}\quad\frac{1}{P_e^*}=\frac{2\upsilon^*}{u\Delta x}=\frac{2\upsilon}{u\Delta x}+\frac{2\bar{\upsilon}}{u\Delta x}=\frac{1}{P_e}+\frac{1}{\bar{P}_e} \tag{7.6.28}$$

where the RHS follows from (7.6.27).

In this, if $P_e^* < 1$, then

$$\frac{2\upsilon}{u\Delta x} + \frac{2\overline{\upsilon}}{u\Delta x} > 1 \quad \text{or} \quad \frac{1}{\overline{P}_e} > 1 - \frac{1}{P_e} \tag{7.6.29}$$

and the solution will not be oscillatory. From (7.6.14), the discrete equation in terms of P_e^* can be written as $(P_e^* - 1)\phi_{j+1} + 2\phi_j - (P_e^* + 1)\phi_{j-1} = 0$.

Recall that the discrete solution is oscillatory when $|\mu| > 1$ (since $\phi_N = \mu^N$). The roots of the modified equations are $\mu^N = c_1$ or $c_2((1+P_e^*)/(1-P_e^*))^N$.

From (7.6.21) and the previous, we obtain the following solution for P_e^*:

$$\left(\frac{1+P_e^*}{1-P_e^*}\right) = e^{\left(\frac{u}{\upsilon}\Delta x\right)} \tag{7.6.30}$$

The RHS of (7.6.30) can be expressed in terms of P_e. That is:

$$\frac{1+P_e^*}{1-P_e^*} = e^{\frac{u}{\upsilon}\Delta x} = e^{2\left(\frac{u\Delta x}{2\upsilon}\right)} = e^{2P_e} \quad \text{or} \quad 1 - e^{2P_e} = -P_e^*(1+e^{2P_e}) \tag{7.6.31}$$

Therefore,

$$P_e^* = \frac{e^{2P_e} - 1}{e^{2P_e} + 1} = \frac{e^{P_e} - e^{-P_e}}{e^{P_e} + e^{-P_e}} \quad \text{or} \quad P_e^* = \tanh(P_e) \tag{7.6.32}$$

Substituting (7.6.32) into (7.6.28) yields

$$\frac{1}{\tanh(P_e)} = \frac{1}{P_e} + \frac{1}{\overline{P}_e} \quad \text{or} \quad \frac{1}{\overline{P}_e} = \coth(P_e) - \frac{1}{P_e} \tag{7.6.33}$$

Equations (7.6.28) and (7.6.33) may be combined to give

$$\overline{P}_e = \frac{u\Delta x}{2\overline{\upsilon}} = \left[\coth(P_e) - \frac{1}{P_e}\right]^{-1} \tag{7.6.34}$$

Using (7.6.34), we can also express $\overline{\upsilon}$ in terms of P_e:

$$\overline{\upsilon} = \frac{1}{2}u\Delta x\left[\coth(P_e) - \frac{1}{P_e}\right] = \alpha\frac{u\Delta x}{2} \tag{7.6.35}$$

Finally, it becomes apparent that we may define the parameter α as:

$$\alpha = \left[\coth(P_e) - \frac{1}{P_e}\right] \tag{7.6.36}$$

Note that when $\alpha = 0$, (7.6.23) is simply a central difference method, and when $\alpha = 1$, it is a full upwind formulation.

To illustrate the spatial instability and resulting nodal oscillations, we consider the one dimensional advection–diffusion equation as given in (7.6.2) with boundary conditions (7.6.3).

Example 7.2 1D Advection–diffusion equation The plots below compare the exact solution with finite element solutions for both no upwinding and full upwinding. In the first example, 10 elements are used. The Peclet number, Pe, is chosen to be 1. Comparing the results with the analytical solution, we can see the nodal exactness. (see Figure 7.4a) In the second example, the Pedet number is 3. From Figure 7.4(b), we can see that the oscillation appears in the Galerkin method while the Petrov–Galerkin method eliminates the instability. From the results shown in Figure 7.4, we can see that the numberical solutions obtained by the Galerkin method are spatially unstable. More oscillations appear when a larger Peclet number is used. However, with upwind Petrov Galerkin, the solutions are stabilized.

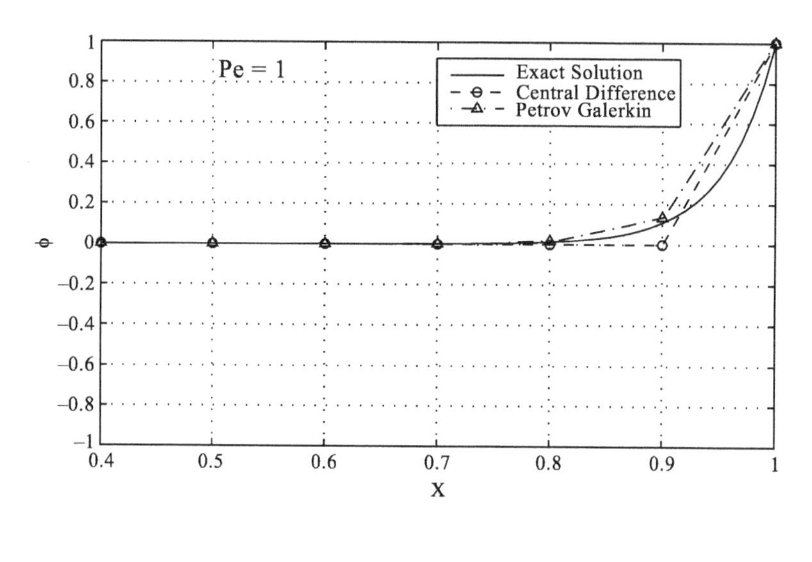

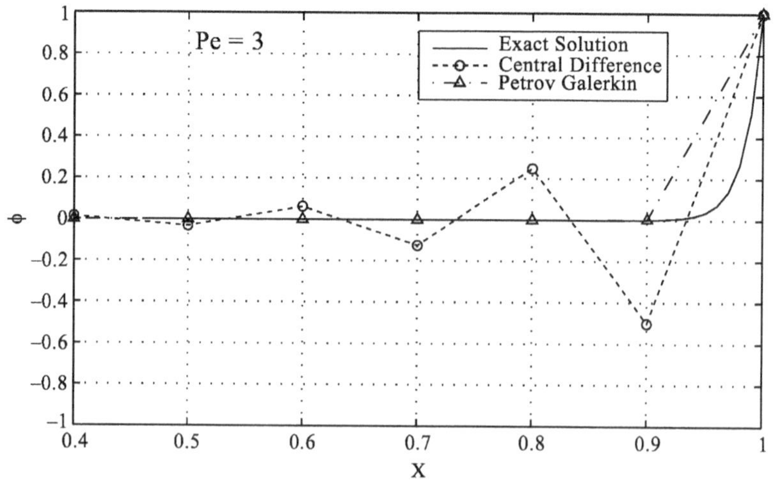

Figure 7.4

7.6.5 SUPG Multiple Dimensions

In the preceding, the fundamental concepts of SUPG have been developed in one dimension. In the following we develop a SUPG discretization of the advection–diffusion equation with the artificial viscosity in the streamline direction.

The advection–diffusion equation in multiple dimensions is

$$\mathbf{u}\cdot\nabla\phi-\upsilon\nabla^2\phi=0 \quad \text{in } \Omega \tag{7.6.37a}$$

where the boundary conditions are

$$\phi=g \quad \text{on} \quad \Gamma_g \qquad \upsilon\nabla\phi\cdot\mathbf{n}=0 \quad \text{on } \Gamma_t \tag{7.6.37b}$$

The weak form of the advection–diffusion equation for a streamline-upwind/Petrov–Galerkin formulation of (7.6.37a) is obtained by multiplying (7.6.37a) by the test function $\tilde{w}$ and integrating over the domain Ω:

$$\int_\Omega \tilde{w}(\mathbf{u}\cdot\nabla\phi-\upsilon\nabla^2\phi)d\Omega=0 \tag{7.6.37c}$$

Similar to the one-dimensional formulation presented above, let the Petrov–Galerkin test functions be defined by

$$\tilde{w}\equiv \underbrace{w}_{\text{Galerkin test function}} + \underbrace{\tau\mathbf{u}\cdot\nabla w}_{\text{discontinuous test function}} \tag{7.6.37d}$$

thus giving

$$0=\int_\Omega (w+\tau\mathbf{u}\cdot\nabla w)(\mathbf{u}\cdot\nabla\phi-\upsilon\nabla^2\phi)d\Omega \tag{7.6.37e}$$

where the stabilization parameter is given by

$$\tau=|\alpha|\frac{h}{\|\mathbf{u}\|} \quad \text{and} \quad h\equiv\Delta x \tag{7.6.37f}$$

where α is given by (7.6.36). For a time-dependent problem, the stabilization parameter can be set by $\tau=|\alpha|\frac{\Delta t}{2}$.

Note that $w\in u_0$ and $\tilde{w}\notin u_0$ as in the one-dimensional case. Applying integration by parts and the divergence theorem, the weak form of (7.6.37a) can be shown to be

$$\begin{aligned}0=&\underbrace{\int_\Omega w\mathbf{u}\cdot\nabla\phi d\Omega+\int_\Omega \upsilon\nabla w\cdot\nabla\phi d\Omega}_{\text{Galerkin terms}}\\&\underbrace{+\sum_{e=1}^{N_e}\int_{\Omega^e}\tau(\mathbf{u}\cdot\nabla w)(\mathbf{u}\cdot\nabla\phi)d\Omega-\sum_{e=1}^{N_e}\int_{\Omega^e}\tau(\mathbf{u}\cdot\nabla w)\upsilon\nabla^2\phi d\Omega}_{\text{streamline upwind stabilization terms}}\end{aligned} \tag{7.6.38}$$

In (7.6.38), the domain integral is now approximated by the open set of the interiors of all elements. This avoids the singularities of $\nabla^2\phi$ on the boundaries. As can be seen from the above equation, the Petrov–Galerkin terms are simply the sum of the standard Galerkin terms plus the streamline upwind stabilization terms. Namely,

$$\text{Petrov–Galerkin} = \text{Galerkin} + \text{Streamline upwind stabilization}$$

It can be seen from the second term in (7.6.38), namely the first stabilization term, that the artificial viscosity acts only in the streamline direction. This can be made more explicit as follows. Let the streamline be in the direction x. Then the integrand is $(u_x w_{,x})(u_x \phi_{,x})$. The other terms vanish. Therefore, the viscosity is just in the direction of the flow, that is, the streamline. Hence the name streamline upwind Petrov–Galerkin.

7.7 Petrov–Galerkin Formulation of Momentum Equation

We now develop a SUPG discretization of the ALE momentum equation, (7.5.2). The Petrov–Galerkin formulation involves the selection of the test function,

$$\delta\tilde{v}_i = \delta v_i + \delta v_i^{PG} \tag{7.7.1}$$

where $\delta v_i = 0$ on Γ_{v_i}, i.e. $\delta v_i \in u_0$. For now, we refrain from imposing any condition on δv_i^{PG} except that $\delta v_i^{PG} \in C^{-1}$. The PG test function δv_i^{PG} is given in Section 7.7.2. For the time being, $\delta\tilde{v}_i \neq 0$ on Γ_{v_i} $\left(\delta v_i^{PG} \neq 0 \text{ on } \Gamma_{v_i}\right)$. Inserting this test function into the weak form of the equation of motion, (7.5.2), yields

$$\int_\Omega \delta v_i \rho \dot{v}_i \, d\Omega - \int_\Omega \delta v_i \sigma_{ji,j} \, d\Omega - \int_\Omega \delta v_i \rho b_i \, d\Omega + \sum_{e=1}^{N_e} \int_\Omega \delta v_i^{PG} \left(\rho \dot{v}_i - \sigma_{ji,j} - \rho b_i\right) d\Omega = 0 \tag{7.7.2}$$

The first three terms are standard Galerkin terms while the last term serves as a stabilization term. Its influence on the formulation will be governed by our choice of stabilization trial functions, δv_i^{PG}.

Going through the usual integration by parts and the divergence theorem on the second term and noting that $\delta v_i \in u_0$, the weak form of the momentum equation becomes

$$\underbrace{\int_\Omega \delta v_i \rho \dot{v}_i \, d\Omega + \int_\Omega \delta v_{i,j} \sigma_{ij} \, d\Omega - \int_\Omega \delta v_i \rho b_i \, d\Omega - \int_{\Gamma_t} \delta v_i \bar{t}_i \, d\Gamma}_{\text{Galerkin terms}}$$
$$+ \underbrace{\sum_{e=1}^{N_e} \int_\Omega \delta v_i^{PG} (\rho \dot{v}_i - \sigma_{ij,j} - \rho b_i) d\Omega = 0}_{\text{Stabilization terms}} \tag{7.7.3}$$

The above equation consists of the standard Galerkin terms and the stabilization term. For the Galerkin terms, $\delta v_i \in u_0$ and $\sigma \in C^{-1}$ are appropriate. However, the stabilization term requires that $\delta v_i^{PG} \in C^{-1}$ and $\sigma \in C^0$. The C^0 continuity requirement for the Cauchy stress is a major drawback of this Petrov–Galerkin formulation since it requires the trial function v_i to be C^1. A remedy is proposed in Section 7.8.

Prior to the development of the matrix formulation, we wish to describe an alternative Petrov–Galerkin formulation which requires $\sigma \in C^{-1}$, but $\delta v_{i,j}^{PG} \in C^{-1}$. This formulation is closely related to the one developed in Section 7.6.3.

7.7.1 Alternative Stabilization Formulation

In this section we describe an alternative stabilization to circumvent some of the difficulties associated with the more conventional Petrov–Galerkin method outlined in the previous section. Our aim is to avoid the computation of the gradient of the Cauchy stress (i.e., $\sigma_{ji,j}$) which, as will be shown in Section 7.8, is often complicated and computationally time-consuming.

From the weak form (7.5.2) and using integration by parts, the divergence theorem on the second term gives

$$\int_{\Omega} \delta\tilde{v}_i \rho \dot{v}_i \, d\Omega + \int_{\Omega} \delta\tilde{v}_{i,j} \sigma_{ji} \, d\Omega - \int_{\Omega} \delta\tilde{v}_i \rho b_i \, d\Omega - \int_{\Omega} \delta\tilde{v}_i \bar{t}_i \, d\Gamma - \int_{\Gamma_v} \delta\tilde{v}_i \sigma_{ji} n_j \, d\Gamma = 0 \tag{7.7.4}$$

where the test function (7.7.1) applies with $\delta v_i \in \Gamma_{v_i}$. It is noted that the first four terms correspond to the Galerkin formulation given in (7.5.2). In deriving (7.5.2), we assume that $\delta\tilde{v}_i = 0$ on Γ_v. If indeed we can pick $\delta\tilde{v}_i = 0$ on Γ_v, the last term of (7.7.4) is zero and hence (7.7.4) reduces to (7.5.2).

In this formulation, we require $\delta v_i = 0$ on Γ_v and hence using (7.7.4):

$$\int_{\Gamma_v} \delta\tilde{v}_i \sigma_{ji} n_j \, d\Gamma = \sum_{e=1}^{N_e} \int_{\Gamma_v} \delta v_i^{PG} \sigma_{ji} n_j \, d\Gamma \tag{7.7.5}$$

This extra term imposes $\sigma_{ji} n_j = \bar{t}_i$ on Γ_v as an unknown and it is weighted by the test function δv_i^{PG}. Substituting (7.7.5) and (7.2.17) back into (7.7.4) gives

$$\int_{\Omega} \delta\tilde{v}_i \rho v_{i,t[x]} \, d\Omega + \int_{\Omega} \delta\tilde{v}_i \rho c_j v_{i,j} \, d\Omega = -\int_{\Omega} \delta\tilde{v}_{i,j} \sigma_{ji} \, d\Omega + \int_{\Omega} \delta\tilde{v}_{i\rho} b_i \, d\Omega + \int_{\Gamma_t} \delta\tilde{v}_i \bar{t}_i \, d\Gamma + \sum_{e=1}^{N_e} \int_{\Gamma_v} \delta v_i^{PG} \sigma_{ji} n_j \, d\Gamma \tag{7.7.6}$$

This form of the momentum equation contains no stress gradients! Instead, the stress gradient integral has been replaced by a boundary integral (7.7.5) because $\delta v_i^{PG} \in C^0$.

Prior to the development of the finite element equations, we describe the test function δv_i^{PG}. Our selection is based on Section 7.6 with an aim of obtaining a streamline upwinding that provides an artificial viscosity in the streamline direction for each component of the momentum equation.

7.7.2 The δv_i^{PG} Test Function

Following the Petrov–Galerkin formulation of the advective–diffusion equation in Section 7.6, the test function δv_i^{PG} is chosen to be

$$\delta v_i^{PG} = \tau c_j \delta v_{i,j} \quad \text{where} \quad \tau = |\alpha| \frac{h}{2\,\|\mathbf{c}\|} \tag{7.7.7}$$

The selection of τ is motivated by (7.6.36).

In the above, h is the characteristic element length and $|\alpha| = 1$. For a discussion of this parameter, see Hughes and Mallet (1986), Hughes and Tezduyar (1984) and Brooks and Hughes (1982).

7.7.3 Finite Element Equation

The discretization of the variational form (7.7.3) closely follows Section 7.5. We use the trial functions, (7.5.7), with the test function

$$\delta\tilde{v}_i = \delta v_{iI} N_I + \delta v_i^{PG}$$

The discretization of (7.7.3) gives (cf. 7.5.18–23)

$$\delta\mathbf{v}^T\left(\mathbf{M}\frac{d\mathbf{v}}{dt} + \mathbf{L}\mathbf{v} + \mathbf{f}^{\text{int}} - \mathbf{f}^{\text{ext}}\right) + \sum_{e=1}^{N_e}\int_\Omega \delta\mathbf{v}^{PG}\cdot\left(\rho\left.\frac{d\mathbf{v}}{dt}\right|_\chi + \rho\mathbf{c}\cdot\nabla\mathbf{v} - \nabla\boldsymbol{\sigma} - \rho\mathbf{b}\right)d\Omega = 0 \qquad (7.7.8)$$

where the mass $\mathbf{M}$, convective $\mathbf{L}$, $\mathbf{f}^{\text{int}}$ and $\mathbf{f}^{\text{ext}}$ matrices are given as in (7.5.19–20) with $\bar{\mathbf{N}}$ set equal to $\mathbf{N}$.

The discretization of the stabilization terms in (7.7.8) is obtained by interpolating δv_i^{PG} as

$$\delta v_i^{PG} = \tau c_j N_{I,j}\delta v_{iI} \qquad (7.7.9)$$

and assuming the $\sigma_{ji,j}$ is C^{-1}. The resulting stabilization matrices are

$$\begin{aligned}&\sum_{e=1}^{N_e}\int_\Omega \delta\mathbf{v}^{PG}\cdot\left(\rho\frac{d\mathbf{v}}{dt}|_\chi + \rho\mathbf{c}\cdot\nabla\mathbf{v} - \nabla\boldsymbol{\sigma} - \rho\mathbf{b}\right)d\Omega \\ &= \int_\Omega \delta\mathbf{v}^T\left(\mathbf{M}_{\text{stab}}\frac{d\mathbf{v}}{dt} + \mathbf{L}_{\text{stab}}\mathbf{v} + \mathbf{f}^{\text{int}}_{\text{stab}} - \mathbf{f}^{\text{ext}}_{\text{stab}}\right)d\Omega\end{aligned} \qquad (7.7.10)$$

where

$$\mathbf{M}_{\text{stab}} = \mathbf{I}[M_{IJ}]_{\text{stab}} = \left(\int_\Omega \rho\tau c_j N_{I,j} N_J\, d\Omega\right)\mathbf{I} \qquad (7.7.11a)$$

$$\mathbf{L}_{\text{stab}} = \mathbf{I}[L_{IJ}]_{\text{stab}} = \left(\int_\Omega \rho\tau c_j N_{I,j} c_i N_{J,i}\, d\Omega\right)\mathbf{I} \qquad (7.7.11b)$$

$$\mathbf{f}^{\text{int}}_{\text{stab}} = \left[f_{iI}^{\text{int}}\right]_{\text{stab}} = -\int_\Omega \tau c_j N_{I,j}\sigma_{ki,k}\, d\Omega \qquad (7.7.11c)$$

$$\mathbf{f}^{\text{ext}}_{\text{stab}} = \left[f_{iI}^{ext}\right]_{\text{stab}} = \int_\Omega \rho\tau c_j N_{I,j} b_i\, d\Omega \qquad (7.7.11d)$$

The discrete Petrov–Galerkin momentum equation is

$$(\mathbf{M} + \mathbf{M}_{\text{stab}})\frac{d\mathbf{v}}{dt} + (\mathbf{L} + \mathbf{L}_{\text{stab}})\mathbf{v} + \left(\mathbf{f}^{\text{int}} + \mathbf{f}^{\text{int}}_{\text{stab}}\right) = \left(\mathbf{f}^{\text{ext}} + \mathbf{f}^{\text{ext}}_{\text{stab}}\right) \qquad (7.7.12)$$

We simply write (7.7.12) as

$$\mathbf{M}\frac{d\mathbf{v}}{dt}+\mathbf{L}\mathbf{v}+\mathbf{f}^{\text{int}}=\mathbf{f}^{\text{ext}} \tag{7.7.13}$$

It is noted that $\mathbf{f}^{\text{ext}}_{\text{stab}}$ does not include the traction and $\mathbf{L}_{\text{stab}}$ is a symmetric matrix ($\mathbf{L}$ is nonsymmetric); $\mathbf{L}_{\text{stab}}$ can be interpreted as an artificial viscosity with a viscosity tensor given by $\rho\tau\mathbf{c}\mathbf{c}^T$. This can be seen by rewriting (7.7.11b) in 3-D:

$$[L_{IJ}]_{\text{stab}}=\int_\Omega \rho\tau\begin{bmatrix}N_{I,x} & N_{I,y} & N_{I,z}\end{bmatrix}\begin{bmatrix}c_1c_1 & c_1c_2 & c_1c_3\\ c_2c_1 & c_2c_2 & c_2c_3\\ c_3c_1 & c_3c_2 & c_3c_3\end{bmatrix}\begin{bmatrix}N_{J,x}\\ N_{J,y}\\ N_{J,z}\end{bmatrix}d\Omega \tag{7.7.14}$$

Examining ((7.7.6) closely, we can show that if we ignore the surface integral, Γ_v, the discretization of ((7.7.6) is identical to that of (7.5.2). Hence, the finite element equations are given by (7.5.18) through (7.5.23). The definition of $\bar{N}_I$ can be obtained by substituting (7.7.7) into (7.7.1):

$$\delta\tilde{v}_i=N_I\delta v_{iI}+\tau c_j N_{I,J}\delta v_{iI}=(N_I+\tau c_j N_{I,j})\delta v_{iI}\equiv\bar{N}_I\delta v_{iI} \tag{7.7.15}$$

The remaining Γ_v surface term can be interpreted as a reaction force term $\mathbf{f}^{\text{react}}$:

$$\left[\mathrm{f}^{\text{react}}_{iI}\right]=\int_{\Gamma_v}\tau c_j N_{I,j}\sigma_{ji}n_j\,d\Gamma \tag{7.7.16}$$

The final expression of the discrete momentum equation becomes

$$\mathbf{M}\frac{d\mathbf{v}}{dt}+\mathbf{L}\mathbf{v}+\left(\mathbf{f}^{\text{int}}-\mathbf{f}^{\text{react}}\right)=\mathbf{f}^{\text{ext}} \tag{7.7.17}$$

where the matrices are defined in (7.5.19) through (7.5.22) with $\bar{N}_I$ given in (7.7.15) and $\mathbf{f}^{\text{react}}$ given in (7.7.16). We prefer to put $\mathbf{f}^{\text{react}}$ on the left hand side since $\mathbf{f}^{\text{react}}$ is a function of $\boldsymbol{\sigma}$ similar to $\mathbf{f}^{\text{int}}$; $\boldsymbol{\sigma}$ is also a function of the velocity.

7.8 Path-Dependent Materials

In this section, an ALE formulation for path-dependent materials is presented. Formulations for a standard Galerkin discretization, Streamline-upwind/Petrov-Galerkin (SUPG) method and operator splitting method are derived. All other path-dependent state variables should be updated similarly. An computational method with explicit integration and a flow-chart are presented. Finally, some elastic and elastic-plastic wave propagation examples are given.

The stress state in a path-dependent material depends on the history of the material point. A path-dependent material can be readily treated in Lagrangian description because quadrature points coincide with material points regardless of the deformation of the continuum. On the other hand, in an ALE description, a quadrature point does not coincide with a material point so that the stress history needs to be convected by the relative velocity **c**, as indicated below in (7.8.1). Note that the spatial derivatives of the stress are involved in the convection term. When C^{-1} functions are used to interpolate the element stresses, the ambiguity of the stress derivatives in the transport terms in (7.8.1) are ambiguous at the element interface.

7.8.1 Strong Form of Stress Update

The Cauchy stress is given by the Jaumann rate constitutive equation as in Chapter 5:

$$\sigma_{ij,t[\chi]} + c_k \sigma_{ij,k} = C_{ijkl}^{\sigma J} D_{kl} + \sigma_{kj} W_{ik} + \sigma_{ki} W_{jk} \tag{7.8.1}$$

In the above equations, D_{ij} and W_{ij} are elements of the rate-of-deformation tensor and the spin tensor, respectively; see Chapter 3. $C_{ijkl}^{\sigma J}$ is the material response tensor which relates any frame-invariant rate of the Cauchy stress to the velocity strain. Both geometric and material nonlinearities will be included.

With the product rule for derivatives, (7.8.1) is equivalent to the following equations:

$$\sigma_{ij,t[\chi]} + y_{ijk,k} - c_{k,k}\sigma_{ij} = C_{ijkl}^{\sigma J} D_{kl} + \sigma_{kj} W_{ik} + \sigma_{ki} W_{jk}; \quad y_{ijk} = \sigma_{ij} c_k \tag{7.8.2}$$

where y_{ijk} is the stress–velocity product. In the following finite element computations, these two equations will replace (7.8.1) in the weak form.

7.8.2 Weak Form of Stress Update

As for the continuity and momentum equations, we may obtain the weak form of the constitutive equations by multiplying (7.8.2) by a test function:

$$\begin{aligned} &\int_\Omega \delta\tilde{\sigma}_{ij}\sigma_{ij,t[\chi]} d\Omega + \int_\Omega \delta\tilde{\sigma}_{ij} y_{ijk,k} d\Omega - \int_\Omega \delta\tilde{\sigma}_{ij} c_{k,k}\sigma_{ij} d\Omega \\ &= \int_\Omega \delta\tilde{\sigma}_{ij} C_{ijkl}^{\sigma J} D_{kl} d\Omega + \int_\Omega \delta\tilde{\sigma}_{ij}\{\sigma_{kj} W_{ik} + \sigma_{ki} W_{jk}\} d\Omega \end{aligned} \tag{7.8.3}$$

and

$$\int_\Omega \delta\tilde{y}_{ijk} y_{ijk} d\Omega = \int_\Omega \delta\tilde{y}_{ijk}\sigma_{ij} c_k d\Omega \tag{7.8.4}$$

7.8.3 Finite Element Discretization

We obtain the discrete form of the constitutive equations by letting $\mathbf{N}^s$ and $\mathbf{N}^y$ as sets of shape functions. Let the test and trial functions be $\sigma_{ij} = N_I^s s_{ijI}$, $y_{ijk} = N_I^y y_{ijkI}$; $\delta\tilde{\sigma}_{ij} = \bar{N}_I^s \delta s_{ijI}$ and $\delta\tilde{y}_{ijk} = \bar{N}_I^y \delta y_{ijkI}$. Note that the test functions and the shape functions for Cauchy stresses are used only in the constitutive equations.

The resulting discrete constitutive equations are

$$\mathbf{M}^s \frac{d\mathbf{s}}{dt} + \sum_{k=1}^{N_{sd}} \mathbf{G}_k^T \mathbf{Y}_k - \mathbf{D}\mathbf{s} = \mathbf{z} \tag{7.8.5}$$

$$\mathbf{M}^y \mathbf{Y}_k = \mathbf{L}_k^y \mathbf{s} \quad \text{(no sum on } k) \tag{7.8.6}$$

where $\mathbf{M}^s$ and $\mathbf{D}$ are the generalized mass and diffusion matrices for stress; $\mathbf{G}_k^T \mathbf{Y}_k$ corresponds to the generalized convective term; $\mathbf{M}^y$ and $\mathbf{L}_k^y$ are the generalized mass and convective matrices for stress–velocity product respectively; and $\mathbf{z}$ is the generalized stress vector. The identity matrix, $\mathbf{I} \in R^{N_\sigma} \times R^{N_\sigma}$. The matrices are given in Box 7.2.

Box 7.2 Matrices for ALE stress update

$$\mathbf{M}^s = \mathbf{I}\left[M_{IJ}^s\right] = \mathbf{I}\int_\Omega \bar{N}_I^s N_J^s \, d\Omega \tag{B7.2.1}$$

$$\mathbf{D} = \mathbf{I}\left[D_{IJ}\right] = \mathbf{I}\int_\Omega \bar{N}_I^s c_{k,k} N_J^s \, d\Omega \tag{B7.2.2}$$

$$\mathbf{z} = \left[\mathbf{z}_I\right] = \int_\Omega \bar{N}_I^s C_{ijkl}^{\sigma J} D_{kl} \, d\Omega + \int_\Omega \bar{N}_I^s \left\{\sigma_{kj} W_{ik} + \sigma_{ki} W_{jk}\right\} d\Omega \tag{B7.2.3}$$

$$\mathbf{M}^y = \mathbf{I}\left[M_{IJ}^y\right] = \mathbf{I}\int_\Omega \bar{N}_I^y N_J^y \, d\Omega \tag{B7.2.4}$$

$$\mathbf{s} = [\mathbf{s}_J] = [s_{ijJ}] \tag{B7.2.5}$$

For each convective velocity component c_k, we define:

$$\mathbf{G}_k^T = \mathbf{I}\left[G_{IJ}^T\right]_k = \mathbf{I}\int_\Omega \bar{N}_I^s N_{J,k}^y \, d\Omega \tag{B7.2.6}$$

$$\mathbf{L}_k^y = \mathbf{I}\left[L_{IJk}^y\right] = \mathbf{I}\int_\Omega \bar{N}_I^y c_k N_J^s \, d\Omega \tag{B7.2.7}$$

$$\mathbf{Y}_k = [\mathbf{y}_{kJ}] = [y_{ijkJ}] \tag{B7.2.8}$$

The stress–velocity product $\mathbf{Y}_k$ is stored at each node as a $n_\sigma \times 1$ column vector for each convective velocity c_k. The diagonal form for $\mathbf{M}^y$ is obtained by locating the numerical integration points at nodes.

7.8.4 *Stress Update Procedures*

7.8.4.1 Stress Update Procedure for Galerkin

The corresponding matrix equations have been given in (7.8.5) and (7.8.6). The stress–velocity product $\mathbf{Y}_k$ can be eliminated by inverting $\mathbf{M}^y$ in (7.8.6) and substituting into (7.8.5):

$$\mathbf{M}^s \frac{d\mathbf{s}}{dt} + \sum_{k=1}^{N_{sd}} \mathbf{G}_k^T (\mathbf{M}^y)^{-1} \mathbf{L}_k^y \mathbf{s} - \mathbf{D}\mathbf{s} = \mathbf{z} \tag{7.8.7}$$

where $\sum_{k=1}^{N_{sd}} \mathbf{G}_k^T (\mathbf{M}^y)^{-1} \mathbf{L}_k^y \mathbf{s}$ can be identified as the convective term; the upwind techniques should be applied to evaluate $\mathbf{L}_k^y \mathbf{s}$. When $\mathbf{c} = \mathbf{0}$, i.e., $\boldsymbol{\chi} = \mathbf{X}$, (7.8.7) degenerates to the usual stress update formula in the Lagrangian description, $\mathbf{M}^s \frac{d\mathbf{s}}{dt} = \mathbf{z}$.

All the path-dependent material properties, such as yield strains, effective plastic strains, yield stresses, and back stresses, should be convected via (7.8.7) with $\mathbf{s}$ replaced by these variables, and with $\mathbf{z}$ appropriately modified.

7.8.4.2 Stress Update Procedure for SUPG Method

In a nonlinear finite element formulation for path-dependent materials, such as elastic–plastic materials, the stress and state variables are available only at quadrature points. In order to establish the nodal values for the stress-velocity product, a weak formulation is needed. Based on the one-dimensional study (Liu, Belytschko and Chang, 1986), in which the upwind procedure is used to define this intermediate variable, the artificial viscosity technique (streamline upwind) is considered here as a generalization of this upwind procedure to multi-dimensional cases.

The relation for the stress–velocity product of (7.8.2) is modified to accommodate the artificial viscosity tensor A_{ijkm} by

$$y_{ijk} = \sigma_{ij} c_k - A_{ijkm,m} \tag{7.8.8}$$

The ingredients of the artificial viscosity tensor consist of a tensorial coefficient multiplied by the stress:

$$A_{ijkm} = \mu_{km} \sigma_{ij} \tag{7.8.9}$$

where the tensorial coefficient is constructed and the scalar $\bar{\mu}$ is given by:

$$\mu_{km} = \bar{\mu} c_k c_m / c_n c_n \quad \text{and} \quad \bar{\mu} = \sum_{i=1}^{N_{SD}} \alpha_i c_i h_i / N_{SD} \tag{7.8.10}$$

Here h_i is the element length in the i-direction; N_{SD} designates the number of space dimensions; and α_i is the artificial viscosity parameter given by

$$\alpha_i = \begin{cases} \dfrac{1}{2} & \text{for } c_i > 0 \\ -\dfrac{1}{2} & \text{for } c_i < 0 \end{cases} \tag{7.8.11}$$

The weak form corresponding to (7.8.8) can be obtained by multiplying the stress-velocity product by the test function over the spatial domain Ω:

$$\int_\Omega \delta y_{ijk} y_{ijk} \, d\Omega = \int_\Omega \delta y_{ijk} \sigma_{ij} c_k \, d\Omega - \int_\Omega \delta y_{ijk} A_{ijkm,m} \, d\Omega \tag{7.8.12}$$

This equation may be written as

$$\int_{\Omega} \delta y_{ijk} y_{ijk} \, d\Omega = \int_{\Omega} \delta y_{ijk} \sigma_{ij} c_k \, d\Omega + \int_{\Omega} \delta y_{ijk,m} A_{ijkm} \, d\Omega \tag{7.8.13}$$

by applying the divergence theorem and assuming no traction associated with the artificial viscosity on the boundary. The expression for A_{ijkm}, (7.8.9–10), can be substituted into this equation to yield

$$\int_{\Omega} \delta y_{ijk} y_{ijk} \, d\Omega = \int_{\Omega} (\delta y_{ijk} + \delta \overline{y}_{ijk}) \sigma_{ij} c_k \, d\Omega \quad \text{where} \quad \delta \overline{y}_{ijk} = \delta y_{ijk,m} \overline{\mu} c_m / c_n c_n \tag{7.8.14}$$

The above can be viewed as a modification of the Galerkin finite element method because of the transport nature of the stress-velocity product.

The shape functions for the stress-velocity product can be chosen to be the standard C^0 functions. The number and position of quadrature points for (7.8.14) should be selected to be the Gauss quadrature points, since the stress histories in (7.8.14) are only available at these points.

Following the procedures given above, the stress–velocity product can be defined at each nodal point and it can be substituted into the constitutive equation of (7.8.5) to calculate the rate of change of stresses with the same procedure as that of without artificial viscosity. Note that the interpolation functions for stresses are integrated over the spatial element domain. The task of selecting the number of quadrature points for the displacement finite element poses another important issue. For example, the volumetric locking phenomenon for fully integrated elements arises when the material becomes incompressible (see Section 8.4). While selective reduced integration can overcome this difficulty, it is just as costly as full quadrature. The non-linear two-quadrature point element (Liu *et al.*, 1988) appears to be a good candidate for large-scale computations because it exhibits nearly the same accuracy as the selective reduced integration element with only one-third of the cost.

Following the procedure described by Liu, Belytschko and Chang (1986), the displacement element is divided into M subdomains, where M denotes the number of quadrature points. Each subdomain is designated by Ω_I ($I = 1, \ldots, M$), which contains the quadrature points $\mathbf{x}_I$, and no two subdomains overlap. Associated with Ω_I, a stress interpolation function N_I^s is assigned and its value is prescribed to be unity only at the quadrature point $\mathbf{x} = \mathbf{x}_I$ such that $N_I^s(\mathbf{x}_J) = \delta_{IJ}$. The test function in Ω_I is chosen to be the Dirac delta function $\overline{N}_I^s = \delta(\mathbf{x} - \mathbf{x}_I)$.

Substitution of these functions into the constitutive equation represents a mathematical requirement that the residual of the weak form vanishes at each collocative quadrature point. Because the collocation point is coincident with the quadrature point, the algebraic equations resulting from (7.8.5) can be easily worked out without numerical integration and given as below.

The general mass matrix is

$$M_{IJ}^s = \int_{\Omega} \overline{N}_I^s N_J^s \, d\Omega \tag{7.8.15}$$

where the subscripts I and J range from 1 to M, the number of stress quadrature points per element. The transpose of the divergence operator matrix reads

$$G_{IJk}^T = \int_{\Omega} \overline{N}_I^s N_{J,k}^y \, d\Omega$$

For a 2D 4-node element, it will be

$$\mathbf{G}_k^T = \begin{bmatrix} N_{1,k}^y(x_1) & N_{2,k}^y(x_1) & \cdots & N_{M,k}^y(x_1) \\ N_{1,k}^y(x_2) & N_{2,k}^y(x_2) & \cdots & N_{M,k}^y(x_2) \\ \vdots & \vdots & \vdots & \vdots \\ N_{1,k}^y(x_M) & N_{2,k}^y(x_M) & \cdots & N_{M,k}^y(x_M) \end{bmatrix}_{M\times M} \tag{7.8.16}$$

The generalized diffusion matrix for stress is

$$D_{IJ} = \int_\Omega \bar{N}_I^s c_{k,k} N_J^s \, d\Omega$$

or

$$\mathbf{D} = \begin{bmatrix} c_{k,k}(x_1) & & & \\ & c_{k,k}(x_2) & & \\ & & \ddots & \\ & & & c_{k,k}(x_M) \end{bmatrix}_{M\times M} \tag{7.8.17}$$

The generalized stress vector is

$$z_{ijI} = \int_\Omega \bar{N}_I^s \left\{ C_{ijkl}^{\sigma J} D_{kl} + \sigma_{kj} W_{ik} + \sigma_{ki} W_{jk} \right\} d\Omega$$

or

$$\mathbf{z} = \begin{bmatrix} \left(C_{ijkl}^{\sigma J} D_{kl} + \sigma_{kj} W_{ik} + \sigma_{ki} W_{jk} \right)_{x_1} \\ \left(C_{ijkl}^{\sigma J} D_{kl} + \sigma_{kj} W_{ik} + \sigma_{ki} W_{jk} \right)_{x_2} \\ \vdots \\ \left(C_{ijkl}^{\sigma J} D_{kl} + \sigma_{kj} W_{ik} + \sigma_{ki} W_{jk} \right)_{x_M} \end{bmatrix}_{M\times 1} \tag{7.8.18}$$

7.8.4.3 Stress Update Procedure for Operator Splitting Method

In addition to the methods shown in the last section for the fully coupled equations, another approach referred to as an operator split is used. Conceptually, the approach is simple. 'Splitting' stands for 'decomposing' a set of PDE operators into several sets of simple PDE operators which will be solved sequentially. An operator split decouples the various physical phenomena in the governing equations to obtain simpler equations which are solved more easily. In exchange for a certain loss of accuracy, the operator split offers a generic advantage: simpler equations lead to simpler and stable algorithms, specifically designed for each decoupled equation according to the different physical characteristics.

In the operator split scheme, the stress update is divided into two steps:

$$\sigma^{\text{trial}} = \sigma_n + \left.\frac{\partial \sigma_n}{\partial t}\right|_{\mathbf{X}} \Delta t \quad \text{(Lagrangian step)} \tag{7.8.19}$$

$$\sigma_{n+1} = \sigma^{\text{trial}} + \left.\frac{\partial \sigma^{\text{trial}}}{\partial t}\right|_{\chi} \Delta t + \frac{1}{2}\left.\frac{\partial^2 \sigma^{\text{trial}}}{\partial t^2}\right|_{\chi} \Delta t^2 \quad \text{(Eulerian step)} \tag{7.8.20}$$

in which $\frac{\partial \sigma_n}{\partial t}\big|_{\mathbf{X}}$ and $\frac{\partial \sigma^{\text{trial}}}{\partial t}\big|_{\chi}$ are computed by the general formula (7.8.21). To illustrate the operator split concept, we consider the transport equation for one component of the Cauchy stress:

$$\sigma,_{t[X]} = \sigma,_{t[\chi]} + c_i \sigma,_i = q \tag{7.8.21}$$

where q is determined by the constitutive law. An example of q is the right-hand side of (7.8.1).

The first phase of the operator split is to solve a Lagrangian step without considering the convective effect as below:

$$\sigma^n,_{t[\mathbf{X}]} = q, \quad c_i = 0 \tag{7.8.22}$$

In this Lagrangian phase, it is integrated in time to update stresses from $\sigma^{(t)}$ (stress at time t) to σ^{trial} (denotation of the Lagrangian updated stress), neglecting the convective terms, which is equivalent to assuming that mesh points $\boldsymbol{\chi}$ move with material particles $\mathbf{X}$, that is $\boldsymbol{\chi} = \mathbf{X}$. Thus this Lagrangian phase can proceed in the same way as the usual updated Lagrangian procedure. In addition, it is well known that the stresses are obtained at Gauss points.

After obtaining $\sigma^n,_{t[\mathbf{X}]}$, we can compute σ^{trial} by (7.8.19). Between the two phases, there is a rezoning procedure, which is essentially a remeshing procedure. We may denote it as $\boldsymbol{\chi}^{n+1} = \hat{\boldsymbol{\Phi}}^{-1}(\mathbf{x}^n, t^{n+1})$. Here $\hat{\boldsymbol{\Phi}}^{-1}$ describes the motion between the spatial domain and referential domain. After the new ALE mesh $\boldsymbol{\chi}^{n+1}$ is established, the Eulerian step begins, which remaps all the state variables from coordinate $\mathbf{x}^n$ to $\boldsymbol{\chi}^{n+1}$ in the same configuration according to the advection algorithm.

The second phase deals with the convective term that has not been taken into account during the Lagrangian phase, where the governing PDE is

$$\sigma,_{t[\chi]} + c_i \sigma,_i = 0 \quad q = 0 \tag{7.8.23}$$

and during which phase the stresses are updated from σ^{trial} to $\sigma^{t_{n+1}}$.

A schematic illustration of the whole procedure is shown in Figure 7.5.

According to the two-phase strategy above, the constitutive equation is split into the parabolic equation of the Lagrangian phase and the hyperbolic equation of the convection phase.

Here, we follow Liu, Belytschko and Chang (1986) to apply explicit methods to integrate the convection equation. To compute gradients of the stress fields, the Lax–Wendroff explicit smoothing is employed.

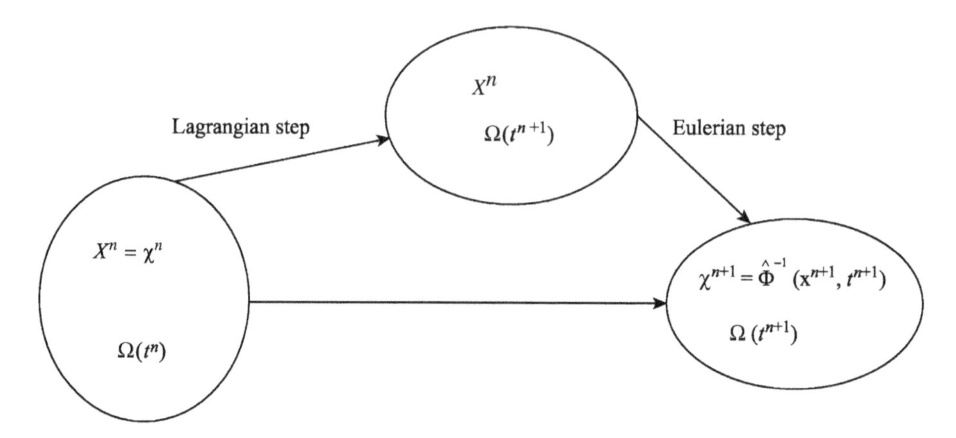

Figure 7.5

Lax–Wendroff update The key point of the Lax–Wendroff method is to replace the time derivatives of dependent variables with spatial derivatives using the governing equations. For the partial differential equations of (7.8.23), we have

$$\sigma_{,t[\chi]}^{\text{trial}} = -c_i^{t_{n+1/2}}\sigma_{,i}^{\text{trial}} \tag{7.8.24}$$

and

$$\begin{aligned}\sigma_{,tt[\chi]}^{\text{trial}} &= \left(\sigma_{,t[\chi]}^{\text{trial}}\right)_{,t} = -\left(c_i^{t_{n+1/2}}\sigma_{,i}^{\text{trial}}\right)_{,t} = -c_i^{t_{n+1/2}}\left(\sigma_{,t[\chi]}^{\text{trial}}\right)_{,i} \\ &= -c_i^{t_{n+1/2}}\left(-c_j^{t_{n+1/2}}\sigma_{,j}^{\text{trial}}\right)_{,i} = c_i^{t_{n+1/2}}c_j^{t_{n+1/2}}\sigma_{,ij}^{\text{trial}}\end{aligned} \tag{7.8.25}$$

After substituting the above two equations into the Taylor series expansion of $\sigma^{t_{n+1}}$ with respect to time:

$$\sigma^{t_{n+1}} = \sigma^{t} + \sigma_{,t[\chi]}^{\text{trial}}\Delta t + \frac{1}{2}\sigma_{,tt[\chi]}^{\text{trial}}\Delta t^2 \tag{7.8.26}$$

we obtain the update equation of the Lax–Wendroff method as

$$\sigma^{t_{n+1}} = \sigma^{t} - c_i^{t_{n+1/2}}\sigma_{,i}^{\text{trial}}\Delta t + \frac{1}{2}c_i^{t_{n+1/2}}c_j^{t_{n+1/2}}\sigma_{,ij}^{\text{trial}}\Delta t^2 \tag{7.8.27}$$

where $c_i^{t_{n+1/2}}$ is the convective velocity evaluated at the mid-step.

In (7.8.27), the gradient of stress $\boldsymbol{\gamma}$ where $\gamma_i = \frac{\partial}{\partial x_i}\sigma^{\text{trial}}$, and its first derivative, $\frac{\partial}{\partial x_i}\gamma$, at each Gauss quadrature point is required. In order to obtain $\boldsymbol{\gamma}$ and its first derivative, a classical least-square projection is employed to obtain a smoothed field of stress gradient, $\boldsymbol{\gamma}^h(x) = \Sigma N_I(x)\boldsymbol{\gamma}_I$ where $\boldsymbol{\gamma}^h(x)$ is a H^1 function. Since initially $(\sigma^{\text{trail}})^h$ is a C^{-1} function, $\boldsymbol{\gamma}$ is not available at nodal points. To obtain $\boldsymbol{\gamma}_I$ by using the divergence theorem, we have

$$\int_\Omega \mathbf{N}_I^\gamma \boldsymbol{\gamma} d\Omega = -\int_\Omega \sigma\nabla\mathbf{N}_I^\gamma d\Omega + \int_\Gamma \mathbf{N}_I^\gamma \sigma\mathbf{n} d\Gamma \tag{7.8.28}$$

where $\mathbf{n}$ is the outward unit normal in the current configuration and $\mathbf{N}^{\gamma}$ is a set of test shape functions. After the assembly, we obtain the linear set of equations:

$$\mathbf{M}^{\gamma}\gamma = \mathbf{\Phi} \quad \text{where} \quad \mathbf{M}^{\gamma} = \left[M_{IJ}^{\gamma}\right] = \int_{\Omega} N_I^{\gamma} N_J^{\gamma} d\Omega \tag{7.8.29}$$

and $\boldsymbol{\gamma}$ is the vector of nodal smoothed values of the stress gradient, and the vector $\mathbf{\Phi}$ is defined as

$$\mathbf{\Phi} = \left[\Phi_I\right] = \sum_e \left[-\int_{\Omega} \sigma \nabla N_I^{\gamma} d\Omega + \int_{\Gamma} N_I^{\gamma} \sigma \mathbf{n} d\Gamma\right] \tag{7.8.30}$$

To make the algorithm explicit, the lumped mass matrix is used instead of the consistent one. After doing so, the solution of the stress gradient of $\boldsymbol{\gamma}$ is straightforward. Then the stress is updated by (7.8.27). To obtain the stress value at Gauss points, the collocation technique can be applied to handle the weak form of (7.8.27).

7.8.5 *Finite Element Implementation of Stress Update Procedures in 1D*

In this section, we will compare various stress updates. For illustrative purposes, we consider the one-dimensional case. In 1D, the shape functions and the corresponding test functions for density, velocity, energy and stress-velocity product may be chosen to be piece wise linear C^0 functions:

$$N_1 = \frac{1}{2}(1-\xi) \quad N_2 = \frac{1}{2}(1+\xi) \tag{7.8.31}$$

where $\xi \in [-1, 1]$. The test and trial functions for all variables are identical. The full upwind method can be applied for all the matrices involving convection effects.

In a uniform mesh, a 1D rod is divided into M segments of equal size h. The elements and the nodes are numbered sequentially from 1 to M and $M+1$, respectively. Let c_m designates the convective velocity at node m and s_m stand for the stress in element m. For simplicity, all the nodal convective velocities are considered to be positive and there is only one quadrature point per element.

7.8.5.1 Application of SUPG in 1D

According to the stress update, (7.8.5), its matrix and vectors can be expressed as follows:

$$\text{generalized mass matrix } \mathbf{M}^s = h\mathbf{I}_{M\times M} \tag{7.8.32}$$

The transpose of the discrete divergence operator matrix is

$$\mathbf{G}_x^T = \begin{bmatrix} -1 & 1 & & & & \\ & \ddots & \ddots & & & \\ & & -1 & 1 & & \\ & & & \ddots & \ddots & \\ & & & -1 & 1 & \end{bmatrix}_{M\times(M+1)} \tag{7.8.33}$$

The matrix $\mathbf{M}^y$ is diagonalized by locating the integration points at the nodes (Lobatto quadrature):

$$\mathrm{diag}(\mathbf{M}^y) = h\left\{\frac{1}{2}, 1, \cdots, 1, \cdots 1, \frac{1}{2}\right\}^T_{(M+1)\times 1}$$

If exact integration is used, the update stress is:

$$\mathbf{L}_k^y\mathbf{s} = \frac{1}{6}h\{(2c_1 + c_2)s_1, \cdots, (c_{m-1} + 2c_m)s_{m-1} + (2c_m + c_{m+1})s_m, \cdots, (c_M + 2c_{M+1})s_M\}^T_{M\times 1} \tag{7.8.34}$$

If full upwind is used, it becomes:

$$\mathbf{L}_k^y\mathbf{s} = \frac{1}{2}h\{0, \cdots, (c_m + c_{m+1})s_m, \cdots, (c_M + c_{M+1})s_M\}^T_{(M+1)\times 1} \tag{7.8.35}$$

The generalized diffusion vector is:

$$\mathbf{D}\mathbf{s} = \{(-c_{1,1} + c_{2,1})s_1, \cdots, (-c_{m,1} + c_{m+1})s_m, \cdots, (-c_{M,1} + c_{M+1,1})s_M\}^T_{M\times 1} \tag{7.8.36}$$

and the rate of change of stress due to material deformation (the rotation of stress vanishes in the 1D case) is

$$\mathbf{z} = h\{\dot{s}_1, \cdots, \dot{s}_m, \cdots, \dot{s}_M\}^T_{M\times 1} \tag{7.8.37}$$

where $\dot{s} = C^{\sigma J}_{1111}v_{(1,1)}$.

By substituting (7.8.32–7.8.37) into (7.8.7), the rate of change of stress in an ALE description can be shown to be as follows.

If exact integration is used:

$$\frac{d\mathbf{s}}{dt} = \begin{Bmatrix} ds_1/dt \\ \vdots \\ ds_m/dt \\ \vdots \\ ds_M/dt \end{Bmatrix}_{M\times 1} = \frac{1}{h}\begin{Bmatrix} (-c_1 + c_2)s_1 \\ \vdots \\ (-c_m + c_{m+1})s_m \\ \vdots \\ (-c_M + c_{M+1})s_M \end{Bmatrix} - \frac{1}{6h}\begin{Bmatrix} (-3c_1)s_1 + (2c_2 + c_3)s_2 \\ \vdots \\ -(c_{m-1} + 2c_m)s_{m-1} \\ -(c_m - c_{m+1})s_m \\ +(2c_{m+1} + c_{m+2})s_{m+1} \\ \vdots \\ -(c_{M-1} + 2c_M)s_{M-1} \\ +3c_{M+1}s_M \end{Bmatrix} + \begin{Bmatrix} s_{1,t[\mathbf{x}]} \\ \vdots \\ s_{m,t[\mathbf{x}]} \\ \vdots \\ s_{M,t[\mathbf{x}]} \end{Bmatrix} \tag{7.8.38}$$

If full upwind is used to evaluate $\mathbf{L}_k^y\mathbf{s}$:

$$\frac{d\mathbf{s}}{dt} = \begin{Bmatrix} ds_1/dt \\ \vdots \\ ds_m/dt \\ \vdots \\ ds_M/dt \end{Bmatrix}_{M\times 1} = \frac{1}{h}\begin{Bmatrix} (-c_1+c_2)s_1 \\ \vdots \\ (-c_m+c_{m+1})s_m \\ \vdots \\ (-c_M+c_{M+1})s_M \end{Bmatrix} - \frac{1}{2h}\begin{Bmatrix} (-c_1+c_2)s_1 \\ \vdots \\ (-c_{m-1}+c_m)s_{m-1} \\ +(c_m+c_{m+1})s_m \\ \vdots \\ -(c_{M-1}+c_M)s_{M-1} \\ +2(c_M+c_{M+1})s_M \end{Bmatrix} + \begin{Bmatrix} s_{1,t[\mathbf{x}]} \\ \vdots \\ s_{m,t[\mathbf{x}]} \\ \vdots \\ s_{M,t[\mathbf{x}]} \end{Bmatrix} \tag{7.8.39}$$

The second bracket on the right-hand side of (7.8.38) shows the central differencing (or simple averaging) effects for the transport of stresses, while (7.8.39) exhibits the donor-cell differencing. This can be further clarified by letting

$$c_1 = \cdots = c_m = \cdots = c_{M+1} = c(\text{constant})$$

If exact integration is used:

$$\frac{d\mathbf{s}}{dt} = \begin{Bmatrix} ds_1/dt \\ \vdots \\ ds_m/dt \\ \vdots \\ ds_M/dt \end{Bmatrix}_{M\times 1} = -\frac{c}{2h}\begin{Bmatrix} -s_1+s_2 \\ \vdots \\ -s_{m-1}+s_{m+1} \\ \vdots \\ -s_{M-1}+s_M \end{Bmatrix} + \begin{Bmatrix} s_{1,t[\mathbf{x}]} \\ \vdots \\ s_{m,t[\mathbf{x}]} \\ \vdots \\ s_{M,t[\mathbf{x}]} \end{Bmatrix} \tag{7.8.40}$$

If full upwind is used:

$$\frac{d\mathbf{s}}{dt} = \begin{Bmatrix} ds_1/dt \\ \vdots \\ ds_m/dt \\ \vdots \\ ds_M/dt \end{Bmatrix}_{M\times 1} = -\frac{c}{h}\begin{Bmatrix} s_1 \\ \vdots \\ -s_{m-1}+s_m \\ \vdots \\ -s_{M-1}+2s_M \end{Bmatrix} + \begin{Bmatrix} s_{1,t[\mathbf{x}]} \\ \vdots \\ s_{m,t[\mathbf{x}]} \\ \vdots \\ s_{M,t[\mathbf{x}]} \end{Bmatrix} \tag{7.8.41}$$

Equation (7.8.40) shows that the transport of the stresses at odd and even elements tends to be decoupled. Therefore physically unrealistic oscillations would be expected when the simple averaging method is employed to evaluate the spatial derivatives of stresses.

7.8.5.2 Application of Operator Split in 1D Case

For illustrative purposes, a 1D rod, which is divided into M segments of equal size h, is considered. Assuming

$$c_1 = c_m = c_{M+1} = c(constant) > 0$$

we can see that this procedure of update will not work for constant stress and linear shape function since the RHS of (7.8.30) will be zero. In addition, we can see that for the Lax–Wendroff update, the shape function of **s**, $\mathbf{N}^s$, must be in the same order as $\mathbf{N}^\gamma$. Assuming both of them are linear shape functions, we can obtain

$$\mathrm{diag}(\mathbf{M}^y) = h\left\{\frac{1}{2}, 1, \cdots, 1, \cdots 1, \frac{1}{2}\right\}^T_{(M+1)\times 1} \tag{7.8.42}$$

$$\boldsymbol{\Phi} = \frac{1}{2}\left[-s_1 + s_2, -s_1 + s_3, -s_2 + s_4, \cdots, -s_{m-1} + s_{m+1}, \cdots - s_{M-1} + s_{M+1}\right]^T_{(M+1)\times 1} \tag{7.8.43}$$

7.8.6 Explicit Time Integration Algorithm

Explicit time integration, the coupled equations (7.5.13 and 7.5.18), will be integrated by a predictor-corrector method (Hughes and Liu, 1978).

This kind of predictor-corrector method is similar to the Newmark algorithm. The major difference is that the former algorithm is explicit, while the latter is implicit.

The continuity equations are

$$\frac{d\boldsymbol{\rho}}{dt} = -\left(\mathbf{M}_n^\rho\right)^{-1}\left(\mathbf{L}_n^\rho\tilde{\boldsymbol{\rho}}_{n+1} + \mathbf{K}_n^\rho\tilde{\boldsymbol{\rho}}_{n+1}\right) \quad \text{where} \quad \tilde{\boldsymbol{\rho}}_{n+1} = \boldsymbol{\rho}_n + (1-\alpha)\Delta t\frac{d\boldsymbol{\rho}_n}{dt} \tag{7.8.44}$$

$$\boldsymbol{\rho}_{n+1} = \tilde{\boldsymbol{\rho}}_{n+1} + \alpha\Delta t\frac{d\boldsymbol{\rho}_{n+1}}{dt} \tag{7.8.45}$$

The momentum equations are

$$\frac{d\mathbf{v}}{dt} = (M_n)^{-1}\left(\mathbf{f}_{n+1}^{\mathrm{ext}} - \mathbf{f}_n^{\mathrm{int}} - \mathbf{L}_n\tilde{\mathbf{v}}_{n+1}\right) \quad \text{where} \quad \tilde{\mathbf{v}}_{n+1} = \mathbf{v}_n + (1-\gamma)\Delta t\frac{d\mathbf{v}_n}{dt} \tag{7.8.46}$$

$$\mathbf{v}_{n+1} = \tilde{\mathbf{v}}_{n+1} + \gamma\Delta t\frac{d\mathbf{v}_{n+1}}{dt} \tag{7.8.47}$$

Equation (7.8.46) needs to be used in conjunction with

$$\tilde{\mathbf{d}}_{n+1} = \mathbf{d}_n + \Delta t\mathbf{v}_n + \left(\frac{1}{2} - \beta\right)\Delta t^2\frac{d\mathbf{v}_n}{dt} \tag{7.8.48}$$

$$\mathbf{d}_{n+1} = \tilde{\mathbf{d}}_{n+1} + \beta\Delta t^2\frac{d\mathbf{v}_{n+1}}{dt} \tag{7.8.49}$$

to calculate the $\mathbf{f}_n^{\mathrm{int}}$.

In these equations, α, β and γ are the computational parameters. For explicit calculations, the following parameters are used:

$$\alpha = 0,\ \beta = 0,\ \gamma \geq \frac{1}{2} \tag{7.8.50}$$

The algorithm is presented in Box 7.3.

Box 7.3 Explicit time integration

1. Initialization. Set $n = 0$, input initial conditions.
2. Time stepping loop, $t \in [0, t_{max}]$.
3. Integrate the mesh velocity to obtain the mesh displacement and spatial coordinates.
4. Calculate the incremental stresses, yield stresses, and back stresses by integration of (7.8.7), which stress update procedures have been discussed in detail in the last section:
   (a) the rate of stresses due to convection,
   (b) the rate of stresses due to rotation,
   (c) the rate of stresses due to deformation.
5. Compute the internal force vector.
6. Compute the acceleration by the equations of motion, (7.8.46).
7. Compute the density by the equation of mass conservation, (7.8.44).
8. Integrate the acceleration to obtain the velocity.
9. If $(n+1)\Delta t > t_{max}$, stop; otherwise, replace n by $n+1$ and go to Step 2.

7.9 Linearization of the Discrete Equations

The derivation of the Jacobian matrices, that is, the tangent stiffness and load stiffness matrices, was given in Chapter 6 for Lagrangian elements. In this section, tangential matrices are modified for the ALE description. This mainly involves accounting for the transport terms. However, the complications start with the time derivatives in the referential domain (Liu *et al.*, 1991; Hu and Liu, 1993).

7.9.1 Internal Nodal Forces

Referring to (4.4.11), the internal nodal force can be written as an integral over the referential (ALE) domain as follows:

$$\begin{aligned} f_{iI}^{\text{int}} &= \int_{\Omega} \frac{\partial N_I}{\partial x_k} \sigma_{ki} \, d\Omega \\ &= \int_{\hat{\Omega}} \frac{\partial N_I}{\partial \chi_k} \frac{\partial \chi_k}{\partial x_m} \sigma_{mi} \hat{J} \, d\hat{\Omega} \end{aligned} \tag{7.9.1}$$

In the second equation we have used the chain rule and the relation $d\Omega = \hat{J} d\hat{\Omega}$ where $\hat{J} = \det \hat{\mathbf{F}}$ and $\hat{F}_{ij} = \frac{\partial x_i}{\partial \chi_j}$. We define the nominal stress on the referential configuration as

$$\hat{P}_{ki} = \hat{J} \frac{\partial \chi_k}{\partial x_m} \sigma_{mi} = \hat{J} \hat{F}_{km}^{-1} \sigma_{mi} \tag{7.9.2}$$

Substituting this expression into (7.9.1), the internal nodal forces can be written as

$$f_{iI}^{\text{int}} = \int_{\hat{\Omega}} \frac{\partial N_I}{\partial \chi_k} \hat{P}_{ki} \, d\hat{\Omega} \tag{7.9.3}$$

Taking the time derivative of the above gives (noting that the domain $\hat{\Omega}$ is fixed and the shape functions are independent of time)

$$\frac{df_{iI}^{\text{int}}}{dt} = \int_{\hat{\Omega}} \frac{\partial N_I}{\partial \chi_k} \hat{P}_{ki,t[\chi]} \, d\hat{\Omega} \tag{7.9.4}$$

Following an analogous procedure to that of Section 6.4.1, note that the nominal stress on $\hat{\Omega}$ can be written as $\hat{\mathbf{P}} = \hat{\mathbf{S}}\hat{\mathbf{F}}^T$ where $\hat{\mathbf{S}}$ is the second Piola–Kirchhoff stress on $\hat{\Omega}$. Analogously to steps (6.4.23) to (6.4.10),

$$\begin{aligned} \hat{\mathbf{P}}_{,t[\chi]} &= \hat{\mathbf{F}}^{-1} \cdot \left(\hat{\mathbf{F}} \cdot \hat{\mathbf{S}}_{,t[\chi]} \cdot \hat{\mathbf{F}}^T + \hat{\mathbf{F}} \cdot \hat{\mathbf{S}} \cdot \hat{\mathbf{F}}^T \cdot \hat{\mathbf{F}}^{-T} \cdot \hat{\mathbf{F}}_{,t[\chi]}{}^T \right) \\ &= \hat{\mathbf{F}}^{-1} \cdot \left(\hat{J} \boldsymbol{\sigma}^{\nabla \hat{T}} + \hat{J} \boldsymbol{\sigma} \cdot \hat{\mathbf{L}}^T \right) \end{aligned} \tag{7.9.5}$$

where $\boldsymbol{\sigma}^{\nabla \hat{T}} = \hat{J}^{-1} \hat{\mathbf{F}} \cdot \hat{\mathbf{S}}_{,t[\chi]} \cdot \hat{\mathbf{F}}^T = \boldsymbol{\sigma}_{,t[\chi]} - \hat{\mathbf{L}} \cdot \boldsymbol{\sigma} - \boldsymbol{\sigma} \cdot \hat{\mathbf{L}}^T + (\text{trace } \hat{\mathbf{L}})\boldsymbol{\sigma}$ is the Truesdell rate of Cauchy stress based on the referential time derivative and the mesh velocity gradient $\hat{\mathbf{L}} = \hat{\mathbf{F}}_{,t[\chi]} \cdot \hat{\mathbf{F}}^{-1}$. Substituting (7.9.5) into (7.9.4) gives

$$\begin{aligned} \frac{df_{iI}^{\text{int}}}{dt} &= \int_{\hat{\Omega}} \frac{\partial N_I}{\partial \chi_k} \hat{F}_{km}^{-1} \left(\sigma_{mi}^{\nabla \hat{T}} + \sigma_{ml} \hat{L}_{il} \right) \hat{J} \, d\hat{\Omega} \\ &= \int_{\Omega} \frac{\partial N_I}{\partial x_k} \left(\sigma_{ki}^{\nabla \hat{T}} + \sigma_{kl} \hat{L}_{il} \right) d\Omega \end{aligned} \tag{7.9.6}$$

Note that $\sigma_{ki}^{\nabla \hat{T}}$ can also be written as

$$\sigma_{ki}^{\nabla \hat{T}} = \sigma_{ki,t[\chi]} - \hat{v}_{k,m} \sigma_{mi} - \sigma_{km} \hat{v}_{i,m} + \hat{v}_{m,m} \sigma_{ki} \tag{7.9.7}$$

analogous to the definition of the Truesdell rate in Box 3.5. The first term on the right-hand side of (7.9.7) can also be expressed by

$$\sigma_{ki,t[\chi]} = \dot{\sigma}_{ki} - c_l \sigma_{ki,l} \quad \text{or} \quad \frac{\partial \sigma_{ki}(\boldsymbol{\chi}, t)}{\partial t} = \frac{\partial \sigma_{ki}(\mathbf{X}, t)}{\partial t} - c_l \frac{\partial \sigma_{ki}}{\partial x_l} \tag{7.9.8a}$$

$$\hat{L}_{km} = L_{km} - c_{k,m}; \quad \hat{v}_{m,m} = v_{m,m} - c_{m,m} \tag{7.9.8b}$$

Using (7.9.8), (7.9.7) becomes

$$\sigma_{ki}^{\nabla \hat{T}} = \dot{\sigma}_{ki} - c_l \sigma_{ki,l} - (L_{km} - c_{k,m})\sigma_{mi} - \sigma_{km}(L_{im} - c_{i,m}) + (v_{m,m} - c_{m,m})\sigma_{ki} \tag{7.9.9}$$

After regrouping the terms (7.9.9) and substituting it into (7.9.6), we get

$$\frac{df_{iI}^{\text{int}}}{dt}=\int_{\Omega} N_{I,k}[\dot{\sigma}_{ki}-\sigma_{mi}L_{km}+v_{m,m}\sigma_{ki}]d\Omega+\int_{\Omega} N_{I,k}[\sigma_{mi}c_{k,m}-c_l\sigma_{ki,l}-c_{m,m}\sigma_{ki}]d\Omega$$

or

$$\frac{df_{iI}^{\text{int}}}{dt}=\int_{\Omega} N_{I,k}\left(\sigma_{ki}^{\nabla\tau}+\sigma_{kl}v_{i,l}\right)d\Omega+\int_{\Omega} N_{I,k}[\sigma_{km}c_{i,m}-c_l\sigma_{ki,l}-c_{m,m}\sigma_{ki}]d\Omega \tag{7.9.10}$$

The first integral is associated with the Lagrangian representation (6.4.20) while the second integral is associated with the ALE convective effect.

After discretizing (7.9.10), df_{iI}^{int}/dt can be described as follows:

$$\frac{d\mathbf{f}_{iI}^{\text{int}}}{dt}=\mathbf{K}^{\text{lag}}\cdot\mathbf{v}+\mathbf{K}^{\text{ale}}\cdot\mathbf{c}\quad\text{or}\quad f_{iI,t[\chi]}^{\text{int}}=\left(K_{iIjJ}^{\text{lag}}v_{jJ}+K_{ijIJ}^{\text{ale}}c_{jJ}\right) \tag{7.9.11}$$

where $\mathbf{v}=[v_{jJ}]$, $\mathbf{c}=[c_{jJ}]$, $\mathbf{K}^{\text{lag}}=\left[K_{iIjJ}^{\text{lag}}\right]$, and $\mathbf{K}^{\text{ale}}=\left[K_{ijIJ}^{\text{ale}}\right]$. $\mathbf{K}^{\text{lag}}$ in (7.9.11) comes from the first integral, which by $\mathbf{K}^{\text{lag}}=\mathbf{K}^{\text{mat}}+\mathbf{K}^{\text{geo}}$. $\mathbf{K}^{\text{mat}}$ and $\mathbf{K}^{\text{geo}}$ are defined in Section 6.4.1.

Further expanding $\mathbf{K}^{\text{ale}}$ gives

$$K_{iIlJ}^{\text{ale}}=\int_{\Omega} N_{I,k}\sigma_{km}N_{J,m}\delta_{il}d\Omega-\int_{\Omega} N_{I,k}\sigma_{ki,l}N_J\,d\Omega-\int_{\Omega} N_{I,k}\sigma_{ki}N_{J,l}\,d\Omega \tag{7.9.12}$$

7.9.2 External Nodal Forces

For the linearization $\mathbf{f}^{\text{ext}}$, we will start by using the time derivative of a surface integral. Nanson's relation between surface elements in the current and referential (ALE) configurations is

$$\mathbf{n}\,d\Gamma=\hat{J}\hat{\mathbf{n}}\cdot\hat{\mathbf{F}}^{-1}d\hat{\Gamma} \tag{7.9.13}$$

where $\hat{\mathbf{n}}$ is the unit outward normal to the surface on $\hat{\Omega}$. Taking the time derivative of (7.9.13) gives

$$\begin{aligned}\frac{d}{dt}(\mathbf{n}\,d\Gamma)&=\hat{J}_{,t[\chi]}\hat{\mathbf{n}}\cdot\hat{\mathbf{F}}^{-1}d\hat{\Gamma}+\hat{J}\hat{\mathbf{n}}\cdot(\hat{\mathbf{F}}^{-1})_{,t[\chi]}d\hat{\Gamma}\\&=\hat{J}\operatorname{div}\hat{\mathbf{v}}\hat{\mathbf{n}}\cdot\hat{\mathbf{F}}^{-1}d\hat{\Gamma}+\hat{J}\hat{\mathbf{n}}\cdot(-\hat{\mathbf{F}}^{-1}\cdot\hat{\mathbf{L}})d\hat{\Gamma}\\&=(\operatorname{div}\widehat{\mathbf{v}\mathbf{I}}-\hat{\mathbf{L}}^T)\hat{J}\hat{\mathbf{n}}\cdot\hat{\mathbf{F}}^{-1}d\hat{\Gamma}\\&=(\operatorname{div}\widehat{\mathbf{v}\mathbf{I}}-\hat{\mathbf{L}}^T)\cdot\mathbf{n}\,d\hat{\Gamma}\end{aligned} \tag{7.9.14}$$

In the second of (7.9.14) we have used the results $\hat{J}_{,t[\chi]} = \hat{J}\text{div}\,\hat{\mathbf{v}}$ and $(\hat{\mathbf{F}}^{-1})_{,t[\chi]} = -\hat{\mathbf{F}}^{-1} \cdot \hat{\mathbf{L}}$. The time derivative of a surface integral is therefore given by

$$\frac{d}{dt}\int_{\Gamma} g\mathbf{n}\,d\Gamma = \int_{\Gamma}\left[\left(g_{,t[\chi]} + g\nabla\cdot\hat{\mathbf{v}}\right)\mathbf{I} - g\hat{\mathbf{L}}^T\right]\cdot\mathbf{n}\,d\Gamma \tag{7.9.15}$$

Therefore, (7.9.15) can also be rewritten as

$$\frac{d}{dt}\int_{\Gamma} g\mathbf{n}\,d\Gamma = \int_{\Gamma}\left[\left(g_{,t[\chi]} + g\nabla\cdot(\mathbf{v}-\mathbf{c})\right)\mathbf{I} - g(\mathbf{L}^T - \nabla\mathbf{c}^T)\right]\cdot\mathbf{n}\,d\Gamma \tag{7.9.16}$$

After substituting the relation $g_{,t[\chi]} = \dot{g} - c_i g_i$ and regrouping the terms with the convective velocity, the equation becomes

$$\frac{d}{dt}\int_{\Gamma} g\mathbf{n}\,d\Gamma = \int_{\Gamma}\left[\left(\dot{g} + g\nabla\cdot\mathbf{v}\right)\mathbf{I} - g\mathbf{L}^T\right]\cdot\mathbf{n}\,d\Gamma + \int_{\Gamma}\left[\left(-g\nabla\cdot\mathbf{c}\right)\mathbf{I} + g\nabla\mathbf{c}^T - \mathbf{c}\cdot\nabla g\right]\cdot\mathbf{n}\,d\Gamma \tag{7.9.17}$$

The first term on the right-hand side of (7.9.17) resembles the original material time derivative in the Lagrangian formulation; and the second term is the extra convected terms for solving the time derivative in the reference domain of the ALE algorithm.

For the particular case of pressure loading let $g = -pN_I$. The time derivative of the external nodal force can be written as

$$\frac{d\mathbf{f}^{\text{ext}}}{dt} = (\mathbf{K}^{\text{ext}})^{\text{lag}}\cdot\mathbf{v} + (\mathbf{K}^{\text{ext}})^{\text{ale}}\cdot\mathbf{c} \quad \text{or} \quad \frac{df_{iI}^{\text{ext}}}{dt} = \left(K_{iIjJ}^{\text{ext}}\right)^{\text{lag}} v_{jJ} + \left(K_{iIjJ}^{\text{ext}}\right)^{\text{ale}} c_{jJ} \tag{7.9.18}$$

where $(\mathbf{K}^{\text{ext}})^{\text{lag}}$ is given in (6.4.32); and $(\mathbf{K}^{\text{ext}})^{\text{ale}}$ is defined as

$$\left(K_{iIjJ}^{\text{ext}}\right)^{\text{ale}} = -\int_{\Gamma}(-pN_I)N_{J,j}\delta_{ki}n_k\,d\Omega + \int_{\Gamma}(-pN_I)N_{J,i}n_j\,d\Omega - \int_{\Gamma}(-pN_I)_{,j}\,N_J n_i\,d\Omega \tag{7.9.19}$$

7.10 Mesh Update Equations

7.10.1 Introduction

The option of arbitrarily moving the mesh in the ALE description offers interesting possibilities. By means of ALE, moving boundaries (which are material surfaces) can be tracked with the accuracy characteristic of Lagrangian methods and the interior mesh can be moved so as to avoid excessive element distortion and entanglement. However, this requires that an effective algorithm for updating the mesh, that is, the mesh velocities $\hat{\mathbf{v}}$, must be prescribed. The mesh should be prescribed so that mesh distortion is avoided and so that boundaries and interfaces remain at least partially Lagrangian.

In this section, we will describe several procedures for updating the mesh. The material and mesh velocities are related by (7.2.16). Once one of them is determined, the other is automatically fixed. It is important to note that, if $\hat{\mathbf{v}}$ is given, $\hat{\mathbf{d}}$ and $\hat{\mathbf{a}}$ can be computed and there is no need to evaluate $\mathbf{w}$. On the other hand, if $\hat{\mathbf{v}}$ is considered the unknown but $\mathbf{w}$ is given, (7.2.16) must be solved to evaluate $\hat{\mathbf{v}}$ before updating the mesh. Finally, mixed reference velocities can be given (i.e., a component of $\hat{\mathbf{v}}$ can be prescribed and $\mathbf{w}$ in the other(s)). Finding the *best* choice for these velocities and an algorithm for updating the mesh constitutes one of the major hurdles in developing an effective implementation of the ALE description. Depending on which velocity ($\hat{\mathbf{v}}$ or $\mathbf{w}$ or mixed) is prescribed, three different cases may be studied.

7.10.2 Mesh Motion Prescribed A Priori

The case where the mesh motion $\hat{\mathbf{v}}$ is given corresponds to an analysis where the domain boundaries are known at every instant. When the boundaries of the fluid domain have a known motion, the mesh movement along this boundary can be prescribed *a priori*.

7.10.3 Lagrange–Euler Matrix Method

The case where the relative velocity $\mathbf{w}$ is arbitrarily defined is discussed by Hughes, Liu and Zimmerman (1981). Let $\mathbf{w}$ be given by

$$w_i = \left.\frac{\partial \chi_i}{\partial t}\right|_X = \left(\delta_{ij} - \alpha_{ij}\right) v_j \tag{7.10.1}$$

where δ_{ij} is the Kronecker delta and α_{ij} is a matrix of parameters called the Lagrange–Euler parameter matrix; $\alpha_{ij} = 0$ if $i \neq j$ and $\alpha_{\underline{ii}}$ real (underlined indices meaning no sum on them). In general, the α's can vary in space and be time-dependent; however α_{ij} is usually taken as time-independent. According to (7.10.1) the relative velocity $\mathbf{w}$ becomes a linear function of the material velocity. It was chosen because, if $\alpha_{ij} = \delta_{ij}$, $\mathbf{w} = \mathbf{0}$ and the Lagrangian description is obtained, whereas if $\alpha_{ij} = 0$, $\mathbf{w} = \mathbf{v}$, yielding the Eulerian formulation. The Lagrange–Euler matrix needs to be given once and for all at each grid point.

Since $\mathbf{w}$ is defined by (7.10.1), the other velocities are determined by (7.2.16), which become, respectively,

$$c_i = \frac{\partial x_i}{\partial \chi_j}\left(\delta_{jk} - \alpha_{jk}\right) v_k \tag{7.10.2}$$

and

$$\hat{v}_i = v_i - \left(\delta_{jk} - \alpha_{jk}\right) v_k \frac{\partial x_i}{\partial \chi_j} \tag{7.10.3}$$

The latter equations must be satisfied in the referential domain along its boundaries. Substituting (7.2.7) into (7.10.3) yields a basic equation for mesh rezoning:

$$\left.\frac{\partial x_i}{\partial t}\right|_\chi + \left(\delta_{jk} - \alpha_{jk}\right) v_k \frac{\partial x_i}{\partial \chi_j} - v_i = 0 \tag{7.10.4}$$

The explicit form of (7.10.4) in 2D is listed:

$$\left.\frac{\partial x_1}{\partial t}\right|_\chi + (1-\alpha_{11})v_1\frac{\partial x_1}{\partial \chi_1} + (1-\alpha_{22})v_2\frac{\partial x_1}{\partial \chi_2} - v_1 = 0 \tag{7.10.5}$$

$$\left.\frac{\partial x_2}{\partial t}\right|_\chi + (1-\alpha_{11})v_1\frac{\partial x_2}{\partial \chi_1} + (1-\alpha_{22})v_2\frac{\partial x_2}{\partial \chi_2} - v_2 = 0 \tag{7.10.6}$$

Equation (7.10.4) differs only in its last term from the one proposed by Hughes, Liu and Zimmerman (1981). This difference is not noticeable if the Lagrange–Euler parameters α_{ij} are chosen equal to zero or one. Moreover, (7.10.4) includes the Jacobian matrix (i.e., $\partial x_i/\partial \chi_j$) that is missing in the Liu and Ma (1982) formulation. Finally, (7.10.4) is a transport equation without any diffusion so the classic numerical difficulties associated with transport equations are expected.

The ALE technique with a mesh update based on the Lagrange–Euler parameters is very useful in surface wave problems. We assume that the free surface is oriented relative to the global coordinates so that it can be written as $x_{3s} = x_{3s}(x_1, x_2, t)$. An Eulerian description is used in the x_1 and x_2 directions (i.e. $x_1 = \chi_1$ and $x_2 = \chi_2$). The free surface is defined by one spatial coordinate which is a continuous and differentiable function of the other two spatial coordinates and time. In this case the Lagrange–Euler matrix has only one nonzero term, α_{33} (usually equal to 1), and the only nontrivial equation in (7.10.4) is

$$\left.\frac{\partial x_{3s}}{\partial t}\right|_\chi + v_1\frac{\partial x_{3s}}{\partial \chi_1} + v_2\frac{\partial x_{3s}}{\partial \chi_2} - v_3 = (\alpha_{33}-1)v_3\frac{\partial x_{3s}}{\partial \chi_3} \tag{7.10.7}$$

The above equation is easily recognized as the kinematic equation of the surface and may be written as

$$\left.\frac{\partial x_{3s}}{\partial t}\right|_\chi + v_i n_i N_s = a(x_1, x_2, x_{3s}, t) \tag{7.10.8}$$

where the components of **n** form the unit normal pointing out from the surface. The components of the normal vector are given by

$$\frac{1}{N_s}\left(\frac{\partial x_{3s}}{\partial \chi_1}, \frac{\partial x_{3s}}{\partial \chi_2}, -1\right) \tag{7.10.9a}$$

with N_s given by

$$N_s = \left[1 + \left(\frac{\partial x_{3s}}{\partial \chi_1}\right)^2 + \left(\frac{\partial x_{3s}}{\partial \chi_2}\right)^2\right]^{\frac{1}{2}} = \left[1 + \left(\frac{\partial x_{3s}}{\partial x_1}\right)^2 + \left(\frac{\partial x_{3s}}{\partial x_2}\right)^2\right]^{\frac{1}{2}} \tag{7.10.9b}$$

where $a(x_1, x_2, x_{3s}, t)$ is the so-called accumulation rate function expressing the gain or loss of mass under the free surface. It can be seen by comparing (7.10.7) and (7.10.8) that the accumulation rate function is

$$a(x_1, x_2, x_{3s}, t) = (\alpha_{33} - 1)v_3 \frac{\partial x_{3s}}{\partial \chi_3} = w_3 \frac{\partial x_{3s}}{\partial \chi_3} \tag{7.10.10}$$

The free surface is a material surface; along the free surface the accumulation rate must be zero; and consequently α_{33} has to be taken equal to one. This can also be deduced by noticing that no particles can cross the free surface, so w_3 must be zero. Although (7.10.4) can be applied to problems where x_1 and/or x_2 are not Eulerian by prescribing nonzero αs in these directions, controlling the element shapes by adjusting the α's is very difficult.

Controlling the mesh by (7.10.1) has some disadvantages. For instance, while $\hat{\mathbf{v}}$ has a clear physical interpretation (i.e., the mesh velocity), $\mathbf{w}$ is much more difficult to visualize (except in the direction perpendicular to material surfaces, where it is identically zero) and therefore it is very difficult to maintain regular-shaped elements inside the fluid domain by just prescribing the αs. Because of this main drawback the mixed formulation introduced by Huerta and Liu (1988), called the deformation gradient method, was developed, and this is discussed next.

7.10.4 Deformation Gradient Formulations

Because of the limitations of the α scheme, a mixed formulation is developed for the resolution of (7.2.16). One of the goals of the ALE method is the accurate tracking of the moving boundaries which are usually material surfaces. Hence, along these surfaces we enforce $\mathbf{w} \cdot \mathbf{n} = 0$ where $\mathbf{n}$ is the exterior normal. The other goal of the ALE technique is to avoid element entanglement and this is better achieved, once the boundaries are known, by prescribing the mesh displacements independently (through the potential equations, for instance), or the velocities, because both $\hat{\mathbf{d}}$ and $\hat{\mathbf{v}}$ directly govern the element shape. Therefore, one can prescribe $\mathbf{w} \cdot \mathbf{n} = 0$ along the domain boundaries while defining the nodal displacements or prescribing the velocities, $\hat{\mathbf{d}}$ or $\hat{\mathbf{v}}$, in the interior.

The system of differential equations defined in (7.2.16) has to be solved along the moving boundaries. Notice first that solving for w_i in terms of $(v_i - \hat{v}_i)$, (7.2.16) can be rewritten as

$$c_j \equiv v_j - \hat{v}_j = F_{ji}^{\chi} w_i \tag{7.10.11}$$

Define the Jacobian matrix of the map between the spatial and ALE coordinates by

$$F_{ij}^{\chi} \equiv \frac{\partial x_i}{\partial \chi_j} \tag{7.10.12}$$

Its inverse is

$$(\mathbf{F}^{\chi})^{-1} = \frac{1}{\hat{J}} \begin{bmatrix} \hat{J}^{11} & -\hat{J}^{12} & \hat{J}^{13} \\ -\hat{J}^{21} & \hat{J}^{22} & -\hat{J}^{23} \\ \hat{J}^{31} & -\hat{J}^{32} & \hat{J}^{33} \end{bmatrix} \equiv \frac{\hat{J}^{ij}}{\hat{J}} \tag{7.10.13}$$

where $\hat{J}^{ij}$ are the cofactors of F_{ij}^{χ}; $\hat{J}$ is the determinent of the Jacobian Matrix. Multiplying the inverse Jacobian matrix on both sides of (7.10.11) and substituting (7.10.13) into (7.10.11) yields

$$\frac{\hat{J}^{ji}}{\hat{J}}(v_j - \hat{v}_j) = w_i \quad \text{or} \quad \hat{J}^{ji}(v_j - \hat{v}_j) = \hat{J}w_i \tag{7.10.14}$$

Dividing $\hat{J}^{\underline{ii}}$ on both sides of (7.10.14) gives

$$\frac{\hat{J}^{ji}}{\hat{J}^{\underline{ii}}}(v_j - \hat{v}_j) = \frac{\hat{J}}{\hat{J}^{\underline{ii}}}w_i = v_i - \left.\frac{\partial x_i}{\partial t}\right|_{\chi} + \sum_{\substack{j=1 \\ j\neq i}}^{nsd} \frac{v_j - \hat{v}_j}{\hat{J}^{\underline{ii}}}\hat{J}^{ji} \tag{7.10.15}$$

This equation is derived using the fact that the first term of the left-hand side of the equation equals 1. When the LHS of (7.10.15) has been simplified using the definition of $\hat{v}_j$, (7.2.7), it can be written as

$$\left.\frac{\partial x_i}{\partial t}\right|_{\chi} - v_i - \sum_{\substack{j=1 \\ j\neq i}}^{NSD} \frac{v_j - \hat{v}_j}{\hat{J}^{\underline{ii}}}\hat{J}^{\underline{ji}} = -\frac{\hat{J}}{\hat{J}^{\underline{ii}}}w_i \tag{7.10.16}$$

Notice that the cofactor $\hat{J}^{\underline{ii}}$ appears in the denominator to account for the motion of the mesh in the plane perpendicular to χ_i because (7.10.16) are verified in the reference domain $\hat{\Omega}$, not in the actual deformed domain Ω.

Examples for (7.10.16) in 1D, 2D and 3D are given in Box 7.4.

Box 7.4 Examples of (7.10.16) in 1D, 2D and 3D

1D

$$\left.\frac{\partial x_1}{\partial t}\right|_{\chi} - v_1 = -\frac{\hat{J}}{\hat{J}^{11}}w_1 \tag{B7.4.1}$$

where $\hat{J}^{11} = 1$.

2D

$$\left.\frac{\partial x_1}{\partial t}\right|_{\chi} - v_1 - \frac{v_2 - \hat{v}_2}{\hat{J}^{11}}\hat{J}^{21} = -\frac{\hat{J}}{\hat{J}^{11}}w_1 \tag{B7.4.2}$$

$$\left.\frac{\partial x_2}{\partial t}\right|_{\chi} - v_2 - \frac{v_1 - \hat{v}_1}{\hat{J}^{22}}\hat{J}^{12} = -\frac{\hat{J}}{\hat{J}^{22}}w_2 \tag{B7.4.3}$$

where $\hat{J}^{11} = \frac{\partial x_2/\partial \chi_2}{\hat{J}}$ and $\hat{J}^{22} = \frac{\partial x_1/\partial \chi_1}{\hat{J}}$.

3D

$$\left.\frac{\partial x_1}{\partial t}\right|_\chi - v_1 - \frac{v_2 - \hat{v}_2}{\hat{J}^{11}}\hat{J}^{21} - \frac{v_3 - \hat{v}_3}{\hat{J}^{11}}\hat{J}^{31} = -\frac{\hat{J}}{\hat{J}^{11}}w_1 \tag{B7.4.4}$$

$$\left.\frac{\partial x_2}{\partial t}\right|_\chi - v_2 - \frac{v_1 - \hat{v}_1}{\hat{J}^{22}}\hat{J}^{12} - \frac{v_3 - \hat{v}_3}{\hat{J}^{22}}\hat{J}^{32} = -\frac{\hat{J}}{\hat{J}^{22}}w_2 \tag{B7.4.5}$$

$$\left.\frac{\partial x_3}{\partial t}\right|_\chi - v_3 - \frac{v_1 - \hat{v}_1}{\hat{J}^{33}}\hat{J}^{13} - \frac{v_2 - \hat{v}_2}{\hat{J}^{33}}\hat{J}^{23} = -\frac{\hat{J}}{\hat{J}^{33}}w_3 \tag{B7.4.6}$$

For purposes of simplification, we assume that the moving free surface is perpendicular to one coordinate axis in the reference domain. Let the free surface be perpendicular to χ_3; the first two equations in (7.10.16) are trivial because the mesh velocity is prescribed in the direction of χ_1 and χ_2. Therefore the mesh motion is known, but the third equation must be solved for $\hat{v}_3$ given w_3, $\hat{v}_1$, and $\hat{v}_2$. It may be written explicitly as

$$\hat{v}_3 - \frac{\hat{J}^{13}}{\hat{J}^{33}}(v_1 - \hat{v}_1) - \frac{\hat{J}^{23}}{\hat{J}^{33}}(v_2 - \hat{v}_2) - v_3 = -\frac{\hat{J}}{\hat{J}^{33}}w_3 \tag{7.10.17}$$

or

$$\begin{aligned}\left.\frac{\partial x_{3s}}{\partial t}\right|_\chi &- \frac{v_1 - \hat{v}_1}{\hat{J}^{33}}\hat{J}^{13}\left(\frac{\partial x_{3s}}{\partial_{\chi_1}}, \frac{\partial x_{3s}}{\partial_{\chi_2}}\right) - \frac{v_2 - \hat{v}_2}{\hat{J}^{33}}\hat{J}^{23}\left(\frac{\partial x_{3s}}{\partial_{\chi_1}}, \frac{\partial x_{3s}}{\partial_{\chi_2}}\right) - v_3 \\ &= -\frac{w_3}{\hat{J}^{33}}\hat{J}\left(\frac{\partial x_{3s}}{\partial_{\chi_1}}, \frac{\partial x_{3s}}{\partial_{\chi_2}}\right)\end{aligned} \tag{7.10.18}$$

where $\hat{v}_3$ has been substituted by $\left.\frac{\partial x_{3s}}{\partial t}\right|_\chi$; $\hat{J}^{13}$, $\hat{J}^{23}$, and the Jacobian $\hat{J}$ are functions of $\partial x_{3s}/\partial_{\chi_1}$ and $\partial x_{3s}/\partial_{\chi_2}$; $\hat{J}^{33}$ is not dependent on x_{3s}; and x_{3s} is the free surface equation. In (7.10.18) x_{3s} is the unknown function, while $\hat{v}_1$, $\hat{v}_2$ and w_3 are known. If $\hat{v}_1 = \hat{v}_2 = 0$ (the Eulerian description is used in χ_1 and χ_2), the kinematic surface equation, (7.10.8), is again obtained. However, with the mixed formulation $\hat{v}_1$ and $\hat{v}_2$ can be prescribed and therefore better numerical results are obtained than by defining in (7.10.7) α_{11} and α_{22}, whose physical interpretation is much more obscure.

7.10.5 Automatic Mesh Generation

The Laplacian method for remeshing is based on updating the position of the nodes by solutions of the Laplace equation space (I, J) into real space (x, y). It is motivated by the fact that the contours of solutions to the Laplace equation are approximately orthogonal.

The determination of positions of the ALE nodes is posed as finding $x(I, J)$ and $y(I, J)$ such that they satisfy the following equations:

$$L^2(x) = -\frac{\partial^2 x}{\partial I^2} + \frac{\partial^2 x}{\partial J^2} = 0; \quad L^2(y) = -\frac{\partial^2 y}{\partial I^2} + \frac{\partial^2 y}{\partial J^2} = 0 \quad \text{in } \Omega \tag{7.10.19}$$

where I and J are taken to be independent variables and $x(I)$ and $y(J)$ are the coordinates of nodes I and J when I and J take on integer values. The boundary conditions in 2D are

$$x(I, J) = \bar{x}(I, J); \quad y(I, J) = \bar{y}(I, J) \quad \text{in } \Gamma \tag{7.10.20}$$

Another useful mesh generation scheme is by solving a fourth-order differential equation:

$$L^4(x) = -\frac{\partial^4 x}{\partial I^2 \partial J^2}; \quad L^4(y) = \frac{\partial^4 y}{\partial I^2 \partial J^2} \tag{7.10.21}$$

Equations (7.10.19) and (7.10.21) can be solved by the finite difference method with a Gauss–Seidel iteration scheme. Meshes generated by the Laplace equation are distorted near the boundary where a high curvature occurs. However, the fourth-order equation gives a better mesh shape because a higher differentiation is employed. An equipotential method regards the mesh lines as two intersecting sets of equipotentials with each set satisfying Laplace's equation in the interior.

7.10.6 Mesh Update Using a Modified Elasticity Equation

In this mesh moving scheme, the mesh 'flow' is governed by the modified equations of linear elasticity equation without body force and prescribed traction. When mesh movement takes place, these equations are solved to determine the internal nodal displacements of the mesh based on the given placements. The method works for any mesh type and for any type of movement. The added generality comes at the cost of solving this additional equation system each time the mesh is deformed.

Consider an elastic body occupying a region $\Omega \subset R^{n_{sd}}$ with boundary Γ. The displacement of the mesh is given by $\hat{\mathbf{d}}(\mathbf{x})$ and is governed by the equilibrium equations of elasticity without body force. The strain tensor ε is related to the displacement gradients by

$$\varepsilon = \frac{1}{2}\left(\nabla\hat{\mathbf{u}} + (\nabla\hat{\mathbf{u}})^T\right) \tag{7.10.22}$$

The stiffness matrix $\mathbf{k}$ in the local coordinate is defined as

$$\mathbf{k}_e = \int_\xi \mathbf{B}^T \mathbf{D} \mathbf{B} J \; d\xi \tag{7.10.23}$$

If the integrand of (7.10.23) is multiplied by J/J, then the stiffness matrix becomes

$$\mathbf{k}_e = \int_\xi \mathbf{B}^T \frac{J\mathbf{D}}{J} \mathbf{B} J \; d\xi \tag{7.10.24}$$

Therefore, $\mathbf{D}$, the material constant, can be redefined as

$$\tilde{\mathbf{D}} = J\mathbf{D} \tag{7.10.25}$$

It is desirable to retain the structure of the mesh in the more refined areas, and have most of the deformation weighted towards the larger element regions of the mesh. To accomplish this, a variable stiffness coefficient is desirable where the small elements are more rigid. We implement this by dividing material constant, $\tilde{\mathbf{D}}$, by the Jacobian of the transformation from the element domain to the physical domain. By doing this, smaller elements which are more susceptible to distortion and located in areas where refinement is important retain their shape better. The element stiffness matrix $\mathbf{k}_e$ is assembled into the global stiffness matrix, $\mathbf{K}$.

7.10.7 Mesh Update Example

As an example of the ALE mesh update using the modified elasticity equation, consider the finite element mesh for fluid flow around a cylinder located in a rectangular fluid domain as shown in Figure 7.6(a). The updated mesh after the cylinder has displaced an amount of 0.25 w in the y-direction is shown in Figure 7.6(b). The boundaries of the rectangular domain remain fixed. As can be seen from the figure, the mesh refinement about the cylinder is retained in the updated mesh and no significant distortion of the elements has occurred.

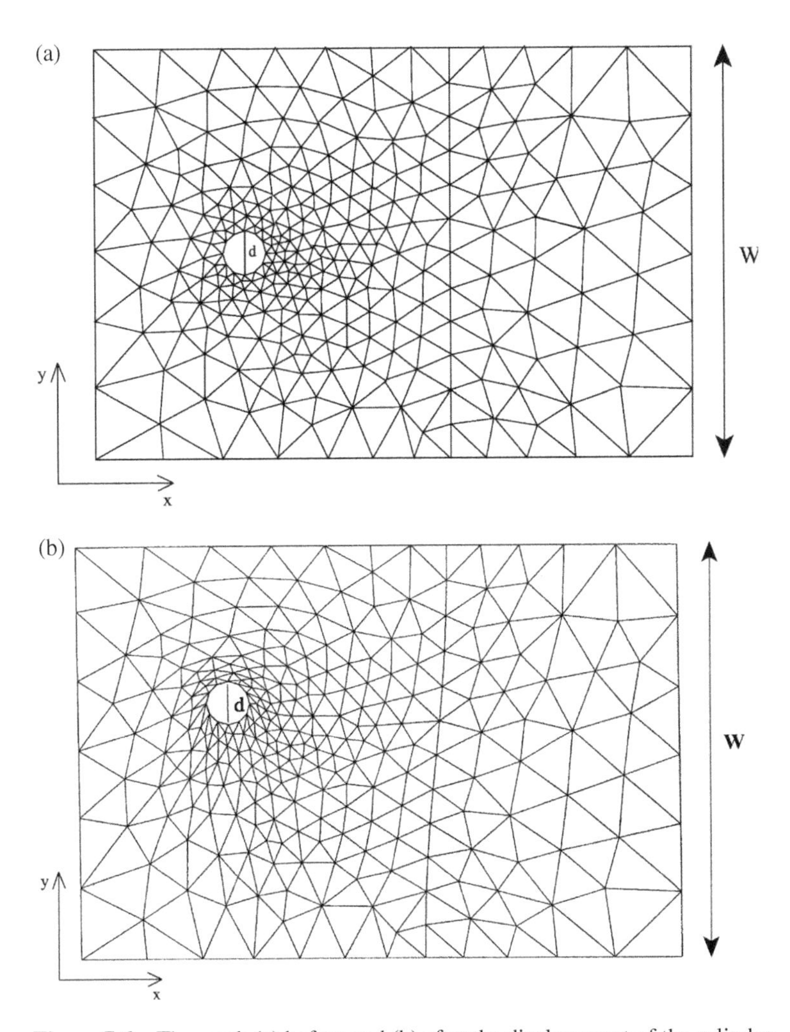

Figure 7.6 The mesh (a) before and (b) after the displacement of the cylinder

7.11 Numerical Example: An Elastic–Plastic Wave Propagation Problem

An elastic–plastic wave propagation problem is used to assess the ALE approach in conjunction with the regular fixed mesh method. The problem statement, given in Figure 7.7, represents a 1D infinitely long, elastic–plastic hardening rod. Constant density and isothermal conditions are assumed to simplify the problem. Thus only the momentum equation and constitutive equation are considered for this problem. It should be noted that this elastic–plastic wave propagation problem does not require an ALE mesh and the problem was selected because it provides a severe test of the stress update procedure and because of the availability of an analytic solution. The problem is solved using 400 elements which are uniformly spaced with a mesh size of 0.1. The mesh is arranged so that no reflected wave will occur during the time interval under consideration. Four stages are involved in this problem:

1. $t \in [0, t_1]$, the mesh is fixed, and a square wave is generated at the origin.
2. $t \in [t_1, t_2]$, the mesh is fixed and the wave travels along the bar.
3. $t \in [t_2, t_3]$, two cases are studied:
   Case A: the mesh is moved uniformly to the left-hand side with a constant speed $-\hat{v}^*$;
   Case B: same as Case A except the mesh is moved to the right.
4. $t = t_3$, the stress is reported as a function of spatial coordinates in Figures 7.7 and 7.8 for Case A and Case B, respectively.

For both cases, the momentum and stress transport are taken into account by employing the full upwind method for elastic and elastic–plastic materials. Material properties and computational parameters are

$$\rho = 1,\ E = 10^4,\ E/E_T = 3,\ \sigma_{y0} = 75,\ \sigma_0 = -100$$

$$\Delta x = \Delta\chi = 0.1,\ \hat{v}^* = 0.25\sqrt{E/\rho},\quad (t_1 = 45,\ t_2 = 240,\ t_3 = 320)\ (\times 10^{-3})$$

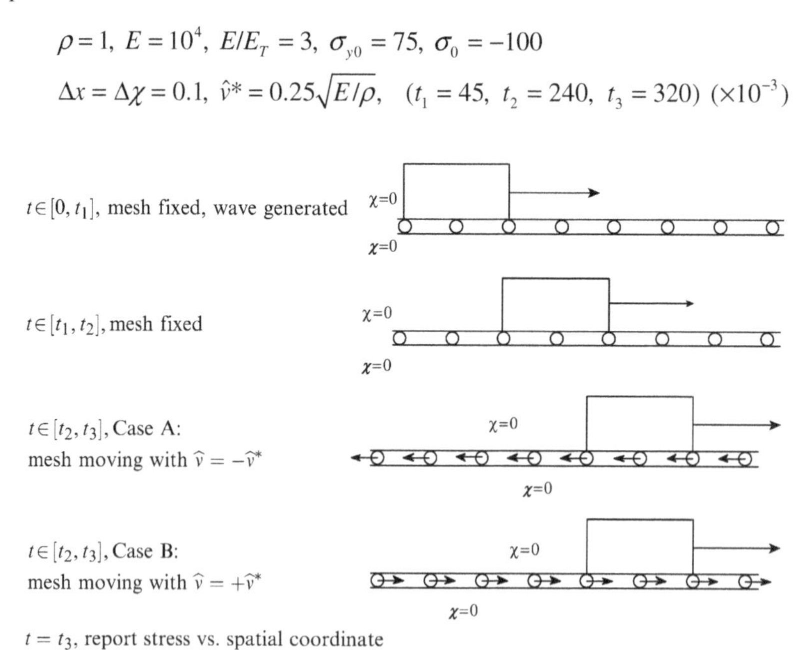

Figure 7.7 Problem statement and computational parameters for 1D wave propagation problem

The results according to several time step sizes are reported in Table 7.2. The wave arrival time for both methods, with or without the upwinding technique, agree with the fixed mesh runs. However, the scheme without the upwinding technique causes severe unrealistic spatial oscillations in Case A because of the significant transport effects. The new method proposed

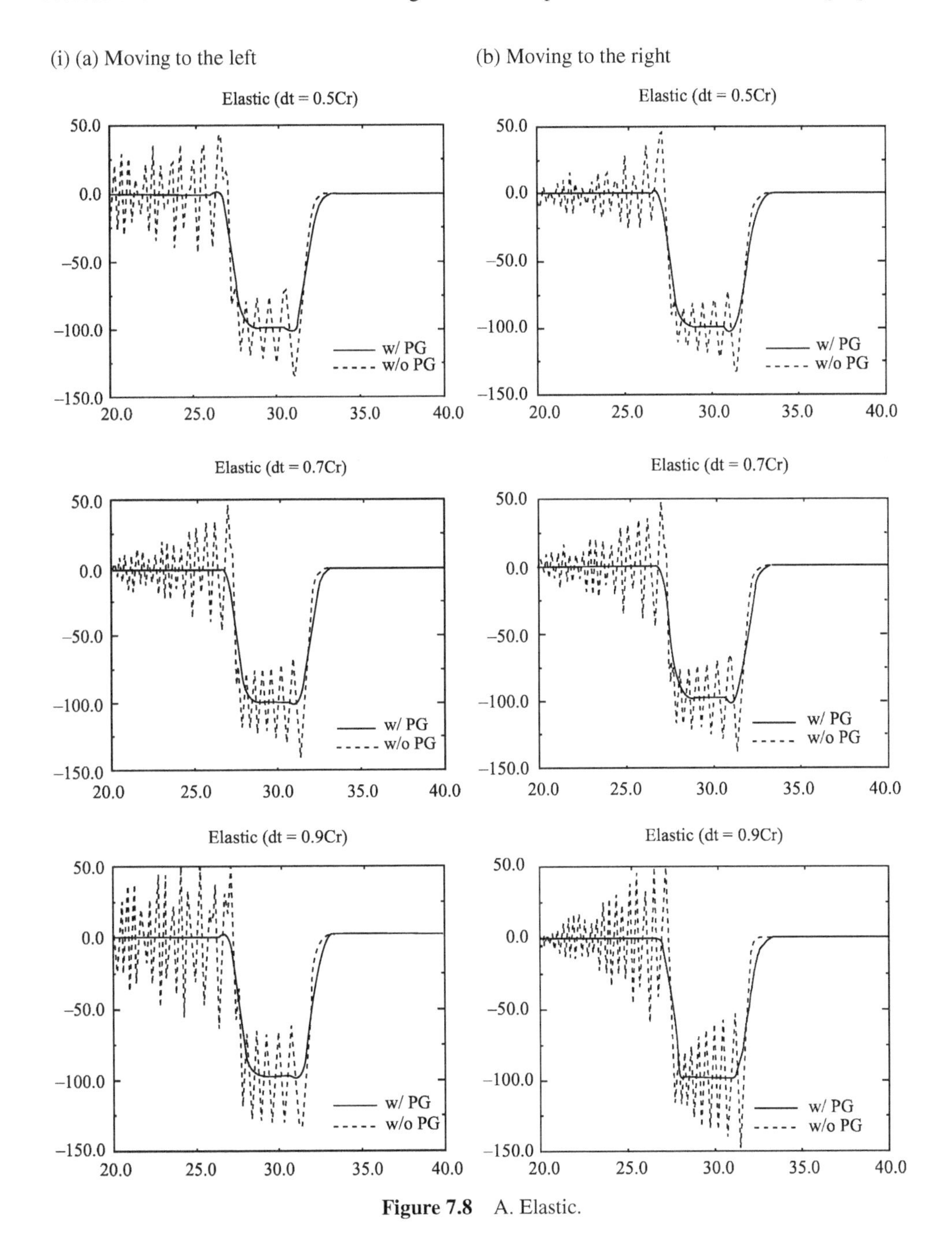

Figure 7.8 A. Elastic.

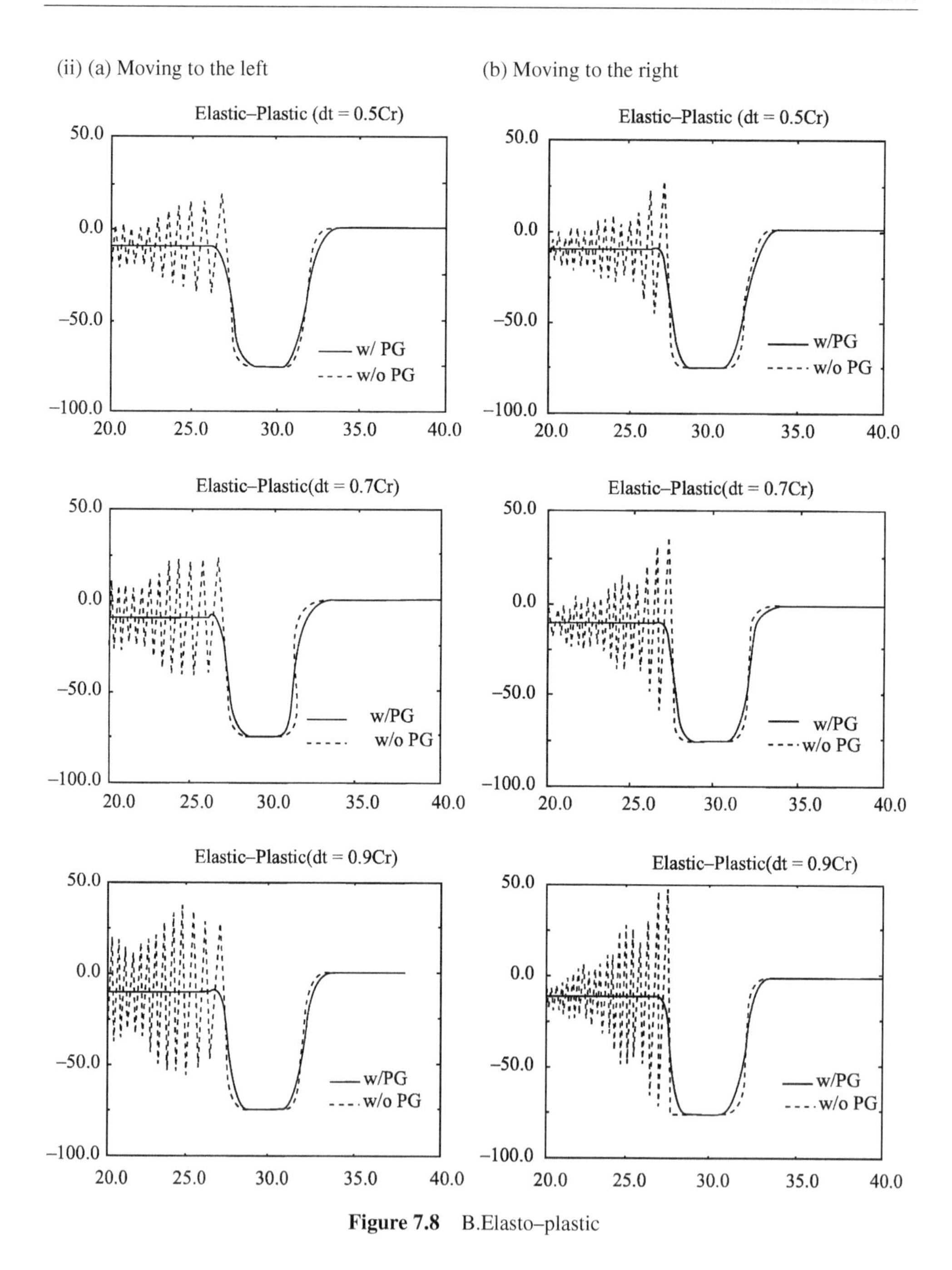

Figure 7.8 B.Elasto–plastic

here eliminates these oscillations completely. Based on these studies, it is also found that the transport of stresses as well as yield stress (and back stresses if kinematic hardening) plays an important role in ALE computations for path-dependent materials, and the proposed update procedure is quite accurate and effective.

Table 7.2 Parameters for elastic–plastic wave propagation problem

Time step (Δt)	$\Delta t/Cr^a$	Number of time steps
0.040×10^{-2}	0.5	400
0.056×10^{-2}	0.7	286
0.072×10^{-2}	0.9	222

$Cr^a = \Delta x/((E/\rho)^{1/2} + |c|)$

7.12 Total ALE Formulations

7.12.1 Total ALE Conservation Laws

In order to establish the integral form of conservation laws in ALE, the material time derivative of a volume integral in referential form is studied. Let $G(t)$ be defined by the volume integral

$$G(t) = \int_{\hat{\Omega}} \hat{f}(\chi, t)\ d\hat{\Omega} = \int_{\Omega} f(\mathbf{x},t) d\boldsymbol{\Omega};\ \hat{f}(\chi, t) = \hat{J} f(\mathbf{x}, t);\ \hat{J} = \det\left(\frac{\partial x_i}{\partial \chi_j}\right) \tag{7.12.1}$$

Recalling the Reynolds transport theorem in Eulerian Description, (3.2.16), and applying it to (7.12.1) gives

$$\frac{DG(t)}{Dt} = \int_{\Omega} \left.\frac{\partial f}{\partial t}\right|_{\mathbf{x}} d\Omega + \int_{\Gamma} v_i n_i f\ d\Gamma \tag{7.12.2}$$

where v_i is the particle velocity observed in the spatial coordinate and n_i is the outward normal to the surface of spatial volume Ω (i.e., Γ). Generalizing the same procedure as used to derive the Reynolds transport theorem in the Eulerian description, we can write a referential form:

$$\frac{DG(t)}{Dt} = \int_{\hat{\Omega}} \left.\frac{\partial \hat{f}}{\partial t}\right|_{\chi} d\hat{\Omega} + \int_{\hat{\Gamma}} w_i \hat{n}_i \hat{f}\ d\hat{\Gamma} \tag{7.12.3}$$

where w_i is defined in (7.2.10), and $\hat{n}_i$ is the outward normal to the surface of $\hat{\Omega}$ (i.e., $\hat{\Gamma}$). Physically, (7.12.3) states that the rate of change of $G(t)$ is equal to the sum of the amount instantaneously created in $\hat{\Omega}$ and the flux through the boundary surface $\hat{\Gamma}$ induced by the relative movement of the reference frame.

7.12.1.1 Conservation of Momentum

The principle of conservation of momentum states that the *total* rate of change of the linear momentum of the medium occupying the referential volume $\hat{\Omega}$ at time t,

$$\frac{D}{Dt} \int_{\hat{\Omega}} \hat{\rho}(\chi, t) \mathbf{v}(\chi, t) d\hat{\Omega} \tag{7.12.4a}$$

is equal to the net force exerted on it:

$$\int_{\hat{\Gamma}} \hat{\mathbf{t}} d\hat{\Gamma} + \int_{\hat{\Omega}} \hat{\rho} \mathbf{b} d\hat{\Omega} \tag{7.12.4b}$$

where $\hat{\rho}(\boldsymbol{\chi}, t) = \hat{J} \rho(\mathbf{x}, t)$; $\hat{\mathbf{t}}$ is the force per unit area acting on the surface of the volume $\hat{\Gamma}$, and $\mathbf{b}$ is the body force per unit of mass acting in $\hat{\Omega}$. The force on the deformed spatial surface per unit of referential area, $\hat{\mathbf{t}}$, may be written as a function of the nominal stress tensor $\hat{\mathbf{P}}$ (see (7.9.2)) and the outward unit normal $\hat{\mathbf{n}}$ on the referential surface as

$$\hat{t}_i = \hat{P}_{ji} \hat{n}_j \tag{7.12.5}$$

Substituting (7.12.5) into (7.12.4) and using the divergence theorem to transform the surface $d\Gamma_0$ integral into a volume integral gives

$$\frac{D}{Dt} \int_{\hat{\Omega}} \hat{\rho} v_i d\hat{\Omega} = \int_{\hat{\Omega}} \left[\frac{\partial \hat{P}_{ji}}{\partial \chi_j} + \hat{\rho} b_i \right] d\hat{\Omega} \tag{7.12.6}$$

Using Reynold's transport theorem and the divergence theorem, the left-hand side in this equation is transformed into

$$\int_{\hat{\Omega}} \left[\left. \frac{\partial (\hat{\rho} v_i)}{\partial t} \right|_{\chi} + \frac{\partial (w_j \hat{\rho} v_i)}{\partial \chi_j} \right] d\hat{\Omega} = \int_{\hat{\Omega}} \left[\frac{\partial \hat{P}_{ji}}{\partial \chi_j} + \hat{\rho} b_i \right] d\hat{\Omega} \tag{7.12.7}$$

which is reduced to

$$\left. \frac{\partial (\hat{\rho} v_i)}{\partial t} \right|_{\chi} + \frac{\partial (w_j \hat{\rho} v_i)}{\partial \chi_j} = \frac{\partial \hat{P}_{ji}}{\partial \chi_j} + \hat{\rho} b_i \quad \text{in } \hat{\Omega} \tag{7.12.8}$$

after noticing that $d\hat{\Omega}$ is arbitrarily chosen. Equation (7.12.8) can be further simplified by using the continuity equation, (7.12.10), given next; that is, the momentum equation may be written in referential form:

$$\hat{\rho} \left. \frac{\partial v_i}{\partial t} \right|_{\chi} + \hat{\rho} w_j \frac{\partial v_i}{\partial \chi_j} = \frac{\partial \hat{P}_{ji}}{\partial \chi_j} + \hat{\rho} b_i \quad \text{in } \hat{\Omega} \tag{7.12.9}$$

7.12.1.2 Conservation of Mass and Conservation of Energy

It is left to the reader as an exercise to show that the equation of continuity is

$$\left. \frac{\partial \hat{\rho}}{\partial t} \right|_{\chi} + \frac{\partial (\hat{\rho} w_i)}{\partial \chi_i} = 0 \quad \text{in } \hat{\Omega} \tag{7.12.10}$$

and the energy equation is given by

$$\left.\frac{\partial(\hat{\rho}E)}{\partial t}\right|_{\chi}+\frac{\partial}{\partial \chi_i}(w_i\hat{\rho}E)=\frac{\partial(v_j\hat{P}_{ij})}{\partial \chi_i}+\hat{\rho}b_i v_i-\frac{\partial q_i}{\partial \chi_i}+\hat{\rho}s \quad \text{in } \hat{\Omega} \tag{7.12.11}$$

Synthesis By defining vectors $\mathbf{V}$, $\hat{\mathbf{E}}_i$, and $\mathbf{F}$ as follows:

$$\mathbf{V}^T=\{1 \quad v_1 \quad v_2 \quad v_3 \quad E\} \tag{7.12.12}$$

$$\hat{\mathbf{E}}_i=\begin{Bmatrix} 0 \\ \hat{P}_{i1} \\ \hat{P}_{i2} \\ \hat{P}_{i3} \\ \hat{P}_{ij}v_j \end{Bmatrix}+\begin{Bmatrix} 0 \\ 0 \\ 0 \\ 0 \\ -q_i \end{Bmatrix} \tag{7.12.13}$$

$$\mathbf{F}=\begin{Bmatrix} 0 \\ b_1 \\ b_2 \\ b_3 \\ b_j v_j+s \end{Bmatrix} \tag{7.12.14}$$

the continuity equation (7.12.10), the momentum equation (7.12.9) and the energy equation (7.12.11) can be written as

$$\left.\frac{\partial(\hat{\rho}\mathbf{V})}{\partial t}\right|_{\chi}+\frac{\partial}{\partial \chi_i}(w_i\hat{\rho}\mathbf{V})=\frac{\partial \hat{\mathbf{E}}_i}{\partial \chi_i}+\hat{\rho}\mathbf{F} \quad \text{in } \hat{\Omega} \tag{7.12.15}$$

This is a quasi-Eulerian form of the conservation laws.

7.12.2 Reduction to Updated ALE Conservation Laws

Multiplying (7.12.15) by $d\hat{\Omega}$ and integrating over the referential volume gives

$$\int_{\hat{\Omega}}\left[\left.\frac{\partial(\hat{\rho}\mathbf{V})}{\partial t}\right|_{\chi}+\frac{\partial}{\partial \chi_i}(w_i\hat{\rho}\mathbf{V})\right]d\hat{\Omega}=\int_{\hat{\Omega}}\left[\frac{\partial \hat{\mathbf{E}}_i}{\partial \chi_i}+\hat{\rho}\mathbf{F}\right]d\hat{\Omega} \tag{7.12.16}$$

Substituting $d\hat{\Omega}=\hat{J}^{-1}d\Omega$ into the LHS of (7.12.16) yields

$$\int_{\Omega}\hat{J}^{-1}\left[\frac{\partial(\hat{\rho}\mathbf{V})}{\partial t}\Big|_{\chi}+\frac{\partial}{\partial \chi_i}(w_i\hat{\rho}\mathbf{V})\right]d\Omega \tag{7.12.17}$$

If there are no shocks or discontinuities, the momentum components can be differentiated via the chain rule to give

$$\int_{\Omega} \hat{J}^{-1} \left[\mathbf{V} \frac{\partial \hat{\rho}}{\partial t}\bigg|_{\chi} + \hat{\rho} \frac{\partial \mathbf{V}}{\partial t}\bigg|_{\chi} + \hat{\rho} w_i \frac{\partial \mathbf{V}}{\partial \chi_i} + \mathbf{V} \frac{\partial (w_i \hat{\rho})}{\partial \chi_i} \right] d\Omega \tag{7.12.18}$$

Regrouping terms results in the following expression:

$$\int_{\Omega} \hat{J}^{-1} \left\{ \mathbf{V} \left[\frac{\partial \hat{\rho}}{\partial t}\bigg|_{\chi} + \frac{\partial (w_i \hat{\rho})}{\partial \chi_i} \right] + \hat{\rho} \left[\frac{\partial \mathbf{V}}{\partial t}\bigg|_{\chi} + w_i \frac{\partial \mathbf{V}}{\partial \chi_i} \right] \right\} d\Omega \tag{7.12.19}$$

Using the continuity equation (7.12.10), and $\rho = \hat{J}^{-1}\hat{\rho}$, (7.12.19) reduces to

$$\int_{\Omega} \rho \left[\frac{\partial \mathbf{V}}{\partial t}\bigg|_{\chi} + w_i \frac{\partial \mathbf{V}}{\partial \chi_i} \right] d\Omega \tag{7.12.20}$$

and applying the chain rule to the second term of (7.12.20), in order to examine this expression further, gives

$$\int_{\Omega} \rho \left[\frac{\partial \mathbf{V}}{\partial t}\bigg|_{\chi} + w_j \frac{\partial \mathbf{V}}{\partial x_i} \frac{\partial x_i}{\partial \chi_j} \right] d\Omega \tag{7.12.21}$$

Applying the definition of the convective velocity, c_i (7.2.16), gives

$$\int_{\Omega} \rho \left[\frac{\partial \mathbf{V}}{\partial t}\bigg|_{\chi} + c_i \frac{\partial \mathbf{V}}{\partial x_i} \right] d\Omega \tag{7.12.22}$$

Similar volume transformations are applied to the RHS of (7.12.16) to give

$$\int_{\hat{\Omega}} \hat{\rho} \mathbf{F} \, d\hat{\Omega} = \int_{\Omega} \rho \mathbf{F} \, d\Omega \tag{7.12.23}$$

One can also show, by analogy to the previous derivation, that

$$\int_{\hat{\Omega}} \frac{\partial \hat{\mathbf{E}}_i}{\partial \chi_i} d\hat{\Omega} = \int_{\Omega} \frac{\partial \mathbf{E}_i}{\partial x_i} \, d\Omega \tag{7.12.24}$$

where $\mathbf{E}_i$ is defined as

$$\mathbf{E}_i = \begin{Bmatrix} 0 \\ \sigma_{i1} \\ \sigma_{i2} \\ \sigma_{i3} \\ \sigma_{ij} v_j \end{Bmatrix} + \begin{Bmatrix} 0 \\ 0 \\ 0 \\ 0 \\ -q_i \end{Bmatrix} \quad \text{in } \Omega \tag{7.12.25}$$

Finally, combining (7.12.22) through (7.12.24) gives

$$\int_{\Omega}\left[\rho\left.\frac{\partial \mathbf{V}}{\partial t}\right|_{\chi}+\rho c_i\frac{\partial \mathbf{V}}{\partial x_i}\right]d\Omega=\int_{\Omega}\left[\frac{\partial \mathbf{E}_i}{\partial x_i}+\rho\mathbf{F}\right]d\Omega \tag{7.12.26}$$

The updated ALE conservation laws given in Section 7.3 can be obtained by taking an arbitrary volume Ω. Hence the conservation laws in vector form are given by

$$\rho\left.\frac{\partial \mathbf{V}}{\partial t}\right|_{\chi}+\rho c_i\frac{\partial \mathbf{V}}{\partial x_i}=\frac{\partial \mathbf{E}_i}{\partial x_i}+\rho\mathbf{F} \tag{7.12.27}$$

The component forms of (7.12.27) are identical to those derived in Section 7.3.

7.13 Exercises

7.1. Develop your own code (MATLAB®, FORTRAN, C, Maple, etc.) to solve the 1D advection-diffusion equation

$$u\frac{d\phi}{dx}=\upsilon\frac{d^2\phi}{dx^2}$$

with Galerkin and SUPG method separately. (See Example 7.2].)

The BCs and parameters are assigned to a real world problem to determine the particles distribution at the steady state. Consider a 1m length segment in a long tube filled with a solution. At steady state, the particle concentration at end A is 5% and that at end B is 20%, that is, $\phi(x=0)=0.05$, $\phi(x=1)=0.2$. The solvent flows in the tube from end A to B under a constant velocity $u = 2$ m/s. The particles' diffusion coefficient in the solvent is $\upsilon = 0.025$ m²/s. Provide the distribution of particles concentration distribution along the tube segment.

Simulate and discuss the following situations:

(a) Mesh the domain with 10, 20, 50, 100 and 200 uniform size elements. What is the element Peclet number P_e for each mesh? Compare the analytical solution, Galerkin and SUPG prediction. In SUPG method, select $\gamma=\frac{\Delta x}{2}\left(\coth(P_e)-\frac{1}{P_e}\right)$ for each case. Discuss the stability and accuracy of the results.

(b) In the mesh with 20 uniform size elements, conduct the SUPG with $\gamma = 10\gamma_0$, $2\gamma_0$, $0.5\gamma_0$ and $0.1\gamma_0$, where $\gamma_0=\frac{\Delta x}{2}\left(\coth(P_e)-\frac{1}{P_e}\right)$. Discuss the influence of γ.

(c) Mesh the domain with a nonuniform mesh. Discuss the following:

c.1) Where should the finer mesh be?

c.2) How to select a proper γ?

8

Element Technology

8.1 Introduction

The objective of element technology is to develop elements with better performance, particularly for large-scale calculations and for incompressible materials. Low-order elements when applied to incompressible materials tend to lock volumetrically. In volumetric locking, the displacements are underpredicted by large factors: it is not uncommon for the displacement to be an order of magnitude too small for otherwise reasonable meshes. Although incompressible materials are quite rare in linear stress analysis, many materials behave in a nearly incompressible manner in the nonlinear regime. For example, the plastic behavior of von Mises elastic–plastic materials is incompressible. Any element that locks volumetrically will not perform well for von Mises materials. Rubbers are also incompressible. Therefore, the ability to treat incompressible materials effectively is important in nonlinear finite elements. However, most elements have shortcomings when applied to incompressible or nearly incompressible materials. An understanding of these shortcomings and their remedies is crucial in the selection of elements for nonlinear analysis.

For large-scale calculations, underintegration is used to achieve faster elements. For three dimensions, cost reductions on the order of 8 have been achieved through underintegration compared to full integration. However, underintegration requires the stabilization of the element. Although stabilization has not been too popular in the academic literature, it is ubiquitous in large-scale calculations in industry. As shown in this chapter, it is theoretically sound and can be combined with multi-field concepts to obtain elements of high accuracy.

Nonlinear Finite Elements for Continua and Structures, Second Edition.
Ted Belytschko, Wing Kam Liu, Brian Moran, and Khalil I. Elkhodary.

Companion Website: www.wiley.com/go/belytschko

To eliminate volumetric locking, two approaches have evolved:

1. Multi-field elements in which the pressures or stress and strain fields are also considered as dependent variables.
2. Reduced integration procedures in which certain terms of the weak form are underintegrated.

Multi-field elements are based on multi-field weak forms or multi-field variational principles; they are also known as mixed elements and hybrid elements. In multi-field elements, in addition to the displacements, variables such as the stresses or strains are considered as dependent variables and interpolated independently of the displacements. This enables the strain or stress fields to be designed so as to avoid volumetric locking. We shall see that the additional variables are actually Lagrange multipliers, and they enable constraints such as incompressibility to be handled more effectively.

In some cases, the strain or stress fields are also designed to achieve better accuracy for beam bending or other specific problems. It should be stressed that mixed elements can improve the capability of an element only for constrained media or specific classes of problems. Mixed methods cannot improve the general performance of an element when there are no constraints such as incompressibility. Many of the papers on mixed methods give the impression that mixed elements are inherently superior to single-field elements, but there is no convincing evidence for this claim. There is instead considerable evidence that the rate of convergence can never exceed that of the corresponding single-field element in the absence of constraints, and we will show some numerical results that support this statement. Thus the only goals that can be achieved by mixed elements is to avoid locking and to improve behavior in a selected class of problems, such as beam bending.

The unfortunate byproduct of using multi-field variational principles is that in many cases the resulting elements possess instabilities in the additional fields. Thus most 4-node quadrilaterals based on multi-field weak forms are subject to a pressure instability. This requires another fix, so that the resulting element can be quite complex. The development of truly robust elements is not easy, particularly for low-order approximations.

We will begin the chapter with an overview of element performance in Section 8.2. This section describes the characteristics of many of the most widely used elements for modeling continua. The description is limited to elements which are based on polynomials of quadratic order or lower, since elements of higher order are seldom used in nonlinear analysis at this time. Terms such as consistency, polynomial completeness and the reproducing conditions are defined. Rates of convergence for various elements in linear problems are given. The implications of these results for nonlinear problems is examined on the basis of the smoothness of solutions. Hierarchical and p-elements are omitted because these elements have been used very little in nonlinear analysis.

The patch tests are described in Section 8.3. These are important, useful tests for the theoretical soundness of an element and the correctness of its implementation. Patch tests can be used to examine whether an element is convergent, whether it avoids locking and whether it is stable. Various forms of the patch test are described which are applicable to both static and explicit programs. The concepts of correct rank and rank deficiency of elements are also presented.

To illustrate element technology, we will focus on the 4-node isoparametric quadrilateral element (Q4). This element is convergent for compressible materials without any modifications, so none of the techniques described in this chapter are needed for compressible materials. On the other hand, for incompressible or nearly incompressible materials, this element locks, as shown in Section 8.4.3.

Section 8.5 describes some of the major multi-field weak forms and their application to element development. The first multi-field variational principle to be discovered was the Hellinger–Reissner variational principle, but it is not considered because it can not readily be used with strain-driven constitutive equations. Therefore, we will confine ourselves to various forms of the three-field weak form, which are related to the Hu–Washizu variational principle, in which the stress, strain measure, and displacement are dependent variables, that is, unknown fields. The elements of Pian and Sumihara (1985) and Simo and Rifai (1990) are also described and extensions to total Lagrangian forms and variational principles are given.

Although the multi-field elements are primarily aimed at circumventing volumetric locking, they apply more generally to what can collectively be called constrained media problems. Another important class of such problems are structural elements, such as shells, where the constraint applies to the motion of the normal to the reference surface. The same techniques described in this chapter will be used in Chapter 9 to develop shell elements.

From a theoretical viewpoint, underintegrated and selective reduced elements are quite similar to elements based on multi-field variational principles; an equivalence was proven by Malkus and Hughes (1978) for certain classes of elements. Both underintegrated and mixed elements suffer from the instabilities in the Lagrange multiplier fields. For incompressible problems, this is the pressure field.

Sections 8.6 and 8.7 will describe quadrilaterals with one-point quadrature. It will be shown that this element is rank-deficient, which results in spurious singular modes known as hourglass modes. Perturbation hourglass control is then described. Stabilization methods based on mixed variational principles are also developed. These are known as physical hourglass control since the hourglass parameters can be expressed in terms of material and geometric parameters. Finally, the extension of these results to the 8-node hexahedron is sketched.

In Section 8.8 some numerical results are presented to demonstrate the performance of various elements. It is shown that the multi-field elements and one-point quadrature elements avoid volumetric locking. The chapter concludes with a brief look at the stability of mixed elements.

8.2 Element Performance

8.2.1 Overview

In this section, an overview of characteristics of several of the most widely used elements will be presented. We will concentrate on elements in two dimensions, since the properties of these elements parallel those in three dimensions. Some of the elements we will discuss are illustrated in Figure 8.1. The overview is limited to continuum elements; shell elements are described in Chapter 9.

In selecting elements, the ease of mesh generation for a particular element should be borne in mind. Triangles and tetrahedral elements are very attractive because they are the easiest to mesh. Mesh generators for quadrilateral elements tend to be less robust. Therefore, triangular and tetrahedral elements are preferable when all other performance characteristics are comparable for the problem at hand.

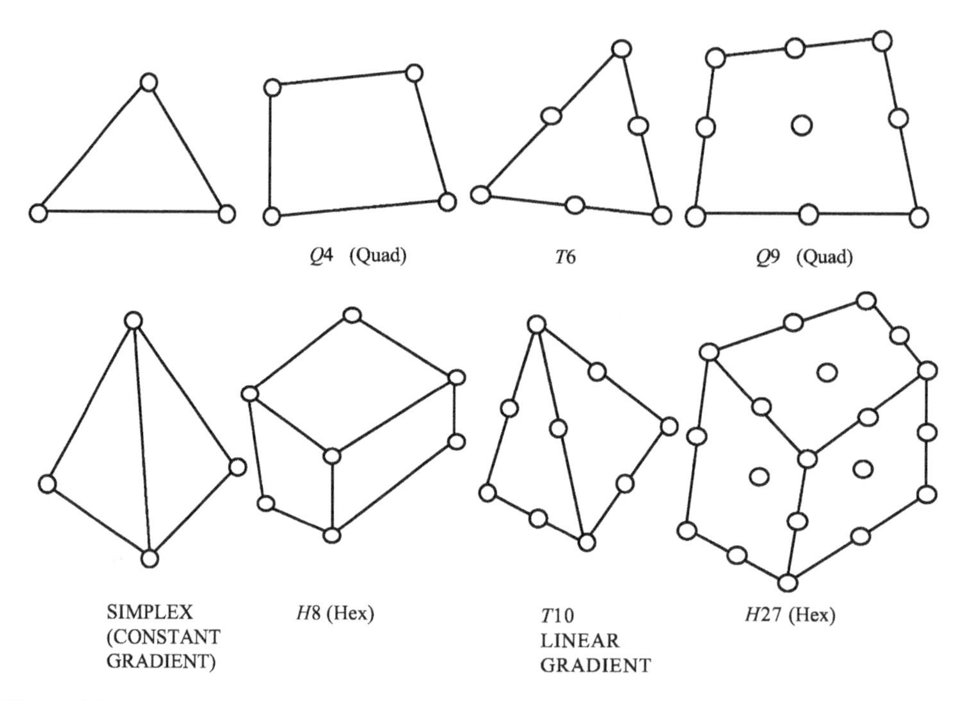

Figure 8.1 Linear and quadratic finite elements in two and three dimensions; hidden nodes are not shown

The most frequently used low-order elements in two dimensions are the 3-node triangle and the 4-node quadrilateral. The corresponding three-dimensional elements are the 4-node tetrahedron and the 8-node hexahedron, respectively. As is well-known to anyone familiar with linear finite element methods, the displacement fields of the triangle and tetrahedron are linear and the gradients of the displacement and velocity fields are constant. The displacement fields of the quadrilateral and hexahedron are bilinear and trilinear, respectively, and the strains are a combination of constant and linear terms; the strains are not linear complete. All of these elements can reproduce a linear displacement field and constant strain field exactly. Consequently they satisfy the standard patch test, which is described in Section 8.3.

The simplest elements are the 3-node triangle in two dimensions and the 4-node tetrahedron in three dimensions. These are also known as simplex elements because a simplex is a set of $n+1$ points in n dimensions. Both simplex elements perform poorly for incompressible materials. The triangular element exhibits severe *volumetric locking in plane strain problems*. The proviso *plane strain* is added because volumetric locking does not occur in plane stress problems, for in plane stress the thickness of the element can change to accommodate incompressible materials. The *tetrahedron locks for incompressible and nearly incompressible materials*.

Volumetric locking can be avoided for simplex elements by using special arrangements of the elements. For example, the cross-diagonal arrangement of triangles shown in Figure 8.2(a) eliminates locking (Nagtegaal, Parks and Rice, 1974). However, meshing in this arrangement of elements is similar to meshing quadrilaterals, so the benefits arising from triangular meshing are lost. Furthermore, cross-diagonal meshes lock when the center node is not exactly on

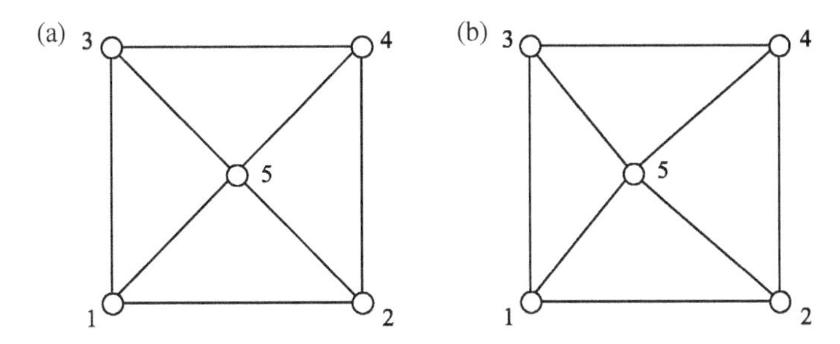

Figure 8.2 The cross-diagonal mesh pattern which avoids volumetric locking; the center node must be precisely on the intersection of diagonals; the mesh on the right locks

the intersection of the diagonals as shown in Figure 8.2(b). In large-displacement problems, such configurations always develop. In addition, cross-diagonal meshes do not satisfy the LBB condition, so pressure oscillations are possible (see Section 8.9 for LBB condition).

Simplex elements also manifest stiff behavior in other situations, such as beam bending. Stiff behavior is convergent, but manifested by poor accuracy for coarse meshes. Stiff behavior is not as deleterious as locking, but it is nevertheless undesirable for it means that very fine meshes are needed to obtain reasonable accuracy.

Four-node quadrilaterals and 8-node hexahedra are generally more accurate than 3-node triangles and tetrahedra, respectively. When fully integrated, that is, 2×2 Gauss quadrature for the quadrilateral or $2 \times 2 \times 2$ quadrature for hexahedra, these elements also lock for incompressible materials. They also tend to be stiff in beam bending when fully integrated.

Volumetric locking can be eliminated in these elements by reduced integration, namely one-point quadrature, or selective-reduced integration, which consists of one-point quadrature on the volumetric terms and 2×2 quadrature on the deviatoric terms; this was described in Section 4.5.4. The resulting elements exhibit good convergence in the displacements for incompressible materials.

The underintegrated, SRI (Selective Reduced Integration), and multi-field versions of the 4-node quadrilateral and 8-node hexahedral elements are all plagued by a major flaw: they exhibit a spatial instability in the pressure field. As a consequence, the pressure will often be oscillatory, with the pattern shown in Figure 8.3. This oscillatory pattern in the pressures is known as checkerboarding. Checkerboarding is sometimes harmless: for example, in materials governed by the von Mises elasto-plastic law the strain rate is independent of the pressure, so pressure oscillations are almost harmless, although they lead to errors in the elastic strains. Checkerboarding can be eliminated by filtering or by viscosities. Nevertheless it is undesirable, and a user of finite elements should be aware of its possibility with these elements. Oscillations in the stresses are possible for most elements based on multi-field variational principles.

The fastest form of the quadrilateral is the underintegrated, one-point quadrature element: it is often three to four times as fast as the selective-reduced quadrature quadrilateral element. In three dimensions, the corresponding speedup is of the order of 6 to 8. The one-point quadrature element also suffers from pressure oscillations, and in addition possesses instabilities in the displacement field. These instabilities are studied in Section 8.7.2. They have various names: hourglassing, keystoning, kinematic modes, spurious zero energy modes and chicken-wiring are some of the appellations for these modes. These modes can be controlled quite

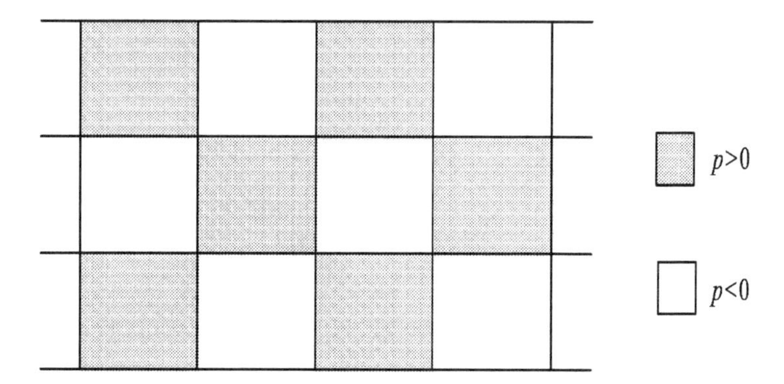

Figure 8.3 The checkerboarding mode, a consequence of a pressure instability

effectively. In fact, the rate of convergence is not decreased by a consistent control of these modes, so for many large-scale calculations, one-point quadrature with hourglass control is very effective. Hourglass control is described in Section 8.7.

The next highest order elements are the 6-node triangle and the 8- and 9-node quadrilaterals. The counterparts in three dimensions are the 10-node tetrahedron and the 20- and 27-node hexahedra. These elements reproduce quadratic and linear fields when the element sides are straight, but only the linear fields when the element sides are curved. These elements cannot reproduce a quadratic displacement field exactly when the sides are not straight. Of course, curved sides are a compelling advantage of finite elements, for they enable boundary conditions to be met for curved boundaries. However, curved element boundaries should only be used for exterior surfaces, since their presence decreases the accuracy of the element. Ciarlet and Raviart (1972), in a landmark paper, proved that the convergence of these elements is second-order when the midside nodes are near the midpoint, though how near is near enough is often an open question.

In nonlinear problems with large deformations, the performance of these elements degrades when the midside nodes move substantially; this has already been discussed in the one-dimensional context in Example 2.5. Element distortion is a vexing drawback of higher-order elements for large-deformation problems: the convergence rate of higher-order elements degrades significantly as they are distorted, and furthermore solution procedures often fail when distortion becomes excessive.

The 6-node triangle fails the LBB condition for incompressible materials. The 9-node quadrilateral when developed by a multi-field variational principle with a linear pressure field satisfies the LBB condition and does not lock. It is the only element we have discussed so far which has flawless behavior for incompressible materials.

The higher-order elements are not well suited to dynamics or large-deformation problems with Lagrangian meshes. It is difficult to develop good diagonal mass matrices for these elements. Furthermore, for higher-order elements, wave propagation solutions tend to be quite noisy because of the presence of optical modes: see Belytschko and Mullen (1978). In large-deformation problems, these elements fail more often and deteriorate in accuracy more rapidly than low-order elements because the Jacobian determinant can easily become negative at quadrature points.

8.2.2 *Completeness, Consistency, and Reproducing Conditions*

We first define the terms completeness, polynomial completeness, reproducing conditions and consistency. The last term is used with a variety of meanings; in this book, we adopt a specific definition based on its original definition in finite difference methods.

8.2.2.1 Completeness

Completeness has a general definition that for any Cauchy sequence in a space, the limit is in the same space. We are interested in what completeness implies on the ability of a set of basis functions to approximate a function. If a set of basis functions $\phi_I(\mathbf{x})$ is complete in H_r, then for any function $f(\mathbf{x}) \in H_r$, it follows that

$$\left\| f(\mathbf{x}) - a_I \phi_I(\mathbf{x})_{H_r} \right\| \to 0 \quad \text{as} \quad n \to \infty \tag{8.2.1}$$

(see Appendix 2 for a description of norms; note the implicit sums on the repeated indices). The appropriate norm H_r depends on the smoothness and regularity of the variable of interest, and what we are interested in. For example, if we are interested in the first derivatives of a variable, the H_1 norm would be the choice.

8.2.2.2 Reproducing Conditions

Reproducing conditions test the ability of an approximation to reproduce a function exactly. For functions of a variable $\mathbf{x}$, these conditions can be stated as follows: a set of approximating functions $N_J(\mathbf{x})$ reproduce $p(\mathbf{x})$ if when $u_J = p(\mathbf{x}_J)$, then

$$N_J(\mathbf{x})u_J = N_J(\mathbf{x})p(\mathbf{x}_J) = p(\mathbf{x}) \tag{8.2.2}$$

that is, when the nodal values of the approximation are given by $p(\mathbf{x}_J)$, then the function $p(\mathbf{x})$ is reproduced exactly by the approximation. This equation is quite subtle and contains more than first meets the eye. When the reproducing condition holds, the shape functions or interpolants are able *to exactly reproduce* the given function $p(\mathbf{x})$. For example, if the shape functions are able to reproduce the constant, then if $u_J = 1$, the approximation should be exactly unity:

$$N_J(\mathbf{x})u_J = \sum_J N_J(\mathbf{x}) = 1 \tag{8.2.3}$$

Isoparametric finite element shape functions meet the constant reproducing condition, and therefore sum to 1. Functions which possess this property are called *partitions of unity*.

Similarly, if the shape functions reproduce linear functions x_i, then if $u_I = x_{iI}$, it follows that $u = x_i$, so the reproducing condition for a linear function is

$$N_J(\mathbf{x})u_J = N_J(\mathbf{x})x_{jJ} = x_j \tag{8.2.4}$$

Any approximation which satisfies the linear reproducing conditions can be shown to be complete in H_1. This is called linear completeness by Hughes (1987), but the term reproducing condition seems more appropriate, since completeness refers to a more general condition described by (8.2.1); functions which satisfy (8.2.1) are complete, but not all complete bases satisfy (8.2.2). For example, Fourier series are complete but do not satisfy (8.2.2). Therefore, when using completeness in this sense we will append an adjective such as linear completeness or quadratic completeness.

8.2.2.3 Consistency

A third definition pertaining to convergence is consistency. Consistency is usually defined in the context of finite difference methods (see, for example, Strikwerda, 1989). A discrete approximation $L^h(u)$ of a partial differential equation $L(u)$ is consistent if the error is of the order of the mesh size, that is, if

$$L(u) - L^h(u) = O(h^n), \quad \text{with } n \geq 1 \tag{8.2.5}$$

This states that the truncation error of a consistent discrete approximation must tend to zero as the element size tends to zero. For time-dependent problems, the discretization error will be a function of the time step and the element size h, and the truncation error of the time and spatial discretizations must tend to zero.

8.2.2.4 Lax Equivalence Theorem

A landmark in finite difference methods is the Lax equivalence theorem. It states that for a well-posed problem, a discretization that is stable and consistent is convergent. Therefore, it is often written that

$$\text{completeness} + \text{stability} \rightarrow \text{convergence}$$

In finite element methods, a corresponding proof is not available. Furthermore, the second leg of a finite element method is not consistency, which is very difficult to establish for an arbitrary mesh. Instead, finite element convergence proofs are based on completeness. Completeness is implied by the reproducing conditions. Stability often appears in proofs of finite elements in the disguise of the coercivity conditions. Therefore, in finite element solutions of equilibrium problems, we can write

$$\text{completeness} + \text{stability} \rightarrow \text{convergence}$$

Completeness plays a central role in the performance of elements. If an element can reproduce polynomials of sufficiently high order and is stable, it will converge (though we do not know of a proof of this generality in finite elements). These notions are implicit in the patch tests, to be described later, which test the reproducing conditions of an element and, in some cases, its stability.

8.2.3 Convergence Results for Linear Problems

The following briefly summarizes some convergence results for finite element solutions to linear, elliptic problems. Elliptic problems, as summarized in Chapter 1, include most equilibrium problems where the material is stable. The convergence results are expressed in terms of the reproducing capabilities of the element: if the finite element solution $\mathbf{u}^h(\mathbf{x})$ is generated by elements which can reproduce polynomials of order k exactly, and if the solution $\mathbf{u}(\mathbf{x})$ is sufficiently smooth and regular for the Hilbert norm H_r to exist, then

$$\left\|\mathbf{u}-\mathbf{u}^h\right\|_{H_m} \leq Ch^{\alpha}\mathbf{u}_{Hr}, \quad \alpha = \min\,(k+1-m,\ r-m) \tag{8.2.6}$$

where h is a measure of element size and C is an arbitrary constant which is independent of h and varies from problem to problem (Strang and Fix, 1973: 107; Hughes, 1987: 269; Oden and Reddy, 1976: 275); note that the last two references give interpolation estimates at the pages cited; these technically are not equivalent to the convergence rate, but are an upper bound on the rate of convergence.

We will now examine the implications of this theorem for various elements for linear problems. The parameter α indicates the rate of convergence of the finite element solution: the greater the value of α, the faster the finite element solution converges to the exact solution and the more accurate the element.

It is important to note that the rate of convergence is limited by the smoothness of the solution. If there are no sharp corners or cracks, a linear equilibrium solution is analytic, that is, infinitely smooth, so r tends to infinity. The second term in the definition for α, $r-m$, plays no role for smooth solutions. However, if the solution is not smooth, for example, if there are discontinuities in the derivatives, then r is finite. For example, if there are discontinuities in the second derivatives, then r is at most 2, and the term $r-m$ governs convergence.

We first examine what (8.2.6) means for accuracy in displacements for smooth elastic solutions for various elements. In that case, we consider the H_0 norm, which is equivalent to the L_2 norm, so $m=0$. The 3-node triangle, the 4-node quadrilateral, the 4-node tetrahedron and the 8-node hexahedron all reproduce linear polynomials exactly, so $k=1$. Therefore, for the elements we have listed that satisfy linear reproducing conditions, we obtain that $\alpha=\min(k+1-m,\ r-m)=\min(1+1-0,\ \infty-0)=2$.

This result is illustrated later in this chapter in Figure 8.12, which shows a log–log plot of the H_0 error norm for a linear–complete element. In the log–log plot, the graph of error in displacements versus element size is a straight line with a slope proportional to the rate of convergence. When the solution converges quadratically, the slope is 2. Equation (8.2.6) is an asymptotic result which holds only as the element size goes to zero, but it agrees very well with numerical results for practical meshes.

We next consider the higher-order elements, namely the 6-node triangle, the 9-node quadrilateral, the 10-node tetrahedron and the 27-node hexahedron with straight edges. In this case $k=2$, and the remaining constants are unchanged. Then $\alpha=3$, so the rate of convergence is cubic in the displacements. This increase of one order in convergence rate is quite significant. A higher-order element provides a tremendous increase in accuracy when the solution is smooth.

The results for the strains, that is, the derivatives of the displacement field, can be estimated by similar arguments. In this case $m=1$ since the error in strains is measured by the H_1-norm. The rates of convergence are then one order lower than that of displacements: $k=1$ for elements with linear completeness, so $\alpha=1$ for the error in the H_1 norm and the rate of convergence in the strains is linear, and $k=2$ for elements with quadratic completeness, so $\alpha=2$.

Similar conclusions can be drawn for parabolic partial differential equations. However, for hyperbolic differential equations, the situation is more complicated and less favorable to higher-order elements. Recall that in hyperbolic differential equations, discontinuities may occur in the derivatives of the solution. Therefore, if the data, that is, the initial and boundary conditions, are not smooth, the solution will not be smooth. Furthermore, it is possible for discontinuities, such as shocks, to develop even for smooth data. Therefore, higher order elements are of advantage in hyperbolic problems only when the data is smooth and the solution is expected to remain smooth.

8.2.4 Convergence in Nonlinear Problems

It is possible to use these results to ascertain the performance of elements for nonlinear problems. The interpolation estimates which form the basis for (8.2.6) apply to nonlinear problems and are always upper bounds on the performance of an element. In other words, it is impossible for an element to converge more rapidly than the estimates (8.2.6).

According to (8.2.6), the performance of elements for nonlinear problems will depend on the smoothness of the solution. This in turn depends primarily on the smoothness of the constitutive equation and the response. For elliptic problems, if the constitutive equation is continuously differentiable, that is, C^1, such as a hyperelastic model for rubber, then the rate of convergence should be the same as for elastic, linear materials. However, for constitutive equations which are C^0, such as elastic–plastic materials, the second term in the definition of α in (8.2.6) governs the rate of convergence. For example, in an elastic–plastic material, the relation between stress and strain is C^0. Therefore the displacements are at most C^1 and $r=2$. Then from (8.2.6) the rate of convergence of the displacements is at most of order 2, i.e. $\alpha=2$. Thus there appears to be no benefit in going to higher-order elements for nonsmooth materials. Similarly, the rate of convergence in the strains is at most of order $\alpha=1$ for elasto-plastic materials.

In summary, for elliptic problems with smooth constitutive equations where smooth solutions are expected, higher-order elements are advantageous because of their higher rate of convergence. If the constitutive equation lacks sufficient smoothness, then there is no advantage to using higher order elements. These results are also relevant for hyperbolic problems: when the data is very smooth, there is some benefit in going to higher-order elements, provided that a consistent mass matrix is used. For data which lack smoothness or when nonlinearities induce roughness, there is little advantage to higher-order elements. In time-dependent problems, the total error depends on the combined effects of time discretization error and space discretization error.

In nonlinear problems, accuracy can be degraded further in Lagrangian meshes by element distortion. This deterioration of element performance with deformation is more severe for the higher-order elements. In Eulerian meshes element distortion is a problem only when the initial mesh is distorted, since an Eulerian element does not change with time. The amount of element distortion should be considered in the choice of an element for nonlinear analysis. For Lagrangian mesh solutions with very large deformations, higher-order elements are less beneficial even when the constitutive equation and response are smooth.

Even linear, elastic solutions have discontinuities in derivatives: at interfaces between different materials, the derivatives of the displacement are discontinuous. However, any reasonable analyst aligns the element edges with the material interfaces. In that case, the full accuracy of higher-order elements can be retained since they can represent discontinuities in derivatives effectively along element edges. In elastic–plastic and hyperbolic problems, on the other hand, discontinuities sweep through the model and, as the problem evolves, they often proliferate. Thus the effects of rough constitutive equations and hyperbolicity in nonlinear problems can be devastating to accuracy.

It should be stressed that the convergence rates (8.2.6) only apply to linear, elliptic problems without singularities. The major impediment to obtaining such convergence results for nonlinear problems is probably the lack of stability of nonlinear solutions. The estimate (8.2.6) is representative of element performance for nonpathological, nonlinear elliptic problems. Furthermore, keep in mind that it is an upper bound on accuracy: a solution cannot be more accurate than the interpolation power of the approximation.

8.3 Element Properties and Patch Tests

8.3.1 Patch Tests

The patch tests are extremely useful for examining the soundness of element formulations and their completeness and stability. The patch test was conceived by Irons and reported in Bazeley *et al.* (1965) to examine the soundness of a nonconforming plate element. In its original form, the patch test was primarily a test for polynomial completeness, that is, the ability to reproduce exactly a polynomial of order k. It was proposed by Strang (1972) that the patch test is equivalent to the conditions needed for finite element convergence. In the cited paper, Strang also says that if the finite element equations are considered as 'a finite difference scheme, then the patch test would be equivalent to the formal consistency of the difference equations with the correct differential equation.' In fact, the finite element equations for irregular meshes in two dimensions do not appear to be consistent in the sense of finite difference equations for irregular meshes, or at least any consistency is very difficult to show. But the myth of equivalence to difference equations has persisted, and many authors, including the authors of this book, often speak of the consistency of Galerkin discretizations. The power of the patch test arises primarily from its demonstration of another property required for convergence, the completeness of the approximation. In the following we describe several variants of the patch test.

8.3.2 Standard Patch Test

We first describe the standard patch test, which checks for polynomial completeness of the displacement field, that is, the ability of the element to reproduce polynomials of a specified order. In addition, the test checks the implementation and the program. Sometimes the element is correct but fails the patch test because of a faulty implementation or a bug in the programming.

In the standard patch test, a patch of elements such as shown in Figure 8.4 is used. The elements should be distorted as shown because rectangular elements may satisfy the patch test when arbitrary quadrilaterals do not. No body forces should be applied, and the material properties should be uniform and linear–elastic. The displacements of the nodes on the periphery

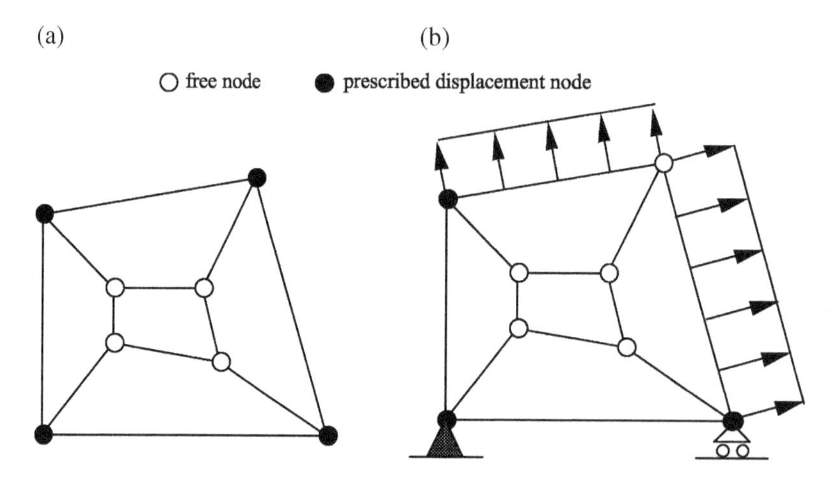

Figure 8.4 Standard patch test (a) with all boundary displacements prescribed and (b) extended patch test for stability

of the patch are prescribed according to the order of the patch test, so $\Gamma_u = \Gamma$. For a test of the linear reproducing conditions, the displacement field on Γ_u is prescribed by

$$\begin{aligned} u_x(\mathbf{x}) &= \alpha_{x0} + \alpha_{x1}x + \alpha_{x2}y \\ u_y(\mathbf{x}) &= \alpha_{y0} + \alpha_{y1}x + \alpha_{y2}y \end{aligned} \quad \text{in 2D,} \qquad u_i(\mathbf{x}) = \alpha_{i0} + \alpha_{ij}x_j \quad \text{in general} \tag{8.3.1}$$

where are constants set by the user; they should all be nonzero to test the reproducing condition completely. This displacement field is prescribed on the nodes of the entire boundary of the patch Γ_u, so the prescribed displacements are

$$\begin{aligned} u_{xI} &= \alpha_{x0} + \alpha_{x1}x_I + \alpha_{x2}y_I \\ u_{yI} &= \alpha_{y0} + \alpha_{y1}x_I + \alpha_{y2}y_I \end{aligned} \quad \text{in 2D,} \qquad u_{iI} = \alpha_{i0} + \alpha_{ij}x_{jI} \quad \text{in general} \tag{8.3.2}$$

To satisfy the patch test, the finite element solution should be given by (8.3.1) throughout the patch: the nodal displacements at the interior nodes should be given by (8.3.2), and the strains should be constant and given by the application of the strain–displacement equations to the displacement in (8.3.1):

$$\begin{aligned} \varepsilon_x &= u_{x,x} = \alpha_{x1} \\ \varepsilon_y &= u_{y,y} = \alpha_{y2} \\ 2\varepsilon_{xy} &= u_{x,y} + u_{y,x} = \alpha_{x2} + \alpha_{y1} \end{aligned} \quad \text{in 2D,} \qquad \varepsilon_{ij} = \frac{1}{2}(\alpha_{ij} + \alpha_{ji}) \quad \text{in general} \tag{8.3.3}$$

The stresses should also be constant. All of these conditions should be met to a high degree of precision, on the order of the precision of the computer. On a machine with eight significant digits, the results should be accurate to at least five digits (the exact number of digits depends on the computer, since the number of digits in the machine arithmetic varies).

The significance of the standard patch test lies in its verification of the reproducing conditions. When an exact solution is in the subspace of finite element approximations, the finite element solution must correspond to the exact solution. That (8.3.1) is the exact solution to the linear-elastic problem can be established as follows: since the strains are constant, and the material properties uniform, the stresses are constant. Therefore, since there are no body forces, the equilibrium equation (3.5.37) is satisfied exactly. Since linear elastic solutions are unique, (8.3.1) is the exact solution.

When the patch test fails, then either the finite element cannot reproduce the linear field exactly, that is, it is not linearly complete, or there is an error in the implementation. Whether the reproducing conditions are satisfied can be checked independently by setting the nodal displacements at all nodes according to (8.3.2) and checking the strains at the quadrature points. This test in fact suffices as a test of the reproducing conditions. The rest of the patch test, including the solution of the linear equations, is a check on the implementation and the programming.

8.3.3 Patch Test in Nonlinear Programs

The patch test as described can be extended to nonlinear programs. The linear field (8.3.1) is applied with large values of α_{ij}. Since the displacement field is linear, the deformation gradient and the Green strain tensor must be constant, and so the PK2 stress and the nominal stress are constant. Therefore, the equilibrium equation (3.6.10) is satisfied in the absence of body forces, and (8.3.1) is a solution. It is not a unique solution, but this is a difficult issue which we skip. If an element meets the reproducing conditions in a linear patch test, it must meet the linear reproducing conditions in a nonlinear patch test. Therefore, adding the nonlinear patch test is more a test of the nonlinear implementation than of the reproducing conditions.

8.3.4 Patch Test in Explicit Programs

The patch test as described previously is not applicable to explicit programs because they cannot solve the equilibrium equations. However, the patch test can be modified for explicit programs, as described in Belytschko, Wong and Chiang (1992). In this patch test, the initial velocities are prescribed by a linear field

$$\begin{aligned} v_x(\mathbf{x}) &= \alpha_{x0} + \alpha_{x1}x + \alpha_{x2}y \\ v_y(\mathbf{x}) &= \alpha_{y0} + \alpha_{y1}x + \alpha_{y2}y \end{aligned} \quad \text{in 2D,} \qquad v_i(\mathbf{x}) = \alpha_{i0} + \alpha_{ij}x_j \quad \text{in general} \tag{8.3.4}$$

where α_{ij} are arbitrary constants; these should be very small because otherwise the geometric nonlinearities are triggered, and this patch test will not work in its entirety. These are applied by setting the initial nodal velocities by

$$\begin{aligned} v_{xI} &= \alpha_{x0} + \alpha_{x1}x_I + \alpha_{x2}y_I \\ v_{yI} &= \alpha_{y0} + \alpha_{y1}x_I + \alpha_{y2}y_I \end{aligned} \quad \text{in 2D,} \qquad v_{iI} = \alpha_{i0} + \alpha_{ij}x_{jI} \quad \text{in general} \tag{8.3.5}$$

No external forces should be applied. The equations of motion are integrated one time step, and the rate-of-deformation and the accelerations at the end of the time step are checked. The rate-of-deformation should have the correct constant values in all of the elements and the accelerations should vanish at all of the interior nodes. The accelerations should vanish for the following reason: if the reproducing conditions are met, the stresses should be constant and then from the momentum equation, (3.5.33), in the absence of body forces, the accelerations should vanish.

The test should be met to a high degree of precision if the constants α_{ij} are small enough. For example, when the constants are of order 10^{-4}, the accelerations should be of order 10^{-7}.

8.3.5 Patch Tests for Stability

Taylor *et al.* (1986) have devised a modified patch test which also checks for spatial stability of the displacement field. It also tests whether the traction boundary conditions are implemented correctly. The main difference from the standard patch test is that the displacements are not prescribed at all nodes of the boundary. Instead, the displacement boundary conditions are the minimum required to preclude rigid body motion, as shown in Figure 8.4(b) (the figure shows discrete boundary conditions; equivalent continuous boundary conditions are difficult to construct).

This test is not an infallible test for detecting spatial instabilities. Noncommunicable spurious singular modes will often not be detected. Furthermore, this test can detect only displacement instabilities, not pressure instabilities. To thoroughly check an element for spatial instabilities, an eigenvalue analysis of a single free element, that is, a completely unconstrained element, and a patch of elements should be made. The number of zero eigenvalues should be equal to the number of rigid body modes. For example, in two dimensions, an element or a patch of elements should possess three zero eigenvalues, corresponding to two translations and one rotation, whereas in three dimensions, an element should possess six zero eigenvalues, three translational and three rotational rigid body modes. If there are more zero eigenvalues, the model has a displacement instability; this is also called rank deficiency of the stiffness matrix, as discussed in Section 8.3.8.

8.3.6 Linear Reproducing Conditions of Isoparametric Elements

It will now be shown that *isoparametric elements of any order reproduce the complete linear velocity* (displacement) field; that is, all isoparametric elements are linear-complete. An arbitrary isoparametric element with n nodes is considered. The map between the current configuration and the parent element is given by

$$x_i(\xi) = N_I(\xi)x_{iI} \tag{8.3.6}$$

For an isoparametric element, the dependent variable u is interpolated by the same shape functions, so

$$u(\xi) = N_I(\xi)u_I \tag{8.3.7}$$

Let the dependent variable be a linear function of the spatial coordinates, so

$$u = \alpha_0 + \alpha_j x_j \tag{8.3.8}$$

where α_0 and α_i are arbitrary parameters. If the nodal values of the field are given by the above, then

$$u_I = \alpha_0 + \alpha_j x_{jI} \qquad \text{or} \qquad \mathbf{u} = \alpha_0 \mathbf{1} + \alpha_i \mathbf{x}_i \tag{8.3.9}$$

where $\mathbf{u}$ is the column matrix of order n, the number of nodes; $\mathbf{1}$ is a column matrix given by $1_J = 1$, $J = 1$ to n, and $\mathbf{x}_i$ are column matrices of nodal coordinates. For a 4-node quadrilateral, these column matrices are

$$\mathbf{u} = [u_1 \quad u_2 \quad u_3 \quad u_4]^T, \quad \mathbf{1} = [1 \quad 1 \quad 1 \quad 1]^T \tag{8.3.10}$$

$$\mathbf{x}_1 \equiv \mathbf{x} = [x_1 \quad x_2 \quad x_3 \quad x_4]^T, \quad \mathbf{x}_2 \equiv \mathbf{y} = [y_1 \quad y_2 \quad y_3 \quad y_4]^T \tag{8.3.11}$$

Substituting the nodal values of the dependent variable given by (8.3.9) into (8.3.7) yields

$$u = u_I N_I(\xi) = (\alpha_0 1_I + \alpha_j x_{jI}) N_I(\xi) \tag{8.3.12}$$

and rearranging the terms gives

$$u = \alpha_0 (1_I N_I(\xi)) + \alpha_i (x_{iI} N_I(\xi)) \tag{8.3.13}$$

It is recognized from (8.3.6) that the coefficients of α_i in the last sum on the right-hand side of this correspond to x_i so

$$u = \alpha_0 1_I N_I(\xi) + \alpha_i x_i = \alpha_0 + \alpha_i x_i \tag{8.3.14}$$

where we have used the reproducing condition for a constant, (8.2.3), in the last step.

This is precisely the linear field (8.3.8) from which the nodal values u_I were defined in (8.3.10). In other words, by prescribing the nodal values by a linear field, the shape functions have exactly reproduced this linear field.

Thus the isoparametric element reproduces the constant and linear fields exactly. As a consequence, the element satisfies the linear patch test. Henceforth we will simply refer to the highest-order field which is reproduced: that is, when an element reproduces linear and constant fields, we will say it reproduces linear fields; when it reproduces quadratic, linear and constant fields we will say it reproduces quadratic fields.

Although the linear reproducing property of isoparametric elements appears at first to be somewhat trivial, it is central to convergence proofs of finite elements. It is not an intrinsic attribute of interpolants, as can be appreciated by noting that it does not hold for all terms in an element interpolation. The quadratic terms in a 3-node one-dimensional element are not reproduced when the nodes are not equispaced, and the bilinear term in a 4-node quadrilateral is not reproduced.

Similarly, higher-order isoparametric elements cannot reproduce all polynomials contained in their isoparametric fields except under special conditions. For example, the 9-node Lagrange element cannot reproduce quadratic fields unless the element is straight-sided with equispaced nodes. When the edges are curved, a quadratic polynomial is not reproduced and the accuracy of the element decreases. The convergence proofs of Ciarlet and Raviart (1972) show that the order of convergence of the displacements in the L_2 norm for the 9-node element is of order h^3 only when the midpoint node is 'close' to the midpoint of the side.

8.3.7 Completeness of Subparametric and Superparametric Elements

In linear finite elements, the terms subparametric and superparametric refer to elements in which the parent to spatial map $\mathbf{x}(\xi)$ is of lower or higher order, respectively, than the interpolant for the dependent variable $u(\xi)$. Subparametric and superparametric elements are defined as follows:

$$\text{subparametric: } \mathbf{x}(\xi) \text{ lower order than } u(\xi)$$

$$\text{superparametric: } \mathbf{x}(\xi) \text{ higher order than } u(\xi)$$

Examples of these elements are illustrated in Figure 8.5. The subparametric element is linear complete, but the superparametric element is not. To demonstrate this for a subparametric, let the dependent variable be $u(\xi)$ interpolated by a Lagrange element with n_u nodes and the current configuration by a map with n_x nodes, so

$$u(\xi,\eta) = \sum_{I=1}^{n_u} u_I N_I^u(\xi,\eta) \tag{8.3.15}$$

$$\mathbf{x} = \sum_{I=1}^{n_x} \mathbf{x}_I N_I^x(\xi,\eta) \quad \text{or} \quad x_i = \sum_{I=1}^{n_x} x_{iI} N_I^x(\xi,\eta) \tag{8.3.16}$$

(In this subsection, the sums are indicated explicitly). The displacement interpolant is distinguished from the spatial interpolant the superscript. We now define a set of n_u nodes $(\bar{x}_I, \bar{y}_I)$ which are obtained by evaluating (x, y) by (8.3.16) at the n_u nodes in the parent element. Then the map between the parent element and the current configuration is

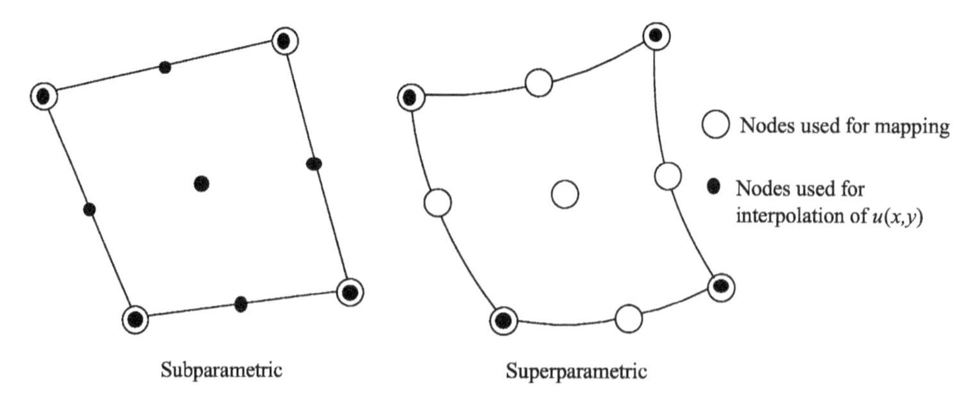

Figure 8.5 Examples of subparametric and superparametric elements based on the 4-node and 9-node Lagrange shape functions

$$\mathbf{x} = \sum_{I=1}^{n_u} \bar{\mathbf{x}}_I N_I^u(\xi,\eta) \tag{8.3.17}$$

The arguments invoked in going from (8.3.8) to (8.3.14) can now be repeated to establish the linear completeness of the subparametric element.

Superparametric elements, in which the map between the initial configuration and the parent element is of higher order than the interpolation of the displacement, cannot reproduce a linear field: the step of inserting nodes we made previously is not feasible.

In summary, isoparametric and subparametric elements reproduce linear fields. Therefore, for these elements, when the dependent variable is a displacement, the correct constant strain state is obtained for a linear displacement field, and the patch test will be satisfied. The element will also represent rigid body translation and rotation exactly since this motion is a linear field.

8.3.8 *Element Rank and Rank Deficiency*

To be stable, an element must have the proper rank. If the rank of an element is deficient, the discretization is unstable and will exhibit spurious oscillations in a dynamic solution. In static problems, rank deficiency manifests itself as singularity or near-singularity of the linearized equations. The instability associated with rank deficiency is a weak instability: the modes associated with a rank deficiency are *zero-frequency modes which grow linearly with time*, similar to a weak instability in numerical integration. In many cases, rank deficiency of elements leads to a system in which the smallest eigenvalue is positive, but much smaller than the correct lowest eigenvalue. The spurious modes then have a very small stiffness, and although they grow slowly, they become large enough to destroy the solution. If the rank of an element exceeds the proper rank, the element will strain in rigid body motion and either fail to converge or converge very slowly.

The proper rank of an element tangent stiffness or linear stiffness matrix is

$$\text{proper rank}\,(\mathbf{K}_e) = \text{order}(\mathbf{K}_e) - \text{n}_{RB} \tag{8.3.18}$$

where n_{RB} is the number of rigid body modes. We use the symbol $\mathbf{K}_e$ for both the tangent and linear stiffness in this chapter. The rank deficiency for a generic stiffness matrix is given by

$$\text{rank deficiency}\,(\mathbf{K}_e) = \text{proper rank}\,(\mathbf{K}_e) - \text{rank}\,(\mathbf{K}_e) \tag{8.3.19}$$

The generic form of the numerically integrated element stiffness is given by (see Box 6.5)

$$\mathbf{K}_e = \sum_{Q=1}^{n_Q} \bar{w}_Q \left(J_\xi \mathbf{B}^T [\mathbf{C}] \mathbf{B} \right)\Big|_{\xi_Q} \tag{8.3.20}$$

The above gives the linear elastic stiffness matrix if $[\mathbf{C}]$ is the linear elastic matrix. The Jacobian determinants J_ξ are assumed to be positive at all quadrature points, that is, the element cannot be excessively distorted. The matrix $[\mathbf{C}]$ is assumed to be positive definite; otherwise, it is possible for the element to be rank-deficient even for a proper element design. For example, if the tangent modulus matrix vanishes at all quadrature points due to loss of material stability, the element stiffness will obviously be of rank zero. The geometric stiffness is not considered in the above because we wish the element rank to be correct for any stress and we do not to consider the decrease in rank due to a geometric instability.

Another way of expressing (8.3.18) is

$$\dim(\ker(\mathbf{K}_e)) = n_{RB} \quad \text{where } \mathbf{z} \in \ker(\mathbf{K}_e) \quad \text{if} \quad \mathbf{K}_e\mathbf{z} = 0 \tag{8.3.21}$$

where the kernel of $\mathbf{K}_e$ is defined on the RHS. Any mode in the kernel of a stiffness matrix is a zero-energy mode, since the energy is $\frac{1}{2}\mathbf{z}^T\mathbf{K}_e\mathbf{z}$. A sufficient condition for the stability of a mesh is that the rank of all elements in the mesh be correct. This can be seen by noting that if all elements have the proper rank, then the only zero-energy mode of any element is a rigid body mode. Then any motion which is not a rigid body motion must have nonzero energy. The preceding is not a necessary condition, because some elements have so-called noncommunicable modes. These noncommunicable zero-energy modes cannot exist in a mesh without deformation energy in some of the elements. Assemblages of those elements will have the proper rank even when the elements do not.

To explain why the proper rank of the element is necessary for stability, consider a linear stability analysis of the system, Section 6.5.4. The frequency μ of the linearized equations should be zero if and only if the model is in a rigid body mode. However, if the element possesses a spurious singular mode that can exist in a global pattern, then the eigenvalue problem (6.5.15) will have a zero root with this nonphysical mode. Therefore this mode will grow linearly with time: $\mathbf{d} = \mathbf{y}t$ where $\mathbf{y}$ is the spurious singular mode. The spurious mode is a weak instability; it does not grow exponentially. Note that the rigid body modes, if they are not eliminated by constraints, also grow linearly with time, but this is a correct solution. The spurious singular mode grows linearly with time and pollutes the solution.

Spurious modes can be detected by an eigenvalue analysis: if the number of zero eigenvalues of an element or an assemblage of elements exceeds the number of rigid body modes, the element possesses spurious singular modes.

8.3.9 *Rank of Numerically Integrated Elements*

We now examine the rank of a numerically integrated stiffness $\mathbf{K}_e$. We assume that the tangent modulus $\mathbf{C}$ is positive-definite and that J_ξ is positive at all quadrature points. The numerically integrated element stiffness (8.3.20) can be rewritten as

$$\mathbf{K}_e = \mathring{\mathbf{B}}^T \mathring{\mathbf{C}} \mathring{\mathbf{B}} \tag{8.3.22}$$

where

$$\mathring{\mathbf{B}} = \begin{bmatrix} \mathbf{B}(\xi_1) \\ \mathbf{B}(\xi_2) \\ \vdots \\ \mathbf{B}(\xi_{n_Q}) \end{bmatrix} \quad \mathring{\mathbf{C}} = \begin{bmatrix} \bar{w}_1 J_\xi [\mathbf{C}]\big|_{\xi_1} & \mathbf{0} & \dots & \mathbf{0} \\ \mathbf{0} & \bar{w}_2 J_\xi [\mathbf{C}]\big|_{\xi_2} & & \mathbf{0} \\ \vdots & & \ddots & \vdots \\ \mathbf{0} & \mathbf{0} & \dots & \bar{w}_{n_Q} J_\xi [\mathbf{C}]\big|_{\xi_{n_Q}} \end{bmatrix} \tag{8.3.23}$$

It is well-known from linear algebra that the rank of a product of two matrices is always less than or equal to the rank of either of matrices (see Noble, 1969):

$$\text{rank}\,\mathbf{K}_e \leq \min\,(\text{rank}\,\mathring{\mathbf{B}},\quad \text{rank}\,\mathring{\mathbf{C}}) \tag{8.3.24}$$

When J_ξ and $\mathbf{C}$ are positive at all quadrature points, the rank of $\mathring{\mathbf{C}}$ is always greater than or equal to the rank of $\mathring{\mathbf{B}}$, so (8.3.24) can be replaced by

$$\text{rank}\,\mathbf{K}_e \leq \text{rank}\,\mathring{\mathbf{B}} \tag{8.3.25}$$

The inequality applies only in rare cases. The rank of $\mathring{\mathbf{B}}$ is bounded by the following:

$$\text{rank}\,\mathring{\mathbf{B}} \leq \dim(\mathbf{D}) \tag{8.3.26}$$

where the dimension of $\mathbf{D}$ is equal to the number of linearly independent functions in $\mathbf{D}$.

The rank-sufficiency of the quadrilateral Q4 will now be examined for various quadrature schemes. Q4 has four nodes with two degrees of freedom at each node, so order $(\mathbf{K}_e) = 8$. The number of rigid body modes is three: translations in the x and y directions and rotation in the (x, y) plane. By (8.3.18), the proper rank of $\mathbf{K}_e = 5$.

The most widely used quadrature scheme for Q4 is 2×2 Gauss quadrature. The number of quadrature joints $n_Q = 4$ and the number of rows in each $\mathbf{B}(\xi_\alpha) = 3$, so the number of rows in $\mathring{\mathbf{B}} = 12$. This exceeds the proper rank. However, for the velocity field (8.4.15) it is easy to show that the rate-of-deformation is

$$\{\mathbf{D}\} = \begin{Bmatrix} \alpha_{x1} + \alpha_{x3}h_{,x} \\ \alpha_{y2} + \alpha_{y3}h_{,y} \\ \alpha_{x2} + \alpha_{y1} + \alpha_{x3}h_{,y} + \alpha_{y3}h_{,x} \end{Bmatrix} \tag{8.3.27}$$

This field consists of five linearly independent vectors:

$$\begin{Bmatrix} \alpha_{x1} \\ 0 \\ 0 \end{Bmatrix}, \begin{Bmatrix} 0 \\ \alpha_{y2} \\ 0 \end{Bmatrix}, \begin{Bmatrix} 0 \\ 0 \\ \alpha_{x2} + \alpha_{y1} \end{Bmatrix}, \begin{Bmatrix} \alpha_{x3}h_{,x} \\ 0 \\ \alpha_{x3}h_{,y} \end{Bmatrix}, \begin{Bmatrix} 0 \\ \alpha_{y3}h_{,y} \\ \alpha_{y3}h_{,x} \end{Bmatrix} \tag{8.3.28}$$

Thus $\dim(\mathbf{D}) = 5$. The number of rows in $\mathring{\mathbf{B}} = 12$ but at most 5 can be linearly independent, so it follows from (8.3.26) that $\text{rank}\,(\mathring{\mathbf{B}}) = 5$. The rank of $\mathring{\mathbf{B}}$ for this element cannot exceed 5 regardless of how many quadrature points are used. The element stiffness then has the proper rank with 2×2 quadrature if at least five of the rows of $\mathring{\mathbf{B}}$ are linearly independent; this can be shown for any nondegenerate element, but it is painful.

The rank of the element stiffness of the quadrilateral for one-point quadrature can be ascertained similarly. In one-point quadrature, $\mathring{\mathbf{B}}$ consists of $\mathbf{B}$ evaluated at a single point:

$$\mathring{\mathbf{B}} = \mathbf{B}(\mathbf{0}) = \begin{bmatrix} \mathbf{b}_x^T(\mathbf{0}) & \mathbf{0} \\ \mathbf{0} & \mathbf{b}_y^T(\mathbf{0}) \\ \mathbf{b}_y^T(\mathbf{0}) & \mathbf{b}_x^T(\mathbf{0}) \end{bmatrix} \tag{8.3.29}$$

where $\mathbf{b}_x$ and $\mathbf{b}_y$ are given later in (8.4.9). Since the $\overset{\circ}{\mathbf{B}}$ matrix has three rows which are linearly independent, unless the element is degenerate, its rank is 3. Therefore the rank of the stiffness matrix $\mathbf{K}_e$ is 3 by (8.3.24), and from (8.3.18–19) we can see that the element has a rank deficiency of 2. This rank deficiency can cause serious difficulties unless it is corrected.

8.4 Q4 and Volumetric Locking

8.4.1 Element Description

Throughout this chapter, various properties of elements will be illustrated in terms of the 4-node quadrilateral. A Lagrangian formulation will be assumed for specificity, but many of the properties also apply to ALE and Eulerian formulations. To avoid excessive repetition of the name 4-node quadrilateral, we will often use the shorter name Q4. This element has been introduced in Example 4.2, but we repeat some of the equations in a notation that will be useful for the analysis of this element.

The motion, that is, the map between current configuration and the parent element, for Q4 is given by (we resume the implied summation notation for repeated subscripts)

$$x_i(\xi,t) = N_I x_{iI} = \mathbf{N}\mathbf{x}_i \tag{8.4.1}$$

where $\mathbf{N}$ is a row matrix consisting of the four isoparametric shape functions $\mathbf{N} = [N_I] = [N_1, N_2, N_3, N_4]$. The shape functions are given in (E4.2.2) and $\mathbf{x}_i$, $i = 1$ to 2, are the column matrices of the nodal coordinates:

$$\mathbf{x}_1 \equiv \mathbf{x} = [x_1, x_2, x_3, x_4]^T, \quad \mathbf{x}_2 \equiv \mathbf{y} = [y_1, y_2, y_3, y_4]^T \tag{8.4.2}$$

The parent element is a biunit square as shown in Figure 8.6, with the nodes numbered counter-clockwise beginning at the lower left-hand corner as shown.

The displacements and velocities are written similarly as

$$u_i = N_I u_{iI} = \mathbf{N}\mathbf{u}_i, \quad v_i = N_I v_{iI} = \mathbf{N}\mathbf{v}_i \tag{8.4.3}$$

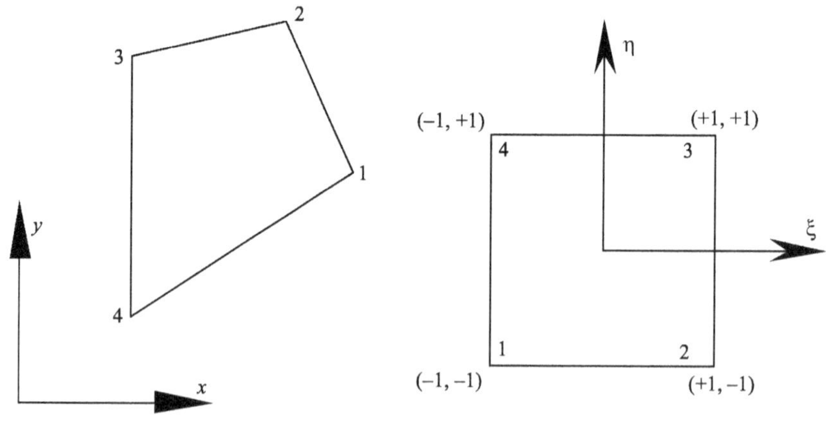

Figure 8.6 Spatial and parent element domains for the 4-node quadrilateral, QUAD4

where $\mathbf{v}_x$ and $\mathbf{v}_y$ are the column matrices of components of the element nodal velocities:

$$\mathbf{v}_x^T = [v_{x1}, v_{x2}, v_{x3}, v_{x4}], \qquad \mathbf{v}_y^T = [v_{y1}, v_{y2}, v_{y3}, v_{y4}] \tag{8.4.4}$$

The rate-of-deformation field has been obtained in Example 4.2. In Voigt form it can be written as

$$\begin{Bmatrix} D_x \\ D_y \\ 2D_{xy} \end{Bmatrix} = \begin{bmatrix} N_{I,x} & 0 \\ 0 & N_{I,y} \\ N_{I,y} & N_{I,x} \end{bmatrix} \begin{Bmatrix} v_{xI} \\ v_{yI} \end{Bmatrix} \quad \text{or} \quad \{\mathbf{D}\} = \begin{bmatrix} \mathbf{N},_x & \mathbf{0} \\ \mathbf{0} & \mathbf{N},_y \\ \mathbf{N},_y & \mathbf{N},_x \end{bmatrix} \begin{Bmatrix} \mathbf{v}_x \\ \mathbf{v}_y \end{Bmatrix} \equiv \mathbf{B}\dot{\mathbf{d}} \tag{8.4.5}$$

The RHS of (8.4.5) is not in standard form: in the $\mathbf{d}$ matrix, the x-components of all nodal velocities appear first, followed by the y-components of all nodal velocities. The form given here simplifies some of the subsequent analysis.

The element Jacobian determinant J_ξ is obtained via (E4.2.9) and the motion:

$$J_\xi = \frac{1}{8}[x_{24}y_{31} + x_{31}y_{42} + (x_{21}y_{34} + x_{34}y_{12})\xi + (x_{14}y_{32} + x_{32}y_{41})\eta] \tag{8.4.6}$$

Note that the bilinear term is absent in the element Jacobian determinant J_ξ. The element Jacobian determinant at the origin of the parent element is given by evaluating of (8.4.6) at $\xi = \eta = 0$, giving

$$J_\xi = \frac{1}{8}(x_{24}y_{31} + x_{31}y_{42}) = \frac{A}{4} \tag{8.4.7}$$

where A is the area of the element. From (8.4.5), it can be seen that the $\mathbf{B}$ matrix at the origin of the parent element coordinate system can be expressed as

$$\mathbf{B}^T(\mathbf{0}) = \begin{bmatrix} \mathbf{b}_x & \mathbf{0} & \mathbf{b}_y \\ \mathbf{0} & \mathbf{b}_y & \mathbf{b}_x \end{bmatrix} \tag{8.4.8}$$

where

$$\mathbf{b}_x^T \equiv \mathbf{b}_1^T = \mathbf{N},_x = \frac{1}{2A}[y_{24}, y_{31}, y_{42}, y_{13}], \quad \mathbf{b}_y^T \equiv \mathbf{b}_2^T = \mathbf{N},_y = \frac{1}{2A}[x_{42}, x_{13}, x_{24}, x_{31}] \tag{8.4.9}$$

8.4.2 Basis Form of Q4 Approximation

We will now develop a form of the velocity approximation for Q4 which makes analysis easier. This form was first given in Belytschko and Bachrach (1986), and the reader is referred to that paper for an alternative, slower development. To develop this velocity field, we define two sets of four column matrices:

$$\mathbf{p}_I = [\mathbf{1} \quad \mathbf{b}_x \quad \mathbf{b}_y \quad \gamma], \qquad \mathbf{q}_J = [\mathbf{1} \quad \mathbf{x} \quad \mathbf{y} \quad \mathbf{h}] \tag{8.4.10}$$

where

$$\mathbf{h}^T = [1 \quad -1 \quad 1 \quad -1] \qquad \mathbf{1}^T = [1 \quad 1 \quad 1 \quad 1]$$
$$\gamma = \frac{1}{4}(\mathbf{h} - (\mathbf{h}^T\mathbf{x})\mathbf{b}_x - (\mathbf{h}^T\mathbf{y})\mathbf{b}_y) \quad \bar{\mathbf{1}} = \frac{1}{4}(\mathbf{1} - (\mathbf{1}^T\mathbf{x})\mathbf{b}_x - (\mathbf{1}^T\mathbf{y})\mathbf{b}_y) \tag{8.4.11}$$

The usefulness of the two sets of vectors in (8.4.10) is that they are biorthogonal:

$$\mathbf{p}_I^T\mathbf{q}_J = \delta_{IJ} \tag{8.4.12}$$

This is easily verified for most of the terms. The biorthogonality of $\mathbf{b}_i$ and $\mathbf{x}_i$ is shown as follows:

$$b_{iI}x_{jI} = \frac{\partial N_I}{\partial x_i}x_{jI} = \frac{\partial x_j}{\partial x_i} = \delta_{ij} \tag{8.4.13}$$

where the second step follows from (8.4.9) and the third step from (8.4.1). Furthermore, for a nondegenerate element, the two sets $\mathbf{p}_I$ and $\mathbf{q}_I$ are linearly independent, so each spans the space R^4. Therefore any vector in R^4 can be expressed as a linear combination of either of these sets of vectors.

Since the bilinear velocity field includes the linear fields, the velocity field can be written as

$$v_i = \alpha_{i0} + \alpha_{i1}x + \alpha_{i2}y + \alpha_{i3}h \quad \text{where} \quad h = \xi\eta, \quad h_I = h(\xi_I, \eta_I) \tag{8.4.14}$$

The nodal velocities can then be expressed as

$$v_{iI} = \alpha_{i0} + \alpha_{i1}x_I + \alpha_{i2}y_I + \alpha_{i3}h_I \quad \text{or} \quad \mathbf{v}_i = \alpha_{i0}\mathbf{1} + \alpha_{i1}\mathbf{x} + \alpha_{i2}\mathbf{y} + \alpha_{i3}\mathbf{h} \tag{8.4.15}$$

where $\mathbf{h}$ is given by (8.4.11). Premultiplying (8.4.15) by $\mathbf{p}_K$ and exploiting the orthogonality (8.4.12), we have

$$\alpha_{iK} = \mathbf{p}_{K+1}^T\mathbf{v}_i \quad \text{for} \quad K = 0 \quad \text{to} \quad 3 \tag{8.4.16}$$

or

$$\alpha_{i0} = \tilde{\mathbf{1}}^T\mathbf{v}_i, \quad \alpha_{i1} = \mathbf{b}_x^T\mathbf{v}_i, \quad \alpha_{i2} = \mathbf{b}_y^T\mathbf{v}_i, \quad \alpha_{i3} = \gamma^T\mathbf{v}_i \tag{8.4.17}$$

So by substituting this into (8.4.14) the velocity field can be written

$$v_i = (\bar{\mathbf{1}}^T\mathbf{v}_i) + (\mathbf{b}_x^T\mathbf{v}_i)x + (\mathbf{b}_y^T\mathbf{v}_i)y + (\gamma^T\mathbf{v}_i)h \tag{8.4.18}$$

This form is very useful for dissecting the properties of the Q4 element. It provides the velocity field in terms of the linear components and the bilinear term, $h = \xi\eta$. With this form, we will see that it is possible to construct assumed strain fields which avoid various types of locking.

The function $h = \xi\eta$ has the interesting and useful property that its derivatives are orthogonal to the constant field:

$$\int_{\Omega e} h_{,x}\, d\Omega = \int_{\Omega e} h_{,y}\, d\Omega = 0 \tag{8.4.19}$$

We will see that as a consequence, the energy of the bilinear velocity can be decoupled from the energy of the linear velocity, and the stiffness matrices can be decoupled similarly.

8.4.3 Locking in Q4

The 4-node quadrilateral locks in plane strain for incompressible and nearly incompressible materials when it is fully integrated. The motion for an incompressible material must be isochoric, that is, the Jacobian $J=1$. In rate form, this implies that $\dot{J}=0$, which by (3.2.25) is equivalent to $v_{i,i}=0$. In the following, we present two explanations of volumetric locking for the 4-node quadrilateral. We first present the arguments for an incompressible material, and then extend them to nearly incompressible materials.

Consider element 1 in Figure 8.7. The only nodal velocities which can be nonzero are

$$v_{x3}=-\beta_1 a \qquad v_{y3}=+\beta_2 b \tag{8.4.20}$$

where β_K are arbitrary (we have chosen this form to simplify subsequent developments). All other nodal velocities of element 1 must be zero to satisfy the boundary conditions. Differentiating (8.4.18) shows that the dilatation rate D_{ii} for an arbitrary motion is

$$D_{ii}=v_{x,x}+v_{y,y}=\mathbf{b}_x^T\mathbf{v}_x+\mathbf{b}_y^T\mathbf{v}_y+\gamma^T\mathbf{v}_x h,_x+\gamma^T\mathbf{v}_y h,_y \tag{8.4.21}$$

For element 1 we can show that $\mathbf{b}_x^T=\frac{1}{2ab}\left[-b \quad b \quad b \quad -b\right]$, $\mathbf{b}_y^T=\frac{1}{2ab}\left[-a \quad -a \quad a \quad a\right]$, $\gamma^T=\frac{1}{4}[1 \quad -1 \quad 1 \quad -1]$. The nodal velocities (8.4.20) give $\mathbf{b}_x^T\mathbf{v}_x=-\beta_1/2, \mathbf{b}_y^T\mathbf{v}_y=\beta_1/2$. So the constant part of the dilatation rate is nonzero unless $\beta_1=\beta_2$, so an isochoric motion requires $\beta_1=\beta_2$. In that case, it can be shown that $A=0$. But when $\beta_1=\beta_2\equiv\beta$,

$$D_{ii}=\frac{1}{4}\beta(bh,_y-ah,_x)=\frac{\beta}{2}(\bar{x}-\bar{y}) \quad \text{where} \quad \bar{x}=x/a, \bar{y}=y/b \tag{8.4.22}$$

This vanishes only along the line $\bar{y}=\bar{x}$! So although the motion of the element is a constant volume motion, the rate of dilatation is nonzero everywhere except on that line. For the motion to be isochoric throughout the element, β must vanish and node 3 cannot move.

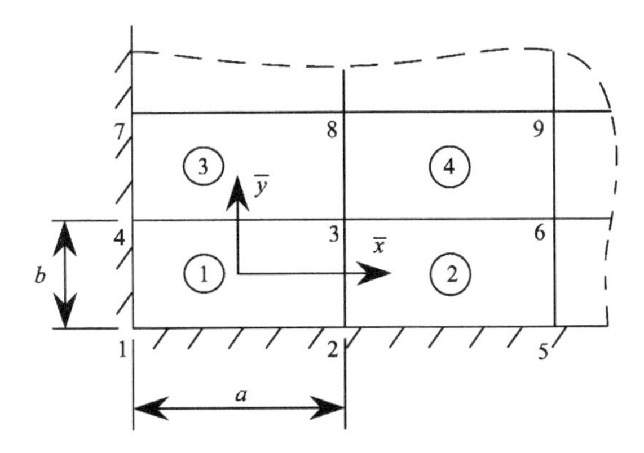

Figure 8.7 Mesh of rectangular elements fixed on two sides; only part of the mesh is shown

If node 3 cannot move, nodes 2 and 3 then provide a rigid boundary for the left-hand side of element 2, and it can be shown by repeating these arguments for element 2 that node 6 cannot move. This argument can then be repeated for all elements of the mesh to show that the velocities of all nodes must vanish, that is, *the finite element model locks*. This argument also applies to skewed elements.

Another way to examine this behavior is to consider an element velocity expressed by (8.4.14). The rate of dilatation is then given by

$$D_{ii} = v_{x,x} + v_{y,y} = \alpha_{x1} + \alpha_{y2} + \alpha_{x3}h,_x + \alpha_{y3}h,_y \tag{8.4.23}$$

We can evaluate the change in area of the element by integrating the rate of dilatation over the element domain:

$$\int_{\Omega e} D_{ii} dA = \int_{\Omega e} (\alpha_{x1} + \alpha_{y2} + \alpha_{x3}h,_x + \alpha_{3y}h,_y)dA \tag{8.4.24}$$

Integrating the RHS of (8.4.24), using (8.4.19) and setting the result to zero to reflect isochoric motion gives

$$\int_{\Omega_e} D_{ii} dA = (\alpha_{x1} + \alpha_{y2})A = 0 \tag{8.4.25}$$

This shows that for any isochoric velocity field, it is necessary that $\alpha_{y2} = -\alpha_{x1}$. The dilatation rate for a motion that keeps the element area constant is then

$$D_{ii} = \alpha_{x3}h,_x + \alpha_{y3}h,_y \tag{8.4.26}$$

This dilatation rate will be nonzero everywhere within the element except along the curve $\alpha_{x3}h,_x = -\alpha_{x3}h,_y$, even when the overall element deformation is volume preserving. Thus the element cannot reproduce an isochoric motion. Note also that the above shows that the portion of the motion that causes difficulties is the hourglass mode, for it preserves the volume but its dilatation rate is nonzero within the element.

These arguments can be easily extended to a nearly incompressible material. For simplicity, we consider a linear material. If we subdivide the linear elastic strain energy into hydrostatic and deviatoric parts, it can be written as

$$W^{\text{int}} = \frac{1}{2}\int_{\Omega^e} \left(K(u_{i,i})^2 + 2\mu\varepsilon_{ij}^{\text{dev}}\varepsilon_{ij}^{\text{dev}} \right) d\Omega \tag{8.4.27}$$

where K is the bulk modulus and μ the shear modulus. In any isochoric motion, the total volume of the element will remain constant. However, the motion must be isochoric throughout the element. Otherwise, when K is very large (a nearly incompressible material), any nonzero volumetric strain will absorb all of the energy.

Thus volumetric locking arises from the inability of the element to exactly represent an isochoric motion. To eliminate locking, the strain field must be designed so that the dilatation in the assumed strain field vanishes throughout the element. This may be restated as follows: *to avoid locking, the strain field must be isochoric throughout the element for any velocity field which preserves the volume of the element.* In particular, for the quadrilateral, the dilatation must vanish in the entire element for the hourglass mode, because this motion is equivoluminal.

8.5 Multi-Field Weak Forms and Elements

8.5.1 Nomenclature

Multi-field weak forms, also called mixed weak forms, are used for the construction of elements which do not lock and possess other performance enhancements. The resulting elements are often called *assumed strain* and *assumed stress* elements, because the strain and/or stress field in these elements is interpolated independently of the displacement field. Other names for these elements are mixed elements and hybrid elements. These techniques are beneficial for constrained media problems, such as incompressible problems, because standard elements with full integration often lock in constrained media problems. Examples of other constrained media problems are beams, shells and irrotational flow. Multi-field weak forms make it possible to design elements which neither lock nor exhibit excessive stiffness for constrained media problems. Mixed or multi-field *variational principles* can be developed for conservative problems.

8.5.2 Hu–Washizu Weak Form

The most general of the multi-field weak forms is the Hu–Washizu variational principle. This variational principle was developed subsequent to Reissner's development of the two-field principle, the Hellinger–Reissner principle, in which the displacements and the stresses are the unknown fields. The two-field principle is seldom used in nonlinear analysis because it is incompatible with strain-driven constitutive models. An interesting anecdote on the three-field principle appears in Reissner (1996: 434). Evidently, after completing his work on the two-field principle, Eric Reissner was visited by Washizu, who told him of his development of the three-field principle. As Reissner relates the story, he,

> ...first objected that since only stresses and displacements would be encountered in the boundary conditions of problems it was not natural to consider strain displacement relations in ways other than defining relations. I was, however, soon persuaded that the three-field theorem which Washizu, and independently Hu, had proposed was a valuable advance which I wished I had thought of myself.

We shall first give a statement of the Hu–Washizu weak form which emanates from this variational principle and show that the corresponding strong form consists of the kinetic equations (conservation of momentum, interior continuity conditions, and traction boundary conditions), the strain–displacement equation (rate-of-deformation in terms of velocities) and constitutive equation.

The Hu–Washizu weak form involves three dependent tensor variables: the velocity $\mathbf{v}(\mathbf{X}, t)$, the rate-of-deformation $\bar{\mathbf{D}}(\mathbf{X},t)$, and the stress $\bar{\boldsymbol{\sigma}}(\mathbf{X}, t)$. The rate-of-deformation $\bar{\mathbf{D}}$ is often called the assumed rate-of-deformation, and $\bar{\boldsymbol{\sigma}}$ the assumed stress, since they are interpolated independently of the velocity field. Superposed bars are placed on all assumed fields except the velocity to distinguish them from fields evaluated in terms of other variables; for example, the velocity strain $\mathbf{D}$ emanates from the velocity via the kinematic relations, while the stress $\boldsymbol{\sigma}$ is evaluated by the constitutive equation from the assumed velocity strain. Thus $\mathbf{D}$ and $\boldsymbol{\sigma}$ represent $\mathbf{D}(\mathbf{v})$ and $\sigma(\bar{\mathbf{D}})$, but the functional dependence is often omitted in the following.

The weak form is

$$0=\delta\Pi^{HW}(\mathbf{v},\bar{\mathbf{D}},\bar{\sigma})=\int_{\Omega}\delta\bar{\mathbf{D}}:\sigma(\bar{\mathbf{D}})d\Omega+\int_{\Omega}\delta[\bar{\sigma}:(\mathbf{D}(\mathbf{v})-\bar{\mathbf{D}})]\,d\Omega-\delta P^{\text{ext}}+\delta P^{\text{kin}} \quad (8.5.1)$$

or

$$0=\delta\Pi^{HW}(\mathbf{v},\bar{\mathbf{D}},\bar{\sigma})=\int_{\Omega}\delta\bar{D}_{ij}\sigma_{ij}(\bar{\mathbf{D}})d\Omega+\int_{\Omega}\delta[\bar{\sigma}_{ij}(D_{ij}(\mathbf{v})-\bar{D}_{ij})]\,d\Omega-\delta P^{\text{ext}}+\delta P^{\text{kin}} \quad (8.5.2)$$

In this, in the second term on the RHS, δ is an operator with the rules of a variational operator, so $\delta(uv)=\delta uv+u\delta v$. The virtual external and intertial power are the same as in the single-field principle, (B4.2.6–7). The stress $\sigma(\bar{\mathbf{D}})$ can be a function of stresses and state variables in addition to being a function of $\bar{\mathbf{D}}$, but we only explicitly indicate the dependence on $\bar{\mathbf{D}}$; note that $\boldsymbol{\sigma}$ is a function of the *assumed velocity strain*.

No derivatives of the assumed velocity strain $\bar{\mathbf{D}}$ or the assumed stress $\bar{\boldsymbol{\sigma}}$ appear in this weak form, so these variables need only be piecewise continuous functions of $\mathbf{X}$, that is C^{-1} functions. The trial functions for the $\bar{\mathbf{D}}$ and $\bar{\boldsymbol{\sigma}}$ need not satisfy any boundary conditions. The first derivative of the velocity appears in the Hu–Washizu weak form, so the velocity must be continuously differentiable, that is, $\mathbf{v}\in C^0$; furthermore $\mathbf{v}$ must satisfy the kinematic boundary conditions, while $\delta\mathbf{v}$ must vanish on kinematic boundaries. The conditions on the test and trial functions for the velocity are thus identical to those of the principle of virtual power, so $\mathbf{v}\in u$ and $\delta\mathbf{v}\in u_0$, where the definition of u_0 are given by (4.3.1). The test functions $\delta\mathbf{v}(\mathbf{X}),\delta\bar{\mathbf{D}}(\mathbf{X}),\delta\bar{\sigma}(\mathbf{X})$, as shown, are independent of time. One subtle point worth noting is that since the derivative of the velocity field appears as a product with the assumed stress, the continuity requirements given here may be stronger than needed for the boundedness of the energy. However, implementations with lesser continuity would be awkward in a finite element method.

The Hu–Washizu weak form can be viewed as a Lagrange multiplier form of the single-field weak form. The constraint is the relationship between $\bar{\mathbf{D}}$ and $\mathbf{D}=\text{sym}\nabla\mathbf{v}$; the Lagrange multiplier is the assumed stress $\bar{\sigma}$.

The statement of the equivalence of the weak and strong forms is:
if at any time $t,\mathbf{v}\in u,\bar{\mathbf{D}}\in C^{-1},\bar{\mathbf{D}}=\bar{\mathbf{D}}^T,\bar{\sigma}\in C^{-1},\bar{\sigma}=\bar{\sigma}^T$ and

$$\delta\Pi^{HW}(\mathbf{v},\bar{\mathbf{D}},\bar{\sigma})=0\quad\forall\delta\mathbf{v}\in u_0,\quad\forall\delta\bar{\mathbf{D}}\in C^{-1},\quad\delta\bar{\mathbf{D}}=\delta\bar{\mathbf{D}}^T;\quad\forall\delta\bar{\sigma}\in C^{-1},\delta\bar{\sigma}=\delta\bar{\sigma}^T \quad (8.5.3)$$

then

$$\frac{\partial\bar{\sigma}_{ji}}{\partial x_j}+\rho b_i=\rho\dot{v}_i \qquad \text{(momentum equation)} \qquad (8.5.4)$$

$$n_i\bar{\sigma}_{ij}=t_j^*\quad\text{on}\quad\Gamma_{t_j} \qquad \text{(traction boundary condition)} \qquad (8.5.5)$$

$$\bar{\sigma}_{ij}=\sigma_{ij}(\bar{\mathbf{D}},\bar{\sigma},\ldots,\text{etc.}) \qquad \text{(constitutive equation)} \qquad (8.5.6)$$

$$\bar{D}_{ij}=D_{ij}(\mathbf{v})=\frac{1}{2}\left(\frac{\partial v_i}{\partial x_j}+\frac{\partial v_j}{\partial x_i}\right) \qquad \text{(strain measure)} \qquad (8.5.7)$$

$$[\![n_i\bar{\sigma}_{ij}]\!]=0\quad\text{on}\quad\Gamma_{\text{int}} \qquad \text{(interior continuity conditions)} \qquad (8.5.8)$$

where $u = \{\mathbf{v}(\mathbf{X},t) \mid \mathbf{v} \in C^0, v_i = \bar{v}_i \text{ on } \Gamma_{v_i}\}$, $u_0 = \{\delta v_i(\mathbf{X}) \mid \delta v_i \in C^0, \delta v_i = 0 \text{ on } \Gamma_{v_i}\}$.

Two of the equations in the strong form can use some explanation. Equation (8.5.6) is the strong form of the constitutive equation: it states that the assumed stress equals the stress computed by the constitutive law. Equation (8.5.7) is the strong form of the definition of the velocity strain: the assumed velocity strain is equal to the symmetric gradient of the velocity field (we use the symbol $\mathbf{D}$ to simplify the notation).

In the following, we will demonstrate that the weak form (8.5.3) implies (8.5.4–8). We will also illustrate that the assumed stress field $\bar{\boldsymbol{\sigma}}$ is a Lagrange multiplier on the relation between the velocity strain and velocity. For this purpose we will replace $\bar{\boldsymbol{\sigma}}$ by a symmetric Lagrange multiplier field $\boldsymbol{\lambda}=\boldsymbol{\lambda}^T$, which is a second-order tensor, and show that this Lagrange multiplier can be identified as the stress at the end of the proof. For simplicity, we consider simple boundaries where all components of either the traction or velocity are prescribed, so $\Gamma_t = \Gamma - \Gamma_v$.

Taking the variations of the second term in (8.5.2), with $\bar{\boldsymbol{\sigma}}$ replaced by $\boldsymbol{\lambda}$ and D_{ij} replaced by $v_{i,j}$ yields

$$\int_\Omega \delta[\lambda_{ij}(v_{i,j} - \bar{D}_{ij})]\, d\Omega = \int_\Omega [\delta\lambda_{ij}(v_{i,j} - \bar{D}_{ij})] + \lambda_{ij}(\delta v_{i,j} - \delta\bar{D}_{ij})]\, d\Omega \tag{8.5.9}$$

in view of the symmetry of λ_{ij}. Next consider the third term on the RHS in (8.5.9):

$$\begin{aligned}\int_\Omega \delta v_{i,j}\lambda_{ij}\, d\Omega &= \int_\Omega [(\delta v_i \lambda_{ij})_{,j} - \delta v_i \lambda_{ij,j}]\, d\Omega \\ &= \int_{\Gamma_t} \delta v_i \lambda_{ij} n_j\, d\Gamma + \int_{\Gamma_{\text{int}}} \delta v_i [\![\lambda_{ij} n_j]\!]\, d\Gamma - \int_\Omega \delta v_i \lambda_{ij,j}\, d\Omega\end{aligned} \tag{8.5.10}$$

where in the second line we have used the Gauss divergence theorem on the first term of the RHS and immediately changed Γ to Γ_t because $\delta v_i = 0$ on $\Gamma_v = \Gamma - \Gamma_t$. Substituting (8.5.10) into (8.5.9) and the result in (8.5.2) gives

$$\begin{aligned}0 = \delta\Pi^{HW} &= \int_\Omega [\delta v_i(-\lambda_{ij,j} - \rho b_i + \rho\dot{v}_i) + \delta\lambda_{ij}(sym\ v_{i,j} - \bar{D}_{ij}) \\ &- \delta\bar{D}_{ij}(\lambda_{ij} - \sigma_{ij}(\bar{\mathbf{D}}))]\, d\Omega + \int_{\Gamma_t} \delta v_i(\lambda_{ij} n_j - t_i^*)\, d\Gamma + \int_{\Gamma_{\text{int}}} \delta v_i [\![\lambda_{ij} n_j]\!]\, d\Gamma\end{aligned} \tag{8.5.11}$$

Using the arbitrariness of the test functions then gives equations identical to (8.5.4–8), except they are expressed in terms of $\boldsymbol{\lambda}$ instead of $\bar{\boldsymbol{\sigma}}$. If we replace $\boldsymbol{\lambda}$ by $\bar{\boldsymbol{\sigma}}$, we obtain the strong form. The strong form consists of the generalized momentum balance (momentum equation, traction boundary conditions and interior stress continuity conditions), the constitutive equation, and the strain-rate-velocity relations. The above also shows that if a Lagrange multiplier constrains the relation between rate-of-deformation and velocity, then the Lagrange multiplier is the stress.

8.5.3 *Alternative Multi-Field Weak Forms*

To cure volumetric locking, it suffices to design the dilatational part of the rate-of-deformation and the pressure field. For this purpose, a weak form involving only these fields is desirable.

This is obtained by splitting the stresses and velocity strains in (8.5.1) into hydrostatic and deviatoric parts and letting $\bar{\mathbf{D}}^{\text{dev}} = \mathbf{D}^{\text{dev}}, \bar{\boldsymbol{\sigma}}^{\text{dev}} = \boldsymbol{\sigma}^{\text{dev}}, \delta\bar{\mathbf{D}}^{\text{dev}} = \delta\mathbf{D}^{\text{dev}}$ and $\delta\bar{\boldsymbol{\sigma}}^{\text{dev}} = \delta\boldsymbol{\sigma}^{\text{dev}}$. The internal virtual power in the Hu–Washizu weak form (8.5.2) can then be simplified to

$$\delta P^{\text{int}} = \int_{\Omega} \left(\delta D_{ij}^{\text{dev}} \sigma_{ij}^{\text{dev}} \left(\bar{D}_{kk}, D_{ij}^{\text{dev}} \right) - \delta \bar{D}_{iip} \left(\bar{D}_{kk}, D_{ij}^{\text{dev}} \right) \right) d\Omega - \int_{\Omega} \delta \left[\bar{p} \left(D_{ii} - \bar{D}_{ii} \right) \right] d\Omega \quad (8.5.12)$$

where D_{ii} is the rate of dilatation obtained from the velocity field, $\bar{D}_{ii}$ is the assumed rate of dilatation and $\bar{p}$ is the assumed pressure field; the external and kinetic powers are unchanged. This multi-field weak form adds only two scalar fields as unknowns: the assumed rate of dilatation and the assumed pressure field. This weak form is applicable to nearly incompressible materials.

For incompressible materials, the assumed dilatation should vanish, so we let $\bar{D}_{ii} = 0$ and $\delta\bar{D}_{ii} = 0$. The internal power (8.5.12) can then be simplified to

$$\delta P^{\text{int}} = \int_{\Omega} \delta D_{ij}^{\text{dev}} \sigma_{ij}^{\text{dev}} (\mathbf{D}^{dev}) d\Omega - \int_{\Omega} \delta [\bar{p} D_{ii}] d\Omega \quad (8.5.13)$$

In the above, the rate incompressibility condition $D_{ii} = v_{i,i} = 0$ is a constraint with the Lagrange multiplier the assumed pressure $\bar{p}$. Note that the deviatoric stress is a function of $\mathbf{D}(\mathbf{v})$. This weak form has only one additional unknown, the assumed pressure.

8.5.4 Total Lagrangian Form of the Hu–Washizu

The weak form (8.5.1) can be expressed in many other ways; Atluri and Cazzani (1995) give a comprehensive account of these variations. We shall illustrate a few of them here. These alternative forms are obtained by using different conjugate measures of stress and strain in the Hu–Washizu three-field weak form, but at times things get tricky.

If we convert (8.5.2) to the conjugate pair consisting of the nominal stress $\mathbf{P}$ and the deformation gradient $\mathbf{F}$, the three-field weak form becomes

$$0 = \delta\Pi^{HW}(\mathbf{u}, \bar{\mathbf{F}}, \bar{\mathbf{P}}) = \int_{\Omega_0} \delta\bar{F}_{ij} P_{ji}(\bar{\mathbf{F}}) d\Omega_0 + \int_{\Omega_0} \delta[\bar{P}_{ji}(F_{ij}(\mathbf{u}) - \bar{F}_{ij}) d\Omega_0 - \delta W^{\text{ext}} + \delta W^{\text{kin}} \quad (8.5.14)$$

where the external and kinetic virtual energies are given in (B4.5.3–4). The internal virtual work for the three-field weak form can also be given in terms of the energy-conjugate pair consisting of the PK2 stress $\mathbf{S}$ and the Green strain $\mathbf{E}$:

$$\delta W^{\text{int}} = \int_{\Omega_0} \delta\bar{E}_{ij} S_{ij}(\bar{\mathbf{E}}) d\Omega_0 + \int_{\Omega_0} \delta[\bar{S}_{ij}(E_{ij}(\mathbf{u}) - \bar{E}_{ij})] d\Omega_0 \quad (8.5.15)$$

The total Lagrangian forms are more difficult to work with. Since the nominal stress is not symmetric, four components have to be interpolated in two dimensions and the balance of angular momentum has to be considered. Often this is done by expressing the constitutive equation in terms of the PK2 stress $\mathbf{S}$ and then letting $\bar{\mathbf{P}} = \overline{\mathbf{S}\mathbf{F}^T}$ (Box 3.2). Other difficulties arise from the fact that $\mathbf{F}$ is not a measure of strain (it does not vanish in rigid body rotation), so it is more difficult to design this field. The Green strain complicates things, since $\bar{\mathbf{E}}$ is quadratic for a linear gradient field; its design is not as simple as the design of the velocity-strain field $\bar{\mathbf{D}}$.

In the total Lagrangian framework, it is possible to write a three-field *variational principle* for static, conservative problems, that is, those in which the loads and materials possess potentials. The three-field form corresponding to the single-field principle in Section 4.9.3 is

$$\Pi^{HW} = (\mathbf{u}, \bar{\mathbf{S}}, \bar{\mathbf{E}}) = \int_{\Omega_0} w(\bar{\mathbf{E}}) d\Omega_0 + \int_{\Omega_0} \bar{\mathbf{S}} : (\mathbf{E} - \bar{\mathbf{E}}) \, d\Omega_0 - W^{ext} \tag{8.5.16}$$

where $w(\bar{\mathbf{E}})$ is the internal potential energy in terms of the assumed strain. This can also be written in terms of the nominal stress and the deformation tensor and other stress–strain pairs which are conjugate in energy.

For incompressible and nearly incompressible materials, the following total Lagrangian form has been investigated by Simo, Taylor and Pister (1985):

$$W(\mathbf{u}, \bar{p}) = \int_{\Omega_0} w\left(\bar{J}^{-\frac{1}{3}} \mathbf{F}(\mathbf{u})\right) + \bar{p}(J(\mathbf{u}) - \bar{J}) \Big) d\Omega_0 - W^{ext} \tag{8.5.17}$$

where $\bar{p}$ is the assumed pressure and $\bar{J}$ the assumed field for the Jacobian determinant. For an incompressible material, this can be reduced to

$$W(\mathbf{u}, \bar{p}) = \int_{\Omega_0} \left(w\left(J^{-\frac{1}{3}} \mathbf{F}(\mathbf{u})\right) + \bar{p}\left(J(\mathbf{u}) - 1\right) \right) d\Omega_0 - W^{ext} \tag{8.5.18}$$

This can also be viewed as the Lagrange multiplier method which enforces the incompressibility condition $J = 1$; see Section 5.4.2, where it is also noted that $J^{-\frac{1}{3}}\mathbf{F}$ is a measure of deviatoric deformation. These weak forms are more difficult to work with than (8.5.13), since the Jacobian J is nonlinear in the displacements. However, this approach enforces the incompressibility condition more accurately than the rate form (8.5.13) for large increments.

8.5.5 *Pressure–Velocity* (p–v) *Implementation*

Discretization of (8.5.13) involves two fields: the velocity $\mathbf{v}(\mathbf{X},t)$ and the assumed pressure $\bar{p}(\mathbf{X},t)$. Two types of pressure approximations are used:

1. globally defined fields $\bar{p}(\mathbf{X},t)$;
2. element-specific fields $\bar{p}(\mathbf{X},t)$ where the pressure depends only on parameters associated with a single element that can be eliminated on an element level prior to assembly of the global equations; this requires the perturbed Lagrangian.

We will begin with globally defined fields and then modify them to obtain the discrete equations for the second approach.

The trial velocity and pressure fields are

$$v_i(\xi,t) = N_{iA}(\xi)\dot{d}_A(t) \text{ or } v = \mathbf{N}\dot{\mathbf{d}} \tag{8.5.19}$$

$$\begin{aligned} \bar{p}(\xi,t) &= N_A^p(\xi) p_A(t) \quad \text{or} \quad \bar{p} = \mathbf{N}^p \mathbf{p} \\ \dot{\mathbf{d}}^T &= [v_{x1}, v_{y1}, v_{x2}, v_{y2}, \ldots], \qquad \mathbf{p}^T = [p_1, p_2, \ldots] \end{aligned} \tag{8.5.20}$$

In these, we have included a component index for the velocity shape functions. We will not use different shape functions for different components; this notation has been chosen to simplify some of the subsequent development. The superscript 'p' is appended to the pressure interpolants since they may differ from the velocity interpolants.

The virtual internal power for an incompressible material (8.5.13) can then be expressed as (see Section 4.5.4)

$$\delta P^{\text{int}} = \delta \dot{d}_B \int_\Omega B_{ijB}^{dev} \sigma_{ij}^{dev} d\Omega + \delta(p_A G_{AB} \dot{d}_B) \quad \text{where } G_{AB} = -\int_\Omega N_A^p N_{Bi,i} d\Omega$$
$$B_{ijA} = \frac{1}{2}(N_{iA,j} + N_{jA,i}), \quad B_{ijA}^{dev} = B_{ijA} - \frac{1}{3} B_{kkA} \delta_{ij} \tag{8.5.21}$$

The external and kinetic power is unchanged from the single field form in Box 4.3, so

$$\delta P^{ext} - \delta P^{\text{kin}} = \delta \dot{d}_A \left(f_A^{ext} - M_{AB} \ddot{d}_B \right) \tag{8.5.22}$$

Taking the variation of the second term in (8.5.21) gives

$$\delta P^{\text{int}} = \delta \dot{d}_B \left[\int_\Omega B_{ijB}^{dev} \sigma_{ji}^{dev} d\Omega + G_{AB} p_A \right] + \delta p_A G_{AB} \dot{d}_B \tag{8.5.23}$$

Combining (8.5.22) with (8.5.23) we obtain

$$M_{AB} \ddot{d}_B + f_A^{\text{int}} = f_A^{ext} \quad \text{where } f_A^{\text{int}} = \int_\Omega B_{ijA}^{dev} \sigma_{ji}^{dev} d\Omega + G_{BA} p_B \tag{8.5.24}$$

$$G_{AB} \dot{d}_B = 0 \tag{8.5.25}$$

Equation (8.5.24) is the semidiscrete equation of motion with the internal nodal forces redefined to account for the pressure approximation. Equation (8.5.25) is the incompressibility condition. It is worthwhile to compare the internal nodal forces obtained by this multi-field approach to the single-field result (B4.3.2). Writing out $\mathbf{G}$ by (8.5.21), (8.5.24) becomes

$$f_A^{\text{int}} = \int_\Omega \left(B_{ijA}^{dev} \sigma_{ji}^{dev} - B_{iiA} N_B^p p_B \right) d\Omega \tag{8.5.26}$$

so the stresses have simply been split into deviatoric and hydrostatic components with the interpolation (8.5.20) for the hydrostatic part.

The equations of motion and the isochoric constraint (8.5.24–25) can be written in Voigt matrix form as

$$\mathbf{M}\ddot{\mathbf{d}} + \mathbf{f}^{\text{int}} = \mathbf{f}^{ext} \quad \text{where } \mathbf{f}^{\text{int}} = \int_\Omega \mathbf{B}_{dev}^{\mathrm{T}} \left\{ \sigma^{dev} \right\} \mathrm{d}\Omega + \mathbf{G}^{\mathrm{T}} \mathbf{p} \tag{8.5.27}$$

$$\mathbf{G}\ddot{\mathbf{d}} = 0 \tag{8.5.28}$$

8.5.5.1 Linearization

The linearized equations for the previous can be obtained by the procedure in Section 6.4. We use the notation in Section 6.5.3 with $\boldsymbol{\delta}\mathbf{d}$ the incremental displacement in the iterative Newton procedure, $\boldsymbol{\Delta}\mathbf{d}$ the step increment. The linearized equations are

$$\begin{bmatrix} \mathbf{A} & \mathbf{G}^T \\ \mathbf{G} & \mathbf{0} \end{bmatrix} \begin{Bmatrix} \delta\mathbf{d} \\ \delta\mathbf{p} \end{Bmatrix} = \begin{Bmatrix} -\mathbf{r}_v - \mathbf{G}^T\mathbf{p}_v \\ -\mathbf{G}\Delta\mathbf{d}_v \end{Bmatrix} \tag{8.5.29}$$

where $\mathbf{A}$ is given by (6.3.23) and

$$\mathbf{K}_{IJ}^{\text{int}} = \int_{\Omega} \left(\mathbf{B}_I^T \left[\mathbf{C}_{dev}^{\sigma T} \right] \mathbf{B}_J + \mathbf{I} \boldsymbol{B}_I^T (\sigma - \bar{p}\mathbf{I})\, \boldsymbol{B}_J \right) d\Omega \tag{8.5.30}$$

This is the standard form of the Lagrange multiplier problem; compare to (6.3.40). In this formulation, the pressure field must be global; it cannot be eliminated at the element level. This multi-field form is useful since only one additional unknown field is added to the single-field form: the pressure. The pressure cannot be eliminated on the element level because the matrix coefficient of the pressure on the LHS of the above is zero.

Thus the multi-field weak forms are basically constrained forms of the single-field principle. For the p–v formulation, the constraint is the incompressibility condition $v_{i,i}=0$ and the Lagrange multiplier is the assumed pressure.

8.5.6 *Element Specific Pressure*

In the perturbed Lagrangian, the pressure can be eliminated at the element level. It is left as an exercise to show that the linearized equations are

$$\begin{bmatrix} \mathbf{A} & \mathbf{G}^T \\ \mathbf{G} & \mathbf{0} \end{bmatrix} \begin{Bmatrix} \delta\mathbf{d} \\ \delta\mathbf{p} \end{Bmatrix} = \begin{Bmatrix} -\mathbf{r}_v - \mathbf{G}^T\mathbf{p}_v \\ -\mathbf{G}\Delta\mathbf{d}_v - \mathbf{H}\mathbf{p}_v \end{Bmatrix} \tag{8.5.31}$$

where

$$\mathbf{H} = \left[H_{AB} \right],\ H_{AB} = \frac{1}{K} \int_{\Omega} N_A^P N_B^P \, d\Omega \tag{8.5.32}$$

where K is the penalty parameter, which is the counterpart of β^{-1} in (6.3.45).

If the pressure is C^{-1}, it can be eliminated at the element level. We replace (8.5.20) by $\bar{p} = \mathbf{N}\mathbf{p}^e$, where $\mathbf{p}^e$ are element pressure variables. The discrete equations are then identical to (8.5.31) except that they pertain to an element:

$$\begin{bmatrix} \mathbf{A}_e & \mathbf{G}_e^T \\ \mathbf{G}_e & \mathbf{H}^e \end{bmatrix} \begin{Bmatrix} \delta\mathbf{d}^e \\ \delta\mathbf{p}^e \end{Bmatrix} = \begin{Bmatrix} \mathbf{f}_e^{\text{int}} \\ \mathbf{0} \end{Bmatrix} \tag{8.5.33}$$

The constraint equations are then the second line in the previous

$$\mathbf{G}^e \delta\mathbf{d}^e + \mathbf{H}^e \delta\mathbf{p}^e = 0 \tag{8.5.34}$$

where $\mathbf{G}^e$ and $\mathbf{H}^e$ are the element counterparts of the global matrices defined above. This can be solved at the element level for $\delta\mathbf{p}^e$, and the result can be substituted into the first of (8.5.33) to give an effective element tangent matrix.

8.5.7 Finite Element Implementation of Hu–Washizu

The development of the finite element equations via the Hu–Washizu principle involves the approximation of three tensor fields. The resulting number of scalar fields is very large: in three dimensions, the number of scalar fields associated with $\bar{\mathbf{D}}(\mathbf{X},t), \bar{\boldsymbol{\sigma}}(\mathbf{X},t)$ and $\mathbf{v}(\mathbf{X},t)$ is 6, 6 and 3, respectively (symmetry has been invoked for $\bar{\mathbf{D}}$ and $\bar{\boldsymbol{\sigma}}$); in two dimensions, the number of scalar fields is 3, 3 and 2, respectively, for a total of eight unknown fields. Therefore, the application of the Hu–Washizu weak form is very costly, since it involves four times as many unknown fields in two dimensions as the single-field weak form, and five times as many in three dimensions. Therefore, the Hu–Washizu is seldom implemented directly. However, it is instructive to examine its implementation.

This implementation of the Hu–Washizu weak form takes advantage of the weaker continuity requirements on $\bar{\mathbf{D}}(\mathbf{X},t)$ and $\bar{\boldsymbol{\sigma}}(\mathbf{X},t)$: these assumed fields are defined on an element level, and then eliminated before assembly. The finite element approximations for the dependent variables are given by

$$v_i(\xi,t) = N_{iA}(\xi)\dot{d}_A(t) \quad \text{or} \quad \mathbf{v} = \mathbf{N}\dot{\mathbf{d}} \tag{8.5.35}$$

$$\bar{D}_{ij}(\xi,t) = N^D_{ijA}(\xi)\alpha^e_A(t) \text{ or } \bar{D}_a = N^D_{aA}\alpha^e_A, \quad \text{or} \quad \{\bar{\mathbf{D}}\} = \mathbf{N}_D\alpha^e \tag{8.5.36}$$

$$\bar{\sigma}_{ij}(\xi,t) = N^\sigma_{ijA}(\xi)\beta^e_A(t) \text{ or } \bar{\sigma}_a = N^\sigma_{aA}\beta^e_A, \quad \text{or} \quad \{\bar{\sigma}\} = \mathbf{N}_\sigma\beta^e \tag{8.5.37}$$

where $N_{iA}(\xi)$ are C^0 interpolants. The Voigt forms, where the indices 'ij' are converted to a single index, are shown on the right. The interpolants $N^D_{ijI}(\xi)$ and $N^\sigma_{ijI}(\xi)$ are C^{-1} and are defined in each element in terms of parameters that pertain only to that element; the interpolants for the assumed stress and velocity strain fields are symmetric in the indices 'ij', i.e. $N^D_{ijA} = N^D_{jiA}$ and $N^\sigma_{ijA} = N^\sigma_{jiA}$. The Hu–Washizu weak form can also be implemented with C^0 fields for the assumed stress and velocity strain, but the cost is great. The test functions are

$$\delta v_i(\xi) = N_{iA}(\xi)\delta\dot{d}_A, \quad \delta\bar{D}_{ij}(\xi) = N^D_{ijC}(\xi)\delta\alpha^e_C, \quad \delta\bar{\sigma}_{ij}(\xi) = N^\sigma_{ijB}(\xi)\,\delta\beta^e_B \tag{8.5.38}$$

Substituting the approximations (8.5.35–38) into (8.5.1), we obtain

$$\begin{aligned} 0 = \delta\Pi^{HW} &= \sum_e \delta\alpha^e_C \int_{\Omega_e} N^D_{ijC}\sigma_{ij}(\bar{\mathbf{D}})d\Omega \\ &+ \sum_e \int_{\Omega_e} \delta\left[\beta^e_B N^\sigma_{ijB}\left(B_{ijA}\dot{d}_A - N^D_{ijC}\alpha^e_C\right)\right]d\Omega - \delta P^{ext} + \delta P^{\mathrm{kin}} \end{aligned} \tag{8.5.39}$$

The velocity strain and stress interpolations do not appear in either δP^{ext} or δP^{kin}. Therefore, these terms are unchanged from the single-field principle of virtual power, so (see Section 4.4):

$$\delta P^{ext} - \delta P^{\mathrm{kin}} = \delta\dot{d}_A\left(f^{ext}_A - M_{AB}\ddot{d}_B\right) \tag{8.5.40}$$

where f_A^{ext} and M_{AB} are the external nodal forces and the mass matrix given in Box 4.3. Only the internal nodal forces differ from the single-field form. For convenience the internal power in (8.5.39) is rewritten as

$$\delta P_{HW}^{\text{int}} = \sum_e \left(\delta\alpha_C^e \, \tilde{\sigma}_C^e + \delta\left(\beta_B^e \tilde{B}_{BA}^e \dot{d}_A - \beta_B^e G_{BC}^e \alpha_C^e \right) \right) \tag{8.5.41}$$

where

$$\tilde{\sigma}_C^e = \int_{\Omega_e} N_{ijC}^D \sigma_{ij}(\bar{\mathbf{D}}) d\Omega = \int_{\Omega_e} N_{aC}^D \{\sigma_a(\bar{\mathbf{D}})\} d\Omega \quad \text{or} \quad \{\tilde{\sigma}_e\} = \int_{\Omega_e} (\mathbf{N}_D)^T \{\sigma(\bar{\mathbf{D}})\} d\Omega \tag{8.5.42}$$

$$\tilde{B}_{BA}^e = \int_{\Omega_e} N_{ijB}^\sigma B_{ijA} \, d\Omega = \int_{\Omega_e} N_{aB}^\sigma B_{aA} \, d\Omega \quad \text{or} \quad \tilde{\mathbf{B}}_e = \int_{\Omega_e} \mathbf{N}_\sigma^T \mathbf{B} \, d\Omega \tag{8.5.43}$$

$$G_{BC}^e = \int_{\Omega_e} N_{ijB}^\sigma N_{ijC}^D \, d\Omega = \int_{\Omega_e} N_{aB}^\sigma N_{aC}^D \, d\Omega \quad \text{or} \quad \mathbf{G}_e = \int_{\Omega_e} \mathbf{N}_\sigma^T \mathbf{N}_D \, d\Omega \tag{8.5.44}$$

Taking the variation of the second term in (8.5.41) and writing out the resulting expression gives

$$\begin{aligned} 0 = \delta\Pi^{HW} &= \sum_e \Big[\delta\alpha_C^e \left(\tilde{\sigma}_C^e - G_{BC}^e \beta_B^e \right) + \delta\beta_B^e \left(\tilde{B}_{BA}^e \dot{d}_A - G_{BC}^e \alpha_C^e \right) \\ &\quad + \delta\dot{d}_A \tilde{B}_{KA} \beta_K^e \Big] - \delta\dot{d}_A \left(f_A^{\text{ext}} - M_{AB} \ddot{d}_B \right) \end{aligned} \tag{8.5.45}$$

Using the connectivity matrix to relate the element nodal velocities to the global nodal velocities by $\dot{d}_A^e = L_{AB}^e \dot{d}_B, \delta\dot{d}_A^e = L_{AB}^e \delta d_B$ (Section 2.5), we obtain from the arbitrariness of δd_A that the above yields

$$\sum_e L_{BA}^e \tilde{B}_{CB}^e \beta_C^e - f_A^{ext} + M_{AB} \ddot{d}_B = 0 \quad \forall A = n_{SD}(I-1) + i, \quad \text{where } (i, I) \notin \Gamma_{v_i} \tag{8.5.46}$$

where the RHS indicates that the degree-of-freedom A is not on a prescribed velocity boundary.

These equations can be written in a form similar to the single-field equations of motion (B4.3.1) if we define

$$f_A^{\text{int},e} = \tilde{B}_{BA}^e \beta_B^e \quad \text{or} \quad \mathbf{f}_e^{\text{int}} = \tilde{B}_e^T \beta_e \tag{8.5.47}$$

The global internal forces can be obtained from the above by standard column matrix assembly. The equations of motion are then:

$$f_A^{\text{int}} - f_A^{ext} + M_{AB} \ddot{d}_B = 0 \quad \text{or} \quad \mathbf{f}^{\text{int}} - \mathbf{f}^{ext} + \mathbf{M}\mathbf{a} = \mathbf{0} \tag{8.5.48}$$

The fields $\bar{\mathbf{D}}(\mathbf{X},t)$ and $\bar{\sigma}(\mathbf{X},t)$ can be obtained as follows. Invoking the arbitrariness of $\delta\alpha_A^e$ and $\delta\beta_A^e$ in (8.5.45) gives, respectively:

$$\tilde{\sigma}_C^e = G_{BC}^e \beta_B^e \quad \text{or} \quad \{\tilde{\boldsymbol{\sigma}}_e\} = \mathbf{G}_e^T \beta_e \tag{8.5.49}$$

$$\tilde{B}_{BA}^e \dot{d}_A^e = G_{BC}^e \alpha_C^e \quad \text{or} \quad \tilde{\mathbf{B}}_e \dot{\mathbf{d}}_e = \mathbf{G}_e \alpha_e \tag{8.5.50}$$

Equation (8.5.49) is the *discrete constitutive equation*, and (8.5.50) is the *discrete equation for the rate-of-deformation*. The assumed rate-of-deformation is obtained by combining (8.5.36) and (8.5.50):

$$\bar{D}_{ij} = N^{D}_{ijA}\left(G^{e}_{AB}\right)^{-1}\tilde{B}^{e}_{BD}\dot{d}_{D} \quad \text{or} \quad \{\bar{\mathbf{D}}(\xi,t)\} = \mathbf{N}^{D}(\xi)\mathbf{G}_{e}^{-1}\tilde{B}_{e}\dot{\mathbf{d}}_{e}(t) \tag{8.5.51}$$

Box 8.1 gives the algorithm for computing the nodal internal forces in the element. It can be seen by comparing Box 8.1 with Box 4.3 that substantially more computations are required than for a single-field element. Therefore, elements based on this multi-field weak form are seldom used in large-scale computations.

Box 8.1 Internal force calculation in mixed element

1. Obtain α_e from $\dot{\mathbf{d}}_e$ by solving (8.5.50).
2. Evaluate the rate-of-deformation $\bar{\mathbf{D}}$ by (8.5.51).
3. Compute the stress $\sigma(\bar{\mathbf{D}})$ by the constitutive equation.
4. Compute $\{\tilde{\sigma}_e\}$ by (8.5.42).
5. Compute $\boldsymbol{\beta}_e$ from (8.5.49): $\beta_e = \mathbf{G}^{-1}\{\tilde{\sigma}_e\}$.
6. Compute nodal internal forces by (8.5.47).

8.5.8 Simo–Hughes B-Bar Method

Simo and Hughes (1986) have presented a technique which significantly simplifies the implementation of assumed strain elements. In this implementation, the assumed velocity strain is expressed directly in terms of the nodal velocities (which is often difficult). The velocity strains are then

$$\{\bar{\mathbf{D}}\} = \bar{\mathbf{B}}\dot{\mathbf{d}} \quad \text{or} \quad \bar{D}_{ij} = \bar{B}_{ijA}\dot{d}_{A} \tag{8.5.52}$$

The $\bar{\mathbf{B}}$ matrix plays a similar role as $\mathbf{N}_D$ in (8.5.36): it is a velocity strain interpolant, but it is expressed in terms of the nodal velocities.

A key step in the development of this method is that the assumed stresses are assumed to be orthogonal to the difference between the assumed velocity strain and the symmetric gradient of the velocity:

$$\int_{\Omega} \bar{\boldsymbol{\sigma}} : (\mathbf{D} - \bar{\mathbf{D}})d\Omega = 0 \quad \text{so} \quad \int_{\Omega} \bar{\boldsymbol{\sigma}} : (\mathbf{B}\dot{\mathbf{d}} - \bar{\mathbf{B}}\dot{\mathbf{d}})d\Omega = 0 \quad \forall \dot{\mathbf{d}} \tag{8.5.53}$$

The internal power in the Hu–Washizu weak form (8.5.1) is then

$$\delta P^{\text{int}}_{HW} = \int_{\Omega} \delta\bar{D}_{ij}\sigma_{ij}(\bar{\mathbf{D}})d\Omega = \delta\dot{d}_{A}\int_{\Omega} \bar{B}_{ijA}\sigma_{ij}(\bar{\mathbf{D}})d\Omega \tag{8.5.54}$$

because the second term in (8.5.2) vanishes by (8.5.53). From the definition of the internal nodal forces, we obtain

$$f_A^{\text{int}} = \int_\Omega \bar{B}_{ijA}\sigma_{ij}(\bar{\mathbf{D}})d\Omega \quad \text{or} \quad \mathbf{f}^{\text{int}} = \int_\Omega \bar{\mathbf{B}}^T\{\sigma(\bar{\mathbf{D}})\}d\Omega \tag{8.5.55}$$

The above formula for the internal nodal forces is identical to that in the single-field method described in Chapter 4, except that the **B** matrix has been replaced by an assumed strain matrix $\bar{\mathbf{B}}$ and the stress is a function of the assumed strain. This implementation, often called the B-bar form, was pioneered by Hughes (1987). It is striking in its simplicity and elegance: all of the equations are identical to the single-field finite element form but the kinematic fields can be designed to avoid locking as with the three-field Hu–Washizu principle. An interesting point about this formulation is that its construction requires an orthogonal stress field $\bar{\boldsymbol{\sigma}}$, but this stress is never used. Because the entire space of continuous function is available for the construction of $\bar{\boldsymbol{\sigma}}$, it can always be constructed. But since it is never used, we do not need to construct it!

For the total Lagrangian formulation, the second term in (8.5.15) is assumed to vanish and the corresponding B-bar equations for the nodal forces and rate of Green strain are

$$f_A^{\text{int}} = \int_{\Omega_0} \bar{B}_{ijA}^0 S_{ij} d\Omega_0 \quad \text{or} \quad \mathbf{f}^{\text{int}} = \int_{\Omega_0} \bar{\mathbf{B}}_0^T\{\mathbf{S}(\bar{\mathbf{E}})\}d\Omega_0 \tag{8.5.56}$$

$$\dot{\bar{E}}_{ij} = \bar{B}_{ijA}^0 \dot{d}_A \quad \text{or} \quad \dot{\bar{\mathbf{E}}} = \bar{\mathbf{B}}_0\dot{\mathbf{d}} \tag{8.5.57}$$

where $\bar{\mathbf{E}}$ is the assumed Green strain and $\bar{\mathbf{B}}_0$ is the assumed strain counterpart of (4.9.20).

8.5.9 Simo–Rifai Formulation

Simo and Rifai (1990) have given an implementation of the three-field weak form which they call an enhanced element formulation. The basic idea is that they modified (or enhanced) the velocity strain field **D** rather than constructing an entirely new assumed velocity strain field. This facilitates the development of higher-order three-field elements. The assumed velocity strain field is

$$\bar{\mathbf{D}} = \mathbf{D} + \mathbf{D}^{\text{enh}} \equiv \mathbf{D} + \tilde{\mathbf{D}} \tag{8.5.58}$$

where $\mathbf{D}^{\text{enh}} \equiv \tilde{\mathbf{D}}$ is called the enhanced velocity strain field. Substituting (8.5.58) into (8.5.1) we obtain the following three-field weak form:

$$0 = \delta\Pi^{HW}(\mathbf{v}, \tilde{\mathbf{D}}, \bar{\sigma}) = \int_\Omega (\delta\mathbf{D} + \delta\tilde{\mathbf{D}}) : \sigma(\mathbf{D} + \tilde{\mathbf{D}})d\Omega - \int_\Omega \delta(\bar{\sigma} : \tilde{\mathbf{D}})d\Omega - \delta P^{ext} + \delta P^{\text{kin}} \tag{8.5.59}$$

It can be shown that the strong form corresponding to the above consists of the generalized momentum balance, (8.5.4–5), and (8.5.8), while (8.5.7) is replaced by $\mathbf{D}^{\text{enh}} \equiv \tilde{\mathbf{D}} = \mathbf{0}$. This outcome is at first puzzling: the enhanced velocity strain vanishes in the strong form! However, in a discretization, it will not vanish; when properly designed, it improves the element.

In the implementation of this weak form, the velocity field is approximated as before by (8.5.35). The enhanced velocity strain field is expressed in terms of unknown parameters α_A by

$$\tilde{D}_{ij}(\xi, t) = N_{ijA}^D(\xi)\alpha_A(t) \tag{8.5.60}$$

with a similar test function. In the discretization procedure, it is assumed that the assumed stresses are orthogonal to the enhanced velocity strains $\tilde{\mathbf{D}}$ as in the Simo–Hughes implementation. Substituting the above and (8.5.35) into (8.5.59) and invoking the orthogonality then gives (for the equilibrium problem)

$$\int_{\Omega}\left(\delta\dot{d}_A B_{ijA} + \delta\alpha_A N_{ijA}^D\right)\sigma_{ij}(\bar{\mathbf{D}})d\Omega - \delta\dot{d}_A f_A^{\text{ext}} = 0 \tag{8.5.61}$$

where f_A^{ext} are the external nodal forces. The resulting discrete equations are then

$$\int_{\Omega} B_{ijA}\sigma_{ij}(\bar{\mathbf{D}})d\Omega - f_A^{\text{ext}} = 0, \quad \int_{\Omega} N_{ijA}^D\sigma_{ij}(\bar{\mathbf{D}})d\Omega = 0 \tag{8.5.62}$$

As can be seen from the previous, the nodal internal forces are identical to the single field form in Box 4.3 except that the stresses are functions of the modified velocity strains. The enhanced field can be element specific, so that the RHS equation in (8.5.62) is element specific. However, the equations for the parameters α_A are nonlinear when the constitutive law is nonlinear. Therefore, finding them involves the solution of nonlinear equations for each element.

The cognomen 'enhanced strain element' is a little misleading. As we will see from the Example 8.1 and the limitation principles, the three-field approach cannot enhance a strain, it can only suppress certain terms in the velocity strain. Moreover, since the higher-order terms in the velocity strain cannot be represented easily, this approach cannot completely suppress the terms that need to be suppressed. For example, in Q4 it is desirable to suppress the nonconstant part of the shear. The Simo–Rifai approach does not suppress this term completely when the element is not rectangular. On the other hand, in elements of order greater than Q4, the Simo–Rifai approach can provide significant savings in computational effort.

Example 8.1 Two-node rod element We consider a 2-node rod element in one dimension with linear velocity and constant velocity strain and stress fields. The element is illustrated in Figure 2.4. The cross-sectional area of the element is A and its length is ℓ.

The rod is assumed to be in a state of uniaxial stress, so the only nonzero components of interest are $v_x(\xi, t)$, $D_{xx}(\xi, t)$ and $\sigma_{xx}(\xi, t)$. Using linear approximations for these variables gives

$$v_x(\xi,t) = [1-\xi, \xi]\begin{Bmatrix} v_{x1}(t) \\ v_{x2}(t) \end{Bmatrix} = \mathbf{N}\dot{\mathbf{d}}^e, \quad \xi = x/\ell = X/\ell_0 \tag{E8.1.1}$$

$$\bar{D}_{xx} = [1, \ \xi]\begin{Bmatrix} \alpha_1 \\ \alpha_2 \end{Bmatrix} = \mathbf{N}_D\alpha, \quad \bar{\sigma}_{xx} = [1, \ \xi]\begin{Bmatrix} \beta_1 \\ \beta_2 \end{Bmatrix} = \mathbf{N}_\sigma\beta \tag{E8.1.2}$$

The superscripts e have been dropped since all matrices pertain to a single element. The $\mathbf{B}$ matrix is identical to that of the single-field element with a linear velocity field:

$$\mathbf{B} = \frac{\partial}{\partial x}[N] = \frac{\partial}{\partial x}[1-\xi, \ \xi] = \frac{1}{\ell}[-1, \ 1] \tag{E8.1.3}$$

The $\tilde{\mathbf{B}}$ and $\mathbf{G}$ matrices are obtained by (8.5.43) and (8.5.44):

$$\tilde{\mathbf{B}} = \int_{\Omega} \mathbf{N}_{\sigma}^{T} \mathbf{B} d\Omega = \int_{0}^{1} \begin{Bmatrix} 1 \\ \xi \end{Bmatrix} \frac{1}{\ell} [-1, \quad 1] A \ell d\xi = \frac{A}{2} \begin{bmatrix} -2 & +2 \\ -1 & +1 \end{bmatrix}$$
$$\mathbf{G} = \int_{\Omega} \mathbf{N}_{\sigma}^{T} \mathbf{N}_{D} d\Omega = \int_{0}^{1} \begin{Bmatrix} 1 \\ \xi \end{Bmatrix} [1, \quad \xi] A \ell d\xi = \frac{A\ell}{6} \begin{bmatrix} 6 & 3 \\ 3 & 2 \end{bmatrix} \tag{E8.1.4}$$

$$\mathbf{G}^{-1} = \frac{2}{A\ell} \begin{bmatrix} 2 & -3 \\ -3 & 6 \end{bmatrix} \tag{E8.1.5}$$

It can be seen from (8.5.50) that

$$\boldsymbol{\alpha} = \mathbf{G}^{-1} \tilde{\mathbf{B}} \dot{\mathbf{d}} \quad \text{or} \quad \begin{Bmatrix} \alpha_1 \\ \alpha_2 \end{Bmatrix} = \frac{1}{\ell} \begin{bmatrix} -1 & +1 \\ 0 & 0 \end{bmatrix} \begin{Bmatrix} v_{x1} \\ v_{x2} \end{Bmatrix} \tag{E8.1.6}$$

so that α_2 always vanishes, that is, the assumed velocity strain is constant. Thus enriching or enhancing the velocity strain with terms which are absent from the gradient of the velocity field has no effect whatsoever.

The nodal internal forces are given by (8.5.47):

$$\mathbf{f}^{\text{int}} = \tilde{\mathbf{B}}^{T} \boldsymbol{\beta} = \tilde{\mathbf{B}}^{T} \mathbf{G}^{-1} \tilde{\boldsymbol{\sigma}} = \frac{1}{\ell} \begin{bmatrix} -1 & 0 \\ +1 & 0 \end{bmatrix} \int_{\Omega} \begin{Bmatrix} 1 \\ \xi \end{Bmatrix} \sigma_{xx}(\bar{D}_{xx}) d\Omega \tag{E8.1.7}$$

Since the second column of this matrix vanishes, β_2 has no effect, that is, the linear term in the stress interpolant has no effect. The nodal internal forces thus depend only on the mean value of the stress. If the stress is constant, then the expression for the internal nodal forces becomes

$$\begin{Bmatrix} f_{x1} \\ f_{x2} \end{Bmatrix}^{\text{int}} = A\sigma_{xx} \begin{Bmatrix} -1 \\ +1 \end{Bmatrix} \tag{E8.1.8}$$

which is identical to the expression developed in Chapter 2.

Thus the element obtained by this multi-field variational approach is identical to that obtained by the single-field weak form. The reason for this lies in the choice of velocity and assumed velocity strain. If the assumed velocity strain includes all terms in the gradient of the velocity field, then the mixed weak form will lead to the same element as the single-field weak form. *Adding terms to the velocity strain approximation beyond the gradient of the velocity has no effect* for a linear constitutive law. This result is proven by Stolarski and Belytschko (1987) and is called a limitation principle. The limitation principle for the two-field variational principle was discovered by Fraeijs de Veubeke (1965); an elegant generalization of limitation principles for multi-field variational principles is given by Alfano and de Sciarra (1996). Because of the limitation principles, it is clear that the benefits of the mixed variational approach are restricted to those elements where something can be gained by removing part of the strain field; adding to the strain field confers no benefits.

8.6 Multi-Field Quadrilaterals

In the following, multi-field quadrilaterals are developed by the assumed strain method. The velocity-strain fields are designed so as to avoid volumetric locking and locking in bending due to shear, often called shear locking. An element which locks is useless, so the elimination of locking is imperative.

8.6.1 *Assumed Velocity Strain to Avoid Volumetric Locking*

In the following, all components are assumed in a corotational system that is aligned with the element coordinate base vectors as described in Section 9.5.6. The velocity-strain field associated with the velocity field (8.4.18) is

$$
\begin{aligned}
\{\mathbf{D}\} &= \left\{\begin{matrix} v_{x,x} \\ v_{y,y} \\ v_{x,y} + v_{y,x} \end{matrix}\right\} = \begin{bmatrix} \mathbf{b}_x^T + h,_x \boldsymbol{\gamma}^T & 0 \\ 0 & \mathbf{b}_y^T + h,_y \boldsymbol{\gamma}^T \\ \mathbf{b}_y^T + h,_y \boldsymbol{\gamma}^T & \mathbf{b}_x^T + h,_x \boldsymbol{\gamma}^T \end{bmatrix} \left\{\begin{matrix} \mathbf{v}_x \\ \mathbf{v}_y \end{matrix}\right\} \\
&= \left\{\begin{matrix} D_{xx}^c + \dot{q}_x h,_x \\ D_{yy}^c + \dot{q}_y h,_y \\ 2D_{xy}^c + \dot{q}_x h,_y + \dot{q}_y h,_x \end{matrix}\right\} \quad \text{where} \quad \dot{q}_i = \boldsymbol{\gamma}^T \mathbf{v}_i
\end{aligned}
\tag{8.6.1}
$$

and the superscript 'c' indicates the constant part of the velocity-strain field.

In Section 8.4.3, we explained why Q4 locks with 2×2 quadrature for incompressible materials. We showed that locking is due to the dilatational field associated with the hourglass modes. From (8.6.1), it can be seen that the hourglass modes induce the nonconstant part of the extensional velocity strains.

In constructing a velocity strain interpolant which will not lock for incompressible materials, we then have two alternatives:

1. The nonconstant terms of the first two rows of (8.6.1) can be dropped.
2. The first two rows can be modified so that no volumetric velocity strains occur in the hourglass modes.

The first alternative leads to the assumed velocity strain

$$
\{\bar{\mathbf{D}}\} = \left\{\begin{matrix} D_{xx}^c \\ D_{yy}^c \\ 2D_{xy}^c + \dot{q}_x h,_y + \dot{q}_y h,_x \end{matrix}\right\}
\tag{8.6.2}
$$

In the second alternative, the assumed velocity strain field is given by

$$\{\bar{\mathbf{D}}\} = \begin{Bmatrix} D^c_{xx} + \dot{q}_x h,_x - \dot{q}_y h,_y \\ D^c_{yy} + \dot{q}_y h,_y - \dot{q}_x h,_x \\ 2D^c_{xy} + \dot{q}_x h,_y + \dot{q}_y h,_x \end{Bmatrix} \tag{8.6.3}$$

In (8.6.3), the higher-order part of the rate of dilatation $\bar{D}_{xx} + \bar{D}_{yy}$ vanishes in the hourglass mode regardless of the value of $\dot{q}_x$ and $\dot{q}_y$: $\bar{D}_{xx} + \bar{D}_{yy} = D^c_{xx} + D^c_{yy}$.

The question then arises as to which of the two alternatives, (8.6.2) or (8.6.3), is preferable. We will see that for elements which involve beam bending, the performance of the element can be improved strikingly by omitting the nonconstant part of the shear. A constant shear cannot be combined with the extensional strains in (8.6.2) because the strain field would then contain only three independent functions, and the element would be rank deficient. Therefore if the higher-order shear terms are to be suppressed, (8.6.3) is the better alternative. This gives

$$\{\bar{\mathbf{D}}\} = \begin{Bmatrix} D^c_{xx} + \dot{q}_x h,_x - \dot{q}_y h,_y \\ D^c_{yy} + \dot{q}_y h,_y - \dot{q}_x h,_x \\ 2D^c_{xy} \end{Bmatrix} \tag{8.6.4}$$

The above velocity-strain field corresponds to the 'Optimal Incompressible' or OI element in Belytschko and Bachrach (1986). This element performs well in beam bending problems when one set of element sides are parallel to the axis of the beam and the elements are not too distorted.

The performance of Q4 in bending can be enhanced even further for isotropic, elastic problems by using a strain field which depends on Poisson's ratio as follows:

$$\{\bar{\mathbf{D}}\} = \begin{Bmatrix} D^c_{xx} + \dot{q}_x h,_x - \bar{v}\dot{q}_y h,_y \\ D^c_{yy} + \dot{q}_y h,_y - \bar{v}\dot{q}_x h,_x \\ 2D^c_{xy} \end{Bmatrix} \quad \text{where } \bar{v} = \begin{cases} v & \text{for plane stress} \\ v/(1-v) & \text{for plane strain} \end{cases} \tag{8.6.5}$$

This is called the 'Quintessential Bending and Incompressible' element, or QBI, in Belytschko and Bachrach (1986). It has excellent accuracy in bending for linear elastic beams. For a rectangle this element corresponds to the incompatible element of Wilson *et al.* (1973).

A more general form of the assumed strain field (8.6.3) is:

$$\{\bar{\mathbf{D}}\} = \bar{\mathbf{B}}\dot{\mathbf{d}} \quad \text{where} \quad \bar{\mathbf{B}} = \begin{Bmatrix} \mathbf{b}^T_x + e_1 h,_x \boldsymbol{\gamma}^T & e_2 h,_y \boldsymbol{\gamma}^T \\ e_2 h,_x \boldsymbol{\gamma}^T & \mathbf{b}^T_y + e_1 h,_y \boldsymbol{\gamma}^T \\ \mathbf{b}^T_y + e_3 h,_y \boldsymbol{\gamma}^T & \mathbf{b}^T_x + e_3 h,_x \boldsymbol{\gamma}^T \end{Bmatrix} \tag{8.6.6}$$

where e_1, e_2, and e_3 are arbitrary constants. The entire family of assumed velocity strain fields for $e_1 = -e_2$ is isochoric.

The constants e_a can be considered unknowns at the element level and determined by (8.5.62). However, this entails the solution of equations at the element level for each element nodal force evaluation. In most cases, these unknowns can be ascertained adequately by inspection. For moderately refined meshes, the differences resulting from changes in e_i are insignificant if the field is isochoric and the shear vanishes.

8.6.2 Shear Locking and its Elimination

The effect of 'parasitic' shear is somewhat different than that of 'parasitic' volumetric strains. When volumetric locking occurs, the results completely fail to converge; with spurious shear, the solutions converge but rather slowly. Thus the term 'excessive shear stiffness' is probably more precise, but the term shear locking is often used.

To explain shear locking and its elimination by projection, we consider a beam in pure bending modeled by a single row of elements as shown in Figure 8.8. In pure bending, the moment is constant so the resultant shear $s=\int\sigma_{xy}dy$ must vanish, since by equilibrium the shear is the derivative of the moment: $s=m_{,x}$. However, since all elements deform in the x-hourglass mode in bending, $\dot{q}_x \neq 0$ and the shear is nonzero by (8.6.1).

To eliminate shear locking, the portion of the shear velocity strain due to the hourglass mode must be eliminated. This can be accomplished by letting $e_3=0$ in (8.6.6). In pure bending, the nodal displacements in the local coordinate systems shown in Figure 8.8 are given by $\mathbf{u}_{\hat{x}} = c\mathbf{h}$, where c is an arbitrary constant. The linear strain energy for arbitrary e_3 is

$$W_{\text{shear}} = \frac{1}{4}\mu e_3^2 c^2 H_{yy} = 0 \tag{8.6.7}$$

where μ is the shear modulus and

$$H_{ij} = \int_{\Omega} h_{,i} h_{,j}\, d\Omega \tag{8.6.8}$$

so it vanishes when $e_3=0$; parasitic shear in bending is thus eliminated by dropping the shear strain associated with the hourglass mode.

Table 8.1 lists the constants for (8.6.6) for the assumed strain elements considered here. Note that the fully integrated Q4 element corresponds to $e_1 = e_3 = 1$, $e_2=0$. The element ADS is an element in which the deviatoric velocity strain is assumed. SRI is the selective-reduced integration element described in Section 5.5. SRI stabilization cannot be derived by the

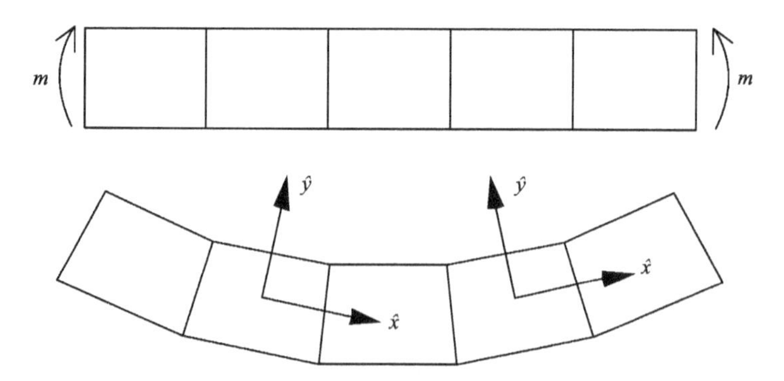

Figure 8.8 A beam in pure bending showing that the deformation is primarily in the hourglass mode

Table 8.1 Constants for assumed velocity-strain stabilization

Element	e_1	e_2	e_3	c_1	c_2	c_3
Q4 (plane strain) 2×2 quadrature	1	0	1	$\frac{1-\nu}{1-2\nu}$	1/2	$\frac{1}{2(1-2\nu)}$
Q4 (plane stress) 2×2 quadrature	1	0	1	$\frac{1}{1-\nu}$	1/2	$\frac{1+\nu}{2(1-\nu)}$
SRI				1	1/2	1/2
ASQBI	1	$-\bar{\nu}$	0	$1+\bar{\nu}$	0	$-\bar{\nu}(1+\bar{\nu})$
ASOI	1	−1	0	2	0	−2
ADS	1/2	−1/2	0	1/2	0	−1/2

assumed strain approach. ADS is another assumed velocity strain element with isochoric higher order fields based on assumed deviatoric velocity strains.

8.6.3 *Stiffness Matrices for Assumed Strain Elements*

The linear stiffness matrix for any of the assumed strain quadrilaterals is

$$\mathbf{K}_e = \mathbf{K}_e^{1pt} + \mathbf{K}_e^{\text{stab}} \tag{8.6.9}$$

where $\mathbf{K}_e^{1pt}$ is the one-point quadrature stiffness, Belytschko and Bindeman (1991). The quadrature point is $\xi=\eta=0$, and $\mathbf{K}_e^{\text{stab}}$ is the rank 2 stabilization stiffness, which is given by

$$\mathbf{K}_e^{\text{stab}} = 2\mu \begin{bmatrix} (c_1 H_{xx} + c_2 H_{yy})\boldsymbol{\gamma\gamma}^T & c_3 H_{xy}\boldsymbol{\gamma\gamma}^T \\ c_3 H_{xy}\boldsymbol{\gamma\gamma}^T & (c_1 H_{yy} + c_2 H_{xx})\boldsymbol{\gamma\gamma}^T \end{bmatrix} \tag{8.6.10}$$

where the constants c_1, c_2, and c_3 are given by Table 8.1. It can be surmised from the above that the Q4 element will lock in plane strain for nearly incompressible materials since c_1 and c_2 get very large as $\nu \rightarrow ½$.

8.6.4 *Other Techniques in Quadrilaterals*

The previous account does not follow a historical path and some of the key developments are not mentioned. The concept of element technology originated in Wilson *et al.* (1973) who was the first to improve the performance of the 4-node quadrilateral by adding incompatible modes. The effect of these modes is identical to suppressing part of the strain fields by multi-field methods. Incompatible modes always pose a pedagogical dilemma, since a teacher first stresses the importance of compatible fields in developing the single-field finite elements and then has to dismantle this pedagogical structure by introducing incompatible modes. Furthermore, incompatible modes pass the patch test for arbitrary meshes only if the quadrature is adjusted. As Strang and Fix (1973) said: ‘two wrongs do make a right in California’ (they were referring to the incorrect incompatible modes combined with a modified quadrature). By the time one gets through with incompatible modes, the element is a Rube Goldberg contraption; the design of elements is much cleaner with assumed strain methods.

An important element which was developed in the two-field format with an assumed strain is the Pian–Sumihara (1985) element. This element uses the following assumed stress field:

$$\sigma_{\xi\xi} = \beta_1 + \beta_2\eta, \quad \sigma_{\eta\eta} = \beta_3 + \beta_4\xi, \quad \sigma_{\xi\eta} = \beta_5$$

where $\sigma_{\xi\xi}$, $\sigma_{\eta\eta}$ and $\sigma_{\xi\eta}$ are the covariant components of the stress; the concept appears earlier in Wempner *et al.* (1982). The element must use a single orientation of the curvilinear coordinate system in transforming to the covariant components; otherwise, it fails the patch test. Since the shear stress field is constant, the element works well in beam bending.

This idea of approximating curvilinear components is appealing to some workers because the element is then frame-invariant: the stiffness is the same regardless of how the element is aligned relative to the coordinate system. If the stresses and strains are expressed in terms of the corotational coordinate system as in Section 9.5, the element is also frame-invariant. But these curvilinear approaches also have disadvantages. Assumed strain or stress fields based on curvilinear components require many transformations between the curvilinear and physical components. The rate of convergence is not improved. Thus the advantages of approximating the curvilinear components are not substantive.

8.7 One-Point Quadrature Elements

8.7.1 Nodal Forces and B-Matrix

We have already seen in Section 8.3 that when one-point quadrature is used with Q4, the element is rank deficient. For large-scale calculations, one-point quadrature elements are attractive because of their speed and accuracy. However, one-point quadrature elements require stabilization. Prior to describing these stabilization procedures, we examine the one-point quadrature element.

The internal nodal forces are given by (4.5.23) with the quadrature point corresponding to the origin of the coordinate system in the reference plane

$$\mathbf{f}^{\text{int}} = 4\boldsymbol{B}^T(\mathbf{0})\boldsymbol{\sigma}(\mathbf{0})J_\xi(\mathbf{0}) = A\boldsymbol{B}^T(\mathbf{0})\boldsymbol{\sigma}(\mathbf{0}) \tag{8.7.1}$$

In Voigt notation

$$\mathbf{f}^{\text{int}} = A\mathbf{B}^T(\mathbf{0})\{\boldsymbol{\sigma}(\mathbf{0})\} = A\begin{bmatrix} \mathbf{b}_x & \mathbf{0} & \mathbf{b}_y \\ 0 & \mathbf{b}_y & \mathbf{b}_x \end{bmatrix}\begin{Bmatrix} \sigma_{xx} \\ \sigma_{yy} \\ \sigma_{xy} \end{Bmatrix} \tag{8.7.2}$$

The assumed rate-of-deformation at the quadrature point is given by

$$\{\mathbf{D}(\mathbf{0})\} = \mathbf{B}(\mathbf{0})\dot{\mathbf{d}} \tag{8.7.3}$$

8.7.2 Spurious Singular Modes (Hourglass)

We next examine the structure of the spurious singular modes of the Q4 element. Any motion that is not a rigid body motion and results in no straining of the element is a spurious singular mode. Consider the nodal velocities

$$(\dot{\mathbf{d}}^{Hx})^T = \left[\mathbf{v}_x^T \quad \mathbf{v}_y^T\right] = [\mathbf{h}^T \, \mathbf{0}] (\dot{\mathbf{d}}^{Hy}) = \left[\mathbf{v}_x^T \quad \mathbf{v}_y^T\right] = [\mathbf{0} \quad \mathbf{h}^T] \tag{8.7.4}$$

where, as before, $\mathbf{h}^T = [+1, -1, +1, -1]$. It can easily be verified that $\mathbf{b}_x^T\mathbf{h} = 0, \mathbf{b}_y^T\mathbf{h} = 0$. Therefore, it follows that $\mathbf{B}(\mathbf{0})\dot{\mathbf{d}}^{Hx} = \mathbf{0}$ and $\mathbf{B}(\mathbf{0})\dot{\mathbf{d}}^{Hy} = \mathbf{0}$ that ist the velocity strain vanishes at the quadrature point for these modes.

Figure 8.9 shows the spurious modes in a rectangular element; the modes are shown individually on the left while the deformation due to both modes is shown at the right. A mesh in hourglass mode is shown in Figure 8.10. As can be seen, a vertical pair of elements in this mode looks like an hourglass, a device for measuring time by the flow of sand from the top to the bottom. For this reason, this spurious singular mode is often called the hourglass mode or hourglassing. The spurious mode in Figure 8.10 is called an x-hourglass because it involves motion only in the x-direction.

The hourglass mode, as can be seen in Figure 8.10, is communicable. This means that every element can go into the hourglass mode without any strain in any element. No energy is absorbed by the mode and the mode spreads like a communicable disease. When the model is constrained by boundary conditions, it is often not possible for the hourglass mode to develop without straining at least a few elements. However, the stiffness of the global hourglass mode then is still very small, and the associated frequency is very low (much lower than the true lowest frequency). Hourglassing is a spatial instability, like the advection–diffusion instability described in Chapter 7.

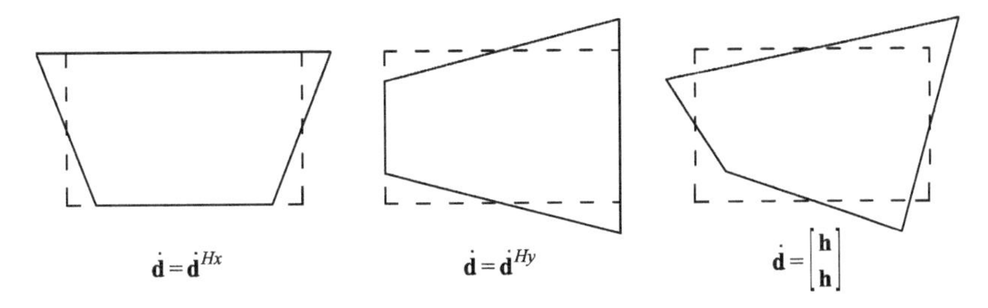

Figure 8.9 Hourglass modes of deformation in a quadrilateral element

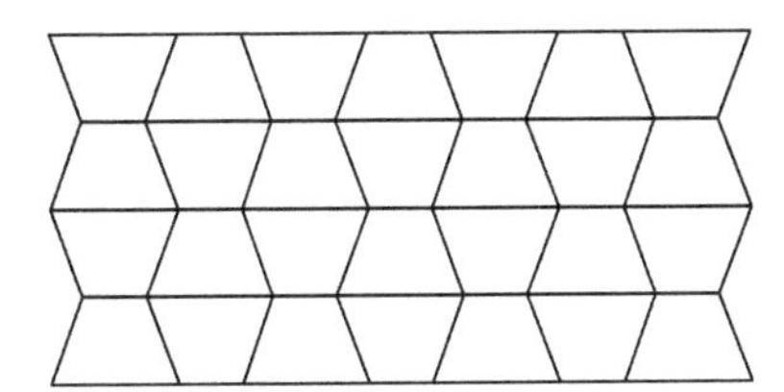

Figure 8.10 Mesh in hourglass mode of deformation

Hourglassing first appeared in finite difference hydrodynamics, where the derivatives were evaluated by transforming them to contour integrals; see Wilkins and Blum (1975). This procedure tacitly assumed that the derivatives are constant in the domain enclosed by each contour. This leads to finite difference equations which are equivalent to quadrilateral finite elements with one-point quadrature. This equivalence was demonstrated by Belytschko, Kennedy and Schoeberle (1975); also see Belytschko (1983). Many *ad hoc* procedures for hourglass control were developed by finite difference researchers, such as control based on the relative rotations of element sides; however, these methods did not to maintain completeness.

This singularity of discrete models due to rank deficiency occurs in many other settings, so a variety of names have evolved. For example, they occur frequently in mixed or hybrid elements, where they are called zero-energy modes or spurious zero-energy modes. Hourglass modes are zero-energy modes, since in these modes the strain vanishes at the quadrature points. Therefore they do no work in the discrete model and $\dot{\mathbf{d}}_{Hx}^T \mathbf{f}^{\text{int}} = \dot{\mathbf{d}}_{Hy}^T \mathbf{f}^{\text{int}} = 0$. In structural analysis, spurious singular modes occur when the redundancy is insufficient, that is, the number of structural members or supports is insufficient to preclude rigid body motion of part of the structure. Such modes often occur in three-dimensional truss models. In structural analysis, they are called kinematic modes, and because of the close relationship between the structural and finite element communities, this name has also been applied to spurious singular modes. Other names which have been applied to this phenomenon are keystoning, chickenwiring, and mesh instability.

For finite element discretizations of partial differential equations, spurious singular mode appears to be the most precise name. For example, the names kinematic modes and zero-energy modes are not appropriate for the Laplace equation. In elements where the spurious singular mode has a distinctive appearance, such as the hourglass pattern in Q4, we shall also use that name. Spurious singular modes are a manifestation of the rank deficiency of the element stiffness.

The evolution of an hourglass mode in a transient problem is shown in Figure 8.11. In this problem, the beam was supported at a single node to facilitate the appearance of the hourglass

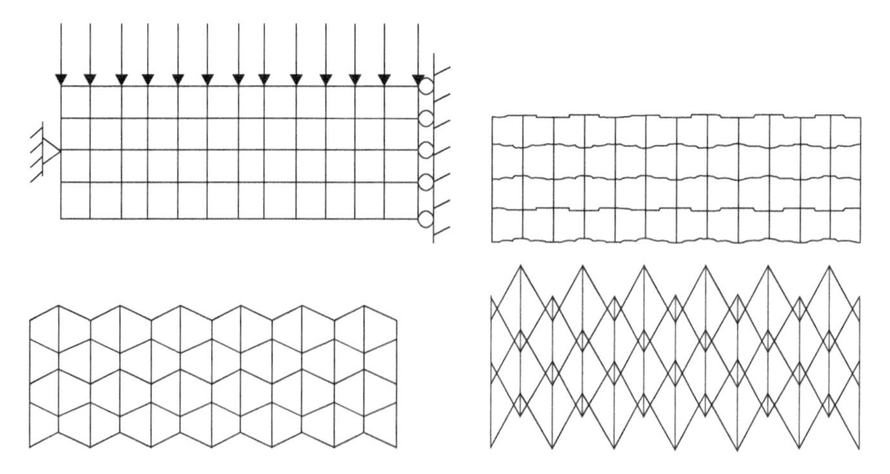

Figure 8.11 Four snapshots of a simply supported beam showing the evolution of the hourglass mode (due to symmetry, only half the beam was modeled)

mode. If all nodes at the left-hand end of the beam were fixed to simulate a clamped support, the hourglass mode would not appear. However, for large meshes and nonlinear materials, they could reappear. Although rank-deficient elements may sometimes appear to be stable, they should not be used without an appropriate stabilization.

8.7.3 Perturbation Hourglass Stabilization

In perturbation hourglass control, a small correction is added to the discrete system to restore the correct rank of the element. It is important to augment the rank without disturbing the linear completeness of the isoparametric element. One approach is to augment the **B** matrix of the one-point quadrature element by two rows which are orthogonal to the other three. The orthogonality ensures that the additional rows are linearly independent of the first three and that the correction terms do not affect the response to linear fields, so the that the linear completeness of **B** is not lost.

The additional two rows of the **B** matrix are chosen to be the $\boldsymbol{\gamma}$ vector given in (8.4.11), which is orthogonal to all linear fields. Adding these two rows corresponds to adding two generalized strains. The element matrices for the one-point quadrature Q4 with a hypoelastic law are then

$$\tilde{\mathbf{B}} = \begin{bmatrix} \mathbf{b}_x^T & \mathbf{0} \\ \mathbf{0} & \mathbf{b}_y^T \\ \mathbf{b}_y^T & \mathbf{b}_x^T \\ \boldsymbol{\gamma}^T & \mathbf{0} \\ \mathbf{0} & \boldsymbol{\gamma}^T \end{bmatrix} [\tilde{\mathbf{C}}^{\sigma T}] = \begin{bmatrix} C_{11} & C_{13} & C_{13} & 0 & 0 \\ C_{21} & C_{22} & C_{23} & 0 & 0 \\ C_{31} & C_{32} & C_{33} & 0 & 0 \\ 0 & 0 & 0 & C^Q & 0 \\ 0 & 0 & 0 & 0 & C^Q \end{bmatrix} \tag{8.7.5}$$

$$\{\boldsymbol{\sigma}\} = [\sigma_x, \sigma_y, \sigma_{xy}, Q_x, Q_y]^T \{\mathbf{D}\} = [D_x, D_y, 2D_{xy}, \dot{q}_x, \dot{q}_y]^T \tag{8.7.6}$$

where $\tilde{\mathbf{C}}^{\sigma T}$ is the two-dimensional constitutive matrix augmented by two rows and columns which relate the generalized hourglass stress–strain pair; or $\tilde{\boldsymbol{\sigma}}$ and $\tilde{\mathbf{D}}$ are the stress and velocity-strain matrices augmented by the stabilization stresses and strains (Q_x, Q_y) and (q_x, q_y), respectively.

To satisfy linear completeness, the stabilization strains should vanish when the nodal velocities emanate from a linear field.

Since the **B(0)** already satisfies the linear reproducing conditions, it is necessary that $\dot{q}_x = \dot{q}_y = 0$ for linear fields. This implies that

$$\boldsymbol{\gamma}^T \mathbf{v}_i^{\text{lin}} = \boldsymbol{\gamma}^T (\alpha_{i0} \mathbf{1} + \alpha_{i1} \mathbf{x} + \alpha_{i2} \mathbf{y}) = 0 \forall \alpha_{il} \tag{8.7.7}$$

The above can be interpreted as an orthogonality condition: $\boldsymbol{\gamma}$ must be orthogonal to all linear fields. In addition, to stabilize the element, $\boldsymbol{\gamma}$ and $\mathbf{b}_i$ must be linearly independent so that the rank of $\tilde{\mathbf{B}}$ is 5. The $\boldsymbol{\gamma}$ matrix given in (8.4.11), $\boldsymbol{\gamma} = \frac{1}{4}[\mathbf{h} - (\mathbf{h}^T \mathbf{x}_i)\mathbf{b}_i]$, satisfies these conditions. The orthogonality properties have been verified with (8.4.12–13). Some of the other conditions are left as exercises.

8.7.4 Stabilization Procedure

The nodal internal forces for the stabilized element are

$$\mathbf{f}^{\text{int}} = A\tilde{\mathbf{B}}^T(\mathbf{0})\{\tilde{\boldsymbol{\sigma}}\} = A\boldsymbol{B}^T(\mathbf{0})\boldsymbol{\sigma}(\mathbf{0}) + A\begin{Bmatrix} Q_x\boldsymbol{\gamma} \\ Q_y\boldsymbol{\gamma} \end{Bmatrix} = A\boldsymbol{B}^T(\mathbf{0})\boldsymbol{\sigma}(\mathbf{0}) + \mathbf{f}^{\text{stab}} \tag{8.7.8}$$

The first term in the internal force is that obtained by one-point quadrature. The second term defines the stabilization nodal force matrix $\mathbf{f}^{\text{stab}}$. The velocity strains and generalized hourglass stresses are obtained by

$$\{\mathbf{D}\} = \mathbf{B}(\mathbf{0})\dot{\mathbf{d}}, \quad \dot{Q}_x = C^Q\dot{q}_x, \quad \dot{Q}_y = C^Q\dot{q}_y \quad \text{where} \quad \dot{q}_x = \boldsymbol{\gamma}^T\mathbf{v}_x, \quad \dot{q}_y = \boldsymbol{\gamma}^T\mathbf{v}_y \quad (8.7.9\text{a, b})$$

One can also use a combination of stiffness and viscous hourglass control

$$Q_i = C^Q q_i + \xi_D C^Q \dot{q}_i \tag{8.7.10}$$

where ξ_D is the fraction of critical damping in the stabilization. Since $\boldsymbol{\gamma}$ is independent of linear fields, $\dot{q}_i = 0$ in rigid body rotation. Note that all of the above computations should be made in a corotational coordinate system. For a total Lagrangian formulation, a corotational system is not needed, though it is best to align the initial coordinate system with the elements' intrinsic coordinates to achieve element invariance. However, as mentioned before, element invariance has no effect on convergence.

For this stabilization, the linear stiffness matrix is:

$$\mathbf{K}_e = \mathbf{K}_e^{1pt} + C^Q A\begin{bmatrix} \boldsymbol{\gamma\gamma}^T & \boldsymbol{\gamma\gamma}^T \\ \boldsymbol{\gamma\gamma}^T & \boldsymbol{\gamma\gamma}^T \end{bmatrix} \tag{8.7.11}$$

The rank of the element stiffness is 5, which is the correct rank for the Q4.

8.7.5 Scaling and Remarks

Since the parameter C^Q in (8.7.9) is not a true material constant, it must be normalized to provide approximately the same degree of stabilization for any geometry and material properties. The objective is to obtain a scaling which perturbs the element sufficiently to ensure the correct rank but not to shift the solution too much from the one-point quadrature element, since it is convergent and does not lock volumetrically.

A procedure for selecting C^Q is given in Belytschko and Bindeman (1991). They select C^Q so the maximum eigenvalue of the stabilization stiffness is scaled to the maximum eigenvalue of the underintegrated stiffness. The hourglass frequencies in a fully integrated element are usually much higher than the frequencies of interest in a solution, and it would be desirable to shift the hourglass frequencies into this part of the spectrum. However, to avoid locking, the stabilization parameters cannot be large.

According to Flanagan and Belytschko (1981), the maximum eigenvalue of $\mathbf{K}_e\mathbf{z} = \lambda\mathbf{M}_e\mathbf{z}$ for the one-point quadrature element for an isotropic material is bounded by (see (6.4.64)):

$$Ac^2\mathbf{b}_i^T\mathbf{b}_i \le 2\lambda_{\max} \le 2Ac^2\mathbf{b}_i^T\mathbf{b}_i \tag{8.7.12}$$

where c is the elastic dilatational wavespeed. The eigenvalue associated with the hourglass mode can be estimated by a Raleigh quotient. We let the estimate of the eigenvector be $\mathbf{z}^T = [\mathbf{h}, \mathbf{0}]$. The corresponding eigenvalue estimate can be obtained by:

$$\lambda^h = \frac{\mathbf{z}^T \mathbf{K}_e \mathbf{z}}{\mathbf{z}^T \mathbf{M}_e \mathbf{z}} = \frac{AC^Q \mathbf{h}^T \boldsymbol{\gamma}\boldsymbol{\gamma}^T \mathbf{h}}{\mathbf{h}^T \mathbf{M}_e \mathbf{h}} = \frac{C^Q}{\rho} \tag{8.7.13}$$

where the second equality follows because $\mathbf{K}_e^{1pt}\mathbf{z} = \mathbf{0}$ and the last term follows from (8.4.11). An identical estimate would be obtained by using the y-hourglass mode in the Raleigh quotient. From (8.7.12) and (8.7.13), we find that the stabilization eigenvalue λ^h is scaled to the lower bound on the maximum eigenvalue of the element if

$$C^Q = \frac{1}{2}\alpha_s c^2 \rho A \mathbf{b}_i^T \mathbf{b}_i \tag{8.7.14}$$

where α_s is a scaling parameter. Recommended values of α_s are about 0.1. For static problems, a stabilization based on eigenvalues of $\mathbf{K}_e\mathbf{z} = \lambda\mathbf{z}$ may be more appropriate.

The recommended value of the stabilization parameter does not preclude the hourglass mode from appearing. When the data is rich in the hourglass mode, it is difficult to suppress the hourglass mode even with large values of the stabilization parameter. For example, point loads often cause hourglassing. Sometimes these nodal forces arise in poorly programmed algorithms. For example, one widely used program was notorious for hourglassing; it was found that the nodal forces in the contact algorithm alternated in sign. In such situations, prevention of the hourglass instability is almost impossible. But there are times when an hourglass mode appears inexplicably, particularly when materials are unstable or close to instability. It is best then to switch to a fully integrated element in those subdomains where it appears. However, if the material is nearly incompressible, a mixed element or selective reduced integration should be used, or you will cure hourglassing by locking the mesh!

The perturbation hourglass parameters should not be too large for nearly incompressible materials, since they then can cause locking. If large values of hourglass control are needed, physical hourglass control is recommended (see the next section). The ratio of the hourglass energy to the total energy should be monitored. If it is large (on the order of 3% or 5%), the results will be in error by the same order of magnitude. If the problem is large, the hourglass energy should be monitored on subdomains, just as we recommended for the energy balance in Section 6.2.3.

There is also the choice of whether to control the modes by viscous forces or by stiffness stabilization: see (8.7.10). Viscous stabilization is often needed when large impulsive loads are applied. For moderate deformations and long duration simulations, stiffness stabilization is better because some of the hourglass energy is recoverable. In many cases, a combination is best.

8.7.6 Physical Stabilization

Hourglass stabilization procedures have been developed on the basis of assumed strain methods. In these procedures, the stabilization parameters are based on the material properties and geometry of the element. This type of stabilization is also called physical hourglass control. These stabilization methods will not lock for incompressible materials even when the stabilization parameters are of order 1. In this section, physical hourglass control is developed for the Q4 element. In developing the physical hourglass control, two assumptions must be made:

1. the spin is constant within the element;
2. the material response is uniform within the element.

The assumed velocity strain fields for stabilization are identical to those in Section 8.6.1. In the remainder of the development, the special case, $e_1 = -e_2 = 1$ which corresponds to the OI element, is considered. The corotational components of the velocity strain are then given by

$$\begin{Bmatrix} \bar{D}_{\hat{x}} \\ \bar{D}_{\hat{y}} \\ 2\bar{D}_{\hat{x}\hat{y}} \end{Bmatrix} = \begin{bmatrix} \mathbf{b}_{\hat{x}}^T + h,_{\hat{x}} \boldsymbol{\gamma}^T & -h,_{\hat{y}} \boldsymbol{\gamma}^T \\ -h,_{\hat{x}} \boldsymbol{\gamma}^T & \mathbf{b}_{\hat{y}}^T + h,_{\hat{y}} \boldsymbol{\gamma}^T \\ \mathbf{b}_{\hat{y}}^T & \mathbf{b}_{\hat{x}}^T \end{bmatrix} \begin{Bmatrix} \mathbf{v}_{\hat{x}} \\ \mathbf{v}_{\hat{y}} \end{Bmatrix} = \bar{\mathbf{B}}\dot{\mathbf{d}} \tag{8.7.15}$$

This defines the $\bar{\mathbf{B}}$ matrix. The nodal internal force is given by (8.5.55)

$$\hat{\mathbf{f}}^{\text{int}} = \int_{\Omega} \bar{\mathbf{B}}^T \{\hat{\boldsymbol{\sigma}}\} d\Omega \tag{8.7.16}$$

For a hypoelastic material, the stress rate is then given by

$$\frac{\{\partial\hat{\boldsymbol{\sigma}}(\xi, t)\}}{\partial t} = [\hat{\mathbf{C}}^{\sigma T}]\{\bar{\mathbf{D}}(\mathbf{0})\}, \begin{Bmatrix} \dot{\hat{Q}}_x \\ \dot{\hat{Q}}_y \end{Bmatrix} = \begin{Bmatrix} (\hat{C}_{11} - \hat{C}_{12})(\dot{q}_{\hat{x}} h,_{\hat{x}} - \dot{q}_{\hat{y}} h,_{\hat{y}}) \\ (\hat{C}_{22} - \hat{C}_{21})(\dot{q}_{\hat{y}} h,_{\hat{y}} - \dot{q}_{\hat{x}} h,_{\hat{x}}) \end{Bmatrix} \tag{8.7.17}$$

where we have separated the stress into the part that is constant and the part which varies within the element; superscripts on $\mathbf{C}$ are dropped. It can be seen from (8.7.17) that the corotational stress rate always has the same distribution within the element, so the stress also has the same form.

Taking advantage of this form of the stress field, the orthogonality of $h,_i$ to the constant field, (8.4.19), and the fact that $\hat{\mathbf{C}}$ is constant in the element gives

$$\hat{\mathbf{f}}^{\text{int}} = A\bar{\mathbf{B}}^T(\mathbf{0})\{\hat{\boldsymbol{\sigma}}(\mathbf{0})\} + \mathbf{f}^{\text{stab}} \tag{8.7.18}$$

where $\hat{\mathbf{f}}^{\text{stab}}$ are the stabilization nodal forces, which are given by

$$\hat{\mathbf{f}}^{\text{stab}} = \begin{Bmatrix} Q_{\hat{x}} \hat{\boldsymbol{\gamma}} \\ Q_{\hat{y}} \hat{\boldsymbol{\gamma}} \end{Bmatrix} \quad \text{where} \quad \begin{Bmatrix} \dot{Q}_{\hat{x}} \\ \dot{Q}_{\hat{y}} \end{Bmatrix} = (e_1)^2 (\hat{C}_{11} - \hat{C}_{12} - \hat{C}_{21} + \hat{C}_{22}) \begin{Bmatrix} H_{\hat{x}\hat{x}} \dot{q}_{\hat{x}} - H_{\hat{x}\hat{y}} \dot{q}_{\hat{y}} \\ H_{\hat{y}\hat{y}} \dot{q}_{\hat{y}} - H_{\hat{x}\hat{y}} \dot{q}_{\hat{x}} \end{Bmatrix} \tag{8.7.19}$$

Note that the rates of the stabilization stresses are expressed in corotational components so that they are frame-invariant. The procedure for the element is summarized in Box 8.2.

In physical hourglass control, no parameters are needed for the hourglass stabilization. By assuming that the material response is homogeneous, it is possible to integrate (8.7.19) in closed form to obtain the stabilization forces. Since these forces are based on an assumed velocity strain field which does not lock, the element is well suited to incompressible materials.

Box 8.2 Element nodal force calculation

1. Update corotational coordinate system.
2. Transform nodal velocities $\mathbf{v}$ and coordinates $\mathbf{x}$ to corotational coordinate system.
3. Compute velocity strain at the quadrature point by (8.7.15).
4. Compute stress rate by constitutive law and update stress.
5. Compute generalized hourglass strain rates by (8.7.9b).
6. Compute the generalized hourglass stress rates by (8.7.10) and update the generalized hourglass stresses by integrating in time.
7. Compute $\hat{\mathbf{f}}^{\text{int}}$ by (8.7.18).
8. Transform $\hat{\mathbf{f}}^{\text{int}}$ to global system and assemble.

A similar element QBI has better performance in bending. The main shortcoming is that in some problems, the assumption of homogeneous material response is not satisfactory.

The following remarks apply. The stress rate will correspond to the Green–Naghdi rate if the rotation of the corotational coordinate system is properly defined. Deviations from the assumptions of constant spin and material properties are proportional to the strength of the hourglass modes for smooth materials; thus in h-adaptive methods, it is advantageous to refine those elements which exhibit substantial hourglass energy, as in Belytschko, Wong and Plaskacz (1989). For rough materials, such as elastic–plastic materials, substantial inhomogeneities in material response can occur even in the absence of an hourglass mode.

8.7.7 Assumed Strain with Multiple Integration Points

One-point quadrature is usually advantageous for speed; however, for nonsmooth stress fields, more integration points are sometimes useful. For example, for elastic beam problems, very accurate solutions can be obtained with only one element through the depth of the beam. However, for elastic–plastic beams, 4 to 10 elements are needed through the depth to obtain an accurate solution, because the stress is not smooth through the depth. The number of integration points can be increased by refining the mesh, or by increasing the number of integration points in each element. The latter has the advantage of increasing accuracy without reducing the stable time step.

The assumed strain fields developed in Section 8.6.1 can be used with any number of integration points without locking since the strain fields have zero dilatational strain throughout the element domain. The internal force vector for multi-point integration with an assumed strain field is

$$\mathbf{f}_e^{\text{int}} = \sum_{Q=1}^{n_Q} \bar{w}_Q J_\xi(\xi_Q) \bar{\mathbf{B}}^T(\xi_Q) \boldsymbol{\sigma}(\xi_Q, t) \tag{8.7.20}$$

where ξ_Q are the quadrature points.

Stabilization forces may be necessary depending on the location of the integration points. If we consider a rectangular $a \times b$ element with a corotational coordinate system, the referential axes are parallel to the corotational axes, so $\xi_{,\hat{x}} = 1/a, \eta_{,\hat{y}} = 1/b, \eta_{,\hat{x}} = \xi_{,\hat{y}} = 0$. From the preceding, it is apparent that $h_{,\hat{x}} = 0$ along the η axis and $h_{,\hat{y}} = 0$ along the ξ axis. Therefore if the integration points are all along one referential axes, stabilization is still needed in that direction to maintain rank sufficiency. Full 2×2 quadrature by (8.7.20) is rank sufficient, but nearly the same results are obtained with this two-point quadrature scheme:

$$\hat{f}_{iI}^{\text{int},e} = 2J_{\xi}(\mathbf{0})\sum_{Q=1}^{2} B_{Ij}(\xi_Q)\hat{\sigma}_{ji}(\xi_Q) \tag{8.7.21}$$

The two integration points are either $\boldsymbol{\xi}_1=(-3^{-1/2}, -3^{-1/2})$, $\boldsymbol{\xi}_2=(3^{-1/2}, 3^{-1/2})$, or $\boldsymbol{\xi}_1=(-3^{-1/2}, 3^{-1/2})$, $\boldsymbol{\xi}_2 = (3^{-1/2}, -3^{-1/2})$. The choice among the integration points makes little difference in the accuracy of the solution, but there will be small differences. This two-point integration scheme is similar to the IPS2 element reported in Liu, Chang and Belytschko (1988). The formulation here differs in that an assumed strain field is used to improve accuracy.

8.7.8 Three-Dimensional Elements

The concepts developed in the previous three sections can be extended to three-dimensional hexahedra, though there are some additional complications. The following development follows Belytschko and Bindeman (1993). The stabilization vectors for elements in 2D and 3D can be written in a general form

$$\gamma_{\alpha I} = \beta(h_{\alpha I} - (h_{\alpha I}x_{jI})b_{jI}) \quad \text{where range of } \alpha = \begin{cases} 1 \text{ in 2D} \\ 4 \text{ in 3D} \end{cases} \tag{8.7.21}$$

The velocity fields are

$$v_i(\xi, t) = \alpha_{i0}(t) + \alpha_{ij}(t)x_j(\xi) + \alpha_{i\alpha+3}(t)h_{\alpha}(\xi) \tag{8.7.23}$$

$$= (\mathbf{1} + \mathbf{b}_j x_j(\xi) + \gamma_{\alpha}h_{\alpha}(\xi))^T \mathbf{v}_i \tag{8.7.24}$$

$$h_{\alpha}(\xi) = [\eta\zeta \quad \zeta\xi \quad \xi\eta \quad \xi\eta\zeta], \quad \mathbf{b}_j = \mathbf{N}_{,j}(\mathbf{0}) \tag{8.7.25}$$

where the repeated Greek index is summed over a range of 4. The assumed velocity strain fields can be constructed using the same arguments as before. For example, an isochoric field with constant shear is given in Belytschko and Bindeman (1993):

$$\bar{\mathbf{B}}^T = \begin{bmatrix} \mathbf{b}_x + \frac{2}{3}\mathbf{g}_x^{1234} & -\frac{1}{3}\mathbf{g}_x^{1234} & -\frac{1}{3}\mathbf{g}_x^{1234} & \mathbf{g}_y^{12} & \mathbf{b}_z + \mathbf{g}_z^{13} & \mathbf{b}_y \\ -\frac{1}{3}\mathbf{g}_y^{1234} & \mathbf{b}_y + \frac{2}{3}\mathbf{g}_y^{1234} & -\frac{1}{3}\mathbf{g}_y^{1234} & \mathbf{b}_z + \mathbf{g}_x^{12} & \mathbf{0} & \mathbf{b}_x + \mathbf{g}_z^{23} \\ -\frac{1}{3}\mathbf{g}_z^{1234} & -\frac{1}{3}\mathbf{g}_z^{1234} & \mathbf{b}_z + \frac{2}{3}\mathbf{g}_z^{1234} & \mathbf{b}_y & \mathbf{b}_x + \mathbf{g}_x^{13} & \mathbf{g}_x^{23} \end{bmatrix} \tag{8.7.26}$$

where

$$\mathbf{g}_i^{1234} = h_{\alpha,i}\gamma_i \tag{8.7.27}$$

and $\mathbf{g}_i^{IJ}$ are defined by omitting the missing indices from the sum. The element is best used with a corotational coordinate system. It is easy to show that the assumed velocity strain field is isochoric for any nodal velocities, but there are many other combinations that lead to an isochoric field. The element structure in three dimensions is much more complex than in two dimensions. A theory as to the best structure has not yet been developed.

8.8 Examples

In the following, we illustrate the performance of assumed strain and fully integrated quadrilaterals in two dimensions. Our intent is to show how the performance of elements is improved by assumed strain fields for incompressible materials and beam bending. We also show that for problems that do not include these features, nothing is gained by an assumed strain element. The elements included in this study are:

1. FB: Flanagan–Belytschko perturbation hourglass control of stiffness type; the stiffness control parameter α_s is specified.
2. ASOI: assumed velocity strain element.
3. ASQBI, assumed quintessential bending.
4. ADS, assume deviatoric assumed strain.

All of the elements except the first are described by (8.6.6) and Table 8.1.

8.8.1 Static Problems

The displacement of the free end is reported in Table 8.2 for rectangular and skewed elements. The convergence of these elements for an elastic beam is shown in Figure 8.12 for an incompressible, isotropic elastic material in plane strain. It can be seen that all of the elements have the same rate of convergence except Q4 = QUAD4 with full 2 × 2 quadrature. The latter locks, as expected. For the nonlocking elements, the rate of convergence of the displacements in the L_2 norm is almost exactly 2.0 and the rate of convergence in the energy norm (which is almost equivalent to the H_1 norm) is 1.0. These are the rates of convergence expected for a linear complete element according to Section 8.2.3. ASQBI has the same accuracy as the Pian–Sumihara element. The Pian–Sumihara element is slightly more accurate than the other elements for the skewed case. Otherwise, the elements are comparable.

Our studies have shown that the ASQBI element with one-point quadrature provides accurate coarse mesh solutions for the elastic beam; however, the accuracy for the elastic–plastic

Table 8.2 Ratio of computed to analytical solution for displacement of elastic cantilever beam

Material	MESH	QUAD4	ASQBI, and Pian–Sumihara		ASOI	ADS
$\nu = 0.25$	rectangular	0.708	0.986		0.862	1.155
$\nu = 0.49999$	rectangular	0.061	0.982		0.982	1.205
$\nu = 0.25$	skewed	0.689	0.948	0.955	0.834	1.112
$\nu = 0.4999$	skewed	0.061	0.957	0.960	0.957	1.170

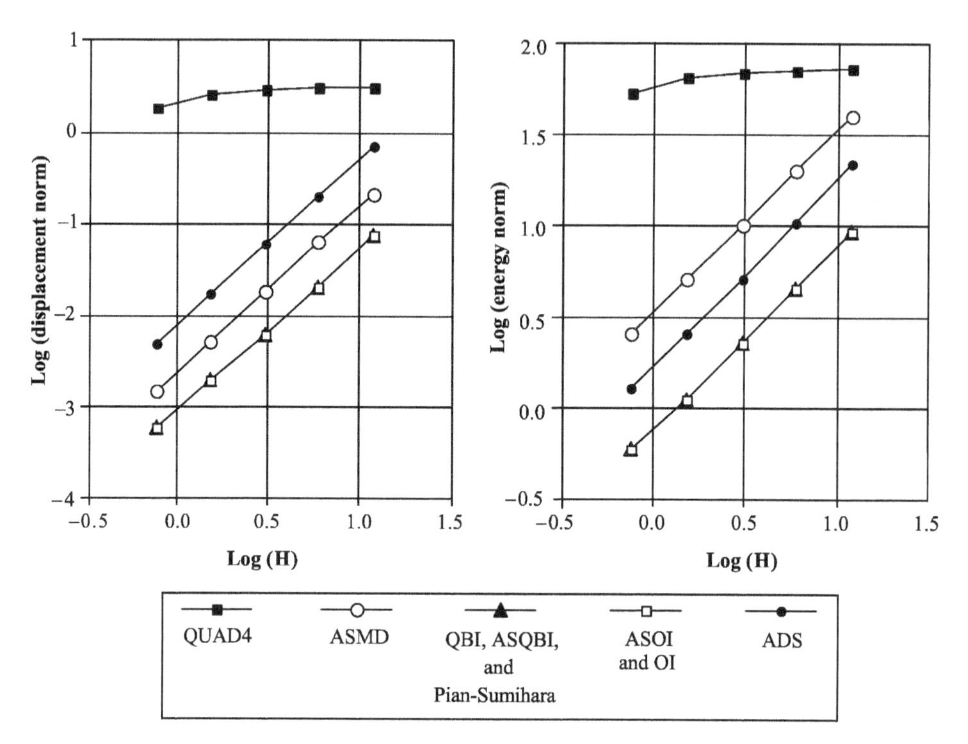

Figure 8.12 Convergence of displacement and energy error norms; Poisson's ratio $\nu=0.4999$, plane strain

beam is poor. The error in the elastic–plastic solution can be attributed to an insufficient number of quadrature points. This not surprising if we recall that the stress in a beam varies linearly through the depth, so yielding first occurs at the top and bottom surfaces and then spreads toward the midline. With one-point quadrature, the stress is sampled only at the center of the element, so the quadrature point does not reflect the stress field in the element adequately.

The two-point integration scheme (8.7.21) improves the accuracy by placing integration points near the edge of the element and increasing the number of sampling points. Even better coarse mesh accuracy can be obtained in bending by placing three to ten quadrature points along the normal to the axis of the beam. However, this requires the element to 'know' the geometry of the beam or the analyst to number the element nodes appropriately.

8.8.2 Dynamic Cantilever Beam

A cantilever beam in plane strain is shown in Figure 8.13 with both elastic and elastic–plastic materials. The material is isotropic with $\nu=0.25$, $E=1\times10^4$, and density $\rho=1$. A von Mises yield function and linear isotropic strain hardening with modulus $H=0.01E$ is used. The yield stress $\sigma_Y=300$. A similar problem is reported in Liu, Chang and Belytschko (1988).

The problem involves very large displacements, of the order of one-third the length of the beam. A solution for a mesh of 32×192 elements serves as a benchmark. The end displacements

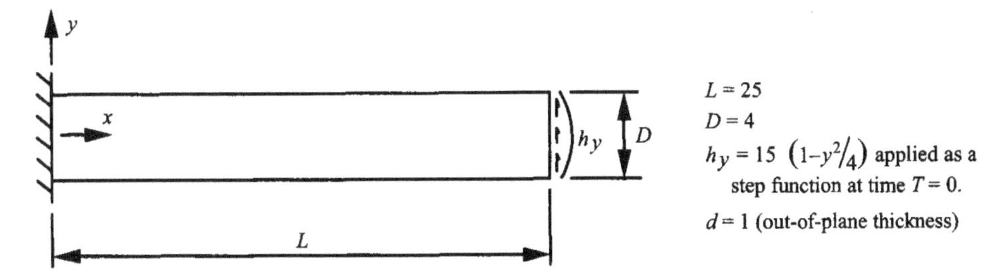

Figure 8.13 Dynamic cantilever beam problem

Table 8.3 Maximum end displacement for dynamic solution of elastic–plastic cantilever

Element	1×6	2×12	4×24	8×48	2×6	4×12
QUAD4 (2×2)	4.69	6.30	7.31	7.85	4.94	6.61
FB (0.1)	15.9	8.39	8.18	8.14	7.22	7.67
FB (0.3)	7.68	7.05	7.59	7.92	5.35	6.69
ASOI	4.78	6.17	7.17	7.76	6.11	7.00
ASQBI	6.89	6.86	7.54	7.90	6.79	7.34
ASQBI (2×2)	6.98	7.52	7.86	8.05	7.27	7.68
ASQBI (2-pt)	7.00	7.53	7.87	8.06	7.28	7.69
ADS	14.2	8.15	8.12	8.12	7.94	7.94
ASSRI	6.05	6.63	7.42	7.86	5.23	6.60

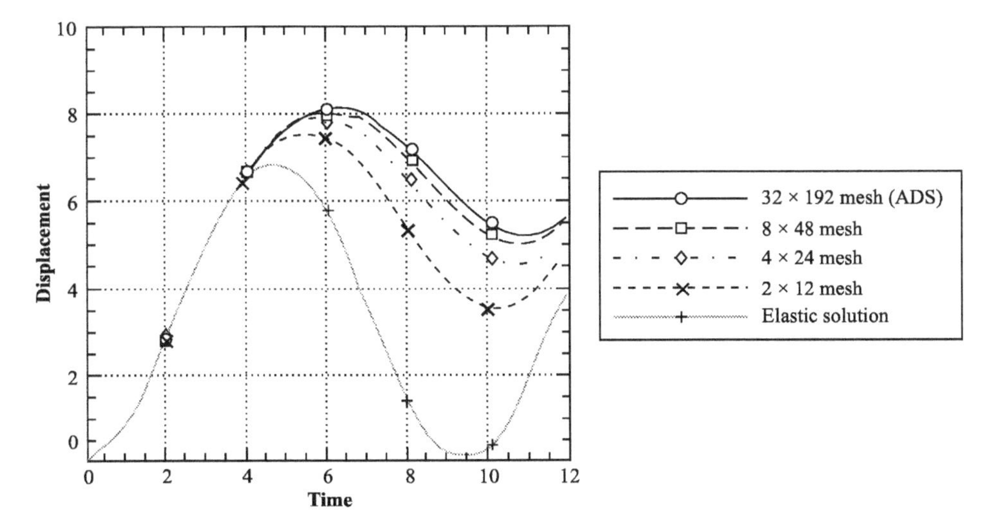

Figure 8.14 End displacement of elastic cantilever; ASQBI (two-point) element

for elastic–plastic beams are listed in Table 8.3. Figure 8.14 shows the time history of the y-component of the displacement at the end of the cantilever. With mesh refinement all elements converge to the same result.

Table 8.4 lists the fraction of internal energy absorbed by the hourglass mode when the end displacement is maximum. As expected, nearly all the strain energy is in the hourglass mode

Table 8.4 Hourglass energy in the mesh when the end displacement is maximum (normalized by total strain energy)

Mesh	FB (0.1)	ASOI	ASQBI	ADS
1×6	0.982	0.975	0.981	0.988
2×12	0.108	0.327	0.247	0.124
4×24	0.033	0.110	0.079	0.036
8×48	0.011	0.035	0.026	0.012

for the coarse (1×6) mesh. As the mesh is refined, the energy in the hourglass mode decreases rapidly. Hourglass control is comparatively ineffective in bending, because as noted before, all of the elements go into an hourglass mode. Therefore, even for relatively fine meshes, a significant part of the energy is absorbed by the hourglass control.

For all of the elements, the onset of plastic deformation is retarded by coarse meshes. This is most evident in the ASQBI elements. The ADS or FB elements are too flexible, which tends to mask the error caused by an insufficient number of quadrature points. In elastic–plastic bending, the number of quadrature points in the direction perpendicular to the axis of the beam is critical.

Each level of mesh refinement slows an explicit run by a factor of 8, while additional integration points slow it by less than 2 for ASQBI (two-point) and 4 for ASQBI (2×2). For this simple constitutive equation, the additional computations for a second stress evaluation are largely offset by the elimination of stabilization, so ASQBI (two-point solutions are only 10% slower than for the stabilized one-point element). The following remarks pertain to element performance:

1. The Q4 element with 2×2 quadrature performs no better than the stabilized one-point elements.
2. The value of the stabilization parameter α_s has a significant effect on the solution of bending problems with perturbation stabilization (FB) when the mesh is coarse.

The desire for high accuracy in beam bending is driven by the ubiquitousness of bending problems in engineering and nature. A quadrilateral with near-perfect accuracy for beams would be delightful. However, most of the quest for further improvement is misguided. MacNeal (1994) has shown that for nonrectangular elements, it is impossible to avoid parasitic shear. Since parasitic shear must be suppressed completely to obtain the perfect results for bending with a single layer of elements, MacNeal's analysis shows that it cannot be done.

8.8.3 Cylindrical Stress Wave

A two-dimensional domain with a circular hole at its center was modeled with 4876 quadrilateral elements. A compressive load was applied to the hole and the dynamic solution obtained by explicit time integration. The model was large enough to prevent the wave from reflecting from the outer boundary. Elastic and elastic–plastic materials were used. Details are given in Belytschko and Bindeman (1991). Errors were estimated by comparing the two-dimensional

Table 8.5 Normalized L_2 norms of error in displacements for axisymmetric wave problem

θ	QUAD4	FB (0.1)	ASMD	ASQBI	ASOI	
0°	0.014	0.014	0.014	0.014	0.013	elastic
45°	0.022	0.022	0.019	0.019	0.012	elastic
0°	0.0063	0.0063	0.0061	0.0061	0.0061	plastic
45°	0.0069	0.0069	0.0086	0.0088	0.0073	plastic

solutions with an axisymmetric one-dimensional computation with a very fine mesh. The normalized L_2 error norms at the end of the run are given in Table 8.5.

It can be seen that the accuracy of all of the elements is comparable. Note that the error includes time integration error, and we do not know its proportion to the spatial error. In any case, the accuracy of the stabilized elements is equivalent to that of the 2×2 quadrature elements, and there is no improvement in going to a multi-field element. We quote these results to support our contention that multi-field elements improve accuracy only when there is locking or when the element is tailored to a specific problem class, such as beam bending.

8.9 Stability

In multi-field methods, instabilities often occur in the assumed strain and stress fields. These instabilities are very common; in fact very few of the mixed or underintegrated (or SRI quadrature) elements posses stable stress fields. It is an irony of mixed methods that in resolving one difficulty, locking they introduce another. The removal of volumetric locking leads to pressure instabilities, and the removal of shear locking by mixed methods in shells leads to shear instabilities. These instabilities are related to the presence of Lagrange multipliers: any constraint treated by a Lagrange multiplier or a perturbed penalty approach must be carefully designed not only to meet the constraint but also to be stable; the latter is usually more difficult.

The pressure instability is a spatial instability, like the instability in the advection–diffusion equation. To study it by the standard stability methods which we have used previously, we need to examine the stability of the discrete equations spatially. Unfortunately, the pressure instability does not appear in one dimension, and the two-dimensional stability analysis entails a large amount of algebra; furthermore, it is feasible only for regular meshes.

The stability properties of the pressure field are related to the LBB condition (the letter L for Ladezhvanskaya is often dropped recently; it appears in Ladezhvanskaya (1968). This condition imposes severe restrictions on the assumed stress and strain fields. A readable account of this theory may be found in Bathe (1996: 301).

Necessary but not sufficient conditions for the stability of the pressure approximation are given by Zienkiewicz and Taylor (1991). We will give a brief outline of their discussion and the major results, except that we pose the problem as a linearized stability problem. The equations for a linearized stability analysis of the Lagrange multiplier problem (8.5.29) are obtained by adding the inertia, yielding

$$\begin{bmatrix} \mathbf{K}^{\text{int}} & \mathbf{G}^T \\ \mathbf{G} & \mathbf{0} \end{bmatrix} \begin{Bmatrix} \mathbf{d} \\ \mathbf{p} \end{Bmatrix} + \begin{bmatrix} \mathbf{M} & \mathbf{0} \\ \mathbf{0} & \mathbf{0} \end{bmatrix} \begin{Bmatrix} \ddot{\mathbf{d}} \\ \ddot{\mathbf{p}} \end{Bmatrix} = 0 \tag{8.9.1}$$

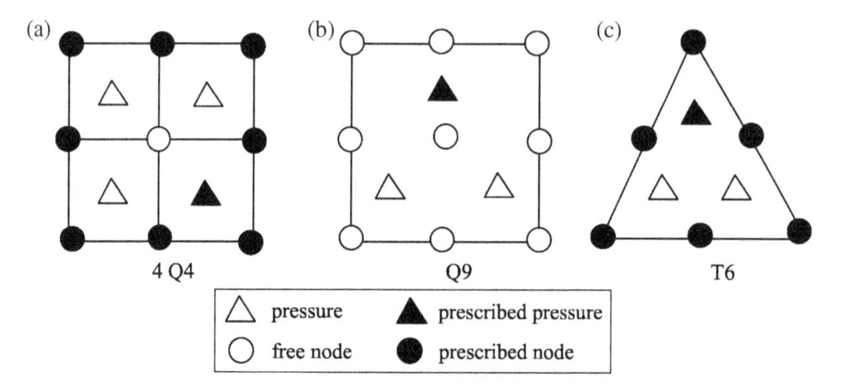

Figure 8.15 Multi-field p–v elements and stability check on pressure elements

To perform a stability analysis of these equations, we let the nodal displacements and pressure solutions be exponential: $\mathbf{d}=\mathbf{a}e^{\mu t}$, $\mathbf{p}=\mathbf{b}e^{\mu t}$. Substituting these nodal displacements and pressures into (8.9.1), we obtain the eigenvalue problem

$$\begin{bmatrix} \mathbf{A} & \mathbf{G}^T \\ \mathbf{G} & \mathbf{0} \end{bmatrix} \begin{Bmatrix} \mathbf{a} \\ \mathbf{b} \end{Bmatrix} = \begin{Bmatrix} \mathbf{0} \\ \mathbf{0} \end{Bmatrix} \text{ where } \mathbf{A} = \mathbf{M} + \mu^2 \mathbf{K}^{\text{int}} \tag{8.9.2}$$

By the Raleigh nesting theorem, the frequencies of the constrained system are all nested between the frequencies of **A**. We note that if we eliminate **a** from the system by using the first equation, we obtain

$$\mathbf{G}\mathbf{A}^{-1}\mathbf{G}^T\mathbf{b} = \mathbf{0} \tag{8.9.3}$$

It is crucial for stability that a solution exists for the pressure solution **b**. Zienkiewicz and Taylor (1991) provide some nice guidelines on this basis. They argue that for **b** to exist, the rank of **A** must be greater than or equal to the rank of **G**. Since the rank of **A** is n_{dof}, where n_{dof} is the number of degrees of freedom, then if n_p is the number of pressure parameters in the model, it follows that a necessary condition for solvability of the system (8.9.3) for the pressure variables is that

$$n_{\text{dof}} \geq n_p \text{ or } \min(n_{\text{dof}}) \geq \max(n_p) \tag{8.9.4}$$

where the second condition follows if we wish the element pressure field to be solvable for any valid combination of boundary conditions. It is stressed that the above are necessary conditions, not sufficient conditions for the solvability of the pressure field, that is, it is possible for the above to be met and the element still be unstable.

Remarkably, the simple condition (8.9.4) is an excellent guide as to when the mixed element provides stable pressure fields. Some widely used elements are given shown in Figure 8.15 in a mesh for the Zienkiewicz–Taylor test. The first is the 4-node quadrilateral with a constant pressure. The displacements have been prescribed around the periphery of the mesh to establish a configuration for which n_{dof} is minimum. One of the pressures is prescribed to eliminate the constant pressure mode. It can be seen that $n_{\text{dof}}=2$, since only one node with two degrees of freedom is free, whereas $n_p = 3$. So (8.9.4) is not met and difficulties are expected with the pressure field. If the element fails the test for any mesh, it is not LBB-stable. On the other hand, in the 9-node element shown in Figure 8.15(b), $n_{\text{dof}} = 2$ and $n_p = 2$, so the condition is met. This element is indeed one of the few elements for which element-specific pressure fields are well-behaved. It can be shown that this element meets the LBB condition.

The character of the difficulties introduced by the multi-field approach can be understood better by a specific example. We consider the mesh in Figure 8.15(a). It is left as an exercise to show that the constraint matrix **G** for this mesh is given by

$$\mathbf{G}^T = \begin{bmatrix} \Delta y & -\Delta y & -\Delta y & \Delta y \\ \Delta x & \Delta x & -\Delta x & -\Delta x \end{bmatrix} \tag{8.9.5}$$

where Δx and Δy are the element lengths along the x and y axes, respectively. There are two pressure vectors in the kernel of $\mathbf{G}^T$: $\mathbf{p} = \mathbf{1}$ and $\mathbf{p} = \mathbf{h}$, where $\mathbf{1}$ and $\mathbf{h}$ are defined in (8.4.11). The first is the constant pressure mode, which is a proper singular mode. The second is the checkerboarding mode (Figure 8.3) alternately positive and negative. Note the similarity of the singular pressure mode to the spurious singular mode of the one-point quadrature quadrilateral: both can be described by the vector **h**. In both cases, the solution grows linearly in time. Like an hourglass mode, the pressure is a weak instability.

The same spurious pressure mode occurs in the penalty method. Recall from Section 6.3 that the penalty method is equivalent to the perturbed Lagrangian. That the spurious pressure mode is a zero frequency mode of the perturbed Lagrangian form is easily verified. Therefore it must also be a mode of the penalty method. Any element that has spurious pressure modes in the Lagrange multiplier formulation will have spurious modes in the penalty formulation.

For robustness, an element must satisfy the LBB condition or the pressure field must be stabilized. To meet the LBB condition, the element must not lock volumetrically. Among the two dimensional, low-order elements with element-specific pressure fields, only the 9-node quadrilateral with a linear pressure field and the 6-node triangle with a constant pressure field satisfy the LBB condition. The 6-node triangle also satisfies the LBB condition for a C^0 linear pressure, but this entails a global pressure approximation. The theoretical basis of the LBB condition, incidentally, has only been developed for linear problems. Again we have used the ubiquitous practice of applying the results of linear theory to nonlinear problems when we do not have a nonlinear theory.

For elements that do not lock volumetrically but have pressure instabilities, such as the 4-node quadrilateral with constant pressures and constant-strain triangles in the cross-diagonal pattern, a pressure stabilization should be used. A pressure stabilization applicable to the Q4 has been given by Franca and Frey (1992).

8.10 Exercises

8.1. Show that when $X_2 \neq \frac{1}{2}(X_1 + X_3)$, then the 3-node element in Example 2.5 does not reproduce the quadratic displacement field. *Hint*: set the node displacements by a quadratic field in X and examine the resulting field.

8.2. Show that the weak form (8.5.12) leads to the following strong form:

$$\bar{p} = p, \bar{D}_{ii} = D_{ii}, \quad \sigma^{\text{dev}}_{ij,j} - \bar{p},_i + \rho b_i = \rho \dot{v}_i$$

$$n_i\left(\sigma^{\text{dev}}_{ij} - \bar{p}\delta_{ij}\right) = \bar{t}_j \text{ on } \Gamma_t, \quad \left[n_i\left(\sigma^{\text{dev}}_{ij} - \bar{p}\delta_{ij}\right)\right] = 0 \text{ on } \Gamma_{\text{int}}$$

8.3. Show that the weak form (8.5.13) leads to the following strong form:

$$\sigma^{\text{dev}}_{ij,j} - \bar{p},_i + \rho b_i = \rho \dot{v}_i, \bar{p} = p, \bar{D}_{ii} = 0$$

$$n_i\left(\sigma^{\text{dev}}_{ij} - \bar{p}\delta_{ij}\right) = \bar{t}_j \text{ on } \Gamma_t, \left[n_i\left(\sigma^{\text{dev}}_{ij} - \bar{p}\delta_{ij}\right)\right] = 0 \text{ on } \Gamma_{\text{int}}$$

8.4. By using the transformation for stresses and letting $\delta\mathbf{D} = \delta\mathbf{F}$, show that (8.5.1) can be transformed to (8.5.14).

9

Beams and Shells

9.1 Introduction

Shell elements and other structural elements are invaluable in the modeling of many engineered components and natural structures. Thin shells appear in many products, such as the sheet metal in an automobile, the fuselage, wings and rudder of an airplane, the housings of products such as cellphones, washing machines, and computers. Modeling these items with continuum elements would require a huge number of elements and lead to extremely expensive computations. As we have seen in Chapter 8, modeling a beam with hexahedral continuum elements requires a minimum of about five elements through the thickness. Thus even a low-order shell element can replace five or more continuum elements, which improves computational efficiency immensely. Furthermore, modeling thin structures with continuum elements often leads to high aspect ratios, which degrades the conditioning of the equations and the accuracy of the solution. In explicit methods, continuum element models of thin structures are restricted to very small time steps by stability requirements. Therefore structural elements are very useful in engineering analysis.

Structural elements are classified as:

1. beams, in which the motion is described as the function of a single independent variable;
2. shells, where the motion is described as a function of two independent variables;
3. plates, which are flat shells loaded normal to their surface.

Plates are usually modeled by shell elements in computer software. Since they are just flat shells, we will not consider plate elements separately. Beams, on the other hand, require some

Nonlinear Finite Elements for Continua and Structures, Second Edition.
Ted Belytschko, Wing Kam Liu, Brian Moran, and Khalil I. Elkhodary.

Companion Website: www.wiley.com/go/belytschko

additional theoretical considerations and provide simple models for learning the fundamentals of structural elements, so we will devote a substantial part of this chapter to beams.

Shell finite elements can be developed in two ways:

1. by using a weak form of the classical shell equations for momentum balance (or equilibrium);
2. by developing the element directly from a continuum element by imposing the structural assumptions; this is called the continuum-based (CB) approach.

The first approach is difficult, particularly for nonlinear shells, since the governing equations for nonlinear shells are very complex and awkward to deal with; they are usually formulated in terms of curvilinear components of tensors, and features such as variations in thickness, junctions and stiffeners are generally difficult to incorporate. There is still disagreement as to what are the best nonlinear shell equations. The CB (continuum-based) approach, on the other hand, is straightforward, yields excellent results, is applicable to arbitrarily large deformations and is widely used in commercial software and research. Therefore we will concentrate on the CB methodology. It is also called the degenerated continuum approach; we prefer the appellation continuum-based, coined by Stanley (1985), since there is nothing degenerate about these elements.

The CB methodology is not only simpler, but also provides an intellectually more appealing framework for developing shell elements than classical shell theories. In most plate and shell theories, the equilibrium or momentum equations are developed by imposing the kinematic assumptions on the motion and then using the principle of virtual work to derive the partial differential equations. The development of a weak form of the momentum equation for the purpose of discretization then entails going back to the principle of virtual work. In the CB approach, the kinematic assumptions are imposed on the test and trial functions in the weak form for continua. Thus the CB shell methodology is a more straightforward way of obtaining the discrete equations for shells and other structures.

In the CB methodology for shells the kinematic assumptions are imposed by two approaches:

1. on the motion of the continuum in the weak form, or
2. on the discrete equations for continua.

We will begin with a description of beams in two dimensions. This will provide a setting for discussing various structural theories and comparing them with CB theory. In contrast to the organization of the previous chapters, we will begin with a description of the implementation where the simplicity and key features of the CB approach are most apparent. We will then examine CB beam elements more thoroughly from a theoretical viewpoint using the first approach.

CB shell elements are developed next. Again, we begin with the implementation, illustrating how many of the techniques developed for continuum elements in the previous chapters can be applied directly to shells. The CB shell theory developed here is a synthesis of various approaches reported in the literature but also incorporates a new treatment of changes in thickness due to large deformations. The methodologies for describing large rotations in three dimensions are also described.

Two of the shortcomings of CB shell elements are then described: shear and membrane locking. These phenomena are examined in the context of beams but the insights gained are applicable to shell elements. Methods for circumventing these difficulties by means of assumed strain fields are described and examples of elements which alleviate shear and membrane locking are given.

We conclude with a description of the 4-node quadrilateral shell elements used in explicit programs, often called one-point quadrature elements. These elements are fast and robust and well suited to large-scale analysis. Several elements of this genre are reviewed and compared.

9.2 Beam Theories

9.2.1 *Assumptions of Beam Theories*

The key feature which distinguishes structures from continua is that assumptions are made about the motion and the state of stress in the element. These assumptions are based on hypotheses verified by experimental observations. The assumptions on the motion of thin shells are called kinematic assumptions, while the assumptions on the stress field are called kinetic assumptions.

The major kinematic assumption concerns the motion of the normals to the midline (also called reference line) of the beam. In linear structural theory, the midline is usually chosen to be the loci of the centroids of the cross-sections of the beam. However, the position of a reference line has no effect on a CB element: any line which corresponds approximately to the shape of the beam may be chosen as the reference line. The position of the reference line affects only the values of the resultant moments; the stresses and the overall response are not affected. We will use the terms reference line and midline interchangeably, noting that even when the term midline is used the precise location of this line relative to the cross-section of the beam is irrelevant in a CB element. The plane defined by the normals to the midline is called the normal plane. Figure 9.1 shows the reference line and normal plane for a beam.

Two types of beam theory are widely used: Euler–Bernoulli beam theory and Timoshenko beam theory. The kinematic assumptions of these theories are:

1. *Euler–Bernoulli beam theory*: the planes normal to the midline are assumed to remain plane and normal; this is also called engineering beam theory and the corresponding shell theory is called the Kirchhoff–Love shell theory.
2. *Timoshenko beam theory*: the planes normal to the midline are assumed to remain plane, but not necessarily normal; this is also called shear beam theory, and the corresponding shell theory is called the Mindlin–Reissner shell theory.

Euler–Bernoulli beams, as we shall see shortly, do not admit any transverse shear, whereas beams governed by the second assumption do admit transverse shear. The motions of an Euler–Bernoulli beam are a subset of the motions allowed by shear beam theory.

For the purpose of describing the consequences of these kinematic assumptions, we consider a straight beam along the x-axis in two dimensions as shown in Figure 9.1. Let the x-axis coincide with the midline and the y-axis with the normal to the midline. We consider only instantaneous motion of a specific current configuration, so the following equations do not constitute a nonlinear theory. We will first express the kinematic assumptions mathematically and develop the rate-of-deformation tensor; the rate-of-deformation will have the same

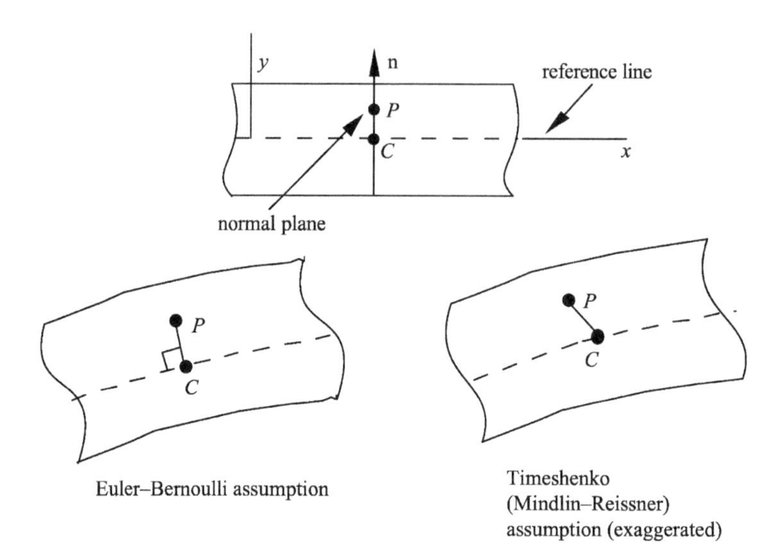

Figure 9.1 Motion in an Euler–Bernoulli beam and a shear (Timoshenko) beam; in the Euler–Bernoulli beam, the normal plane remains plane and normal, whereas in the shear beam the normal plane remains plane but not normal

properties as the linear strain since the equations for the rate-of-deformation can be obtained by replacing displacements by velocities in the linear strain–displacement relations. The aim of the following is to illustrate the consequences of the kinematic assumptions on the strain field, not to construct a theory which is worth implementing.

9.2.2 *Timoshenko (Shear Beam) Theory*

We first describe the Timoshenko beam theory. The major kinematic assumptions of this theory are that the normal planes remain plane, that is, flat, and that no deformation occurs within that plane. Thus the planes normal to the midline rotate as rigid bodies. Consider the motion of a point P whose orthogonal projection on the midline is point C. If the normal plane rotates as a rigid body, the velocity of point P relative to the velocity of point C is given by

$$\mathbf{v}_{PC} = \boldsymbol{\omega} \times \mathbf{r} \tag{9.2.1}$$

where $\boldsymbol{\omega}$ is the angular velocity of the plane and $\mathbf{r}$ is the vector from C to P. In two dimensions, the only nonzero component of the angular velocity is the z-component, so $\boldsymbol{\omega} = \dot{\theta}\mathbf{e}_z \equiv \omega\mathbf{e}_z$ where $\dot{\theta}(x,t)$ is the angular velocity of the normal. Since $\mathbf{r} = y\mathbf{e}_y$, the relative velocity is

$$\mathbf{v}_{PC} = \boldsymbol{\omega} \times \mathbf{r} = -y\omega\mathbf{e}_x \tag{9.2.2}$$

The velocity of any point on the midline is a function of x and time t, so $\mathbf{v}^M(x,t) = v_x^M\mathbf{e}_x + v_y^M\mathbf{e}_y$. The velocity of any point is then the sum of the relative velocity (9.2.1) and the midline velocity

$$\mathbf{v} = \mathbf{v}^M + \boldsymbol{\omega} \times \mathbf{r} = \left(v_x^M - y\omega\right)\mathbf{e}_x + v_y^M \mathbf{e}_y \tag{9.2.3a}$$

$$v_x(x, y, t) = v_x^M(x,t) - y\omega(x,t), \quad v_y(x,y,t) = v_y^M(x,t) \tag{9.2.3b}$$

Applying the definition of the rate-of-deformation, $D_{ij} = \text{sym}(v_{i,j})$ (see Section 3.3.2) gives

$$D_{xx} = v_{x,x}^M - y\omega_{,x}, \quad D_{yy} = 0, \quad D_{xy} = \frac{1}{2}\left(v_{y,x}^M - \omega\right) \tag{9.2.4a-c}$$

It can be seen that the only nonzero components of the rate-of-deformation are the axial component, D_{xx}, and the shear component, D_{xy}; the latter is called the *transverse shear*.

It can be seen immediately from (9.2.4) that the dependent variables v_i^M and ω need only be C^0 for the rate-of-deformation to be finite throughout the beam. Thus the standard isoparametric shape functions can be used in the construction of shear beam elements. Theories for which the interpolants need only be C^0 are often called C^0 structural theories.

9.2.3 Euler–Bernoulli Theory

In the Euler–Bernoulli (engineering beam) theory, the kinematic assumption is that the normal plane remains plane and normal. Therefore the angular velocity of the normal is given by the rate of change of the slope of the midline

$$\omega = v_{y,x}^M \tag{9.2.5}$$

By examining (9.2.4c) it can be seen that this is equivalent to requiring the shear D_{xy} to vanish, which implies that the angle between the normal and the midline does not change, that is, the normal remains normal. The axial displacement is then given by

$$v_x(x, y, t) = v_x^M(x, t) - y v_{y,x}^M(x, t) \tag{9.2.6}$$

The rate-of-deformation in Euler–Bernouili beam theory is given by

$$D_{xx} = v_{x,x}^M - y v_{y,xx}^M, \qquad D_{yy} = 0, \qquad D_{xy} = 0 \tag{9.2.7}$$

Two features are noteworthy in the previous:

1. the transverse shear vanishes:
2. the second derivative of the velocity appears in the expression for the rate-of-deformation, so the velocity field must be C^1.

Whereas the Timoshenko beam has two dependent variables (unknown), only a single dependent variable appears in the Euler–Bernoull beam. Similar reductions take place in the corresponding shell theories: a Kirchhoff–Love shell theory has only three dependent variables, whereas a Mindlin–Reissner theory has five dependent variables (six are often used in practice; this is discussed in Section 9.5).

Euler–Bernoulli beam theory is often called a C^1 theory because of the need for C^1 approximations. This requirement is the biggest disadvantage of Euler–Bernoulli and Kirchhoff–Love theories, since a C^1 approximation is difficult to construct in multi-dimensions. For this reason, C^1 structural theories are seldom used in software except for beams. Beam elements are often based on Euler–Bernoulli theory because C^1 interpolants are easily constructed in one dimension.

Transverse shear is of significance only in thick beams. However, Timoshenko beams and Mindlin–Reissner shells are frequently used even when transverse shear has little effect on the response. When a beam is thin, the transverse shear energy in a Timoshenko beam model will tend to zero in well-behaved elements. Thus the normality hypothesis, which implies that transverse shear vanishes for thin beams and experiments, is also observed in numerical solutions.

These assumptions are based primarily on experimental evidence: predictions of this theory agree with measurements. For elastic materials, closed-form analytic solutions for beams also support this theory. For arbitrary nonlinear materials, the error due to the structural assumptions has not been ascertained analytically.

9.2.4 Discrete Kirchhoff and Mindlin–Reissner Theories

A third approach, which is used only in numerical methods, is the discrete Kirchhoff theories. In the discrete Kirchhoff theory, the Kirchhoff–Love assumption is only applied discretely, that is, at a finite number of points, usually the quadrature points. Transverse shear then develops at other points in the element but is ignored. Similarly, discrete Mindlin–Reissner elements are formulated by imposing these assumptions discretely.

9.3 Continuum-Based Beam

In the following, the continuum-based (CB) formulation for a beam in two dimensions is developed. The governing equations for structures are identical to those for continua:

1. Conservation of mass
2. Conservation of linear and angular momentum
3. Conservation of energy
4. Constitutive equations
5. Strain–displacement equations.

To specialize these equations to beams, the assumptions of beam theory are imposed on the motion and the state of stress.

In this section, we will impose the kinematic constraints of CB beam theory on the discrete equations, that is, the continuum finite element will be modified so that it behaves like a beam. In the next section, we will develop the CB beam by imposing the kinematic assumption on the motion before developing the weak form and the discrete equations. These two sections, 9.3 and 9.4, will introduce many of the concepts and techniques which are used in CB shell elements. The elements to be developed are applicable to nonlinear materials and geometrical nonlinearities. Either an updated Lagrangian or a total Lagrangian formulation can be used, but we will emphasize the former. Lagrangian elements are almost always used for shells and structures because they consist of closely separated surfaces which are difficult to treat with Eulerian elements.

We will not go through the steps followed in Chapters 2, 4, and 7 of developing a weak form for the momentum equation and showing the equivalence to the strong form. Instead, we will start with the discrete equations for continua which were developed in Chapter 4.

9.3.1 Definitions and Nomenclature

A CB beam element is shown in Figure 9.2. The parent element is also shown. As can be seen in Figure 9.2, the continuum element has nodes only on the top and bottom, for the motion must be linear in η. These nodes are called *slave nodes*, A 6-node quadrilateral is shown here as the underlying continuum element, but any other continuum element with nodes on the top and bottom surfaces can also be used. The reference line is coincident with the line $\eta = 0$.

The lines of constant ξ are called *fibers*. The unit vectors along fibers are called *directors*, denoted by $\mathbf{p}(\xi, t)$; they are also called *pseudonormals*. The directors play the same role in the CB theory as normals in the classical Mindlin–Reisser theory, hence the alternative name pseudonormals. Lines of constant η are called *laminae*.

Master nodes are introduced at the intersections of the fibers connecting slave nodes with the reference line. The degrees-of-freedom of these nodes describe the motion of the beam. The equations of motion will be formulated in terms of the generalized forces and velocities of the master nodes. Each master node is associated with a pair of slave nodes along a common fiber; see Figure 9.2. The slave nodes are identified either by superscript asterisks or by superscript plus and minus signs on the node numbers: thus node I^+ and I^- are slave nodes associated with master node I and lie on the top (+) and bottom (−) surfaces of the beam; I^* are the

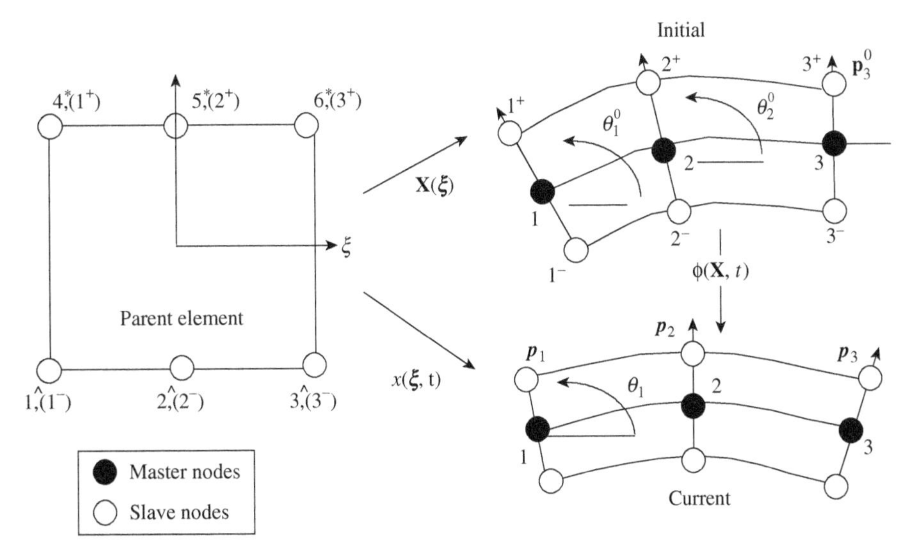

Figure 9.2 A three-node CB beam element and the underlying 6-node continuum element: the two notations for slave nodes of the underlying continuum element are shown

alternate node numbers of the continuum element. Each triplet of nodes I^-, I, and I^+, is collinear and lies on the same fiber. The appellations 'top' and 'bottom' have no exact definition; either surface of the beam can be designated as the 'top' surface.

The two sets of node numbers for the continuum element are related by

$$\begin{aligned} I^- &= I^* & \text{for} \quad I^* &\le n_N \\ I^+ &= I^* - n_N & \text{for} \quad I^* &> n_N \end{aligned} \tag{9.3.1}$$

The master node associated with any slave node can be obtained by converting the slave node numbers I^* to the node numbers with + and – superscripts using the above rule; the integer value then gives the master node number.

For each point in the beam, a *corotational coordinate system* is defined with $\hat{\mathbf{e}}_x$ tangent to the lamina; $\hat{\mathbf{e}}_y$ is normal to the lamina and its direction may change through the thickness of the beam; it is not necessarily collinear with the director. This coordinate system is also called a *laminar coordinate system* because one of the axes is tangent to the lamina.

9.3.2 Assumptions

The following assumptions are made on the motion and stress state:

1. The fibers remain straight.
2. The transverse normal stress is negligible, which is called the plane stress condition or zero normal stress condition:

$$\hat{\sigma}_{yy} = 0 \tag{9.3.2}$$

3. The fibers are inextensible.

The first assumption will be called the *modified Mindlin–Reissner assumption* in this book. It differs from the classical Mindlin–Reissner assumption, which requires the *normal to remain straight*. In the C^0 elements to be developed, the fibers are usually not normal to the midline, so constraining the motion of a fiber is not equivalent to constraining the motion of the normal. The resulting theory is similar to a single director Cosserat theory. We will use the appellation 'modified Mindlin–Reissner' for these assumptions for both beams and shells; for beams it could also be called a modified Timoshenko assumption.

If the CB beam element is to approximate Timoshenko beams, it is necessary for the fibers to be aligned as closely as possible with the normal to the midline. This can be achieved by specifying the initial positions of the slave nodes so that the fibers are close to the normal. Otherwise the behavior of the CB beam element may deviate substantially from Timoshenko beam behavior and may not agree with the observed behavior of beams. Exercise 9.1 shows that it is impossible to align the fibers with the normal exactly along the entire length of a C^0 element.

It should be noted that the inextensibility of fibers applies only to the motion. Inextensibility contradicts the plane stress assumption: the fibers are usually close to the $\hat{y}$ direction and so if $\hat{\sigma}_{yy}= 0$, the velocity strain $\hat{D}_{yy}$ generally cannot vanish. The contradiction is reconciled by not using the motion to compute $\hat{D}_{yy}$. Instead, $\hat{D}_{yy}$ is obtained from the constitutive equation by the requirement that $\hat{\sigma}_{yy}=0$, and the change in thickness is computed from $\hat{D}_{yy}$. This is equivalent to obtaining the thickness from conservation of matter since the plane stress constitutive equation incorporates the conservation of matter. The nodal internal forces are then modified to reflect changes in the thickness. Thus the inextensibility assumption applies only to the kinematics.

We have not given the plane stress condition in terms of the PK2 stress or nominal stress, for unless simplifying assumptions are made, they are more complex than (9.3.2): the plane stress condition requires that the $\hat{y}$-component of the physical stress vanish, which is not equivalent to requiring the corresponding component of the PK2 stress to vanish.

9.3.3 *Motion*

The motion of the beam is described by the translations of the master nodes, $x_I(t)$, $y_I(t)$, and the rotations of the nodal directors $\theta_I(t)$; see Figure 9.2. The angle $\theta_I(t)$ is positive counterclockwise from the x-axis. The motion is given by the standard isoparametric map for continuum elements in terms of the slave node motions:

$$\mathbf{x}(\boldsymbol{\xi},t)=\sum_{I^+=1}^{n_N}\mathbf{x}_{I^+}(t)N_{I^+}(\xi,\eta)+\sum_{I^-=1}^{n_N}\mathbf{x}_{I^-}(t)N_{I^-}(\xi,\eta)=\sum_{I^*=1}^{2n_N}\mathbf{x}_{I^*}(t)N_{I^*}(\xi,\eta) \tag{9.3.3}$$

In the above $N_{I*}(\xi,\eta)$ are the *standard shape functions for continua* (indicated by asterisks or superscripts '+' and '–' signs on nodal indices).

The *shape functions of the underlying continuum element must be linear in* η for the above motion to be consistent with the modified Mindlin–Reissner assumption. Therefore the parent element can have only two nodes along the η direction, that is, there can be only two slave nodes along a fiber.

The velocity field is obtained by taking the material time derivative of the previous, which gives

$$\mathbf{v}(\boldsymbol{\xi},t)=\sum_{I^+=1}^{n_N}\mathbf{v}_{I^+}(t)N_{I^+}(\xi,\eta)+\sum_{I^-=1}^{n_N}\mathbf{v}_{I^-}(t)N_{I^-}(\xi,\eta)=\sum_{I^*=1}^{2n_N}\mathbf{v}_{I*}(t)N_{I^*}(\xi,\eta) \tag{9.3.4}$$

We now impose the inextensibility and the modified Mindlin–Reissner assumptions on the motion of the slave nodes:

$$\mathbf{x}_{I^+}(t)=\mathbf{x}_I(t)+\frac{1}{2}h_I^0\mathbf{p}_I(t),\qquad \mathbf{x}_{I^-}(t)=\mathbf{x}_I(t)-\frac{1}{2}h_I^0\mathbf{p}_I(t) \tag{9.3.5a,b}$$

where $\mathbf{p}_I(t)$ is the director at master node I, and h_I^0 is the initial thickness of the beam at node I (or more precisely a pseudo-thickness since it is the distance between the top and bottom surfaces along a fiber). This is the crucial step in the conversion of a continuum element to a CB beam.

The director at node I *is* a unit vector along the fiber (I^-, I^+), so the current nodal directors are given by

$$\mathbf{p}_I(t)=\frac{1}{h_I^0}(\mathbf{x}_{I^+}(t)-\mathbf{x}_{I^-}(t))=\mathbf{e}_x\ \cos\theta_I+\mathbf{e}_y\ \sin\theta_I \tag{9.3.6}$$

where the initial thickness is given by

$$h_I^0 = \left\| \mathbf{X}_{I^+} - \mathbf{X}_{I^-} \right\| \tag{9.3.7}$$

and $\mathbf{e}_x$ and $\mathbf{e}_y$ are the global base vectors. The above can also be derived by subtracting (9.3.5b) from (9.3.5a). The initial nodal directors are

$$\mathbf{p}_I^0(t) = \frac{1}{h_I^0}\left(\mathbf{X}_{I^+} - \mathbf{X}_{I^-}\right) = \mathbf{e}_x \cos\theta_I^0 + \mathbf{e}_y \sin\theta_I^0 \tag{9.3.8}$$

where θ_I^0 is the initial angle of the director at node I. It is easily shown that the motion (9.3.5) meets the inextensibility requirement on the fibers through node I. It will be shown in Section 9.4 that all fibers remain constant in length according to CB beam theory. However, this result does not hold for CB finite elements: see Example 9.1.

The velocities of the slave nodes are the material time derivative of (9.3.5), yielding

$$\mathbf{v}_{I^+}(t) = \mathbf{v}_I(t) + \frac{1}{2}h_I^0\omega_I(t)\times\mathbf{p}_I(t), \qquad \mathbf{v}_{I^-}(t) = \mathbf{v}_I(t) - \frac{1}{2}h_I^0\boldsymbol{\omega}_I(t)\times\mathbf{p}_I(t) \tag{9.3.9}$$

where we have used (9.2.1) to express the nodal velocities in terms of the angular velocities, noting that the vectors from the master node to the slave nodes are $\frac{1}{2}h_I^0\mathbf{p}_I(t)$ and $-\frac{1}{2}h_I^0\mathbf{p}_I(t)$ for the top and bottom slave nodes, respectively. Since the model is two-dimensional, $\boldsymbol{\omega} = \omega\mathbf{e}_z \equiv \dot{\theta}\mathbf{e}_z$ and the slave node velocity can be written by using (9.3.6), (9.3.7), and (9.3.9) as:

$$\mathbf{v}_{I^+} = \mathbf{v}_I - \omega_{zI}((y_{I^+} - y_I)\mathbf{e}_x - (x_{I^+} - x_I)\mathbf{e}_y) = \mathbf{v}_I - \frac{1}{2}\omega_{zI}h_I^0(\mathbf{e}_x \sin\theta_I - \mathbf{e}_y \cos\theta_I) \tag{9.3.10}$$

$$\mathbf{v}_{I^-} = \mathbf{v}_I - \omega_{zI}((y_{I^-} - y_I)\mathbf{e}_x - (x_{I^-} - x_I)\mathbf{e}_y) = \mathbf{v}_I + \frac{1}{2}\omega_{zI}h_I^0(\mathbf{e}_x \sin\theta_I - \mathbf{e}_y \cos\theta_I) \tag{9.3.11}$$

The motion of the master nodes is described by three degrees of freedom per node:

$$\mathbf{d}_I(t) \equiv \mathbf{d}_I^{\text{mast}} = \begin{bmatrix} u_{xI}^M & u_{yI}^M & \theta_I \end{bmatrix}^T, \quad \dot{\mathbf{d}}_I(t) = \begin{bmatrix} v_{xI}^M & v_{yI}^M & \omega_I \end{bmatrix}^T \tag{9.3.12}$$

Equations (9.3.10–11) can be written in matrix form as

$$\begin{Bmatrix} \mathbf{v}_{I^-} \\ \mathbf{v}_{I^+} \end{Bmatrix}^{\text{slave}} = \{v_{xI^-}, v_{yI^-}, v_{xI^+}, v_{yI^+}\}^T = \mathbf{T}_I\,\dot{\mathbf{d}}_I^{\text{mast}} \text{ (no sum on } I) \tag{9.3.13}$$

where we have added the superscripts 'slave' and 'mast' to emphasize that the continuum nodes are the slave nodes and the beam nodes the master nodes. From a comparison of (9.3.13) and (9.3.10–11) we can see that

$$\mathbf{T}_I = \begin{bmatrix} 1 & 0 & y_I - y_{I^-} \\ 0 & 1 & x_{I^-} - y_I \\ 1 & 0 & y_I - y_{I^+} \\ 0 & 1 & x_{I^+} - x_I \end{bmatrix} \text{and define } \mathbf{T} = \begin{bmatrix} \mathbf{T}_1 & \mathbf{0} & \cdot & \mathbf{0} \\ \mathbf{0} & \mathbf{T}_2 & \cdot & \mathbf{0} \\ \cdot & \cdot & \cdot & \cdot \\ \mathbf{0} & \mathbf{0} & \cdot & \mathbf{T}_n \end{bmatrix} \tag{9.3.14}$$

9.3.4 Nodal Forces

The master nodal internal forces are related to the slave nodal internal forces by the transformation rule given in Section 4.5.5, (4.5.35). Since the slave nodal velocities are related to the master nodal velocities by (9.3.13–14), the nodal forces are related by

$$\mathbf{f}_I^{\text{mast}} = \begin{bmatrix} f_{xI} \\ f_{yI} \\ m_I \end{bmatrix}^{\text{mast}} = \mathbf{T}_I^T \begin{Bmatrix} \mathbf{f}_{I^-} \\ \mathbf{f}_{I^+} \end{Bmatrix}^{\text{slave}} = \begin{bmatrix} 1 & 0 & 1 & 0 \\ 0 & 1 & 0 & 1 \\ y_I - y_{I^-} & x_{I^-} - x_I & y_I - y_{I^+} & x_{I^+} - x_I \end{bmatrix} \begin{Bmatrix} f_{xI^-} \\ f_{yI^-} \\ f_{xI^+} \\ f_{yI^+} \end{Bmatrix} \tag{9.3.15}$$

The external nodal forces at the master modes can be obtained from the slave external forces by the same transformation. The column matrix of nodal forces consists of the two force components f_{xI} and f_{yI} and the moment m_I. The nodal forces are conjugate in power to the velocities of the master nodes, i.e. the power of the forces at node I *is* given by $\dot{\mathbf{d}}_I^T \mathbf{f}_I$ (no sum on I). The superscripts 'mast' and 'slave' will be dropped from now on.

9.3.5 Constitutive Update

To convert the standard continuum element to a CB beam, the plane stress assumption (9.3.2) must be enforced. For this purpose, it is convenient to use laminar components of the stress and velocity strain. The laminar base vectors $\hat{\mathrm{e}}_i$ are constructed so that $\hat{\mathbf{e}}_x$ *is* tangent to the lamina and $\hat{\mathbf{e}}_y$ is normal to the lamina (see Figure 9.3):

$$\hat{\mathbf{e}}_x = \frac{\mathbf{x}_{,\xi}}{\|\mathbf{x}_{,\xi}\|} = \frac{x_{,\xi}\,\mathbf{e}_x + y_{,\xi}\,\mathbf{e}_y}{\left(x_{,\xi}^2 + y_{,\xi}^2\right)^{1/2}}, \quad \hat{\mathbf{e}}_y = \frac{-y_{,\xi}\,\mathbf{e}_x + x_{,\xi}\,\mathbf{e}_y}{\left(x_{,\xi}^2 + y_{,\xi}^2\right)^{1/2}} \tag{9.3.16}$$

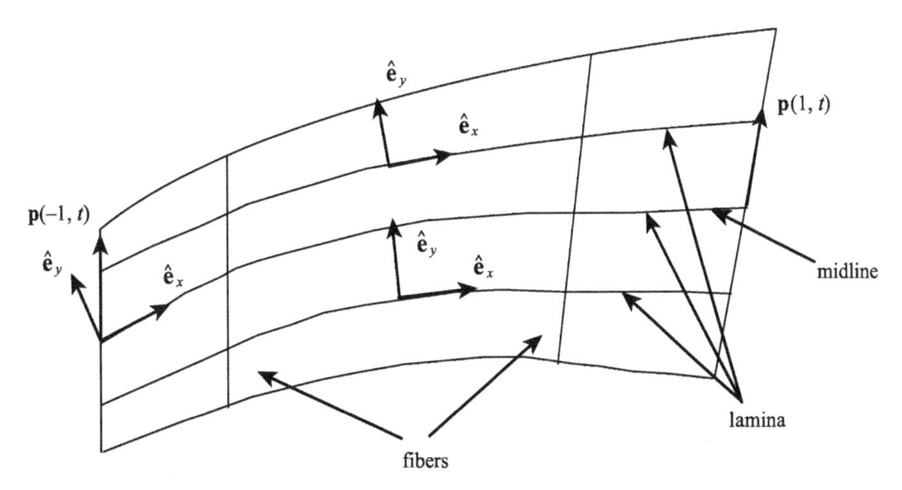

Figure 9.3 Schematic of CB beam showing the laminae, the corotational unit vectors $\hat{\mathbf{e}}_x, \hat{\mathbf{e}}_y$ and the director $\mathbf{p}(\xi, t)$ at the ends; note $\mathbf{p}$ usually does not coincide with $\hat{\mathbf{e}}_y$

$$x_{,\xi} = \sum_{I^*} x_{I^*} N_{I^*,\xi}(\xi,\eta), \quad y_{,\xi} = \sum_{I^*} y_{I^*} N_{I^*,\xi}(\xi,\eta) \tag{9.3.17}$$

We have superposed a hat on the laminar components because they rotate with the material and hence can be considered corotational. Unless the normals to the reference line remain normal, the angular velocity of this system is not precisely $\mathbf{W}$ or $\mathbf{\Omega}$ and the rotation of the lamina is not exactly $\mathbf{R}$. However, in most shell problems, the shear is small so that the differences are minute. If you consider the differences important, you can set up another coordinate system rotated by the polar decomposition $\mathbf{R}$, but bear in mind that the plane stress condition should be imposed in the laminar system. The laminar components of the rate-of-deformation are given by (3.4.16):

$$\hat{\mathbf{D}} = \mathbf{R}_{\text{lam}}^T \mathbf{D} \mathbf{R}_{\text{lam}} \quad \text{where} \quad \mathbf{R}_{\text{lam}} = \begin{bmatrix} \mathbf{e}_x \cdot \hat{\mathbf{e}}_x & \mathbf{e}_x \cdot \hat{\mathbf{e}}_y \\ \mathbf{e}_y \cdot \hat{\mathbf{e}}_x & \mathbf{e}_y \cdot \hat{\mathbf{e}}_y \end{bmatrix} \tag{9.3.18}$$

In the evaluation of the stress, the plane stress constraint $\hat{\sigma}_{yy} = 0$ must be observed. If the constitutive equation is in rate form, the constraint is $D\hat{\sigma}_{yy}/Dt = 0$. For example, for an isotropic hypoelastic material, the stress rate components are given by

$$\frac{D}{Dt}\{\hat{\boldsymbol{\sigma}}\} = \frac{D}{Dt}\begin{Bmatrix} \hat{\sigma}_{xx} \\ \hat{\sigma}_{xy} \\ \hat{\sigma}_{yy} \end{Bmatrix} = \frac{D}{Dt}\begin{Bmatrix} \hat{\sigma}_{xx} \\ \hat{\sigma}_{xy} \\ 0 \end{Bmatrix} = \frac{E}{1-\nu^2}\begin{bmatrix} 1 & 0 & \nu \\ 0 & \frac{1}{2}(1-\nu) & 0 \\ \nu & 0 & 1 \end{bmatrix}\begin{Bmatrix} \hat{D}_{xx} \\ 2\hat{D}_{xy} \\ \hat{D}_{yy} \end{Bmatrix} \tag{9.3.19}$$

The stress and velocity strain components in the Voigt form have been rearranged so that the 'yy' component is last. Solving the last line for the $\hat{D}_{yy}$ gives $\hat{D}_{yy} = -\nu\hat{D}_{xx}$. Substituting the preceding into (9.3.19) gives

$$\frac{D\hat{\sigma}_{xx}}{Dt} = E\hat{D}_{xx}, \quad \frac{D\hat{\sigma}_{xy}}{Dt} = \frac{E}{(1+\nu)}\hat{D}_{xy} \tag{9.3.20}$$

For more general materials (including laws which lack symmetry in the moduli, such as nonassociated plasticity) the constitutive rate relation can be written as

$$\frac{D}{Dt}\begin{Bmatrix} \hat{\sigma}_{xx} \\ \hat{\sigma}_{xy} \\ 0 \end{Bmatrix} = \begin{bmatrix} \hat{C}_{11} & \hat{C}_{13} & \hat{C}_{12} \\ \hat{C}_{31} & \hat{C}_{33} & \hat{C}_{32} \\ \hat{C}_{21} & \hat{C}_{23} & \hat{C}_{22} \end{bmatrix}^{\text{lam}} \begin{Bmatrix} \hat{D}_{xx} \\ 2\hat{D}_{xy} \\ \hat{D}_{yy} \end{Bmatrix} \tag{9.3.21}$$

where $\hat{\mathbf{C}}^{\text{lam}}$ are the tangent moduli and the last equation enforces the plane stress condition. The last line is used to solve for $\hat{D}_{yy}$. A correction for rotation is not needed. Examples of tangent moduli for hypoelastic and elastic–plastic materials are given in Chapter 5. $\mathbf{\Omega}$ closely approximates the spin of the laminar corotational triad, so the Green–Naghdi tangent moduli closely approximate $\hat{\mathbf{C}}^{\text{lam}}$. In fact, for thin beams, where the normal remains normal, the two are identical.

9.3.6 Continuum Nodal Internal Forces

Except for the imposition of the plane stress condition, the computation of the slave internal nodal forces is identical to that in the underlying continuum element. The integral (E4.2.12) is evaluated by numerical quadrature. Neither full quadrature (4.5.23) nor the selective-reduced quadrature (4.5.33) can be used in a CB beam. Both quadrature schemes result in shear locking, to be described in Section 9.7. Shear locking can be avoided in the 2-node element by using a single stack of quadrature points at $\xi = 0$. This quadrature scheme is also called selective-reduced integration. It integrates the axial forces exactly (if the underlying continuum element is rectangular) but underintegrates the transverse shear stresses: see Hughes (1987). The number of quadrature points in the η-direction depends on the material law and the accuracy desired. For a smooth hyperelastic material law, three Gauss quadrature points are often adequate. For an elastic–plastic law, a minimum of about five quadrature points is needed since the stress distribution is not continuously differentiable. Elastic–plastic stress distributions have discontinuous derivatives, such as shown in Figure 9.4. Gauss quadrature is not optimal for elastic–plastic laws since these quadrature schemes are based on higher-order polynomial interpolation, which tacitly assumes smoothness in the data. The trapezoidal rule is often used because it is more effective for less smooth functions.

To illustrate the selective-reduced integration procedure that circumvents shear locking, we consider a 2 node beam element based on a 4-node quadrilateral continuum element. The nodal forces are obtained by integration with a single stack of quadrature points at $\xi = 0$ (see Section 4.5.4):

$$\left[f_{xI^*}, f_{yI^*}\right]^{\text{int}} = \frac{h}{h^0}\sum_{Q=1}^{n_Q}\left(\left[N_{I^*,x} \quad N_{I^*,y}\right]\begin{bmatrix}\sigma_{xx} & \sigma_{xy}\\ \sigma_{xy} & \sigma_{yy}\end{bmatrix}\bar{w}_Q a J_\xi\right)\Bigg|_{(0,\eta_Q)} \tag{9.3.22}$$

where η_Q are the η_Q quadrature points through the thickness of the beam, $\bar{w}_Q$ are the quadrature weights, a is the dimension of the beam in the z-direction and J_ξ is the Jacobian determinant with respect to the parent element coordinates, (4.4.43). The only difference from the continuum element relation (E4.2.14) is the factor h/h^0 which accounts approximately for the

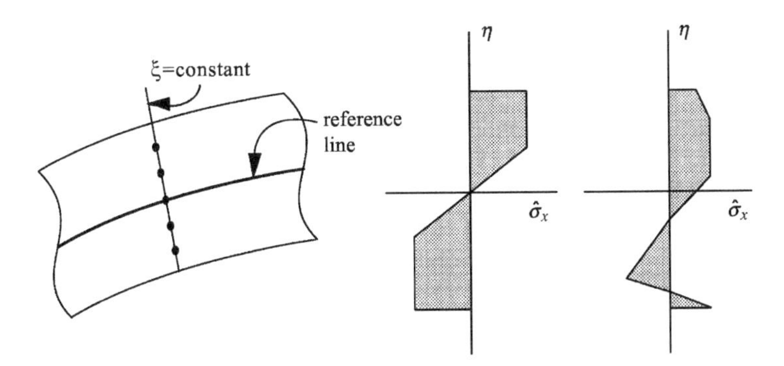

Figure 9.4 A stack of quadrature points and examples of axial stress distribution for an elastic–plastic material

change in thickness. The stresses as computed by (9.3.20–21) must be rotated back to the global system prior to evaluating the nodal internal forces by (9.3.22).

The nodal internal forces can also be computed in terms of the corotational components of the stress by (4.6.10):

$$\left[f_{xI^*}, f_{yI^*}\right]^{\text{int}} = \frac{h}{h^0}\sum_{Q=1}^{n_Q}\left(\left[N_{I^*,\hat{x}} N_{I^*,\hat{y}}\right]\begin{bmatrix}\hat{\sigma}_{xx} & \hat{\sigma}_{xy}\\ \hat{\sigma}_{xy} & 0\end{bmatrix}\begin{bmatrix}R_{x\hat{x}} & R_{y\hat{x}}\\ R_{x\hat{y}} & R_{y\hat{y}}\end{bmatrix}\bar{w}_Q a J_{\xi}\right)\Bigg|_{(0,\eta_Q)} \tag{9.3.23}$$

The stress component $\hat{\sigma}_{yy}$ vanishes in (9.3.23) because of the zero normal stress condition. In the above, the corotational laminar coordinate system generally differs for each quadrature point.

An algorithm for computing the nodal forces is given in Box 9.1. The algorithm employs an updated Lagrangian formulation with the constitutive update in terms of corotational, laminar components.

Box 9.1 Algorithm for CB beam element

1. The positions and velocities of the slave nodes are computed by (9.3.6), (9.3.5) and (9.3.9):

$$\mathbf{p}_I(t) = \mathbf{e}_x \cos\,\theta_I + \mathbf{e}_y\,\sin\theta_I \tag{B9.1.1}$$

$$\mathbf{x}_{I^+}(t) = \mathbf{x}_I(t) + \frac{1}{2}h_I^0\mathbf{p}_I(t) \quad \mathbf{x}_{I^-}(t) = \mathbf{x}_I(t) - \frac{1}{2}h_I^0\mathbf{p}_I(t) \tag{B9.1.2}$$

$$\mathbf{v}_{I^+}(t) = \mathbf{v}_I(t) + \frac{1}{2}h_I^0\boldsymbol{\omega}_I(t)\times\mathbf{p}_I(t) \quad \mathbf{v}_{I^-}(t) = \mathbf{v}_I(t) - \frac{1}{2}h_I^0\boldsymbol{\omega}_I(t)\times\mathbf{p}_I(t) \tag{B9.1.3}$$

(no sum on I in the above).

2. At quadrature points, $Q = 1$ to n_Q:
   i. set up laminar base vectors: $\hat{\mathbf{e}}_x, \hat{\mathbf{e}}_y, \mathrm{R}_{\text{lam}}$ by (9.3.16–18)
   ii. compute laminar components: $\hat{\mathbf{v}}_I = \mathbf{R}_{\text{lam}}^T\mathbf{v}_I, \hat{\mathbf{x}}_{I^*} = R_{\text{lam}}^T\mathbf{x}_{I^*}$
   iii. compute shape function gradients:

$$N_{I,\hat{x}}^T = N_{I,\xi}^T\hat{\mathbf{x}}_{,\xi}^{-1} \tag{B9.1.4}$$

   iv. compute the velocity gradient and velocity strain:

$$\hat{\mathbf{L}} = \hat{\mathbf{v}}_I\hat{\mathbf{B}}_I^T = \hat{\mathbf{v}}_I N_{I,\hat{x}}^T, \quad \hat{\mathbf{D}} = \frac{1}{2}(\hat{\mathbf{L}} + \hat{\mathbf{L}}^T) \tag{B9.1.5}$$

   v. update the stress $\hat{\sigma}$ by constitutive equation (Section 9.3.5, Chapter 5)
   vi. add stress contribution to slave nodal forces by (9.3.23)

   END loop

3. Compute master nodal forces by (9.3.15).

9.3.7 Mass Matrix

The mass matrix of the CB beam element can be obtained by the transformation (4.5.38):

$$\mathbf{M} = \mathbf{T}^T \hat{\mathbf{M}} \mathbf{T} \tag{9.3.24}$$

where $\hat{\mathbf{M}}$ is the mass matrix for the underlying continuum element. The mass matrix is time dependent, which is unusual for a Lagrangian element and is due to the kinematic constraints of CB theory. $\hat{\mathbf{M}}$ can be either the consistent or the lumped mass of the continuum element. Equation (9.3.24) does not yield a diagonal matrix even when the diagonal mass matrix of the continuum element is used. Two techniques are used to obtain diagonal matrices:

1. the row sum technique,
2. physical lumping.

For a CB beam based on a rectangular 4-node continuum element, the second procedure yields

$$\mathbf{M} = \frac{\rho_0 h_0 \ell_0 a_0}{420} \begin{bmatrix} 210 & 0 & 0 & 0 & 0 & 0 \\ 0 & 210 & 0 & 0 & 0 & 0 \\ 0 & 0 & \alpha h_0^2 & 0 & 0 & 0 \\ 0 & 0 & 0 & 210 & 0 & 0 \\ 0 & 0 & 0 & 0 & 210 & 0 \\ 0 & 0 & 0 & 0 & 0 & \alpha h_0^2 \end{bmatrix} \tag{9.3.25}$$

where α is a scale factor for the rotational inertia. The scale factor is chosen in explicit codes so that the critical time step depends only on the translational degrees of freedom, that is, so that the time step avoids the ℓ^2 dependence seen in Table 6.1. This was proposed by Key and Beisinger (1971).

The above lumped mass matrix does not account for the time dependence of the T matrix. If we account for the time dependence of $\mathbf{T}$, the inertial force according to (4.5.41) is given by

$$\mathbf{f}^{\text{kin}} = \mathbf{T}^T \hat{\mathbf{M}} \mathbf{T} \ddot{\mathbf{d}} + \mathbf{T}^T \hat{\mathbf{M}} \dot{\mathbf{T}} \dot{\mathbf{d}} \tag{9.3.26}$$

where $\hat{\mathbf{M}}$ is given in Example 4.2 and $\mathbf{T}_I$ is given by (9.3.14). The matrix $\dot{\mathbf{T}}$ is obtained by taking a time derivative of (9.3.14) which gives

$$\dot{\mathbf{T}}_I = \frac{d\mathbf{T}_I}{dt} = \omega_I \begin{bmatrix} 0 & 0 & x_I - x_{I^-} \\ 0 & 0 & y_I - y_{I^-} \\ 0 & 0 & x_I - x_{I^+} \\ 0 & 0 & y_I - y_{I^+} \end{bmatrix} \tag{9.3.27}$$

Thus the acceleration will include a term proportional to the square of the angular velocity, and the inertial terms in the semidiscrete equations are no longer linear in the velocities.

The time integration of the equations of motion is then more complex. Furthermore, this term is often small, so it is usually neglected.

9.3.8 Equations of Motion

The equations of motion at a master node are given by

$$\mathbf{M}_{IJ}\ddot{\mathbf{d}}_J + \mathbf{f}_I^{\text{int}} = \mathbf{f}_I^{\text{ext}} \tag{9.3.28}$$

where the nodal forces and nodal velocities are

$$\mathbf{f}_I = \begin{Bmatrix} f_{xI} \\ f_{yI} \\ m_I \end{Bmatrix} \qquad \dot{\mathbf{d}}_I = \begin{Bmatrix} v_{xI} \\ v_{yI} \\ \omega_I \end{Bmatrix} \tag{9.3.29}$$

For a diagonal mass matrix the equations of motion at a node are

$$\begin{bmatrix} M_{11} & 0 & 0 \\ 0 & M_{22} & 0 \\ 0 & 0 & M_{33} \end{bmatrix}_I \begin{Bmatrix} \dot{v}_{xI} \\ \dot{v}_{yI} \\ \dot{\omega}_I \end{Bmatrix} + \begin{Bmatrix} f_{xI} \\ f_{yI} \\ m_I \end{Bmatrix}^{\text{int}} = \begin{Bmatrix} f_{xI} \\ f_{yI} \\ m_I \end{Bmatrix}^{\text{ext}} \tag{9.3.30}$$

where M_{ii}, $i = 1$ to 3, are the assembled diagonal masses at node I. Although we have not derived these equations explicitly, they follow from (4.4.24) since we have only transformed the variables by (9.3.13) and (9.3.15). For equilibrium processes, the inertial term is dropped.

9.3.9 Tangent Stiffness

The tangential and load stiffnesses are obtained from the corresponding matrices for the underlying continuum element by the transformation (4.5.42). These matrices do not need to be rederived for CB beams. The tangent stiffness matrix is based on Example 6.3:

$$\hat{\mathbf{K}}_{I^*J^*}^{\text{int}} = \int_\Omega \hat{\mathbf{B}}_{I^*}^T \left[\hat{\mathbf{C}}_P^{\text{lam}}\right] \hat{\mathrm{B}}_{J^*}\, d\Omega + \mathbf{I} \int_\Omega \boldsymbol{B}_{I^*}^T \boldsymbol{\sigma} B_{J^*}\, d\Omega \tag{9.3.31}$$

$$\left[\hat{\mathbf{C}}_P^{\text{lam}}\right] = \hat{\mathbf{C}}_{aa} - \hat{\mathbf{C}}_{ab}\hat{\mathbf{C}}_{bb}^{-1}\hat{\mathbf{C}}_{ba}, \qquad \hat{\mathbf{B}}_{I^*} = \begin{bmatrix} N_{I^*,\hat{x}} & 0 \\ N_{I^*,\hat{y}} & N_{I^*,\hat{x}} \end{bmatrix} \tag{9.3.32}$$

where $\hat{\mathbf{C}}_{aa}$, $\hat{\mathbf{C}}_{ab}$, $\hat{\mathbf{C}}_{ba}$ and $\hat{\mathbf{C}}_{bb}$ are the submatrices of $[\hat{\mathbf{C}}^{\text{lam}} + \hat{\mathbf{C}}^*]$ needed to eliminate the last row. Note that the correction matrix $\hat{\mathbf{C}}^*$ depends on the spin which is chosen. The tangent stiffness can be expressed in terms of the master nodes by the stiffness transformation (4.5.42), $\mathbf{K} = \mathbf{T}^T\hat{\mathbf{K}}\mathbf{T}$, with the transformation matrix T given by (9.3.14). The load stiffness is similarly obtained from the continuum element load stiffness.

9.4 Analysis of the CB Beam

9.4.1 Motion

In order to obtain a better understanding of the CB beam, it is worthwhile to examine its motion from a viewpoint which more closely parallels classical beam theory. The analysis in this section leads to discrete equations which are identical to those described in the previous section. It is more pleasing conceptually but working in this framework is more burdensome, since many of the entities needed for a standard implementation, such as the tangent stiffness and the mass matrix, have to be developed from scratch, whereas in the previous approach they are inherited from a continuum element.

We start with a description of the motion. To satisfy the modified Mindlin–Reissner assumption, the motion must be linear in η, that is, through the thickness of the beam. Consequently we can describe the motion of the CB beam by

$$\mathbf{x}(\xi,\eta,t) = \mathbf{x}^M(\xi,t) + \frac{1}{2}\eta h^0(\xi)\mathbf{p}(\xi,t) \tag{9.4.1}$$

where $\mathbf{x}^M(\xi, t)$ is the current configuration of the reference line and $\mathbf{p}(\xi, t)$ is the director field along the midline. Another expression for the motion equivalent to the previous is

$$\mathbf{x}(\xi,\eta,t) = \mathbf{x}^M(\xi,t) + \eta\mathbf{x}^B(\xi,t) \tag{9.4.2}$$

where $\mathbf{x}^B(\xi, t)$ is the bending part of the motion given by

$$\mathbf{x}^B = \frac{1}{2}h^0\mathbf{p} \tag{9.4.3}$$

The variables ξ and η are curvilinear coordinates. Note that although we use the same nomenclature for the parent element coordinates, (9.4.1) need not refer to a parent element. The top and bottom surfaces of the beam are given by $\eta = 1$ and $\eta = -1$, respectively, and $\eta = 0$ corresponds to the midline. The initial configuration is given by (9.4.1) at the initial time:

$$\mathbf{X}(\xi,\eta) = \mathbf{X}^M(\xi) + \eta\frac{h^0}{2}\mathbf{p}_0(\xi) \tag{9.4.4}$$

where $\mathbf{p}_0(\xi)$ is the initial director and $\mathbf{X}^M(\xi)$ describes the initial midline.

In this form of the motion, it is straightforward to show that all fibers are inextensible. The length of a fiber is the distance between the top and bottom surfaces along the fiber, that is, the distance between the points at $\eta = -1$ and $\eta = 1$ for a constant value of ξ. Using (9.4.1) it follows that the length of any fiber in the deformed configuration is given by

$$\begin{aligned} h(\xi,t) &= \|\mathbf{x}(\xi,1,t) - \mathbf{x}(\xi,-1,t)\| = \left\| \left(\mathbf{x}^M + \frac{h^0}{2}\mathbf{p}\right) - \left(\mathbf{x}^M - \frac{h^0}{2}\mathbf{p}\right) \right\| \\ &= \|h^0\mathbf{p}\| = h^0 \end{aligned} \tag{9.4.5}$$

where the last step follows from the fact that the director **p** is a unit vector. Hence the length of a fiber is always $h^0(\xi)$. This property does not hold for a finite element approximation; this will be shown in Example.

The displacement is obtained by subtracting (9.4.4) from (9.4.1), which gives

$$\mathbf{u}(\xi,\eta,t) = \mathbf{u}^M(\xi,t) + \eta\frac{h^0}{2}(\mathbf{p}(\xi,t)-\mathbf{p}_0(\xi)) = \mathbf{u}^M(\xi,t)+\eta\mathbf{u}^B(\xi,t) \tag{9.4.6}$$

where $\mathbf{u}^B(\xi, t)$ is called the bending displacement. Since the directors are unit vectors, the second term on the **RHS** of the above is a function of a *single dependent variable*, the angle $\theta(\xi, t)$, which is measured counterclockwise from the x-axis as shown in Figure 9.2. The dependence of the bending displacement on a single dependent variable can be clarified by expressing the second term of (9.4.6) in terms of the global base vectors:

$$\mathbf{u} = \mathbf{u}^M + \eta\frac{h^0}{2}(\mathbf{e}_x(\cos\,\theta - \cos\theta_0) + \mathbf{e}_y(\sin\theta - \sin\theta_0)) \tag{9.4.7}$$

where $\theta_0(\xi)$ is the initial angle of the director. The velocity is the material time derivative of the displacement (9.4.7):

$$\mathbf{v}(\xi,\eta,t) = \mathbf{v}^M(\xi,t) + \eta\frac{h^0}{2}\dot{\mathbf{p}}(\xi,t) = \mathbf{v}^M(\xi,t)+\eta\mathbf{v}^B(\xi,t) \tag{9.4.8}$$

where the bending velocity $\mathbf{v}^B(\xi, t)$ is defined from the previous as

$$\mathbf{v}^B = \frac{h^0}{2}\dot{\mathbf{p}} \tag{9.4.9}$$

Using (9.2.1), (9.4.8) can be written

$$\mathbf{v} = \mathbf{v}^M + \eta\frac{h^0}{2}\boldsymbol{\omega}\times\mathbf{p} = \mathbf{v}^M + \eta\frac{h^0}{2}\omega\mathbf{q} \tag{9.4.10}$$

where $\omega = \dot{\theta}$ and $\boldsymbol{\omega} = \omega(\xi, t)\mathbf{e}_z$ is the angular velocity of the director and

$$\mathbf{q} = \mathbf{e}_z \times \mathbf{p} = -\hat{\mathbf{e}}_x \cos\bar{\theta} + \hat{\mathbf{e}}_y \;\sin\bar{\theta} \tag{9.4.11}$$

where $\bar{\theta}$ is the angle between the normal to the midline and the director, as shown in Figure 9.5.

We will now compare the above velocity with that in Timoshenko beam theory, (9.2.2). For this purpose, we express the vectors in terms of the corotational base vectors:

$$\mathbf{p} = \hat{\mathbf{e}}_x \sin\bar{\theta} + \hat{\mathbf{e}}_y \;\cos\bar{\theta}, \qquad \mathbf{v}^M = \hat{v}_x^M\hat{\mathbf{e}}_x + \hat{v}_y^M\hat{\mathbf{e}}_y \tag{9.4.12}$$

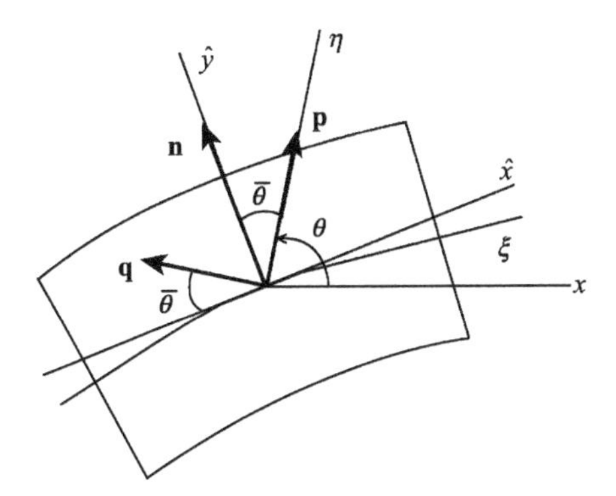

Figure 9.5 Nomenclature for CB beam in two dimensions showing director **p** and normal **n**

The velocity can then be written in terms of the corotational base vectors as

$$\mathbf{v} = \hat{v}_x^M \hat{\mathbf{e}}_x + \hat{v}_y^M \hat{\mathbf{e}}_y + \eta \frac{h^0}{2} \omega(-\hat{\mathbf{e}}_x \cos\bar{\theta} + \hat{\mathbf{e}}_y \sin\bar{\theta}) \tag{9.4.13}$$

From (9.4.2) and Figure 9.5, it can be seen that

$$\frac{h^0}{2}\eta \cos\bar{\theta} = \hat{y} \quad \text{so} \quad \frac{h^0}{2}\eta = \frac{\hat{y}}{\cos\bar{\theta}} \tag{9.4.14}$$

Substituting the above into (9.4.13) and writing the velocity vectors as column matrices gives

$$\begin{Bmatrix} \hat{v}_x \\ \hat{v}_y \end{Bmatrix} = \begin{Bmatrix} \hat{v}_x^M \\ \hat{v}_y^M \end{Bmatrix} + \omega\hat{y} \begin{Bmatrix} -1 \\ \tan\bar{\theta} \end{Bmatrix} \tag{9.4.15}$$

Comparing this to (9.2.3) we see that when $\bar{\theta} = 0$, the above corresponds exactly to the velocity field of classical Timoshenko theory, and as long as $\bar{\theta}$ is small, it is a good approximation. However, analysts often let $\bar{\theta}$ take on large values, like $\pi/4$, by placing the slave nodes so that the director is not aligned with the normal. When the angle between the director and the normal is large, the velocity field differs substantially from that of classical Timoshenko theory.

The acceleration is given by the material time derivative of the velocity:

$$\dot{\mathbf{v}} = \dot{\mathbf{v}}^M + \eta \frac{h^0}{2}(\dot{\boldsymbol{\omega}} \times \mathbf{p} + \boldsymbol{\omega} \times (\boldsymbol{\omega} \times \mathbf{p})) \tag{9.4.16}$$

so (4.4.16) gives

$$\begin{aligned} \delta v_{iI} f_{iI}^{\text{kin}} &= \int_\Omega \delta \mathbf{v} \cdot \rho \left(\dot{\mathbf{v}}^M + \eta \frac{h^0}{2}(\dot{\boldsymbol{\omega}} \times \mathbf{p} + \boldsymbol{\omega} \times (\boldsymbol{\omega} \times \mathbf{p})) \right) d\Omega \\ &= \int_\Omega \delta \mathbf{v} \cdot \rho \left(\dot{\mathbf{v}}^M + \eta \frac{h^0}{2}(\dot{\omega}\mathbf{q} - \omega^2 \mathbf{p}) \right) d\Omega \end{aligned} \tag{9.4.17}$$

Thus the inertial force depends on ω^2, as seen earlier in (9.3.26).

The dependent variables for the beam are the two components of the midline velocity $\mathbf{v}^M(\xi, t)$ and the angular velocity $\omega(\xi, t)$; alternatively one can let the midline displacement $\mathbf{u}^M(\xi, t)$ and the current angle of the director $\theta(\xi, t)$ be the dependent variables. The constraints of CB beam theory replace the two translational velocity components of a two-dimensional continuum by the two translational components of the midline and the rotation of the director. However, the new dependent variables are functions of a single space variable, ξ, whereas the independent variables of the continuum are functions of two space variables. This reduction in the dimensionality of the problem is one of the benefits of structural theories.

9.4.2 *Velocity Strains*

We examine next the velocity strains in the CB beam. This is accomplished by a series expansion of the velocity strain through the thickness. The results for $\bar{\theta} = 0$ is:

$$D_{\hat{x}\hat{x}} = v^M_{\hat{x},\hat{x}} - \eta\frac{h^0}{2}(\omega,_{\hat{x}} + p_{\hat{x},\hat{x}} v^M_{\hat{x},\hat{x}}) + O\left(\frac{\eta h^0}{R}\right)^2 \tag{9.4.18}$$

$$\hat{D}_{xy} = \frac{1}{2}(-\omega + v^M_{\hat{y},\hat{x}}) + O\left(\frac{\eta h^0}{R}\right) \tag{9.4.19}$$

The axial velocity strain $\hat{D}_{xx}$ varies linearly through the depth of the beam. It consists of three parts:

1. $v^M_{\hat{x},\hat{x}}$, the stretching of the midline; because this term is in the corotational coordinates, it also couples bending to axial strain;
2. $\frac{\eta h^0}{2}\omega,\hat{x}$, the bending velocity strain, which varies linearly with η;
3. $\frac{\eta h^0}{2} p_{\hat{x},\hat{x}} v^M_{\hat{x},\hat{x}}$, the bending velocity strain that results from stretching; this couples stretching and bending but tends to be less important than the first term.

The transverse shear component $D_{\hat{x}\hat{y}}$ also varies linearly through the depth, but the constant term is dominant. This linear variation is not consistent with observed transverse shears. However, the distribution of transverse shear plays a minor role in the overall response of most beams. For thin homogeneous beams, it becomes insignificant and the shear energy is just a penalty that enforces the Euler–Bernoulli hypothesis that normals remain normal. Thus the exact form of the shear is not important in thin homogeneous beams. For composites and thick beams, corrections of the transverse shear are often required.

The above equations are generally not used for the calculation of the velocity strains in a CB beam. It is only sensible to use formulas such as (9.4.18–19) when speed is critical or when the constitutive equations are expressed in terms of resultant stresses. Otherwise the standard continuum expressions given in Chapter 4 are best.

9.4.3 Resultant Stresses and Internal Power

In classical beam and shell theories, the stresses are treated in terms of their integrals, known as resultant stresses. The predominance of the constant and linear parts of the stresses is the basis for the replacing the stresses by resultant stresses. It should be pointed that resultant stresses are not necessary in computations, and are computationally efficient only for linear materials.

In the following, we examine the resultant stresses for CB beam theory. To make the development more manageable, we assume the director to be normal to the reference line, that is, that $\bar{\theta} = 0$. We consider a curved beam in two dimensions with the reference line parametrized by r; $0 \le r \le L$, where r has physical dimensions of length, in contrast to the curvilinear coordinate ξ, which is nondimensional. To define the resultant stresses, we will express the virtual internal power (4.6.12) in terms of corotational components of the Cauchy stress. We omit the power due to $\hat{\sigma}_{yy}$, which vanishes due to the plane stress assumption. This gives

$$\delta P^{\text{int}} = \int_0^L \int_A (\delta \hat{D}_{xx} \hat{\sigma}_{xx} + 2\delta \hat{D}_{xy} \hat{\sigma}_{xy}) dA\ dr \tag{9.4.20a}$$

In the previous, the three-dimensional domain integral has been changed to an area integral and a line integral. This integral is a good approximation to the integral over the volume if the directors at the endpoints are normal to the reference line and the thickness to radius ratio is small in comparison to unity. We next invoke the kinematics of a CB beam. Substituting (9.4.18–19) into (9.4.20a) gives

$$\delta P^{\text{int}} = \int_0^L \int_A (\delta v^M_{\hat{x},\hat{x}} \hat{\sigma}_{xx} - \hat{y}(\delta\omega,_{\hat{x}} + p_{\hat{x},\hat{x}} \delta v^M_{\hat{x},\hat{x}}) \hat{\sigma}_{xx} + (-\delta\omega + \delta v^M_{\hat{y},\hat{x}}) \hat{\sigma}_{xy})\ dA\ dr \tag{9.4.20b}$$

The following area integrals of the stresses (also known as zeroth and first moments) are defined:

$$\begin{aligned}
&\text{membrane force} \quad && n = \int_A \hat{\sigma}_{xx} dA \\
&\text{moment} && m = -\int_A \hat{y} \hat{\sigma}_{xy} dA \\
&\text{shear force} && s = \int_A \hat{\sigma}_{xy} dA
\end{aligned} \tag{9.4.21}$$

The above are known as resultant stresses or generalized stresses; they are shown in Figure 9.6 with their positive sign conventions. The resultant n is the normal force, also called the membrane force or axial force. This is the net force tangent to the midline due to the axial stresses. It is also the zeroth moment of the axial stresses. The moment m is the first moment of the axial stresses about the reference line. The shear force s is the net resultant (zeroth moment) of the transverse shear stresses. These definitions correspond with the customary definitions in structural mechanics.

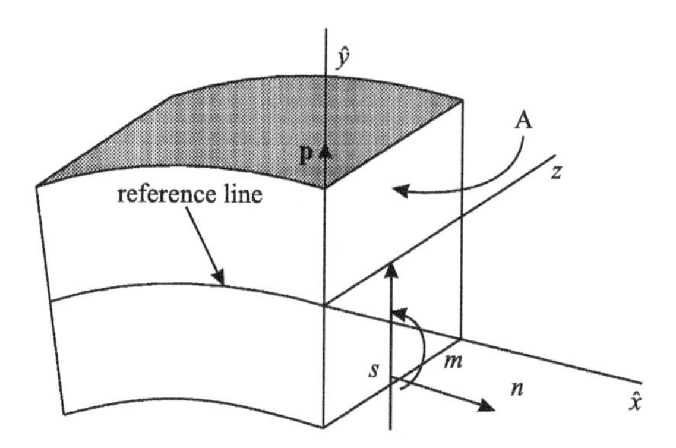

Figure 9.6 Resultant stresses in 2D beam; cross-sectional area A is also shown

With these definitions, the internal virtual power (9.4.20) becomes

$$\delta P^{\text{int}} = \int_0^L \Big(\underbrace{\delta v^M_{\hat{x},\hat{x}} n}_{\text{axial}} + \underbrace{\left(\delta\omega_{,\hat{x}} + p_{\hat{x},\hat{x}} \delta v^M_{\hat{x},\hat{x}}\right) m}_{\text{bending}} + \underbrace{\left(-\delta\omega + \delta v^M_{\hat{y},\hat{x}}\right) s}_{\text{shear}} \Big) dr \tag{9.4.22}$$

The physical names of the various powers are indicated. The axial or membrane power is the power expended in stretching the beam, the bending power in bending the beam. The transverse shear power arises also from bending. Thus the power integral has been reduced to a one-dimensional integral. This is a salient feature of structural theories: the dimension is reduced by one by the imposition of the constraints on the motion.

9.4.4 *Resultant External Forces*

The development of generalized external forces parallels the development of the generalized internal forces we have just completed. We start with the expression for a continuum and then impose the constraints on the motion. Due to the constraint, the generalized external forces involve integrals over the thickness of the CB beam. Assume that $\mathbf{p}$ is coincident with $\hat{y}$ at the ends of the beam. The top and bottom surfaces of the beam are denoted by Γ_{tb} and the surfaces on the ends by Γ. We start with the virtual external power for continua, given by (B4.2.5):

$$\delta P^{\text{ext}} = \int_{\Gamma_{tb}\cup\Gamma} \left(\delta\hat{v}_x \hat{t}^*_x + \delta\hat{v}_y \hat{t}^*_y\right) d\Gamma + \int_\Omega \left(\delta\hat{v}_x \hat{b}_x + \delta\hat{v}_y \hat{b}_y\right) d\Omega \tag{9.4.23}$$

In the above and the remainder of this chapter, we use asterisks to designate prescribed tractions. We now restrict the motion by imposing the CB assumptions as given by (9.4.15). Substituting (9.4.15) into (9.4.23) (with $\bar{\theta} = 0$) yields

$$\delta P^{\text{ext}} = \int_{\Gamma_{tb}\cup\Gamma} \left(\left(\delta\hat{v}^M_x - \delta\omega\hat{y}\right)\hat{t}^*_x + \delta\hat{v}^M_y \hat{t}^*_y\right) d\Gamma + \int_\Omega \left(\left(\delta\hat{v}^M_x - \delta\omega\hat{y}\right)\hat{b}_x + \delta\hat{v}^M_y \hat{b}_y\right) d\Omega \tag{9.4.24}$$

The generalized external forces are now defined similarly to the resultant stresses by taking the zeroth and first moments of the tractions:

$$n^* = \int_A \hat{t}_x^* \, dA, \quad m^* = -\int_A \hat{y}\hat{t}_x^* \, dA, \quad s^* = \int_A \hat{t}_y^* \, dA \tag{9.4.25}$$

The tractions and body force between the endpoints become generalized body forces in beam theory. They are defined by

$$\bar{f}_x = \int_{\Gamma_{tb}} \hat{t}_x^* \, d\Gamma + \int_A \hat{b}_x \, dA, \quad \bar{f}_y = \int_{\Gamma_{tb}} \hat{t}_y^* \, d\Gamma + \int_A \hat{b}_y \, dA, \quad M = -\int_{\Gamma_{tb}} \hat{y}\hat{t}_x^* \, d\Gamma + \int_A \hat{y}\hat{b}_y \, dA \tag{9.4.26}$$

Since the dependent variables have been changed from $v_i(x, y)$ to $v_i^M(r)$ and $\omega(r)$ by the Mindlin–Reissner assumption, the definitions of boundaries are changed accordingly: the boundaries become the endpoints of the beam. The endpoints are the intersections of the midline with Γ. The external virtual power (9.4.23), in light of the definitions (9.4.25–26), becomes

$$\delta P^{\text{ext}} = \int_0^L \left(\delta\hat{v}_x \bar{f}_x + \delta\hat{v}_y \bar{f}_y + \delta\omega M\right) dr + \delta\hat{v}_x n^* \Big|_{\Gamma_n} + \delta\hat{v}_y s^* \Big|_{\Gamma_s} + \delta\omega_m^* \Big|_{\Gamma_m} \tag{9.4.27}$$

where Γ_n, Γ_m and Γ_s are the endpoints of the beam at which the normal (axial) force, moment, and shear force are prescribed, respectively. A boundary on which the traction t_x is prescribed becomes a prescribed normal force boundary and moment boundary, Γ_n and Γ_m. The transformation of a continuum to a CB beam thus changes the nature of the traction boundary conditions. The traction boundary conditions are weakened: only the zeroth and first moments of the tractions can be prescribed. It is possible in beam theory to have a boundary Γ which is a Γ_m boundary but not a Γ_n boundary. An example is a boundary which is fixed in the x-direction but connected to a coil spring with its axis aligned with $\mathbf{e}_z$. These subtleties are due to the change in dependent variables which results from the kinematic constraint.

9.4.5 *Boundary Conditions*

The boundary conditions are subdivided into natural and essential boundary conditions. The velocity (displacement) boundary conditions are essential and given by

$$\hat{v}_x^M = \hat{v}_x^{M*} \text{ on } \Gamma_{\hat{v}_x}, \qquad \hat{v}_y^M = \hat{v}_y^{M*} \text{ on } \Gamma_{\hat{v}_y}, \qquad \omega = \omega^* \text{on } \Gamma_\omega \tag{9.4.28}$$

where the subscript on Γ indicates which velocity is prescribed. The angular velocity is independent of the orientation of the coordinate system so we have not superposed a hat on it.

The generalized traction boundary conditions are:

$$n = n^* \text{ on } \Gamma_n, \qquad s = s^* \text{ on } \Gamma_s, \qquad m = m^* \text{ on } \Gamma_m \tag{9.4.29}$$

Note that (9.4.28) and (9.4.29) are boundary conditions on kinematic and kinetic variables which are conjugate in power. Variables which are conjugate in power cannot be prescribed on the same boundary, but one of the pair must be prescribed on any boundary, so it follows that

$$\begin{aligned} &\Gamma_n \cup \Gamma_{\hat{v}_x} = \Gamma, \quad \Gamma_n \cap \Gamma_{\hat{v}_x} = 0, \\ &\Gamma_s \cup \Gamma_{\hat{v}_y} = \Gamma, \quad \Gamma_s \cap \Gamma_{\hat{v}_y} = 0 \\ &\Gamma_m \cup \Gamma_\omega = \Gamma, \quad \Gamma_m \cap \Gamma_\omega = 0 \end{aligned} \tag{9.4.30}$$

9.4.6 *Weak Form*

The weak form of the momentum equation for a beam is given by

$$\delta P^{\text{kin}} = \delta P^{\text{int}} = \delta P^{\text{ext}} \quad \forall(\delta v_{x,} \delta v_{y,} \delta\omega) \in U_0 \tag{9.4.31}$$

where U_0 is the space of piecewise differentiable functions, that is, C^0 functions, which vanish on the corresponding prescribed displacement boundaries. The functions need only be C^0 since only the first derivatives of the dependent variables appear in the virtual power expressions. Note that the weak form has the same structure as the weak form for continua.

9.4.7 *Strong Form*

We will not derive the strong form equivalent to (9.4.31) for an arbitrary geometry. This can be done, see Simo and Fox (1989), but it is awkward without curvilinear tensors. Instead, we will develop the strong form for small strain theory for a straight beam of uniform cross-section which lies along the x-axis, with inertia and applied body moments neglected. For the above simplifications, using the definitions (9.4.22) and (9.4.27), (9.4.31) can be reduced to

$$\begin{aligned} &\int_0^L (\delta v_{x,x} n + \delta\omega,_x m + (\delta v_{y,x} - \delta\omega)s - \delta v_x \bar{f}_x - \delta v_y \bar{f}_y)dx \\ &\quad -(\delta v_x n^*)\big|_{\Gamma_n} - (\delta\omega m^*)\big|_{\Gamma_m} - (\delta v_y s^*)\big|_{\Gamma_s} = 0 \end{aligned} \tag{9.4.32}$$

The hats have been dropped since the local coordinate system coincides with the global system at all points. The procedure for finding the equivalent strong form then parallels the procedure used in Section 4.3. The idea is to remove all derivatives of test functions in the weak form, so that the above can be written as products of the test functions with a function of the resultant forces and their derivatives. This is accomplished by using integration by parts, which is sketched next for each term in the weak form:

$$\int_0^L \delta v_{x,x} n \, dx = -\int_0^L \delta v_x n,_x \, dx + (\delta v_x n)\big|_{\Gamma_n} - \sum_i [\![\delta v_x n_{\Gamma_i}]\!] \tag{9.4.33}$$

$$\int_0^L \delta\omega,_x m dx = -\int_0^L \delta\omega m,_x \, dx + (\delta\omega m)\big|_{\Gamma_m} - \sum_i [\![\delta\omega m_{\Gamma_i}]\!] \tag{9.4.34}$$

$$\int_0^L \delta v_{y,x} s \, dx = -\int_0^L \delta v_y s_{,x} \, dx + (\delta v_y s)\Big|_{\Gamma_s} - \sum_i [\![\delta v_y s_{\Gamma_i}]\!] \tag{9.4.35}$$

where Γ_i are the points of discontinuity. In each of the above we have applied the fundamental theorem of calculus as given in Chapter 2 for a piecewise continuously differentiable function. We have also used the fact that the test functions vanish on the prescribed displacement boundaries, so the boundary term only applies to the traction boundary points. Substituting (9.4.33–35) into (9.4.32) gives (after a change of sign):

$$\begin{aligned}&\int_0^L (\delta v_x (n_{,x} + f_x) + \delta\omega(m_{,x} + s) + \delta v_y (s_{,x} + f_y))dx \\ &\quad + \sum_i (\delta v_x [\![n]\!] + \delta v_y [\![s]\!] + \delta\omega [\![m]\!])_{\Gamma_i} \\ &\quad - \delta v_x (n^* - n)\Big|_{\Gamma_n} + \delta\omega(m^* - m)\Big|_{\Gamma_m} + \delta v_y (s^* - s)\Big|_{\Gamma_s} = 0\end{aligned} \tag{9.4.36}$$

Using the density theorem as given in Section 4.3 then gives the following strong form:

$$n_{,x} + f_x = 0, \quad s_{,x} + f_y = 0, \quad m_{,x} + s = 0, \tag{9.4.37}$$

$$[\![n]\!] = 0, \qquad [\![s]\!] = 0, \qquad [\![m]\!] = 0 \text{ on } \Gamma_i \tag{9.4.38}$$

$$n = n^* \text{ on } \Gamma_n, \quad s = s^* \text{ on } \Gamma_s, \quad m = m^* \text{ on } \Gamma_m \tag{9.4.39}$$

which are respectively, the equations of equilibrium, the internal continuity conditions, and the generalized traction (natural) boundary conditions.

These equilibrium equations are well known in structural mechanics. These equilibrium equations are not equivalent to the continuum equilibrium equations, $\sigma_{ij,j} + b_i = 0$. Instead, they can be considered *weak forms of the continuum equilibrium equations*, since they hold for integrals of the stresses, namely the moment, shear and normal force. These equilibrium equations are a direct consequence of constraining the motion by the Timoshenko assumptions. By constraining the test functions, the equilibrium equations are weakened.

9.4.8 *Finite Element Approximation*

The finite element approximation to (9.4.1) is constructed by means of one-dimensional shape functions $N_I(\xi)$:

$$\mathbf{x}(\xi, \eta, t) = \left(\mathbf{x}_I^M(t) + \eta \frac{h^0}{2} \mathbf{p}_I(t) \right) N_I(\xi) \tag{9.4.40}$$

where repeated upper case indices are summed to n_N. As indicated above, the *product of the thickness with the director is interpolated.* If the thickness and the director are interpolated independently, the second term in the above is quadratic in the shape functions and the motion differs from that of the underlying continuum element, (9.3.3). It follows immediately from the above that the initial configuration of the element is given by

$$\mathbf{X}(\xi,\eta)=\left(\mathbf{X}_I^M+\eta\frac{h^0}{2}\mathbf{p}_I^0\right)N_I(\xi) \tag{9.4.41}$$

The displacement is obtained by taking the difference of (9.4.40) and (9.4.41), which gives

$$\mathbf{u}(\xi,\eta,t)=\left(\mathbf{u}_I^M(t)+\eta\frac{h^0}{2}(\mathbf{p}_I(t)-\mathbf{p}_I^0)\right)N_I(\xi) \tag{9.4.42}$$

Taking the material time derivative of the above gives the velocity

$$\mathbf{v}(\xi,\eta,t)=\left(\mathbf{v}_I^M(t)+\eta\frac{h^0}{2}(\omega\mathbf{e}_z\times\mathbf{p}_I(t))\right)N_I(\xi) \tag{9.4.43}$$

This velocity field is identical to the velocity field generated by substituting (9.3.9) into (9.3.4): see Example 9.1. Thus any element generated by this approach will be identical to an element implemented as described in Section 9.3. Therefore we will not pursue this approach further.

Example 9.1 Two-node beam element The CB beam theory is used to formulate a 2-node CB beam element based on a 4-node, continuum quadrilateral. The element is shown in Figure 9.7. We place the reference line (midline) midway between the top and bottom surfaces; the line coincides with $\eta = 0$ in the parent domain; although this placement is not necessary it is convenient. The master nodes are placed at the intersections of the

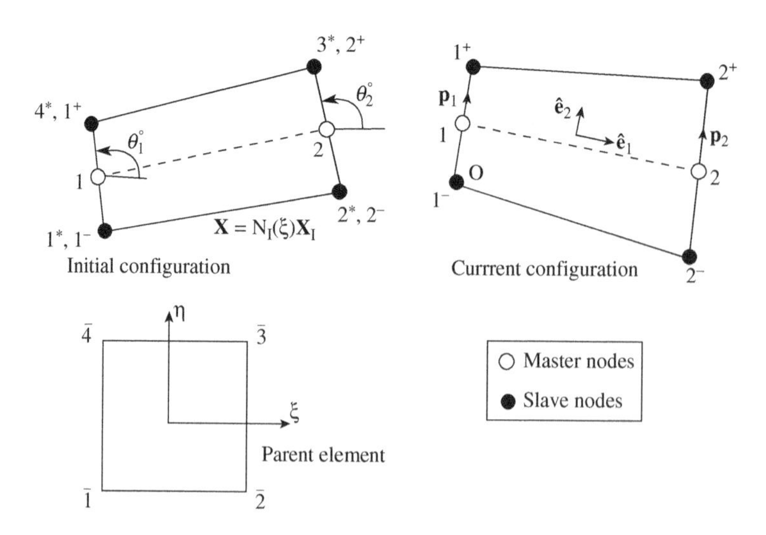

Figure 9.7 Two-node CB beam element based on 4-node quadrilateral continuum element

reference line with the edges of the element. The slave nodes are the corner nodes and are labeled by the two numbering schemes described previously.

The motion of the 4-node continuum element is

$$\mathbf{x} = \mathbf{x}_{I^*}(t)N_{I^*}(\xi,\eta) = \mathbf{x}_{1^*}N_{I^*} + \mathbf{x}_{2^*}N_{2^*} + \mathbf{x}_{3^*}N_{3^*} + \mathbf{x}_{4^*}N_{4^*} \tag{E9.1.1}$$

where $N_{I*}(\xi, \eta)$ are the standard 4-node isoparametric shape functions

$$N_{I^*}(\xi,\eta) = \frac{1}{4}(1+\xi_{I^*}\xi)(1+\eta_{I^*}\eta) \qquad (\text{no sum on } I^*) \tag{E9.1.2}$$

Writing out this motion gives

$$\begin{aligned}\mathbf{x} = \mathbf{x}_{I^*}N_{I^*} &= \frac{1}{4}\mathbf{x}_{1^*}(1-\xi)(1-\eta) + \frac{1}{4}\mathbf{x}_{2^*}(1+\xi)(1-\eta)\\ &+ \frac{1}{4}\mathbf{x}_{3^*}(1+\xi)(1+\eta) + \frac{1}{4}\mathbf{x}_{4^*}(1-\xi)(1+\eta)\\ &= \frac{1}{4}(\mathbf{x}_{1^*}+\mathbf{x}_{4^*})(1-\xi) + \frac{1}{4}(\mathbf{x}_{4^*}-\mathbf{x}_{1^*})\eta(1-\xi)\\ &+ \frac{1}{4}(\mathbf{x}_{2^*}+\mathbf{x}_{3^*})(1+\xi) + \frac{1}{4}(\mathbf{x}_{3^*}-\mathbf{x}_{2^*})\eta(1+\xi)\end{aligned} \tag{E9.1.3}$$

Let

$$\mathbf{x}_1(t) = \frac{1}{2}(\mathbf{x}_{1^*}+\mathbf{x}_{4^*}) \equiv \frac{1}{2}(\mathbf{x}_{1^-}+\mathbf{x}_{1^+}), \quad \mathbf{x}_2(t) \equiv \frac{1}{2}(\mathbf{x}_{2^*}+\mathbf{x}_{3^*}) = \frac{1}{2}(\mathbf{x}_{2^-}+\mathbf{x}_{2^+}) \tag{E9.1.4}$$

$$\|\mathbf{x}_{4^*}-\mathbf{x}_{1^*}\| = h_1^0, \quad \|\mathbf{x}_{3^*}-\mathbf{x}_{2^*}\| = h_2^0, \quad \mathbf{p}_1 = \frac{\mathbf{x}_{4^*}-\mathbf{x}_{1^*}}{h_1^0}, \qquad \mathbf{p}_2 = \frac{\mathbf{x}_{3^*}-\mathbf{x}_{2^*}}{h_2^0}$$

Substituting the above into (E9.1.1–2) gives

$$\mathbf{x} = \frac{1}{2}\mathbf{x}_1(t)(1-\xi) + \frac{1}{2}\mathbf{x}_2(t)(1+\xi) + \eta\frac{h_1^0}{4}\mathbf{p}_1(t)(1-\xi) + \eta\frac{h_2^0}{4}\mathbf{p}_2(t)(1+\xi) \tag{E9.1.5}$$

which corresponds to the motion (9.4.40). Thus (E9.1.1) and (E9.1.5) are different representations of the same motion.

Inextensibility of all fibers Although the nodal fibers are inextensible, the other fibers in an element may change in length. This can easily be seen without any equations by considering the specific situation shown in Figure 9.8: the fiber at the midpoint obviously shortens.

Nodal forces The master nodal forces are given by (9.11.15):

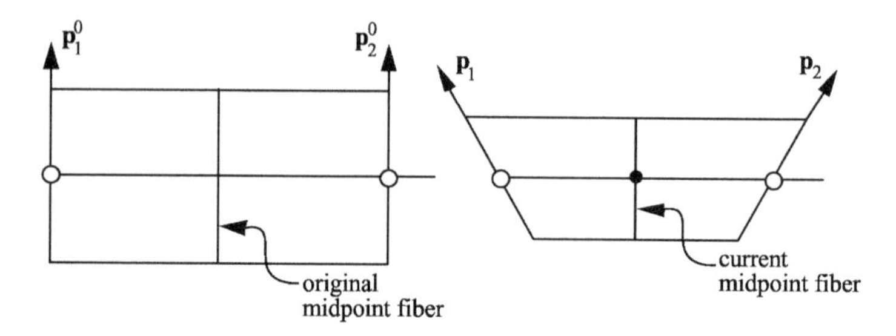

Figure 9.8 Deformation of a 2-node CB beam showing shortening of midpoint fiber

$$\begin{Bmatrix} f_{xI} \\ f_{yI} \\ m_I \end{Bmatrix} = \mathbf{T}_I^T \begin{Bmatrix} f_{xI^-} \\ f_{yI^-} \\ f_{xI^+} \\ f_{yI^+} \end{Bmatrix} \text{where} \quad \mathbf{T}_I = \begin{bmatrix} 1 & 0 & y_I - y_{I^-} \\ 0 & 1 & x_{I^-} - x_I \\ 1 & 0 & y_I - y_{I^+} \\ 0 & 1 & x_{I^+} - x_I \end{bmatrix} \tag{E9.1.6}$$

Evaluating this gives

$$f_{xI} = f_{xI^+} + f_{xI^-} \qquad f_{yI} = f_{yI^+} + f_{yI^-} \tag{E9.1.7}$$

$$m_I = (y_I - y_{I^-})f_{xI^-} + (x_{I^-} - x_I)f_{yI^-} + (y_I - y_{I^+})f_{xI^+} + (x_{I^+} - x_I)f_{yI^+} \tag{E9.1.8}$$

So the transformation gives what is expected from equilibrium: the master nodal force is the sum of the slave nodal forces and the master nodal moment is the moment of the slave nodal forces about the master node.

Green strain This element formulation can be applied to constitutive equations in terms of the PK2 stress and the Green strain. The computation of the Green strain requires the knowledge of θ_I and x_I. The director in the initial and current configurations is given by

$$p_{xI}^0 = \cos\theta_I^0, \quad p_{yI}^0 = \sin\theta_I^0 \quad p_{xI} = \cos\theta_I, \quad p_{yI} = \sin\theta_I \tag{E9.1.9}$$

The positions of the slave nodes can then be computed by specializing (9.4.1) to the nodes, which gives

$$\begin{aligned}
X_{1^+} &= X_1 + \frac{h_0}{2}p_{x1}^0, \quad Y_{1^+} = Y_1 + \frac{h_0}{2}p_{y1}^0 \quad x_{1^+} = x_1 + \frac{h}{2}p_{x1}, \quad y_{1^+} = y_1 + \frac{h}{2}p_{y1} \\
X_{1^-} &= X_1 - \frac{h_0}{2}p_{x1}^0, \quad Y_{1^-} = Y_1 - \frac{h_0}{2}p_{y1}^0 \quad x_{1^-} = x_1 - \frac{h}{2}p_{x1}, \quad y_{1^-} = y_1 - \frac{h}{2}p_{y1} \\
X_{2^-} &= X_2 - \frac{h_0}{2}p_{x2}^0, \quad Y_{2^-} = Y_2 - \frac{h_0}{2}p_{y2}^0 \quad x_{2^-} = x_2 - \frac{h}{2}p_{x2}, \quad y_{2^-} = y_2 - \frac{h}{2}p_{y2} \\
X_{2^+} &= X_2 + \frac{h_0}{2}p_{x2}^0, \quad Y_{2^+} = Y_2 + \frac{h_0}{2}p_{y2}^0 \quad x_{2^+} = x_2 + \frac{h}{2}p_{x2}, \quad y_{2^+} = y_2 + \frac{h}{2}p_{y2}
\end{aligned} \tag{E9.1.10}$$

The displacement of the slave nodes is then obtained by taking the difference of the nodal coordinates. The displacement of any point can then be obtained by the continuum displacement field by $\mathbf{u} = \mathbf{u}_I^* N_I^*$. The Green strain can then be computed by (3.3.6) and the PK2 stress by the constitutive law. The above is a simplistic way to perform the computation. To minimize roundoff error, computations should be done in terms of displacements as in Box 4.6.

Velocity strains for rectangular element When the underlying continuum element is rectangular and the midline of the beam is along the x-axis, the velocity field (9.4.15) is

$$\mathbf{v} = \mathbf{v}^M - y\omega\mathbf{e}_x \tag{E9.1.11}$$

because the directors are in the y-direction and $\bar{\theta} = 0$. Writing out the components of the previous with one-dimensional, linear shape functions gives

$$v_x = v_{x1}^M \frac{1}{2}(1-\xi) + v_{x2}^M \frac{1}{2}(1+\xi) - y\left(\omega_1 \frac{1}{2}(1-\xi) + \omega_2 \frac{1}{2}(1+\xi)\right) \tag{E9.1.12}$$

$$v_y = v_{y1}^M \frac{1}{2}(1-\xi) + v_{y2}^M \frac{1}{2}(1+\xi) \tag{E9.1.13}$$

where $\xi \in [-1, 1]$. The velocity strain components are then given by:

$$D_{xx} = \frac{\partial v_x}{\partial x} = \frac{1}{\ell}(v_{x2}^M - v_{x1}^M) - \frac{y}{\ell}(\omega_2 - \omega_1) \tag{E9.1.14}$$

$$2D_{xy} = \frac{\partial v_y}{\partial x} + \frac{\partial v_x}{\partial y} = \frac{1}{\ell}(v_{y2}^M - v_{y1}^M) - \left(\omega_1 \frac{1}{2}(1-\xi) + \omega_2 \frac{1}{2}(1+\xi)\right) \tag{E9.1.15}$$

The component D_{yy} is computed by the plane stress condition.

9.5 Continuum-Based Shell Implementation

In this section, continuum-based (CB) shell finite elements are developed. This approach was pioneered by Ahmad, Irons and Zienkiewicz (1970); a nonlinear version of this theory was presented by Hughes and Liu (1981a,b). Extensions and generalizations are given in Buechter and Ramm (1992) and Simo and Fox (1989). In the CB shell implementation, as for CB beams, it is not necessary to repeat all of the steps followed in the discretization of the continuum, that is, developing a weak form, discretizing the problem by using finite element interpolants, and so on. Instead the shell element is developed in this section by imposing the constraints of the shell theory on a continuum element. Subsequently, we will examine CB shells from a more theoretical viewpoint by imposing the constraints on the test and trial motions prior to the finite element discretization.

9.5.1 Assumptions in Classical Shell Theories

To describe the kinematic assumptions for shells, we need to define a reference surface, often called a midsurface. The reference surface, as the second name implies, is generally placed midway between the initial top and bottom surfaces of the shell. As in nonlinear CB beams, the exact placement of the reference surface in nonlinear shells is irrelevant.

Before developing the CB shell theory, we briefly review the kinematic assumptions of classical shell theories. As in beams, there are two types of kinematic assumptions, those that admit transverse shear and those that do not. The theories which admit transverse shear are called Mindlin–Reissner theories, whereas the theories which do not admit transverse shear are called Kirchhoff–Love theories. The kinematic assumptions in these two shell theories are:

1. *Kirchhoff–Love theory*: the normal to the midsurface remains straight and normal.
2. *Mindlin–Reissner theory*: the normal to the midsurface remains straight.

Experimental results show that the Kirchhoff–Love assumptions are met by thin shells. For thicker shells or composites, the Mindlin–Reissner assumptions are more appropriate because transverse shear effects become important. Transverse shear effects are particularly important in composites. Mindlin–Reissner theory can also be used for thin shells: in that case the normal will remain approximately normal and the transverse shears will almost vanish.

One point which needs to be made is that Mindlin–Reissner theory was originally developed for small deformation problems, and most of their experimental verification has been made for small deformations. Once the strains are large, it is not clear whether it is better to assume that the *current normal remains straight* or that the *initial normal remains straight*. Currently, in most theoretical work, the initial normal is assumed to remain straight. This choice is probably made because it leads to a cleaner theory. We know of no experiments that show that this assumption is better than the assumption that the current normal remains instantaneously straight.

9.5.2 Coordinates and Definitions

In the implementation and theory of CB shell elements, the shell is modeled by a single layer of three-dimensional elements. The motion is then constrained to reflect the modified Mindlin–Reissner assumptions.

A CB element with nine master nodes and the associated three-dimensional continuum element are shown in Figure 9.9. The parent element coordinates are ξ^i, $i = 1$ to 3; we also use the notation $\xi^1 \equiv \xi$, $\xi^2 \equiv \eta$, and $\xi^3 \equiv \zeta$; the coordinates ξ^i are curvilinear coordinates. Each surface of constant ζ is called a *lamina*. The reference surface corresponds to $\zeta = 0$. The reference surface is parametrized by the two curvilinear coordinates (ξ, η), ξ^α in indicial notation (Greek letters are used for indices with a range of 2). Lines along the ζ axis are called *fibers*, and the unit vector along a fiber is called a *director*. These definitions are analogous to the corresponding definitions for CB beams given previously. The thickness is defined as follows. The distance along the fiber between the bottom surface and the reference surface is denoted by $h^-(\xi, \eta, t)$, and the distance along the fiber between the top surface and the reference surface is denoted by h^+. The thickness is then given by $h = h^+ + h^-$. This is not the customary definition for the

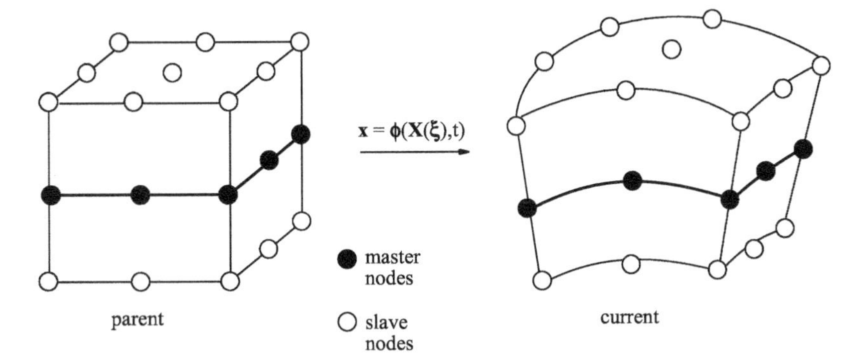

Figure 9.9 Nine-node CB shell element based on 18-node continuum element

thickness of a shell, which is usually defined as the distance between the top and bottom surfaces along the normal, but is common in CB shell theories.

9.5.3 Assumptions

In the CB shell theory, the major assumptions are:

1. fibers remain straight (modified Mindlin–Reissner assumption);
2. the stress normal to the midsurface vanishes (also called the plane stress condition);
3. the momentum due to the extension of the fiber and momentum balance in the direction of the fiber are neglected.

The first assumption differs from classical Mindlin–Reissner theory in that the fibers are constrained to remain straight, not the normal. As in beams, the classical Mindlin–Reissner kinematic assumption cannot be imposed exactly in a CB element with C^0 interpolants. Nodes should be placed so that the fiber direction is as close as possible to the normal. While assumption 1 can be viewed as an approximation to the classical Mindlin–Reissner assumption, we prefer to call it a modified Mindlin–Reissner assumption.

It is often assumed in CB shell theories that the fibers are inextensional. However, this assumption is applied only as stated above and does not hold throughout the formulation. Since the thickness changes for large deformations, some of the effects of the extensibility of the fibers must be included.

9.5.4 Coordinate Systems

Three coordinate systems are defined:

1. The global Cartesian system, (x, y, z) with base vectors $\mathbf{e}_i$.
2. The corotational laminar coordinates $(\hat{x}, \hat{y}, \hat{z})$ with base vectors $\hat{\mathbf{e}}_i$, often called laminar coordinates. These are constructed at each point so that the plane defined by $\hat{\mathbf{e}}_1$ and $\hat{\mathbf{e}}_2$ is

tangent to the lamina at that point. The laminar base vectors vary from point to point, but are viewed as a global Cartesian system; see Section 3.4.3. The construction of the laminar systems is described later.

3. Nodal coordinate systems associated with the master nodes; the associated orthogonal base vectors are identified by superposed bars, as in $\bar{\mathbf{e}}_I(t)$, and the subscript identifies the node. For a lumped mass matrix, the base vectors are coincident with the principal coordinates of the mass matrix. They can also be defined by

$$\bar{\mathbf{e}}_{zI}(t) = \mathbf{p}_I(t) \tag{9.5.1}$$

The orientation of the two other base vectors is arbitrary.

The laminar coordinates do not rotate exactly by the **R** from the polar decomposition theorem. However, particularly if the transverse shears are small, the difference is minute and it is worthwhile to think of them as corotational.

9.5.5 *Finite Element Approximation of Motion*

The underlying continuum element for a CB shell is a three-dimensional isoparametric element with $2n_N$ nodes, with n_N nodes on the top and bottom surfaces as shown in Figure 9.9. To observe the modified Mindlin–Reissner assumption that the motion is linear in ζ, the continuum element may have at most two slave nodes along any fiber. The mesh is Lagrangian and either an updated or a total Lagrangian formulation can be employed. We will describe the updated Lagrangian formulation, but remind the reader that the updated Lagrangian formulation can be transformed to a total Lagrangian formulation as shown in Chapter 4.

At the intersections of the fibers connecting a pair of slave nodes with the reference surface, we define n_N master nodes as shown in Figure 9.9. As in the beam, two notations are used for the slave nodes; the two numbering schemes are related by (9.3.1).

The formulation may have either five or six degrees of freedom per master node. We will emphasize the six-degree-of-freedom formulations and discuss the five-degree-of-freedom formulation and the relative merits later. The nodal velocities and forces at a master node in the six-degree-of-freedom formulation are

$$\dot{\mathbf{d}}_I = \begin{bmatrix} v_{xI} & v_{yI} & v_{zI} & \omega_{xI} & \omega_{yI} & \omega_{zI} \end{bmatrix}^T \tag{9.5.2a}$$

$$\mathbf{f}_I = \begin{bmatrix} f_{xI} & f_{yI} & f_{zI} & m_{xI} & m_{yI} & m_{zI} \end{bmatrix}^T \tag{9.5.2b}$$

where ω_{iI} are the components of the angular velocity and m_{iI} are the components of the moment at node I.

The finite element approximation to the motion in terms of the slave nodes is

$$\mathbf{x}(\xi, t) \equiv \phi(\xi, t) = \sum_{I^-=1}^{nN} \mathbf{x}_{I^-}(t) N_{I^-}(\xi) + \sum_{I^+=1}^{nN} \mathbf{x}_{I^+}(t) N_{I^+}(\xi) = \sum_{I^*=1}^{2nN} \mathbf{x}_{I^*}(t) N_{I^*}(\xi) \tag{9.5.3}$$

where $N_I^*(\xi)$ are standard isoparametric, three-dimensional shape functions and ξ are the parent element coordinates. The notation for slave node numbers is identical to that given for beams in Section 9.3.1. Recall that in a Lagrangian element, the element coordinates can be used as surrogate material coordinates. The velocity field of the underlying continuum element is given by

$$\mathbf{v}(\boldsymbol{\xi}, t) = \sum_{I^*=1}^{2nN} \dot{\mathbf{x}}_{I^*}(t) N_{I^*}(\boldsymbol{\xi}) \equiv \sum_{I^*=1}^{2n_N} \mathbf{v}_{I^*}(t) N_{I^*}(\boldsymbol{\xi}) \tag{9.5.4}$$

where $\dot{\mathbf{x}}_{I^*} \equiv \mathbf{v}_{I^*}$ is the velocity of slave node I^*. Since the motion of the CB shell is linear in ζ, it can be written as

$$\mathbf{x} = \mathbf{x}^M + \zeta \mathbf{x}^B \equiv \mathbf{x}^M + \bar{\zeta}\mathbf{p} \tag{9.5.5a}$$

$$\mathbf{x}^B = h^+\mathbf{p} \text{ and } \bar{\zeta} = \zeta h^+ \quad \text{when } \zeta > 0, \quad \mathbf{x}^B = h^-\mathbf{p} \text{ and } \bar{\zeta} = \zeta h^- \text{ when } \quad \zeta < 0 \tag{9.5.5b}$$

Equations (9.5.3) and (9.5.5a) are two alternative descriptions of the same motion. The former is a function of three independent spatial variables, that is, a continuum representation, whereas the latter is a function of two independent variables, the curvilinear coordinates on the reference surface.

The velocity field is obtained by taking material time derivatives of (9.5.5):

$$\mathbf{v} = \mathbf{v}^M + \zeta \mathbf{v}^B \equiv \mathbf{v}^M + \bar{\zeta}\dot{\mathbf{p}} + \dot{\bar{\zeta}}\,\mathbf{p} \tag{9.5.6}$$

The velocities of the slave nodes are next expressed in terms of the translational velocities of the master nodes $\mathbf{v}_I^M = [v_{xI}^M, v_{yI}^M, v_{zI}^M]^T$ and the angular velocities of the director $\omega_I = [\bar{\omega}_{xI}, \bar{\omega}_{yI}, \bar{\omega}_{zI}]^T$. Writing (9.5.6) at the nodes and using (9.2.1) gives

$$\mathbf{v}_{I^+} = \mathbf{v}_I^M + h_I^+ \boldsymbol{\omega}_I \times \mathbf{p}_I + \dot{h}_I^+ \mathbf{p}_I, \quad \mathbf{v}_{I^-} = \mathbf{v}_I^M - h_I^- \boldsymbol{\omega}_I \times \mathbf{p}_I - \dot{h}_I^- \mathbf{p}_I \tag{9.5.7}$$

where $\dot{h}_I^+$ and $\dot{h}_I^-$ are the rates of change of the thickness.

As stated in the third assumption, momentum balance (or equilibrium in statics) is not enforced for the relative motion in the direction $\mathbf{p}$. So the terms involving $\dot{h}_I^+$ and $\dot{h}_I^-$ in the expressions for the velocity strain are dropped in constructing the equations of motion. They are also omitted in the strain rate computation because the thickness is obtained from the constitutive equations by the plane stress condition. In fact it is often said that in CB shell theory the fibers are inextensional. This is confusing, since while inextensibility is used to omit momentum balance due to relative motion in the $\mathbf{p}$ direction, *the change of thickness is not neglected* in the computation of the nodal internal forces.

To obtain the relation between the nodal velocities for a triplet of nodes along a fiber, we express the cross-product in (9.5.7) as $h^+\boldsymbol{\omega}_I \times \mathbf{p}_I = \boldsymbol{\Lambda}^+\boldsymbol{\omega}_I$, where $\boldsymbol{\Lambda}^+$ is the skew-symmetric tensor given by $\Lambda_{ij}^+ = h^+ e_{ijk} p_k$: see (3.2.44). Then using a similar relation for $\mathbf{v}_{I^-}$, we can relate the slave nodal velocities to the master nodal velocities by

$$\begin{Bmatrix} \mathbf{v}_{I^-} \\ \mathbf{v}_{I^+} \end{Bmatrix} = \mathbf{T}_I \dot{\mathbf{d}}_I \text{ (no sum on } I) \tag{9.5.8}$$

$$\mathbf{v}_{I^-} = [v_{xI^-}, v_{yI^-}, v_{zI^-}]^T, \quad \mathbf{v}_{I^+} = [v_{xI^+}, v_{yI^+}, v_{zI^+}]^T \tag{9.5.9}$$

$$\mathbf{T}_I = \begin{bmatrix} \mathbf{I} & \mathbf{\Lambda}^- \\ \mathbf{I} & \mathbf{\Lambda}^+ \end{bmatrix} \tag{9.5.10}$$

$$\mathbf{\Lambda}^- = -h_I^- \begin{bmatrix} 0 & p_z & -p_y \\ -p_z & 0 & p_x \\ p_y & -px & 0 \end{bmatrix} = \begin{bmatrix} 0 & z_{I^-} - z_I & y_I - y_{I^-} \\ z_I - z_{I^-} & 0 & x_{I^-} - x_I \\ y_{I^-} - y_I & x_I - x_{I^-} & 0 \end{bmatrix} \tag{9.5.11}$$

$$\mathbf{\Lambda}^+ = h_I^+ \begin{bmatrix} 0 & p_z & -p_y \\ -p_z & 0 & p_x \\ p_y & -p_x & 0 \end{bmatrix} = \begin{bmatrix} 0 & z_{I^+} - z_I & y_I - y_{I^+} \\ z_I - z_{I^+} & 0 & x_{I^+} - x_I \\ y_{I^+} - y_I & x_I - x_{I^+} & 0 \end{bmatrix} \tag{9.5.12}$$

It can be seen from (9.5.11) and (9.5.12) that the current thicknesses are used in the master nodal force computation, so accounting for the extensibility of the fibers.

9.5.6 Local Coordinates

The slave nodal forces for the CB shell, that is, the forces at the nodes of the underlying continuum element, are obtained by the usual procedures for continuum elements as given in Chapter 4. Of course, the plane stress assumption and computation of the thickness change must be considered. For this purpose, a corotational laminar coordinate system with base vectors $\hat{\mathbf{e}}_i$ is set up at each quadrature point, and the constitutive update is made in this coordinate system.

Several methods are available for setting up the base vectors $\hat{\mathbf{e}}_i$. The method given next is from Hughes (1987: 386). The objective is to find an orthogonal set of base vectors $\hat{\mathbf{e}}_i$ as close as possible to the covariant base vectors $\mathbf{g}_\alpha$ given by

$$\mathbf{g}_\alpha = \frac{\partial \mathbf{x}}{\partial \xi^\alpha} \tag{9.5.13}$$

The vectors $\mathbf{g}_\alpha$ define a plane tangent to the lamina, and the base vectors $\hat{\mathbf{e}}_i$ will also lie in this plane. The base vector normal to this plane is given by

$$\hat{\mathbf{e}}_z = \frac{\mathbf{g}_1 \times \mathbf{g}_2}{\|\mathbf{g}_1 \times \mathbf{g}_2\|} \tag{9.5.14}$$

The base vectors are developed in two steps. First an auxiliary set of vectors are defined by

$$\mathbf{a} = \frac{\mathbf{g}_1}{\|\mathbf{g}_1\|} + \frac{\mathbf{g}_2}{\|\mathbf{g}_2\|}, \qquad \mathbf{b} = \hat{\mathbf{e}}_z \times \mathbf{a} \tag{9.5.15}$$

The new base vectors are then given by

$$\hat{\mathbf{e}}_x = \frac{\mathbf{a}-\mathbf{b}}{\|\mathbf{a}-\mathbf{b}\|}, \quad \hat{\mathbf{e}}_y = \frac{\mathbf{a}+\mathbf{b}}{\|\mathbf{a}+\mathbf{b}\|} \tag{9.5.16}$$

The laminar components can be evaluated in two ways:

1. Compute **D** as in Example 4.3, (E4.3.5–8), and transform the components by

$$\hat{\mathbf{D}} = \mathbf{R}^T_{\text{lam}}\mathbf{D}\mathbf{R}_{\text{lam}}, \quad (R_{ij})_{\text{lam}} = \mathbf{e}_i \cdot \hat{\mathbf{e}}_j \tag{9.5.17}$$

2. At each point compute the velocity and then the rate-of-deformation in the laminar coordinate system using

$$\hat{\mathbf{v}}_I = \mathbf{R}^T_{\text{lam}}\mathbf{v}_I, \quad \hat{\mathbf{x}}_I = \mathbf{R}^T_{\text{lam}}\mathbf{x}_I, \quad \hat{\mathbf{L}} = \hat{\mathbf{v}}_I N_{I,\hat{\mathbf{x}}} = \hat{\mathbf{v}}_I N_{I,\xi}\hat{\mathbf{x}}_{,\xi}^{-1}, \quad \hat{\mathbf{D}} = \frac{1}{2}(\hat{\mathbf{L}} + \hat{\mathbf{L}}^T) \tag{9.5.18}$$

The standard procedures for three-dimensional Lagrangian continuum elements as described in Chapter 4 are used to evaluate all the components except $\hat{D}_{zz}$; the through-the-thickness component $\hat{D}_{zz}$ is computed from the plane stress condition $\hat{\sigma}_{zz} = 0$.

9.5.7 Constitutive Equation

Any of the continuum material laws described in Chapter 5 can be used for CB shells. However, the plane stress condition must be enforced. The methods for imposing constraints described in Section 6.3.8, such as Lagrange multiplier methods and penalty methods, may be used for imposing this constraint.

For a material model in rate form, the plane stress constraint can be applied as follows. The rate update equations are written in Voigt form:

$$\begin{Bmatrix} \hat{\sigma}_{xx} \\ \hat{\sigma}_{yy} \\ \hat{\sigma}_{xy} \\ \hat{\sigma}_{xz} \\ \hat{\sigma}_{yz} \\ \hat{\sigma}_{zz} \end{Bmatrix}^{n+1} = \begin{Bmatrix} \hat{\sigma}_{xx} \\ \hat{\sigma}_{yy} \\ \hat{\sigma}_{xy} \\ \hat{\sigma}_{xz} \\ \hat{\sigma}_{yz} \\ 0 \end{Bmatrix}^{n+1} = \begin{Bmatrix} \hat{\sigma}_{xx} \\ \hat{\sigma}_{yy} \\ \hat{\sigma}_{xy} \\ \hat{\sigma}_{xz} \\ \hat{\sigma}_{yz} \\ 0 \end{Bmatrix}^{n} + \Delta t \begin{bmatrix} \hat{\mathbf{C}}_{aa} & \hat{\mathbf{C}}_{ab} \\ \hat{\mathbf{C}}^T_{ab} & \hat{\mathbf{C}}_{bb} \end{bmatrix}^{\text{lam}} \begin{Bmatrix} \hat{D}_{xx} \\ \hat{D}_{yy} \\ 2\hat{D}_{xy} \\ 2\hat{D}_{xz} \\ 2\hat{D}_{yz} \\ \hat{D}_{zz} \end{Bmatrix}^{n+\frac{1}{2}} \tag{9.5.19}$$

By letting $\hat{\sigma}_{zz} = 0$ in the previous, the plane stress condition has been imposed. The stress and velocity strain components have been reordered from standard Voigt form so that the thickness component is last. The matrices $\hat{\mathbf{C}}_{aa}$ and $\hat{\mathbf{C}}_{ab}$ are, respectively, 5 × 5 and 5 × 1 submatrices of the tangent modulus matrix, $\hat{\mathbf{C}}^{\text{lam}}$. A modified matrix $\mathbf{C}_{aa}$ which relates the nonzero stress increments can easily be obtained by eliminating the sixth equation, yielding

$$\tilde{\mathbf{C}}^P_{aa} = \hat{\mathbf{C}}_{aa} - \hat{\mathbf{C}}_{ab}\hat{\mathbf{C}}^{-1}_{bb}\hat{\mathbf{C}}^T_{ab} \tag{9.5.20}$$

The rate-of-deformation component $\hat{D}_{zz}$ is obtained from the last line of (9.5.19) and is used to obtain the thickness change as described next.

9.5.8 *Thickness*

The thickness can be obtained either in rate form or directly. The thickness at any time is given by

$$h^+ = \int_0^1 h_0^+ F_{\zeta\zeta}(+\zeta)d\zeta, \qquad h^- = \int_0^1 h_0^- F_{\zeta\zeta}(-\zeta)d\zeta \tag{9.5.21}$$

where $F_{\zeta\zeta}$ is obtained by $F_{\zeta\zeta} = F_{ij}(\mathbf{e}_i \cdot \mathbf{p})(\mathbf{e}_j \cdot \mathbf{p})$. Note that the thickness as defined here differs somewhat from the customary condition of thickness: it is the distance between the top and bottom surfaces along a fiber. The rates of h^+ and h^- are given by

$$\dot{h}^+ = \int_0^1 h^+ D_{\zeta\zeta}(+\zeta)d\zeta, \quad \dot{h}^- = \int_0^1 h^- D_{\zeta\zeta}(-\zeta)d\zeta \tag{9.5.22}$$

where $D_{\zeta\zeta}$ is obtained by $D_{\zeta\zeta} = \hat{D}_{ij}(\hat{\mathbf{e}}_i \cdot \mathbf{p})(\hat{\mathbf{e}}_j \cdot \mathbf{p})$. The thickness update given here provides a two-parameter approximation to the thickness. Since the deformation gradient is approximately linear through the thickness in isoparametric CB elements, this usually suffices. Often, one-parameter forms are used which account only for the average change of thickness. The two-parameter form is more accurate because when bending is superimposed on stretch, the thickness change on the compressive side and the tensile side vary. A more accurate alternative is to compute the new locations of all quadrature points, but this is usually not needed.

9.5.9 *Master Nodal Forces*

The nodal internal and external forces at the master nodes can be obtained from the slave nodal forces by (4.3.35), that is,

$$\mathbf{f}_I = \mathbf{T}_I^T \begin{Bmatrix} \mathbf{f}_{I^-} \\ \mathbf{f}_{I^+} \end{Bmatrix} \text{(no sum on } I\text{)} \tag{9.5.23}$$

where $\mathbf{f}_I$ is given by (9.5.2b) and $\mathbf{T}_I$ is given by (9.5.10). The slave nodal forces are computed by the procedures for a continuum element: see Section 4.5 and Example 4.3.

9.5.10 Mass Matrix

The mass matrix of the CB shell element can be obtained by the transformation (4.5.38), with $\hat{\mathbf{M}}$ the mass matrix for the underlying continuum element. The 6 × 6 submatrices of the mass matrix are then given by

$$\mathbf{M}_{IJ} = \mathbf{T}_I^T \hat{\mathbf{M}}_{IJ} \mathbf{T}_J \text{(no Sum on } I \text{ or } J\text{)} \tag{9.5.24}$$

where $\mathbf{M}_{IJ}$ is the 6 × 6 submatrix of the mass matrix associated with nodes I and J.

In explicit programs and for low-order elements, diagonal mass matrices are frequently employed. The diagonal submatrices of a diagonal mass matrix are given by

$$\mathbf{M}_{II} = \begin{bmatrix} \mathbf{M}_{tI} & 0 \\ 0 & \mathbf{M}_{rI} \end{bmatrix} = \begin{bmatrix} M_{tI} & 0 & 0 & 0 & 0 & 0 \\ 0 & M_{tI} & 0 & 0 & 0 & 0 \\ 0 & 0 & M_{tI} & 0 & 0 & 0 \\ 0 & 0 & 0 & \bar{M}_{xxI} & 0 & 0 \\ 0 & 0 & 0 & 0 & \bar{M}_{yyI} & 0 \\ 0 & 0 & 0 & 0 & 0 & \bar{M}_{zzI} \end{bmatrix} \tag{9.5.25}$$

where $\mathbf{M}_{tI} = [M_{tI}]$ is the translational mass and $\mathbf{M}_{rI} = [\bar{M}]$ are the rotational inertias, which correspond to the products of inertia about the node. As indicated previously, the components of the rotational mass are customarily expressed in terms of the nodal coordinate system. An isotropic, diagonal mass matrix has the property $\bar{M}_r = M_{xx} = M_{yy} = M_{zz}$, so its components are identical in any coordinate system. The Key–Beisinger (1971) scaling is frequently used in explicit codes to avoid limiting the stable time step by the rotational behavior; see Section 9.3.7.

9.5.11 Discrete Momentum Equation

For the diagonal mass matrix given above, the translational equations of motion at a node are given by

$$\mathbf{M}_{tI}\dot{\mathbf{v}}_I + \mathbf{f}_I^{\text{int}} = \mathbf{f}_I^{\text{ext}} \left(\text{no sum on } I\right) \tag{9.5.26}$$

where the nodal forces and nodal velocities are

$$\mathbf{f}_I = \begin{Bmatrix} f_{xI} \\ f_{yI} \\ f_{zI} \end{Bmatrix}, \mathbf{v}_I = \begin{Bmatrix} v_{xI} \\ v_{yI} \\ v_{zI} \end{Bmatrix} \tag{9.5.27}$$

The rotational motion of each node is described by the triad of nodal base vectors $\bar{e}_{iI}$, $i = 1$ to 3. The orientation of the triad of node I is coincident with the principal coordinates of the moment of inertia tensor M_{ijI} of the node. The rotational equations of motion are expressed in the nodal coordinate system, since the rotational mass matrix is invariant in these coordinates. The equations of motion for an anisotropic diagonal mass are given by

$$M_{xxI}\,\dot{\bar{\omega}}_{xI} + (\bar{M}_{zzI} - \bar{M}_{yyI})\bar{\omega}_{yI}\bar{\omega}_{zI} + \bar{m}_{xI}^{\text{int}} = \bar{m}_{xI}^{\text{ext}} \tag{9.5.28}$$

$$M_{yyI}\,\dot{\bar{\omega}}_{yI} + (\bar{M}_{xxI} - \bar{M}_{zzI})\bar{\omega}_{xI}\bar{\omega}_{zI} + \bar{m}_{yI}^{\text{int}} = \bar{m}_{yI}^{\text{ext}} \tag{9.5.29}$$

$$M_{zzI}\,\dot{\bar{\omega}}_{zI} + (\bar{M}_{yyI} - \bar{M}_{xxI})\bar{\omega}_{xI}\bar{\omega}_{yI} + \bar{m}_{zI}^{\text{int}} = \bar{m}_{zI}^{\text{ext}} \tag{9.5.30}$$

where the bars indicate the components in the nodal system. The above are the well-known Euler's equations of motion. They are nonlinear in the angular velocities but the quadratic terms vanish for an isotropic rotational mass matrix.

9.5.12 Tangent Stiffness

The tangent stiffness and load stiffness matrices can be obtained from that of the underlying continuum element by the standard transformation for stiffness matrices,

$$\mathbf{K}_{IJ} = \mathbf{T}_I^T \bar{\mathbf{K}}_{IJ} \mathbf{T}_J \,(\text{no sum on } I \text{ or } J) \tag{9.5.31}$$

where $\bar{\mathbf{K}}_{IJ}$ is the tangent stiffness matrix for the continuum element and $\mathbf{T}_I$ is given in (9.5.10). The tangential and load stiffness matrices for continuum elements are given in Section 6.4.

9.5.13 Five Degree-of-Freedom Formulation

As will be seen in the next section, the motion of a shell in the absence of kinks and junctions can be treated with five degrees-of-freedom per node. In this case, the nodal velocities at a master node are

$$\mathbf{v}_I = [v_{xI}, v_{yI}, v_{zI}, \bar{\omega}_{xI}, \bar{\omega}_{yI}]^T \tag{9.5.32}$$

where the angular velocity component $\bar{\omega}_{zI}$ has been omitted. In a five-degree-of-freedom formulation, the angular velocity components must be expressed in the nodal system which rotates with the node. The nodal forces are conjugate to the nodal velocities in power and are given by

$$\mathbf{f}_I = [f_{xI},\ f_{yI},\ f_{zI},\ \bar{m}_{xI},\ \bar{m}_{yI}]^T \tag{9.5.33}$$

The relationship between the slave and master nodal velocities for each triplet of nodes along a fiber for a five-degree-of-freedom formulation can be written in matrix form similar to (9.5.8–12)

$$\begin{Bmatrix} \bar{\mathbf{v}}_{I^-} \\ \bar{\mathbf{v}}_{I^+} \end{Bmatrix} = \mathbf{T}_I \dot{\mathbf{d}}_I \left(\text{no sum on } I\right) \tag{9.5.34}$$

where the nodal velocities have been expressed in the nodal coordinate system of the master node for convenience. In the previous,

$$\bar{\mathbf{v}}_{I^-} = [\bar{v}_{xI^-},\ \bar{v}_{yI^-},\ \bar{v}_{zI^-}]^T,\quad \bar{\mathbf{v}}_{I^+} = [\bar{v}_{xI^+},\ \bar{v}_{yI^+},\ \bar{v}_{zI^+}]^T \tag{9.5.35}$$

$$\dot{\bar{\mathbf{d}}}_I = [\bar{v}_{xI},\ \bar{v}_{yI},\ \bar{v}_{zI},\ \bar{\omega}_{xI},\ \bar{\omega}_{yI}]^T \tag{9.5.36}$$

$$\mathbf{T}_I = \begin{bmatrix} \mathbf{I}_{3\times3} & \boldsymbol{\Lambda}^- \\ \mathbf{I}_{3\times3} & \boldsymbol{\Lambda}^+ \end{bmatrix},\quad \boldsymbol{\Lambda}^- = -h_I^- \begin{bmatrix} 0 & 1 \\ -1 & 0 \\ 0 & 0 \end{bmatrix},\quad \boldsymbol{\Lambda}^+ = h_I^+ \begin{bmatrix} 0 & 1 \\ -1 & 0 \\ 0 & 0 \end{bmatrix}, \tag{9.5.37}$$

The transformation of slave nodal forces to master nodal forces is given by (9.5.23).

The five-degree-of-freedom formulation fits better with the theory of CB shells than do six-degree-of-freedom formulations, as will become clear in the next section. In fact, when the shell is flat, the stiffness is singular for a six-degree-of-freedom formulation. On the other hand, a five-degree-of-freedom formulation must be modified at corners and to incorporate structural features such as stiffeners. For software with a variable number of degree of freedom at a node, using six degrees of freedom only at the nodes where the additional degrees of freedom are required is probably best.

9.5.14 *Large Rotations*

The treatment of large rotations in three dimensions is described in the following. This topic has been extensively explored in the literature on large displacement finite element methods and multi-body dynamics; for an account, see Crisfield, (1991: 183) or Shabana (1998). Large rotations are usually treated by Euler angles in classical dynamics texts. However, Euler angles are nonunique for certain orientations and lead to awkward equations of motion. Therefore alternative techniques which are cleaner and more robust are usually employed.

9.5.15 *Euler's Theorem*

A fundamental concept in large rotations is the theorem of Euler. This theorem states that in any rigid body rotation, there exists a line which remains fixed; the body rotates about this line. From this theorem general formulas for the rotation matrix can be developed.

Consider the rotation of vector $\mathbf{r}$ by an angle θ about the axis defined by the unit vector $\mathbf{e} = \mathbf{e}_1$. The vector after the rotation is denoted by $\mathbf{r}'$ as shown in Figure 9.10. The rotation matrix $\mathbf{R}$ relates $\mathbf{r}'$ to $\mathbf{r}$ by

$$\mathbf{r}' = \mathbf{R}\mathbf{r} \tag{9.5.38}$$

where $\mathbf{R}$ is to be determined. We will first derive the formula

$$\mathbf{r}' = \mathbf{r} + \sin\theta\mathbf{e}\times\mathbf{r} + (1-\cos\theta)\mathbf{e}\times(\mathbf{e}\times\mathbf{r}) \tag{9.5.39}$$

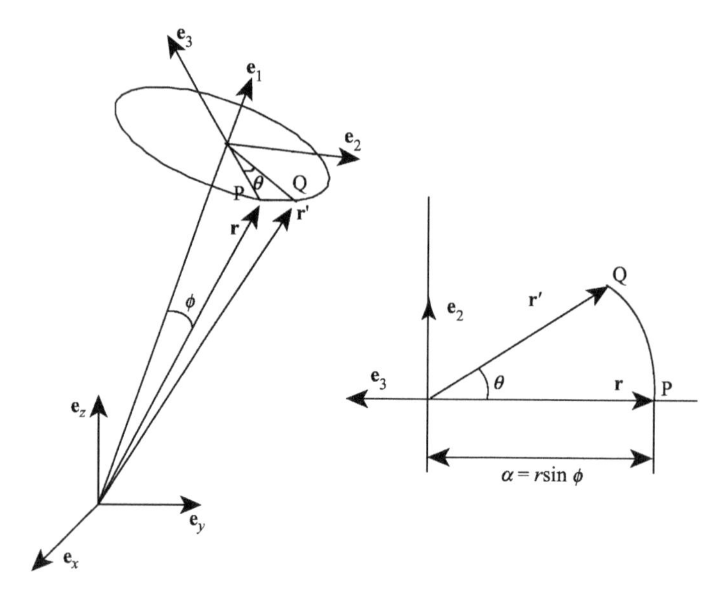

Figure 9.10 Rotation of a vector **r** viewed as a rotation about a fixed axis $\boldsymbol{\theta} = \theta\mathbf{e}$ according to Euler's theorem; on the right a top view along the $\boldsymbol{\theta}$ axis is shown

where **e** is a unit vector along the axis of rotation, whose existence we know by Euler's theorem. The schematic on the right of Figure 9.10 shows the body as viewed along the **e** axis. It can be seen from this schematic that

$$\mathbf{r}' = \mathbf{r} + \mathbf{r}_{PQ} = r + \alpha \ \sin\ \theta\mathbf{e}_2\ +\ \alpha(1-\cos\ \theta)\mathbf{e}_3 \text{ where } \alpha = r\sin\phi \tag{9.5.40}$$

From the definition of the cross-product it follows that

$$\alpha\mathbf{e}_2 = r\ \sin\phi\mathbf{e}_2 = \mathbf{e}\times\mathbf{r},\quad \alpha\mathbf{e}_3 = r\ \sin\ \phi\mathbf{e}_3 = \mathbf{e}\times(\mathbf{e}\times\mathbf{r}) \tag{9.5.41}$$

Substituting (9.5.41), $\alpha\mathbf{e}_2 = \mathbf{e}\times\mathbf{r}$ and $\alpha\mathbf{e}_3 = \mathbf{e}\times(\mathbf{e}\times\mathbf{r})$, into (9.5.40) yields (9.5.39).

We next rewrite (9.5.39) in matrix form. Recall from (3.2.44–46) that a skew-symmetric tensor can be defined by $\Omega_{ij}(\mathbf{v}) = -e_{ijk}v_k$ where e_{ijk} is the permutation symbol, so that

$$\mathbf{v}\times\mathbf{r} = \boldsymbol{\Omega}(\mathbf{v})\mathbf{r} \quad \text{where} \quad \boldsymbol{\Omega}(\mathbf{v}) = \begin{bmatrix} 0 & -v_3 & v_2 \\ v_3 & 0 & -v_1 \\ -v_2 & v_1 & 0 \end{bmatrix} \tag{9.5.42}$$

for any vector **v**. From the definition of $\boldsymbol{\Omega}(\mathbf{v})$ it follows that

$$\boldsymbol{\Omega}(\mathbf{e})\mathbf{r} = \mathbf{e}\times\mathbf{r},\quad \boldsymbol{\Omega}^2(\mathbf{e})\mathbf{r} = \boldsymbol{\Omega}(\mathbf{e})\boldsymbol{\Omega}(\mathbf{e})\mathbf{r} = \mathbf{e}\times(\mathbf{e}\times\mathbf{r}) \tag{9.5.43}$$

This is illustrated in Figure 9.11 along with higher powers of $\boldsymbol{\Omega}(\mathbf{v})$.

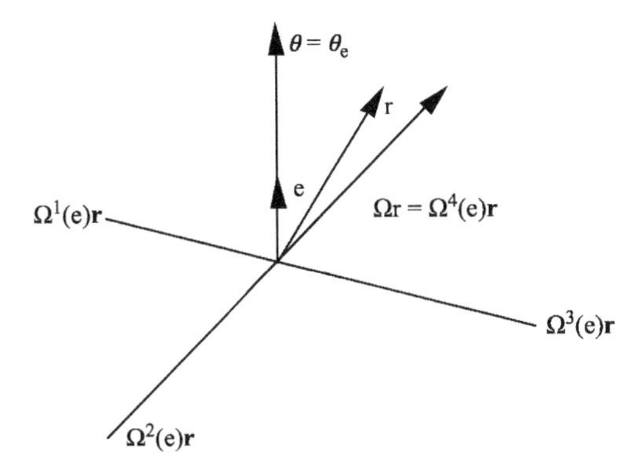

Figure 9.11 The powers of $\mathbf{\Omega}(\mathbf{e})$

Replacing the cross-products in (9.5.39) with the vector products yields

$$\mathbf{r}' = \mathbf{r} + \sin\theta\mathbf{\Omega}(\mathrm{e})\mathbf{r} + (1-\cos\theta)\mathbf{\Omega}^2(\mathbf{e})\mathbf{r} \tag{9.5.44}$$

Comparison of (9.5.44) with (9.5.38) yields

$$\mathbf{R} = \mathbf{I} + \sin\theta\mathbf{\Omega}(\mathbf{e}) + (1-\cos\theta)\mathbf{\Omega}^2(\mathbf{e}) \tag{9.5.45}$$

In writing the rotation matrix, it is often useful to define a column matrix $\boldsymbol{\theta}$ given by $\boldsymbol{\theta} = \theta\mathbf{e}$. In terms of $\mathbf{\Omega}(\boldsymbol{\theta})$, (9.5.45) can be written as

$$\mathbf{R} = \mathbf{I} + \frac{\sin\theta}{\theta}\mathbf{\Omega}(\boldsymbol{\theta}) + \frac{1-\cos\theta}{\theta^2}\mathbf{\Omega}^2(\boldsymbol{\theta}) \tag{9.5.46}$$

The column matrix $\boldsymbol{\theta}$ is called a pseudovector because it does not possess the properties of a vector. In particular, the operation of addition is not commutative: if the pseudovector $\boldsymbol{\theta}_{12}$ corresponds to the rotation $\boldsymbol{\theta}_1$ followed by the rotation $\boldsymbol{\theta}_2$, and if $\boldsymbol{\theta}_{21}$ corresponds to the rotation $\boldsymbol{\theta}_2$ followed by the rotation $\boldsymbol{\theta}_1$, then $\boldsymbol{\theta}_{12} \neq \boldsymbol{\theta}_{21}$. A column matrix which lacks any of the properties of a vector cannot be considered a vector. The lack of commutativity in addition of rotations is often illustrated in introductory physics texts by rotating a book 90° about its spine followed by a 90° rotation about its bottom, and comparing this with a 90° rotation about the bottom followed by a 90° rotation about the spine. The configurations at the conclusion of these two processes are not the same, which means that the addition of rotations is not commutative.

9.5.16 Exponential Map

A powerful way to describe rotation is the *exponential map*, which gives the rotation matrix **R** by

$$\mathbf{R} = \exp(\mathbf{\Omega})(\boldsymbol{\theta}) = \sum_{n=1}^{\infty}\frac{\mathbf{\Omega}^n(\boldsymbol{\theta})}{n!} = \mathbf{I} + \mathbf{\Omega}(\boldsymbol{\theta}) + \frac{\mathbf{\Omega}^2(\boldsymbol{\theta})}{2} + \frac{\mathbf{\Omega}^3(\boldsymbol{\theta})}{6} + \cdots \tag{9.5.47}$$

This form can be used to obtain approximations to the rotation matrix. To develop the exponential map we note that the matrix $\boldsymbol{\Omega}(\boldsymbol{\theta})$ satisfies the following recurrence relation:

$$\boldsymbol{\Omega}^{n+2}(\boldsymbol{\theta}) = -\theta^2\boldsymbol{\Omega}^n(\boldsymbol{\theta}), \quad \text{or} \quad \boldsymbol{\Omega}^{n+2}(\mathbf{e}) = -\boldsymbol{\Omega}^n(\mathbf{e}) \tag{9.5.48}$$

which is illustrated in Figure 9.11. This can be obtained by (9.5.43):

$$\boldsymbol{\Omega}^{n+2}(\mathbf{e})\mathbf{r} = \mathbf{e}\times(\mathbf{e}\times\boldsymbol{\Omega}^n(\mathbf{e})\mathbf{r} = -\boldsymbol{\Omega}^n(\mathbf{e})\mathbf{r} \tag{9.5.49}$$

for any vector $\mathbf{r}$. The trigonometric functions $\sin\theta$ and $\cos\theta$ can be expanded in Taylor's series

$$\sin\theta = \theta - \frac{\theta^3}{3!} + \frac{\theta^5}{5!} - \cdots, \quad \cos\theta = 1 - \frac{\theta^2}{2!} + \frac{\theta^2}{4!} - \cdots \tag{9.5.50}$$

Substituting (9.24.50) into (9.24.45) gives

$$\mathbf{R} = \mathbf{I} + \left(\theta - \frac{\theta^3}{3!} + \cdots\right)\boldsymbol{\Omega}(\mathbf{e}) + \left(\frac{\theta^2}{2!} - \frac{\theta^4}{4!} + \cdots\right)\boldsymbol{\Omega}^2(\mathbf{e}) \tag{9.5.51}$$

Using (9.5.48), we can rewrite this as (note the change to $\boldsymbol{\Omega}(\boldsymbol{\theta}) = \theta\boldsymbol{\Omega}(\mathbf{e})$)

$$\mathbf{R} = \mathbf{I} + \left(\boldsymbol{\Omega}(\boldsymbol{\theta}) + \frac{1}{3!}\boldsymbol{\Omega}^3(\boldsymbol{\theta}) - \cdots\right) + \left(\frac{1}{2}\boldsymbol{\Omega}^2(\boldsymbol{\theta}) - \frac{1}{4!}\boldsymbol{\Omega}^4(\boldsymbol{\theta}) + \cdots\right) \tag{9.5.52}$$

from which (9.5.47) follows immediately.

9.5.17 First- and Second-Order Updates

The exponential map provides a simple framework for developing first- and second-order updates of the nodal triad $\bar{\mathbf{e}}_i$. Since these are usually used with incremental displacements, we write the following developments in terms of $\Delta\boldsymbol{\theta}$. The first-order update is obtained by retaining only linear terms in the exponential map (9.5.47), which gives

$$\bar{\mathbf{e}}_i^{\text{new}} = (\mathbf{I} + \boldsymbol{\Omega}(\Delta\boldsymbol{\theta}))\bar{\mathbf{e}}_i^{\text{old}} \equiv \mathbf{R}_{(1)}\bar{\mathbf{e}}_i^{\text{old}} \tag{9.5.53}$$

where $\mathbf{R}_{(1)}$ is a first-order approximation to the rotation tensor. The above update is too inaccurate for most purposes.

The second-order update is obtained by including quadratic terms of (9.5.47):

$$\bar{\mathbf{e}}_i^{\text{new}} = \left(\mathbf{I} + \boldsymbol{\Omega}(\Delta\boldsymbol{\theta}) + \frac{1}{2}\boldsymbol{\Omega}^2(\Delta\boldsymbol{\theta})\right)\bar{\mathbf{e}}_i^{\text{old}} \equiv \mathbf{R}_{(2)}\bar{\mathbf{e}}_i^{\text{old}} \tag{9.5.54}$$

where $\mathbf{R}_{(2)}$ is a second-order approximation $\mathbf{R}$. Without modification, the above is not useful since the triad does not remain orthonormal. By requiring $\mathbf{R}$ in (9.5.54) to satisfy $\mathbf{R}^T\mathbf{R} = \mathbf{I}$, another form of a second-order update can be obtained:

$$\mathbf{R} = \mathbf{I} + \frac{1}{1+\frac{1}{4}\theta^2}\left(\mathbf{\Omega}(\Delta\mathbf{\theta}) + \frac{1}{2}\mathbf{\Omega}^2(\Delta\mathbf{\theta})\right) \tag{9.5.55}$$

The above can be thought of as a radial return of the update so that the triad remains of unit length.

9.5.18 *Hughes–Winget Update*

A second-order accurate update of the triad based on the first-order rotation tensor can be also constructed by using the midpoint rule. The rotation matrix is approximated by

$$\mathbf{R} = \left(\mathbf{I} - \frac{1}{2}\mathbf{\Omega}(\Delta\mathbf{\theta})\right)^{-1}\left(\mathbf{I} + \frac{1}{2}\mathbf{\Omega}(\Delta\mathbf{\theta})\right) \tag{9.5.56}$$

It can be shown that $\mathbf{R}^T\mathbf{R} = \mathbf{I}$, so the triad remains orthogonal and of unit length. This update is described in Hughes and Winget (1980); it also appears in Frajeis de Veubeke (1965). The above is obtained as follows. We first write a central difference approximation to (3.2.42):

$$\mathbf{\Omega} = \dot{\mathbf{R}}\mathbf{R}^T \rightarrow \mathbf{\Omega}\mathbf{R} = \dot{\mathbf{R}} \rightarrow \frac{1}{2}\mathbf{\Omega}^{n+½}(\mathbf{R}^{n+1} + \mathbf{R}^n) = \frac{\mathbf{R}^{n+1} - \mathbf{R}^n}{\Delta t} \tag{9.5.57}$$

This can be rearranged as

$$\left(\mathbf{I} - \frac{\Delta t}{2}\mathbf{\Omega}^{n+½}\right)\mathbf{R}^{n+1} = \left(\mathbf{I} + \frac{\Delta t}{2}\mathbf{\Omega}^{n+½}\right)\mathbf{R}^n \tag{9.5.58}$$

Letting $\mathbf{\Omega}(\Delta\mathbf{\theta}) = \frac{\Delta t}{2}\mathbf{\Omega}^{n+½}$ and $\mathbf{R}^n = \mathbf{I}$ in the above gives (9.5.56).

9.5.19 *Quaternions*

In this approach, the rotation matrix is treated as a function of four parameters:

$$q_0 = \cos\frac{\theta}{2}, \qquad q_i = e_i \sin\frac{\theta}{2}, \qquad i = 1 \text{ to } 3 \tag{9.5.59}$$

The parameters q_i are called quaternions. To express the rotation matrix in terms of the quaternions, we first note that

$$\mathbf{\Omega}^2(\mathbf{\theta})\mathbf{r} = \mathbf{\theta}\times(\mathbf{\theta}\times\mathbf{r}) = (\mathbf{\theta}\otimes\mathbf{\theta} - \theta^2\mathbf{I})\mathbf{r} \quad \text{or} \quad \mathbf{\Omega}^2(\mathbf{\theta}) = \mathbf{\theta}\otimes\mathbf{\theta} - \theta^2\mathbf{I} \tag{9.5.60}$$

Substituting the half-angle formulas $\sin\theta = 2\sin\frac{\theta}{2}\cos\frac{\theta}{2}$ and $\cos\theta = 2\cos^2\frac{\theta}{2} - 1$ and (9.5.60) into (9.5.46) gives

$$\begin{aligned}\mathbf{R} &= \left(2\cos^2\frac{\theta}{2} - 1\right)\mathbf{I} + \frac{2}{\theta}\cos\frac{\theta}{2}\sin\frac{\theta}{2}\boldsymbol{\Omega}(\boldsymbol{\theta}) + \frac{2}{\theta^2}\sin^2\frac{\theta}{2}\boldsymbol{\theta}\otimes\boldsymbol{\theta} \\ &= \left(2\cos^2\frac{\theta}{2} - 1\right)\mathbf{I} + 2\cos\frac{\theta}{2}\sin\frac{\theta}{2}\boldsymbol{\Omega}(\mathbf{e}) + 2\sin^2\frac{\theta}{2}\mathbf{e}\otimes\mathbf{e}\end{aligned}$$

where we have used the definition $\boldsymbol{\theta} = \theta\mathbf{e}$ in the last line. Now expressing the above in terms of the quaternions by (9.5.59) gives

$$\mathbf{R} = 2\left(q_0^2 - \frac{1}{2}\right)\mathbf{I} + 2q_0\boldsymbol{\Omega}(\mathbf{q}) + 2\mathbf{q}\otimes\mathbf{q} \equiv 2\left(q_0^2 - \frac{1}{2}\right)\mathbf{I} + 2q_0\boldsymbol{\Omega}(\mathbf{q}) + 2\mathbf{q}\mathbf{q}^T \tag{9.5.61}$$

where $q^T = [q_1, q_2, q_3]$. Writing out the above gives

$$\mathbf{R} = 2\begin{bmatrix} q_0^2 + q_1^2 - \frac{1}{2} & q_1q_2 - q_0q_3 & q_1q_3 + q_0q_2 \\ q_1q_2 + q_0q_3 & q_0^2 + q_2^2 - \frac{1}{2} & q_2q_3 + q_0q_1 \\ q_1q_3 + q_0q_2 & q_2q_3 + q_0q_1 & q_0^2 + q_3^2 - \frac{1}{2} \end{bmatrix} \tag{9.5.62}$$

9.5.20 Implementation

Any of the previous formulas can be used to update the triad of nodal base vectors $\bar{\mathbf{e}}_I$. In dynamics, the equations of motion yield the nodal angular acceleration, which are integrated to give the nodal angular velocities $\bar{\omega}_{iI}$. The matrix $\Delta\boldsymbol{\theta}$ is then given by

$$\Delta\boldsymbol{\theta}_I = \boldsymbol{\omega}_I\Delta t \tag{9.5.63}$$

In statics, the increments $\Delta\theta_{iI}$ are customarily chosen as the degrees of freedom to describe the rotation. In both cases the triad is updated with $\mathbf{R}(\Delta\boldsymbol{\theta}_I)$.

9.6 CB Shell Theory

9.6.1 Motion

We now examine the behavior of the CB shell from a more theoretical point of view. The objective is to examine the motion as a function of two independent variables, ξ and η, and the resulting velocity strain field. In the following, ξ and η are curvilinear coordinates on the surfaces of the shell. They play the same role as the corresponding element coordinates, but in this section they provide a parametrization of the reference surface of the shell. The reference surface of the shell is a two-dimensional manifold in three-dimensional space.

According to the modified Mindlin–Reissner assumption, the fibers remain straight, so the motion must be linear in ζ:

$$\mathbf{x}(\xi,\eta,\zeta,t)=\mathbf{x}^{M}(\xi,\eta,t)+\zeta h^{-}(\xi,\eta,t)\mathbf{p}(\xi,\eta,t)\ \text{for}\ \zeta<0 \tag{9.6.1a}$$

$$\mathbf{x}(\xi,\eta,\zeta,t)=\mathbf{x}^{M}(\xi,\eta,t)+\zeta h^{+}(\xi,\eta,t)\mathbf{p}(\xi,\eta,t)\ \text{for}\ \zeta>0 \tag{9.6.1b}$$

where $h^{-}(\xi,\eta,t)$ and $h^{+}(\xi,\eta,t)$ are the distances from the reference surface to the top and bottom surfaces along the director, respectively. The initial midsurface is assumed to be midway between the top and bottom surfaces, so $h_0^{-}=h_0^{+}=h_0/2$. This can also be written in the compact form (9.5.5a):

$$\mathbf{x}=\mathbf{x}^{M}+\zeta\mathbf{x}^{B}\equiv\mathbf{x}^{M}+\bar{\zeta}\mathbf{p} \tag{9.6.2}$$

where $\mathbf{x}^{B}$ is the motion due to bending as defined in (9.5.5b) and $\bar{\zeta}$ is also defined in (9.5.5b). The coordinates of the shell in the original configuration are obtained by evaluating (9.6.2) at the initial time:

$$\mathbf{X}(\xi,\eta,\zeta)=\mathbf{X}^{M}(\xi,\eta)+\bar{\zeta}_0\mathbf{p}_0(\zeta,\eta)=\mathbf{X}^{M}+\zeta\mathbf{X}^{B} \tag{9.6.3}$$

where $\mathbf{p}_0=\mathbf{p}(\xi,\eta,0)$ and $\bar{\zeta}_0$ are defined in terms of the initial configuration. The displacement field is obtained by taking the difference of (9.6.2) and (9.6.3):

$$\mathbf{u}(\xi,\eta,\zeta,t)=\mathbf{u}^{M}+\zeta\mathbf{u}^{B}=\mathbf{u}^{M}+\bar{\zeta}_0(\mathbf{p}-\mathbf{p}_0)+\Delta\bar{\zeta}\mathbf{p}=\mathbf{u}^{M}+\zeta\mathbf{u}^{B}+\Delta\bar{\zeta}\mathbf{p} \tag{9.6.4}$$

where

$$\mathbf{u}^{M}=\mathbf{x}^{M}-\mathbf{X}^{M},\quad \mathbf{u}^{B}=\mathbf{x}^{B}-\mathbf{X}^{B}=\frac{h_0}{2}(\mathbf{p}-\mathbf{p}_0)$$
$$\Delta\bar{\zeta}=\bar{\zeta}-\bar{\zeta}_0=\zeta(h^{+}-h_0^{+})\quad\text{for}\quad\zeta>0,\quad \Delta\bar{\zeta}=\bar{\zeta}-\bar{\zeta}_0=\zeta(h^{-}-h_0^{-})\quad\text{for}\quad\zeta<0 \tag{9.6.5}$$

As can be seen from the previous, the bending displacement $\mathbf{u}^{B}$ depends on the difference between two unit vectors, $\mathbf{p}-\mathbf{p}_0$. Therefore the bending displacement $\mathbf{u}^{B}$ can be described by two dependent variables. The motion can then be described by five dependent variables: the three translations of the midsurface, $\mathbf{u}^{M}=\left[u_x^{M},\ u_y^{M},\ u_z^{M}\right]$, and the two which describe the bending displacement, $\mathbf{u}^{B}$.

The velocity field is obtained by taking the material time derivative of the displacement or motion, using (9.2.1) to write the rate of the director:

$$\mathbf{v}(\zeta,\eta,\zeta,t)=\mathbf{v}^{M}(\xi,\eta,t)+\bar{\zeta}\boldsymbol{\omega}(\xi,\eta,t)\times\mathbf{p}+\dot{\bar{\zeta}}\mathbf{p} \tag{9.6.6}$$

The last term in (9.6.6) represents the rate of change in the thickness. The above velocity field can also be written as

$$\mathbf{v}(\xi,\eta,\xi,t)=\mathbf{v}^{M}+\zeta\mathbf{v}^{B}+\dot{\bar{\zeta}}\mathbf{p}\ \text{where}\ \begin{cases}\mathbf{v}^{B}=h^{+}\boldsymbol{\omega}\times\mathbf{p} & \text{for}\ \zeta>0\\ \mathbf{v}^{B}=h^{-}\boldsymbol{\omega}\times\mathbf{p} & \text{for}\ \zeta<0\end{cases} \tag{9.6.7}$$

As can be seen from the above, the velocity of any point in the shell consists of the sum of the velocity of the reference plane $\mathbf{v}^M$, the bending velocity $\mathbf{v}^B$, and the velocity due to the change in thickness. The bending velocity depends on the angular velocity of the director. The component of the angular velocity parallel to the director $\mathbf{p}$ is irrelevant since it causes no change in $\mathbf{p}$. This component is called the *drilling component* or the *drill* for short. Since the drill has no effect on deformation, it is apparent that a five-degree-of-freedom formulation is more consistent with CB shells than a six-degree-of-freedom formulation.

9.6.2 Velocity Strains

We will next examine the velocity strains for CB shell theory. For simplicity, we consider a shell with the reference surface at the midsurface and neglect the change in thickness. The velocity strains will be studied by taking a series expansion about the midsurface. The resulting formulas are not recommended for calculations except when resultant stresses are used. The analysis is based on Belytschko, Wong and Stolarski (1989). For simplicity, it is assumed that the director is collinear with the normal and that the reference plane is the midplane, so $h^+ = h^- = h/2$.

Consider the derivative of a generic function $f(\mathbf{x})$. Our objective is to expand the expression for the velocity strain in powers of $\tilde{\zeta} = \zeta h/2R$, which are small for a shell. Therefore the terms of higher order than quadratic can be dropped. We start with the three-dimensional form of the chain rule (note that we have written the transpose of our usual form):

$$\begin{Bmatrix} f_{,\xi} \\ f_{,\eta} \\ f_{,\zeta} \end{Bmatrix} = \hat{\mathbf{x}}_{\xi}^{T} f_{,\hat{\mathbf{x}}} = \begin{bmatrix} \hat{x}_{,\xi} & \hat{y}_{,\xi} & \hat{z}_{,\xi} \\ \hat{x}_{,\eta} & \hat{y}_{,\eta} & \hat{x}_{,\eta} \\ \hat{x}_{,\zeta} & \hat{y}_{,\zeta} & \hat{z}_{,\zeta} \end{bmatrix} \begin{Bmatrix} f_{,\hat{x}} \\ f_{,\hat{y}} \\ f_{,\hat{z}} \end{Bmatrix}$$
$$= \begin{bmatrix} \hat{x}_{,\xi}^{M} + \zeta \hat{x}_{,\xi}^{B} & \hat{y}_{,\xi}^{M} + \zeta \hat{y}_{,\xi}^{B} & \hat{z}_{,\xi}^{M} + \zeta \hat{z}_{,\xi}^{B} \\ \hat{x}_{,\eta}^{M} + \zeta \hat{x}_{,\eta}^{B} & \hat{y}_{,\eta}^{M} + \zeta \hat{y}_{,\eta}^{B} & \hat{z}_{,\eta}^{M} + \zeta \hat{z}_{,\eta}^{B} \\ \hat{x}^{B} & \hat{y}^{B} & \hat{z}^{B} \end{bmatrix} \begin{Bmatrix} f_{,\hat{x}} \\ f_{,\hat{y}} \\ f_{,\hat{z}} \end{Bmatrix} \tag{9.6.8}$$

We consider only the x and y derivatives. Since the coordinate system is corotational, on the reference plane $\hat{z}_{,\zeta} \cong h/2$, $\hat{z}_{,\eta} = \hat{z}_{,\xi} = 0$. Inverting the above in closed form for the submatrix associated with the x and y derivatives yields

$$\hat{\mathbf{x}}_{,\xi}^{-1} = \frac{1}{J_{\xi}} \begin{bmatrix} \hat{y}_{,\eta}^{M} + \zeta \hat{y}_{,\eta}^{B} & -\left(\hat{x}_{,\eta}^{M} + \zeta \hat{x}_{,\eta}^{B}\right) \\ -\left(\hat{y}_{,\xi}^{M} + \zeta \hat{y}_{,\xi}^{B}\right) & \hat{x}_{,\eta}^{M} + \zeta \hat{x}_{,\xi}^{B} \end{bmatrix} \tag{9.6.9}$$

where

$$J_{\xi} = J_0 \left(1 + \frac{J_1}{J_0} \zeta \right) + O(\zeta^2), \quad J_0 = \hat{x}_{,\xi}^{M} \hat{y}_{,\eta}^{M} - \hat{x}_{,\eta}^{M} \hat{y}_{,\xi}^{M} \tag{9.6.10}$$

Expressing the bending motion $\mathbf{x}^B$ in terms of the director components in (9.6.9) gives

$$\hat{\mathbf{x}},_{\xi}^{-1} = \frac{1}{J_{\xi}}\underbrace{\begin{bmatrix} \hat{y},_{\eta}^{M} & -\hat{x},_{\eta}^{M} \\ -\hat{y},_{\xi}^{M} & \hat{x}_{\xi}^{M} \end{bmatrix}}_{\mathbf{A}} + \zeta\frac{h}{2J_{\xi}}\underbrace{\begin{bmatrix} p_{\hat{y},\eta} & -p_{\hat{x},\eta} \\ -p_{\hat{y},\xi} & p_{x,\xi} \end{bmatrix}}_{\mathbf{B}} \tag{9.6.11}$$

We note that the radius of curvature R is given by

$$\frac{1}{R} = \max\left(\left(p_{\hat{x},\xi}^2 + p_{\hat{y},\xi}^2\right)^{1/2}, \left(p_{\hat{x},\eta}^2 + p_{\hat{y},\eta}^2\right)^{1/2}\right) \tag{9.6.12}$$

So $\mathbf{B}$ in (9.6.11) is of order R^{-1} and we can rewrite (9.6.11) as

$$\hat{\mathbf{x}},_{\xi}^{-1} = \frac{1}{J_0(1+b\tilde{\zeta})}(\mathbf{A} + \tilde{\zeta}R\mathbf{B}) \quad \text{where} \quad b = \frac{2RJ_1}{hJ_0} \tag{9.6.13}$$

We now assume that

$$\tilde{\zeta} \equiv \zeta\frac{h}{2R} << 1 \quad \text{or} \quad \frac{h}{2R} << 1 \tag{9.6.14}$$

where R is the radius of curvature. The second equation above follows from the first since $|\zeta| \leq 1$. This condition on the ratio of thickness to radius of curvature is an important requirement for the applicability of shell theory. When it is not met, shell theory is not applicable.

Multiplying both the numerator and denominator of (9.6.13) by $1 - b\tilde{\zeta}$ and dropping terms of quadratic order or higher in $\tilde{\zeta}$ then gives (after replacing $\tilde{\zeta}$ by ζ via (9.6.14)):

$$\mathbf{x},_{\xi}^{-1} = \frac{1}{J_0}\underbrace{\begin{bmatrix} \hat{y},_{\eta}^{M} & -\hat{x},_{\eta}^{M} \\ -\hat{y},_{\xi}^{M} & \hat{x}_{\xi}^{M} \end{bmatrix}}_{\mathbf{x}_{M,\xi}^{-1}} + \frac{\zeta h}{2J_0}\underbrace{\begin{bmatrix} p_{\hat{y},\eta} - c\hat{y},_{\eta}^{M} & \left(p_{\hat{x},\eta} - c\hat{x},_{\eta}^{M}\right) \\ -\left(p_{\hat{y},\xi} - c\hat{y},_{\xi}^{M}\right) & \left(p_{\hat{x},\xi} - c\hat{x},_{\xi}^{M}\right) \end{bmatrix}}_{\mathbf{x}_{B,\xi}^{-1}} \tag{9.6.15}$$

where $c = \frac{2J_1}{hJ_0}$ Note that the denominator is $J_0(1 - \tilde{\zeta}^2 + O(\tilde{\zeta}^3))$ after the multiplication, so the only term retained in the numerator is J_0. The velocity gradient $\mathbf{L}$ is then given by $\mathbf{L} = \mathbf{v},_{\xi}\, \tilde{\mathbf{x}},_{\xi}^{-1}$. The rate-of-deformation $\hat{D}_{\alpha\beta}$ is the symmetric part of $\mathbf{L}$, so it is also linear through the thickness. This corresponds to the through-the-thickness distribution of strains in classical shell theories. Note that the *applicability* of classical shell theory and the *equivalence of classical shell theory and CB shell theory both hinge on the inequality* (9.6.14).

9.6.3 Resultant Stresses

The internal power is given by

$$\delta P^{\text{int}} = \int_{\Omega} (\delta\hat{D}_{\alpha\beta}\hat{\sigma}_{\alpha\beta} + 2\delta\hat{D}_{3\alpha}\hat{\sigma}_{3\alpha})d\Omega \tag{9.6.16}$$

where $\hat{\sigma}_{33}$= o by the plane stress condition. If we note that $\delta\hat{D}_{\alpha\beta} = \delta\hat{D}^M_{\alpha\beta} + \bar{\zeta}\delta\hat{D}^B_{\alpha\beta}$ and define resultants by

$$\hat{n}_{\alpha\beta} = \int_{-1}^{1} \hat{\sigma}_{\alpha\beta}\frac{h}{2}d\zeta, \quad \hat{m}_{\alpha\beta} = \int_{-1}^{1} \zeta\hat{\sigma}_{\alpha\beta}\frac{h}{2}d\zeta, \quad \hat{s}_{\alpha} = \int_{-1}^{1} \hat{\sigma}_{\alpha 3}\, d\zeta, \tag{9.6.17}$$

The internal power is given by

$$\delta P^{\text{int}} = \int_S (\delta\hat{D}_{\alpha\beta}\hat{n}_{\alpha\beta} + 2\delta\hat{D}_{3\alpha}\hat{s}_{\alpha} + \delta\hat{\kappa}_{\alpha\beta}\hat{m}_{\alpha\beta})dS \tag{9.6.18}$$

where S is the reference surface. The above powers are associated with membrane, shear and bending, respectively.

9.6.4 Boundary Conditions

The boundary is a contour C that is the intersection of the lateral surface of the underlying solid with the reference surface of the shell, as shown in Figure 9.12. Tractions applied to the top and bottom surfaces are treated like body forces. The boundary conditions are expressed in a local coordinate system with base vectors $\hat{\mathbf{e}}_x, \hat{\mathbf{e}}_y$, and $\hat{\mathbf{e}}_z$, where $\hat{\mathbf{e}}_y$ is tangent to C, $\hat{\mathbf{e}}_z$ is normal to the reference surface and $\hat{\mathbf{e}}_x = \hat{\mathbf{e}}_y \times \hat{\mathbf{e}}_z$. The tractions on C are subdivided into zeroth and first-order moments:

$$\hat{f}_i^* = \int_{-1}^{1} \hat{t}_i \frac{h}{2}d\zeta, \quad \hat{m}_i^* = \int_{-1}^{1} e_{ijk}\hat{x}_j\hat{t}_k \frac{h}{2}d\zeta \tag{9.6.19}$$

Only the first two moments have any effect on the response of the CB shell.

The external power is given by

$$\delta P^{\text{ext}} = \int_C (\delta\hat{v}_i\hat{f}_i^* + \delta\hat{\omega}_i\hat{m}_i^*)ds \tag{9.6.20}$$

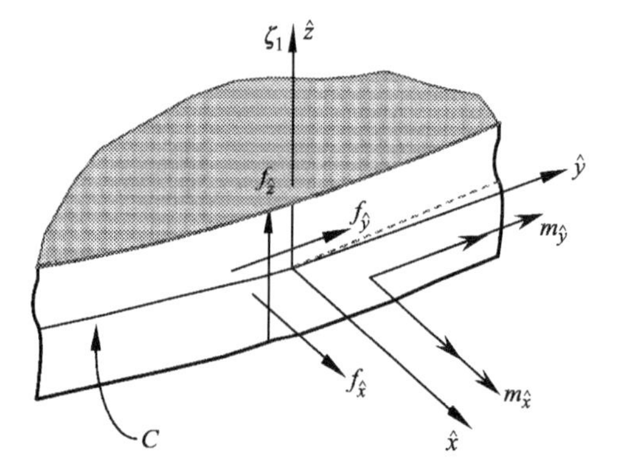

Figure 9.12 Applied tractions and resultant external forces and moments

where ds is the differential of the arc length along C. When the boundary lies in a plane, the nodal external forces resulting from nonzero applied fractions are given by

$$\hat{f}_{i\hat{I}}^{\text{ext}} = \int_C N_I(\xi)\hat{f}_i^* ds \tag{9.6.21}$$

$$\hat{m}_{i\hat{I}}^{\text{ext}} = \int_C N_I(\xi)\hat{m}_i^* ds \quad i = 1 \text{ to } 3 \text{ for 6 d.o.f.}, i = 1 \text{ to } 2 \text{ for 5 d.o.f.} \tag{9.6.22}$$

where $N_I(\xi)$ is the projection of the CB shape function with $\zeta = 0$ onto the edge.

The boundary conditions for Mindlin–Reissner shells are then

$$\hat{v}_i = \hat{v}_i^* \text{ on } C_{\hat{v}_i}, \quad \hat{f}_i = \hat{f}_i^* \text{ on } C_{\hat{f}_i}, \quad C_{\hat{v}_i} \cap C_{\hat{f}_i} = 0, \quad C_{\hat{v}_i} \cup C_{\hat{f}_i} = C \tag{9.6.23}$$

$$\hat{\omega}_\alpha = \hat{\omega}_\alpha^* \text{ on } C_{\hat{\omega}_\alpha}, \quad \hat{m}_\alpha = \hat{m}_\alpha^* \text{ on } C_{\hat{m}_\alpha}, \quad C_{\hat{\omega}_\alpha} \cap C_{\hat{m}_\alpha} = 0, \quad C_{\hat{\omega}_\alpha} \cup C_{\hat{m}_\alpha} = C \tag{9.6.24}$$

In the classical shell theory and in five-degree-of-freedom implementation, only two components of the moment are prescribed along a boundary because the drilling degree-of-freedom is not a dependent variable. However, when a six-degree-of-freedom formulation is used, all three must be prescribed. These boundary conditions are exact only when the boundary C lies in a plane. For general boundaries, the generalized moments should be defined in directly in terms of the tractions

$$f_{iI}^{ext} = \int_C \int_{-1}^{+1} N_I(\xi,\eta,\zeta) t_i^* \frac{h}{2} d\zeta ds, \quad m_{iI}^{ext} = \int_C \int_{-1}^{+1} N_I(\xi,\eta,\zeta) e_{ijk} \hat{x}_j t_k^* \frac{h}{2} d\zeta ds \tag{9.6.25}$$

9.6.5 *Inconsistencies and Idiosyncrasies of Structural Theories*

The Mindlin–Reissner and Kirchhoff–Love assumptions introduce several inconsistencies. In the Mindlin–Reissner theory, the shear stresses $\hat{\sigma}_{xz}$ and $\hat{\sigma}_{yz}$ are constant through the depth of the shell. However, unless a shear traction is applied to the top or bottom surfaces, the transverse shear must vanish at these surfaces because of the symmetry of the stress tensor. An analysis of equilibrium an elastic beam shows that the transverse shear stress should be quadratic through the depth of a beam and vanish at the top and bottom surfaces. Therefore a constant shear stress distribution overestimates the shear energy. A correction factor, known as a shear correction, is often used to reduce the energy associated with the transverse shear, and accurate estimates of this factor can be made for elastic beams and shells. For nonlinear materials, however, it is difficult to estimate a shear correction factor.

The inconsistency of Kirchhoff–Love theory is even more severe, since the kinematic assumption results in a vanishing transverse shear. In a beam, it is well known in structural theories that the shear must be nonzero if the moment is not constant. Thus the Kirchhoff–Love kinematic assumption is inconsistent with equilibrium. Nevertheless, comparison with experiments shows that it is quite accurate, and for thin, homogenous shells it is just as accurate as the Mindlin–Reissner theory. Transverse shear does not play an important role in the deformation of thin structures, so its inclusion has little effect. Because of their simplicity, Mindlin–Reissner elements are used even when transverse shear effects are negligible.

The modified Mindlin–Reissner CB models provide additional possibilities for errors. If the directors are not normal to the midsurface, the motion deviates markedly from the motion which is observed experimentally.

The zero normal stress assumption is inconsistent when a normal traction is applied to either surface of the shell. Obviously, the normal stress must equal the applied normal traction for equilibrium. However, they are neglected in structural theories because they are much smaller than the axial stresses; only a small fraction of the energy is absorbed by the normal stress and it has little effect on the deformation.

The analyst should also be aware of severe boundary effects in shells. Certain boundary conditions result in edge effects where the behavior changes dramatically in a narrow boundary layer. For certain boundary conditions, singularities can occur in corners of the boundary.

One reason for using the structural kinematic assumptions is that they improve the conditioning of the discrete equations. If a shell is modeled with three-dimensional continuum elements, the degrees of freedom are the translations at all of the nodes. The natural modes associated with through-the-thickness strains have very large eigenvalues. As a consequence, the conditioning of the linearized equilibrium equations or linearized equations for an implicit update can be very poor. The conditioning of shell equations is also not as good as that of standard continuum models, but it is substantially better than that of continuum models of thin shells. In explicit methods, continuum models of thin structures have very small critical time steps due to the large eigenvalues of the through-the-thickness mode. CB shell models can provide much larger critical time steps.

9.7 Shear and Membrane Locking

9.7.1 Description and Definitions

Among the most troublesome characteristics of shell elements are shear and membrane locking. Shear locking results from the spurious appearance of transverse shear. More precisely, it emanates from the inability of many elements to represent deformations in which the transverse shear should vanish. Since the shear stiffness is often significantly greater than the bending stiffness, the spurious shear absorbs a large part of the energy imparted by the external forces and the predicted deflections and strains are much too small, hence the name *shear locking*.

The observed behavior of thin beams and shells indicates that the normals to the midline remain straight and normal, and that hence the transverse shears vanish. This behavior can be viewed as a constraint on the motion of the continuum. While the normality constraint is not exactly enforced in the shear-beam or CB shell theories, the normality constraint appears as a shear energy, that is, a penalty term in the energy. The penalty factor increases as the thickness decreases, so as the thickness decreases, shear locking becomes more prominent. Shear locking does not appear in C^1 elements, since in C^1 elements the motion is defined so that the normals remain normal. In C^0 elements, the normal can rotate relative to the midline, so spurious transverse shear and locking can appear.

Membrane locking results from the inability of shell finite elements to represent inextensional modes of deformation. Shells bend without stretching: take a piece of paper and see how easily you can bend it; this is called inextensional bending. However, stretching a piece of paper by hand is almost impossible. Shells behave similarly: their bending stiffness is small

Table 9.1 Analogy of locking phenomena

Constraint	Shortcoming of finite element motion	Locking type
Incompressibility, isochoric motion, J = constant, $v_{i,i} = 0$	Volumetric strain appears in element	Volumetric locking
Kirchhoff–Love constraint, $\hat{D}_{xz} = \hat{D}_{yz} = 0$	Transverse shear strain appears in pure bending	Shear locking
Inextensibility constraint	Membrane strain appears in inextensional bending mode	Membrane locking

but their membrane stiffness is large. When the finite element cannot bend without stretching, the energy is incorrectly shifted to membrane energy, resulting in underprediction of displacements and strains. Membrane locking is particularly important in simulation of buckling since many buckling modes are completely or nearly inextensional.

Shear and membrane locking are similar to the volumetric locking described in Chapter 8: when a finite element approximation to motion cannot satisfy a constraint, the constrained mode is much stiffer than the stiffness of the correct motion. In the case of volumetric locking, the constraint is incompressibility, while for shear and membrane locking the constraints are the Kirchhoff–Love normality constraint and the inextensibility constraint in bending (which has nothing to do with the inextensibility of fibers). The analogies are summarized in Table 9.1. It should be noted that the shear-free behavior of thin shells is not an exact constraint. For thicker shells and beams, some transverse shear is expected, but just as elements that lock volumetrically perform poorly for nearly incompressible materials, shell elements which look in shear perform poorly for moderately thick shells even when transverse shear appears.

9.7.2 Shear Locking

In the following, we will use linear strain displacement equations, which are valid for only small strains and rotation. This description of shear and membrane locking closely follows Stolarski, Belytschko and Lee (1994). To examine the causes of shear locking, we consider the two-node beam element described in Example 9.1. For simplicity, let the element lie along the x-axis and consider the linear response, so we replace D_{ij} by the linear strain ε_{ij} and the velocity by the displacement in the kinematic relations. The transverse shear strain is given by the counterpart of (E9.1.15):

$$2\varepsilon_{xy} = \frac{1}{\ell}\left(u_{x2}^{M} - u_{x1}^{M}\right) - \theta_1 \frac{1}{2}(1-\xi) - \theta_2 \frac{1}{2}(1+\xi) \quad \text{where } \xi \in [-1, +1] \qquad (9.7.1)$$

We now consider the element in a state of pure bending: with $u_{x1} = u_{x2} = 0$, $\theta_1 = -\theta_2 = \alpha$. For these nodal displacements, (9.7.1) gives

$$2\varepsilon_{xy} = \alpha\xi \qquad (9.7.2)$$

From the equilibrium equation, (9.4.37), the shear $s(x)$ should vanish when the moment is constant. However, from (9.7.2) we see that the transverse shear strain, and hence the transverse shear stress $\sigma_{xy} = 2G\varepsilon_{xy}$, are nonzero in most of the element. In fact they are nonzero everywhere except at $\xi = 0$. The transverse shear which appears in a state of pure bending is often called *parasitic shear*.

This parasitic transverse shear has a large effect on the behavior of the element. To explain the severity of the effect, the energies associated with the bending and shear strains are examined for a linear, elastic beam of unit depth and rectangular cross-section. The bending energy for this nodal displacements is given by

$$W_{\text{bend}} = \frac{E}{2}\int_{\Omega} y^2 \theta_{,x}^2 \, d\Omega = \frac{Eh^3}{24}\int_0^{\ell} \theta_{,x}^2 \, dx = \frac{Eh^3}{24\ell}(\theta_2 - \theta_1)^2 = \frac{Eh^3\alpha^2}{6\ell} \tag{9.7.3}$$

where the rotations associated with the bending mode $\theta_1 = -\theta_2 = \alpha$ have been used in the last expression. The shear energy for the beam is given by

$$W_{\text{shear}} = \frac{E}{(1+\nu)}\int_{\Omega} \varepsilon_{xy}^2 \, d\Omega = \frac{Eh}{(1+\nu)}\int_0^{\ell} (\theta - u_{y,x})^2 \, dx = \frac{Eh\ell\alpha^2}{3(1+\nu)} \tag{9.7.4}$$

The ratio of these two energies is $W_{\text{shear}}/W_{\text{bend}}$ is proportional to $(\ell/h)^2$. Thus the shear energy is significantly greater than the bending energy when $\ell > h$. Since the shear energy should vanish in pure bending, this parasitic shear energy absorbs a large part of the available energy. The result is a significant underprediction of the total displacement. However, in contrast to volumetric locking, where no convergence is observed with refinement, elements that lock in shear converge to the correct solution, but very slowly.

Equation (9.7.2) immediately suggests why underintegration can alleviate shear locking in this element: note that the transverse shear vanishes at $\xi = 0$, which corresponds to the quadrature point in one-point quadrature. Thus, the spurious transverse shear is eliminated by underintegration of the shear-related terms.

Shear locking in the three-node beam with quadratic interpolants is less obvious than for the two-node beam. Consider a three-node beam element of length ℓ with parent coordinate $\xi = 2x/\ell$, $-1 \le \xi \le 1$. The shear strain in this element is given by

$$\begin{aligned} 2\varepsilon_{xy} = u_{\hat{y},x} - \theta = \frac{1}{\ell}\left[(2\xi - 1)u_{y1} - 4\xi u_{y2} + (2\xi + 1)u_{y3}\right] \\ -\frac{1}{2}(\xi^2 - \xi)\theta_1 - (1 - \xi^2)\theta_2 - \frac{1}{2}(\xi^2 + \xi)\theta_3 \end{aligned} \tag{9.7.5}$$

Consider a state of pure bending, $\theta_1 = -\theta_3 = \alpha$, $\theta_2 = 0$, $u_{y1} = u_{y3} = 0$ and $u_{y2} = \alpha\ell/4$. Substituting these nodal displacements into (9.7.5) shows that the transverse shear vanishes throughout the element. There is no reason to expect locking based on this result. However, consider another deformation, $u_y = \alpha\xi^3$, $\theta = u_y, x = 6\alpha\xi^2/\ell$. The shear should vanish since the normals remain normal. However, the transverse shear given by (9.7.5) for the nodal displacements corresponding to this deformation is

$$2\varepsilon_{xy} = \frac{2\alpha}{\ell}(1 - 3\xi^2) \tag{9.7.6}$$

so the finite element approximation gives nonzero shear everywhere except at $\xi = \pm 1/\sqrt{3}$. Therefore a large amount of transverse shear will occur in this element for the shear-free mode, and it will not be effective in modeling thin beams.

9.7.3 *Membrane Locking*

To illustrate membrane locking we use the Marguerre shallow beam equations:

$$\varepsilon_{xx} = u^M_{x,x} + w,^0_x u_y, x - y\theta,_x, \qquad 2\varepsilon_{xy} = u_y, x - \theta \tag{9.7.7a,b}$$

where w^0 is the initial displacement in the z-direction of the midline from the chord of the beam, that is, the x-axis. The variable w^0 reflects the curvature of the beam: for a straight beam $w^0 = 0$. It should be stressed that while these kinematic relations differ from the CB beam equations, they closely approximate the linear CB equations for shallow beams, that is, when $w^0(x)$ is small.

Consider a three-node beam element of length ℓ with parent coordinate $\xi = 2x/\ell$, $-1 \le \xi \le 1$. In an inextensional mode, the membrane strain ε_{xx} must vanish. Integrating the expression for $u^M_{x,x}$ in (9.7.7a) and letting $\varepsilon_{xx} = 0$ for $y = 0$ gives

$$u^M_{x3} - u^M_{x1} = -\int_0^\ell w^0,_x u_{y,x}\, dx \tag{9.7.8}$$

Consider a beam in a pure bending mode so $\theta_1 = -\theta_3 = \alpha$ and $u_{z1} = u_{z3} = 0$. In the absence of transverse shear it follows from (9.7.7b) that

$$u_{y2} = \int_0^{\ell/2} \theta(x) dx = \frac{\alpha\ell^2}{4} \tag{9.7.9}$$

In an initially symmetric configuration, $\theta_1^0 = \theta_3^0 = \theta_0, \theta_2^0 = 0. =$ Then (9.7.8) is satisfied if $u_{x1} = -u_{x3} = \dfrac{\theta_0 \alpha\ell}{6}$, $u_{x2} = 0$. Evaluating the membrane strain via (9.7.7a) gives

$$\varepsilon_{xx} = \alpha\theta_0\left(\frac{1}{3} - \xi^2\right) \tag{9.7.10}$$

Thus, in this particular inextensional mode of deformation, the extensional strain is nonzero everywhere except at $\xi = \pm 1/\sqrt{3}$. If the element includes quadrature points where the extensional strain does not vanish, the element will exhibit membrane locking.

Membrane locking can also be explained by examining the orders of the displacement fields. The variables u_x, u_y, and w^0 are quadratic in x, and these quadratic fields are actuated in a pure bending mode. Since $u_{x,x}$ is only linear, the membrane strain (9.7.7a) cannot vanish uniformly throughout the element in a pure bending mode if w^0 is nonzero. Thus membrane locking can be seen to originate from the inability of the finite element interpolant to represent inextensional motions. Shear locking can be explained similarly as the inability of finite element interpolants to represent pure bending modes.

From the preceding, an obvious remedy for membrane and shear locking would be to match the order of the interpolants of different components of the motion. For example, a cubic field u_x would improve the representation of an inextensional mode for quadratic u_y. However, this is incompatible with the framework of CB isoparametric elements: different order interpolations for different components are awkward to program and impair the element's ability to represent rigid body motion exactly, which is crucial for convergence.

If the element is rectilinear, w^0 vanishes and membrane locking will not occur; for a straight element, bending does not generate membrane strains, see (9.7.7a). Membrane locking also does not occur in flat shell elements. Thus, the 2-node beam element never exhibits membrane locking and the 4-node quadrilateral shell element manifests membrane locking only in warped configurations.

Although this model for membrane locking is based on the Maguerre shallow beam equations, it correctly predicts the performance of elements developed by other beam and shell theories and of CB shell elements. The mechanical behavior of shell elements is almost independent of the shell theory as long as the element is shallow. Moreover, as meshes are refined, elements increasingly conform to the shallow shell hypothesis. However, the extension of these analyses to general shell elements is quite difficult, particularly when the element is not rectangular.

9.7.4 *Elimination of Locking*

We have already mentioned in Section 9.7.2 how shear locking can be avoided by underintegrating the shear energy with the quadrature point $\xi = 0$. Restricting the sampling of the shear energy to this point avoids the parasitic shear and hence unlocks the element. Locking can also be circumvented by the multi-field methods described in Chapter 8 by designing the appropriate strain fields. For example, if the Hu–Washizu weak form is used, shear locking can be circumvented by making the transverse shear constant. The transverse velocity shear and shear stress fields are:

$$\bar{D}_{xy} = \alpha_1, \quad \bar{\sigma}_{xy} = \beta_1 \tag{9.7.11}$$

where α_1 and β_1 are determined by the discrete compatibility and constitutive equations.

Assumed strain methods, described in Section 8.5.8, are also used to avoid locking. The essence of the assumed strain approach is to design transverse shear fields and membrane strain fields so that parasitic shear and membrane locking are minimized. Assumed strain fields must be designed so that the correct rank of the stiffness matrix is retained. For the 2-node beam, the assumed strain field should be constant and should vanish in pure bending. We can achieve these aims if

$$\bar{D}_{xy} = D_{xy}(0) \tag{9.7.12}$$

which is a constant field equal to D_{xy} at the midpoint. For this field in pure bending, the assumed strain rate will vanish throughout the element.

For the 3-node CB beam the analyses in Section 9.7.2 also provide guidance as to how to overcome both shear and membrane locking. Remarkably, both the shear in (9.7.6) and the membrane strains in (9.7.10) vanish at the points $\xi = \pm 1/\sqrt{3}$, the Gauss quadrature points for two-point quadrature. These are often called Barlow points because Barlow (1976) first pointed out that if the nodal displacements of an 8-node isoparametric element are set by a cubic field, the stresses at these quadrature points correspond to those obtained from a cubic displacement field. He concluded that 'if the element is used to represent a general cubic displacement field, the stresses at the 2 × 2 Gauss points will have the same degree of accuracy as the nodal displacements'. This discovery has proved very helpful in designing effective shell elements. For example, it explains the success of the reduced integration introduced by Zienkiewicz, Taylor, and Too (1971). So reduced integration on the shear and membrane powers with two-point Gauss quadrature will eliminate shear and membrane locking.

To avoid shear locking by a multi-field approach, let the transverse shear and membrane velocity strains be linear:

$$\bar{D}_{xy} = \alpha_1 + \alpha_2 \xi, \quad \bar{\sigma}_{xy} = \beta_1 + \beta_2 \xi \tag{9.7.13}$$

The Hu–Washizu weak form projects the velocity strains obtained from the velocities onto these linear fields. For computations, it is convenient if the assumed velocity strain field in terms of velocity strains computed from velocities. An assumed strain field of this type can be constructed by taking advantage of the property of the Barlow points we have just described. The fields are

$$\begin{aligned} \bar{D}_{xy} &= D_{xy}\left(-\bar{\xi}\right)\frac{\xi-\bar{\xi}}{-2\bar{\xi}} + D_{xy}\left(\bar{\xi}\right)\frac{\xi+\bar{\xi}}{2\bar{\xi}}, \\ \bar{D}_{xx} &= D_{xx}\left(-\bar{\xi}\right)\frac{\xi-\bar{\xi}}{-2\bar{\xi}} + D_{xx}\left(\bar{\xi}\right)\frac{\xi+\bar{\xi}}{2\bar{\xi}} \end{aligned} \tag{9.7.14}$$

where D_{xy} and D_{xx} are obtained from the velocity field and $\bar{\xi} = 1/\sqrt{3}$. The transverse shear will vanish in the motions considered and the membrane field will vanish in inextensional deformation. Thus the parasitic energies are avoided and the element will not lock.

9.8 Assumed Strain Elements

Shear and membrane locking in shell elements can also be circumvented by assumed strain methods and selective-reduced integration. However, the design of these schemes for shells is more difficult than for beams or continua. For example, in the quadrilateral 4-node plate element with selective-reduced integration described in Hughes (1987: 327) and Hughes, Cohen and Haroun (1978), the element still possesses a spurious singular mode, the w-hourglass mode. Thus while selective-reduced integration provides robust elements for continua, it is not as successful for shells.

9.8.1 Assumed Strain 4-Node Quadrilateral

The construction of the transverse shear field is motivated by (9.7.2). From these equations we can deduce that if the transverse shear distribution for a beam in bending is linear but vanishes

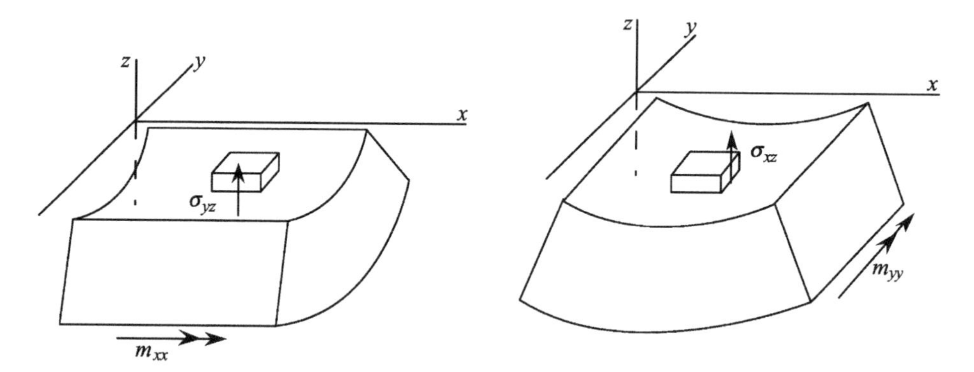

Figure 9.13 Rectangular element under pure bending showing the transverse shear which is activated, if not suppressed, by assumed strain methods. For the deformation shown m_{xx} and m_{yy} are negative

at the center, its projection on the constant field also vanishes. We first consider a planar, rectangular shell element. A rectangular shell element behaves similarly to a beam: when a bending moment is applied to the two ends as shown in Figure 9.13, the transverse shear σ_{xz} should vanish. When the material is isotropic, the transverse shear $\bar{D}_{xz}$ should also vanish. These conditions can be met by making the shear constant, that is, by letting $\bar{D}_{xz} = \alpha_1$ where α_1 is a constant, and using the Hu–Washizu weak form to evaluate α_1. This assumed transverse shear vanishes for a constant moment. However, a constant transverse shear leads to rank deficiency and hence an unstable element. To restore stability, a linear dependence on y is added, so the assumed transverse shear is

$$\bar{D}_{xz} = \alpha_1 + \alpha_2 y \tag{9.8.1}$$

The linear term has no effect on the behavior under the bending moment m_{yy}, so the unlocking is not disturbed. Similarly the transverse shear $\bar{D}_{yz}$ is taken to be

$$\bar{D}_{yz} = \beta_1 + \beta_2 x \tag{9.8.2}$$

This concept is an extended to quadrilaterals as follows. The transverse shear $\bar{D}_{\xi\hat{z}}$ is assumed constant in the ξ direction to avoid parasitic shear, linear in the η direction to stabilize the element. A similar argument for $\bar{D}_{\eta\hat{z}}$ gives

$$\bar{D}_{\xi\hat{z}}(\xi, \eta, \zeta, t) = \alpha_1 + \alpha_2\eta, \quad \bar{D}_{\eta\hat{z}}(\xi, \eta, \zeta, t) = \beta_1 + \beta_2\xi \tag{9.8.3}$$

where α_i and β_i are arbitrary parameters.

In the application of the Hu–Washizu weak form, the parameters α_i and β_i are found by the discrete compatibility equations. However, this complicates the implementation. Instead, the assumed rate-of-deformation $\bar{\mathbf{D}}$ can be interpolated in terms of $\mathbf{D}$ at selected points. The midpoints of the edges are chosen as interpolation points. At these points the transverse shear vanishes for a rectangular element, as illustrated for a beam element in Section 9.7.1. Remarkably, this property of the shear vanishing at the midpoints of the edges for a constant

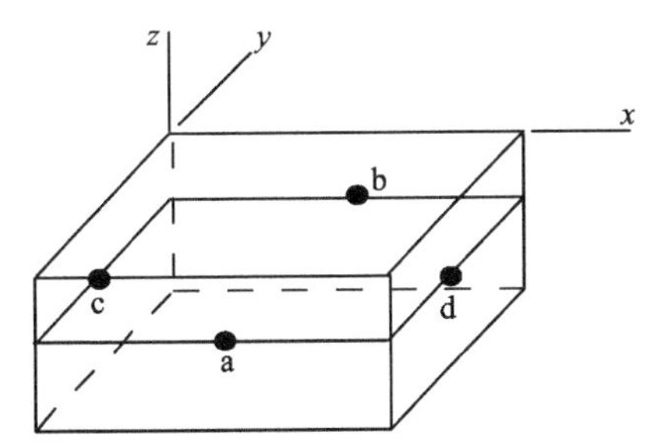

Figure 9.14 Interpolation points for shear in 4-node quadrilateral

moment holds for arbitrary quadrilaterals. We can take advantage of this by letting the assumed rates of transverse velocity shear be

$$\bar{D}_{\xi\zeta} = \frac{1}{2}\left(D_{\xi\zeta}(\xi_a, t) + D_{\xi\zeta}(\xi_b, t)\right) + \frac{1}{2}\left(D_{\xi\zeta}(\xi_a, t) - D_{\xi\zeta}(\xi_b, t)\right)\eta \tag{9.8.4}$$

$$\bar{D}_{\eta\zeta} = \frac{1}{2}\left(D_{\eta\zeta}(\xi_c, t) + D_{\eta\zeta}(\xi_d, t)\right) + \frac{1}{2}\left(D_{\eta\zeta}(\xi_a, t) - D_{\eta\zeta}(\xi_d, t)\right)\xi \tag{9.8.5}$$

where $\boldsymbol{\xi}_a = (0, -1, 0)$, $\boldsymbol{\xi}_b = (0, 1, 0)$, $\boldsymbol{\xi}_c = (-1, 0, 0)$, $\boldsymbol{\xi}_d = (1, 0, 0)$. The interpolation points are shown in Figure 9.14. The $\xi\zeta$ component is interpolated in the previous instead of the $\zeta\hat{z}$ component. The rate-of-deformation at the interpolation points is computed from the velocity field. This assumed strain field was first constructed on the basis of physical arguments by MacNeal (1982) and Hughes and Tezduyar (1981); the referential interpolation was given by Wempner, Talaslidis and Hwang (1982), who used the three-field Hu–Washizu weak form. Dvorkin and Bathe (1984) developed the previous interpolated strain field.

9.8.2 *Rank of Element*

The rank sufficiency of the above element can be checked by the methods given in Section 8.3.9. We illustrate this for a flat shell element in the x-y plane. Only the bending behavior is considered. Each node then has three relevant degrees of freedom: v_{zI}, ω_{xI}, ω_{yI}. Since there are four nodes, the element has 12 degrees of freedom. Three of these are rigid body motions: rotations about the x- and y-axes and translation in the z-direction. The proper rank of the element is then 9. The bending field has the same structure in terms of ω_x and ω_y as the plane field examined in (8.3.28), so it possesses five linearly independent fields. The two transverse shears from (9.8.3) possess four linearly independent fields. Thus the total of linearly independent fields is nine, which suffices to provide the proper rank for the element.

9.8.3 *Nine-Node Quadrilateral*

Assumed strain fields for the nine-node shell that avoid membrane and shear locking have been given by Huang and Hinton (1986) and Bucalem and Bathe (1993). In the latter, the assumed velocity strains are interpolated by the points shown in Figure 9.15(a):

$$\bar{D}_{\xi\xi} = \sum_{I=1}^{2}\sum_{J=1}^{3} D_{\xi\xi}\left(\xi_I, \eta_J, \zeta, t\right) N_{IJ}^{(1,2)}(\xi, \eta) \tag{9.8.6}$$

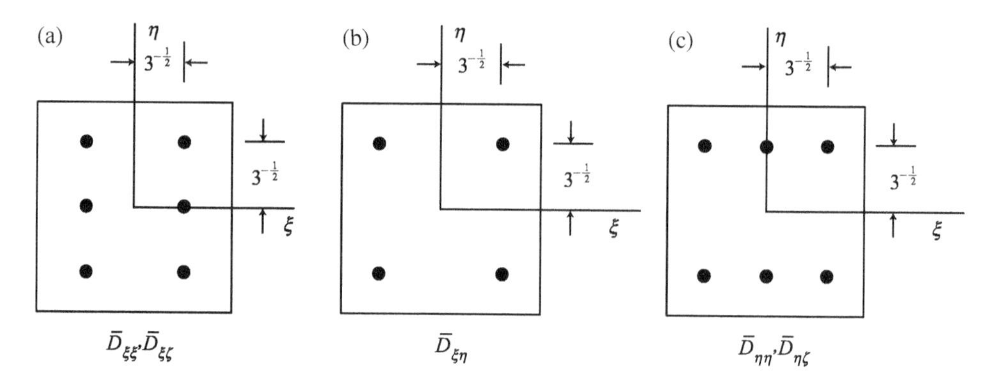

Figure 9.15 Interpolation points for assumed velocity strains in 9-node shell element

$$\bar{D}_{\xi\xi} = \sum_{I=1}^{2}\sum_{J=1}^{3} D_{\xi\xi}\left(\xi_I, \eta_J, \xi, t\right) N_{IJ}^{(1,2)}(\xi, \eta) \tag{9.8.7}$$

where I and J refer to the usual numbering and $\xi_I = (-3^{-1/2}, 3^{-1/2})$, $\eta_J = (-3^{-1/2}, 0, 3^{-1/2})$, and $N_{IJ}^{(a,b)}(\xi, \eta)$ is given by

$$n_{IJ}^{(a,b)}(\xi\eta) = N_I^{(a)}(\xi) N_J^{(b)}(\eta) \tag{9.8.8}$$

where $N_I^{(a)}(\xi)$ are the one-dimensional Lagrange interpolants of order a. The beam example in the previous section sheds some light on the rationale for the above: at the Gauss quadrature points, the transverse shear vanishes in bending and the membrane strain vanishes in inextensional bending. Thus the element should not exhibit parasitic transverse shears or membrane strains. The higher-order terms η^2 and $\eta^2\xi$ in the $\bar{D}_{\xi\xi}$ and $\bar{D}_{\xi\xi}$ fields provide stability. The velocity strains $\bar{D}_{\eta\eta}$ and $\bar{D}_{\eta\xi}$ are interpolated with the points shown in Figure 9.15(c). The shear component $\bar{D}_{\xi\eta}$ is interpolated with the points shown in Figure 9.15(b) using $N_{IJ}^{(1,1)}(\xi, \eta)$.

9.9 One-Point Quadrature Elements

In explicit software, the most widely used shell elements are 4-node quadrilaterals with one-point quadrature. Here one-point quadrature refers to the number of quadrature points in the reference plane: actually, anywhere from three to 30 or more quadrature points are used through the thickness, depending on the complexity of the nonlinear material response. Therefore, we often refer to one stack of quadrature points. These elements are commonly used in large-scale industrial analysis because they work well with diagonal mass matrices and are extremely robust. Higher-order elements, such as those based on quadratic isoparametric interpolants, converge more rapidly to smooth solutions. However, most large-scale problems involve nonsmooth phenomena, such as elasto-plasticity and contact-impact, so the greater approximation power of higher-order elements is not realizable in these problems; see Section 8.2.

Table 9.2 Four-node quadrilateral shell elements

Element	Acronym	Passes patch test	Correct in twist	Cost	Robustness
Belytschko–Tsay (1983)	BT	No	No	1.0	High
Hughes–Liu (1981a, b)	HL	No	Yes		High*
Belytschko–Wong–Chiang (1992)	BWC	No	Yes	1.2	Moderate
Belytschko–Leviathan (1994b)	BL	Yes	Yes	2.0	Moderate to low
Englemann–Whirley (1990)	YASE	No	No	–	Moderate
Full quadrature MacNeal–Wempner (Dvorkin–Bathe, 1984)	DB	Yes	Yes	3.5	Moderate to low

We will summarize the elements which have been most frequently used in software. These elements are listed in Table 9.2, along with some of their features and drawbacks. We then describe two of these elements in more detail, drawing on the material which precedes this to abbreviate the description. The earliest one-point quadrature shell element is the Belytschko–Tsay (BT) element (Belytschko and Tsay, 1983; Belytschko, Lin and Tsay, 1984). It is constructed by combining a flat, 4-node element with a plane quadrilateral 4-node membrane element. As indicated in Table 9.2, it does not respond correctly when its configuration is warped (this shortcoming manifests itself primarily when one or two lines of elements are used to model twisted beams).

The Hughes–Liu (HL) element, partially described in Hughes and Liu (1981a, b), is based on the CB shell theory. In explicit codes, it is used with a single stack of quadrature points, so it also requires hourglass control; the techniques developed in Belytschko, Lin and Tsay (1984) are used. It is significantly slower than the BT element.

The BWC element (see Table 9.2) corrects the twist, that is, the warped configuration defect in the BT element. Otherwise, it is quite similar. In the BL element, the physical hourglass control described in Chapter 8 is implemented. This hourglass control is based on a multi-field variational principle and the Dvorkin–Bathe strain approximation, (9.8.4–5). For elastic materials, it reproduces that element. However, in practice the inhomogeneity of the strain and stress state prevent accurate physical hourglass control. Nevertheless, this form of hourglass control provides a substantial advantage; it can be increased to moderately large values without locking, whereas in the BT element high values of the hourglass control parameters result in shear locking.

Both the BL element and any fully integrated element are afflicted with another shortcoming. In problems with large distortions, these elements fail suddenly and dramatically, aborting the simulation. The BT element, on the other hand, is very robust under severe distortion and seldom aborts a computation. This is highly valued in industrial settings. So the advantage of single quadrature point elements does not reside only in their superior speed; in addition, they tend to be more robust in problems where severe distortions are expected, such as car crash simulation.

The YASE element (the acronym for 'yet another shell element') incorporates the Pian–Sumihara (1985) membrane field for improved membrane response in beam bending, that is, for improved flexural performance, see Section 8.6.4. Otherwise, it is identical to the BT element.

The BT, BWC, and BL elements are based on a discrete Mindlin–Reissner theory; they are not continuum-based. 'Discrete' refers to the fact that the Mindlin–Reissner assumption is applied to the motion only at the quadrature point. The motion is constrained by requiring the current normal to remain straight. This can be viewed as another modification of the Mindlin–Reissner assumption; rather than requiring the *initial* normal to remain straight, the *current* normal is required to remain straight. A corotational formulation is used. Although in the original papers the corotational coordinate system was aligned with $\hat{\mathbf{e}}_x$ along $\mathbf{x}_{,\xi}$, this can lead to difficulties, so the technique described in Section 8.9 is recommended.

The velocity field is given by:

$$\mathbf{v}(\xi,t) = \mathbf{v}^M(\xi,\eta,t) + \bar{\zeta}(\boldsymbol{\omega}(\xi,\eta,t)\times\tilde{\mathbf{p}}(\xi,\eta,t)) \tag{9.9.1}$$

where a tilde is superimposed on the director $\tilde{\mathbf{p}}$ to indicate that it may differ from the director as defined in Section 9.6; it is the current normal to the reference surface. The finite element approximation to the motion is

$$\mathbf{v}(\xi,t) = \sum_{I=1}^{4}(\mathbf{v}_I(t) + \bar{\zeta}\boldsymbol{\omega}_I(t)\times\tilde{\mathbf{p}}_I)N_I(\xi,\eta) \tag{9.9.2}$$

Converting the cross-product to a matrix product, this can be written

$$\mathbf{v}(\xi,t) = \sum_{I=1}^{4}(\mathbf{v}_I(t) + \bar{\zeta}\,\Omega(\boldsymbol{\omega}_I)\tilde{\mathbf{p}}_I)N_I(\xi,\eta) \tag{9.9.3}$$

where N_I are the four-node isoparametric shape functions and $\boldsymbol{\Omega}(\boldsymbol{\omega}_I)$ is defined in (9.5.42). The corotational rate-of-deformation at the quadrature point $\xi = \eta = 0$ is given by

$$\hat{D}_{\alpha\beta} = \hat{D}^M_{\alpha\beta} + \bar{\zeta}\hat{\kappa}_{\alpha\beta} \tag{9.9.4}$$

where $\hat{\kappa}_{\alpha\beta}$ are the curvature rates. The membrane strains and membrane hourglass control are computed in the corotational coordinate system by the procedure and equations given in Section 8.7, Box 8.2. The curvature rates at the quadrature point are given by

$$\hat{\kappa}_{xx} = \frac{1}{2A}\left(\hat{y}_{24}\hat{\omega}_{y13} + \hat{y}_{31}\hat{\omega}_{y42}\right) + \frac{2z_\gamma}{A^2}\left(\hat{x}_{13}\hat{v}_{x13} + \hat{x}_{42}\hat{v}_{x24}\right) \tag{9.9.5}$$

$$\hat{\kappa}_{yy} = -\frac{1}{2A}\left(\hat{x}_{42}\hat{\omega}_{x13} + \hat{x}_{13}\hat{\omega}_{x24}\right) + \frac{2z_\gamma}{A^2}\left(\hat{y}_{13}\hat{v}_{y13} + \hat{y}_{42}\hat{v}_{y24}\right) \tag{9.9.6}$$

$$\begin{aligned} 2\hat{\kappa}_{xy} &= \frac{1}{2A}\left(\hat{x}_{42}\hat{\omega}_{y13} + \hat{x}_{31}\hat{\omega}_{y24} - \hat{y}_{24}\hat{\omega}_{x13} + \hat{y}_{31}\hat{\omega}_{x24}\right) \\ &+ \frac{2z_\gamma}{A^2}\left(\hat{x}_{13}\hat{v}_{y13} + \hat{x}_{42}\hat{v}_{y24} + \hat{y}_{13}\hat{v}_{x13} + \hat{y}_{42}\hat{v}_{x24}\right) \end{aligned} \tag{9.9.7}$$

where $z_\gamma = \hat{\gamma}^T\mathbf{z}$ and where $\hat{\gamma}$ is given in (8.4.11). The last terms in the curvature rate expressions do not vanish for a rigid body rotation in an arbitrary coordinate system. In the corotational system, the nodal velocities $\hat{v}_x$ and $\hat{v}_y$ are proportional to $z_y\mathbf{h}$ in rigid body rotation and it can be shown that the curvature rates vanish for rigid body rotation.

Since only one stack of quadrature points is used, the element is rank-deficient without stabilization. This can easily be seen by comparison with the rank analysis in Section 9.8.2. Since the element lacks the linear terms in the transverse shear and in the curvature rate, the rank of the bending part is 5 for one-point quadrature: the rate-of-deformation field consists of three constant moments and two constant shears. Thus the rank deficiency of the bending part is four. The spurious singular modes are shown in Hughes (1987: 333). Three of the modes are communicable, while one, the in-plane twist mode, is not. The three communicable modes are controlled by hourglass control:

$$\dot{q}_\alpha^B = \hat{\gamma}_I \hat{\omega}_{\alpha I}, \quad \dot{q}_3^B = \hat{\gamma}_I \hat{v}_{zI}, \quad \dot{Q}_\alpha^B = C_1^{QB} \dot{q}_\alpha^B, \quad \dot{Q}_3^B = C_2^{QB} \dot{q}_3^B \tag{9.9.8}$$

$$C_1^{QB} = \frac{1}{192} r_\theta (Eh^3 A) \mathbf{b}_\alpha^T \mathbf{b}_\alpha, \; C_2^{QB} = \frac{1}{12} r_w (Gh^3) \mathbf{b}_\alpha^T \mathbf{b}_\alpha \tag{9.9.9}$$

where E and G are Young's modulus and the shear modulus, A is the area of the element, and $\mathbf{b}_\alpha$ are defined in (8.4.9). The parameters r_θ and r_w are set by the user and should range from 0.01 to 0.05. These generalized hourglass strain rates are not orthogonal to rigid body rotation for warped configurations, so projections which eliminate the rigid body effects are necessary – see Belytschko and Leviathan (1994b). In addition, there are two hourglass modes associated with the membrane response; these and their control have been described in Section 8.7. All of the elements except BL use perturbation hourglass control.

Since the formulation is in a corotational laminar coordinate system, the stress rate corresponds closely to the Green–Naghdi rate. The formulation thus requires a constitutive law which relates the Green–Naghdi rate to the corotational rate-of-deformation tensor. The plane stress condition must be enforced as in Section 9.5.7. Under these conditions, the formulation is valid for arbitrarily large strains.

9.10 Exercises

9.1. Consider the three-node CB element shown in Figure 9.16. The shape functions are quadratic in ξ. Develop the velocity field and the rate-of-deformation in the corotational coordinates. Give an expression for the nodal forces. Develop an expression for the angle between the pseudonormal $\boldsymbol{p}$ and the true normal to the midline.

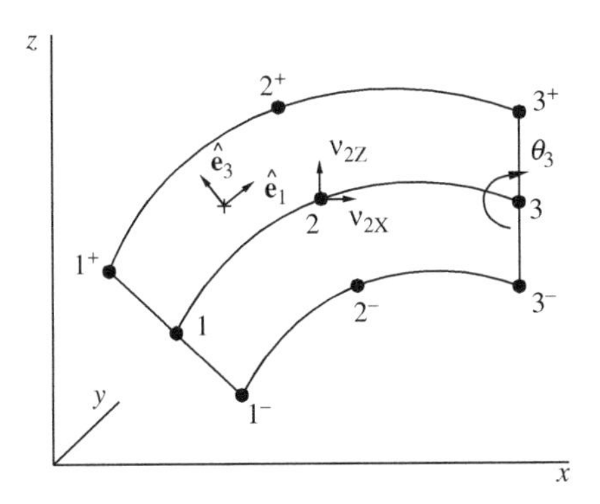

Figure 9.16 Three-node CB beam

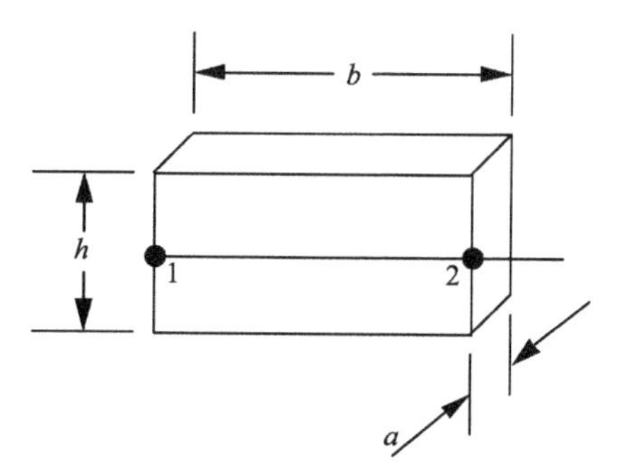

Figure 9.17 Two-node CB-beam

9.2. Consider a plate (a flat shell) in the x–y plane governed by the Mindlin–Reissner theory. Show that the rate-of-deformation is given by

$$D_{xx} = \frac{\partial v_x^M}{\partial x} + z\frac{\partial \omega_y}{\partial x}, \quad D_{yy} = \frac{\partial v_x^M}{\partial y} - z\frac{\partial \omega_x}{\partial y}, \quad D_{xy} = \frac{1}{2}\left(\frac{\partial v_x^M}{\partial y} + \frac{\partial v_x^M}{\partial x}\right) + \frac{z}{2}\left(\frac{\partial \omega_y}{\partial y} - \frac{\partial \omega_x}{\partial x}\right)$$

$$D_{xz} = \frac{1}{2}\left(\omega_y + \frac{\partial v_z^M}{\partial x}\right), \quad D_{yz} = \frac{1}{2}\left(-\omega_x + \frac{\partial v_z^M}{\partial y}\right)$$

9.3. Consider the lumped mass for a rectangular CB-beam element (Figure 9.17), $\hat{\mathrm{M}} = \frac{1}{8}m\mathrm{I}$, $m = \rho_0 a_0 b_0 h_0$ where ρ_0, a_0, b_0, and h_0 are the initial density and dimensions of the rectangular continuum element underlying the beam element. Using the transformation (9.3.24), develop a mass matrix for the 2-node CB element and diagonalize the result with the row-sum technique.

9.4. Develop the consistent mass matrix for a rectangular continuum element.
  (a) develop a consistent mass for the CB beam using (9.3.24), that is, $\mathbf{M} = \mathbf{T}^T\hat{\mathbf{M}}\mathbf{T}$ for a beam element lying along the x-axis as shown in Figure 9.17;
  (b) develop the complete inertia term including the time-dependent term in (9.3.26).

10

Contact-Impact

All solvable problems are trivial, all nontrivial problems are unsolvable.

Santayana

10.1 Introduction

This chapter introduces modeling of contact and impact. Many problems in the simulation of prototype tests and manufacturing processes involve contact and impact. For example, in the simulation of a drop test, the components must be separated by so-called sliding interfaces which can model contact, sliding and separation. In the simulation of manufacturing processes, sliding interfaces are also important: the modeling of the surfaces between the die and workpiece in sheet metal forming, the modeling of the tool–workpiece interface in machining, and the modeling of extrusion are some examples of where sliding interfaces are needed. In crash simulation of automobiles, many components, including the engine, wheels, radiator, and so on, can contact during the crash and their surfaces are treated as sliding interfaces. The treatment of impact always requires a subsequent treatment of contact, since bodies which impact will stay in contact until rarefaction waves result in release.

This chapter presents the governing equations and finite element procedures for contact-impact for Lagrangian meshes; the modeling of contact with Eulerian meshes is not considered here. The governing equations for bodies in contact are identical to the equations introduced previously, but it is necessary to add the kinetic and kinematic conditions on the contact interface. The key condition is the *condition of impenetrability*: namely, the condition that two

Nonlinear Finite Elements for Continua and Structures, Second Edition.
Ted Belytschko, Wing Kam Liu, Brian Moran, and Khalil I. Elkhodary.

Companion Website: www.wiley.com/go/belytschko

bodies cannot interpenetrate. Impenetrability cannot be expressed as a simple equation, so several simplified approaches have evolved. We will consider two of these: a rate form which is useful for explicit dynamics and a form based on closest point projection. The latter is primarily useful for implicit methods and equilibrium solutions. Both the classical Coulomb friction models and interface constitutive models are described.

Next, the weak forms of the governing equations are developed. Four approaches to treating the contact surface constraints are considered:

1. The Lagrange multiplier method
2. The penalty method
3. The augmented Lagrangian method
4. The perturbed Lagrangian method.

These weak forms incorporate the various ways of imposing constraints which were explored in Chapters 6 and 8.

The Lagrange multiplier weak form for contact-impact differs from the weak form for single bodies in that it is an inequality; it is often called a weak inequality and the corresponding variational principles are called variational inequalities. In the discretization of contact problems by Lagrange multiplier methods, the multipliers must be approximated on the contact interface. The multiplier must observe the constraint that the normal tractions be compressive. In penalty methods, the traction inequalities emerge from the Heaviside step function which is embedded in the penalty force.

Contact-impact problems are among the most difficult nonlinear problems because the response in contact-impact problems is not smooth. The velocities normal to the contact interface are discontinuous in time when impact occurs. For Coulomb friction models, the tangential velocities along the interface are discontinuous when stick-slip behavior is encountered. These characteristics of contact-impact introduce significant difficulties in the time integration of the discrete equations and impair the performance of Newton algorithms. The appropriate choice of methodologies and algorithms is crucial to success, and regularization techniques are highly useful in obtaining robust solution procedures.

10.2 Contact Interface Equations

10.2.1 Notation and Preliminaries

Contact-impact algorithms in general purpose software can treat the interaction of many bodies, but multi-body contact consists of the interaction of pairs of bodies. So we consider the two-body problem shown in Figure 10.1. We have denoted the current configurations of the two bodies by Ω^A and Ω^B and denote the union of the two bodies by Ω. The boundaries of the bodies are denoted by Γ^A and Γ^B. Although the two bodies are interchangeable with respect to their mechanics, in some equations and algorithms the bodies are distinguished as master and slave: body A is designated as the master, body B as the slave. When we wish to distinguish field variables that are associated with a particular body, we append a superscript A or B; when neither of these superscripts appears, the field variable applies to the union of the two bodies. Thus the velocity field $\mathbf{v}(\mathbf{X}, t)$ refers to the velocity field in both bodies, whereas $\mathbf{v}^A(\mathbf{X}, t)$ refers to the velocity in body A.

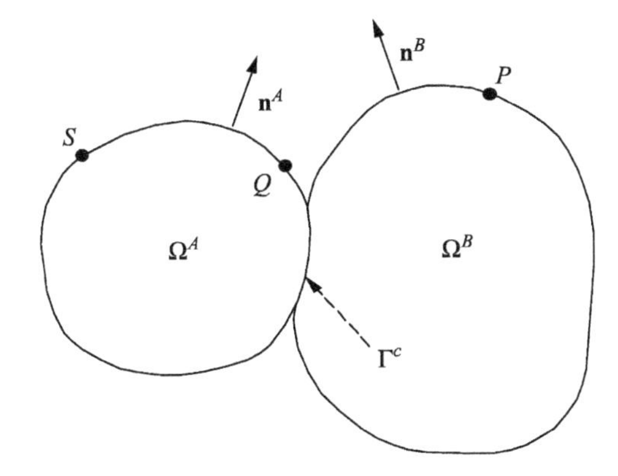

Figure 10.1 Model problem for contact-impact showing notation

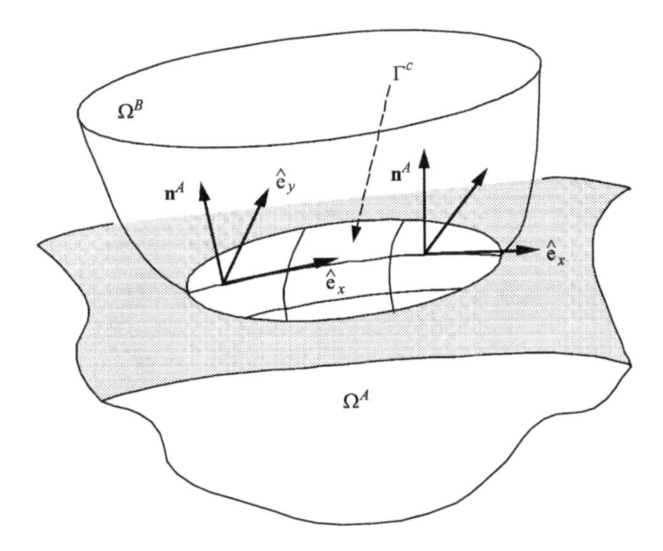

Figure 10.2 Contact interface showing local unit vectors referred to master surface A

The contact interface consists of the intersection of the surfaces of the two bodies and is denoted by Γ^c:

$$\Gamma^c = \Gamma^A \cap \Gamma^B \tag{10.2.1}$$

This contact interface consists of the two physical surfaces of the two bodies which are in contact, but since they are coincident we refer to a single interface Γ^c. In numerical solutions, the two surfaces will usually not be coincident. In those cases, Γ^c refers to the master surface. Moreover, although the two bodies may be in contact on several disjoint interfaces, we designate their union by a single symbol Γ^c. The contact interface is a function of time, and its determination is an important part of the solution of the contact-impact problem.

In constructing the interface equations, it is convenient to express vectors in terms of local components of the contact surface. A local coordinate system is set up at each point of the master contact surface as shown in Figure 10.2. At each point, we can construct unit vectors tangent to the surface of the master body $\hat{\mathbf{e}}_1^A \equiv \hat{\mathbf{e}}_x^A$ and $\hat{\mathbf{e}}_2^A \equiv \hat{\mathbf{e}}_y^A$. The procedure for obtaining these unit vectors is identical to that used in shell elements: see Chapter 9. The normal for body A is given by

$$\mathbf{n}^A = \hat{\mathbf{e}}_1^A \times \hat{\mathbf{e}}_2^A \tag{10.2.2}$$

On the contact surface

$$\mathbf{n}^A = -\mathbf{n}^B \tag{10.2.3}$$

that is, the normals of the two bodies are in opposite directions. The velocity fields are expressed in terms of local components by

$$\mathbf{v}^A = v_N^A \mathbf{n}^A + \hat{v}_\alpha^A \hat{\mathbf{e}}_\alpha^A = v_N^A \mathbf{n}^A + \mathbf{v}_T^A \tag{10.2.4}$$

$$\mathbf{v}^B = v_N^B \mathbf{n}^A + \hat{v}_\alpha^B \hat{\mathbf{e}}_\alpha^A = -v_N^B \mathbf{n}^B + \mathbf{v}_T^B \tag{10.2.5}$$

where the range of Greek subscripts is 2 in three-dimensional problems; when the problem is two-dimensional, the contact surface becomes a line, so we have a single unit vector $\hat{\mathbf{e}}_1 \equiv \hat{\mathbf{e}}_x$ tangent to this line; the range of the Greek subscripts in (10.2.4) is then one. As can be seen in the above, the components are in the local coordinate system of the master surface A. The normal velocities are given by

$$v_N^A = \mathbf{v}^A \cdot \mathbf{n}^A, \quad v_N^B = \mathbf{v}^B \cdot \mathbf{n}^A \tag{10.2.6}$$

which is obtained by taking the dot product of the expressions in (10.2.4–5) with $\mathbf{n}^A$ and using the fact that the normal is orthogonal to the unit vectors tangent to the plane.

The bodies are governed by the standard field equations given in Boxes 4.1 and 5.1: conservation of mass, momentum and energy, a strain measure, and the constitutive equations. Contact adds the following conditions: the bodies cannot interpenetrate and the tractions must satisfy momentum conservation on the interface. Furthermore, the normal traction across the contact interface cannot be tensile. We classify the requirements on the displacements and velocities as kinematic conditions and the requirements on the tractions as kinetic conditions.

10.2.2 Impenetrability Condition

In a multi-body problem, the bodies must observe the impenetrability condition. The impenetrability condition for a pair of bodies can be stated as

$$\Omega^A \cap \Omega^B = 0 \tag{10.2.7}$$

that is, the intersection of the two bodies is the null set. In other words, the two bodies are not allowed to overlap, which can also be viewed as a compatibility condition. The impenetrability condition is highly nonlinear for large displacement problems, and in general cannot be expressed as an algebraic or differential equation in terms of the displacements. The difficulty arises because in an arbitrary motion it is impossible to anticipate which points of the two bodies will contact. For example, in Figure 10.1, if the bodies are spinning, it is possible for

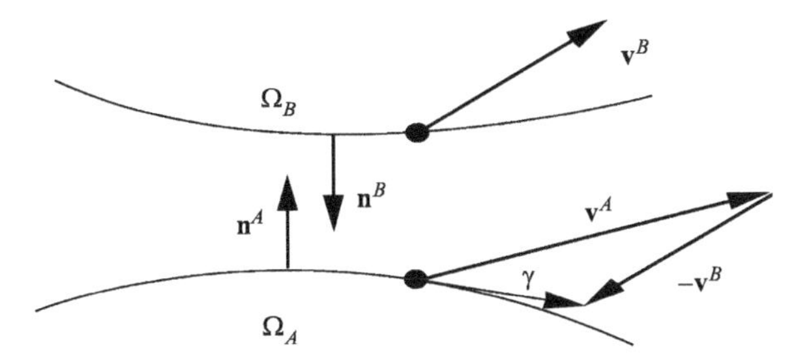

Figure 10.3 Nomenclature for velocities on contact surface; the same nomenclature and relations hold for incremental displacements $\Delta\mathbf{u}$ or variations $\delta\mathbf{u}$ or $\delta\mathbf{v}$; the contacting surfaces are shown separated for clarity

point P to contact point Q, whereas a different relative motion can result in contact of point P with point S. Consequently, an equation which expresses the fact that point P does not penetrate body A cannot be written except in general terms such as (10.2.7).

Because it is not feasible to express (10.2.7) in terms of the displacements, it is convenient to express the impenetrability equations in rate form or incremental form in each stage of the process. The rate form of the impenetrability condition is applied to those portions of bodies A and B which are already in contact, that is, to those points which are on the contact surface Γ^c. It can be written as

$$\gamma_N = \mathbf{v}^A \cdot \mathbf{n}^A + \mathbf{v}^B \cdot \mathbf{n}^B = (\mathbf{v}^A - \mathbf{v}^B)\cdot \mathbf{n}^A \equiv v_N^A - v_N^B \leq 0 \quad \text{on} \quad \Gamma^c \tag{10.2.8}$$

where v_N^A and v_N^B are defined in (10.2.6). Here $\gamma_N(\mathbf{X}, t)$ is the rate of interpenetration of the two bodies; see Figure 10.3. The impenetrability condition (10.2.8) restricts the interpenetration rate for any points on the contact surface to be negative, that is, it expresses the fact that when the two bodies are in contact, then they must either remain in contact ($\gamma_N = 0$) or they must separate ($\gamma_N < 0$). When (10.2.8) is met for all points which are in contact, the impenetrability condition is met exactly. However, (10.2.8) and (10.2.7) are not equivalent. When (10.2.8) is only observed at instants in time as in most numerical methods, interpenetration is possible for points which are closely separated but not in contact. Equation (10.2.8) is applicable only to point-pairs that are in contact. The integral of γ_N depends on the path and is not recommended as a measure of interpenetration.

Imposing (10.2.8) introduces discontinuities in the velocity time histories. Prior to contact, the normal velocities are not equal, whereas subsequent to impact, the normal velocity components must observe (10.2.8). These discontinuities in time complicate the time integration of the discrete equations.

Many authors use the quantity $-\gamma_N$ to characterize the interaction of the two bodies and call it the gap rate. *The gap rate is the negative of the interpenetration rate.* It may appear inconsistent to speak of an interpenetration rate when impenetrability is a fundamental condition on the solution. However, in many numerical methods, a small amount of interpenetration is allowed, and inequality (10.2.8) will not be observed exactly.

The relative tangential velocity is given by

$$\gamma_T = \hat{\gamma}_{Tx}\hat{\mathbf{e}}_x + \hat{\gamma}_{Ty}\hat{\mathbf{e}}_y = \mathbf{v}_T^A - \mathbf{v}_T^B \tag{10.2.9}$$

The middle term is included to illustrate that the relative tangential velocity in three dimensions is a two-component vector. As can be seen from (10.2.9), the expression for the relative tangential velocity is similar to the expression for the normal relative velocities, (10.2.8).

10.2.3 Traction Conditions

The tractions must observe the balance of momentum across the contact interface. Since the interface has no mass, this requires that the sum of the tractions on the two bodies vanishes:

$$\mathbf{t}^A + \mathbf{t}^B = \mathbf{0} \tag{10.2.10}$$

The tractions on the surfaces of the two bodies are defined by Cauchy's law:

$$\mathbf{t}^A = \boldsymbol{\sigma}^A \cdot \mathbf{n}^A \quad \text{or} \quad t_i^A = \sigma_{ij}^A n_j^A \tag{10.2.11}$$

$$\mathbf{t}^B = \boldsymbol{\sigma}^B \cdot \mathbf{n}^B \quad \text{or} \quad t_i^B = \sigma_{ij}^B n_j^B \tag{10.2.12}$$

The normal tractions are defined by

$$t_N^A = \mathbf{t}^A \cdot \mathbf{n}^A \quad \text{or} \quad t_N^A = t_j^A n_j^A \tag{10.2.13}$$

$$t_N^B = \mathbf{t}^B \cdot \mathbf{n}^A \quad \text{or} \quad t_N^B = t_j^B n_j^A \tag{10.2.14}$$

Note that the normal components refer to the master body. The normal component of momentum balance can be obtained by taking a dot product of (10.2.10) with the normal vector $\mathbf{n}^A$, which gives

$$t_N^A + t_N^B = 0 \tag{10.2.15}$$

We do not consider any adhesion between the contact surfaces in the normal direction, so the normal tractions cannot be tensile. The condition that the normal tractions cannot be tensile, that is, that they are compressive, is stated as

$$t_N \equiv t_N^A(\mathbf{x},t) = -t_N^B(\mathbf{x},t) \leq 0 \tag{10.2.16}$$

so this condition requires t_N^B to be positive since t_N^B is the projection of the traction on body B onto the unit normal of A, which points into body B. Note that this expression is not symmetric with respect to bodies A and B. The normal of one of the bodies is selected to define the normal traction, and the sign of the normal traction of the bodies will depend on this choice of normal.

The tangential tractions are defined by

$$\mathbf{t}_T^A = \mathbf{t}^A - t_N^A \mathbf{n}^A, \quad \mathbf{t}_T^B = \mathbf{t}^B - t_N^B \mathbf{n}^A \tag{10.2.17}$$

so the tangential tractions are the total tractions projected on the master contact surface. Momentum balance requires that

$$\mathbf{t}_T^A + \mathbf{t}_T^B = \mathbf{0} \tag{10.2.18}$$

The above equation can be obtained by substituting (10.2.17) into (10.2.10) and using (10.2.15).

When a frictionless model of contact is used, the tangential tractions vanish:

$$\mathbf{t}_T^A = \mathbf{t}_T^B = \mathbf{0} \tag{10.2.19}$$

We have used the phrase 'frictionless model of contact' to emphasize that it is not implied that friction is absent, but rather that friction is neglected in the model because it is deemed unimportant. Subsequently we shall just say frictionless contact, but it should be understood that friction never vanishes in reality.

Although one of the bodies has been chosen as the master body in developing the preceding contact interface equations, these equations are symmetric with respect to the bodies when the two contact surfaces are coincident and (10.2.3) is observed. Thus it does not matter which body is chosen as the master body. However, when the two surfaces are not coincident, as in most numerical solutions, then the choice of the master body changes the results somewhat.

10.2.4 *Unitary Contact Condition*

Conditions (10.2.8) and (10.2.16) can be combined into a single equation

$$t_N \gamma_N = 0 \tag{10.2.20}$$

which is called the *unitary contact condition*. This equation also expresses the fact that the normal components of the contact forces do no work. That this condition must hold on the contact surface can be seen as follows: when the bodies are in contact and remain in contact, $\gamma_N = 0$, whereas when contact ceases, $\gamma_N \le 0$ but the normal traction vanishes, so the product always vanishes.

10.2.5 *Surface Description*

The surfaces of the contacting bodies are described by curvilinear coordinates $\zeta^A = \left[\zeta_1^A, \zeta_2^A\right]$ and $\zeta^B = \left[\zeta_1^B, \zeta_2^B\right]$, where the superscripts designate the bodies. In two dimensions, the contact surfaces are lines which are parameterized by ζ^A and ζ^B.

Points on the contact surface can be specified by the reference coordinates of either body, but it is conventional to choose one body as the master and use its reference coordinates. Body A is the master and the motion of the contact interface is described by $\mathbf{x}(\boldsymbol{\zeta}^A, t) = \boldsymbol{\phi}^A(\boldsymbol{\zeta}^A, t)$. The covariant base vectors of the contact surface of body A are given by

$$\mathbf{a}_\alpha = \frac{\partial \boldsymbol{\phi}^A}{\partial \zeta^\alpha} \equiv \boldsymbol{\phi},_\alpha^A \equiv \mathbf{x},_\alpha^A \tag{10.2.21}$$

In an implementation, the covariant base vectors are usually replaced by Cartesian base vectors defined as described in Section 9.5.

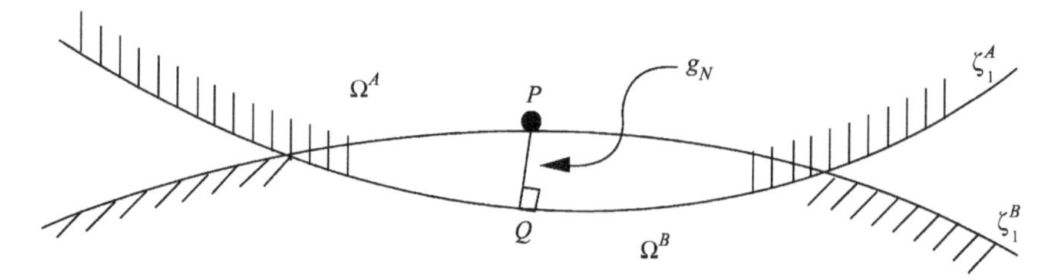

Figure 10.4 Point Q is the closest point on body A to P on body B: it is the orthogonal projection of P on A

10.2.6 Interpenetration Measure

In the following, an expression is developed for the interpenetration. This derivation is quite complicated and not readily applied to finite elements, so the next sections can be skipped on a first reading. We follow Wriggers (1995) and Wriggers and Miehe (1994). Consider the situation shown in Figure 10.4, where point P has penetrated body A. We wish to find the magnitude of interpenetration, which is denoted by $g_N(\boldsymbol{\zeta}^B, t)$.

The penetration of a point P on body B which is inside body A is defined as the minimum distance to any point on the surface of body A. The distance between point P with coordinates $x^B(\zeta^B, t)$ and any point on the surface of A is given by

$$\ell_{AB} = \left\|\mathbf{x}^B(\zeta^B, t) - \mathbf{x}^A(\zeta^A, t)\right\| \equiv \left[(x^B - x^A)^2 + (y^B - y^A)^2 + (z^B - z^A)^2\right]^{\frac{1}{2}} \tag{10.2.22}$$

The interpenetration $g_N(\boldsymbol{\zeta}^B, t)$ is the minimum of the above and is considered nonzero only when point P is inside body A. The latter condition can be checked by examining the projection of the normal to body A on $\mathbf{x}^B - \mathbf{x}^A$: when the projection is negative, point P is inside body A and has therefore interpenetrated, otherwise P is not inside body A and there is no interpenetration. So the definition of the interpenetration is

$$g_N(\boldsymbol{\zeta}^B, t) = \min_{\zeta^A} \alpha \ell_{AB}, \quad \alpha = \begin{cases} 1 & \text{if}\,(\mathbf{x}^B - \mathbf{x}^A)\cdot\mathbf{n}^A \le 0 \\ 0 & \text{if}\,(\mathbf{x}^B - \mathbf{x}^A)\cdot\mathbf{n}^A > 0 \end{cases} \tag{10.2.23}$$

The coordinate $\bar{\zeta} \equiv \zeta^A$ which minimizes $g_N(\boldsymbol{\zeta}^B, t)$, i.e. the point $\mathbf{x}^A(\bar{\zeta}, t)$ which minimizes ℓ_{AB} is found by setting the derivative of ℓ_{AB} with respect to $\bar{\zeta}$ to zero:

$$\frac{\partial \ell_{AB}}{\partial \bar{\zeta}^\alpha} = \frac{\partial}{\partial \bar{\zeta}^\alpha}\left\|\mathbf{x}^B - \mathbf{x}^A\right\| = \frac{\mathbf{x}^B - \mathbf{x}^A}{\left\|\mathbf{x}^B - \mathbf{x}^A\right\|}\cdot\left(\frac{-\partial \mathbf{x}^A}{\partial \bar{\zeta}^\alpha}\right) \equiv -\mathbf{e}\cdot\mathbf{a}_\alpha = 0 \tag{10.2.24}$$

where $\mathbf{a}_\alpha$ is given by (10.2.21) and $\mathbf{e} = (\mathbf{x}^B - \mathbf{x}^A)/\|\mathbf{x}^B - \mathbf{x}^A\|$, so $\mathbf{e}$ is a unit vector from body A to body B. Since $\mathbf{e}$ is orthogonal to the tangent vectors $\mathbf{a}_\alpha$ according to (10.2.24), it is normal to the surface A. So ℓ_{AB} is minimum when $\mathbf{e}$ is normal to the surface of A; $\mathbf{x}^A(\bar{\zeta}, t)$ is called the

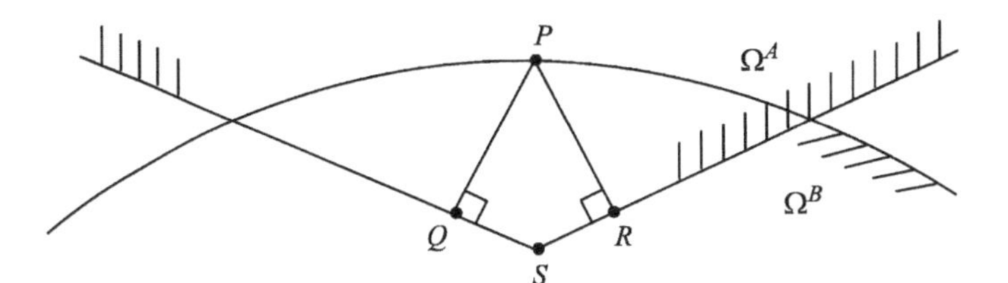

Figure 10.5 Penetration by a surface with a kink showing the resulting nonuniqueness of the point of orthogonal projection

orthogonal projection of the point P onto the surface A. This is a result that permeates mathematics: the shortest distance from a point to a space or manifold is the orthogonal projection. The result is illustrated in Figure 10.4 in two dimensions. Note that when the bodies have interpenetrated, $\mathbf{e}$ is opposite to the direction of the outward pointing normal, so $\mathbf{e} = -\mathbf{n}^A$.

The minimizer $\bar{\zeta}$ is obtained by solving the nonlinear algebraic equations (10.2.24). In three dimensions (10.2.24) involves two equations in two unknowns; in two dimensions it involves a single equation. Once $\bar{\zeta}$ is determined, the interpenetration g_N can be found by (10.2.23).

This approach to defining the interpenetration poses difficulties when the two bodies are not smooth or not locally convex. For example, in the situation shown in Figure 10.5 the minimizer of ℓ_{AB} is not unique: there are two points which are orthogonal projections of the point P. In these situations, it is difficult to develop schemes which lead to a uniquely defined measure of the interpenetration.

10.2.7 Path-Independent Interpenetration Rate

In this section, the rate of interpenetration will be developed from $g_N(\boldsymbol{\zeta}, t)$ given in (10.2.23); the integral of $\dot{g}_N(\zeta, t)$ is path-independent. We define the rate of interpenetration $g_N(\boldsymbol{\zeta}, t)$, as given in (10.2.23), only when $\alpha \neq 0$;

$$\dot{g}_N = \frac{\partial g_N(\zeta, t)}{\partial t} = \frac{\mathbf{x}^B(\zeta, t) - \mathbf{x}^A(\bar{\zeta}, t)}{\left\|\mathbf{x}^B(\zeta, t) - \mathbf{x}^A(\zeta, t)\right\|} \cdot \left(\frac{\partial \mathbf{x}^B(\zeta, t)}{\partial t} - \frac{\partial \mathbf{x}^A(\bar{\zeta}, t)}{\partial t}\right) \tag{10.2.25}$$

Based on the discussion following (10.2.24), we know that the minimum is attained when $(\mathbf{x}^B - \mathbf{x}^A)/\|\mathbf{x}^B - \mathbf{x}^A\|$ corresponds to the normal to body A. Using this and the fact that $\mathbf{v}^B = \partial \mathbf{x}^B(\boldsymbol{\zeta}, t)/\partial t$, this can be rewritten as

$$\dot{g}_N = \mathbf{n}^B \cdot \left(\mathbf{v}^B - \frac{\partial \mathbf{x}^A(\bar{\zeta}, t)}{\partial t}\right) \tag{10.2.26}$$

It is important to observe that $\bar{\zeta}$ is not a material coordinate, because in order to remain the closest point projection, this point moves independently of the material. Thus the second term in the parentheses of the RHS of (10.2.26) is not a material derivative. The point can be considered an ALE point: it is neither fixed in space nor coincident with a material point. Using ALE derivatives from Chapter 7 (or simply the chain rule), it follows that

$$\mathbf{v}^A = \frac{\partial \mathbf{x}^A(\zeta,t)}{\partial t} = \frac{\partial \mathbf{x}^A}{\partial t}(\bar{\zeta},t) + \frac{\partial \mathbf{x}^A}{\partial \bar{\zeta}^\alpha}\frac{\partial \bar{\zeta}^\alpha}{\partial t}$$

$$\text{so} \quad \frac{\partial \mathbf{x}^A(\bar{\zeta},t)}{\partial t} = v^A - \frac{\partial \mathbf{x}^A}{\partial \bar{\zeta}^\alpha}\frac{\partial \bar{\zeta}^\alpha}{\partial t} \equiv \mathbf{v}^A - \mathbf{x}_{,\alpha}^A \frac{\partial \bar{\zeta}^\alpha}{\partial t} \tag{10.2.27}$$

Substituting (10.2.27) into (10.2.26), and using (10.2.3) it follows that

$$\dot{g}_N = \mathbf{n}^B \cdot \left(\mathbf{v}^B - \mathbf{v}^A + \mathbf{x}_{,\alpha}^A \frac{\partial \bar{\zeta}^\alpha}{\partial t} \right) = \mathbf{n}^A \cdot \mathbf{v}^A - \mathbf{n}^A \cdot \mathbf{v}^B - \mathbf{n}^A \cdot \mathbf{x}_{,\alpha}^A \frac{\partial \bar{\zeta}^\alpha}{\partial t} \tag{10.2.28}$$

Comparing (10.2.8) and (10.2.28), it can be seen that the normal interpenetration rate differs from the normal projection of the relative velocities γ_N unless $\bar{\zeta}_{,t} = \mathbf{0}$. Whenever the two surfaces of the contacting bodies are coincident $\bar{\zeta}_{,t} = \mathbf{0}$, so

$$\gamma_N = \dot{g}_N \tag{10.2.29}$$

The above development of interpenetration rate requires that the bodies be continuously differentiable, that is, C^1. Otherwise, in situations such as shown in Figure 10.5, $\bar{\zeta}$ is not a continuous function of time, as when for example the closest point shifts from Q to R.

10.2.8 Tangential Relative Velocity for Interpenetrated Bodies

If the bodies have interpenetrated, (10.2.9) does not give the relative tangential velocities of two points on the contact surfaces; (10.2.9) is exact only when the two bodies are in contact but have not interpenetrated. To obtain a relation for the tangential velocities which holds for interpenetrated bodies, we follow Wriggers (1995). In this approach, the relative tangential velocity is defined in terms of the velocities of a point P on body B and its closest point projection. The relative tangential velocity is defined by

$$\dot{\mathbf{g}}_T = \bar{\zeta}_{,t}^{\alpha} \mathbf{a}_\alpha \tag{10.2.30}$$

which involves the rate $\bar{\zeta}_{,t}^{\alpha}$ which appears in (10.2.27); $\bar{\zeta}_{,t}^{\alpha}$ is obtained from (10.2.24) as follows. Since (10.2.24) always pertains to the closest point, the time derivative of the RHS of (10.2.24) must vanish. So multiplying (10.2.24) by $\|\mathbf{x}^B - \mathbf{x}^A\|$ and using (10.2.21), $\mathbf{a}_\alpha = \partial \mathbf{x}^A / \partial \zeta^\alpha$, we have

$$\frac{\partial}{\partial t}\left[(\mathbf{x}^B(\zeta,t) - x^A(\bar{\zeta},t)) \cdot \mathbf{a}_\alpha \right] = 0 \tag{10.2.31}$$

where ζ is fixed. From (10.2.21)

$$\frac{\partial \mathbf{a}_\alpha}{\partial t} = \frac{\partial}{\partial t}\left(\frac{\partial \mathbf{x}^A}{\partial \zeta^\alpha}\right) = \frac{\partial}{\partial \zeta^\alpha}\left(\frac{\partial \mathbf{x}^A}{\partial t} + \frac{\partial \mathbf{x}^A}{\partial \zeta^\beta}\frac{\partial \bar{\zeta}^\beta}{\partial t}\right) = \mathbf{v},_\alpha^A + \mathbf{x},_{\alpha\beta}^A \bar{\zeta},_t^\beta \tag{10.2.32}$$

The remaining steps are as follows (the independent variables are suppressed when convenient):

Derivative of product in (10.2.31):

$$\left(\mathbf{x},_t^B(\zeta, t) - \mathbf{x},_t^A(\bar{\zeta}, t)\right)\cdot \mathbf{a}_\alpha + (\mathbf{x}^B - \mathbf{x}^A)\cdot \mathbf{a}_{\alpha,t} = 0 \tag{10.2.33}$$

Using $\mathbf{v}^{BA} \equiv \mathbf{v}^B - \mathbf{v}^A$, (10.2.27), $\mathbf{x}^{BA} \equiv \mathbf{x}^B - \mathbf{x}^A$, (10.2.32) for $\mathbf{a}_{\alpha,t}$:

$$\left(\mathbf{v}^{BA} + \mathbf{x},_\beta^A \bar{\zeta},_t^\beta\right)\cdot \mathbf{a}_\alpha + \mathbf{x}^{BA}\cdot\left(\mathbf{v},_\alpha^A + \mathbf{x},_{\alpha\beta}^A \bar{\zeta},_t^\beta\right) = 0 \tag{10.2.34}$$

Using $\mathbf{x},_\beta^A = \mathbf{a}_\beta$, and rearranging (10.2.34) gives

$$\left(-\mathbf{a}_\alpha \cdot \mathbf{a}_\beta - \mathbf{x}^{BA}\cdot \mathbf{x},_{\alpha\beta}^A\right)\bar{\zeta},_t^\beta = \mathbf{x}^{BA}\cdot \mathbf{v},_\alpha^A + \mathbf{v}^{BA}\cdot \mathbf{a}_\alpha \tag{10.2.35}$$

This is a system of two linear algebraic equations in the two unknowns $\bar{\zeta},_t^\beta$; all terms on the right-hand side are known. Once the time derivatives $\bar{\zeta},_t^\beta$ have been obtained, $\dot{\mathbf{g}}_T$ can be determined from (10.2.30).

When $\mathbf{x}^{BA} = 0$, (10.2.35) can be simplified to

$$\mathbf{a}_\alpha \cdot \mathbf{a}_\beta \bar{\zeta},_t^\beta = (\mathbf{v}^A - \mathbf{v}^B)\cdot \mathbf{a}_\alpha \tag{10.2.36}$$

The RHS by (10.2.9) are the components of $\boldsymbol{\gamma}_T$, while the LHS of the above are the components of $\dot{\mathbf{g}}_T$; so we can see that when the surfaces are coincident, $\dot{\mathbf{g}}_T = \gamma_T$.

Thus the displacement-based definition of relative tangential velocity, (10.2.30), is consistent with the tangential velocity defined in (10.2.9) in the absence of interpenetration. The kinetic and kinematic contact interface equations are summarized in Box 10.1.

Box 10.1 Contact interface conditions

Kinetic conditions:

$$\mathbf{t}^A + \mathbf{t}^B = \mathbf{0} \tag{B10.1.1}$$

$$\text{normal}: t_N^A + t_N^B = 0,\ t_N^A \equiv \mathbf{t}^A\cdot\mathbf{n}^A, t_N^B \equiv \mathbf{t}^B\cdot\mathbf{n}^A, t_N \equiv t_N^A \le 0 \tag{B10.1.2}$$

$$\text{tangential}: \mathbf{t}_T^A + \mathbf{t}_T^B = \mathbf{0},\quad \mathbf{t}_T^A \equiv \mathbf{t}^A - t_N^A\mathbf{n}^A,\quad \mathbf{t}_T^B \equiv \mathbf{t}^B - t_N^B\mathbf{n}^A \tag{B10.1.3}$$

Kinematic conditions in velocity form:

$$\gamma \equiv \gamma_N = (\mathbf{v}^A - \mathbf{v}^B)\cdot \mathbf{n}^A \equiv v_N^A - v_N^B \leq 0 \quad \text{(B10.1.4)}$$

$$\boldsymbol{\gamma}_T = \mathbf{v}_T^A - \mathbf{v}_T^B = \mathbf{v}^A - \mathbf{v}^B - \mathbf{n}^A(\mathbf{v}^A - \mathbf{v}^B)\cdot \mathbf{n}^A \quad \text{(B10.1.5)}$$

Unitary contact condition:

$$t_N \gamma_N = 0 \quad \text{(B10.1.6)}$$

Kinematic conditions and definitions in displacement form:

$$g \equiv g_N = \min_{\bar{\zeta}} \left\| \mathbf{x}^B(\zeta,t) - \mathbf{x}^A(\bar{\zeta},t) \right\| \quad \text{if} \quad \left[\mathbf{x}^B(\zeta,t) - \mathbf{x}^A(\bar{\zeta},t)\right]\cdot \mathbf{n}^A \leq 0 \quad \text{(B10.1.7)}$$

$$\dot{g}_N = \mathbf{n}^A \cdot \mathbf{v}^A + \mathbf{n}^B \cdot \mathbf{v}^B - \mathbf{n}^A \cdot \mathbf{x}_{,\alpha}^A \bar{\zeta}_{,t}^{\alpha} \quad \text{(B10.1.8)}$$

Example 10.1 Consider two surfaces which have partially interpenetrated. The master body is a 9-node isoparametric element, so the three nodes of surface A are defined by a quadratic mapping:

$$\begin{Bmatrix} x \\ y \end{Bmatrix}^A = (1-r^2)\begin{Bmatrix} 2 \\ 1 \end{Bmatrix} + \frac{1}{2}r(1+r)\begin{Bmatrix} 3 \\ 3 \end{Bmatrix} \quad \text{where} \quad r \equiv \zeta^A, -1 \leq r \leq 1 \quad \text{(E10.1.1)}$$

The surface of the slave body B is a horizontal line given by

$$\begin{Bmatrix} x \\ y \end{Bmatrix}^B = \begin{Bmatrix} 4s \\ 1.5 \end{Bmatrix}, \quad s \equiv \zeta^B, \quad 0 \leq s \leq 1 \quad \text{(E10.1.2)}$$

The interpenetration in the example has been exaggerated. Note that $\mathbf{n}^B \neq -\mathbf{n}^A$ along the interface. For the point P on surface B with coordinates (1,1.5), we next find the interpenetration. The orthogonal projection of the point Q minimizes ℓ_{PQ}:

$$\begin{aligned} \ell_{PQ} &= \left\| \mathbf{x}^B(\zeta^B) - \mathbf{x}^A(\zeta^A) \right\| = \left((x^B - x^A)^2 + (y^B - y^A)^2 \right)^{1/2} \\ &= \left\{ \left[1 - \left(2(1-r^2) + \frac{3}{2}r(1+r) \right) \right]^2 + \left[\frac{3}{2} - \left((1-r^2) + \frac{3}{2}r(1+r) \right) \right]^2 \right\}^{1/2} \end{aligned} \quad \text{(E10.1.3)}$$

The minimizer is given by

$$0 = \frac{d\ell_{PQ}}{dr} = \frac{1}{\ell_{PQ}}\left(r^3 + 3r + \frac{3}{4} \right) \quad \text{(E10.1.4)}$$

The root is found numerically to be $r = -0.2451$, so $(x_Q, y_Q) = (1.6023, 0.6624)$.

10.3 Friction Models

10.3.1 Classification

The models for the tangential tractions are collectively called friction models. There are basically three types of friction models:

1. Coulomb friction models, which are based on the classical theories of friction commonly taught in undergraduate mechanics and physics courses.
2. Interface constitutive equations, which give the tangential forces by equations similar to constitutive equations used for materials.
3. Asperity-lubricant models, which model the behavior of the physical characteristics of the interface, often on a microscale.

The demarcations between these classes are not sharp: some models adopt features of more than one of these three.

10.3.2 Coulomb Friction

Coulomb friction models originate from friction models for rigid bodies. In the adaptation of Coulomb friction models to continua, they are applied at each point of the contact interface, which gives:

if A and B are in contact at $\mathbf{x}$, then

$$\text{(a) if } \left\|\mathbf{t}_T(\mathbf{x},t)\right\| < -\mu_F t_N(\mathbf{x},t), \qquad \gamma_T(\mathbf{x},t) = \mathbf{0} \tag{10.3.1}$$

$$\text{(b) if } \left\|\mathbf{t}_T(\mathbf{x},t)\right\| = -\mu_F t_N(\mathbf{x},t), \qquad \gamma_T(\mathbf{x},t) = -\kappa(\mathbf{x},t)\mathbf{t}_T(\mathbf{x},t), \quad \kappa \geq 0 \tag{10.3.2}$$

where κ is a variable which is determined from the solution of the momentum equation. The condition that the two bodies are in contact at a point implies that the normal traction $t_N \leq 0$, so the RHS of the two expressions, $-\mu_F t_N$, is always positive. Condition (a) is known as the stick condition: when the tangential traction at a point is less than the critical value, no relative tangential motion is permitted, that is, *the two bodies stick*. Condition (b) corresponds to sliding, and the second part of that equation expresses the fact that the tangential friction must be in the direction opposite to the relative tangential velocity.

Coulomb friction closely resembles a rigid plastic material. If the tangential velocity γ_T is interpreted as a strain and the tangential tractions are interpreted as stresses, the first relation in (10.3.1) can be interpreted as a yield function. According to (10.3.1), when the yield criterion is not met, the tangential velocity vanishes. Once the yield function is satisfied, the tangential velocity is in the direction given in (10.3.2). These attributes parallel a rigid plastic material model.

There are several alternative ways of stating Coulomb's law which are equivalent to the above. For example, Demkowicz and Oden (1981) give Coulomb's law as (the spatial dependence of the variables has been dropped for simplicity):

if A and B are in contact at X, then

$$\left\|\mathbf{t}_T\right\| \leq -\mu_F t_N \quad \text{and} \quad \mathbf{t}_T \cdot \boldsymbol{\gamma}_T + \mu_F \left|t_N\right| \left|\boldsymbol{\gamma}_T\right| = 0 \tag{10.3.3}$$

The stick condition of Coulomb friction is its most troublesome characteristic, since it induces discontinuities in the time history of the relative tangential velocity. When the motion of a point changes from slip to stick, the relative tangential velocity γ_T discontinuously jumps to zero. Thus the tangential velocities are not smooth, which makes numerical procedures difficult.

10.3.3 Interface Constitutive Equations

A different approach to defining interface laws was pioneered by Michalowski and Mroz (1978) and Curnier (1984). This approach is motivated by the theory of plasticity and the analogy between Coulomb friction and elasto-plasticity we alluded to previously. Constitutive models of interface behavior are motivated by surface roughness due to asperities, such as shown in Figure 10.6, which occur at the microscale on even the smoothest surface. Friction is generated by the interaction of the asperities during sliding. Sliding initially causes elastic deformations of these asperities, so a true stick condition cannot occur in nature, that is, the stick condition is an idealization of observed behavior. Elastic deformation of the asperities is followed by 'grinding' down of the asperities as the sliding proceeds. The elastic deformations of the asperities are reversible, whereas the grinding down is irreversible, so ascribing an elastic character to the initial sliding and a plastic character to subsequent sliding is natural.

As an example of an interface constitutive law we describe an adaptation of Curnier's (1984) plasticity theory for friction. This model contains all of the ingredients of a plasticity theory for continua: a decomposition of deformation into reversible and irreversible components, a yield function and a flow law. In this description of the Curnier model, we have replaced relative displacements by relative velocities, so the following applies to problems involving arbitrary time histories and large relative sliding.

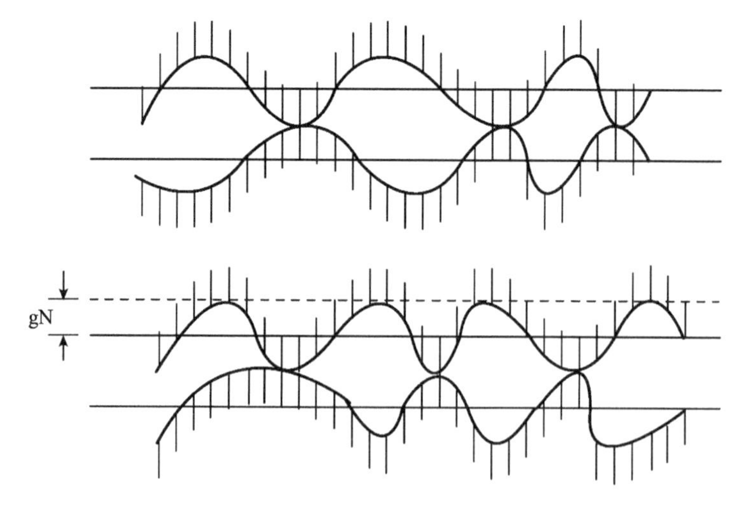

Figure 10.6 Asperities on contacting surface

The relative tangential velocities $\boldsymbol{\gamma}_T$ are subdivided into an adherence, which is the elastic deformation of the asperities, and a slip, the grinding down of the asperities:

$$\boldsymbol{\gamma}_T = \boldsymbol{\gamma}_T^{\text{adh}} + \boldsymbol{\gamma}_T^{\text{slip}} \tag{10.3.4}$$

Here $\boldsymbol{\gamma}^{\text{adh}}$ is the reversible part, $\boldsymbol{\gamma}^{\text{slip}}$ is the irreversible part. A wear function is defined by

$$D^c = \int_0^t \left(\boldsymbol{\gamma}_T^{\text{slip}} \cdot \boldsymbol{\gamma}_T^{\text{slip}}\right)^{\frac{1}{2}} dt \tag{10.3.5}$$

which is reminiscent of the definition of effective plastic strain.

Two functions are defined to construct the plastic interface law as a function of the tractions $\mathbf{t}$:

1. a yield function, $f(\mathbf{t})$;
2. a potential function for the flow law, $h(\mathbf{t})$.

The yield function determines the onset of plastic response, the potential function the relationship between the slip (plastic strain rate) and the tangential tractions.

The theory is similar to the nonassociative plasticity theories given in Section 5.6. Therefore, we will only sketch the steps and point out the need for nonassociative plasticity. The yield function for Coulomb-type behavior corresponds to the Coulomb friction condition:

$$f(t_N, \mathbf{t}_T) = \|\mathbf{t}_T\| + \mu_F t_N = 0 \tag{10.3.6}$$

Note the similarity to (10.3.1). In two dimensions this yield function takes the form shown in Figure 10.7: $\mathbf{t}_T = t_T \hat{\mathbf{e}}_x$ in two dimensions, so the yield function consists of two lines with slopes $\pm\mu_F$ as shown. In three dimensions, $\mathbf{t}_T = \hat{t}_\alpha \hat{\mathbf{e}}_\alpha = \hat{t}_x \hat{\mathbf{e}}_x + \hat{t}_y \hat{\mathbf{e}}_y$, and (10.3.6) becomes

$$f(t_N, \mathbf{t}_T) = \left(\hat{t}_x^2 + \hat{t}_y^2\right)^{1/2} + \mu_F t_N = 0 \tag{10.3.7}$$

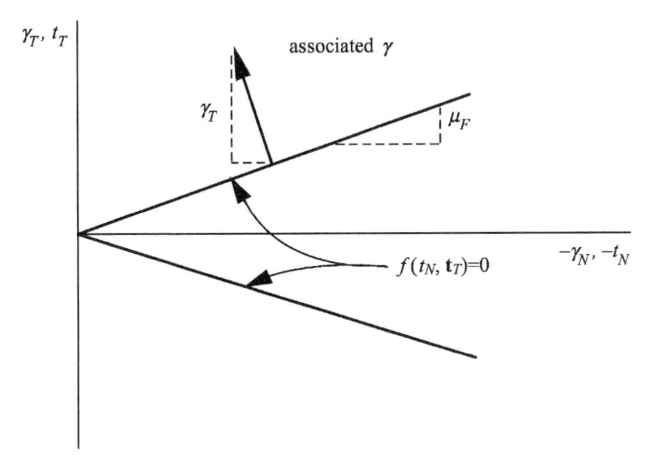

Figure 10.7 Coulomb yield surface in two dimensions

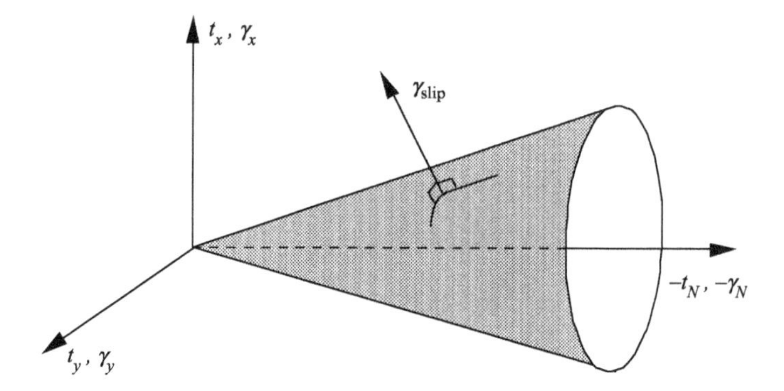

Figure 10.8 Coulomb surface for contact in three dimensions and the associated slip

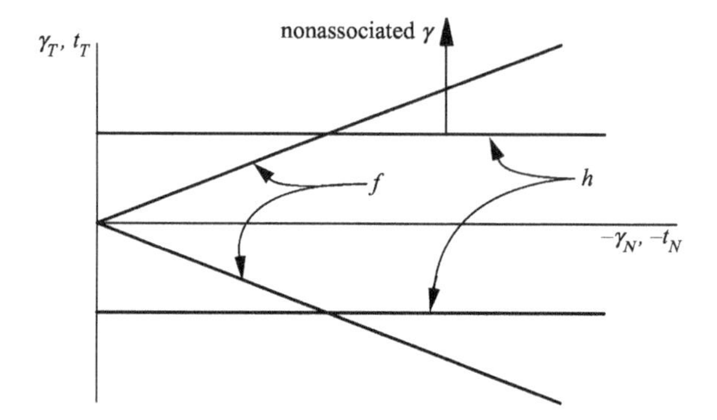

Figure 10.9 Non-associated flow law in two dimensions showing the yield function f and potential function h

so the yield function is a cone as shown in Figure 10.8.

In a nonassociative plasticity, the potential function for the slip differs from the yield function. One possible potential function for a nonassociative theory is

$$h(t_N, \mathbf{t}_T) = \|\mathbf{t}_T\| - \beta = 0 \tag{10.3.8}$$

where β is a constant whose magnitude is irrelevant. This potential function is shown in Figure 10.9.

To write the complete relations for a plasticity theory of friction in two and three dimensions, it is convenient to define

$$\boldsymbol{\gamma} = \begin{Bmatrix} \gamma_N \\ \gamma_T \end{Bmatrix} \text{ in 2D,} \quad \boldsymbol{\gamma} = \begin{Bmatrix} \gamma_N \\ \boldsymbol{\gamma}_T \end{Bmatrix} = \begin{Bmatrix} \gamma_N \\ \hat{\gamma}_x \\ \hat{\gamma}_y \end{Bmatrix} \text{ in 3D} \tag{10.3.9}$$

$$\mathbf{Q} = \begin{Bmatrix} t_N \\ t_T \end{Bmatrix} \text{ in 2D,} \quad \mathbf{Q} = \begin{Bmatrix} t_N \\ \mathbf{t}_T \end{Bmatrix} = \begin{Bmatrix} t_N \\ \hat{t}_x \\ \hat{t}_y \end{Bmatrix} \text{ in 3D} \tag{10.3.10}$$

The adhesive strains are then related to the stresses by

$$\mathbf{Q}^{\nabla} = \mathbf{C}^{Q}\boldsymbol{\gamma}^{\text{adh}} \quad \text{or} \quad Q_i^{\nabla} = C_{ij}^{Q}\gamma_j^{\text{adh}} \tag{10.3.11}$$

which is the counterpart of the hypoelastic law for continua. Usually $\mathbf{C}^Q$ is diagonal since little experimental information is available on the coupling between different components of the frictional traction and the relative motion.

The adhesive slip rates are given by the nonassociative flow law. We consider perfectly plastic sliding, in which there is no increase in the tractions with the accumulation of slip:

$$\boldsymbol{\gamma}^{\text{slip}} = \alpha\frac{\partial h}{\partial \mathbf{Q}} \quad \text{or} \quad \gamma_i^{\text{slip}} = \alpha\frac{\partial h}{\partial Q_i} \tag{10.3.12}$$

We define

$$\mathbf{f} \equiv \frac{\partial f}{\partial \mathbf{Q}}, \quad \mathbf{h} \equiv \frac{\partial h}{\partial \mathbf{Q}} \tag{10.3.13}$$

The steps for developing the constitutive equation for the frictional surface are then:

$$\mathbf{f}^T\dot{\mathbf{Q}} = 0 \qquad \text{consistency} \tag{10.3.14}$$

$$\mathbf{Q}^{\nabla} = \mathbf{C}^{Q}(\boldsymbol{\gamma} - \boldsymbol{\gamma}^{\text{slip}}) \qquad \text{(10.3.11) and (10.3.4)} \tag{10.3.15}$$

$$\mathbf{f}^T\mathbf{C}^{Q}(\boldsymbol{\gamma} - \alpha\mathbf{h}) = 0 \qquad \text{(10.3.12) and (10.3.14) into (10.3.15)} \tag{10.3.16}$$

$$\alpha = \frac{\mathbf{f}^T C^{Q}\boldsymbol{\gamma}}{\mathbf{f}^T\mathbf{C}^{Q}\mathbf{h}} \qquad \text{solve (10.3.16) for } \alpha \tag{10.3.17}$$

$$\mathbf{Q}^{\nabla} = \mathbf{C}^{Q}\left(\boldsymbol{\gamma} - \frac{\mathbf{f}^T\mathbf{C}^{Q}\boldsymbol{\gamma}}{\mathbf{f}^T\mathbf{C}^{Q}\mathbf{h}}\mathbf{h}\right) \qquad \text{(10.3.17) and (10.3.12) into (10.3.15)} \tag{10.3.18}$$

The objective (i.e. frame-invariant) rate is related to the material rate by

$$\frac{\partial \mathbf{Q}(\zeta, t)}{\partial t} = \mathbf{Q}^{\nabla} + \mathbf{Q}\cdot\mathbf{W} \tag{10.3.19}$$

where $\mathbf{W}$ is the projection of the spin given by (3.3.11) onto the surface. Update procedures analogous to that in elasto-plasticity, Section 5.6, can be used.

The reason for choosing a nonassociative flow law can be explained by examining sliding in two-dimensions. For an associated flow law, the irreversible slips are given by $\gamma_N^{\text{slip}} = \alpha\partial f/\partial t_N = \alpha\mu_F$ and $\gamma_T^{\text{slip}} = \alpha\partial f/\partial t_T = \alpha\,\text{sign}(t_T)$. Since $\alpha \geq 0$, and $t_N < 0$ in contact, this implies that $\gamma_N^{\text{slip}} < 0$ so the bodies would separate after the onset of slip (recall γ_N is positive in interpenetration). If the slips are given by the potential flow laws using the nonassociated potential (10.3.8), the slips in two dimensions can be written as

$$\gamma_N^{\text{slip}} = \alpha \frac{\partial h}{\partial t_N} = 0, \quad \gamma_T^{\text{slip}} = \alpha \frac{\partial h}{\partial t_T} = \alpha \,\text{sign}(t_T) \tag{10.3.20}$$

Thus the normal slip vanishes, that is, no irreversible normal separation occurs during sliding with the nonassociative law.

Hardening can also be incorporated as in elasto-plasticity: see Section 5.6. When the normal tractions are large, the asperities are often ground down, and irreversible 'strains' γ_N^{slip} develop. This can be modeled by a cap model, see DiMaggio and Sandler (1971).

10.4 Weak Forms

10.4.1 Notation and Preliminaries

The weak form of the momentum equation and the contact interface conditions will be developed for a Lagrangian mesh. This development is also applicable to an ALE mesh when the contact surface is treated as Lagrangian. For simplicity, we start with frictionless contact and defer the treatment of tangential tractions to the last part of this section. We restrict the following developments to the case where all traction or velocity components are prescribed on a traction or displacement boundary, respectively.

The contact surface is neither a traction nor a displacement boundary. The total boundary of body A is given by

$$\Gamma^A = \Gamma_t^A \cup \Gamma_u^A \cup \Gamma^c \tag{10.4.1}$$

and we note that

$$\Gamma_t^A \cap \Gamma_u^A = 0, \quad \Gamma_t^A \cap \Gamma^c = 0, \quad \Gamma_u^A \cap \Gamma^c = 0 \tag{10.4.2}$$

$$\Gamma_t = \Gamma_t^A \cup \Gamma_t^B, \quad \Gamma_u = \Gamma_u^A \cup \Gamma_u^B \tag{10.4.3}$$

Identical relations hold for body B.

The trial solutions are in the space of *kinematically admissible velocities*, and as in Chapter 4 we choose the velocities to be the cardinal dependent variable; the displacements can be obtained by time integration. The trial solution is $\mathbf{v}(\mathbf{X}, t) \in U$ where the space of trial functions is defined by

$$u = \{\mathbf{v}(\mathbf{X},t) | \mathbf{v} \in C^0(\Omega^A), \quad \mathbf{v} \in C^0(\Omega^B), \quad \mathbf{v} = \bar{\mathbf{v}} \text{ on } \Gamma_u\} \tag{10.4.4}$$

This space of trial functions is similar to that for the single-body problem, but the velocities are separately approximated in the two bodies; the velocity fields in U are not continuous across the contact interface. The admissible velocity fields are here given as C^0, that is, in H^1 but for purposes of convergence analysis in linear elastostatics the displacements should be in $H^{1/2}$: see Kikuchi and Oden (1988). This is similar to the function space that is used in fracture mechanics problems to handle the singular stresses at the crack tip. In contact problems, singularities occur at the edge. However, unlike in fracture mechanics, the singularities in contact problems do not appear to be of any engineering significance, since the roughness of surfaces appears to eliminate the appearance of even near-singular behavior in the stresses.

The space of test functions is defined by

$$U_0 = \{U \text{ with } \overline{\mathbf{v}} = 0\} \tag{10.4.5}$$

which parallels the definition in Section 4.3.

10.4.2 Lagrange Multiplier Weak Form

A common approach to imposing the contact constraints is by means of Lagrange multipliers. We will follow the description given by Belytschko and Neal (1991). Let the Lagrange multiplier trial function be $\lambda(\boldsymbol{\zeta}, t)$ and the corresponding test functions be $\delta\lambda\,(\boldsymbol{\zeta}, t)$. These functions reside in the following spaces:

$$\lambda(\zeta,t) \in J^+, J^+ = \{\lambda(\zeta,t) \mid \lambda \in C^{-1}, |\lambda \geq 0 \text{ on } \Gamma^c\} \tag{10.4.6}$$

$$\delta\lambda(\zeta) \in J^-, J^- = \{\delta\lambda(\zeta) \mid \delta\lambda \in C^{-1}, \delta\lambda \leq 0 \quad \text{on} \quad \Gamma^c\} \tag{10.4.7}$$

The weak form is

$$\delta P_L(\mathbf{v}, \delta\mathbf{v}, \lambda, \delta\lambda) \equiv \delta P + \delta G_L \geq 0 \quad \forall \delta\mathbf{v} \in U_0, \quad \forall \delta\lambda \in J^- \tag{10.4.8}$$

where

$$\delta G_L = \int_{\Gamma^c} \delta(\lambda \gamma_N)\, d\Gamma \tag{10.4.9}$$

In the previous, δP is defined in Box 4.2 and $\mathbf{v} \in U$, $\lambda \in J^+$. This weak form is equivalent to the momentum equation, the traction boundary conditions, the interior continuity conditions (generalized momentum balance) and the following contact interface conditions: impenetrability (10.2.8), momentum balance on normal tractions (10.2.15) and the frictionless condition (10.2.19). The Lagrange multiplier field need only be C^{-1} because its derivatives do not appear in the weak form. The requirement that the normal interface traction be compressive is a constraint on the trial space for the Lagrange multipliers. Note that the above *weak form is an inequality*.

The above is a standard way of appending a constraint to a weak form by means of Lagrange multipliers: compare to the Hu–Washizu variational principle. The only difference from the Hu–Washizu form is that the constraint is an inequality.

The equivalence of the weak form to the momentum equation, the traction boundary conditions and the contact conditions is shown by a procedure that parallels that in Section 4.2. (The interior continuity conditions are omitted by assuming sufficient smoothness). Recall that δP is given in Box 4.2 as

$$\delta P = \int_{\Omega} [\delta v_{i,j} \sigma_{ji} - \delta v_i (\rho b_i - \rho \dot{v}_i)]\, d\Omega - \int_{\Gamma_t} \delta v_i \bar{t}_i\, d\Gamma \tag{10.4.10}$$

where we have used commas to denote derivatives with respect to the spatial variables and a superposed dot to denote the material time derivative. All integrals in (10.4.10) apply to the union of both bodies, that is, $\Omega = \Omega^A \cup \Omega^B$, $\Gamma_t = \Gamma_t^A \cup \Gamma_t^B$ and so on. The first step is to integrate the internal virtual power by parts and apply Gauss's theorem:

$$\int_{\Omega} (\delta v_i \sigma_{ji})_{,j} \, d\Omega = \int_{\Gamma_t} \delta v_i \sigma_{ji} n_j \, d\Gamma + \int_{\Gamma^c} \left(\delta v_i^A t_i^A + \delta v_i^B t_i^B \right) d\Gamma \tag{10.4.11}$$

We have used the fact that the integral on the displacement boundary Γ_u vanishes because $\delta v_i = 0$ on Γ_u and Cauchy's law (B3.1.1) has been applied to obtain the expressions in the last integral. The first integral on the RHS of (10.4.11) applies to both bodies, as can be seen from the definition (10.4.3). A contact surface integral results on each body when Gauss's theorem is applied, so to express the result as a single integral over the contact surface, the identity of the body to which the variables pertain is indicated by the superscripts A and B.

The integrand of the second integral on the RHS of (10.4.11) is now subdivided into components normal and tangential to the contact surface. In indicial notation this gives

$$\delta v_i^A t_i^A = \delta v_N^A t_N^A + \delta \hat{v}_\alpha^A \hat{t}_\alpha^A \tag{10.4.12}$$

where the range of alpha is 1 for two-dimensional problems and 2 for three-dimensional problems. A similar relationship can be written for body B. This is clearer to some people in vector notation, where

$$\delta \mathbf{v}^A \cdot \mathbf{t}^A = \left(\delta v_N^A \mathbf{n}^A + \delta \mathbf{v}_T^A \right) \cdot \left(t_N^A \mathbf{n}^A + \mathbf{t}_T^A \right) = \delta v_N^A t_N^A + \delta \mathbf{v}_T^A \cdot \mathbf{t}_T^A \tag{10.4.13}$$

The simplification to the second line is obtained by noting that $\mathbf{n}$ is normal to the tangent vectors $\mathbf{t}_T$ and $\mathbf{v}_T$. The second term in (10.4.13) is an alternative expression for $\delta \hat{v}_\alpha \hat{t}_\alpha$.

Substituting (10.4.11) and (10.4.12) into (10.4.10) gives

$$\begin{aligned} \delta P = & \int_{\Omega} \delta \dot{v}_i \left(\rho \dot{v}_i - b_i - \sigma_{ij,j} \right) d\Omega + \int_{\Gamma_t} \delta v_i \left(\sigma_{ji} n_j - \bar{t}_i \right) d\Gamma \\ & + \int_{\Gamma^c} \left(\delta v_N^A t_N^A + \delta v_N^B t_N^B + \delta \hat{v}_\alpha^A \hat{t}_\alpha^A + \delta \hat{v}_\alpha^B \hat{t}_\alpha^B \right) d\Gamma \end{aligned} \tag{10.4.14}$$

Now consider (10.4.9):

$$\delta G_L = \int_{\Gamma^c} \delta(\lambda \gamma_N) d\Gamma = \int_{\Gamma^c} (\delta \lambda \gamma_N + \delta \gamma_N \lambda) d\Gamma \tag{10.4.15}$$

Substituting (10.2.8) into (10.4.15) gives

$$\delta G_L = \int_{\Gamma^c} \left(\delta \lambda \gamma_N + \lambda \left(\delta v_N^A - \delta v_N^B \right) \right) d\Gamma \tag{10.4.16}$$

Combining (10.4.14) and (10.4.16) yields

$$0 \le \delta P_L = \int_{\Omega} \delta v_i \left(\sigma_{ji,j} - \rho b_i - \rho \dot{v}_i\right) d\Omega + \int_{\Gamma_t} \delta v_i \left(\sigma_{ji} n_j - \bar{t}_i\right) d\Gamma$$
$$+ \int_{\Gamma^c} \left[\delta v_N^A \left(t_N^A + \lambda\right) + \delta v_N^B \left(t_N^B - \lambda\right) + \left(\delta \hat{v}_\alpha^A \hat{t}_\alpha^A + \delta \hat{v}_\alpha^B \hat{t}_\alpha^B\right) + \delta\lambda \gamma_N \right] d\Gamma \tag{10.4.17}$$

Extracting the strong form from the weak inequality is similar to the procedure described in Section 4.3.2 but we must consider the inequalities on the test functions. Whenever the test function is unconstrained, there is no restriction on the sign of the term which multiplies the test function and the term must vanish by the density theorem. It follows from the first two integrals in the above that

$$\sigma_{ji,j} - \rho b_i = \rho \dot{v}_i \text{ in } \Omega, \quad \sigma_{ji} n_j = \bar{t}_i \text{ on } \Gamma_t \tag{10.4.18}$$

that is, that the momentum equation and the natural boundary conditions are satisfied in bodies A and B. In all terms of the contact surface integrand except the last, the test function is unconstrained, so we obtain the equalities

$$\hat{t}_\alpha^A = 0 \text{ and } \hat{t}_\alpha^B = 0 \text{ on } \Gamma^c, \text{ or } \mathbf{t}_T^A = \mathbf{t}_T^B = \mathbf{0} \text{ on } \Gamma^c \tag{10.4.19a}$$

$$\lambda = -t_N^A \text{ and } \lambda = t_N^B \text{ on } \Gamma^c \tag{10.4.19b}$$

By eliminating λ from (10.4.20) we obtain the momentum balance condition on the normal tractions:

$$t_N^A + t_N^B = 0 \quad \text{on } \Gamma^c \tag{10.4.20}$$

The test function $\delta\lambda$, in the last term of the integrand of (10.4.17) is negative by (10.4.7). Therefore, γ_N does not necessarily vanish. However, it can be deduced that γ_N must be nonpositive, that is, that the weak *inequality* implies

$$\gamma_N \le 0 \quad \text{on } \Gamma^c \tag{10.4.21}$$

which is the interpenetration inequality (10.2.8).

Equations (10.4.18–21) constitute the strong form corresponding to the weak form (10.4.8). This set includes the momentum equation, the interior continuity conditions, and the traction (natural) boundary conditions on both bodies. On the contact surface, the strong form includes momentum balance of the normal tractions and the inequality on the interpenetration rate. The compressive character of the normal tractions follows from the restriction on the Lagrange multiplier test function (10.4.6).

10.4.3 Contribution of Virtual Power to Contact Surface

At this point, for the purpose of simplifying subsequent proofs, we observe that the only contribution of δP to the conditions on the contact interface is

$$\delta P_1(\Gamma^c) = \int_{\Gamma^c} \left(\delta v_i^A t_i^A + \delta v_i^B t_i^B\right) d\Gamma = \int_{\Gamma^c} \left(\delta v_N^A t_N^A + \delta v_N^B t_N^B + \delta \mathbf{v}_T^A \cdot \mathbf{t}_T^A + \delta \mathbf{v}_T^B \cdot \mathbf{t}_T^B\right) d\Gamma \quad (10.4.22)$$

The remaining terms in δP are equivalent to the momentum equation and traction boundary conditions on the surfaces not in contact. So replacing δP by δP_1 is equivalent to the preceding.

If the contact surface is frictionless, then the last two terms in (10.4.22) vanish, so the contribution of δP to the contact interface is

$$\delta P_2\left(\Gamma^c\right) \equiv \int_{\Gamma^c} \left(\delta v_N^A t_N^A + \delta v_N^B t_N^B\right) d\Gamma \quad (10.4.23)$$

Replacing δP by δP_2 implies the momentum equation, the traction boundary conditions, and the frictionless condition (10.2.19). These results will be used in the proofs which follow.

10.4.4 Rate-Dependent Penalty

In the penalty method, the impenetrability constraint is imposed as a penalty normal traction along the contact surface. In contrast to the Lagrange multiplier method, the penalty method allows some interpenetration. However, it is easier to implement and is quite widely used. We consider two forms of the penalty method:

1. A penalty proportional to the square of the interpenetration rate γ_N;
2. A penalty which is an arbitrary function of the interpenetration and its rate.

The second is more useful for applications in nonlinear problems, because a strictly velocity-dependent penalty allows too much interpenetration.

In the penalty methods, the test and trial functions are identical to those in the Lagrange multiplier method, (10.4.4–5). The equivalence of the weak form to the strong form for the penalty method can be stated as follows:

if $\quad \mathbf{v} \in u \text{ and } \delta P_p(\mathbf{v}, \delta\mathbf{v}) = \delta P + \delta G_p = 0 \; \forall \delta\mathbf{v} \in U_0 \quad (10.4.24)$

where

$$\delta G_p = \int_{\Gamma^c} \frac{\beta}{2} \delta\left(\gamma_N^2\right) H(\gamma_N) d\Gamma \quad (10.4.25)$$

then the momentum equation and natural boundary conditions are satisfied in the two bodies, the normal tractions on Γ^c satisfy momentum balance and are compressive, and the tangential tractions vanish on Γ^c.

In the above $H(\gamma_N)$ is the Heaviside step function,

$$H\left(\gamma_N\right) = \begin{cases} 1 & \text{if } \gamma_N > 0 \\ 0 & \text{if } \gamma_N < 0 \end{cases} \quad (10.4.26)$$

The functional δP is defined in (10.4.10) and β is the *penalty parameter*. The penalty parameter can be a function of the spatial coordinates. The weak form associated with the penalty method is not an inequality; the discontinuous nature of the contact-impact problem is introduced by the Heaviside step function that appears in (10.4.25). This weak form does not imply the impenetrability condition, which is satisfied only approximately in the penalty method.

To show that the weak form implies the strong form, we begin by taking the variation of δG_P, which gives

$$\delta G_P = \int_{\Gamma^c} \beta \gamma_N \delta \gamma_N H(\gamma_N) d\Gamma \tag{10.4.27}$$

Using (10.2.8) in (10.4.27) gives

$$\delta G_P = \int_{\Gamma^c} \beta \gamma_N^+ \left(\delta v_N^A - \delta v_N^B \right) d\Gamma \quad \text{where } \gamma_N^+ = \gamma_N H(\gamma_N) \tag{10.4.28a,b}$$

We then combine the above with $\delta P_2(\Gamma^c)$, given in (10.4.23). This yields

$$\delta P_P = \int_{\Gamma^c} \left[\delta v_N^A \left(t_N^A + \beta \gamma_N^+ \right) + \delta v_N^B \left(t_N^B - \beta \gamma_N^+ \right) \right] d\Gamma = 0 \tag{10.4.29}$$

The arbitrariness of δv_N^A and δv_N^B on Γ^c then yields

$$t_N^A + \beta \gamma_N^+ = 0 \quad \text{on } \Gamma^c \tag{10.4.30}$$

$$t_N^B - \beta \gamma_N^+ = 0 \quad \text{on } \Gamma^c \tag{10.4.31}$$

Combining these two equations gives

$$t_N^A = -t_N^B = -\beta \gamma_N^+ \le 0 \tag{10.4.32}$$

where the inequality follows from the fact that $\gamma_N^+ \ge 0$ when the penalty is active because of violation of the impenetrability constraint. Thus the weak form implies that the normal fractions satisfy momentum balance and are compressive. The momentum equation, traction boundary conditions and frictionless condition were implied by using (10.4.23) in (10.4.29).

The penalty weak form, unlike the Lagrange multiplier weak form, does not enforce the continuity of the velocities across the contact interface; in fact, the velocities will be discontinuous across the interface. The magnitude of the discontinuity can be obtained from (10.4.28b) and (10.4.32), which gives

$$\gamma_N^+ = \left(v_N^A - v_N^B \right) H(\gamma_N) = -t_N^A / \beta$$

Thus the discontinuity in the relative normal velocity component is inversely proportional to the penalty parameter β; as β is increased, the discontinuity in the velocities will decrease.

10.4.5 Interpenetration-Dependent Penalty

The previously shown form of the penalty method often performs quite poorly since it may allow excessive interpenetration. The normal traction is nonzero only when the relative velocities lead to continued interpenetration. As soon as the relative velocities of contiguous points of the two surfaces become equal or negative, the normal traction vanishes. Substantial interpenetration may consequently persist in the solution. Therefore, in penalty methods, it is recommended that the normal traction also be a function of the interpenetration as defined in (10.2.23). For this purpose, we define an interface pressure by $p = \bar{p}(g_N, \gamma_N)H(\bar{p})$ where g_N is defined in (10.2.23). The weak form is then given by (10.4.24) with

$$\delta G_p = \int_{\Gamma^c} \delta\gamma_N p \, d\Gamma \tag{10.4.33}$$

The same procedure as before then gives

$$t_N^A + p = 0 \quad \text{and} \quad t_N^B - p = 0 \quad \text{on } \Gamma^c \tag{10.4.34}$$

Combining these above equations gives

$$t_N^A = -t_N^B = -p = -\bar{p}(g_N, \gamma_N)H(\bar{p}) \tag{10.4.35}$$

Thus the tractions are always compressive and satisfy momentum balance. The tractions are functions of the interpenetration and rate of interpenetration. An example of a penalty function is

$$\bar{p} = (\beta_1 g_N + \beta_2 \gamma_N) \tag{10.4.36}$$

where β_1 and β_2 are penalty parameters. An alternate expression is

$$\bar{p} = \beta_1 g_N H(g_N) + \beta_2 \gamma_N H(\gamma_N) \tag{10.4.37}$$

10.4.6 Perturbed Lagrangian Weak Form

The perturbed Lagrangian weak form is

$$\mathbf{v} \in U, \lambda \in C^{-1} \text{ and } \delta P_{PL} = \delta P + \delta G_{PL} = 0 \quad \forall \delta\mathbf{v} \in U_0, \quad \forall \delta\lambda \in C^{-1} \tag{10.4.38}$$

In the previous

$$\delta G_{PL} = \int_{\Gamma^c} \delta\left(\lambda \gamma_N^+ - \frac{1}{2\beta}\lambda^2 \right) d\Gamma \tag{10.4.39}$$

where γ_N^+ is defined by (10.4.28b) and (10.2.8) and β is a large constant, that is, a penalty parameter. It can be seen that the second term in the above integrand is a perturbation of the Lagrange multiplier weak form, (10.4.8); $\lambda^2/2\beta$ is small since β is large.

In this weak form, the test and trial functions for the Lagrange multiplier are unconstrained. This weak form is equivalent to generalized momentum balance and the traction inequalities

(10.2.16) on the contact interface. It will be shown that as in the penalty method, the impenetrability condition (10.2.8) is met only approximately.

The equivalence to the strong form is shown as follows. From (10.4.39),

$$\delta G_{PL} = \int_{\Gamma^c} \left(\delta\lambda \gamma_N^+ + \lambda \delta\gamma_N^+ - \frac{1}{\beta} \lambda \delta\lambda \right) d\Gamma \tag{10.4.40}$$

Combining δG_{PL} with the terms that emerge from δP once the momentum equation, traction boundary conditions and frictionless interface conditions are met, δP_2 (Γ^c) in (10.4.22), yields

$$\begin{aligned} 0 = \delta G_{PL} + \delta P_2 &= \int_{\Gamma^c} \delta\lambda \left(\gamma_N^+ - \frac{\lambda}{\beta} \right) d\Gamma \\ &+ \int_{\Gamma^c} \delta v_N^A \left(t_N^A + \lambda H(\gamma_N) \right) + \delta v_N^B \left(t_N^B - \lambda H(\gamma_N) \right) d\Gamma \end{aligned} \tag{10.4.41}$$

Since the test functions δv_N^A and δv_N^B are arbitrary, it follows that

$$t_N^A = -\lambda H(\gamma_N) \quad \text{on } \Gamma^c \tag{10.4.42}$$

$$t_N^B = \lambda H(\gamma_N) \quad \text{on } \Gamma^c \tag{10.4.43}$$

The test function $\delta\lambda$ is unconstrained, so (10.4.41) yields

$$\lambda = \beta \gamma_N^+ \quad \text{on } \Gamma^c \tag{10.4.44}$$

Combining these yields

$$t_N^A = -t_N^B = -\beta \gamma_N^+ = -\beta \left(v_N^A - v_N^B \right) H(\gamma_N) \quad \text{on } \Gamma^c \tag{10.4.45}$$

So the tractions satisfy momentum balance and are compressive on the contact interface.

The above strong form of the contact surface conditions is almost identical to those which emanate from the penalty method. This similarity is also found in the discrete equations: the perturbed Lagrangian weak form is a penalty weak form in disguise.

10.4.7 Augmented Lagrangian

The augmented Lagrangian formulation has been developed to exploit improved methods for solving the Lagrange multiplier problem (cf. Bertsekas, 1984). The weak form is given by

$$\delta P_{AL}(\mathbf{v}, \delta\mathbf{v}, \lambda, \delta\lambda) = \delta P + \delta G_{AL} \geq 0 \quad \forall \delta\mathbf{v} \in U_0,\ \delta\lambda \in J^- \tag{10.4.46}$$

$$\delta G_{AL} = \int_{\Gamma^c} \delta \left[\lambda \gamma_N(\mathbf{v}) + \frac{\alpha}{2} \gamma_N^2(\mathbf{v}) \right] d\Gamma \tag{10.4.47}$$

where $\mathbf{v} \in U, \lambda \in J^{+}(\Gamma^{c})$; $\gamma_{N}(\mathbf{v})$ is defined by (10.2.8) and α is a positive parameter determined as part of the solution procedure.

The equivalence of this weak form to the strong form is shown in the following. Expanding the integrand in (10.4.47) gives

$$\delta G_{AL} = \int_{\Gamma^{c}} \left[\delta\lambda\gamma_{N} + \lambda\left(\delta v_{N}^{A} - \delta v_{N}^{B}\right) + \alpha\gamma_{N}\left(\delta v_{N}^{A} - \delta v_{N}^{B}\right) \right] d\Gamma \tag{10.4.48}$$

where (10.2.8) has been used for $\delta\gamma_{N}$. Combining the above with (10.4.23) gives

$$\int_{\Gamma^{c}} \left[\delta\lambda\gamma_{N} + \delta v_{N}^{A}\left(\lambda + \alpha\gamma_{N} + t_{N}^{A}\right) - \delta v_{N}^{B}\left(\lambda + \alpha\gamma_{N} - t_{N}^{B}\right) \right] d\Gamma \geq 0 \tag{10.4.49}$$

Since all of the variations are arbitrary, we obtain that on Γ^{c}

$$\delta\lambda: \quad \gamma_{N} = v_{N}^{A} - v_{N}^{B} \leq 0 \tag{10.4.50}$$

$$\delta v_{N}^{A}: \quad \lambda = -\alpha\gamma_{N} - t_{N}^{A} \tag{10.4.51}$$

$$\delta v_{N}^{B}: \quad \lambda = -\alpha\gamma_{N} + t_{N}^{B} \tag{10.4.52}$$

Equations (10.4.51) and (10.4.52) can be combined to yield

$$t_{N}^{A} = -t_{N}^{B} = -\lambda - \alpha\gamma_{N} \tag{10.4.53}$$

Thus the normal interface traction satisfies momentum balance.

10.4.8 Tangential Tractions by Lagrange Multipliers

All of the previous formulations can be modified to handle interface friction by adding a term to the weak form to enforce continuity of the tangential tractions (see Box 10.2 for weak forms). We let

$$\delta P_{C} = \delta P + \delta G_{N} + \delta G_{T} \tag{10.4.54}$$

The weak form is an inequality for the Lagrange and augmented Lagrange methods:

$$\delta P_{C} \geq 0 \quad \text{when} \quad \delta G_{N} = \delta G_{L} \quad \text{or} \quad \delta G_{AL} \tag{10.4.55}$$

The weak form is an equality for the penalty and perturbed Lagrange methods:

$$\delta P_{C} = 0 \quad \text{when} \quad \delta G_{N} = \delta G_{P} \quad \text{or} \quad \delta G_{PL} \tag{10.4.56}$$

In both cases

$$\delta G_{T} = \int_{\Gamma^{c}} \delta\boldsymbol{\gamma}_{T} \cdot \mathbf{t}_{T}\, d\Gamma \equiv \int_{\Gamma^{c}} \delta\hat{\gamma}_{\alpha}\hat{t}_{\alpha}\, d\Gamma \tag{10.4.57}$$

where $\mathbf{t}_T$ is a traction tangent to the contact interface which is computed by a friction model. We have put hats on the expressions in indicial notation to indicate that these components are in the local coordinates of the tangent plane of the contact interface.

To obtain the strong form, we take what remains from δP after extracting the momentum equation, the traction boundary conditions, and contact conditions along the normal as before:

$$0 = \int_{\Gamma^c} \left(\delta \mathbf{v}_T^A \cdot \mathbf{t}_T^A + \delta \mathbf{v}_T^B \cdot \mathbf{t}_T^B + \delta \boldsymbol{\gamma}_T \cdot \mathbf{t}_T \right) d\Gamma \tag{10.4.58}$$

Note that $\mathbf{t}_T$ differs from $\mathbf{t}_T^A$ and $\mathbf{t}_T^B$; $\mathbf{t}_T$ is the tangential traction given by the interface constitutive equation, whereas $\mathbf{t}_T^A$ and $\mathbf{t}_T^B$ are the tractions on the interface which result from the stresses in bodies A and B, respectively. By the definition of γ_T, (10.2.9), we can write $\delta \boldsymbol{\gamma}_T = \delta \mathbf{v}_T^A - \delta \mathbf{v}_T^B$. Substituting this into the above we have, after rearranging the terms,

$$0 = \int_{\Gamma^c} \left[\delta \mathbf{v}_T^A \cdot \left(\mathbf{t}_T^A + \mathbf{t}_T \right) + \delta \mathbf{v}_T^B \cdot \left(\mathbf{t}_T^B - \mathbf{t}_T \right) \right] d\Gamma \tag{10.4.59}$$

From this we can extract

$$\mathbf{t}_T^A = -\mathbf{t}_T \quad \text{and} \quad \mathbf{t}_T^B = \mathbf{t}_T \quad \text{on } \Gamma^c \tag{10.4.60}$$

Eliminating $\mathbf{t}_T$ from this we have

$$\mathbf{t}_T^A + \mathbf{t}_T^B = \mathbf{0} \quad \text{or} \quad \hat{t}_\alpha^A + \hat{t}_\alpha^B = 0 \quad \text{on } \Gamma^c \tag{10.4.61}$$

Thus the additional term δG_T in the weak form corresponds to the momentum balance of the tangential tractions on the contact interface. Without this term in the weak form, the tangential tractions vanish, that is, the interface is frictionless.

When the stick condition applies to a part of the contact interface, the constraint of no tangential slip can be imposed by Lagrange multipliers. For simplicity we consider the entire contact surface to be in a stick condition. So we add a Lagrange multiplier term, as in Section 8.4, to enforce the stick condition. This term is denoted by δG_{TS} which is given by

$$\delta G_{TS} = \int_{\Gamma^c} \delta(\boldsymbol{\gamma}_T \cdot \boldsymbol{\lambda}_T) d\Gamma \equiv \int_{\Gamma^c} \delta(\hat{\gamma}_\alpha \hat{\lambda}_\alpha) d\Gamma \tag{10.4.62}$$

The strong forms corresponding to $\delta P_C = \delta P + \delta G_N + \delta G_{TS} = 0$ with δG_{TS} given by (10.4.62) are generalized momentum balance, normal traction balance, tangential traction balance $\mathbf{t}_T^A = -\boldsymbol{\lambda}$, $\mathbf{t}_T^B = \boldsymbol{\lambda}$, and the stick condition $\boldsymbol{\gamma}_T = 0$ on Γ^c. The Lagrange multiplier can be eliminated from the preceding relations between the tractions and the Lagrange multiplier to give (10.4.61).

Box 10.2 Weak forms

$$\delta P_C = \delta P + \delta G + \delta G_T \tag{B10.2.1}$$

Tangential tractions: $\delta G_T = \int_{\Gamma^c} \delta \boldsymbol{\gamma}_T \cdot \mathbf{t}_T \, d\Gamma \equiv \int_{\Gamma^c} \delta \hat{\gamma}_\alpha \hat{t}_\alpha \, d\Gamma$ see (10.4.57) (B10.2.2)

Lagrangian: $\delta G = \delta G_L = \int_{\Gamma^c} \delta(\lambda \gamma_N) d\Gamma, \quad \delta P_C \geq 0$ (B10.2.3)

Penalty: $\delta G = \delta G_P = \int_{\Gamma^c} \frac{1}{2} \beta \delta(\gamma_N^2) d\Gamma, \quad \delta P_C = 0$ (B10.2.4)

Augmented Lagrangian: $\delta G = \delta G_{AL} = \int_{\Gamma^c} \delta\left(\lambda \gamma_N + \frac{\alpha}{2} \gamma_N^2\right) d\Gamma, \quad \delta P_C \geq 0$ (B10.2.5)

Perturbed Lagrangian: $\delta G_N = \delta G_{PL} = \int_{\Gamma^c} \delta\left(\lambda \gamma_N - \frac{1}{2\beta} \lambda^2\right) d\Gamma, \quad \delta P_C = 0$ (B10.2.6)

10.5 Finite Element Discretization

10.5.1 Overview

The finite element equations for the various treatments of contact-impact are developed. The weak statements for all the approaches to the contact-impact problem (penalty, Lagrange multiplier, etc.) involve a sum of the standard virtual power and a contribution from the contact interface. The standard virtual power is discretized exactly as in the absence of contact, so we will use the results developed in Chapter 4. This section concentrates on the discretization of the various contact interface weak forms.

The developments that follow here are applicable to Lagrangian meshes with both updated and total Lagrangian formulations. However, in total Lagrangian formulations, the contact interface conditions must be imposed in terms of the deformed surface. The following discretizations are also applicable to ALE formulations as long as the nodes on the contact surface are Lagrangian. *They are not directly applicable to Eulerian formulations* since we assume that we have at our disposal a referential coordinate that describes the contact surface. Such a coordinate system cannot easily be defined in an Eulerian mesh. In a Lagrangian mesh, the contact surface corresponds to a subset of the boundary of the mesh.

We will first develop the FEM discretization for the Lagrange multiplier method in indicial notation. Indicial notation enables us to go through some subtle steps which will subsequently be skipped in the matrix derivations; anyone who wishes to replicate these steps for other formulations can derive these in indicial notation.

10.5.2 Lagrange Multiplier Method

The velocity $\mathbf{v}$ ($\mathbf{X}$, t) in each body is approximated by C^0 interpolants as in the single body problem; as can be seen from (10.4.4), the velocities of the two bodies need not be continuous across the contact interface; the interpenetration condition will emanate from the discretization of the weak form. As in Chapter 4, we note that the approximation of the velocity field also defines the approximation of the displacement field.

The finite element approximation for the velocity field is expressed in terms of the material coordinates since we are dealing with a Lagrangian mesh. It can alternatively be written in terms of the element coordinates, since as pointed out in Chapter 4 the two sets of coordinates are equivalent. To clarify certain issues, we will initially drop the summation convention on repeated nodal indices and indicate sums explicitly. The velocity fields are

$$v_i^A(\mathbf{X},t) = \sum_{I\in\Omega^A} N_I(\mathbf{X})v_{iI}^A(t) \tag{10.5.1}$$

$$v_i^B(\mathbf{X},t) = \sum_{I\in\Omega^B} N_I(\mathbf{X})v_{iI}^B(t) \tag{10.5.2}$$

If the node numbers of bodies A and B are distinct, then the two velocity fields can be written as a single expression:

$$v_i(\mathbf{X},t) = N_I(\mathbf{X})v_{iI}(t) \tag{10.5.3}$$

where an implicit sum over all nodes is implied on the repeated nodal indices.

The Lagrange multiplier field $\lambda(\zeta, t)$, as can be seen from (10.4.6), is approximated by a C^{-1} field on the contact surface:

$$\lambda(\zeta,t) = \sum_{I\in\Gamma^c} \Lambda_I(\zeta)\lambda_I(t) \equiv \Lambda_I(\zeta)\lambda_I(t), \quad \lambda(\zeta,t) \geq 0 \tag{10.5.4}$$

where $\Lambda_I(\zeta)$ are C^{-1} shape functions. The shape functions for the Lagrange multiplier field often differ from those for the velocities, so different symbols have been used for the two approximations. When the nodes of bodies A and B are not coincident, the mesh for the Lagrange multiplier may differ from that for the velocity field. The need for a different nodal structure for the Lagrange multipliers is discussed later.

The test functions are

$$\delta v_i(\mathbf{X}) = N_I(\mathbf{X})\delta v_{iI}, \quad \delta\lambda(\zeta) = \Lambda_I(\zeta)\delta\lambda_I, \quad \delta\lambda(\zeta) \leq 0 \tag{10.5.5}$$

To develop the semidiscrete equations, the above approximations for the velocity and Lagrange multiplier fields and the test functions are substituted into the weak form, (B10.2.1). The terms emerging from δP are identical to the nodal forces developed in Chapter 4, so they will not be rederived; the results are given in Box 4.3. From (B4.3.1) it follows that

$$\delta P = \delta v_{iI}\left(f_{iI}^{\text{int}} - f_{iI}^{\text{ext}} + M_{ijIJ}\dot{v}_{jJ}\right) \equiv \delta\dot{\mathbf{d}}^T\left(\mathbf{f}^{\text{int}} - \mathbf{f}^{\text{ext}} + \mathbf{M}\ddot{\mathbf{d}}\right) \equiv \delta\mathbf{v}^T\mathbf{r} \tag{10.5.6}$$

where $\mathbf{v}$ designates the nodal velocities henceforth. The interpenetration rate can be expressed in terms of the nodal velocities via (10.2.8) and (10.5.1):

$$\gamma_N = \sum_{I\in\Gamma^c\cap\Gamma^A} N_I v_{iI}^A n_i^A + \sum_{I\in\Gamma^c\cap\Gamma^B} N_I v_{iI}^B n_i^B \tag{10.5.7}$$

where the first sum, as indicated, is over the nodes of body A which are on the contact interface, and the second sum is over the nodes of body B which are on the contact interface. If we assign these nodes distinct node numbers, we can eliminate the distinction between nodes of body A and B and express (10.5.7) as

$$\gamma_N = N_I v_{NI} \tag{10.5.8}$$

The range of the sum on the repeated index I is given in (10.5.7). The normal components are defined via (10.2.6) by

$$v_{NI} = v_{iI}^A n_i^A \quad \text{if } I \text{ in } A, \quad v_{NI} = v_{iI}^B n_i^B \quad \text{if } I \text{ in } B \tag{10.5.9}$$

This approximates the product of the normal and the velocities by the shape functions; a more exact form is given in (10.5.17). Then using the approximations (10.5.1) to (10.5.3), it follows that

$$\int_{\Gamma^c} \delta(\lambda\gamma_N)d\Gamma = \delta v_{NI}\hat{G}_{IJ}^T\lambda_J + \delta\lambda_I\hat{G}_{IJ}v_{NJ} \quad \text{where} \quad \hat{G}_{IJ} = \int_{\Gamma^c}\Lambda_I N_J\, d\Gamma \tag{10.5.10}$$

A superposed hat has been placed on $\hat{G}_{IJ}$ to indicate that it pertains to the velocities in the local coordinate system of the contact interface. Combining (10.5.6) and (10.5.10), we can write the discrete weak form as

$$\sum_{I\in\Omega}\delta v_{iI}r_{iI} + \sum_{I\in\Gamma_\lambda^c}\delta v_{NI}\hat{G}_{IJ}^T\lambda_J + \sum_{I\in\Gamma_\lambda^c}\delta\lambda_I\hat{G}_{IJ}v_{NJ} \ge 0 \tag{10.5.11}$$

where the implicit sum on the index J holds, but the sums on the index I are explicitly stated to indicate the relevant nodes.

The governing equations must be extracted carefully because of the inequalities. The equations for nodes which are not on the contact interface can be directly extracted from the first sum since the test nodal velocities are arbitrary, which yields the standard nodal equations of motion:

$$r_{iI} = 0 \quad \text{or} \quad M_{IJ}\dot{v}_{iJ} = f_{iI}^{\text{ext}} - f_{iI}^{\text{int}} \quad \text{for} \quad I \in \Omega - \Gamma^c - \Gamma_u \tag{10.5.12}$$

To obtain the equations on the contact interface, what remains of the first sum after extracting (10.5.12) is rewritten in the local coordinate systems of the contact interface and combined with the second sum, giving

$$\sum_{I\in\Gamma^c}\left(\delta v_{NI}r_{NI} + \delta\hat{v}_{\alpha I}\hat{r}_{\alpha I} + \delta v_{NI}\hat{G}_{IJ}^T\lambda_J\right) + \sum_{I\in\Gamma_\lambda^c}\delta\lambda_I\hat{G}_{IJ}v_{NJ} \ge 0 \tag{10.5.13}$$

Since the tangential nodal velocities are unconstrained, the weak inequality yields an equality for the coefficients of the nodal velocities. First we set the coefficient of $\delta\hat{v}_{\alpha I}$ to zero, which gives

$$\hat{r}_{\alpha I} = 0 \quad \text{or} \quad M_{IJ}\,\dot{\hat{v}}_{\alpha J} = \hat{f}_{\alpha I}^{\text{ext}} - \hat{f}_{\alpha I}^{\text{int}} \quad \text{for} \quad I \in \Gamma^c \tag{10.5.14}$$

The equation for the normal component at the contact interface nodes (10.5.13) for a frictionless interface gives

$$r_{NI} + \hat{G}^T_{IJ}\lambda_J = 0 \quad \text{or} \quad M_{IJ}\dot{v}_{NJ} + f^{\text{ext}}_{NI} - f^{\text{int}}_{NI} + \hat{G}^T_{IJ}\lambda_J = 0 \quad \text{for} \quad \text{I} \in \Gamma^c \tag{10.5.15}$$

To extract the equations associated with the Lagrange multipliers, we note that $\delta\lambda_I \leq 0$, so the inequality (10.5.13) implies

$$\hat{G}_{IJ} v_{NJ} \leq 0 \tag{10.5.16}$$

In addition, from (10.4.6) the trial Lagrange multiplier field must be positive: $\lambda(\boldsymbol{\xi}, t) \geq 0$. This inequality is difficult to enforce. For elements with piecewise linear edge displacements, this condition is enforced only at the nodes by $\lambda_I \geq 0$, since all minima of $\lambda(\boldsymbol{\xi})$ occur at nodes. For higher-order approximations, the condition must be checked more exhaustively.

The above equations, in conjunction with the strain-displacement equations and the constitutive equation, comprise the complete system of equations for the semidiscrete model. The semidiscrete equations consist of the equations of motion and the contact interface conditions. The equations of motion for nodes not on the contact interface are unchanged from the unconstrained case. On the contact interface, additional forces $\hat{G}_{IJ}\lambda_J$ which represent the normal contact tractions appear. In addition, the impenetrability constraint in the weak form (10.5.16) must be imposed. Like the equations without contact, the semidiscrete equations are ordinary differential equations, but the variables are subject to algebraic inequality constraints on the velocities and the Lagrange multipliers. These inequality constraints substantially complicate the time integration, since the smoothness which is implicitly assumed in most time integration procedures is absent.

For purposes of implementation, it is convenient to write the above equations in matrix form in global components. Let the interpenetration rate be defined in terms of the nodal velocities by

$$\gamma_N = \Phi_{iI}(\boldsymbol{\xi}) v_{iI}(t) \quad \text{where} \quad \Phi_{iI}(\boldsymbol{\xi}) = \begin{cases} N_I(\boldsymbol{\xi}) n_i^A(\boldsymbol{\xi}) & \text{if } I \text{ on } A \\ N_I(\boldsymbol{\xi}) n_i^B(\boldsymbol{\xi}) & \text{if } I \text{ on } B \end{cases} \tag{10.5.17}$$

The contact weak term is then given by

$$G_L = \int_{\Gamma^c} \lambda \gamma_N \, d\Gamma = \int_{\Gamma^c} \lambda_I \Lambda_I \Phi_{jJ} v_{jJ} \, d\Gamma = \boldsymbol{\lambda}^T \mathbf{G} \mathbf{v} \tag{10.5.18}$$

where

$$G_{IjJ} = \int_{\Gamma^c} \Lambda_I \Phi_{jJ} \, d\Gamma, \quad \mathbf{G} = \int_{\Gamma^c} \Lambda^{\mathrm{T}} \boldsymbol{\Phi} \, d\Gamma \tag{10.5.19}$$

where jJ has been converted to a single index by the Voigt column matrix rule to form the matrix expression on the right.

The equations of motion can be written in matrix form by combining this form with matrix forms of the internal, external and inertial power, which gives

$$\delta \mathbf{v}^T(\mathbf{f}^{\text{int}} - \mathbf{f}^{ext} + \mathbf{M}\ddot{\mathbf{d}}) + \delta(\mathbf{v}^T \mathbf{G}^T \boldsymbol{\lambda}) \geq 0 \quad \forall \delta v_{iI} \notin \Gamma_u \quad \text{and} \quad \forall \delta \lambda_I \leq 0 \tag{10.5.20}$$

We will skip the steps represented by (10.5.7–17) and invoke the arbitrariness of $\delta\mathbf{v}$ and $\delta\boldsymbol{\lambda}$ which yields equations of motion and the interpenetration condition

$$\mathbf{M}\ddot{\mathbf{d}} + \mathbf{f}^{\text{int}} - \mathbf{f}^{\text{ext}} + \mathbf{G}^T \boldsymbol{\lambda} = \mathbf{0} \tag{10.5.21}$$

$$\mathbf{G}\mathbf{v} \leq \mathbf{0} \tag{10.5.22}$$

10.5.2.1 Lagrange Multiplier Mesh

The construction of the mesh for the Lagrange multipliers poses some difficulties. In general, the nodes of the two contacting bodies are not coincident, as shown in Figure 10.10(a). Therefore it is necessary to develop a scheme to deal with noncontiguous nodes. One possibility is indicated in Figure 10.10(b), where the nodes for the Lagrange multiplier field are chosen to be the contacting nodes of the master body. This simple scheme is not effective when the mesh of one body is much finer than the mesh of the other. A coarse mesh for the Lagrange multipliers will then lead to interpenetration. An alternative is to place Lagrange multiplier nodes wherever a node appears in either body A or B, as shown in Figure 10.10(b). The disadvantage of that scheme is that when nodes of A and B are close, some Lagrange multiplier elements are very small. This can lead to ill-conditioning of the equations. In three dimensions, such schemes are not viable. For general applications, a separate mesh must be made for the Lagrange multipliers which is independent of either mesh but is at least as fine as the finer of the two.

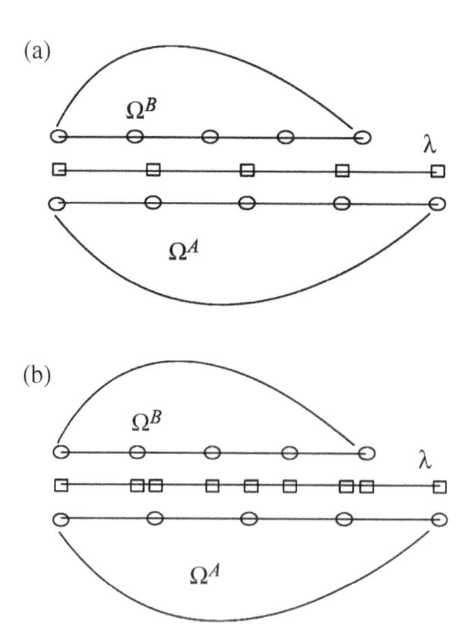

Figure 10.10 Nodal arrangements for two contacting bodies with noncontiguous nodes, showing (a) a Lagrange strain multiplier mesh based on the master body, and (b) an independent Lagrange multiplier mesh with nodes wherever they occur in either body

10.5.3 Assembly of Interface Matrix

The $\mathbf{G}$ matrix can be assembled from 'element' matrices like any other global matrix. To illustrate the assembly procedure, let the column matrices of element e be expressed in terms of the global matrices by

$$\mathbf{v}_e = \mathbf{L}_e\mathbf{v}, \quad \boldsymbol{\lambda}_e = \mathbf{L}_e^{\lambda}\boldsymbol{\lambda}, \tag{10.5.23}$$

where $\mathbf{L}_e$ is defined in Section 2.5. Substituting the above into (10.5.18) gives

$$\boldsymbol{\lambda}^{\mathrm{T}}\mathbf{G}\mathbf{v} = \int_{\Gamma^c} \lambda\gamma_N \, d\Gamma = \sum_e \int_{\Gamma_e^c} \boldsymbol{\lambda}\gamma_N \, d\Gamma = \boldsymbol{\lambda}^T \sum_e \left(\mathbf{L}_e^{\lambda}\right)^T \int_{\Gamma_e^c} \Lambda^T \Phi \, d\Gamma \quad \mathbf{L}_e\mathbf{v} \tag{10.5.24}$$

Since the above must hold for arbitrary $\mathbf{v}$ and $\boldsymbol{\lambda}$ it can be seen by comparing the first and last term of (10.5.24) that

$$\mathbf{G} = \sum_e \left(\mathbf{L}_e^{\lambda}\right)^T \mathbf{G}_e\mathbf{L}_e, \quad \mathbf{G}_e = \int_{\Gamma_e^c} \Lambda^T \boldsymbol{\Phi} \, d\Gamma \tag{10.5.25}$$

Thus the assembly of $\mathbf{G}$ from $\mathbf{G}_e$ is identical to assembly of global matrices such as the stiffness matrix.

10.5.4 Lagrange Multipliers for Small-Displacement Elastostatics

Small-displacement analysis of continua with linear elastic materials is here called small-displacement elastostatics. We use this nomenclature rather than linear elastostatics because due to the inequality constraint on the displacements which arises from the contact condition, these problems are not linear. For small-displacement elastostatics, the discrete impenetrability constraint can be obtained from (10.5.22) by replacing the velocities by the displacements. Thus (10.2.8) and (10.5.17) are replaced by

$$g_N = (\mathbf{u}^A - \mathbf{u}^B) \cdot \mathbf{n}^A \leq 0 \ \text{ on } \ \Gamma^c, \quad g_N = \boldsymbol{\Phi}\mathbf{d} \tag{10.5.26}$$

The discretization procedure is then identical to the above except for replacing velocities by displacements and omitting the inertia, giving

$$\delta\mathbf{d}^T(\mathbf{f}^{\mathrm{int}} - \mathbf{f}^{\mathrm{ext}}) + \delta(\mathbf{d}^T\mathbf{G}\boldsymbol{\lambda}) \geq 0 \quad \forall \delta d_{iI} \notin \Gamma_{u_i} \quad \text{and} \quad \forall \delta\lambda_I \leq 0 \tag{10.5.27}$$

Since the internal nodal forces are not affected by contact, for small-displacement elastostatic problems they can be expressed in terms of the stiffness matrix by

$$\mathbf{f}^{\mathrm{int}} = \mathbf{K}\mathbf{d} \tag{10.5.28}$$

The resulting discrete equations are then

$$\begin{bmatrix} \mathbf{K} & \mathbf{G}^T \\ \mathbf{G} & \mathbf{0} \end{bmatrix} \begin{Bmatrix} \mathbf{d} \\ \boldsymbol{\lambda} \end{Bmatrix} \begin{matrix} = \\ \leq \end{matrix} \begin{Bmatrix} \mathbf{f}^{\mathrm{ext}} \\ \mathbf{0} \end{Bmatrix} \tag{10.5.29}$$

This is the standard form of the Lagrange multiplier problem – see (6.3.41) – except that an inequality appears in the second matrix equation.

Several remarks pertain to the previous, as to other Lagrange multiplier discretizations:

1. The system of linear algebraic equations is no longer positive-definite.
2. The equations as given above are not banded and it is difficult to find an arrangement of unknowns to restore bandedness.
3. The number of unknowns is increased as compared to the system without the contact constraints.

In addition, the solution of the contact problem is complicated by the inequalities. These are very difficult to deal with and often the small-displacement, elastostatic problem is posed as a quadratic programming problem. These difficulties also arise in the implicit time integration of contact problems.

A major disadvantage of the Lagrange multiplier method is the need for a mesh for the Lagrange multipliers. As we have seen in the simple two-dimensional example, this can introduce complications even in two dimensions. In three dimensions, this task is far more complicated. The mesh must change with time as the contact interfaces evolve. In penalty methods there is no need to set up an additional mesh.

In comparison to the penalty method, the advantage of the Lagrange multiplier method is that there are no user-set parameters and the contact constraint can be met almost exactly when the nodes are contiguous. When the nodes are not contiguous, impenetrability may be violated slightly, but not as much as in penalty methods. However, for high velocity impact, Lagrange multipliers often result in very noisy solutions. Therefore, Lagrange multiplier methods are most suited for static and low velocity problems.

10.5.5 Penalty Method for Nonlinear Frictionless Contact

The discrete equations are developed only for the penetration-dependent form of the penalty method. In the penalty method only the velocity field needs to be approximated. Again, the velocity field is C^0 within each body. Continuity between bodies is not stipulated, but is enforced by the penalty method. We will develop only the discrete form of the penalty term δG_p given in (10.4.33). The rest is unchanged from the unconstrained problem. Substituting (10.5.17) into (10.4.33) gives

$$\delta G_P = \delta \mathbf{v}^T \int_{\Gamma^c} \mathbf{\Phi}^T p d\Gamma \equiv \delta \mathbf{v}^T \mathbf{f}^c \quad \text{where} \quad \mathbf{f}^c = \int_{\Gamma^c} \mathbf{\Phi}^T p d\Gamma \tag{10.5.30}$$

Using (10.5.30) and (10.5.6) in the weak form (10.4.24) gives

$$\delta P_P = \delta \mathbf{v}^T \mathbf{r} + \delta \mathbf{v}^T \mathbf{f}^c \tag{10.5.31}$$

So the arbitrariness of $\delta \mathbf{v}$ and the definition of $\mathbf{r}$ in (10.5.6) yields

$$\mathbf{f}^{\text{int}} - \mathbf{f}^{\text{ext}} + \mathbf{Ma} + \mathbf{f}^c = \mathbf{0} \tag{10.5.32}$$

Thus in the penalty method the number of equations is unchanged from the unconstrained problem. The inequalities do not appear explicitly among the discrete equations but are enforced by the step function in the contact penalty forces.

10.5.6 Penalty Method for Small-Displacement Elastostatics

For small-displacement elastostatics, we replace velocities by displacements as previously. Equation (10.4.37) with $\beta_2 = 0$ and (10.5.17) give

$$\bar{p} = \beta_1 g_N = \beta_1 \mathbf{\Phi d} \tag{10.5.33}$$

Substituting this into (10.5.30) gives

$$\mathbf{f}^c = \int_{\Gamma^c} \mathbf{\Phi}^T \bar{p}(g_N) H(g_N) d\Gamma = \int_{\Gamma^c} \beta_1 \mathbf{\Phi}^T \mathbf{\Phi} H(g_N) d\Gamma \mathbf{d} \tag{10.5.34}$$

or

$$\mathbf{f}^c = \mathbf{P}_c \mathbf{d}, \quad \mathbf{P}_c = \int_{\Gamma^c} \beta_1 \mathbf{\Phi}^T \mathbf{\Phi} H(g_N) d\Gamma \tag{10.5.35}$$

Substituting (10.5.35) and (10.5.28) into (10.5.32) and dropping the inertial term gives

$$(\mathbf{K} + \mathbf{P}_c)\mathbf{d} = \mathbf{f}^{\text{ext}} \tag{10.5.36}$$

This is a system of algebraic equations of the same order as the problem without contact. The contact constraints are enforced through the penalty forces $\mathbf{P}_c\mathbf{d}$. The algebraic equations are not linear because, as can be seen from (10.5.35), the matrix $\mathbf{P}_c$ involves the Heaviside step function of the gap, which depends on the displacements.

In contrast to the Lagrange multiplier methods it can be seen that:

1. the number of unknowns is not increased by the enforcement of the contact constraints.
2. the system equations remain positive-definite since **G** is positive-definite.

The disadvantage of the penalty approach is that the enforcement of the impenetrability condition is only approximate and its effectiveness depends on the selection of the penalty parameters. If the penalty parameters are too small, excessive interpenetration occurs. In impact problems, small penalty parameters reduce the maximum computed stresses. Picking the correct penalty parameter is a challenge.

10.5.7 Augmented Lagrangian

In the augmented Lagrangian method, the weak contact term is

$$\delta G_{AL} = \int_{\Gamma^c} \delta\left(\lambda \gamma_N + \frac{\alpha}{2}\gamma_N^2\right) d\Gamma \tag{10.5.37}$$

Using the approximation for the velocity and the Lagrange multiplier, (10.5.17) and (10.5.4), gives

$$\delta G_{AL} = \int_{\Gamma^c} \delta\left(\boldsymbol{\lambda}^T \Lambda^T \mathbf{\Phi v} + \frac{\alpha}{2} \mathbf{v}^T \mathbf{\Phi}^T \mathbf{\Phi v}\right) d\Gamma \tag{10.5.38}$$

Taking the variations gives

$$\delta G_{AL} = \delta\boldsymbol{\lambda}^T\mathbf{G}\mathbf{v} + \delta\mathbf{v}^T\mathbf{G}^T\boldsymbol{\lambda} + \delta\mathbf{v}^T\mathbf{P}_c\mathbf{v} \tag{10.5.39}$$

where $\mathbf{P}_c$ is defined in Box 10.3. Writing out the weak form $\delta P_{AL} = \delta P + \delta G_{AL} \geq 0$ using (10.5.37–39) then gives

$$\mathbf{f}^{\text{int}} - \mathbf{f}^{\text{ext}} + \mathbf{M}\mathbf{a} + \mathbf{G}^T\boldsymbol{\lambda} + \mathbf{P}_c\mathbf{v} = \mathbf{0} \quad \text{and} \quad \mathbf{G}\mathbf{v} \leq \mathbf{0} \tag{10.5.40}$$

Comparing (10.5.40) with (B10.3.1–2) we can see that the contact forces in the augmented Lagrangian method are the sum of those in the Lagrangian and penalty methods. The impenetrability constraint is identical to that in the Lagrange multiplier method.

Box 10.3 Semidiscrete equations for nonlinear contact

Lagrange multiplier:

$$\mathbf{M}\mathbf{a} - \mathbf{f} + \mathbf{G}^T\boldsymbol{\lambda} = \mathbf{0}, \quad \mathbf{G}\mathbf{v} \leq \mathbf{0}, \quad \boldsymbol{\lambda}(\mathbf{x}) \geq 0, \quad \mathbf{f} = \mathbf{f}^{\text{ext}} - \mathbf{f}^{\text{int}} \tag{B10.3.1}$$

Penalty:

$$\mathbf{M}\mathbf{a} - \mathbf{f} + \mathbf{f}^c = \mathbf{0}, \quad \mathbf{f}^c = \int_{\Gamma^c} \boldsymbol{\Phi}^T p(g_N) H(g_N)\, d\Gamma \tag{B10.3.2}$$

Augmented Lagrangian:

$$\mathbf{M}\mathbf{a} - \mathbf{f} + \mathbf{G}^T\boldsymbol{\lambda} + \mathbf{P}_c\mathbf{v} = \mathbf{0}, \quad \mathbf{G}\mathbf{v} \leq \mathbf{0} \tag{B10.3.3}$$

Perturbed Lagrangian:

$$\mathbf{M}\mathbf{a} - \mathbf{f} + \mathbf{G}^T\boldsymbol{\lambda} = \mathbf{0}, \quad \mathbf{G}\mathbf{v} - \mathbf{H}\boldsymbol{\lambda} = \mathbf{0} \tag{B10.3.4}$$

$$\mathbf{G} = \int_{\Gamma^c} \boldsymbol{\Lambda}^T\boldsymbol{\Phi}\, d\Gamma, \quad \mathbf{H} = \frac{1}{\beta}\int_{\Gamma^c} \boldsymbol{\Lambda}^T\boldsymbol{\Lambda}\, d\Gamma, \quad \mathbf{P}_c = \int_{\Gamma^c} \alpha\boldsymbol{\Phi}^T\boldsymbol{\Phi}\, d\Gamma$$

For small-displacement elastostatics, we use the same procedure as before. We replace the nodal velocities by nodal displacements, and using (10.5.28) gives

$$\begin{bmatrix} \mathbf{K} + \mathbf{P}_c & \mathbf{G}^T \\ \mathbf{G} & \mathbf{0} \end{bmatrix} \begin{Bmatrix} \mathbf{d} \\ \boldsymbol{\lambda} \end{Bmatrix} \begin{matrix} = \\ \leq \end{matrix} \begin{Bmatrix} \mathbf{f}^{\text{ext}} \\ \mathbf{0} \end{Bmatrix} \tag{10.5.41}$$

This further illustrates that the augmented Lagrangian method is a synthesis of penalty and Lagrange multiplier methods, (10.5.29) and (10.5.36).

10.5.8 Perturbed Lagrangian

The semidiscretization of the perturbed Lagrangian formulation is obtained from (10.4.38) with velocity and Lagrange multiplier approximations (10.5.3–5). We will not go through the steps, since they are identical to the previous discretizations. The discrete equations are

$$\mathbf{f}^{\text{int}} - \mathbf{f}^{\text{ext}} + \mathbf{Ma} + \mathbf{G}^T\boldsymbol{\lambda} = \mathbf{0} \tag{10.5.42}$$

$$\mathbf{Gv} - \mathbf{H}\boldsymbol{\lambda} = \mathbf{0} \tag{10.5.43}$$

The above are the momentum equation and the impenetrability condition. The matrix **G** is defined by (10.5.19) and

$$\mathbf{H} = \int_{\Gamma_c} \frac{1}{\beta} \Lambda^T \Lambda d\Gamma \tag{10.5.44}$$

The constraint equations (10.5.43) can be used to eliminate $\boldsymbol{\lambda}$:

$$\mathbf{f}^{\text{int}} - \mathbf{f}^{\text{ext}} + \mathbf{Ma} + \mathbf{G}^T\mathbf{H}^{-1}\mathbf{Gv} = \mathbf{0} \tag{10.5.45}$$

This is similar to the discrete penalty equation (10.5.32) with the penalty parameter β appearing through **H**. The last term $\mathbf{G}^T\mathbf{H}^{-1}\mathbf{Gv}$ in (10.5.45) represents the contact forces.

The semidiscrete equations for small-displacement elastostatics for the perturbed Lagrangian methods are

$$\begin{bmatrix} \mathbf{K} & \mathbf{G}^T \\ \mathbf{G} & -\mathbf{H} \end{bmatrix} \begin{Bmatrix} \mathbf{d} \\ \boldsymbol{\lambda} \end{Bmatrix} = \begin{Bmatrix} \mathbf{f}^{\text{ext}} \\ \mathbf{0} \end{Bmatrix} \tag{10.5.46}$$

Comparing the above to the Lagrangian method, (10.5.29), we can see that it differs only in the lower right submatrix, which is **0** in the Lagrangian multiplier method.

Example 10.2 Finite element equations for one-dimensional contact-impact Consider the two rods shown in Figure 10.11. We consider a rod of unit cross-sectional area. The contact interface consists of the nodes at the ends of the rods, which are numbered 1 and 2. The unit normals, as shown in Figure 10.11, are $n_x^A = 1$, $n_x^B = -1$. The contact interface in one-dimensional problems is rather odd since it consists of a single point. The velocity fields in the two elements which border the contact interface are given by

$$v(\xi,t) = \mathbf{N}(\xi,t)\dot{\mathbf{d}} = [\xi^A, 1-\xi^B, \xi^B]\dot{\mathbf{d}} \quad \text{where} \quad \dot{\mathbf{d}}^T = [v_1, v_2, v_3] \tag{E10.2.1}$$

The **G** matrix is given by (10.5.19); in a one-dimensional problem, the integral is replaced by a single function value, the function evaluated at the contact point:

$$\begin{aligned} \mathbf{G} &= \left[\xi^A n_x^A, \left(1-\xi^B\right) n_x^B, \xi^B\right]\Big|_{\xi^A=1,\xi^B=0} \\ &= [(1)(1), 1(-1), 0] = [1, -1, 0] \end{aligned} \tag{E10.2.2}$$

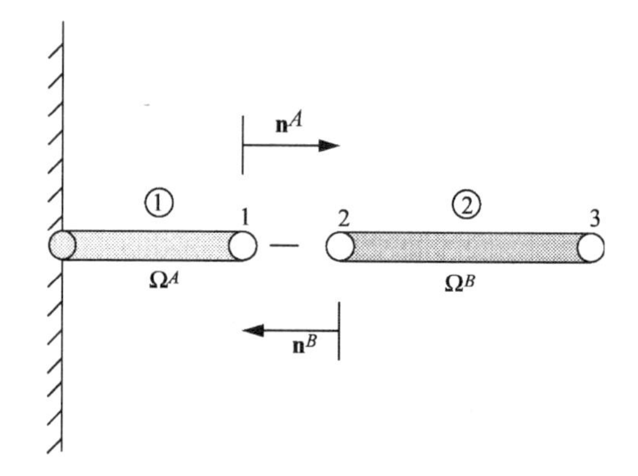

Figure 10.11 One-dimensional example of contact: see Example 10.2

The impenetrability condition in rate form, (10.5.22), is given by

$$\mathbf{G}\dot{\mathbf{d}} \le 0 \quad \text{or} \quad [1,-1,0]\dot{\mathbf{d}} = v_1 - v_2 \le 0 \tag{E10.2.3}$$

Lagrange multiplier method The previous contact condition, $v_1 - v_2 \le 0$, can be obtained by inspection: when the two nodes are in contact, the velocity of node 1 must be less than or equal to the velocity of node 2. If the two nodal velocities are equal, the nodes remain in contact; whereas when the inequality holds, they release. These conditions are not sufficient to check for initial contact, which should be checked in terms of the nodal displacements: $x_1 - x_2 \ge 0$ indicates contact has occurred during the previous time step.

Since there is only one point of contact, only a single Lagrange multiplier is needed to enforce the contact constraint. The equations of motion, (B10.3.1), are

$$\begin{bmatrix} M_{11} & M_{12} & M_{13} \\ M_{21} & M_{22} & M_{23} \\ M_{31} & M_{32} & M_{33} \end{bmatrix} \begin{Bmatrix} \ddot{d}_1 \\ \ddot{d}_2 \\ \ddot{d}_3 \end{Bmatrix} - \begin{Bmatrix} f_1 \\ f_2 \\ f_3 \end{Bmatrix} + \begin{Bmatrix} 1 \\ -1 \\ 0 \end{Bmatrix} \lambda_1 = 0 \quad \text{and} \quad \lambda_1 \ge 0 \tag{E10.2.4}$$

The last terms on the LHS in (E10.2.4) are the nodal forces resulting from contact between nodes 1 and 2. These forces are equal and opposite and vanish when the Lagrange multiplier vanishes. The equations of motion are identical to the equations for an unconstrained finite element mesh except at the nodes that are in contact. The equations for a diagonal mass matrix with unit area can be written as

$$M_1 a_1 - f_1 + \lambda_1 = 0, \quad M_2 a_2 - f_2 - \lambda_1 = 0, \quad M_3 a_3 - f_3 = 0 \tag{E10.2.5}$$

where $a_I = \ddot{d}_I$; the fourth equation is (E10.2.3).

The equations for small-displacement elastostatics, (10.5.29), are obtained by combining the $\mathbf{G}$ matrix, (E10.2.2), with the assembled stiffness:

$$\begin{bmatrix} k_1 & 0 & 0 & 1 \\ 0 & k_2 & -k_2 & -1 \\ 0 & -k_2 & k_2 & -1 \\ 1 & -1 & 0 & 0 \end{bmatrix} \begin{Bmatrix} d_1 \\ d_2 \\ d_3 \\ \lambda_1 \end{Bmatrix} \begin{matrix} \\ = \\ \\ \geq \end{matrix} \begin{Bmatrix} f_1 \\ f_2 \\ f_3 \\ 0 \end{Bmatrix}^{\text{ext}} \tag{E10.2.6}$$

where k_I is the stiffness of element I. The assembled stiffness matrix in the absence of contact, that is, the upper left-hand 3×3 matrix, is singular, but with the addition of the contact interface conditions, the complete 4×4 matrix becomes regular.

Penalty method To write the equation for the penalty method, we will use the penalty law $p=\beta g=\beta(x_1-x_2)H(g)=\beta(X_1-X_2+u_1-u_2)H(g)$, $g\equiv g_N$. Then evaluating (10.5.30) gives

$$\mathbf{f}^c = \int_{\Gamma^c} \mathbf{\Phi}^T p d\Gamma = \begin{Bmatrix} 1 \\ -1 \\ 0 \end{Bmatrix} \beta g \tag{E10.2.7}$$

In this integral, the integrand is evaluated at the interface point (Γ^c is a point). Equations (B10.3.2) for a diagonal mass are then

$$M_1 a_1 - f_1 + \beta g = 0, \quad M_2 a_2 - f_2 - \beta g = 0, \quad M_3 a_3 - f_3 = 0 \tag{E10.2.8}$$

These equations are similar to that for the Lagrange multiplier method, (E10.2.5) except that the Lagrange multiplier is replaced by the penalty force and (E10.2.3) is absent.

To construct the small-displacement, elastostatic equations for the penalty method, we first evaluate $\mathbf{P}_c$ by (10.5.35):

$$\mathbf{P}_c = \int_{\Gamma^c} \bar{\beta} \mathbf{\Phi}^T \mathbf{\Phi} d\Gamma = \bar{\beta} \begin{bmatrix} +1 \\ -1 \\ 0 \end{bmatrix} \begin{bmatrix} +1 & -1 & 0 \end{bmatrix} = \bar{\beta} \begin{bmatrix} +1 & -1 & 0 \\ -1 & +1 & 0 \\ 0 & 0 & 0 \end{bmatrix} \tag{E10.2.9}$$

where $\bar{\beta} = \beta_1 H(g)$. Adding $\mathbf{P}_c$ to the linear stiffness yields the following equations for statics

$$\begin{bmatrix} k_1 + \bar{\beta} & -\bar{\beta} & 0 \\ -\bar{\beta} & k_2 + \bar{\beta} & -k_2 \\ 0 & -k_2 & k_2 \end{bmatrix} \begin{Bmatrix} d_1 \\ d_2 \\ d_3 \end{Bmatrix} = \begin{Bmatrix} f_1 \\ f_2 \\ f_3 \end{Bmatrix}^{\text{ext}} \tag{E10.2.10}$$

It can be seen from (E10.2.10) that the penalty method adds a spring with a stiffness $\bar{\beta}$ between nodes 1 and 2. The above equation is nonlinear since $\bar{\beta}$ is a nonlinear function of $g=u_1-u_2$.

Example 10.3 Two-dimensional example Figure 10.12 shows two bodies in contact modeled by 4-node quadrilaterals. The velocity field on the contact line is written in terms of the edge coordinates (the element coordinates projected on the contact line) of body A:

$$\begin{Bmatrix} v_x(\zeta,t) \\ v_y(\zeta,t) \end{Bmatrix} = \begin{bmatrix} N_1^A & 0 & N_2^A & 0 & N_3^B & 0 & N_4^B & 0 \\ 0 & N_1^A & 0 & N_2^A & 0 & N_3^B & 0 & N_4^B \end{bmatrix} \dot{\mathbf{d}} \tag{E10.3.1}$$

where

$$\dot{\mathbf{d}}^T = [v_{x1} \quad v_{y1} \quad v_{x2} \quad v_{y2} \quad v_{x3} \quad v_{y3} \quad v_{x4} \quad v_{y4}] \tag{E10.3.2}$$

$$N_1^A = N_3^B = 1-\zeta, \quad N_2^A = N_4^B = \zeta, \quad \zeta = x/\ell \tag{E10.3.3}$$

The unit normals, as shown in Figure 10.12, are given by $\mathbf{n}^A = \left[n_x^A,\ n_y^A\right]^T = [0,-1]^T$, $\mathbf{n}^B = [0,\ 1]^T$. The $\mathbf{\Phi}$ matrix is given by (10.5.17):

$$\begin{aligned} \mathbf{\Phi} &= \left[N_1 n_x^A \quad N_1 n_y^A \quad N_2 n_x^A \quad N_2 n_y^A \quad N_3 n_x^B \quad N_3 n_y^B \quad N_4 n_x^B \quad N_4 n_y^B\right] \\ &= \left[0 \quad -N_1^A \quad 0 \quad -N_2^A \quad 0 \quad N_3^B \quad 0 \quad N_4^B\right] \end{aligned} \tag{E10.3.4}$$

The Lagrange multiplier is approximated by a linear field:

$$\lambda(\zeta,t) = \mathbf{\Lambda}\boldsymbol{\lambda} = [1-\zeta \quad \zeta]\begin{Bmatrix} \lambda_1 \\ \lambda_2 \end{Bmatrix} \tag{E10.3.5}$$

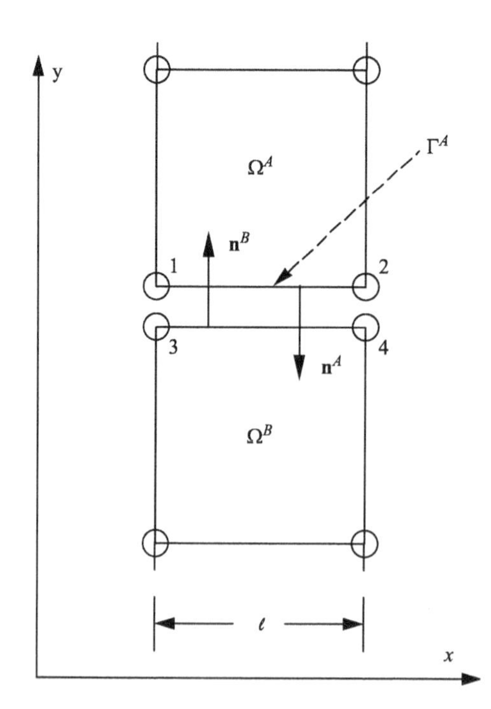

Figure 10.12 Two impacting quadrilaterals; the two edges of the contact line have been drawn separately for clarity: see Example 10.3

The **G** matrix (10.5.19) is given by

$$\mathbf{G} = \int_0^1 \mathbf{\Lambda}^T \mathbf{\Phi} \ell d\zeta = \frac{\ell}{6}\begin{bmatrix} 0 & -2 & 0 & -1 & 0 & 2 & 0 & 1 \\ 0 & -1 & 0 & -2 & 0 & 1 & 0 & 2 \end{bmatrix} \tag{E10.3.6}$$

This matrix resembles a consistent mass matrix for a rod: contact at node 1 results in forces at node 2, and vice versa. The contact nodal forces are strictly in the y-direction; all x-components of contact nodal forces vanish since the odd columns of the **G** matrix vanish. This is expected, since the normal to the contact edge is in the y-direction and the contact interface is frictionless.

10.5.9 Regularization

In a regularization procedure, a feature of the solution which is difficult to handle because it causes discontinuities or singularities is replaced by an artifact which smoothes and regularizes the solution. The classic example of regularization is von Neumann's addition of artificial viscosity to the Euler fluid equations to smooth shocks. Without this artificial viscosity, central difference solutions of the Euler equations in the vicinity of shocks are very oscillatory. As shown by von Neumann, that regularization conserves momentum, so the conservation properties are not impaired by the regularization.

The penalty method plays a similar role in impact. In the Lagrange multiplier method, the velocities at the time of impact on the contact interface are discontinuous in time as shown in Figure 10.13(a). These discontinuities propagate through the body and result in considerable

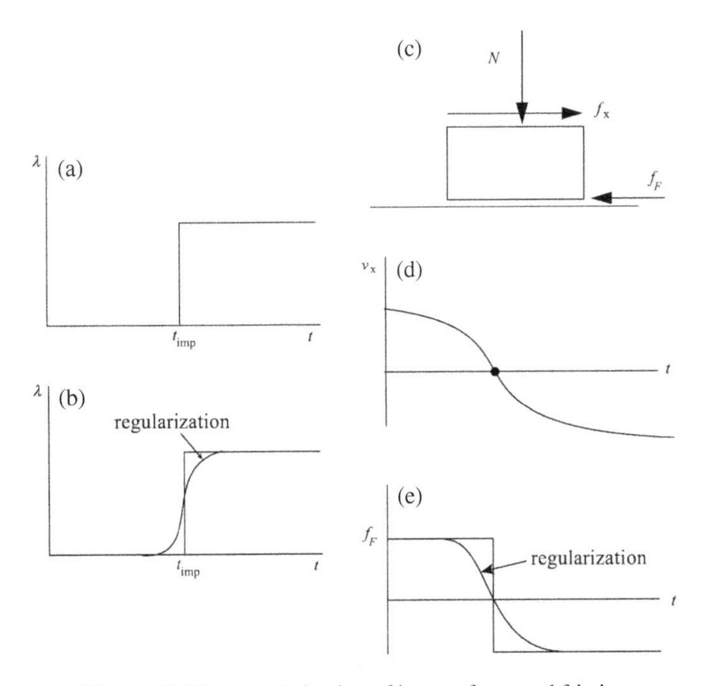

Figure 10.13 Regularization of impact force and friction

noise. The penalty method may be thought of as a regularization of the contact interface conditions: it smoothes the discontinuous velocities as shown in Figure 10.13(b) and preserves momentum conservation. It only relaxes one condition, the impenetrability condition, by allowing some overlap of the two bodies. That is a small price to pay for smoother solutions.

The Curnier–Mroz friction model can also be considered a regularization: the discontinuous Coulomb friction laws are replaced by smoother models. The discontinuous nature of Coulomb friction can be appreciated from a simple illustration. Consider an element on a rigid surface, with interface tractions modeled by Coulomb friction. A vertical force and a horizontal force are applied as shown in Figure 10.13(c); we neglect the deformability of the element. The vertical force is kept constant while the horizontal force yields a horizontal velocity with the time history shown in Figure 10.13(d). The Coulomb friction law then gives the discontinuous friction force shown in Figure 10.13(e). A Curnier–Mroz plastic model will give a smoother force, such as the one labeled 'regularization' in Figure 10.13(e).

Regularization of Coulomb friction differs from regularization of interpenetration because it smoothes the response by introducing additional mechanics, namely the behavior of the asperities, whereas the relaxation of the interpenetrability condition is usually *ad hoc* and not motivated by physical arguments. In fact, one can also attribute some interpenetration in contact-impact problems to compression of asperities. However, the penalty parameters for the normal force are generally not based on mechanics, but instead are chosen so that they eliminate or reduce frequencies above a certain threshold.

10.6 On Explicit Methods

10.6.1 Explicit Methods

In this section we describe the procedures for treating contact impact with explicit time integration. Explicit time integration is well suited to dynamic contact-impact problems for the following reasons.

1. The time steps are small because of stability requirements, so the discontinuities due to contact-impact wreak less havoc.
2. Neither linearization nor a Newton solver is needed, so the deleterious effects of discontinuities on Newton solvers are avoided.

The large time steps made possible by unconditionally stable implicit methods are not effective for discontinuous response. Furthermore, contact-impact introduces discontinuities in the Jacobian, which impedes the convergence of Newton methods.

In explicit algorithms, in each time step, the bodies are first integrated completely independently, as if they were not in contact. This uncoupled update correctly indicates which parts of the body will be in contact at the end of the time step. The contact conditions are then imposed; no iterations are needed to establish the contact interface.

We will describe implementations of Lagrange multiplier and penalty contact-impact algorithms in explicit methods. The issues to be discussed include:

1. structure of the algorithm;
2. effects of contact-impact methods on numerical stability;
3. the correctness of the uncoupled update for predicting the contact interface.

We will also describe certain characteristics of explicit solutions of contact-impact problems which arise from physics and numerics.

10.6.2 Contact in One Dimension

In order to illustrate the characteristics of contact-impact in a simple setting, we first consider a one-dimensional problem. The one-dimensional example is shown in Figure 10.14. We first consider the premise that uncoupled updates of bodies A and B followed by modifications of the interpenetrating nodes for contact-impact lead to correct solutions. For nodes 1 and 2, there are four possibilities during the time step:

1. Nodes 1 and 2 are not in contact and do not contact during the time step.
2. Nodes 1 and 2 are not in contact but impact during the time step.
3. Nodes 1 and 2 are in contact and remain in contact.
4. Nodes 1 and 2 are in contact and separate during a time step, often known as *release*.

For case 3, the statement 'remain in contact' does not imply that the two points must remain contiguous in two- or three-dimensional problems, because relative tangential motion, or sliding, is possible. When two bodies remain in contact, only the normal component of the velocity is considered to be continuous.

All of these possibilities can be correctly treated by an uncoupled update followed by adjustment of the velocities and the displacements of the nodes which have interpenetrated during the time step. The possibilities which need to be explained are cases 2, 3 and 4.

The discrete momentum equations for nodes 1 and 2 are given by (E10.2.5). We will show that when the velocities from the uncoupled update predict interpenetration, then the Lagrange multiplier $\lambda \geq 0$. The results of the uncoupled update are called *trial variables* and are designated by *superposed bars*. The *trial accelerations and velocities* of nodes 1 and 2 are

$$M_1\bar{a}_1 - f_1 = 0, \quad M_2\bar{a}_2 - f_2 = 0, \quad \bar{v}_1^+ = v_1^- + \Delta t\bar{a}_1, \quad \bar{v}_2^+ = v_2^- + \Delta t\bar{a}_2 \tag{10.6.1}$$

The correct velocities by a central difference update of (E10.2.5) are:

$$M_1v_1^+ - M_1v_1^- - \Delta tf_1 + \Delta t\lambda = 0, \qquad M_2v_2^+ - M_2v_2^- - \Delta tf_2 - \Delta t\lambda = 0 \tag{10.6.2}$$

where $(\bullet)^+ \equiv (\bullet)^{n+½}$, $(\bullet)^- \equiv (\bullet)^{n-½}$; all unlabeled variables are at time step n. When the bodies contact during the time step, these equations must hold with the subsidiary condition $v_1^+ = v_2^+$.

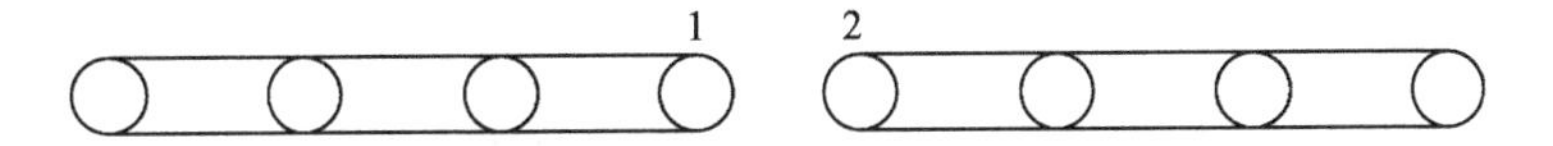

Figure 10.14 Model of two impacting rods

Eliminating λ from the previous equations by using the contact constraint $v_1^+ = v_2^+$ gives the corrected velocities

$$v_1^+ = v_2^+ = \frac{M_1 v_1^- + M_2 v_2^- + \Delta t\left(f_1 + f_2\right)}{M_1 + M_2} \tag{10.6.3}$$

if the nodes have interpenetrated during the time step. This is the well-known equation of conservation of momentum for plastic impact of rigid bodies; more will be said on this later.

We will now show that whenever the trial velocities interpenetrate, then the Lagrange multiplier will be positive, that is, the interface force will be compressive. In other words, the Lagrange multipliers will have the correct sign at any nodes which interpenetrate. This corresponds to showing that

$$\text{if } \bar{v}_1^+ \geq \bar{v}_2^+, \quad \text{then} \quad \lambda \geq 0 \tag{10.6.4}$$

Multiplying (10.6.2a) by M_2 and (10.6.2b) by M_1 and taking the difference of the two equations gives

$$M_1 M_2\left(v_1^- - v_2^-\right) + \Delta t(M_2 f_1 - M_1 f_2) = \lambda \Delta t(M_1 + M_2) \tag{10.6.5}$$

Substituting the expressions for f_1 and f_2 from (10.6.1) into the above and rearranging gives

$$\frac{\Delta t\left(M_1 + M_2\right)}{M_1 M_2}\lambda = \left(v_1^- - v_2^-\right) + \Delta t\left(\bar{a}_1 - \bar{a}_2\right) = \bar{v}_1^+ - \bar{v}_2^+ \tag{10.6.6}$$

where the last equality is obtained by using the central difference updates for the two bodies: $\bar{v}_I^+ = \bar{v}_I + \Delta \bar{a}_I$. The coefficient of λ is positive, so the sign of the RHS gives the sign of λ. Thus (10.6.4) has been demonstrated.

To examine this in more detail, we now consider three of the cases listed earlier (case 1 is trivial since it requires no modification of the nodal velocities):

1. Not in contact/contacts during Δt: then $\bar{v}_1^+ > \bar{v}_2^+$ and $\lambda \geq 0$ by (10.6.6)
2. In contact/remains in contact: then $\bar{v}_1^+ > \bar{v}_2^+$ and $\lambda \geq 0$ by (10.6.6)
3. In contact/release during Δt: then $\bar{v}_1^+ < \bar{v}_2^+$ and $\lambda < 0$ by (10.6.6)

Thus the velocities obtained by the uncoupled updates correctly predict the sign of the Lagrange multiplier λ.

Two other interesting properties of explicit integration of contact-impact can be learned from this example:

1. release cannot occur in the same time step as initial contact, that is, impact;
2. energy is dissipated during impact.

The first statement rests on the fact that the Lagrange multiplier at time step n is computed so that the velocities at time step $n + 1/2$ match. Hence there is no mechanism in an explicit method for release during the time step in which impact occurs. This property is consistent with the mechanics of wave propagation. In impacting bodies, release is caused by rarefaction waves which develop when the compressive waves due to impact reflect from a free surface and reach the point of contact. When the magnitude of this rarefaction is sufficient to cause tension across the contact interface, release occurs. Therefore the minimum time required for release subsequent to impact is two traversals to the nearest free surface (unless a rarefaction wave from some other event reaches the contact surface). The stable time step, you may recall, allows any wave to traverse at most one element in a time step. In explicit time integration, there is insufficient time in a stable time step for the waves to traverse twice the distance to the nearest free surface, so release cannot occur in the same time step as impact.

The second statement can be explained by (10.6.3) which can be recognized as the plastic impact conditions. These always dissipate energy. The dissipation decreases with the refinement of the mesh. In the continuous impact problem, that is, the solution to the PDE, the same conditions hold for the interface but no energy is dissipated because the condition of equal velocities after impact is limited to the impact surfaces. A surface is a set of measure zero, so a change of energy over the surface has no effect on the total energy (for one-dimensional problems the impact surface is a point, which is also a set of measure zero.) In a discrete model, the impacting nodes represent a material layer of thickness $h/2$ adjacent to the contact surface. Therefore, the energy dissipation in a discrete model is always finite. It should be stressed that these arguments do not apply to multi-body models with beams, shells, or rods, where the stiffness through the thickness direction is not modeled. The release and impact conditions are then more complex.

10.6.3 Penalty Method

The discrete equation at the impacting nodes for the two-body problem are (E10.2.8):

$$M_1 a_1 - f_1 + f_1^c = 0, \quad M_2 a_2 - f_2 - f_2^c = 0 \tag{10.6.7}$$

When the nodes are initially coincident, then $x_1 = x_2$ and the interface normal traction can be written as

$$f^c = p = \beta_1 g + \beta_2 \dot{g} = \beta_1 (u_1 - u_2) H(g) + \beta_2 (v_1 - v_2) H(g) \tag{10.6.8}$$

The unitary condition is now violated since the normal traction is positive while the interpenetration rate is positive, so its product no longer vanishes. The post-impact velocities depend on the penalty parameter. The velocities of the two nodes are not equal since the penalty method enforces the impenetrability constraint only approximately. As the penalty parameter is increased, the condition of impenetrability is observed more closely. However, the penalty parameter cannot be arbitrarily large.

In the penalty method, impact and release may occur in the same time step. If the penalty force is very large, it is possible for the relative nodal velocity to reverse in the time step of impact. This anomaly can be eliminated by placing an upper bound on the penalty force, so that the impact is at most perfectly plastic. In other words, the penalty force should be bounded so that the velocities at the end of the impact time step are given by (10.6.3). This yields the following upper bound on the contact force:

$$f^c \leq \frac{M_1 M_2 \left(v_1^- - v_2^-\right)}{\Delta t(M_1 + M_2)} \tag{10.6.9}$$

This bound can be very useful since it provides an upper bound on the penalty.

In contrast to the Lagrange multiplier method, the penalty method always decreases the stable time step. The stable time step can be estimated by applying the eigenvalue element inequality to a linearized model: a group of elements consisting of the penalty spring and the two surrounding elements should be considered, since the penalty element has no mass and therefore has an infinite frequency. This analysis shows that the stable time step for an interpenetration-dependent penalty is given by (Belytschko and Neal, 1991)

$$\Delta t_{\text{crit}} = \sqrt{2}\frac{h}{c}\left(1 + \beta_1 + \sqrt{1 + \beta_1^2}\right)^{-\frac{1}{2}} \tag{10.6.10}$$

for a stiffness penalty when the rate-dependent penalty vanishes ($\beta_2 = 0$).

The decrease in the critical time step varies inversely with the stiffness of the penalty spring β_1: as the β_1 is increased, the stable time step decreases. This estimate of the stable time step is not conservative, even though it is based on the element eigenvalue inequality, that is, it does not necessarily overestimate the stable time step. The bounding character of the estimate is lost because contact-impact is very nonlinear, while the analysis presumes linear behavior.

10.6.4 Explicit Algorithm

The contact-impact conditions are enforced immediately after the boundary conditions; see Box 6.1. Prior to the contact-impact step, all nodes in the model have been updated as if they were not in contact, including the nodes which were in contact in the previous time step.

Some difficulties may occur because the contact-impact modifications are made after the boundary conditions are enforced. For example, for a pair of contacting nodes on a plane of symmetry, it is possible for the contact-impact modifications to result in violation of the symmetry condition. Therefore, boundary conditions sometimes have to be reimposed at contact nodes after the modifications.

For low-order elements, the maximum interpenetration always occurs at the nodes. It is then only necessary to check nodes for interpenetration into elements of another body. This is quite challenging. In a large model, on the order of 10^5 nodes may need to be checked against penetration into a similar number of elements. Obviously a brute force approach to this task is not going to work.

11

EXtended Finite Element Method (XFEM)

11.1 Introduction

For problems characterized by discontinuities, singularities, localized deformations, and complex geometries, the standard finite element mesh becomes cumbersome to work with as it needs to conform to the morphology of these features. We will focus on the modeling of discontinuities by the extended finite element method (XFEM) in this chapter. Discontinuities can be broken down into two main categories: weak and strong, both of which, will be discussed in this chapter. For a more comprehensive overview of XFEM, the readers are referred to Fries and Belytschko (2010).

11.1.1 Strong Discontinuity

The most common example of a strong discontinuity is fracture, which we will focus on in this chapter. The methods described for cracks can also be applied to other strong discontinuities such as dislocations and unbounded inclusions. A brief description of other methods for modeling strong discontinuities is given next.

In standard finite element methods, fracture can be modeled by remeshing (similar to mesh update procedures introduced in Chapter 7) whereby element edges track the crack front and faces. This approach has certain disadvantages: (1) feature-conforming mesh generation is more demanding on the mesh generator, (2) considerable computational resources are spent on remeshing, (3) projection schemes can introduce errors into solution, and (4) the ever-changing mesh poses difficulties for post-processing and solution interpretation. Figure 11.1

Nonlinear Finite Elements for Continua and Structures, Second Edition.
Ted Belytschko, Wing Kam Liu, Brian Moran, and Khalil I. Elkhodary.

Companion Website: www.wiley.com/go/belytschko

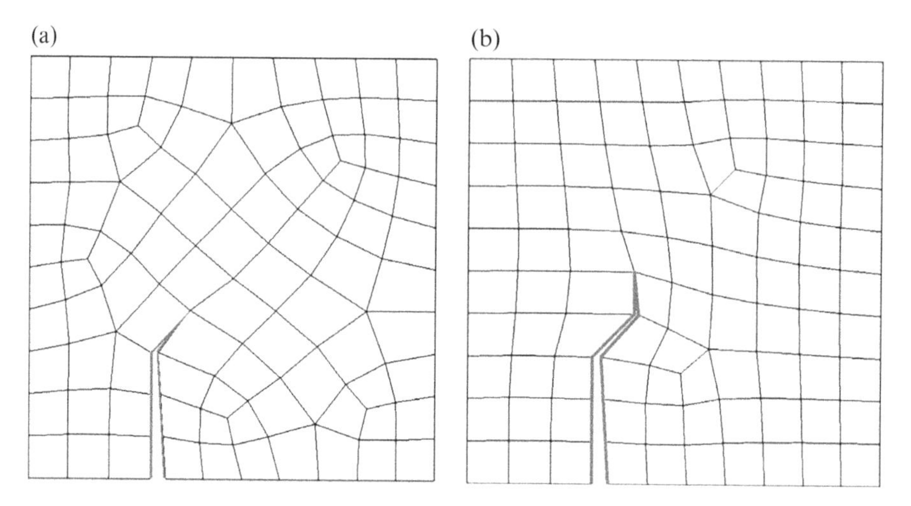

Figure 11.1 (a) Crack-conforming mesh (b) mesh regenerated to track crack evolution

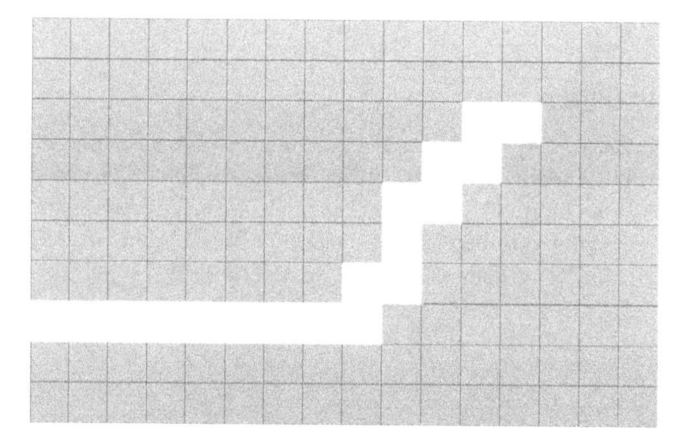

Figure 11.2 Deleted elements track (interpreted) crack path

shows remeshing of a plate with a crack propagating in it. Admittedly, better meshing algorithms and more user intervention can minimize the difference between the two meshes.

The element deletion method is also widely used to simulate fracture problems within the classical FEM framework, and is remarkably easy to incorporate into existing software as it can be simply implemented in constitutive equations as a switch based on certain failure criteria. It models a crack as a set of deleted elements with zero stress, that is, no load bearing capacity. However, as energy is lost when elements are deleted, it's imperative to adjust the constitutive law with respect to the element size lest spurious mesh dependence of the fracture energy be introduced. However, even with this scaling of the constitutive law, the element deletion method is still heavily mesh-dependent. Figure 11.2 shows how deleted elements in a structured mesh track the crack path.

Another way of modeling discontinuities is the inter-element crack method. Xu and Needleman (1994) prepared all element edges for separation from the beginning of the

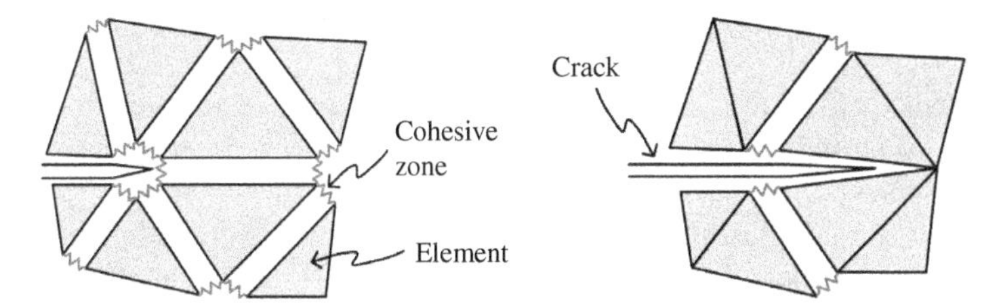

Figure 11.3 (a) All potential cracks with cohesive forces present from the beginning (Xu and Needleman, 1994) (b) cracks with cohesive forces injected as needed (Camacho and Ortiz, 1996). Reproduced from Song JH, Wang H and Belytschko T (2008) A comparative study on finite element methods for dynamic fracture. *Computational Mechanics*, **42**(2), 239–250. Copyright © 2008, Springer

simulation by inserting a cohesive law (which behaved like springs between element edges) between all elements. A more flexible method has been presented by Camacho and Ortiz (1996) in which a fracture criterion is used to determine if a pair of elements is allowed to separate along a shared edge. If the criterion is met, a discontinuity with cohesive force is injected along element edges. Because the crack path has to conform to the element edges, prediction of the path is dependent on the mesh locally, and accurate tracking of the crack path requires a fine mesh. Figure 11.3 shows schematics of these two variations on the inter-element method.

11.1.2 Weak Discontinuity

Weak discontinuities are kinks in the displacement field which result in a jump in the strains. A weak discontinuity has C^0 continuity which is the same as the finite element approximation of the displacement field, however, modeling these weak discontinuities can still be challenging. For example, Chessa *et al.* (2002) developed a method for discontinuities in derivatives in solidification problems.

A common weak discontinuity occurs at the interface between two different materials. A wide range of natural and manufactured materials contain multiple phases, defects, inclusions and/or other internal boundaries, which have significant influences on the material durability and structural integrity. For example, material reinforced by different constituents, such as fiber reinforced composite or particle reinforced metal, usually exhibits stronger macroscopic mechanical properties (these will be discussed in Chapter 12). The improvement of mechanical properties largely depends on the interactions at the interface. Defects, voids, and inclusions are usually unavoidable in industrial manufacturing process, and have a large impact on overall material behavior. Hence the accurate modeling of material interfaces is of much interest to both theorists and practitioners.

In traditional finite element methods, the mesh is required to conform to internal boundaries of the model. Although relatively robust mesh generation schemes exist for these tasks, meshing these defects and inclusions of arbitrary number and distribution is still time-consuming and can results in poorly shaped elements. Additionally if the interface evolves with time, such as in phase transitions, then remeshing must occur.

11.1.3 XFEM for Discontinuities

While these methods for both weak and strong discontinuities require the mesh topology to conform to the discontinuities (conforming mesh), XFEM allows for the mesh to be separated from them (non-conforming mesh). The difference between these meshing requirements can be seen in Figure 11.4, and a further discussion of this topic can be found in Tian *et al.* (2011). The complete independence of the mesh from these entities is the main advantage of XFEM, which dramatically simplifies the solution of problems, such as:

1. the propagation of cracks,
2. the evolution of dislocations,
3. the modeling of grain boundaries, and
4. the evolution of phase boundaries.

The original applications of XFEM to fracture can be seen in Belytschko and Black (1999) and Moës *et al.* (1999), while the evolution of dislocations in complex geometry can be seen in Oswald *et al.* (2009). Wagner *et al.* (2001, 2003) used XFEM to model solids inside a Stokes fluid and Duddu *et al.* (2008) combined XFEM and level set methods to model the discontinuity in the normal derivative of the substrate concentration field at a biofilm-fluid interface. An alternative approach for the treatment of discontinuity has been discussed by Lu *et al.* (2005).

XFEM utilizes extrinsic enrichments structurally similar to other enriched methods, first among them being the partition of unity method (PUM) and the generalized finite element method (GFEM). The enrichments in these methods are realized through the partition of unity (PU) concept, which was proposed by Melenk and Babuška (1996). One important distinction between XFEM and the early works of the other methods is that the domain is enriched only locally; this is achieved by enriching a subset of the nodes.

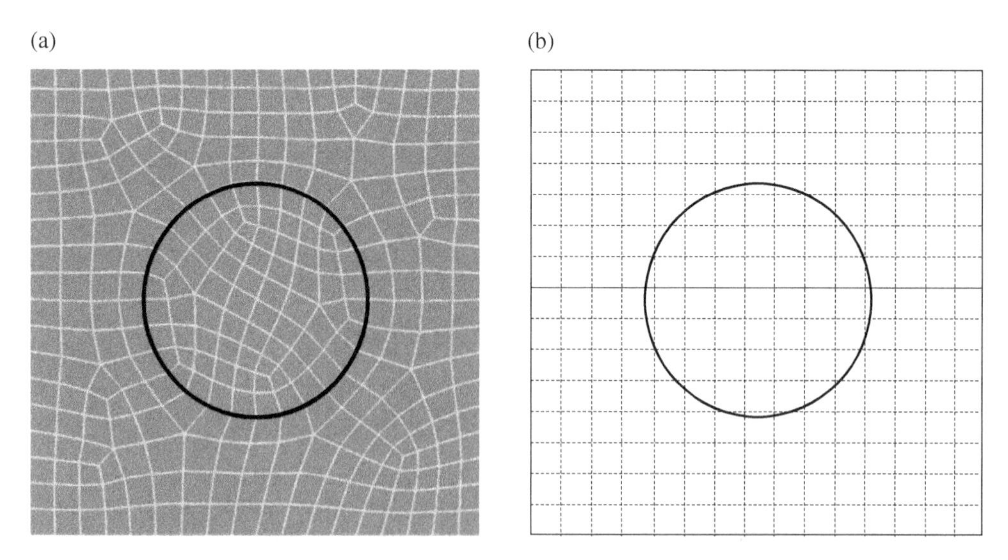

Figure 11.4 (a) Conforming mesh required by classical FEM and (b) non-conforming mesh permissible by XFEM

11.2 Partition of Unity and Enrichments

For convergence and passing the patch test, a finite element approximation has to be able to represent rigid body translation exactly. All Lagrangian finite element shape functions satisfy this requirement because they have the partition of unity property, which states

$$\sum_{\forall I}\varphi_I(\mathbf{X})=1,\ \forall\mathbf{X}\in\Omega_0 \tag{11.2.1}$$

The fact that any function $\Psi(\mathbf{x})$ can be reproduced by a product of functions that satisfy the partition of unity with $\Psi(\mathbf{x})$ (i.e., $\Psi(\mathbf{x})\cdot 1=\Psi(\mathbf{x})$) is exploited in XFEM. It is also known as the reproducing condition, introduced by Liu *et al.* (1995a, b), in the meshfree literature. On a side note, the reproducing kernel particle method by Liu *et al.* (1997) and reproducing kernel element method (Liu *et al.*, 2004; Li *et al.*, 2004; Lu *et al.*, 2004; Simkins *et al.*, 2004) are all based on this condition.

Under this generalized view, we can recast the standard finite element approximation as

$$\mathbf{u}^h(\mathbf{X})=\sum_{\forall I}N_I(\mathbf{X})\Psi(\mathbf{X})\mathbf{u}_I \tag{11.2.2}$$

where N_I are the standard FEM shape functions and $\mathbf{u}_I$ are the standard nodal degrees of freedom, the $\Psi(\mathbf{X})$ here is simply the function $1(\mathbf{X})=1$. The crucial step of arriving at XFEM is realizing the possibility of substituting in more interesting forms of $\Psi(\mathbf{X})$, for example, a discontinuous function, for the vanilla $1(\mathbf{X})$. Additionally this substitution need only be done locally in regions of primary interest such as at cracks, grain boundaries or dislocations, and other parts of the domain can still retain the standard approximation. This leads to the enriched form

$$\mathbf{u}^h(\mathbf{X})=\underbrace{\sum_{\forall I}N_I(\mathbf{X})1(\mathbf{X})\mathbf{u}_I}_{\mathbf{u}^{\mathrm{FE}}}+\underbrace{\sum_{\forall I}\varphi_I(\mathbf{X})\Psi(\mathbf{X})\mathbf{q}_I}_{\mathbf{u}^{\mathrm{enr}}} \tag{11.2.3}$$

where we keep the $1(\mathbf{X})$ intact to emphasize the similarity of the two summation terms. The first part of the right-hand side is the standard finite element approximation ($\mathbf{u}^{\mathrm{FE}}$), whereas the second part is the partition of unity enrichment ($\mathbf{u}^{\mathrm{enr}}$). The nodal values $\mathbf{q}_I$ are unknown parameters that adjust the enrichment so that it best approximates the solution at hand. Thus the enrichment need not be precisely the local solution for the problem at hand; it needs only to capture the characteristics of the local feature. Another advantage of this approximation structure is that when the functions $\varphi_I(\mathbf{X})$ have compact support (i.e., are only nonzero over a small subdomain of the problem), then the discrete equations for the system will be sparse. By contrast, directly adding a global enrichment function to the approximation would lead to non-sparse discrete equations, which are computationally more expensive. However, it should be noted that the displacement field has been decomposed into a combination of the traditional degrees of freedom $\mathbf{u}_I$ and the additional degrees of freedom $\mathbf{q}_I$, so that both $\mathbf{u}_I$ and $\mathbf{q}_I$ have lost their kinematic meanings on their own.

Note that by the partition of unity property, when $\mathbf{q}_I=1$ and $\mathbf{u}_I=0$, the function $\Psi(\mathbf{X})$ is *reproduced* exactly by the approximation. It should be pointed out that the shape functions for the standard approximation and the enrichment need not be the same functions, as indicated in Equation (11.2.3), but generally the same functions are used, that is, generally $\varphi_I(\mathbf{X})=N_I(\mathbf{X})$.

11.3 One-Dimensional XFEM

11.3.1 Strong Discontinuity

We will first consider fracture in one dimension. While a 1D crack cannot illustrate some interesting features present in 2D and 3D fracture, since the crack is a single stationary point, and renders the structure disjointed, the simplicity and representative nature of a 1D problem does make it a good example to familiarize readers with the basic concepts of XFEM.

In order to model a strong discontinuity (crack) the function $\Psi(\mathbf{x})$ should incorporate a strong discontinuity. A common choice is the Heaviside function as it embeds a strong discontinuity, and it is defined as

$$H(\alpha)=\begin{cases}1, & \alpha\geq 0\\ 0, & \text{otherwise}\end{cases} \tag{11.3.1}$$

The Heaviside function is shown in Figure 11.5.

To describe the crack we will use a signed distance function, also called a level set. (For more details on level sets, see Section 11.7: however, full understanding is not critical to this description.) The signed distance function can be expressed in this case as:

$$\phi(X)=X-X_c \tag{11.3.2}$$

where X_c is the location of the fracture. This can then be combined with Equation (11.3.1) to give an equation that is discontinuous at the crack face. Combining this with (11.2.3) we get the displacement for an element cut by a strong discontinuity:

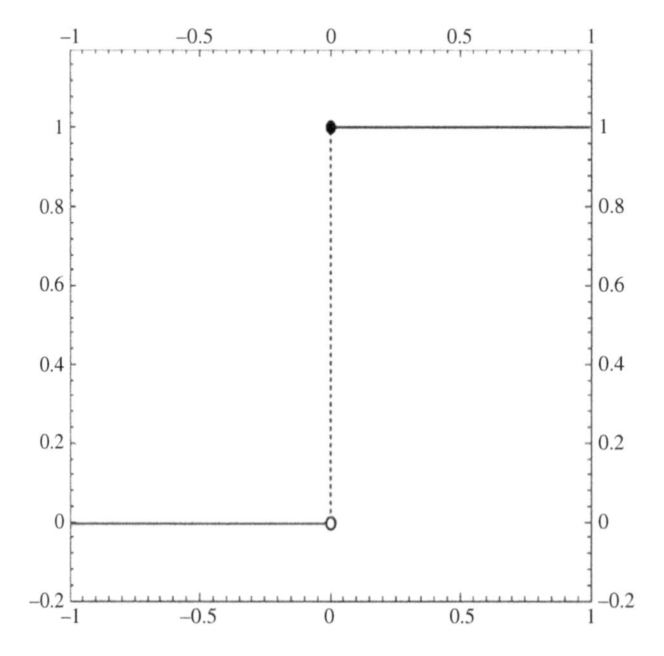

Figure 11.5 Heaviside function for modeling a strong discontinuity

$$u^h(X) = \sum N_I(X)u_I + \sum N_J(X)H(\phi(X))q_J \tag{11.3.3}$$

While the enrichment given in Equation (11.3.3) works well for a single element, it can easily be seen that the neighboring element using only the unenriched displacement field would lead to a incompatibility at the shared nodes. To overcome this issue we will introduce shifting, the preferred method of removing this element boundary incompatibility. The goal of shifting is to remove the effects of q_i on the nodes. To accomplish this we can construct $\Psi(\mathbf{X})$ to be the following:

$$\Psi_J(X) = H(\phi(X)) - H(\phi(X_J)) \tag{11.3.4}$$

where X_J is the value of X at the node J.

It is important to understand that XFEM solutions using different forms of enrichment functions (think of the pair of Heaviside enrichment functions with or without shifting) can model the same discontinuous displacement field. But the solutions will end up with different values of u_I and q_I, since these coefficients correspond to different basis functions. What this means is that u_I and q_I have lost kinematic meaning of their own, only the combination of them can give the full description of the solution.

Example 11.1 A cracked rod modeled with XFEM Consider a 1D cracked rod as shown in Figure 11.6. The rod is cut by a crack at X_c, and there is no jump across this crack initially. The rod is originally of length ℓ_0 and constant cross-sectional area A_0. At any subsequent time t, the two end points have displacement u_1 and u_2, respectively; this motion also caused the crack to open gradually, and the two surfaces of the crack have displacement u_c^- and u_c^+, respectively. The superscripts denote the left-hand side (LHS) and the right-hand side (RHS) of the crack. Note the dependence of these displacements on time t is not explicitly written but assumed. Model this rod using XFEM with Heaviside enrichment and XFEM with *shifted* Heaviside enrichment. Compare the differences of them. The reader is also encouraged to model this rod with standard conforming FEM and compare it with the XFEM models.

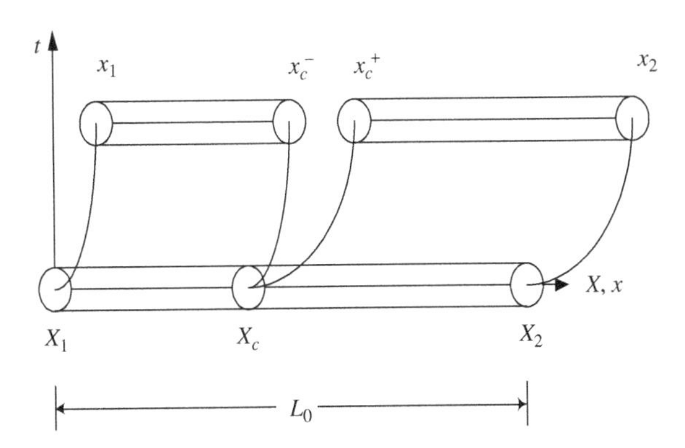

Figure 11.6 1D cracked rod shown in initial configuration and current configuration

Displacement field with Heaviside enrichment We assume linear displacement for the segments formed by the crack, consequently the motion is completely determined by the four displacements specified. We call these the original degrees of freedom since they are what we started with. Intuitively, the displacements of the two crack surfaces are easy to visualize and work with, but this makes our derivation explicitly dependent on the position of the crack. The mesh-dependence is not a problem in this toy problem but becomes quite cumbersome in higher dimensions, especially when the crack is also growing at the same time. So instead of relying on these displacements of the crack surfaces to describe the motion, we get rid of the mesh-dependence by modeling this rod using 1 XFEM element, with the two end points being nodes, each of which have two degrees of freedom, u_i^H and q_i^H, $i = 1, 2$. The displacement field is given by the enriched form in terms of material coordinates:

$$u(X, t) = \frac{1}{\ell_0}\begin{Bmatrix} X_2 - X \\ X - X_1 \\ (X_2 - X)H(X - X_c) \\ (X - X_1)H(X - X_c) \end{Bmatrix}^{\mathrm{T}} \begin{Bmatrix} u_1^H(t) \\ u_2^H(t) \\ q_1^H(t) \\ q_2^H(t) \end{Bmatrix} \tag{11.3.5}$$

The shape functions of the XFEM element enriched with the Heaviside function is outlined in Figure 11.7.

To get a handle on how these degrees of freedom also fully describe the same motion, we express the original displacements in terms of them.

$$\begin{aligned}
u_1 &= \frac{1}{\ell_0}\left[(X_2 - X_1)\cdot u_1^H + (X_1 - X_1)\cdot u_2^H + 0\cdot q_1^H + 0\cdot q_2^H\right] \\
u_c^- &= \frac{1}{\ell_0}\left[(X_2 - X_c)\cdot u_1^H + (X_c - X_1)\cdot u_2^H + 0\cdot q_1^H + 0\cdot q_2^H\right] \\
u_c^+ &= \frac{1}{\ell_0}\left[(X_2 - X_c)\cdot u_1^H + (X_c - X_1)\cdot u_2^H + (X_2 - X_c)\cdot q_1^H + (X_c - X_1)\cdot q_2^H\right] \\
u_2 &= \frac{1}{\ell_0}\left[(X_2 - X_2)\cdot u_1^H + (X_2 - X_1)\cdot u_2^H + (X_2 - X_2)\cdot q_1^H + (X_2 - X_1)\cdot q_2^H\right]
\end{aligned} \tag{11.3.6}$$

We can solve for the degrees of freedom with Heaviside enrichment in terms of the original ones:

$$\begin{aligned}
u_1^H &= u_1 \\
u_2^H &= \frac{(u_1 - u_c^-)X_2 + X_1 u_c^- - X_c u_1}{X_1 - X_c} \\
q_1^H &= \frac{(u_2 - u_c^+)X_1 - (u_1 - u_c^+)X_2 + (u_1 - u_2)X_c}{X_2 - X_c} \\
q_2^H &= \frac{(u_2 - u_c^-)X_1 - (u_1 - u_c^-)X_2 + (u_1 - u_2)X_c}{X_1 - X_c}
\end{aligned} \tag{11.3.7}$$

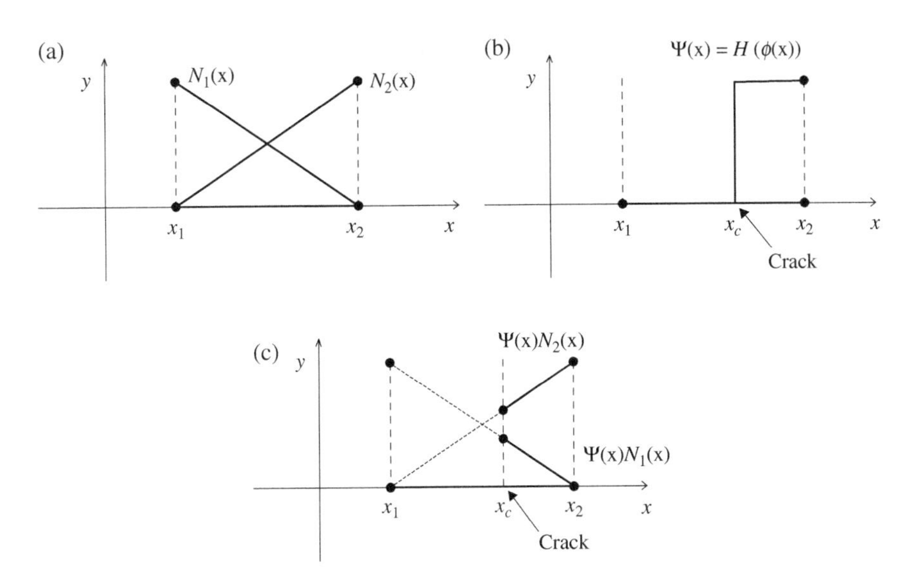

Figure 11.7 The shape functions of the XFEM element with Heaviside enrichment, where $\phi(X)=X-X_c$

It shows clearly that except for u_1^H, all the other displacements do not have an obvious kinematic interpretation.

Displacement field with shifted Heaviside enrichment Now we can do the same with the shifted Heaviside enrichment function

$$u(X,\,t)=\frac{1}{\ell_0}\begin{Bmatrix} X_2-X \\ X-X_1 \\ (X_2-X)[H(X-X_c)-H(X_1-X_c)] \\ (X-X_1)[H(X-X_c)-H(X_2-X_c)] \end{Bmatrix}^{\mathrm{T}}\begin{Bmatrix} u_1^S(t) \\ u_2^S(t) \\ q_1^S(t) \\ q_2^S(t) \end{Bmatrix} \tag{11.3.8}$$

where the superscript S means shifted Heaviside enrichment is used, and the new shape functions of the XFEM element are plotted in Figure 11.8. Similar to what we did in (11.3.6), we can solve for the degrees of freedom in terms of the original ones:

$$\begin{aligned} u_1^S &= u_1 \\ u_2^S &= u_2 \\ q_1^S &= \frac{\left(u_2-u_c^+\right)X_1-\left(u_1-u_c^+\right)X_2+(u_1-u_2)X_c}{X_2-X_c} \\ q_2^S &= \frac{\left(u_2-u_c^-\right)X_1-\left(u_1-u_c^-\right)X_2+(u_1-u_2)X_c}{X_1-X_2} \end{aligned} \tag{11.3.9}$$

We can see that with shifted enrichment, the conventional degrees of freedom retain their own kinematic meanings. Note q_1^S and q_1^H are the same, but q_2^S is scaled when compare to q_2^H.

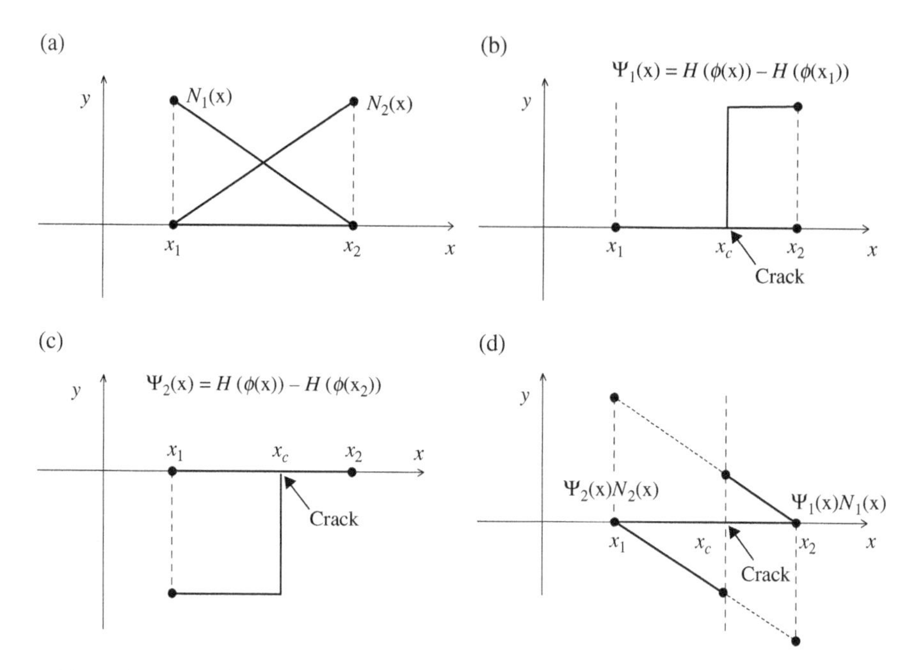

Figure 11.8 The shape functions of the XFEM element with shifted Heaviside enrichment, where $\phi(X) = X - X_c$

Comparison of displacement fields To further illustrate the differences caused by using different enrichment functions, we plot the displacement fields using *the same* numeric values for the degrees of freedom. Bear in mind in this case, we are plotting two different displacement fields, even though their degrees of freedom have identical numeric values. The values we used are listed here:

$$u_1 = 0, \quad u_2 = 1, \quad q_1 = 2, \quad q_2 = 1 \tag{11.3.10}$$

Note no superscripts are used, since they are used for both forms of enrichment functions. The displacement fields are shown in Figure 11.9 and Figure 11.10.

11.3.2 Weak Discontinuity

In order to model a weak discontinuity; $\Psi(X)$ need to be designed as a continuous function, with its derivatives being discontinuous at the interface. The following form of enrichment function has been proposed by Sukumar (2001):

$$\Psi(X) = \left| \sum_{I=1}^{n} \phi_I N_I(X) \right| = |\phi(X)| \tag{11.3.11}$$

where $\phi(X)$ is the signed distance from the weak discontinuity (e.g., material interface) (x_m) to the material point in the element similar to that in Equation (11.3.2). $\sum_{I=1}^{n} \phi_I N_I(X)$ is the discretized form of the signed distance function.

$$\phi_I = \phi(X_I) \tag{11.3.12}$$

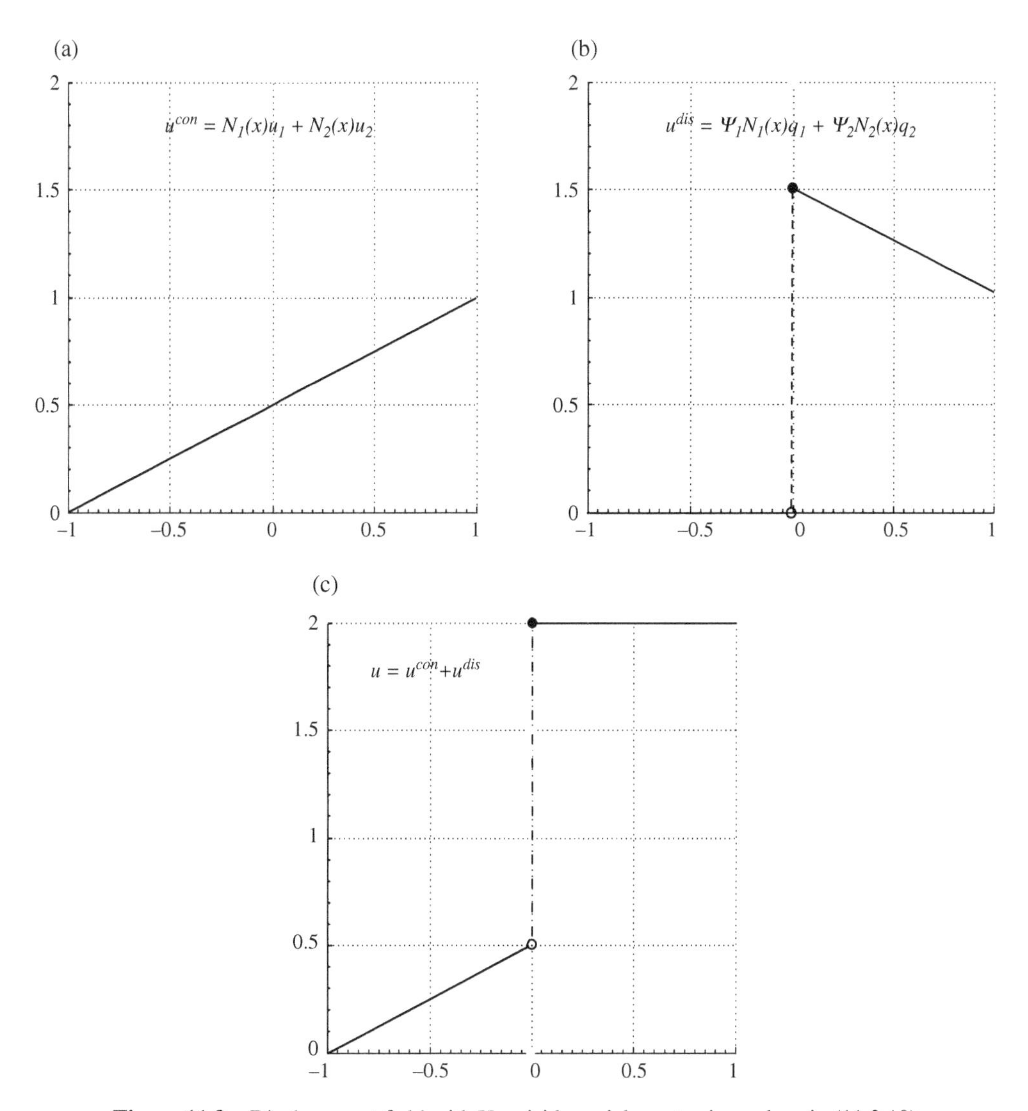

Figure 11.9 Displacement field with Heaviside enrichment using values in (11.3.10)

An alternative enrichment function to (11.3.11) for weak discontinuity can be defined as (Moës, 2003):

$$\Psi(X) = \sum_{i=1}^{n} |\phi_i| N_i(X) - \left| \sum_{i=1}^{n} \varphi_i N_i(X) \right| \tag{11.3.13}$$

Consider a one-dimensional element with material interface at $x = x_c$. We can use the notation x here as the assumption of small strains and displacements is made. The coordinates of the two nodes are x_1 and x_2. The enrichment function (11.3.11) is shown in Figure 11.11. The enrichment function (11.3.13) is shown in Figure 11.12.

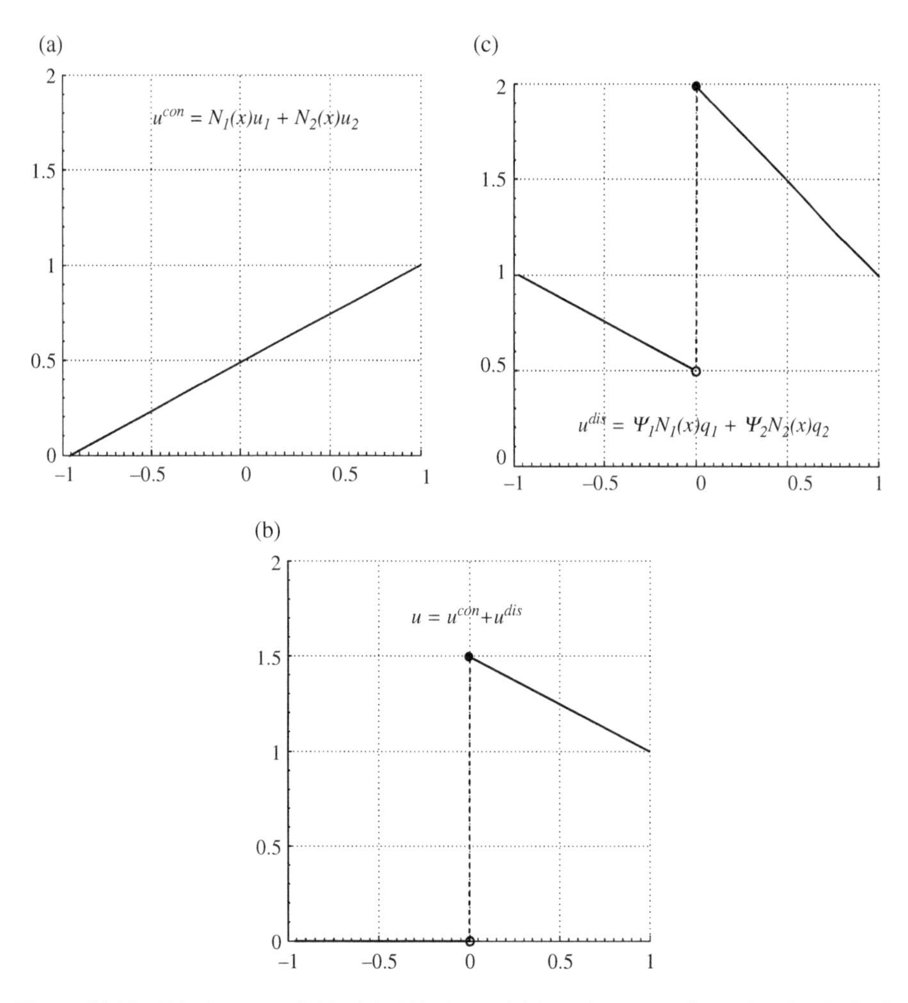

Figure 11.10 Displacement field with shifted Heaviside enrichment using values in (11.3.10)

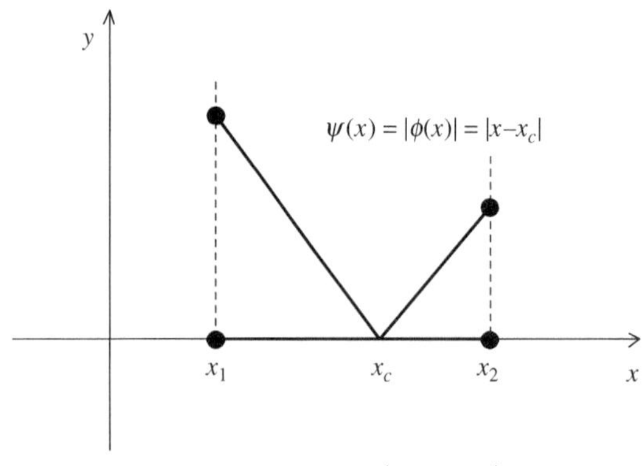

Figure 11.11 Enrichment function $\Psi(x) = \left|\sum_{I=1}^{n} \phi_I N_I(x)\right| = |\phi(x)|$ for weak discontinuity

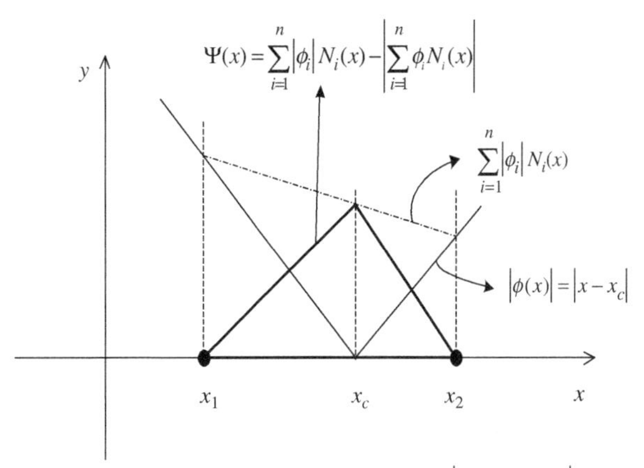

Figure 11.12 Enrichment function $\Psi(x)=\sum_{i=1}^{n}|\phi_i|N_i(x)-\left|\sum_{i=1}^{n}\phi_i N_i(x)\right|$ for weak discontinuity

The two forms of enrichment functions for weak discontinuities are given as (11.3.11) and (11.3.13). Both of these are continuous, but have discontinuous derivatives at the interface. A good exercise is to draw the actual interpolation functions $N_I(X)\Psi(X)$ for $I=1,2$. The major advantage of the second form over the first form of enrichment functions is that it is formulated in such a way that the traditional degrees of freedom u_i maintain their kinematic meanings, that is, displacement, at each node, because the function $\Psi(X)$ vanishes at the nodes, as seen in:

$$\Psi(x_1)=0, \quad \Psi(x_2)=0. \tag{11.3.14}$$

Then, for example, at node 1

$$\begin{aligned} u(x_1) &= \sum_{i=1}^{2}N_i(x_1)u_i+\sum_{j=1}^{2}N_j(x_1)\psi(x_1)q_j \\ &= N_1(x_1)u_1+N_2(x_1)u_2 \\ &= u_1 \end{aligned} \tag{11.3.15}$$

11.3.3 Mass Matrix

The consistent mass matrix for the enriched element can be written as:

$$\mathbf{M}=\begin{bmatrix}\mathbf{M}_{uu} & \mathbf{M}_{uq}\\ \mathbf{M}_{uq}^{T} & \mathbf{M}_{qq}\end{bmatrix} \tag{11.3.16}$$

where $\mathbf{M}_{uu}$ and $\mathbf{M}_{qq}$ are the mass matrices for the traditional degrees of freedom and additional degrees of freedom respectively. $\mathbf{M}_{uq}$ is the coupled term. For dynamic crack problems, explicit time integration is generally the preferred time integration method. In explicit methods a diagonalized mass matrix, that is, lumped mass matrix, is generally used. Multiple lumping techniques are available in literature (Menouillard, 2006, 2008). One option of mass lumping procedure, which keeps the simplicity of implementation, consists of two steps. Firstly, the coupled terms in mass matrix are neglected.

$$\mathbf{M}=\begin{bmatrix}\mathbf{M}_{uu} & 0\\ 0 & \mathbf{M}_{qq}\end{bmatrix}=\rho\int_{\Omega}\begin{bmatrix}N_1N_1 & N_1N_2 & 0 & 0\\ N_1N_2 & N_2N_2 & 0 & 0\\ 0 & 0 & N_1\Psi_1N_1\Psi_1 & N_1\Psi_1N_2\Psi_2\\ 0 & 0 & N_1\Psi_1N_2\Psi_2 & N_2\Psi_2N_2\Psi_2\end{bmatrix}d\Omega \quad (11.3.17)$$

Then the row-sum technique is performed on the remaining part of the mass matrix. It needs to be noted that shifted Heaviside enrichment function is assumed when writing the expression in (11.3.17). In cases where the enrichment function is the same at each node, for example, the weak discontinuities given in this chapter or non-shifted Heaviside enrichment, then in (11.3.17)

$$\Psi(x)=\Psi_1(x)=\Psi_2(x) \quad (11.3.18)$$

Thus, after the row summation procedure, the mass matrix for the additional degrees of freedom can be written as

$$\mathbf{M}_{qq}=\rho\int_{\Omega}\begin{bmatrix}N_1\Psi^2 & 0\\ 0 & N_2\Psi^2\end{bmatrix}d\Omega \quad (11.3.19)$$

11.4 Multi-Dimension XFEM

Now that we have covered a simple version of XFEM we will expand it to multiple dimensions. In multiple dimensions fracture becomes much more interesting with evolving crack tips or fronts, in 2D and 3D respectively. For example, Sukumar *et al.* (2000) and Duan *et al.* (2009) modeled dynamic fractures in 3D for several problems. Modeling of crack tip fields using XFEM requires the ability to model high gradients associated with stress concentrations or singular stress fields.

To deal with cases of high gradients and other similar phenomena in XFEM, multiple enrichments are added. For m enrichment terms, the approximation becomes

$$\mathbf{u}^h(\mathbf{x})=\sum_{\forall I}N_I(\mathbf{x})\mathbf{u}_I+\sum_{j=1}^{m}\sum_{i\in I_j}\varphi_I(\mathbf{x})[\Psi^j(\mathbf{x})-\Psi^j(\mathbf{x}_i)]\mathbf{q}_i^j \quad (11.4.1)$$

where I_j and Ψ^j are the nodal subsets of the enriched nodes and the corresponding enrichment functions, respectively. Note the subtraction of the nodal values in the bracketed term; this is called *shifting*, and what it does is recover the Kronecker-δ property for the enrichment terms (see Section 11.3 for example). Special care is needed to ensure that the resulting approximation space is continuous in displacement along element boundaries (Fries, 2008).

11.4.1 Crack Modeling

Consider a 2D finite element model of a cracked body, as shown in Figure 11.13. Let the set of all nodes in the finite element mesh be denoted by S, the set of nodes of elements around the crack tip (or crack front in three dimensions) be denoted by S_C and the set of nodes of elements cut by the crack, that is, the discontinuity, but not in S_C, be denoted by S_H. The set of

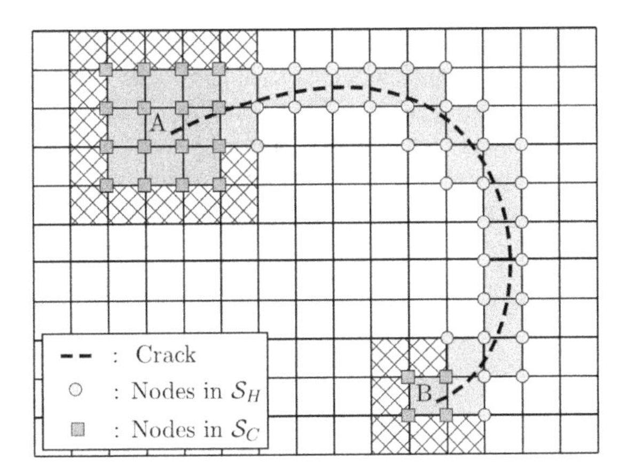

Figure 11.13 An arbitrary crack line (dashed line) in a structured mesh with step enriched and tip enriched elements. Nodes in sets S_C and S_H are denoted by squares and circles, respectively. Cross hatched elements are blended elements

elements with nodes in S_C can be selected by the user. Usually a single element, as shown at crack tip B in Figure 11.13, suffices, but some improvements in accuracy can be obtained by using several elements, as shown at crack tip A. Nodes in S_C and S_H will be referred to as tip enriched and step enriched nodes, respectively, and collectively as enriched nodes.

Let the crack surface be given by an implicit function description, that is, a level set, $\phi(\mathbf{x})=0$ and let $\phi(\mathbf{x})$ have opposite signs on the two sides of the crack. In most applications, $\varphi_I = N_I$ and we will use this here. Then, the XFEM displacement field in the vicinity of a crack is

$$\begin{aligned}\mathbf{u}^h(\mathbf{X}) &= \sum_{\forall I} N_I(\mathbf{X})\mathbf{u}_I + \sum_{J\in S_H} N_J(\mathbf{X})[H(\phi(\mathbf{X})) - H(\phi(\mathbf{X}_J))]\mathbf{q}_J^0 \\ &+ \sum_j \sum_{K\in S_C} N_K(\mathbf{X})[\Psi^{(j)}(\mathbf{X}) - \Psi^{(j)}(\mathbf{X}_K)]\mathbf{q}_K^{(j)}\end{aligned} \tag{11.4.2}$$

where H(·) is the Heaviside step function, and $\Psi^{(j)}$ is a set of enrichment functions which approximate the near-tip behavior, $\mathbf{q}_K^{(j)}$ are the enrichment coefficients which are additional unknowns at the nodes, and $\mathbf{X}_J$ is the position of node J.

It should be stressed that the previous approximation is a local partition of unity, that is, the enrichment is added only where it is useful. This substantially improves the computational efficiency because in general far fewer unknowns are introduced by the enrichment than in a global partition of unity. There are some difficulties in the blending elements around the tip enriched elements. The blending elements are indicated by cross-hatching in Figure 11.13 and can be accounted for by including the enriched displacement field inside of these elements, with the values of $\boldsymbol{q}_k$ set to 0 for nodes not in *Sc*.

Note how the discontinuity is introduced by this approximation. Since $\phi(\mathbf{X})$ defines the crack and since ϕ(X) changes sign across the crack, it follows that H($\phi(\mathbf{X})$) is discontinuous along the crack. We can furthermore show that for elements cut by the crack, the jump in the displacement field across the crack Γ_c is

$$\left[\!\left[\mathbf{u}^h(\mathbf{X})_{\Gamma_c^0}\right]\!\right] = \sum_{J\in S_H} N_J(\mathbf{X})\mathbf{q}_J^0, \qquad \mathbf{X}\in\Gamma_c^0 \tag{11.4.3}$$

Thus the magnitude of the crack-opening displacement directly depends on $\mathbf{q}_J^0$.

Note that the enrichment function is shifted so that the product of the shape function N_I and the enrichment function vanish at each node. Consequently, the step function enrichment for the discontinuity vanishes at the edges of all elements not crossed by the discontinuity and $u^h(\mathbf{x}_J)=\mathbf{u}_J$. Therefore, only those elements that are crossed by the discontinuity need to be treated differently. In addition, this shifting simplifies the blending of enriched elements with standard elements. It is sometimes stated that this shifting enables one to satisfy displacement boundary conditions, just as in the standard finite element method, by setting the nodal displacements appropriately, but this is not strictly true: the displacement boundary conditions will only be met at the nodes when this is done.

11.4.2 Tip Enrichment

The fracture process zone ahead of a crack can be modeled using nonlinear constitutive and damage relations. If the process zone is small with respect to characteristic dimensions of the specimen (or cracked body) and the material behavior is linearly elastic outside the process zone the conditions for Linear Elastic Fracture Mechanics (LEFM) may hold. In LEFM, the material is modeled as linearly elastic throughout and the crack tip stress and strain fields (sometimes called the asymptotic crack tip fields) take on the so-called square root singular form with respect to radial distance from the tip. Readers who are interested in knowing more about LEFM are encouraged to refer to books on the topic, for example, Hertzberg (2012). Here we give the crack tip enrichment functions $\Psi^{(j)}$ that are based on this asymptotic solution. Effective near-tip enrichment can be constructed from the following functions

$$\{\Psi_i\}_{i=1}^{4} = \sqrt{r}\{\sin(\theta/2), \cos(\theta/2), \sin(\theta/2)\sin(\theta), \cos(\theta/2)\sin(\theta)\} \tag{11.4.4}$$

where r and θ are a polar coordinate system with the origin at the crack tip and $\theta=0$ tangent to the crack tip. These functions are shown in Figure 11.14.

For ductile materials the square root form of (11.4.4) does not hold. For elastic–plastic fracture mechanics (EPFM) analytical solutions for the crack tip fields are not available in general, though some special cases have been solved.

Hutchinson (1968) and Rice and Rosengren (1968) developed the asymptotic fields for a power law hardening material near the crack tip. These crack tip fields are generally referred to as the HRR fields. The stresses as given by Rice and Rosengren are:

$$\sigma_{\theta\theta} = r^{-\frac{N}{1+N}} h(\theta) \tag{11.4.5}$$

$$\sigma_{r\theta} = \frac{1+N}{2+N} r^{-\frac{N}{1+N}} h'(\theta) \tag{11.4.6}$$

$$\sigma_{rr} = (1+N) r^{-\frac{N}{1+N}} \left[h(\theta) + \frac{1+N}{2+N} h''(\theta) \right] \tag{11.4.7}$$

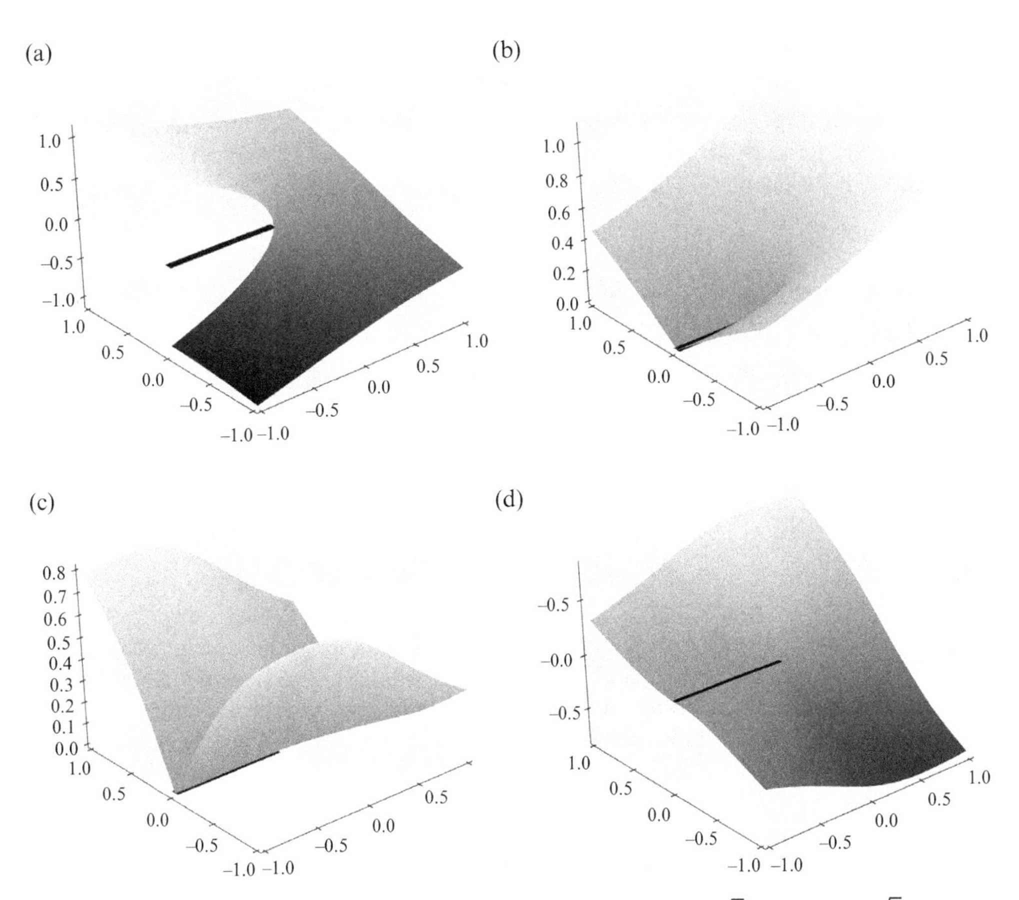

Figure 11.14 The often used tip enrichment functions plotted: (a)$\sqrt{r}\sin(\theta/2)$, (b)$\sqrt{r}\cos(\theta/2)$, (c)$\sqrt{r}\sin(\theta/2)\sin(\theta)$, and (d)$\sqrt{r}\cos(\theta/2)\sin(\theta)$. The polar coordinate system at the crack tip is also drawn

and the displacements are

$$u_r = r^{-\frac{N}{1+N}} g'(\theta) \tag{11.4.8}$$

$$u_\theta = \frac{1+2N}{1+N} r^{-\frac{N}{1+N}} g(\theta) \tag{11.4.9}$$

where N is the hardening exponent and $h(\theta)$ and $g(\theta)$ are non-independent and must depend on hardening law and the stress–strain relation.

Elguedj *et al.* (2006) adapted the HRR field to be modeled by XFEM. They determined through a Fourier analysis that the tip enrichment$\Psi^{(j)}$ can then be expressed as:

$$\{\Psi_i\}_{i=1}^{6} = r^{\frac{N}{1+N}} \left\{ \sin\left(\frac{\theta}{2}\right), \cos\left(\frac{\theta}{2}\right), \sin\left(\frac{\theta}{2}\right)\sin(\theta), \cos\left(\frac{\theta}{2}\right)\sin(\theta), \sin\left(\frac{\theta}{2}\right)\sin(3\theta), \cos\left(\frac{\theta}{2}\right)\sin(3\theta) \right\} \tag{11.4.10}$$

Although the first four terms share the trimetric part of the function, there are two additional terms and the radius term is raised to the $\frac{N}{1+N}$ power as opposed to the $\frac{1}{2}$ power. These tip enrichment functions greatly improve the accuracy of the XFEM tip enrichment for ductile materials. The surfaces for the six tip enrichments can be seen in Figure 11.15. We would recommend the reader consult literature on ductile fracture for more information about EPFM.

11.4.3 Enrichment in a Local Coordinate System

Sometimes, instead of writing the enrichment in component form with respect to the global axes, they are written in terms of a local coordinate system, for example, a system with axes tangent and normal to the discontinuity. For example, in 2D:

$$\mathbf{u}^h(\mathbf{X}) = \sum_{\forall I} N_I(\mathbf{X})\mathbf{u}_I + \sum_{i\in I^*} \varphi_I(\mathbf{X})\cdot\mathbf{t}\left[\Psi(\mathbf{X}) - \Psi(\mathbf{X}_i)\right]q_i^t + \sum_{i\in I^*} \varphi_I(\mathbf{X})\cdot\mathbf{n}\left[\Psi(\mathbf{X}) - \Psi(\mathbf{X}_i)\right]q_i^n \tag{11.4.11}$$

where $\mathbf{t}, \mathbf{n} \in \mathbb{R}^d$ and are the unit tangent vector along the interface and unit normal vector, respectively. Note the two sets of XFEM unknowns q_i^t and q_i^n allow the discontinuity to be different in the tangential and normal directions. In some cases, the special solution properties of a vector field only take place tangent to the interface. Then, an enrichment of only the tangential component of the displacement field is appropriate, for example, in sliding interfaces and dislocations. Then, the displacement equation becomes:

$$\mathbf{u}^h(\mathbf{X}) = \sum_{\forall I} N_I(\mathbf{X})\mathbf{u}_I + \sum_{i\in I^*} \varphi_I(\mathbf{X})\cdot\mathbf{t}[\Psi(\mathbf{X}) - \Psi(\mathbf{X}_i)]q_i^t \tag{11.4.12}$$

where $\mathbf{t} \in \mathbb{R}^d$ is the unit tangent vector along the interface. Note that in contrast to the situation where all components are discontinuous, only *one* dimension of additional unknowns (q_i^t) is needed. Similarly, in three dimensions, two tangent vectors $\mathbf{t}_1$ and $\mathbf{t}_2$ and two sets of XFEM unknowns q_i^{t1} and q_i^{t2} are required.

11.5 Weak and Strong Forms

We shall briefly discuss the strong and weak forms of XFEM, emphasizing the differences from Section 4.3.

We first define the trial and test function spaces as:

$$U = \{\mathbf{v}(\mathbf{X},t) \mid \mathbf{v}(\mathbf{X},t) \in C^0, \mathbf{v}(\mathbf{X},t) = \bar{\mathbf{v}}(t) \text{ on } \Gamma_v, \mathbf{v} \text{ discontinuous on crack surfaces}\} \tag{11.5.1}$$

$$U_0 = \{\delta\mathbf{v}(\mathbf{X}) \mid \mathbf{v}(\mathbf{X}) \in C^0, \delta\mathbf{v}(\mathbf{X}) = 0 \text{ on } \Gamma_v, \delta\mathbf{v} \text{ discontinuous on crack surfaces}\} \tag{11.5.2}$$

The strong form, or the generalized momentum balance is:

$$\frac{\partial \sigma_{ji}}{\partial x_j} + \rho b_i = \rho \ddot{u}_i \quad \text{in } \Omega \tag{11.5.3}$$

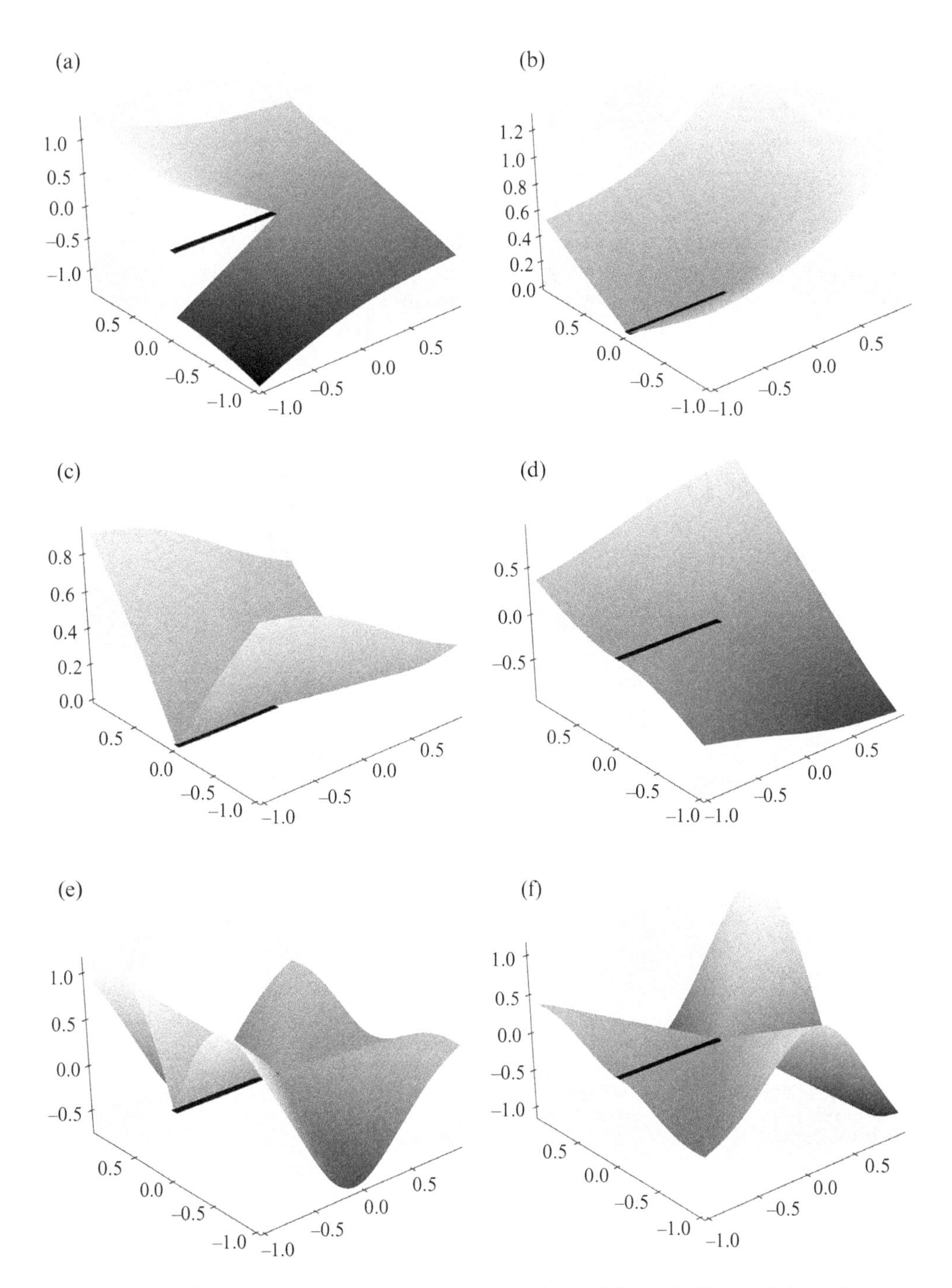

Figure 11.15 Tip enrichment displacement fields based on HRR field. The crack in all images is located where the discontinuity is seen in (a). The hardening exponent for all fields is $N = 39$. The fields are labeled (a–f) in the same order as shown in Equation (11.4.10)

$$n_j \sigma_{ji} = \bar{t}_i \quad \text{on} \quad \Gamma_{t_i} \tag{11.5.4}$$

$$n_j \sigma^+_{ji} = -n_j \sigma^-_{ji} = \tau_i \left([\![u_i]\!] \right) \quad \text{on} \quad \Gamma_c \tag{11.5.5}$$

$$[\![n_j \sigma_{ji}]\!] = 0 \quad \text{on} \quad \Gamma_w \tag{11.5.6}$$

where τ_i is the cohesive force on strong discontinuities Γ_c(e.g., cracks), and the interior traction continuity condition requires net zero traction on weak discontinuities Γ_w(e.g., material interfaces). Note the superscript plus and minus signs refer to the two sides of the strong discontinuity, the crack surface can be considered part of the boundary Γ, and can have prescribed forces and displacements applied to it. Similar procedures can be carried out on the strong form as we did in Section 4.3, and we arrive at the weak form:

$$\delta P^{kin} = \delta P^{ext} - \delta P^{\text{int}} - \delta P^{coh}, \quad \forall \delta \mathbf{v}(\mathbf{X}) \in U_0 \tag{11.5.7}$$

where $\boldsymbol{\delta P^{kin}}$ is the kinetic power associated with inertia, $\boldsymbol{\delta P^{ext}}$ is the external power associated with applied external loads, $\boldsymbol{\delta P^{int}}$ is the internal power, and $\boldsymbol{\delta P^{coh}}$ is the power associated with the cohesive traction on the crack surface $\Gamma_{\mathbf{c}}$. These terms are defined as

$$\delta P^{kin} = \int_\Omega \delta \mathbf{v} \cdot \rho \dot{\mathbf{v}} d\Omega \tag{11.5.8}$$

$$\delta P^{ext} = \int_\Omega \delta \mathbf{v} \cdot \rho \mathbf{b} d\Omega + \int_{\Gamma_t} \delta \mathbf{v} \cdot \bar{\mathbf{t}} d\Gamma_t \tag{11.5.9}$$

$$\delta P^{\text{int}} = \int_\Omega \frac{\partial \delta \mathbf{v}}{\partial \mathbf{x}} : \sigma d\Omega \tag{11.5.10}$$

$$\delta P^{coh} = -\int_{\Gamma_c} \delta [\![\mathbf{u}]\!] \cdot \tau^c d\Gamma_c \tag{11.5.11}$$

The readers are encouraged to carry out a detailed proof of the equivalence between the weak and strong forms.

11.6 Discrete Equations

The discrete XFEM equations are obtained by the substitution of enriched approximation (11.4.2) into the principle of virtual work. Assuming that the system is static and linear leads to the following system of discrete equations

$$\begin{bmatrix} \mathbf{K}^{uu} & \mathbf{K}^{u0} & \cdots & \mathbf{K}^{u4} \\ \mathbf{K}^{u0^{\mathrm{T}}} & \mathbf{K}^{00} & \cdots & \mathbf{K}^{04} \\ \vdots & \vdots & \ddots & \vdots \\ \mathbf{K}^{u4^{\mathrm{T}}} & \mathbf{K}^{14^{\mathrm{T}}} & \cdots & \mathbf{K}^{44} \end{bmatrix} \begin{Bmatrix} \mathbf{d}^u \\ \mathbf{q}^0 \\ \vdots \\ \mathbf{q}^4 \end{Bmatrix} = \begin{Bmatrix} \mathbf{f}^{\text{ext}} \\ \mathbf{Q}^0 \\ \vdots \\ \mathbf{Q}^4 \end{Bmatrix} \text{ or } \begin{bmatrix} \mathbf{K}^{uu} & \mathbf{K}^{uq} \\ \mathbf{K}^{uq^{\mathrm{T}}} & \mathbf{K}^{qq} \end{bmatrix} \begin{Bmatrix} \mathbf{d}^u \\ \mathbf{q} \end{Bmatrix} = \begin{Bmatrix} \mathbf{f}^{\text{ext}} \\ \mathbf{Q} \end{Bmatrix} \tag{11.6.1}$$

where the vector of standard finite element degrees of freedom is $\mathbf{d}^u = \{\mathbf{u}_1, \ldots, \mathbf{u}_n\}^T$ and the vectors of enriched degrees of freedom are $\mathbf{q}^0 = \{\mathbf{q}_1^0, \ldots, \mathbf{q}_{n_H}^0\}^T$ and $\mathbf{q}^i = \{\mathbf{q}_1^i, \ldots, \mathbf{q}_{n_C}^i\}^T$. The scalars n, n_H and n_C are the number of nodes in S, S_H and S_C, respectively. The stiffness matrices are given by

$$\mathbf{K}_{IJ}^{uu} = \int_\Omega \mathbf{B}_I \mathbf{C} \mathbf{B}_J d\Omega \tag{11.6.2}$$

$$\mathbf{K}_{IJ}^{uj} = \int_\Omega \mathbf{B}_I \mathbf{C} \mathcal{B}_J^{(j)} d\Omega, \qquad j \in \{0, 1, 2, 3, 4\} \tag{11.6.3}$$

$$\mathbf{K}_{IJ}^{ij} = \int_\Omega \mathcal{B}_J^{(i)} \mathbf{C} \mathcal{B}_J^{(j)} d\Omega, \qquad i, j \in \{0, 1, 2, 3, 4\} \tag{11.6.4}$$

where $\mathbf{C}$ is the elasticity matrix and $\mathbf{B}_I$ is the standard finite element strain–displacement matrix, which in 2D is

$$\mathbf{B}_I = \begin{bmatrix} N_{I,x} & 0 \\ 0 & N_{I,y} \\ N_{I,y} & N_{I,x} \end{bmatrix}, \qquad \forall I \tag{11.6.5}$$

where a comma denotes differentiation. The first part of Equation (11.6.1) can be written in a more compact form as the second part, where the definitions of the terms in the second part can be easily seen from comparing the two forms. We also note there is a direct correspondence between the layout of the sub-matrices in (11.6.1) and the decomposition of the displacement field into a standard part and enriched parts (11.4.2). The enriched strain–displacement matrices associated with the enriched part of the displacement approximation are

$$\mathcal{B}_I^0 = \begin{bmatrix} (N_I H(\phi(\mathbf{x})) - H(\phi(\mathbf{x}_I)))_{,x} & 0 \\ 0 & (N_I H(\phi(\mathbf{x})) - H(\phi(\mathbf{x}_I)))_{,y} \\ (N_I H(\phi(\mathbf{x})) - H(\phi(\mathbf{x}_I)))_{,y} & (N_I H(\phi(\mathbf{x})) - H(\phi(\mathbf{x}_I)))_{,x} \end{bmatrix}, \qquad \forall I \in S_H \tag{11.6.6}$$

And for $j \in \{1,2,3,4\}$

$$\mathcal{B}_I^j = \begin{bmatrix} (N_I \Psi_j(\phi(\mathbf{x})) - \Psi_j(\phi(\mathbf{x}_I)))_{,x} & 0 \\ 0 & (N_I \Psi_j(\phi(\mathbf{x})) - \Psi_j(\phi(\mathbf{x}_I)))_{,y} \\ (N_I \Psi_j(\phi(\mathbf{x})) - \Psi_j(\phi(\mathbf{x}_I)))_{,y} & (N_I \Psi_j(\phi(\mathbf{x})) - \Psi_j(\phi(\mathbf{x}_I)))_{,x} \end{bmatrix}, \qquad \forall I \in S_C \tag{11.6.7}$$

The Cauchy stress is then

$$\sigma = \mathbf{C}\left(\sum_{\forall I} \mathbf{B}_I \mathbf{u}_I + \sum_{J \in S_H} \mathcal{B}_J^0 \mathbf{q}_J^0 + \sum_{K \in S_C} \sum_{j=1}^{4} \mathcal{B}_K^j \mathbf{q}_K^j \right) \tag{11.6.8}$$

In the absence of body forces, the force vectors due to external loads are

$$\mathbf{f}_I^{\text{ext}} = \int_{\Gamma_t} N_I \bar{\mathbf{t}} d\Gamma, \qquad \forall I \tag{11.6.9}$$

$$\mathbf{Q}_I^0 = \int_{\Gamma_t} (N_I H(\phi(\mathbf{x})) - H(\phi(\mathbf{x}_I)))\bar{\mathbf{t}} d\Gamma, \qquad \forall I \in S_H \tag{11.6.10}$$

$$\mathbf{Q}_I^j = \int_{\Gamma_t} (N_I \Psi_j(\phi(\mathbf{x})) - \Psi_j(\phi(\mathbf{x}_I)))\bar{\mathbf{t}} d\Gamma, \qquad \forall I \in S_C \text{ and } j \in \{1,2,3,4\} \tag{11.6.11}$$

where $\bar{\mathbf{t}}$ are the applied tractions to the domain boundary Γ_t.

Two important characteristics of the discrete equations are worth noting:

1. The B-matrices associated with the enriched degrees of freedom are discontinuous within elements cut by the crack. Therefore the integrands of the stiffness matrices and the force vectors are discontinuous. Furthermore, the B-matrix is singular at the crack tip (crack front in three dimensions). So, standard Gauss quadrature is inadequate for evaluating these integrals.
2. The sets S_H and S_C are small subsets of the total number of nodes. Therefore, the integrals are only nonzero over a small number of elements.

Example 11.2 Derive B matrix for a 1D element with crack Consider a 1D bar element with a crack at position of x_c (refer to Figure 11.6). Derive the **B** matrix for:

1. Material point A at x_A, where $x_A < x_c$
2. Material point B at x_A, where $x_A > x_c$

For a one-dimensional linear element, shape functions take the form of

$$N_1 = \frac{x_2 - x}{x_2 - x_1}, N_2 = \frac{x - x_1}{x_2 - x_1} \tag{E11.2.1}$$

$$\frac{\partial N_1}{\partial x} = \frac{-1}{x_2 - x_1}, \qquad \frac{\partial N_2}{\partial x} = \frac{1}{x_2 - x_1} \tag{E11.2.2}$$

For a linear element, the strain is always constant. Signed distance (level set) values for the two nodes in this element are

$$\phi_1 = \phi(x_1) = x_1 - x_c < 0 \tag{E11.2.3}$$

$$\phi_2 = \phi(x_2) = x_2 - x_c > 0 \tag{E11.2.4}$$

At material points A and B, the level set function, and enrichment functions are

$$\begin{aligned} \phi(x_B) &= x_B - x_c > 0 & \phi(x_A) &= x_A - x_c < 0 \\ \Psi_1(x_B) &= H(\phi_B) - H(\phi_1) & \Psi_1(x_A) &= H(\phi_A) - H(\phi_1) \\ &= H(x_B - x_c) - H(x_1 - x_c) & &= H(x_A - x_c) - H(x_1 - x_c) \\ &= 1 & &= 0 \\ \Psi_2(x_B) &= H(\phi_B) - H(\phi_2) & \Psi_2(x_A) &= H(\phi_A) - H(\phi_2) \\ &= H(x_B - x_c) - H(x_2 - x_c) & &= H(x_A - x_c) - H(x_2 - x_c) \\ &= 0 & &= -1 \end{aligned} \tag{E11.2.5}$$

The **B** matrix for a 1D element is

$$\mathbf{B} = \left[\underbrace{\frac{\partial N_1}{\partial x}, \frac{\partial N_1}{\partial x}\Psi_1}_{\mathbf{B}_1}, \underbrace{\frac{\partial N_2}{\partial x}, \frac{\partial N_2}{\partial x}\Psi_2}_{\mathbf{B}_2} \right] \tag{E11.2.6}$$

For material point A

$$\mathbf{B}(x_A) = [-1,\ 0, 1, -1]\ \frac{1}{x_2 - x_1} \tag{E11.2.7}$$

For material point B

$$\mathbf{B}(x_B) = [-1,\ -1, 1, 0]\frac{1}{x_2 - x_1} \tag{E11.2.8}$$

11.6.1 Strain–Displacement Matrix for Weak Discontinuity

The strain–displacement matrix (assuming small deformations) is given by:

$$\frac{\partial \mathbf{u}}{\partial \mathbf{x}} = \sum_{I=1}^{n} \frac{\partial N_I(\mathbf{x})}{\partial \mathbf{x}} \mathbf{u}_I + \sum_{J=1}^{n} \frac{\partial (N_J(\mathbf{x})\Psi(\mathbf{x}))}{\partial \mathbf{x}} \mathbf{q}_J \tag{11.6.12}$$

The first part of the right-hand side of (11.6.12) is usually denoted as the $\mathbf{B}_I$ matrix.

In the XFEM implementation, the traditional degrees of freedom give rise to this standard form of the **B** matrix. For the additional degrees of freedom, the **B** matrix is defined as

$$\mathcal{B}_J = \begin{bmatrix} \frac{\partial N_J}{\partial x}\Psi + \frac{\partial \Psi}{\partial x} N_J & 0 \\ 0 & \frac{\partial N_J}{\partial y}\Psi + \frac{\partial \Psi}{\partial y} N_J \\ \frac{\partial N_J}{\partial y}\Psi + \frac{\partial \Psi}{\partial y} N_J & \frac{\partial N_J}{\partial x}\Psi + \frac{\partial \Psi}{\partial x} N_J \end{bmatrix}, \quad \forall J \tag{11.6.13}$$

For each node I of the enriched element, the final derivative of displacement field takes the form of

$$\mathbf{B}_I\mathbf{u} = \begin{bmatrix} \frac{\partial N_I}{\partial x} & 0 & \frac{\partial N_I}{\partial x}\Psi + \frac{\partial \Psi}{\partial x}N_I & 0 \\ 0 & \frac{\partial N_I}{\partial y} & 0 & \frac{\partial N_I}{\partial y}\Psi + \frac{\partial \Psi}{\partial y}N_I \\ \frac{\partial N_I}{\partial y} & \frac{\partial N_I}{\partial x} & \frac{\partial N_I}{\partial y}\Psi + \frac{\partial \Psi}{\partial y}N_I & \frac{\partial N_I}{\partial x}\Psi + \frac{\partial \Psi}{\partial x}N_I \end{bmatrix} \cdot \begin{bmatrix} u_{Ix} \\ u_{Iy} \\ q_{Ix} \\ q_{Iy} \end{bmatrix}, \quad \forall I \qquad (11.6.14)$$

where the derivative of the enrichment function for a weak discontinuity is

$$\begin{aligned} \frac{\partial \Psi(\mathbf{x})}{\partial \mathbf{x}} &= \frac{\partial\left(\sum_{I=1}^{n}|\phi_I|N_I(\mathbf{x}) - \left|\sum_{I=1}^{n}\phi_I N_I(\mathbf{x})\right|\right)}{\partial \mathbf{x}} \\ &= \sum_{I=1}^{n}|\phi_I|\frac{\partial N_I(\mathbf{x})}{\partial \mathbf{x}} - \frac{\partial|\phi|}{\partial \mathbf{x}} \end{aligned} \qquad (11.6.15)$$

Bear in mind that the dimensions of the **B** matrix have expanded due to the introduction of the additional degrees of freedom. However, the dimensions of the strain tensor defined as $\mathbf{B}_I\mathbf{u}$ and the stress tensor do not change.

Example 11.3 Derive the B matrix for a 1D element with weak discontinuity Consider a one-dimensional element with the material interface at position x_c. Derive the **B** matrix for material point x_A, where $x_A < x_c$.

At material point A, the level set function, enrichment function, as well as derivative of enrichment function can be written out:

$$\begin{aligned} \phi(x_A) &= x_A - x_c < 0 \\ \Psi(x_A) &= |\phi_1|N_1(x_A) + |\phi_2|N_2(x_A) - |\phi(x_A)| \\ &= |x_1 - x_c|\frac{x_2 - x_A}{x_2 - x_1} + |x_2 - x_c|\frac{x_A - x_1}{x_2 - x_1} + x_A - x_c \\ &= \frac{2(x_A - x_1)(x_2 - x_c)}{x_2 - x_1} \\ \frac{\partial \Psi(x_A)}{\partial x} &= |\phi_1|\frac{\partial N_1(x_A)}{\partial x} + |\phi_2|\frac{\partial N_2(x_A)}{\partial x} - \frac{\partial|\phi(x_A)|}{\partial x} \\ &= -(x_1 - x_c)\cdot\left(-\frac{1}{x_2 - x_1}\right) + (x_2 - x_c)\cdot\frac{1}{x_2 - x_1} + 1 \\ &= \frac{2x_2 - 2x_c}{x_2 - x_1} \end{aligned} \qquad (E11.3.1)$$

The **B** matrix for a 1D element is

$$\mathbf{B}=\left[\underbrace{\frac{\partial N_1}{\partial x},\ \frac{\partial N_1}{\partial x}\Psi+\frac{\partial \Psi}{\partial x}N_1}_{\mathbf{B}_1},\underbrace{\frac{\partial N_2}{\partial x},\ \frac{\partial N_2}{\partial x}\Psi+\frac{\partial \Psi}{\partial x}N_2}_{\mathbf{B}_2}\right] \tag{E11.3.2}$$

For material point A:

$$\begin{aligned}
\mathbf{B}_1(x_A)&=\left[\frac{-1}{x_2-x_1},\ \frac{-1}{x_2-x_1}\cdot\frac{2(x_A-x_1)(x_2-x_c)}{x_2-x_1}+\frac{2x_2-2x_c}{x_2-x_1}\frac{x_2-x_A}{x_2-x_1}\right]\\
&=\left[\frac{-1}{x_2-x_1},\ \frac{2(x_2-x_c)(-2x_A+x_1+x_2)}{(x_2-x_1)^2}\right]\\
\mathbf{B}_2(x_A)&=\left[\frac{1}{x_2-x_1},\ \frac{1}{x_2-x_1}\frac{2(x_A-x_1)(x_2-x_c)}{x_2-x_1}+\frac{2x_2-2x_c}{x_2-x_1}\frac{x_A-x_1}{x_2-x_1}\right]\\
&=\left[\frac{1}{x_2-x_1},\ \frac{4(x_2-x_c)(x_A-x_1)}{(x_2-x_1)^2}\right]\\
\mathbf{B}(x_A)&=[\mathbf{B}_1(x_A),\ \mathbf{B}_2(x_A)]
\end{aligned} \tag{E11.3.3}$$

Deriving the **B** matrix at material point B is left as an exercise for the reader. The implementation of internal nodal force computation is shown in Box 11.1.

Box 11.1 Internal nodal force computation for enriched element

1. $\mathbf{f}^{\text{int}}=0$.
2. Calculate value of sign distance function (ϕ, the level set) for each node.
3. For all quadrature points $\boldsymbol{\xi}_Q$
   i. Compute $[B_{iI}]=[\partial N_I(\boldsymbol{\xi}_Q)/\partial x_i]$ for the conventional degrees of freedom for all nodes I
   ii. Computer $[\mathcal{B}_{iI}]=[\Psi_I(\xi_Q)\partial N_I(\xi_Q)/\partial x_i]$ for the enriched degrees of freedom all nodes I
   iii. $\mathbf{H}=\mathbf{B}_I\mathbf{u}_I+\mathcal{B}_I\mathbf{q}_I$; $H_{ij}=\partial u_i/\partial x_j$
   iv. $\mathbf{F}=\mathbf{I}+\mathbf{H}$
   v. Compute Cauchy stress (σ) or PK2 stress ($\mathbf{S}$) based on constitutive equation.
   vi. $\mathbf{f}^{\text{int}}\leftarrow\mathbf{f}^{\text{int}}+[\mathbf{B}_I\ \ \mathcal{B}_I]^T\boldsymbol{\sigma}J_\xi\bar{w}_Q$

   end loop

Note: Usually in the enriched element, Gaussian quadrature is performed individually on the two subdomains separated by the crack to obtain better accuracy. Thus the coordinates of quadrature points are chosen according to the isoparametric mapping of each subdomain.

11.7 Level Set Method

The level set method (LSM) is a numerical technique for tracking moving interfaces and has proven to be a useful addition to XFEM. We will highlight some key aspects of level sets often used in XFEM implementations; however, this is not a complete guide to level sets and we would encourage the reader to look to external sources for a more complete treatment. Some example applications of XFEM combined with level set are to be found in Sukumar *et al.* (2001) and Bordas *et al.* (2006). The main idea is to represent an interface as a level set curve defined by a function $\phi(\mathbf{x}, t)$, which is one dimension higher than the dimension of the interface itself.

11.7.1 Level Set in 1D

To demonstrate a level set in 1D, we will consider a line segment of length 1, $x \in [0,1]$. Now we consider a point, which serves as an interface on this line segment, $\Gamma(t) \in (0,1)$. Note that we are also considering the dependence of the interface on time. One scenario, which can be described by this simple model, is a thermometer. The location where mercury touches the vacuum is the interface.

This interface can then be formulated as the level set curve by the function $\phi : [0, 1] \times \mathbb{R} \rightarrow \mathbb{R}$, where

$$\Gamma(t) = \{x \in (0, 1) : \phi(x, t) = 0\} \tag{11.7.1}$$

One such function (ϕ) is the signed distance function; in one dimension, it is simply described as:

$$\phi(x, t) = x - \Gamma(t) \tag{11.7.2}$$

11.7.2 Level Set in 2D

For a 2D level set we will consider a circular hole growing with time on a plane surface. The center of the hole is located at x_c. The radius, r(t), changes with time. The moving interface of the hole can then be formulated as the level set curve of a function $\phi : \mathbb{R}^2 \times \mathbb{R} \rightarrow \mathbb{R}$, where

$$\Gamma(t) = \{\mathbf{x} \in \mathbb{R}^2 : \phi(\mathbf{x}, t) = 0\} \tag{11.7.3}$$

The level set function can be chosen as the signed distance to the circumference of the hole

$$\phi(\mathbf{x}, t) = \|\mathbf{x} - \mathbf{x}_c\| - r(t) \tag{11.7.4}$$

As shown here, the geometry of an interface is represented by the zero level set curve, $\phi \equiv \phi(\mathbf{x}, t) = 0$. In essence, the physical description of the interface is converted to a functional representation. In joining LSM and XFEM, a discrete representation of ϕ is created at a fixed set of points $\mathbf{X}_I$ (nodes). These geometric degrees of freedom determine ϕ, and hence the location of the interface. We can associate each finite element node in a mesh with a geometric DOF for the level set function, and use standard shape functions to interpolate ϕ at any point in the domain

$$\phi(\mathbf{X}, t) = \sum_I N_I(\mathbf{X}) \phi_I(t) \tag{11.7.5}$$

where the summation is over all nodes of the element that contains **X**, $N_I(\mathbf{X})$ are standard finite element shape functions, and $\phi_I(t)$ are the nodal values of the level set function which can evolve with time. The interface interpolated in this way is exactly linear for a 3-node triangular element, bilinear (approximately linear) for a 4-node quadrilateral element, and trilinear for an 8 node hexahedral element.

As the level set can evolve with time giving the freedom to track a moving discontinuity. In order to update the level set in time a hyperbolic governing equation must be solved. The governing equation, can be expressed as

$$\frac{\partial \phi}{\partial u} + \mathbf{v}(\mathbf{x}) \cdot \nabla \phi = 0 \tag{11.7.6}$$

where **v** is the velocity of the level set field.

11.7.3 Dynamic Fracture Growth Using Level Set Updates

To describe a crack in 2D or 3D two level sets are used. First ϕ^c tracks the crack itself and ϕ^t tracks the position of the crack tip or crack front in 3D. The crack tip/front moves at velocity **v**, forming an angle θ_c with the tangent to the crack at the tip. These two kinematic quantities can be estimated using fracture mechanics based on the loadings, material properties and dynamic effects and so on, but these are beyond the scope of this chapter. What we do want to emphasize is that this speed is decomposed into the speeds of the two level sets, which are then updated using (11.7.6). This is depicted in Figure 11.16.

The level set method can greatly simplify and speed up the geometric computations in XFEM simulations. However, level sets are originally designed to track moving interfaces, to which crack surfaces do not really belong. So while ϕ^t and the corresponding crack tip evolve with time, ϕ^c on the cracked side of ϕ^t cannot change. However, ϕ^c can change on the uncracked side of ϕ^t (in reference configuration). Ventura *et al.* (2002) developed a geometric view of the level set representation, which better suits the irreversibility of crack surfaces, and it also eliminates the need to solve the evolution differential equation (11.7.6).

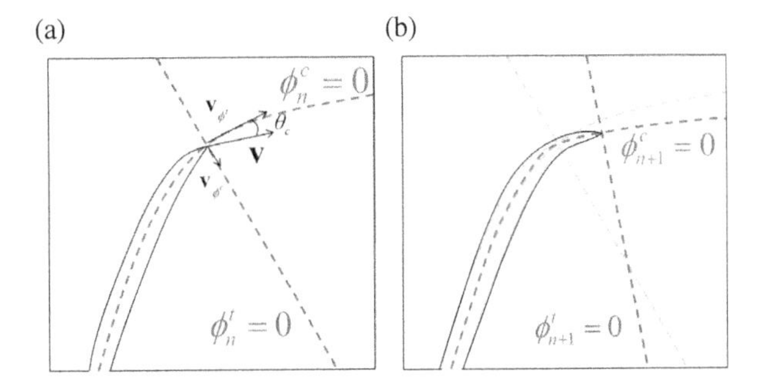

Figure 11.16 A growing crack in 2D at time t_n (a) and at time t_{n+1} (b), respectively

11.8 The Phantom Node Method

There are three major differences between a XFEM and a classical FE code: (1) The quadrature has to consider the special character of the enrichment, (2) the enrichment functions have to be implemented, and (3) the code must be able to deal with a variable number of degrees of freedom per node which holds on the element-level (the element matrices have different dimensions) as well as for the overall system matrix. Furthermore, for the visualization of results with inner-element discontinuities, adjustments in the post-processing tools may be advisable. In particular, the third aspect mentioned previously – the variable number of degrees of freedom per node – may lead to significant problems if XFEM is to be incorporated into an existing FE code and may require substantial background knowledge of the user. (This can be avoided by assigning the enrichment unknowns to additional nodes. For example, a 4-node bilinear quadrilateral with a discontinuous enrichment would be implemented as an 8-node bilinear quadrilateral with the extra four nodes storing the enrichment). Variable number of degrees of freedom per node may also be awkward with respect to parallelization of the code. However, for implementations which consider the needs of XFEM from the beginning, the computational effort scales exactly like a classical FE simulation.

Song *et al.* (2006) developed a XFEM formulation called phantom node method, which is particularly suited to explicit time integration and low-order elements, particularly the one-point quadrature elements. The formalism is similar to that of Hansbo and Hansbo (2004). In the phantom node method one discontinuous element is replaced by two elements with additional phantom nodes or phantom degrees of freedom, thus fewer modifications are needed in implementing this method in an existing finite element code.

11.8.1 Element Decomposition in 1D

The phantom node method only considers the Heaviside enrichment and hence strong discontinuities. We start with the standard XFEM description of the discontinuous field with shifting for 1D

$$\mathbf{u}(X,t)=\sum_{I=1}^{2}N_I(X)\{\mathbf{u}_I(t)+\mathbf{q}_I[H(X-a)-H(X_I-a)]\} \qquad (11.8.1)$$

We will now transform this to a superposed element formulation. Assume node 1 is to the left of the discontinuity. Writing out the equation, we have

$$u=u_1N_1+u_2N_2+q_1N_1H+q_2N_2(H-1) \qquad (11.8.2)$$

where $H_a=H(X-a)$. Using $N_1=N_1H_a+N_1(1-H_a)$ and $N_2=N_2H_a+N_2(1-H_a)$, we can rewrite the equation as

$$u=(u_1+q_1)N_1H_a+u_1N_1(1-H_a)+(u_2-q_2)N_2(1-H_a)+u_2N_2H_a \qquad (11.8.3)$$

Now, if we define

$$\text{element 1}\begin{cases} u_1^{(1)}\equiv u_1 \\ u_2^{(1)}\equiv u_2-q_2 \end{cases} \qquad \text{element 2}\begin{cases} u_1^{(2)}\equiv u_1+q_1 \\ u_2^{(2)}\equiv u_2 \end{cases} \qquad (11.8.4)$$

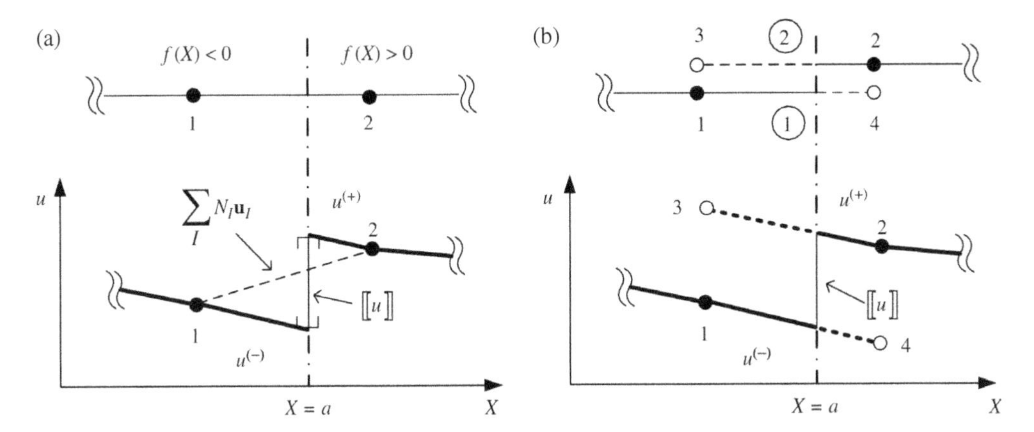

Figure 11.17 The representation of a discontinuity in a 1D model for: (a) standard XFEM; and (b) phantom node method; solid circles denote real nodes and hollow circles denote phantom nodes. Reproduced with permission from Song JH, Areias P, and Belytschko T (2006) A method for dynamic crack and shear band propagation with phantom nodes. *International Journal for Numerical Methods in Engineering*, **67**(6), 868–893. Copyright © 2006, John Wiley & Sons, Ltd

where superscripts and subscripts denote the element and node numbers, respectively. With these definitions, we get

$$u = u_1^{(1)} N_1 (1 - H_a) + u_2^{(1)} N_2 (1 - H_a) + u_1^{(2)} N_1 H_a + u_2^{(2)} N_2 H_a \tag{11.8.5}$$

In this way, we can consider the displacement field to be composed of the displacement fields of two elements: element 1, which is only active for $X<a$, because of the terms $(1-H_a)$ and element 2, which is only active for $X>a$ because of the term H_a. We can see that the discontinuous field can be constructed by adding an extra element, element 2 in this case, as shown in Figure 11.17. Then two phantom nodes are added: in this case they are $u_2^{(1)}$ and $u_1^{(2)}$. The two parts of the model are completely disjoint except for a cohesive law that relates the traction across the discontinuity to the jump in the displacement. The displacement jump across the crack is

$$\begin{aligned} [\![u_{X=a}]\!] &= \lim_{\varepsilon \to 0} [u(X+\varepsilon) - u(X-\varepsilon)]_{X=a} \\ &= (N_1(a) u_1^{(2)} + N_2(a) u_2^{(2)}) - (N_1(a) u_1^{(1)} + N_2(a) u_2^{(1)}) \\ &= N_1(a)(u_1^{(2)} - u_1^{(1)}) + N_2(a)\left(u_2^{(2)} - u_2^{(1)}\right) \\ &= q_1 N_1(a) + q_2 N_2(a) \end{aligned} \tag{11.8.6}$$

11.8.2 Element Decomposition in Multi-Dimensions

Now we can develop similarly the two-element displacement field form for a multi-node element in two or three dimensions that is completely cut by a crack. We start with the conventional XFEM displacement field

$$\mathbf{u}(\mathbf{X}, t) = \sum_{I=1}^{n^N} N_I(\mathbf{X}) \{ \mathbf{u}_I(t) + \mathbf{q}_I [H(\phi(\mathbf{X})) - H(\phi(\mathbf{X}_I))] \} \tag{11.8.7}$$

Expanding (11.8.7) as we did for the one-dimensional case by subdividing each term into parts that are associated with $\phi(\mathbf{X})<0$ and $\phi(\mathbf{X})>0$, we have

$$\mathbf{u}=\sum_{I=1}^{n^N}[\mathbf{u}_I N_I(1-H)+\mathbf{u}_I N_I H+\mathbf{q}_I N_I(H-H_I)] \tag{11.8.8}$$

where $H=H(\phi(\mathbf{X}))$. We now further expand both fields by duplicating them with the multipliers $H_I^-=H(-\phi(\mathbf{X}_I))$ and $H_I^+=H(\phi(\mathbf{X}_I))$, which does not change the fields and make use of the fact that

$$\begin{aligned} H-H_I &= H-1 \text{ when } H_I^+\neq 0 \\ H-H_I &= H \quad \text{ when } H_I^-\neq 0 \end{aligned} \tag{11.8.9}$$

$$\begin{aligned} \mathbf{u}=\sum_{I=1}^{n^N}[&\mathbf{u}_I N_I(1-H)H_I^+ +\mathbf{u}_I N_I(1-H)H_I^- +\mathbf{u}_I N_I H H_I^+ \\ &+\mathbf{u}_I N_I H H_I^- +\mathbf{q}_I N_I(1-H)H_I^+ +\mathbf{q}_I N_I H H_I^-] \end{aligned} \tag{11.8.10}$$

We then rewrite (11.87) as

$$\mathbf{u}=\sum_{I=1}^{n^N}\left[(\mathbf{u}_I-\mathbf{q}_I)N_I(1-H)H_I^+ +\mathbf{u}_I N_I(1-H)H_I^- +\mathbf{u}_I N_I H H_I^+ +(\mathbf{u}_I+\mathbf{q}_I)N_I H H_I^-\right] \tag{11.8.11}$$

If we then let

$$\begin{aligned} \mathbf{u}_I^1 &= \begin{cases} \mathbf{u}_I & \text{if } \phi(\mathbf{X}_I)<0 \\ \mathbf{u}_I-\mathbf{q}_I & \text{if } \phi(\mathbf{X}_I)>0 \end{cases} \\ \mathbf{u}_I^2 &= \begin{cases} \mathbf{u}_I+\mathbf{q}_I & \text{if } \phi(\mathbf{X}_I)<0 \\ \mathbf{u}_I & \text{if } \phi(\mathbf{X}_I)>0 \end{cases} \end{aligned} \tag{11.8.12}$$

we get the two-element form of the displacement field as

$$\mathbf{u}(\mathbf{X},t)=\sum_{I\in S_1}\underbrace{\mathbf{u}_I^1(t)N_I(\mathbf{X})}_{\mathbf{u}^1(\mathbf{X},t)}H(-\phi(\mathbf{X}))+\sum_{I\in S_2}\underbrace{\mathbf{u}_I^2(t)N_I(\mathbf{X})}_{\mathbf{u}^2(\mathbf{X},t)}H(\phi(\mathbf{X})) \tag{11.8.13}$$

Thus, the XFEM field for a completely cut element can be written as the sum of two element fields; one, $\mathbf{u}_1(\mathbf{X}, t)$, which holds for $\phi(\mathbf{X})<0$ and the other, $\mathbf{u}_2(\mathbf{X}, t)$, which holds for $\phi(\mathbf{X})>0$. This decomposition is shown in Figure 11.18. This form corresponds to the concept proposed by Hansbo and Hansbo (2004), though they did not present it in this form. Areias and Belytschko (2006) pointed out that the Hansbo and Hansbo formulation is another form of the XFEM displacement field.

Note that this equivalence holds for any element, that is, 3-node triangles, 8-node quadrilaterals, and so on. Recasting the discontinuous field in this form simplifies the implementation of the element in existing finite element codes. It is only necessary to add an extra element (i.e., element 2 in the cases here) and phantom nodes and modify the element quadrature. The phantom nodes are defined by

$$\text{Node I is a phantom node in}\begin{cases} \text{element 1 if } \phi(\mathbf{X}_I)>0 \\ \text{element 2 if } \phi(\mathbf{X}_I)<0 \end{cases} \tag{11.8.14}$$

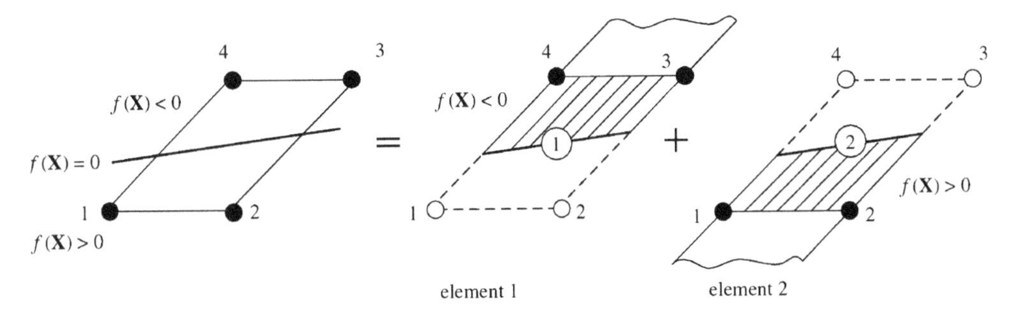

Figure 11.18 Decomposition of a cracked element into two elements; solid and hollow circles denote original nodes and added *phantom* nodes, respectively

11.9 Integration

The extension of the approximation space by enrichment functions $\Psi(\mathbf{x})$ with special properties, for example, discontinuities, complicates the integration of the weak form. Standard Gauss quadrature, which is frequently used in the classical FEM, requires a smooth integrand and a finite order polynomial. Classical FEM generally satisfy these requirements inside of each element. In the presence of jumps or kinks within elements, the accuracy of Gauss quadrature and other methods that similarly assume smoothness is drastically decreased. Another way to see the deficiency of Gauss quadrature is that the number of integration points required for exact summation increases linearly with the polynomial order of the integrand, and a discontinuity can only be expressed approximately by polynomials of finite order. Therefore, for discontinuous enrichments, special procedures are required for the accurate integration of the weak form.

11.9.1 Integration for Discontinuous Enrichments

A popular approach of circumnavigating the discontinuity is a decomposition of the elements into subelements (integration elements) that align with the discontinuity. This was proposed in the early works on XFEM. While the subelements are conforming, using them inside of XFEM still has significant advantages to remeshing. These subelements do not introduce new degrees of freedom, affect the critical time step, or require specific aspect ratios.

To show one example of how this decomposition is done, let us assume that the interface is described implicitly by a discretized level set function that is interpolated by classical shape functions, so that the zero level is given by

$$\phi^h(\mathbf{X}) = \sum_{i \in I} N_i(\mathbf{X})\phi_i = 0 \tag{11.9.1}$$

For simplicity, we exclude the situation where the level-set function is zero at a node, that is, $\phi_i \neq 0$ for all $i \in I$. The interface is, in general, *curved* in the current element geometry, as it results from finding the roots in the *reference* element and projecting these points into the current element geometry (e.g., by an isoparametric mapping). The interface is planar inside the elements only for linear interpolants (e.g., 3-node triangular and 4-node tetrahedral

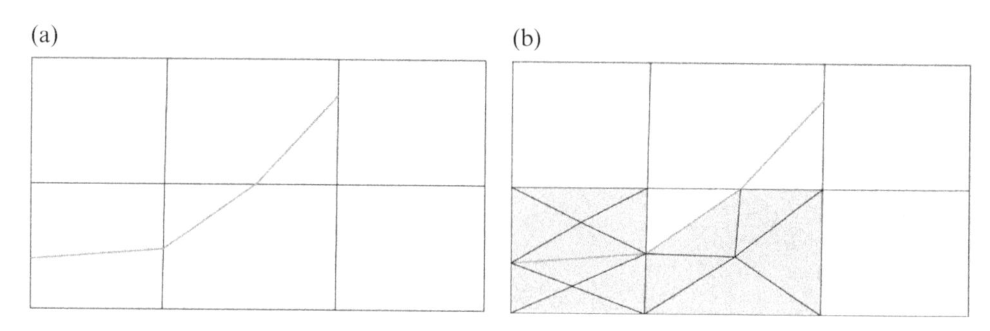

Figure 11.19 (a) polygons formed when elements cut by a crack (b) triangulate polygons to create subdomains (shaded) for accurate quadrature of the weak form (two cases depicted, i.e., triangulating quadrilaterals and pentagons)

elements). Planar interfaces enable decomposition of the element into polygons in two dimensions and polyhedra in three dimensions. After decomposing the polygons into quadrilaterals and triangles, Gauss quadrature can be employed enabling the exact integration of polynomials.

One issue with using subelements is that data must be projected to their integration points which can incur some error. Thus, the 'exact integration' property for the reference element is lost. Additionally when breaking a 4-node quadrilateral element into subelements of a linear or bilinear nature the correct mapping of the cut surface is not possible as it is bilinear and subelements are linear along their edges. For quadrilateral elements, it can be justified to replace the curved interface by a straight line, which is determined from the intersections of the interface with the element edges. It is often preferable to decompose cut quadrilateral reference elements into two triangles and use linear interpolation in each triangle to determine the interface. The interface is then always piecewise straight and polygonal subelements for integration purposes are easily obtained. An example is shown in Figure 11.19.

An alternative approach proposed by Ventura *et al.* (2005) for the quadrature of integrands with jumps or kinks inside elements does not require the decomposition of elements. Suppose that the integrand consists of an arbitrary polynomial $\mathcal{P}$ multiplied by a (strongly or weakly) discontinuous function $\mathcal{D}$,

$$\int_{\Omega_e} \mathcal{P} \cdot \mathcal{D} \, d\Omega \tag{11.9.2}$$

where Ω_e is the element domain. Then, the idea is to replace the discontinuous function $\mathcal{D}$ by a polynomial $\tilde{\mathcal{D}}$ such that the result is exactly the same as that obtained by the decomposition approach described previously. The function $\tilde{\mathcal{D}}$ is called the *equivalent polynomial*. The order of the integrand is increased; however, the important advantage is that standard Gauss quadrature over the entire element can be deployed. A drawback of this approach is that the resulting definitions of the equivalent polynomials depend on the enrichment and element type. Furthermore, for a given weak form, there may be different polynomials $\mathcal{P}$ that are multiplied by the discontinuous term $\mathcal{D}$. Then, for each case an individual equivalent polynomial $\tilde{\mathcal{D}}$ needs to be determined.

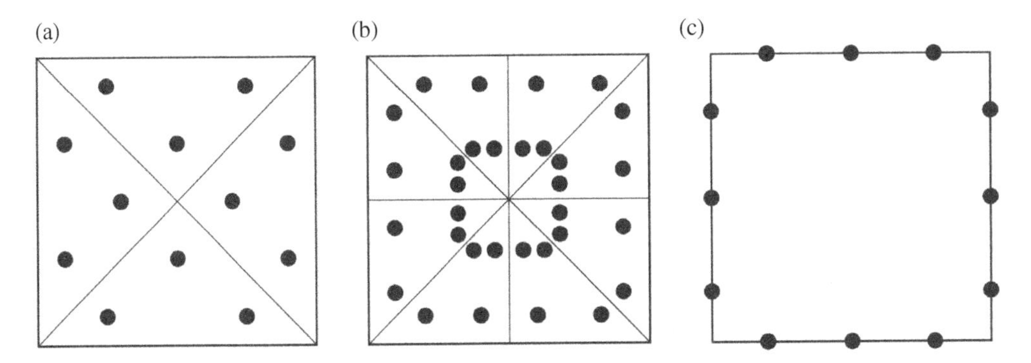

Figure 11.20 Comparison of different quadrature schemes (Ventura *et al.*, 2009): (a) subdomain integration (b) The nearly polar integration method with eight subdomains and a mapped second-order biunit square quadrature rule (c) The boundary integration method with a fifth-order-exact quadrature rule along each edge. Reproduced with permission from Ventura G, Gracie R, and Belytschko T (2009) Fast integration and weight function blending in the extended finite element method. *International Journal for Numerical Methods in Engineering*, **77**(1), 1–29. Copyright © 2009, John Wiley & Sons, Ltd

11.9.2 Integration for Singular Enrichments

Special quadrature rules are also recommended for enrichment functions that are singular within elements. We do not consider the situation where the singularity coincides with an element node because this has also been studied within the classical FEM framework. We also note that the integration of singular integrands is standard in the boundary element method.

Due to high gradients near the singularity, a concentration of integration points in the vicinity of the singularity improves the results significantly. This can be achieved by using polar integration approaches as described in Laborde *et al.* (2005) and Béchet *et al.* (2005) for two dimensions. This is shown in Figure 11.20(b). Laborde *et al.* have shown that this approach eliminates the singular term from the quadrature. The idea is to decompose the element containing the crack tip into triangles, so that each triangle has one node at the singularity ('singularity node'). Tensor-product type Gauss points in a quadrilateral reference element are mapped into each triangle such that two nodes of the quadrilateral coincide at the singularity node of each triangle. This approach is well suited for point singularities; however, the extension to three dimensions where a singularity may be present along a front is not straightforward.

A transformation of the integration domain containing the singularity into a contour integral is proposed by Ventura *et al.* (2009), for crack and dislocation enrichments in the framework of linear elasticity. This is shown in Figure 11.20(c). Contour integrals are evaluated with much less computational effort than domain integrals. These methods can also be used to reduce domain integrals to surface integrals in three-dimensional problems.

11.10 An Example of XFEM Simulation

These simulations concern an experiment reported by Kalthoff and Winkler (1988) in which a maraging steel 18Ni1900 plate with two initial edge notches is impacted by a projectile as shown in Figure 11.21. The material properties are $\rho = 8000\,\text{kg/m}^3$, $E = 190\,\text{GPa}$ and $\nu = 0.30$.

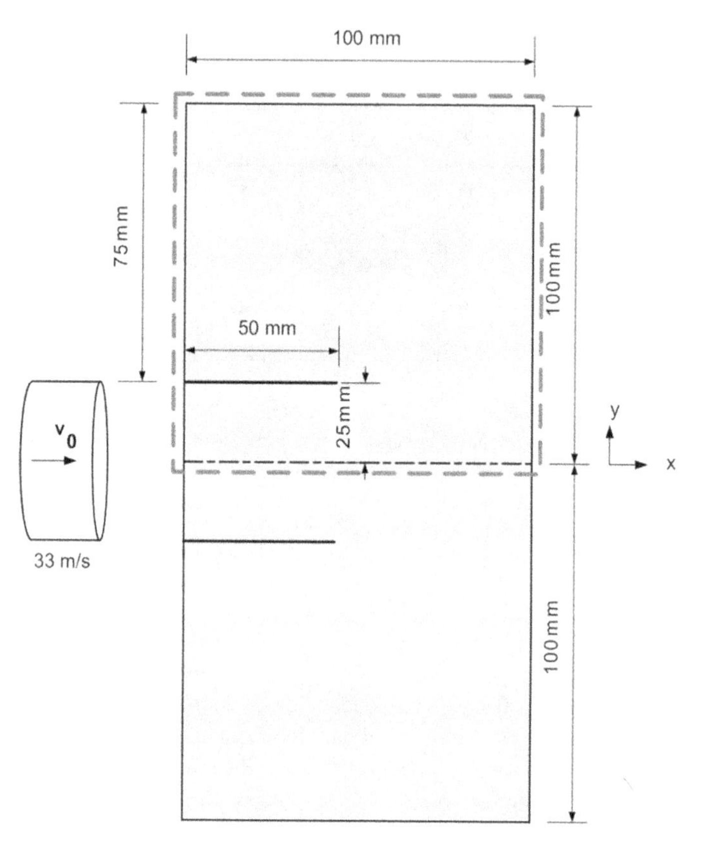

Figure 11.21 Setup for Kalthoff's experiments. The dashed box represents the half symetry examined numerically

In the experiment, two different failure modes are observed by modifying the projectile speed v_0. At high impact velocities, a shear band is observed to emanate from the notch at an angle of $-10°$ with respect to initial notch; at lower strain rates, brittle failure with a crack propagation angle of about 70° is observed. We focus on the brittle failure mode.

Horizontal symmetry is used in modeling the problem. A symmetry condition is applied on the bottom edge of the model ($u_y = 0$), and a step velocity on the cracked edge for $0 \le y \le 25\,\text{mm}$; the other edges are traction-free. We assume that the projectile has the same elastic impedance as the plate section that is impacted, so the impact velocity is approximately one-half of the projectile speed (Lee and Freund, 1990). The impact velocity is chosen to be $v = 16.5\,\text{m/s}$. In this example, the loss of hyperbolicity criterion (Belytschko, 2003) is used to advance the crack. The hyperbolicity condition means that the equation of motion keeps the form of a hyperbolic partial differential equation, which can be viewed as a condition for material stability. Therefore the violation of this condition can serve as the fracture criterion for crack propagation. This condition is identical to the condition of ellipticity of the equations of equilibrium.

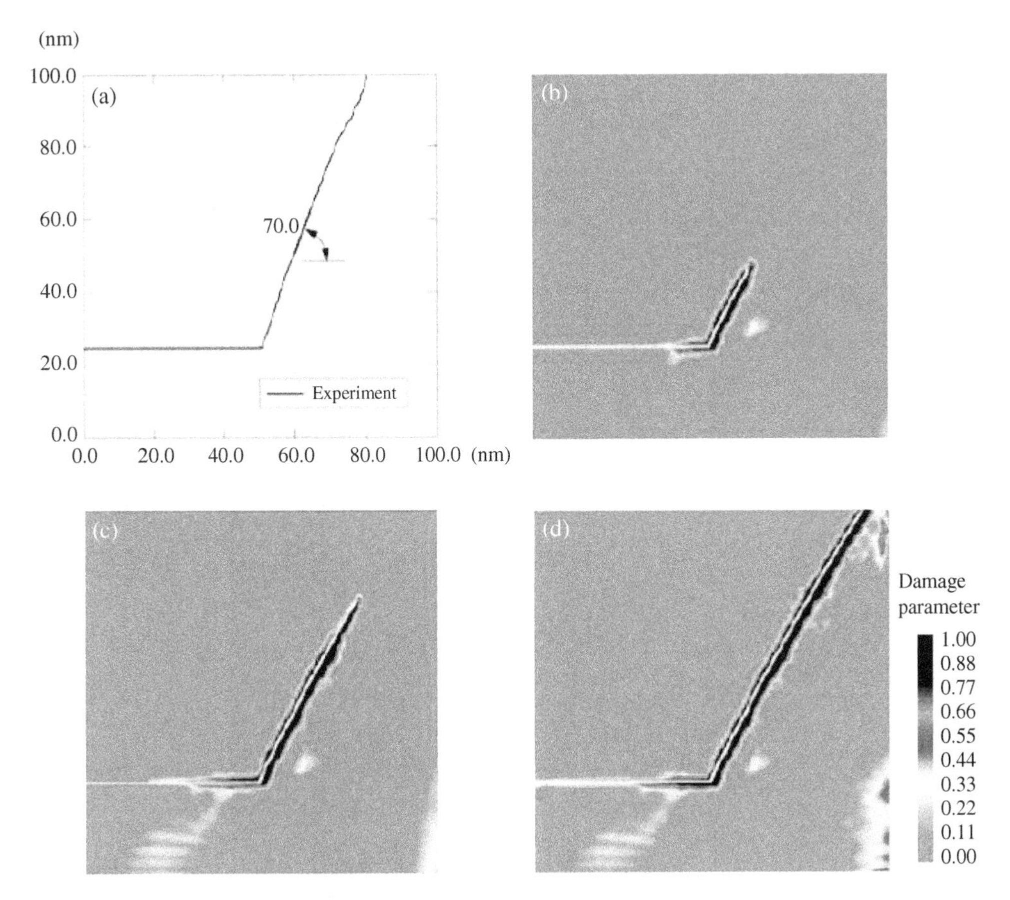

Figure 11.22 Comparison between XFEM results and experiments: (a) the experimental crack path reported by Kalthoff and Winkler (1988); XFEM crack path using loss of hyperbolicity criterion with a 100×100 quadrilateral mesh at: (b) t=42.64 μs; (c) t=53.58 μs; (d) t=88.58 μs. Reproduced from Song JH, Wang H and Belytschko T (2008). A comparative study on finite element methods for dynamic fracture. *Computational Mechanics*, **42**(2), 239–250. Copyright © 2008, Springer

The remaining details are summarized for the sake of completeness. The cohesive energy is taken to be mode I energy release rate $G_f = 2.213 \times 10^4\,\text{J/m}^2$, the critical crack opening displacement $\delta_{max} = 5.378 \times 10^{-5}\,\text{m}$. Lemaitre's damage model (Lemaitre, 1971) is used with the following damage parameters: A=1.0 B=200 and the damage threshold $\varepsilon_{D_0} = 3 \times 10^{-3}$. A 100×100 structured mesh is used.

The contour plot in Figure 11.22 represents the field of the damage parameters. A comparison of final crack path for the experiment and different numerical schemes is given in Figure 11.23. The XFEM crack propagation results show an almost straight crack path. The overall crack angle is about 58°, which agrees reasonably well with the experimental value of 70°.

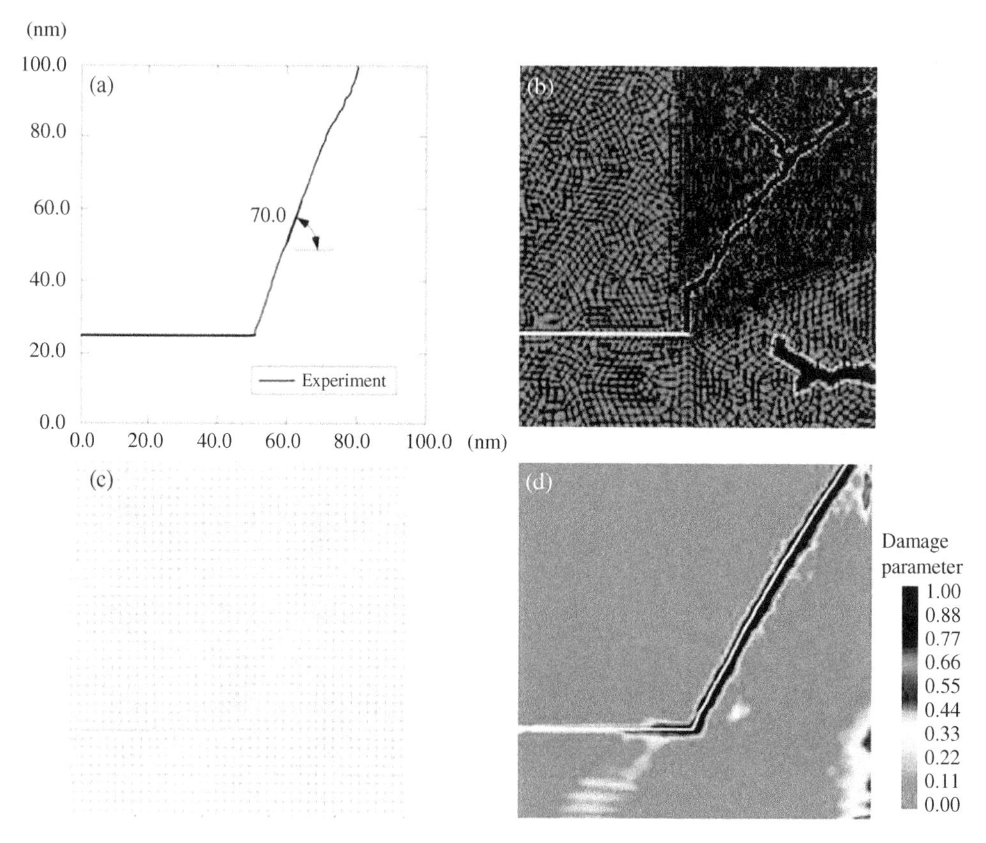

Figure 11.23 Comparison between XFEM results and other numerical methods on Kalthoff's experiments (a) Experimental results (Kalthoff and Winkler, 1988) (b) Element deletion method with unstructured mesh (Song, 2008) (c) Needleman's inter-element method with 40 × 40 structured cross triangular mesh (Song, 2008) (d) XFEM with 100 × 100 quadrilateral mesh (Song, 2008). Reproduced from Song JH, Wang H and Belytschko T (2008) A comparative study on finite element methods for dynamic fracture. *Computational Mechanics*, **42**(2), 239–250. Copyright © 2008, Springer

11.11 Exercise

11.1. Consider a 1D bar which has a total length of 20 mm with its left end at −10 mm and right end at 10 mm and a discontinuity surface at $x_c = 0$ mm. The bar is stretched by an external load (P = 1 MPa) on the left end and fixed on the right. Assuming that the bar is made of linear elastic material with material properties E = 200 GPa, $\rho = 7.83\,\text{g/cm}^3$, $\nu = 0.3$.

Write a 1D code to solve this problem using explicit formulation with standard Verlet time integrator. Implement XFEM to model the crack surface. To simplify the solution, no cohesive force on the crack surface needs to be considered. One of two ways of implementing XFEM can be used: the original XFEM or the phantom node method. The hints that follow are given for implementing the original XFEM.

Hints:

(a) Calculate the level set values (signed distance function to the crack surface) for each node. Note that only one element cut by the crack will be enriched in this case. It will have both positive and negative nodal level set values.

(b) Derive the **B** matrix for the enriched element and use the shifted enrichment function introduced in this chapter for strong discontinuity. Keep in mind that the dimension of the **B** matrix has changed, since it now has twice the degrees of freedom as before. However, the dimension of the stress shouldn't change. Keep the standard form for the **B** matrix for all the other unenriched elements.

(c) Integrate $\mathbf{B}^{T}\boldsymbol{\sigma}$ over the enriched element carefully to get the internal force. Since **B** is now a discontinuous function over the enriched domain, using the same number of gauss integration points as the unenriched elements will give poor results. There are two ways around: (i) Use a lot of integration points. (ii) Integrate the two parts of the enriched element formed by the crack separately.

12

Introduction to Multiresolution Theory

12.1 Motivation: Materials are Structured Continua

Most applications in science and engineering are macroscopic. The macroscale mechanical behavior of materials is therefore of primary interest. At the macroscale, the appearance of materials is continuous, a fact that has been used to justify the direct application of the theories of continuum mechanics to model their engineering behavior. Obviously, different materials *exhibit* different engineering properties, a fact accounted for in continuum mechanics via the development of a plurality of constitutive theories that aim to match subsets of materials to respectively observed properties, such as yield strength increase, ductility to failure, strain-rate sensitivity, or susceptibility to thermal expansion. However, questions as to why materials behave differently, or even why materials differ amongst themselves in the first place, receive only little attention in the classical constitutive theories of continuum mechanics, which are usually satisfied with a mere *description* of macroscopically observed phenomena. Such constitutive theories are thus called phenomenological constitutive theories.

Materials scientists, on the other hand, have long known that macroscale material performance is, to a great extent, determined by *microstructure*, that is, those structural features revealed by the various techniques of microscopy, which compose the material and often span multiple length scales (see Polmear, 2006).

It was soon recognized that microstructure can be controlled through material synthesis and processing parameters, that is, addition of alloying elements, raising working temperature, controlling induced plastic strain (e.g., area reduction per rolling pass) and strain rate (e.g., production rate). The general aim, to which the theories of mechanics can only contribute in part, is to create

Nonlinear Finite Elements for Continua and Structures, Second Edition.
Ted Belytschko, Wing Kam Liu, Brian Moran, and Khalil I. Elkhodary.

Companion Website: www.wiley.com/go/belytschko

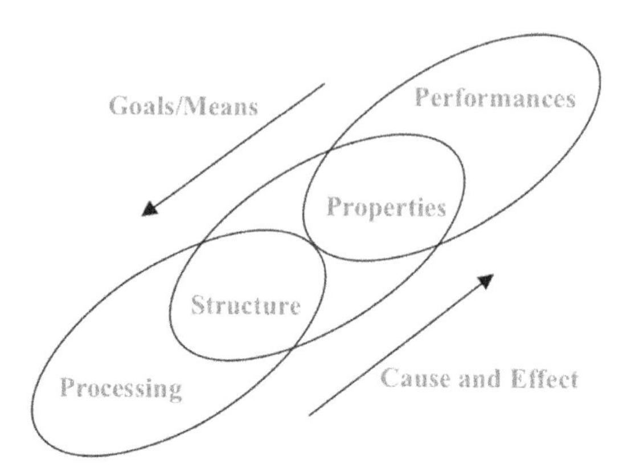

Figure 12.1 Materials structure, properties, and relationships. Redrawn from Olson, G.B., Computational design of hierarchically structured materials. *Science*, 1997

and optimize microstructures which exhibit properties (e.g., strength, toughness) that meet specific performance and design criteria, reliably and at minimal cost. The general procedure to achieve this aim may be understood as follows, as summarized in Figure 12.1 (Olson, 1997): (1) linking microstructure to processing parameters at each stage of processing, (2) developing models that relate material properties to their structure, and (3) predicting as a function of structure the performance of the material, then testing experimentally to verify the predictions. If performance criteria are not met, new microstructures are then synthesized and processed in a manner thought to yield improvement in performance, usually based on experiments and experience.

Consider, for example, metals, as represented by the schematic in Figure 12.2, which often exhibit a multiplicity of microstructural features. Upon mild magnification, the metal reveals a structure that in two dimensions looks like a jigsaw puzzle, each piece of which is called a grain or crystal. Grains are volumes of material whose atoms are arranged according to a single crystalline orientation, formed during solidification, and separated by grain boundaries, which are the surfaces that delineate orientation mismatches between neighboring grains. The orientation of a grain is defined during solidification by the ordering and consolidation of atoms along preferred directions that act as a basis for an atomic lattice that spans the grain. A lattice, and its orientation, can only be revealed by advanced microscopy, which requires rather high magnification. The different ways in which atoms can order themselves in a lattice may be revealed by yet higher magnification, to classify what are known as crystal structures (Kelly, Groves *et al.*, 2000; De Graef and McHenry, 2007). This *nesting* of material structures, as revealed by repeated magnification (e.g., specimen, grain, atomic lattice, crystal structure), emphasizes the 'multi-level' nature of engineering materials (e.g., Xiao and Belytschko, 2004; Liu, Qian *et al.*, 2010).

Continuing the example of metals, tailoring their chemistry by the addition of alloying elements, even in trace amounts, can substantially affect the energetics of the material system during solidification, favoring the segregation of heterogeneous clusters of atoms or the interspersion of secondary crystals across the grains (Polmear, 2006). Consider Figure 12.3. Suppose that at a high temperature, element B is added to element A, such that it contributes 5% of the alloy's mass, and that the molten mixture AB is cooled from liquid state to room temperature. It is first observed that the solidification of crystals of a single phase (called γ) occurs, which is a uniform solid solution of B in A. Upon further cooling, the solubility of B in A decreases, since solubility depends on temperature, so that some of

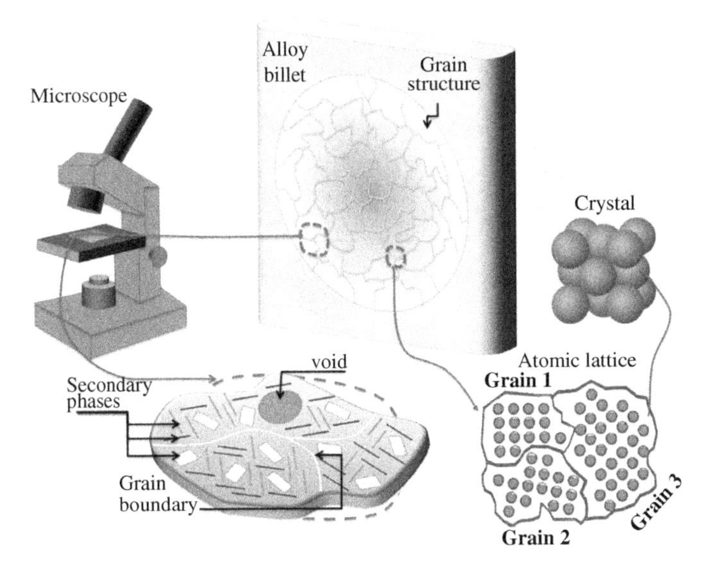

Figure 12.2 Complexity of material structure revealed upon successive magnification

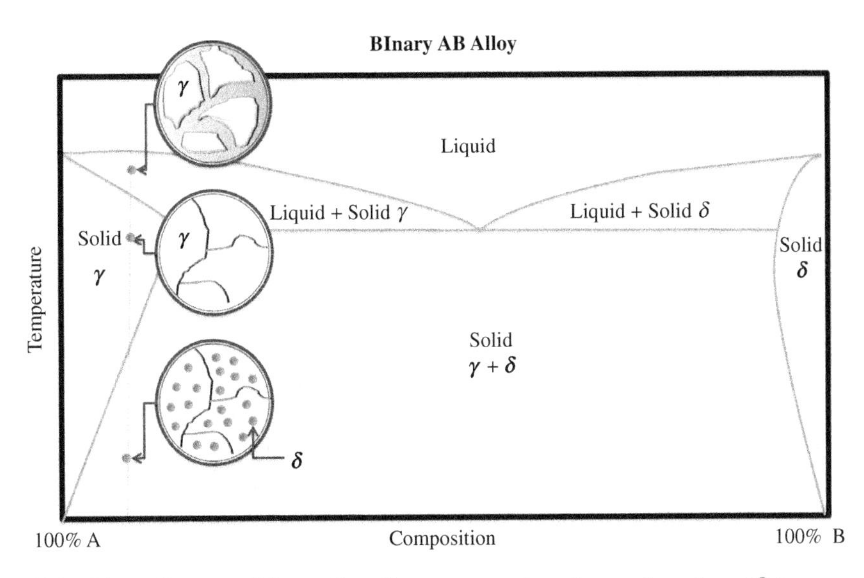

Figure 12.3 Phase diagram of binary alloy. Shows segregation of secondary phase (δ) by precipitation

B must segregate from phase γ to form a secondary phase δ, which may or may not be crystalline in structure.

In general, the ratio and configuration of the phases composing an alloy may be controlled by variation of alloy composition, as well as processing temperature. An experimental determination of diagrams such as Figure 12.3, called *phase diagrams*, can guide metallurgists to

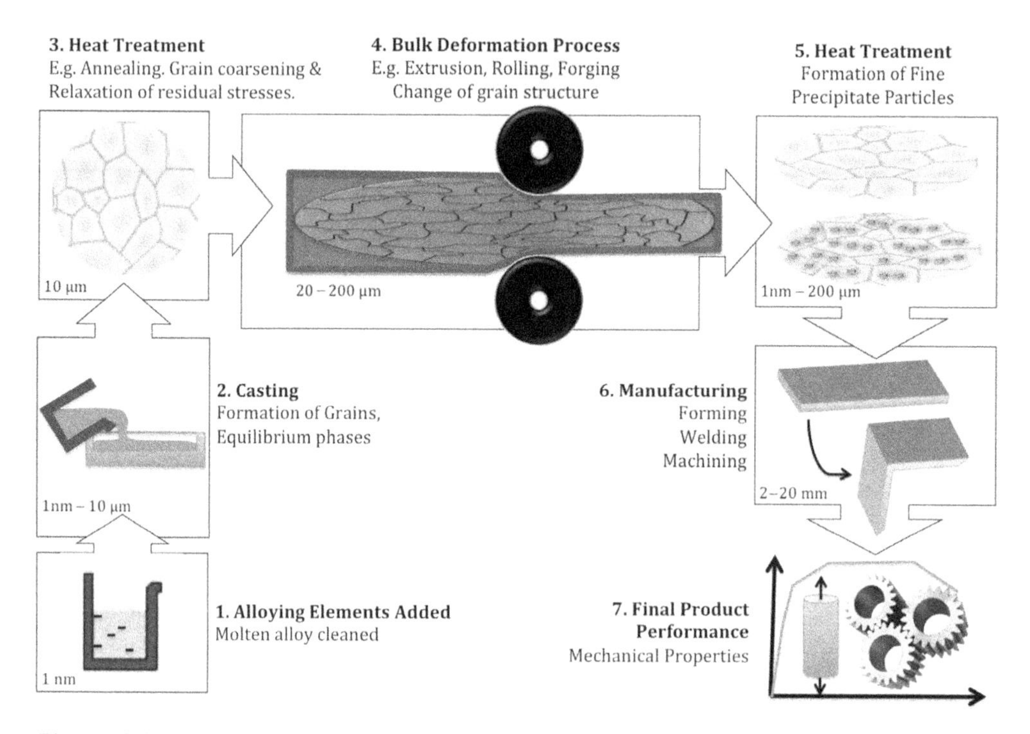

Figure 12.4 Processing of alloy to control its microstructure and ultimately its mechanical properties

the synthesis of new alloys, that is, alloys with modified microstructures. Further, controlling subsequent steps of production of an alloy, as summarized in Figure 12.4, can induce different alloy properties, for the same composition. That is, activating different reactions by application of thermomechanical treatment will alter the resultant energetically favorable microstructure.

For example, an alloy is first formed by mixing elements in their molten state and then cast into ingots, to solidify and lock in the grain structure. Subsequent heating of the solid (annealing) permits grain coarsening, by releasing residual stresses, and the general softening of the alloy to allow for its easy forming by extrusion, rolling, or other bulk deformation processes which shape the ingot into intermediate products. Ensuing heating and controlled cooling can induce in some alloys the precipitation (segregation) of fine particulates. In general, such particulates act as the main strengthening agents for many alloys, due to their ability to block *dislocation* movement (for more on dislocations, see Chapter 13).

Control over the microstructural features of an alloy, for example, grain size, secondary phases, and so on, is desirable and in large part achievable through synthesis and processing. Such control can greatly enhance alloy strength, toughness and ductility (Hull and Bacon, 2006; Polmear, 2006), which is the basic motivation for alloying. More recent advances in materials science, beyond metals and alloys, can yield even more impressive control over material structure (e.g., nano-fabrication), but remain throttled by the expense of experimentation, which undermines the rate at which new materials may be deployed in expanding world markets. To accelerate the cycle through which new materials may reach their markets, predicting and testing their performance must be made less expensive.

There has been a rising trend in mechanics to develop theories that can supply materials scientists with predictions for the properties of materials from knowledge of their microstructure, with only minimal resort to experimentation. Such efforts have led to numerous formulations of a *generalized mechanics*, which incorporates structural information more directly in augmented balance laws. Generalized mechanics is associated with *mechanism-based* approaches to constitutive modeling, which take into consideration statistical descriptions of local and non-local deformation physics. The complexity of the resulting mathematics, in particular the nonlinearity of the governing physical laws and the wealth of parameters that can influence deformation, has been a driving force for mechanicians to integrate their theories with finite element methods, often tailored for high-performance computing. As such, many modern theories in computational mechanics aim to produce rapid, low cost, reliable and general-purpose finite element codes for the virtual performance testing of materials. These theories relate material structure to their properties with high fidelity and in a manner that can be modeled by finite element methods.

Multiresolution continuum theory (MCT) is one such theory (McVeigh, Vernerey *et al.*, 2006; Vernerey, Liu *et al.*, 2007; McVeigh and Liu, 2008; McVeigh and Liu, 2009; Liu, Qian *et al.*, 2010; McVeigh and Liu, 2010; Tian, Chan *et al.*, 2010; Greene, Liu *et al.*, 2011); it is based on a multilevel view of microstructures, which provides for a continuum description of all embedded microstructural features, so that they may contribute, either separately or coupled, to the evolving mechanical behavior of a material at the engineering scales of observation. Each structural level is thus assumed to exhibit different material properties and a different deformation state. The reason is that in general, the micro-response of a material point to a local load deviates considerably from the macro-response of a specimen to applied loads and boundary conditions, which tends to be the aggregate of micro-responses averaged over a larger (active) volume. Thus, MCT permits a separate strain, stress and mechanism-based material law to be associated with each level of microstructure represented by a material point of a macroscopic specimen. To understand the MCT approach, we first review the basic description of bulk deformation and mechanics of microstructured continua.

12.2 Bulk Deformation of Microstructured Continua

Thus far it has been emphasized that microstructure plays a dominant role in determining the mechanical properties of materials. Moreover, in most engineering applications the scale of observation of a material and its structures is large enough for features to appear continuous, that is, discrete atomic lattices or molecular structures cannot be discerned at the levels of required performance. In the same vein, the deformation undergone by material microstructures can be described by continuous functions, in approximation to the collective motion of the atoms. The advantage of resorting to continuous approximation to deformation is twofold. On the one hand, retention of atomic detail is computationally prohibitive at engineering scales of observation (~1 cm or greater, ~1 s or longer). On the other hand atomic detail is often unnecessary, since neighboring atoms tend to follow macroscopic trends in motion, so that their positions could be thought to approximately coincide with macroscopically permissible modes of deformation that are determined more efficiently by continuum methods.

12.3 Generalizing Mechanics to Bulk Microstructured Continua

12.3.1 The Need for a Generalized Mechanics

Given that we model the bulk deformation using a continuum-based approach, there remains the matter of capturing *local* microstructure; consider Figures 12.5 and 12.6.

Figure 12.5 shows a schematic of a small slice taken from a heterogeneous microstructure with multiple secondary phases and grain boundaries, before and after deformation. It can be seen from the figure that each phase deforms differently, and that a crack could open up between grains, or elsewhere in general. What happens to the surrounding material is not shown, but it may be assumed that the slice shown remains attached to the specimen from which it was taken. Figure 12.6(a) on the other hand shows a schematic for an idealized specimen undergoing ductile deformation to failure. In this figure, the secondary phases and grains cannot be discerned, and the formation of the micro-crack in Figure 12.5 cannot be seen at this scale of observation. Instead, necking is predicted to occur over the central portion of the specimen, shrinking the load-bearing zone until rupture. Figure 12.5 and Figure 12.6(a) look very different, in spite of being characteristic of deformation of the same specimen, albeit at different scales. Furthermore, the boundary conditions of Figure 12.5 require compatibility of deformation between the

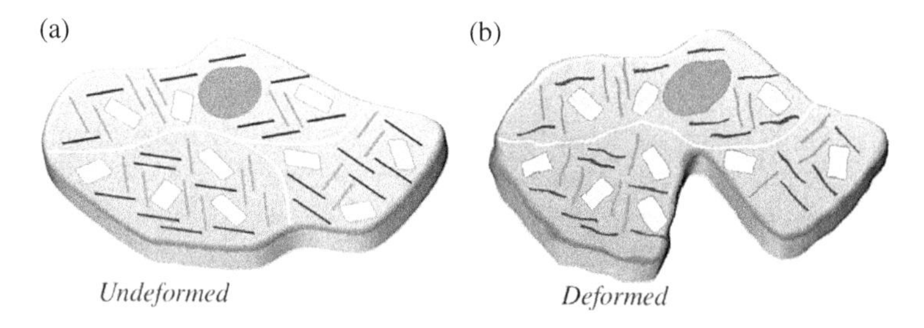

Figure 12.5 Local view of deformation and fracture of microstructure

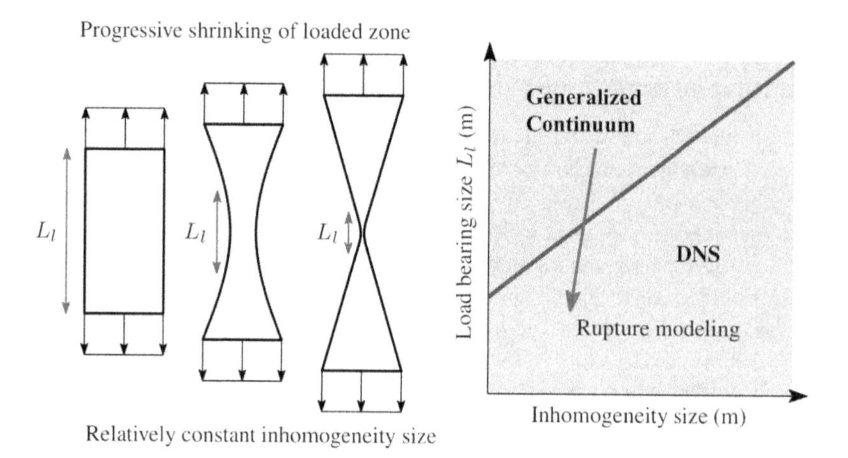

Figure 12.6 (a) Global view of deformation and fracture of specimen, (b) domain of applicability of generalized continuum theory

material slice and its surroundings, which evolve with deformation. In Figure 12.6(a), however, the top and bottom surfaces are constrained to move at a fixed velocity, the sides are free, and these boundary conditions do not necessarily change with deformation.

To reconcile the mechanics leading to Figure 12.5 and Figure 12.6(a), knowing that the scale of interest to engineering applications usually corresponds to Figure 12.6(a), or larger, two approaches may spring to mind. The first approach is to take the slice in Figure 12.5 to span the entire the specimen with all its structural details, and to model the structure by *direct numerical simulation* (DNS) with realistic loads and boundary conditions. This approach is often computationally prohibitive, as there could be millions of secondary phases, grains, and so on to include in the models of complex microstructures. The second approach is to model the entire specimen by means of *generalized continuum* theories which incorporate additional terms into the balance laws to account for microstructural effects on macro-deformation.

Typically, generalized continuum methods are suitable for applications where DNS is not feasible, due to computational limitations, or unnecessary. They are characterized by their ability to capture effectively the interactions between embedded structures and excitation energies that are of comparable spatial frequency. This is especially true when load-bearing zones may contract to characteristic length scales (e.g., spacing between particles) leading to localized deformation of a microstructure. The applicability of generalized continuum theories is summarized schematically in Figure 12.6(b). Thus, generalized continuum theories are effective when each of two key conditions is present:

1. Deformation may be continuously approximated at the scale of observation,
2. Sub-scale interactions of microstructure and load cannot be ignored and can contribute to local inhomogeneous modes of deformation.

To capture inhomogeneous deformation, the response of each material *point* (where a point represents any portion of microstructure such as Figure 12.5), is conditioned by that of its neighbors, so that information about the *neighborhood* must be included in the balance law for a proper characterization of the evolving mechanics across microstructural length scales.

Note that if local inhomogeneous deformation is not likely to occur, effective microstructural properties may be assigned to each material point by the methods of micromechanics (Mura, 1987; Nemat-Nasser, and Hori 1999), and classical continuum methods discretized by standard finite element procedures would suffice.

12.3.2 Major Ideas for a Generalized Mechanics

As indicated by Figure 12.7, information about the neighborhood of a material point may be introduced via two main approaches, depending on whether or not it may be derived from the original degrees of freedom (DOF) at a material point. If derivable from the original DOF, higher gradients of the same DOF are taken, thus generating higher order continuum theories (e.g., Mindlin, 1965; Ahmadi and Firoozbakhsh, 1975; Muhlhaus and Aifantis, 1991a, b; Fleck, Muller *et al.*, 1994; Fleck and Hutchinson, 1997; Borst, Pamin *et al.*, 1999; Chambon, Caillerie *et al.*, 2001; Geers, Kouznetsova *et al.*, 2001; Gurtin, 2002; Kouznetsova, Geers *et al.*, 2002; Wagner and Liu, 2003; Chambon, Caillerie *et al.*, 2004; Georgiadis, Vardoulakis *et al.*, 2004; Kouznetsova, Geers *et al.*, 2004; Gurtin and Anand, 2005; Engelen, Fleck *et al.*, 2006; Fleck and

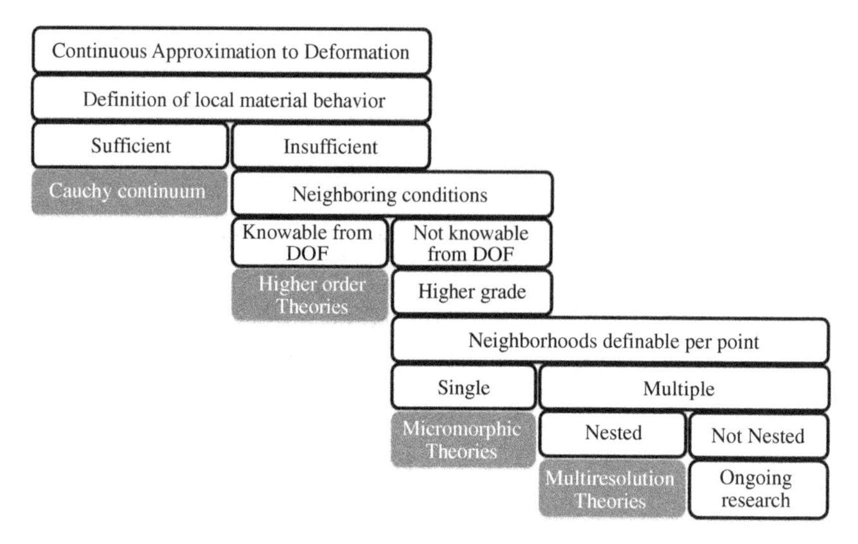

Figure 12.7 Important ideas in generalized mechanics, leading up to multiresolution continuum theory

Willis, 2009a, b; Luscher, McDowell *et al.*, 2010). If not, then additional DOF need be appended to each material point to inform it of its neighboring conditions and thus generate *higher-grade* continuum theories (e.g., Cosserat and Cosserat, 1909; Mindlin, 1964; Germain, 1973; Eringen, 1999; Nappa, 2001; Iesan, 2002; Chen and Lee, 2003a, b; Lee and Chen, 2005; Forest and Sievert, 2006; Zeng, Chen *et al.*, 2006; Vernerey, Liu *et al.*, 2007; Forest, 2009; Jänicke, Diebels *et al.*, 2009; Regueiro, 2009). As will be discussed in this chapter, MCT extends higher-grade continuum theories to situations where multiple neighborhoods per material point need be defined due to the complexity of deformation of rich microstructures in modern materials.

12.3.3 Higher-Order Approach

Simple examples of gradient phenomena are bending and diffusion, as shown by schematic in Figure 12.8. The physics of bending depends on curvature, which is computable from the first gradient of strain, while the physics of diffusion depends on the gradient of the concentration of a diffusing species. From both these examples it is clear that a gradient phenomenon requires a finite span of microstructure over which to evolve and is therefore naturally associated with neighborhoods and characteristic length scales of a microstructure.

Curvature is also the *second* gradient of displacement, and as such is only worth retaining in the formulations of continuum mechanics when deformation at a material point is sufficiently non-uniform, for example, locally inhomogeneous. To understand this, consider the following expansion of a non-uniformly deformed element $d\mathbf{x}$ (e.g., bending in Figure 12.8)

$$d\mathbf{x} = \mathbf{F} \cdot d\mathbf{X} + \frac{1}{2} d\mathbf{X} \cdot \nabla_0 \mathbf{F} \cdot d\mathbf{X} + \dots \tag{12.3.1}$$

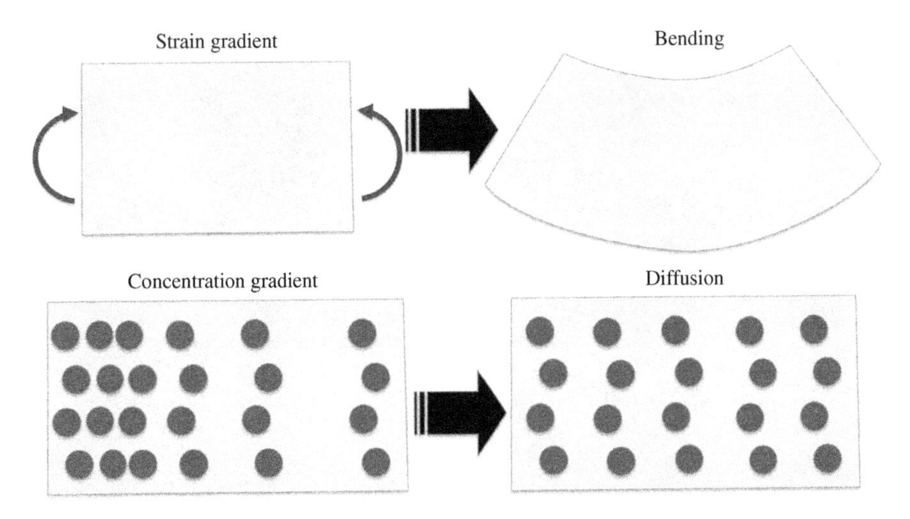

Figure 12.8 Two examples where gradients naturally arise

where $\mathbf{F}=\nabla_0\phi$ is the deformation gradient of any nonlinear motion $\phi(\mathbf{X},t)$. If deformation is locally uniform, only the first term of the expansion of $\phi(\mathbf{X},t)$ needs be retained. When $\nabla_0\mathbf{F}$ is large with respect to a material element $d\mathbf{X}$ and the microstructural length scales it represents, then it is important to include gradients of strain in the formulations of continuum mechanics. In the case of rate formulations, gradients of velocity gradients should instead be considered. For example, the internal power density may be written as (e.g., Mindlin and Tiersten, 1962),

$$p_{\text{int}} \equiv \boldsymbol{\sigma} : \mathbf{L} + \boldsymbol{\sigma\sigma} \vdots \nabla\mathbf{L} \tag{12.3.2}$$

where $\nabla\mathbf{L}=L_{ij,k}$ is the spatial gradient of the velocity gradient, that is, the velocity second gradient, and $\boldsymbol{\sigma\sigma}$ is the energy-conjugate kinetic measure known as *double stress*, which generalizes the concept of bending moment per unit area. The triple-dot indicates contraction on all three indices of the tensors, that is, $\boldsymbol{\sigma\sigma} \vdots \nabla\mathbf{L}=\sigma\sigma_{ijk}L_{ij,k}$.

12.3.4 Higher-Grade Approach

A typical higher-grade theory applied to microstructures is the *micromorphic* theory (Eringen and Suhubi, 1964; Mindlin, 1964; Germain, 1973; Eringen, 1999; Chen and Lee, 2003a,b,c), often with some adaptation (McVeigh, Vernerey *et al.*, 2006; Vernerey, Liu *et al.*, 2007; McVeigh and Liu, 2008; McVeigh and Liu, 2009; McVeigh and Liu, 2010), where a material point of finite size is defined by considering nested particles in the microstructure, see Figure 12.9. The idea is that the deformation inside a material point is not uniform, so capturing its variation is of interest. In general, the dominant phase away from the nested structure is assumed to deform according to distortion map $\mathbf{F}_0$, while the nested phase deforms by $\mathbf{F}_1$; deformation in between would vary from $\mathbf{F}_0$ to $\mathbf{F}_1$.

The displacement $\mathbf{u}_1(\mathbf{X})$of the nested phase (Figure 12.9) can act as a basic set of DOFs. The dominant phase on the other hand exhibits a far-field displacement $\mathbf{u}_0(\mathbf{X})$, see Figure 12.10.

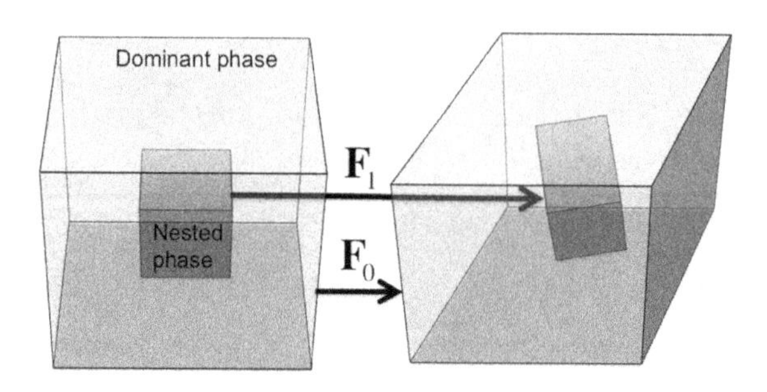

Figure 12.9 Embedded sub-structure in a material point

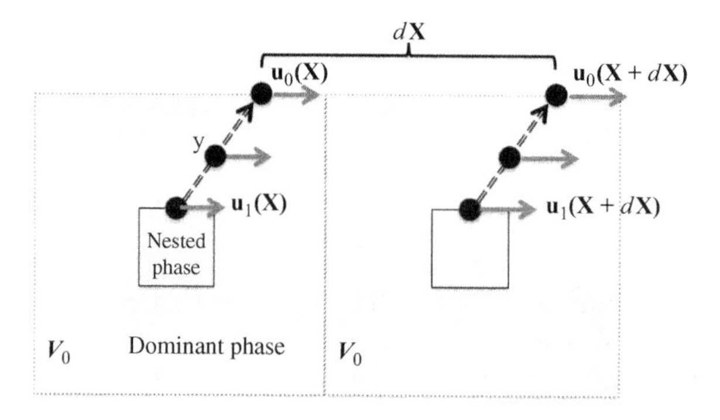

Figure 12.10 Two neighboring micromorphic material points in the reference configuration

The relative displacement or velocity between and the nested phase and the far field can be used to derive additional DOFs.

It is possible to express the deformation gradient anywhere within the micromorphic material point (Figure 12.10) in terms of $\mathbf{F}_1$ and the interior position vector, $\mathbf{y}$. To this end, the displacement at any point $\mathbf{y}$ inside a material point is first assumed as of the form,

$$\mathbf{u}(\mathbf{X},\mathbf{y}) = \mathbf{u}_1(\mathbf{X}) + \mathbf{f}_1 \cdot \mathbf{y}, \tag{12.3.3}$$

where $\mathbf{f}_1 \equiv \mathbf{I} + \dfrac{\partial \mathbf{u}}{\partial \mathbf{y}}$ is a constant second-order *relative* micro-distortion tensor, which is treated as the set of additional DOF that characterizes micro-deformability. Note that we here use a lowercase bold symbol $\mathbf{f}$ for the micro-distortion tensor to avoid confusion with macro-deformation tensor $\mathbf{F}$.

Variants of the micromorphic theory can be generated by tailoring $\mathbf{f}_1$ so as to permit only certain micro-deformabilities. For example in the *micro-polar* theory (Eringen, 1965; Yang and Lakes, 1982; Kennedy and Kim, 1993; Kadowaki and Liu, 2005; Yan, Larsson *et al.*, 2006), only micro-rotation is allowed, that is, $\mathbf{f}_1 = \mathbf{R}_1$ where $\mathbf{R}_1$ is a rotation tensor. Similarly, for *micro-strain* theory (Forest and Sievert, 2006), only rotation-free strain is permitted, that is, $\mathbf{f}_1 = \mathbf{U}_1$, where $\mathbf{U}_1$ is the micro-strain tensor. Yet another variant is to assume micro-rotation is followed by expansion or contraction, that is, $\mathbf{f}_1 = \lambda \mathbf{R}_1$, where λ is the stretch, to yield the *micro-stretch* theory (Eringen, 1990).

The macro-gradient of deformation at a point $\mathbf{y}$ within a micromorphic material point $\mathbf{X}$ is in general given by (see Figure 12.10)

$$\mathbf{F}(\mathbf{X},\mathbf{y}) = \mathbf{I} + \frac{\partial \mathbf{u}(\mathbf{X},\mathbf{y})}{\partial \mathbf{X}} = \mathbf{I} + \frac{\partial (\mathbf{u}_1(\mathbf{X}) + \mathbf{f}_1 \cdot \mathbf{y})}{\partial \mathbf{X}}. \tag{12.3.4}$$

Thus in terms of $\mathbf{F}_1$, which is equal to $\mathbf{I}+\frac{\partial \mathbf{u}_1}{\partial \mathbf{X}}$, the macro-deformation gradient is,

$$\mathbf{F}(\mathbf{X},\mathbf{y})=\mathbf{F}_1(\mathbf{X})+\nabla_X\mathbf{f}_1\cdot\mathbf{y} \tag{12.3.5}$$

The gradient operator is here given a sub-script ∇_X to avoid confusion between material point coordinates $\mathbf{X}$ and imbedded point coordinates $\mathbf{y}$. An interpretation of these equations is given in Figure 12.11(a,b), each of which shows by schematic four contiguous micromorphic material points, two in the horizontal direction and two in the vertical direction. Figure 12.11(a) shows that in spite of locally amplified strain at each material point in the regions near the right interfaces between dominant phase (e.g., matrix) and nested feature (e.g., inclusion), we find that $\mathbf{f}_1(\mathbf{X},\mathbf{y}^*)=\mathbf{f}_1(\mathbf{X}+d\mathbf{X},\mathbf{y}^*)$ for any two corresponding points $\mathbf{y}^*$ at a distance $d\mathbf{X}$ apart. That is, the micro-mechanisms driving the local, or sub-scale, amplification of deformation are equally active in contiguous micromorphic material points, so that $\nabla_X\mathbf{f}_1=0$. Compare this behavior with that in Figure 12.11(b). The two material points in the left column exhibit more active sub-scale mechanisms than those in the right column, so that $\nabla_X\mathbf{f}_1\neq 0$. The micromorphic material points on the left of Figure 12.11(b) are thus predicted to localize macroscopic deformation (hence the darker color), since by Equation (12.3.5) they will deform more than their right neighbors. Note that the cause of localization in this micromorphic approach derives from sub-scale activity, which will be modeled by distinct (but coupled) constitutive laws. Also note that, as in the example of Figure 12.11(a), when $\mathbf{f}_1\neq 0$ but $\nabla_X\mathbf{f}_1=0$, an additional internal power from the activation of subscale mechanisms should be considered, as these mechanisms will store or dissipate energy, even when localization of macro-deformation does not occur.

It is common to formulate micromorphic theory in terms of the velocity field $\mathbf{v}(\mathbf{x})$, expressed in the current configuration, to simplify the description of large deformation mechanics (Germain, 1973), which is more easily treated in a rate form. The macro-velocity gradient $\mathbf{L}$ at any embedded point $\mathbf{y}$ is thus expressed as,

$$\mathbf{L}(\mathbf{x},\mathbf{y})=\mathbf{L}_1(\mathbf{x})+\nabla_x\mathbf{l}_1(\mathbf{x})\cdot\mathbf{y}, \tag{12.3.6}$$

where $\mathbf{l}_1\equiv\frac{\partial \mathbf{v}}{\partial \mathbf{y}}$ is the relative micro-velocity gradient, which acts as the set of additional DOFs. Here ∇_x indicates the gradient with respect to the spatial coordinates $\mathbf{x}$ of a material point, again to avoid confusion with imbedded point coordinates.

The contribution of these additional DOFs to the mechanics of deformation may be considered by modifying the balance laws. For example, the principle of virtual power, which balances momentum, would be enhanced with these additional degrees of freedom and conjugate stress measures, as will be discussed in the following section.

12.3.5 Reinterpretation of Micromorphism for Bulk Microstructured Materials

12.3.5.1 Internal Power

To prepare the ground for our discussion of the *multiresolution continuum theory*, which aims to capture the internal power of complex microstructures, let us depart from the micromorphic picture described in the previous section and reinterpret the kinematic quantities of Equation (12.3.6) by rewriting the velocity gradient as follows,

(a)

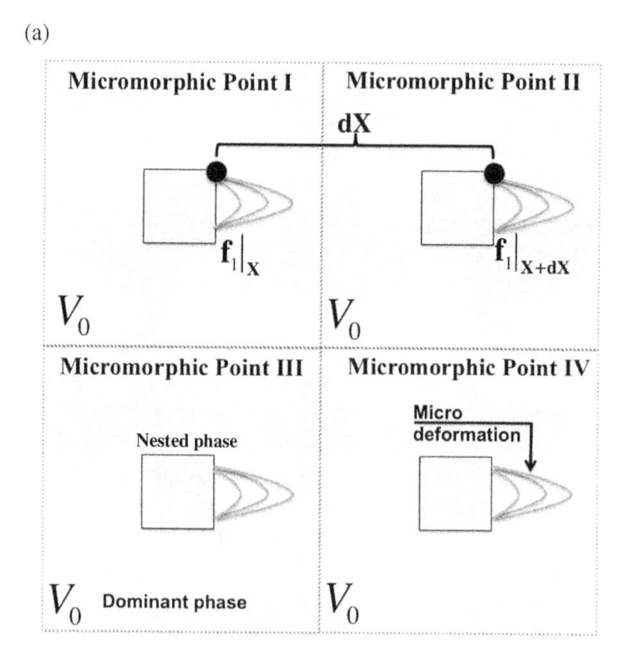

(b)

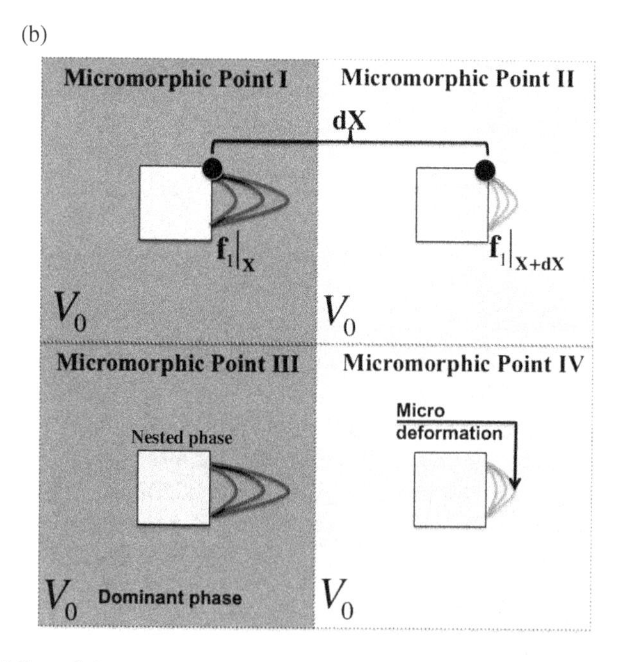

Figure 12.11 (a) Micro-deformation is the same across multiple micromorphic material points. (b) Micro-deformation exhibits a gradient across micromorphic material points in the x-direction

$$\mathbf{L}(\mathbf{x},\mathbf{y}) = \mathbf{L}_0(\mathbf{x}) + (\mathbf{L}_1(\mathbf{x}) - \mathbf{L}_0(\mathbf{x})) + \nabla_x \mathbf{l}_1(\mathbf{x})\cdot\mathbf{y}, \tag{12.3.7}$$

where $\mathbf{L}_0$ is some reference velocity gradient for material point $\mathbf{x}$. The relative micro-velocity gradient was redefined as $\mathbf{l}_1 \equiv \mathbf{L}_1 - \mathbf{L}_0$. Its spatial macro-gradient $\nabla_x \mathbf{l}_1 \equiv \nabla_x \mathbf{L}_1$, that is, $\nabla_x \mathbf{L}_0 \rightarrow \mathbf{0}$, or the reference velocity gradient $\mathbf{L}_0$ is assumed constant at the material point. Thus,

$$\mathbf{L}(\mathbf{x},\mathbf{y}) = \mathbf{L}_0 + \mathbf{l}_1 + \nabla_x \mathbf{L}_1 \cdot \mathbf{y}. \tag{12.3.8}$$

The internal power density would thus be defined as (in the current configuration),

$$p_{\text{int}} = \boldsymbol{\sigma}_0 : \mathbf{L}_0 + \mathbf{s}_1 : \mathbf{l}_1 + \mathbf{ss}_1 \,\vdots\, (\nabla_x \mathbf{L}_1) \tag{12.3.9}$$

$\boldsymbol{\sigma}_0$ is a reference Cauchy stress conjugate to the reference velocity gradient $\mathbf{L}_0$. The reference $\mathbf{L}_0$ thus replaces the far-field value of the micromorphic picture in Figure 12.11. $\mathbf{s}_1$ is the stress that penalizes the relative micro-deformation $\mathbf{l}_1$ (higher-grade degrees of freedom), and $\mathbf{ss}_1$ is the double stress that penalizes the gradient in micro-deformation.

The physical implications of this reinterpretation are now presented. The internal power defined in Equation (12.3.9) can be rewritten as,

$$p_{\text{int}} = p_{\text{hom}} + p_{\text{inhom}}, \tag{12.3.10}$$

where $p_{\text{hom}} \equiv \boldsymbol{\sigma}_0 : \mathbf{L}_0$ comes from the *homogeneous* component of the material point deformation in the current configuration, and $p_{\text{inhom}} \equiv \mathbf{s}_1 : \mathbf{l}_1 + \mathbf{ss}_1 \vdots \nabla_x \mathbf{L}_1$ comes from the *inhomogeneous* part, which arises due to the local activation of sub-scale mechanisms that drive deformation away from the homogeneous pattern; see Figure 12.12.

$\mathbf{s}_1$ is defined from a volume average (McVeigh, Vernerey *et al.*, 2006; Vernerey, Liu *et al.*, 2007) taken in the current configuration,

$$\mathbf{s}_1 = \frac{V_0}{V_1^a}\left(\frac{1}{V_0}\int_{V_1^a} \boldsymbol{\sigma}_p dV\right) = \frac{1}{V_1^a}\int_{V_1^a} \boldsymbol{\sigma}_p dV, \tag{12.3.11}$$

where $V_1^a(\mathbf{x}) \subset V_0(\mathbf{x})$ is the sub-volume of a material point $\mathbf{x}$ (whose volume is $V_0(\mathbf{x})$) wherein a given sub-scale mechanism (labeled by sub-script 1) is activated. Superscript a in V_1^a stands for *active*. An example of V_1^a would be, as shown in Figure 12.12, one of the quadrilaterals taken from the grid dividing a material point of volume V_0 in the current configuration for the first inhomogeneous mode.

$\boldsymbol{\sigma}_p$ is the Cauchy penalty stress acting on each embedded point $\mathbf{y}$ of a micromorphic material point. $\boldsymbol{\sigma}_p$ penalizes the deviation of the local inhomogeneous deformation from the surrounding homogeneous pattern. The stress *density* created in the micromorphic material point as a result of the embedded inhomogeneous mechanism corresponds with the sum (i.e., integral) of all stresses it generates within the active volume, divided by the total volume of the material point. There are approximately $\frac{V_0}{V_1^a}$ such sub-volumes spanning the material point, each of which is assumed active (for simplicity) and contributes to the internal power at the same rate $\mathbf{l}_1$, so that Equation (12.3.11) results. The double stress $\mathbf{ss}_1$, is similarly defined by the volume average,

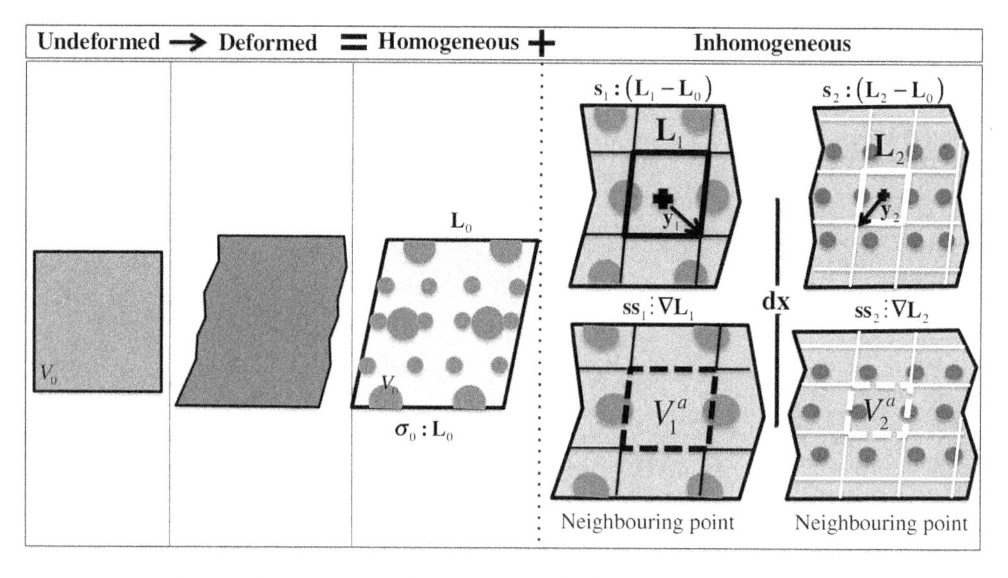

Figure 12.12 Multiresolution decomposition of material point deformation into homogeneous and inhomogeneous components. Each inhomogeneous component is also augmented with a macro-gradient, which helps quantify the state of an active micro-mechanism in neighboring material points

$$\mathbf{ss}_1 = \frac{1}{V_1^a} \int_{V_1^a} \boldsymbol{\sigma}_m \otimes \mathbf{y} dV, \tag{12.3.12}$$

where $\boldsymbol{\sigma}_m$ is the Cauchy stress in the microstructure at a point within the same active volume, defined in the current configuration. Subscript m refers to *local* fields within the microstructure.

12.3.5.2 A Link to RVE Modeling

To compute stresses in complex microstructures, rarely are integrals of Equations (12.3.11) and (12.3.12) computed analytically; instead *homogenization* is applied computationally. The homogeneous power density at any point in the continuum may be approximated from the average power density of a superimposed representative volume element (RVE) of the actual microstructure, subjected to similar loads. p_{hom} is thus given in terms of the local stress $\boldsymbol{\sigma}_m$ and velocity gradient $\mathbf{L}_m$ within the RVE, where subscript m refers to *local* fields within the RVE (microstructural model). That is,

$$p_{\text{hom}} = \frac{1}{V_{RVE}} \int_{V_{RVE}} \boldsymbol{\sigma}_m : \mathbf{L}_m dV \tag{12.3.13}$$

Using the Hill–Mandel lemma (McVeigh, 2007), one can equate the average power in a volume to the power arising from the average Cauchy stress and the average velocity gradient,

$$p_{\text{hom}}(\mathbf{x}) = \boldsymbol{\sigma}_0 : \mathbf{L}_0 \tag{12.3.14}$$

where,

$$\boldsymbol{\sigma}_0(\mathbf{x}) = \frac{1}{V_{RVE}} \int_{V_{RVE}} \boldsymbol{\sigma}_m dV, \quad \text{and} \quad \mathbf{L}_0(\mathbf{x}) = \frac{1}{V_{RVE}} \int_{V_{RVE}} \mathbf{L}_m dV. \tag{12.3.15a,b}$$

The Hill–Mandel lemma assumes that the boundary of the RVE satisfies the conditions that, (1) the prescribed velocity $\mathbf{v}=\mathbf{L}_0 \cdot \mathbf{x}$ for any point $\mathbf{x}$ on the Dirichlet boundary, and (2) the traction $\mathbf{t}=\boldsymbol{\sigma}_0 \cdot \mathbf{n}$ on the Neumann boundary.

This conventional homogenization approach would seem incapable of representing the inhomogeneous component of deformation, since the homogenized solution is limited by the resolution of the averaging operation performed at the RVE scale. To tackle this difficulty, the Hill–Mandel approach should be extended to account for both homogeneous and inhomogeneous contributions. In this framework, the inhomogeneous contribution to the internal power is defined as the difference between the average power density in the active volume $V^a{}_1$ wherein the subscale mechanism evolves and the power density averaged over the entire RVE volume V_{RVE}, which represents a material point of volume V_0, that is,

$$p_{\text{inhom}}(\mathbf{x}) = \frac{1}{V_1^a} \int_{V_1^a} p_m dV - \frac{1}{V_{RVE}} \int_{V_{RVE}} p_m dV, \quad \text{where} \quad V_1^a \subset V_{RVE}. \tag{12.3.16}$$

With the aim to match the kinematic quantities of Equation (12.3.8), the local internal power p_m at any point $\mathbf{y}$ of the microstructure where inhomogeneous deformation evolves will be assumed to depend on the velocity gradient and second gradient, that is, $p_m \equiv f(\mathbf{L}_m, \nabla_{\mathbf{x}} \mathbf{L}_m)$. To fully define this dependency, it is further assumed that conjugate to the velocity gradient is the stress $\boldsymbol{\sigma}_m$ computed from the RVE model, and that conjugate to the second gradient of the velocity will be the tensor product $\boldsymbol{\sigma}_m \otimes \mathbf{y}$. Using these assumptions, Equation (12.3.13) and Equation (12.3.14) and plugging into Equation (12.3.16), we obtain

$$\begin{aligned} p_{\text{inhom}}(\mathbf{x}) &= \frac{1}{V_1^a} \int_{V_1^a} (\boldsymbol{\sigma}_m : \mathbf{L}_m + \boldsymbol{\sigma}_m \otimes \mathbf{y} \vdots \nabla_{\mathbf{x}} \mathbf{L}_m) dV - \boldsymbol{\sigma}_0 : \mathbf{L}_0 \\ &= \frac{1}{V_1^a} \int_{V_1^a} (\boldsymbol{\sigma}_m : \mathbf{L}_m - \boldsymbol{\sigma}_0 : \mathbf{L}_0 + \boldsymbol{\sigma}_m \otimes \mathbf{y} \vdots \nabla_{\mathbf{x}} \mathbf{L}_m) dV \end{aligned}, \tag{12.3.17}$$

We can then write inhomogeneous power density in terms of the local *relative* velocity gradient $\mathbf{l}_m \equiv \mathbf{L}_m - \mathbf{L}_0$ and invoke power equivalence for the penalty stress $\boldsymbol{\sigma}_p$, that is, $\boldsymbol{\sigma}_p : (\mathbf{L}_m - \mathbf{L}_0) = \boldsymbol{\sigma}_m : \mathbf{L}_m - \boldsymbol{\sigma}_0 : \mathbf{L}_0$, to obtain

$$p_{\text{inhom}}(\mathbf{x}) = \frac{1}{V^a{}_1} \int_{V^a{}_1} (\boldsymbol{\sigma}_p : (\mathbf{L}_m - \mathbf{L}_0) + \boldsymbol{\sigma}_m \otimes \mathbf{y} \vdots \nabla_{\mathbf{x}} \mathbf{L}_m(\mathbf{x})) dV \tag{12.3.18}$$

The final step to recover from the RVE simulation corresponding continuum quantities, is to assume that,

$$\mathbf{L}_1(\mathbf{x}) = \frac{1}{V_1^a} \int_{V_1^a} \mathbf{L}_m dV, \quad \nabla_{\mathbf{x}} \mathbf{L}_1(\mathbf{x}) = \frac{1}{V_1^a} \int_{V_1^a} \nabla_{\mathbf{x}} \mathbf{L}_m dV. \tag{12.3.19a,b}$$

Substituting Equation (12.3.19) into Equation (12.3.18), the inhomogeneous virtual internal power density can now be written as, again using the Hill–Mandel lemma (McVeigh and Liu, 2008),

$$p_{\text{inhom}} = \mathbf{s}_1 : (\mathbf{L}_1 - \mathbf{L}_0) + \mathbf{ss}_1 \vdots \nabla_x \mathbf{L}_1 \tag{12.3.20}$$

where $\mathbf{s}_1$ and $\mathbf{ss}_1$ are defined by Equations (12.3.11) and (12.3.12) respectively. These volume averages are interpreted as *continuum* micro-stresses, resolved by computational RVE modeling of the active mechanisms in $V_1^a \subset V_{RVE}$, where V_{RVE} in this approach represents material point volume V_0.

For the cases where $\mathbf{L}_1 - \mathbf{L}_0 \cong \mathbf{0}$ component wise, the corresponding power term in Equation (12.3.20) would be ignored. When $\|\mathbf{L}_1| \gg |\mathbf{L}_0\|$, the local mechanism dominates and the penalty stress $\boldsymbol{\sigma}_p \cong \boldsymbol{\sigma}_m$, the microstructural stress, component wise. Thus, $\mathbf{s}_1 = \frac{1}{V_1^a} \int_{V_1^a} \boldsymbol{\sigma}_m dV$, which is easy to compute from the RVE model. For cases where the difference between $\mathbf{L}_1$ and $\mathbf{L}_0$ is intermediate, more elaborate methods are needed to compute $\boldsymbol{\sigma}_p$, a discussion of which is postponed to Section 12.6.

This reinterpretation of the micromorphic approach is more general than the gradient approach, since it includes in the balance law both the gradient of deformation and additional degrees of freedom for relative deformation. In fact, if $\mathbf{s}_1$ is interpreted as a Lagrange multiplier, this formulation reduces to the second-gradient approach, which may be seen from Equation (12.3.20) by noticing that the relative deformation becomes negligible and $\mathbf{L}_1 = \mathbf{L}_0$.

12.4 Multiscale Microstructures and the Multiresolution Continuum Theory

A difficulty to modeling microstructures by a direct application of the reinterpreted micromorphic approach of Section 12.3 is the fact that microstructures possess a multiplicity of embedded length scales. For example, Figure 12.11 with an inclusion at the centroid of the material point is not usually representative of actual materials. The extension to N-scales, see Figure 12.12, to capture multiple *nested* neighborhoods in real microstructures, has therefore been proposed to tackle this difficulty, and lays the foundation of the *multiresolution* continuum theory (MCT). For the case where neighborhoods of a microstructural material point cannot be viewed as nested, similar ideas have also been proposed (Elkhodary, Greene *et al.*, 2013), but will not be discussed here.

In a rate formulation of multiscale mechanics, linear superposition of the evolving N-scale active mechanisms is assumed. Micro-stress and micro-double-stress can be introduced for each scale of inhomogeneous deformation such that its contribution to inhomogeneous internal power density may be captured in a summation as follows,

$$p_{\text{int}} = \boldsymbol{\sigma}_0 : \mathbf{L}_0 + \sum_{n=1}^{N} (\mathbf{s}_n : \mathbf{l}_n + \mathbf{ss}_n \vdots \nabla_x \mathbf{L}_n) \tag{12.4.1}$$

where $\mathbf{l}_n \equiv \mathbf{L}_n - \mathbf{L}_0$ is the nth relative micro-deformation rate, taken with respect to the homogeneous deformation rate $\mathbf{L}_0$. $\mathbf{L}_n$ is thus the nth micro-deformation rate and represents the average deformation rate in active volume V_n^a.

In Figure 12.12, the perturbation in material point deformation induced by the nth inhomogeneous mechanism is assumed to be positive over some portions of the material point and negative in equally many portions, such that the sum of deviations due to the nth inhomogeneous mode is zero. As such, $\mathbf{L}_0$ could be used as a reference state for all inhomogeneous modes of deformation, and justifies using $\mathbf{l}_n = \mathbf{L}_n - \mathbf{L}_0$ as the quantity to be penalized in the power statement (12.4.1). This assumption of using $\mathbf{L}_0$ as a reference for all inhomogeneous modes of deformation produces direct coupling between penalty stresses from all scales in the final strong form, as will be seen in Section 12.5.4. However, this assumption has been a source of contention, since it is can lead to a double-count of energies (Luscher, McDowell *et al.*, 2010) when $n \geq 2$ and $\mathbf{L}_0$ can no longer be reasonably assumed as a reference for all inhomogeneous modes. This difficulty has motivated the redefinition $\mathbf{l}_n \equiv \mathbf{L}_n - \mathbf{L}_{n-1}$ in more recent MCT formulations, where the nth micro-deformation is considered active only after the onset of the n–1st mechanism. However, when $\mathbf{l}_n = \mathbf{L}_n - \mathbf{L}_{n-1}$ is assumed, concurrent coupling between *all* scales is not represented explicitly in the strong form of MCT governing equations. In the following, we adhere to the original definition $\mathbf{l}_n \equiv \mathbf{L}_n - \mathbf{L}_0$, assuming the appropriate conditions hold.

The power contribution of a given inhomogeneous deformation mode (see Figure 12.12), which causes portions of the material point to lean forward and others equally backward with respect to the homogeneous mode, could be thought of as a sum to which each of these portions contributes the same average power value. Therefore, in the internal power density definition all $\frac{V_0}{V_n^a}$ copies of the nth micro-deformation sub-volume must be considered, so that conjugate kinetic measures are defined by,

$$\begin{aligned} \mathbf{s}_n &\equiv \frac{1}{V_n^a} \int_{V_n^a} \boldsymbol{\sigma}_p^n(\lambda_1, \ldots, \lambda_{n-1}, \phi_n, .., \phi_N) dV \\ \mathbf{ss}_n &\equiv \frac{1}{V_n^a} \int_{V_n^a} \boldsymbol{\sigma}_p^n(\lambda_1, \ldots, \lambda_{n-1}, \phi_n, .., \phi_N) \otimes \mathbf{y}_n dV \end{aligned} \quad . \qquad (12.4.2a,b)$$

The purpose of $\mathbf{s}_n$ in the power statement Equation (12.4.1) is to penalize the inhomogeneous mode of deformation driven by an nth scale mechanism when it evolves above and beyond the homogeneous (i.e., coarser) mode of deformation. The purpose of $\mathbf{ss}_n$ is to penalize the gradients in micro-deformation. λ_i is a set of localization variables communicated to the nth scale from its ith neighborhood, for $i \in \{1, \ldots, n-1\}$, and ϕ_j is a set of internal state variables evolving at the jth embedded scale. Typically, localization variables for a given scale evolve from functions of internal variables at coarser scales, that is, $\lambda_i \in \{f_j(\phi_{i-1}, .., \phi_1), j \in \mathbb{N}\}$. As such, the nesting of neighborhoods required by this multiresolution approach is simply an assumption that organizes the definition of stresses $\mathbf{s}_n$ and double-stresses $\mathbf{ss}_n$ in a multilevel scheme, but is not a requirement of the power density statement (12.4.1).

In this approach, it is thought that the local inhomogeneous mechanisms activated at the nth scale and give rise to $\mathbf{s}_n$ are not the same mechanisms that govern the relatively homogeneous deformation of the coarser $n-1$ scales, albeit they may influence the coarser scales and their stresses, by means of evolving internal variables they communicate thereto. It should also be remarked that this approach considers no additional conditions of *compatibility* between the deform-

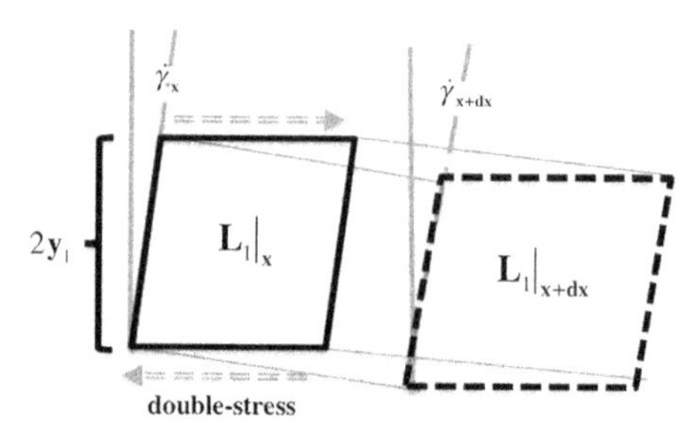

Figure 12.13 Interpretation of the macro-gradient of the velocity micro-gradient and double stress in multiresolution continuum theory

Table 12.1 Continuum measures and their origin within the RVE

Continuum Point Tensor…	*…is a Volume Average of the Field…*	*…at Scale*
$\boldsymbol{\sigma}_0$	$\boldsymbol{\sigma}_m$	V_0
$\mathbf{s}_n$	$\boldsymbol{\sigma}_p^n$	V_n
$\mathbf{ss}_n$	$\boldsymbol{\sigma}_p^n \otimes \mathbf{y}_n$	V_0
$\mathbf{L}_0$	$\mathbf{L}_m$	V_0
$\mathbf{L}_n$	$\mathbf{L}_m$	V_n
$\nabla_x \mathbf{L}_n$	$\nabla_x \mathbf{L}_m$	V_n
Power Equivalence	$\boldsymbol{\sigma}_p^n : (\mathbf{L}_n - \mathbf{L}_0) = \boldsymbol{\sigma}_m : \mathbf{L}_m - \boldsymbol{\sigma}_0 : \mathbf{L}_0$	V_n
Internal Power Density	$p_{\text{int}} = \boldsymbol{\sigma}_0 : \mathbf{L}_0 + \sum_{n=1}^{N} (\mathbf{s}_n : \mathbf{l}_n + \mathbf{ss}_n \vdots \nabla_x \mathbf{L}_n)$	V_0

ing $\frac{V_0}{V_n^a}$ sub-volumes at a given scale n, so that a direct sum of their contributions may be assumed in the internal power statement for simplicity, leading to the definitions of Equation (12.4.2).

Consider the highlighted sub-volumes in Figure 12.12, one of which is extracted and depicted in Figure 12.13. Two corresponding sub-volumes of finite span $2\mathbf{y}_1$ which are embedded in contiguous material points at a distance $d\mathbf{x}$ apart may in general deform at different rates $\mathbf{L}_1(\mathbf{x})$ and $\mathbf{L}_1(\mathbf{x}+d\mathbf{x})$. It may then be supposed that these two sub-volumes are connected linearly, so that the quantity $\nabla_x \mathbf{L}_1(\mathbf{x})$ represents the local gradient in micro-deformation. The non-uniformity of micro-deformation, characterized by this local gradient, will be counteracted by double-stresses evolving over the finite span of the sub-volume. In the simple illustration of Figure 12.13, the gradient in the illustrated shear rate component, $\partial_x \dot{\gamma}$, is assumed to be countered by a shear stress-couple, acting across an arm $2\mathbf{y}_1$.

Notice that as $\mathbf{L}_2$ and $\mathbf{L}_1$ are independent modes of inhomogeneous deformation, each attributable to a different microstructural mechanism, the double-stresses penalizing these gradients are also assumed independent. In general therefore, the penalty stress $\boldsymbol{\sigma}_p$ arising in Equation (12.4.2) will not be directly equivalent to the local microstructural stress $\boldsymbol{\sigma}_m$ in an RVE model, except under specific simplifying assumptions. More careful considerations for deriving penalty stresses from microstructural quantities predicted in RVE models are presented in Section (12.6).

The key RVE relationships to MCT are summarized in Table 12.1 for the case in which micro-stresses are used, where each scale corresponds to a sub-volume V_n, and typically, $V_n \subset V_{n-1} \subset \ldots \subset V_0$. All volumes are expressed in the current configuration.

12.5 Governing Equations for MCT

The complete governing equations for MCT, suitable for large dynamic deformation, are derivable from the balance law of mechanical power, which states that any existing difference between externally applied power p_{ext} and internally generated power p_{int} must be compensated by motion of the system, whose inertial forces give rise to kinetic power p_{kin}. Hence,

$$p_{ext} - p_{int} = p_{kin}, \qquad (12.5.1)$$

or defining the Lagrangian density π,

$$\pi = p_{kin} - (p_{ext} - p_{int}) = 0. \qquad (12.5.2)$$

The trajectory of any material point in the MCT model must be made to obey this law at every instant of time, that is, π=0 for $\forall t$. In simulations where generalized displacement fields are controlled, p_{ext} and p_{kin} are given at a time instant and the internal power p_{int} is generated to bring the system into balance. Any trajectory that is near to but not a solution of the MCT model will exhibit a $\pi \neq 0$. Hence, to find a solution for the MCT model, π may be thought of as the functional which is to be minimized or maximized, by the principle of *virtual* power, which searches all trajectories in the vicinity of the solution and finds the right one by setting,

$$\delta \int_{\Omega} \pi d\Omega = 0. \qquad (12.5.3)$$

To find the governing MCT equations it is thus required to find the first variations of p_{int}, p_{ext}, and p_{kin}, all integrated over the entire body volume Ω.

12.5.1 Virtual Internal Power

The first variation of internal power given by Equation (12.4.1) is the sum of the homogeneous and inhomogeneous parts. Thus,

$$\delta P_{int} = \int_{\Omega} \left(\sigma_0 : \delta \mathbf{L}_0 + \sum_{n=1}^{N} (\mathbf{s}_n : \delta \mathbf{l}_n + \mathbf{ss}_n \vdots \delta \nabla_{\mathbf{x}} \mathbf{L}_n) \right) d\Omega. \qquad (12.5.4)$$

12.5.2 Virtual External Power

The virtual external power is found by extending the Cauchy-type expression to,

$$\begin{aligned} \delta P_{ext} = & \int_{\Omega} \left(\mathbf{b} \cdot \delta \mathbf{v}_0 + \sum_{n=1}^{N} \mathbf{B}_n : \delta \mathbf{l}_n \right) d\Omega \\ & + \int_{\Gamma} \left(\mathbf{t} \cdot \delta \mathbf{v}_0 + \sum_{n=1}^{N} \mathbf{T}_n : \delta \mathbf{l}_n \right) d\Gamma \end{aligned}, \qquad (12.5.5)$$

where it is now written in terms of the boundary (Γ) macro-traction force-density $\mathbf{t}$ and the corresponding penalty micro double traction $\mathbf{T}_n$, in addition to the body (Ω) macro-force-density $\mathbf{b}$

and the corresponding penalty micro body double force $\mathbf{B}_n$. $\mathbf{v}_0$ is the velocity of the MCT material point as it undergoes homogeneous deformation.

12.5.3 Virtual Kinetic Power

The virtual kinetic power extends an expression developed by Mindlin (1964) to multiple scales (Vernerey, Liu *et al.*, 2007). First, the variation in MCT kinetic energy density must be defined as

$$\delta e_{\text{kin}} = \bar{\rho}_0 \mathbf{v}_0 \cdot \delta \mathbf{v}_0 + \sum_n \left(\frac{1}{V_n^a} \int_{V_n^a} (\bar{\rho}_n - \bar{\rho}_{n-1})(\mathbf{v}_0 + \boldsymbol{v}_n) \cdot \delta(\mathbf{v}_0 + \boldsymbol{v}_n)\, d\text{V} \right) \tag{12.5.6}$$

where $\boldsymbol{v}_{(n)} = \mathbf{l}_{(n)} \cdot \mathbf{y}_{(n)}$, where $\mathbf{l}_n$ the relative micro-velocity gradient and $\mathbf{y}_n$ is the length-scale of the nth inhomogeneous mechanism evolving in active sub-volume V_n^a. $\bar{\rho}_0$ is the average MCT material point mass density, and $\bar{\rho}_n$ is the average mass density within active sub-volume V_n^a. Taking the material time derivative of the volume integral of Equation (12.5.6) and applying Reynold's transport theorem and the conservation of mass, the resulting power variation can simplify to,

$$\delta \text{P}_{\text{kin}} = \int_{\Omega} \left(\rho \dot{\mathbf{v}}_0 \cdot \delta \mathbf{v}_0 + \sum_{n=1}^{N} \boldsymbol{\alpha}_n \cdot \text{P}_n : \delta \mathbf{l}_n \right) d\Omega, \tag{12.5.7}$$

where, $\rho = \bar{\rho} + \sum_n (\bar{\rho}_n - \bar{\rho}_{n-1})$. $\alpha_{(n)} \equiv \dot{\mathbf{l}}_{(n)} + \mathbf{l}_{(n)} \cdot \mathbf{l}_{(n)}$ defines the relative micro-acceleration of the nth scale sub-volumes, and $\text{P}_{(n)} \equiv \frac{1}{V_n^a} \int_{V_n^a} (\bar{\rho}_n - \bar{\rho}_{n-1}) \mathbf{y}_{(n)} \otimes \mathbf{y}_{(n)} d\text{V}$ defines the relative second moment of mass density local to sub-volume V_n^a, and represents inertia of the sub-volume. Derivation of Equation (12.5.7) from Equation (12.5.6) is left as an exercise (see Exercise 12.1).

12.5.4 Strong Form of MCT Equations

In deriving the strong form for typical MCT applications, which have been focused on modeling localization of deformation, it may be assumed that $(\mathbf{L}_n - \mathbf{L}_0) = \mathbf{l}_n \cong \mathbf{L}_n$ component wise. Thus, variations of p_{ext} and p_{kin} with respect to $\mathbf{l}_n$ in the principle of virtual power could be replaced by direct variations in $\mathbf{L}_n$. However, variation of p_{int} is written in terms of $(\mathbf{L}_n - \mathbf{L}_0)$ to retain in the strong form direct stress-coupling across scales. $\mathbf{L}_n$ thus serves as the set of higher-grade degrees of freedom for the nth scale, in lieu of $\mathbf{l}_n$.

With this assumption in mind, by applying the divergence theorem to the variation of principle of virtual power (Equation 12.5.1) and making use of the arbitrariness of the variations $\delta\mathbf{v}$ and $\delta\mathbf{L}_n$, the resulting equilibrium equations and boundary conditions for MCT theory become,

$$\begin{aligned} \left(\boldsymbol{\sigma} - \sum_{n=1}^{N} \mathbf{s}_n \right) \cdot \nabla_{\mathbf{x}} + \mathbf{b} = \dot{\mathbf{v}}_0 \rho \;\; \text{in}\, \Omega \\ \mathbf{ss}_n \cdot \nabla_{\mathbf{x}} - \mathbf{s}_n + \mathbf{B}_n = \alpha_n \cdot \text{P}_n \;\; \text{in}\, \Omega \end{aligned} \tag{12.5.8a,b}$$

with boundary conditions,

$$\mathbf{t}-\left(\boldsymbol{\sigma}-\sum_{n=1}^{N}\mathbf{s}_n\right)\cdot\mathbf{n}=0 \text{ on } \Gamma_t$$
$$\mathbf{T}_n-\mathbf{ss}_n\cdot\mathbf{n}=0 \text{ on } \Gamma_{T_n} \tag{12.5.9a,b}$$

Derivation of Equation (12.5.8) and Equation (12.5.9) is left as an exercise for the reader (see Exercise 12.2).

Additional simplifying assumptions could be made. The penalty micro-stresses can be identified for simplicity with local stresses from RVE models when localization dominates homogeneous deformation, that is, $\boldsymbol{\sigma}_p^n \cong \boldsymbol{\sigma}_m$ component wise. This assumption greatly simplifies constitutive modeling of the sub-scale inhomogeneous modes. It is also possible to identify $\mathbf{T}_n$ with $\frac{1}{V_n^a}\int_{V_n^a}\mathbf{t}\otimes\mathbf{y}_n dV$ and $\mathbf{B}_n$ with $\frac{1}{V_n^a}\int_{V_n^a}\mathbf{b}\otimes\mathbf{y}_n dV$, so as to simplify the definition of the boundary conditions and body forces in a typical MCT simulation. For such an identification, the micro-stresses $\mathbf{T}_n$ and $\mathbf{B}_n$ are restricted to represent the couples which the macro-fields $\mathbf{t}$ and $\mathbf{b}$ create over the finite sub-span $\mathbf{y}_n$ of the material point.

12.6 Constructing MCT Constitutive Relationships

To solve the MCT governing equations, the material time derivatives of stress are required,

$$\dot{\boldsymbol{\sigma}}=\dot{\boldsymbol{\sigma}}(\mathbf{L}_0),\ \dot{\mathbf{s}}_n=\dot{\mathbf{s}}_n(\mathbf{L}_n-\mathbf{L}_0), \text{ and } \dot{\mathbf{ss}}_n=\dot{\mathbf{ss}}_n(\nabla_x\mathbf{L}_n). \tag{12.6.1}$$

The objective components of these rates require constitutive definitions, which of course depend on the microstructure of a material and the modeling strategy adopted. Typically, to deal with the nth embedded constitutive relationship, only the inhomogeneous rate of deformation $\mathbf{D}_n-\mathbf{D}_0$ is considered (i.e., the symmetric part of the velocity gradient, $\mathbf{D}=sym(\mathbf{L})$). Each of these constitutive models can be developed from RVE computational models of the microstructure by averaging at various scales V_n. Typically, constitutive models in the rate-form are developed according to the strategy in Box 12.1.

Box 12.1 Constitutive modeling strategy in MCT

1. An RVE is defined and loaded under a given boundary condition.
2. The average stress $\boldsymbol{\sigma}_0$ and velocity gradient $\mathbf{L}_0$ associated with the deforming RVE are recorded to *calibrate* a macroscale constitutive model, that is, they are used to determine material parameters in a *predefined* constitutive model.
3. The RVE model is then examined to identify regions where deformation is strongly inhomogeneous. Inhomogeneous deformation generally arises due to interactions between microstructural features.

4. An *appropriate* averaging volume V_n^a is identified for each inhomogeneous deformation mechanism. The averaging volume should be sufficiently large to capture a linear variation in the inhomogeneous deformation fields, as gradients need be computed.
5. Relative velocity gradient $(\mathbf{L}_n - \mathbf{L}_0)$ is then computed as the difference between the average strain in the inhomogeneously deforming region V_n^a and the average RVE velocity gradient from step 2.
6. Penalty micro-stress $\boldsymbol{\sigma}_p^n$ within V_n^a is computed via a *power equivalence relationship* (Equation 12.6.7). This field is then averaged over V_n^a for the continuum micro-stress $\mathbf{s}_n$ (Equation 12.4.2a and Table 12.1).
7. Principal values of the continuum micro-stress $\mathbf{s}_n$ are then computed. This step should be repeated for multiple loading conditions to generate an inhomogeneous mechanism plastic potential ϕ_n.
8. Plastic potential ϕ_n, with *mechanism-based* hardening parameters and internal variables, is constructed for the nth inhomogeneous mechanism from the data collected in Step 7. The form of the plastic potential ϕ_n may be, (a) pre-defined when the physics is easy to interpret in terms of a penalty stress (e.g., internal friction, or interface activity), or (b) by general curve-fitting of the data obtained in Step 7.
9. Elastic properties for the nth inhomogeneous mechanism, which are needed to update stress rates from hypo-elastic relations (Equation 12.6.3), may be constructed from simple micromechanical principles applied to the active sub-volume V_n^a, such as the *rule of mixtures* (Kubin and Mortensen, 2003), for example.
10. Average macro-velocity second gradient $\nabla_x \mathbf{L}_n$ is computed in V_n^a (Table 12.1).
11. Continuum double micro-stress $\mathbf{ss}_n$ is computed in V_n^a according to Equation (12.4.2b).
12. Plastic potential $\phi\phi_n$ for double micro-stress $\mathbf{ss}_n$ is then assumed and calibrated against RVE results. Often, however, a single plastic potential is used to combine $\mathbf{s}_n$ and $\mathbf{ss}_n$ so as to couple first degree and higher degree mechanisms.

In more detail, let us consider the nth sub-scale mechanism, where an additive elastic–plastic decomposition of the deformation rate and its gradient is assumed,

$$\mathbf{D}_n = \mathbf{D}_n^e + \mathbf{D}_n^p \text{ and } \nabla_x \mathbf{D}_n = \nabla_x \mathbf{D}_n^e + \nabla_x \mathbf{D}_n^p. \tag{12.6.2a,b}$$

The corresponding objective stress and double stress rate may be computed from the elastic parts of the rate of deformation according to a generalized Hooke's law such that,

$$\overset{\nabla}{\mathbf{s}}_n \equiv \mathbf{C}_n : (\mathbf{D}_n^e - \mathbf{D}_0^e),\ \overset{\nabla}{\mathbf{ss}}_n \equiv \mathbf{CC}_n \,\vdots\, \nabla_x \mathbf{D}_n^e \tag{12.6.3a,b}$$

where $\mathbf{C}_n$ is the elasticity tensor characterizing the properties of the microstructure of the nth sub-volume, and could be constructed by micromechanical laws, such as the rule of mixtures. $\mathbf{CC}_n$ is a sixth-order elastic tensor, typically approximated by $\mathbf{CC}_n = \frac{1}{V_n^a} \int_{V_n^a} \mathbf{y}_n \otimes \mathbf{C}_n \otimes \mathbf{y}_n dV$, so that no additional material constants need be defined beyond the components of $\mathbf{C}_n$ in the description of gradient phenomena. Note that to update the stress increments in a finite element approach, material derivatives of the stress and double stress are required. These may be obtained for example from Jaumann objective rates $\left(\overset{\nabla J}{\mathbf{s}}_n, \overset{\nabla J}{\mathbf{ss}}_n\right)$ as follows (where the tangent moduli in Equation (12.6.3) need be appropriately defined as discussed in Chapter 5),

$$(\dot{\mathbf{s}}_n)_{ij} = \left(\overset{\nabla J}{\mathbf{s}}_n\right)_{ij} + (\mathbf{W})_{ik}(\mathbf{s}_n)_{kj} + (\mathbf{s}_n)_{ik}(\mathbf{W})_{jk}$$
$$(\dot{\mathbf{ss}}_n)_{ijk} = \left(\overset{\nabla J}{\mathbf{ss}}_n\right)_{ijk} + (\mathbf{ss}_n)_{ojk}(\mathbf{W})_{io} + (\mathbf{ss}_n)_{iok}(\mathbf{W})_{jo} + (\mathbf{ss}_n)_{ijo}(\mathbf{W})_{ko} \tag{12.6.4a,b}$$

In order to compute the objective stress rate (Equation 12.6.3) it is required to compute the plastic part of the deformation rate $\mathbf{D}_n^p$ and its gradient $\nabla_x \mathbf{D}_n^p$, which will subtracted from the total rate and its gradient respectively, according to Equation (12.6.2). These plastic parts are typically computed from plastic potentials ϕ_n and $\phi\phi_n$ with an associative flow rule. Thus,

$$\mathbf{D}_n^p - \mathbf{D}_0^p = \lambda \frac{\partial \phi_n}{\partial \mathbf{s}_n}, \text{ and } \nabla_{\mathbf{x}} \mathbf{D}_n^p = \lambda \frac{\partial \phi\phi_n}{\partial \mathbf{ss}_n} \tag{12.6.5a,b}$$

where λ is a plastic multiplier, as discussed in Chapter 5. The remaining task associated with the development of constitutive models for sub-scale mechanisms is to find an appropriate mechanism-based expression for the plastic potentials ϕ_n and $\phi\phi_n$. ϕ_n in particular may not be easy to visualize in many modeling situations, since it is a potential for a penalty stress, which is conjugate to a relative deformation, rather than an absolute deformation; unlike $\phi\phi_n$ which is conjugate to an absolute deformation, since by assumption $\nabla_{\mathbf{x}} \mathbf{D}_0^p \to \mathbf{0}$.

Need for a systematic approach to construct ϕ_n from RVEs was thus immediately recognized, and a solution was first proposed by (McVeigh and Liu, 2008). The approach herein presented extends some aspects of that work. In general, local *plastic* power equivalence is defined in V_n^a,

$$\boldsymbol{\sigma}_p^n : \left(\mathbf{D}_m^p - \mathbf{D}_0^p\right) \equiv \boldsymbol{\sigma}_m : \mathbf{D}_m^p - \boldsymbol{\sigma}_0 : \mathbf{D}_0^p \tag{12.6.6}$$

where $\mathbf{D}_m^p$ is the local plastic rate of deformation within the nth sub-volume and $\mathbf{D}_0^p$ is the homogeneous plastic rate of deformation. It is possible to define a penalty stress that satisfies Equation (12.6.6) via the relation,

$$\boldsymbol{\sigma}_p^n \equiv \frac{1}{3} p_n \left(\mathbf{D}_m^p - \mathbf{D}_0^p\right)^{-1}, \tag{12.6.7}$$

where $p_n \equiv \left(\boldsymbol{\sigma}_m : \mathbf{D}_m^p - \boldsymbol{\sigma}_0 : \mathbf{D}_0^p\right)$ in V_n^a, provided substantial inhomogeneous deformation has evolved and $\left(\mathbf{D}_m^p - \mathbf{D}_0^p\right)$ is invertible. These conditions are typically satisfied when modeling large deformation of heterogeneous materials. When $\mathbf{D}_m^p$ is component wise almost equal to $\mathbf{D}_0^p$, it is reasonable to assume $\boldsymbol{\sigma}_p^n \equiv 0$. Since p_n is scalar, the eigenvectors of $\boldsymbol{\sigma}_p^n$ will coincide with those of the inhomogeneous mode of deformation in this assumption. Moreover, it is often helpful to define the equivalent relative plastic deformation-rate measure δ_n as follows,

$$\delta_n = \sqrt{\frac{2}{3}\left(\mathbf{D}_m^p - \mathbf{D}_0^p\right) : \left(\mathbf{D}_m^p - \mathbf{D}_0^p\right)}. \tag{12.6.8}$$

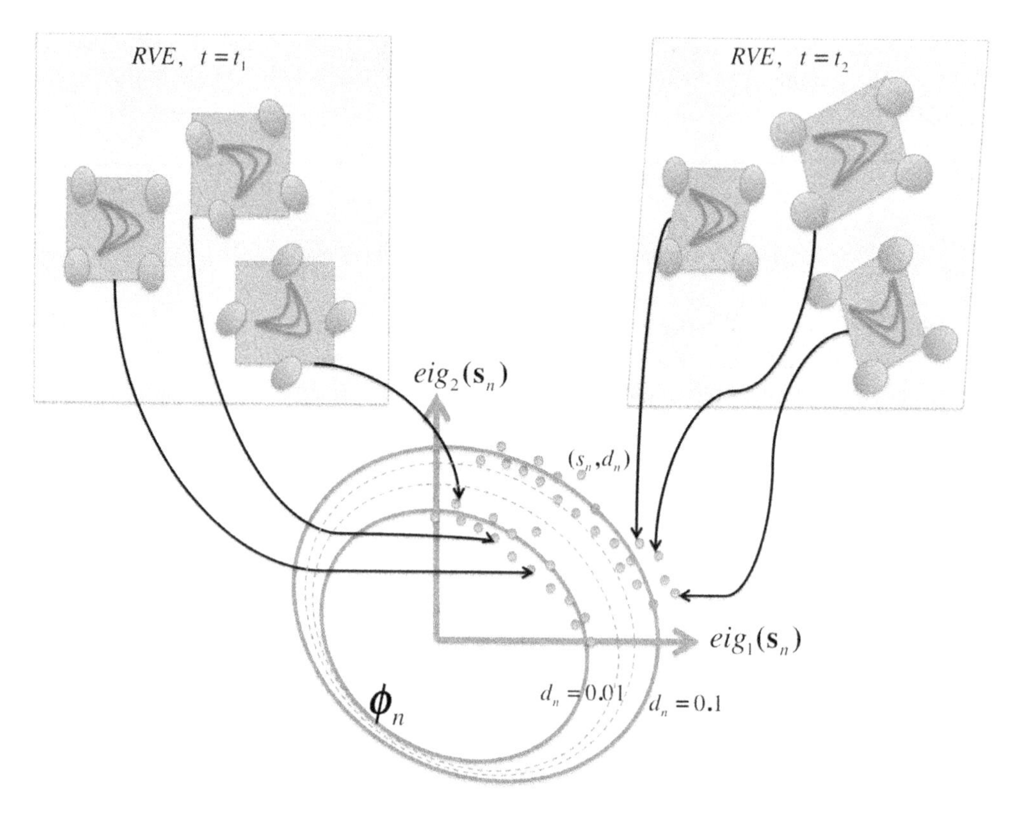

Figure 12.14 Finding the plastic potential for the nth scale inhomogeneous mechanism

The effective relative deformation over V_n^a is accordingly defined by,

$$d_n = \frac{1}{V_n^a} \int_{V_n^a} \delta_n dV. \tag{12.6.9}$$

Pairing every d_n to a corresponding equivalent penalty stress s_n, generates a point on the potential surface. The equivalent penalty stress is defined by,

$$s_n = \frac{1}{V_n^a} \int_{V_n^a} \sqrt{\frac{3}{2} \boldsymbol{\sigma}_p^n : \boldsymbol{\sigma}_p^n} dV, \tag{12.6.10}$$

where $\boldsymbol{\sigma}_p^n$ is defined by Equation (12.6.7). For a given (d_n, s_n) pair, one should find the principal values of the penalty stress $\mathbf{s}_n$, defined by Equation (12.4.2). Next, multiple stress points corresponding to various (d_n, s_n) pairs are graphed in stress space with the principal axes of $\mathbf{s}_n$ as coordinates, see Figure 12.14. The stress points should be sampled from different regions of interest in the RVE, and/or by running multiple simulations with different loading conditions, for example, torsion, pure shear and biaxial tension, and/or by sampling multiple regions of the RVE at different times. If multiple stress points of similar d_n values are grouped, a yield

Table 12.2 Homogeneous and inhomogeneous quantities used in MCT modeling

RVE Quantity	*Description*
V_n^a	Volume to average active nth scale mechanism
$\boldsymbol{\sigma}_p^n$	$\boldsymbol{\sigma}_p^n \equiv \frac{1}{3} p_n \left(\mathbf{D}_m^p - \mathbf{D}_0^p\right)^{-1}$
s_n	$s_n = \frac{1}{V_n^a} \int_{V_n^a} \sqrt{\frac{3}{2} \boldsymbol{\sigma}_p^n : \boldsymbol{\sigma}_p^n} dV$
$\mathbf{s}_n$	$\mathbf{s}_n = \frac{1}{V_n^a} \int_{V_n^a} \boldsymbol{\sigma}_p^n dV$
d_n	$d_n = \frac{1}{V_n^a} \int_{V_n^a} \delta_n dV$
Φ_n	Any set of history variables used to fit the plastic potential
$\phi, \phi\phi_n$	Plastic potential

surface ϕ_n may be passed through them by curve fitting. When d_n values are sorted from small to large, evolution of the yield surface ϕ_n may be modeled according to some fitting history variables. Fuller development of this approach would combine elements of statistical analysis that are beyond the scope of this chapter, but remains the subject of active research.

A simpler approach to constitutive modeling from RVEs is more frequently adopted (McVeigh, Vernerey *et al.*, 2006; Liu and McVeigh, 2008; McVeigh and Liu, 2008; McVeigh and Liu, 2009; McVeigh and Liu, 2010). It can often be assumed that the yield surface ϕ_n is of a given form, based on the knowledge of the physics to be modeled, so that only parameter fitting based on RVE results is needed (e.g., McVeigh and Liu, 2008). Moreover, a plot obtained from the RVE of the effective Euler–Almansi plastic strain ε_n^{eff} versus the effective Cauchy penalty-stress Σ_n^{eff} may be used to fit a pre-defined hardening law that accompanies an associative flow rule (e.g., McVeigh and Liu 2008). The effective stress Σ_n^{eff} would be computed from the plastic power equivalence,

$$\Sigma_n^{eff} \equiv p_n / \left\| \mathbf{D}_m^p - \mathbf{D}_0^p \right\|. \tag{12.6.11}$$

It should be noted here that were the assumption $\nabla_{\mathbf{x}} \mathbf{D}_0^p \to \mathbf{0}$ not suitable for the higher-order phenomena, then a similar approach to constructing $\phi\phi_n$ (which would now characterize relative gradients instead of absolute ones) from RVE modeling would be necessary. The higher dimensionality of the double-stress tensors would render such an approach more complicated and time consuming, which is thus typically avoided.

The quantities to be computed for MCT models based on RVEs are summarized in Table 12.2.

12.7 Basic Guidelines for RVE Modeling

An MCT material point, see Figure 12.12, represents the effective behavior of a finite volume of heterogeneous microstructure. The link between the continuum material point of infinitesimal size $d\mathbf{x}$ and the microstructure it represents is given through a set of constitutive relations as discussed in Section 12.6. Each of these constitutive relations is characterized by a mathematical model, the form of which is predetermined by the physics of deformation. The mathematical models potentially involve a multiplicity of parameters that need be calibrated to match the specific material deformation mode being modeled. This can be achieved by correlating the mathematical constitutive model to either, (1) experimentally observed material response, or (2) the *average* response taken from a direct numerical simulation (DNS) of a *representative* material volume, which must be sufficiently large to capture the evolving inhomogeneous deformation mode of interest. Such a material volume is called a *representative volume element* (RVE) as described in Section 12.6. RVEs are particularly useful to find constitutive relations for difficult-to-measure phenomena, such as sub-surface damage, adiabatic temperature rise, or dynamic crack propagation for instance.

There are three basic considerations to account for in a typical RVE simulation performed by finite element methods. The first is what constitutive model to assign to each microstructural constituent of the DNS model. Chapter 5 presents various constitutive models from which to select, based on the phenomena to be captured in the simulation (e.g., hyper-elastic, visco-elastic, plastic, etc.). On the other hand, an example of a more elaborate mechanism-based constitutive model for crystalline materials that is often used in RVE simulations is outlined in Chapter 13.

The second consideration is the size of the RVE model, that is, how much microstructure is needed to capture the inhomogeneous mode of interest. The third basic consideration is concerning the boundary conditions that should be applied to the RVE computational model. In the remainder of this section we discuss briefly these latter two considerations.

12.7.1 Determining RVE Cell Size

To select proper RVE model sizes, we must first consider three characteristic lengths (McVeigh, 2007) (see, Figure 12.15). The first, called L, is the scale at which deformation and its variation across a body are considered macroscopic. The second, called λ, is the approximate wavelength of macro-variation in a state-quantity of interest (e.g., macro-strain ε) about its average $\bar{\varepsilon}$. Thus, as can be seen from Figure 12.15, λ corresponds with the span of microstructure that can be reasonably represented by a homogeneous deformation $\bar{\varepsilon}$. The third scale, called a, is the span over which a material's underlying properties undergo considerable variation about their mean value, which corresponds with a microstructural length scale (e.g., spacing between particulates). Multiscale microstructures would thus exhibit a set of such length scales, and A would then be used to represent the entire set. To model a particular inhomogeneous mode, it is often possible to include only

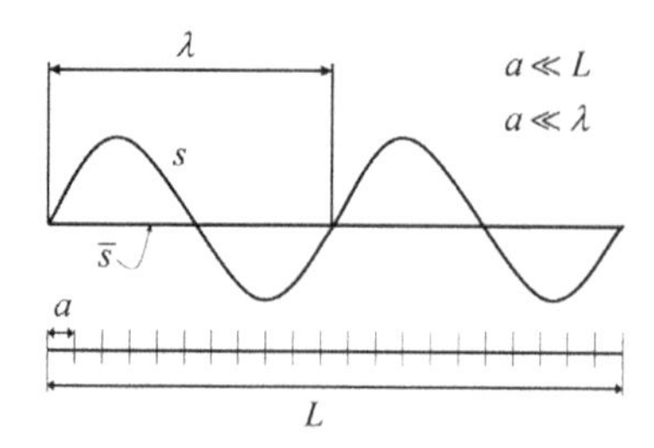

Figure 12.15 RVE cell size is much smaller than the structural size and macro-deformation variation. Reproduced from McVeigh, C. J. (2007). *Linking Properties to Microstructure through Multiresolution Mechanics*. Doctor of Philosophy thesis, Northwestern University

a subset of the microstructural length scales $\alpha \subset A$ in the RVE model, if the governing physics is known to be dominated only by those length scales in α. a in Figure 12.15 would then be defined as $a = \max(\alpha)$.

The RVE cell size can then be identified with λ, which can be determined according to the following criteria.

Criterion (1): *L and λ must be large compared to a*. This is the separation of scales principle (Auriault, 1991); specifically, the homogeneous strain $\bar{\varepsilon}$ should reasonably represent the macroscopic average material response over a span $\lambda >> a$.

Criterion (2): *The properties of interest of the microstructure in a volume element a^3 are the same as those in any other region of the body*. This ensures that the microstructure modeled in the RVE is similar throughout the body so that it is indeed representative of the material response at any point of the body.

12.7.2 RVE Boundary Conditions

Boundary conditions need be applied to induce deformation in the RVE. The common types of boundary conditions applied to computational RVEs are described here.

12.7.2.1 Traction Boundary Conditions

A homogeneous stress $\boldsymbol{\sigma}_0$ can be induced in the RVE through an applied traction $\mathbf{t}$ on the RVE surface S_0.

$$\mathbf{t} = \boldsymbol{\sigma}_0 \cdot \mathbf{n} \quad \text{on} \quad S_0 \tag{12.7.1}$$

In practice, traction boundary conditions become unstable if the average stress decreases due to softening or a material instability. For this reason, velocity based boundary conditions are often preferred.

12.7.2.2 Velocity Boundary Conditions

Velocity boundary conditions $\mathbf{v}$ applied on the RVE surface S_0 can replicate a homogeneous deformation rate $\mathbf{D}_0$ in the RVE, by satisfying the equation,

$$\mathbf{v} = \mathbf{D}_0 \cdot \mathbf{x} \quad \text{on} \quad S_0 \tag{12.7.2}$$

where $\mathbf{x}$ has its origin at the center of the RVE. In this case the boundaries remain flat and rigid, and the RVE tends to be over constrained could lead to unphysical pressure build ups.

12.7.2.3 Periodic Boundary Conditions

Boundary conditions that represent periodic kinematic fields can be applied by identifying the velocities on one surface of the RVE model with those on the opposite surface. That is,

$$\mathbf{v}(\mathbf{x}) = \mathbf{v}(\mathbf{x} + \mathbf{r}), \tag{12.7.3}$$

where $\mathbf{x} \in S_0$ and $(\mathbf{x} + \mathbf{r}) \in S_1$ are matching points on opposite surfaces (S_0, S_1) of the RVE.

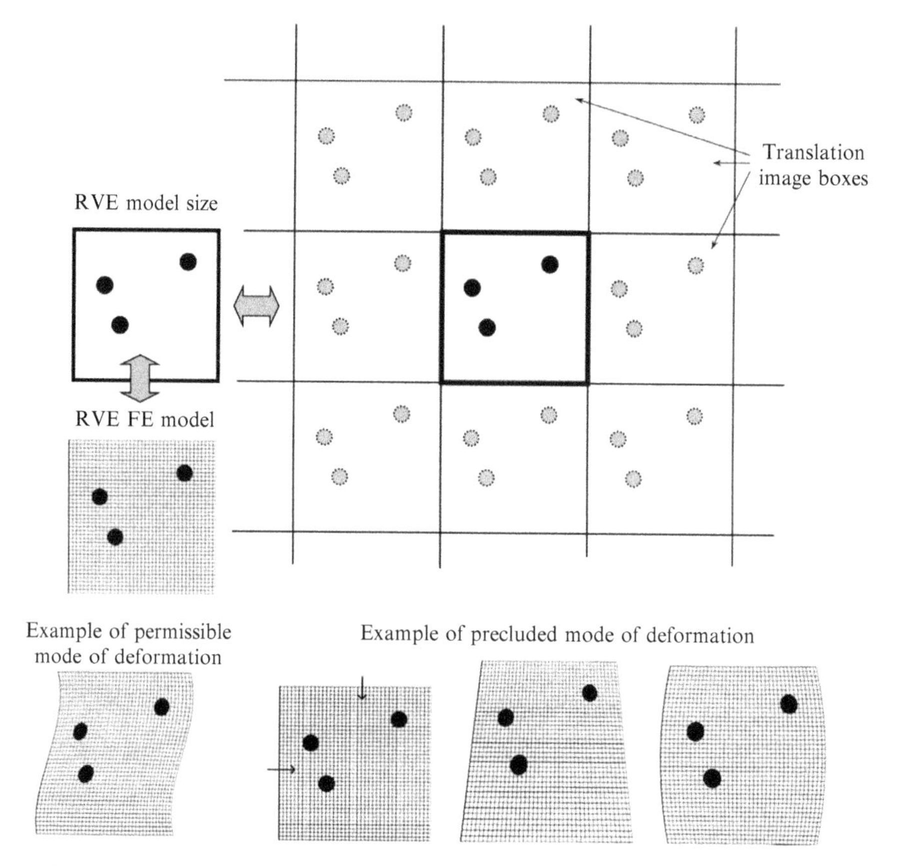

Figure 12.16 Periodic boundary conditions and their implications on permissible deformation modes

Figure 12.16 gives an example of periodic boundary conditions and their basic implications. Essentially, the RVE model and its deformation represent a repeat unit in a periodic structure. As such, modes of deformation that preserve distance between corresponding points on periodic surfaces are permissible, while modes of deformation that do not are not permissible, examples of which are shown in Figure 12.16. Research in the area of periodic boundary conditions is still ongoing, for example, (Mesarovic and Padbidri, 2005).

Periodic boundary conditions can be applied in finite element software by defining 'tie constraints' (Hibbitt, Karlson *et al.*, 2007) between opposite surfaces (e.g., periodic boundaries). For tie constraints, a slave surface is defined so that all nodes on it are constrained to follow matching *points* on a master surface. If a matching point on the master surface is not a node, then the slave node motion is interpolated from the neighboring nodes of the matching point on the master surface. In this manner, the degrees of freedom on a slave surface are eliminated.

12.7.2.4 General Linear Constraints

As a less restrictive boundary condition, it is possible to constrain the motion of points on opposite surfaces by linear equations. Consider four points on the RVE surface as shown in Figure 12.17. Linear constraints can be applied to the nodal degrees of freedom, for example the velocities $\mathbf{v}$,

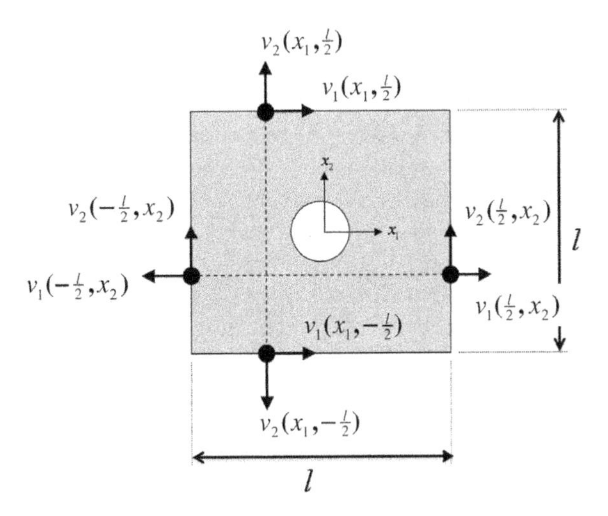

Figure 12.17 Four nodes to define linear constraints in an RVE. Reproduced from Auriault J.L., Is an equivalent macroscopic description possible? *International Journal of Science and Engineering*, 1991. **29**(7), 785–795. Copyright © 1991, Elsevier

$$v_1\left(\frac{l}{2}, x_2\right) - v_1\left(-\frac{l}{2}, x_2\right) = d_1, \quad v_2\left(x_1, \frac{l}{2}\right) - v_2\left(x_1, -\frac{l}{2}\right) = d_2$$
$$v_1\left(x_1, \frac{l}{2}\right) - v_1\left(x_1, -\frac{l}{2}\right) = d_3, \quad v_2\left(\frac{l}{2}, x_2\right) - v_2\left(-\frac{l}{2}, x_2\right) = d_4 \tag{12.7.4}$$

d_1, d_2,d_3 and d_4 are constant velocities which define the homogeneous velocity gradient $(\mathbf{L}_0)_{11}$, $(\mathbf{L}_0)_{22}$, $(\mathbf{L}_0)_{12}$ and $(\mathbf{L}_0)_{12}$ as,

$$(\mathbf{L}_0)_{11} = \log\left(1+\frac{d_1}{l}\right), \quad (\mathbf{L}_0)_{22} = \log\left(1+\frac{d_2}{l}\right)$$
$$(\mathbf{L}_0)_{12} = \log\left(1+\frac{d_3}{l}\right), \quad (\mathbf{L}_0)_{21} = \log\left(1+\frac{d_4}{l}\right) \tag{12.7.5}$$

Hence, only four velocities are required to define an average deformation rate $\mathbf{L}_0$ in two-dimensional RVE models.

It should be noted that velocities of corresponding points on opposite boundaries in Figure 12.17 will differ by a constant, but corresponding velocity gradients will be the same, so that periodic patterns in deformation result in the RVE.

12.7.2.5 Other Boundary Conditions

Many other boundary conditions can be applied. For example, maintaining the state of stress triaxiality, by continually updating the applied lateral pressure on the RVE (Socrate, 1995). Triaxiality is particularly interesting when studying inhomogeneous mechanisms in the RVE associated with void nucleation and growth.

Another example particular to dynamic simulations is that of non-reflecting boundaries (Keller and Givoli, 1989; Givoli, 1991). When waves propagate from an RVE surface or emanate from an inhomogeneity in its interior they are bound to hit the RVE boundaries after some time has elapsed. Excitation energies from the reflections of these waves back into the RVE tend to pollute the results of a simulation, since RVE boundaries are typically non-physical. As such, *infinite* elements (Bettess, 1977; Zienkiewicz, Emson *et al.*, 1983) have been introduced, among other schemes, to absorb incident waves on RVE boundaries, and can be used to define RVE boundary conditions.

Sub-modeling strategies (Hibbitt, Karlson *et al.*, 2007) have also been adopted, where coarse-scale simulations are first conducted to obtain global solution fields (e.g., velocities, displacements) at a structural level. These solution fields are then extracted from a set of interior nodes, specifically nodes that trace the boundary of a region that becomes an RVE driven by these nodal solutions. For the RVE simulation, it is possible to add interior details (i.e., microstructure) to the mesh, for a higher resolution of the localized physics of deformation. This approach can be repeated for any number of zoom-ins as desired, where each RVE can serve as a coarse-scale simulation for a subsequent smaller scale RVE.

These and more boundary conditions (e.g., contact, see Chapter 10) can be applied using commercial finite element software with relative ease, in combination with a variety of material models; an observation which situates RVE modeling in an attractive light when considering MCT constitutive model development.

12.8 Finite Element Implementation of MCT

The presented MCT formulation allows a C^0 continuous discretization within the finite element framework. We first define a vector $\mathbf{d}_\alpha$, which holds the DOF at node α,

$$\mathbf{d}_\alpha = \begin{bmatrix} \mathbf{v}_\alpha \\ \mathbf{L}_\alpha^1 \\ \vdots \\ \mathbf{L}_\alpha^{N-1} \end{bmatrix} \tag{12.8.1}$$

Let $\mathbf{d}^e$ represent the vector of DOF in the element. For a quadrilateral element, therefore,

$$\mathbf{d}^e = \left[\mathbf{d}_1^T, \mathbf{d}_2^T, \mathbf{d}_3^T, \mathbf{d}_4^T\right]^T. \tag{12.8.2}$$

If $\mathbf{N}^e$ represents the matrix of nodal interpolation functions, then

$$\mathbf{d}(\mathbf{x}) \equiv \begin{bmatrix} \mathbf{v}(\mathbf{x}) \\ \mathbf{L}_1(\mathbf{x}) \\ \vdots \\ \mathbf{L}_{N-1}(\mathbf{x}) \end{bmatrix} = \mathbf{N}^e(\mathbf{x})\mathbf{d}^e \tag{12.8.3}$$

Likewise, if $\mathbf{B}^e$ is the matrix of spatial derivatives of $\mathbf{N}^e$, then the vector of macro-strain and gradients of micro-strain will be,

$$\mathbf{E}(\mathbf{x}) \equiv \begin{bmatrix} \mathbf{L}_0(\mathbf{x}) \\ \nabla_x \mathbf{L}_1(\mathbf{x}) \\ \vdots \\ \nabla_x \mathbf{L}_{N-1}(\mathbf{x}) \end{bmatrix} = \mathbf{B}^e(\mathbf{x})\mathbf{d}^e. \tag{12.8.4}$$

Generalized body forces and surface tractions may be represented respectively as,

$$\boldsymbol{\beta}^e = [\mathbf{b}, \mathbf{B}_1, \ldots, \mathbf{B}_{N-1}] \quad \text{and} \quad \boldsymbol{\tau}^e = [\mathbf{t}, \mathbf{T}_1, \ldots, \mathbf{T}_{N-1}]. \tag{12.8.5}$$

The inertia matrix would be, after necessary simplifications,

$$\mu^e = \begin{bmatrix} \rho\mathbf{I} & & & \\ & \mathrm{P}_1 & & \\ & & \ddots & \\ & & & \mathrm{P}_{N-1} \end{bmatrix}, \tag{12.8.6}$$

where density ρ is defined in Equation (12.5.7), I is the identity matrix, and $\mathbf{P}_i$ is the second moment of mass density in sub-volume i. The material stress rates from all N scales will be solved for from corresponding material subroutines, and may then be grouped in a vector $\dot{\Sigma}$ as,

$$\left(\dot{\Sigma}\right)^T = [V(\dot{\boldsymbol{\sigma}}), V(\dot{\mathbf{s}}_1), V(\dot{\mathbf{ss}}_1), \ldots, V(\dot{\mathbf{s}}_{N-1}), V(\dot{\mathbf{ss}}_{N-1})], \tag{12.8.7}$$

where $V(\cdot)$ indicates writing the each tensor out according to the Voigt notation. The stress vector would thus be updated from the first order update,

$$\sum\nolimits_{t+\Delta t} = \sum\nolimits_t + \Delta t \dot{\sum}. \tag{12.8.8}$$

Making use of the foregoing definitions, the resulting MCT finite element discretization becomes: for the internal power variation,

$$\delta P_{\text{int}} = \sum_{e=1}^{e=e_{\max}} \left\{ \int_{\Omega_e} (\delta \mathbf{d}^e)^T (\mathbf{B}^e)^T \sum\nolimits_{t+\Delta t} d\Omega \right\}, \tag{12.8.9}$$

for the external power variation,

$$\delta P_{\text{ext}} = \sum_{e=1}^{e=e_{\max}} \left\{ \int_{\Omega_e} (\delta \mathbf{d}^e)^T (\mathbf{N}^e)^T \boldsymbol{\beta}^e d\Omega \right\} + \sum_{e=1}^{e=e_{\max}} \left\{ \int_{\Gamma_e} (\delta \mathbf{d}^e)^T (\mathbf{N}^e)^T \tau^e d\Gamma \right\}, \tag{12.8.10}$$

and for the kinetic power variation,

$$\delta P_{\text{kin}} = \sum_{e=1}^{e=e_{\max}} \left\{ \int_{\Omega_e} (\delta \mathbf{d}^e)^T (\mathbf{N}^e)^T \mu^e \mathbf{N}^e d\Omega \right\}. \tag{12.8.11}$$

In a dynamic explicit environment an internal force vector $\mathbf{f}_{\text{int}}$ may be formed, and these powers may be recast as,

$$\delta P_{\text{int}} = \sum_{e=1}^{e=e_{\text{max}}} \left\{ (\delta \mathbf{d})^T (\mathbf{X}^e)^T \mathbf{f}^e_{\text{int}} \right\}, \tag{12.8.12a}$$

$$\delta P_{\text{ext}} = \sum_{e=1}^{e=e_{\text{max}}} \{ (\delta \mathbf{d})^T (\mathbf{X}^e)^T \mathbf{f}^e_{\text{ext}} \}, \tag{12.8.12b}$$

$$\delta P_{\text{kin}} = \sum_{e=1}^{e=e_{\text{max}}} \{ (\delta \mathbf{d})^T (\mathbf{X}^e)^T \mathbf{M}^e \mathbf{X}^e \ddot{\mathbf{d}} \}, \tag{12.8.12c}$$

where $\mathbf{d}$ represents the vector holding all increments in DOF of the MCT system defined over a solid body Ω, $\mathbf{X}^e$ is the connectivity matrix, and having defined the internal force vector $\mathbf{f}^e_{\text{int}}$ for an element,

$$\mathbf{f}^e_{\text{int}} = \int_{\Omega_e} (\mathbf{B}^e)^T \sum\nolimits_{t+\Delta t} d\Omega, \tag{12.8.13}$$

the external force vector $\mathbf{f}^e_{\text{ext}}$ for an element,

$$\mathbf{f}^e_{\text{ext}} = \int_{\Omega_e} (\mathbf{N}^e)^T \beta^e d\Omega + \int_{\Gamma_e} (\mathbf{N}^e)^T \boldsymbol{\tau}^e d\Gamma, \tag{12.8.14}$$

and the mass matrix for an element,

$$\mathbf{M}^e = \int_{\Omega_e} (\mathbf{N}^e)^T \mu^e \mathbf{N}^e d\Omega. \tag{12.8.15}$$

12.9 Numerical Example

12.9.1 *Void-Sheet Mechanism in High-Strength Alloy*

The MCT computational strategy introduced in this chapter is now applied to high strength alloys. In particular, the proposed hierarchical material model accounts for the void-sheet mechanisms shown in Figure 12.18.

The alloy initially deforms plastically in a homogeneous pattern, until voids nucleate at large secondary particles as shown in (2) of Figure 12.18 (Vernerey, 2006). As these voids grow, localized shear deformation, that is, inhomogeneous deformation, evolves in between these particles, as shown in (3). Soon after, a multitude of voids may nucleate at the interfaces of the numerous small secondary particles, as shown in (4), creating a *void sheet*. All voids will then grow and coalesce to induce failure of the MCT material point, as shown in (5) and (6). The mechanism shown in (2) to (6) is called a void-sheet mechanism, and is responsible for ductile modes of failure at the macroscale in high-strength alloys (Rogers, 1960).

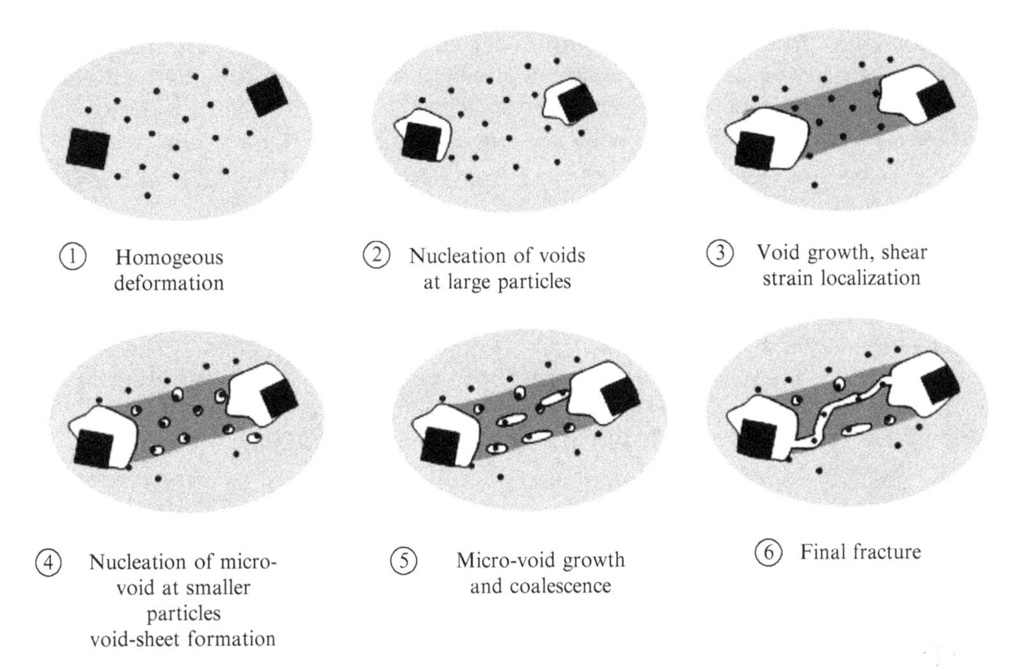

Figure 12.18 Void sheet mechanism as the inhomogeneous deformation mechanism to be captured by MCT modeling. Reproduced from Vernerey, F. J. (2006). *Multi-scale Continuum Theory for Materials with Microstructure*. Doctor of Philosophy thesis, Northwestern University

12.9.2 MCT Multiscale Constitutive Modeling Outline

Constitutive modeling starts by the introduction of RVE Ω_1 for the void-sheeting mechanisms, and Ω, of size larger than Ω_1, for the homogeneous deformation mode. Homogenization of results is then conducted. For the scale of void-sheeting, an elasticity tensor $\mathbf{C}_1$, using the rule of mixtures, is thus constructed. A plastic potential ϕ_1, using the approach outlined in Section 12.6, would then be determined for the void-sheet mechanism. In this example, however, it is assumed that $\mathbf{l}_1 \cong \mathbf{L}_1$ component wise for a void-sheeting mechanism, since the mechanism is expected to localize deformation well beyond the homogeneous level. Moreover, only the symmetric part $\mathbf{D}_1$ of the velocity gradient will be retained in this numerical example; see Exercise 12.4. The plastic potential ϕ_1 is identified with one characterizing void-sheeting, instead of a potential that penalizes deviation from a homogeneous deformation. A potential of the form,

$$\phi_1(\mathbf{s}_1, \mathbf{ss}_1) = \phi_1(\boldsymbol{\sigma}_1, \boldsymbol{\sigma\sigma}_1) = s_1^{eq} - s_1^{y} = 0, \tag{12.9.1}$$

is assumed, following (Fleck and Hutchinson, 1993; Fleck and Hutchinson, 1997; Fleck and Hutchinson, 2001), where

$$s_1^{eq} \equiv \sqrt{\frac{3}{2}\boldsymbol{\sigma}_1^{dev} : \boldsymbol{\sigma}_1^{dev} + \left(\frac{a_1}{l_1}\right)^2 \boldsymbol{\sigma\sigma}_1^{dev} \vdots \boldsymbol{\sigma\sigma}_1^{dev}}, \tag{12.9.2}$$

is the equivalent micro-stress where void-sheeting emerges, averaged over Ω_1, and s_1^y is the corresponding yield value. l_1 is a microstructural length scale characteristic of the span of double stress, and a_1 is a parameter to be adjusted numerically. No gradient plasticity is assumed, and the elasticity tensor $\mathbf{C}_1$ constructed from the RVE model suffices to fully characterize elastic gradient behavior defined by,

$$\mathbf{CC}_1 = \frac{1}{V_1^a} \int_{V_1^a} \mathbf{y}_1 \otimes \mathbf{C}_1 \otimes \mathbf{y}_1 \, dV, \tag{12.9.3}$$

where $\mathbf{y}_1$ represents half the span of Ω_1.

Homogeneous deformation is then described by an elasticity tensor $\mathbf{C}_0$, again using the rule of mixtures, but over the RVE volume Ω. A corresponding macro-plastic potential ϕ is constructed using a Gurson-type yield surface, for example see (Vernerey, Liu *et al.*, 2007),

$$\phi = \left(\frac{\sigma_0^{eff}}{\sigma_y} \right)^2 + 2q_0^1 f_0 \cosh\left(\frac{3q_0^2 p_1}{2\sigma_y} \right) - \left(1 + q_0^3 (f_0)^2\right) = 0, \tag{12.9.4}$$

where q_0^1, q_0^2, q_0^3 are constants determined by fitting to the homogenized RVE response in Ω, and f_0 evolves with void growth, nucleation, and coalescence, averaged over Ω, which characterizes void-sheeting. Different functional forms for f_1 may be specified; the one here used is outlined in (Vernerey, Liu *et al.*, 2007).

12.9.3 Finite Element Problem Setup for a Two-Dimensional Tensile Specimen

A two dimensional plane-strain model of a specimen in tension was investigated. A two-scale MCT model is compared with a direct numerical simulation (DNS). The geometry of the specimen, boundary conditions and finite element discretization are depicted in Figure 12.19.

A four-node quadrilateral element with 2 × 2 Gaussian integration was developed. For the two-scale MCT model, the degrees of freedom vector per element may be given as,

$$\mathbf{v}^e = \left[v_1(1)\, v_2(1)\, D_{11}^1(1)\, D_{22}^1(1)\, D_{12}^1(1) \cdots v_1(4)\, v_2(4)\, D_{11}^1(4)\, D_{22}^1(4)\, D_{12}^1(4) \right]^{eT}$$

For each component, the number in the parentheses designates node number, the superscript the scale of inhomogeneous deformation, and subscripts identify the component of velocity or velocity micro-strain. The corresponding shape function matrix may be expressed as,

$$\mathbf{N}^e(\mathbf{x}) = \begin{bmatrix} N_1^1 & 0 & 0 & 0 & 0 & N_1^2 & 0 & 0 & 0 & 0 & N_1^3 & 0 & 0 & 0 & 0 & N_1^4 & 0 & 0 & 0 & 0 \\ 0 & N_2^1 & 0 & 0 & 0 & 0 & N_2^2 & 0 & 0 & 0 & 0 & N_2^3 & 0 & 0 & 0 & 0 & N_2^4 & 0 & 0 & 0 \\ 0 & 0 & N_3^1 & 0 & 0 & 0 & 0 & N_3^2 & 0 & 0 & 0 & 0 & N_3^3 & 0 & 0 & 0 & 0 & N_3^4 & 0 & 0 \\ 0 & 0 & 0 & N_4^1 & 0 & 0 & 0 & 0 & N_4^2 & 0 & 0 & 0 & 0 & N_4^3 & 0 & 0 & 0 & 0 & N_4^4 & 0 \\ 0 & 0 & 0 & 0 & N_5^1 & 0 & 0 & 0 & 0 & N_5^2 & 0 & 0 & 0 & 0 & N_5^3 & 0 & 0 & 0 & 0 & N_5^4 \end{bmatrix}_{[5\times 20]}$$

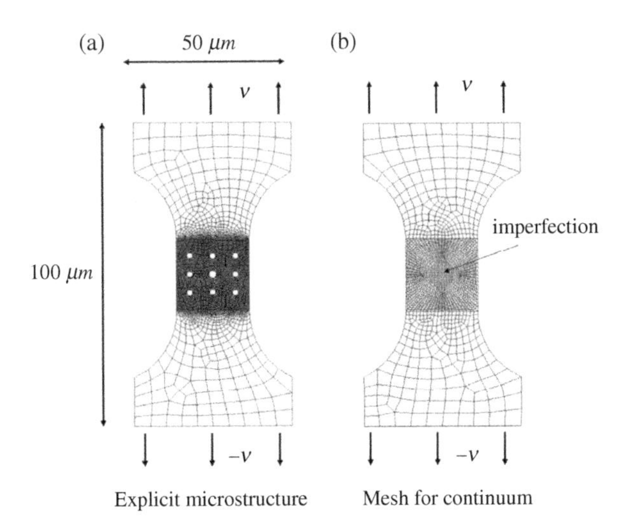

Figure 12.19 Meshes used in DNS and MCT simulations. Reproduced from Vernerey, F. J. (2006). *Multi-scale Continuum Theory for Materials with Microstructure.* Doctor of Philosophy thesis, Northwestern University

All degrees of freedom ($I = 1, \ldots, 5$) will be interpolated by bilinear functions, so that $N_1^I = N_2^I = N_3^I = N_4^I = N_5^I \equiv N_I(\mathbf{x})$ of Example 4.2, and the shape function matrix becomes,

$$
\mathbf{N}^e(\mathbf{x}) = \begin{bmatrix}
N_1 & 0 & 0 & 0 & 0 & N_2 & 0 & 0 & 0 & 0 & N_3 & 0 & 0 & 0 & 0 & N_4 & 0 & 0 & 0 & 0 \\
0 & N_1 & 0 & 0 & 0 & 0 & N_2 & 0 & 0 & 0 & 0 & N_3 & 0 & 0 & 0 & 0 & N_4 & 0 & 0 & 0 \\
0 & 0 & N_1 & 0 & 0 & 0 & 0 & N_2 & 0 & 0 & 0 & 0 & N_3 & 0 & 0 & 0 & 0 & N_4 & 0 & 0 \\
0 & 0 & 0 & N_1 & 0 & 0 & 0 & 0 & N_2 & 0 & 0 & 0 & 0 & N_3 & 0 & 0 & 0 & 0 & N_4 & 0 \\
0 & 0 & 0 & 0 & N_1 & 0 & 0 & 0 & 0 & N_2 & 0 & 0 & 0 & 0 & N_3 & 0 & 0 & 0 & 0 & N_4
\end{bmatrix}_{[5\times 20]}
$$

The MCT mass matrix is then computed, using Equation (12.8.15) and $\mathbf{N}^e(\mathbf{x})$. A corresponding generalized strain interpolation matrix $\mathbf{B}^e(\mathbf{x})$ may be expressed as,

$$
\mathrm{B}^e = \begin{bmatrix}
N_{1,x} & 0 & 0 & 0 & 0 & N_{2,x} & 0 & 0 & 0 & 0 & N_{3,x} & 0 & 0 & 0 & 0 & N_{4,x} & 0 & 0 & 0 & 0 \\
0 & N_{1,y} & 0 & 0 & 0 & 0 & N_{2,y} & 0 & 0 & 0 & 0 & N_{3,y} & 0 & 0 & 0 & 0 & N_{4,y} & 0 & 0 & 0 \\
N_{1,y} & N_{1,x} & 0 & 0 & 0 & N_{2,y} & N_{2,x} & 0 & 0 & 0 & N_{3,y} & N_{3,x} & 0 & 0 & 0 & N_{4,y} & N_{4,x} & 0 & 0 & 0 \\
0 & 0 & N_{1,x} & 0 & 0 & 0 & 0 & N_{2,x} & 0 & 0 & 0 & 0 & N_{3,x} & 0 & 0 & 0 & 0 & N_{4,x} & 0 & 0 \\
0 & 0 & N_{1,y} & 0 & 0 & 0 & 0 & N_{2,y} & 0 & 0 & 0 & 0 & N_{3,y} & 0 & 0 & 0 & 0 & N_{4,y} & 0 & 0 \\
0 & 0 & 0 & N_{1,x} & 0 & 0 & 0 & 0 & N_{2,x} & 0 & 0 & 0 & 0 & N_{3,x} & 0 & 0 & 0 & 0 & N_{4,x} & 0 \\
0 & 0 & 0 & N_{1,y} & 0 & 0 & 0 & 0 & N_{2,y} & 0 & 0 & 0 & 0 & N_{3,y} & 0 & 0 & 0 & 0 & N_{4,y} & 0 \\
0 & 0 & 0 & 0 & N_{1,x} & 0 & 0 & 0 & 0 & N_{2,x} & 0 & 0 & 0 & 0 & N_{3,x} & 0 & 0 & 0 & 0 & N_{4,x} \\
0 & 0 & 0 & 0 & N_{1,y} & 0 & 0 & 0 & 0 & N_{2,y} & 0 & 0 & 0 & 0 & N_{3,y} & 0 & 0 & 0 & 0 & N_{4,y}
\end{bmatrix}_{[9\times 20]}
$$

This yields the MCT generalized velocity-strain vector,

$$
E(\mathbf{x}) = \left[\varepsilon_{11}, \varepsilon_{22}, \gamma_{12}, D^1{}_{11,1}, D^1{}_{11,2}, D^1{}_{22,1}, D^1{}_{22,2}, D^1{}_{12,1}, D^1{}_{12,2} \right]^T .
$$

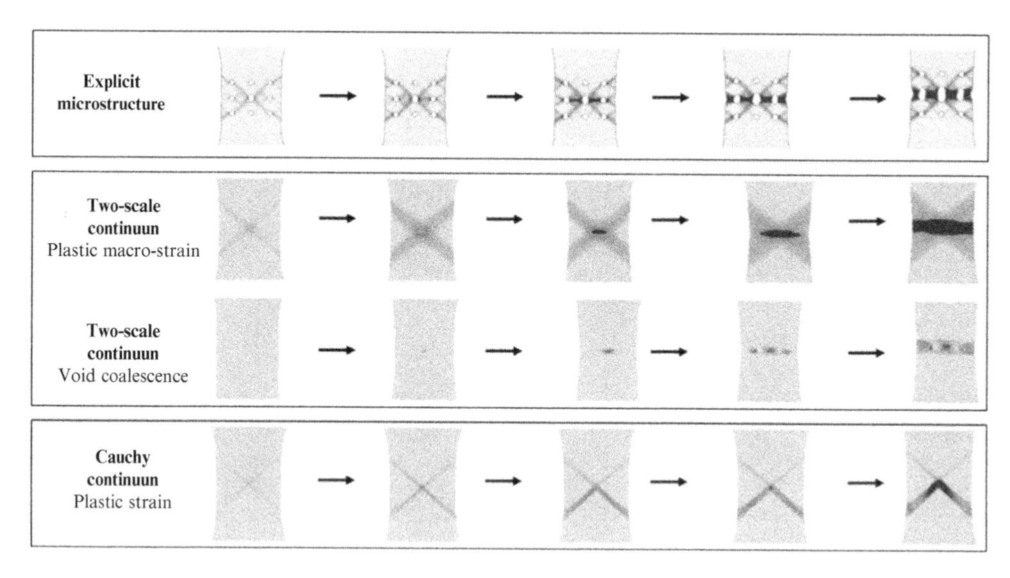

Figure 12.20 Comparing evolving plastic strain in DNS, MCT and the Cauchy-type continuum Reproduced from Vernerey, F. J. (2006). *Multi-scale Continuum Theory for Materials with Microstructure*. Doctor of Philosophy thesis, Northwestern University

For the two-scale MCT simulation, a finite element discretization that is fine enough to capture size effects at the scale of voids and imperfection is used. A higher initial porosity is specified in the constitutive model at the center of the specimen, to initiate localization.

The DNS (explicit) microstructure is represented by a periodic distribution of circular voids of diameter 1 μm, see Figure 12.19, with a 3% volume fraction. An imperfection is introduced by setting the void at the center slightly larger than others. As localization is expected to take place in the vicinity of imperfections, voids are only modeled in the central region of the specimen.

12.9.4 Results

A comparison of the macroscopic plastic strain distribution for the DNS and the two-scale material is shown in Figure 12.20. The results are displayed at four different times t_1, t_2, t_3 and t_4. At time $t=t_1$, the material response reaches a macroscopic instability point (softening of the macroscopic behavior due to void growth), and deformation begins to localize in shear band oriented at 45° with respect to the tensile direction. A stable void growth is observed at time $t=t_2$. In the two-scale continuum, this is accounted for by the appearance of a couple-micro-stress (not shown in the figure). The onset of void coalescence occurs at time $t=t_3$, corresponding to a yielding of the micro-stresses that decreases the resistance of the material to micro-deformation. Consequently, the development of a region of intense plastic strain at the central portion of the sample is seen at time t_4. By comparing these results with the strain distribution in the model of explicit microstructure, this example shows that the different stages of deformation are very well captured by the two-scale MCT model. In Figure 12.20, the evolution of the homogeneous plastic strain E_p and the inhomogeneous plastic strain E_p^1 are shown. It can be clearly seen in the figure that as

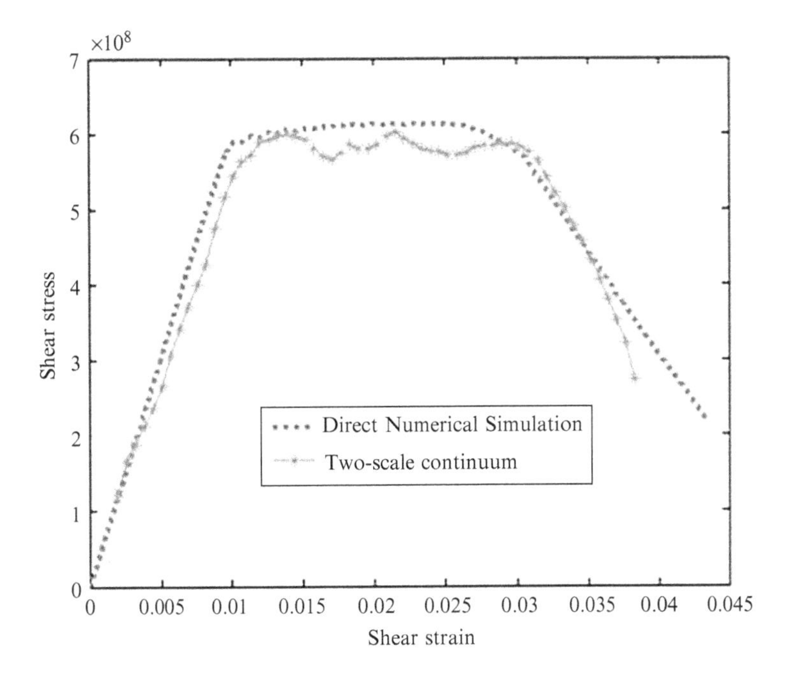

Figure 12.21 Explicit (DNS) and MCT shear stress prediction. Reproduced from Vernerey, F. J. (2006). *Multi-scale Continuum Theory for Materials with Microstructure*. Doctor of Philosophy thesis, Northwestern University

deformation localizes horizontal band forms due to the emergence of microscopic plastic strains, that is, the void-sheet mechanism.

The macroscopic shear stress response of the specimen is plotted for the DNS and the two scale MCT model in Figure 12.21. Good agreement is observed between the two curves in the first stages of deformation up to the softening point. Only in the very last stages of failure, softening of the two-scale material becomes more severe. This is attributed to a loss of size effects after void coalescence.

A simulation using a Cauchy-type continuum, that is, classical continuum, was also performed for comparison with MCT as shown in Figure 12.20, and again plotted at $t=t_4$ in Figure 12.22. The solution given by the classical continuum cannot predict the void sheet behavior after the onset of macroscopic instability. In this case, the solution is mesh dependent and localization occurs in thin shear bands at 45° to the tensile axis, behaving quite differently from the actual solution with the explicit microstructure model, while the two-scale MCT model compares well with the behavior of the explicit microstructure. The other main advantage of the two-scale MCT model resides in the gain of computational time. For instance, the typical element size for the two-scale MCT model is 10 times larger than that used in the DNS (1 μm and 0.1 μm respectively). Furthermore, no (explicit) geometric modeling of the microstructure was needed. This implies that finite element discretization of the specimen need not be modified when the microstructure (i.e., void size, spacing in this case) is changed. These features are essentially what make multi-scale MCT modeling an attractive framework for parametric studies of the nonlinear deformation of heterogeneous microstructures.

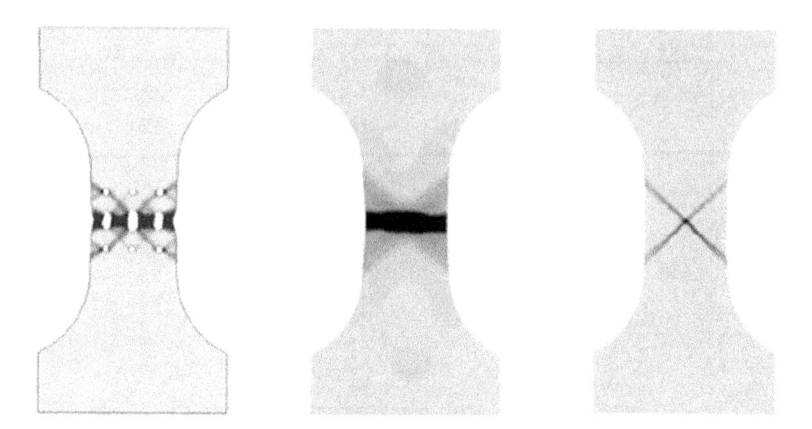

Figure 12.22 Plastic strain patterns in: *left*, explicit (DNS), *center*, MCT, and *right*, Cauchy-type continuum. Reproduced from Vernerey, F. J. (2006). *Multi-scale Continuum Theory for Materials with Microstructure*. Doctor of Philosophy thesis, Northwestern University

12.10 Future Research Directions of MCT Modeling

Application of MCT to a variety of materials, in addition to the metals and alloys herein emphasized, has been the focus of much recent research. For instance, the modeling of ceramic matrix composites (McVeigh and Liu, 2009) by MCT successfully captured the inhomogeneous shear modes that arise at the embedded interfaces, which lead to failure. Also, modeling of elastomeric nano-composites with carbon filler by MCT (Tang, Greene *et al.*, 2012) succeeded in distinguishing the different roles of cross-linked polymer chains from free chains in determining the viscoelasticity of the nano-composite in the vicinity of crack tips.

As suggested in Figure 12.7, recent efforts have also aimed at extending the multiresolution approach to situations where length scales are *not* nested (Elkhodary, Greene *et al.*, 2013), adopting a strategy that partitions a material point into contiguous neighborhoods, reminiscent of a domain decomposition strategy, and whose mechanical information is sampled along some fiber. Relaxing the assumption that neighborhoods are nested aims to facilitate the multiscale modeling of complex mesostructures that exhibit a plurality of comparable length scales.

Linking of the MCT constitutive modeling approach to rigorous statistical analysis, the need for which was suggested in Section 12.6, is also being currently studied (Greene, Liu *et al.*, 2011), in particular with an aim to quantify uncertainty in predicted bulk material properties as a function of stochastic constitutive descriptions of microstructures.

Finally, with the rapidly expanding computational capacity of modern super-computing, for example, *exascale* computing (Kogge, Bergman *et al.*, 2008), there is current research into the concurrent coupling of variants of the MCT framework to high-resolution material modeling methods, for example, crystal plasticity (see Chapter 13), phase field theory (Chen, 2002) and molecular dynamics (Yamakov, Wolf *et al.*, 2002), to better capture the localized mechanisms of inhomogeneous deformation.

12.11 Exercises

12.1. In Section 12.5.3 the kinetic power is defined for MCT theory. Beginning with the definition of kinetic energy density Equation (12.5.6), show that Equation (12.5.7) results for the kinetic power. (*Hint*: See derivation in Vernerey, Liu *et al.*, 2007).

12.2. Using the assumptions outlined in Section 12.5.4, derive the MCT strong form Equation (12.5.8) and Equation (12.5.9). (*Hint*: See derivation in Vernerey, Liu *et al.*, 2007).

12.3. Show that the stress rate $\overset{\nabla}{\mathbf{s}}_n$ and double stress rate $\overset{\nabla}{\mathbf{ss}}_n$ in Equation (12.6.4) are objective.

12.4. (a) What condition on the stresses and double stresses would permit rewriting the variation of MCT internal power as,

$$\delta P_{\text{int}} = \int_{\Omega} \left(\boldsymbol{\sigma}_0 : \delta \mathbf{D}_0 + \sum_{n=1}^{N} \left(\mathbf{s}_n : \delta \mathbf{d}_n + \mathbf{ss}_n \vdots \delta \nabla_x \mathbf{D}_n \right) \right) d\Omega,$$

where $\mathbf{d}_n = \mathbf{D}_n - \mathbf{D}_0$.

(b) What savings in degrees of freedom per node could be achieved as a result?

12.5. *Computer project*: Write a code for the finite element implementation a of two-scale MCT model, applied to a one-dimensional rod, using explicit time stepping. Each scale is to be defined as elasto-plastic. To induce localized deformation, set the yield strength and corresponding strain in the middle element of your mesh to half the values assigned to the other elements. Plot the stress distribution across the rod for different elasto-plastic properties of the imbedded scale, holding the macro-properties constant.

13

Single-Crystal Plasticity

13.1 Introduction

In this chapter, the focus will be on the computational modeling of *crystalline* materials, that is, materials where the microstructure is mostly composed of *crystals* (Hull and Bacon, 2006; Polmear, 2006). Understanding the properties of crystallinity to be captured by an appropriate constitutive model is central to the modeling of large plastic deformation.

As can be seen from Figure 13.1, it is an experimental fact that the stress-strain behavior of a large polycrystalline aggregate is different from that of a single crystal, which will define the local response of a material point in crystalline plasticity models. The local plastic response of a single-crystal is generally characterized by three distinct stages: (1) easy glide, (2) hardening, and (3) recovery, which are governed by the motion of crystalline defects (Hull and Bacon, 2006), that is, *dislocations*, as will be outlined in the coming sections of this chapter. These three stages cannot in general be discerned at the macroscale (Figure 13.1a), which may be thought of as a homogenization of all local responses in a polycrystalline aggregate.

Local and global rates of deformation also differ, the global rate being the appropriate volume average of local rates. When dislocation evolution is stable, the local plastic response will be averaged over a uniformly deforming specimen. When dislocation evolution becomes unstable at some material point, only part of the crystal becomes plastically active, and localized deformation for the single crystal will ensue (Estrin and Kubin, 1986). For specimens composed of a multitude of randomly oriented crystals, their plastic response may appear isotropic. However, single-crystals are strongly anisotropic in their plasticity. This discrepancy has been explained in terms of the preference of dislocations to nucleate in only specific

Nonlinear Finite Elements for Continua and Structures, Second Edition.
Ted Belytschko, Wing Kam Liu, Brian Moran, and Khalil I. Elkhodary.

Companion Website: www.wiley.com/go/belytschko

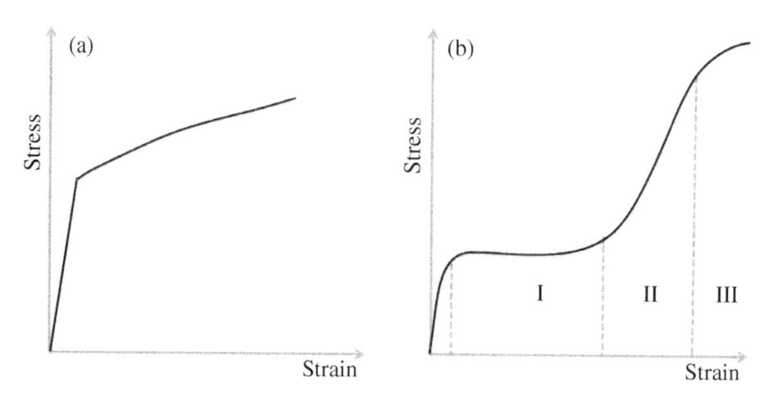

Figure 13.1 Stress-strain in (a) polycrystalline aggregate, (b) single crystal

planes of crystals, and to propagate only along certain directions contained in those planes, as will be seen later in this chapter.

In summary, while macroscopic plastic deformation may appear homogeneous and in some cases isotropic, the microscale origins of plasticity are neither homogeneous nor isotropic. Inhomogeneity of plastic deformation is an important aspect of large deformation mechanics that phenomenological constitutive laws are unable to capture properly. For instance, in applications of metal forming such as deep drawing, effects such as *spring-back* and *earing* are important to model and control, and were shown (see review in Roters, Eisenlohr *et al.*, 2010) to be dependent on the evolving plastic inhomogeneity which remains difficult for phenomenological plasticity models to capture.

Plasticity, in its atomic origins, involves the passage of atoms between equilibrium positions in the crystal (Bulatov and Cai, 2006; Hull and Bacon, 2006), so that keeping track of patterns of atomic movement is necessary for its proper description. To efficiently model plastic deformation, however, due to the large scale of engineering applications, it is not feasible to retain an atomic resolution of crystals, so a continuum approach becomes necessary. To this end, the theory of single-crystal plasticity (Rice and Asaro, 1977) has been developed, which offers unique theoretical insight into the microstructural mechanisms that govern the mechanical behavior of crystals in general, and metals and alloys in particular.

Crystalline plasticity recognizes that plastic deformation occurs along preferred planes and directions in crystals, so that only specific orientations at a material point are used to construct the plastic deformation tensors. In this manner, plastic anisotropy is automatically built into the constitutive model.

The possibility of inhomogeneous plastic deformation was shown to be dependent on the evolution of densities of dislocations (Estrin and Kubin, 1986), which are in fact the bearers of plasticity in crystals. Thus, explicit connection to dislocation densities, as internal variables, needs be made in crystal-plasticity formulations, so as to reproduce of the mechanisms of glide, hardening, and recovery that physically govern plasticity and its patterns of inhomogeneity.

Presenting single-crystal constitutive formulations that accurately characterize the micromechanics of crystals, within a finite element framework, is the aim of this chapter. Constructing appropriate homogenization techniques for polycrystals (see, for example, Van Houtte, 1987; Van Houtte, Li *et al.*, 2005) to obtain polycrystalline aggregate responses would be a natural follow-up. However, additional complications to modeling the homogenized response arise; for example, with polycrystals it is more challenging to determine active slip directions (Boyce, Weber *et al.*, 1989). In this chapter, therefore, focus will be on computational modeling of single-crystals in RVEs. The role of RVE modeling has already been outlined in Chapter 12.

The way this chapter is written is as follows. First, an extended discussion of crystals, their structures, Burgers vectors and slip systems, is presented in Sections 13.2–13.4. The discussion aims to provide a fairly detailed outline of the necessary crystalline properties that serve as *input data* for single-crystal plasticity models. In particular, the fact that not all crystals possess orthogonal bases is cause for some additional pre-processing of slip-related data, as typically presented in literature, such as the Miller indices of slip systems. A summary of the pre-processing procedure is then given in Box 13.1.

The chapter then shifts focus to the theory behind single crystal plasticity. The kinematics of slip in single crystals is first discussed in Section 13.5, along with the simplifying assumptions adopted by the crystal plasticity algorithm presented at the end of this chapter. Next, the kinetics of single crystal plasticity is considered in a bottom-up approach. The change of internal structure of crystals due to slip, represented by evolving dislocation density mechanisms, is described in Section 13.6. The associated hardening of crystals based on the evolving dislocation densities is subsequently outlined in Section 13.7. The dislocation density theory outlined in Sections 13.6 and 13.7 is then linked to the stressing of single crystal in Section 13.8 to complete the kinetic formulation.

The final algorithm for dislocation density based single crystal plasticity, combining the developments of the first eight sections, is put together in Box 13.2 in Section 13.9. Section 13.10 concludes the chapter with a numerical example.

13.2 Crystallographic Description of Cubic and Non-Cubic Crystals

In this section crystals and their different structures modeled by crystal plasticity are introduced, and a method to identify planes and directions in crystals is presented. The aim is to lay the foundation for the definition of slip systems, which define plasticity in crystals.

Crystalline materials are characterized by the fact their atoms tend to order themselves with definiteness to form a *space lattice* (Azaroff, 1984; Kelly, Groves *et al.*, 2000; Giacovazzo, Monaco *et al.*, 2002), that is, an infinite array of atoms stretching out in three-dimensional space. Each repeat-unit within the space lattice is termed a *unit cell*. By applying discrete translations to these unit cells in 3D space the lattice is generated. It has been shown that only seven *lattice systems* exist (Azaroff, 1984), defined by the shape of the unit cells, which are illustrated in Figure 13.2. The shape of a unit cell is defined by three axes, **a**, **b** and **c**, and the three angles between them, α (between **b** and **c**), β (between **a** and **c**) and γ (between **a** and **b**).

Distributed within these seven lattice systems are 14 possible *Bravais lattices*, defined by whether the unit cell is: (1) *primitive* (P), that is, having only corner *lattice points*, (2) *side-centered* (C, or B, or A) having lattice points centered on a pair of faces along an axis (**c**, **b**, or **a** respectively), in addition to the corner points, (3) *body-centered* (I), having a lattice point in the center of the unit cell in addition to the corner points, or (4) *face centered* (F), having a lattice point centered on each face, in addition to the corner points. Not every lattice system can exhibit P, C, I and F unit cells, so those that remain possible are the 14

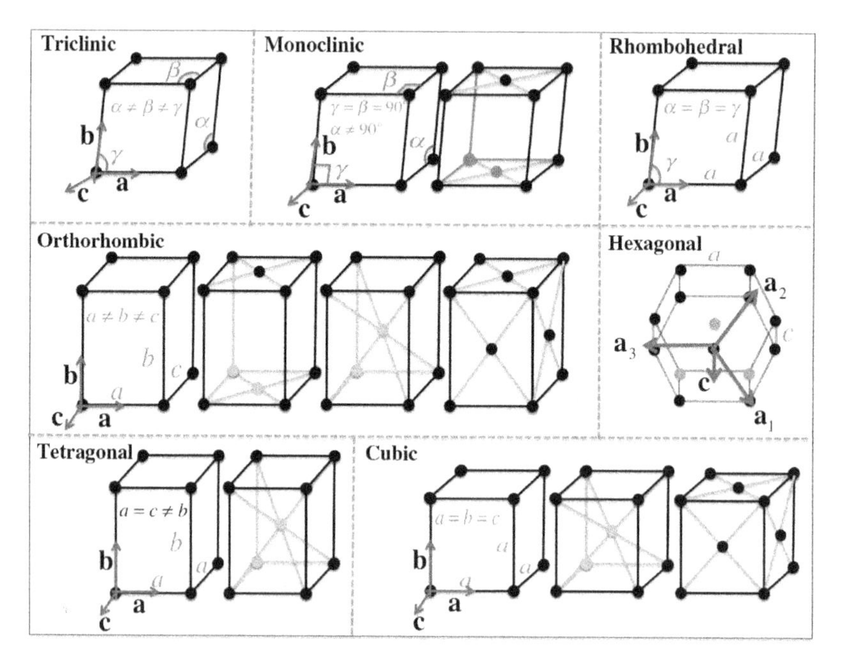

Figure 13.2 Possible unit cells in crystals

Bravais lattices shown in Figure 13.2. In metals and alloys, all points shown in the unit cells (whether on the vertices, the faces, or in the cell interior) correspond to atomic locations; however, in more general materials, these points would simply correspond with *point groups* (Azaroff, 1984). To label directions and planes of interest in these unit cells, *Miller indices* are introduced (Kelly, Groves *et al.*, 2000).

13.2.1 Specifying Directions

Miller indices for directions in the Bravais lattice can be obtained from the *fractional* coordinates of the first atom (more generally, lattice point) to intercept the desired direction vector. The coordinates of the atom are measured via a position vector pointing from an origin fixed at some corner of the unit cell (e.g., Figure 13.3). The position vector may be written in terms of the Bravais lattice *basis*, a set of three primitive vectors (call them [**x**,**y**,**z**]) that define crystallographic axes of a unit cell,

$$\mathbf{r}_{uvw} = u\mathbf{x} + v\mathbf{y} + w\mathbf{z}, \tag{13.2.1}$$

where u, v and w are respectively the **x**, **y** and **z** fractional coordinates of the atom whose position is sought. Once the position vector is defined, its coefficients are grouped in square brackets, that is, $[uvw]$, and any fractions are cleared to yield the Miller indices for that direction, written $[hkl]$ where h, k and l are integers. For example, a point of coordinates $\left(-1, \frac{3}{4}, \frac{1}{2}\right)$ will generate a direction with Miller indices $[\bar{4}32]$. By convention, negative numbers are written with an over-bar instead of a minus sign.

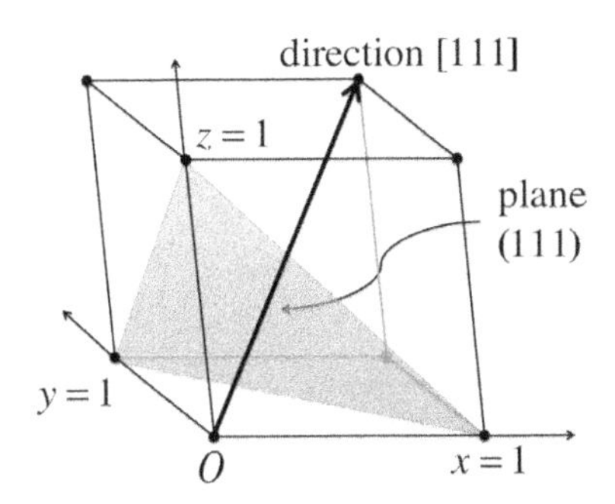

Figure 13.3 Planes and directions of a unit cell

13.2.2 Specifying Planes

Before we introduce Miller indices for planes, it is beneficial to outline the concept of a *reciprocal lattice* first (Azaroff, 1984; Kelly, Groves *et al.*, 2000; Giacovazzo, Monaco *et al.*, 2002). Having chosen for the Bravais lattice a set of three primitive vectors as a basis, call them $[\mathbf{a}_1,\mathbf{a}_2,\mathbf{a}_3]$, it is possible to define a reciprocal lattice with basis $[\mathbf{b}^1,\mathbf{b}^2,\mathbf{b}^3]$ as follows,

$$[\mathbf{b}^1,\mathbf{b}^2,\mathbf{b}^3]^T = [\mathbf{a}_1,\mathbf{a}_2,\mathbf{a}_3]^{-1}, \tag{13.2.2}$$

where each row of the inverse matrix corresponds to a basis vector in the reciprocal lattice.

More explicitly,

$$\mathbf{b}^1 = \frac{\mathbf{a}_2 \times \mathbf{a}_3}{det[\mathbf{a}_1,\mathbf{a}_2,\mathbf{a}_3]}, \mathbf{b}^2 = \frac{\mathbf{a}_3 \times \mathbf{a}_1}{det[\mathbf{a}_1,\mathbf{a}_2,\mathbf{a}_3]}, \mathbf{b}^3 = \frac{\mathbf{a}_1 \times \mathbf{a}_2}{det[\mathbf{a}_1,\mathbf{a}_2,\mathbf{a}_3]}. \tag{13.2.3}$$

It follows from the definition of the cross product that each vector $\mathbf{b}^i$ in the reciprocal lattice is *normal* to a plane defined by two corresponding basis vectors $(\mathbf{a}_j, \mathbf{a}_k)$ from the Bravais lattice. Hence, the reciprocal lattice basis and Bravais lattice basis are duals, and

$$\mathbf{b}^i \cdot \mathbf{a}_j = \delta^i_j, 1 \le i, j \le 3 \tag{13.2.4}$$

where δ^i_j is the Kronecker delta.

Any vector in the reciprocal lattice is but a linear combination of its basis vectors $\mathbf{b}^i$. As crystal planes may be identified by their normals, a natural method by which to label planes of atoms is to associate with a vector from the reciprocal lattice a corresponding plane in the Bravais lattice, as shown in Figure 13.4. Specifically, a plane of *Miller index* (*hkl*), where *h*, *k*, and *l* are integers, will be defined as the plane whose normal is the reciprocal lattice vector $\mathbf{g}^{hkl}$, given by

$$\mathbf{g}^{hkl} = h\mathbf{b}^1 + k\mathbf{b}^2 + l\mathbf{b}^3, \tag{13.2.5}$$

To visualize the corresponding plane inside the Bravais lattice note that any vector (say, $\mathbf{r}$) contained in plane (*hkl*) must satisfy the relation,

$$\mathbf{r} \cdot \mathbf{g}^{hkl} = hr_1 + kr_2 + lr_3 = 0. \tag{13.2.6}$$

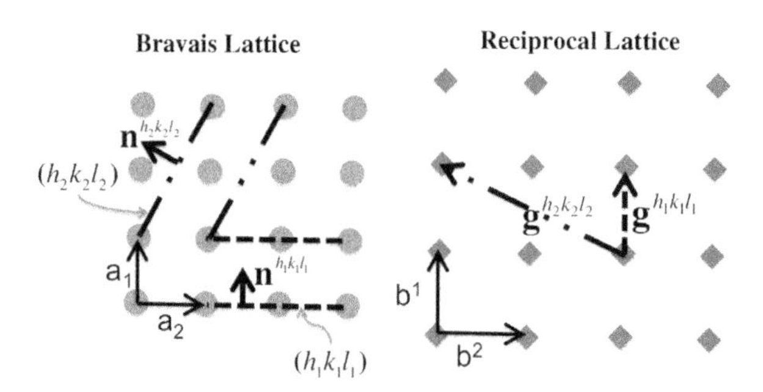

Figure 13.4 (a) Direct Bravais lattice, (b) Reciprocal lattice

This equation can be interpreted as a plane passing through the origin of the Bravais lattice. The orientation of the plane can be found from its intercepts with the crystal axes if the plane is shifted parallel to itself to set the right-hand side of Equation (13.2.6) to unity, that is, when

$$hr_1 + kr_2 + lr_3 = 1. \tag{13.2.7}$$

Then, it is immediate that the plane of Miller indices (hkl) is but one whose intercepts are $\left(\frac{1}{h}, \frac{1}{k}, \frac{1}{l}\right)$ in the Bravais lattice, as shown in Figure 13.3 for the plane (111). Conversely, given a plane of intercepts $\left(\frac{1}{h}, \frac{1}{k}, \frac{1}{l}\right)$ in a unit cell, its Miller index would be (hkl) and the corresponding unit normal is simply,

$$\mathbf{n}^{hkl} = \left|\mathbf{g}^{hkl}\right|^{-1} \mathbf{g}^{hkl}. \tag{13.2.8}$$

It is easy to verify that for cubic crystals the normal to a plane (hkl) points in the direction $[hkl]$, and it is left as an exercise (see Exercise 13.1). This observation, however, does not hold for non-cubic crystals, and calculation of reciprocal lattice vectors as outlined above becomes a requirement for proper implementation of single-crystal plasticity.

Correctly labeling planes and directions in crystals by Miller indices is especially important in materials engineering due to the directional dependency, that is, *anisotropy*, of various material properties, such as optical properties, reactivity, surface tension and dislocation motion. Anisotropic motion of dislocations, along preferential directions and planes, is fundamental to the following developments of crystal plasticity (Rice and Asaro, 1977).

13.3 Atomic Origins of Plasticity and the Burgers Vector in Single Crystals

In this section a brief discussion of the atomic origins of plasticity in crystals is presented, with an aim to define the *Burgers vector*, which is essential to the formulation of plastic shear strain in crystals. Plastic deformation in metals is *isochoric*, that is, volume-preserving or incompressible (Section 5.4.2), so that plastic shear remains the mechanism of interest, and whose modeling lays the foundation for crystalline plasticity.

If shearing forces are applied on a *perfect* crystal lattice, where all atoms are in full registry, the shear stress needed to slide a layer of atoms over another as a function of displacement x is given by (Hull and Bacon, 2006),

$$\tau = \frac{Gc}{2\pi a} \sin \frac{2\pi x}{c}, \tag{13.3.1}$$

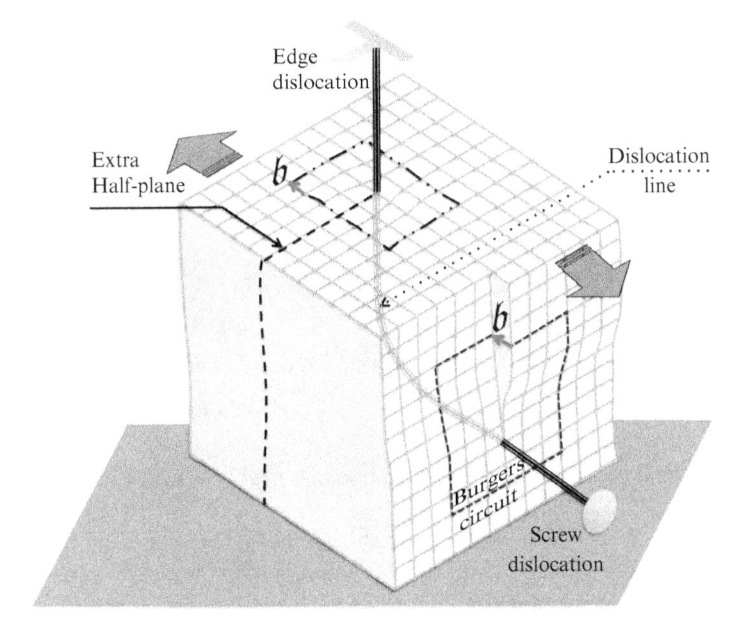

Figure 13.5 Dislocation line in a crystal

Where a is the vertical spacing between atoms, c is the horizontal spacing, and G is the shear modulus. It could thus be concluded that the theoretical shear strength (τ_{yield}) of a perfect crystal is on the order of $\frac{1}{2\pi}G$. Better estimates set the theoretical strength of perfect crystals at $\frac{1}{30}G$ (Hull and Bacon, 2006). However, experimental measurements find that the shear strength of a crystal is well below these estimates, and usually in the range (Hull and Bacon, 2006),

$$1\times10^{-8}G \le \tau_{yield} \le 1\times10^{-4}G. \tag{13.3.2}$$

To explain this difference, the *dislocation* theory was put forward (Orowan, 1934; Taylor, 1934). A dislocation is a *topological* defect (Anthony and Azirhi, 1995; Bulatov and Cai, 2006), that is, a defect in the way certain atoms in a lattice connect to their neighbors. Specifically, a dislocation is a *line defect*. It may be thought of as the boundary between slipped and un-slipped material. Two basic types of dislocations exist in a crystal: *edge* and *screw* (Hirth and Lothe, 1982), as illustrated in Figure 13.5.

An edge dislocation is caused by the termination of a plane of atoms in the middle of a crystal. It can be visualized as the edge of an extra half-plane of atoms as shown in Figure 13.6(c) and on the top surface of Figure 13.5. A screw dislocation may be visualized by slicing a crystal partway along some plane, say a vertical plane. Slide the right half against the left, such that the atoms on right come into full registry with new atoms on the left in their new positions (thus maintaining crystallinity), and after this distortion it is found that the boundary line at the tip of the slice defines the screw dislocation line. In a real crystal dislocations are *mixed*, in the sense that they could typically be resolved into edge and screw components. A mixed

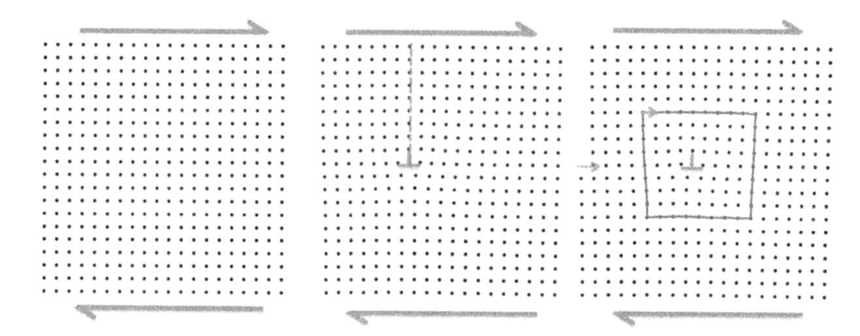

Figure 13.6 (a) square lattice under shear, (b) square lattice with dislocation under shear, (c) dislocation moves

dislocation line can be visualized as a curved line that begins as a screw and ends as an edge, as shown in Figure 13.5.

Both screw and edge dislocations induce localized residual shear stresses around their cores. In addition to shear stress, edge dislocations induce normal stress because of their natural asymmetry. On the side of the extra half-plane compression arises, and on the side of the 'missing' plane there is tension (Hull and Bacon, 2006). Dislocations, by virtue of the distortions and residual stresses they introduce in a lattice, make it easier for slip to occur in atomic layers containing them, and the energy barrier associated with their motion is low enough to explain the experimentally measured crystal shear strengths (Hull and Bacon, 2006).

Figure 13.6(a) shows the side view of a square lattice subjected to shear forces. The edge of the extra half-plane (indicated by an upside down 'T' in the figure) corresponds with the side view of the dislocation line. The topological defect created by a dislocation may be inferred from Figure 13.6(b) from the fact that atoms on a dislocation line have five nearest neighbors, instead of six as in a perfect simple cubic lattice in three dimensions.

It is customary to quantify the lattice defect a dislocation introduces by means of a *Burgers vector*. A Burgers vector measures *closure failure* in any circuit drawn through the lattice when circumscribing a dislocation (Hull and Bacon, 2006). That is, suppose a closed rectangular circuit is placed by marching through lattice points in a perfect portion of a crystal. Then, if the same circuit is placed around a dislocation, it is found that the start and end points of the circuit no longer coincide. Closure failure can arise from, for instance, the extra half plane that the circuit needs cover as shown in Figure 13.6(c). To maintain crystallinity, the magnitude of a Burgers vector must correspond to a *lattice vector*, that is, the distance between two lattice points. The magnitude of a Burgers vector ($\boldsymbol{b}$) is given by (De Graef and McHenry, 2007),

$$|\boldsymbol{b}| = \sqrt{g_{ij}\Delta s^j \Delta s^i}, \text{ (sum on } i \text{ and } j) \tag{13.3.3}$$

where $|\Delta \mathbf{s}| = |\mathbf{p}_m - \mathbf{p}_n|$ is the *repeat distance* along slip direction $\mathbf{s}$, that is, the distance between two consecutive lattice points m and n located at positions $\mathbf{p}_m$ and $\mathbf{p}_n$ respectively. g_{ij} are the components of a *metric* tensor, which helps define distances in a crystal that may exhibit non-orthonormal bases, and are given by the dot product of the lattice basis vectors, that is, $g_{ij} = \mathbf{a}_i \cdot \mathbf{a}_j$. In its general form, thus applicable to crystals of the lowest symmetry, that is, *triclinic*, to those of highest symmetry, that is, *cubic*, the metric tensor would be (De Graef and McHenry, 2007),

$$\mathbf{g} = \begin{bmatrix} a^2 & ab\cos\gamma & ac\cos\beta \\ ab\cos\gamma & b^2 & bc\cos\alpha \\ ac\cos\beta & bc\cos\alpha & c^2 \end{bmatrix}, \tag{13.3.4}$$

where lengths a, b, c and angles α, β, γ, are as defined in Figure 13.1. Its inverse $\mathbf{G}$ represents the metric in reciprocal space, and may be expressed as (De Graef and McHenry, 2007),

$$\mathbf{G} = \frac{1}{V^2}\begin{bmatrix} b^2c^2\sin^2\alpha & abc^2\Phi(\alpha,\beta,\gamma) & abc^2\Phi(\gamma,\alpha,\beta) \\ abc^2\Phi(\alpha,\beta,\gamma) & a^2c^2\sin^2\beta & abc^2\Phi(\beta,\gamma,\alpha) \\ abc^2\Phi(\gamma,\alpha,\beta) & abc^2\Phi(\beta,\gamma,\alpha) & a^2b^2\sin^2\gamma \end{bmatrix} \tag{13.3.5}$$

where $V^2 = a^2b^2c^2(1 - \cos^2\alpha - \cos^2\beta - \cos^2\gamma + 2\cos\alpha\cos\beta\cos\gamma)$ is the determinant of $\mathbf{g}$, and the function of ordered angles $\Phi(A,B,C) = \cos A\cos B - \cos C$.

For an edge dislocation, the Burgers vector is perpendicular to the dislocation line, while for the screw dislocation, the Burgers vector is parallel to the dislocation line, as can be gathered from Figure 13.5. In a mixed dislocation the Burgers vector is neither normal nor parallel to the dislocation line. A dislocation retains its identity as it moves when a crystal deforms plastically, and in that sense may be thought of as a *soliton* (Bulatov and Cai, 2006), so that its Burgers vector is conserved.

In continuum mechanics it is posited that the Burgers vector be defined by (Kroner, 1981; Bulatov and Cai, 2006; Gurtin, 2006),

$$\boldsymbol{b} = \oint \mathbf{F}^e d\mathbf{X}, \tag{13.3.6}$$

where $\mathbf{F}^e$ is the elastic part of the deformation gradient, obtainable from the multiplicative decomposition $\mathbf{F} = \mathbf{F}^e\mathbf{F}^p$ (see Section 5.7.1). This formula yields what is called the *Volterra* dislocation (Bulatov and Cai, 2006), because of the continuous topology it assumes of a crystal in lieu of a discrete lattice structure.

When deformation is entirely elastic $\mathbf{F}^e = \mathbf{F} = \nabla_0\mathbf{u}$, that is, elastic deformation becomes the gradient of the displacement vector field. Elastic deformation may thus be seen as a *conservative* field, whose potential is displacement. The closed-path integral of Equation (13.3.6) must therefore be zero, so that a zero Burgers vector results. In the case of plasticity $\mathbf{F}^e \neq \nabla_0\mathbf{u}$, which is made clear by the multiplicative decomposition of $\mathbf{F} = \mathbf{F}^e\mathbf{F}^p$. It follows that elastic deformation is not a conservative field, and the closed-path integral does not vanish, so a Burgers vector results. In this sense, a Volterra dislocation quantifies the extent to which elastic deformation fails to be conservative, which ultimately relates to the *evolving* dislocation content of a deforming crystal.

13.4 Defining Slip Planes and Directions in General Single Crystals

In this section we define slip planes and directions in crystals, which are essential to the definition of the kinematics of single crystal plasticity, and conclude with Box 13.1, which presents all the pre-processing steps performed on crystallographic data which is to be fed into the single crystal plasticity algorithm defined in the remainder of this chapter.

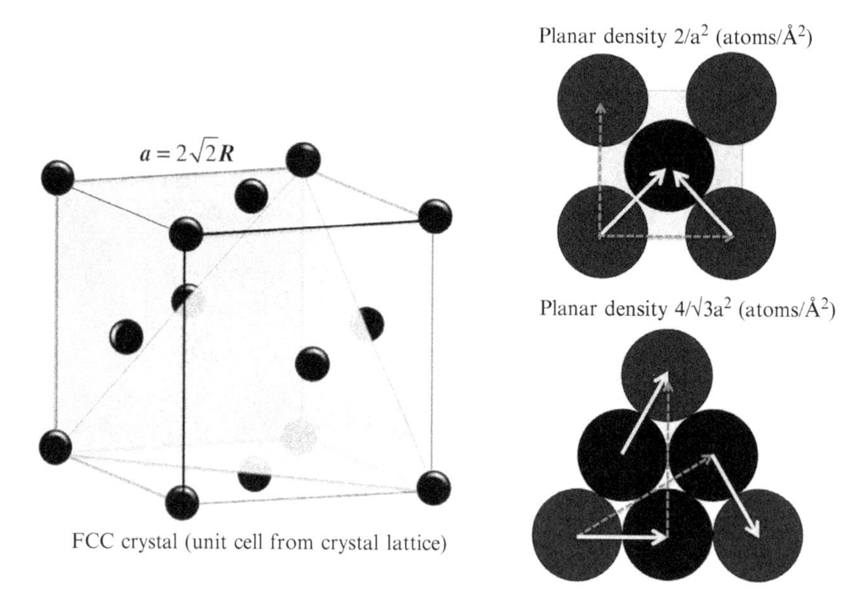

Figure 13.7 (a) FCC crystal, (b) (001) and (111) planes

The *motion* of dislocations typically characterizes plasticity in crystalline materials. As indicated by the arrows in Figure 13.6(c), application of shear activates the motion of a dislocation line, and leaves behind a step of magnitude $|\boldsymbol{b}|$ along the direction of motion, in the plane containing the dislocation, and defines a *unit* of slip. During plastic deformation of metals and alloys, multitudes of dislocations move and interact; however, it should be noted that the crystalline material between them remains undistorted (except for typically negligible elastic lattice distortions).

In general, dislocations move within the most *closely packed* crystal planes and along the most closely packed crystal directions. Figure 13.7 illustrates a face centered cubic (FCC) crystal. The (001) and (111) planes have been singled out to illustrate the difference in atomic packing between them, where it is seen that the (111) plane has a greater planar atomic density than the (001) plane. The figure also indicates in each plane by solid arrows possible slip directions (most closely packed) and by dotted arrows directions that are not closely packed. Dislocation motion can occur, for example, in the (111) plane and along the $[0\bar{1}1]$, $[\bar{1}01]$ and $[\bar{1}10]$ directions (Kelly, Groves *et al.*, 2000). These directions are members of a *family*. A family of directions is designated by angle brackets, for example, $\langle 1\bar{1}0\rangle$.

Three other distinct planes crystallographically equivalent to (111) exist in an FCC crystal, namely $(\bar{1}11)$, $(\bar{1}\bar{1}1)$, and $(1\bar{1}1)$. These four planes also belongs to a family. In general, a family of planes is designated by curly brackets, such as $\{111\}$. An example of families of planes and directions is shown in Figure 13.8.

Examples of slip planes and directions in three basic crystals are given in Table 13.1, namely FCC, body centered cubic (BCC) and hexagonal close packed crystals (HCP) (Kelly, Groves *et al.*, 2000). Notice that materials scientists find it convenient to use four indices to define planes and directions in HCP crystals, where three indices are used to describe the hexagonal base (along $\mathbf{a}_1$, $\mathbf{a}_2$ and $\mathbf{a}_3$ in Figure 13.2) and one index defines the height of the unit cell

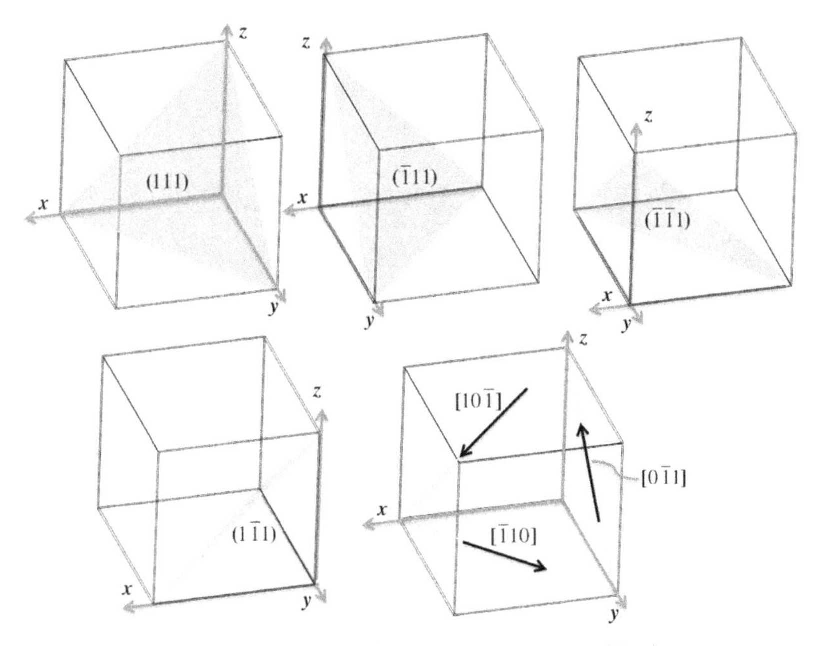

Figure 13.8 {111} family of planes in a cubic crystal, and $\langle\bar{1}10\rangle$ family of directions

Table 13.1 Slip systems

Crystal	Slip systems	Bravais Lattice (Section 13.2)
FCC metals: Al, Cu,...	$\langle 1\bar{1}0\rangle\{11$	F (Face centered)
BCC metals: Fe, W, Na...	$\langle 1\bar{1}1\rangle\{110\}$	I (Body centered)
HCP metals: Zn, Mg, Cd...	$\langle 11\bar{2}0\rangle(0001), \langle 11\bar{2}0\rangle\{10\bar{1}1\}$ $\langle 11\bar{2}0\rangle\{10\bar{1}0\}, \langle 11\bar{2}0\rangle\{11\bar{2}2\}$	P (Primitive)

(along the **c**-axis). To recover the three-index notation, the first and third index should be added, that is: $(a_1,a_2,a_3,d)\rightarrow(a_2,(a_1+a_3),d)$ in HCP crystals. This three-index notation is better suited to the crystal-plasticity algorithm herein developed.

In general, a pair of vectors is always needed to characterize dislocation motion: slip plane normal **n** and slip direction **s**. This pair of vectors defines a *slip system* (Table 13.1) and ultimately defines plasticity in crystalline materials. The pair of vectors must satisfy the relation $\mathbf{n}\cdot\mathbf{s}=0$ to form a slip system. Thus [110] and (111) do not form a slip system, since the dot product has a value of 2. On the other hand, the pair [$1\bar{1}0$] and (111), or [$\bar{1}10$] and (111), or [110] and ($\bar{1}11$), and so on, can form a slip system, since the dot product is zero.

In many applications, the slip systems of a crystal have a predetermined orientation relation with respect to the loading axis. For example, the cold rolling of metal billets may often induce preferential crystallographic orientation of grains, which is referred to as *texture*

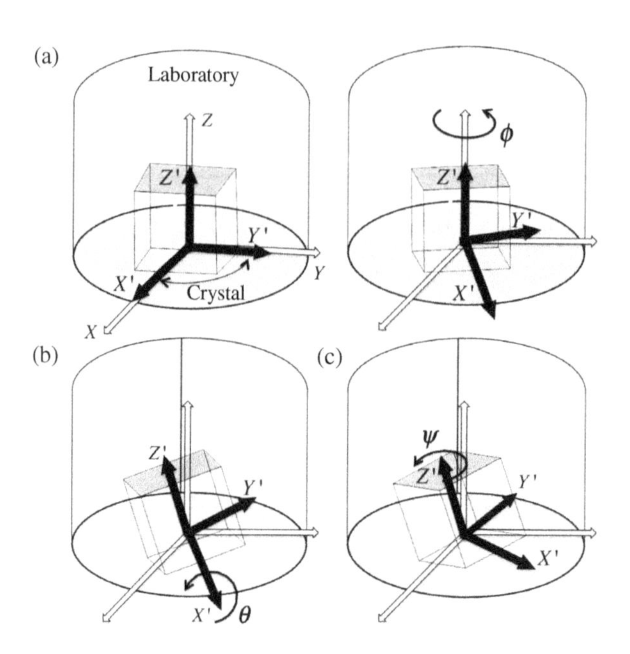

Figure 13.9 Euler angles that specify crystal orientation

(Van Houtte, Li *et al.*, 2005). When loading a cold-rolled specimen, therefore, crystal orientations will have specific distributions with respect to the loading axis, a phenomenon that greatly contributes to anisotropic plastic deformation (Roters, Eisenlohr *et al.*, 2010). This orientation effect can be easily captured in single-crystal plasticity by applying the following transformation to the slip directions and normals of slip system α of the crystal $(\mathbf{s}_c^\alpha, \mathbf{n}_c^\alpha)$ to map them to the laboratory frame $(\mathbf{s}_l^\alpha, \mathbf{n}_l^\alpha)$,

$$\left[\mathbf{s}_l^\alpha, \mathbf{n}_l^\alpha\right] = \mathbf{T}_l^c\left[\mathbf{s}_c^\alpha, \mathbf{n}_c^\alpha\right], \tag{13.4.1}$$

where $\mathbf{T}_c^l$ is the 3D rotation matrix defined by Euler angle triplets (ϕ, θ, ψ). There are multiple conventions that determine Euler angles, depending on the sequence of rotations. A common choice for crystallographers is the *Bunge* convention (Randle and Engler, 2009). A 3D rotation is in this convention assumed to arise from the sequence shown in Figure 13.9: (1) a rotation ϕ about the Z-axis, which will change the orientation of vectors in the X-Y plane (2) a rotation θ about the X-axis, which is in now in a new position due to application of (1), and (3) a rotation ψ about the Z-axis in its new position after application of (1) and (2). As such, the transformation matrix for a 3D rotation may be given by,

$$\mathbf{T}_c^l = \begin{bmatrix} c_1c_3 - c_2s_1s_3 & -c_1s_3 - c_2c_3s_1 & s_1s_2 \\ c_3s_1 + c_1c_2s_3 & c_1c_2c_3 - s_1s_3 & -c_1s_2 \\ s_2s_3 & c_3s_2 & c_2 \end{bmatrix} \tag{13.4.2}$$

where $c_i = \mathbf{cos}(\delta_i)$, $s_i = \mathbf{sin}(\delta_i)$ and δ_i takes the value of ϕ, θ, or ψ when $i = 1, 2$ or 3, respectively. Euler angles may be read off experimentally obtained *pole figures* (Randle and Engler, 2009). Pole figures are two-dimensional plots of orientation distributions of crystal planes in a material sample; they are briefly discussed in Appendix 4.

A summary is given in Box 13.1 of the main equations in Sections 13.2–13.4. The box generates the slip-related input data for single crystal plasticity material subroutines. The plasticity algorithm using this input data is developed in the remainder of this chapter.

Box 13.1 Pre-processing: determining initial slip directions and normals

1. Determine three basis vectors for the crystal to be modeled, $\{\mathbf{a}_1, \mathbf{a}_2, \mathbf{a}_3\}$
2. Normalize the basis vectors, $\mathbf{a}_i = \mathbf{a}_i \Big/ \sqrt{a_{1i}^2 + a_{2i}^2 + a_{3i}^2}$, $i \in \{1, 2, 3\}$
3. Get the Miller indices for all slip planes $\{(hkl)^\alpha, 1 \le \alpha \le nss\}$ and slip directions $\{(uvw)^\alpha, 1 \le \alpha \le nss\}$ of the crystal. This information may be found from literature, or by selecting the densest planes and directions of the crystal in absence of experimental data.
4. Define and normalize each of the slip direction vectors as follows,

$$\mathbf{s}^{uvw} = (u^2 + v^2 + w^2)^{-1/2}[u\mathbf{a}_1 + v\mathbf{a}_2 + w\mathbf{a}_3]$$

5. Pair each slip direction vector with the appropriate Burgers vector magnitude. Information on Burgers vectors for specific crystals may be found in literature, or by calculating the repeat distance of the lattice along the corresponding slip direction using,

$$|\boldsymbol{b}| = \sqrt{g_{ij} \Delta s^j \Delta s^i}$$

where the metric $\mathbf{g}$ is defined by Eq. 13.3.4, and $|\Delta\mathbf{s}| = |\mathbf{p}_m - \mathbf{p}_n|$ is the repeat distance.

6. Transform Miller indices of planes to normal vectors using the reciprocal lattice construct,

For cubic crystals: $\mathbf{n}^{hkl} = (h^2 + k^2 + l^2)^{-1/2}[h\mathbf{e}_1 + k\mathbf{e}_2 + l\mathbf{e}_3]$

For non-cubic crystals: $\mathbf{n}^{hkl} = |\mathbf{g}^{hkl}|^{-1}\mathbf{g}^{hkl}$, where $\mathbf{g}^{hkl}$ is defined by Equation (13.2.5).

7. Align the slip plane normal $\mathbf{n}^{hkl}$ and slip direction $\mathbf{s}^{uvw}$ of a crystal with respect to a pre-defined loading axis, or according to sample texture, by applying the transformation,

$$\left[\mathbf{s}_l^\alpha, \mathbf{n}_l^\alpha\right] = \mathbf{T}_l^c \left[\mathbf{s}_c^\alpha, \mathbf{n}_c^\alpha\right]$$

with $\mathbf{T}_l^c$ as defined in Equation (13.4.2).

As a further remark, there are situations where a crystal's spatial orientation must follow that of its surrounding material. For example, precipitate crystals may nucleate and grow along energetically favorable planes (*habit planes*) and in specific directions of the surrounding matrix crystal, resulting in what is known as *rational orientation relations*; see examples in (Wang and Starink, 2005). Such a case can also be modeled by single-crystal plasticity, provided the right sequence of plane and direction transformations is first applied; see, for example, Elkhodary, Lee *et al.* (2011).

Example 13.1 Determine the slip planes and normals of an FCC crystal having Euler angles (0°, 60°, 0°) with respect to the sample axes

Solution From Table 13.1, the family of directions and planes of interest in FCC crystals is the $\langle 1\bar{1}0\rangle\{111\}$. Only the pairs satisfying the relation $\mathbf{n}\cdot\mathbf{s}=0$ need be considered. There are only 12 independent such pairs, if we identify positive slip and negative slip directions with

Table E13.1 FCC directions and normals, unrotated and with Euler angles (0°, 60°, 0°)

Slip system number	Direction			Plane		
	Miller indices	Vector $\mathbf{s}_c$	Transformed Vector ($\mathbf{s}_l$)	Miller indices	Vector $\mathbf{n}_c$	Transformed Vector ($\mathbf{n}_l$)
1	$[\bar{1}01]$	$\frac{1}{\sqrt{2}}[-1,0,1]$	[−0.7071 −0.6124 0.3536]	(111)	$\frac{1}{\sqrt{3}}[1,1,1]$	[0.5774 −0.2113 0.7887]
2	$[\bar{1}10]$	$\frac{1}{\sqrt{2}}[-1,1,0]$	[−0.7071 0.3536 0.6124]	(111)	$\frac{1}{\sqrt{3}}[1,1,1]$	[0.5774 −0.2113 0.7887]
3	$[0\bar{1}1]$	$\frac{1}{\sqrt{2}}[0,-1,1]$	[0 −0.9659 −0.2588]	(111)	$\frac{1}{\sqrt{3}}[1,1,1]$	[0.5774 −0.2113 0.7887]
4	[011]	$\frac{1}{\sqrt{2}}[0,1,1]$	[0 −0.2588 0.9659]	$(\bar{1}\bar{1}1)$	$\frac{1}{\sqrt{3}}[-1,-1,1]$	[−0.5774 −0.7887 −0.2113]
5	$[\bar{1}10]$	$\frac{1}{\sqrt{2}}[-1,1,0]$	[−0.7071 0.3536 0.6124]	$(\bar{1}\bar{1}1)$	$\frac{1}{\sqrt{3}}[-1,-1,1]$	[−0.5774 −0.7887 −0.2113]
6	[101]	$\frac{1}{\sqrt{2}}[1,0,1]$	[0.7071 −0.6124 0.3536]	$(\bar{1}\bar{1}1)$	$\frac{1}{\sqrt{3}}[-1,-1,1]$	[−0.5774 −0.7887 −0.2113]
7	[101]	$\frac{1}{\sqrt{2}}[1,0,1]$	[0.7071 −0.6124 0.3536]	$(\bar{1}11)$	$\frac{1}{\sqrt{3}}[-1,1,1]$	[−0.5774 −0.2113 0.7887]
8	[110]	$\frac{1}{\sqrt{2}}[1,1,0]$	[0.7071 0.3536 0.6124]	$(\bar{1}11)$	$\frac{1}{\sqrt{3}}[-1,1,1]$	[−0.5774 −0.2113 0.7887]
9	$[0\bar{1}1]$	$\frac{1}{\sqrt{2}}[0,-1,1]$	[0 −0.9659 −0.2588]	$(\bar{1}11)$	$\frac{1}{\sqrt{3}}[-1,1,1]$	[−0.5774 −0.2113 0.7887]
10	[011]	$\frac{1}{\sqrt{2}}[0,1,1]$	[0 −0.2588 0.9659]	$(1\bar{1}1)$	$\frac{1}{\sqrt{3}}[1,-1,1]$	[0.5774 −0.7887 −0.2113]
11	[110]	$\frac{1}{\sqrt{2}}[1,1,0]$	[0.7071 0.3536 0.6124]	$(1\bar{1}1)$	$\frac{1}{\sqrt{3}}[1,-1,1]$	[0.5774 −0.7887 −0.2113]
12	$[\bar{1}01]$	$\frac{1}{\sqrt{2}}[-1,0,1]$	[−0.7071 −0.6124 0.3536]	$(1\bar{1}1)$	$\frac{1}{\sqrt{3}}[1,-1,1]$	[0.5774 −0.7887 −0.2113]

the same system. FCC slip systems can be represented by the Miller indices in Table E13.1. The slip vectors ($\mathbf{s}_c$) are obtained by simply normalizing the direction Miller indices to unity, as shown in Table E13.1. Slip normals ($\mathbf{n}_c$) in the table are obtained for FCC, that is, cubic, crystals by applying the equation: $\mathbf{n}^{hkl} = (h^2 + k^2 + l^2)^{-1/2}[h\mathbf{e}_1 + k\mathbf{e}_2 + l\mathbf{e}_3]$, according to Step 6 in Box 13.1. Using $\mathbf{T}_l^C$ as defined in Equation (13.4.2) with Euler angles (0, π/3,0) we find,

$$\mathbf{T}_l^c = \begin{bmatrix} 1 & 0 & 0 \\ 0 & 0.5 & -0.866 \\ 0 & 0.866 & 0.5 \end{bmatrix}.$$

Finally, by Step 7 in Box 13.1, $\mathbf{s}_l$ and $\mathbf{n}_l$ in Table E13.1 are obtained.

13.5 Kinematics of Single Crystal Plasticity

In this section we outline the details of the kinematic formulation necessary for the implementation of the single crystal plasticity and conclude with expedient simplifications adopted in the algorithm developed in this chapter along with some final remarks.

13.5.1 Relating the Intermediate Configuration to Crystalline Mechanics

As indicated in Figure 13.10, the structure of a crystalline solid at the mesoscale may be viewed as an aggregate of volume elements, or *lattice blocks*, which are undeformed parts of the lattice, potentially delineated by dislocations. For crystalline solids subjected to a continuous deformation gradient $\mathbf{F}$, two basic mechanisms of *deformation* are assumed to take place

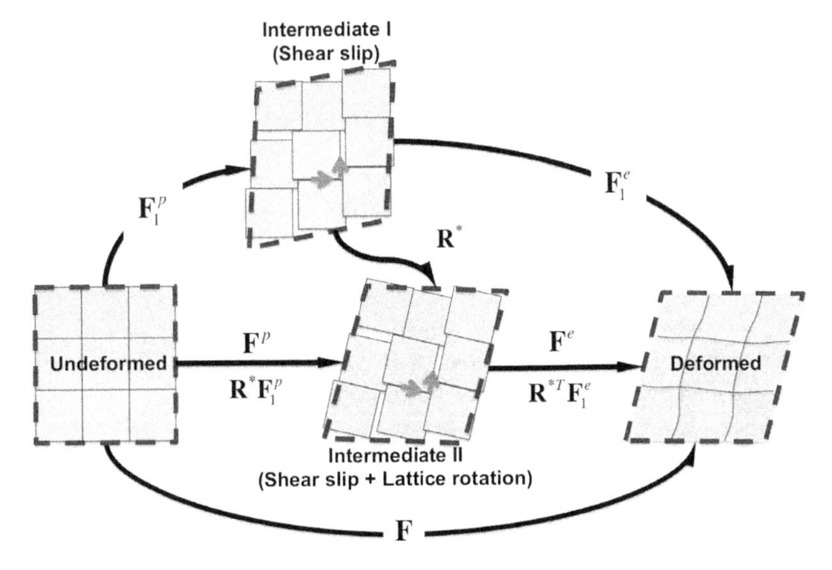

Figure 13.10 Kinematics of a continuously deformed crystal

concurrently, (1) slip between lattice blocks, followed by elastic lattice distortions, and (2), lattice *rotations* also arise under a general deformation gradient **F** and account for observable phenomena such as the *geometric softening* of crystals (Asaro, 1979). Decomposition of the gradient **F** into these deformation and rotation mechanisms would thus facilitate the constitutive description of single crystals. Two decompositions of **F** have been typically adopted, differing by whether to first introduce lattice rotations in the deformed configuration (Asaro, 1983) or instead the intermediate configuration (Onat, 1982), as shown in Figure 13.10.

In the first approach, the intermediate configuration is plastically deformed, stress-free and its lattice remains aligned with that in the undeformed configuration. Such a configuration is achieved by a deformation designated as $\mathbf{F}_1^p$. To arrive at the deformed configuration, which is stressed and whose lattice has undergone the necessary rotations, an elastic deformation designated by $\mathbf{F}_1^e$ needs be applied. This leads to a multiplicative decomposition of the deformation gradient into elastic and plastic parts (Section 5.7.1),

$$\mathbf{F} = \mathbf{F}_1^e \mathbf{F}_1^p. \tag{13.5.1a}$$

In the second approach, the intermediate configuration is also plastically deformed, and stress-free, but the lattice is rotated to its final orientation by $\mathbf{R}^*$. This intermediate configuration is reached by a plastic distortion $\mathbf{F}^p$. The deformed configuration is then reached from the intermediate by application of a symmetric elastic distortion $\mathbf{F}^e$, which accounts for lattice stretching. The second approach leads to the decomposition,

$$\mathbf{F} = \mathbf{F}^e \mathbf{F}^p. \tag{13.5.1b}$$

The two approaches are related. As shown in Figure 13.10, intermediate configuration II follows from intermediate configuration I by a lattice rotation, so that $\mathbf{F}^p = \mathbf{R}^* \mathbf{F}_1^p$. Similarly, the elastic distortion of the second approach relates to the first, since $\mathbf{F}^e = \mathbf{R}^{*T} \mathbf{F}_1^e$. As $\mathbf{F}^e$ is required to be symmetric, it can be shown that $\mathbf{F}^e = \mathbf{V}_1^e$, the left stretch tensor obtained from the polar decomposition $\mathbf{F}_1^e = \mathbf{V}_1^e \mathbf{R}_1$, and $\mathbf{R}_1 = \mathbf{R}^*$, the rotation tensor from the same polar decomposition equals the lattice rotation (see Exercise 13.2). The two approaches have been shown to yield identical results, so that the choice of one or the other is mostly a matter of convenience (Boyce, Weber *et al.*, 1989). Tensors that are defined over intermediate configuration II will be denoted with an over-bar, while those in intermediate configuration I will use a tilde.

The velocity gradient ($\mathbf{L} = \dot{\mathbf{F}}\mathbf{F}^{-1}$) is similarly decomposed into a an elastic part and a plastic part by plugging Equation (13.5.1b) into its definition (see Equation 5.7.8),

$$\mathbf{L} = \dot{\mathbf{F}}^e (\mathbf{F}^e)^{-1} + \mathbf{F}^e \dot{\mathbf{F}}^p (\mathbf{F}^p)^{-1} (\mathbf{F}^e)^{-1} \tag{13.5.2}$$

Denoting by $\bar{\mathbf{L}}^p$ the product $\dot{\mathbf{F}}^p (\mathbf{F}^p)^{-1}$, which designates the plastic part of the velocity gradient viewed in intermediate configuration II, we can further define the plastic rate of deformation and spin by the additive decomposition,

$$\bar{\mathbf{L}}^p = \bar{\mathbf{D}}^p + \bar{\mathbf{W}}^p \tag{13.5.3}$$

where $\bar{\mathbf{D}}^p = sym(\bar{\mathbf{L}}^p)$ and $\bar{\mathbf{W}}^p = skew(\bar{\mathbf{L}}^p)$. Note that in intermediate configuration II the symmetric and skew symmetric parts of $\bar{\mathbf{L}}^p$ need be written in covariant form (see Section 5.10), so that they may be identified with the pullbacks of their spatial counterparts. Thus, the tensor $\bar{\mathbf{C}}^e = \mathbf{F}^{eT}\mathbf{F}^e$ must be used in these definitions of *sym*() and *skew*(), as in Equation (5.7.12).

Due to the symmetry imposed on $\mathbf{F}^e$, it can be shown that $\bar{\mathbf{W}}^p$ is fully defined by the algebraic relation (Onat, 1982; Boyce, Weber *et al.*, 1989),

$$\bar{\mathbf{W}}^p = \mathbf{W} - \mathcal{W} : (\mathbf{D} + \bar{\mathbf{D}}^p), \tag{13.5.4}$$

where $\mathcal{W}$ is a fourth order tensor with components of the order of elastic strain (see Exercise 13.5). A constitutive law, on the other hand is required for the definition of $\bar{\mathbf{D}}^p$, which will be discussed presently. First, recall that intermediate configuration II undergoes both shear slip and lattice rotation. The plastic deformation $\mathbf{F}^p$ in Equation (13.5.2) can therefore be decomposed as $\mathbf{F}^p = \mathbf{R}^*\mathbf{F}^{slip}$, where $\mathbf{F}^{slip} = \mathbf{F}_1^p$ in Figure 13.10 accounts for the plastic deformation in the lattice, and $\mathbf{R}^*$ is the lattice rotation. Plugging this decomposition into the definition of $\bar{\mathbf{L}}^p$ (i.e. $\dot{\mathbf{F}}^p(\mathbf{F}^p)^{-1}$) we find,

$$\bar{\mathbf{L}}^p = \dot{\mathbf{R}}^*\mathbf{R}^{*T} + \mathbf{R}^*\dot{\mathbf{F}}_1^p\left(\mathbf{F}_1^p\right)^{-1}\mathbf{R}^{*T}. \tag{13.5.5}$$

From Equations 13.5.5 and 13.5.3 we find that $\bar{\mathbf{D}}^p = \mathbf{R}^*\tilde{\mathbf{D}}^p\mathbf{R}^{*T}$, where $\tilde{\mathbf{D}}^p \equiv sym\left(\dot{\mathbf{F}}_1^p\left(\mathbf{F}_1^p\right)^{-1}\right)$ is the rate of plastic deformation in the un-rotated lattice, that is, in intermediate configuration I in Figure 13.10. $\bar{\mathbf{D}}^p$ is thus the push-forward of $\tilde{\mathbf{D}}^p$ from intermediate configuration I to II. $\bar{\mathbf{W}}^p = \dot{\mathbf{R}}^*\mathbf{R}^{*T} + \mathbf{R}^*\tilde{\mathbf{W}}^p\mathbf{R}^{*T}$, where $\tilde{\mathbf{W}}^p \equiv skew\left(\dot{\mathbf{F}}_1^p\left(\mathbf{F}_1^p\right)^{-1}\right)$ is the spin in the un-rotated lattice. Thus, $\bar{\mathbf{W}}^p$ combines the push-forward of $\tilde{\mathbf{W}}^p$ from intermediate configuration I to II and the lattice spin, which is designated henceforth as $\mathbf{\Omega}^* \equiv \dot{\mathbf{R}}^*\mathbf{R}^{*T}$. Note that in configuration I tensor $\tilde{\mathbf{C}}^e = \mathbf{F}_1^{eT}\mathbf{F}_1^e$ should be used in the definitions of *sym*() and *skew*().

13.5.2 Constitutive Definitions of the Plastic Parts of Deformation Rate and Spin

In real crystals, multiple slip systems may admit dislocation motion at any one time. Such a situation gives rise to *poly-slip*, that is, concurrent slip in multiple systems, and depends on a crystal's orientation with respect to the load. Now, in intermediate configuration I, where the lattice is elastically undistorted and un-rotated, the plastic parts of deformation rate $\tilde{\mathbf{D}}^p$ and spin $\tilde{\mathbf{W}}^p$ can be related to the instantaneous crystallographic shear slip-rates in all active slip systems through the definition,

$$\tilde{\mathbf{D}}^p = \sum_\alpha \tilde{\mathbf{P}}^\alpha\dot{\gamma}^\alpha, \quad \tilde{\mathbf{W}}^p = \sum_\alpha \tilde{\mathbf{Q}}^\alpha\dot{\gamma}^\alpha, \tag{13.5.6}$$

where $\tilde{\mathbf{P}}^\alpha = sym(\tilde{\mathbf{S}}^\alpha)$, $\tilde{\mathbf{Q}}^\alpha = skew(\tilde{\mathbf{S}}^\alpha)$ and $\tilde{\mathbf{S}}^{(\alpha)} = \tilde{\mathbf{s}}^{(\alpha)} \otimes \tilde{\mathbf{n}}^{(\alpha)}$ is the *Schmidt tensor* (Rice and Asaro, 1977) defining slip system α, that is, the dyadic product defined by slip-direction vector $\tilde{\mathbf{s}}^\alpha$ and slip-plane normal $\tilde{\mathbf{n}}^\alpha$, viewed in intermediate configuration I. Again, to obtain covariant components, *sym*() and *skew*() should use $\tilde{\mathbf{C}}^e$ in these definitions. Deformation rate $\tilde{\mathbf{D}}^p$ and spin $\tilde{\mathbf{W}}^p$ are therefore expressible as linear combinations of shear slip evolving in all slip systems, which collectively span all possible orientations along which a crystal can plastically deform. It should be emphasized that, according to Equation (13.5.6), a crystal can only deform plastically along special planes and directions, unlike the homogeneous continuum material point, which is classically assumed capable to strain in any direction.

Note that plastic deformation of single crystals is essentially incompressible by the kinematic formulation of Equation (13.5.6), which can be understood from the following consideration. The plastic dilatation rate may be computed from (compare Equation 5.7.40),

$$\tilde{\mathbf{C}}^{e-1} : \tilde{\mathbf{D}}^{p} = \sum_{\alpha} \tilde{\mathbf{C}}^{e-1} : \tilde{\mathbf{P}}^{\alpha} \dot{\gamma}^{\alpha} = 0, \tag{13.5.7}$$

using the orthogonality of slip direction $\tilde{\mathbf{s}}^{\alpha}$ and slip plane normal $\tilde{\mathbf{n}}^{\alpha}$ (see Exercise 13.3).

As seen in Figure 13.10, intermediate configuration II requires rotating the lattice by $\mathbf{R}^{*}$. Thus, $\bar{\mathbf{P}}^{\alpha}$ and $\bar{\mathbf{Q}}^{\alpha}$, which are push-forwards of those in Equation (13.5.6), need be defined as,

$$\begin{aligned} \bar{\mathbf{P}}^{\alpha} &\equiv \mathbf{R}^{*} \tilde{\mathbf{P}}^{\alpha} \mathbf{R}^{*T} \\ \bar{\mathbf{Q}}^{\alpha} &\equiv \mathbf{R}^{*} \tilde{\mathbf{Q}}^{\alpha} \mathbf{R}^{*T} \end{aligned} \tag{13.5.8}$$

Hence, with respect to Equation (13.5.3), the plastic rate of deformation and spin are given by,

$$\bar{\mathbf{D}}^{p} = \sum_{\alpha} \bar{\mathbf{P}}^{\alpha} \dot{\gamma}^{\alpha}, \quad \bar{\mathbf{W}}^{p} = \Omega^{*} + \sum_{\alpha} \bar{\mathbf{Q}}^{\alpha} \dot{\gamma}^{\alpha}. \tag{13.5.9}$$

$\bar{\mathbf{P}}^{\alpha}$ and $\bar{\mathbf{Q}}^{\alpha}$ need be stored and updated as tensorial internal variables in a crystal plasticity algorithm to correctly specify lattice orientation in intermediate configuration II, as $\mathbf{R}^{*}$ evolves with time in a simulation. The corotational rates of $\bar{\mathbf{P}}^{\alpha}$ and $\bar{\mathbf{Q}}^{\alpha}$ are given as,

$$\begin{aligned} \overset{\nabla}{\bar{\mathbf{P}}^{\alpha}} &= \dot{\bar{\mathbf{P}}}^{\alpha} - \Omega^{*} \bar{\mathbf{P}}^{\alpha} + \bar{\mathbf{P}}^{\alpha} \Omega^{*} = \mathbf{0} \\ \overset{\nabla}{\bar{\mathbf{Q}}^{\alpha}} &= \dot{\bar{\mathbf{Q}}}^{\alpha} - \Omega^{*} \bar{\mathbf{Q}}^{\alpha} + \bar{\mathbf{Q}}^{\alpha} \Omega^{*} = \mathbf{0}. \end{aligned} \tag{13.5.10}$$

Any rotation $\mathbf{R}$ of intermediate configuration II between time t and $t + \Delta t$ needs satisfy the differential equation,

$$\dot{\mathbf{R}} = \Omega^{*} \mathbf{R}, \tag{13.5.11}$$

during a given time step of a simulation, with initial condition $\mathbf{R}(t) = \mathbf{I}$. Solving for $\mathbf{R}(t + \Delta t)$ from Equation (13.5.11), for example, using the Hughes–Winget update in Section 9.5.18, it is then possible to update $\bar{\mathbf{P}}^{\alpha}$ and $\bar{\mathbf{Q}}^{\alpha}$ using,

$$\begin{aligned} \bar{\mathbf{P}}^{\alpha}_{t+\Delta t} &= \mathbf{R}_{t+\Delta t} \bar{\mathbf{P}}^{\alpha}_{t} \mathbf{R}^{T}_{t+\Delta t} \\ \bar{\mathbf{Q}}^{\alpha}_{t+\Delta t} &= \mathbf{R}_{t+\Delta t} \bar{\mathbf{Q}}^{\alpha}_{t} \mathbf{R}^{T}_{t+\Delta t} \end{aligned} \tag{13.5.12}$$

13.5.3 Simplification of the Kinematics by Restriction to Small Elastic Strain

It is possible to simplify the foregoing kinematic formulations when elastic stretching of the lattice is ignored. For the large deformation of metal crystals, such an assumption is particularly well suited. In this case, $\mathbf{F}^{e} = \mathbf{V}^{e}_{1} = \mathbf{I}$. Then, intermediate configuration II becomes practically indistinguishable from the deformed configuration in Figure 13.10. This simplification will be central to the algorithm herein developed. It greatly contributes to expediency of computation, and the approximation maintains validity for a wide class of crystalline materials that undergo large deformation (e.g., typical metals and alloys).

Plastic deformation will thus be considered to evolve directly in the deformed configuration, using the relation $\mathbf{F}^e = \mathbf{I}$. Though $\mathbf{F}^e$ is here taken as a constant (identity), the *rate* of elastic deformation $\mathbf{D}^e$ will not be ignored, since it provides an efficient vehicle for the definition of Cauchy stress. Hence, with $\mathbf{D}^p = \bar{\mathbf{D}}^p$ in this simplification, the elastic rate of deformation $\mathbf{D}^e$ is obtained from $\mathbf{D}^e \equiv \mathbf{D} - \mathbf{D}^p$. Cauchy stresses are then easily computed from hypo-elastic relations (see Chapter 5). The kinematic quantities of interest for the rate-dependent algorithm herein developed for single-crystal plasticity will thus be given by,

$$\begin{aligned}
\mathbf{D}^p &= \sum_\alpha \mathbf{P}^\alpha \dot{\gamma}^\alpha \\
\mathbf{D}^e &= \mathbf{D} - \mathbf{D}^p \\
\mathbf{W}^p &= \sum_\alpha \mathbf{Q}^\alpha \dot{\gamma}^\alpha \\
\mathbf{\Omega}^* &= \mathbf{W} - \mathbf{W}^p \; (\text{since } \mathbf{W} = \bar{\mathbf{W}}^p \text{ when } \boldsymbol{W} \to \mathbf{0}),
\end{aligned} \tag{13.5.13}$$

where $\mathbf{P}^\alpha = \mathbf{I}\bar{\mathbf{P}}^\alpha\mathbf{I} = \bar{\mathbf{P}}^\alpha$ and $\mathbf{Q}^\alpha = \mathbf{I}\bar{\mathbf{Q}}^\alpha\mathbf{I} = \bar{\mathbf{Q}}^\alpha$ are the simplified push forwards to the deformed configuration of the symmetric and anti-symmetric Schmidt tensors from intermediate configuration II.

13.5.4 Final Remarks

It is noted that if, as shown in Figure 13.10, slip occurs in multiple planes, small gaps, overlaps or displacement discontinuities, will arise between lattice blocks, which result in a *non-compact* intermediate configuration (Kroner, 1981), that is, one that is not simply connected (Section 5.7.1), so that $\mathbf{F}^p$ cannot be a gradient of a plastic displacement field. The deformation map $\mathbf{F}^p$ is thus generally *incompatible* with the existence of a unique displacement field (Kroner, 1981). $\mathbf{F}^e$ maps from the non-compact intermediate configuration to the current configuration and is consequently also incompatible. As with any continuum, however, it is required that the compact reference configuration be mapped by a compatible deformation $\mathbf{F}$ to a compact current configuration, that is, one without holes, overlaps or discontinuities. Hence, $\mathbf{F}^p$ and $\mathbf{F}^e$ must evolve in a consistent manner, such that their product $\mathbf{F}^e\mathbf{F}^p$ be compatible. This compatibility condition may be expressed for any closed path taken around the deforming material point as (Kroner, 1981),

$$\delta \equiv \oint_C \mathbf{F}^e\mathbf{F}^p \mathbf{dX} = \mathbf{0}. \tag{13.5.14}$$

The algorithm developed in this chapter, using kinematic quantities as defined in Equation (13.5.13) in a finite element implementation, guarantees satisfaction of Equation (13.5.14).

13.6 Dislocation Density Evolution

In this section we introduce dislocation evolution, which ultimately serves for a bottom-up Cauchy stress update at any material point that uses single crystal plasticity. That is, by evolving internal variables at a material point that represent the dislocation structure, it is possible to track the evolution of crystal strength and the corresponding stresses.

There are on the order of 10^6–10^{12} dislocations in every cm^2 of metal crystals, depending on the processing history of the material, such as the degree of deformation it has undergone and the various microstructural features formed during heat treatment (Hull and Bacon, 2006; Polmear, 2006). To predict engineering performance of materials it is convenient to smear out individual dislocations into *densities*, so as to study their evolution (causing plastic deformation) in a continuum approximation. A dislocation density, ρ, is defined as the total dislocation line length of all dislocations per unit volume $[\mathrm{m\ m^{-3}}] = [\mathrm{m^{-2}}]$.

It may be assumed that during any increment of strain exceeding the elastic limit a change ensues in the dislocation structure within multiple slip systems, which defines overall plastic activity of a crystal. The dislocation structure in each slip system is further taken to be potentially composed of *mobile* (glissile) and *immobile* (sessile) components. That is, for a given state of the material, the dislocation structure in slip system α, represented by total dislocation density ρ^α, is additively decomposed into a *mobile* density (ρ_m^α) and an *immobile* density (ρ_{im}^α) as (Estrin, Krausz *et al.*, 1996),

$$\rho^\alpha = \rho_m^\alpha + \rho_{im}^\alpha \tag{13.6.1}$$

This decomposition follows from the view that dislocation motion accounts for plastic deformation and therefore *ductility*, while immobile dislocations contribute to the *strengthening* or *hardening* of crystals. Note that a single dislocation line may have both mobile and immobile segments at any one time. All plastic mechanisms activated by local straining of a material point will then be seen as resulting from the generation or annihilation of mobile and immobile dislocation densities in their respective slip systems, as well as their interaction across slip systems. Furthermore, during deformation, mobile dislocations can be immobilized and immobile dislocations released, which suggests evolutionary laws for mobile and immobile dislocations need to be coupled. For a large class of crystalline materials, dislocation density evolutionary laws could thus take the form (Estrin and Kubin, 1986),

$$\begin{aligned} \dot{\rho}_m^\alpha &= \dot{\rho}_{\text{generation}}^\alpha - \dot{\rho}_{\text{interaction}}^{\alpha-} \\ \dot{\rho}_{im}^\alpha &= -\dot{\rho}_{\text{annihilation}}^\alpha + \dot{\rho}_{\text{interaction}}^{\alpha+} \end{aligned} \tag{13.6.2}$$

To completely define the terms in Equation (13.6.2), often material specific assumptions need be made. However, fairly general evolutionary laws that apply to metals of different crystal structures have been derived in (Estrin and Kubin, 1986) and enhanced by subsequent research (see for example Kameda and Zikry, 1996; Rezvanian, Zikry *et al.*, 2007; Shi and Zikry, 2009; Shanthraj and Zikry, 2011).

A material point, which is at the mesoscale, is always assumed to represent sufficient crystalline structure such that the generation, annihilation and interaction mechanisms of dislocation densities may arise. An example of evolutionary laws, which capture the terms of Equation (13.6.2) at a material point, is the following linear superposition of various mobile and immobile mechanisms (see Estrin, Krausz *et al.*, 1996; Shanthraj and Zikry, 2011),

$$\dot{\rho}_{m}^{(\alpha)} = \left|\dot{\gamma}^{(\alpha)}\right|\left(g_{s}^{(\alpha)}\rho_{im}^{(\alpha)}\left(\rho_{m}^{(\alpha)}\right)^{-1} - g^{(\alpha)}{}_{m0}\rho_{m}^{(\alpha)} - g^{(\alpha)}{}_{im0}\sqrt{\rho_{im}^{(\alpha)}}\right) \quad (13.6.3a)$$

$$\dot{\rho}_{im}^{(\alpha)} = \left|\dot{\gamma}^{(\alpha)}\right|\left(g_{m1}^{(\alpha)}\rho_{m}^{(\alpha)} + g_{im1}^{(\alpha)}\sqrt{\rho_{im}^{(\alpha)}} - g_{r}^{(\alpha)}\rho_{im}^{(\alpha)}\exp\left(\frac{-\Delta H_{0}}{kT}\left[1-\sqrt{\frac{\rho_{im}^{(\alpha)}}{\rho_{im}^{\text{sat}}}}\right]\right)\right). \quad (13.6.3b)$$

ΔH_0 is an activation enthalpy for dislocation mechanisms, k is the Boltzmann constant, T the current absolute temperature, and ρ_{sat}^{im} is an experimentally determined saturation value for immobile dislocation densities.

A set of coefficients in Equation (13.6.3) was specified for the material-dependence of the degree of activity of specific dislocation mechanisms: g_s, for the generation of mobile dislocation densities; g_{m0}, for the annihilation of mobile dislocations by mobile dislocation self-interactions; g_{m1}, for dislocation loop and debris formation of result of self-interaction of mobile dislocations; g_r, for the recovery of immobile dislocations; and g_{im1} and g_{im0}, the immobilization of mobile dislocations due to immobile dislocation interaction with mobile dislocations (Estrin and Kubin, 1986). Explicit expressions for these coefficients in terms of the generation, annihilation and interaction mechanisms are outlined in Table 13.2, along with typical values for metals (Estrin and Kubin, 1986). For the finite element algorithm developed in this chapter, utilization of these typical values is deemed sufficient; however, in Appendix 5 a derivation of these coefficients and an explanation of the different terms and symbols appearing in Table 13.2 is presented for the interested reader.

Preference for using the typical coefficients in Table 13.2 to specify the degree of activity of specific dislocation mechanisms is due to the fact that they greatly simplify computational

Table 13.2 Coefficients of dislocation density evolutionary equations

Coefficient	Equation	Typical value
g_s^α	$\frac{\phi^{(\alpha)}}{\left\lvert \boldsymbol{b}^{(\alpha)}\right\rvert}\cdot\sum_\eta c_{\alpha\eta}(\rho_{im}^\eta)^{1/2}$	$\frac{1}{\left\lvert \boldsymbol{b}^{(\alpha)}\right\rvert}\times 2.76\text{e}^{-5}$
g_{m0}^α	$\phi^{(\alpha)}\cdot l_c\cdot\sum_\eta\left(\sqrt{a_{(\alpha)\eta}}\left[\frac{\rho_m^\eta}{\rho_m^{(\alpha)}\left\lvert \boldsymbol{b}^{(\alpha)}\right\rvert}+\frac{\dot{\gamma}^\eta}{\dot{\gamma}^{(\alpha)}\left\lvert \boldsymbol{b}^\eta\right\rvert}\right]\right)$	$\frac{1}{\left\lvert \boldsymbol{b}^{(\alpha)}\right\rvert}\times 5.53$
g_{im0}^α	$\frac{\phi^{(\alpha)}\cdot l_c}{(\rho_{im}^{(\alpha)})^{1/2}}\cdot\sum_\eta\left(\sqrt{a_{(\alpha)\eta}}\cdot\rho_{im}^\eta\right)$	$\frac{1}{\left\lvert \boldsymbol{b}^{(\alpha)}\right\rvert}\times 0.0127$
g_{m1}^α	$\frac{\phi^{(\alpha)}\cdot l_c}{\left\lvert \boldsymbol{b}^{(\alpha)}\right\rvert\dot{\gamma}^{(\alpha)}\rho_m^{(\alpha)}}\cdot\sum_{\eta,\kappa\le\eta}\left(z_{(\alpha)}^{\eta\kappa}\sqrt{a_{\eta\kappa}}\left[\frac{\rho_m^\eta\dot{\gamma}^\kappa}{\left\lvert \boldsymbol{b}^\kappa\right\rvert}+\frac{\rho_m^\kappa\dot{\gamma}^\eta}{\left\lvert \boldsymbol{b}^\eta\right\rvert}\right]\right)$	$\frac{1}{\left\lvert \boldsymbol{b}^{(\alpha)}\right\rvert}\times 5.53$
g_{im1}^α	$\frac{\phi^{(\alpha)}\cdot l_c}{\left\lvert \boldsymbol{b}^{(\alpha)}\right\rvert\dot{\gamma}^{(\alpha)}(\rho_{im}^{(\alpha)})^{1/2}}\cdot\sum_{\eta,\kappa\le\eta}\left(z_{(\alpha)}^{\eta\kappa}\sqrt{a_{\eta\kappa}}\cdot\rho_m^\kappa\dot{\gamma}^\eta\right)$	$\frac{1}{\left\lvert \boldsymbol{b}^{(\alpha)}\right\rvert}\times 0.0127$
g_r^α	$\frac{\phi^{(\alpha)}\cdot l_c}{\dot{\gamma}^{(\alpha)}}\cdot\sum_\eta\left(\sqrt{a_{(\alpha)\eta}}\cdot\frac{\dot{\gamma}^\eta}{\left\lvert \boldsymbol{b}^\eta\right\rvert}\right)$	6.69e[5]

implementation, as well as enable the straightforward application of nonlinear stability analysis techniques to predict the conditions for the onset of non-uniform deformation in order to explain important localization phenomena, such as the formation of shear localization bands (Estrin and Kubin, 1986). It should be emphasized that Equation (13.6.3) and these coefficients are only an example of possible dislocation density evolutionary laws; various other laws could be found in literature, depending on material and plastic phenomenon of interest.

13.7 Stress Required for Dislocation Motion

This section uses the dislocation densities evolved in Section 13.6 to determine crystal strength, which is an essential component in the stress update algorithm of single crystals.

As was mentioned earlier, around each dislocation core there must exist a stress field associated with the distortion it brings to crystalline order. Thus, if a mobile dislocation is to move in a crystal that already contains a number of immobile dislocations, it must interact with and overcome the additional stress fields surrounding these dislocations. Immobile dislocations in this sense act as obstacles to mobile dislocations and their density in large part determines the threshold stress required for dislocation motion (plasticity) within a slip system. The threshold value is known as the *critical resolved shear stress* ($\overline{\tau}^{\alpha}_{ref}$) and defines the *strength* of a crystal, while its rate of change characterizes the hardening of the crystal, and determines the shape of the stress strain curve (Figure 13.1b). If $\overline{\tau}^{\alpha}_{ref}$ increases the crystal is said to *harden*; if it decreases the crystal is said to *soften*. The resolved shear stress $\overline{\tau}^{\alpha}$, which is the component of the deviatoric stress tensor seen in slip system α in intermediate configuration II of Figure 13.10, must always be compared to the $\overline{\tau}^{\alpha}_{ref}$, and the condition for slip to occur could be stated as,

$$\left|\overline{\tau}^{\alpha}\right| > \overline{\tau}^{\alpha}_{ref}(\Phi). \tag{13.7.1}$$

The value of $\overline{\tau}^{\alpha}_{ref}$ generally depends on, in addition to the density of immobile dislocations, the crystal structure, the point defects that interact with slip systems to hinder dislocation motion, as well as temperature and other *internal variables*, (see Hull and Bacon, 2006). Φ in Equation (13.7.1) acts as a placeholder for the list of possible internal variables governing the evolution of the critical shear stress in a crystal. Taylor (1934) showed that the necessary $\overline{\tau}^{\alpha}_{ref}$ for *single-slip* to occur may be given by,

$$\overline{\tau}^{(\alpha)}_{ref} = G\left|\boldsymbol{b}^{(\alpha)}\right| / d^{(\alpha)} = G\left|\boldsymbol{b}^{(\alpha)}\right|\sqrt{\rho^{(\alpha)}_{im}}, \tag{13.7.2}$$

where $\rho^{(\alpha)}_{im}$ is the density of *immobile* dislocations in slip system α, and $d^{(\alpha)}$ is the average dislocation spacing in the slip system. This relation elucidates the need for Equation (13.6.3) to define the evolving strength of a crystal and the corresponding shape of the stress-strain curve, and determines the connection between crystal plasticity theory and dislocation theory. Various modifications to Eq. 13.7.2 have been proposed to better account for dislocation interaction in the case of poly-slip, as well as thermal effects. For example, (Elkhodary, Lee *et al.*, 2011)

$$\overline{\tau}^{\alpha}_{ref} = \left(\overline{\tau}^{\alpha}_{y} + G\sum_{\beta=1}^{nss}\left|\boldsymbol{b}^{\beta}\right|\sqrt{a_{\alpha\beta}\rho^{\beta}_{im}}\right)\left(\frac{T}{T_0}\right)^{-\xi} \tag{13.7.3}$$

where $\overline{\tau}_y^{\alpha}$ is the static yield stress on slip system (α), *nss* is the number of slip systems, $|\boldsymbol{b}^{\beta}|$ is the magnitude of the Burgers vector, and the Taylor coefficients $a_{\alpha\beta}$ reflect the strength of interaction between slip-systems, which vary between 0 and 1, and could be approximated from the relation $a_{\alpha\beta} = 2\sqrt{P_{ij}^{\alpha} P_{ij}^{\beta}}$ for simplicity (Elkhodary, Lee *et al.*, 2011). T is the current temperature, T_0 is a reference temperature, and ξ is the thermal softening exponent.

13.8 Stress Update in Rate-Dependent Single-Crystal Plasticity

In this section, updating the resolved shear stress, needed in the Cauchy stress update, is discussed to complete the single crystal plasticity formulation. The complete formulation is summarized in the next section as an algorithm to be implemented as a material subroutine within the finite element framework.

13.8.1 *The Resolved Shear Stress*

The resolved shear stress $\overline{\tau}^{\alpha}$ is defined by pulling back the deviatoric Cauchy stress σ^{dev} to intermediate configuration II (Figure 13.10) and projecting onto slip system α (Havner, 1992),

$$\overline{\tau}^{\alpha} = |\mathbf{F}^e|((\mathbf{F}^e)^{-1}\boldsymbol{\sigma}^{dev}(\mathbf{F}^e)^{-T} : (\overline{\mathbf{s}}^{\alpha} \otimes \overline{\mathbf{n}}^{\alpha})). \quad (13.8.1)$$

As was discussed in Section 13.5, in the algorithm to be developed in this chapter we make the simplifying assumption that $\mathbf{F}^e = \mathbf{I}$, that is, the elastic lattice distortions are negligible for plastically deforming crystals. In this approximation, resolved shear stress $\overline{\tau}^{\alpha}$ may be identified with a shear stress τ^{α} in the *deformed* configuration and computed from σ^{dev} as,

$$\tau^{\alpha} = \boldsymbol{\sigma}^{dev} : (\mathbf{s}^{\alpha} \otimes \mathbf{n}^{\alpha}) = \boldsymbol{\sigma}^{dev} : \mathbf{P}^{\alpha}, \quad (13.8.2)$$

having used the symmetry of $\boldsymbol{\sigma}^{dev}$.

13.8.2 *The Resolved Shear Stress Rate*

It is required to update the resolved shear stress τ^{α} in any single crystal plasticity algorithm. This requires an integration of its material rate,

$$\dot{\tau}^{(\alpha)} = \dot{\boldsymbol{\sigma}}^{dev} : \mathbf{P}^{\alpha} + \boldsymbol{\sigma}^{dev} : \dot{\mathbf{P}}^{\alpha} \quad (13.8.3)$$

In the present approximation, the deviatoric Cauchy stress is identifiable with its pull-back on intermediate configuration II. That is, the stress tensor should rotate with the crystal lattice, so that its material rate may be given by,

$$\dot{\boldsymbol{\sigma}}^{dev} = \overset{\nabla}{\boldsymbol{\sigma}}^{dev} + \boldsymbol{\Omega}^{*}\boldsymbol{\sigma}^{dev} - \boldsymbol{\sigma}^{dev}\boldsymbol{\Omega}^{*}. \quad (13.8.4)$$

Combining the first of Equation (13.5.10) (using $\mathbf{P}^\alpha$ instead of $\bar{\mathbf{P}}^\alpha$) with Equation (13.8.4) and substituting into Equation (13.8.3) results in the material rate,

$$\dot{\tau}^{(\alpha)} = \left(\overset{\nabla}{\boldsymbol{\sigma}}{}^{dev} + \boldsymbol{\Omega}^*\boldsymbol{\sigma}^{dev} - \boldsymbol{\sigma}^{dev}\boldsymbol{\Omega}^*\right) : \mathbf{P}^\alpha + \boldsymbol{\sigma}^{dev} : (\boldsymbol{\Omega}^*\mathbf{P}^\alpha - \mathbf{P}^\alpha\boldsymbol{\Omega}^*)$$
$$\dot{\tau}^{(\alpha)} = \overset{\nabla}{\boldsymbol{\sigma}}{}^{dev} : \mathbf{P}^\alpha, \tag{13.8.5}$$

The last part of Equation (13.8.5) is a result of the skew-symmetry of the lattice spin tensor Ω^*, the proof of which is left as an exercise (see Exercise 13.4). The deviatoric Cauchy stress rate is defined from a hypo-elastic constitutive law,

$$\overset{\nabla}{\boldsymbol{\sigma}}{}^{dev} \equiv \mathbf{C}^* : (\mathbf{D}^e)^{dev}, \tag{13.8.6}$$

where $\mathbf{C}^*$ is the elasticity tensor. If the elasticity tensor is assumed isotropic, only the shear modulus G needs be retained. While crystals are elastically anisotropic, errors introduced by this simplification are typically small when large inelastic deformations are being modeled. In this case the material rate of the resolved shear stress may be recast as,

$$\dot{\tau}^{(\alpha)} = \mathbf{n}^{(\alpha)} \cdot 2G(\mathbf{D}^{dev} - \mathbf{D}^p) \cdot \mathbf{s}^{(\alpha)} \tag{13.8.7}$$

13.8.3 Updating Resolved Shear Stress in Rate-Dependent Materials

For rate-dependent materials a well-justified empirical power law may relate shear rate in a slip system to the applied shear stress (Kocks, 1987), hence the constitutive equation,

$$\dot{\gamma}^{(\alpha)} = \dot{\gamma}^{(\alpha)}_{ref} \left(\frac{\tau^{(\alpha)}}{\tau^{(\alpha)}_{ref}}\right)^{1/m}, \tag{13.8.8}$$

where $\dot{\gamma}^{\alpha}_{ref}$ is a reference shear strain-rate which corresponds to the critical resolved shear stress τ^{α}_{ref}, and m is the rate-sensitivity parameter. This power law is applicable at strain rates below a critical value of $\dot{\gamma}_{cr}$, where phonon drag becomes active (which occurs only under extreme dynamic loading).

Substituting into Equation (13.8.7) the kinematic relation Equation (13.5.9) (with $\bar{\mathbf{D}}^p = \mathbf{D}^p$) and Equation (13.8.8) for the power law, the material rate of the resolved shear stress may be given by,

$$\dot{\tau}^{(\alpha)} = \mathbf{n}^{(\alpha)} \cdot 2G\left(\mathbf{D}^{dev} - \sum_\eta \mathbf{P}^\eta \left(\dot{\gamma}^\eta_{ref}\left(\frac{\tau^\eta}{\tau^\eta_{ref}}\right)^{1/m}\right)\right) \cdot \mathbf{s}^{(\alpha)}, \quad \forall\alpha. \tag{13.8.9}$$

This is a coupled system of stiff nonlinear ordinary differential equations (Zikry, 1994). The system relates the material rate of shear stress in slip system α to: the shear stress (τ^η) in every active slip system η of orientation $\mathbf{P}^\eta$, shear modulus G, the applied deviatoric deformation rate tensor ($\mathbf{D}^{dev}$) at a material point, and current orientation of slip system α ($\mathbf{n}^\alpha$,$\mathbf{s}^\alpha$).

Different schemes have been proposed to integrate this system of equations, such as ones which rely on explicit Runge–Kutta methods for speed, but shift to backward Euler integration whenever stiffness is detected (Zikry, 1994). More recently, various stiff ODE solvers featuring multiple algorithms have become readily available in FORTRAN and C, such as ODEPACK from the Lawrence Livermore National Lab, and IntelODE from Intel, and can be called directly in material subroutines, to compute the resolved shear stress.

13.8.4 Updating the Cauchy Stress

Once all the resolved shear stresses have been updated, they may be plugged back into the power law (Equation 13.8.8) to yield the current slip rates for all slip systems. These may then be plugged into Equation (13.5.9) to compute the plastic part of the deformation rate $\mathbf{D}^p$. The deviatoric elastic part is then computed from $(\mathbf{D}^e)^{dev} = \mathbf{D}^{dev} - \mathbf{D}^p$. Finally, the Cauchy stresses at the material point may be updated by the first order formula,

$$\begin{aligned}\boldsymbol{\sigma}_{t+\Delta t} &= \boldsymbol{\sigma}_t + \Delta\boldsymbol{\sigma} \\ &= \boldsymbol{\sigma}_t + \left(\kappa \mathbf{I}\dot{e} + 2G(\mathbf{D}^{dev})^e - \boldsymbol{\sigma}_t\boldsymbol{\Omega}^* + \boldsymbol{\Omega}^*\boldsymbol{\sigma}_t\right)\Delta t,\end{aligned} \tag{13.8.10}$$

where σ_t is the Cauchy stress at the beginning of the time step, κ is the bulk modulus, $\dot{e}$ is the volumetric strain rate, given by the trace of the deformation rate tensor $\mathbf{D}$, and Δt is the stable time step for an explicit time-stepping algorithm.

13.8.5 Adiabatic Temperature Update

As a final remark, for high strain rate applications the *Fourier modulus*, the ratio of the rate of heat conduction to the rate of heat storage, is typically small (Zikry and Nemat-Nasser, 1990), so that adiabatic conditions may be assumed. From the balance of energy, in absence heat conduction, the local rate of change of temperature of a material point is,

$$\dot{T} = \chi(\rho c_p)^{-1}\boldsymbol{\sigma} : \mathbf{D}^p, \tag{13.8.11}$$

where χ is the fraction of plastic work converted to heat, typically $\chi = 0.9$–1.0, ρ is the mass density of the material point, and c_p is the specific heat at constant pressure. This simplified energy equation may be integrated for temperature, which is then plugged into Equation (13.7.3) for the next time step, and will capture the effect of thermal softening of single crystals when high strain rate conditions apply.

13.9 Algorithm for Rate-Dependent Dislocation-Density Based Crystal Plasticity

In this section, we present the necessary equations needed to implement rate-dependent dislocation-density based single crystal plasticity as a material subroutine in a displacement-controlled finite element framework. Box 13.2 contains the procedure that implements dislocation-density based single-crystal plasticity. It is a semi-implicit algorithm for ease of implementation. That is, when solving for the resolved shear stresses (τ^α), the algorithm presented utilizes slip normals and directions ($\mathbf{n}^\alpha, \mathbf{s}^\alpha$) at the beginning of the time increment (i.e., at time t), as well as the dislocation densities ($\rho_{im}^{\alpha}, \rho_{m}^{\alpha}$) at time t to calculate crystal strength ($\tau_{\text{ref}}^{\alpha} = \bar{\tau}_{\text{ref}}^{\alpha}$ in Equation 13.7.3), and the local temperature (T) at time $t+\Delta t$. ($\mathbf{n}^{\alpha}, \mathbf{s}^{\alpha}$) and ($\rho_{im}^{\alpha} \rho_{m}^{\alpha}$) are subsequently updated to time ($t+\Delta t$). In this algorithm it is further assumed that $\Omega^{*} = \mathbf{W} - \sum_{\alpha} \mathbf{Q}^{\alpha} \dot{\gamma}^{\alpha} \equiv \mathbf{W}^{e}$. That is, lattice rotation rate corresponds to elastic spin, for simplicity.

Box 13.2 Dislocation-density based single-crystal plasticity algorithm

1. Compute material rate of deformation gradient
$$\dot{F}_{ij}^{t+\Delta t} = \Delta t^{-1}\left(F_{ij}^{t+\Delta t} - F_{ij}^{t}\right)$$
2. Compute velocity gradient
$$L_{ij} = \dot{F}_{ik}^{t+\Delta t}\left(F_{kj}^{t+\Delta t}\right)^{-1}$$
3. Compute deformation rate
$$D_{ij} = \frac{1}{2}(L_{ij} + L_{ji})$$
4. Compute spin
$$W_{ij} = \frac{1}{2}(L_{ij} - L_{ji})$$
5. Assemble the Schmidt tensor (Symmetric and Skew) at beginning of increment (refer to Equation 13.5.6 and Equation 13.5.9)
$$P_{ij}^{(\alpha)} = \frac{1}{2}\left(n_i^{(\alpha)} s_j^{(\alpha)} + n_j^{(\alpha)} s_i^{(\alpha)}\right)\Big|_t ,$$
$$Q_{ij}^{(\alpha)} = \frac{1}{2}\left(n_i^{(\alpha)} s_j^{(\alpha)} - n_j^{(\alpha)} s_i^{(\alpha)}\right)\Big|_t$$
6. Setup and solve coupled ODE system for resolved shear stresses ($\tau_{t+\Delta t}^{\alpha}$), for all α, from nonlinear ODE system. (Note: $\tau_{\text{ref}}^{\alpha} = \bar{\tau}_{\text{ref}}^{\alpha}$ in Equation (13.7.3) as the deformed configuration is identified with intermediate configuration II)
$$\begin{aligned} \dot{\tau}^{(1)} &= \mathbf{n}_t^{(1)} \cdot \mathbf{MAT} \cdot \mathbf{s}_t^{(1)} \\ \dot{\tau}^{(2)} &= \mathbf{n}_t^{(2)} \cdot \mathbf{MAT} \cdot \mathbf{s}_t^{(2)} \\ &\vdots \\ \dot{\tau}^{(nss)} &= \mathbf{n}_t^{(nss)} \cdot \mathbf{MAT} \cdot \mathbf{s}_t^{(nss)} \end{aligned}$$
$$\mathbf{MAT} = 2G\left(\mathbf{D}_{t+\Delta t}^{dev} - \sum_{\eta} \mathbf{P}_t^{\eta}\left(\dot{\gamma}_{ref}^{\eta}\left(\frac{\tau_t^{\eta}}{\tau_{ref}^{\eta}\big|_t}\right)^{1/m}\right)\right)$$
7. Update crystallographic shear slip rates for all slip systems (rate dependent materials)
$$\dot{\gamma}^{(\alpha)} = \dot{\gamma}_{ref}^{(\alpha)}\left(\frac{\tau^{(\alpha)}\big|_{t+\Delta t}}{\tau_{ref}^{(\alpha)}\big|_t}\right)^{1/m}$$
8. Update plastic parts of deformation rate and spin
$$D_{ij}^{p} = \sum_{\alpha} P_{ij}^{\alpha} \dot{\gamma}^{\alpha}$$
9. Compute corresponding elastic parts of deformation rate and spin
$$D_{ij}^{e} = D_{ij} - D_{ij}^{p}, \; W_{ij}^{e} = W_{ij} - \sum_{\alpha} Q_{ij}^{\alpha} \dot{\gamma}^{\alpha}$$
10. Update the Cauchy stress
$$\boldsymbol{\sigma}_{t+\Delta t} = \boldsymbol{\sigma}_t + \left(\kappa \mathbf{I}\dot{e} + 2G \cdot dev(\mathbf{D}^{e}) - \boldsymbol{\sigma}_t \mathbf{W}^{e} + \mathbf{W}^{e} \boldsymbol{\sigma}_t\right)\Delta t$$

11. Update crystal slip system orientations (for the next time step; used in Steps 5, 6 and 8). This step replaces Equations. (13.5.10–12).	$\mathbf{n}^{\alpha}_{t+\Delta t} = \mathbf{n}^{\alpha}_{t}(\mathbf{I}+\mathbf{W}^{e}\Delta t), \mathbf{s}^{\alpha}_{t+\Delta t} = \mathbf{s}^{\alpha}_{t}(\mathbf{I}+\mathbf{W}^{e}\Delta t),$
12. Setup and solve coupled ODE system for dislocation densities (for next time step). If coefficients in ODEs are taken constant, each slip system may be updated separately. Loop over α to obtain $\rho^{\alpha}_{m}\big\|_{t+\Delta t}$ and $\rho^{\alpha}_{im}\big\|_{t+\Delta t}$	$\dot{\rho}^{(\alpha)}_{m} = \left\|\dot{\gamma}^{(\alpha)}\right\|\left(g^{(\alpha)}_{s}\rho^{(\alpha)}_{im}\left(\rho^{(\alpha)}_{m}\right)^{-1} - g^{(\alpha)}_{m0}\rho^{(\alpha)}_{m} - g^{(\alpha)}_{im0}\sqrt{\rho^{(\alpha)}_{im}}\right)$ $\dot{\rho}^{(\alpha)}_{im} = \left\|\dot{\gamma}^{(\alpha)}\right\|\left(g^{(\alpha)}_{m1}\rho^{(\alpha)}_{m} + g^{(\alpha)}_{im1}\sqrt{\rho^{(\alpha)}_{im}} - g^{(\alpha)}_{r}\rho^{(\alpha)}_{im}\right)$
13. Update the temperature. For high strain rate applications, adiabatic conditions assumed.	$T\big\|_{t+\Delta t} = T\big\|_{t} + \chi(\rho c_{v})^{-1}\boldsymbol{\sigma}:\mathbf{D}^{p}\Delta t,$
14. Update the shear strength of the crystal (for the next time step; used in Step 6)	$\tau^{(\alpha)}_{ref}\big\|_{t+\Delta t} = \left(\tau^{(\alpha)}_{y} + G\sum_{\beta=1}^{nss}\left\|b^{\beta}\right\|\sqrt{a_{(\alpha)\beta}\,\rho^{\beta}_{im}\big\|_{t+\Delta t}}\right)\left(\frac{T\big\|_{t+\Delta t}}{T_{0}}\right)^{-\xi}$
15. Update the accumulated plastic shear slip in crystal (for post processing)	$\gamma^{p}_{t+\Delta t} = \gamma^{p}_{t} + \Delta t\sqrt{\frac{2}{3}D^{p}_{ij}D^{p}_{ij}}$
16. Update lattice rotation matrix (for post processing)	$\psi^{t+\Delta t}_{ij} = \psi^{t}_{ij} + \Delta t W^{e}_{ij}$

13.10 Numerical Example: Localized Shear and Inhomogeneous Deformation

The poly-slip dislocation-density based crystal plasticity formulation outlined in this chapter was implemented in the explicit dynamic finite element software ABAQUS\Explicit by means of the user-defined material subroutine VUMAT (Hibbitt, Karlson *et al.*, 2007). The aim was to investigate the behavior of a single crystal of FCC aluminum subjected to a nominal tensile strain of 30%, by pulling on the top and bottom faces at a rate of 5 ms^{-1}, consistent with dynamic loading conditions (Figure 13.11).

The crystal modeled is 15 μm in diameter × 15 μm in length. A cylindrical shape was chosen to avoid localization of deformation at corner points. Free boundary conditions were selected for the curved surface. Euler angles of (0°, 0°, 0°) were used to align the crystal with the model axes. The material properties used in this example are summarized in Table 13.3. The mesh contains 2028 8-node trilinear hexahedral elements (called C3D8R in ABAQUS), with reduced integration and physical (assumed-strain) hourglass control (see Sections 8.7.6 and 8.7.8).

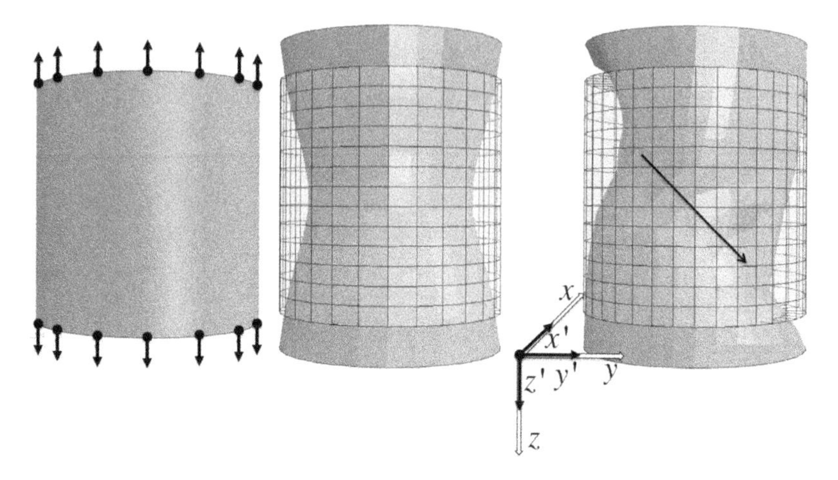

Figure 13.11 *Left*: Single crystal, *Middle*: J_2 plasticity, *Right*: Single crystal plasticity

Table 13.3 Material properties for aluminum

		Aluminum	
Property	Description	Value	Ref.
E (GPa)	Young's modulus	69	(Zhu, Shiflet *et al.*, 2006)
ν	Poisson's ratio	0.34	
τ_y (MPa)	Static yield stress	35	–
ρ (g/cm^3)	Mass density	2.70	–
c_p (J/kgK)	Specific heat	902	(Smithells, 2004)
$\Delta H/k$ (K)	Ration of activation enthalpy to Boltzmann constant	2500	(Ali, Podus *et al.*, 1979)
$\dot{\gamma}_{ref}$ (s^{-1})	Reference strain rate	0.001	(Zikry and Kao, 1997)
$\dot{\gamma}_{crit}$ (s^{-1})	Critical strain rate	3×10^4	
$\lvert b\rvert$ (m)	Burgers vector (same in all slip systems)	3×10^{-10}	
ρ^0_{im} (m^{-2})	Initial immobile dislocation density	10^{12}	–
ρ^{sat}_{im} (m^{-2})	Saturation value for immobile dislocation density	10^{16}	–
ρ^0_{mo} (m^{-2})	Initial mobile dislocation density	10^{10}	–
T_0 (K)	Reference temperature	293	–
M	Strain rate sensitivity	0.02	(Zikry, 1994)
ξ	Thermal softening exponent	0.5	
χ	Fraction of plastic dissipation to heat	0.9	

As can be seen from Figure 13.11, single crystal plasticity predicts deformation localizes along a 45° incline. Simple J_2 plasticity, however, could not predict this behavior, but instead predicts necking in all directions in the central region, much like in macroscopic polycrystalline aggregates. Resulting logarithmic strain iso-surfaces are plotted in Figure 13.12, which

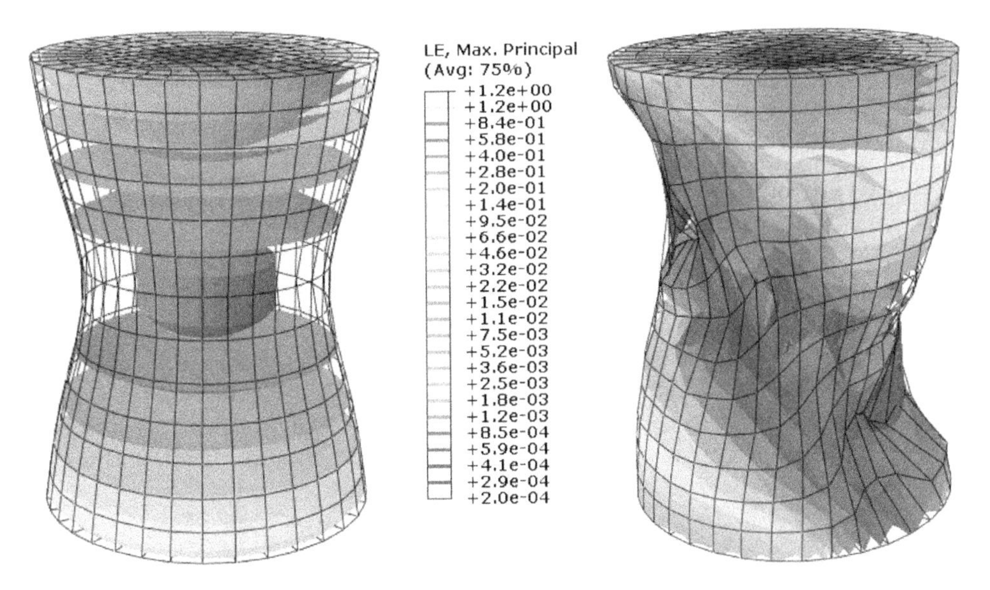

Figure 13.12 Logarithmic strain compared in single crystal plasticity and J_2 plasticity

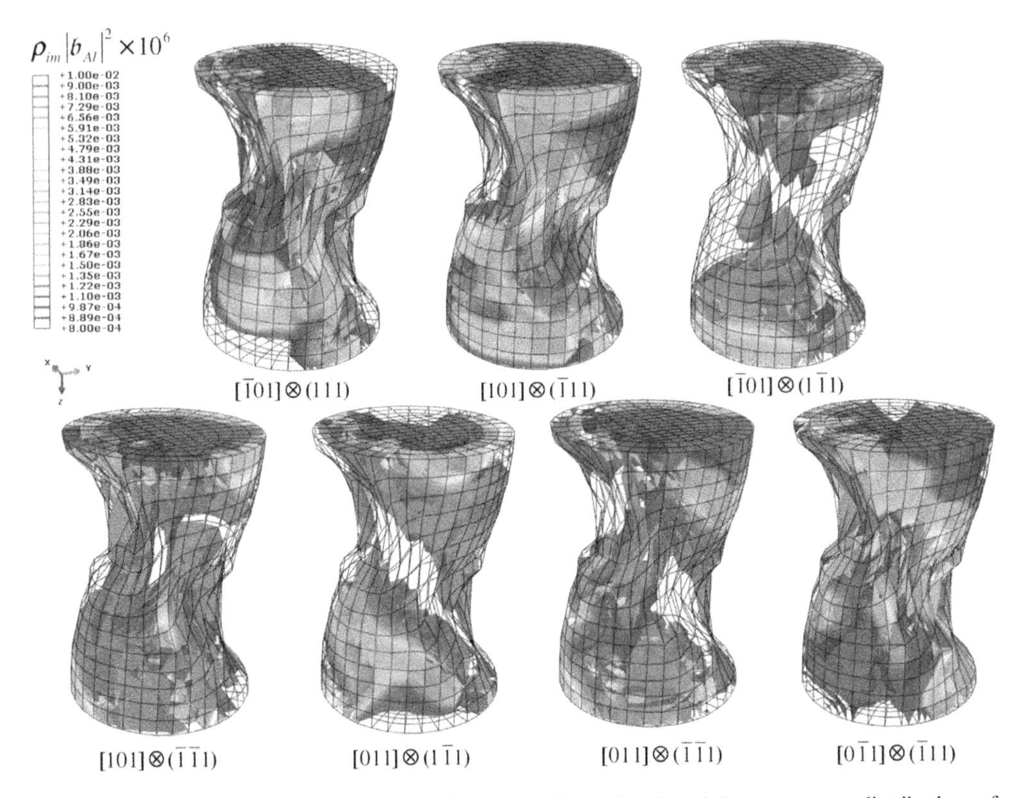

Figure 13.13 Immobile dislocation density iso-surfaces showing inhomogeneous distribution of plastic activity across the cylindrical sample. Shear slip is anisotropic, since plastic activity (represented by dislocation evolution) varies with slip system

clearly indicate the localization of deformation, reaching a maximum value of 120% strain, whereas the nominal strain is 30%.

This simple simulation illustrates the inhomogeneity and anisotropy of plasticity in single crystals that are missed by typical phenomenological constitutive models, in this example J_2 plasticity. In particular, recalling Equation (13.5.9) $\left(\bar{\mathbf{D}}^p = \sum_\alpha \bar{\mathbf{P}}^\alpha \dot{\gamma}^\alpha\right)$, plastic shear for single crystals is forced to evolve along specific orientations, as encoded by the Schmidt tensor $\bar{\mathbf{P}}^\alpha = sym(\bar{\mathbf{s}}^\alpha \otimes \bar{\mathbf{n}}^\alpha)$. However, no such constraint exists in J_2 plasticity, which permits material points to strain in any direction.

Furthermore, the evolved immobile dislocation densities (ρ_{im}^α, Equation 13.6.3) for the seven most active slip systems are shown in Figure 13.13. These are needed in the update of crystal strength (τ_{ref}^α, Equation 13.7.2). Crystal strength is used to compute the slip rate ($\dot{\gamma}^\alpha$, Equation 13.8.8), which completes the definition of $\bar{\mathbf{D}}^p$. From the figure it is seen that plastic activity changes with slip system in every element, so that orientation dependence of shear in single crystals is confirmed. In J_2 plasticity, this orientation dependence, that is, anisotropy, is not considered and necking in all directions ensues as seen in Figure 13.12.

13.11 Exercises

13.1. Show that for cubic crystals the unit normal vector for plane (hkl)

$$\mathbf{n}^{hkl} = (h^2+k^2+l^2)^{-1/2}(h\mathbf{e}_1 + k\mathbf{e}_2 + l\mathbf{e}_3),\text{ where } \mathbf{e}_i \text{ are Cartesian unit basis vectors.}$$

13.2. Show that according to Figure 13.10 $\mathbf{F}^e = \mathbf{V}_1^e$ and $\mathbf{R}_1 = \mathbf{R}^*$, where $\mathbf{F}_1^e = \mathbf{V}_1^e \mathbf{R}_1$.

13.3. Show that crystalline plasticity is incompressible (*Hint*: use 13.5.7).

13.4. Show that the second of Equation (13.8.5) results from the skew symmetry of $\mathbf{\Omega}^*$.

13.5. Find the explicit expression for $\boldsymbol{W}$, in Equation (13.5.4), in component form.
Hint: See derivation in (Boyce, Weber *et al.*, 1989.)

13.6. Computer problem.
(a) Write a poly-slip, rate-dependent, single crystal plasticity subroutine, based on Box 13.2 and Table 13.3, to reproduce the results in Figure 13.11–Figure 13.13.
(b) Change the Euler angles to (0°, 60°, 0°) and compare the deformation pattern at 30% nominal strain. Use Box 13.1 to help re-define your slip directions and normals correctly, or use Table E13.1.

Appendix 1

Voigt Notation

Voigt notation In finite element implementations, symmetric second-order tensors are often written as column matrices. We will call this and any other conversion of higher-order tensors to column matrices Voigt notation. The procedure for converting symmetric second-order tensors to column matrices is called the Voigt rule.

Kinetic Voigt rule The Voigt rule depends on whether a tensor is a kinetic quantity, such as a stress, or a kinematic quantity, such as a strain. The Voigt rule for kinetic tensors, such as the symmetric tensor $\boldsymbol{\sigma}$, is

$$\text{tensor} \rightarrow \text{Voigt}$$

$$\text{2D (Table A1.1)}: \boldsymbol{\sigma} \equiv \begin{bmatrix} \sigma_{11} & \sigma_{12} \\ \sigma_{21} & \sigma_{22} \end{bmatrix} \rightarrow \begin{Bmatrix} \sigma_{11} \\ \sigma_{22} \\ \sigma_{12} \end{Bmatrix} = \begin{Bmatrix} \sigma_1 \\ \sigma_2 \\ \sigma_3 \end{Bmatrix} \equiv \{\boldsymbol{\sigma}\} \tag{A.1.1}$$

$$\text{3D (Table A1.2)}: \boldsymbol{\sigma} \equiv \begin{bmatrix} \sigma_{11} & \sigma_{12} & \sigma_{13} \\ \sigma_{21} & \sigma_{22} & \sigma_{23} \\ \sigma_{31} & \sigma_{32} & \sigma_{33} \end{bmatrix} \rightarrow \begin{Bmatrix} \sigma_{11} \\ \sigma_{22} \\ \sigma_{33} \\ \sigma_{23} \\ \sigma_{13} \\ \sigma_{12} \end{Bmatrix} = \begin{Bmatrix} \sigma_1 \\ \sigma_2 \\ \sigma_3 \\ \sigma_4 \\ \sigma_5 \\ \sigma_6 \end{Bmatrix} \equiv \{\boldsymbol{\sigma}\} \tag{A.1.2}$$

Nonlinear Finite Elements for Continua and Structures, Second Edition.
Ted Belytschko, Wing Kam Liu, Brian Moran, and Khalil I. Elkhodary.

Companion Website: www.wiley.com/go/belytschko

Table A1.1 2D Voigt rule

σ_{ij}		σ_a
i	j	a
1	1	1
2	2	2
1	2	3

Table A1.2 3D Voigt rule

σ_{ij}		σ_a
i	j	a
1	1	1
2	2	2
3	3	3
2	3	4
1	3	5
1	2	6

The correspondence between the indices of the second-order tensor and the indices of the column matrix is given in Table A1.1 (in two dimensions) and in Table A1.2 (three dimensions). As shown, the order of the terms in the column matrix can be remembered by drawing a line down the main diagonal of the tensor, then up the last column, and back across the first row (if there are any elements left). Any tensor or matrix converted by the Voigt rule is said to be in Voigt form, and is enclosed by brackets as shown above.

When indices are used for a tensor in Voigt form, we use subscripts beginning with letters a to g. Thus σ_{ij} is replaced by σ_a in going from tensor to Voigt form.

Kinematic Voigt rule The Voigt rule for a second-order, kinematic tensor, such as the strain ε_{ij}, is also as given in Table A1.1. However, the shear strains, that is, the components with unequal indices, are multiplied by 2. Thus the Voigt rule for the strains is

$$\text{tensor} \rightarrow \text{Voigt}$$

$$\text{in 2D}: \boldsymbol{\varepsilon} \equiv \begin{bmatrix} \varepsilon_{11} & \varepsilon_{12} \\ \varepsilon_{21} & \varepsilon_{22} \end{bmatrix} \rightarrow \begin{Bmatrix} \varepsilon_{11} \\ \varepsilon_{22} \\ 2\varepsilon_{12} \end{Bmatrix} = \begin{Bmatrix} \varepsilon_1 \\ \varepsilon_2 \\ \varepsilon_3 \end{Bmatrix} \equiv \{\boldsymbol{\varepsilon}\} \tag{A.1.3}$$

$$\text{in 2D}: \boldsymbol{\varepsilon} \equiv \begin{bmatrix} \varepsilon_{11} & \varepsilon_{12} & \varepsilon_{13} \\ & \varepsilon_{22} & \varepsilon_{23} \\ \text{sym} & & \varepsilon_{33} \end{bmatrix} \rightarrow \begin{Bmatrix} \varepsilon_{11} \\ \varepsilon_{22} \\ \varepsilon_{33} \\ 2\varepsilon_{23} \\ 2\varepsilon_{13} \\ 2\varepsilon_{12} \end{Bmatrix} \equiv \{\boldsymbol{\varepsilon}\} \tag{A.1.4}$$

The factor of 2 on the shear strains results from the requirement that the expressions for the energy be equivalent in Voigt notation and indicial notation. It is easy to verify that the following expressions for the increment in energy are all equal:

$$\rho dw^{\text{int}} = d\varepsilon_{ij}\sigma_{ij} = d\boldsymbol{\varepsilon} : \boldsymbol{\sigma} = \{d\varepsilon\}^T\{\sigma\} \tag{A.1.5}$$

The factor of 2 in the shear strains can be remembered by observing that the strains in Voigt notation are the engineering shear strains.

Vectorization of matrices Other matrices are also often translated to column matrices, which in computer science are called vectors (note the difference in meaning from mechanics, where a vector is a first-order tensor); sometimes we call this vectorization. This term should not be confused with vectorization in programming. We will denote nodal first-order tensors by double subscripts, such as u_{iI}, where i is the component index and I is the node number. The component index is always lower case, the node number index always upper case; sometimes their order is interchanged.

The following rule is used for converting a matrix such as f_{iI} to column matrices:

$$\text{elements of } \mathbf{f} \text{ are } f_a = f_{iI} \text{ where } a = (I-1)n_{SD} + i \text{ and } n_{SD} \text{ is the number of space dimensions} \tag{A1.6}$$

Voigt rule applied to higher-order tensors The Voigt rule is particularly useful for converting fourth-order tensors, which are awkward to implement in programming, to second-order matrices. For example, the linear elastic law in indicial notation involves the fourth-order tensor C_{ijkl}:

$$\sigma_{ij} = C_{ijkl}\varepsilon_{kl} \quad \text{or in tensor notation } \sigma = \mathbf{C}:\varepsilon \tag{A.1.7}$$

The Voigt matrix form of (A1.7) is

$$\{\boldsymbol{\sigma}\} = [\mathbf{C}]\{\varepsilon\} \quad \text{or} \quad \sigma_a = C_{ab}\varepsilon_b \tag{A.1.8}$$

where $a \leftarrow ij$ and $b \leftarrow kl$ as in Table A1.1 for two dimensions and Table A1.2 for three dimensions. When writing the Voigt expression in matrix indicial form, indices at the beginning of the alphabet are used.

For example, the Voigt matrix form of the elastic constitutive matrix in plane strain is

$$[\mathbf{C}] = \begin{bmatrix} C_{11} & C_{12} & C_{13} \\ C_{21} & C_{22} & C_{23} \\ C_{31} & C_{32} & C_{33} \end{bmatrix} = \begin{bmatrix} C_{1111} & C_{1122} & C_{1112} \\ C_{2211} & C_{2222} & C_{2212} \\ C_{1211} & C_{1222} & C_{1212} \end{bmatrix} \tag{A.1.9}$$

The first matrix refers to the elastic coefficients in Voigt notation, the second in tensor notation; the number of subscripts specifies whether the matrix is expressed in Voigt or tensor notation. To verify this conversion, note, for example, the expression for σ_{12} from (A1.7):

$$\sigma_{12} = C_{1211}\varepsilon_{11} + C_{1212}\varepsilon_{12} + C_{1221}\varepsilon_{21} + C_{1222}\varepsilon_{22} \tag{A.1.10}$$

The translation of the above in Voigt notation is

$$\sigma_3 = C_{31}\varepsilon_1 + C_{33}\varepsilon_3 + C_{32}\varepsilon_2 \tag{A.1.11}$$

which can be shown to be equivalent to (A1.10) if we use $\varepsilon_3 = \varepsilon_{12} + \varepsilon_{21} = 2\varepsilon_{12}$ and the minor symmetry of **C**, namely $C_{1212} = C_{1221}$. The modification of the above for plane stress is given in most texts on linear elasticity.

The Voigt rule and vectorization are sometimes combined. For example, the higher-order matrix B_{ijkK} is often used to relate strains to nodal displacements by

$$\varepsilon_{ij} = B_{ijkK}u_{kK} \tag{A.1.12}$$

where the indices i, j pertain to a kinematic tensor. To translate this to Voigt notation, the kinematic Voigt rule is used for ε_{ij} and the first two indices of B_{ijkK} and vectorization is used for the second pair of indices of B_{ijkK} and the indices of u_{kK}. Thus

$$\text{elements of } [\mathbf{B}] \text{ are } B_{ab} \text{ where } (i, j) \rightarrow a \text{ by the Voigt rule} \tag{A.1.13}$$

and

$$b = (K-1)n_{SD} + k \tag{A.1.14}$$

In two dimensions, the matrix counterpart of B_{ijKk} is then written as

$$[\mathbf{B}]_K = \begin{bmatrix} B_{111K} & B_{112K} \\ B_{221K} & B_{222K} \\ 2B_{121K} & 2B_{122K} \end{bmatrix} = \begin{bmatrix} B_{11xK} & B_{11yK} \\ B_{22xK} & B_{22yK} \\ 2B_{12xK} & 2B_{12yK} \end{bmatrix} \tag{A.1.15}$$

where we have replaced the third numerical indices with a Latin index in the second expressions; we will use both forms. If the range of K is 3, the $[\mathbf{B}]$ matrix is

$$[\mathbf{B}] = \begin{bmatrix} B_{xxx1} & B_{xxy1} & B_{xxx2} & B_{xxy2} & B_{xxx3} & B_{xxy3} \\ B_{yyx1} & B_{yyy1} & B_{yyx2} & B_{yyy2} & B_{yyx3} & B_{yyy3} \\ 2B_{xyx1} & 2B_{xyy1} & 2B_{xyx2} & 2B_{xyy2} & 2B_{xyx3} & 2B_{xyy3} \end{bmatrix} \tag{A.1.16}$$

where the first two indices have been replaced by the corresponding letters. The expression corresponding to (A1.12) can then be written as

$$\varepsilon_a = B_{ab}u_b \quad \text{or} \quad \{\varepsilon\} = [\mathbf{B}]\mathbf{d} \tag{A.1.17}$$

Note that we do not put brackets around the **d**, because the Voigt rule has not been used on **d**; we enclose a variable in brackets only when the Voigt rule has been applied to it.

The Voigt rule is particularly useful in the implementation of stiffness matrices. In indicial notation, the stiffness matrix is

$$K_{rIsJ} = \int_{\Omega} B_{ijrI} C_{ijkl} B_{klsJ} \, d\Omega \tag{A.1.18}$$

Converting the indices in the above matrices by the kinematic Voigt rule and vectorization gives

$$K_{ab} = \int_{\Omega} B_{ae} C_{ef} B_{fg} d\Omega \rightarrow [\mathbf{K}] = \int_{\Omega} [\mathbf{B}]^T [\mathbf{C}][\mathbf{B}] d\Omega = \int_{\Omega} \mathbf{B}^T [\mathbf{C}] \mathbf{B} d\Omega \tag{A.1.19}$$

where the indices '*rI*' and '*sJ*' have been converted to '*a*' and '*b*', *respectively, by vectorization and the indices* '*ij*' and '*kl*' have been converted to '*e*' and '*f*', respectively, by the kinematic Voigt rule. In this last expression, the brackets are dropped on **B**, which we do frequently when the form is obvious from the context. Another useful form of the stiffness matrix is obtained by retaining the nodal indices, which gives

$$[\mathbf{K}]_{IJ} = \int_{\Omega} [\mathbf{B}]_I^T [\mathbf{C}][\mathbf{B}]_J \, d\Omega \tag{A.1.20}$$

where **[B]*I*** is given in (A1.15).

Appendix 2

Norms

Norms are used in this book primarily for simplifying the notation. No proofs are given that rely on the properties of normed spaces, so the student need only learn the definitions of the norms as given below. It is also worthwhile to learn an interpretation of a norm as a distance. This is easily grasped by first learning the norms ℓ_n, which are vector norms. The extension to norms in function spaces such as the Hilbert spaces and the space of Lebesque integrable functions, L_2 (often we refer to this as the space 'el-two'), is then straightforward.

We begin with the norm ℓ_2, which is simply Euclidean distance. If we consider an n-dimensional vector $\mathbf{a}$, often written as $\mathbf{a} \in R^n$, then the ℓ_2 norm is given by

$$\|\mathbf{a}\|_2 \equiv \|\mathbf{a}\|\ell_2 = \left(\sum_{i=1}^{n} a_i^2\right)^{\frac{1}{2}} \tag{A.2.1}$$

In (A2.1), the symbol $\|\cdot\|$ indicates a norm and the subscript 2 in combination with the fact that the enclosed variable is a vector indicates that we are referring to the ℓ_2 norm. For $n = 2$ or 3, respectively, the ℓ_2 norm is simply the length of the enclosed vector. The distance between two points, or the difference between two vectors, is written as

$$\|\mathbf{a} - \mathbf{b}\|_2 = \left(\sum_{i=1}^{n} (a_i - b_i)^2\right)^{\frac{1}{2}} \tag{A.2.2}$$

Nonlinear Finite Elements for Continua and Structures, Second Edition.
Ted Belytschko, Wing Kam Liu, Brian Moran, and Khalil I. Elkhodary.

Companion Website: www.wiley.com/go/belytschko

Fundamental properties of thc ℓ_2 norm are that:

1. it is positive;
2. it satisfies the triangle inequality;
3. it is linear.

The ℓ_k norms are generalizations for the above definition to arbitrary $k>1$ as follows:

$$\|\mathbf{a}\|_k = \left(\sum_{i=1}^{n} |a_i|^k \right)^{\frac{1}{k}} \tag{A.2.3}$$

Norms for $k \neq 2$ are seldom used except for $k=\infty$, which is called the maximum norm. The infinity norm gives the component of the vector with the maximum absolute value:

$$\|\mathbf{a}\|_\infty = \max_i |a_i| \tag{A.2.4}$$

One of the principal applications of these norms is to define errors. Thus if we have an approximate solution $\mathbf{d}^{\text{app}}$ to a set of discrete equations and the exact solution is $\mathbf{d}^{\text{exact}}$, then a measure of the error is

$$\text{error} = \left\| \mathbf{d}^{\text{app}} - \mathbf{d}^{\text{exact}} \right\|_2 \tag{A.2.5}$$

If you are concerned with the maximum error in any component of the solution, then you should select the infinity norm. The idea is to use norms to achieve what you need: they are not immutable. In using norms to assess errors in solutions, it is recommended that the error be normalized, for example,

$$\text{error} = \frac{\left\| \mathbf{d}^{\text{app}} - \mathbf{d}^{\text{exact}} \right\|_2}{\left\| \mathbf{d}^{\text{app}} \right\|_2} \tag{A.2.6}$$

Norms of functions are defined analogously to the above: a function can be thought of as an infinite-dimensional vector. Thus the norm in function space that corresponds to ℓ_2 is given by

$$\|a(x)\|_{L_2} = \left(\sum_{i=1}^{n} a^2(x_i)\Delta x \right)^{\frac{1}{2}} = \left(\int_0^1 a^2(x)dx \right)^{\frac{1}{2}} \tag{A.2.7}$$

This norm is called the L_2 norm, and the space of functions for which this norm is well defined and bounded is called the space of L_2 functions; often just the subscript '2' is given by the norm, as in $\|\cdot\|_2$. This space is the set of all functions which are square integrable, and it includes the space of all functions which are piecewise continuous, that is, C^{o}.

The Dirac delta function $\delta(x-y)$ is defined by

$$f(x) = \int_{-\infty}^{+\infty} f(y)\delta(x-y)dy \tag{A.2.8}$$

The Dirac delta function is not square integrable! It can be thought of as a function which is infinite at $x=y$ but vanishes everywhere else. The mathematical definition of this function is the topic of the theory of Schwartz distributions. This theory is essential for a good understanding of convergence theory but not for the topics in this book.

The space of functions L_2 is a special case of a more general group of spaces called Hilbert spaces. The norm in the Hilbert space H_1 is defined by

$$\|a(x)\|_{H_1} = \left(\int_0^1 (a^2(x) + a_{,x}^2(x))dx \right)^{\frac{1}{2}} \tag{A.2.9}$$

Just as for vector norms, the major utility of these norms is in measuring errors in functions. Thus if the finite element solution for the displacement in a one-dimensional problem is denoted by $u^h(x)$ and the exact solution is $u(x)$, then the error in the displacement can be measured by

$$\text{error} = \|u^h(x) - u(x)\|_{L_2} \tag{A.2.10}$$

The error in the strain, that is, the first derivative of the displacement, can be measured by the H_1 norm. While this norm also includes the error in the function itself, the error in the derivative almost always dominates. On the other hand, you could measure the error in the strain by the L_2 norm of the first derivative. This is not a true norm in mathematics, because it can vanish for a nonzero function (just take a constant), so it is called a seminorm.

These norms can be generalized to arbitrary domains in multi-dimensional space and to vectors and tensors by just changing the integrals and integrands. Thus the L_2 norm of the displacement on a domain is given by

$$\|\mathbf{u}(\mathbf{x})\|_{L_2} = \left(\int_\Omega u_i(\mathbf{x}) u_i(\mathbf{x}) d\Omega \right)^{\frac{1}{2}} \tag{A.2.11}$$

The definition of the H_1 norm is

$$\|\mathbf{u}(\mathbf{x})\|_{H_1} = \left(\int_\Omega u_i(\mathbf{x}) u_i(\mathbf{x}) + u_{i,j}(\mathbf{x}) u_{i,j}(\mathbf{x}) d\Omega \right)^{\frac{1}{2}} \tag{A.2.12}$$

In general, the precise space to which a norm pertains is not given. Usually only an integer appears by the norm sign. The norm must then be inferred from the context.

In linear stress analysis, the energy norm is often used to measure error. It is given by

$$\text{energy norm} = \left(\int_\Omega \varepsilon_{ij}(\mathbf{x}) C_{ijkl} \varepsilon_{kl}(\mathbf{x}) d\Omega \right)^{\frac{1}{2}} \tag{A.2.13}$$

Its behavior is similar to that of the H_1 norm.

Appendix 3

Element Shape Functions

Three-node triangle

Consider the parent 3-node triangle shown in Figure A3.1 with parent (also called area or barycentric) coordinates ξ_1, ξ_2 and ξ_3. The area coordinates satisfy the relation $\xi_1 + \xi_2 + \xi_3 = 1$. We take ξ_1 and ξ_2 as the independent parent coordinates. The parametrization ξ_3 = constant therefore consists of a family of straight lines parallel to the edge 1–2. The line $\xi_3 = 1$ coincides with the edge 1–2 and the line $\xi_3 = 0$ passes through the origin. The shape functions N_I for the linear (constant strain) triangle are equivalent to the area coordinates, i.e., $N_I = \xi_I$ where I denotes node number.

The area coordinates are given by

$$\xi = \begin{Bmatrix} \xi_1 \\ \xi_2 \\ \xi_3 \end{Bmatrix} = \frac{1}{2A} \begin{bmatrix} y_{23} & y_{32} & x_2 y_3 - x_3 y_2 \\ y_{31} & x_{13} & x_3 y_1 - x_1 y_3 \\ y_{12} & x_{21} & x_1 y_2 - x_2 y_1 \end{bmatrix} \begin{Bmatrix} x \\ y \\ 1 \end{Bmatrix}, \qquad x_{IJ} = x_I - x_J, \quad y_{IJ} = y_I - y_J \tag{A.3.1}$$

where $2A = (x_{32} y_{12} - x_{12} y_{32})$ is twice the area of the triangle. Expressions for the derivatives of the area coordinates with respect to x and y are obtained by differentiating the above, that is, by the first and second columns of the square matrix, respectively. A useful formula for carrying out integration over elements is

$$\int_A \xi_1^i \xi_2^j \xi_3^k \, dA = \frac{(i!\ j!\ k!)}{(i+j+k+2)!} 2A \tag{A.3.2}$$

Nonlinear Finite Elements for Continua and Structures, Second Edition.
Ted Belytschko, Wing Kam Liu, Brian Moran, and Khalil I. Elkhodary.

Companion Website: www.wiley.com/go/belytschko

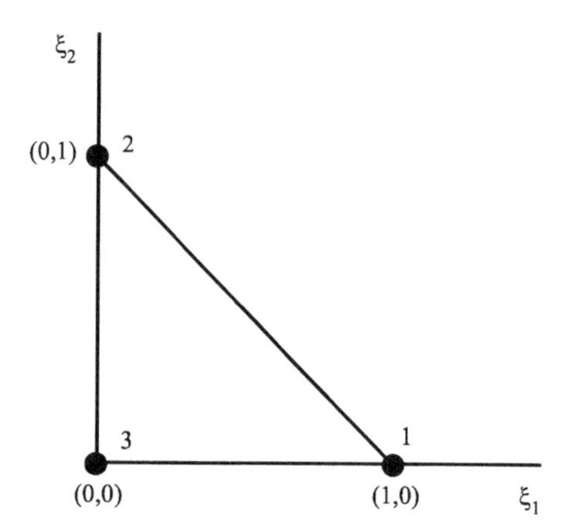

Figure A3.1 Three-node triangle

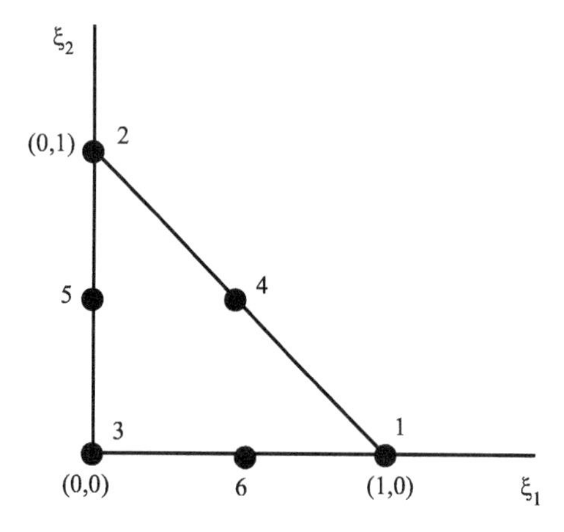

Figure A3.2 Six-node triangle

Six-node triangle

For the 6-node (quadratic displacement) triangle, nodes are added at the midpoints of the sides as shown in Figure A3.2. The shape functions are given by

$$\begin{aligned} &N_I = \xi_I(2\xi_I - 1), \quad I = 1,3 \\ &N_4 = 4\xi_1\xi_2, \quad N_5 = 4\xi_2\xi_3, \quad N_6 = 4\xi_1\xi_3 \end{aligned} \tag{A.3.3}$$

Four-node tetrahedron

The shape functions for the 4-node tetrahedral element can be specified as follows (see Figure A3.3). Define the matrix $\mathbf{A}$:

$$\mathbf{A} = \begin{bmatrix} 1 & x_1 & y_1 & z_1 \\ 1 & x_2 & y_2 & z_2 \\ 1 & x_3 & y_3 & z_3 \\ 1 & x_4 & y_4 & z_4 \end{bmatrix} \tag{A.3.4}$$

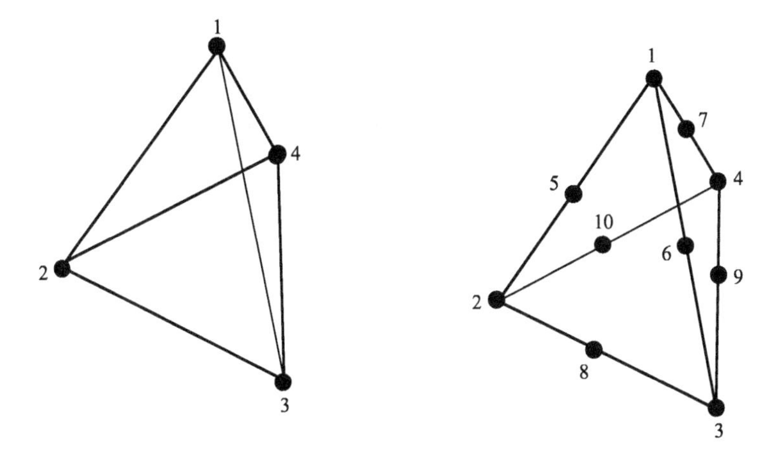

Figure A3.3 Four-node and ten-node tetrahedral elements; nodes 5 through 10 are at the midpoints of the sides

where (x_I, y_I, z_I) are the nodal coordinates ($I = 1, 4$) and the local node numbering is defined by choosing the first node and then numbering the remaining three nodes in a counterclockwise direction as seen from the first. The shape functions are given by

$$N_I(x, y, z) = m_{1I} + m_{2I}x + m_{3I}y + m_{4I}z \tag{A.3.5}$$

where

$$m_{IJ}\frac{1}{6V}(-1)^{(I+J)}\hat{A}_{IJ} \tag{A.3.6}$$

and where $\hat{A}_{IJ}$ are the minors of **A**, that is, the determinant of the matrix obtained by deleting column I and row J of **A**. The element volume is given by $V = \det \mathbf{A}/6$.

The shape functions may also be regarded as volume coordinates, ξ_I (analogous to area coordinates for the triangle). For node I, for example,

$$\xi_I = N_I(x, y, z) = \frac{\text{volume } pJKL}{V} \tag{A.3.7}$$

where $_pJKL$ is the tetrahedron formed by the point p located at (x, y, z) and the remaining three nodes, J, K, L. The following integration formula is useful for element formulation:

$$\int_V \xi_1^i \xi_2^j \xi_3^k \xi_4^m \, dV = \frac{(i!\, j\,!k!\; m!)}{(i+j+k+m+3)!}6V \tag{A.3.8}$$

Ten-node tetrahedron

The shape functions for the 10-node tetrahedron (Figure a3.3) are given by

$$\begin{aligned} &N_1 = \xi_I(2\xi_I - 1), && I = 1,4 \\ &N_5 = 4\xi_1\xi_2, && N_6 = 4\xi_1\xi_3, \quad N_7 = 4\xi_1\xi_4 \\ &N_8 = 4\xi_2\xi_3, && N_9 = 4\xi_3\xi_4, \quad N_{10} = 4\xi_2\xi_4 \end{aligned} \tag{A.3.9}$$

Four-node quadrilateral

The domain of a straight-edged quadrilateral element is defined by the location of its four nodal points $\mathbf{x}_I$, $I = 1, 4$, in the R^2 plane. We assume the nodal points are labeled in ascending order corresponding to the counterclockwise direction: see Figure A3.4.

The nodal coordinates are given by Table A3.1, and the shape functions are given by

$$N_I(\xi) = N_I(\xi,\eta) = \frac{1}{4}(1+\xi_I\xi)(1+\eta_I\eta), \qquad I = 1,4 \tag{A.3.10}$$

Nine-node isoparametric element

The shape functions for the 9-node isoparametric element (Figure A3.5) are given by

$$\begin{aligned} N_I &= \frac{1}{4}(\xi_I\xi+\xi^2)(\eta_I\eta+\eta^2), \quad I = 1.4 \\ N_5 &= \frac{1}{2}(1-\xi^2)(\eta^2-\eta), \quad N_6 = \frac{1}{2}(\xi^2+\xi)(1-\eta^2) \\ N_7 &= \frac{1}{2}(1-\xi^2)(\eta^2+\eta), \quad N_8 = \frac{1}{2}(\xi^2-\xi)(1-\eta^2) \\ N_9 &= (1-\xi^2)(1-\eta^2) \end{aligned} \tag{A.3.11}$$

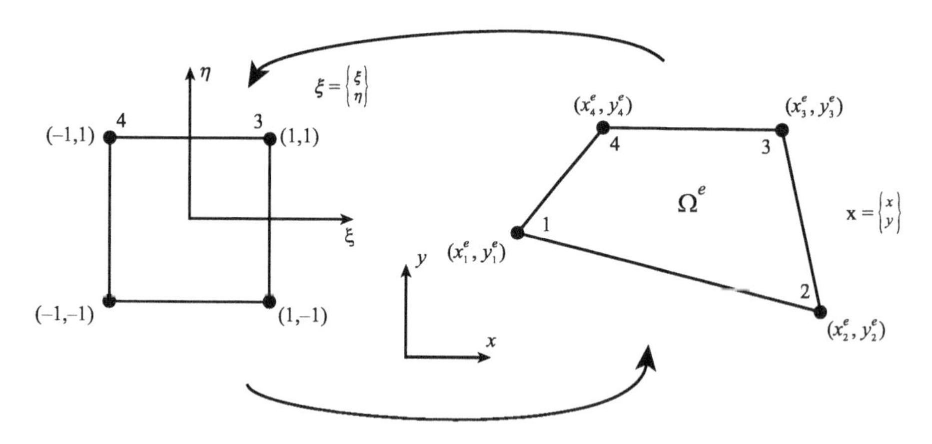

Figure A3.4 Bilinear quadrilateral element domain and local node ordering

Table A3.1 Nodal parent coordinates for 4-node quadrilateral

I	ξ_I	η_I
1	−1	−1
2	1	−1
3	1	1
4	−1	1

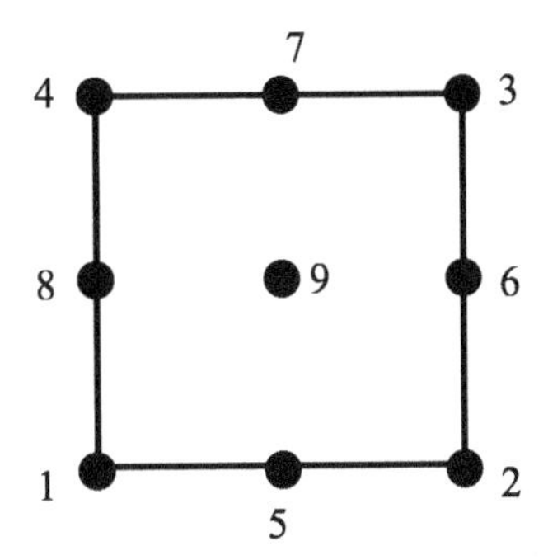

Figure A3.5 Nine-node isoparametric element. The parent domain is the same as for the 4-node quadrilateral. Nodes 5 through 9 are at the midpoints of the sides

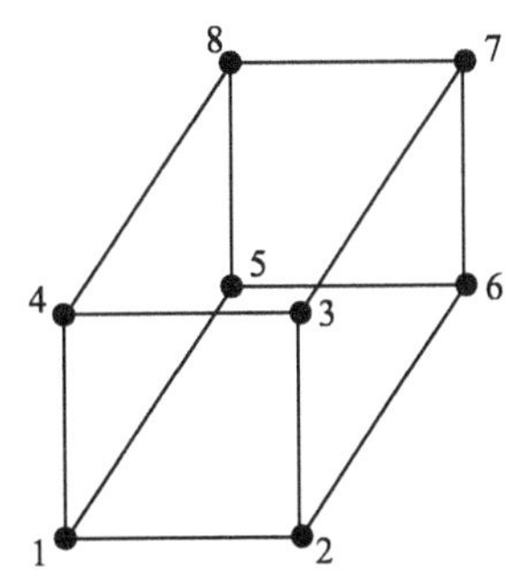

Figure A3.6 Eight-node hexahedral element

Table A3.2 Nodal parent coordinates for 8-node hexahedral element

I	ξ_I	η_I	ζ_I
1	−1	−1	−1
2	1	−1	−1
3	1	1	−1
4	−1	1	−1
5	−1	−1	1
6	1	−1	1
7	1	1	1
8	−1	1	1

Eight-node hexahedral element

For an 8-node trilinear hexahedral element (Figure A3.6), the nodal coordinates are given by Table A3.2 and the shape functions are given by

$$N_I(\xi) = N_I(\xi, \eta, \varsigma) = \frac{1}{8}(1 + \xi_I \xi)(1 + \eta_I \eta)(1 + \varsigma_I \varsigma), \quad I = 1,8 \tag{A.3.12}$$

Table A3.3 Gauss points and weights (p is the order of polynomial which is exactly reproduced by the quadrature scheme)

n_Q	ξ_i	w_i	$p = 2n_Q - 1$
1	0	2	1
2	$\pm 1/\sqrt{3}$	1	3
3	0	8/9	5
	$\pm\sqrt{3/5}$	5/9	
4	$\pm\sqrt{\dfrac{3-2\sqrt{6/5}}{7}}$	$\dfrac{1}{2}+\dfrac{1}{6\sqrt{6/5}}$	
	$\pm\sqrt{\dfrac{3-2\sqrt{6/5}}{7}}$	$\dfrac{1}{2}-\dfrac{1}{6\sqrt{6/5}}$	7

Gaussian quadrature

Gauss points and weights for one-dimensional quadrature over the interval [−1, 1] are given in Table A3.3. For quadrilateral and hexahedral elements, this table can be used in conjunction with the multi-dimensional quadrature formulas given in Chapter 4, Equations (4.5.21–24).

Appendix 4

Euler Angles from Pole Figures

In this appendix we briefly outline the utility of pole figures and the way by which to determine Euler angles from them. For a more extended discussion see (Kelly, *et al.* 2000; Cullity and Stock, 2001).

As indicated in Figure A4.1, for a polycrystalline aggregate each grain possesses a different crystallographic orientation. To plot these different crystal orientations on a two-dimensional plot, *stereographic projection* is used. To identify a given plane, say the (001) plane, in each crystal, draw a unit normal to that plane. Since the plane could be oriented along any direction in three-dimensional space, its unit normal could intersect any point on a unit sphere, as shown in (b–d) in Figure A4.1. From that intersection point a line is drawn to the south pole of the sphere, designated by an **S** in the figure. Then, the intersection of that line with the equatorial plane, designated by a black spot, would represent the crystal's orientation on what becomes called a *pole figure* (i.e., the equatorial circle with the black spots in it, Figure A4.1e–f).

For normals (*poles*) that point beneath the equator, a line from the sphere's surface to the north pole (**N**) is drawn instead of the south pole, and a separate pole figure of the lower hemisphere is thus generated. The pole figure plotted in Figure A4.1(e) includes (as gray circles) the (001) planes of crystals 1, 2 and 3 shown in Figure A4.1(b–d). It further shows a uniform distribution of spots, indicating no preferential orientation in space for a hypothetical polycrystalline aggregate. Such a case is referred to as *no texture*. Figure A4.1(f), on the other hand, shows clustering of spots in specific regions of the pole figure, which indicates preferential orientation in space for crystals in a specimen, that is, *texture*. The latter is common when metals are processed by rolling, for example.

It is possible to extract the distribution of Euler angles for the corresponding planes from a pole figure, which is required in Box 13.1 to correctly align slip directions and normals in the

Nonlinear Finite Elements for Continua and Structures, Second Edition.
Ted Belytschko, Wing Kam Liu, Brian Moran, and Khalil I. Elkhodary.

Companion Website: www.wiley.com/go/belytschko

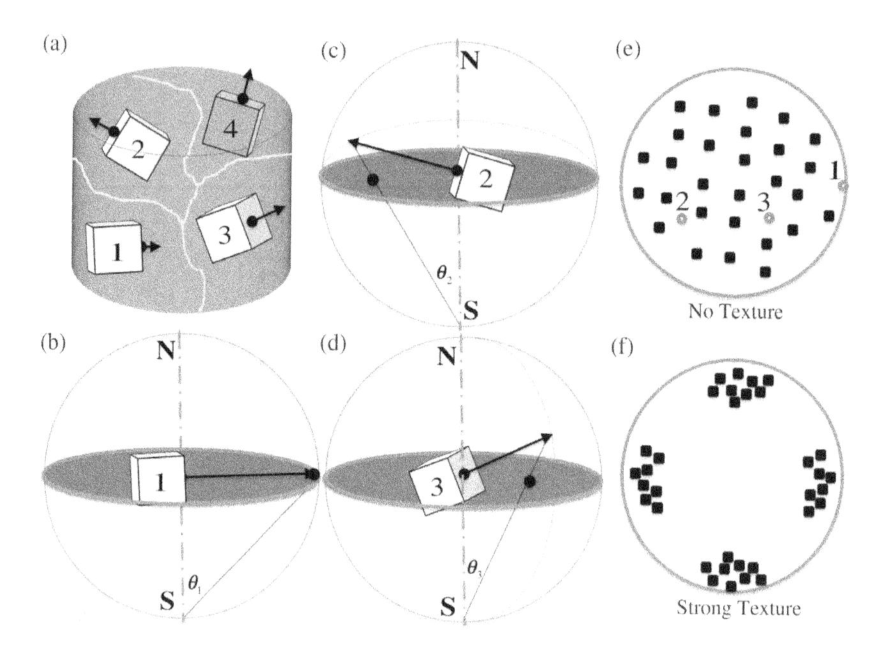

Figure A4.1 (a) Polycrystalline aggregate showing four grains, (b–d) finding the stereographic projections of grains 1–3, (e) a pole figure showing no texture, (f) a pole figure showing strong texture

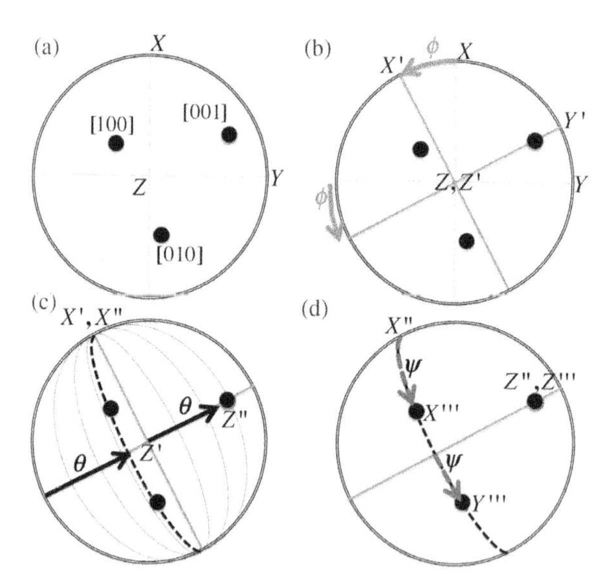

Figure A4.2 Finding Euler angles from a pole figure

single crystal plasticity algorithm explained in Box 13.2 (see Chapter 13). A representative procedure is shown in Figure A4.2.

Crystallographically equivalent plane normals are identified, [001], [010] and [100] for this example. First, a rotation ϕ about the Z-axis is applied, so as to make the Y-axis pass through the [001] pole, as shown in Figure A4.2(b), thereby creating new axes, X′, Y′ and Z′. Second, a rotation θ is applied about the X′-axis to bring the Z′-axis in coincidence with the [001] pole. The Z′-axis is here renamed Z″ and the X′-axis X″, Figure A4.2(c). The plane whose normal is the Z″ axis is then identified with the major circle found from the outer diameter by the same angle θ, as shown in Figure A4.2(c). Finally, a rotation ψ is applied about the Z″-axis to bring the X″-axis in coincidence with the [100] pole, which is now labeled X‴. At this point, the crystal's orientation is fully determined by these three Euler angles measured from the given pole figure, and the [010] pole may be identified with the Y‴ axis and the [001] pole with the Z‴-axis.

Appendix 5

Example of Dislocation-Density Evolutionary Equations

In this appendix an explanation and derivation are presented of the specific dislocation density evolutionary laws selected for Equation (13.6.3), that is,

$$\dot{\rho}_m^{(\alpha)} = \left|\dot{\gamma}^{(\alpha)}\right| \left(g_s^{(\alpha)} \rho_{im}^{(\alpha)} \left(\rho_m^{(\alpha)}\right)^{-1} - g_{m0}^{(\alpha)} \rho_m^{(\alpha)} - g_{im0}^{(\alpha)} \sqrt{\rho_{im}^{(\alpha)}} \right) \tag{A.5.1a}$$

$$\dot{\rho}_{im}^{(\alpha)} = \left|\dot{\gamma}^{(\alpha)}\right| \left(g_{m1}^{(\alpha)} \rho_m^{(\alpha)} + g_{im1}^{(\alpha)} \sqrt{\rho_{im}^{(\alpha)}} - g_r^{(\alpha)} \rho_{im}^{(\alpha)} \exp\left(\frac{-\Delta H_0}{kT} \left[1 - \sqrt{\frac{\rho_{im}^{(\alpha)}}{\rho_{im}^{\text{sat}}}} \right] \right) \right). \tag{A.5.1b}$$

These laws are not unique in the literature, and the discussion herein is merely to illustrate a possible choice of evolutionary laws, with a fair level of generality and that capture the spirit of the approach of dislocation density modeling. First, however, a brief discussion of dislocation mobility is in order.

A.5.1 Dislocation Mobility

Dislocation motion is generally of two types: *conservative* and/or *non-conservative*, depending on whether the number of atoms in the neighborhood of a dislocation core changes as a result of the motion (Bulatov and Cai, 2006). As shown in Figure A5.1, an edge dislocation (a) can move conservatively by *gliding* in the slip direction (b), as was discussed earlier, or non-conservatively by *climbing* in a direction normal to the slip plane (c). For an edge dislocation to climb upward, a line of atoms must continually be emitted from above the dislocation core; whereas for climb to proceed downward as shown in the figure, a line of atoms must continually be absorbed under the dislocation core. Such a process may occur in real

Nonlinear Finite Elements for Continua and Structures, Second Edition.
Ted Belytschko, Wing Kam Liu, Brian Moran, and Khalil I. Elkhodary.

Companion Website: www.wiley.com/go/belytschko

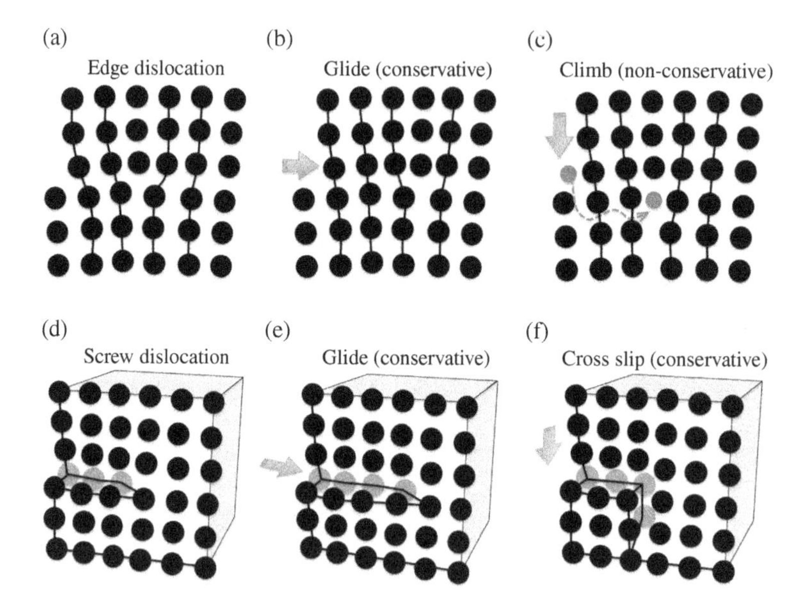

Figure A5.1 Conservative and non-conservative dislocation motion

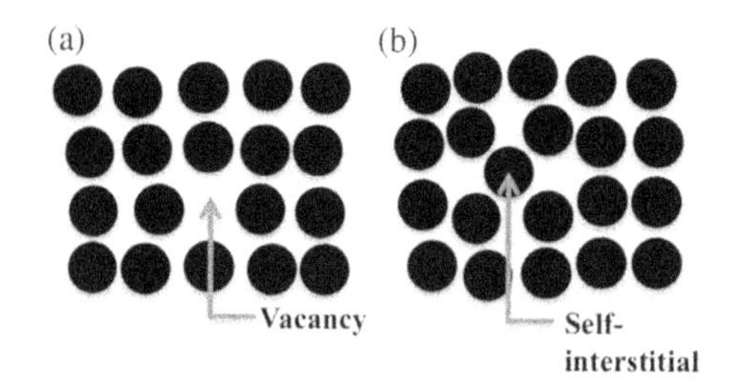

Figure A5.2 (a) vacancy, (b) interstitial. Adapted from Bulatov and Cai, 2006

crystals when atomic motion is assisted by substantial thermal agitation (i.e., deformation proceeds at high temperature), and when the crystal possesses a population of *point defects*, that is, *interstitials* and *vacancies* (see Figure A5.2), which may become attracted to dislocation cores by virtue of their interacting stress fields (Bulatov and Cai, 2006), as shown in Figure A5.1(c).

Conversely, screw dislocations can migrate from one slip system to another, which shares the same slip direction, by the mechanism of *cross-slip*, as shown in Figure A5.1(d–f). This process may be understood from the fact that screw dislocations are defined by parallel dislocation line and burgers vectors, such that no unique slip plane normal exists. For many crystals therefore (e.g., BCC structures), multiple planes can accommodate screw dislocation motion, and cross-slip may be observed to evolve along the different slip systems for the same applied load as shown in Figure A5.1(e–f). As no atoms need be added or removed by this process, cross slip and screw dislocation motion in general are conservative.

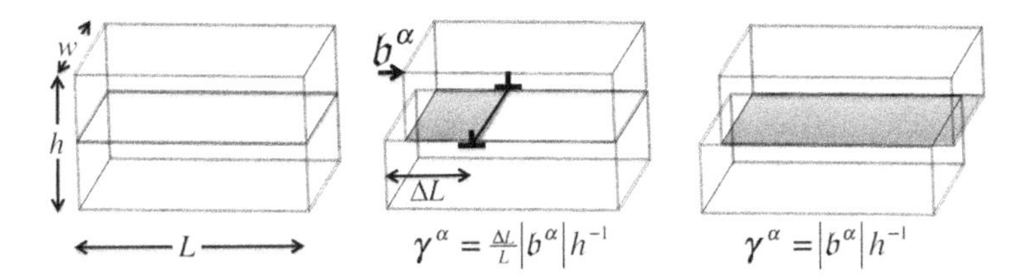

Figure A5.3 Mechanism behind the Orowan rule

Thus, both the screw and edge components of mixed dislocations can move during deformation. In general, their motion evolves at a rate that depends on the exact mechanism (glide, climb, cross-slip). However, Orowan's rule summarizes this result by relating mesoscale plastic shear rate ($\dot{\gamma}^{\alpha}$) to the density of mobile dislocations (ρ_m^{α}), the Burgers vector ($b^{(\alpha)}$), and the *effective* dislocation speed v^{α} by (Hull and Bacon, 2006),

$$\dot{\gamma}^{(\alpha)} = \rho_m^{(\alpha)} \left| \boldsymbol{b}^{(\alpha)} \right| v^{(\alpha)} \tag{A.5.2}$$

This relation may be understood by considering Figure A5.3. For a crystal of volume $V = Lwh$, when a single dislocation traverses the entire crystal, an average shear strain $\gamma^{\alpha} = |\boldsymbol{b}^{\alpha}| h^{-1}$ is created (Figure A5.3, right). For a dislocation that traveled only partway across the crystal the average shear strain would rescale to $\gamma^{\alpha} = \Delta L \cdot L^{-1} \cdot |\boldsymbol{b}^{\alpha}| h^{-1}$ (Figure A5.3, middle). For N such dislocations in the slip system, the resulting average shear strain would be $\gamma^{(\alpha)} = N^{(\alpha)} \Delta L \cdot L^{-1} \cdot |\boldsymbol{b}^{(\alpha)}| h^{-1}$. We can rehash this equation as $\gamma^{(\alpha)} = N^{(\alpha)} w V^{-1} \cdot |\boldsymbol{b}^{(\alpha)}| \Delta L$, where V is the volume and $wV^{-1} = L^{-1} h^{-1}$. Recognizing that $\rho^{\alpha} = N^{\alpha} w V^{-1}$, i.e. dislocation density is the total length of dislocations per unit volume, we find that $\gamma^{(\alpha)} = \rho^{(\alpha)} |\boldsymbol{b}^{(\alpha)}| \Delta L$. Differentiating in time yields Equation (A5.2).

There are various equations that characterize dislocation velocity v^{α}, which must be determined in Equation (A5.2), depending on the details of the evolving dislocation mechanism, see for example (Caillard and Martin, 2003). Quite generally, the speed v^{α} can be described by a relation of the form (Caillard and Martin, 2003),

$$v^{\alpha} = v_0 \left(\frac{\overline{\tau}^{*\alpha}}{\overline{\tau}_{ref}} \right)^n \exp\left(-\frac{\Delta E}{kT} \right), \tag{A.5.3}$$

where $\overline{\tau}^{*\alpha} = \overline{\tau}^{\alpha} - a\sqrt{\rho^{\alpha}}$ is the effective *resolved* shear stress, with $\overline{\tau}^{(\alpha)} \equiv \overline{\mathbf{s}}^{T(\alpha)} \overline{\boldsymbol{\sigma}}^{dev} \overline{\mathbf{n}}^{(\alpha)}$ being the projection of the deviatoric part of the Cauchy stress tensor $\overline{\boldsymbol{\sigma}}^{dev}$ in slip system $\boldsymbol{\alpha}$, pulled back to the intermediate configuration, and a is a constant with units of MPa$\cdot m$. The parentheses indicate 'no sum' on the index. Superscript α runs through all active slip systems of a crystal. n is the stress exponent, and ΔE is the activation energy for dislocation motion. k is the Boltzmann constant, and T the current absolute temperature. For conservative and non-conservative motion different values for n, a, ΔE and v_0 would be needed. Relation Equation (A5.2) is central to different developments of the evolutionary laws for dislocation densities. In the following, development of the different terms of Equation (A5.1) will be consistent with (Shanthraj and Zikry, 2011).

A.5.2 Dislocation Generation

Dislocation densities tend to increase during any deformation beyond the elastic limit, as they are the bearers of plasticity in a crystal. Generation of dislocations proceeds from various sources in a crystal and takes the form (e.g., Estrin and Kubin, 1986; Estrin, Krausz *et al.*, 1996; Shanthraj and Zikry, 2011),

$$\dot{\rho}^{(\alpha)}_{generation} \equiv \rho^{(\alpha)}_{source} \frac{v^{(\alpha)}}{l_c}, \tag{A.5.4}$$

where $\rho^{(\alpha)}_{source}$ is the local density of dislocation generation sites in slip system α, $v^{(\alpha)}$ is the average speed at which dislocations are emitted from the source, and l_c is the characteristic spacing between sources.

It is rather difficult to determine $\rho^{(\alpha)}_{source}$, since multiple sites for dislocation generation may arise under different loading conditions and for different microstructures. Commonly, immobile dislocations form a connected network, called a *Frank net*, which extends throughout the crystal (Gurtin, 2006). In slip system α, segments of mobile dislocation lines will inevitably get pinned and immobilized by this net. These immobilized segments often act as sources to new dislocations by means of a *Frank–Read* mechanism (Estrin and Kubin, 1986; Gurtin, 2006).

As shown in Figure A5.4, a dislocation segment immobilized and pinned at both ends by obstacles (e.g., Frank net) can bow out under an applied shear stress to form a dislocation loop that is emitted away from the obstacles. In the process, the immobile segment reforms and remains pinned so that the dislocation loop formation and emission mechanism may be sustained. For such a case, it may be assumed that $\rho^{(\alpha)}_{source} = \phi^{(\alpha)} \cdot \rho^{(\alpha)}_{im}$, where ϕ^{α} is the probability immobile dislocations in slip system α are conducive to the activation of a Frank–Read mechanism. Emission speed $v^{(\alpha)}$ used in Equation (A5.4) may be computed from Orowan's relation, while distance l_c for a Frank net can be determined from $l_c = \left(\sum_{\eta} \left(\rho^{\eta}_{im}\right)^{1/2} \right)^{-1}$.

A.5.3 Dislocation Annihilation

Dislocation annihilation proceeds as a result of a process known as *dynamic recovery* (Estrin and Kubin, 1986; Hull and Bacon, 2006), which is a thermally activated mechanism that evolves with large deformation, whereby immobile dislocations are released to immediately

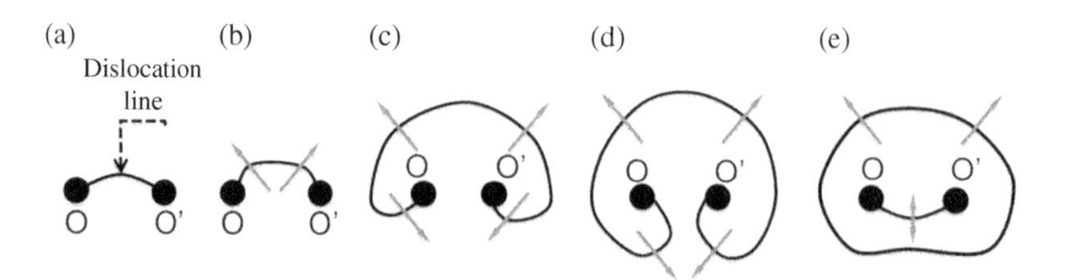

Figure A5.4 Frank–Read mechanism

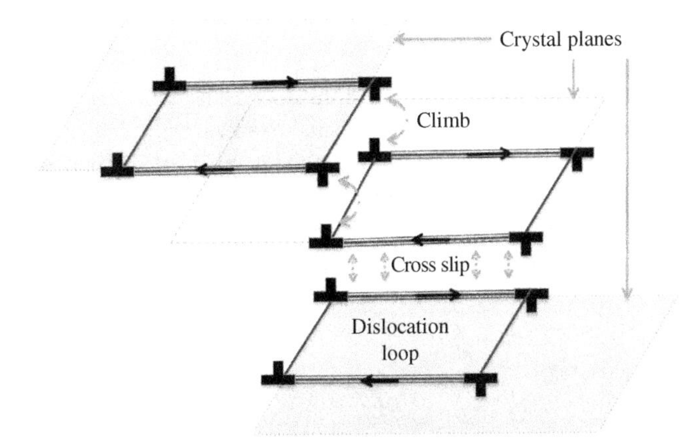

Figure A5.5 Climb and cross-slip. Adapted from J. P. Poirier, On the symmetrical role of cross-slip of screw dislocations and climb of edge dislocations as recovery processes controlling high temperature creep, Vol. 11, 1976. Revue De Physique Applique'e

cancel out with neighbors of opposite signs. Together with dislocation generation, annihilation accounts for the experimental observation that dislocation densities tend to saturate at large strain. It can be shown that recovery obeys an Arrhenius type equation of the form (Shanthraj and Zikry, 2011),

$$\dot{\rho}^{(\alpha)}_{annihilation} = \dot{A}^{(\alpha)} \rho^{(\alpha)}_{im} \exp(-\Delta H^{(\alpha)} / kT). \qquad \text{(A.5.5)}$$

$\dot{A}^{\alpha}$ is the *attempt* rate of dislocation annihilation by its fundamental mechanisms of climb and cross-slip, as indicated in Figure A5.5 for idealized rectangular dislocation loops composed of two edge components (thin lines) and two *split* screw components (thick lines). $\dot{A}^{\alpha}$ is thus measured in s^{-1} and reflects: (1) the rate by which segments of mobile edge dislocations of opposite sign climb to annihilate one another, and (2) the rate by which segments of mobile screw dislocations of opposite sign cross-slip to annihilate one another. The attempt rate is therefore related to the Orowan velocities for both non-conservative and conservative motions, and may be heuristically given by (Shanthraj and Zikry, 2011),

$$\dot{A}^{(\alpha)} = \phi^{(\alpha)} \sum_{\eta} \left(\sqrt{a_{(\alpha)\eta}}\, l_c \dot{\gamma}^{\eta} \left| \boldsymbol{b}^{\eta} \right|^{-1} \right), \qquad \text{(A.5.6)}$$

where ϕ^{α} is the probability interacting dislocations will annihilate at a given temperature. ΔH in Equation (A5.5) is the enthalpy of activation of the climb and cross-slip mechanisms responsible for annihilation. It may be taken to decrease from a reference value ΔH_0 for a crystal to zero, in proportion to the ratio of immobile dislocation densities to their observed saturation value, such that the process of annihilation reach its maximum rate as dislocation densities saturate, consistent with experimental observation. That is (Shanthraj and Zikry, 2011),

$$\Delta H^{\alpha} = \Delta H_0 \left(1 - \sqrt{\rho^{\alpha}_{im} \left(\rho^{sat}_{im} \right)^{-1}} \right). \qquad \text{(A.5.7)}$$

A.5.4 General Dislocation Interaction

As mobile dislocations move, they interact with themselves as well as with immobile dislocations, so that the resulting dislocation structure changes in ways other than by means of the generation and the thermally activated annihilation mechanisms described previously. The net effect of these general interactions is usually immobilization of a fraction of the mobile densities, which on the one hand decreases mobile dislocation densities, and on the other hand adds to the immobile dislocation junctions. Reduction of mobile dislocation densities may be expressed as a result of mobile-mobile and mobile-immobile interactions as (Shanthraj and Zikry, 2011),

$$\dot{\rho}^{(\alpha)-}_{\text{interaction}} = \phi^{(\alpha)} \cdot l_c \sum_{\eta} \rho^{(\alpha)}_m \left(\rho^{\eta}_m \overline{v}_{\alpha\eta} + \rho^{\eta}_{im} v_{\alpha} \right), \tag{A.5.8}$$

Where velocity $\overline{v}_{\alpha\eta} = \frac{1}{2}(v_\alpha + v_\eta)$ measures the average mobility of dislocations in slip systems α and η, and may be determined from the Orowan relations described earlier.

To determine the rate at which stable dislocation junctions are formed, Frank's energetic criterion needs be applied to all interaction combinations. The criterion specifies whether it is energetically favorable for intersecting dislocations from slip systems β and γ to form a junction in system α. Favorability is measured by a set of binary coefficients $z^{\beta\gamma}_{\alpha} \in [0, 1]$ that take a value of zero if formation of a junction is not favorable, and unity when favorable. The test for favorability may be stated as (Shanthraj and Zikry, 2011),

$$G\,|\boldsymbol{b}^{\alpha}|^2 < G\left(|\boldsymbol{b}^{\gamma}|^2 + G\,|\boldsymbol{b}^{\beta}|^2\right) \text{ and } \boldsymbol{b}^{\alpha} = \boldsymbol{b}^{\gamma} + \boldsymbol{b}^{\beta}, \tag{A.5.9}$$

where it is recognized that strain energy scales with $G|\boldsymbol{b}^{\eta}|^2$. Once the set $z^{\beta\gamma}_{\alpha}$ has been determined, dislocation junction formation rate in slip system α may be given by (Shanthraj and Zikry, 2011),

$$\dot{\rho}^{(\alpha)+}_{\text{interaction}} = \phi^{(\alpha)} \cdot l_c \sum_{\beta,\gamma \le \beta} z^{\beta\gamma}_{\alpha} \sqrt{a_{\beta\gamma}} \left(\rho^{\beta}_m \rho^{\gamma}_m \overline{v}_{\beta\gamma} + \rho^{\beta}_m \rho^{\gamma}_{im} v_{\beta} + \rho^{\gamma}_m \rho^{\beta}_{im} v_{\gamma} \right) \tag{A.5.10}$$

A.5.5 Final Remarks

Finally, combination and manipulation of the equations herein derived for dislocation density generation, annihilation and interaction would yield the terms in Equation (A5.1), and the coefficients defined in Chapter 13, Table 13.2.

As dislocation densities tend to saturate under large deformations, the coefficients of Table 13.2 may be taken as roughly constant over most of the strain range. For simplicity, it may be further assumed that $g_{m1} = g_{m0}$ and $g_{im1} = g_{im0}$ (e.g., Estrin and Kubin, 1986; Estrin, Krausz *et al.*, 1996), which lead to the constant values in Table 13.2.

Glossary

Symbols

$\dot{f}$	For a field, the superposed dot denotes the material time derivative, i.e. $\dot{f}(\mathbf{X}, t) = \partial f(\mathbf{X}, t)/\partial t$; for a function of time only, it is the ordinary time derivative, i.e. $\dot{f}(t) = df(t)/dt$
$f_{,x}$	Derivative with respect to the variable x; when the comma is followed by an index, such as i, j, k, to s, it is the derivative with respect to the corresponding spatial coordinate, i.e. $f_{,i} = \partial f/\partial x_i$
ϕ^*	Pull-back of a tensor from spatial to reference configurations using the deformation gradient. The precise operation depends on the context. For strain rate, for example, $\dot{\mathbf{E}} = \varphi^*\mathbf{D} = \mathbf{F}^T \cdot \mathbf{D} \cdot \mathbf{F}$; for stress $\mathbf{S} = \varphi^*\tau = \mathbf{F}^{-1} \cdot \tau \cdot \mathbf{F}$
ϕ_*	Push-forward of a tensor from reference to spatial configurations, e.g. $\mathbf{D} = \varphi^*\mathbf{E} = \mathbf{F}^{-T} \cdot \dot{\mathbf{E}} \cdot \mathbf{F}^{-1}$, $\tau = \varphi^*\mathbf{S} = \mathbf{F} \cdot \mathbf{S} \cdot \mathbf{F}^T$
ϕ_e^*, ϕ_*^e	Pull-back and push-forward, respectively, using the elastic part of the deformation gradient
$L_v\tau$	Lie derivative – the push-forward of the material time derivative of the pull-back of a quantity (also called the convected rate), as in $$L_v\tau = \phi_*\left(\frac{D}{Dt}(\phi^*\tau)\right)$$
$\cdot$ as in $\mathbf{a} \cdot \mathbf{b}$	Contraction of inner indices; for vectors, $\mathbf{a} \cdot \mathbf{b}$ is the scalar product $a_i b_i$; if one or more of the variables are tensors of second order or higher, the contraction is on the inner indices, i.e. $\mathbf{A} \cdot \mathbf{B}$ represents $A_{ij}B_{jk}$, $\mathbf{A} \cdot \mathbf{a}$ represents $A_{ij}a_j$

Nonlinear Finite Elements for Continua and Structures, Second Edition.
Ted Belytschko, Wing Kam Liu, Brian Moran, and Khalil I. Elkhodary.

Companion Website: www.wiley.com/go/belytschko

: as in **A**: **B**	Double contraction of inner indices: **A**: **B** is given by $A_{ij}B_{ij}$, **C: D** is $\mathbf{C}_{ijkl}D_{kl}$; note the order of the indices! Note also that if **A** or **B** is symmetric **A**: **B** = $A_{ij}B_{ji}$
×	As in **a** × **b** indicates cross-product; in indicial notation, $\mathbf{a} \times \mathbf{b} \rightarrow e_{ijk}a_jb_k$
⊗	As in **a** ⊗ **b** indicates vector product; in indicial notation, $\mathbf{a} \otimes \mathbf{b} \rightarrow a_ib_j$; in matrix notation, $\mathbf{a} \otimes \mathbf{b} \rightarrow \{a\}\{b\}^T$

Variables

A	Jacobian matrix of the system $\mathbf{r} = \mathbf{0}$; used for other purposes in chapters other than chapter 6
$A^{(i)}$	First through fourth elasticity tensors for $i = 1, 4$ respectively.
$\boldsymbol{B}_I, \boldsymbol{B}$	$\boldsymbol{B}$ is the column matrix of spatial derivatives of shape functions I, given by $\boldsymbol{B}_{iI} = \partial N_I/\partial x_i$; $\boldsymbol{B}$ is a rectangular matrix formed by $[\boldsymbol{B}_I, \boldsymbol{B}_2, \ldots, \boldsymbol{B}_n]$
B_{0I}/B_0	Column matrix of material derivatives of shape functions I, given by $B_{0iI} = \partial N_I/\partial X_i$; B_0 is a rectangular matrix $[B_{01}, B_{02}, \ldots, B_{0n}]$
$\mathbf{B}_I, \mathbf{B}$	Matrix of spatial derivatives of shape functions in Voigt notation arranged so that $\{\mathbf{D}\} = \mathbf{B}_I\dot{\mathbf{d}}_I$; **B** is a rectangular matrix $[\mathbf{B}_1, \mathbf{B}_2, \ldots, \mathbf{B}_n]$
$\mathbf{B}_{0I}, \mathbf{B}_0$	Matrix of material derivatives of shape functions in Voigt notation arranged so that $\{\dot{\mathbf{E}}\} = \mathbf{B}_{0I}\dot{\mathbf{d}}_I$; $\mathbf{B}_0$ is a rectangular matrix $[\mathbf{B}_{01}, \mathbf{B}_{02}, \ldots, \mathbf{B}_{0n}]$
C	Cauchy Green tensor, $\mathbf{C} = \mathbf{F}^T \cdot \mathbf{F}$; it is distinguished from the material response matrices which follow by absence of a superscript
$\mathbf{C}^{SE}, C^{SE}_{ijkl}, [\mathbf{C}^{SE}]$	Material tangent moduli relating $\dot{\mathbf{S}}$ to $\dot{\mathbf{E}}$
$\mathbf{C}^{\tau}, C^{\tau}_{ijkl}, [\mathbf{C}^{\tau}]$	Material tangent moduli relating convected rate of Kirchhoff stress $\tau^{\nabla c}$ to **D**
$\mathbf{C}^{\sigma J}, C^{\sigma J}_{ijkl}, [\mathbf{C}^{\sigma J}]$	Material tangent moduli relating Jaumann rate of Cauchy stress $\sigma^{\nabla J}$ to **D**
$\mathbf{C}^{\sigma T}, C^{\sigma T}_{ijkl}, [\mathbf{C}^{\sigma T}]$	Material tangent moduli relating Truesdell rate of Cauchy stress $\sigma^{\nabla T}$ to **D**
$\mathbf{C}^{\mathrm{alg}}, C^{\mathrm{alg}}_{ijkl}, [\mathbf{C}^{\mathrm{alg}}]$	Algorithmic moduli relating Truesdell rate of Cauchy stress $\sigma^{\nabla T}$ to **D** (where rates are based on finite increments)
$\mathbf{D}, D_{ij}, \{\mathbf{D}\}$	Rate of deformation, velocity strain, $\mathbf{D} = \mathrm{sym}(\nabla \mathbf{v})$
$\mathbf{E}, E_{ij}$	Green strain tensor, $\mathbf{E} = \frac{1}{2}(\mathbf{F}^T \cdot \mathbf{F} - \mathbf{I})$
$\mathbf{F}, F_{ij}$	Deformation gradient, $F_{ij} = \partial x_i/\partial X_j$
$\mathbf{F}^e, \mathbf{F}^p$	Elastic and plastic parts of the deformation gradient, $\mathbf{F} = \mathbf{F}^e \cdot \mathbf{F}^p$
J	Determinant of Jacobian between spatial and material coordinates, $J = \det[\partial x_i/\partial X_j]$
$\hat{J}_{ij}$	Cofactors of F^{χ}_{ij}.
J_{ξ}	Determinant of Jacobian between spatial and element coordinates, $J_{\xi} = \det[\partial x_i/\partial \xi_j]$
J^0_{ξ}	Determinant of Jacobian between material and element coordinates, $J^0_{\xi} = \det[\partial X_i/\partial \xi_j]$
K	Linear stiffness matrix
$\mathbf{K}^{\mathrm{int}}, \mathbf{K}^{\mathrm{ext}}$	Tangent stiffness matrices for internal and external nodal forces, $\mathbf{K}^{\mathrm{int}} = \partial \mathbf{f}^{\mathrm{int}}/\partial \mathbf{d}$, $\mathbf{K}^{\mathrm{ext}} = \partial \mathbf{f}^{\mathrm{ext}}/\partial \mathbf{d}$
$\mathbf{K}^{\mathrm{mat}}, \mathbf{K}^{\mathrm{geo}}$	Material and geometric tangent stiffness, respectively
$\mathbf{K}^{\mathrm{ale}}$	Stiffness matrix account for the ALE part for the momentum equation
$\mathbf{L}, L_{ij}$	Spatial gradient of velocity field: see (3.3.18)
$\mathbf{L}_e$	Connectivity matrix
$\mathbf{M}, \mathbf{M}_{IJ}, M_{ijIJ}$	Mass matrix: see Sections 4.4.3 and 4.4.9
N_I	Shape functions

$\mathbf{P}, P_{ij}$	Nominal stress (transpose of first Piola–Kirchhoff stress)
$\mathbf{Q}$	Rotation tensor/matrix used in frame invariance and material symmetry
$\mathbf{R}, R_{ij}$	Rotation matrix: see Section 3.2.8.
$\mathbf{S}, S_{ij}$	Second Piola–Kirchhoff (PK2) stress
$\mathbf{U}, U_{ij}$	Right stretch tensor
u, u_0	Space of kinematically admissible displacements and velocities; u_0 is the space u_0 with the functions vanishing where they are prescribed: see (4.3.1–2)
W	Work
$W^{int}, W^{ext}, W^{inert}$	Internal, external, and inertial work
$\mathbf{X}$	Material (Lagrangian) coordinates
$X_{iI}, \mathbf{X}_I$	$\mathbf{X}_I = [X, Y, Z]$ = nodal material coordinates
$\boldsymbol{\chi}$	Referential coordinates (ALE formulations)
$\mathbf{b}, b_i$	Body force: see Section 3.5.5
$\mathbf{d}$	Nodal displacements stored in Voigt form
$\mathbf{e}_i, \mathbf{e}_i$	$[\mathbf{e}_x, \mathbf{e}_y, \mathbf{e}_z]$, base vectors of coordinates
f	Yield function in elastic–plastic constitutive models
$\mathbf{f}, \mathbf{f}_I, f_{iI}$	Nodal forces
$\mathbf{f}^{int}, \mathbf{f}_I^{int}, f_{iI}^{int}$	Internal nodal forces
$\mathbf{f}^{ext}, \mathbf{f}_I^{ext}, f_{iI}^{ext}$	External nodal forces
$\mathbf{h}$	Plastic moduli in elastic–plastic constitutive models
$\mathbf{n}, n_i, \mathbf{n}_0, n_i^0$	Unit normal in current (deformed) and initial (reference, undeformed) configurations
$\mathbf{q}, q_i$	Heat flux, also collection of internal variables in constitutive models
$\mathbf{t}, t_i$	Surface tractions (see Section 3.5.5)
$\mathbf{u}, u_i$	Displacement field
$\mathbf{u}_I, u_{iI}$	Matrix of components of displacement at node I
$\mathbf{v}_I, v_{iI}$	Matrix of components of velocity at node I
$\mathbf{v}, v_i$	Velocity field
$w, \bar{w}$	Hyperelastic potential on reference and intermediate configurations, respectively, e.g. $\mathbf{S} = \partial w/\partial \mathbf{E}$
$\mathbf{x}_{IJ} \equiv \mathbf{x}_I - \mathbf{x}_J$	Difference in nodal coordinates
$\mathbf{x}, x_i$	Spatial (Eulerian) coordinates
$x_{iI}, \mathbf{x}_I$	$\mathbf{x}_I = [x_1, y_1, z_1]$ = nodal spatial coordinates
Γ, Γ_0	Boundary of body in current (deformed) and initial (reference, undeformed) configurations
Γ_{int}	Surfaces of interior discontinuities
ξ, ξ_i	$[\xi, \eta, \zeta]$ are parent element coordinates, also used as curvilinear coordinates
ρ, ρ_0	Current and original density
$\sigma, \sigma_{ij}, \{\sigma\}$	Cauchy (physical) stress tensor
$\tau, \tau_{ij}, \{\tau\}$	Kirchhoff stress tensor
$\phi(\mathbf{X}, t)$	Mapping from the initial configuration Ω_0 to the current or spatial configuration Ω
ϕ	Viscoplastic overstress function
$\hat{\phi}(\boldsymbol{\chi}, t)$	Mapping from the referential configuration $\hat{\Omega}$ to the spatial configuration Ω
$\psi(\mathbf{X}, t)$	Mapping from the initial configuration Ω_0 to the referential configuration Ω
$\psi, \bar{\psi}$	Hyperelastic potential on reference or intermediate configurations, respectively, e.g. $\mathbf{S} = 2\partial\psi/\partial\mathbf{C}$
$\Omega, \Omega_0, \hat{\Omega}$	Domain of current (deformed), initial (undeformed) and referential configurations

References

Abeyaratne R and Knowles JK (1988) Unstable elastic materials and the viscoelastic response of bars in tension, *J. Appl. Mech.*, **55**, 491–492.

Ahmad S, Irons BB and Zienkiewicz OC (1970) Analysis of thick and thin shell structures by curved finite elements, *International Journal for Numerical Methods in Engineering*, **2**, 419–151.

Ahmadi G and Firoozbakhsh K (1975) First strain gradient theory of thermoelasticity. *International Journal of Solids and Structures* **11**(3), 339–345.

Aifantis EC (1984) On the structural origin of certain inelastic models, *J. Engrg. Mat. Tech.*, **106**, 326–330.

Alfano G and de Sciarra FM (1996) Mixed finite element formulations and related limitation principles; a general treatment, *Computer Methods in Applied Mechanics and Engineering*, **104**, 105–130.

Ali AA, Podus GN, Sirenko AF *et al.* (1979) Determining the thermal activation parameters of plastic deformation of metals from data on the kinetics of creep and relaxation of mechanical stresses, *Strength of Materials*, **11**(5), 496.

Anthony K-H and Azirhi A (1995) Dislocation dynamics by means of Lagrange formalism of irreversible processes – complex fields and deformation Processes, *International Journal of Engineering Science*, **33**(15), 2137–2148.

Areias P and Belytschko T (2006) A comment on the article 'A finite element method for simulation of strong and weak discontinuities in solid mechanics' by A. Hansbo and P. Hansbo [*Comput. Methods Appl. Mech. Eng.*, **193**(2004) 3523–3540], *Computer Methods in Applied Mechanics and Engineering*, **195**, 1275–1276.

Argyris JH (1965), Elasto-plastic matrix displacement analysis of three-dimensional continua, *J. Royal Aeronautical Society*, **69**, 633–635.

Asaro RJ (1979) Geometrical effects in the inhomogeneous deformation of ductile single crystals, *Acta Metallurgica*, **27**(3), 445–453.

Asaro RJ (1983) *Micromechanics of Crystals and Polycrystals*, Advances in Applied Mechanics, Vol. **23**, Academic Press, New York.

Asaro RJ (1983) Crystal plasticity, *Journal of Applied Mechanics*, **50**, 921–934.

Asaro RJ and Rice JR (1977) Strain localization in ductile single crystals, *J. Mech. Phys. Solids*, **25**, 309–338.

Atluri SN and Cazzani A (1995) Rotations in computational solid mechanics, *Arch. Comp. Mech.*, **2**, 49–138.

Auriault JL (1991) Is an equivalent macroscopic description possible? *International Journal of Science and Engineering* **29**(7), 785–795.

Azaroff LV (1984) *Introduction to Solids*, Tata McGraw-Hill Education.

Nonlinear Finite Elements for Continua and Structures, Second Edition.
Ted Belytschko, Wing Kam Liu, Brian Moran, and Khalil I. Elkhodary.

Companion Website: www.wiley.com/go/belytschko

Barlow J (1976) Optimal stress locations in finite element models, *International Journal for Numerical Methods in Engineering*, **10**, 243–251.

Bathe KJ (1996) *Finite Element Procedures*, Prentice-Hall, Englewood Cliffs, NJ.

Bayliss A, Belytschko T, Kulkarni M and Lott-Crumpler DA (1994) On the dynamics and the role of imperfections for localization in thermoviscoplastic materials, *Modeling and Simulation in Materials Science and Engineering*, **2**, 941–964.

Bazant ZP and Belytschko T (1985) Wave propagation in strain softening bar: exact solution, *J. Engrg. Mech. ASCE*, **111**, 381–389.

Bazant ZP and Cedolin L (1991) *Stability of Structures*, Oxford University Press, Oxford.

Bazant ZP, Belytschko T and Chang TP (1984) Continuum theory for strain softening, *J. Engrg. Mech., ASCE*, **110** (3), 1666–1692.

Bazeley GP, Cheung YK, Irons BM and Zienkiewicz OC (1965) *Triangular elements in plate bending*, *Proc. First Conf. on Matrix Methods in Structural Mechanics*, Wright-Patterson AFB, Ohio.

Béchet E, Minnebo H, Moës N and Burgardt B (2005) Improved implementation and robustness study of the X-FEM for stress analysis around cracks, *International Journal for Numerical Methods in Engineering*, **64**(8), 1033–1056.

Belytschko T (1976) Methods and programs for analysis of fluid–structure systems, *Nucl. Engrg. Design*, **42**, 41–52.

Belytschko T (1983) Overview of semidiscretization, in *Computational Methods for Transient Analysis*, T Belytschko and TJR Hughes (eds), North-Holland, Amsterdam.

Belytschko T and Bachrach WE (1986) Efficient implementation of quadrilaterals with high coarse-mesh accuracy, *Computer Methods in Applied Mechanics and Engineering*, **54**, 279–301.

Belytschko T, Bazant ZP, Hyun YW and Chang TP (1986) Strain softening materials and finite element solutions, *Comput. Struct.*, **23** (2), 163–180.

Belytschko T and Bindeman LP (1991) Assumed strain stabilization of the 4-node quadrilateral with 1-point quadrature for nonlinear problems, *Computer Methods in Applied Mechanics and Engineering*, **88**, 311–340.

Belytschko T and Bindeman LP (1993) Assumed strain stabilization of the eight node hexahedral element, *Computer Methods in Applied Mechanics and Engineering*, **105**, 225–260.

Belytschko T and Black T (1999) Elastic crack growth in finite elements with minimal remeshing. *International Journal for Numerical Methods in Engineering*, **45**(5), 601–620.

Belytschko T, Chen H, Xu J and Zi G (2003) Dynamic crack propagation based on loss of hyperbolicity and a new discontinuous enrichment, *International Journal for Numerical Methods in Engineering*, **58**(12), 1873–1905.

Belytschko T, Chiang HY and Plaskacz E (1994) High resolution two-dimensional shear band computations: imperfections and mesh dependence, *Computer Methods in Applied Mechanics and Engineering*, **119**, 1–15.

Belytschko T, Fish J and Englemann B (1988) A finite element method with embedded localization zones, *Computer Methods in Applied Mechanics and Engineering*, **70**, 59–90.

Belytschko T and Hsieh BJ (1973) Nonlinear transient finite element analysis with convected coordinates, *International Journal for Numerical Methods in Engineering*, **7**, 255–271.

Belytschko T and Hughes TJR (1983) *Computational Methods for Transient Analysis*, North-Holland, Amsterdam.

Belytschko T and Kennedy JM (1978) Computer models for subassembly simulation, *Nucl. Engrg. Design*, **49**, 17–38.

Belytschko T, Kennedy JM and Schoeberle DF (1975) On finite element and finite difference formulations of transient fluid structure problems, *Proc. Comp. Methods in Nucl. Engrg.*, American Nuclear Society, Savannah, GA, **2**, IV: 39–43.

Belytschko T and Leviathan I (1994a) Physical stabilization of the 4-node shell element with one-point quadrature, *Computer Methods in Applied Mechanics and Engineering*, **113**, 321–350.

Belytschko T and Leviathan I (1994b) Projection scheme for one-point quadrature shell elements, *Computer Methods in Applied Mechanics and Engineering*, **115**, 277–286.

Belytschko T, Lin JI and Tsay CS (1984) Explicit algorithms for the nonlinear dynamics of shells, *Computer Methods in Applied Mechanics and Engineering*, **42**, 225–251.

Belytschko T and Liu WK (1985) Computer methods for transient fluid–structure analysis of nuclear reactors, *Nuclear Safety*, **26**, 14–31.

Belytschko T, Moran B and Kulkarni M (1990) On the crucial role of imperfections in quasistatic viscoplastic solutions, *Appl. Mech. Review*, **43** (5), 251–256.

Belytschko T and Mullen R (1978) On dispersive properties of finite element solutions, in *Modern Problems in Wave Propagation*, J Achenbach and J Miklowitz (eds), John Wiley & Sons, Inc., New York, 67–82.

Belytschko T and Neal MO (1991) Contact-impact by the pinball algorithm with penalty and Lagrangian methods, *International Journal for Numerical Methods in Engineering*, **31**, 547–572.

Belytschko T and Schoeberle DF (1975) On the unconditional stability of an implicit algorithm for structural dynamics, *J. Appl. Mech.*, **42**, 865–869.

Belytschko T, Smolinski P and Liu WK (1985a) Stability of multi-time step partitioned integrators for first order finite element systems, *Computer Methods in Applied Mechanics and Engineering*, **49**, 281–297.

Belytschko T, Stolarski H, Liu WK, Carpenter N and Ong JSJ (1985b) Stress projection for membrane and shear locking in shell finite elements, *Computer Methods in Applied Mechanics and Engineering*, **51**, 221–258.

Belytschko T and Tsay CS (1983) A stabilization procedure for the quadrilateral plate element with one-point quadrature, *International Journal for Numerical Methods in Engineering*, **19**, 405–419.

Belytschko T, Wong BL and Chiang HY (1992) Advances in one-point quadrature shell elements, *Computer Methods in Applied Mechanics and Engineering*, **96**, 93–107.

Belytschko T, Wong BL and Stolarski H (1989) Assumed strain stabilization procedure for the 9-node Lagrange shell element, *International Journal for Numerical Methods in Engineering*, **28**, 385–114.

Belytschko T, Yen HJ and Mullen R (1979) Mixed methods in time integration, *Computer Methods in Applied Mechanics and Engineering*, **17/18**, 259–275.

Benson DJ (1989) An efficient, accurate simple ALE method for nonlinear finite element programs, *Computer Methods in Applied Mechanics and Engineering*, **72**, 205–350. Bertsekas DP (1984) *Constrained Optimization and Lagrange Multiplier Methods*, Academic Press, New York.

Bettess P (1977) Infinite elements. *International Journal for Numerical Methods in Engineering* **11**(1), 53–64.

Bonet J and Wood RD (1997) *Nonlinear Continuum Mechanics for Finite Element Analysis*, Cambridge University Press, New York.

Bordas S and Moran B (2006) Enriched finite elements and level sets for damage tolerance assessment of complex structures, *Engineering Fracture Mechanics*, **73**(9), 1176–1201.

Borst Rd, Pamin J and Marc GC (1999) On coupled gradient-dependent plasticity and damage theories with a view to localization analysis. *European Journal of Mechanics – A/Solids*, **18**, 939–962.

Boyce MC, Parks DM and Argon AS (1988) Large inelastic deformation of glassy polymers; Part 1: Rate dependent constitutive model, *Mech. Mater.*, **7**, 15–33.

Boyce MC, Weber G, Parks DM *et al.* (1989) On the kinematics of finite strain plasticity, *Journal of the Mechanics and Physics of Solids*, **37**(5), 647–665.

Bridgman P (1949) *The Physics of High Pressure*, Bell and Sons, London.

Brooks AN and Hughes TJR (1982) Streamline upwind/Petrov–Galerkin formulations for convection dominated flows with particular emphasis on the incompressible Navier–Stokes equations, *Computer Methods in Applied Mechanics and Engineering*, **32**, 199–259.

Bucalem M and Bathe KJ (1993) Higher order MITC general shell elements, *International Journal for Numerical Methods in Engineering*, **36**, 3729–3754.

Buechter N and Ramm E (1992) Shell theory versus degeneration – a comparison of large rotation finite element analysis, *International Journal for Numerical Methods in Engineering*, **34**, 39–59.

Bulatov V and Cai W (2006) *Computer Simulations of Dislocations*, Oxford University Press, Oxford.

Caillard D and Martin J-L (2003) *Thermally Activated Mechanisms in Crystal Plasticity*, Elsevier Science, Amsterdam.

Camacho GT and Ortiz M (1996) Computational modeling of impact damage in brittle materials, *International Journal of Solids and Structures*, **33**(20), 2899–2938.

Chambon R, Caillerie D and Matsuchima T (2001) Plastic continuum with microstructure, local second gradient theories for geomaterials: localization studies. *International Journal of Solids and Structures*, **38**(46–47), 8503–8527.

Chambon R, Caillerie D and Tamangnini C (2004) A strain space gradient plasticity theory for finite strain. *Computer Methods in Applied Mechanics and Engineering.* **193**(27–29), 2797–2826.

Chandrasekharaiah DD and Debnath L (1994) *Continuum Mechanics*, Academic Press, Boston, MA.

Chen LQ (2002) Phase-field models for microstructure evolution. *Annual Review of Materials Research*, **32**(1), 113–140.

Chen Y and Lee JD (2003a) Connecting molecular dynamics to micromorphic theory. (I) Instantaneous and averaged mechanical variables. *Physica A: Statistical Mechanics and its Applications*, **322**, 359–376.

Chen Y and Lee JD (2003b) Connecting molecular dynamics to micromorphic theory. (II) Balance laws. *Physica A: Statistical Mechanics and its Applications*, **322**, 377–392.

Chen Y and Lee JD (2003c) Determining material constants in micromorphic theory through phonon dispersion relations. *International Journal of Engineering Science*, **41**, 871–886.

Chessa J, Smolinski, P and Belytschko T (2002) The extended finite element method (XFEM) for solidification problems, *International Journal for Numerical Methods in Engineering*, **53**(8), 1959–1977.

Chung L and Hulbert G (1993) A time integration algorithm for structural dynamics with improved numerical dissipation; the generalized α-method, *J. Appl. Mech.*, **60**, 371–375.

Ciarlet PG and Raviart PA (1972) Interpolation theory over curved elements, *Computer Methods in Applied Mechanics and Engineering*, **1**, 217–249.

Coleman BD and Noll W (1961) Foundations of linear viscoelasticity, *Rev. Modern Phys.*, **33**, 239–249.

Cook RD, Malkus DS and Plesha ME (1989) *Concepts and Applications of Finite Element Analysis*, 3rd edn, John Wiley & Sons, Ltd, Chichester.

Cosserat EMP and Cosserat F (1909) *Theorie des Corps Deformables*. A. Hermann et Fils, Paris.

Costantino CJ (1967) Finite element approach to stress wave problems, *Journal of Engineering Mechanics Division, ASCE*, **93**, 153–166.

Courant R, Friedrichs KO and Lewy H (1928) Über die partiellen Differenzensleichungen der Mathematischen Physik, *Math. Ann.*, **100**, 32.

Crisfield MA (1980) A fast incremental/iterative solution procedure that handles 'snap-through', *Comput. Struct.*, **13**, 55–62.

Crisfield MA (1991) *Non-linear Finite Element Analysis of Solids and Structures*, Vol. **1**, John Wiley & Sons, Inc., New York.

Cuitino A and Ortiz M (1992) A material-independent method for extending stress update algorithms from small-strain plasticity to finite plasticity with multiplicative kinematics, *Engrg. Comput.*, **9**, 437–451.

Cullity BD and Stock SR (2001) *Elements of X-Ray Diffraction*, Prentice Hall, Upper Saddle River, NJ.

Curnier A (1984) A theory of friction, *Int. J. Sol. Struct.*, **20**, 637–647.

Daniel WJT (1997) Analysis and implementation of a new constant acceleration subcycling algorithm, *International Journal for Numerical Methods in Engineering*, **40**, 2841–2855.

Daniel WJT (1998) A study of the stability of subcycling algorithms in structural dynamics, *Computer Methods in Applied Mechanics and Engineering*, **156**, 1–13.

Dashner PA (1986) Invariance considerations in large strain elasto-plasticity, *J. Appl. Mech.*, **53**, 55–60.

de Borst R (1987) Computation of post-bifurcation and post-failure behaviour of strain softening solids, *Comput. Struct.*, **25** (2), 211–224.

de Borst R and Mulhaus HB (1993) Gradient-dependent plasticity-formulation and algorithmic aspects, *International Journal for Numerical Methods in Engineering*, **35**, 521–539.

De Graef M and McHenry ME (2007) *Structure of Materials: An Introduction to Crystallography, Diffraction and Symmetry*, Cambridge University Press, Cambridge.

Demkowicz L and Oden JT (1981) On some existence and uniqueness results in contact problems with nonlocal friction, Report 81–13, Texas Institute of Computational Mechanics (TICOM), University of Texas at Austin.

Dennis JE and Schnabel RB (1983) Numerical Methods for Unconstrained Optimization and Nonlinear Equations, Prentice-Hall, Englewood Cliffs, NJ.

Dhatt G and Touzot G (1984) *The Finite Element Method Displayed*, John Wiley & Sons, Ltd, Chichester.

Dienes JK (1979) On the analysis of rotation and stress rate in deforming bodies, *Acta Mechanica*, **32**, 217–232.

DiMaggio FL and Sandler IS (1971) Material model for granular soils, *Journal of Engineering Mechanics Division* ASCE, 935–950.

Dobovsek I and Moran B (1996) Material instabilities in rate in dependent solids, *European Journal of Mechanics and Solids*, **15** (2), 267–294.

Doyle TC and Ericksen JL (1956) *Nonlinear Elasticity, Advances in Applied Mechanics*, Vol. **4**, Academic Press, New York.

Duan Q, Song JH, Menouillard T and Belytschko T (2009) Element-local level set method for three-dimensional dynamic crack growth, *International Journal for Numerical Methods in Engineering*, **80**(12), 1520–1543.

Duddu R, Bordas S, Chopp D and Moran B (2008) A combined extended finite element and level set method for biofilm growth, *International Journal for Numerical Methods in Engineering*, **74**(5), 848–870.

Dvorkin EN and Bathe KJ (1984) A continuum mechanics based four-node shell element for general nonlinear analysis, *Engrg. Comput.*, **1**, 77–88.

Elguedj, T, Gravouil A and Combescure A (2006) Appropriate extended functions for X-FEM simulation of plastic fracture mechanics, *Computer Methods in Applied Mechanics and Engineering*, **195**(7–8) 501–515.

Elkhodary K, Greene S, Tang S and Liu WK (2013) The archetype-blending continuum theory. *Comp. Meth. Appl. Mech. Eng.*, **254**, 309–333.

Elkhodary K, Lee W, Suna LP, *et al.* (2011) Deformation mechanisms of an Ω precipitate in a high-strength aluminum alloy subjected to high strain rates, *Journal of Materials Research*, **26**(04), 487–497.

Engelen RAB, Fleck NA, Peerlings RHJ and Geers MGD (2006) An evaluation of higher-order plasticity theories for predicting size effects and localisation. *International Journal of Solids and Structures*, **43**(7–8), 1857–1877.

Englemann BE and Whirley RG (1990) *A new elastoplastic shell element formulation for DYNA3D, Report UCRL-JC-104826*, Lawrence Livermore National Laboratory, CA.

Eringen AC (1965) *Linear theory of micropolar elasticity*, DTIC Document.

Eringen AC (1990) Theory of thermo-microstretch elastic solids. *International Journal of Engineering Science*, **28**(12), 1291–1301.

Eringen AC (1999) *Microcontinuum Field Theories I: Foundations and Solids*. Springer, New York.

Eringen AC and Suhubi ES (1964) Nonlinear theory of simple microelastic solids. *International Journal of Engineering and Science*, **2**, 189–203, 389–404.

Estrin Y and Kubin L (1986) Local strain hardening and nonuniformity of plastic deformation, *Acta Metallurgica*, **34**(12), 2455–2464.

Estrin Y, Krausz A, *et al.* (1996) *Unified Constitutive Laws of Plastic Deformation*. Academic Press, New York, p. 69.

Estrin Y, Krausz AS and Krausz K (1996) Dislocation-density related constitutive modelling, in *Unified Constitutive Laws of Plastic Deformation*, (eds) AS Krausz and K Krausz, Academic Press, New York, pp. 69–106.

Flanagan DP and Belytschko T (1981) A uniform strain hexahedron and quadrilateral with orthogonal hourglass control, *International Journal for Numerical Methods in Engineering*, **17**, 679–706.

Fleck NA and Hutchinson JW (1993) A phenomenological theory for strain gradient effects in plasticity. *Journal of the Mechanics and Physics of Solids*, **41**(12), 1825–1857.

Fleck NA and Hutchinson JW (1994) A phenomenological theory for strain gradient effects in plasticity, *J. Mech. Physics of Solids*, **41**, 1825–1857.

Fleck NA and Hutchinson JW (1997) Strain gradient plasticity. In *Advances in Applied Mechanics, Volume 33*, WH John and YW Theodore (eds), Elsevier. pp. 295–361.

Fleck NA and Hutchinson JW (2001) A reformulation of strain gradient plasticity. *Journal of the Mechanics and Physics of Solids*, **49**, 2245–2271.

Fleck NA and Willis JR (2009a) A mathematical basis for strain-gradient plasticity theory - Part I: Scalar plastic multiplier. *Journal of the Mechanics and Physics of Solids*, **57**(1), 161–177.

Fleck NA and Willis JR (2009b) A mathematical basis for strain-gradient plasticity theory. Part II: Tensorial plastic multiplier. *Journal of the Mechanics and Physics of Solids*, **57**(7), 1045–1057.

Fleck NA, Muller GM, Ashby MF and Hutchinson JW (1994) Strain gradient plasticity: Theory and experiment. *Acta Metallurgica et Materialia*, **42**(2), 475–487.

Forest S (2009) Micromorphic approach for gradient elasticity, viscoplasticity, and damage. *Journal of Engineering Mechanics*, **135**(3), 117–131.

Forest S and Sievert R (2006) Nonlinear microstrain theories. *International Journal of Solids and Structures*, **43**(24), 7224–7245.

Fraeijs de Veubeke, F (1965) Displacement equilibrium models in the finite element method, in *Stress Analysis*, OC Zienkiewicz and GS Holister (eds), John Wiley & Sons, Ltd, London, 145–196.

Franca LP and Frey SL (1992) Stabilized finite element methods: II The incompressible Navier–Stokes equations, *Computer Methods in Applied Mechanics and Engineering*, **99**, 209–233.

Fries TP (2008) A corrected XFEM approximation without problems in blending elements, *International Journal for Numerical Methods in Engineering*, **75**(5), 503–532.

Fries TP and Belytschko T (2010) The extended/generalized finite element method: An overview of the method and its applications, *International Journal for Numerical Methods in Engineering*, **84**(3), 253–304.

Geers M, Kouznetsova VG and Brekelmans WAM (2001) Gradient-enhanced computational homogenization for the micro–macro scale transition. *J. Phys. IV France*, **11**(PR5), Pr5-145–Pr145-152.

Georgiadis HG, Vardoulakis I and Velgaki EG (2004) Dispersive Rayleigh-wave propagation in microstructured solids characterized by dipolar gradient elasticity. *Journal of Elasticity*, **74**(1), 17–45.

Germain P (1973) The method of virtual power in continuum mechanics. Part 2: microstructure. *SIAM Journal on Applied Mathematics*, **25**(3), 556–575.

Giacovazzo C, Monaco HL, Artioli G *et al.* (2002) *Fundamentals of Crystallography. Part of International Union of Crystallography Monographs on Crystallography Series*. Oxford University Press, New York, pp. 74–76.

Givoli D (1991) Non-reflecting boundary conditions. *Journal of Computational Physics*, **94**(1), 1–29.

Green AE and Rivlin RS (1957) The mechanics of non-linear materials with memory: Part I, *Arch. Rat. Mech. Anal.*, **1**, 1–21.

Greene MS, Liu Y, Chen W and Liu WK (2011) Computational uncertainty analysis in multiresolution materials via stochastic constitutive theory. *Computer Methods in Applied Mechanics and Engineering*, **200**, 309–325.

Gurson AL (1977) Continuum theory of ductile rupture by void nucleation and growth: Part I – Yield criteria and flow rules for porous ductile media, *Journal of Engineering Materials and Technology*, **99**, 2–15.

Gurtin ME (2002) A gradient theory of single-crystal viscoplasticity that accounts for geometrically necessary dislocations. *Journal of the Mechanics and Physics of Solids*, **50**(1), 5–32.

Gurtin ME (2006) The Burgers vector and the flow of screw and edge dislocations in finite-deformation single-crystal plasticity, *Journal of the Mechanics and Physics of Solids*, **54**, 1882–1898.

Gurtin ME and Anand L (2005) A theory of strain-gradient plasticity for isotropic, plastically irrotational materials. Part II: Finite deformations. *International Journal of Plasticity*, **21**(12), 2297–2318.

Hallquist JO (1994) *LS-DYNA Theoretical Manual*.

Hallquist JO and Whirley RG (1989) *DYNA3D Users' manual: nonlinear dynamic analysis of structures in three dimensions, UCID-19592, Rev. 5*, Lawrence Livermore National Laboratory, CA.

Hansbo A and Hansbo P (2004) A finite element method for the simulation of strong and weak discontinuities in solid mechanics, *Computer Methods in Applied Mechanics and Engineering*, **193**(33), 3523–3540.

Harren SV, Lowe T, Asaro RJ and Needleman A (1989) Analysis of large-strain shear in rate-dependent face-centred cubic polycrystals: correlation of micro-and macromechanics, *Phil. Trans. R. Soc. Lond.*, **A328**, 433–500.

Havner KS (1992) *Finite Plastic Deformation of Crystalline Solids*, Cambridge University Press, New York.

Hertzberg RW, Vinci RP and Hertzberg JL (2012) *Deformation and Fracture Mechanics of Engineering Materials*, John Wiley & Sons, Inc., Hoboken, NJ.

Hibbitt D, Karlsson B, Sorensen P *et al.* (2007) *ABAQUS Analysis User's Manual*. ABAQUS Inc., Providence, RI.

Hilber HM, Hughes TJR and Taylor RL (1977) Improved numerical dissipation for time integration algorithms in structural dynamics, *Earthquake Engineering and Structural Dynamics*, **5**, 282–292.

Hill R (1950) *The Mathematical Theory of Plasticity*, Oxford University Press, Oxford.

Hill R (1962) Acceleration waves in solids, *J. Mech. Phys. Solids*, **10**, 1–16.

Hill R (1975) Aspects of Invariance in Solid Mechanics, *Advances in Applied Mechanics*, **18**, Academic Press, New York.

Hillerborg A, Modeer M and Peterson PE (1976) Analysis of crack formation and crack growth in concrete by means of fracture mechanics and finite elements. *Cement Concrete Res.*, **6**, 773–782.

Hirth JP and Lothe J (1982) *Theory of Dislocations*, John Wiley & Sons, Ltd, Chichester.

Hodge PG (1970) *Continuum Mechanics*, McGraw-Hill, New York.

Hu YK and Liu WK (1993) An ALE hydrodynamic lubrication finite element method with application to strip rolling, *International Journal for Numerical Methods in Engineering*, **36**, 855–880.

Huang HC and Hinton E (1986) A new nine-node degenerated shell element with enhanced membrane and shear interpolants, *International Journal for Numerical Methods in Engineering*, **22**, 73–92.

Huerta A and Liu WK (1988) Viscous flow with large free surface motion, *Computer Methods in Applied Mechanics and Engineering*, **69**, 277–324.

Hughes TJR (1984) Numerical Implementation of Constitutive Models: Rate-Independent Deviatoric Plasticity, *Theoretical Foundation for Large-Scale Computations of Nonlinear Material Behavior*, S Nemat-Nemat-Nasser, RJ Asaro and GA Hegemier (eds), Martinus Nijhoff Publishers, Dordrecht, pp. 29–57.

Hughes TJR (1987) *The Finite Element Method*, Linear Static and Dynamic Finite Element Analysis, Prentice-Hall, Englewood Cliffs, NJ.

Hughes TJR (1996) personal communication.

Hughes TJR, Cohen M and Haroun M (1978) Reduced and selective integration techniques in the finite element analysis of plates, *Nuclear Engineering and Design*, **46**, 203–222.

Hughes TJR and Liu WK (1978) Implicit – explicit finite elements in transient analysis, *J. Appl. Mech.*, **45**, 371–378.

Hughes TJR and Liu WK (1981a) Nonlinear finite element analysis of shells: Part 1, Two-dimensional shells, *Computer Methods in Applied Mechanics and Engineering*, **26**, 167–181.

Hughes TJR and Liu WK (1981b) Nonlinear finite element analysis of shells: Part 2, Three-dimensional shells, *Computer Methods in Applied Mechanics and Engineering*, **26**, 331–362.

Hughes TJR, Liu WK and Zimmerman TK (1981) Lagrangian–Eulerian finite element formulation for incompressible viscous flows, *Computer Methods in Applied Mechanics and Engineering*, **29**, 329–349.

Hughes TJR and Mallet M (1986) A new finite element formulation for computational fluid dynamics: iii The generalized streamline operator for multidimensional advective-diffusive systems, *Computer Methods in Applied Mechanics and Engineering*, **58**, 305–328.

Hughes TJR and Pister KS (1978) Consistent Linearization in Mechanics of Solids and Structures, *Computers and Structures*, **8**, 391–397.

Hughes TJR, Taylor RL and Kanoknukulchai W (1977) A simple and efficient element for plate bending, *International Journal for Numerical Methods in Engineering*, **11**, 1529–1543.

Hughes TJR and Tezduyar TE (1981) Finite elements based upon Mindlin plate theory with particular reference to the four-node isoparametric element, *J. Appl. Mech.*, **58**, 587–596.

Hughes TJR and Tezduyar TE (1984) Finite element methods for first-order hyperbolic systems with particular emphasis on the compressible Euler equations, *Computer Methods in Applied Mechanics and Engineering*, **45**, 217–284.

Hughes TJR and Winget J (1980) Finite rotation effects in numerical integration of rate-constitutive equations arising in large-deformation analysis, *International Journal for Numerical Methods in Engineering*, **15**, 1862–1867.

Hull D and Bacon DJ (2006) *Introduction to Dislocations*. Butterworth-Heinemann, Oxford.

Hutchinson JW (1968) Singular behavior at the end of a tensile crack in a hardening material, *Journal of the Mechanics and Physics of Solids*, **16**(1), 13–31.

Iesan D (2002) On the micromorphic thermoelasticity. *International Journal of Engineering Science*, **40**(5), 549–567.

Jänicke R and Diebels S (2009) Two-scale modelling of micromorphic continua. Continuum *Mechanics and Thermodynamics*, **21**(4), 297–315.

Johnson G and Bammann DJ (1984) A discussion of stress rates in finite deformation problems, *International Journal for Numerical Methods in Engineering*, **20**, 735–737.

Kadowaki H and Liu WK (2005) A multiscale approach for the micropolar continuum model. *Computer Modelling in Engineering Sciences*, **7**(3), 269–282.

Kalthoff JF and Winkler S (1988) Failure mode transition at high rates of shear loading, DGM Informationsgesellschaft mbH, *Impact Loading and Dynamic Behavior of Materials*, **1**, 185–195.

Kameda T and Zikry MA (1996) Three dimensional dislocation-based crystalline constitutive formulation for ordered intermetallics, *Scripta Materialia*, **38**(4), 631–636.

Keller JB and Givoli D (1989) Exact non-reflecting boundary conditions. *Journal of Computational Physics*, **82**(1), 172–192.

Kelly A, Groves GW and Kidd P (2000) *Crystallography and Crystal Defects*. John Wiley & Sons, Ltd, Chichester.

Kennedy TC and Kim JB (1993) Dynamic analysis of cracks in micropolar elastic materials. *Engineering Fracture Mechanics*, **44**(2), 207–216.

Key SW and Beisinger ZE (1971) The transient dynamic analysis of thin shells by the finite element method, *Proc. 3rd Conf. on Matrix Methods in Structural Analysis*, Wright-Patterson AFB, Ohio.

Khan AS and Huang S (1995) *Continuum Theory of Plasticity*, John Wiley & Sons, Inc., New York.

Kikuchi N and Oden JT (1988) *Contact Problems in Elasticity: A Study of Variational Inequalities and Finite Element Methods*, SIAM, Philadelphia, PA.

Kleiber M (1989) *Incremental Finite Element Modelling in Non-linear Solid Mechanics*, Ellis Horwood, Chichester.

Knowles JK and Sternberg E (1977) On the failure of ellipticity of the equations for finite elastostatics in plane strain, *Arch. Rat. Mech. Anal.*, **63**, 321–336.

Kocks U (1987) Constitutive behaviour based on crystal plasticity, In *Constitutive Equations for Creep and Plasticity*, AK Miller (ed.), Elsevier, Amsterdam.

Kogge P, Bergman K, Borkar S, *et al.* (2008) *Exascale Computing Study: Technology Challenges in Achieving Exascale Systems. DARPA*, available online at http://users.ece.gatech.edu/mrichard/ExascaleComputingStudy Reports/exascale_final_report_100208.pdf (accessed June 13, 2013)

Kouznetsova VG, Geers MGD and Brekelmans WAM (2002) Multi-scale constitutive modelling of heterogeneous materials with a gradient-enhanced computational homogenization scheme. *International Journal for Numerical Methods in Engineering*, **54**(8), 1235–1260.

Kouznetsova VG, Geers MGD and Brekelmans WAM (2004) Multi-scale second-order computational homogenization of multi-phase materials: a nested finite element solution strategy. *Computer Methods in Applied Mechanics and Engineering*, **193**(48–51), 5525–5550.

Krajcinovic D (1996) *Damage Mechanics*, North-Holland, Amsterdam and New York.

Krieg RD and Key SW (1976) Implementation of a time dependent plasticity theory into structural computer programs, in *Constitutive Equations in Viscoplasticity: Computational and Engineering Aspects*, JA Stricklin and KJ Saczalski (eds), ASME, New York.

Kroner E (1981) Continuum theory of defects. In *Les Houches, Session XXXV, 1980-Physics of Defects*. R Balian, M Kleman and JP Poitier (eds), North-Holland, Amsterdam, pp. 219–315.

Kubin LP and Mortensen A (2003) Geometrically necessary dislocations and strain-gradient plasticity: a few critical issues. *Scripta Materialia*, **48**, 119–125.

Kulkarni M, Belytschko T and Bayliss A (1995) Stability and error analysis for time integrators applied to strain-softening materials, *Computer Methods in Applied Mechanics and Engineering*, **124**, 335–363.

Laborde P, Pommier J, Renard Y and Salaün M (2005) High-order extended finite element method for cracked domains, *International Journal for Numerical Methods in Engineering*, **64**(3), 354–381.

Ladyzhesnkaya OA (1968) *Linear and Quasilinear Elliptic Equations*, Academic Press, New York.

Lasry D and Belytschko T (1988) Localization limiters in transient problems, *International Journal of Solids and Structures*, **24**(6), 581–597.

Lee EH (1969) Elastic–plastic deformation at finite strains, *J. Appl. Mech.*, **36**, 1–6.

Lee JD and Chen Y (2005) Material forces in micromorphic thermoelastic solids. *Philosophical Magazine*, **85**(33), 3897–3910.

Lee YJ and Freund LB (1990) Fracture initiation due to asymmetric impact loading of an edge cracked plate. *Journal of Applied Mechanics*, **57**(1), 104–111.

Lemaitre J (1971) Evaluation of dissipation and damage in metal submitted to dynamic loading. *Proceedings of ICM 1, Kyoto, Japan*.

Lemaitre J and Chaboche JL (1990) *Mechanics of Solid Materials*, Cambridge University Press, Cambridge.

Li S, Lu H, Han W, Liu WK and Simkins DC (2004) Reproducing kernel element method Part II: Globally conforming Im/Cn hierarchies, *Computer Methods in Applied Mechanics and Engineering*, **193**(12), 953–987.

Lin JI (1991) Bounds on eigenvalues of finite element systems, *International Journal for Numerical Methods in Engineering*, **32**, 957–967.

Liu WK (1981) Finite element procedures for fluid–structure interactions with application to liquid storage tanks, *Nucl. Engrg. Design*, **65**, 221–238.

Liu WK, Belytschko T and Chang H (1986) An arbitrary Lagrangian–Eulerian finite element method for path-dependent materials, *Computer Methods in Applied Mechanics and Engineering*, **58**, 227–246.

Liu WK and Chang HG (1984) Efficient computational procedures for long-time duration fluid–structure interaction problems, *J. Pressure Vessel Tech.*, **106**, 317–322.

Liu WK and Chang HG (1985) A method of computation for fluid structure interactions, *Comput. Struct.*, **20**, 311–320.

Liu WK, Chang H and Belytschko T (1988) Arbitrary Lagrangian and Eulerian Petrov–Galerkin finite elements for nonlinear continua, *Computer Methods in Applied Mechanics and Engineering*, **68**, 259–310.

Liu WK, Chen JS, Belytschko T and Zhang YF (1991) Adaptive ALE finite elements with particular reference to external work rate on frictional interface, *Computer Methods in Applied Mechanics and Engineering*, **93**, 189–216.

Liu WK, Han W, Lu H, Li S and Cao J (2004) Reproducing kernel element method. Part I: Theoretical formulation. *Computer Methods in Applied Mechanics and Engineering*, **193**(12), 933–951.

Liu WK, Jun S and Zhang YF (1995a) Reproducing kernel particle methods, *International Journal for Numerical Methods in Fluids*, **20**(8–9), 1081–1106.

Liu WK, Jun S, Li S, Adee J and Belytschko T (1995b) Reproducing kernel particle methods for structural dynamics, *International Journal for Numerical Methods in Engineering*, **38**(10), 1655–1679.

Liu WK and Ma DC (1982) Computer implementation aspects for fluid–structure interaction problems, *Computer Methods in Applied Mechanics and Engineering*, **31**, 129–148.

Liu WK and McVeigh C (2008) Predictive multiscale theory for design of heterogeneous materials. *Computational Mechanics*, **42**(2), 147–170.

Liu WK, Qian D, Gonella S et al. (2010) Multiscale methods for mechanical science of complex materials: bridging from quantum to stochastic multiresolution continuum. *International Journal for Numerical Methods in Engineering*, **83**, 1039–1080.

Liu WK, Uras RA and Chen Y (1997) Enrichment of the finite element method with the reproducing kernel particle method, *Journal of Applied Mechanics*, **64**, 861–870.

Losi GU and Knauss WG (1992) Free volume theory and nonlinear thermo-viscoelasticity, *Polym. Sci. Engng.*, **32** (9), 542–557.

Lu H, Kim DW and Liu WK (2005) Treatment of discontinuity in the reproducing kernel element method. *International Journal for Numerical Methods in Engineering*, **63**(2), 241–255.

Lu H, Li S, Simkins Jr DC, Kam Liu W and Cao, J (2004) Reproducing kernel element method Part III: Generalized enrichment and applications, *Computer Methods in Applied Mechanics and Engineering*, **193**(12), 989–1011.

Lubliner L (1990) *Plasticity Theory*, Macmillan, New York.

Luscher DJ, McDowell DL and Bronkhorst CA (2010) A second gradient theoretical framework for hierarchical multiscale modeling of materials. *International Journal of Plasticity*, **26**(8), 1248–1275.

MacNeal RH (1982) Derivation of element stiffness matrices by assumed strain distributions, *Nucl. Engrg. Design*, **33**, 1049–1058.

MacNeal RH (1994) *Finite Elements: Their Design and Performance*, Marcel Dekker, New York.

Malkus D and Hughes TJR (1978) Mixed finite element methods – reduced and selective integration techniques: a unification of concepts, *Computer Methods in Applied Mechanics and Engineering*, **15**, 63–81.

Malvern LE (1969) *Introduction to the Mechanics of a Continuous Medium*, Prentice-Hall, Englewood Cliffs, NJ.

Marcal PV and King IP (1967) Elastic–plastic analysis of two dimensional stress systems by the finite element method, *Int. J. Mechanical Sciences*, **9**, 143–155.

Marin EB and McDowell DL (1997) A semi-implicit integration scheme for rate-dependent and rate-independent plasticity, *Comput. Struct.*, **63** (3), 579–600.

Marsden JE and Hughes TJR (1983) *Mathematical Foundations of Elasticity*, Prentice-Hall, Englewood Cliffs, NJ.

Mase GF and Mase GT (1992) *Continuum Mechanics for Engineers*, CRC Press, Boca Raton, FL.

McVeigh CJ (2007) *Linking Properties to Microstructure through Multiresolution Mechanics*. Doctor of Philosophy, Northwestern University.

McVeigh CJ and Liu WK (2008) Linking microstructure and properties through a predictive multiresolution continuum. *Computer Methods in Applied Mechanics and Engineering*, **197**(41–42), 3268–3290.

McVeigh CJ and Liu WK (2009) Multiresolution modeling of ductile reinforced brittle composites. *Journal of the Mechanics and Physics of Solids*, **57**(2), 244–267.

McVeigh CJ and Liu WK (2010) Multiresolution continuum modeling of micro-void assisted dynamic adiabatic shear band propagation. *Journal of the Mechanics and Physics of Solids*, **58**(2), 187–205.

McVeigh CJ, Vernerey F, Liu WK et al. (2006) Multiresolution analysis for material design. *Computer Methods in Applied Mechanics and Engineering*, **195**(37–40), 5053–5076.

Mehrabadi MM and Nemat-Nasser S (1987) Some basic kinematical relations for finite deformations of continua, *Mech. Mater.*, **6**, 127–138. Michalowski R and Mroz Z (1978) Associated and non-associated sliding rules in contact friction problems, *Arch. Mech.*, **30**, 259–276.

Melenk JM and Babuška I (1996) The partition of unity finite element method: basic theory and applications. *Computer Methods in Applied Mechanics and Engineering*, **139**(1), 289–314.

Menouillard T, Réthoré J, Combescure A and Bung H (2006) Efficient explicit time stepping for the eXtended Finite Element Method (X-FEM), *International Journal for Numerical Methods in Engineering*, **68**(9), 911–939.

Menouillard T, Réthoré J, Moës N, Combescure A and Bung H (2008) Mass lumping strategies for X-FEM explicit dynamics: Application to crack propagation, *International Journal for Numerical Methods in Engineering*, **74**(3), 447–474.

Mesarovic SD and Padbidri J (2005) Minimal kinematic boundary conditions for simulations of disordered microstructures. *Philosophical Magazine* **85**(1), 65–78.

Miehe C (1994) Aspects of the formulation and finite element implementation of large strain isotropic elasticity, *International Journal for Numerical Methods in Engineering*, **37**, 1981–2004.

Mindlin RD (1964) Micro-structure in linear elasticity. *Archive for Rational Mechanics and Analysis*, **16**(1), 51–78.

Mindlin RD (1965) Second gradient of strain and surface-tension in linear elasticity. *International Journal of Solids and Structures*, **1**(4), 417–438.

Mindlin RD and Tiersten HF (1962) Effects of couple-stresses in linear elasticity. *Archive for Rational Mechanics and Analysis*, **11**(1), 415–448.

Moës N, Cloirec M, Cartraud P and Remacle JF (2003) A computational approach to handle complex microstructure geometries, *Computer Methods in Applied Mechanics and Engineering*, **192**(28), 3163–3177.

Moës N, Dolbow J and Belytschko T (1999) A finite element method for crack growth without remeshing, *International Journal for Numerical Methods in Engineering*, **46**(1), 131–150.

Molinari A and Clifton RJ (1987) Analytical characterization of shear localization in thermoviscoplastic materials, *J. Appl. Mech.*, **54**, 806–812.

Moran B (1987) A finite element formulation for transient analysis of viscoplastic solids with application to stress wave propagation problems, *Comput. Struct.*, **27**, 241–247.

Moran B, Ortiz M and Shih CF (1990) Formulation of implicit finite element methods for multiplicative finite deformation plasticity, *International Journal for Numerical Methods in Engineering*, **29**, 483–514.

Mühlhaus HB and Aifantis E (1991a) The influence of microstructure-induced gradients on the localization of deformation in viscoplastic materials. *Acta Mechanica*, **89**(1), 217–231.

Mühlhaus HB and Alfantis EC (1991b) A variational principle for gradient plasticity. *International Journal of Solids and Structures*, **28**(7), 845–857.

Mülhaus HB and Vardoulakis I (1987) The thickness of shear bands in granular materials, *Geotechnique*, **37**, 271–283.

Mura T (1987) *Micromechanics of Defects in Solids*. Kluwer Academic Publishers, Amsterdam.

Nagtegaal JD and DeJong JE (1981) Some computational aspects of elastic–plastic large strain analysis, *International Journal for Numerical Methods in Engineering*, **17**, 15–41.

Nagtegaal JD, Parks DM and Rice JR (1974) On numerically accurate finite element solutions in the fully plastic range, *Computer Methods in Applied Mechanics and Engineering*, **4**, 153–178.

Nappa L (2001) Variational principles in micromorphic thermoelasticity. *Mechanics Research* Communications, **28**(4), 405–412.

Narasimhan R, Rosakis AJ and Moran B (1992) A three dimensional numerical investigation of fracture initiation by ductile failure mechanisms in 4340 steel, *Int. J. Fracture*, **56**, 1–24.

Needleman A (1982) Finite elements for finite strain plasticity problems, in *Plasticity of Metals at Finite Strain: Theory, Computation and Experiment*, EH Lee and RL Mallett (eds), pp. 387–436.

Needleman A (1988) Material rate dependence and mesh sensitivity in localization problems, *Computer Methods in Applied Mechanics and Engineering*, **67**, 69–85.

Needleman A and Tvergaard V (1984) An analysis of ductile rupture in notched bars, *J. Mech. Phys. Solids*, **32**, 461–490.

Nemat-Nasser S and Hori M (1999) *Micromechanics: Overall Properties of Heterogeneous Materials*. Elsevier, Amsterdam.

Noble B (1969) *Applied Linear Algebra*, Prentice-Hall, Englewood Cliffs, NJ.

Nocedal J and Wright SJ (1999) *Numerical Optimization*, Springer, New York.

Nye JF (1985) *Physical Properties of Crystals*, Oxford University Press, Oxford.

O'Dowd NP and Knauss WG (1995) Time dependent large principal deformation of polymers, *J. Mech. Phys. Solids*, **43 (5)**, 771–792.

Oden JT (1972) *Finite Elements of Nonlinear Continua*, McGraw-Hill, New York.

Oden JT and Reddy JN (1976) *An Introduction to the Mathematical Theory of Finite Elements*, John Wiley & Sons, Inc., New York.

Ogden RW (1984) *Non-linear Elastic Deformations*, Ellis Horwood, Chichester.

Olson GB (1997) Computational design of hierarchically structured materials. *Science*, **277**(5330), 1237–1242.

Onat ET (1982) Representation of inelastic behaviour in the presence of anisotropy and finite deformations. In: *Recent Advances in Creep and Fracture of Engineering Materials and Structures*, DRJ Owen and B Wilshire (eds). Pineridge Press, Swansea, pp. 231–264.

Orowan E (1934) Plasticity of crystals, *Z. Phys*, **89**(9–10), 605–659.

Ortiz M and Martin JB (1989) Symmetry-preserving return mapping algorithms and incrementally extremal paths: a unification of concepts, *International Journal for Numerical Methods in Engineering*, **28**, 1839–1853.

Ortiz M, Leroy Y and Needleman A (1987) A finite element for localized failure analysis, *Computer Methods in Applied Mechanics and Engineering*, **61**, 189–214.

Oswald J, Gracie R, Khare R and Belytschko T (2009) An extended finite element method for dislocations in complex geometries: Thin films and nanotubes, *Computer Methods in Applied Mechanics and Engineering*, **198**(21–26), 1872–1886.

Pan J, Saje M and Needleman A (1983) Localization of deformation in rate-sensitive porous plastic solids, *Int. J. Fracture*, **21**, 261–278.

Park KC and Stanley GM (1987) An assumed covariant strain based nine-node shell element, *J. Appl. Mech.*, **53**, 278–290.

Peirce D, Shih CF and Needleman A (1984) A tangent modulus method for rate dependent solids, *Comput. Struct.*, **18**, 875–887.

Perzyna P (1971) *Thermodynamic Theory of Viscoplasticity*, Advances in Applied Mechanics, Vol. **11**, Academic Press, New York.

Pian THH and Sumihara K (1985) Rational approach for assumed stress elements, *International Journal for Numerical Methods in Engineering*, **20**, 1685–1695.

Pijauder-Cabot G and Bazant ZP (1987) Nonlocal damage theory, *J. Engrg. Mech.*, **113**, 1512–1533.

Polmear IJ (2006) *Light Alloys: From Traditional Alloys to Nanocrystals.* Elsevier/Butterworth-Heinemann, Burlington, MA.

Prager W (1945) Strain hardening under combined stress, *J. Appl. Phys.*, **16**, 837–840.

Prager W (1961) *Introduction to Mechanics of Continua*, Ginn & Co., New York.

Randle V and Engler O (2009) *Introduction to Texture Analysis: Macrotexture, Microtexture, and Orientation Mapping*, CRC Press, Boca Raton.

Rashid MM (1993) Incremental kinematics for finite element applications, *International Journal for Numerical Methods in Engineering*, **36**, 3937–3956.

Regueiro RA (2009) Finite strain micromorphic pressure-sensitive plasticity. *Journal of Engineering Mechanics*, **135**(3), 178–191.

Reissner E (1996) *Selected Works in Applied Mechanics and Mathematics*, Jones and Bartlett, Boston, MA.

Rezvanian O, Zikry MA and Rajendran AM (2007) Statistically stored, geometrically necessary and grain boundary dislocation densities: microstructural representation and modelling, *Proceedings of the Royal Society*, **463**, 2833–2853.

Rice JR (1971) Inelastic constitutive relations for solids: internal-variable theory and its application to metal plasticity, *J. Mech. Phys. Solids*, **19**, 443–455.

Rice JR and Asaro R (1977) Strain localization in ductile single crystals, *Journal of Physics and Solids*, **25**, 309–338.

Rice JR and Rosengren GF (1968) Plane strain deformation near a crack tip in a power-law hardening material, *Journal of the Mechanics and Physics of Solids*, **16**(1), 1–12.

Richtmeyer RD and Morton KW (1967) *Difference Methods for Initial Value Problems*, John Wiley & Sons, Inc., New York.

Riks E (1972) The application of Newton's method to the problem of elastic stability, *J. Appl.* Mech., **39**, 1060–1066.

Rivlin RS and Saunders DW (1951) Large elastic deformations of isotropic materials: VII Experiments on the deformation of rubber, *Phil. Trans. Roy. Soc. Lond.*, **A243**, 251–288.

Rogers HC (1960) The tensile fracture of ductile metals. *AIME Trans.*, **218**(3), 498–506.

Roters F, Eisenlohr P, Hantcherli L, *et al.* (2010) Overview of constitutive laws, kinematics, homogenization and multiscale methods in crystal plasticity finite-element modeling: Theory, experiments, applications, *Acta Materialia*, **58**(4), 1152–1211.

Rudnicki JW and Rice JR (1975) Conditions for localization in pressure sensitive dilatant materials, *J. Mech. Phys. Solids*, **23**, 371–394.

Saje M, Pan J and Needleman A (1982) Void nucleation effects on shear localization in porous plastic solids, *Int. J. Fracture*, **19**, 163–182.

Schreyer HL and Chen Z (1986) One dimensional softening with localization, *J. Appl. Mech.*, **53**, 791–797.

Seydel R (1994) *Practical Bifurcation and Stability Analysis: From Equilibrium to Chaos*, Springer-Verlag, Heidelberg.

Shabana AA (1998) *Dynamics of Multi-Body Systems*, Cambridge University Press, Cambridge.

Shanthraj P and Zikry M (2011) Dislocation density evolution and interactions in crystalline materials, *Acta Materialia*, **59**(20), 7695–7702.

Shi J and Zikry MA (2009) Grain-boundary interactions and orientation effects on crack behavior in polycrystalline aggregates, *Int. J. Solids and Structures*, **46**, 3914–3925.

Simkins Jr DC, Li S, Lu H and Kam Liu W (2004) Reproducing kernel element method. Part IV: globally compatible Cn (n⊠1) triangular hierarchy, *Computer Methods in Applied Mechanics and Engineering*, **193**(12), 1013–1034.

Simo JC and Fox DD (1989) On a stress resultant geometrically exact shell model, Part I: Formulation and optimal parametrization, *Computer Methods in Applied Mechanics and Engineering*, **72**, 267–304.

Simo JC and Hughes TJR (1986) On the variational foundations of assumed strain methods, *J. Appl. Mech.*, **53**, 1685–1695.

Simo JC and Hughes TJR (1998) *Computational Inelasticity*, Springer-Verlag, New York.

Simo JC and Ortiz M (1985) A unified approach to finite deformation plasticity based on the use of hyperelastic constitutive equations, *Computer Methods in Applied Mechanics and Engineering*, **49**, 221.

Simo JC and Rifai MS (1990) A class of mixed assumed strain methods and the method of incompatible modes, *International Journal for Numerical Methods in Engineer*, **29**, 1595–1638.

Simo JC and Taylor RL (1985) Consistent tangent operators for rate independent elastoplasticity, *Computer Methods in Applied Mechanics and Engineering*, **48**, 101–119.

Simo JC, Oliver J and Armero F (1993) An analysis of strong discontinuities induced by strain-softening in rate-independent inelastic solids, *Computational Mechanics*, **12**, 277–296.

Simo JC, Taylor RL and Pister KS (1985) Variational and projection methods for the volume constraint in finite deformation elastoplasticity, *Computer Methods in Applied Mechanics and Engineering*, **51**, 177–208.

Smithells CJ (2004) *Smithells Metals Reference Book*. Elsevier/Butterworth-Heinemann, Oxford, Burlington, MA.

Smolinski T, Sleith S and Belytschko T (1996) Explicit-explicit subcycling with non-integer time step ratios for linear structural dynamics systems, *Comp. Mech.*, **18**, 236–244.

Socrate S (1995) *Mechanics of Microvoid Nucleation and Growth in High-Strength Metastable Austenitic Steels*. MIT Press.

Song JH, Areias P and Belytschko T (2006) A method for dynamic crack and shear band propagation with phantom nodes, *International Journal for Numerical Methods in Engineering*, **67**(6), 868–893.

Song JH, Wang H and Belytschko T (2008) A comparative study on finite element methods for dynamic fracture, *Computational Mechanics*, **42**(2), 239–250.

Spivak M (1965) *Calculus on Manifolds*, Benjamin, New York.

Stanley GM (1985) *Continum-based shell elements*, PhD thesis, Stanford University, CA.

Stolarski H and Belytschko T (1982) Membrane locking and reduced integration for curved elements, *J. Appl. Mech.*, **49**, 172–177.

Stolarski H and Belytschko T (1983) Shear and membrane locking in curved elements, *Computer Methods in Applied Mechanics and Engineering*, **41**, 279–296.

Stolarski H and Belytschko T (1987) Limitation principles for mixed finite elements based on the Hu–Washizu variational formulation, *Computer Methods in Applied Mechanics and Engineering*, **60**, 195–216.

Stolarski H, Belytschko T and Lee SH (1994) A review of shell finite elements and corotational theories, *Comp. Mech. Adv.*, **2**, 125–212.

Strang G (1972) Variational crimes in the finite element method, in *The Mathematical Foundations of the Finite Element Method with Applications to Partial Differential Equations*, AK Aziz (ed.), Academic Press, New York, pp. 689–710.

Strang G and Fix GJ (1973) *An Analysis of the Finite Element Method*, Prentice-Hall, Englewood Cliffs, NJ.

Strikwerda JC (1989) *Finite Difference Schemes and Partial Differential Equations*, Wadsworth, Belmont, CA.

Sukumar N, Chopp DL and Moran B (2003) Extended finite element method and fast marching method for three-dimensional fatigue crack propagation, *Engineering Fracture Mechanics*, **70**(1), 29–48.

Sukumar N, Chopp DL, Moës N and Belytschko T (2001) Modeling holes and inclusions by level sets in the extended finite-element method, *Computer Methods in Applied Mechanics and Engineering*, **190**(46), 6183–6200.

Sukumar N, Moës N, Moran B and Belytschko, T (2000) Extended finite element method for three-dimensional crack modelling, *International Journal for Numerical Methods in Engineering*, **48**(11), 1549–1570.

Tang S, Greene MS and Liu WK (2012) Two-scale mechanism-based theory of nonlinear viscoelasticity. *J. Mech. Phys. Solids*. **60**, 199–226.

Taylor G (1934) The mechanism of plastic deformation of crystals. Part I. Theoretical, Proceedings of the Royal Society of London. Series A, *Containing Papers of a Mathematical and Physical Character*, **145**(855), 362–387.

Taylor RL, Simo JC, Zienkiewicz OC and Chan AHC (1986) The patch test: a condition for assessing FEM convergence, *International Journal for Numerical Methods in Engineering*, **22**, 39–62.

Thompson JMT and Hunt GW (1984) *Elastic Instability Phenomena*, John Wiley & Sons, Ltd, Chichester.

Tian R, Chan S, Tang S, *et al.* (2010) A multiresolution continuum simulation of the ductile fracture process. *Journal of the Mechanics and Physics of Solids*, **58**, doi: 1681\961700.

Tian R, To, AC and Liu WK (2011) Conforming local meshfree method, International *Journal for Numerical Methods in Engineering*, **86**(3), 335–357.

Truesdell C and Noll W (1965) The non-linear field theories of mechanics, in *Encyclopedia of Physics*, S Flugge (ed.), 3/3, Springer-Verlag, Berlin.

Turner MR, Clough R, Martin H and Topp L (1956) Stiffness and deflection analysis of complex structures, *J. Aero. Sci.*, **23** (9), 805–823.

Tvergaard V (1981) Influence of voids on shear band instabilities under plane strain conditions, *Int. J. Fracture*, **17**, 389–407.

Tvergaard V and Needleman A (1984) Analysis of the cup-cone fracture in a round tensile bar, *Acta Metallurgica*, **32**, 157–169.

Van Houtte P (1987) The Taylor and the relaxed Taylor theory, *Textures and Microstructures*, **7**, 29–72.

Van Houtte P, Li S, Seefeldt M, *et al.* (2005) Deformation texture prediction: from the Taylor model to the advanced Lamel model, *International Journal of Plasticity*, **21**(3), 589–624.

Ventura G, Gracie R and Belytschko T (2009) Fast integration and weight function blending in the extended finite element method, *International Journal for Numerical Methods in Engineering*, **77**(1), 1–29.

Ventura G, Moran B and Belytschko T (2005) Dislocations by partition of unity, *International Journal for Numerical Methods in Engineering*, **62**(11), 1463–1487.

Ventura G, Xu JX and Belytschko T (2002) A vector level set method and new discontinuity approximations for crack growth by EFG, *International Journal for Numerical Methods in Engineering*, **54**(6), 923–944.

Vernerey FJ (2006) *Multi-scale Continuum Theory for Materials with Microstructure*. Doctor of Philosophy, Northwestern University.

Vernerey FJ, Liu WK and Moran B (2007) Multi-scale micromorphic theory for hierarchical materials. *Journal of the Mechanics and Physics of Solids*, **55**, 2603–2651.

Wagner GJ, Ghosal S and Liu WK (2003) Particulate flow simulations using lubrication theory solution enrichment. *International Journal for Numerical Methods in Engineering*, **56**(9), 1261–1289.

Wagner GJ and Liu WK (2003) Coupling of atomistic and continuum simulations using a bridging scale decomposition. *Journal of Computational Physics*, **190**(1), 249–274.

Wagner GJ, Moës N, Liu WK and Belytschko T (2001) The extended finite element method for rigid particles in Stokes flow. *International Journal for Numerical Methods in Engineering*, **51**(3), 293–313.

Wang SC and Starink MJ (2005) Precipitates and Intermetallic phases in precipitation hardening Al-Cu-Mg-(Li) based alloys, *International Materials Review*, **50**, 193–215

Wempner GA (1969) Finite elements, finite rotations and small strains, *Int. J. Sol. Struct.*, **5**, 117–153.

Wempner GA, Talaslidis D and Hwang CM (1982) A simple and efficient approximation of shells via finite quadrilateral elements, *J. Appl. Mech.*, **49**, 331–362.

Wilkins ML (1964) Calculation of elastic–plastic flow, in *Methods of Computational Physics*, Vol. **3**, B Alder, S Fernbach and M Rotenberg (eds), Academic Press, New York.

Wilson EL, Taylor RL, Doherty WP and Ghaboussi J (1973) Incompatible displacement models, in *Numerical and Computer Models in Structural Mechanics*, SJ Fenves *et al.* (eds), Academic Press, New York.

Wriggers P (1995) Finite element algorithms for contact problems, *Arch. Comp. Meth. Engrg.*, **2** (4), 1–49.

Wriggers P and Miehe C (1994) Contact constraints within coupled thermomechanical analysis: a finite element model, *Computer Methods in Applied Mechanics and Engineering*, **113**, 301–319.

Wright TW and Walter JW (1987) On stress collapse in adiabatic shear, *J. Mech. Phys. Solids*, **35** (6), 205–212.

Xiao SP and Belytschko T (2004) A bridging domain method for coupling continua with molecular dynamics. *Computer Methods in Applied Mechanics and Engineering*, **193**(17–20), 1645–1669.

Xu XP and Needleman A (1994) Numerical simulations of fast crack growth in brittle solids, *Journal of the Mechanics and Physics of Solids*, **42**(9), 1397–1434.

Yamakov V, Wolf D, Phillpot SR, et al. (2002) Dislocation processes in the deformation of nanocrystalline aluminium by molecular-dynamics simulation. *Nature Materials*, **1**(1), 45–49.

Yan Z, Larsson R, Jing-yu F, *et al.* (2006) Homogenization model based on micropolar theory for the interconnection layer in microsystem packaging. *IEEE Xplore, High Density Microsystem Design and Packaging and Component Failure Analysis, HDP 2006.*

Yang JFC and Lakes RS (1982) Experimental study of micropolar and couple stress elasticity in compact bone in bending. *Journal of Biomechanics*, **15**(2), 91–98.

Zeng X, Chen Y and Lee JD (2006) Determining material constants in nonlocal micromorphic theory through phonon dispersion relations. *International Journal of Engineering Science*, **44**(18–19), 1334–1345.

Zhong ZH (1993) *Finite Element Procedures for Contact-Impact Problems*, Oxford University Press, New York.

Zhu AW, Shiflet GJ and Stark EA (2006) First principles calculations for alloy design of moderate temperature age-hardenable Al alloys, *Materials Science Forum*, **519–521**, 35–43.

Ziegler H (1950) A modification of Prager's hardening rule, *Quart. J. Appl. Math.*, **17**, 55–65.

Zienkiewicz OC, Emson C and Bettess P (1983) A novel boundary infinite element. *International Journal for Numerical Methods in Engineering*, **19**(3), 393–404.

Zienkiewicz OC and Taylor RL (1991) *The Finite Element Method*, McGraw-Hill, New York.

Zienkiewicz OC, Taylor RL and Too JM (1971) Reduced integration techniques in general analysis of plates and shells, *International Journal for Numerical Methods in Engineering*, **3**, 275–290.

Zikry M (1994) An accurate and stable algorithm for high strain-rate finite strain plasticity, *Computers & Structures*, **50**(3), 14.

Zikry M and Kao M (1997) Inelastic microstructural failure modes in crystalline materials: The S33A and S11 high angle grain boundaries, *International Journal of Plasticity*, **13**(4), 31.

Zikry M and Nemat-Nasser S (1990) High strain-rate localization and failure of crystalline materials, *Mechanics of Materials*, **10**(3), 215–237.

Index

Note: Page numbers in **bold** refer to Boxes.

Nonlinear Finite Elements for Continua and Structures, Second Edition.
Ted Belytschko, Wing Kam Liu, Brian Moran, and Khalil I. Elkhodary.

Companion Website: www.wiley.com/go/belytschko

CPSIA information can be obtained
at www.ICGtesting.com
Printed in the USA
BVHW010552191119
564228BV00016B/170/P